COLLEGE BOTANY

[Vol. I]

COLLEGE BOTANY

[Vol. I]

Hirendra Chandra Gangulee M.Sc., D.Sc., F.N.A.

Formerly Professor of Botany, Darjeeling Govt. College
Presidency College, Kolkata
University College of Science, Kolkata

Kumud Shankar Das M.Sc.

Formerly Professor of Botany, Surendranath College, Kolkata

AND

Chittatosh Dutta M.Sc., Ph.D. [Lond.], D.I.C. [Lond.]

Formerly Assistant Professor of Botany, Presidency College, Kolkata
Lecturer in Plant Physiology, University College of Science, Kolkata

PLANT PHYSIOLOGY AND ECOLOGY

Revised by

Shyamapada Sen M.Sc., D.Phil., F.N.A.

Formerly Professor of Botany, Kalyani University

New Central Book Agency (P) Ltd

Co-Published by

Bijay Publications

E-mail : bijaypublications@gmail.com

Exclusively Distributed by

Bijay Pustak Bhandar (P) Ltd

90/6A M. G. Road, Bhawani Dutta Lane, Kolkata-700007

✆ : (033) 4605-7856, 8777898576, 8274916104

E-mail : bijaypustakbhandar@gmail.com

NCBA

REGD OFFICE
8/1 Chintamoni Das Lane, Kolkata 700 009, India
email: ncbakolkata@gmail.com

OVERSEAS
NCBA (UK) Ltd, 149 Park Avenue North, Northampton, NN3 2HY, UK
email: ncbauk@yahoo.com

CORPORATE OFFICE
212 Shahpur Jat, New Delhi 110 049
email: ncbaexports@ncbapvtltd.com

BIJAY PUBLICATIONS
90/6A M. G. Road, Bhawani Dutta Lane, Kolkata-7
E-mail : bijaypublications@gmail.com

BIJAY PUSTAK BHANDAR (P) LTD
90/6A M. G. Road, Bhawani Dutta Lane, Kolk
E-mail : bijaypustakbhandar@gmail.com

COLLEGE BOTANY [Vol. I] • H. C. Gangulee, K. S. Das, C. Dutta and S. Sen

First Published: 1959
Fifth Edition: 1984
Sixth Edition: 1988
Reprinted: 1991, 1993, 1994, 1997, 1998, 1999, 2001, 2002, 2005, 2006, 2007, 2011, 2012, 2014, 2015, 2016, 2017, 2018, 2019, 2023

PUBLISHER • TYPESETTER • PRINTER
New Central Book Agency (P) Ltd
8/1 Chintamoni Das Lane, Kolkata 700 009

CO-PUBLISHED BY
Bijay Publications
90/6A M G Road, Bhawani Dutta Lane, Kolkata 700007

ISBN: 978 81 7381 028 2

Price ₹ 975.00

PREFACE TO THE FIFTH EDITION

It is a painful reality that of the three original authors, two died in course of the last two years and I am now the only one left. But, it is very hopeful that younger and promising teachers are likely to take over. Dr. S. P. Sen has already revised the Plant Physiology & Ecology Parts and others may follow in future.

Thanks are due to all friends, well-wishers and students who are supporting this publication.

Akshaya Tritiya,
4th May, 1984
Calcutta

H. C. GANGULEE

PREFACE TO THE FIRST EDITION

Teachers in undergraduate Botany classes in India find themselves handicapped as they can recommend only foreign books to their students as Indian books of that standard are not available. For sometimes past, it had been the ambition of the authors to present such a book to the students and teachers, as foreign books, however good, suffer from the drawback of citations from plants which are mostly unfamiliar to our students. The first volume of this venture is, at last, out. We have tried to make it as up-to-date as possible, although we do not claim any originality of approach as we have followed the generally accepted lines. But, we are aware that, as in all first ventures, there must be many shortcomings in this book. We apologise for these as well as for the printing errors. At the same time we expect that our fellow teachers and specialists will co-öperate with us by sending in their suggestions and criticisms which will be thankfully received and will help us in improving the next edition.

The book is primarily meant for the students of degree classes of all Universities in India and the neighbouring countries, but we hope that it may be helpful to students of more advanced courses and also to students of Agriculture and allied subjects.

We are glad to record our thanks to our colleagues at Presidency College and Surendranath College and also to a wide circle of friends of other Institutions for their continued interest. We are particularly grateful to our teachers—Dr. G. P. Majumdar, former Professor of Botany, Presidency College, Calcutta and Dacca University; Dr. I. Banerji, Head of the Department of Botany, Calcutta University and Dr. J. C. Sen Gupta, Chief Botanist, Botanical Survey of India and former Principal, Presidency College, Calcutta and also to Dr. P. N. Bhaduri, Head of the Department of Botany, Presidency College, Calcutta for the kind encouragement we have received. Our thanks are also due to Sri L. K. Roy who has taken great pain in preparing the diagrams and to Sri Jatindranath Sen for undertaking this costly publication.

Basant Panchami
12th February, 1959

H. C. GANGULEE
K. S. DAS
C. T. DATTA

CONTENTS

COLLEGE BOTANY

VOLUME ONE

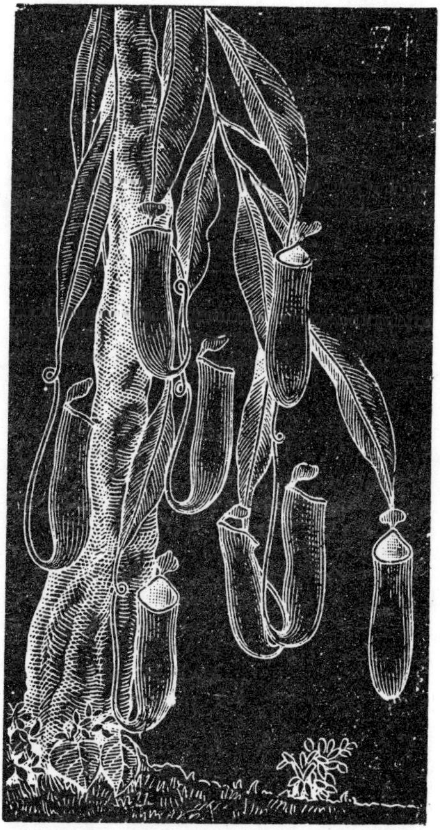

1 [one]

GENERAL CONSIDERATIONS

The word **Botany** is derived from the Greek noun *botane* which means herb. It is a branch of biology which deals with the study of plants. **Biology** (*bios*=life, *logos*=discourse) is the science of life. The scope of this natural science is vast, as it covers the study of life in plants, in animals, as well as in men. For practical purposes biology is subdivided into two sister sciences—botany, dealing with plants, and zoology, with animals.

Before taking up discussion on botany it is desirable to make an attempt to know something about life and living matters. One can hardly give a satisfactory answer to the question 'what is life?' Life is abstract. It has no objective reality apart from the living body. Hence it evades definition. It may be compared to the flame of a candle which is produced by the burning of the candle substance. Flame has no separate existence apart from the candle substance, and it persists as long as the burning or oxidation goes on. Life similarly finds expression in the various activities of plants and animals and it has no real existence outside the living body. The flame flickers out when the candle substance supplying energy to it is exhausted. Life ceases to exist when the store of energy in the living body is totally spent up. Life, thus, involves a continuous expense of energy just comparable to the constant loss of heat and light by the candle flame. Activities of living things are manifested in the ceaseless conversion of energy—the potential energy into the kinetic form. It finds expression in movement, production of heat, light, and electricity which are utilised by the organism for its own benefit. One is familiar with the locomotion found in the animals. They swim, fly, or run about either in search of food or to avoid the enemy. The lotus opens its petals at sunrise to close them up again at sunset. The two sides of a leaf of Venus' fly-trap, an insectivorous plant, close up when touched by an insect, when the latter is entrapped in a fatal grip. Heat is evolved from the germinating seeds and from the body of a busy bee. The glow-worm emits light energy to find its way in the darkness of the night. The deep-sea fishes have phosphorescent patches along their sides glowing with light like port-holes of an ocean-going liner, which help them in moving in the depth of the ocean—a cheerless dark place where sun's rays do not penetrate. The electric eels of the Amazon basin and the electric rays of the Indian ocean possess specialised electric organs for electrocuting their enemies.

CHARACTERISTICS OF LIVING THINGS

The concept of life is vague, but its manifestations in the living body are rather distinct. So the living bodies possess some distinctive characteristics by which they can be distinguished from the non-living.

A living body has its own kind of size and form which may vary within narrow limits, whereas the inanimate objects have no definite size and form. Water, for instance, may exist as a drop of rain or as an ocean. It may be moulded into a mass of cloud or an iceberg or an icicle. But it is in all the cases the same H_2O, though varying in physical form and size to such a wide extent.

All living things are composed of the essential living substance, **protoplasm,** which has been very aptly described by the famous biologist T. H. Huxley in the nineteenth century as 'the physical basis of life'. Protoplasm is a soft, jelly-like substance which is always in a state of motion. Bits of protoplasm occur in small units known as **cells.** A number of cells destined to perform a particular function constitutes **a tissue.**

Tissues are aggregated into **tissue-systems,** and the tissue-systems together form an organ. The organs are interdependent and work together harmoniously to make the life of the organism possible. Thus a living body exhibits a wonderful **organization.**

It has been stated that protoplasm undergoes a series of chemical changes, and that cessation of activities ends in disorganization and death. The sum-total of the changes is collectively referred to as **metabolism.** Some changes are constructive and some destructive. The formative or constructive phase is called *anabolism* in which process the organism gains energy in potential form and reserve it for future use; whereas the destructive phase, known as *katabolism*, involves breaking down of the organic substances, so that the potential energy is converted into the kinetic form. The two processes of building up and breaking down must go on simultaneously. The real mystery of life centres round this peculiar dual process of waste and repair the protoplasm constantly undergoes. Metabolic changes find expression in *nutrition*, *respiration*, *secretion* and *excretion*.

Nutrition is the process of taking in of food just comparable to the fuelling of a machine. The energy locked up in the food molecules is used by the living body for its own purpose. It actually occurs in several successively running stages. The food is first taken in or *injested*. The complex matters are broken down to simpler forms by secretion of digestive juices called enzymes, or in other words food is *digested*. The digested food then diffuses in the living body and is carried to the regions where they are mostly needed. That is how food is *absorbed*. The absorbed food is either immediately utilised for th repair of the body as a source of energy or it may be stored up for future use. The food thus *assimilated*. The residue left during these changes is thrown out or *ejected*.

Respiration implies much more than mechanical breathing involving taking in of oxygen and giving out of carbon dioxide. It is really the motive force of converting the potential energy stored up in the living body during anabolic changes into an utilisable kinetic form. Respiration is a process of biological oxidation—a katabolic process, when organic materials of the living body are broken down to inorganic forms with liberation of energy.

During metabolic changes protoplasm produces many new substances. Some of them are positively useful to the organism. They are known as *secretions*. Enzymes, the digestive juices, and hormones, which have profound influence on the growth of the body and mind, are the examples. The waste products, also called *excretions*, are useless and usually accumulate in special reservoirs or dead inactive tissues. They may be thrown out during fall of leaves, and with bark and seeds periodically.

Anabolism usually gains over katabolism and the result is **growth.** In a living body growth is not a matter of simple *accretion* or deposition of similar substance, as found in case of a sugar crystal which enlarges when put in concentrated sugar solution, but it is a biological process regulated by protoplasm. In fact, new particles are intercalated here between the already existing ones. This process is referred to as growth by *intussusception*.

Movement is another character of the living things. Automatic or spontaneous movement is noticed in the protoplasm of aquatic plants like *Vallisneria*. In higher animals movement is more prominent as they develop special organs for the purpose, e.g., fins for swimming, wings for flying, and legs for walking. *Irritability* or power of responding to external stimuli is an inherent character of living bodies. The stimuli may be physical like heat or light, mechanical such as touch or contact, and even chemical, such as presence of acids and bases. A moth flies into the flame under the influence of light, and the leaves of the sensitive plant shrink down modestly to touch. The roots of the plants go down under the influence of gravity and a muscle contracts when stimulated by an electric shock. The organism responds to external stimuli with the definite purpose of maintaining its own life as well as for perpetuation of the whole race. Purposive behaviour is a feature of the living body.

Reproduction is a unique property of the living things which can produce young ones like themselves after having attained maturity, and can thus maintain continuity of life. There are different methods of reproduction, but the young ones always resemble the parents in the long run—what is referred to as the *law of biogenesis*.

Vital activities such as nutrition, respiration, etc., exhibit *rhythmicity* or periodicity. A stage of intense activity is always followed by a distinct pause, however small that may be. In the mean time protoplasm gathers its vital force necessary for the next active phase.

Every organism shows a distinct **life-cycle.** In the young stage it takes food and grows actively. Having attained maturity it reproduces its own kinds. In course of time it grows old or *senescence* sets in when protoplasmic activities gradually slow down. Finally it dies and *death* ends all.

PLANTS AND ANIMALS

Man has always been impressed by the vastness and variety of living things. The plants and animals are the main living things. Though some small microscopic organisms have been claimed as plants by the botanists and as animals by the zoologists, the higher plants may be readily distinguished from the higher animals. The main points of difference are being discussed here.

Majority of the plants are green in colour, as they possess the essential green pigment, *chlorophyll.* The green plants absorb water with dissolved mineral matters from the soil and carbon dioxide gas from the outer atmosphere. With the aid of chlorophyll they can synthesise complex carbohydrate food out of the two raw materials in presence of sunlight. This unique process of manufacture of carbohydrate food, known as *photosynthesis*, is an absolute monopoly of the green plants. The carbohydrates form the ground substance in the manufacture of proteins and fats. Thus green plants are self-supporting or *autotrophic* bodies, and their mode of nutrition is referred to as *holophytic.* The animals are *heterotrophic*, having *holozoic* mode of nutrition, as they cannot manufacture food, and are perforce dependent on the plants directly or indirectly for their entire supply of food.

Plants absorb the raw materials in liquid or gaseous forms. They cannot injest any solid, nor any solid moves inside the plants. The insectivorous plants are the exceptions, as they get part of their nitrogen requirement from the bodies of the insects they catch. The animals, excepting a few parasites, can injest solid food.

Metabolism in the plants is predominantly anabolic. Growth is localised to certain regions and continues till the last days of their lives. Most of the plants are freely branched and have an open form of growth. The animals have more active habits and metabolism is predominantly katabolic here. They have more compact forms, and growth ceases long before death.

Specialised excretory passages for getting rid of waste products are absent in the plants. In fact, the waste products are sent to the bark of the trees, to the leaves before their fall and are thus periodically thrown out. Animals have well-developed excretory systems for eliminating the waste products which are formed in much larger quantities.

Most of the plants remain anchored to the soil by roots and root-like organs and have thus no power of locomotion. They can, however, move their parts in response to external stimuli The animals in general have the power of locomotion, as they have to move about in search of food. Response to external stimuli is more quick in animals, due to the fact that they possess distinct nervous systems.

The plant cells usually remain surrounded by rigid walls made of complex insoluble carbohydrate cellulose. In some cases protoplasm disappears in course of time and the wall is left behind. These are obviously dead cells. Cellulose cell wall is never found in the animal cells which really remain surrounded by their delicate protoplasmic membranes. Distinct *vacuoles,* which are intercellular cavities containing watery cell-sap, are present in the plant cells, whereas they are scarce in animal cells. A star-shaped *central body* usually occurs near the nucleus in an animal cell and plays an important part during cell division It is of rare occurrence in plant cells.

SUBDIVISIONS OF BOTANY

Botany is the science of plant life. As such it includes everything which has reference to plants. It considers the external appearance of the plant organs, their internal structure and organization, the various vital activities like nutrition, respiration, growth, movement and reproduction carried on by plants, as well as a systematic classification based on resemblances and differences, their life-histories, their adaptations to varying environmental conditions, relationships amongst themselves and with other living things, their distribution in the world in space and time and their economic values specially to mankind.

As botany is a vast science embracing everything having connection with plants, it is usually divided into a few branches or subdivisions for the convenience of study.

Morphology (*morphe*=form, *logos*=discourse) includes the study of the form and structure of the plant organs. The gross external features that can be examined with the naked eye, come under external morphology. Internal morphology or **anatomy** (*ana*=asunder, *temnein*=to cut) deals with the internal structures. The aid of the magnifying apparatus, microscope, is indispensable for the study of internal structures and other minute details. **Histology** (*histos*=cobweb) is the study of minute structures. So it forms a part of anatomy. **Cytology** is the special study of the cell, the unit of structure and function. **Physiology** (*physis*=nature or life) deals with the vital functions performed by the plants. It seeks to explain the processes like metabolism, growth, movement and reproduction in terms of chemistry and physics as far as possible. **Taxonomy** (*taxis*=arrangement, *nomos*=law) or systematic botany is concerned with the classification of plants or division of the plant kingdom into smaller and smaller groups in a systematic manner, and with the naming of the plants for identification. **Ecology** (*oikos*=house) considers the influence of environment or surroundings on plants and plant communities and the various adaptations exhibited by them according to the situations where they grow. **Plant geography** is the study of the distribution of plants in the different parts of the world and the factors responsible for it. **Paleobotany** considers the distribution of plants in time. This study is based on the fossil remains of the plants that existed in the dim past. **Genetics** is a modern science dealing with study of heredity in plants.

Like other natural sciences botany may also be studied from two aspects—*pure* and *applied.* The subdivisions stated above belong to pure botany which are concerned with the fundamental knowledge of the subject regardless of their practical application. Applied branches are those which are particularly related to the well-being of mankind and thus form part of **economic botany.** *Agriculture* dealing with the study of crop-plants, *Horticulture* concerned with the study of garden plants, *Forestry*, with the forest trees, *Pharmacognosy*, with the drug plants, *Plant pathology*, with the plant diseases with relevant suggestions about prevention and treatment, and *Plant-breeding* are the applied branches of botany. Besides these, special investigations of certain groups of plants are also referred to as sub-sciences, viz., *Algology*, the study of algae; *Mycology*, of fungi; *Bacteriology*, of bacteria; and so on.

IMPORTANCE OF PLANTS AND SCOPE OF BOTANY

We are surrounded by a world of plants and have to depend on them for almost all our daily necessities of life. The principal items of our food are derived directly or indirectly from plants. The green plants have the unique power of synthesising complex organic food matters from simple substances in presence of sunlight. That process involves taking in of injurious carbon dioxide gas from the atmosphere and giving out of oxygen. In this way, the plants, apart from playing the role of universal food producer for animals including men, purify the air and thus make possible all life on earth. Plants, in fact, make the earth's crust habitable for animals.

All the important food cereals like rice, wheat, maize, millets, barley, oat, bazra are derived directly from the plants. Fats and oils produced by the plants are used as food, and also in other industries like manufacture of soap, as lubricants and also in paints and varnishes. Different types of spices; beverages like tea, coffee, cocoa and wines and liquors are also plant products. Timber, wood, bamboos and canes are extensively used as building materials and also for furniture. Plant fibres are the chief materials for the manufacture of clothings. Plants are the sources of a large number of medicines indispensable to mankind. The wonder drugs of the present age—the antibiotics, penicillin, streptomycin, aureomycin, etc., which have revolutionised medical science, are extracted from some of the lower forms of plants. Coal, an indispensable commodity in the civilized world, is nothing but the remains of buried plants that existed long long before—in the past geological ages. Tar, petroleum, aniline dyes and saccharine, are produced from coal. Many other essential substances like paper, rubber, resin, cork, are also plant products.

The roots of the plants go down and form an intricate and extensive system in the soil. That is how the soil particles are held in position and the soil becomes porous. As a result rain water is considerably absorbed and provides a source of pure water for wells and reservoirs. Otherwise, there would be an excessive run off after heavy rains which would result in disastrous floods and erosion of fertile top soil. A cover of vegetation also prevents erosion caused by wind storms.

Further, the aesthetic value of the plants is by no means meagre. The world of plants has been the source of inspiration to many artists, poets, litterateures and musicians from times immemorial.

Knowledge of Botany is an essential pre-requisite for the pursuit of many applied sciences like Agriculture, Horticulture, Sericulture, Animal Husbandry, Forestry, Pharmacology and Medicine.

THE PLANT KINGDOM

Quite a large number of plants exist in this world of nature. More than 3,50,000 plants have been identified, described and named by the botanists and a pretty large number still remain unknown. Thus they are not only large in number, but are equally varied in nature. They inhabit all the conceivable places in the world and can often withstand extreme unfavourable conditions. The smallest bacteria, which occur everywhere and some of which are the causes of many diseases we suffer from, are plants and form one extreme, and the gigantic forest trees form the other. Transitional stages between the two extremes are numerous. The problem is really a stupendous one and it cannot be tackled unless a suitable and systematic plan of dividing the plant kingdom into smaller and smaller groups is devised. Attempts to classify plants were made even by the early ancients like Theophrastus in Greece and the Indian sages in the ages of the Upanishads. Since the sixteenth century, different systems of classification have been proposed. Modern classification is truly systematic, as it is based on the interrelationships amongst the plant groups.

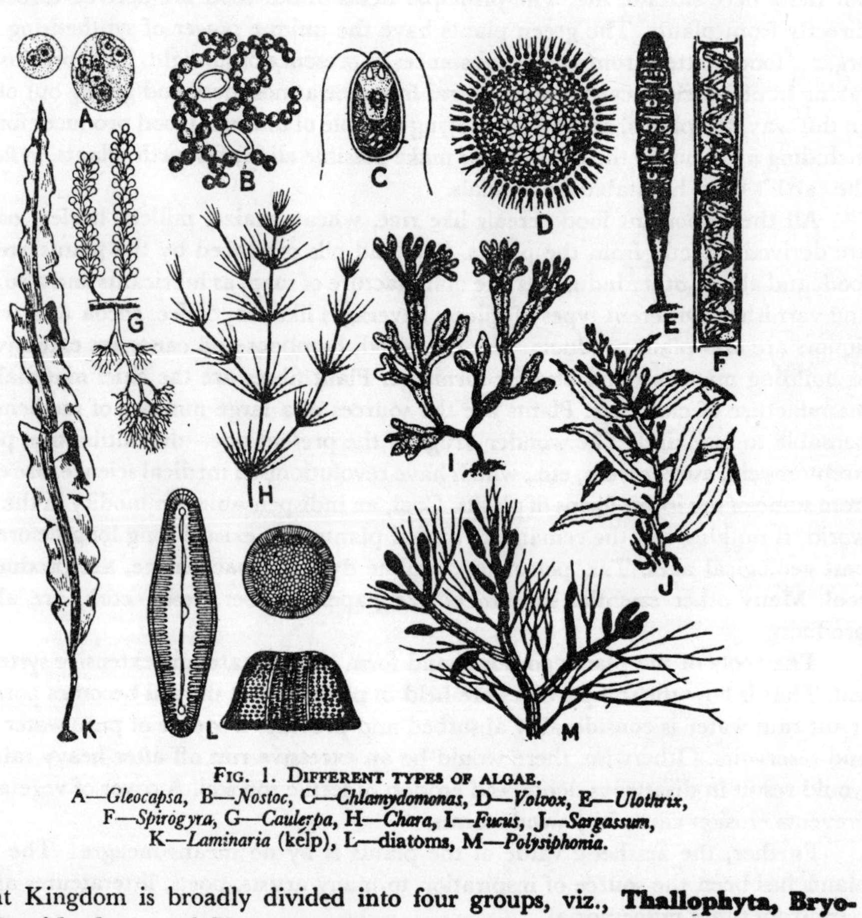

FIG. 1. DIFFERENT TYPES OF ALGAE.
A—*Gleocapsa*, B—*Nostoc*, C—*Chlamydomonas*, D—*Volvox*, E—*Ulothrix*, F—*Spirogyra*, G—*Caulerpa*, H—*Chara*, I—*Fucus*, J—*Sargassum*, K—*Laminaria* (kelp), L—diatoms, M—*Polysiphonia*.

Plant Kingdom is broadly divided into four groups, viz., **Thallophyta, Bryophyta, Pteridophyta** and **Spermatophyta.** Of the four groups the first three never bear flowers as they are popularly known and, hence, are also known as *flowerless* plants or **Cryptogams.** The last group includes the plants which bear flowers and are also known as *flowering* plants or **Phanerogams.** It is considered now that pteridophyta is not a homogeneous group and a clear line of demarcation hardly exists between pteridophyta and the next higher group spermatophyta. In view of above consideration which is based on a sound study of fossils and living plants, it is advocated that all vascular plants, naturally including pteridophytes and spermatophytes, should be put under one group **Tracheophyta.**

I. **Thallophyta.** Plants belonging to this group are the simplest and most primitive ones. Their bodies are not differentiated into organs like root, stem and leaf. In fact, the plant body of a thallophyte is an undifferentiated mass of cells, known as a *thallus*. A thalloid body may be unicellular or a colony of cells which does not exhibit any division of labour. The sex organs are usually unicellular.

The green thallophytes possessing chlorophyll are known as **Algae** (Fig. 1). They usually grow in water or in moist situations. Some marine algae like the sea-weeds and kelps are fairly large in size. In addition to chlorophyll, other pigments may also be present in algae. The fresh-water algae are generally green or blue-green in colour, whereas the marine ones are red or brown. These are *autotrophic* plants, as they can manufacture their own food.

Non-green thallophytes characterised by total absence of chlorophyll are called **Fungi** (Fig. 2). They grow either on dead, rotten organic matters as *saprophytes* or live as *parasites* on other living bodies, which are referred to as *hosts*. Moulds and mushrooms are the familiar examples of saprophytic fungi. Parasitic fungi infect a pretty large number of economic plants as well as animals and often cause considerable damage.

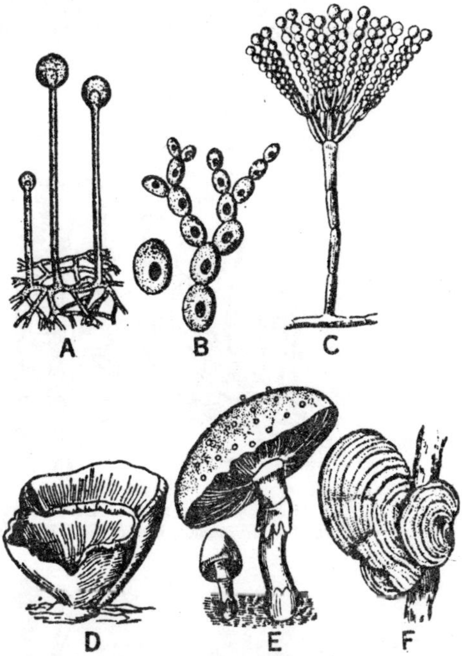

FIG. 2. DIFFERENT TYPES OF FUNGI.
A—*Mucor*, B—*Yeast*, C—*Penicillium*, D—*Peziza*, E—*Agaricus*, F—*Polyporus*.

Bacteria (Fig. 3) form a group of infinitely small non-chlorophyllous micro-organisms which lead saprophytic or parasitic existence. Many of them are *pathogenic;* they are causes of some plant and animal diseases. Saprophytic bacteria are rather beneficial. They work as 'scavengers' of the soil and many of them are used in industries.

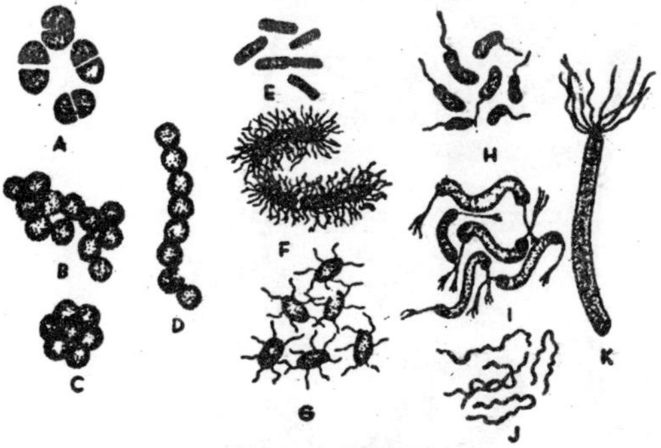

FIG. 3. DIFFERENT TYPES OF BACTERIA.
A to D—Coccus forms, E to G—Bacillus forms, H to K—Spirillum forms.

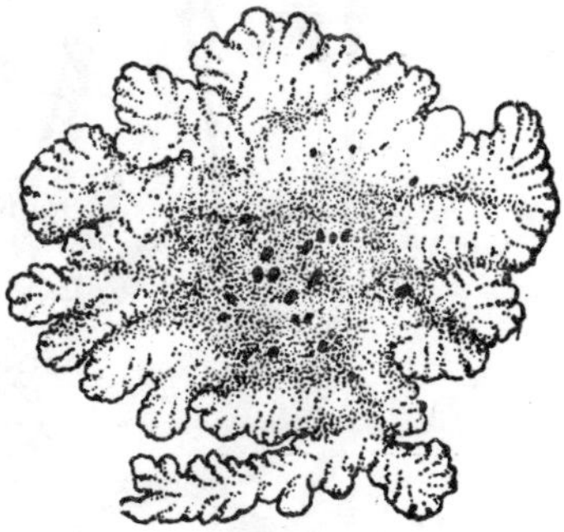

FIG. 4. Lichen—a foliose type.

Lichens (Fig. 4) constitute an extra-ordinary group of greyish green plants which grow on rocks, tree-trunks, dead wood, etc. A lichen is a peculiar combination of an alga and a fungus—the two live deriving mutual benefit. The alga manufactures

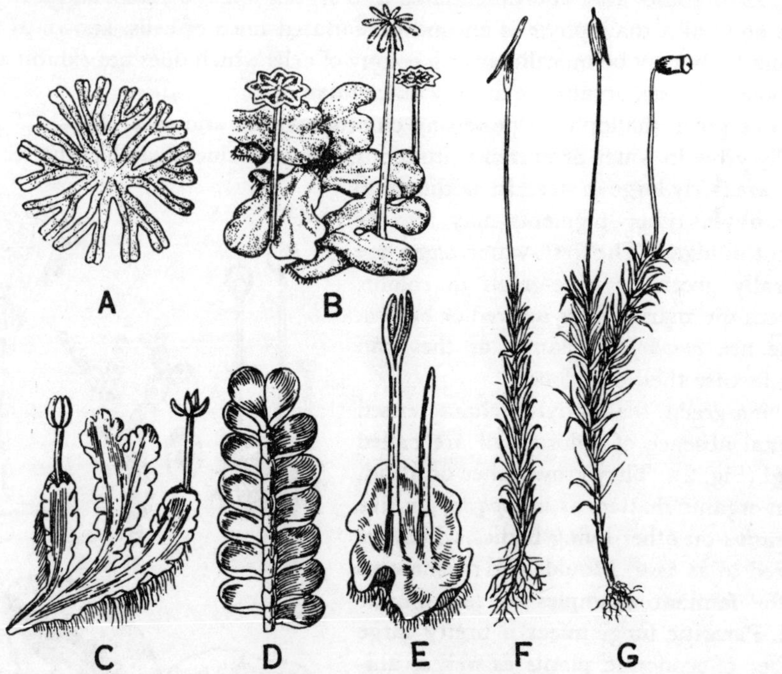

FIG. 5. DIFFERENT TYPES OF BRYOPHYTES.
A—*Riccia*, B—*Marchantia*, C—*Pellia*, D—*Lejeunea*, E—*Anthoceros*, F—*Barbula*, G—*Polytrichum*.

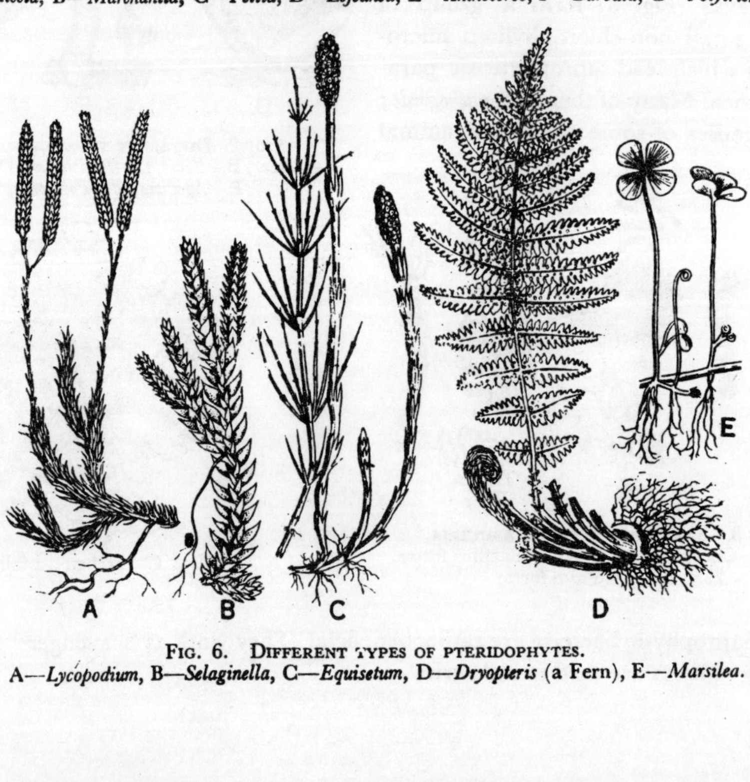

FIG. 6. DIFFERENT TYPES OF PTERIDOPHYTES.
A—*Lycopodium*, B—*Selaginella*, C—*Equisetum*, D—*Dryopteris* (a Fern), E—*Marsilea*.

carbohydrate food which becomes available to the fungus, and the latter absorbs and retains water and thus keep the algal cells moist. So it is a nice example of *symbiosis*.

II. **Bryophyta.** This group consists of plants (Fig. 5) which are more advanced than the thallophytes. They usually grow in moist places. Some bryophytes like the liverworts have thalloid bodies, whereas the mosses show slight differentiation of plant body. In fact, in a moss the plant body is differentiated into a small stem and simple leaves, but true roots are absent. The sex organs are multicellular and the gametes always remain surrounded by a jacket of sterile cells.

III. **Pteridophyta.** This group includes the *vascular cryptogams* like club-mosses, horse-tails and ferns (Fig. 6) which are universally distributed all over the world. Most of them are terrestrial plants flourishing well in moist and shady places, and some of them are aquatic. The pteridophytes have well-differentiated plant bodies, consisting of roots, stems and leaves. Moreover, they possess vascular bundles. The sex organs are multi-cellular and the gametes remain surrounded by sterile cells.

IV. **Spermatophyta.** The so-called higher plants belong to this group. Apart from distinct differentiation of the plant body like the pteridophytes, development of typical flowers and consequent reproduction through seeds are the outstanding features of this group. Spermatophytes are again divided into two groups, viz., **Gymnosperms** and **Angiosperms.**

Gymnosperms (*gymnos*=naked, *sperma*=seed) are the naked-seeded plants (Fig. 7). They have very simple flowers without accessory whorls and the microsporophylls (stamens) and megasporophylls (carpels) remain aggregated in cones. Ovules are present on

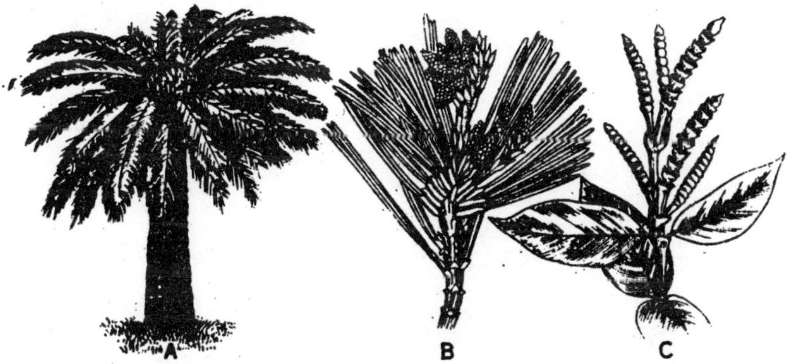

FIG. 7. GYMNOSPERMS.
A—*Cycas*, B—twig of *Pinus*, C—twig of *Gnetum*.

the surface of the megasporophylls and are directly pollinated by the pollen grains. There is nothing like ovary, style and stigma, and naturally there is no fruit. Gymnosperms constitute a group intermediate between pteridophytes and angiosperms. They formed dense vegetation in the past geological ages but many of them are extinct now.

Angiosperms (*angeion*=a case) are the *closed-seeded* plants. These are the most highly developed plants which bear flowers having conspicuous accessory and essential whorls. Carpels have the ovary, style and stigma. With the stimulus of fertilisation the ovary usually develops into the fruit and the ovules into seeds. Thus the seeds remain within the fruits. Angiosperms exhibit wide diversities as regards their form and structure, ranging from the smallest duck-weeds to the huge forest trees; and they are capable of growing in all types of situations. They are further put into two subdivisions; **dicotyledons** and **monocotyledons,** depending on the number of cotyledons in the embryo.

The above classification proves beyond any shadow of doubt that in the plant kingdom there is a gradual ascent in the complexity of structure from the lowest thallophytes to the highest angiosperms. Gradual adaptibility to terrestrial habitat from aquatic is also a notable feature. It is interesting to note that some plants like the mosses are amphibious, like the amphibians of animal kingdom, in the sense that they live partly in water and partly on land.

PART ONE
The External Morphology of Angiosperms

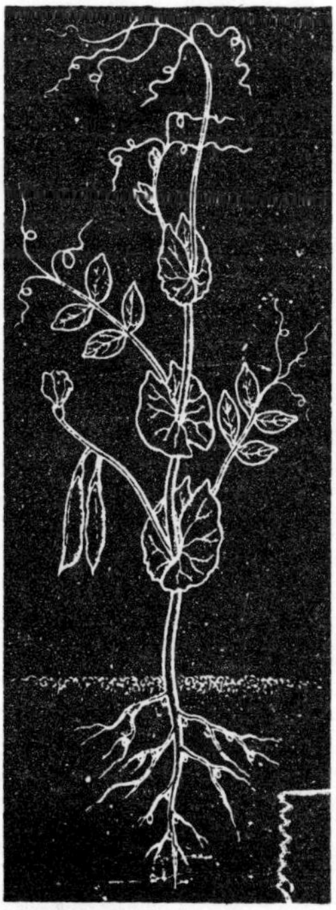

CHAPTER I

INTRODUCTION

A critical study of the **external morphology** or the external forms of higher plants is necessary primarily in order to be able to describe the plants accurately and to enable one to distinguish between similar-looking plants. This is a point essential in the study of Systematic Botany as well as in studying *variations* in connection with Evolution and Genetics. Out of the general science of Morphology other sciences like **Experimental Morphology, Morphogenesis,** etc., have sprung up which try to find out why and how each organ grows. The developmental study of an organ gives a true insight into the nature of an organ. **Embryology** (which is more internal than external morphology) or the study of the development of embryos has developed so much that it is sometimes treated as a separate science.

While speaking about the general morphology of Angiosperms one means the morphology of typical forms, that is, the ordinary plants we see around us like the mango tree, the rose shrub or the grasses. Such plants grow on soil (*terrestrial*), live in a moderate environment and are completely independent for their food supply (*autotrophic*). But one should have some idea of the exceptional forms one usually meets.

According to the mode of living, plants may be classified as follows:

I. **Autotrophic plants** or **Autophytes:** These are the normal plants living independently, building their own food out of inorganic raw material. There may be different types of them according to their *ecological environments*. Different environments cause differences in their morphology. Thus, we find special adaptations in aquatic plants, terrestrial plants, xerophytes, mangrove plants, etc. Among them may be considered the

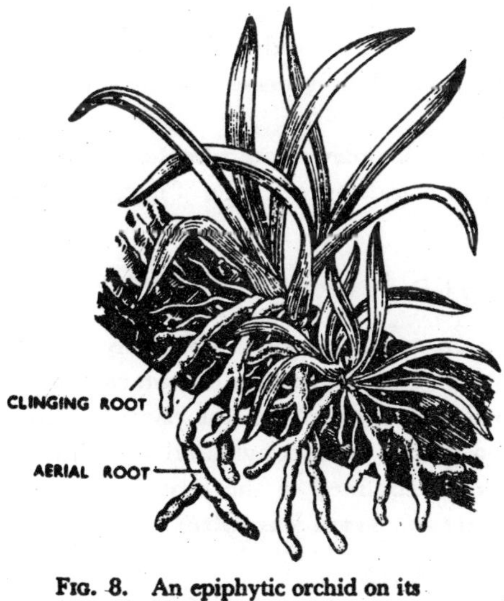

FIG. 8. An epiphytic orchid on its perch. Note two types of roots.

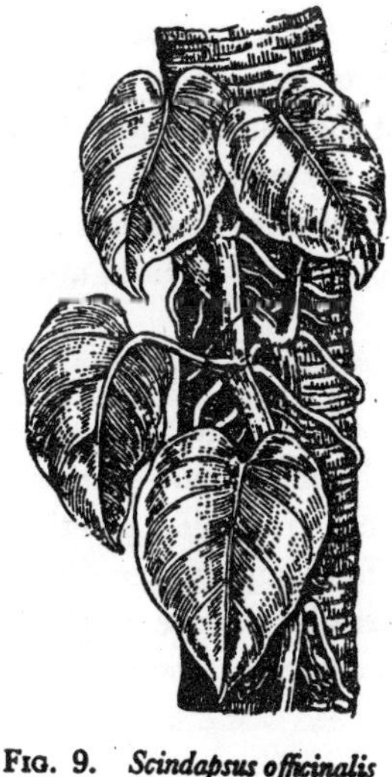

FIG. 9. *Scindapsus officinalis*—an epiphyte with climbing roots.

Epiphytes as, although they are peculiar, they are completely autotrophic. These plants usually grow on the top of other plants but are not parasitic on them. They have two types of roots. Some are *clinging roots* which are rather small and creep into the cracks in the bark of the tree on which the epiphyte grows. These clinging roots not only fix the epiphyte to its support but also absorb the little nutriment that is found within the debris accumulating on the bark. The other roots are the *aerial roots* which are more prominent and hang about in the air. These aerial roots are usually green and covered by a special fibrous tissue called *velamen* which is sponge-like in absorbing the moisture contained in air as well as rain water. Most orchids (Figs. 8 & 147) are epiphytes. There are also instances of epiphytes among some Pteridophytes like ferns. *Ficus* trees like banyan or peepul often begin their lives as epiphytes. Some seed may be deposited (say, with the excreta of a bird) on the top of a date palm or on the roof of a house where the seedling grows as an epiphyte. But, later on, these epiphytic *Ficus* plants send their roots down to the soil. Once the plant gets a foothold on the soil it becomes strong and gradually smothers its support or destroys the building. Quite reverse is the case of *Scindapsus officinalis* (*gajapipul*) of Araceae (Fig. 9) which begins its life as a terrestrial plant but, after climbing up, gets detached from the soil and becomes an epiphyte.

II. **Heterotrophic plants** obtain their nutrition by irregular methods. The nutrients are of organic origin and, therefore, derived from living or dead organisms. Heterotrophic plants may entirely depend on this irregular nutrition or substitute it by regular inorganic nutrients. These are of different types according to the special modes of nutrition. When an organism takes its position on another living organism and sucks the nutriments prepared by the latter for its own use, it is called a **parasite.** The second plant or animal, at whose expense the parasite is growing, is called the *host.* If the dependent organism lives, not on a living host, but on its dead body or on its decayed products (*e.g.*, rotten leaves, humus or dung) then it is called a **saprophyte.** It is not always possible to demarcate between parasites and saprophytes as a parasite growing on a living host may continue to grow after the host is dead.

Sometimes, two plants may live together in organic association without causing harm to either of them. The plant A may prepare one type of nutrition (say, nitrogenous) and pass a part of it to the plant B while the plant B may produce another type of nutriments (say, carbohydrates) and pass some of it to A. This is a case of mutual helping and not parasitism and is known as **symbiosis.**

Of special interest are the **Insectivorous** or **Carnivorous plants** which get their nitrogenous nutrients from small animals like insects. This is a case of feeding on animal protein and so does not come under parasitism.

PARASITES

A parasite may depend for its nutrition on its host totally or partially. When partially dependent, it retains its green colour with which it prepares its own carbohydrates. These are called **partial parasites** or **semiparasites.** When the parasite obtains all its food supply from its host it does not even possess chlorophyll and is called a **total parasite.** Some angiospermic parasites remain confined to the root or to the shoot. Accordingly, they are named as **root parasites** or **stem parasites.**

PARTIAL PARASITES

Among **partial stem parasites** the Loranthaceae is an important family. The mistletoe (*Viscum album*—Fig. 10) and *Loranthus* sp., specially the latter genus, are quite common on large trees like mango and often cause serious damage. They possess thick green leaves and remain firmly perched on branches by means of *haustoria.* The seeds of these plants are sticky and, when left by birds on the twigs of plants, germinate there with the radicle coming out as a sucker. Germination does not proceed far if a suitable host be not obtained. The sucker fixes the plant on the host. In *Viscum* this sucker, after penetrating the host tissues, gives out branches which form an internal sucking system. The *Loranthus* sucker, on the other hand, grows externally along the bark and gives out haustorial discs here and there. The haustorial discs or suckers connect the vascular systems of the two plants so that the xylem stream from the host is drawn into the parasite. *Cassytha filiformis* (Fig. 11) of Lauraceae is another partial stem parasite found in the

FIG. 10

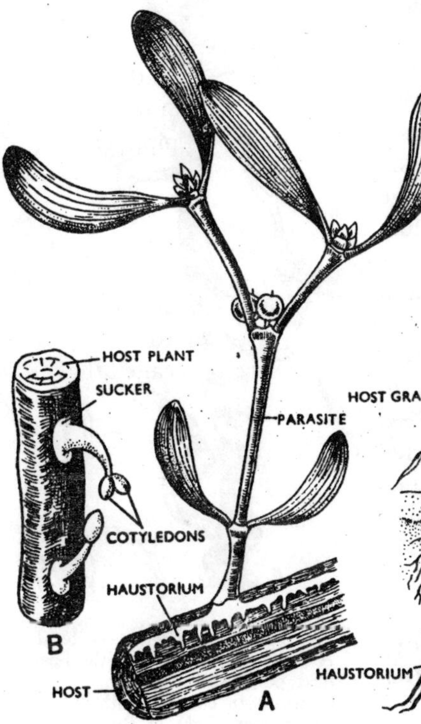

FIG. 11

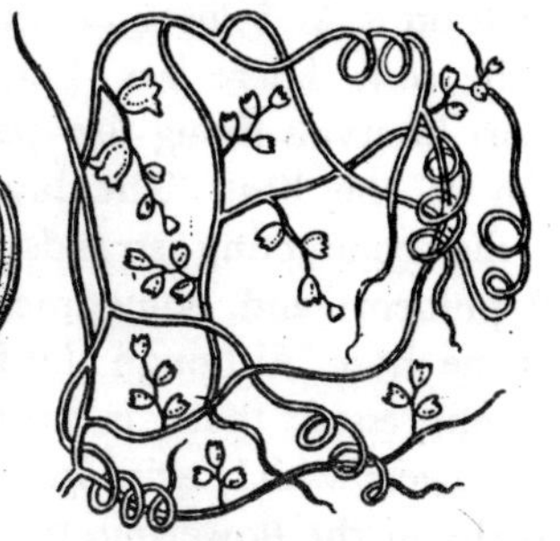

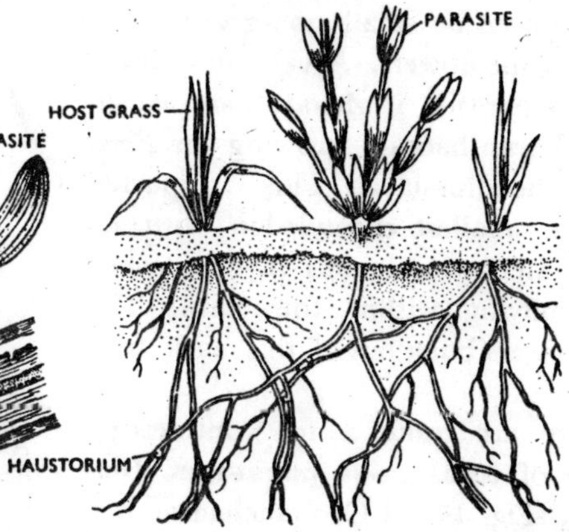

FIG. 12

PARTIAL PARASITES. FIG. 10. *Viscum album*—a partial stem parasite. A. The plant showing the haustoria exposed. B. Stages of germination in two plantlets attached to the host. FIG. 11. *Cassytha filiformis*—another partial stem parasite. FIG. 12. *Thesium alpinum*—a partial root parasite with haustorial suckers attached to grass roots.

tropics. The plant is a leafless wiry stem devoid of normal roots which entwines the host plant and comes in contact with the conducting tissues of the host by means of haustorical discs here and there. *Cassytha* is a partial parasite because the stem is greenish and is capable of some carbon assimilation.

Partial root parasites are not as common as the stem parasites. The sandal wood tree (*Santalum album* of Santalaceae), grown in a commercial scale in Mysore, develops as a root parasite. Some of its roots penetrate their sucking roots into the roots of nearby plants. Although it possesses independent roots as well, it does not grow luxuriantly if the host plants are not grown side by side. In cooler climates there are several plants which obtain their nutrition through roots attached to grasses and other plants. *Thesium* (Fig. 12) and *Rhinanthus* are such examples. They often seriously damage pastures.

TOTAL PARASITES

Total parasites are completely dependent on their hosts and are devoid of functional roots.

Among **total stem parasites** the golden yellow coloured twiner dodder (*Cuscuta reflexa* of Convolvulaceae—Figs. 13 & 231) is very common. The plant is formed solely of the twining wiry stem devoid of leaves. Even the embryo is devoid of cotyledons. The seed germinates as a terrestrial plant but, as soon as it gets its host, it forms a dense entanglement on it and loses all connection with the soil. There are haustorial discs at short intervals fixing the parasite on the stem of the host. The haustoria penetrate the conducting strands (both xylem and phloem) and, sometimes even the pith of the host. Although the foliage leaves are suppressed, there is no reduction in the flowers which arise in clusters from the nodes in the flowering time and there may even be small non-green bracts subtending the flowers. A very interesting total stem parasite is *Arceuthobium minutissimum* (Loranthaceae) growing on *Pinus excelsa* on the Himalayas. This is reported to be the smallest dicotyledon with no stem but an inconspicuous mycelium-like stock ramifying within the bark. The minute flowers burst out of this bark here and there.

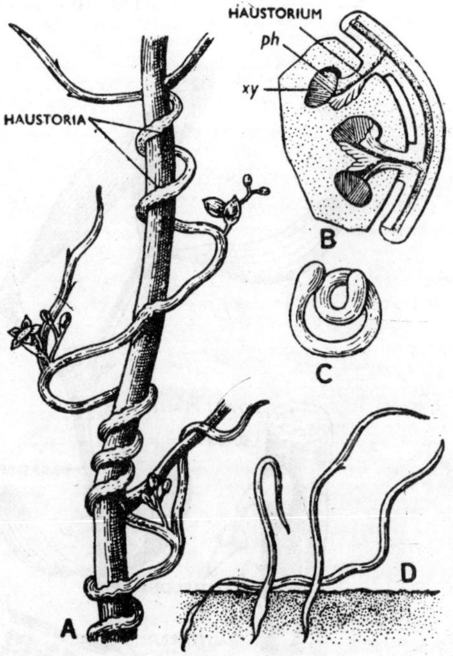

FIG. 13. *Cuscuta reflexa*—a total stem parasite. A. The parasite twining the host. B. T.s. of the host plant showing parasite in l.s. Haustoria in intimate contact with the xylem (*xy*) and the phloem (*ph*) of the host. C. An embryo of a *Cuscuta* sp. D. Stages of germination of *Cuscuta* in soil.

There are quite a few interesting examples of **total root parasites.** *Orobanche* sp. (Fig. 14) of Orobanchaceae is a leafless parasite growing on the roots of a number of horticultural crops like brinjal, tobacco, tomato, potato, mustard, etc. The seed germinates only on contact with root of the host and forms an underground tuber. Then it gives rise to an aerial inflorescence of coloured flowers with brown bracts. Another plant of similar habits is *Aeginetia acaulis*, also belonging to Orobanchaceae, which is common in the Orissa forests being a parasite on the roots of *khus-khus (Vetiveria zizanioides)* and weeds in the shade of *shal* (*Shorea robusta*) forests. Quite a similar example is *Balanophora dioica* (Fig. 15) of Balanophoraceae found on tree roots at an elevation of about 6000 ft. in Assam. This also is represented by leafless inflorescences. The most interesting example among total root parasites is that of *Rafflesia arnoldi* (Fig. 16) found in the East Indies. Parasitism has developed to the extreme here. The plant body is a tissue forming a branched filamentous body ramifying *inside* the tissue of the roots (sometimes stem also) of its host (*Cissus*, *Vitis*, etc.). The vegetative body has, therefore, been reduced almost to the level of the fungi. But, when flowering time comes, large buds begin to develop on this tissue and ultimately burst out of the roots. The buds open out to form some of the largest flowers known, sometimes reaching one metre in diameter and 15 Kg in weight.

The flower has a fleshy colour and the smell of rotten meat which attracts carrion flies for pollination. *Sapria himalayana,* a relative of *Rafflesia* and of similar habits grows in the extreme Eastern Himalaya (Abor Hills).

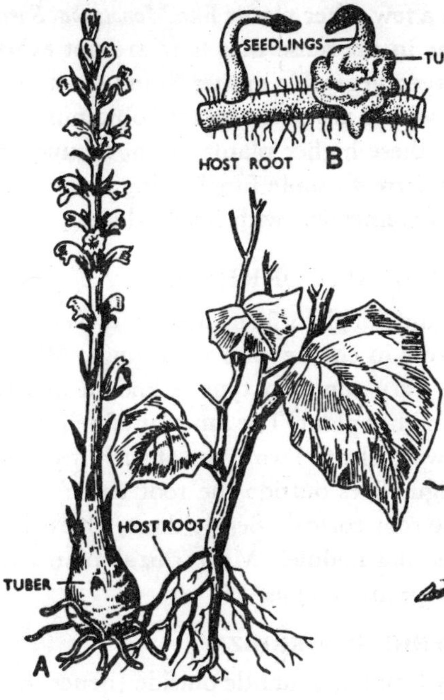

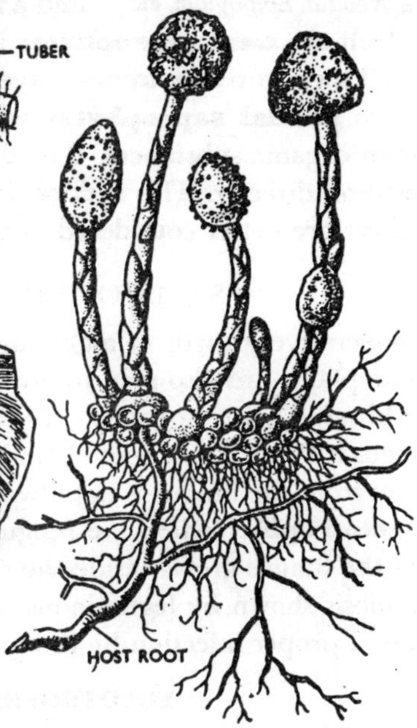

FIG. 14 FIG. 15

TOTAL ROOT PARASITES. FIG. 14. *Orobanche.* A. An *Orobanche* plant in flower growing on brinjal root. B. Germination of *Orobanche* showing two seedlings developing leafless tuberous stems. FIG. 15. *Balanophora dioica* developing on host root.

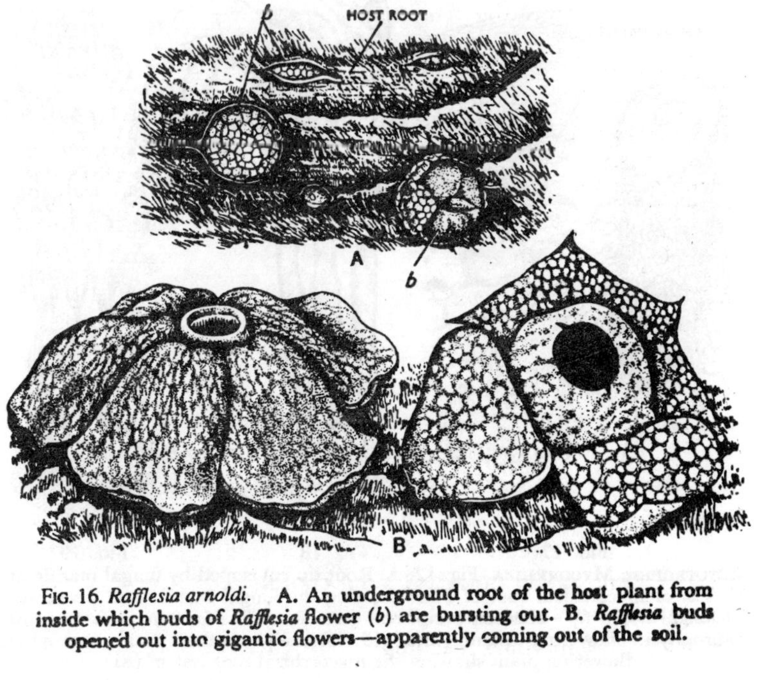

FIG. 16. *Rafflesia arnoldi.* A. An underground root of the host plant from inside which buds of *Rafflesia* flower (*b*) are bursting out. B. *Rafflesia* buds opened out into gigantic flowers—apparently coming out of the soil.

SAPROPHYTES

Saprophytes live on dead organisms or decayed organic products. They are quite common among fungi and bacteria. Among the higher plants there are some orchids (*Corallorhiza, Neottia, Epipogium*, etc.) and a few other plants like *Monotropa*, *Sarcodes*, etc., of *Pyrolaceae* which are **total saprophytes** in the sense that they are not even green and they obtain all their nutrition from decayed vegetable matter (humus) on the forest floor. There are also **partial saprophytes** which are green plants obtaining part of their nutrition from organic substances. But, these higher plants are not found to utilise the organic nutrition directly. This is done through some fungi living in their roots. These plants are therefore better considered in connection with Mycorrhiza.

SYMBIOSIS AND MYCORRHIZA

The cooperative growth of algae and fungi to form lichens is well known. Some Angiospermic plants are also seen to grow in conjunction with fungi. All of them come under **symbiosis** (living together). A very common type is known as **mycorrhiza,** where a fungus is found in the root of a higher plant. The amount of help that the higher plant receives from the fungus is not always clear, Two types of such mycorrhiza are recognized—**ectotrophic** (when the fungus lives outside the root tissue) and **endotrophic** (when the fungus grows inside the root cortex). Beside these, there is the special type of symbiosis shown by leguminous root nodules. Mycorrhizal plants often fail to grow without a proper infection by the specific fungus.

ECTOTROPHIC MYCORRHIZA

Fungi are known to occur regularly forming a mantle outside (hence, *ectotrophic*) the roots of trees of Fagaceae (oak, beech, etc.), Betulaceae (birch) and some Coniferae (pines —Fig. 17). The hyphi penetrate only between the outer cells and replace the normal

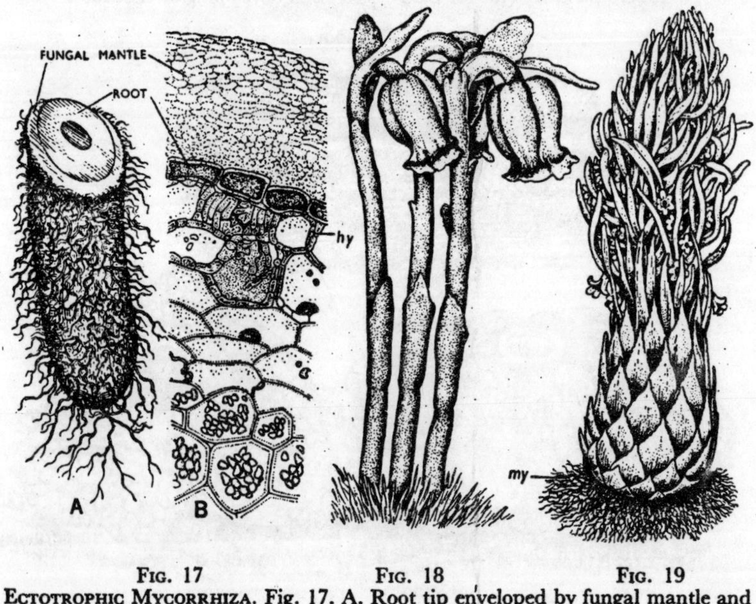

FIG. 17 FIG. 18 FIG. 19

ECTOTROPHIC MYCORRHIZA. Fig. 17. A. Root tip enveloped by fungal mantle and B. T.s. of mycorrhizal root of *Pinus* showing the fungal mantle and penetration of fungal hyphi (*hy*) between peripheral cells. FIG. 18. *Monotropa uniflora*—a total saprophytic flowering plant. FIG. 19. *Sarcodes*—another total saprophytic flowering plant showing the mycorrhizal root system (*my*).

root-hairs. In extreme cases we get the **total saprophytes,** where the roots are completely covered by the mantle of fungal hyphi so that the plant has to obtain all its nutrition through these fungi and do not even bear any chlorophyll. Such total saprophytic mycorrhiza is seen in *Monotropa* (the Indian pipe—Fig. 18), *Sarcodes* (Snow plant—Fig. 19), etc.

ENDOTROPHIC MYCORRHIZA

In these cases the fungi occur internal to the root tissue. In the case of ordinary green orchids the seed does not germinate properly if it does not get infected with some specific fungus at an early stage. Possibly, the young seedling gets some of its nutrition through

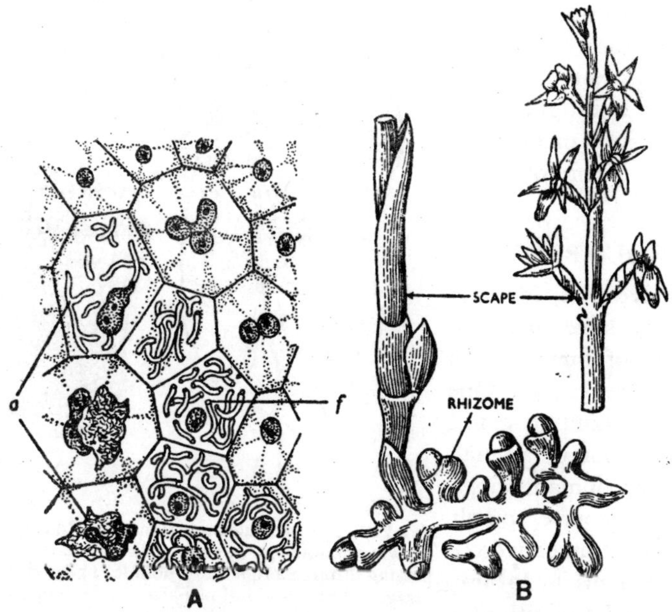

Fig. 20. Endotrophic Mycorrhiza. A. Section through the mycorrhizal region of an Orchid tuber showing fungal hyphi (*f*) within host cells. Some cells (*a*) show such hyphal mass in course of digestion. B. The Orchid *Corallorhiza innata* showing rootless mycorrhizal rhizome and flowering scape.

this fungus. But, even in these cases, the fungus becomes unnecessary when the plant matures. On the other hand, there are certain non-green orchids (*Corallorhiza, Neottia, Epipogium,* etc.), which are total saprophytes. *Corallorhiza* (Fig. 20) rhizome branches freely under the rich humus in Pine forests and bears no root. A flowering scape bears colourless scale leaves. The rhizome is heavily infected with fungus. Heather, ling (*Calluna*) and other members of Ericaceae also get infected with some mycorrhizal fungus at the seedling stage. Occurrence of such mycorrhizal fungus is widespread among the Pteridophytes.

PLANTS WITH ROOT-NODULES

Leguminous plants are well known for the presence of nodules on roots (Fig. 21) which harbour some bacterium (e.g., *Bacillus radicicola* or *Pseudomonas radicicola*) which is able to fix atmospheric nitrogen and supply it to the host plant in the form of amino compounds. This fact is of great importance in agriculture and in the maintenance of the nitrogen cycle. These bacteria are usually present in the soil and when the leguminous plants are grown, the bacteria enter the roots through the root hairs. Thereafter, their presence causes the development of the nodules or tubercles as pathological growth. The nodule cells are full of the bacteria. If, accidentally, the soil is sterile, the leguminous plants do not develop well and when growing leguminous crops on such soil, inoculation with the necessary strain of bacterium may be necessary. It is also found that the nodules do not develop well if the soil be too rich in nitrates.

This is a true case of *symbiosis* as the bacteria do not harm the leguminous plants but supply them with nitrogenous substances while they receive other forms of nutrients in return.

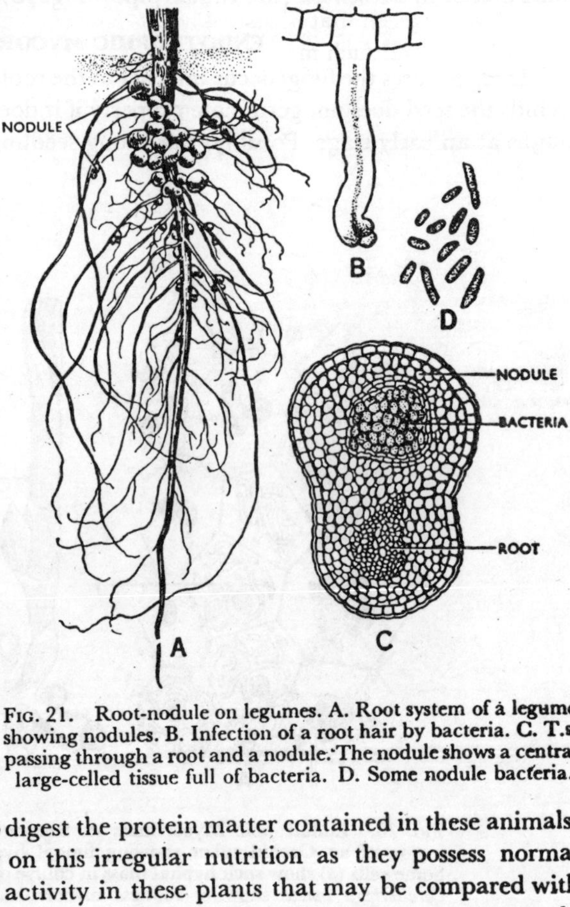

FIG. 21. Root-nodule on legumes. A. Root system of a legume showing nodules. B. Infection of a root hair by bacteria. C. T.s. passing through a root and a nodule. The nodule shows a central large-celled tissue full of bacteria. D. Some nodule bacteria.

INSECTIVOROUS PLANTS

The **insectivorous** (or **carnivorous**) plants have attracted the imagination of literary men ever since they were known. One may read stories of hunting plants which prey upon large animals like dogs and even human beings. Such stories are, of course, purely imaginary. The insectivorous plants have mechanisms to trap animals no bigger than butterflies or grasshoppers and are adapted to digest the protein matter contained in these animals. These plants depend only partially on this irregular nutrition as they possess normal leaves and roots. There is no such activity in these plants that may be compared with the beasts of prey. Following are the types of insectivorous plants that are commonly known in India and outside:

FAMILY DROSERACEAE

There are a number of well-known plants belonging to this family:

Drosera or **Sundew** (Fig. 22): Some four species of *Drosera* are well known. Of these *Drosera indica*, *D. burmanni* and *D. peltata* occur in India while *D. rotundifolia* is found in British peats and bogs where absorption by roots is difficult. The species vary mainly in the shape of leaves which is narrow and long in *D. indica*, spathulate in *D. burmanni*, peltate in *D. peltata* and round in *D. rotundifolia*. All the species are small herbs a few inches high. *D. burmanni* is seen all over Eastern India (often in rice fields) as small plants with rosettes of radical leaves which look like small reddish patches from a distance. *D. peltata* from Khasia Hills, the Western Himalayas and the Ghats has sometimes both radical and cauline leaves or only cauline leaves. In all Droseras the peculiarity is that the upper surface of each leaf blade is covered with glandular hairs called *tentacles* which secrete a sticky fluid. This drop of fluid shines in the sun like dew drops hence the name 'Sundew'. Insects are attracted by the glistening drops and as one alights on the leaf surface it gets entangled in the sticky fluid and all the tentacles, which are protein sensitive, bend towards it so that the insect is trapped. It is soon killed while the glands secrete an enzyme for digesting the protein substance and the tentacles remain in a curved position until digestion is complete. The movement of the tentacles is caused by chemical stimulus as it is seen that if a small bit of meat is placed on the leaf, the tentacles react in the same way and such movement cannot be induced by any other means. The digested protein is absorbed by the leaves.

FIG. 22

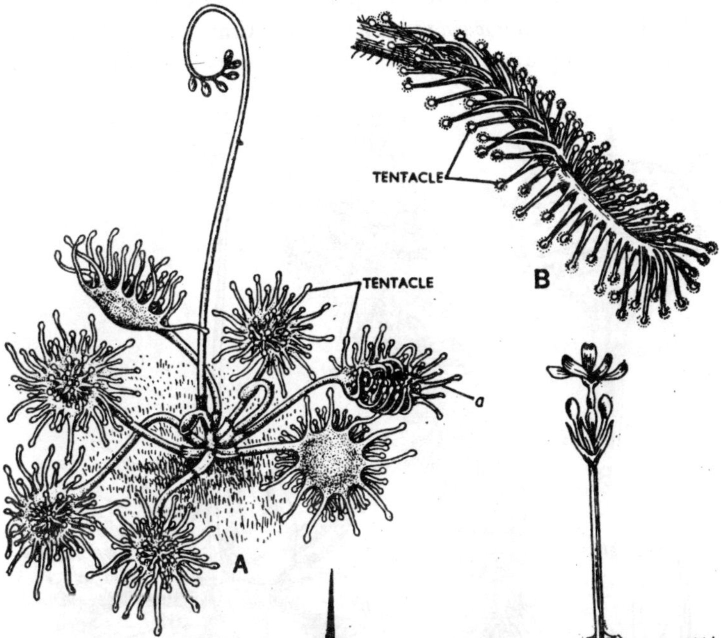

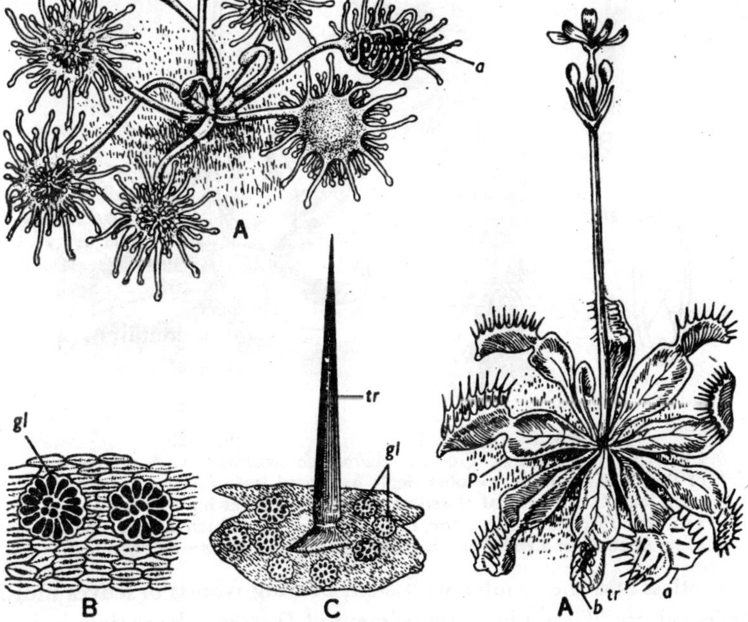

FIG. 23

INSECTIVOROUS PLANTS. FIG. 22. *Drosera rotundifolia*. A. An entire plant with flowering scape. Note leaves with erect tentacles excepting the one on extreme right (*a*) where the tentacles are bent over a trapped insect. B. side view of a leaf. FIG. 23. *Dionoea muscipula*. A. An entire plant with flowering scape. *p*=winged petiole; *a*=leaf in open position; *tr*=trigger hairs; *b*=leaf trapping an insect. B. Glands (*gl*) secreting digestive juice. C. One trigger hair (*tr*) and glands (*gl*).

Dionoea or **Venus' Fly-trap** (Fig. 23): *Dionoea muscipula* is a well-known insectivorous plant from the North American bogs. The radical leaves occur in a rosette and act like rat traps. The petiole of the simple leaf is winged and the lamina is deeply notched at the apex. The leaf margin is slightly incurved and provided with long pointed teeth. The two halves of the leaf are capable of movement with the midrib acting as a hinge. When the leaf folds, the marginal teeth closely fit into one another like the interlocking device of a rat trap. The halves of the leaf normally remain slightly inclined forming an angle at the midrib. On the upper surface of the leaf are six (three on each half) extremely sensitive *trigger hairs* and a number of reddish glands. As a trigger hair is touched by an insect the leaf closes like a trap and the glands secrete copiously a digestive enzyme which digests the protein matter. The leaf reopens only after digestion is complete.

Aldrovanda (Fig. 24): *Aldrovanda vesiculosa*, sometimes called *Malacca jhangi*, used to be found as a free-floating aquatic weed in the salt marshes to the south of Calcutta.

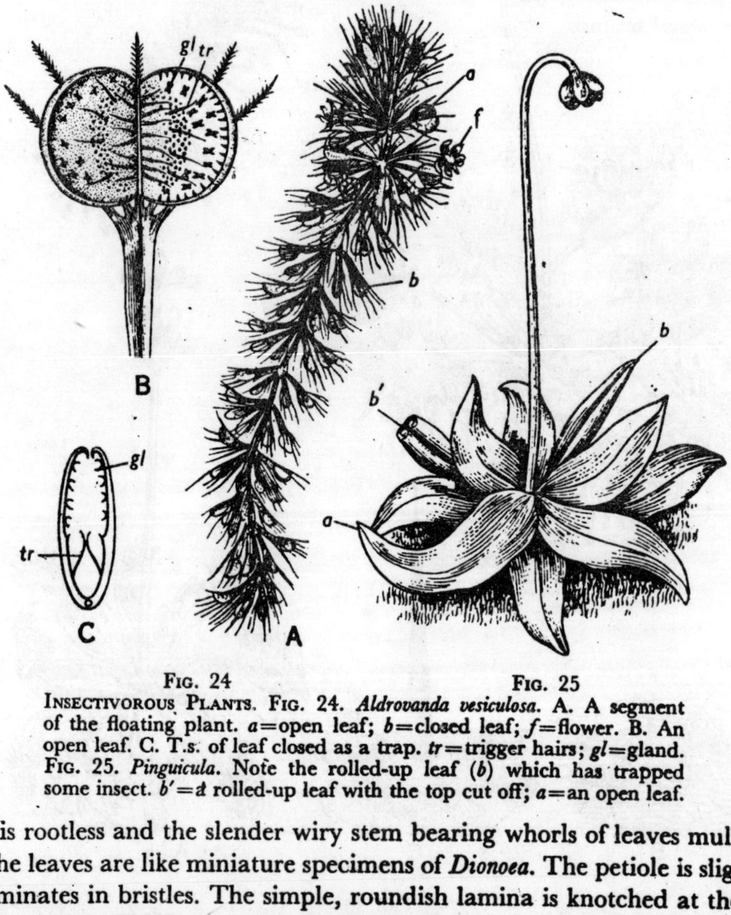

FIG. 24 FIG. 25

INSECTIVOROUS PLANTS. FIG. 24. *Aldrovanda vesiculosa*. A. A segment of the floating plant. *a*=open leaf; *b*=closed leaf; *f*=flower. B. An open leaf. C. T.s. of leaf closed as a trap. *tr*=trigger hairs; *gl*=gland. FIG. 25. *Pinguicula*. Note the rolled-up leaf (*b*) which has trapped some insect. *b'*=a rolled-up leaf with the top cut off; *a*=an open leaf.

The plant is rootless and the slender wiry stem bearing whorls of leaves multiply vegetatively. The leaves are like miniature specimens of *Dionoea*. The petiole is slightly winged and terminates in bristles. The simple, roundish lamina is knotched at the top. The margin of the leaf is slightly curved inwards and bears minute teeth. The two halves of lamina are slightly inclined inwards forming an angle at the midrib. On the upper surface of the leaf, specially close to the midrib, there are a number (more than six) of sensitive *trigger hairs* besides the digestive glands which are distributed all over the surface. As soon as very small animals touch the hairs, the two halves close about the midrib effectively trapping the victims inside forming something like a temporary stomach. The digestive glands then secrete the necessary enzyme and the leaf remains closed until digestion and absorption of the juice is complete.

Pinguicula or **Butterwort** (Fig. 25): Different species of *Pinguicula* grow in the European bogs. One species, *Pinguicula alpina*, is found in the Alpine Himalayas (11,000 to 13,000 ft.). These are small herbs like *Drosera* with scanty development of root. The leaves similarly occur in a rosette but are sessile and do not bear tentacles. On the surface of the leaf are short, stalked hairs secreting a sticky substance and sessile glandular hairs which secrete an enzyme. When an insect alights or crawls on the leaf surface it gets stuck up in the sticky substance and the presence of the nitrogenous substance causes the margins of the leaf to roll upwards and inwards thus trapping the victims. Secretion and digestion follows. The leaf unrolls after digestion and absorption is complete.

FAMILY UTRICULARIACEAE

This is another family of insectivorous plants of which **Utricularia** or **Bladderwort** (Fig. 26) is very common. The genus is widely distributed in tropical and temperate regions with more than 200 species. About 20 species are found in ponds, ditches and marshes or growing among wet mosses in India. The commoner species are free floating,

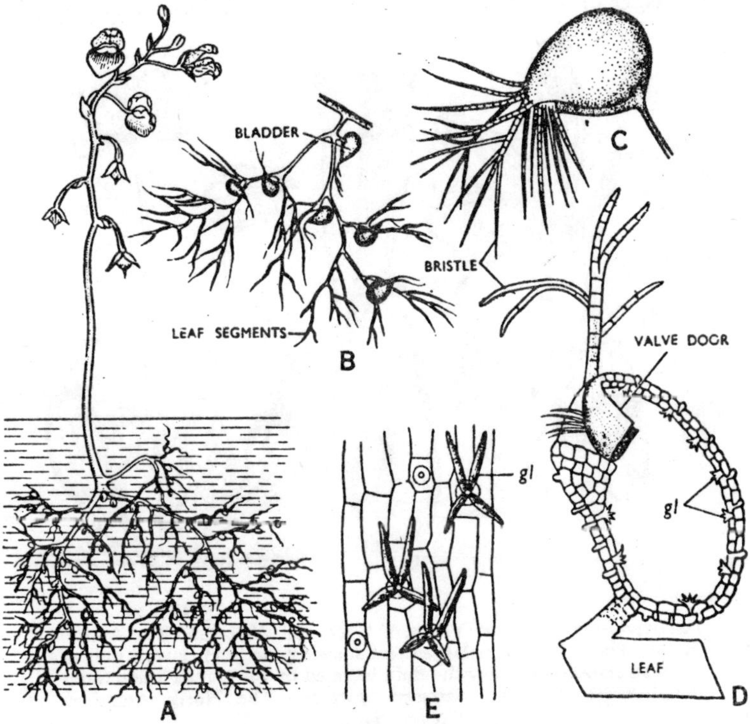

Fig. 26. Insectivorous plant *Utricularia*. A. A plant in flower, floating in water. B. A portion of leaf showing leaf segments and bladders. C. A magnified bladder. D. Section of a bladder showing the valve door and the digestive glands (*gl*). E. Inner wall of a bladder showing glands (*gl*).

aquatic, rootless weeds just like *Aldrovanda* and are found in similar surroundings. Such a plant shows a slender elongated stem multiplying vegetatively which bears finely dissected submerged leaves looking like green roots. Numerous segments of the leaf are modified into *bladders* or *utricles*. The bladders vary in size, some larger ones being as much as one-eighth of an inch in diameter. The inner wall of the bladder is studded with

3 or 4-pronged digestive glands. There is an opening of the bladder which is provided with a valve door which can be pushed in but cannot be pushed out. There are long erect branched hairs or bristles on the valve door as well as around the opening. The bladders are normally filled in with water which is absorbed by the wall leaving it empty. If minute water animals (which can be barely seen with the naked eyes) now push the bristles, the valve door opens and some water carrying the organisms is sucked in. The pressure of this inside water prevents the opening of the valve door. Death, decay and digestion follows and the bladder is again ready for its fresh prey only after the liquid is absorbed. Sometimes hard and indigestible remnants of such organisms are found within the bladders.

PITCHER PLANTS

A number of plants of a few allied families (Nepenthaceae, Sarraceniaceae, etc.) show peculiar pitcher-like modifications of leaves which act as insect-catching traps.

Nepenthes (Fig. 27 and wrapper picture): This is a South Asiatic epiphytic genus of many species of which one is found in Ceylon and one in Assam. By the name

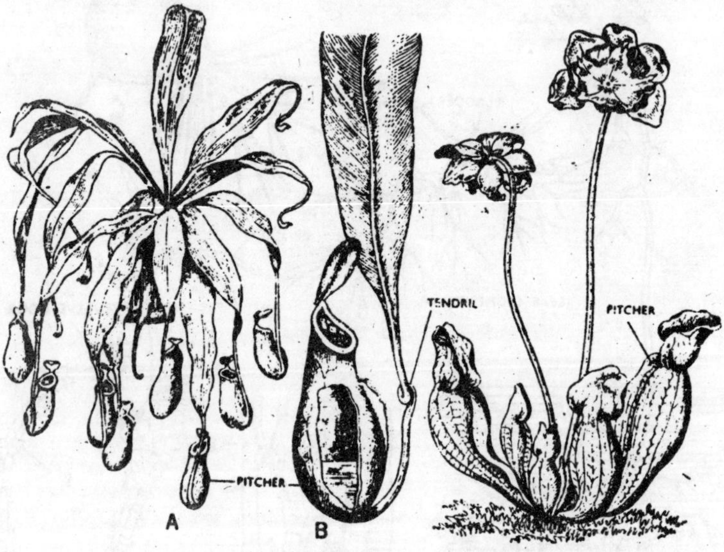

Fig. 27 Fig. 28

INSECTIVOROUS PITCHER PLANTS. FIG. 27. A. *Nepenthes*—The Asiatic Pitcher Plant. B. One enlarged leaf showing the pitcher with part of the wall removed to show the fluid secreted by the glands inside (after Strasburger). FIG. 28. *Sarracenia purpurea*—The North American Pitcher Plant.

pitcher plant usually this genus is signified. The lower part of the petiole is winged and the upper forms a tendrillar structure which may coil round a support to bear the weight of the pitcher-like structure above (which may be modified leaf lamina) with an immovable lid. The lid may be coloured to attract insects and keeps the pitcher closed during immaturity. There is some dispute as to the real morphology of the different parts of the pitcher. Some hold that the winged structure is the leafbase, some others consider the winged part, the tendrillar part and the pitcher as different regions of the petiole. The lid is considered as either the whole lamina or the apex of it. One thing is certain —the pitcher is a part of the leaf, either the petiole or the lamina. The pitcher is 1 to 10 or more inches long. There is no active mechanism for catching insects. The inner walls of a pitcher are studded with button-shaped glands which secrete a fluid partly filling up the vessel. The edge of the pitcher is so slippery and curved in such a manner that insects trying to have a perch on it slip down and cannot climb up the steep wall. There are also some downwards pointed hairs below the rim which prevent the moving up of the insects. These get drowned and are digested by the enzyme secreted from glands. Gradually, the pitcher gets filled up with horny remnants of ants, grasshoppers, cockroaches, etc. The wrapper shows a coloured photograph of the pitcher plant as it grows in its natural surroundings in the Garo Hills in Eastern India.

Sarracenia (Fig. 28): Found in the North American bogs, these pitcher plants have pitchers forming a rosette. The pitchers are similar to those of *Nepenthes* but sessile.

There are a few other types of pitchers like *Cephalotus*, *Heliamphora*, etc. All such pitchers are modifications of leaves and act as insect traps in the same way.

CHAPTER II

THE SEED AND THE SEEDLING

THE SEED

It is customary to begin the study of Morphology with a study of the seed simply because the plant grows out of the seed. Beginning Morphology with the seed is merely a convention and is not to be regarded as any attempt to solve the eternal question—whether the hen or the egg came first. It is found convenient to begin with a description of the seed, then to study the seedling, then to study the different vegetative parts followed by the reproductive organs and finally to end with the formation of the fruit and the seed which brings us to the next generation and completes the life cycle of a flowering plant.

The *Phanerogams* or *Spermatophytes* are endowed with this peculiar but very common detachable vehicle of reproduction carrying the germ of a new individual which is known as the **seed.** Botanically speaking, a seed is to be defined as a *mature integumented megasporangium.* We do not get any integumented megasporangium among the *Cryptogams*, the highest development among which is exemplified by heterosporous plants like *Selaginella* and some more complex fossil forms. Among the *Gymnosperms* the megasporangium is usually covered by a single integument and this is known as the *ovule*. This ovule is naked, i.e., not enclosed in a fruit or the carpellary vessel, hence the name *Gymnosperm*. In the *Angiosperms*, the ovule or ovules are generally endowed with two integuments and are enclosed within a case formed by the carpel or carpels which is known as the *fruit*. When these ovules become mature after fertilisation and some post-fertilisation changes (e.g., hardening and drying up of the whole structure), it becomes the *seed*. It contains an *embryo* which passes a long or short period of *dormancy* and *germinates* when proper conditions are fulfilled. If the embryo within a seed gets killed, it fails to germinate under all conditions and the seed is said to have become *inviable*. *Viability test* is, therefore, very important in Agriculture and Horticulture as a test for assessing the usefulness of any seed sample.

In describing the parts of a seed, one has to proceed a little mechanically at this stage as the real nature of the different parts will become clear only when one has studied the development of the seed which is described towards the end of Morphology. Every seed (general plan—Fig. 29) has a **seedcoat** which is really the two

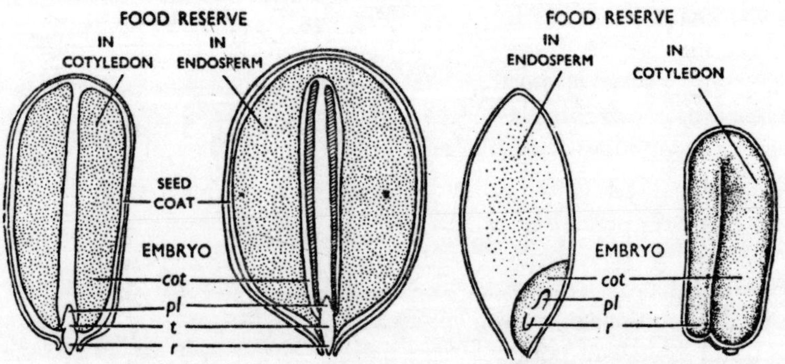

FIG. 29. General plan of the four types of seed. A. Dicotyledonous exalbuminous; B. Dicotyledonous albuminous; C. Monocotyledonous albuminous; D. Monocotyledonous exalbuminous.
Embryo = cotyledon (*cot*) + radicle (*r*) + plumule (*pl*) + tigellum (*t*).

dried up integuments, the **testa** (outer integument) and the **tegmen** (inner integument). The testa and the tegmen may sometimes get so much fused together that they cannot be separated. Inside the seedcoat is the **kernel** of the seed which contains the **embryo** or the rudimentary plant which is the most important part of the seed and may or may not be accompanied by some nutritive tissue (**endosperm** and **perisperm**).

An important part of the embryo is the **cotyledon** or the seed leaf which is a special type of leaf (discussed in connection with Leaf). Angiosperms are divided into the *Monocotyledons* and the *Dicotyledons* according to the number of cotyledons. *Polycotyledonous* types are known among Gymnosperms. Some Angiospermic seeds do not show a proper differentiation of cotyledons but other characteristics put them along with either Monocotyledons or Dicotyledons. Both Dicotyledonous and Monocotyledonous plants may be **Albuminous (Endospermic)** or **Exalbuminous (Exendospermic)** according as whether the endosperm is present or not.

DICOTYLEDONOUS EXALBUMINOUS SEEDS

A typical example of this type is found in the common **pea** (*Pisum sativum*—Fig. 30). On carefully opening a mature green pod along the *dorsal suture* the placental tissue is seen to spread along the *ventral suture* and the roundish seeds are seen arranged in two rows along the length of the pod. Each seed is attached to the *placental tissue* on the fruit suture by a stalk called the *funicle*. The funicle is narrow at the placental end but widens into a disc where it joins the seed. When the mature seed is detached the broad end of the funicle leaves a scar on the seed called the *hilum*. Next to the hilum is a pinhole opening on the seedcoat which is the *micropyle*. If the seed is soaked, wiped and then squeezed, water is seen to ooze out of this micropyle. The seed is covered by the tough *seedcoat* of a light colour. This tough seedcoat is the *testa*. The *tegmen*, which is delicate and completely adherent to the inner side of the testa, is not distinguishable in the mature seed.

On opening the seedcoat the *kernel* is obtained. The kernel in dicotyledonous exalbuminous seeds is the *embryo*. In it the two fleshy *cotyledons* are very conspicuous. These are fleshy as all the nutrients required by the growing seedling in the first few days are kept stored here. The two cotyledons are hinged to an *axis* (*tigellum*) so that they open out like a book. The tigellum represents the axis of the future plant. One end of the tigellum is pointed and protrudes out of the cotyledons. This lies next to the micropyle and is the *radicle* or the rudimentary root. The protruding radicle lies under the pouch-like expansion of the seedcoat and is thus visible even when the seedcoat is not removed. The other end of the tigellum is the feathery *plumule* end which is the first apical bud of the future plant and develops into the shoot. The plumule lies in a groove inside the cotyledons. The point of attachment of the cotyledons to the tigellum is the first node on the axis and careful observation shows the presence of the first lateral buds in the axils of the cotyledons. The portion of the tigellum just *below the cotyledonary node* (i.e., between radicle and node) is called the *hypocotyl* and the portion just *above* (i.e., between node and plumule) the node is the *epicotyl*.

All the dicotyledonous exalbuminous seeds conform to the above plan, though there may be variations in details. In **gram** (*Cicer arietinum*—Fig. 31), the seed is broad at one end and somewhat pointed at the other. On the seedcoat, below the *hilum*, there is another more prominent scar, the *strophiole*, which is a scar left by a funicular outgrowth. The brown seedcoat is the *testa* but on its inner side a papery white membranous layer may be distinguished as the *tegmen*. The embryo does not differ from that of pea except in shape.

There are various types of **bean** seeds of which *Dolichos lablab* (Fig. 32) is very common. The seeds occur in the pod as in peas. They are larger and more or less oval. The *seedcoat* (mainly *testa* with a thin fused *tegmen*) is very hard and black, brown or red in colour. The *funicle* is extended into a long *raphe* which is seen above the hilum. The *embryo* is as usual.

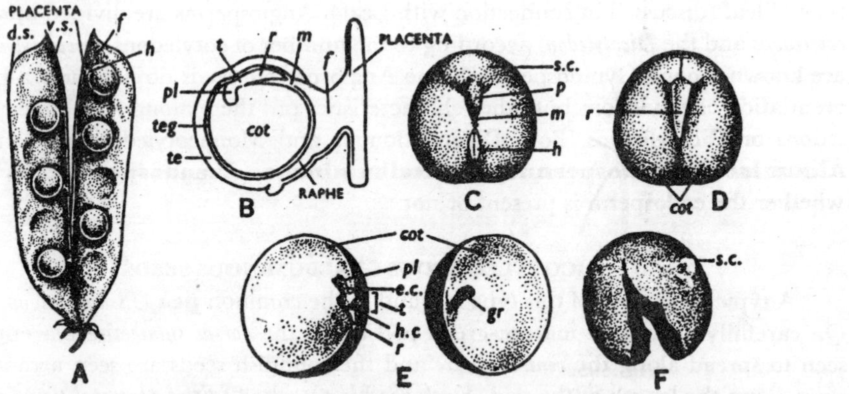

FIG. 30

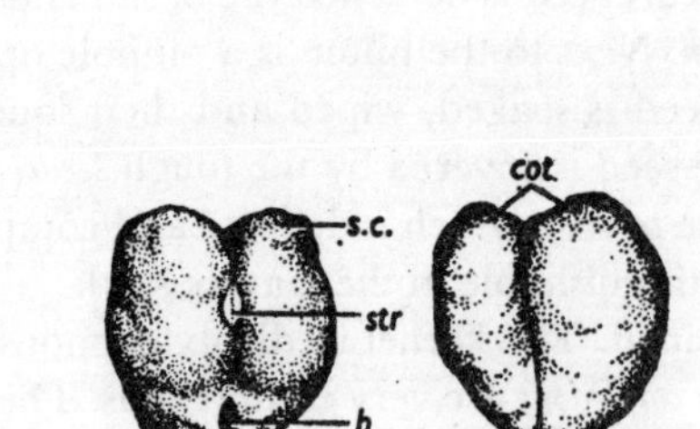

FIG. 31

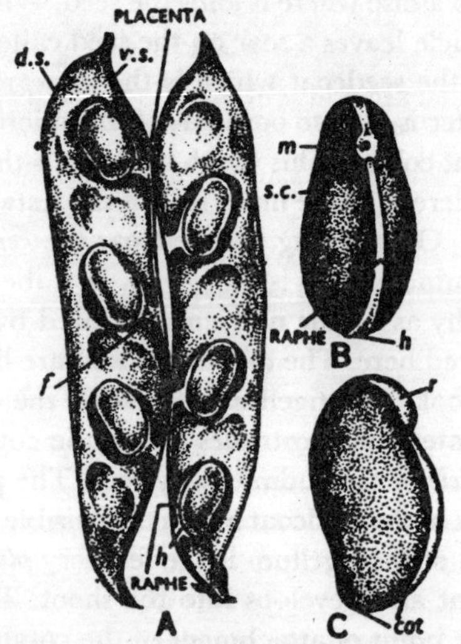

FIG. 32

DICOTYLEDONOUS EXALBUMINOUS SEEDS, FIG. 30. The pea seed. A. Pea pod opened to show positions of seeds. B. Diagrammatic section through seed and placenta. C. Whole seed. D. The kernel after removal of seedcoat. E. Kernel with cotyledons dissected. F. Removed seedcoat. FIG. 31. Gram seed. A. Whole seed. B. The kernel after removal of seedcoat. C. The two dissected cotyledons. D. Peeled seedcoat. FIG. 32. Lablab Bean (*Dolichos lablab*) seed. A. Open pod. B. Whole seed with prominent raphe. C. Kernel after removal of seedcoat. *d.s.*=dorsal suture; *v.s.*=ventral suture; *f*=funicle; *h*=hilum; *te*=testa; *teg*=tegmen; *m*=micropyle; *cot*=cotyledon; *t*=tigellum; *pl*=plumule; *r*=radicle; *s.c.*=seedcoat; *p*=pouch containing radicle; *e.c.*=epicotyl; *h.c.*=hypocotyl; *str*=strophiole; *gr*=groove for plumule

There are quite a few other dicotyledonous exalbuminous seeds that we commonly come across. The **tamarind** (*Tamarindus indica*) seed is rectangular and covered by a very hard shell-like seedcoat. The hard covering on the seed of **mango** (*Mangifera indica*)

is actually the innermost layer of the fruit (*endocarp*) which is not a part of the seed. Inside, the two very large cotyledons are covered by a papery seedcoat. In **jack-fruit** (*Artocarpus heterophyllus* — Fig. 33) the two cotyledons are of irregular and unequal size. The seedcoat shows both testa and tegmen. The cotyledons contain *latex* and, of course, lot of starch. In the **cucurbits** (gourd, cucumber, etc., of Cucurbitaceae) the seed is rather flat. The testa is horny and free from the thin tegmen inside. The cotyledons also

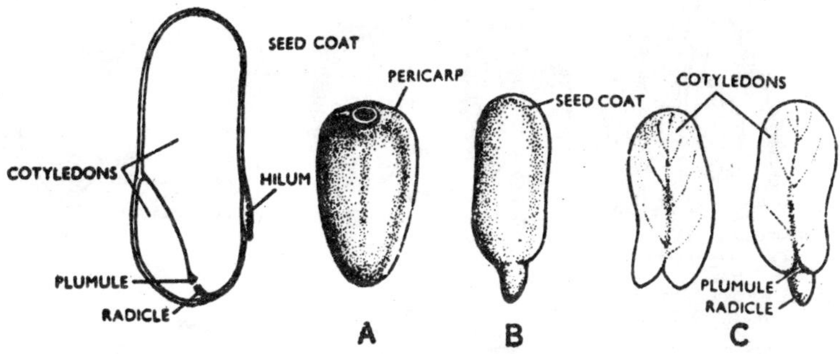

FIG. 33 FIG. 34
DICOTYLEDONOUS EXALBUMINOUS SEEDS. FIG. 33. L.s. of jack-fruit seed. Note unequal cotyledons. FIG. 34. Sunflower seed. A. Fruit enclosing single seed. B. Kernel on breaking open the pericarp. C. Dissected embryo.

are flat although they contain a good amount of food matter including oil. They show vein markings although white. The **sunflower** (*Helianthus annuus*) seed (Fig. 34) is actually its fruit enclosing a single seed. The hard covering is the fruit-wall or the *pericarp*. Inside, the oily *embryo* is found covered by a brownish membranous seedcoat. **Mustard, groundnut** (peanut), **sesame** and **flax** (linseed) seeds contain lot of oil in the cotyledons. In **orange** and **lemon** the seed is sometimes *polyembryonic*, i.e., it contains more than one embryo. The **oak** acorn is a fruit. The testa of the **pomegranate** (*Punica granatum*) is juicy and forms the edible part.

DICOTYLEDONOUS ALBUMINOUS SEEDS

In this type of seeds (ground plan Fig. 29) the food is not stored in the two cotyledons of the embryo but in the **endosperm** external to the embryo. It will be seen later that all embryos get their food from the endosperm which in its turn gets its food from the *nucellus* of the ovule. In the exalbuminous type of seeds the embryo completely consumes the endosperm and nucellus so that they are no longer seen while the food is kept stored in the cotyledons which become swollen. In the albuminous type the endosperm is still present and the cotyledons are thin—acting only as food-sucking organs.

A very common example is the **castor** bean (*Ricinus communis*—Fig. 35) where the fruit is not a bean but a three-chambered capsule. Here, the seedcoat is a hard shell of a mottled black or brown colour. The hilum is almost hidden by an outgrowth, the *caruncle*. The caruncle is spongy and absorbs water readily so that it may be of some use in germination. There is a distinct *raphe* running longitudinally down the seed from the hilum. On breaking open the shell a white mass is found covered by a papery white membrane. This membrane is sometimes supposed to be the tegmen but has been found not to be a part of the seedcoat. It is a remnant of the nucellus which has not been completely used up. Such a remnant is called the **perisperm.** Inside the membrane the whole whitish, fleshy, slightly flattened and oval mass is the *endosperm* which contains much oil. On cutting open the endosperm the *embryo* is found to remain embedded inside. There are two thin, white *cotyledons*. The cotyledons show distinct vein markings like leaves. The veins leave an impression on the endosperm. The two cotyledons are hinged to the *tigellum* which shows the protruding radicle with a short *hypocotyl* behind it and the plumule hidden between the cotyledons.

FIG. 35

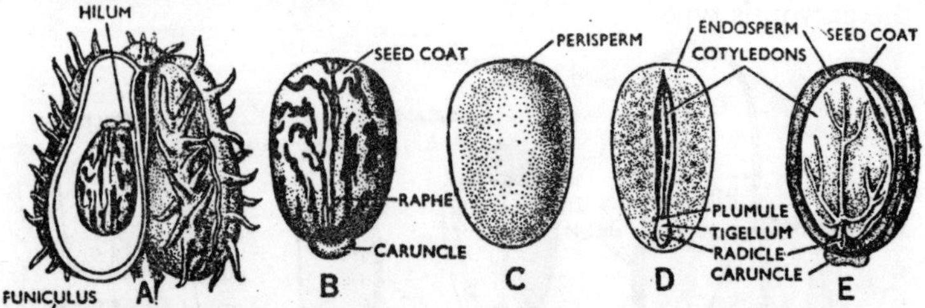

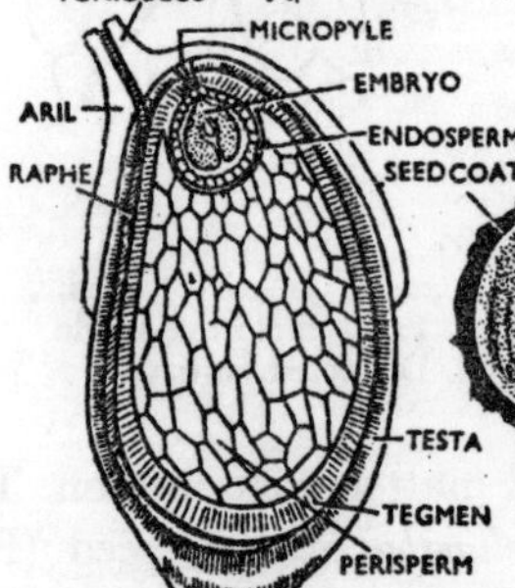

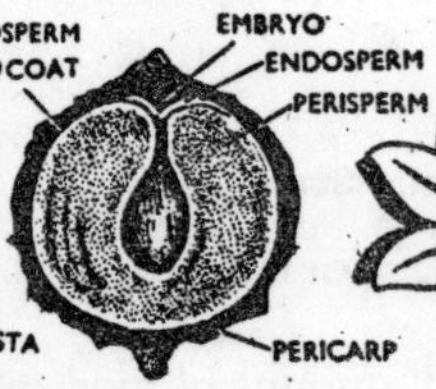

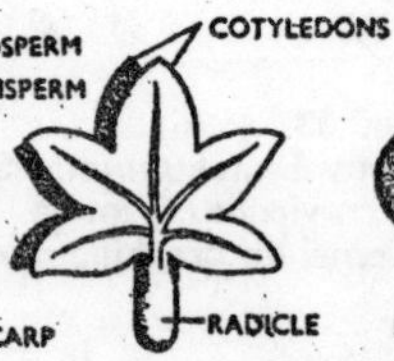

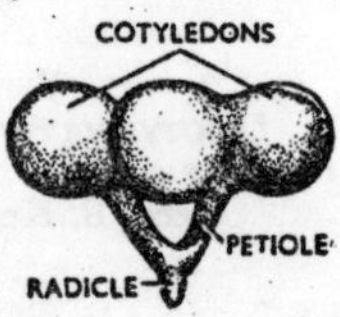

FIG. 36 FIG. 37 FIG. 38 FIG. 39

DICOTYLEDONOUS ALBUMINOUS SEEDS. FIG. 35. Castor seed. A. A capsule showing position of seed. B. Whole seed. C. Kernel on removal of seedcoat. D. L.s. of kernel showing position of embryo. E. Whole seed cut into half showing position of embryo. FIG. 36. L.s. of seed of water-lily (*Nymphaea alba*). FIG. 37. L.s. of single seeded fruit of *Piper nigrum*. FIG. 38. Embryo of *Tilia europoea* showing palmate cotyledons. FIG. 39. Embryo of *Geranium molle* showing lobed cotyledons provided with petioles.

Among common plants, the dicotyledonous albuminous type of seed is also found in **papaw** (*Carica papaya*), **jute, cotton, Mirabilis,** etc. Some whole families show this type of seed, e.g., Magnoliaceae, Annonaceae (custard apple shows a *ruminated endosperm;* other common plants are the mast tree or *Polyalthia longifolia* and *Artabotrys hexapetalus*), **Papaveraceae** (the poppies). The family **Nymphaeaceae** shows *perisperm* and *aril* (an outgrowth like caruncle) in addition to the endosperm. In the **water-lily** (*Nymphaea*—Fig. 36) of this family, the testa and tegmen surround a mass which is the *perisperm.* Inside it, on the pointed end is the minute *embryo* embedded in a small *endosperm.* The *raphe* is prominent. The seed has a third covering called the *aril* which is spongy and helps the seed to float. The family **Piperaceae** shows both endosperm and perisperm. The common **black pepper** (*Piper nigrum*—Fig. 37) of this family is a fruit with a single seed. The shrivelled pericarp encloses a seed with a cellulose perisperm having an embedded endosperm and a tiny embryo. In **coffee** (*Coffea arabica*), the small embryo with two cotyledons is embedded in a mass of horny endosperm formed mostly of hemicellulose. In the lime tree (*Tilia europoea*—Fig. 38) the endosperm embeds an embryo in which the cotyledons are thin, palmately lobed and veined like leaves. *Geranium molle* (Fig. 39) shows a similar embryo within endosperm having cotyledons which are not only lobed but are also stalked.

MONOCOTYLEDONOUS ALBUMINOUS SEEDS

Most of the common monocotyledonous seeds are albuminous. The large endosperm of the cereals is the most important source of starch, the principal food of all people.

Rice (paddy), **wheat** and **maize** may be taken as the type seeds of this class. They are the most important cereal crops of the world. In all these, the grains are actually fruits of the caryopsis type. The rice grain is tightly covered by the husks, in wheat the

FIG. 40

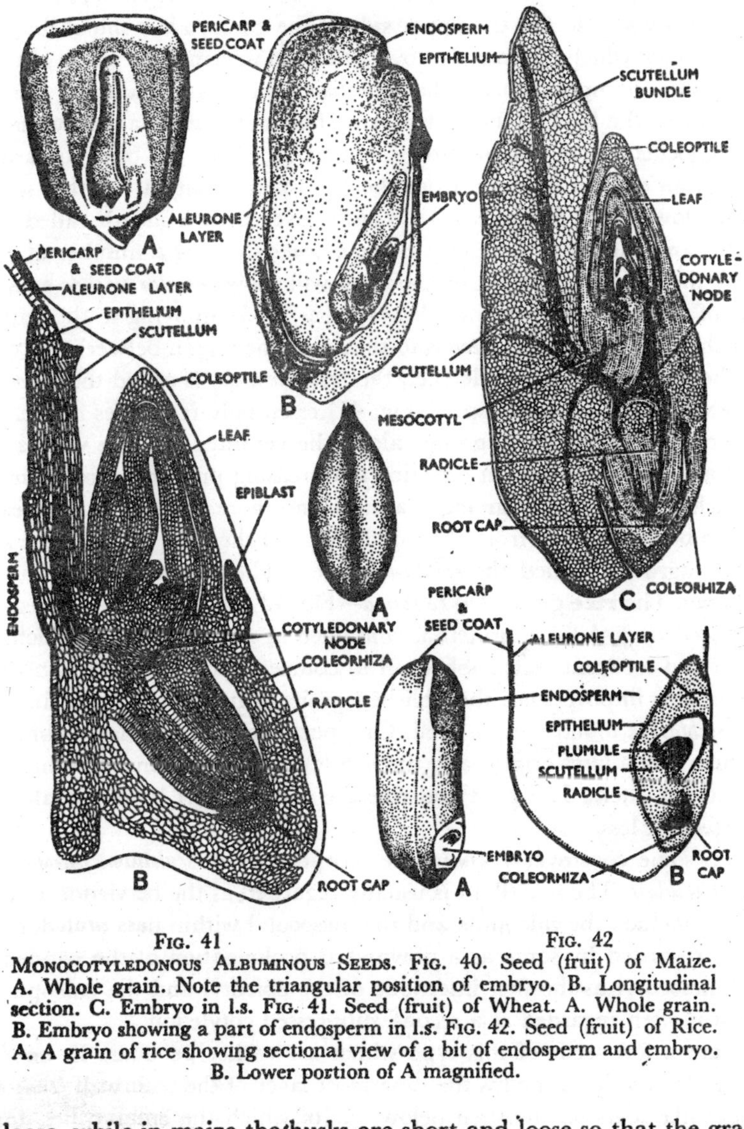

FIG. 41 FIG. 42

MONOCOTYLEDONOUS ALBUMINOUS SEEDS. FIG. 40. Seed (fruit) of Maize. A. Whole grain. Note the triangular position of embryo. B. Longitudinal section. C. Embryo in l.s. FIG. 41. Seed (fruit) of Wheat. A. Whole grain. B. Embryo showing a part of endosperm in l.s. FIG. 42. Seed (fruit) of Rice. A. A grain of rice showing sectional view of a bit of endosperm and embryo. B. Lower portion of A magnified.

husks are loose, while in maize the husks are short and loose so that the grains are exposed. In all the three, the outer coating of the grain is formed by the fusion of the *pericarp* and the *seedcoat*. The micropyle and the hilum cannot be found because of the pericarp covering. Inside, a large *endosperm* forms the bulk of the grain, while a small *embryo* occupies a comparatively small space on one side of the base. The outermost layer of the endosperm is the *aleurone layer* which contains mainly protein.

Of these three, **maize** (*Zea mays*—Fig. 40) has the largest embryo and is the easiest to examine and dissect. Maize grains are flattened and more or less oblong. On the flat face the *embryo* can be seen even from the outside as a whitish deltoid area. The wall (*pericarp+seedcoat*) is yellowish or red and so firmly adherent that it cannot be separated from the kernel. If the seed be cut longitudinally into two and the cut face stained with iodine, the endosperm part (deep blue because of starch) and the embryo part (yellowish) can be easily demarcated. Usually, the upper and outer part of the endosperm is horny (more protein) while the inner part and the part nearest to the embryo is mealy white as it contains mainly starch. On the upper side of the embryo is a shield-shaped structure called the *scutellum* which completely covers the endosperm. The tissue of the scutellum abutting on the endosperm forms the *epithelium* which is a glandular tissue secreting enzymes to digest the food in the endosperm. The scutellum is actually the single cotyledon of monocots whose leaf nature is not at all clear. Its only apparent function is the absorption of food material from the endosperm. Lateral to this scutellum is a short *axis* the lower part of which is the *radicle* covered by a sheath called the *coleorhiza* (root-sheath) and the upper part of it is the *plumule* showing a number of young leaves sheathed by the *coleoptile*. A good preparation will show *vascular strands* emanating from the middle of the axis (the first or *cotyledonary node*) and ramifying in the scutellum. This is the path through which food travels to the axis. The region between the cotyledonary node and the base of the coleoptile (i.e., the *first internode*) is called the *mesocotyl.*

The **wheat** grain (*Triticum* spp.—Fig. 41) essentially resembles maize. The grain is oval and there is a longitudinal groove along the ventral side. The wall is light brown ('wheat colour'). The endosperm contains a decreasing proportion of *gluten* (a protein) as it approaches the scutellum giving it a more and more mealy white appearance. The embryo resembles maize but from the middle part of the axis, opposite the scutellum, a tongue-like outgrowth, called the *epiblast*, is seen. The epiblast is also found in many other *Gramineae*. The **rice** grain (*Oryza sativa*—Fig. 42) shows a brown or red-brown wall (pericarp+seedcoat). A lot of protein is contained in the *aleurone layer* which, along with the pericarp and seedcoat, gets rubbed off in course of polishing rice. The embryo also gets broken. This impoverishes the polished grain in its vitamin, protein and oil contents. The *epiblast* is found here as well. One peculiarity of the rice embryo is that the *plumule* (which is of a later origin) and the *radicle* do not lie on one straight line but form an angle between them. In wheat and rice the embryo part is very small—only about 60% of the seed or less.

There is some controversy about the morphology of *scutellum*, *coleoptile*, *coleorhiza*, *mesocotyl* and *epiblast*. The scutellum is usually regarded as the cotyledon but some botanists prefer to include the coleoptile and the mesocotyl within the cotyledon. The coleorhiza also is regarded by some as a downward prolongation of the cotyledon. Others regard the mesocotyl as the first internode of the epicotyl and the coleoptile as a leaf. The epiblast is often regarded as a second suppressed cotyledon.

There are a few seeds of this type which are rather peculiar. In **coconut** (*Cocos nucifera*—Fig. 43) the hard shell is the innermost layer of the fruit-wall (*endocarp*). There are three eye-like scars on the shell below one of which the embryo lies. On breaking open the shell, the seed is found covered by a dark-brown *seedcoat* which is adherent to the kernel. The white kernel, including the milk within it, is the *endosperm.* The *embryo* lies embedded on the 'top' (actually the base of the fruit) of the kernel below one scar. In **date palm** (*Phoenix sylvestris*—Fig. 44), **palmyra palm** (*Borassus flabellifer*—Fig. 45) and **betel-nut** (*Areca catechu*—Fig. 46) seeds the arrangement is similar but the endosperm is more or less solid and horny containing reserve cellulose. The stony seed of date is covered by a brown testa. The hard endosperm has a longitudinal groove on one side and on the centre of the other side the *embryo* is placed in a small pit. The embryo is

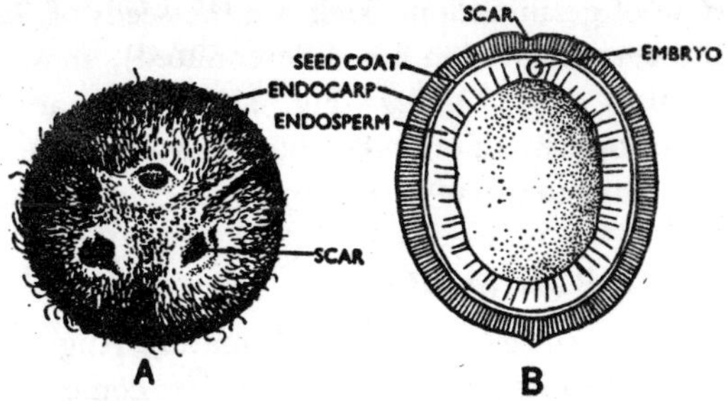

FIG. 43

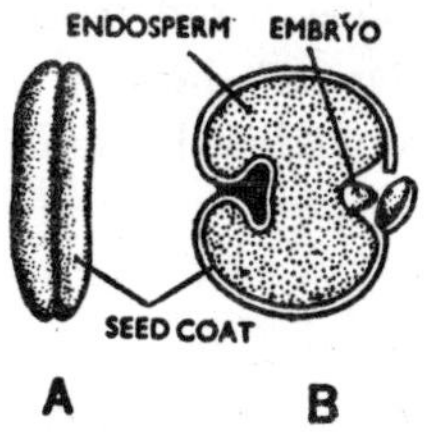

FIG. 44

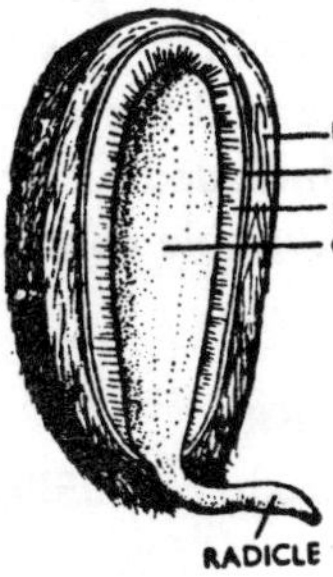

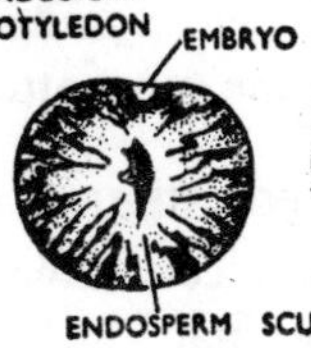

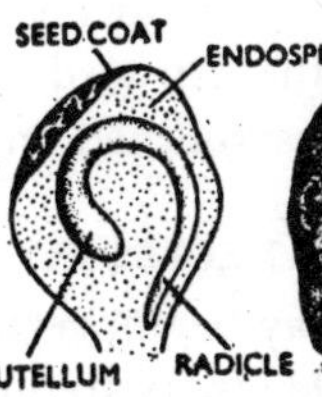

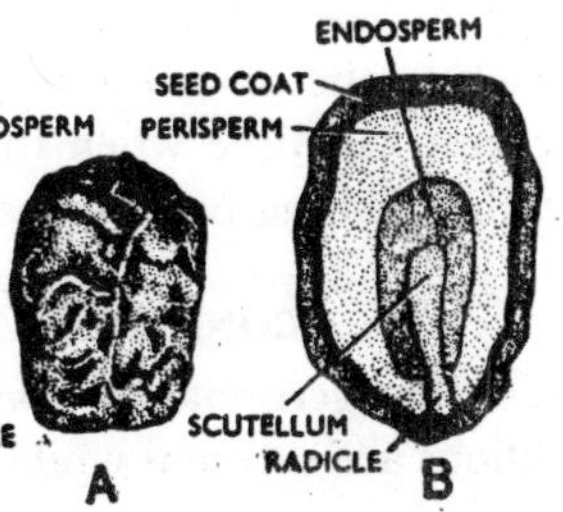

FIG. 45 FIG. 46 FIG. 47 FIG. 48

MONOCOTYLEDONOUS ALBUMINOUS SEEDS. FIG. 43. Coconut. A. Top of hanging shell. B. L.s. showing seed inside shell. FIG. 44. Date palm seed. A. Whole seed. B. T.s. of seed showing embryo in a pit covered by a lid. FIG. 45. L.s. of palmyra palm stone showing seed in the first stage of germination. FIG. 46. L.s. of betel-nut seed showing ruminated endosperm. FIG. 47. L.s. of seed of onion. FIG. 48. Seed of cardamom. A. Whole seed. B. Longitudinal section.

similarly placed in betel-nut and palmyra palm. In betel-nut the endosperm is *ruminated* as in Annonaceae.

The small black seed of **onion** (Fig. 17) is covered by a seedcoat. The inside is filled with a tough semitransparent endosperm within which a curved embryo is embedded. The embryo shows a radicle on one end and a scutellum on the other. The plumule is not seen at this stage. It develops later.

Many seeds belonging to the Scitamineae show the presence of *both endosperm* and *perisperm*. This is the case in **cardamom** (*Amomum*) (Fig. 48) where the endosperm and perisperm are oily and in **Canna** (*Canna indica*).

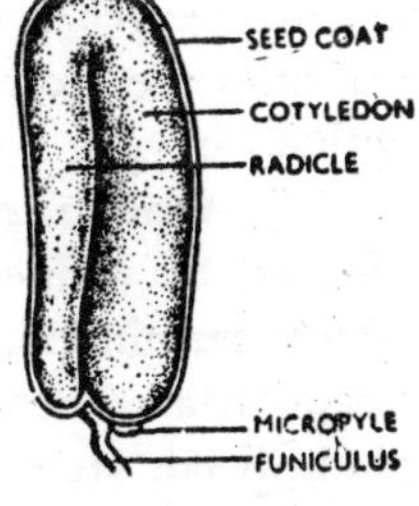

MONOCOTYLEDONOUS EXALBUMINOUS SEED: FIG. 49. Seed of *Alisma plantago* in longitudinal view.

MONOCOTYLEDONOUS EXALBUMINOUS SEEDS

Although all the common monocotyledonous seeds are albuminous there are a few of the exalbuminous type. This type of seed is found in the Aroideae (e.g., *Pathos* and *Amorphophallus campanulatus*) and also widely in the families Hydrocharitaceae (e.g., *Vallisneria*), Alismaceae (e.g., *Alisma plantago*), Naiadaceae, etc. The seed of **Alisma** (Fig. 49) shows a curved embryo with a radicle and a cotyledon as shown in the figure.

ABNORMAL EMBRYOS

The various types of seeds described above are the typical forms. Beside them, various abnormalities are known. Some seeds do not show complete differentiation in the embryo part even at the time of germination. Such are the seeds of *Juncus glaucus*, most orchids (no endosperm and even cotyledon not differentiated), most saprophytes (e.g., *Monotropa*) and most parasites (e.g., *Cuscuta*—Fig. 13). There are dicots (e.g., *Corydales, Abronia*) showing the development of only one cotyledon and monocots showing a second cotyledon. Loranthaceae parasites may show two to six cotyledons.

GERMINATION

Germination is the awakening of the dormant embryo. In all mature Angiospermic seeds the embryo lies in a dormant state when its physiological activities come to a minimum. Even its respiration is so slow as to be detectable only by sensitive instruments. As soon as the necessary conditions are satisfied this *dormancy* is broken and the phenomenon of *germination* begins. Like all growth processes, the process of germination is irreversible, i.e., when germination has once commenced it cannot go back and the seed cannot be brought back to the dormant state. Under germination are included all changes that take place from the time when the dry seed is placed under suitable conditions to the time when the seedling becomes established on the substratum.

CONDITIONS NECESSARY FOR GERMINATION

In order that germination may begin, certain conditions are to be fulfilled. Some of these conditions are external while others are internal.

(A) EXTERNAL FACTORS

(1) **Moisture:** Water is of primary importance in germination. No seed can germinate unless it is thoroughly moistened. Actual submergence under water is not necessary and may even be harmful as it may choke oxygen supply. Water absorbed by the protoplasm enables the resumption of vigorous physiological activities. Digestion, respiration or conduction cannot go on without water. The swelling of the embryo enables it to burst through the seedcoats which again, are softened by water absorption. Oxygen cannot get through the testa unless it is moist.

(2) **Oxygen:** Oxygen is necessary for the seed as for any other living organ. The need of oxygen is even greater during germination as respiration and all physiological activities are more vigorous at this stage. Germination can go on for some time even without oxygen but it is soon checked.

(3) **Optimum temperature:** Like all physiological activities germination is affected by temperate. There is a certain *minimum* and a certain *maximum* beyond which germination cannot take place. Within this range there is a certain *optimum* temperature where germination is most satisfactory. This range varies from plant to plant. Seeds are not usually expected to germinate below 0°C and above 50°C and the optimum often lies between 25–30°C. Some seeds do not germinate well in any given temperature but prefers a fluctuation of temperature during the germinating period.

(4) **Light:** Light is not considered as an essential factor since germination takes place even without light. But, recent experiments have shown it to be of greater importance than what was hitherto thought. Its pronounced effect on germination cannot be minimised. Light affects the germination of different seeds in different ways. It was previously thought that light retards germination in most cases. But experiments by

Kinzel (Germany) with about one thousand species of plants showed that the germination of 70% of seeds is *favoured* by light, in about 26% light inhibits germination while about 4% of the seeds are indifferent to light. Tobacco, *Rumex*, mistletoe (*Viscum album*) seeds do not germinate in darkness while *Datura* and tomato seeds do not germinate well in light. Sometimes, some special treatment of seed (e.g., removal of seedcoat) may do away with this influence of light.

(B) INTERNAL FACTORS

A normal seed is expected to be internally capable of germination. The following factors are of importance in determining this internal capacity:

(1) **Food and auxins:** All normal seeds contain a supply of food which is necessary for the growing embryo and the young seedling. It has already been seen that this food may be contained in the cotyledon (mainly protein and starch in pulses; oil in mustard, groundnut, sesame, linseed, cucurbits, sunflower, etc.), endosperm (starch and a little protein in cereals; oil in castor, coconut, etc.; cellulose in date, areca-nut, etc.; mucilage in mallow or *Malva* sp., etc.), perisperm (black pepper, cardamom, etc.) or testa (pomegranate). *Auxins* are growth-promoting substances whose presence is essential for growth during germination.

(2) **Completion of resting period (dormancy):** Many Angiospermic seeds cannot germinate as soon as they are formed. They have to undergo a *period of dormancy* or *resting period.* The period of dormancy varies from plant to plant. It may be a few days or some months. Most cereals are capable of germination immediately after harvesting while some other seeds do not germinate till after a year. Some plants do not need any resting period. The period of rest may be necessary for various reasons. In some seeds the seedcoat is so hard that it takes time to wither. In this case the period of dormancy may be cut short by breaking open the seedcoat. In others, the embryo may take time to be fully differentiated or some after-ripening may be necessary for the seed. Different treatments may hasten this 'after-ripening'.

(3) **Viability:** Seeds retain their viability (capacity to germinate) for a definite period of time after which the embryo becomes dead for all practical purposes. *Viability test* is necessary to ascertain the germinating capacity of any seed. Conditions of storage (temperature, humidity, etc.) and circumstances in which the seed matured often determine the period for which the seed remains viable. Seeds usually keep well when kept dry, cold and free from insects or fungi. Proper drying of seed is extremely important in the retention of viability although there are a few seeds (e.g., willow, poplar, maple) which fail to germinate if too dry. Weak or immature seeds lose their viability quickly. Hard-coated seeds often remain viable for a long time. Even when every thing is satisfactory, it is found that the period during which the seed remains viable varies in different plants. The period may vary from one growing season to many years. The longest authentic record of this period of viability is that of some lotus (*Nelumbo nucifera*) seed found within peat at the dried up bottom of a lake in Manchuria whose age has been calculated to be some eight hundred years. Stories of 5,000 year old wheat or other cereal grains found in tombs in Egypt or Mohenjodaro proving viable are without any foundation. Seeds that old, are found to be in a carbonised condition.

When viability is apparently lost, it may sometimes be restored by different treatments. This shows that viability may be lost even when the embryo is not really dead.

CHANGES DURING GERMINATION

When all the necessary conditions are satisfied, the first change noticed is *swelling* of the seed by rapid *imbibition* and *osmosis* of water. This may cause a *bursting* of the seed-coat. Absorption of water causes a vigorous *resumption of physiological activities* by the protoplasm. There is rapid *respiration* and *copious secretion of enzymes* which causes *digestion* of stored food. Insoluble food is rendered soluble and complex food made simple. This simple food solution is diluted by water and *conducted* towards the growing epicotyl, hypocotyl, radicle and plumule. Food is *translocated* from perisperm to endosperm, from endosperm to cotyledon and from cotyledon to the growing organs according as which of them are present in the seed. *Assimilation* of this food by the growing organ enables *growth* and the *seedling* soon assumes its ultimate shape. When the seed is placed in the soil and *growth* has become vigorous the radicle is the first organ to grow vigorously. It comes out through the micropyle and fixes the seed to the soil. After this, either the *hypocotyl* or the *epicotyl* begins to grow. When the *hypocotyl* grows first, it pushes the cotyledonary node and all other parts of the seed (with or without the seedcoat) out of the soil and the mode of germination is called **epigeal** or **epigeous.** When the *epicotyl* grows first only the plumule is pushed out of the soil while the cotyledonary node, cotyledons and all other parts remain under the soil. This type of germination is called **hypogeal** or **hypogeous.** In Monocots, however, the hypocotyl does not grow but the epigeous and hypogeous nature is determined by the growth of the cotyledon itself.

EPIGEAL GERMINATION

Epigeal germination is shown by some dicotyledonous plants and a few monocots. Common examples of this type of germination are found in:

Dicotyledonous exalbuminous: Cucurbits, mustard, tamarind, French bean (*Phaseolus vulgaris*), Lablab (*Dolichos lablab*), sunflower.

Dicotyledonous albuminous: Castor.

Monocotyledonous albuminous: rare, found in onion.

Monocotyledonous exalbuminous: Alisma plantago.

(1) **Gourd** (*Cucurbita maxima*—Fig. 50): As germination begins, the straight radicle comes out and fixes the seed to the soil with the secondary roots coming out of the radicle. The *hypocotyl* next grows so quickly that it forms a loop which comes out of the soil and pulls out the rest of the seed. Frequently, the seedcoat gets caught to a peg-like projection at the base of the hypocotyl so that it is cast off easily and the cotyledons are brought out in the air. Next, the cotyledons open out like two leaves, become green, large and thin so that they look and behave like ordinary leaves in every way though differing in form from normal cucurbita leaves. The plumule within the cotyledons becomes exposed and soon grows into the aerial shoot.

(2) **Mustard** (*Brassica* spp.—Fig. 51): The seedcoat is thinner, the two cotyledons are much oily and the radicle is curved in the tiny seed. The stages of germination are essentially the same as in the cucurbits.

(3) **Tamarind** (*Tamarindus indica*—Fig. 52): The testa in this case is very hard. Nevertheless, the radicle comes out first after the testa is burst and fixes the seed by forming the root system. The *hypocotyl* now grows fast and soon pulls out the two large and thick *cotyledons*. The plumule then grows out into the aerial shoot. The cotyledons turn greenish, gradually shrivel up and finally drop off as the food matter within them is used up. The cotyledons, although turning greenish, never look like ordinary leaves as they do in *Cucurbita*.

(4) & (5) **Lablab** (*Dolichos lablab*) and **French bean** (*Phaseolus vulgaris*—Fig. 53): Both of them germinate like tamarind. The fleshy cotyledons do not become leafy but behave like tamarind cotyledons. Lablab is the common flat bean of the plains.

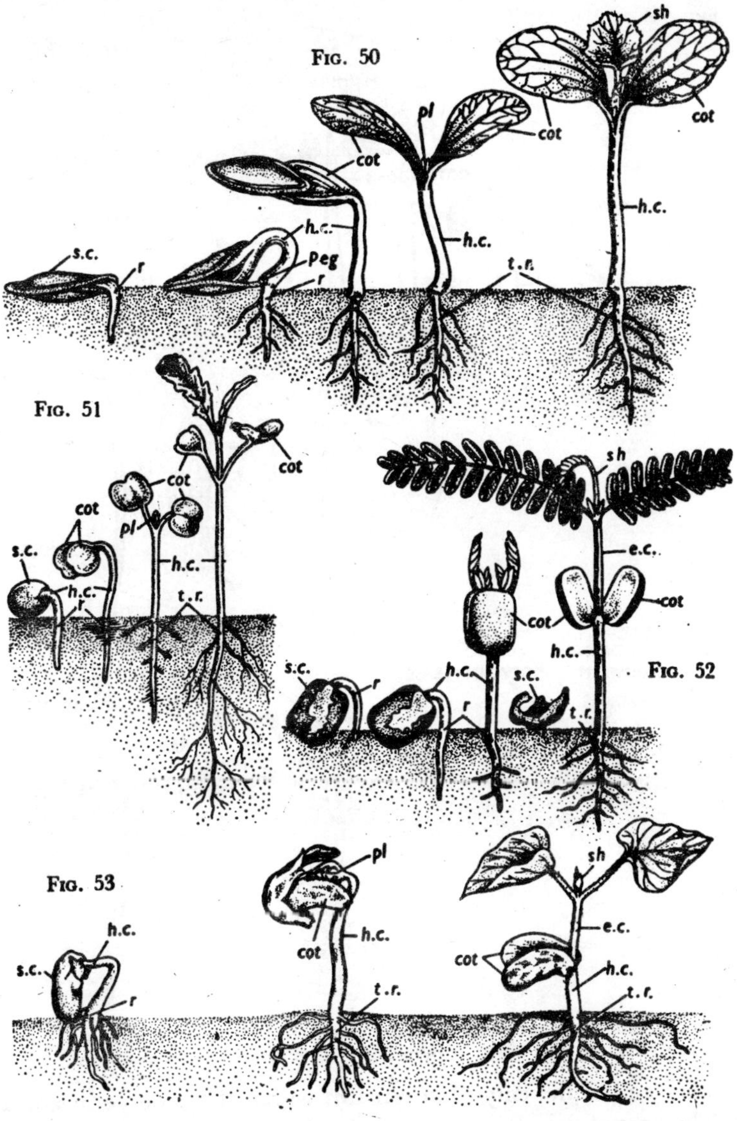

EPIGEOUS GERMINATION OF DICOTYLEDONOUS EXALBUMINOUS SEEDS: Successive stages in gourd (FIG. 50), mustard (FIG. 51), tamarind (FIG. 52) and French bean (FIG. 53). *s.c.*=seedcoat; *r*=radicle; *h.c.*=hypocotyl; *cot*=cotyledon; *pl*=plumule; *t.r.*=taproot; *sh*=shoot; *e.c.*=epicotyl.

(6) **Castor** (*Ricinus communis*—Fig. 54): The shell-like testa first bursts near the caruncle and the radicle grows out. Subsequent growth of the *hypocotyl* pulls out of the soil the two thin cotyledons enclosed in the endosperm. The testa is cracked and is soon cast off but the cotyledons do not come out of the endosperm until the latter is almost consumed by the former. The cotyledons then open up and become green and leafy while the plumule slowly develops into the leafy shoot. The remnant of the endosperm withers and drops off.

(7) **Onion** (*Allium cepa*—Fig. 55): Epigeous germination is very rare among the Monocots. Onion is one of the very few examples. In this case the radicle as well as the base of the curved cotyledon (scutellum) grows out of the seed. The radicle penetrates the

Fig. 54

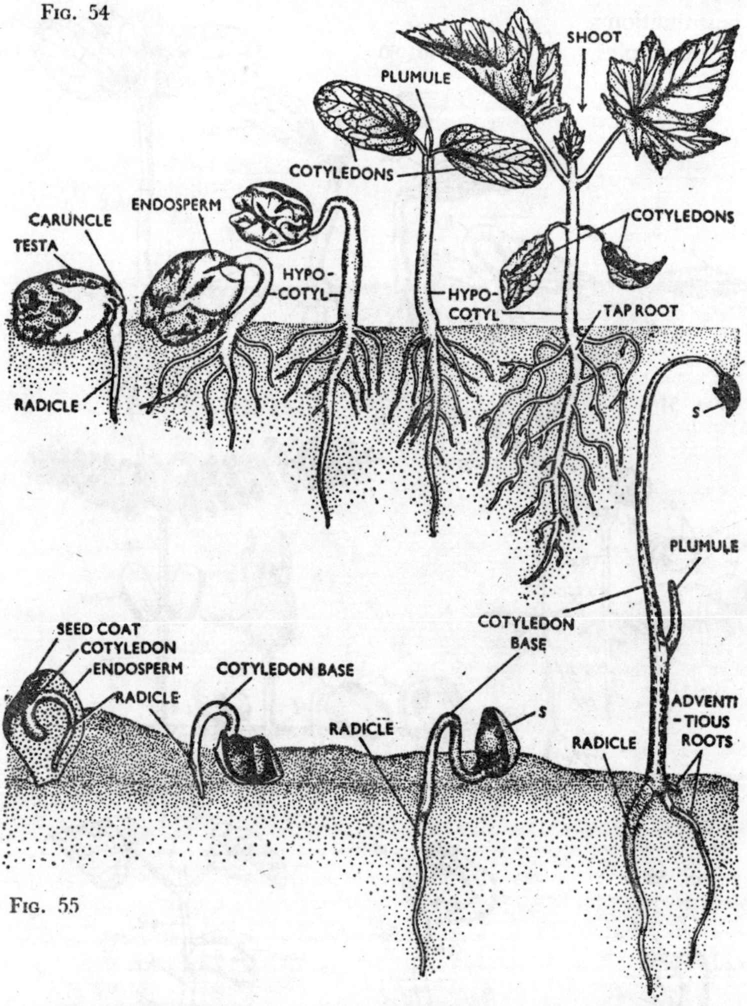

Fig. 55

Epigeous Germination of Albuminous Seeds. Successive stages in castor, a dicot (Fig. 54) and onion, a monocot (Fig. 55). *s*=shrivelled seedcoat.

soil while the other end of the cotyledon remains within the endosperm and sucks the food material. The base of the cotyledon grows further, turns green and pushes the seed out of the soil. The plumule is not visible so long, as it is covered by the base of the cotyledon in the form of a sheath just above the radicle. The plumule now pierces this cotyledon sheath and forms the first cylindrical foliage leaf. Meanwhile, adventitious roots develop from above the radicle forming a fibrous root system which is characteristic of monocots.

It should be noted in this case that the seed is pushed out of the soil not by the growth of the hypocotyl as in the other cases but by that of the base of the cotyledon itself.

(8) *Peperomia peruviana:* This is a dicot (Piperaceae) with endosperm and perisperm showing a peculiar type of germination. During germination one of the cotyledons

remains hypogeal within the endosperm sucking the food material from the latter as well as from the perisperm while the other cotyledon becomes epigeal and green.

HYPOGEAL GERMINATION

Hypogeal germination is shown by some dicotyledons and by most of the monocotyledons. Common examples are:

Dicotyledonous exalbuminous: Pea, gram, broad bean (*Vicia faba*), scarlet runner bean (*Phaseolus multiflorus*), mango, jack-fruit.

FIG. 56

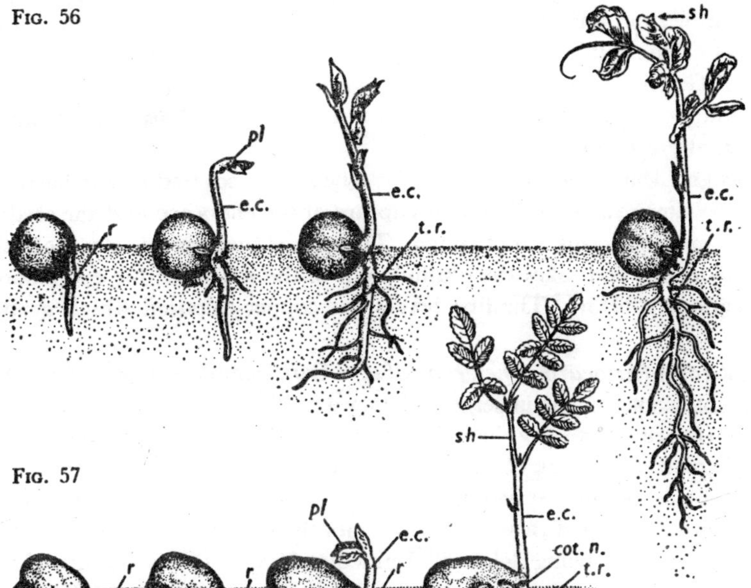

FIG. 57

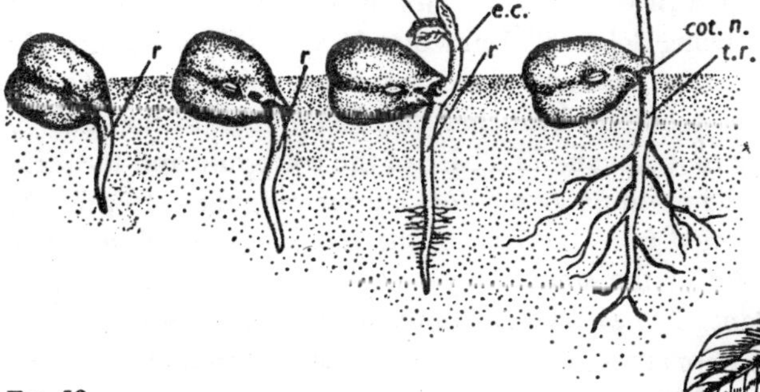

FIG. 58

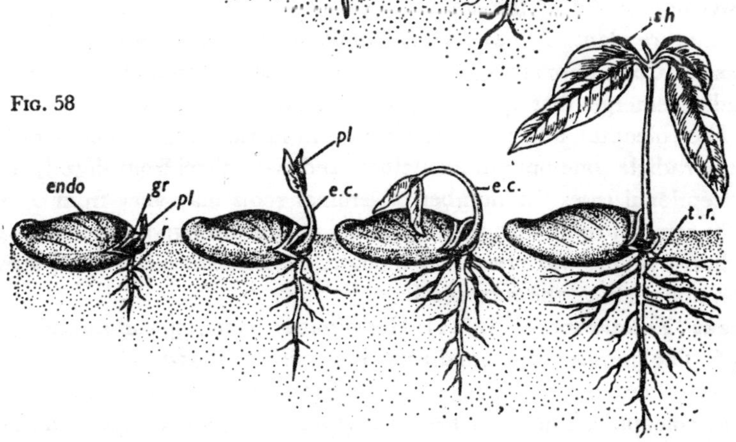

HYPOGEOUS GERMINATION IN DICOTYLEDONOUS EXALBUMINOUS SEEDS. Successive stages in pea (FIG. 56), gram (FIG. 57) and mango (FIG. 58). *r*=radicle; *pl*=plumule; *e.c.*=epicotyl; *t.r.*=taproot; *sh*=shoot; *cot. n.*=cotyledonary node; *endo*=endocarp; *gr*=groove through which radicle and plumule come out.

N.B. In pea and runner bean sometimes a tendency to epigeous germination is noticed.

Monocotyledonous albuminous: Rice, maize, wheat, coconut, date, areca-nut, fan (palmyra) palm.

Dicotyledons

(1) **Pea** (*Pisum sativum*—Fig. 56): The radicle comes out first, penetrates the soil and forms a root system by giving out secondary branches. It is the ***epicotyl*** which grows first here. It arches out and carries the plumule above ground. The plumule soon forms the aerial shoot. The cotyledons, therefore, remain under soil throughout.

(2) & (3) **Gram** (*Cicer arietinum*—Fig. 57) and **Broad Bean** (*Vicia faba*): The mode of germination is essentially the same as in pea. The cotyledons remain underground and are gradually used up.

(4) **Mango** (*Mangifera indica*—Fig. 58): The seed is covered by the hard endocarp. Absorption of water causes swelling and rupture of the endocarp and the seedcoat. The radicle comes out and forms a root system. The ***epicotyl*** then grows, comes out through a slit in the cotyledons and takes the plumule out of the soil while the cotyledons remain within the endocarp below. The first leaves are copper-coloured which gradually become green.

(5) **Jack-fruit** (*Artocarpus heterophyllus*) : Germination is hypogeous. Although the two unequal cotyledons remain underground, they develop a green colour.

(6) **Rice** (*Oryza sativa*—Fig. 59): A day or two after the seed is placed in moist soil, the coleorhiza pierces the base of the fruit and appears as a glistening knob. The radicle next penetrates the soil after splitting the coleorhiza. The coleoptile comes out now. (If the seed remains submerged in water, sometimes the coleoptile may come out first, the emergence of the radicle being delayed.) Immediately after the emergence of the radicle two other roots grow from its base and these are called *seminal roots*. The radicle and seminal roots give rise to secondary branches but, as opposed to the dicotyledons, the radicle does not form the root system. The base of the coleoptile and the mesocotyl now lengthen somewhat and the plumule soon comes out piercing the coleoptile. Meanwhile, adventitious roots are formed from the base of the plumule (top of the mesocotyl) or from the lowermost nodes of the stem. These *adventitious* roots form the fibrous root system of the mature plant. The pierced coleoptile soon withers away.

(7) **Maize** or **Corn** (*Zea mays*—Fig. 60): The mode of germination is essentially the same. The radicle emerges first by piercing the fruit wall and the coleorhiza. The coleoptile follows. The coleoptile and plumule develop as in rice. Three *seminal roots* develop from above the radicle (one opposite scutellum and two others from slightly above that point). In exceptional cases the number of seminal roots may vary from 0 to 10. The radicle and the seminal roots with their branches persist throughout the life of the plant and are not short-living as was previously thought. The *adventitious* roots are formed from the lowermost nodes above the mesocotyl.

(8) **Wheat** (*Triticum* spp.—Fig. 61): Germination as in rice and maize. *Seminal roots* number 4 to 5. The number of seminal roots is sometimes found to be variable. *Adventitious roots*, as in others, develop from above the mesocotyl though some of them may develop on the coleoptile. The branched radicle and the seminal roots probably persist throughout the life of the plant along with adventitious roots higher up.

(9) **Coconut** (*Cocos nucifera*—Fig. 62): The small embryo below one eye on the shell on the top of the endosperm is undifferentiated at first. During germination the lower

(actually the upper end as the fruit remains in a hanging upside down position) end of the embryo forms the cotyledon which begins to grow as a spongy structure inside the endosperm. This spongy cotyledon increases in size as it absorbs the food material stored within the endosperm. The upper end of the embryo develops through the eye carrying the radicle and the plumule. The plumule pierces the fibrous pericarp and emerges like a horn. This develops the aerial shoot even before the roots have come in contact with the soil. The radicle fails to develop any further but several adventitious roots grow from the base of the plumule. The seedling becomes established when the adventitious roots penetrate the soil.

FIG. 59

FIG. 60

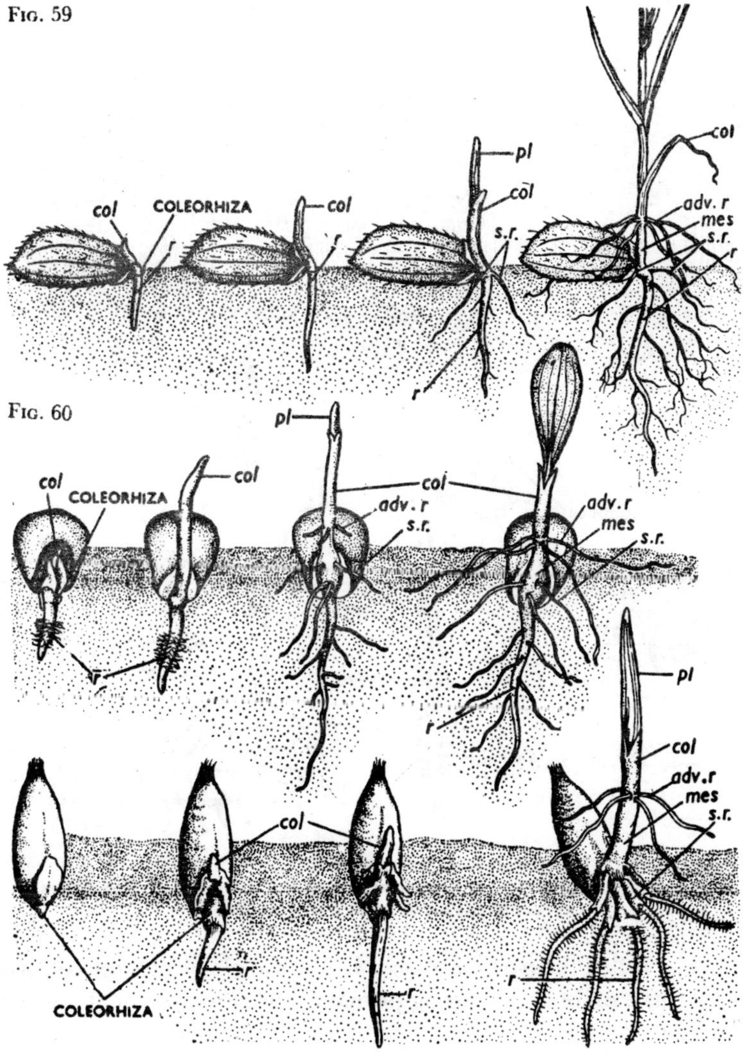

FIG. 61

HYPOGEOUS GERMINATION IN MONOCOTYLEDONOUS SEEDS. Successive stages in rice (FIG. 59), maize (FIG. 60), and wheat (FIG. 61). *col*=coleoptile; *r*=radicle; *pl*=plumule; *s.r.*=seminal roots; *adv. r*=adventitious roots; *mes*=mesocotyl.

(10) **Date** (*Phoenix sylvestris*—Fig. 63)): The stony seed is formed mainly of a cellulose endosperm on one side of which is the small embryo. During germination the cotyledon

begins to grow. The base (sheath and stalk) of the cotyledon grows out by forcing open the soft tissue above the embryo (which often comes out in the form of a lid) while the upper part remains inside the endosperm gradually increasing in size and absorbing more and more of the reserve cellulose transforming the latter into sugar. The base of the cotyledon which penetrates the soil is like a sheath enclosing the axis at its extremity.

FIG. 62

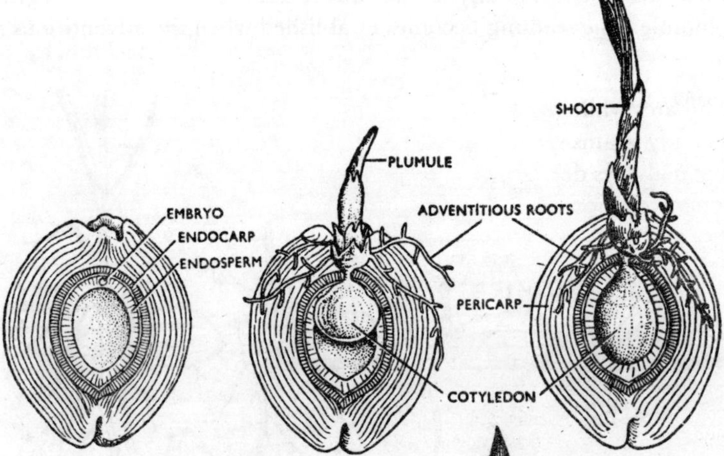

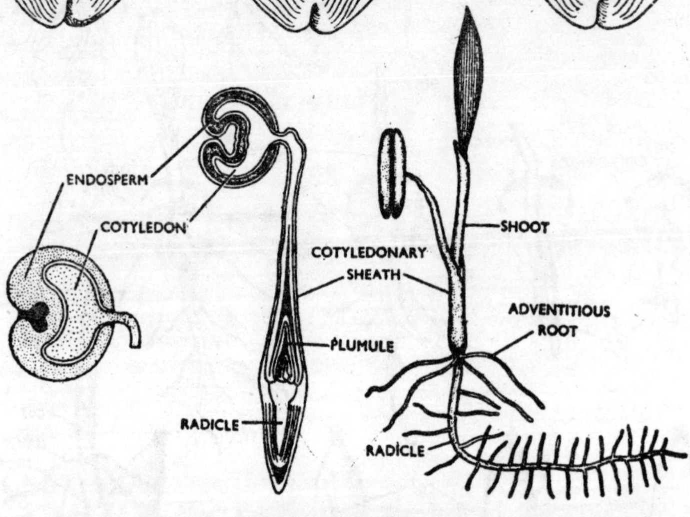

FIG. 63

HYPOGEOUS GERMINATION IN MONOCOTYLEDONOUS SEEDS. Successive stages in coconut (FIG. 62) and date (FIG. 63).

This sheath may be called the *cotyledonary sheath.* The radicle pierces the coleorhiza at the lower end and forms a more or less strong primary root system in the soil which is stronger than in other monocots although it does not form the main root system. Next, the plumule bursts out from one side of the sheath and develops aerial leaves. The upper part of the cotyledonary sheath acts as a stalk for the part of the cotyledon which is still inside the endosperm sucking the nutrients. Adventitious roots are later given out from the base of the aerial shoot and form the main root system of the growing plant.

(11) & (12) **Palmyra** or **Fan Palm** (*Borassus flabellifer*—Fig. 45) and **Betel-** or **Areca-nut** (*Areca catechu*—Fig. 46): Both of these have got hard cellulose endosperms.

The betel-nut endosperm is also ruminated. The manner of germination is practically the same as in date palm.

VIVIPARY

Vivipary means germination of the seed within the fruit while still attached to the mother plant. In the animal world, mammals are viviparous as the embryo differentiates into the young individual while still in the mother's womb. Birds lay eggs and ordinary plants develop seeds and not seedlings on the parent body, so they are not viviparous. Vivipary, however, is found in a number of plants. In the vegetable *Sechium edule* of Cucurbitaceae (locally called *squash*), common in Indian hill towns, it is often seen that the seed germinates while inside the fruit still attached to the mother plant. So is the case of coconut. Paddy grains germinate on the mother plant if they get sufficient moisture. Such vivipary is always dependent on excessive moisture in the atmosphere or within the fruit (e.g., lemons and oranges, tomatoes, melons) and on the absence of any period

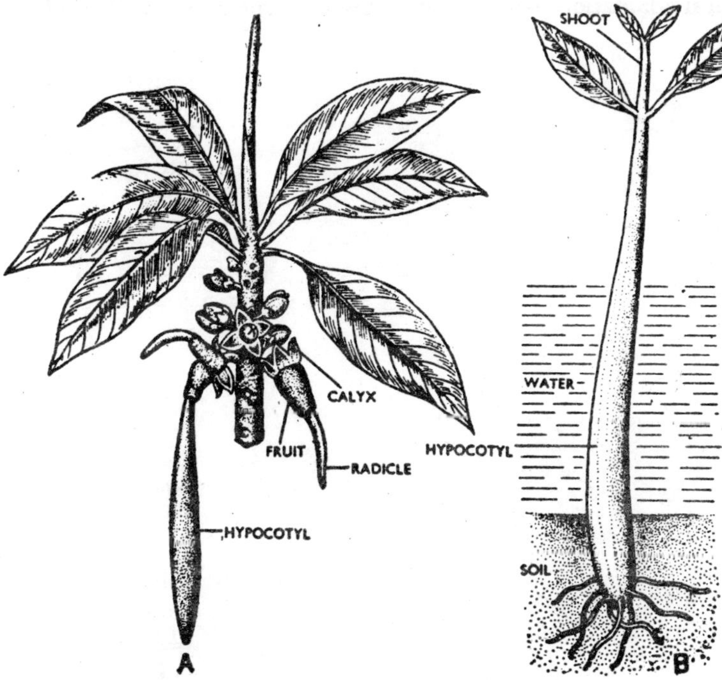

FIG. 64. VIVIPARY. A. A twig of *Rhizophora* showing viviparous germination. B. A seedling has become established on tidal soil.

of dormancy. Vivipary may also take place through vegetative organs, e.g., bulbils of *Agave* (discussed in connection with 'bud'—Fig. 79).

Beside the above, a special kind of vivipary is noticed in the mangrove plants found in estuarine tidal shores of the tropics, e.g., the Sundarbans in the Gangetic delta and similar mangrove formations in the estuaries of different Indo-Burmese rivers. The seed embryos in these plants do not have any resting stage and continue to grow uninterruptedly inside the fruit. The radicle first comes out of the fruit and then the hypocotyl begins to grow very vigorously so that it looks like a club which is usually 2 to 9 inches long and may sometimes attain 18 inches. The length varies according to the species of plant and may be determined by the depth of the water below. The plumule also grows somewhat while the cotyledons remain inside the fruit acting as haustoria. Some fast growing mangrove plants (e.g., *Sonneratia caseolaris* of Sonneratiaceae) do not show any vivipary while in a few others, e.g., *Avicennia* (Verbenaceae) and *Aegiceras* (Myrsinaceae), the type of vivipary is rudimentary. In markedly viviparous plants the process of germination is slow. Common examples are *Rhizophora mucronata, Ceriops decandra, Bruguiera gymnorhiza* and *Kandelia candel* all belonging to the family Rhizophoraceae. Of these, *Rhizophora* grows in the deepest water and shows the largest hypocotyl while *Ceriops* comes next. When the hypocotyl grows very heavy the fruit gets detached from the plant or, in some cases, the axis (i.e., radicle, hypocotyl and plumule) gets detached from the cotyledons and, because of the heaviness and the shape of the hypocotyl, it falls vertically down like a dart so that the radicle penetrates the soil below the shallow water. It soon forms a root system while the plumule grows safely above the water level. Sometimes, if the water be too deep, the viviparous fruit may float with the hypocotyl hanging down and get rooted where the depth of water is just right enabling the radicle to get fixed to the soil. This is a peculiar natural adaptation particularly suiting this type of plants (Fig. 4.4).

CHAPTER III

THE PLANT BODY—THE SHOOT

THE BUD & THE STEM

Given the requisite conditions, the seedling soon develops a luxuriant plant. What one sees above is the **shoot system** or the ascending axis developed out of the plumule while below, covered by the soil, is an equally important portion—the **root system,**

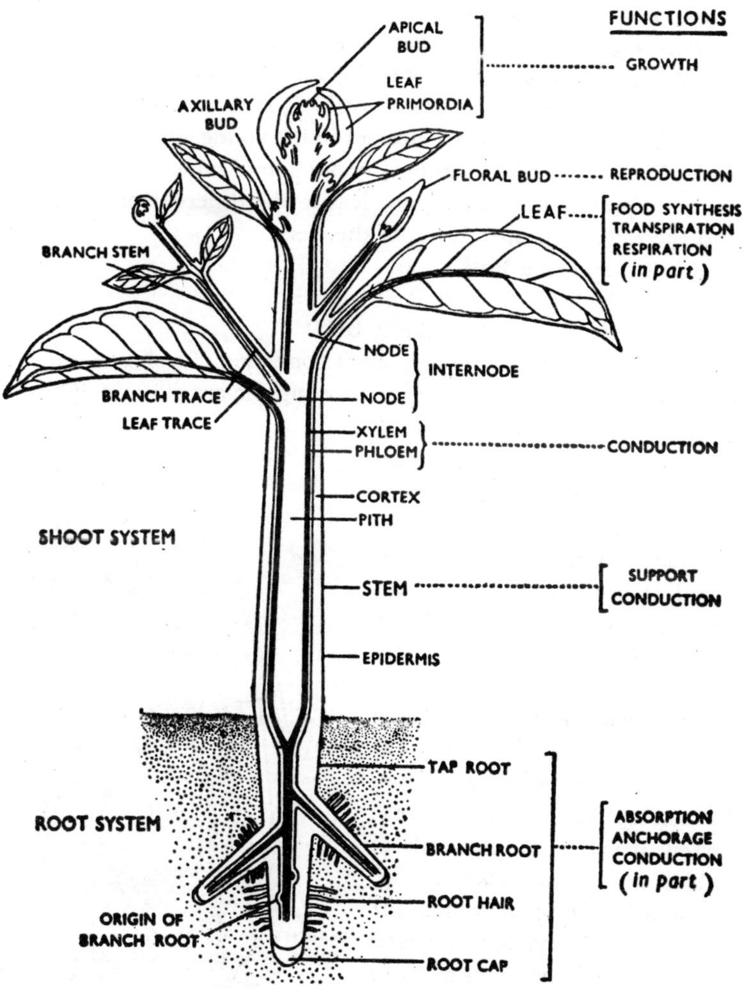

FIG. 65. Ground plan of a plant.

The shoot is the emblem of greenness because of the green leaves while the root, growing in darkness, is never green. Fig. 65 shows the ground plan of a plant the details of which are going to be discussed in the following chapters.

The skeleton of the shoot is formed by the **stem.** The main stem is developed as a direct prolongation of the *tigellum* of the embryo and grows vertically up while its branches grow horizontally or may be somewhat inclined. On the stem are borne the similar branch stems, the leafy appendages that are dissimilar to the stem and the reproductive modifications of shoot which are commonly called *flowers*. The young stem is green but later on it may change its appearance because of secondary growth and the formation of bark. A stem may bear multicellular hairs and branches on it develop exogenously. As opposed to the root, the *nodes* divide the stem into *internodes*, the internodal regions being bare while leaves and branches come out only from the nodal regions. If the nodes are very prominently developed, as in some grasses, the stem has a *jointed* or *articulated* appearance. A stem is usually cylindrical but it may be triangular (*Cyperus rotundas*), square (Labiatae), ribbed with ridges and furrows (*Cucurbita*), flat (Cactaceae), etc. It is usually solid but the pith may disorganise rendering it hollow or *fistular* as in many grasses. The surface of the stem may be smooth or *glabrous*, hairy, prickled or spiny, etc.

The normal stem supports its own branches and the leaves, which latter carry on the important physiological functions of photosynthesis and transpiration. By virtue of the chlorophyll and the stomata which are present in the young stem it can also carry on the functions of a leaf but to a limited extent. Conduction of mineral solutions from the root to the leaves and of prepared food from the leaf to the different parts goes on through the xylem and the phloem, respectively, of the stem.

A shoot has its origin in a **bud.**

THE BUD

A bud (Fig. 66) is a compact young shoot consisting of a short stem overlapped by crowded immature leaves which effectively cover up the growing tip. Thus, the growing stem tips are as well protected as the root tips covered by root caps. *Flower buds* also are buds but they are modified to carry on the special function of reproduction. Sometimes, a bud may be merely a mass of undifferentiated meristematic tissue not showing the leaf-primordia. The plumule is the first bud of the plant. It grows vertically up opening the leaves from below and the *apical bud* always remains at the apex. While the stem is growing, *lateral buds* grow at every leaf axil in an acropetal succession. These are the *axillary buds*. The first axillary buds often develop at the axils of cotyledons which are the first leaves. Fig. 67 shows a potato seedling in which the buds at the axils of the cotyledons develop the first tubers. The axillary buds grow to develop the branches. Each branch, as it will grow, will have its apical bud and will give rise to lateral axillary buds in the same way as the main stem. Each axillary bud at the axil of every leaf does not grow as no plant needs so many branches. Most buds remain *dormant*, i.e., they remain in very rudimentary forms. Buds may also remain dormant because of bad nutrition. Dormant buds may resume growth at a later stage when conditions for their development are

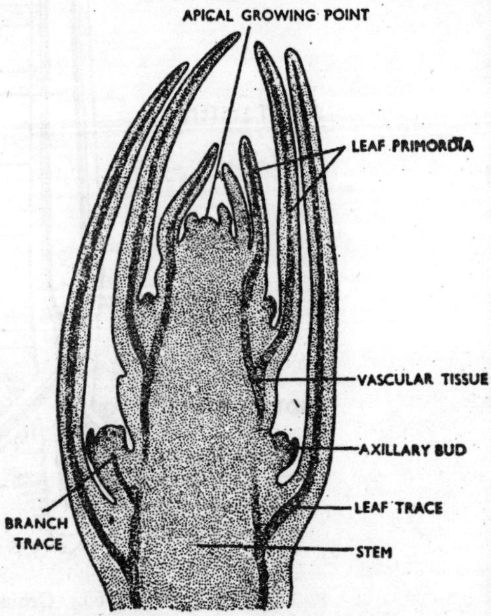

FIG. 66. L.s. of a vegetative bud.

fulfilled, e.g., when the main shoot is cut off. In that case the nutrient sap which would have gone to the main shoot would be diverted towards the dormant buds and would cause their development into new, strong branches. Such branches are *deferred branches* which look out of position as they do not fit into the acropetal order. Some buds fall off after formation so that they may be called *deciduous buds*. The disposition of the buds determine the final shape of the shoot, e.g., as the lateral buds do not develop the palm stem assumes the typical unbranched form.

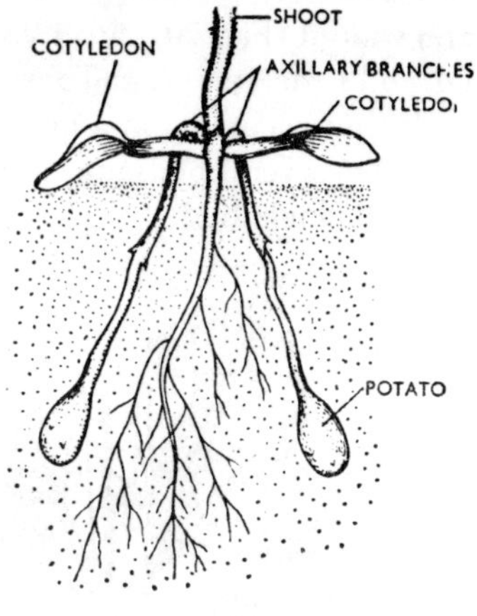

FIG. 67. A potato seedling showing development of buds at cotyledon axils.

DISPLACEMENT OF BUDS

When a bud develops from a usual position (i.e., terminal or axillary) it is a *normal* bud. In some cases, although the bud arises from the normal axillary position, during its growth it is carried to some distance along the lamina of a leaf or along the internode as the tissues of the two organs get fused together so that the bud appears to have developed at the middle of a lamina or at some such extra-axillary position. Organs developed from such a displaced bud also look displaced (Fig. 68). Such displacement of a bud by fusion may be called *adnation*.

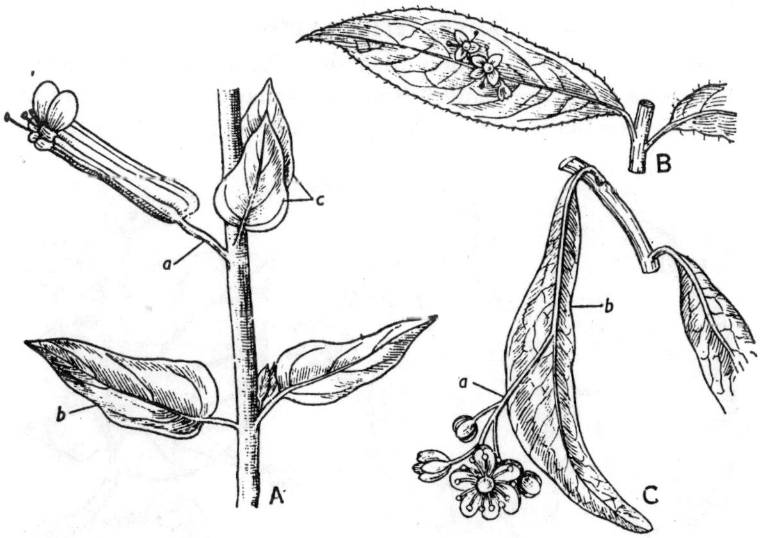

FIG. 68. Displacement of bud by adnation (after Strasburger). A. *Cuphea* of *Lythraceae*. Flower *a* is axillary to leaf *b* in origin but has been shifted to a higher position almost reaching the next higher pair of leaves *c*. B. *Helwingia* inflorescence adherent to foliage leaf giving the false appearance of origin from leaf lamina. C. Inflorescence of *Tilia ulmifolia* (*a*) adherent to its bract (*b*) as in B.

ACCESSORY BUDS

While only one bud is found in a normal position, sometimes there may be more than one. The apical bud is single in practically all Angiosperms. But, in some lower plants, it splits into two giving rise to two branch shoots. This is known as *dichotomy*. It is more common to find more than one bud at the axil. The additional axillary bud is then called an **accessory bud** as found in the common hedge plant *Duranta* (Fig. 69). Accessory bud may be placed one above another (*superposed* or *serial* buds) as in *Aristolochia* and walnut (Fig. 70), side by side (*collateral*) as in many Liliaceae and Solanaceae (brinjal, chilli, etc.) plants and also in maple (Fig. 71), or in a cluster. Sometimes, a number of buds thus occurring in a group may fuse to form a composite *fasciated branch.* In *Cuscuta* (dodder) several buds may appear to be in the same axil because of the branching of the original single bud.

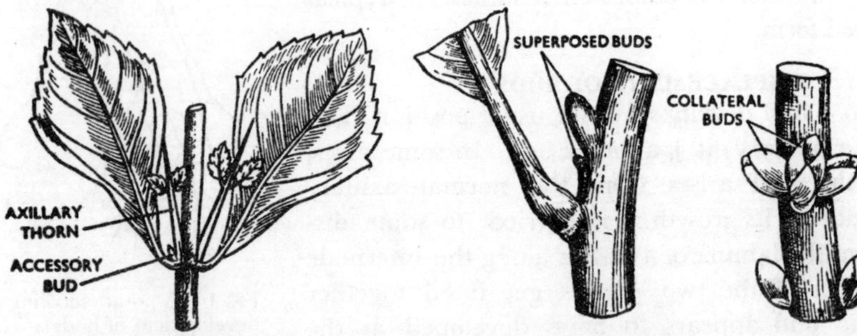

FIG. 69 FIG. 70 FIG. 71

ACCESSORY BUDS. FIG. 69. *Accessory* bud in *Duranta*. FIG. 70. *Superposed* buds in walnut. A number of accessory buds are placed one above another of which the topmost one has developed most. Fig 71. Collateral buds in maple. Three buds occur side by side.

ADVENTITIOUS BUDS

When a bud grows from a position other than the normal (i.e., apex or axil) then it is an **adventitious bud.** Some adventitious buds may arise from extra-axillary positions on the stem (*cauline* buds), from roots (*radical* buds) or from leaves (*epiphyllous* buds).

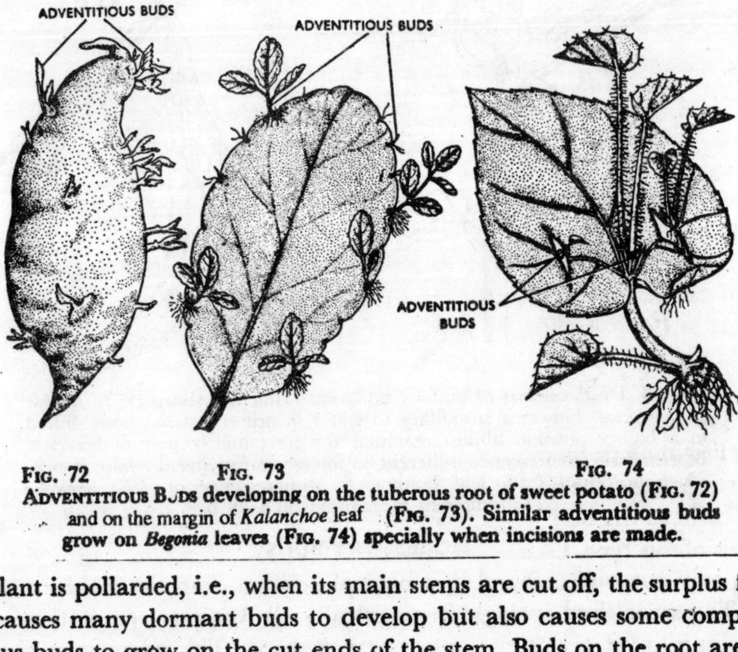

FIG. 72 FIG. 73 FIG. 74

ADVENTITIOUS BUDS developing on the tuberous root of sweet potato (FIG. 72) and on the margin of *Kalanchoe* leaf (FIG. 73). Similar adventitious buds grow on *Begonia* leaves (FIG. 74) specially when incisions are made.

When a plant is pollarded, i.e., when its main stems are cut off, the surplus flow of sap not only causes many dormant buds to develop but also causes some completely new adventitious buds to grow on the cut ends of the stem. Buds on the root are known in many plants. When the trunk is cut off, some plants like woodapple (*Aegle marmelos*) and poplar grow buds on the roots. Root buds are largely used for the propagation of some agricultural crops like sweet potato (*Ipomoea batatas*—Fig. 72) or palwal (*Trichosanthes dioica*) where root cuttings are used as 'seed'. Similar buds with roots are known to grow on the margin or the surface of leaves of *Begonia, Kalanchoe, Gesnera, Gloxinia*, etc. These again are greatly utilised by horticulturists. Young plantlings are obtained at the angles of the crenate margin of mature (Fig. 73) and *Kalanchoe* leaves when they are placed on moist soil. Similar plantlings grow near incisions made on *Begonia* leaves (Fig. 74).

PROTECTION OF BUDS

Buds protect the delicate growing points of the stem. Sometimes this protection has to be accentuated. This may be done by a covering of hairs (e.g., *Zizyphus*), resinous secretions, etc.; by the leaf bases themselves or by stipules and, more conspicuously, by the **bud-scales** (Fig. 75). These bud-scales are prominently seen to cover the buds of jack-fruit, different *Ficus* and *Magnolia* species, etc. The *scales* are tough, modified

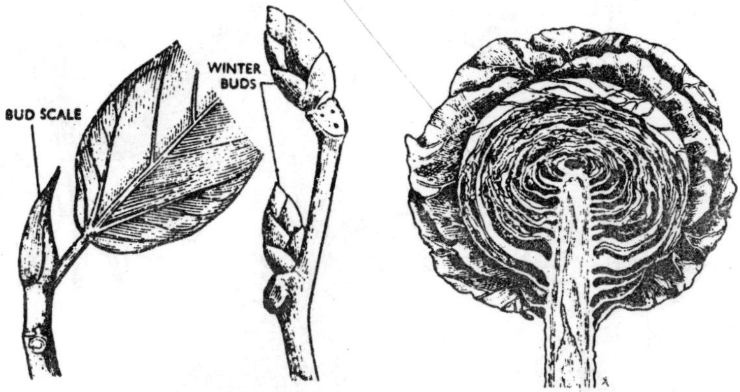

FIG. 75 FIG. 76 FIG. 77
FIG. 75. *Bud-scale* on *scaly bud* of a *Ficus*. FIG. 76. Winter buds.
FIG. 77. L.s. of a cabbage showing its bud nature.

leaves of the nature of *cataphylls* (discussed under types of leaves). They may also be modified leaf bases, petioles or stipules. The true morphology of bud-scales (i.e., which part of leaf they represent) varies from plant to plant and even on the same plant where transitional forms may be seen. Bud-scales do not serve as foliage leaves but fall off as soon as the bud opens. Buds with scales are sometimes called **scaly buds.** Most plants in tropical climates bear buds that are not covered at all, i.e., *naked buds.* On the other hand, plants in very cold countries bear specially protected buds called **winter buds** (Fig. 76). During winter the leaves fall off but the abscission layers do not touch the axillary buds. These winter buds remain well covered by a series of scale leaves and lie almost dormant during the winter. During spring time these buds burst out of their coverings and the new leaves open out. Some of the winter vegetables of our country remain in the bud stage during winter. Cabbage, cauliflower and Brussels sprouts are buds of this type. One whole cabbage (Fig. 77) is nothing but a gigantic leaf bud and a cauliflower is similarly a reproductive bud. Brussels sprouts are formed of leafy buds. All these buds open out at the close of the winter giving rise to leaves and flowering shoots.

MODIFICATION OF BUDS

While leaf axils are normally occupied by buds, sometimes other structures like thorns or tendrils occupy their positions. Such thorns or tendrils are then modifications of the buds or, perhaps more correctly, of stem. Most peculiar, however, is the modification of whole buds into swollen structures known as **bulbils** which serve for the propagation of these plants. The bulbils borne on the axils of *Dioscorea bulbifera* (yam—Fig. 78) are swollen structures which may serve as famine food. When these get detached

FIG. 78

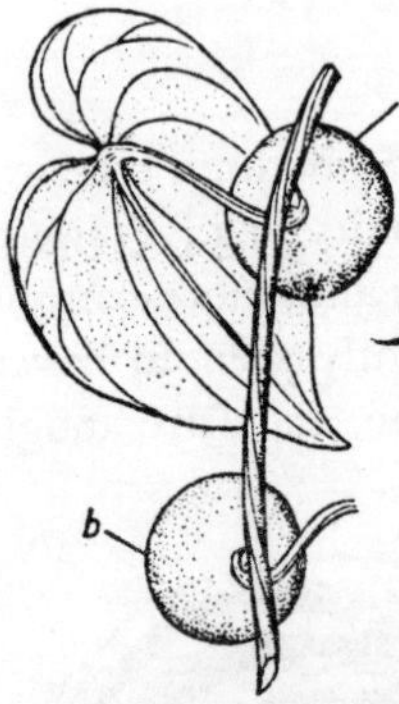

FIG. 79

FIG. 80

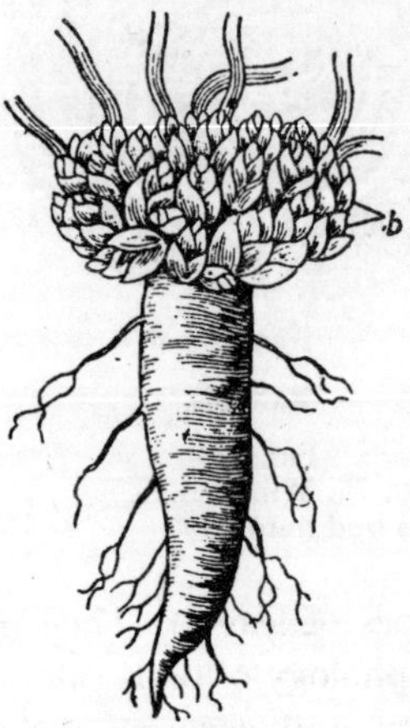

FIG. 81

FIG. 82

BULBILS. FIG. 78. Axillary bulbils (*b*) of *Dioscorea bulbifera*. FIG. 79. Some flowers in an inflorescence of *Agave* have turned into viviparous bulbils (*b*). *f*=normal flower buds. FIG. 80. *Globba bulbifera* showing similar transformation of flowers (*f*) into bulbils (*b*). A normal flower and a bulbil are shown separately. FIG. 81. Bulbils (*b*) developing on top of swollen root of *Oxalis*. FIG. 82. Bulbils (*b*) round the crown and the base of pineapple.

from the parent plant and fall on the ground they germinate into new plants and thus serve as excellent means of vegetative propagation. *Lilium bulbiferum* also shows similar bulbils. The large inflorescence of *Agave* (Century plant) shows many of its floral buds transformed into bulbils (Fig. 79) which germinate while still on the inflorescence—thus showing *vivipary*. Similar transformation of flower buds is seen in the inflorescences of *Globba bulbifera* (Fig. 80), *Allium sativum* (garlic), *Crassula* and other plants. The common *Oxalis* shows bulbils occurring like adventitious buds (Fig. 81) just above the swollen root.

Such adventitious bulbils are also seen round the crown and the base of some pineapples (*Ananas*—Fig. 82). The type of modified stem known as *bulb* (discussed under 'stem') may also be considered as a modification of buds.

PREFOLIATION

The arrangement of the young leaves in the bud prior to opening is known as **prefoliation.** It is also called *vernation.* It can be studied from two aspects—(a) the manner how each individual leaf is folded or rolled—known as **ptyxis,** and (b) the arrangement of all the leaves in relation to one another—**vernation proper.**

(a) **Ptyxis** (Fig. 83) (Arrangement of individual leaves independent of others):

Leaves may be **folded** in different ways, e.g., (1) *reclinate* or *inflexed*—upper half on lower half as in some garden croton; (2) *conduplicate*—folded lengthwise along the midrib as in guava, custard apple, magnolia, *Bauhinia*, etc.; (3) *plicate*—folded longitudinally several times along some prominent veins as in palms. In other cases the leaves are **rolled** in different ways, viz.; (4) *circinate*—leaf rolled up from apex to base like the tail of a dog

Fig. 83

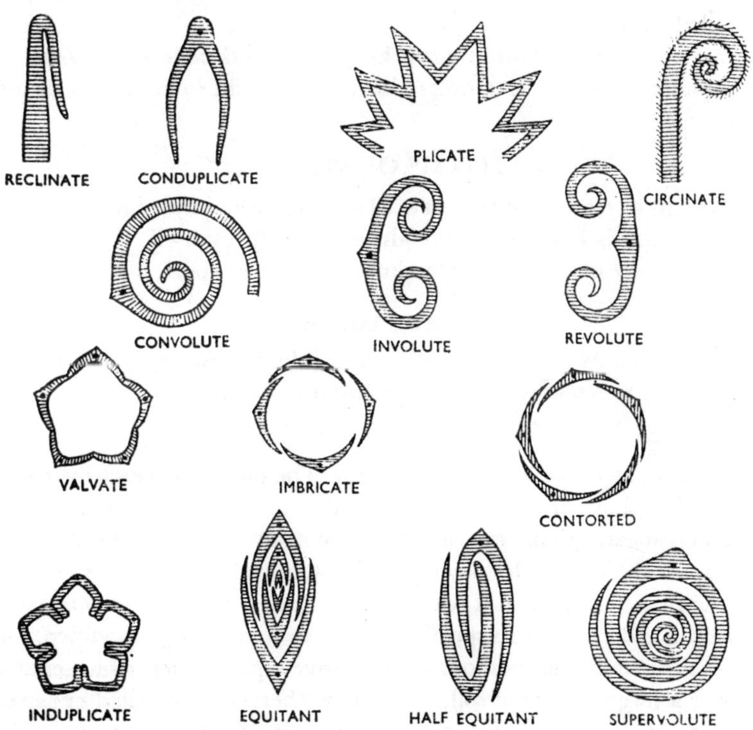

Fig. 84

Fig. 83. Ptyxis types (first two rows). Fig. 84. Types of vernation proper (last two rows).

as in ferns where it is a characteristic feature; (5) *convolute*—leaf rolled up from one margin to the other like the rolling up of a sheet of paper as in banana and *Canna;* (6) *involute*—the two margins (or round margin in peltate leaves) are rolled *up* towards the midrib or the centre of the leaf as in *Antigonon,* lotus, water-lily, etc.; (7) *revolute*—the leaf is similarly rolled *down* towards its lower surface as in *Nerium* (oleander). Finally, (8) the leaf may be *crumpled* in an irregular manner as in cabbage.

It should be remembered that, besides all these, the bud leaf may lie completely *flat* without showing any rolling or folding whatsoever.

(b) **Vernation proper** (Fig. 84) (Prefoliation arrangement of all the leaves in the bud as a whole):

When the individual leaves are flat or slightly convex the vernation may be (1) *valvate*—when the leaves are arranged almost in a circle while their margins may just touch one another without overlapping; (2) *imbricate*—when the margins overlap one another but not in any particular order; (3) *contorted* or twisted or spiral—such overlapping is regular in one direction so that one margin of every leaf overlaps the leaf on one side while the other margin is overlapped by the leaf on the other side thus giving a twisted appearance to the bud; (4) *induplicate* is a variety of valvate where the edges of the leaves instead of slightly touching one another are bent inwards for a little distance. When the leaf is folded or rolled the arrangement may be (5) *equitant*—conduplicate leaves standing face to face and one completely overlapping the other by both the margins; (6) *half-equitant* or *obvolute*—when such overlapping is incomplete so that only one-half of a blade is inside the other; (7) *supervolute*—when a convolute leaf encloses another presenting a completely rolled up bud.

The vernation proper of floral buds is known as **aestivation** or *prefoliation* and will be studied in that connection: Examples also will be found to be more common among flowers.

FORMS OF STEMS

The stem form of plants is either **erect (strong)** when the plant can stand erect or **weak** when the plant is not strong enough to stand up by itself. In the latter case the plant has to trail along the ground or to climb up some support.

ERECT FORMS

A majority of plants have the erect habit. These again may be subdivided into **herbs, shrubs** and **trees** according to the strength of the stem.

HERBS

The herbaceous stem is extremely succulent. The plant is necessarily very small and may not even rise above the surface of the soil. Very often (specially among Monocots) it is subterranean giving rise to some leaves at soil level. At the reproductive season the **scape** (Fig. 85) or the flowering shoot, which is a short stem bearing the inflorescence, springs forth. Such a plant may be called *acaulescent* (*caulis*=stem) as it is apparently stemless. Because of its special nature the stem is often modified. Grasses and some other herbs have special methods of vegetative reproduction which enable them to creep and cover a large surface of soil. They may, therefore, be called **creepers.** Plants of the grass type are sometimes stronger when they give rise to erect stems called **culms** which are more or less hollow at the internodes but solid at the nodes. A culmination of this sort of development is met with in bamboo (Fig. 86) which is really a tall grass but can no longer be called a herb.

According to the duration of its life herbs may be : (1) **Ephemerals**—when the plants live only for a few weeks. This happens when the growing season is very short as near deserts or in very cold countries. The plant has to complete its life cycle within that short time. (2) **Annuals**—They last for one season every year after which they flower and die. There are a large number of annuals in our country represented by the seasonal flowers, vegetables and crops. (3) **Biannuals** or **biennials** require two years to complete their life cycle. Many annuals of our country like the cabbage and the radish are biannuals in cold countries as the temperature there does not allow sufficient growth in one year. (4) **Perennials** grow for a longer time. This perennation may not be in one stage but in annual stages as in plants growing by rhizomes, etc. In these, the aerial shoot dies every year but the underground parts (rhizome, etc.) form the link year after year. Bananas are true perennials as are all the shrubs and trees.

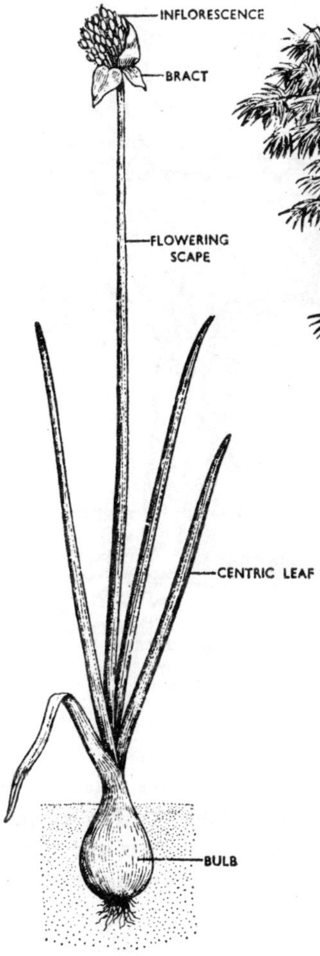

FIG. 85. Flowering scape of onion.

FIG. 86. Culm of bamboo.

Some plants are stronger than herbs but still not strong enough to attain the level of shrubs. Vegetables like brinjals and chillies come under these. These have got slightly woody stems. They may be called **woody herbs** or **undershrubs.**

SHRUBS

Shrubs are larger than herbs and usually give rise to woody branches from the level of the soil forming a bushy plant without any main trunk. The woody branches are never as strong as the trunk of a tree and the plants are much smaller than trees. Shrubs are very common among flowering plants as rose, china-rose, etc.

TREES

Trees are the biggest plants with prominent woody trunks which usually give rise to branches above. Among them (1) the **caudex** (Fig. 87) usually does not branch at all. The lateral buds on the tall columnar trunk are dormant or dead. On the top there is a crown of leaves. This is the habit of palms. They may branch only abnormally. An unbranched stem like this may also be called *columnar*. (2) In the **excurrent** (Fig. 88) trees

FIG. 87 FIG. 88 FIG. 89

TREE FORMS. FIG. 87. Coconut. FIG. 88. Mast tree (*Polyalthia*). FIG. 89. A *Ficus* tree (shown defoliated to clarify branching habit).

the main stem grows indefinitely and the side branches develop in a strict acropetal order as in the Mast tree (*Polyalthia longifolia*), *Casuarina* and Pine. (3) The apical bud of a **deliquescent** (Fig. 89) tree is weaker than the strong lateral buds and very often, it is destroyed at some phase of its life so that the tree has a spreading habit as in banyans and many common trees.

WEAK FORMS

Weak plant forms are not capable of growing erect without some help. These may be **creepers, trailers** or **climbers.**

CREEPERS (FIG. 90)

Creeping plants are actually herbaceous plants. The plants themselves are erect but they give rise to small horizontal branches which are weak and after growing for some length get rooted and give rise to small daughter plants which again increase in a like manner. Thus, a plant radiates in all directions and soon covers up a big area. This is the habit of grasses (e.g., *Cynodon dactylon* or *dub*), *Oxalis*, sweet potato (*Ipomoea batatas*), etc. The most usual propagation is by means of *runners* (*Centella asiatica* — Fig. 90) but may also be by thin rhizomes, suckers, etc. Creepers actually represent not single plants but a series of plants of successive vegetative generations—mother plant, daughter plants, grand-daughter plants, and so on. Sometimes the creepers are regarded as a special group of trailers.

TRAILERS

Trailers are *prostrate* plants having weak, long and slender stems which trail along the surface and do not try to climb up any support. These are different from creepers in being single plants which do not root from the nodes and thus do not form daughter plants. As the plants usually lie flat they are called *procumbent* (Fig. 91). A very common example is *Basella.* In some cases the branch, after growing horizontally for some length, straightens up the growing head like an erect plant. Such a trailer as *Portulaca* or *Lindenbergia* is termed *decumbent* (Fig. 92). Branches of some trailers grow profusely in all directions as in *Boerhaavia* (*Punarnava*) when it may be termed *diffuse.*

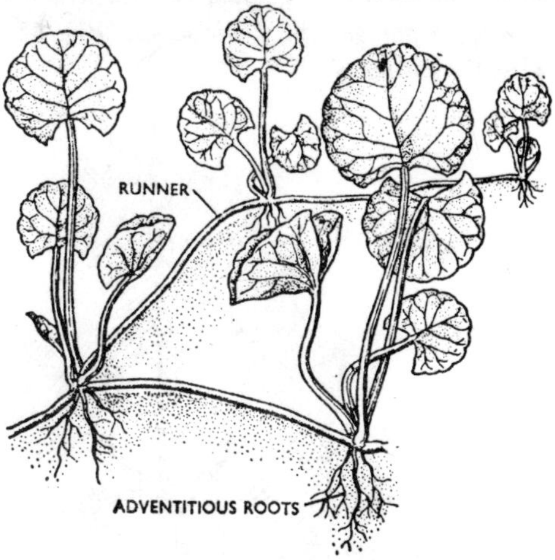

FIG. 90. *Centella asiatica* creeping by means of runners.

FIG. 91 FIG. 92
FIG. 91. *Procumbent* trailers of *Basella.* FIG. 92. *Decumbent* trailing habit of *Lindenbergia urticifolia.*

CLIMBERS

The climbers also are weak plants but they usually grow besides some support and climb up the latter so as to expose their leaves fully to light. They are thus biologically well adapted to grow in forests in the shade of tall trees as they are able to utilise the available light while the expenditure of material in forming a strong stem is avoided. The comparatively soft weak stem is thus able to bear large leaves, flowers and fruits.

The actual methods of climbing are various. Darwin, in his classical work on *Climbing Plants* recognised four groups of climbers—(1) *twiners;* (2) *tendril climbers;* (3) *root climbers* and (4) *scramblers.* To these may be added (5) *adhesive climbers* and (6) *lianas.*

(1) **Twiners or stem climbers**

The weak stems of some plants have the habit of twining round the supports by a special type of growth movement which makes the stem coil round a support. The growing stem apex moves in a circle (*nutation*) and coils round the support if it be available.

The coiling is now known not to be caused by contact but due to the action of gravity. It is specific for the type of plant whether it will coil upwards to the right (anti-clockwise from the top or *sinistrorse*) or to the left (clockwise or *dextrorse*). Among common twiners *Ipomoea quamoclit* (Fig. 93), *Clitoria ternatea*, *Convolvulus* sp. are sinistrorse while *Mikania scandens*, *Asparagus* sp., *Dolichos* sp. (Fig. 94), *Phaseolus* sp. are dextrorse. A few plants may twine the either way.

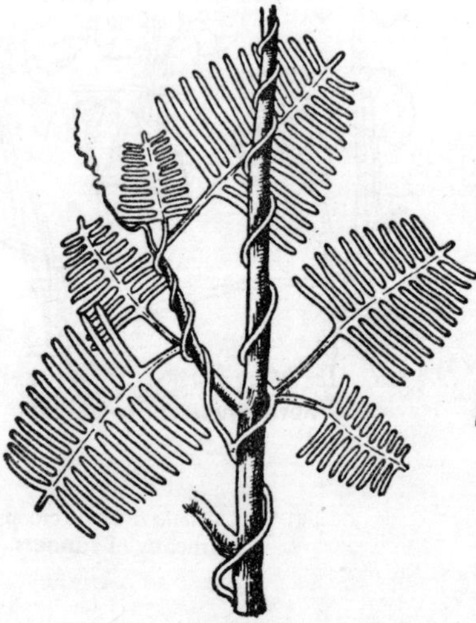

FIG. 93

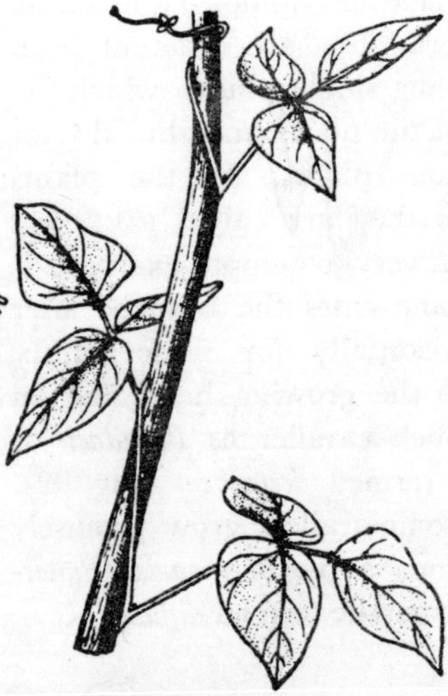

FIG. 94

FIG. 93. Sinistrorse twining of *Ipomoea quamoclit.*
FIG. 94. Dextrorse twining of *Dolichos lablab.*

(2) **Tendril climbers**

The climbing of *tendril climbers* is effected by specialised climbing organs called *tendrils* which are sensitive and capable of growing spirally and encircling a support when it comes in contact with the latter. Different organs of the plant are seen to be modified into tendrils; viz., modified axillary stem (in *Passiflora* or Passion flower—Fig. 95); modified stem apex (in *Vitis* sp.—Figs. 96 & 146); modified leaf (in *Lathyrus* sp. or wild pea—Fig. 97); modified leaflet (in *Pisum* sp. or pea—Fig. 98); modified petiole (in *Clematis* sp. —Fig. 99 and also, possibly, *Nepenthes*—Fig. 27); modified leaf tip (in *Gloriosa*—Fig. 100); modified stipule (in *Smilax*—Fig. 101); modified inflorescence axis (in *Antigonon* sp.—Fig. 102 and *Cardiospermum halicacabum*—Fig. 465C). The tendril of Cucurbitaceae (Fig. 103) is very common but its exact nature is uncertain. It may be a prophyll.

(3) **Root climbers**

These plants give out adventitious roots from their nodes which act as holdfasts in fastening the climber to the support. *Piper betle* or betel vine (Fig. 104), *Piper longum* and other members of Piperaceae; *Scindapsus* of Aroideae (Fig. 9) are common examples. Dodder (*Cuscuta*—Fig. 13) is a twining parasite which is further helped in climbing by the sucking roots (haustoria).

(4) **Scramblers, ramblers** or **struggling plants** and **hook climbers**

A number of weak plants do not actually climb but slowly grow over other bushes and rest there. They are helped in retaining this position by means of superficial outgrowths of the nature of prickles. Examples of such scramblers are found in rose (having

FIG. 95 FIG. 96 FIG. 97

FIG. 98 FIG. 99 FIG. 100

FIG. 101 FIG. 102 FIG. 103

TENDRIL CLIMBERS. FIG. 95. Axillary stem tendril of Passion flower. FIG. 96. Apical stem tendril of *Vitis* sp. FIG. 97. Leaf tendril of *Lathyrus aphaca*. FIG. 98. Leaflet tendril of pea. FIG. 99. Leaf petiole tendril of *Clematis* sp. FIG. 100. Leaf apex tendril of *Gloriosa superba*. FIG. 101. Stipular tendril of *Smilax*. FIG. 102. Inflorescence stalk tendril of *Antigonon*. FIG. 103. Tendril of *Cucurbita*.

curved prickles—Fig. 105), cane (*Calamus*—Fig. 106—scrambling by prickles on leaf and by means of long slender whip like stalk produced from the leaf sheath which is completely covered by curved hooks and helps the plant to grow luxuriantly in tropical forests), *Artabotrys* (curved hooks on flowering peduncles—Fig. 107), *Lantana* (prickles on stem—Fig. 108), *Zizyphus* (spinous stipules which persist after leaves are shed—Fig. 109), and so on.

(5) **Adhesive climbers**

These show extreme modifications where they are provided with organs that serve as adhesive discs which may adhere to flat surfaces. Thus these plants may even grow along flat walls. *Ampelopsis* of Vitaceae (Fig. 110) is such a plant growing adhesive discs from the tendrils. In some *Ficus* sp. (e.g., *Ficus repens*) and in Ivy (*Hedera helix*) such adhesive discs grow from climbing roots so that the plants are able to form carpet-like growths covering steep walls.

(6) **Lianes** or **lianas**

This is a type of perennial climbers that is

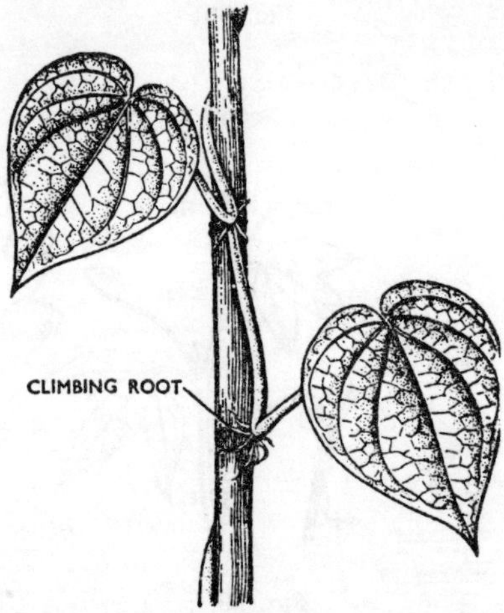

Fig. 104. Root climbing of *Piper betle*.

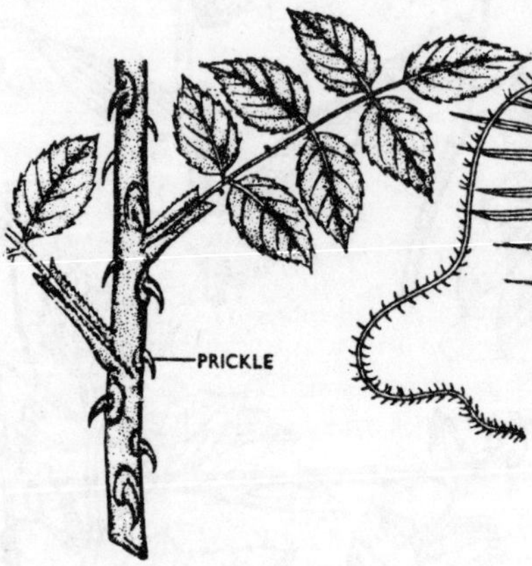

Fig. 105

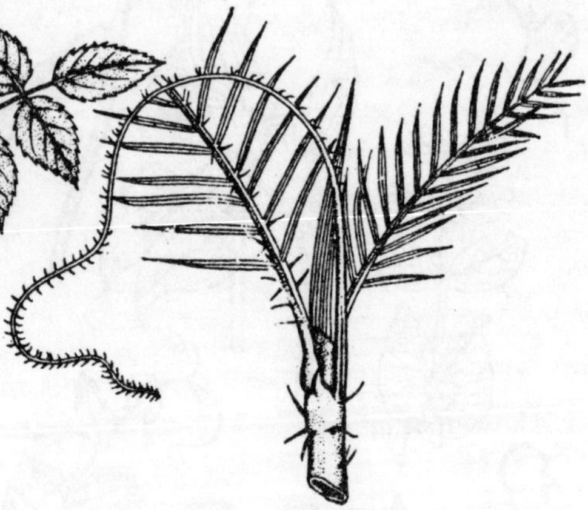

Fig. 106

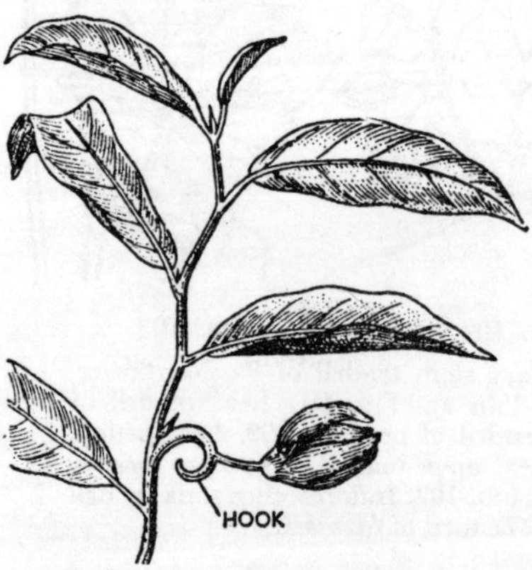

Fig. 107

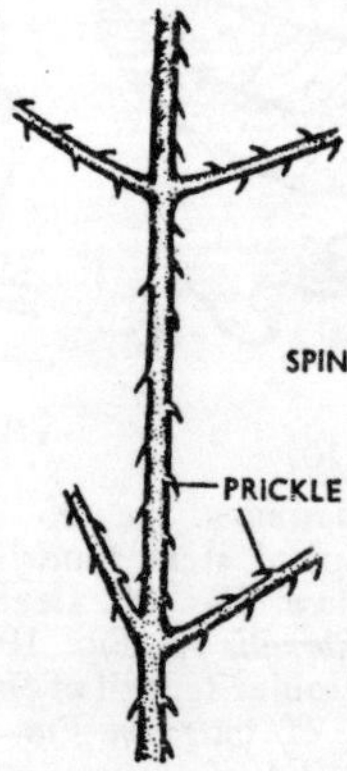

Fig. 108

Fig. 109

Scramblers. Fig. 105. A portion of a climbing rose showing the curved prickles which help in climbing. Fig. 106. *Calamus* sp. (cane) showing prickles which help in its scrambling. Fig. 107. A twig of *Artabotrys hexapetalus* showing hook on flower peduncle. Fig. 108. A piece of stem of *Lantana* showing prickles. Fig. 109. A piece of old stem of *Zizyphus* from which the leaves have been shed leaving the spinous stipules which sometimes help in scrambling.

found in deep forests. The plants at first grow like ordinary twiners growing up tall trees. Once the top leaves have reached sunlight the weak stem becomes stronger and stouter until it becomes woody almost like a slender tree. Such woody climbing stems are then found to hang from tall trees and to join different trees like very stout ropes rendering the forest impenetrable. Common examples are *Hiptage bengalensis, Tinospora*, species of *Bauhinia, Ficus*, etc.

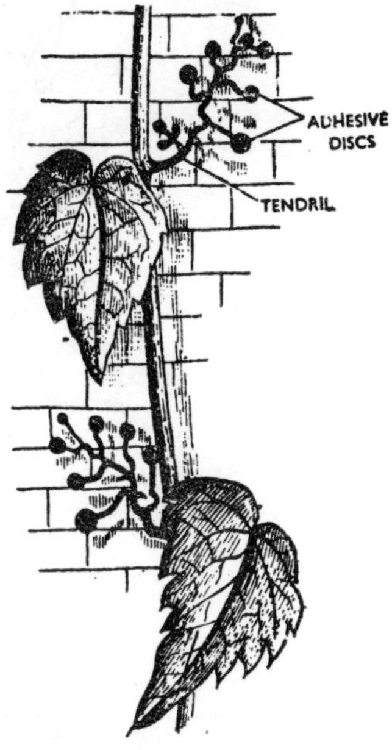

FIG. 110. *Ampelopsis veitchii* climbing by adhesive discs growing from tendrils. (After Strasburger).

BRANCHING OF STEMS

It has been seen that the axillary or lateral buds develop into branch shoots which are exact but smaller copies of the original shoot. The branch shoots again develop lateral buds on them forming branches of the secondary or tertiary order. The same pattern goes on forming branches of higher orders giving rise to the branch-system of the plant. The pattern underlying this branch system is known as **branching.** The large branches are usually called *boughs* while the smallest undeveloped ones are called *twigs*.

The shape of the plant is determined by the type of branching. Thus, in *caudex* the main stem is unbranched, in *excurrent* the apical bud grows indefinitely and in *deliquescent* the apical bud is somehow checked while the lateral buds develop more vigorously (Figs. 87-89).

Branching among the Angiosperms is always caused by the growth of the lateral buds while most Cryptogams show a type of branching by the splitting of apical buds. The former is known as *lateral branching* while the latter is the *dichotomous* type.

A. LATERAL OR AXILLARY BRANCHING

Branching from *lateral* buds is the rule among flowering plants. As the lateral buds are usually axillary, this may also be called *axillary* branching. Branching is **racemose** or **cymose** according as the lateral buds are less vigorous or more vigorous than the apical bud.

(A) RACEMOSE OR MONOPODIAL BRANCHING

Very often the apical bud of the plant grows indefinitely giving rise to a straight stem axis or *podium* on which the lateral buds grow in an aeropetal order. Since there is a single axis for the shoot, the type of branching is **monopodial.** Other names for this type are **racemose** and **indefinite**—the last name because in this type the possibility of vertical growth of the plant is unlimited. This type of branching is the commonest among young Phanerogamic plants. But, as the plant grows older, the apical bud often gets lost and some side branches become very strong so that the plant loses its monopodial character. Typical monopodial branching (Fig. 111) can be seen in the excurrent trees like the conifers or the mast tree (*Polyalthia longifolia*—Fig. 88), *Casuarina*, etc. If, instead of there being a single leaf at each node there be a whorl of leaves, the branches at each node will also be whorled as in *Alstonia scholaris* (Fig. 112).

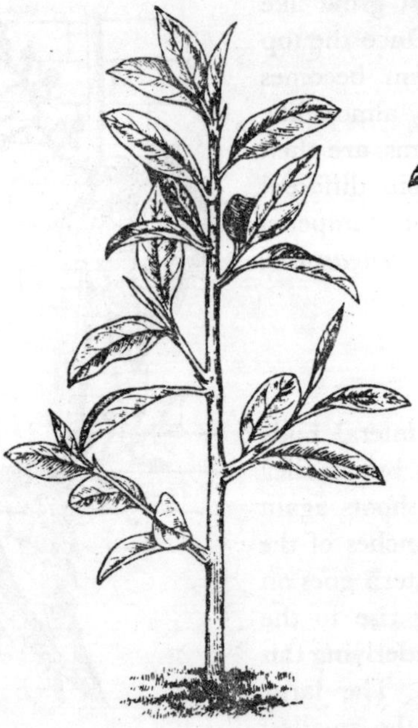

FIG. 111

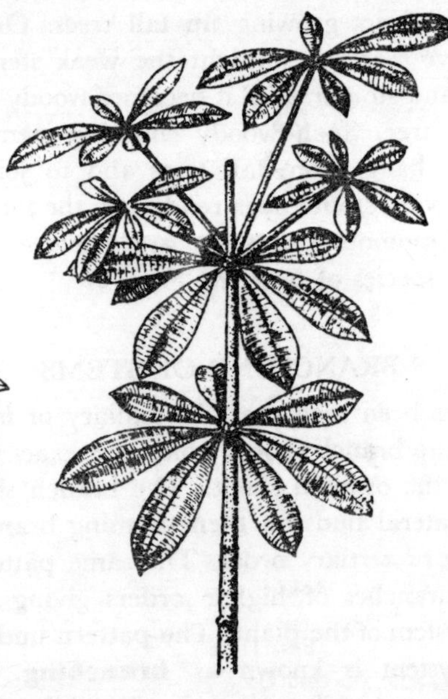

FIG. 112

MONOPODIAL BRANCHING. FIG. 111. Typical form. FIG. 112. Racemose whorled branching in *Alstonia scholaris* having whorled leaves.

(B) CYMOSE BRANCHING

In many tropical trees, as the plant develops, the apical bud gets lost and the side branches develop forming a spreading tree. This gives a deliquescent form (Fig. 89) found special mong trees growing in the open, e.g., mango or jack-fruit. In forests, however, the lower side branches cannot develop because of the absence of light. The plants here are always tall like *Shorea robusta.*

As opposed to the above, in some plants it is the rule that the apical bud does not develop from the early stage and the branching of the plant is dependent solely on the lateral branches immediately below the apex. This is called **cymose** branching. This may also be called the **definite** type of branching as the vertical growth of the plants is limited. These plants cannot be very tall. There may be the following types of cymose branching according as one, two, or more strong branches develop below the apex:

(i) Uniparous cymose

Only one strong branch develops below the stem or the branch apex at each branching. The leaves are arranged spirally so that there is only one axil just below the apex. Among uniparous types again two patterns are distinct:

(*i*) In some, the successive branches below branch tips are alternately to the right and the left. This will give a zigzag form to the stem (Fig. 113A). In nature, however, the plant does not remain zigzag but straightens up (Fig. 113B) forming a false axis or *pseudopodium.* The pseudopodium is also called *sympodium* as the axis is formed by joining up successive branches. This is a **scorpioid cyme.** This is very prominently seen in the vines of different species of *Vitis* (Figs. 96 & 146). In these, the apical buds are transformed into tendrils which are pushed to a side by the straightening up of the sympodium. That is why the leaves are apparently placed opposite the tendrils.

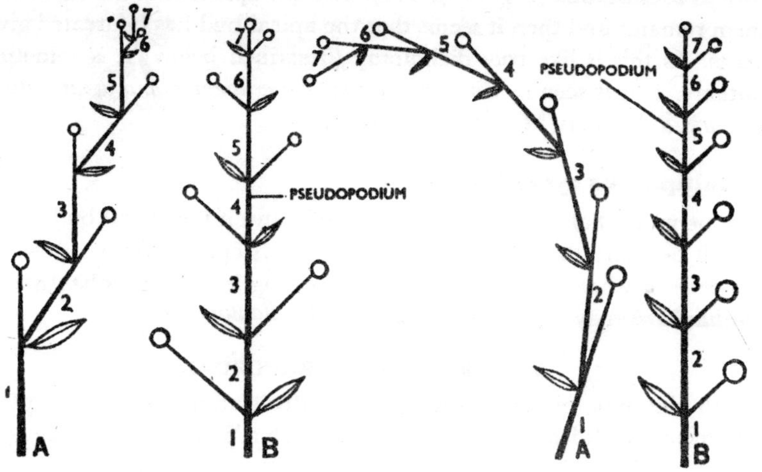

FIG. 113 FIG. 114

PLANS ILLUSTRATING UNIPAROUS CYMOSE BRANCHING. FIG. 113. Scorpioid cyme. A. Showing scorpioid development of successive branches marked 1, 2, 3, 4, ... B. The pseudopodium after straightening up FIG. 114. Helicoid cyme. A. Helicoid development of successive branches marked 1, 2, 3, 4, ... B. Pseudopodium after straightening up.

(*ii*) In the other type of uniparous cyme the branches are successively on one side. This should give rise to a spiral appearance (Fig. 114A), hence the name **helicoid cyme.** But, as in the other case, the pseudopodium straightens up (Fig. 114B) and the real nature is understood only by observing that all the leaves are on one side while all the branches lie opposite. This is found in *Saraca asoca.* The scorpioid cyme again may be **rhipidium** or **cincinnus** and the helicoid may be **drepanium** or **bostryx** according as all the lateral branches are in the same plane or not.

(ii) Biparous cymose

Here, instead of a single lateral shoot developing at each branching, two such

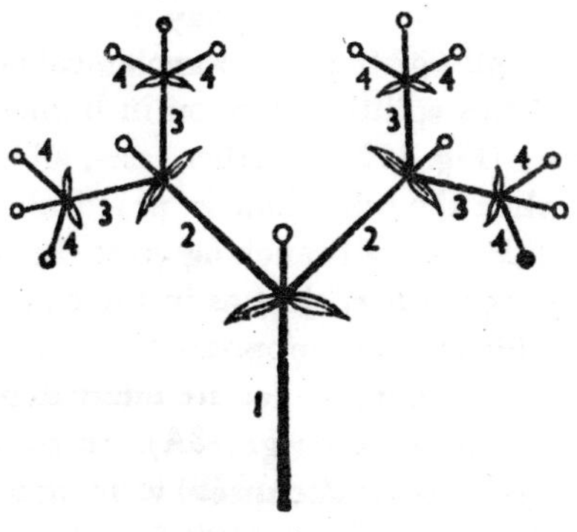

FIG. 115 FIG. 116

FIG. 115. Plan to illustrate dichasium. 1, 2, 3, 4, etc., are the successive branches.
FIG. 116. A twig of *Nerium odorum* showing trichasial branching.

branches develop. This usually happens when there are two opposite leaves at each node and, therefore, there are two opposite axillary buds. This gives rise to a symmetrical shoot known as **dichasium** (Fig. 115). Very often the apical buds die off early so that no trace of them remains and then it seems that the apical bud has bifurcated giving rise to two branches. As this is like true dichotomy (described below) it is sometimes called **false dichotomy.** This is seen in *Mirabilis jalapa, Tabernoemontana divaricata, Plumeria acuminata, Carissa carandas,* Mistletoe (Fig. 10), etc.

(iii) Multiparous cyme

If there be more than two leaves at each node and if the apical bud does not develop, there will be more than two branches at every step. In *Nerium odorum* (oleander) leaves are borne in whorls of three and the branching is of the **trichasium** (Fig. 116) type. **Polychasium** may be seen in *Croton bonplandianum.*

B. DICHOTOMOUS BRANCHING

Although this type of branching is practically confined to the Cryptogams it is necessary to have some idea of it. The growth of such a plant is by a single apical cell or

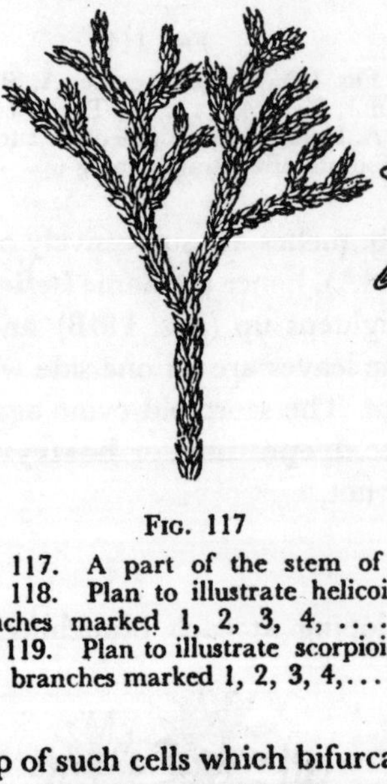

Fig. 117

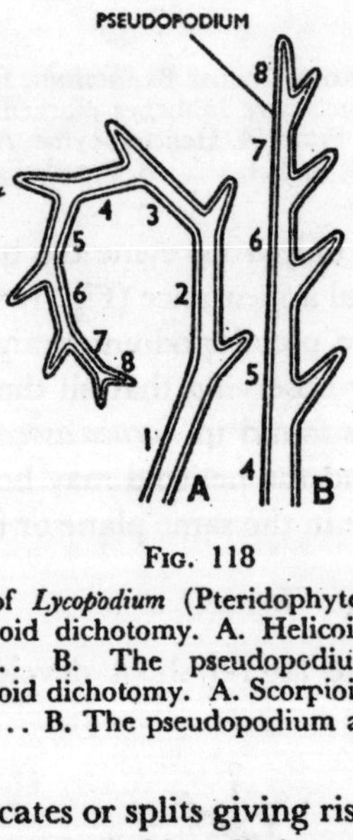

Fig. 118

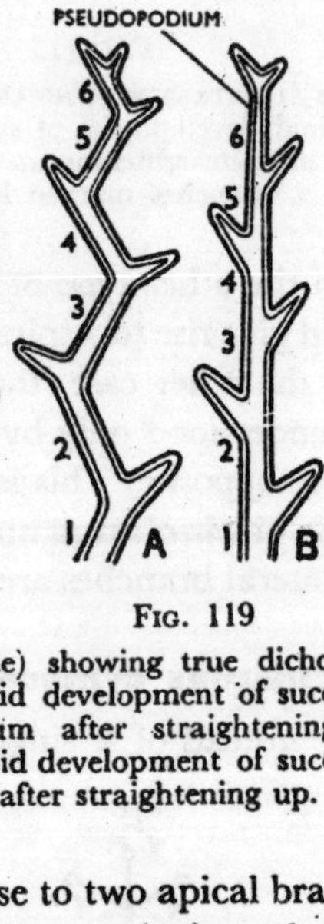

Fig. 119

Fig. 117. A part of the stem of *Lycopodium* (Pteridophyte) showing true dichotomy.
Fig. 118. Plan to illustrate helicoid dichotomy. A. Helicoid development of successive branches marked 1, 2, 3, 4, ... B. The pseudopodium after straightening up.
Fig. 119. Plan to illustrate scorpioid dichotomy. A. Scorpioid development of successive branches marked 1, 2, 3, 4,.... B. The pseudopodium after straightening up.

by a group of such cells which bifurcates or splits giving rise to two apical branches. The tips of these branches again bifurcate and this splitting goes on in branches of higher orders. This is **normal** or **true dichotomy** (Fig. 117). In other cases, after the bifurcation of the tip, one half grows normally while the other half is practically suppressed. This is called **sympodial dichotomy** as this type of branching gives rise to a false axis formed of the bases of successive branches (*sympodium*) just as in the case of uniparous cymose branching. If the branches on one side only are suppressed the plant should take up a coiled structure (Fig. 118A) and the branching is called **helicoid dichotomy.** On the other hand, if alternate branches on the left and the right are suppressed, the zigzag appearance (Fig. 119A) causes the name **scorpioid dichotomy.** In nature, however, these curved stems straighten up forming false axes (Figs. 118B & 119B) or sympodia.

The types of dichotomous branching should be compared with similar cymose forms.

MODIFICATIONS OF STEM

A. UNDERGROUND MODIFICATIONS FOR PERENNATION, VEGETATIVE PROPAGATION AND STORAGE

A number of herbaceous plants find it impossible to survive above ground during unfavourable seasons of the year, usually the winter. Thus, they are short living annuals. Many such plants store up food material in the underground parts. When the aerial shoot withers away during the unfavourable season the underground parts remain safe. Next spring, as the season again becomes favourable, aerial shoots come out which ultimately flower and again die next year after storing up food underground. Perennation is ensured this way by vegetative reproduction. These underground structures, although not green, are recognised to be stem structures because of (1) the presence of nodes and internodes; (2) scale leaves and adventitious roots coming out from the nodes; (3) buds at the axils of scale leaves, at the apices of these underground structures and on the apices of their branches; and (4) their anatomical structures which resemble stems. There are several very common types of them:

(1) **Rhizome**

Rhizomes are underground dorsiventral stems or branches growing horizontally under the surface of the soil. These are brownish in colour so that they are easily mistaken for roots. But, they are divided into nodes and internodes, there are brown scale leaves at the nodes and there are axillary as well as apical buds. Adventitious roots develop on the lower surfaces of the nodes.

The more common types of rhizomes are rather fleshy due to the storage of much food material. An apical bud gives rise to the annual shoot which dies at the close of the season leaving a scar while the rhizome continues its growth by a lateral bud (Fig. 120). The rhizome is, therefore, usually of a *sympodial* nature. Common examples of such rhizomes are ginger (*Zingiber officinale*—Fig. 121) and turmeric (*Curcuma domestica*) which are much branched. The rhizome of *Canna* is not branched so profusely. Rarely, *monopodial* rhizomes may be found as in the case of ferns like *Pteris*. Here, the internodes are rather long and the rhizome grows horizontally by the apical bud which does not terminate in a shoot. Aerial leaves are developed laterally from the nodes.

A special type of rhizome is the **rootstock** which is a rhizome growing up vertically instead of horizontally. Very common examples of rootstock are found in bananas and many aroids (e.g., *Alocasia indica*—Fig. 122).

The **sobole** (Fig. 130) may be considered either as a thin rhizome or as a *runner*. They are described later under runners.

Rhizomes and other kinds of underground stems sometimes bear a special type of roots called *contractile* or *pull roots* (Figs. 126 & 238). These roots contract after formation and pull the growing rhizome or other organ deeper down into the soil. These are illustrated in connection with the root modifications.

(2) **Stem-tuber**

Tuber is a general term applied to any fleshy part of the plant which may store food. These may be stems (underground as in the stem-tubers under discussion and aerial in some bulbils and pseudobulbs) or roots (root-tubers). They are largely used as food and the most common of them is potato (*Solanum tuberosum*—Figs. 67 & 123). When potato is grown from eye buds, after the plant is rooted to the soil, the lower part of its stem is

covered by earth. Axillary or adventitious branches grow from this underground part of the stem (Fig. 123A). These branches swell up at their apices into tubers due to the accumulation of food substances (mostly starch) and the arrest of growth. Each tuber (Fig. 123B) develops slowly. It is a swollen bit of stem on which many internodes are present but not distinct because of the great swelling. The nodal points are noted by the presence of scale leaves at an early stage and *eyes* or rudimentary buds at the axils (marked

Fig. 120

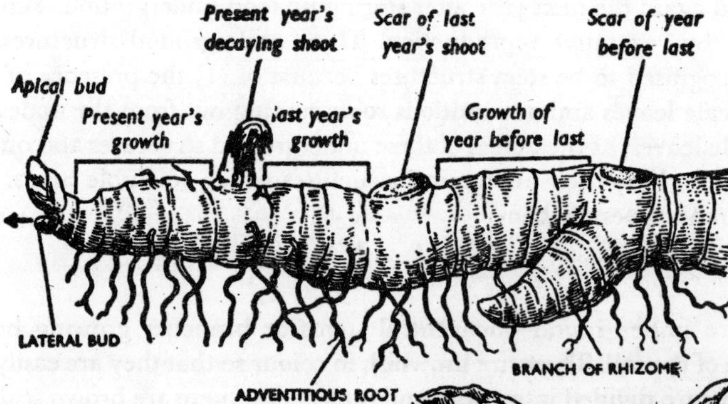

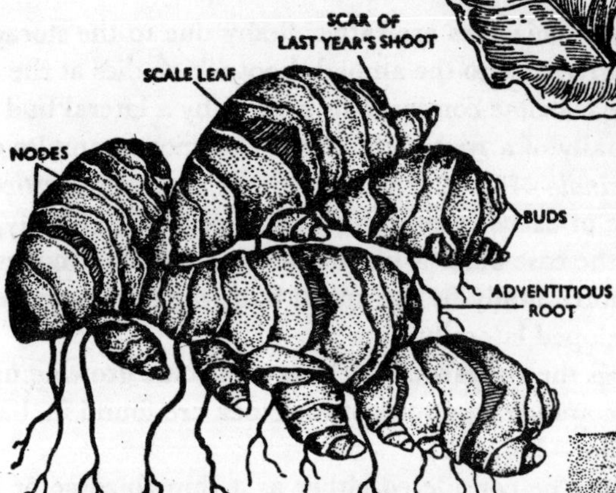

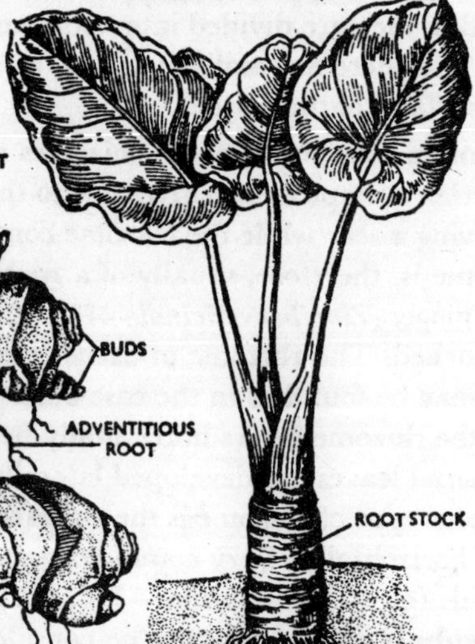

Fig. 121 Fig. 122

Rhizome. Fig. 120. A four-year-old rhizome of a wild plant. The apical bud will develop next year's shoot while growth will be continued by the lateral bud in the direction of the arrow. The stem is, therefore, sympodial. Fig. 121. Rhizome of ginger. Fig. 122. Rootstock (vertical rhizome) of *Alocasia indica*.

by small depressions) of these scale leaves. These *eyes* are arranged spirally on the tuber and are more crowded towards the distal end (called 'rose end') which is terminated by an apical bud. The point of attachment of the tuber to the stem is marked by a scar which is opposite the apical bud. The tuber is protected by a corky skin. Adventitious roots are absent on the tuber at this stage.

Tubers are also found in some other common plants as the weed *Cyperus rotundas* and the vegetable artichokes (*Helianthus tuberosus* or Jerusalem artichoke storing inulin and *Stachys tuberifera* or Chinese artichoke storing stachyose). In *Helianthus tuberosus* (Fig. 124) the scale leaves and nodes are much more prominent.

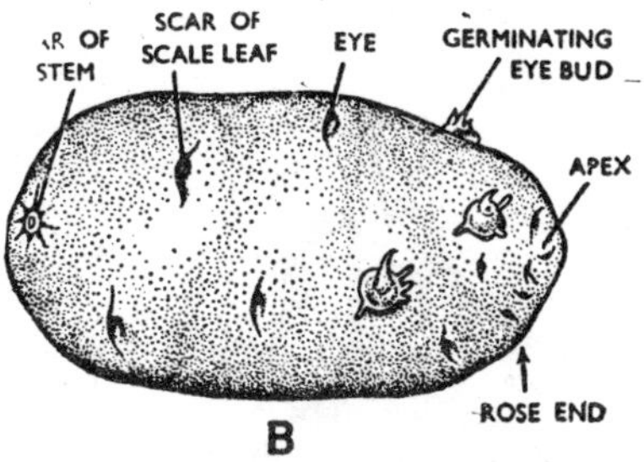

TUBER. FIG. 123. Potato. A. Base of a potato plant showing the development of potato tubers from underground stem branches. B. A single tuber. Some *eyes* towards the *rose end* have commenced development.

When the tubers are supplied with conditions, the eye buds develop into small shoots from the bases of which adventitious roots grow.

(3) **Bulb**

What is commonly known as *bulb* is actually an underground modified bud with a very much reduced stem **(disc)** and *fleshy scale leaves* growing from it. The disc has a convex or conical shape with very much compressed internodes from the nodes of which fleshy scale leaves arise. At the apex there is a bud which develops into a flowering scape at the proper season (Fig. 85). Some daughter bulbs develop from the axillary buds which may also develop branch flowering scapes. Much food is stored up in the fleshy scale leaves which may be true scale leaves (*cataphylls*—described in connection with leaf) or the bases of the green leaves of the last season. From the undersurface of the disc a tuft of adventitious roots grow. In some onions new bulbs develop in a peculiar way. Some miniature bulbs or 'bulbils' are found to replace some of the flowers in its inflorescence in the same way as bulbils develop in *Agave*.

FIG. 124. Stem-tubers of *Helianthus tuberosus*.

Two types of bulbs are distinguished according to the arrangement of the fleshy scale leaves:

(1) In the **tunicated bulb,** as in onion (*Allium cepa*—Fig. 125), tulip (*Tulipa*), hyacinth (*Hyacinthus*) and tuberose (*Polyanthes*), the scale leaves are arranged in a concentric manner as may be seen in a cross-section. The whole bulb is covered by some dry membranous scale leaves which form the *tunic*.

(2) In **scaly** or **imbricate bulbs,** the scale leaves are not concentric but are arranged loosely like the petals of a flower possibly overlapping one another only at the

margins. Such a bulb is not a compact body as in the other case and is usually not covered by any common tunic. This type of bulb is not so very common but may be found in some garlics (*Allium sativum*) and more typically in the lilies (*Lilium* sp.) which also show *contractile roots* (Fig. 126A). In garlic (*Allium sativum*) a number of cylindrical but separate (i.e., not concentric) fleshy scales, called *cloves*, are arranged as in a naked bulb but a group of such 'cloves' are enclosed by a whitish, skinny tunic.

Fig. 125

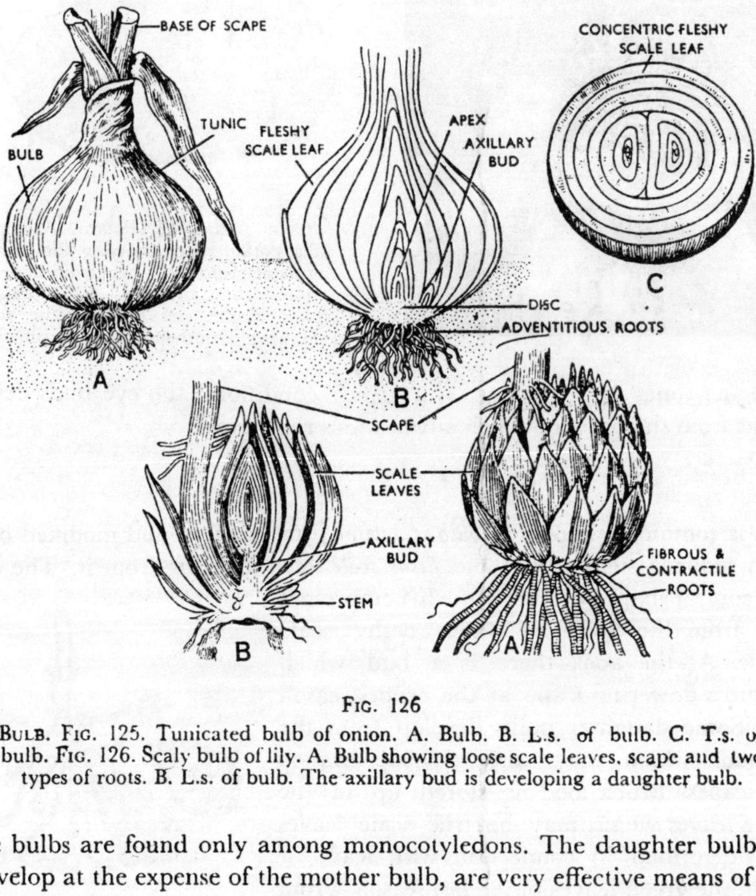

Fig. 126

Bulb. Fig. 125. Tunicated bulb of onion. A. Bulb. B. L.s. of bulb. C. T.s. of bulb. Fig. 126. Scaly bulb of lily. A. Bulb showing loose scale leaves, scape and two types of roots. B. L.s. of bulb. The axillary bud is developing a daughter bulb.

True bulbs are found only among monocotyledons. The daughter bulbs or *cloves*, which develop at the expense of the mother bulb, are very effective means of vegetative propagation.

(4) **Corm**

The *corm*, sometimes called a *solid bulb*, is really a solid very much condensed vertical rootstock with a large apical bud. There are some scale leaves on its body and adventitious roots grow either from its base or all over its body.

In *Amorphophallus campanulatus* (Fig. 127) the corm is a huge condensed single internode with numerous adventitious buds and roots on its body. The apical bud develops the leaves of the year and gradually consumes the food material within the corm so that the latter shrivels up. At the end of the season a new daughter corm is formed by accumulation of food material at the base of the shoot forming a new (second) internode on the top of the shrivelling last year's corm. Adventitious buds also develop into daughter corms. The same phenomenon may be seen in *Crocus sativus* (saffron of Kashmir—Fig. 128) where the corm is covered by some more or less sheathing scale leaves bearing axillary buds. One (sometimes more) axillary bud on the top of the corm forms the aerial shoot which consumes the old corm and the new corm is formed by the swelling of its base towards the close of the season when the aerial shoot dies. Next year, another top

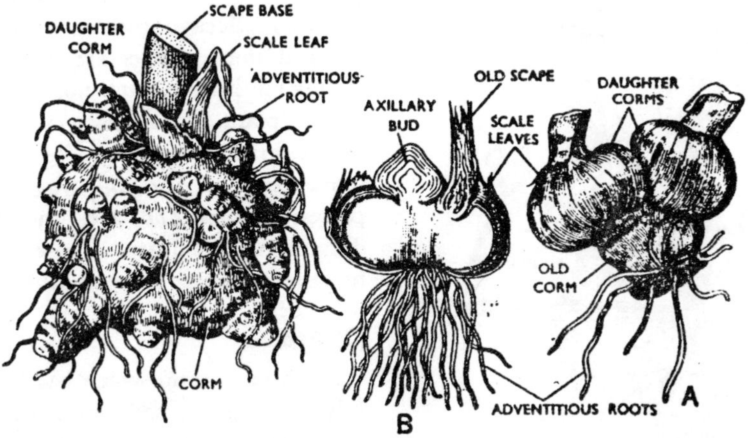

FIG. 127 FIG. 128

CORM. FIG. 127. Corm of *Amorphophallus campanulatus*. FIG. 128. Corms of *Crocus sativus*. A. Two daughter corms arising from an old shrivelled corm. B. L.s. of a corm showing the development of an axillary bud which will form the new aerial shoot.

axillary bud develops on this new corm as the aerial shoot. The development of this type of corm is, therefore, sympodial. In *Colchicum autumnale* (meadow saffron, also from Kashmir) the axillary bud that develops into the aerial shoot is towards the base of the mother corm. Thus, in this case, the new corm is on the side and not on the top of the old corm. The common garden *Gladiolus* lily is also an example of typical corm.

B. SUBAERIAL MODIFICATIONS FOR VEGETATIVE PROPAGATION AND PERENNATION

Certain plants, generally of weak herbaceous habit, propagate quickly by means of special branch stems which give rise to small daughter plants. Thus, although the individual plants are short-living, the colony of plants is perennial. These modified branches either grow along the surface of the soil (water in aquatic plants) or may be partially or wholly underground.

(1) **Runner** and **sobole**

In certain plants showing the creeping habit like the common *dub* grass (*Cynodon dactylon*), *Oxalis* (Fig. 129A), *Centella asiatica* (Fig. 90), strawberry, etc., the axillary buds from the lowest leaves give rise to slightly modified branches. The basal internodes of the bud elongate horizontally and trail along the soil carrying the bud to a distance from the mother plant where it gets fixed to the soil by means of adventitious roots and develops a new daughter plant. This special branch carrying the bud is called a *runner*. It may be of several internodes and have scale leaves on the nodes. Such runners grow in all directions from the mother plant and the daughter plants again propagate in all directions in a similar manner. Thus a single plant soon covers a large area by its progeny. In *Colocasia* (common arum) the runners are somewhat thicker and may be partly or wholly underground (Fig. 129B).

A special type of underground runner or creeping stem is the **sobole** (Fig. 130). It is sometimes included within rhizomes as it may be considered as one of them, only it is thin and does not accumulate any food matter. But, as the term *rhizome* is usually applied to fleshy structures storing food, *soboles* are better considered separately. These are found in the creeping grasses. These grasses grow these underground stems in all directions giving rise to innumerable offsprings. This is a very effective means of propagation. Even if these soboles are cut into pieces during cultivation, each node gives rise to a new plant. Most grassy weeds like the couch grass (*Agropyron*), *kans* (*Saccharum spontaneum*) or *ulu* (*Imperata cylindrica*)—possess soboles and are, therefore, very difficult to eradicate. These weeds are, however, of some value as sand binders.

(2) **Stolon**

This is a term which has been variously used. It is sometimes taken to be synonymous with runner. More commonly, it is applied to a special type of runners which do not grow horizontally from the beginning. At first they grow upwards like ordinary branches and then arch down to meet the soil where daughter plants are formed after striking root. In the horticultural method of *layering* in fruit trees, etc., branches are suitably lowered down to meet the soil where they are made to strike root (Fig. 131). Natural stolons like this are found in *Mentha* sp. and some plants of the Rosaceae (Fig. 132).

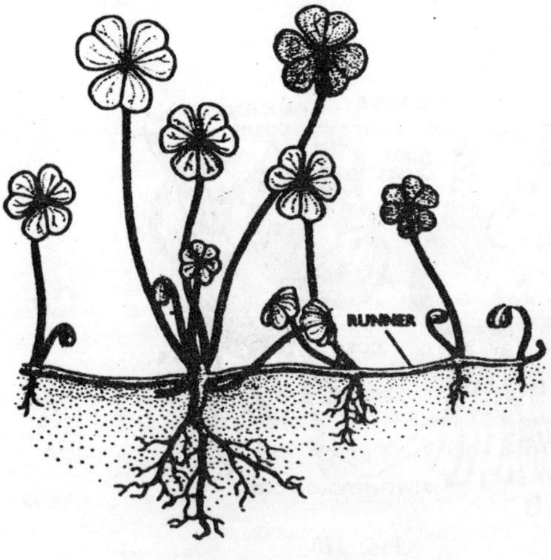

A

B

RUNNER. FIG. 129. A. *Oxalis corniculata* propagating by runners. B. Thick runner of *Colocasia esculenta.*

FIG. 130. The sobole of *Agropyron*.

(3) Sucker

Sucker is an underground runner which soon goes up and forms a daughter plant after striking root. Sucker plants are also formed from axillary buds at the nodes of this underground runner. Plants developed from other underground parts (e.g., roots of some shrubs and trees) are also called suckers. Typical suckers are seen in garden chrysanthemum (Fig. 133), *Mentha arvensis*, etc.

(4) Offset

This is found in aquatic plants and is just like the runner, only it is shorter and thicker. Common instances are the water hyacinth (*Eichhornia crassipes*—Fig. 134) and pistia (*Pistia stratiotes*).

FIG. 131

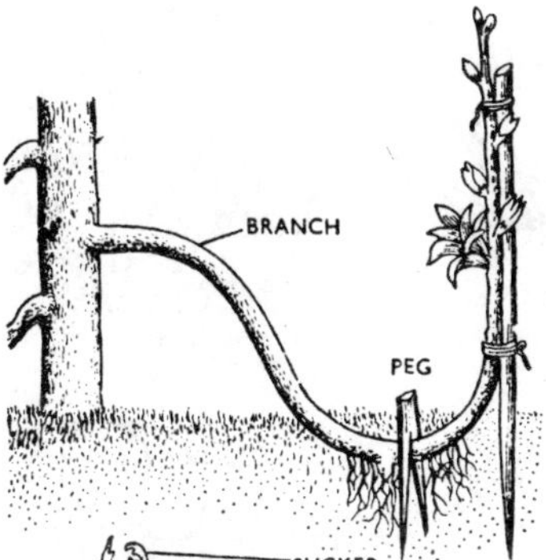

FIG. 132

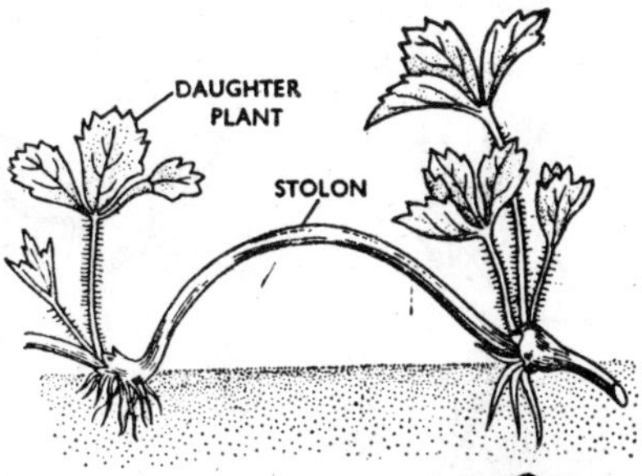

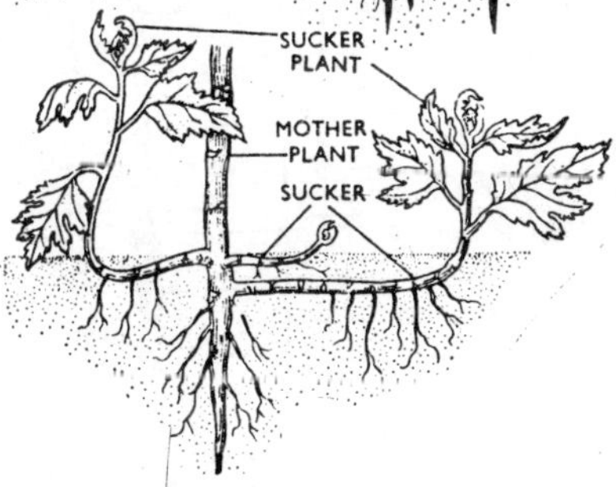

FIG. 133

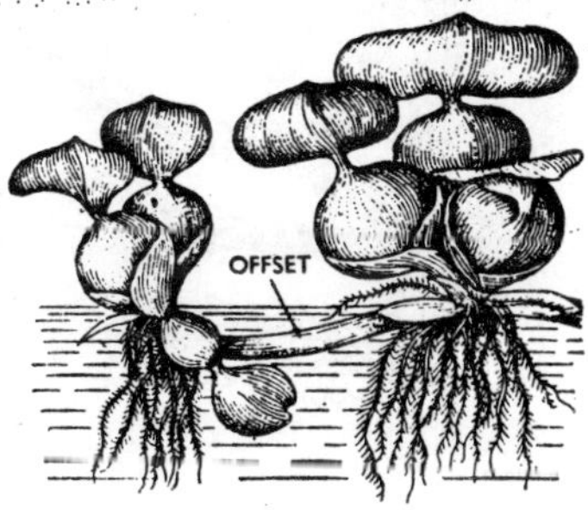

FIG. 134

FIG. 131. Artificial propagation of fruit plants by *layering*. A branch has been bent and forcibly kept down by a peg. The portion covered by a layer of soil has taken root producing an artificial *stolon*. FIG. 132. Natural stolon of *Mentha*. FIG. 133. Suckers of *Chrysanthemum*. FIG. 134. Offset of water hyacinth.

C. AERIAL MODIFICATIONS

The aerial stem is found to be modified for various purposes. Most of these come under **metamorphosed** stems as the modifications are so extreme that the stem nature of these structures is very difficult to make out, except for the origin and the position. Thus, one may say that the stems have been completely changed or *metamorphosed*.

(1) Phylloclade or Cladophyll and Cladode

This is a very interesting modification met with in the xerophytic plants which have to greatly cut off all water loss. This they do by checking the growth of leaves which are mainly responsible for transpiration. But, the stems become flattened or swollen and green so that they may take up the photosynthetic function of the leaves. Such flattened or swollen structures are called *phylloclades* or *cladophylls*. These are often mistaken for leaves. The phylloclade of *Opuntia* (Cactaceae) is very common (Fig. 135). This phylloclade consists of many fleshy flattened internodes similar to underground stem-tubers and the leaves at the nodal positions are metamorphosed into spines or small scaly leaves. Flowers are also borne on the nodes. It develops a strong cuticle and contains much mucilage which helps to retain water. Such phylloclades are common in the families

FIG. 135

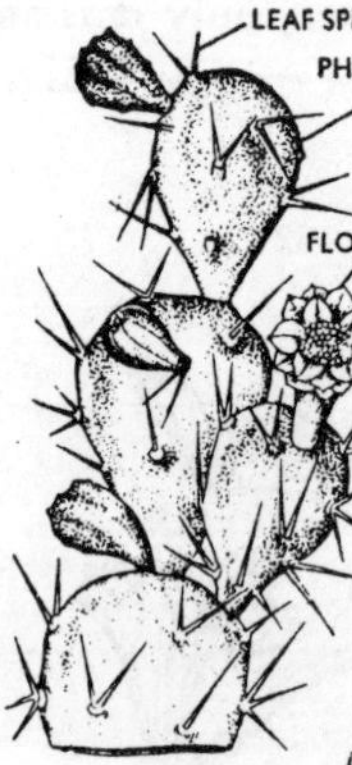

FIG. 136

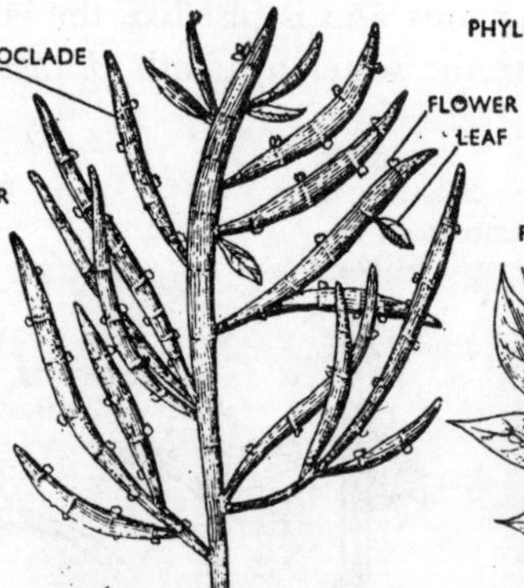

FIG. 137

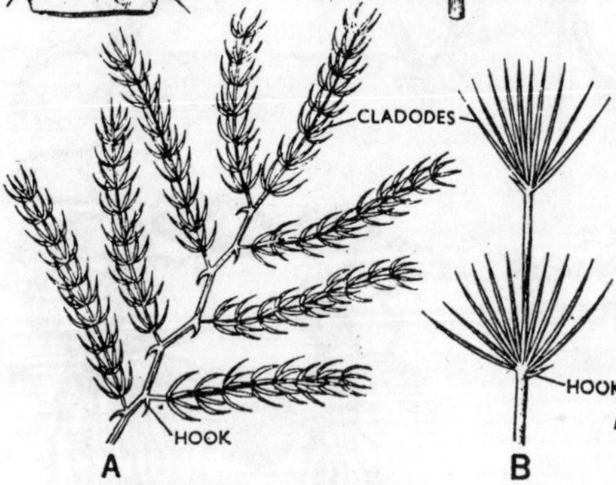

FIG. 138

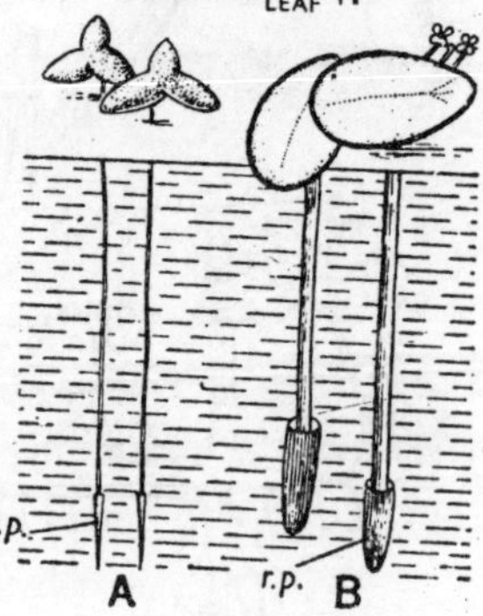

FIG. 139

PHYLLOCLADE. FIG. 135. Phylloclade of *Opuntia monacantha*. Nodes may be located from the positions of the spines (metamorphosed leaves). FIG. 136. Phylloclade of *Muehlenbeckia platiclados*. FIG. 137. Leaf-like phylloclades of *Ruscus aculeatus*. The phylloclades are axillary to real leaves and flowers originate at nodal positions on the phylloclades. FIG. 138. A. Cladodes of *Asparagus racemosus*. The leaves are metamorphosed into hooks. B. shows a magnified bit of the stem. FIG. 139. Young (A) and old (B) plants of *Lemna trisulca* showing flowers developing on it and roots with root-pockets (*r.p.*).

Cactaceae and Euphorbiaceae and in *Muehlenbeckia platiclados* (=*Coccoloba platiclada*) of Polygonaceae (Fig. 136), *Casuarina*, etc. The phylloclades of *Ruscus aculeatus* of Liliaceae (Fig. 137) look exactly like leaves with veins but the flowers show the nodal positions.

Cladode is a type of phylloclade consisting of one internode only. A very beautiful example of this is *Asparagus racemosus* (Fig. 138) of Liliaceae. The ultimate branchlets are of one internode only. These are cylindrical (not fleshy) and are mistaken for leaves. They are cladodes while the leaves are reduced into prickles. Some varieties of *Muehlenbeckia* may also show phylloclades of single internodes which are flattened structures and more leaf-like.

The case of the small aquatic weed *Lemna* (Fig. 139) is rather problematic. A green thalloid leaf-like structure lives freely, multiples vegetatively and develops flowers and roots on it without, apparently, showing any stem. One opinion is that this is an abnormal free-living leaf. A second view is that the plant body is a flattened leaf-like stem without any real leaf. If the second view is correct then this is only another instance of phylloclades.

In some ordinary plants like the Cucurbits the stem sometimes assumes an abnormal flattened appearance. But this is not a phylloclade. It is usually formed by the fusion of a number of buds. This type of abnormal development is known as *fasciation* (*cf.* p. 50).

(4) Thorn

Axillary buds are sometimes arrested and transformed into these hard structures. *Thorns* should be distinguished from *spines* and *prickles* as they are very deep-seated structures with vascular connections, being direct prolongations of the stem. *Prickles* are mere superficial outgrowths on the epidermis. *Spines* are intermediate forms when some other plant organs, e.g., the tips of date palm leaf segments or the stipules of *Acacia*, become hardened sharp structures. The development of such sharp and hard structures is a characteristic of xerophytes.

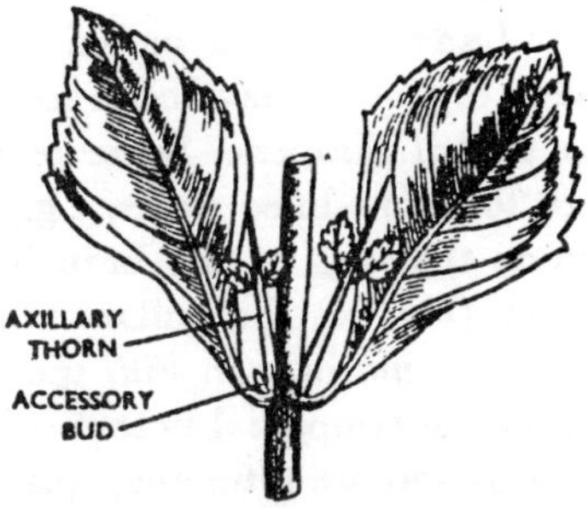

FIG. 140. *Duranta* axillary branch modified into thorn.

FIG. 141

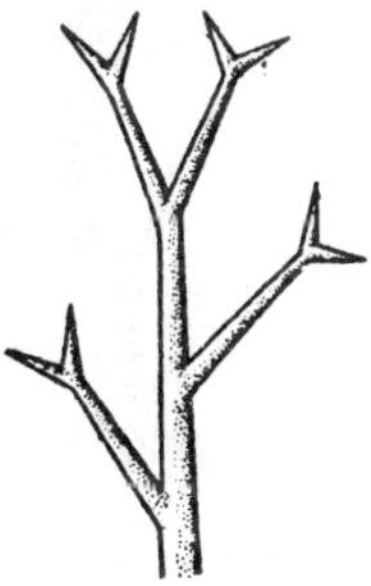

FIG. 142

FIG. 143

FIG. 144

FIG. 145

THORNS. FIG. 141. Branched thorn of *Carissa carandas* (axillary in origin). FIG. 142. Branched thorns of *Flacourtia jangomas*. Some of these are not real thorns but mere subepidermal outgrowths. FIG. 143. Axillary thorns of *Hygrophila auriculata*. FIG. 144. Thorns of prune (*Prunus*) bearing flowers. FIG. 145. Paired "thorns" of *Aegle marmelos*.

Thorns are commonly seen in the garden hedge plant *Duranta* (Fig. 140) where sometimes leaves are found to develop on the thorn showing its stem nature very clearly. Sometimes, the thorn is branched as in *Carissa carandas* (Fig. 141) and *Flacourtia* (Fig. 142). Thorns are found in many other common plants, e.g., *Hygrophila auriculata* (Fig. 143). In prunes the thorn (Fig. 144) bears flowers like an inflorescence axis. The 'thorn' of lemons (Fig. 208) and the double 'thorns' of the common wood-apple (*Aegle marmelos*—Fig. 145) are of problematic origin. They are not considered as true modified stems and are discussed in connection with leaf modifications.

(4) **Stem-tendril**

Tendrils arise as modifications of different plant organs as have been seen when discussing the tendril climbers. Some tendrils are clearly seen to be modified stems, e.g., when they develop on the axils or the branch tips. In passion flower (Fig. 95) the axillary branches are modified into tendrils. In *Vitis* species (Fig. 96) the tips of the sympodial branches are modified into tendrils and are ultimately placed opposite the leaves. *Cissus quadrangularis* (Fig. 146) is another such example. Inflorescence axes, which are also of stem nature, may be modified into tendrils as in *Antigonon* (Fig. 102) and *Cardiospermum* (Fig. 465). The nature of the *Cucurbita* tendril (Fig. 103) is doubtful being, possibly, of prophyll nature.

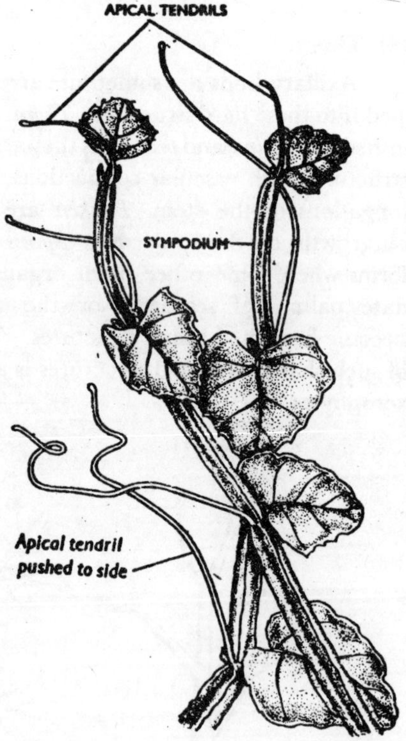

FIG. 146. Twigs of *Vitis quadrangularis* showing apical tendrils which later become lateral (opposing leaves) due to sympodial growth of stem.

(4) **Bulbil**

These are modifications of axillary vegetative or floral buds and have been discussed under the modifications of buds (Figs. 78-82).

(5) **Thalamus**

The *thalamus* or the floral axis is a modified stem and is discussed in connection with the floral organs.

(6) **Pseudobulb**

In many aerial orchids one (rarely, more than one) internode of stem becomes fleshy and tuberous. This structure stores up a good quantity of moisture enabling the plant to survive unfavourable seasons of drought, i.e., the winter. Such structures are known as *pseudobulbs* (Fig. 147).

(7) **Dwarf-shoot** or **spur** or **brachyplast**

While normally all foliage leaves are borne on the nodes of an elongated stem, in some plants most leaves are concentrated on special secondary stems located at axillary positions. These secondary stems are very short with very close internodes so that the leaves appear to be in clusters. They are called *dwar, ots* or *spurs* or *brachyplasts* and are very common among Gymnosperms (e.g., pine). Such dwarf-shoots are also found in certain Angiosperms like *Berberis* (Fig. 148).

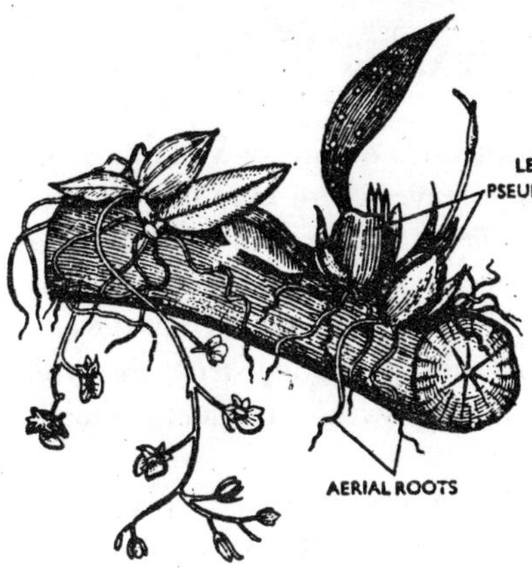

FIG. 147

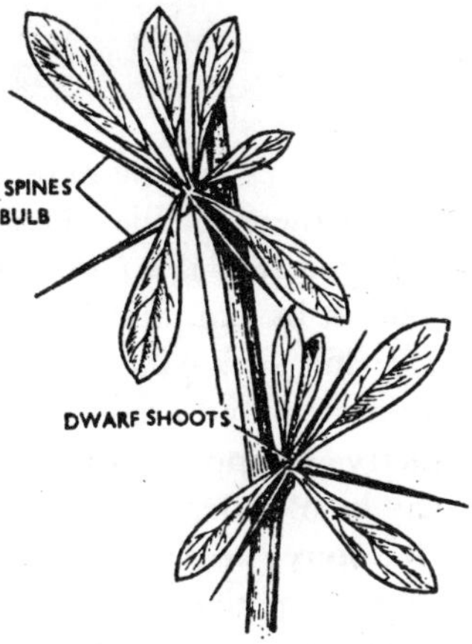

FIG. 148

FIG. 147. An orchid with pseudobulbs. FIG. 148. A twig of *Berberis* showing dwarf-shoots arising from the axils of spinous leaves.

CHAPTER IV

THE SHOOT (Contd.):

THE LEAF

Green foliage leaves are the most conspicuous organs of a plant. The stem forms a skeleton to support these. Besides the foliage leaves there may be leaves of other types. Generally speaking, a **leaf** is a *lateral dissimilar appendage of the stem* borne at a node and subtending a bud at its axil. Such leaves develop from the swollen exogenous protuberances on the growing apex called the *leaf primordia* (Fig. 66). They develop in an acropetal order on the stem and are usually flat, thin, expanded structures while the stem is of radial symmetry. As opposed to the stem the growth of the leaf is limited.

Leaves may be of different kinds:

(1) **Cotyledonary leaf:** The first leaves on the shoot are represented by the cotyledons. The node on the tigellum where the cotyledons are attached in a dicotyledonous seed represents the first node of the shoot. The cotyledons may bear buds in their axils as in potato (Fig. 149). In pea or bean the cotyledons are fleshy because of the storage of food material while in castor it is thin with vein markings. After germination the cotyledon may turn green like the foliage leaves. The cotyledons are sometimes lobed like mature leaves as in *Tilia* and *Geranium* (Figs. 38 & 39). In monocotyledons the scutellum represents the cotyledon and its function is mainly the suction of food material from the endosperm.

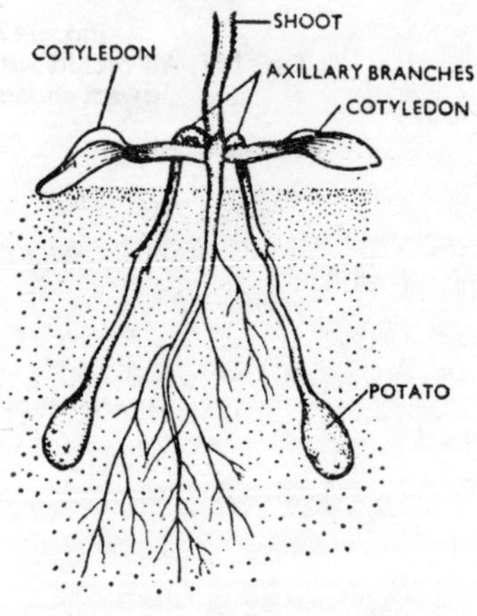

FIG. 149. Potato seedling showing cotyledonary leaves with axillary buds.

(2) **Cataphylls** or **scale leaves:** Different types of scale leaves have already been described in connection with underground stems. They are either dry and papery or fleshy. Such scale leaves may also be found on aerial parts as on bamboos, *Casuarina*, *Asparagus* (Fig. 150), etc., or as bud scales (Fig. 75). Bud scales may represent different parts of a leaf axis as explained while discussing *Buds*.

(3) **Hypsophylls** or **Bract leaves:** These are special leaves which bear floral buds at their axils. They are discussed in connection with the study of reproductive organs.

(4) **Prophylls:** The first few leaves of a branch often differ from the other leaves and then they are termed *prophylls*. Usually there is one prophyll in monocots and two in dicots. In wood-apple (*Aegle marmelos*—Fig. 145) prophylls are represented by spines. The same is the case with the single prophyll of lemons and oranges (*Citrus* spp.—Fig. 208). In many flowering branches they are just the first few leaves which are very small and do not subtend floral buds. In winter buds the outer bud scales may be prophylls. The tendril of *Cucurbita* (Fig. 103) also is, possibly, a modified prophyll.

(5) **Floral leaves** and **sporophylls:** The floral leaves, i.e., sepals, petals, stamens and carpels, are nothing but specialised leaves. Of these, the stamens and carpels may also be termed *sporophylls* in common with the lower plants as they are the spore-bearing organs in the Angiosperms. The leaf nature of these is explained later.

(6). **Foliage leaves:** They are the common green leaves which are discussed in detail in this chapter.

THE FOLIAGE LEAF

Foliage leaves are essential for the plant as leaf is the seat of several very important physiological processes like photosynthesis (which is related to the green colour), transpiration and respiration. Besides these important functions, leaves also protect the buds in their axils and, in special cases, may help in climbing, store food or water (as in *Aloe indica*), serve for vegetative propagation as in the leaf of *Kalanchoe* absorb food material from water as in some aquatic plants and absorb protein as in some insectivorous plants.

Nevertheless, there are a number of plants which are completely or partially leafless. In some of these, the function of leaves is taken up by the stem which remains green and sometimes becomes flat and leaf-like (*phylloclades*). Others become saprophytic or parasitic so that this organ for food manufacture becomes unnecessary. As against these leafless plants, we may consider the case of *Lemna* (Fig. 139) where, according to some, the whole plant is a leaf, there being no stem.

The **duration** of leaves varies. It is *caducous* or *fugdcious* when the leaves fall off as soon as they are formed as in some *Opuntia*. Many trees are characterised by the fact that the leaves fall off at the end of a particular season. These leaves are *deciduous* or *annual*. Some trees shed *all* their leaves during winter, then the trees become bare. New leaves cover the plant on the advent of spring. In most tropical trees, however, the leaves last for a long time and are termed *persistent*. Such trees are *evergreen*.

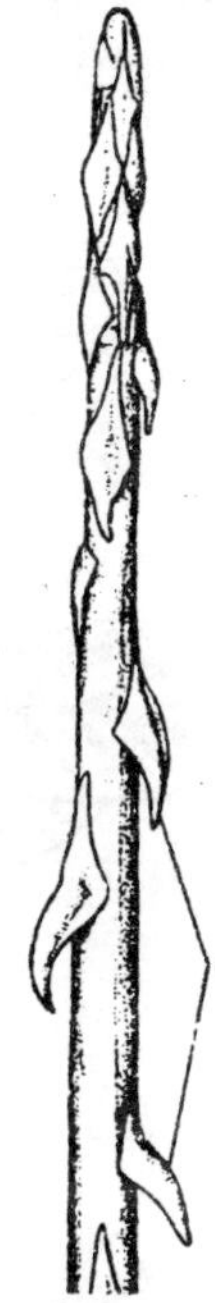

FIG. 150. A growing shoot of *Asparagus* bearing cataphylls.

PHYLLOTAXY

Phyllotaxy is the mode of arrangement of leaves on the stem. It is surprising how regular and strictly mathematical this is. There is a definite law according to which the leaves are arranged on the stem of a particular species. The arrangement is always regular and leaves are never placed on the stem in a haphazard manner.

The foliage leaves are usually spread about on the stem with long or short internodes between them. This arrangement is called cauline. In some cases, however, the leaves arise as a cluster or a rosette from the very short stem just on the top of the root. This arrangement is called *radical* although the leaves do not actually arise from the root but from a very much condensed stem as in the case of radish.

In *cauline* leaves there may be one, two, three or more leaves at each node. When there is only one leaf the arrangement is **spiral** or **alternate** or **acyclic.** This is by far the commonest type of phyllotaxy. When there are two or more leaves at each node the phyllotaxy is **cyclic.**

A. CYCLIC PHYLLOTAXY

In the *cyclic* type of phyllotaxy the leaves at each node form a whorl with the leaves placed on a circle in which the angles between adjacent leaves are the same. Thus, if there be two leaves in a whorl, the two will be placed *opposite* (i.e., at an angular distance of two right angles) one another. If there be three leaves, the angle between leaves in the same whorl is 120° (i.e., one third of a circle), if four, it is 90° and so on. When there are more than two leaves in a whorl the phyllotaxy is called *verticillate.*

(A) OPPOSITE PHYLLOTAXY

In *opposite* phyllotaxy the two leaves at each node are always opposite one another. If the successive pairs of leaves be placed at right angles to one another, the arrangement is termed **opposite decussate** (Fig. 151). In this case, when looked from the top, all the leaves will be found to be arranged along four vertical rows. This is found in

FIG. 151

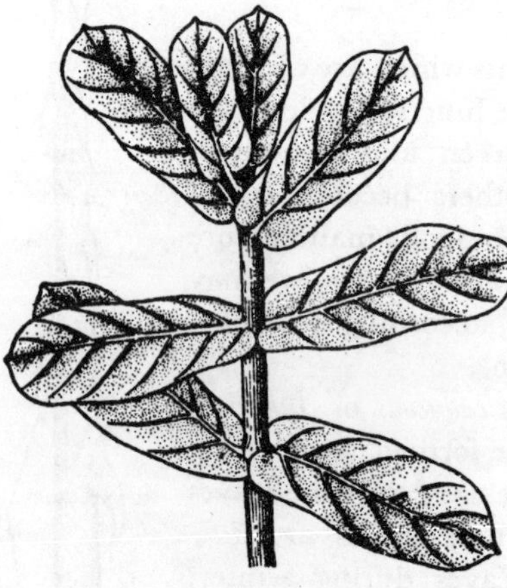

FIG. 152

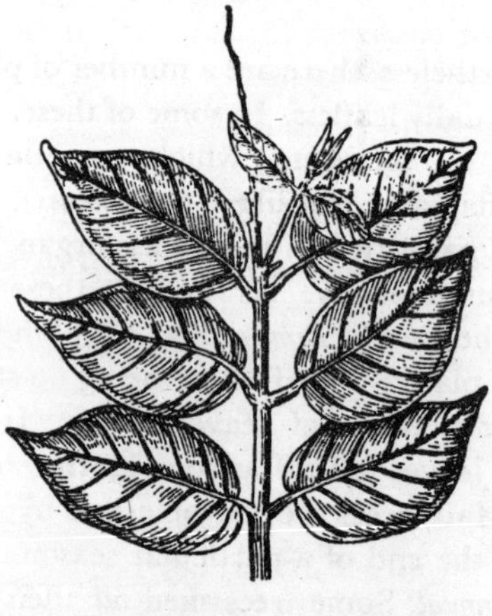

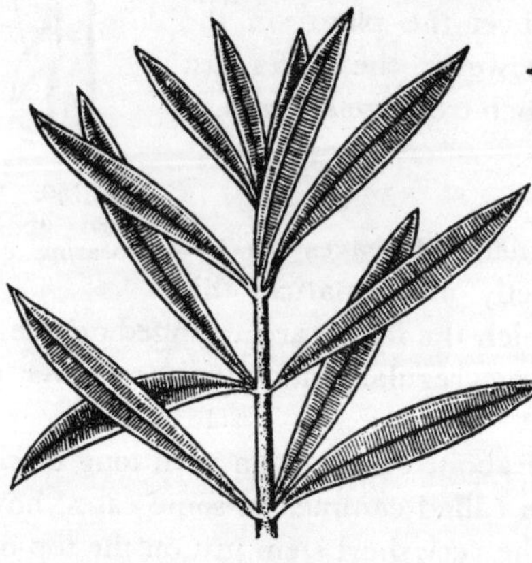

FIG. 153

FIG. 154

PHYLLOTAXY. FIG. 151. Opposite decussate in *Calotropis*. FIG. 152. Opposite superposed in *Quisqualis*. FIG. 153. Verticillate (whorls of three) in *Nerium odorum*. FIG. 154. Spiral (*pentastichous*) in china-rose.

Calotropis, Gardenia, Ixora, etc. In some other plants it is found that the successive pairs are placed exactly on top of one another so that all the leaves lie in one plane and when viewed from above all the leaves are found to lie in two vertical rows. This type of phyllotaxy is known as **opposite superposed** (Fig. 152). It should be noted that while the superposed type is found in many plants like *Psidium guyava* (guava), *Quisqualis,* etc., in many instances the phyllotaxy is initially decussate which later becomes superposed by the flattening out of the twig.

(*B*) *VERTICILLATE PHYLLOTAXY*

Nerium odorum (Fig. 153) shows three leaves forming a whorl at each node while *Alstonia scholaris* (Fig. 112) shows five or more leaves. These are instances of the verticillate type. Sometimes verticillate phyllotaxy is designated as **whorled** phyllotaxy.

B. SPIRAL or ALTERNATE or ACYCLIC PHYLLOTAXY

This is the commonest type (Fig. 154) of phyllotaxy and the mathematical regularity of the arrangement is astonishing. One can easily verify that (1) the angular distance **(angular divergence)** between any two consecutive leaves is constant; (2) when looked from the top (Fig. 155), all the leaves are found to lie in a fixed number of vertical rows or **orthostichies;** (3) the orthostichies are evenly dispersed on a circle, the angle between adjacent orthostichies being constant; (4) although leaves look scattered, a close examination shows that leaves are evenly dispersed on all sides of the stem.

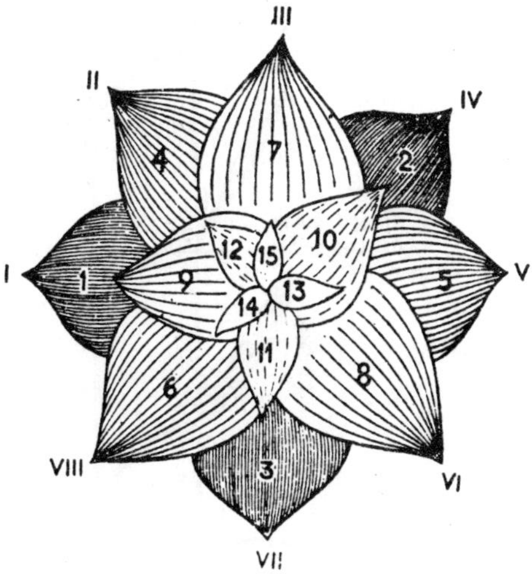

Fig. 155. A twig showing *octastichous* phyllotaxy viewed from top. All leaves fall in eight orthostichies (orthostichy I — leaves 1, 9, 17, etc.; orthostichy II — leaves 4, 12, 20, etc.; and so on).

To correctly understand the plan of a spiral phyllotaxy one has to imagine a line touching the bases of successive leaves. It will be found that this line forms a spiral (hence the name *spiral phyllotaxy*) on the stem. This spiral is called the **genetic spiral** (Fig. 156). This *genetic spiral* can easily be projected on a flat surface to form a *flat spiral* (Fig. 157) and the position of the leaves may be marked on this spiral. The angle subtended at the centre by two consecutive leaves is the *angular divergence.* In practice, the *angular divergence* may be denoted by finding out a leaf which is exactly above a particular leaf, then by counting the number of complete circles covered by the genetic spiral and the number of leaves beginning from the first to the one just before the last. The latter number of leaves also gives the number of orthostichies. Thus,

$$\text{the angular divergence} = \frac{\text{number of circles}}{\text{number of leaves or orthostichies}} \text{ of a circle, i.e., } 360^\circ.$$

Sometimes, only the above fraction is given without calculating the actual divergence in degrees.

A few instances will clarify the issue:

(1) **Distichous or $\frac{1}{2}$ phyllotaxy:** In the family Gramineae (e.g., *Cynodon*, rice) the leaf on the second node is just opposite the leaf on the first node, the third leaf is above the first leaf, the fourth above the second and so on (Figs. 156A & 157A). Thus, the angular divergence is clearly 180° (straight angle) and all the leaves fall in two opposite orthostichies (hence, the name *distichous*). The genetic spiral covers only one circle and there are two leaves on the way in coming to the leaf exactly above the first leaf. Hence, the angular divergence $= \frac{1 \text{ (circle)}}{2 \text{ (orthostichies)}}$ of $360^\circ = 180^\circ$. As only the fraction is usually given, the phyllotaxy is also called $\frac{1}{2}$ phyllotaxy.

(2) **Tristichous or $\frac{1}{3}$ phyllotaxy:** In the common sedge (*Cyperus rotundus*) the leaves are three-ranked (three orthostichies); the fourth leaf is above the first leaf, fifth above second, sixth above third and so on (Figs. 156B & 157B). Three leaves lie in one circle of the genetic spiral. Hence, angular divergence $=\frac{1}{3}$ of $360° = 120°$ or $\frac{1}{3}$ phyllotaxy.

FIG. 156

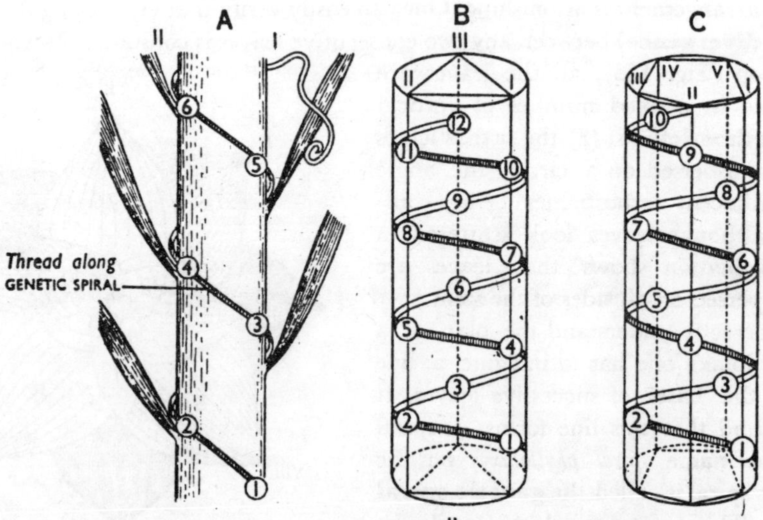

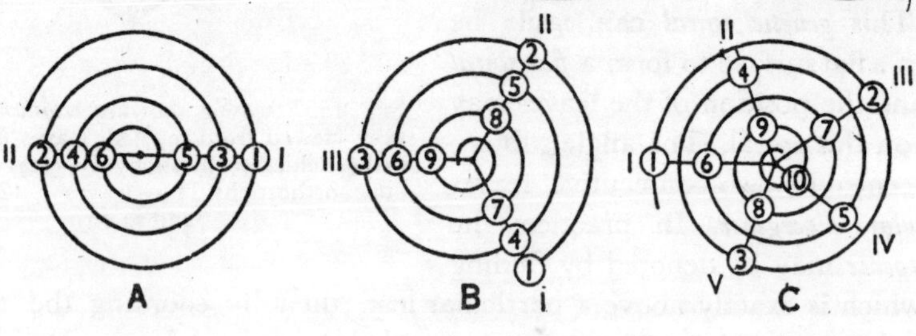

FIG. 157

FIG. 156. The genetic spiral. A. shows a thread tracing the genetic spiral. Nos. 1, 2, 3, ..., etc., are the consecutive leaves. A. *Distichous* or two-ranked (I & II) phyllotaxy. B. *Tristichous* or three-ranked (I, II & III) phyllotaxy. C. *Pentastichous* or five-ranked (I–V) phyllotaxy. FIG. 157. Flat projections of genetic spirals. A. Distichous showing two *orthostichies* (I & II). B. Tristichous showing three *orthostichies* (I, II & III). C. Pentastichous showing five *orthostichies* (I–V).

(3) **Pentastichous or $\frac{2}{5}$ phyllotaxy:** In many common plants like the china-rose and banyan, the 6th leaf is found above the 1st leaf, 7th above the 2nd and so on. The genetic spiral completes two circles in passing these five leaves or five orthostichies (Figs. 156C & 157C). Hence, angular divergence $= \frac{2}{5}$ of $360° = 144°$ or $\frac{2}{5}$ phyllotaxy.

(4) **Octastichous or $\frac{3}{8}$ phyllotaxy:** In this case the ninth leaf is found above the first leaf and the genetic spiral completes three circles in this distance as in the case of the papaw (*Carica papaya*). Thus, there are eight orthostichies (Fig. 155). The angular divergence is $\frac{3}{8}$ of $360° = 135°$.

It is rather interesting to note that in the common types of phyllotaxy the fractions of angular divergence lie in a series in which the numerator and the denominator in each case are obtained by adding up the numerators and denominators of the two preceding phyllotaxies. Thus,

$$\frac{1}{2},\ \frac{1}{3},\ \frac{1+1}{2+3}=\frac{2}{5},\ \frac{1+2}{3+5}=\frac{3}{8},\ \frac{2+3}{5+8}=\frac{5}{13},\ \frac{3+5}{8+13}=\frac{8}{21},\ \cdots$$

This series is called the **fibonacci series** or the **Schimper-Brown series** of divergence.

When the divergence becomes a big fraction it is not possible to count the orthostichies. In plants showing these phyllotaxies, usually the internodes are very short so that the leaves are much crowded together. On date and other palms with persistent leaf

Fig. 158

Fig. 159

Fig. 158. Parastichous arrangement of leaf bases on date palm trunk.
Fig. 159. A pine cone showing parastichous arrangement of sporophylls.

bases showing such complex phyllotaxy one very close spiral of leaf bases is observed. This type of phyllotaxy is called **parastichous** (Fig. 158). The sporophylls on a pine cone are also arranged in the same way and the parastichous arrangement can be seen clearly when the cone is viewed from above (Fig. 159).

Besides the *fibonacci series*, which is by far the commonest, there are possibly some other series of phyllotaxy. Two such rare series are $\frac{1}{4}, \frac{1}{5}, \frac{2}{9}, \frac{3}{14}, \cdots$ and $\frac{1}{2}, \frac{2}{3}, \frac{3}{5}, \frac{5}{8}, \frac{8}{13}, \cdots$

The laws of phyllotaxy may sometimes be disturbed by some torsion of the stem or by incomplete development as towards the apex of a twig.

LEAF MOSAIC

The aim of phyllotaxy is a proper display of the leaves so that all the leaves get the maximum amount of light and the chance of a leaf shading one below be reduced to the minimum. Sometimes the same effect is obtained by a differential growth of the stalks of

Fig. 160. Leaf mosaic in *Acalypha*. The lower leaves have longer petioles so that all leaves are brought to the same level and are not shaded by one another.

the different leaves on a herb or a twig so that all the leaves are brought to the same level presenting a flat *mosaic* surface. The leaf stalks are unequal and leaves originating lower down have longer stalks. Plants in a bush or undergrowths show such leaf mosaic. Leaves twist and turn to fill up all the gaps where light is available. One can see this in *Oxalis*, *Acalypha* (Fig. 160) or *Nasturtium* plants or in a twig of Maple.

HETEROPHYLLY AND ANISOPHYLLY

Usually all the leaves on a plant are more or less of the same shape when mature. Some plants may have different kinds of leaves, e.g., scale leaves and foliage leaves. But, there are only a few plants which show different types of foliage leaves. This is found specially in some plants which remain partly submerged and partly in air. The submerged leaves are usually dissected while the aerial leaves are more or less entire. This is the case

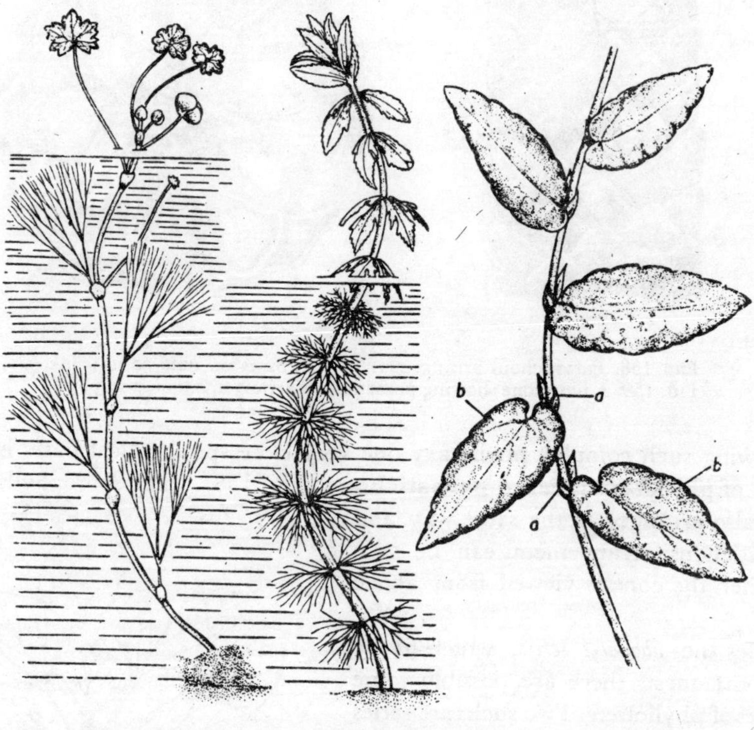

FIG. 161 FIG. 162 FIG. 163
HETEROPHYLLY in the aquatic plants *Ranunculus aquatilis* (Fig. 161) and *Limnophila heterophylla* (FIG. 162) in which the submerged and aerial leaves have different appearances. FIG. 163. ANISOPHYLLY in *Goldfussia glomerata*. *a* and *b* are opposite leaves which are very unequal (after Goebel).

in water chestnut (*Trapa*), *Ranunculus aquatilis* (Fig. 161), *Ranunculus multifidus*, *Myriophyllum indicum*, *Limnophila heterophylla* (Fig. 162), *Sagittaria*, etc. This phenomenon is known as **heterophylly.** Heterophylly is also noticed in some land plants like coriander (*Coriandrum sativum*), *Pterospermum*, *Artocarpus chaplasa*, *Ficus heterophylla*, etc.

A different phenomenon is noticed in some plants with opposite phyllotaxy. Usually opposite leaves on a plant are of the same size but in certain rare cases (e.g., in *Goldfussia glomerata* of Acanthaceae—Fig. 163 and *Pellionia daveanana* of Urticaceae) they are of unequal size. This is known as **anisophylly.**

DEVELOPMENT AND PARTS OF FOLIAGE LEAF

A typical foliage leaf (Fig. 164) has got three parts—(a) **Hypopodium** or the *leaf-base,* (b) **Mesopodium** or the *petiole,* i.e., the stalk of the leaf, and (c) **Epipodium** or the *leaf lamina* or *blade.* The entire leaf axis is called the **phyllopodium.**

In the development of the leaf, the leaf primordium first develops into the leaf-base and then a rudimentary lamina develops on it. Next, the leaf-base completes its development and any outgrowth on it (e.g., *stipule*) is formed. Simultaneously, the lamina expands to form a flat structure. The petiole develops last as a constriction between the base and the lamina and in many cases it does not develop at all. In some rare cases, the development of the leaf may be checked after the formation of the leaf-base.

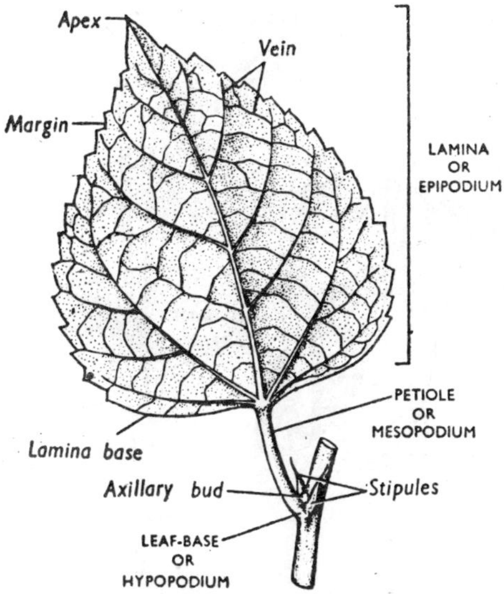

FIG. 164. Parts of a typical leaf.

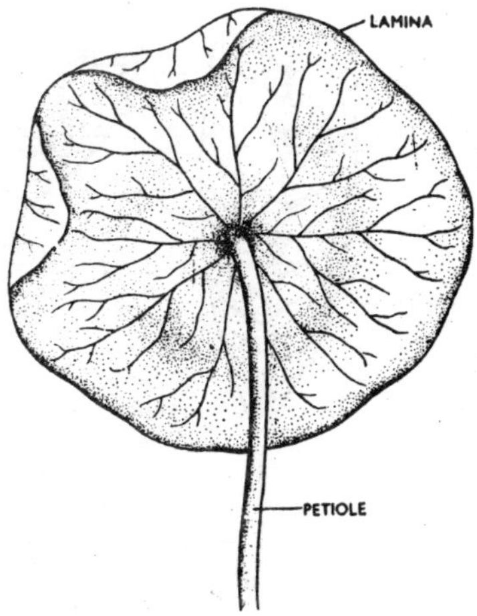

FIG. 165. Peltate leaf of water-lily.

Ordinarily, the leaf blade lies in the same plane as the petiole, the lamina-axis being a prolongation of the petiole. But, in some cases, the petiole is attached to the centre of the lower surface of the lamina, the latter being placed at right angles to the former as in lotus, water-lily (Fig. 165) or garden nasturtium. This type of leaf insertion is called **peltate.**

THE LEAF-BASE (HYPOPODIUM)

The *leaf-base* is the point of attachment of the leaf to the stem. Usually, it protects the small bud as its axil. The base is a prolongation of the tissue of the stem and cannot be demarcated from the petiole and the stem unless it is specially developed. In many instances, as in the leguminous plants (e.g., the sensitive plant *Mimosa pudica* of Mimoseae or *Dolichos* of Papilionaceae—Fig. 179) or in plants like mango (Fig. 166), etc., the leaf-base has a swollen appearance forming a sort of a cushion and is called the **pulvinus.** Such a leaf can be easily plucked off and leaves a scar on the stem.

In most monocotyledons and in a few dicotyledons the leaf-base becomes very prominent and winged. In that case it clasps and sheaths the stem forming a **sheathing leaf-base.** Sometimes the sheathing leaf-base does not clasp a stem inside but the sheath of the next younger leaf and so on. This is the case in banana (Fig. 167) where the apparent stem is only a series of sheathing leaf-bases, one inside another. The sheathing leaf-base may clasp and surround the stem completely when it is termed *amplexicaul* (e.g., grasses, *Polygonum orientale*, *Aethusa* of Umbelliferae—Fig. 168) or such clasping may be incomplete (as in coriander, palms, banana, etc.) when it is called *semi-amplexicaul*.

In a few cases, the petiole as well as the leaf-base are winged. These wings extend down the stem which also appears to be winged and all these wings extend down to the lower node. This is known as a **decurrent** leaf-base as found in *Laggera* of Compositae and *Symphytum* of Boraginaceae (Fig. 169).

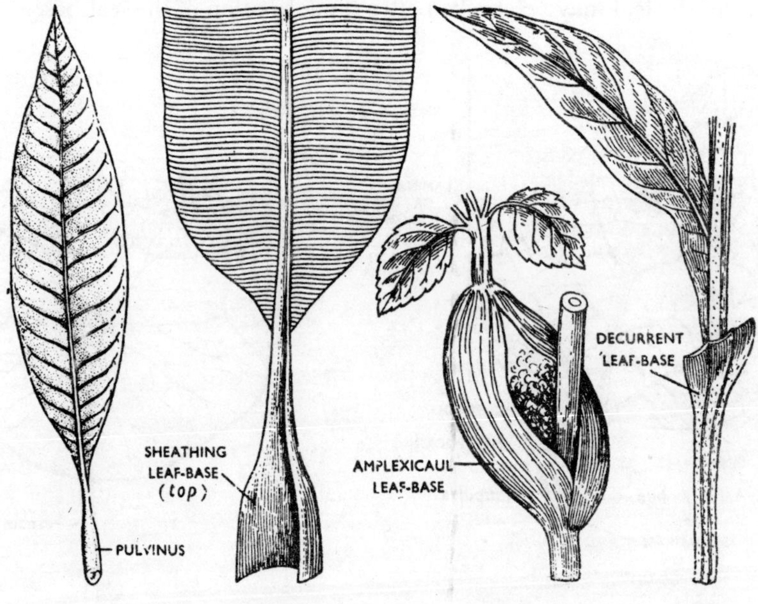

FIG. 166 FIG. 167 FIG. 168 FIG. 169

LEAF-BASE: FIG. 166. Mango leaf showing pulvinus. FIG. 167. Sheathing leaf-base of banana. (Only the top of the leaf-base and the base of the leaf lamina are shown). FIG. 168. Amplexicaul sheathing leaf-base of *Aethusa cynapium*. FIG. 169. Decurrent leaf-base of *Symphytum officinale*.

The leaf-base frequently bears a pair of lateral appendages or outgrowths called **stipules.**

STIPULES

Stipules are outgrowths from the leaf-base which usually protect the young axillary buds. The appearance of a stipule is changed by cohesion, adhesion or by the assumption of special functions. While they are very common among the dicotyledons, they are rather rare among the monocotyledons. When a leaf has got the stipular appendages it is called *stipulate*, when stipules are absent it is *exstipulate* (e.g., mango). The duration of the stipules may be *caducous* (shed very early as in *Michelia champaca*, *Dipterocarpus scabes*, etc.), *deciduous* (lasting one season as in *Dillenia indica*, *Cassia tora*, etc.) or *persistent* (in pea, rose, etc.) as in foliage leaves. Stipules may be highly branched as in *Ipomoea quamoclit*. The following types of stipules are commonly met with:

(1) **Free-lateral:** These stipules are of the simplest type. They are two soft filiform stipules hanging freely on the two sides of the leaf-base as in the china-rose (Fig. 170), cotton, etc. (plants of the family Malvaceae) and also in some other plants.

The four following stipules are modified because of *adhesion* (i.e., union between dissimilar organs, e.g., between stipule and petiole) or *cohesion* (union between different organs of the same type, e.g., stipule and stipule, petal and petal):

(2) **Adnate** or **Adherent:** Here the two stipules are adherent to the sides of the petiole to a certain distance so that the base of the petiole appears to be winged. This is seen in the compound leaf of rose (Fig. 171).

(3) **Intrapetiolar:** Sometimes the two stipules are coherent by their inner margins when a single fused stipule appears to be present *at the axil* of a leaf (*intra* means within). Examples are found in *Gardenia* (Fig. 172A—with opposite leaves), *Houttuynia* (Fig. 172B), etc.

FIG. 170 FIG. 171 FIG. 172A

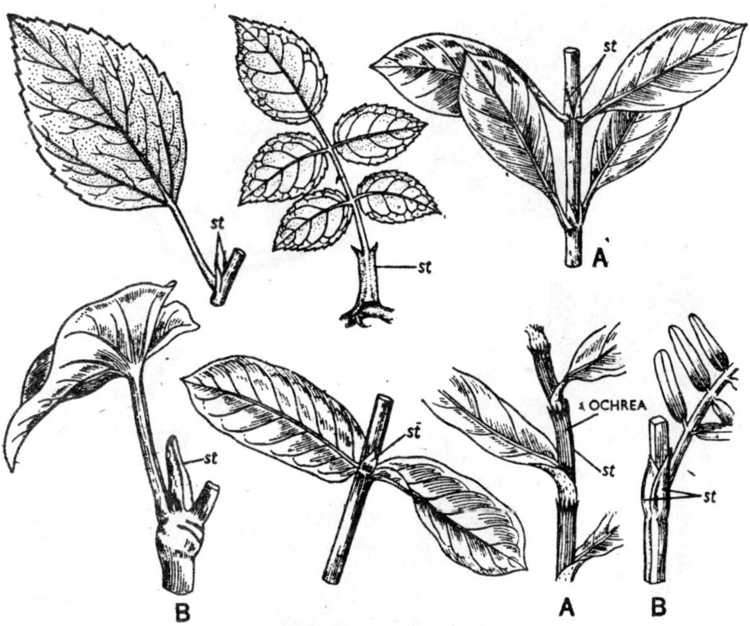

FIG. 172B FIG. 173 FIG. 174

STIPULES. FIG. 170. *Free-lateral* in china-rose. FIG. 171. *Adnate* in rose. FIG. 172A. *Intrapetiolar* in *Gardenia* with opposite leaves. FIG. 172B. *Intrapetiolar* in *Houttuynia*. FIG. 173. *Interpetiolar* in *Ixora* with opposite leaves. FIG. 174. *Ochreate*. A. in *Polygonum*. B. in *Astragalus* with open ochrea. *st*=stipule.

(4) **Interpetiolar:** In certain plants showing opposite phyllotaxy each of the two leaves at a node has stipules from its base. Cohesion takes places between stipules of opposite leaves by their outer margins so that the four stipules are transformed into two fused stipules which lie on two sides of the stem *between* (*inter-*) the petioles of the two opposite leaves. This is clearly seen in *Ixora* (Fig. 173), *Anthocephalus*, *Cephalanthus*, etc.

(5) **Ochreate:** The two stipules of leaf-base fuse to form a tube-like structure covering the internode up to a certain height. The tubular covering is called an *ochrea*. Ochrea is found in different species of *Polygonum* (Fig. 174A). In *Astragalus* (Fig. 174B) the ochrea remains open at one side.

Stipules are sometimes modified to carry on special functions as in the following cases:

(6) **Foliaceous:** In peas the stipules of the compound leaf (Fig. 175) are modified into large leaf-like green structures which take up all the functions of foliage leaves. The

true leaf lamina (the leaflets) is sometimes absent altogether (Fig. 97) being replaced by tendrils.

(7) **Tendrillar:** The stipules are modified into tendrils which help in climbing as in *Smilax* (Fig. 176). (*N.B.* The stipular nature of *Smilax* tendrils is doubtful. It has been shown recently that it rather originates from the petiole.)

(8) **Spiny:** Stipules may be modified into two spines as in the Acacias (Fig. 177A) or in *Zizyphus mauritiana* (Fig. 177B). Such spines serve as defensive armature.

FIG. 175

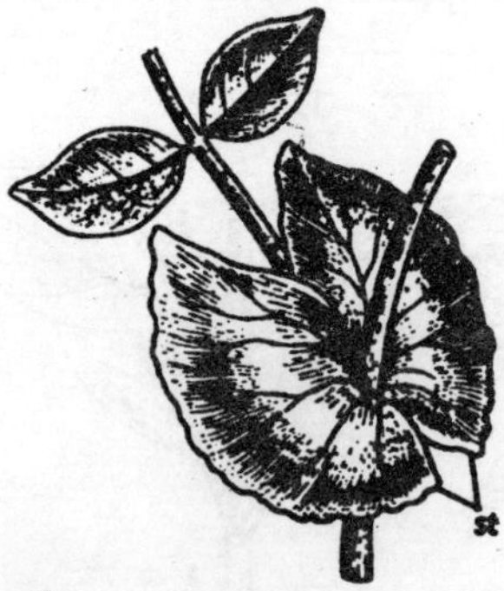

FIG. 176

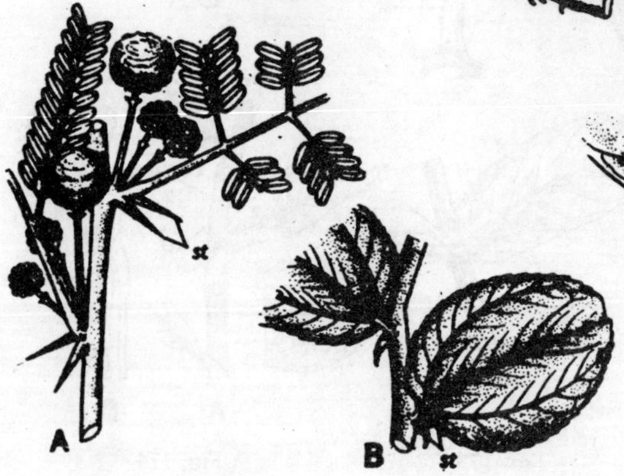

FIG. 177

FIG. 178

STIPULES. FIG. 175. Foliaceous in pea. FIG. 176. *Tendrillar* in *Smilax*.
FIG. 177. *Spiny* stipules. A. *Acacia nilotica*. B. *Zizyphus mauritiana*.
FIG. 178. *Convolute (Bud-scale)* in *Artocarpus*.

(9) **Convolute** or **Bud-scales:** *Bud-scales* which protect the buds (discussed in connection with *protection of buds*) are sometimes modified stipules. The bud-scales fall off as the bud opens. Such bud-scales may be seen in the jack-fruit (*Artocarpus*—Fig. 178).

STIPELS

It is explained later that the leaf lamina is broken up into small bits (leaflets) in compound leaves. The stipules are always borne on the leaf-base of the whole compound leaf and so no such stipule is expected at the bases of small leaflets. But, sometimes, such stipule-like appendages are found to grow on the two sides of the leaflet bases. These appendages are called **stipels** and are regarded by some as rudimentary leaflets which failed to develop. Stipels are present in many leguminous plants like *Dolichos lablab* (Fig. 179), *Desmodium*, *Phaseolus*, *Clitoria*. *Vicia faba*, etc.

THE PETIOLE (MESOPODIUM)

The petiole develops last in the leaf and sometimes may not develop at all. A leaf with a petiole is *petiolate* (mango, etc.) and one without any is *sessile* (*Calotropis*, etc.).

A petiole or leaf-stalk is usually solid and cylindrical but may be more or less flattened, grooved and spongy as in banana or it may be hollow as in papaw (*Carica papaya*). Sometimes, however, the petiole is modified to a much greater extent.

(1) In lemon and shaddock (*Citrus* spp.—Fig. 180) the petiole becomes **winged** and resembles the leaf lamina.

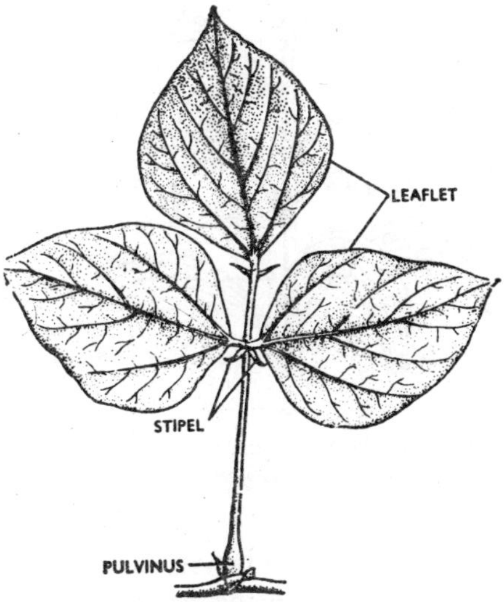

FIG. 179. Trifoliate imparipinnate leaf of *Dolichos lablab* showing stipels and pulvinus.

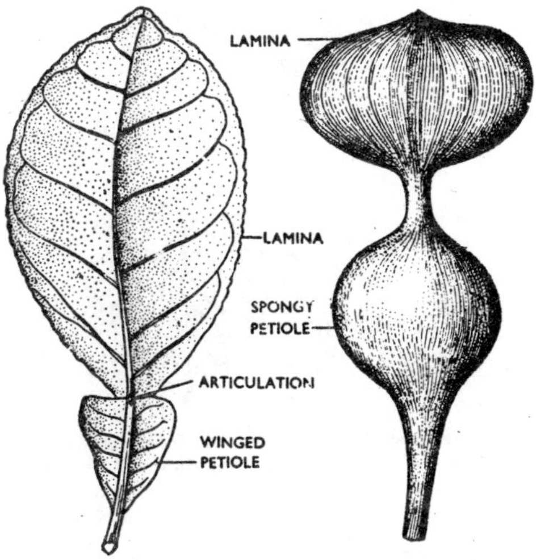

FIG. 180 FIG. 181

FIG. 180. Winged petiole of shaddock (*Citrus grandis*). FIG. 181. Swollen spongy petiole of water hyacinth.

(2) In *Nepenthes* (Fig. 27) the petiole is partly winged and partly tendrillar. (*N.B.* There is some difference of opinion as to the true nature of these organs as discussed in connection with the pitcher plants).

(3) In *Clematis* (Fig. 99) the petioles are **tendrillar.**

(4) In water hyacinth (Fig. 181) and *Trapa bispinosa* the petiole is **swollen** and **spongy.** This type of petiole encloses much air and helps the plant to float.

(5) In *Quisqualis* the leaf blades sometimes fall off and the petioles become **hard spines.**

(6) The most interesting modification of the petiole is the **phyllode.** These are found in Australian Acacias of which there are several species. In *Acacia moniliformis* the seedling plants (Fig. 182A) bear normal compound leaves. As these leaves become mature, the leaflets drop off and the petioles (or sometimes any part of the rachis) become flattened and leaf-like. These flattened petioles which look like ordinary leaves are called *phyllodes.* In mature *A. moniliformis* plants (Fig. 182B) the leaflets and rachis do not develop at all. In *Acacia recurva* even the mature plant shows leaflets on young leaves but these soon drop off and the petioles flatten into phyllodes.

A phyllode is usually borne vertically so that both the surfaces are equally lighted. These plants are of xerophytic habit.

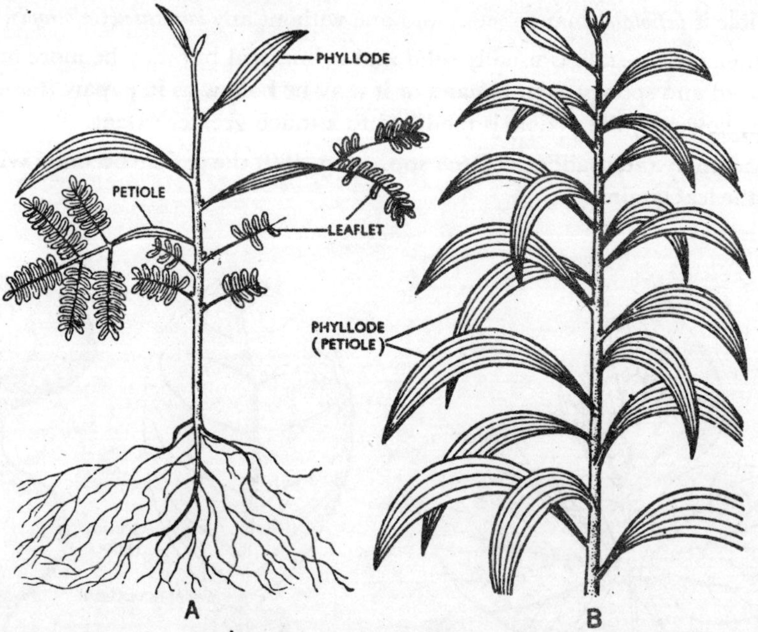

FIG. 182. PHYLLODE of *Acacia moniliformis*. A. A seedling plant showing transformation of petiole into phyllode. B. A twig of a mature plant showing phyllodes only.

LIGULE AND AURICLE

In many plants of the family Gramineae where there is no petiole some peculiar outgrowths grow at the junction of the sheathing leaf-base* and the lamina. This is a tongue-like membranous structure on the inner surface and is called the **ligule.** The ligule may sometimes be incised and may even be a hairy growth. The ligule of *Oryza sativa* (Fig. 183) is strengthened by a vascular supply. The ligule completely closes the sheath so that the growing organs within the sheath are better protected. A ligule is also developed in the palms which covers up the growing parts like a mat.

In rice and some other plants, the ligule is further strengthened by two outgrowths from the two sides of the junction between the sheath and the lamina, growing like two ears. These are called **auricles.** The auricles clasp the younger parts inside the sheath affording even better protection.

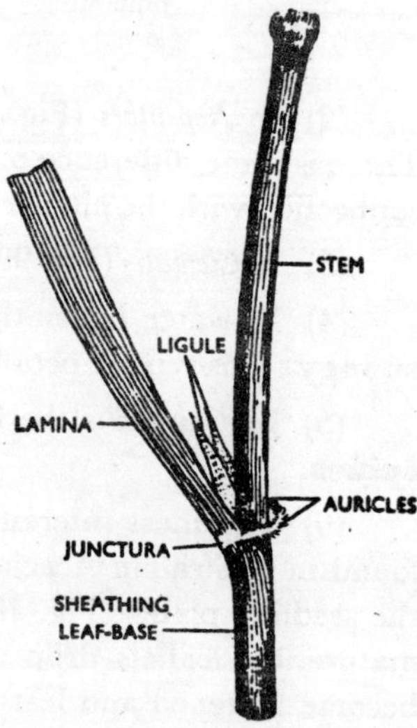

FIG. 183. Ligule and auricles of rice (*Oryza sativa*).

THE LAMINA (EPIPODIUM)

The lamina is the most important part of the leaf and that is the place where all the functions of leaf are carried on. It is the seat of gaseous interchange, starch formation and so on.

*The sheathing leaf-base is considered as a sheathing petiole by some botanists.

The leaf lamina is normally a flat structure. If the *upper* surface (i.e., the surface facing the stem, also called *adaxial* or *ventral surface*) differs in structure from the *lower* (*abaxial* or *dorsal*) surface, the leaves are called **dorsiventral.** In many monocots and shade plants the leaf is placed in such a manner that both the surfaces receive equal light and there is no difference between the two surfaces. Such leaves are called **isobilateral.** In exceptional cases (e.g., the leaf of onion—Fig. 85) the leaves are found to be cylindrical or **centric** when there is no question of two surfaces.

Leaves vary widely in the appearance of the lamina and to describe a plant the lamina must be correctly described. A leaf lamina should be studied from the following aspects—(1) *shape;* (2) *base;* (3) *margin;* (4) *apex;* (5) *surface;* (6) *texture;* (7) *colour;* (8) *odour;* (9) *taste;* (10) *venation;* (11) *leaf incision* and (12) whether *simple* or *compound.*

SHAPE

The shape or outline of the lamina is merely a description of its form. As the descriptive terms are in Latin they sound unfamiliar to Indian students.

The lamina shape may be (Fig. 184):

(*a*) *Narrow throughout*

(1) **Acicular**—needle-shaped as in pine.

(2) **Linear**—longer and slightly broader as in many grasses, tuberose, etc.

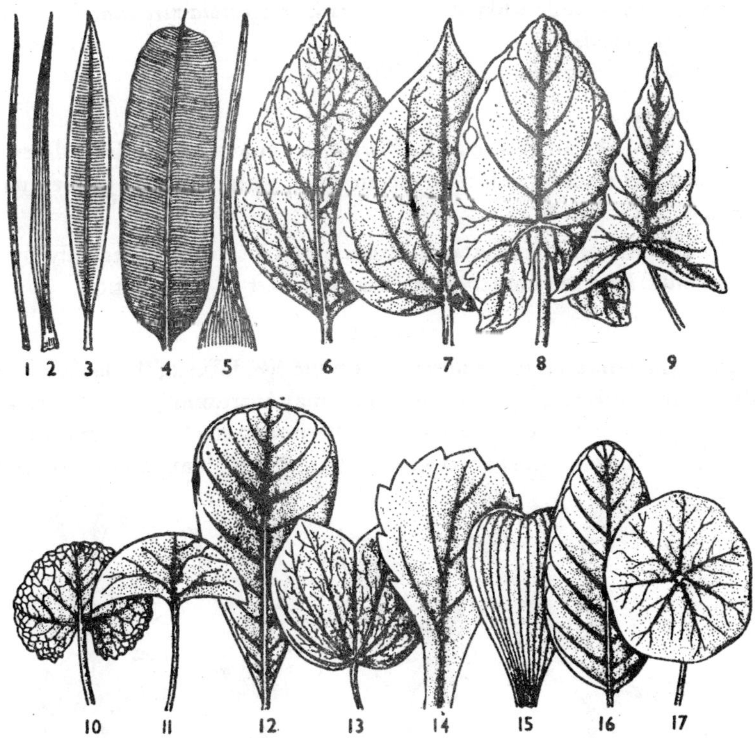

FIG. 184. LAMINA SHAPES. 1. *Acicular* in pine. 2. *Linear* in grass. 3. *Lanceolate* in *Nerium*. 4. *Oblong* in banana. 5. *Subulate* in *Isoetes*. 6. *Ovate* in china-rose. 7. *Cordate* in betel vine. 8. *Sagittate* in *Sagittaria*. 9. *Hastate* in *Ipomoea*. 10. *Reniform* in *Centella*. 11. *Lunate* in *Passiflora lunata*. 12. *Obovate* in jack-fruit. 13. *Obcordate* in *Bauhinia*. 14. *Spathulate* in *Lippia*. 15. *Cuneate* in *Pistia*. 16. *Elliptical* in guava. 17. *Rotund* in water-lily.

(*b*) *Wider but more or less of the same width at base and apex*

(3) **Lanceolate**—shaped like a lance as in *Nerium* and *Polyalthia.*

(4) **Oblong**—more or less rectangular as in banana (*Musa* sp.).

(*c*) *Wide at the base and narrow towards the apex*

(5) **Subulate** or awl-shaped—long and narrow, tapering gradually from base to apex as in *Salsola kali* and *Isoetes* (a pteridophyte).

(6) **Ovate** or egg-shaped—as in china-rose and banyan.

(7) **Cordate** or heart-shaped—with a deep notch at the base as in betel vine (*Piper betle*).

(8) **Sagittate**—shaped like an arrowhead with the two basal lobes pointing towards the base as in *Sagittaria sagittifolia* and in arum.

(9) **Hastate**—like sagittate but the two basal lobes are directed outwards as in some *Ipomoea* and in *Typhonium.*

(10) **Reniform** or kidney-shaped—rounded above with a deep notch at the base as in *Centella asiatica.*

(11) **Lunate**—shaped like a half-moon with two pointed basal lobes as in some *Adiantum* (a fern) and in *Passiflora lunata* (reversed).

(*d*) *Wider at the apex*

(12) **Obovate**—reverse of ovate as in jack-fruit or *Terminalia catappa.*

(13) **Obcordate**—reverse of cordate with an apical notch as in *Bauhinia.*

(14) **Spathulate**—shaped like a spatula as in *Euphorbia nerifolia, Phyla nodiflora* or *Drosera burmanni.*

(15) **Cuneate** or wedge-shaped as in *Pistia stratiotes.*

(*e*) *Symmetrical lamina*

(16) **Elliptical**—like an ellipse as in *Vinca rosea,* guava and India-rubber.

(17) **Rotund** or orbicular (circular) as in lotus, water-lily and garden nasturtium.

(*f*) *Incised forms*

Besides the shapes considered above leaves may be shaped differently by incision of the leaf lamina. This phenomenon is discussed later in this chapter.

BASE

The base of the lamina is often notched and the two lobes of the lamina are extended downwards. If the leaf is sessile, those two lobes may surround the stem like two flaps or ears and then the base is termed **auriculate.** This is seen in *Calotropis procera* (Fig. 185A). The term *amplexicaul* is sometimes used when the auriculate lamina base completely clasps

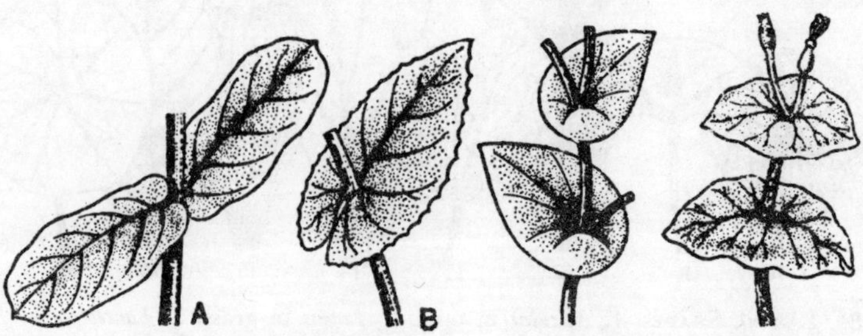

FIG. 185 FIG. 186 FIG. 187

FIG. 185. *Auriculate* lamina base. A. In *Calotropis*. B. The amplexicaul type. FIG. 186. *Perfoliate* base in *Bupleurum*. FIG. 187. *Connate* base in *Canscora*.

the stem (Fig. 185B). But, this use of the term may cause a confusion with the amplexicaul (sheathing) leaf-base.

If the basal lobes fuse together after completely clasping the stem it seems that the stem has perforated the leaf. This is called **perfoliate** (Fig. 186) and is found in *Aloe perfoliata, Pladera perfoliata, Bupleurum,* etc.

Sometimes the bases of opposite leaf laminas become fused together when the two leaves look like one through the centre of which the stem passes (Fig. 187). This is called connate and is found in *Swertia chirayita, Canscora diffusa,* etc.

It has already been seen that in the *decurrent type* (Fig. 169) the petiole as well as the leaf-base and a part of the stem are winged and united. The base of the lamina also is united and continuous with this so that the beginning of the lamina cannot be clearly demarcated.

MARGIN

The margin of the lamina may be (Fig. 188):

(1) **Entire**—when the margin is smooth as in mango.

(2) **Repand**—margin wavy as in *Polyalthia.*

(3) **Serrate**—margin with teeth pointed upwards as in a saw, e.g., china-rose, rose, etc.

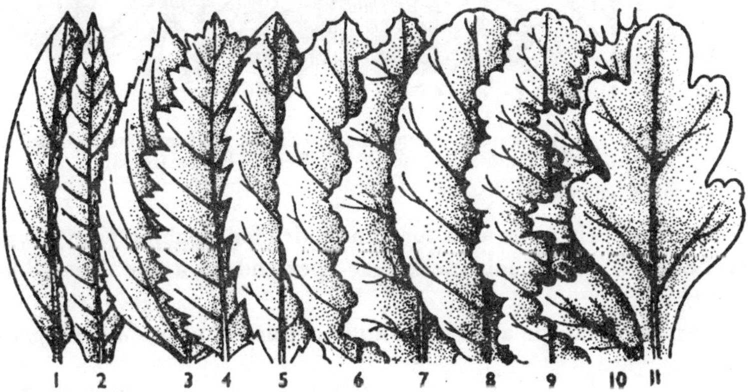

FIG. 188. LEAF MARGIN. 1. Entire. 2. Repand. 3. Serrate. 4. Bi-serrate. 5. Retroserrate. 6. Dentate, 7. Bi-dentate. 8. Crenate. 9. Bi-crenate. 10. Spiny. 11. Lobed.

(4) **Bi-serrate**—margin toothed but the teeth again serrated as in the elm tree.

(5) **Retroserrate**—the teeth are pointed downwards.

(6) **Dentate**—margin toothed, the teeth are pointed outwards, i.e., at right angles, e.g., water-lily.

(7) **Bi-dentate**—margin toothed and the teeth are again dentate.

(8) **Crenate**—margin toothed and the teeth are rounded as in *Centella* and *Kalanchoe.*

(9) **Bi-crenate**—margin toothed and the teeth are again crenate.

(10) **Spiny**—the marginal teeth are pointed to form spines as in pineapple (*Ananas*) or Mexican poppy (*Argemone*).

(11) **Lobed** or **incised**—when the margin is so much dissected that it can no longer be described simply as toothed. There are various types of lobing or incision which are discussed later in connection with *leaf incision.*

APEX

The apex of the lamina may be (Fig. 189):

(1) **Acute**—when pointed and narrow as in mango.
(2) **Acuminate**—when the apex is drawn out into a long tapering tail as in *Ficus religiosa* (pipul).
(3) **Obtuse**—when the apex is broad angled and blunt as in banyan.
(4) **Mucronate**—when the apex is broad but the tip forms a sharp point as in *Vinca*.
(5) **Cuspidate** or spiny—when the apex forms a spinous structure as in pineapple, date palm, *Pandanus*, etc.

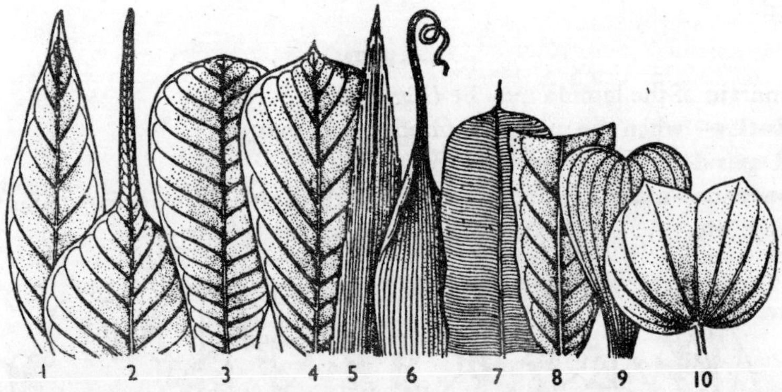

FIG. 189. LEAF APICES. 1. Acute. 2. Acuminate. 3. Obtuse. 4. Mucronate. 5. Cuspidate. 6. Tendrillar. 7. Cirrhose. 8. Truncate. 9. Retuse. 10. Emarginate.

(6) **Tendrillar**—when the apex forms a tendril for climbing as in *Gloriosa*.
(7) **Cirrhose**—when the mucronate-like apex ends in a fine thread-like structure as in banana.
(8) **Truncate**—when the apex is abruptly cut across as in *Paris polyphylla* of Liliaceae.
(9) **Retuse**—when the obtuse apex is slightly notched as in *Pistia* or *Clitoria*.
(10) **Emarginate**—when the obtuse apex is deeply notched as in *Bauhinia*.

SURFACE

The feel of the surface of the leaf may be:

(1) **Glabrous**—when smooth and without any hair, etc., as in mango.
(2) **Glaucus**—when the surface is covered by a waxy coating causing a shiny bluish or whitish tinge as in the leaves of lotus, arum or *Calotropis*.
(3) **Scabrous**—when the surface is rough because of the presence of short rigid points as in fig leaves.
(4) **Viscose**—when the surface is sticky because of some excretion as in *Cleome viscosa*. A similar case is the leaf of *Nicotiana tabacum* which is also sticky and is sometimes described as *glutinous*.
(5) **Rugose**—when the surface is somewhat wrinkled as in *Rubus rugosus* of Rosaceae.
(6) **Gland-dotted**—when glands are found on the leaf surface as in lemons.

(7) **Hairy**—when the surface is covered with hairs. These again may be of many types: (*a*) **pubescent**—when the hair is soft and woolly as in tomato; (*b*) **pilose**—when the hairs are long, distinct and scattered as in *Grewia flavescens* ; (*c*) **villose**—when hairs are long, soft and closely arranged as in *Leucas aspera;* (*d*) **tomentose**—when hairs are short, dense and cottony as in *Terminalia tomentosa, Calotropis procera*, etc.; (*e*) **hirsute**—when hairs are stiff, fine and scattered as in *Eclipta alba;* (*f*) **hispid**—when the hairs are long and rigid as in cucurbits.

(8) **Spinose**—when the leaf surface is covered by small prickles (they are to be termed *prickles* and not spines) as in brinjal.

TEXTURE

The texture of the leaf is:

(1) **Herbaceous**—when the leaf is thin and membranous as in china-rose and rose.

(2) **Coriaceous**—when it is firm and leathery as in mango.

(3) **Succulent**—when soft and juicy as in *Kalanchoe*. The leaves are more or less brittle.

(4) **Gland-dotted**—the presence and nature of glands (e.g., on lemon leaves) may be considered in connection with texture as well as in connection with lamina surface.

COLOUR, ODOUR & TASTE

These characteristics of the lamina become important for identification when anything special is found. Such specialities are to be described in such cases.

VENATION

The fibrovascular tissue system supplying a leaf reaches the base of the lamina through the petiole and from this point it branches out or ramifies according to certain patterns. The ramifications are called **veins** and their arrangement is known as **venation.** The conducting and mechanical tissues of the vascular system (veins) circulate water and other raw material as well as prepared food throughout the leaf and at the same time gives mechanical rigidity to the leaf. The veins, therefore, serve as a circulatory system as well as like a skeleton of the leaf. Venation is very clear on the lower surface of thin leaves.

Venation follows certain basic patterns. In lower plants like the Pteridophytes, the branching of the veins is dichotomous. The Angiosperm leaf shows two principal types of venation. In the first type the strong main veins give rise to finer branches which again branch repeatedly running into one another and forming a *reticulum* or network all over the leaf. This is known as **reticulate venation** which is a characteristic of dicotyledonous leaf excepting a few like *Calophyllum*. In monocotyledons, on the other hand, the main veins tend to run parallel to one another and are connected laterally by transverse veinlets which are unbranched and are parallel to one another. This is known as **parallel** or **striate venation.** A few monocotyledons like some aroids, *Smilax*, *Dioscorea*, etc., however, show reticulate venation. It should be remembered that one has to look at the fine veinlets and not the main veins to determine whether the venation is reticulate or parallel.

A. RETICULATE VENATION (FIG. 190)

The reticulate type of venation shows two variations:

(a) **Unicostate** or **Pinnate**

This type of venation may be likened to racemose branching. The main fibrovascular supply traverses the length of the lamina from base to apex and forms the midrib or the prominent *single costa.* The midrib gives out secondary branch veins resembling the general plan of a feather (hence the name *pinnate*) and these, in their turn, give rise to

FIG. 190

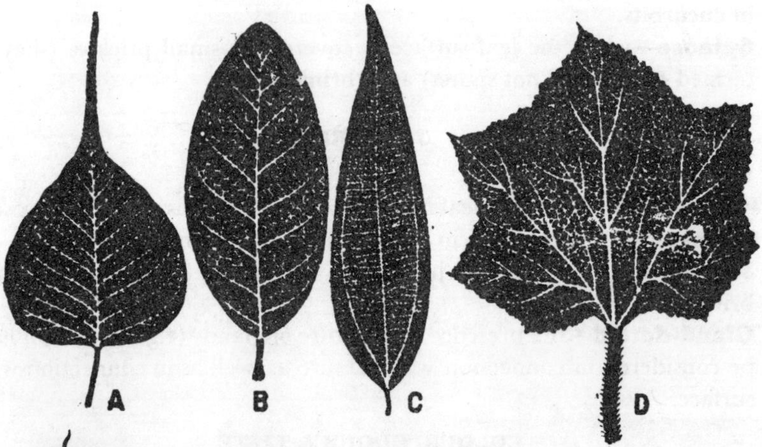

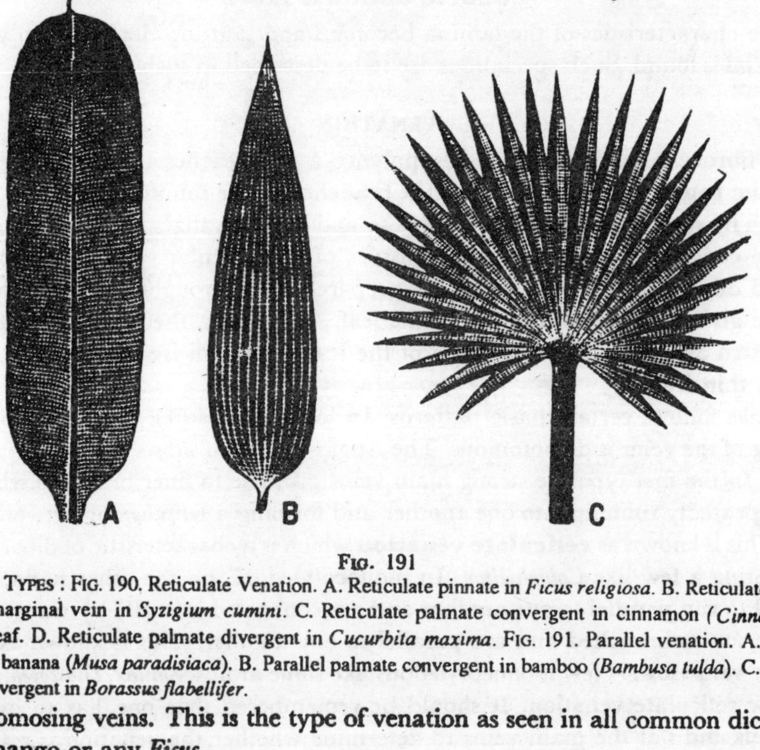

FIG. 191

VENATION TYPES : FIG. 190. Reticulate Venation. A. Reticulate pinnate in *Ficus religiosa*. B. Reticulate pinnate with submarginal vein in *Syzigium cumini*. C. Reticulate palmate convergent in cinnamon (*Cinnamomum tamala*) leaf. D. Reticulate palmate divergent in *Cucurbita maxima*. FIG. 191. Parallel venation. A. Parallel pinnate in banana (*Musa paradisiaca*). B. Parallel palmate convergent in bamboo (*Bambusa tulda*). C. Parallel palmate divergent in *Borassus flabellifer*.

the anastomosing veins. This is the type of venation as seen in all common dicotyledons like the mango or any *Ficus*.

The secondary veins often do not reach the margin but before reaching the edge either fizzle out or curve and run parallel to the margin for a short length. These extremities of the secondary veins often run together and sometimes form a strong

submarginal vein running parallel to the leaf margin as seen in *Plumeria acutifolia* and *Syzigium cumini* of Myrtaceae (Fig. 190B).

(b) **Multicostate** or **Palmate**

In this type the vascular supply, after reaching the base of the lamina, breaks up into a number of equally strong veins or *costas*. Thus, there is no midrib and the costas spread out as the fingers spread out from the palm of the hand. Hence the name *palmate*. Palmate venation, again, may be:

(i) **Convergent**—when the prominent veins (costas) after spreading out from the base again converge towards the tip of the lamina as in *Zizyphus jujuba* of Rhamnaceae and in the cinnamon leaf (*Cinnamomum tamala*—Fig. 190C).

(ii) **Divergent**—when the *costas* spread out farther and farther apart as in the cucurbits and china-rose (Fig. 190D).

B. PARALLEL OR STRIATE VENATION (Fig. 191)

Parallel venation, similarly, may be *unicostate* or *multicostate*.

(a) **Unicostate** or **Pinnate**

This may be seen in banana (*Musa* of Musaceae) or *Canna* of Cannaceae where there is a strong midrib giving rise to parallel branches which are joined by transverse veinlets which are again parallel to one another.

(b) **Multicostate** or **Palmate**

Showing a number of strong veins (costas) which are

(i) **Convergent** in bamboos and grasses.

(ii) **Divergent** in fan palm (*Borassus flabellifer* of Palmaceae).

LEAF INCISION

Simple & Compound Leaf

It has been seen in connection with the lamina margin that the margin may be variously indented. This indentation is sometimes so prominent that the leaf is said to be

Fig. 192

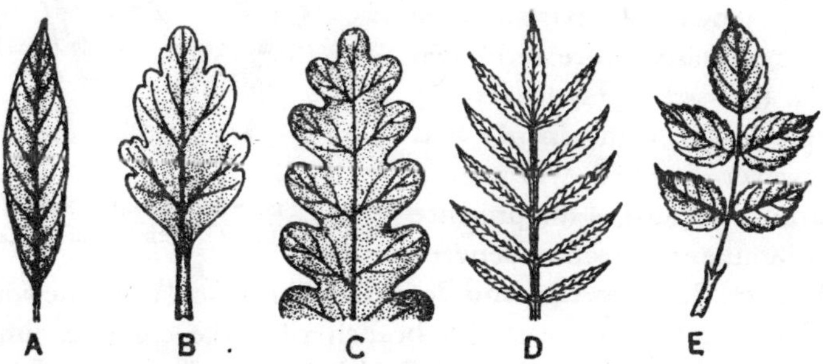

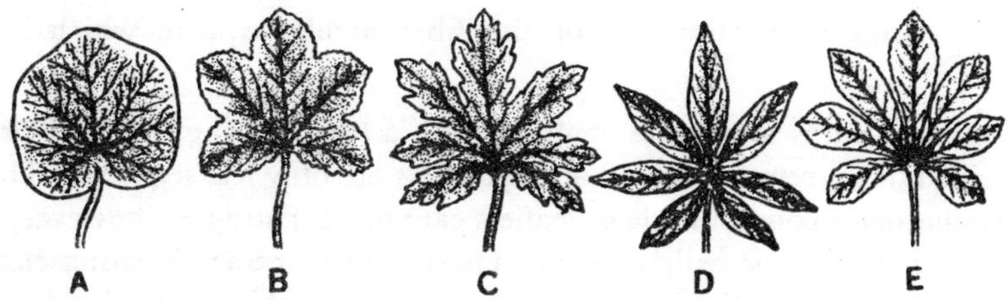

Fig. 193

Leaf Incision. Fig. 192. Incision of pinnate type. A. An entire leaf with pinnate venation.⟶B. Pinnatifid.⟶C. Pinnatipartite.⟶D. Pinnatisect⟶E. A pinnate compound leaf. **Fig. 193. Incision of palmate type. A. An entire leaf with palmate venation.⟶B. Palmatifid. ⟶C. Palmatipartite.⟶D. Palmatisect.⟶E. A palmate compound leaf.**

incised or lobed. The pattern of **leaf incision** depends on the type of venation. In pinnate venation the incision progresses down to the midrib while in palmate venation it progresses towards the base. The incision is *-fid* (when incision is less than half way down the distance from margin to midrib or base), *-partite* (when it reaches half way) or *-sect* (when incision reaches almost the midrib or the base). Thus, according to the *pinnate* (Fig. 192) or the *palmate* (Fig. 193) type of venation the incision may be (1) **pinnatifid** (e.g., poppy or chrysanthemum leaf); (2) **pinnatipartite** (e.g., *Argemone mexicana*); (3) **pinnatisect** (e.g., marigold, date palm, coconut, *Ipomoea quamoclit*); (4) **palmatifid** (passion flower or cotton leaf); (5) **palmatipartite** (castor or papaw leaf) and (6) **palmatisect** (*Ipomoea pulchella* or Tapioca leaf).

Among the pinnately incised types some leaves, like those of radish or mustard, show lobes which are irregularly incised. Usually the lobes towards the apex are larger. These are called **lyrate** (Fig. 194) leaves. Similarly, some palmately lobed leaves show the lobes arranged like the claw of a bird as in *Cayratia pedata*. These are called **pedate** (Fig. 195) leaves.

In extreme cases of leaf incision (e.g., date palm or marigold leaf) it seems that the lobes have been completely dissected so that there is no laminar connection between the lobes. A closer observation, however, shows that a very narrow lamina connects the lobes. If dissection goes on still further even this connection disappears so that the lobes become completely independent of one another. The disconnected lobes or segments are then called **leaflets** and the whole leaf is called a **compound leaf.** The compound leaf again may be **pinnate** (Fig. 192E) or **palmate** (Fig. 193E). As opposed to these compound leaves showing an extreme type of dissection, the ordinary entire or incised (including even the *-sect* types) leaves are known as simple leaves. In a **simple leaf** the lamina, however dissected, will present a single appearance because of the presence of some sort of laminar connection between the dissected lobes. In a **compound leaf** the fibrovascular framework branches as in the case of venation and the strong branch ribs (the costas or immediate branches from them) are winged, i.e., provided with lamina, forming *leaflets*. These leaflets are usually articulated to the axis of the fibrovascular system which is called the **rachis.**

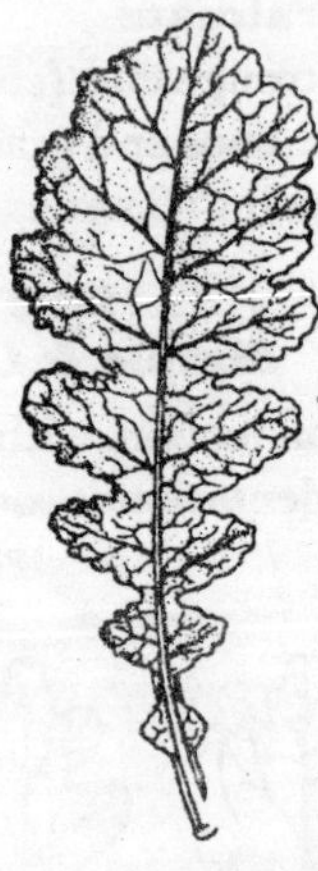

FIG. 194

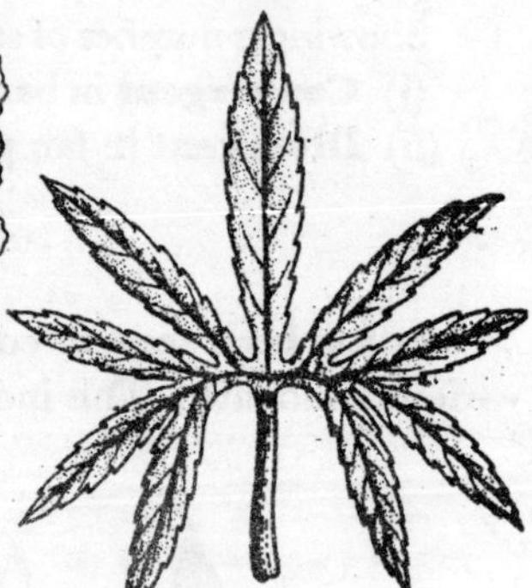

FIG. 195

FIG. 194. Lyrate leaf of radish.
FIG. 195. Pedate leaf of *Cayratia pedata.*

A compound leaf is sometimes confused with a twig bearing leaves. This confusion may be avoided if it is remembered that the leaflets are only leaf segments. There can be no bud terminating a compound leaf, leaflets cannot bear stipules (however, there may be *stipels* as in Fig. 179) or axillary buds. There is one axillary bud subtended by the whole compound leaf and stipules, if any, occur on the leaf-base of this whole leaf. The whole compound leaf can never arise from the axil of another leaf which would have been the case if it were a twig.

A compound leaf, just like venation, may be *pinnate* or *palmate.*

A. PINNATE COMPOUND

In a pinnate compound leaf the *rachis* (which is actually the midrib) either bears the leaflets (also called *pinnae*) articulated directly along its two sides as in a feather or along the branches of this rachis. The patterns may be as follows:

(1) **Simple pinnate** or **Unipinnate:** Leaflets (usually in pairs) are borne directly on rachis.

(*a*) **Paripinnate:** Leaflets are borne in pairs as in *Tamarindus indica* (Fig. 196), different species of *Cassia*, *Swietenia mahogoni*, etc.

FIG. 196

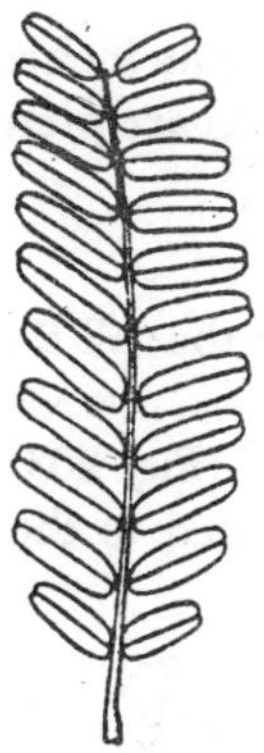

FIG. 197

FIG. 198

FIG. 199

FIG. 200

FIG. 201

FIG. 202

PINNATE COMPOUND LEAF. FIG. 196. Paripinnate in *Tamarindus indica*. FIG. 197. Imparipinnate in rose. FIG. 198. Interruptedly pinnate (imparipinnate) in potato. FIG. 199. Bipinnate in *Acacia nilotica*. FIG. 200. Tripinnate in *Moringa oleifera*. FIG. 201. A portion of decompound leaf of anise. FIG. 202. A portion of decompound leaf with practically no development of lamina.

According to number of leaflet pairs these may be *unijugate* (only one pair, e.g., *Zornia diphylla*), *bijugate* (ground-nut), *trijugate*, etc.

(*b*) **Imparipinnate:** The rachis is terminated by an unpaired odd leaflet as in rose (Fig. 197), *Clitoria* (Papilionaceae), *Murraya exotica* (Rutaceae), *Melia azadirachta* (Meliaceae), etc.

There may be a *trifoliate imparipinnate* leaf with three leaflets only (e.g., *Dolichos* —Fig. 179 or *Crotalaria trifoliatum*) which is distinguished from *palmate trifoliate* leaf by the presence of an elongated rachis. In *Pisum* and *Lathyrus* (Figs. 97 & 98) the terminal and other leaflets may be transformed into tendrils. In potato (Fig. 198), the pairs of leaflets and the terminal one are of unequal sizes and may be described as *interruptedly pinnate.*

(2) **Bipinnate:** The pinnae are dissected again into *pinnules* so that the leaflets (pinnules) are borne not on the rachis but on its branches as in *Acacia nilotica* (Fig. 199), *Mimosa pudica, Caesalpinia pulcherrima,* etc.

(3) **Tripinnate:** Dissection goes to the third order so that the leaflets are found on secondary branches of the rachis as in *Moringa oleifera* of Moringaceae (Fig. 200).

(4) **Decompound:** When incision is of a higher order than in tripinnate as in carrot, anise, fennel (*Foeniculum vulgare*) and other plants of Umbelliferae (Figs. 201 & 202).

In some plants of Umbelliferae (Fig. 202) the leaf is highly incised but there is practically no lamina formation so that only green stalks are present. This is the case in many aquatic plants as well.

B. PALMATE COMPOUND

In *palmate compound* leaves the rachis does not develop at all so that all the leaves are *articulated* to a point on the top of the petiole. These may be:

(1) **Unifoliate:** This is the case in lemon, orange, shaddock (Fig. 203) and other *Citrus* plants (Rutaceae). Only a single leaflet is articulated to the top of the

FIG. 203

FIG. 204

FIG. 205

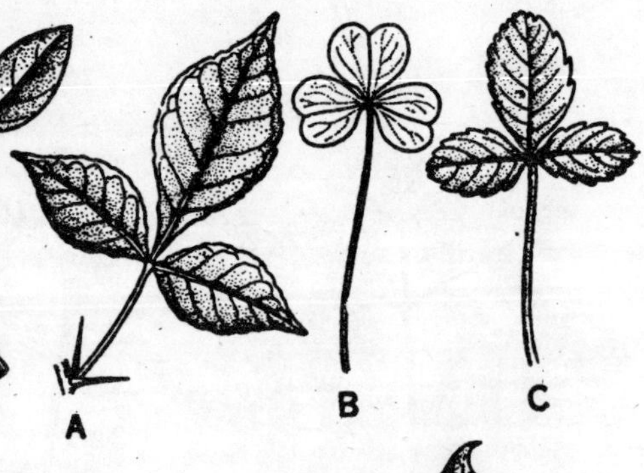

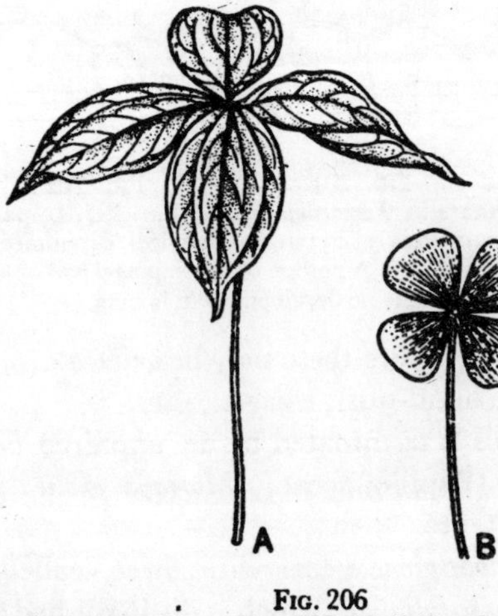

FIG. 206

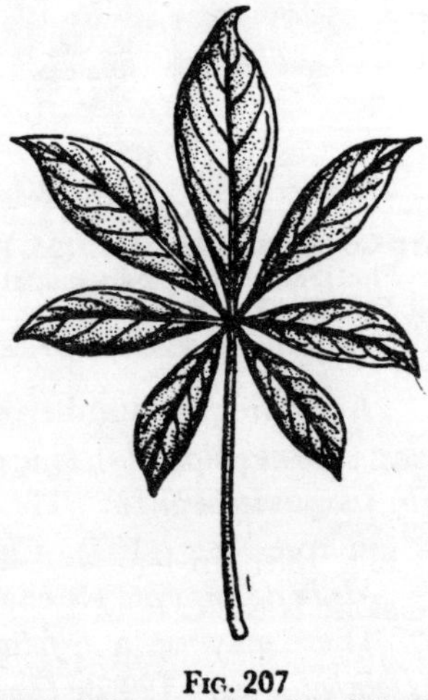

FIG. 207

PALMATE COMPOUND LEAF. FIG. 203. Unifoliate in shaddock (*Citrus grandis*), FIG. 204. Bifoliate in *Hardwickia.* FIG. 205. Trifoliate in A. *Aegle marmelos,* B. *Oxalis* and C. clover. FIG. 206. Quadrifoliate in A. *Paris quadrifolia* and B. *Marsilea.* FIG. 207. Digitate in silk-cotton.

winged petiole. The articulation shows that the leaf is not a simple one but compound.

(2) **Bifoliate palmate** or **binnate:** This type with only two leaflets articulated (Fig. 204) to the rachis is rather rare. This is found in *Balanites aegyptiaca* of Simarubaceae, *Hardwickia binnata* of Papilionaceae, *Campsis grandiflora* of Bignoniaceae, etc.

(3) **Trifoliate** or **ternate** (Fig. 205): Three leaflets are articulated in *Aegle marmelos*, *Oxalis corniculata*, the genus *Trifolium* (clover), etc.

(4) **Quadrifoliate palmate:** This is another rare type found in the Paris plant (*Paris quadrifolia*—Fig. 206A). The very common Pteridophyte *Marsilea quadrifoliata* presents a quadrifoliate appearance (Fig. 206B) although the actual morphology of the four leaflets is not so simple.

(5) **Digitate** or **multifoliate:** If there be more than four leaflets as in the silk-cotton tree (*Bombax ceiba* of Bombacaceae—Fig. 207), *Cleome viscosa and Cleome gynandra* of Capparidaceae, horse chestnut (*Aesculus hippocastanum*), etc.,

MODIFICATIONS OF THE LEAF LAMINA

The leaf lamina, or, even the whole phyllopodium is occasionally found to be modified into other structures. Prominent instances are:

(1) Leaf spines

Spines sometimes arise as modifications of leaf apices or the apices of marginal lobes. These spines may be comparatively small as on the leaves of *Argemone mexicana*, Pineapple, *Agave*, *Aloe*, etc. In the date-palm this spine is quite a hard structure. Development of spinous structures is a feature of the xerophytes. The whole phyllopodium is often transformed into spine as on the phylloclade of *Opuntia* (Fig. 135) and the cladode of

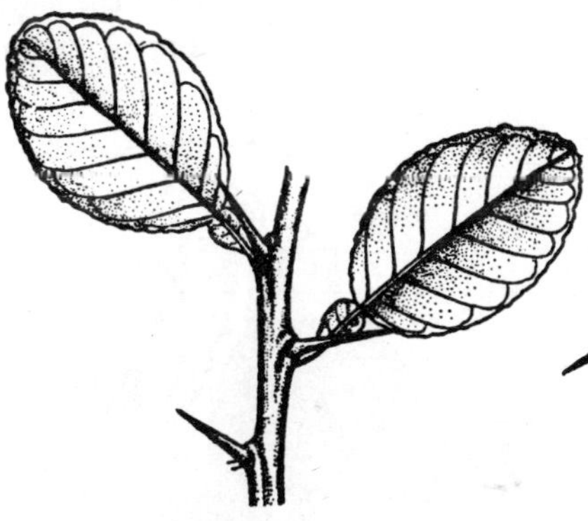

FIG. 208

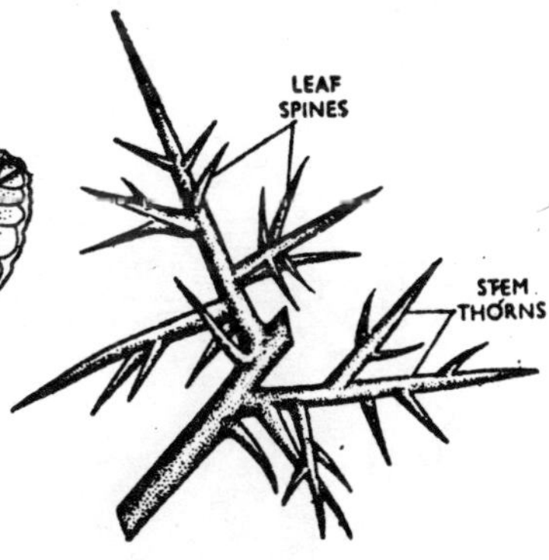

FIG. 209

FIG. 208. Spinous prophyll of *Citrus*. FIG. 209. Spines of *Ulex*. The larger thorns are modified stems while the smaller spines are modified leaves.

Asparagus (Fig. 138). In lemons and oranges (*Citrus* spp.—Fig. 208), the prophyll is a spine. The same is the case with the prophylls (in pairs here) of woodapple (*Aegle marmelos*—Fig. 205A). In *Berberis* (Fig. 148) and *Ulex* (Fig. 209) the leaves are modified into spines and stages of transition from ordinary leaves to spines are seen, specially, in the seedling stage. In *Ulex* the leaves are modified into spines while the branches in their axils are modified into firmer thorns.

(2) **Leaf tendrils**

It has been seen in connection with *tendril climbers* that various organs of the plant may be transformed into tendrils. The entire lamina becomes a tendril in *Lathyrus* (Fig. 97) while the terminal leaflets are so transformed in *Pisum* (Fig. 98), *Naravelia zeylanica* (Ranunculaceae) and *Bignonia venusta*. The leaf apex becomes a tendril in *Gloriosa superba* (Fig. 100). The tendril of Cucurbits may be a prophyll (Fig. 103).

(3) **Hooks**

Climbing hooks are sometimes modified leaves. In *Macfadena unguiscati* from Assam, the three terminal leaflets become claw-like hooks (Fig. 210). The leaf spines on the main branches of *Asparagus* also act as hooks.

(4) **Fleshy leaves**

Some leaves, specially in xerophytes and halophytes, become fleshy because of the storage of water, mucilage and food matter. Such leaves contain a special storage tissue. Common examples are *Portulaca oleracea* (Portulacaceae), *Basella rubra, Suaeda maritima* and *Salsola kali* of Chenopodiaceae, *Aloe* and *Agaves, Kalanchoe, Sedum acre* (stone crop), etc.

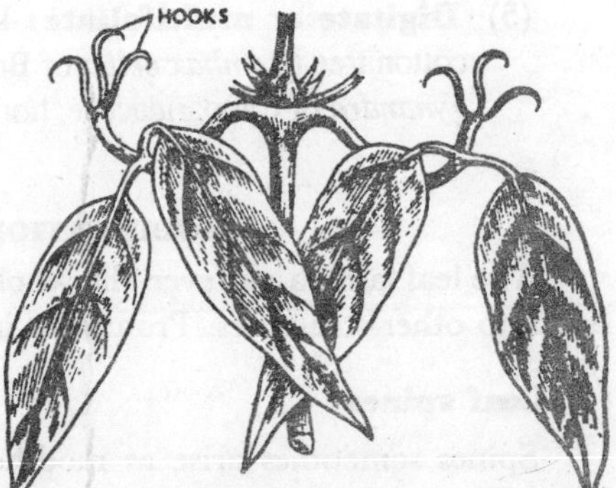

FIG. 210. *Macfadena unguiscati*. Terminal leaflets are modified into hooks.

(5) **Pitcher**

The pitchers of insectivorous pitcher plants, which are wholly or partially modified leaf lamini are described in chapter I (Figs. 27, 28 and coloured photograph on wrapper). Another such pitcher is found in the epiphytic climber *Dischidia rafflesiana* (Asclepiadaceae) from Assam. These pitchers also are modifications of leaves and they

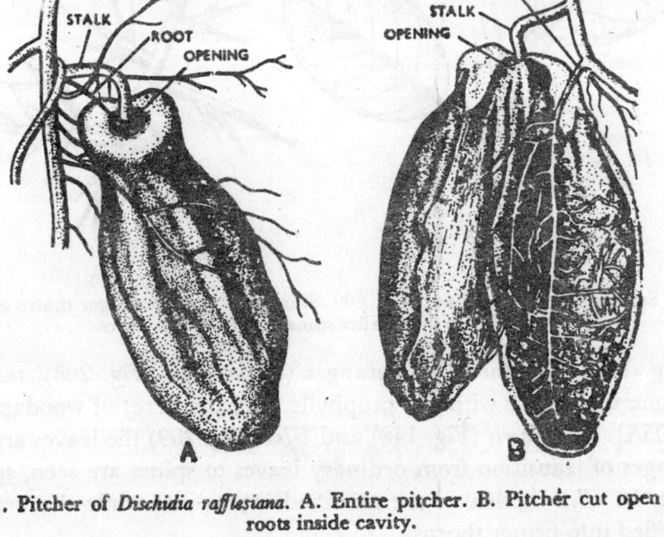

FIG. 211. Pitcher of *Dischidia rafflesiana*. A. Entire pitcher. B. Pitcher cut open to show roots inside cavity.

are provided with openings at the bases but they have nothing to do with insect catching. Rain water as well as debris accumulate within these vessels and this water is absorbed by adventitious roots which grow out from the stem nodes and ramify within the cavities (Fig. 211).

(6) **Bladder**

The bladders which are the insect-catching traps of *Utricularia* have been described in chapter I (Fig. 26). These are modified lobes of leaves.

(7) **Absorbing organs**

The submerged leaves of many aquatic plants (Figs. 152 & 162) are finely dissected and take up the absorptive functions of roots. These plants are usually rootless.

(8) **Free-living leaf**

The peculiar case of *Lemna* (Fig. 139) has already been discussed. Whlle some consider it as a phylloclade, others consider it as a free-living leaf. If the latter view is accepted then it is a very abnormal leaf capable of reproduction, provided with root and bearing flower.

CHAPTER V

THE ROOT

The root is the descending portion of the plant axis. As opposed to the stem, it is positively geotropic, negatively phototropic and positively hydrotropic. The root surface is non-green and it is not divided into nodes and internodes. In stems the branches arise from the nodes so that there is some regularity in their arrangement while this is not so in roots because of the absence of nodes. Nevertheless, the branching in roots also

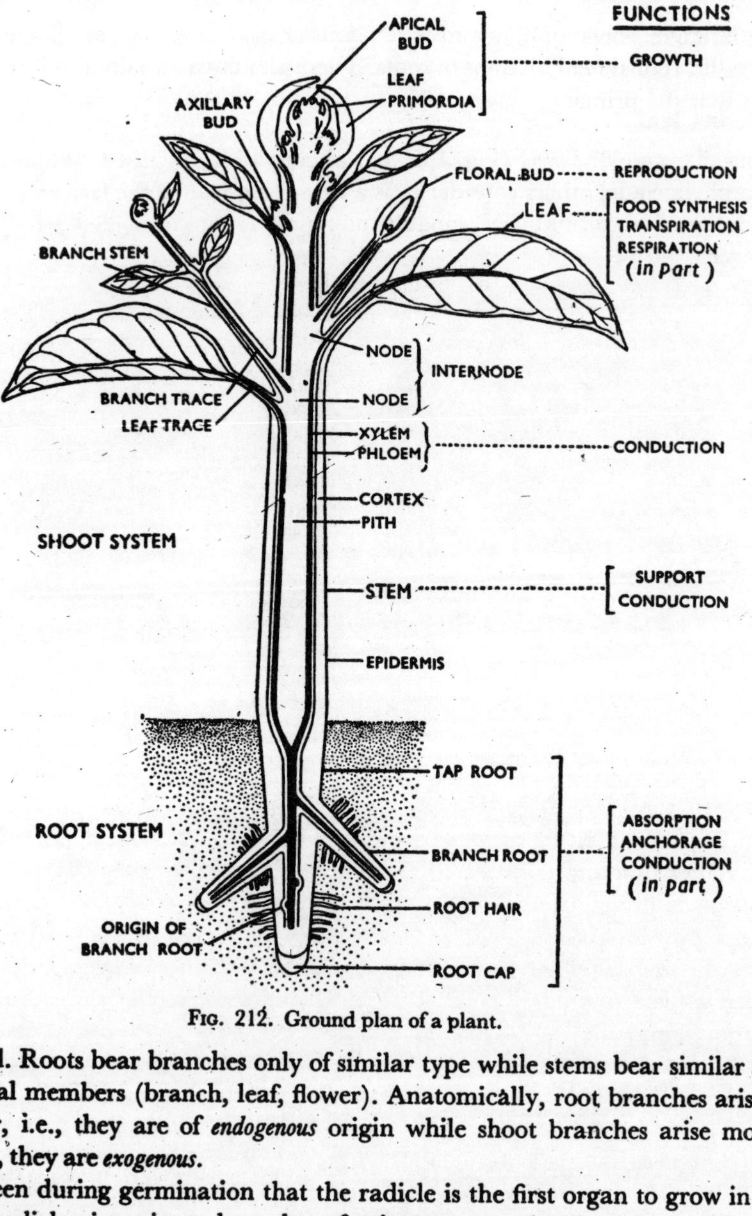

FIG. 212. Ground plan of a plant.

is acropetal. Roots bear branches only of similar type while stems bear similar and dissimilar lateral members (branch, leaf, flower). Anatomically, root branches arise from an inner layer, i.e., they are of *endogenous* origin while shoot branches arise more superficially, i.e., they are *exogenous*.

It is seen during germination that the radicle is the first organ to grow in the seedling. This radicle gives rise to branches of primary, secondary or higher orders forming what is known as the **primary** or **taproot system** (Fig. 213), the main root being known as the *taproot* which is a direct prolongation of the radicle while the branches are often designated by the general expression *secondary roots*. In most dicots the root system ramifies more and more in the soil and anchors the plant very firmly. The ultimate root branches of large trees ramify through a very large and deep area, often occupying no less a space than what the tree occupies above. In most monocots, however, this primary root system does not develop so strongly and does not form the main root system. New roots grow from near the base of the radicle and they are called **seminal roots.** These seminal roots grow and behave in the same manner as the primary root from the radicle. Later on, some more roots grow from the base of the plumule or from its lower nodes. These roots are like fibres and they form a **fibrous root system** (Fig. 214) which is usually the main root system of monocots. In many monocots, at least in the cereals, it is now known that the primary and seminal roots do not die quickly and usually persist.

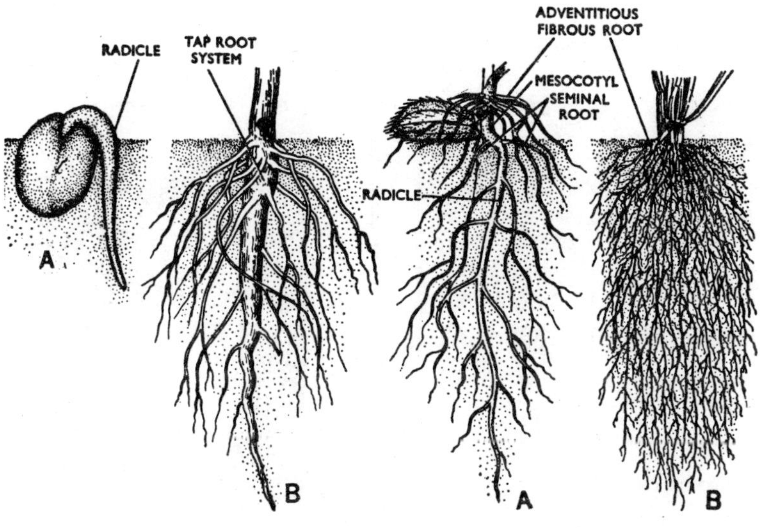

Fig. 213 Fig. 214

Fig. 213. Development of the TAPROOT SYSTEM in dicots. A. Germinating dicot seed with growing radicle. B. Taproot system developed out of a radicle. Fig. 214. Development of the ADVENTITIOUS FIBROUS ROOT SYSTEM in monocots. A. Rice seedling showing radicle roots, seminal roots, adventitious fibrous roots and mesocotyl. B. Rice plant with extensive fibrous root system.

But even in this case, it is the fibrous root system which does the real work. Since the rootlets are rather thin and do not branch enormously as in the case of dicots, fibrous roots cannot be as strong for anchorage as the primary root system. But, there are very few trees among the monocots.

A root system developing from the radicle (with its branches) is known as a **normal root system** while a root developing from anywhere excepting the radicle is known as **adventitious.** Thus, the root system is usually normal in dicots and adventitious in monocots. *Adventitious roots* also arise to serve special functions. Such adventitious roots are seen to develop on various organs like underground stems, stem cuttings, etc. These are described later.

Whether a root belongs to a normal root system or it is adventitious, it shows the same different regions. The **zones** or **regions** of a root are best studied from the tip of a rootlet. As the root remains under the soil and it is difficult to get the fine root endings of large plants, it is easier to study the regions from some germinating seed (Fig. 215), e.g., mustard, gram or pea. It will be seen that the tip of the root is protected by a fine cap-like structure known as the *root-cap* or *calyptra.* This region may be called (1) the **root-cap region.** This *root-cap* serves for the protection of the delicate root-tip while the root is growing down and down into the soil. Friction with soil particles continually wears out the root-cap while it is being constantly renewed by new growth from its base. While most plants show such a simple root-cap (Fig. 216), there are some which show some variation. In *Pandanus* (screw-pine—Fig. 217) the root-cap is *multiple.* The aquatic plants (*Lemna*—Fig. 139 and *Pistia stratiotes*—Fig. 218) usually do not have root-caps but their root apices are protected by what are known as **root-pockets.** Root-pockets are like loose thimbles which keep the root apex covered but, these are not regenerated from the base (which is the case in true root-caps), being of different origin.

Just behind the root-cap is (2) the very short **growing region.** This is the part of the root which actually grows. This region may be further subdivided into the *zones* of

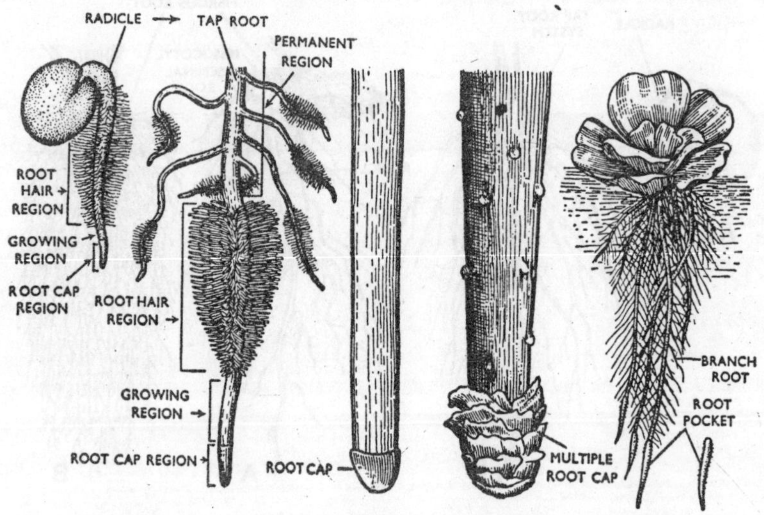

FIG. 215 FIG. 216 FIG. 217 FIG. 218

FIG. 215. A germinating seed and a mature plant showing the four different regions of root. FIG. 216. Simple root-cap of a hanging *Ficus* aerial prop root. FIG. 217. Multiple root-cap of *Pandanus.* FIG. 218. *Pistia stratiotes* plant showing root-pockets.

division and *elongation* according to the stage of growth of the cells. Behind this growing region is (3) the **root-hair** or **piliferous region.** Here the root surface is covered by fine unicellular root-hairs which do the actual *absorption* of solutes. Friction with soil is continually wearing out and destroying the root-hairs while the region of growth, as it extends downwards by growth, is developing new root-hairs forming new root-hair regions. Aquatic plants (Fig. 218) do not have root-hairs and they do not need them. Absorption goes on along the root surface in these plants.

The first three regions take up little space on the tips of the final root branches. All the length of the root behind and above the root-hairs is occupied by (4) the **permanent region.** Most of the root and all its older branches up to the base of the stem are occupied by the permanent region. All root-hairs have been killed and all growth has stopped in this permanent region. No absorption takes place through this region but what has been already absorbed is conducted to the stem above.

Normally, the root performs the **functions** of *anchorage, absorption* and *conduction.*

Anchorage of the plant to the soil is a *mechanical function* while absorption of solutes and their conduction upwards are *physiological functions.* Besides these *normal functions*, roots may sometimes carry on some *special functions* according to which even the structure of a root gets modified. These are the **modifications** of roots.

MODIFICATIONS OF ROOTS

Roots may be modified in order to carry on special functions. Such special functions may be either physiological or mechanical. Either taproots or adventitious roots may undergo such modifications.

(A) *MODIFICATIONS TO CARRY ON SPECIAL PHYSIOLOGICAL FUNCTIONS*

1. **Storage roots** become fleshy by the accumulation of stored up food. Very often the aerial parts of these plants die out during unfavourable seasons and, when the season becomes favourable again, new buds sprout either from the underground fleshy root itself or from a small bit of stem above. Thus, they serve for perennation almost as

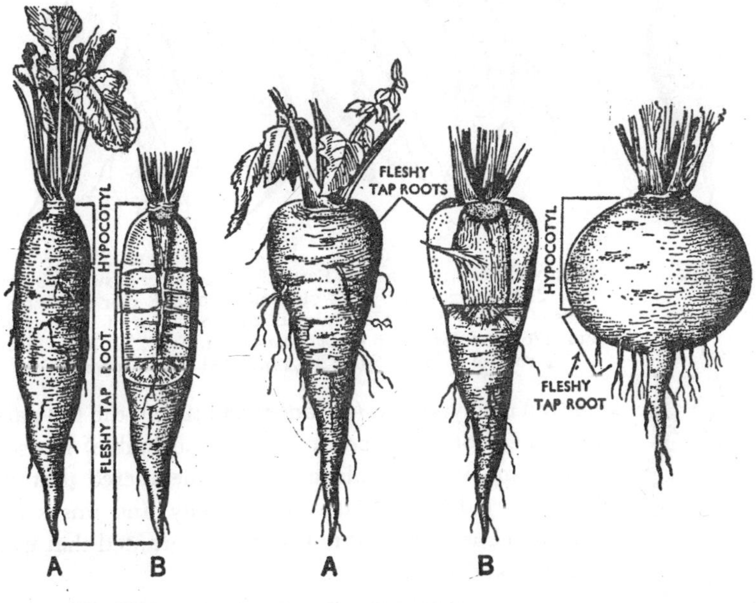

FIG. 219 FIG. 220 FIG. 221

FLESHY TAPROOTS. FIG. 219. Fusiform in radish. A. Whole radish. B. The top part cut off longitudinally. FIG. 220. Conical in carrot. A. Whole carrot. B. The top part cut off longitudinally showing practically no hypocotyl. FIG. 221. Napiform in turnip.

well as the underground stems. In other cases, specially in the tropics, the reserve food is utilised by the plant during flowering and fruiting. These swollen roots serve as important vegetables and food reserve for the human race.

This swelling and storage may involve the taproot (with or without its branches) or adventitious roots. Many common vegetables are instances of *storage in taproots.* Such swollen taproots are named according to the shape assumed. It is **fusiform** (swollen in the middle and gradually tapering towards both ends) in radish (*Raphanus sativus*—Fig. 219), **conical** (broadest on top and gradually tapering towards the lower end) in carrot (*Daucus carota*—Fig. 220) and **napiform** (very much swollen above, abruptly tapering towards the lower end) in turnip (*Brassica campestris* var.

rapa—Fig. 221) or beet (*Beta vulgaris*). It should be noted in this connection that the underground swollen part of these vegetables is not usually formed by the root

FIG. 222 FIG. 223 FIG. 224 FIG. 225

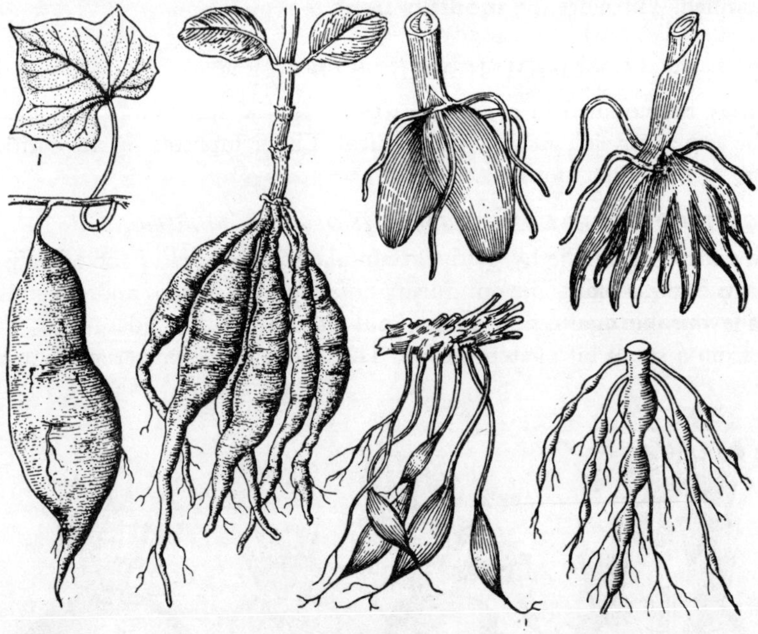

FIG. 226 FIG. 227

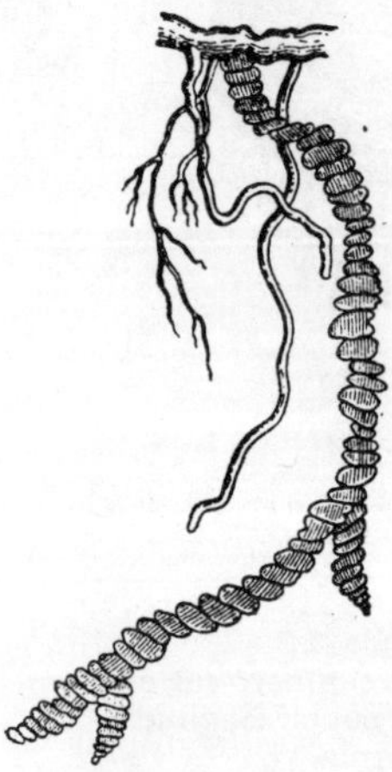

FIG. 228

FLESHY ADVENTITIOUS ROOTS. FIG. 222. Root-tuber of sweet potato. FIG. 223. Fasciculated root of *Dahlia*. FIG. 224. Pair of tuberous roots of *Orchis*. FIG. 225. Palmate tuberous root of an orchid. FIG. 226. Nodulose root of *Curcuma amada*. FIG. 227. Moniliform root of a grass. FIG. 228. Annulated root of ipecac.

alone but the hypocotyl (in radish and turnip the top part is hypocotyl, in carrot the hypocotyl part is negligible) is also incorporated within it. It should also be remembered that the terminologies are for particular *shapes* only and this is not always uniform. Thus, one must not take it for granted that every carrot in the market is *conical*.

Similar swellings are found in the *adventitious roots* of many plants. In sweet potato (*Ipomoea batatas*—Fig. 222) adventitious roots grow from the nodes of the running stem. These become swollen and form the edible **root-tubers.** Sweet potato tubers do even give rise to adventitious buds (Fig. 72) but should be distinguished from the ordinary potato which is a stem-tuber. Tuberous roots are to be found in many common plants like *Dahlia*, Tapioca (*Manihot esculenta*), *Ruellia tuberosa*, etc., and may involve both adventitious and taproots. In some Dahlias (Fig. 223), *Asparagus*, etc., the adventitious roots occur in clusters and all these are swollen. These are called **fasciculated roots.** An interesting tuberous root is that of some orchids. In *Orchis* (Fig. 224) there is a pair of succulent tuberous roots one of which perishes every year while another new one is formed by its side. Such orchid roots may sometimes be of a palmate shape when they may be described as *palmate tuberous* (Fig. 225). In the mango-ginger (*Curcuma amada*—

Fig. 226) and in *Costus speciosus,* only the apices of the adventitious roots become swollen like single beads. This condition is termed as **nodulose.** In *Dioscorea alata,* many grasses and sedges (Fig. 227) and in some *Vitis,* some roots are alternately swollen and constricted presenting a beaded **(moniliform)** appearance. In the valuable medicinal plant ipecac—*Cephaelis* (=*Psychotria*) *ipecacuanha*—(Fig. 228) yielding emetine and now planted on the Himalayas, the roots look as if formed by a number of discs placed one above another. This is described as **annulated.**

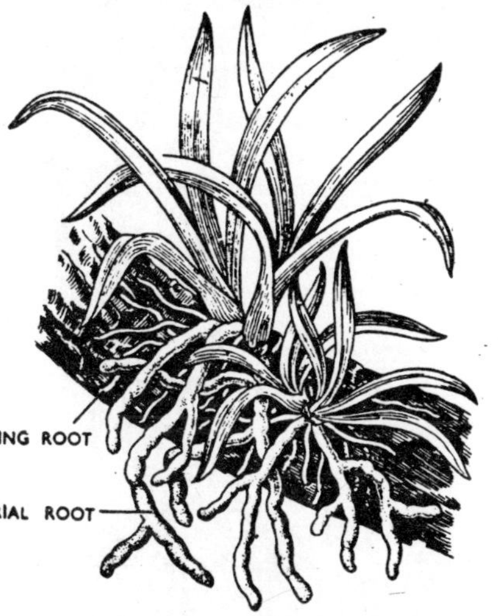

FIG. 229. Orchid showing *aerial* and *clinging* roots.

2. **Epiphytic** or **Aerial absorbing roots** are the peculiar common roots seen hanging from the orchids (Fig. 229). Epiphytes are provided with these *aerial* roots in addition to the clinging roots (Chapter I). The aerial roots are greenish and covered with a spongy *velamen* tissue with which they absorb atmospheric moisture. Their function is, therefore, *absorptive.* They also carry on a little carbon assimilation because of the chlorophyll. Such roots are also found in some aroids.
3. **Assimilatory roots** are green while ordinary roots are not so. In some trees, when urface roots get exposed, they turn green by the development of chlorophyll and carry on some carbon assimilation. Epiphytic roots also are assimilatory. But, certain plants like *Tinospora cordifolia* (Menispermaceae) have long green aerial roots whose main function is carbon assimilation. The peculiar family of plants, Podostemaceae (Fig. 240), which is quite common in places like Cherrapunji, grows in shallow water on the hills. The roots cannot penetrate rocks and are exposed. They are green and assimilatory as well as shoot-forming. These roots often assume a flattened thalloid structure and remain fixed on the rock by means of holdfasts called *haptera* resembling some algae.
4. **Reproductive roots:** Adventitious buds sometimes grow on roots and serve as means of reproduction. Plantlings arise from the base of many garden plants in this way. Root cuttings are the main mode of reproduction in *Trichosanthes dioica,* sweet potato (Fig. 72) and ipecac. Podostemaceae (Fig. 240) propagates through roots. Such roots are, therefore, *reproductive.*
5. **Respiratory roots** or **Pneumatophores:** The roots of some trees and shrubs growing in saline marshes (*mangrove plants*) suffer from the lack of oxygen. To cope with this situation some root branches grow vertically up (i.e., negatively geotropic) into the air (Fig. 230) from the horizontal secondary roots. Atmospheric air enters these roots through the minute pores of special lenticels (*pneumathodes*) on the exposed root tips. They are, therefore, *respiratory* roots. Common examples of plants showing respiratory roots are *Heritiera* (Bengali—*sundri* which has given the name to the Sunderbans), *Rhizophora,* etc.
6. **Sucking roots** or **Haustoria:** These are the sucking roots of parasites discussed in Chapter I. The parasites penetrate the small adventitious rootlets into the tissues of

the host so that there is connection between the conducting systems of the two plants (Fig. 231). This is the channel through which food is sucked in by the parasite from

FIG. 230

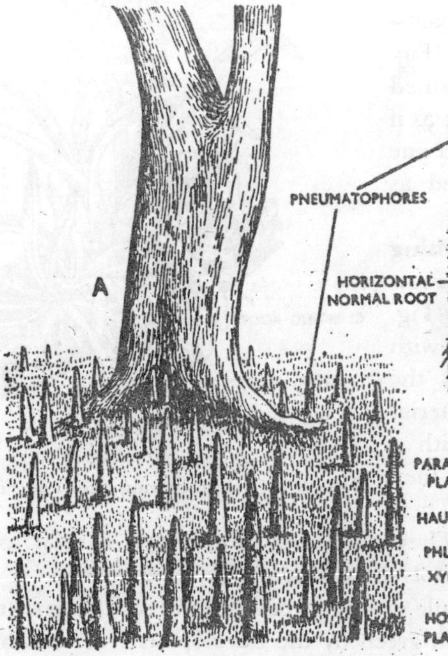

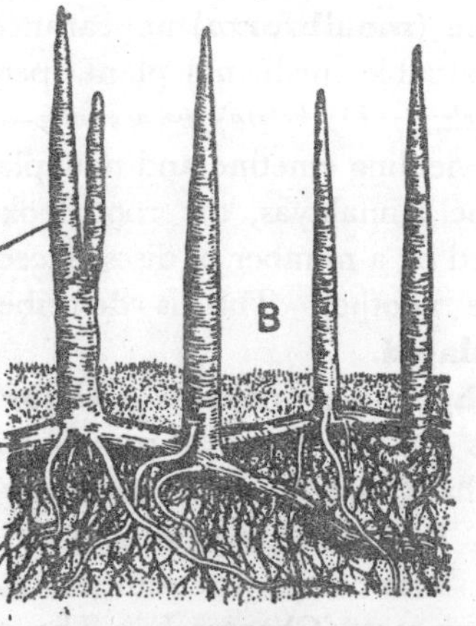

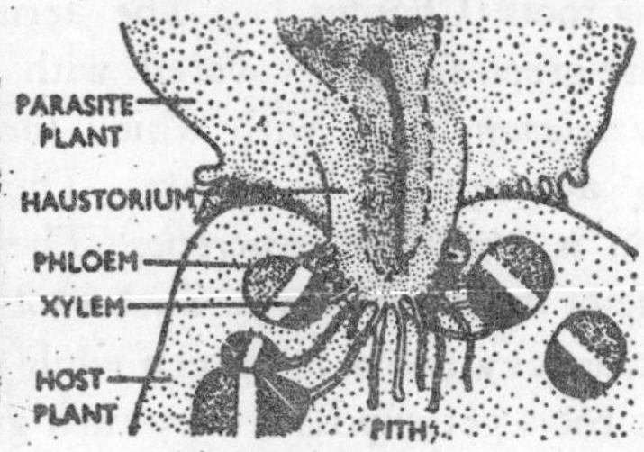

FIG. 231

FIG. 230. Pneumatophores. A. Pneumatopho.es at the base of a mangrove tree. B. Section through soil showing origin of pneumatophores from normal horizontal roots. FIG. 231. Section through host plant and parasite showing how the parasite sucks food material from conducting tissues of the host plant and even from pith.

the host. Such sucking roots or *haustoria* are found in many total and partial parasites (Figs. 10 to 14).

7. **Mycorrhizal roots:** These are the roots of the mycorrhizal saprophytes, discussed in Chapter I. They usually grow in humus and roots are infested with some fungal mycelia. The mycelia form a mantle on the root (Fig. 17) which absorbs food solutions (mostly organic) utilised both by the host plant and the mycorrhizal fungus. The growth of this beneficial fungus on the root atrophies the root which becomes somewhat stunted and often looks abnormal (Fig. 232). The tips do not grow and root-hairs are not formed. Such *mycorrhizal roots* are to be found in trees like pine and birch and in saprophytes like *Monotropa*, etc. (Figs. 17–20).

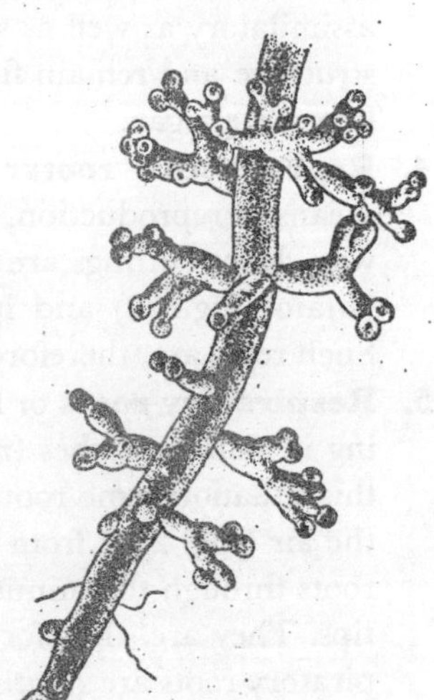

FIG. 232. A mycorrhizal root of pine.

(*B*) *MODIFICATIONS TO CARRY ON SPECIAL MECHANICAL FUNCTIONS*

1. **Prop roots:** In some tropical trees like *Ficus benghalensis* (banyan), the horizontal aerial branches

give rise to aerial roots which are provided with root-caps (Fig. 216) and hang vertically down from the boughs like so many strings. These grow down and, on reaching the soil and becoming anchored, they begin to stouten and ultimately become almost as strong as the main trunk. They support or *prop up* the horizontal branches like so many pillars (Fig. 233). Ultimately, the main trunk may die

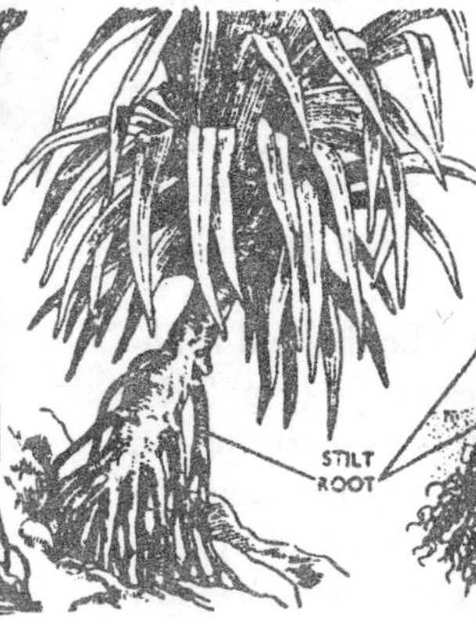

FIG. 233 FIG. 234 FIG. 235

FIG. 233. Prop root of banyan. FIG. 234. Stilt root of *Pandanus*. FIG. 235. Stilt root of maize.

when the prop roots fully replace it. Long living banyan trees cover large areas by their spreading branches supported on prop roots. The banyan tree at the Indian Botanical Garden, Shibpur, Calcutta is about 200 years old, covers a large area and has produced more than 900 prop roots. A similar tree in the Theosophical Society compound Adyar, Madras and another in the Buitenzorg Botanical Garden, Java, are equally remarkable.

Stilt roots: Certain shrubs and small trees like screw-pine (*Pandanus foetidus*—Fig. 234) grow on the edges of tanks, marshes, etc., where the anchorage is not very strong. In these cases, short roots grow obliquely downwards from near the base of the stem and act like *stilts* providing additional support as well as anchorage to the stem. The adventitious roots growing from the lower nodes of maize plants (Fig. 235) act in a similar way. Such stilt roots are also seen in mangrove plants like *Rhizophora*.

FIG. 236. Root buttress.

Root buttresses: In some large trees, instead of stilt roots, there are great plank-like roots radiating from the base of the tree like wings (Fig. 236). These are called *buttresses* and are actually partly root and partly stem. They may be seen in old trees of *Bombax ceiba, Terminalia catappa, Ficus* sp., etc.

Climbing roots: Some climbers can climb up their supports as adventitious roots growing from the nodes of the frail climber twine round and clasp the support as if the climber has been tied to the support at those points. Common examples are to be found in the aroid *Scindapsus officinalis* (Fig. 9), the betel vine (*Piper betle*—Fig. 237), etc.

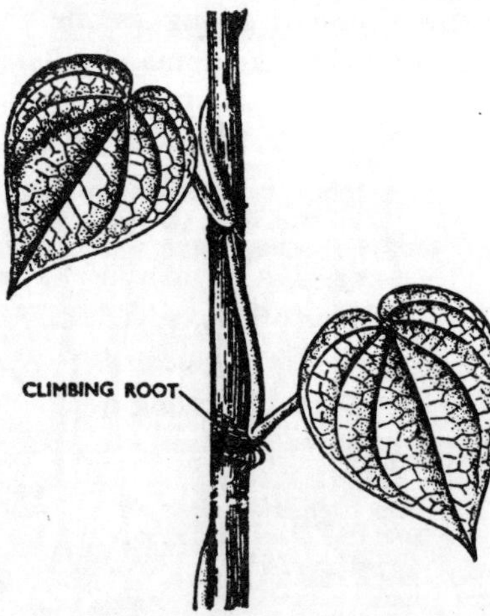

FIG. 237. Climbing root of betel vine.

5. **Clinging roots:** Epiphytes like orchids cling to their support by means of special *clinging roots* which enter the crevices of the support and fix the epiphyte (Fig. 229). Clinging roots may carry on some absorption but are different from the specialised aerial absorbing roots.
6. **Haptera** or **holdfasts:** The thalloid branching root of Podostemaceae (Fig. 240), which is their main plant body carrying on assimilation as well as reproduction, is stuck up to the rock by means of special root processes which may be called *holdfasts* or *haptera* as in some algae. The *adhesive discs* of adhesive climbers like *Ficus repens* and Ivy described on page 60 are also types of holdfasts.

N.B. It may be noted that all the three types of roots, *climbing*, *clinging* and *holdfasts*, may behave in the same way by becoming flattened and disc-like and sticking to their supports or to the substratum.

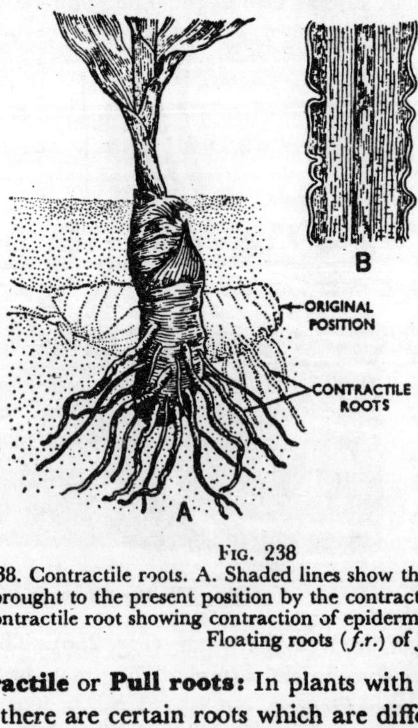

FIG. 238

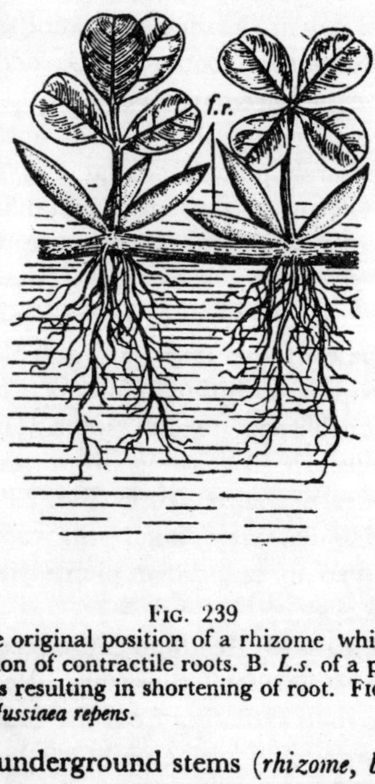

FIG. 239

FIG. 238. Contractile roots. A. Shaded lines show the original position of a rhizome which has been brought to the present position by the contraction of contractile roots. B. *L.s.* of a portion of a contractile root showing contraction of epidermis resulting in shortening of root. FIG. 239. Floating roots (*f.r.*) of *Jussiaea repens*.

7. **Contractile** or **Pull roots:** In plants with underground stems (*rhizome*, *bulb*, *tuber*, *corm*), there are certain roots which are different from the others (Figs. 238 & 126) as they are found to contract or swell so that the aerial shoot or the underground

parts are kept at a proper level in the soil. These roots are, therefore, called *pull* or *contractile roots*. Those may be seen in *Canna*, *Crocus*, *Allium*, *Lilium*, etc.

8. **Floating roots:** Certain aquatic floating plants like *Jussiaea repens* (Onagraceae) have special roots (Fig. 239) growing from the nodes in addition to the ordinary adventitious roots. These roots look like masses of white cotton and are extremely spongy in nature. Their buoyancy helps to keep the plant floating. If taken out of water they dry and shrivel up quickly. The large amount of air contained in the floating roots may also be of some use in the aeration ('breathing') of the submerged organs.

9. **Root-thorns:** In many palms (e.g., *Acanthorhiza* and *Iriartea*), some aroids (*Pothos armatus*) and others, adventitious roots from the base of the stem become hard pointed and thorn-like. These are called *root-thorns*. They are of indirect help as an armature to the plants in driving out marauding animals. Old adventitious roots of coconut, etc., also serve the same purpose.

ROOTLESS PLANTS AND PLANTS THAT ARE ALL ROOTS

While all normal Angiosperms are provided with roots, even among them there are some instances of plants with no roots. They are to be found among aquatic plants and among saprophytes. In aquatic plants like *Utricularia* (Fig. 26) some leaves are finely dissected and these submerged segments carry on the function of roots. In *Wolffia arrhiza*, which is one of the smallest known flowering plants, there is only a very small free floating green structure (leaf or phylloclade as in *Lemna*—Fig. 139). There is no root. Same is the case in *Epipogium*.

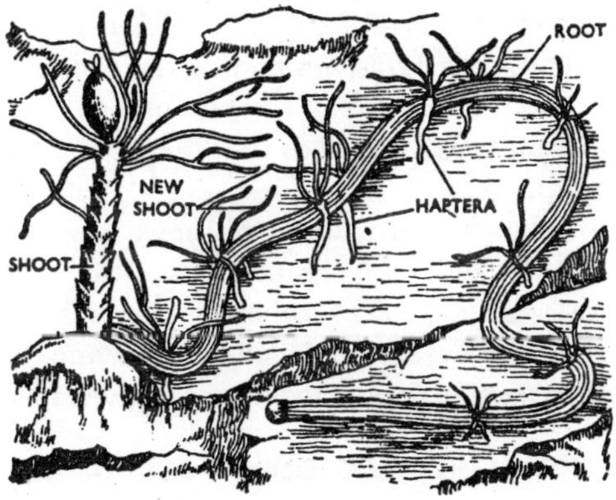

FIG. 240. *Podostemon* plant growing along rock surface. The main plant is a thalloid assimilatory root clinging to the rock by *haptera*. Short-living shoots grow from this root.

On the other extreme we find plants in which the main plant body is a root. In Podostemaceae the main plant body is a thalloid root which carries on assimilation as well as reproduction and grows freely in the shallow water on rock surface. *Podostemon* is quite common in the Cherrapunji area of Meghalaya. The small flowering shoots grow on this root from time to time (Fig. 240). In the parasites *Arceuthobium*, *Rafflesia* (Chapter I, Fig. 16) and *Sapria* the plant bodies are filamentous masses living within the tissue of the host stem or root without showing any differentiation. At the proper time buds develop from this filamentous mass which burst out of the host stem or root and give rise to minute (*Arceuthobium*) or gigantic (*Rafflesia*) flowers. A seed of the saprophyte *Monotropa* (Fig. 18) produces only a free-living root on germination. This branches abundantly in the humus forming a perennial structure. Occasionally, aerial flowering shoots are formed on this root.

CHAPTER VI

HOMOLOGY AND ANALOGY:

DEFENSIVE DEVICES OF PLANTS

HOMOLOGY AND ANALOGY

The study of the **homology** of an organ means the study of its morphological origin, development and position. Organs, however different in their outward appearance, are **homologous** if they are identical in morphological origin. **Analogy,** on the other hand, refers to the outward appearance and the physiological behaviour of an organ. Organs are **analogous** if they are outwardly the same and perform the same function, however different their origins may be. The tendril of pea is a modified leaf or leaflet while that of passion-flower is a modified axillary bud. So, these two tendrils are analogous (similar in outward appearance and behaviour) but not homologous (similar in origin). On the other hand, an axillary thorn and an axillary tendril are dissimilar in appearance but similar in origin and position. Hence, they are homologous but not analogous.

When we study the modifications of some plant organ, we study *homologous structures*. Thus roots and all forms of modified roots are homologous as all of them are roots by origin. Similarly, leaves and all their modifications are homologous.

On the other hand, when one studies particular types of organs, e.g., tendrils of different types, one finds that they are not all homologous. But, all tendrils are *analogous* organs.

The general plan of this book as well as that of most books on Morphology, is to study the different organs (stem, leaf, root, etc.) and their modifications, i.e., *organs homologous to them*. Morphology may also be studied from another angle—by grouping together different analogous organs and then studying their homology. Some such instances of grouping together are given below:

1. HOMOLOGY OF CLIMBING ORGANS WHICH ARE ALL ANALOGOUS TO TENDRILS

Under this head one groups together all types of tendrils and studies the homology of each of them.

(1) Tendrils of *Vitis* are modified apical buds (Figs. 146 & 241B), i.e., homologous to them.

(2) Tendrils of *Passiflora* (Fig. 241A) are homologous to axillary buds.

(3) Tendrils of *Lathyrus aphaca* (Fig. 241C) are homologous to whole leaves.

(4) Tendrils of *Pisum sativum* (Fig. 241D) are homologous to leaflets.

(5) Tendrils of *Clematis* (Fig. 241E) are modified leaf petioles.

(6) Tendrils of *Gloriosa* (Fig. 241F) are homologous to leaf apices.

(7) Tendrils of *Smilax* (Fig. 241G, are homologous to stipules.

(8) Tendrils of *Antigonon* (Fig. 241H) and *Cardiospermum* (Fig. 465C) are modified inflorescence axes.

(9) In the family *Bignoniaceae* tendrils are found which are homologous to rootlets.

(10) The homology of the tendrils of *Cucurbits* (Fig. 241I) is not clear and has been supposed to be different by different authors. These tendrils are sometimes simple and sometimes branched. Goebel's opinion is that simple tendrils are modified prophylls of

axillary shoots, while branched tendrils are modified extra-axillary buds in which the lower portions are modified lateral branches while the upper tendril branches are modified leaves.

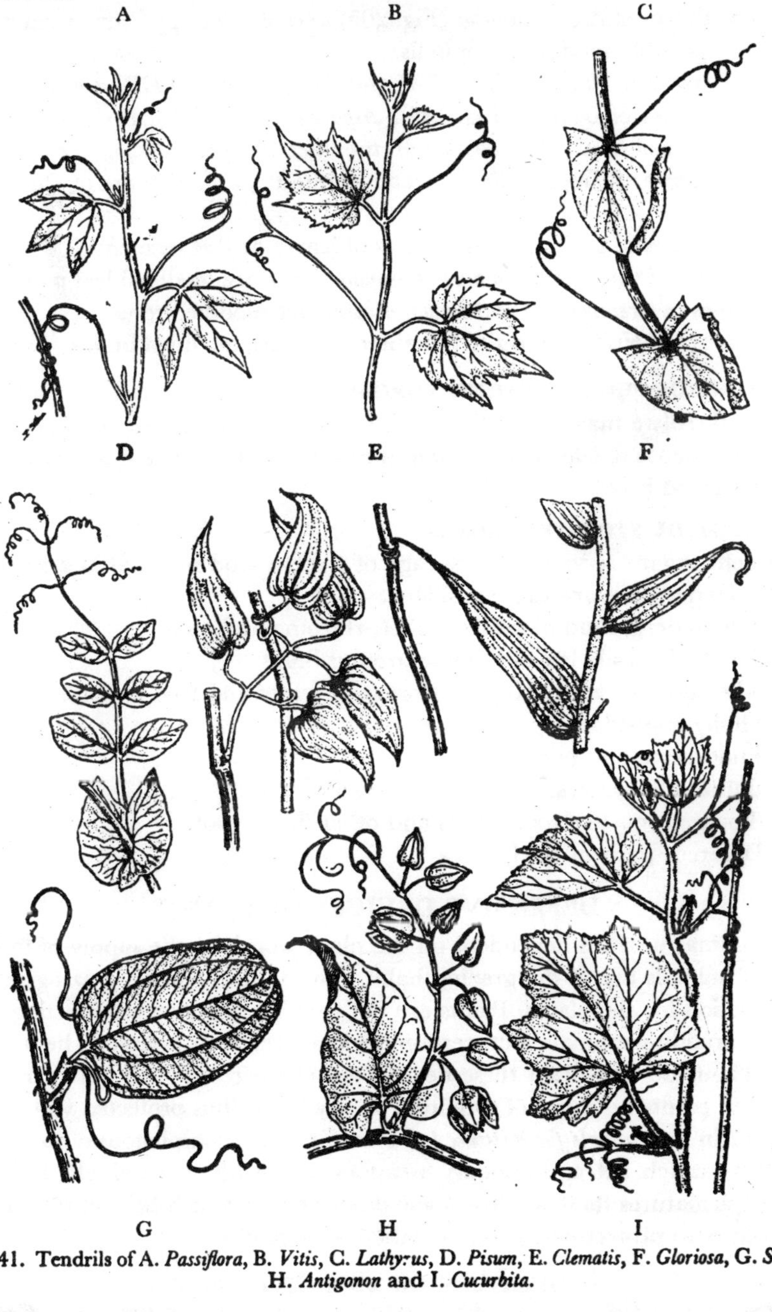

Fig. 241. Tendrils of A. *Passiflora*, B. *Vitis*, C. *Lathyrus*, D. *Pisum*, E. *Clematis*, F. *Gloriosa*, G. *Smilax*, H. *Antigonon* and I. *Cucurbita*.

2. HOMOLOGY OF DEFENSIVE ORGANS WHICH ARE ANALOGOUS TO THORNS AND SPINES

Thorns and spines are different—the former being stronger structures. But all spines and thorns may be treated as analogous as they look similar and perform the same function.

(1) The thorns of *Duranta* (Fig. 140), *Bougainvillaea*, *Carissa* (Fig. 141), *Flacourtia* (Fig. 142), *Hygrophila* (Fig. 143), prune (Fig. 144) and the bigger thorns of *Ulex* (Fig. 209) are all modified axillary buds or stems.

(2) The thorns of *Aegle marmelos* (Fig. 205) and of *Citrus* spp. (lemons and oranges—Fig. 208) are possibly modified prophylls.

(3) The spines of *Acacias* (Fig. 177A) and *Zizyphus* (Fig. 177B) are stipules.

(4) The spines of *Opuntia* (Fig. 135), *Asparagus* (Fig. 138), *Berberis* (Fig. 148) and the smaller thorns of *Ulex* (Fig. 209) are all modified leaves.

(5) The spines of *Trapa bispinosa* (water-chestnut—Fig. 319) are modified persistent sepals.

(6) The spines of date-palm are apices of leaves or leaf segments.

(7) Spines of *Punica granatum* and *Quisqualis indica* are modified leaf petioles.

(8) Spines of *Acanthorhiza* and *Pothos armatus* are modified roots.

(9) Spines of *Spinifex littoralis* (Chapter LIV) are modified bracts.

3. HOMOLOGY OF PHOTOSYNTHETIC ORGANS

Under these we may consider the *phylloclades* (including *cladodes*) and the *phyllodes* which are analogous to leaves performing the same function. The *foliaceous stipules* should also be considered here.

4. HOMOLOGY OF STORAGE ORGANS

Different organs serve for the storage of reserve food material and so they are all analogous. Under these one may consider:

(1) The underground stems like tuber, rhizome and corm.

(2) The fleshy scale leaves of underground bulbs.

(3) The fleshy leaves serving for storage of water and food material.

(4) Fleshy cotyledons.

(5) Endosperm and perisperm.

(6) Bulbils of *Dioscorea*.

(7) Storage roots like root-tubers and other fleshy roots.

(8) Different parts of the fruit.

DEFENSIVE DEVICES OF PLANTS

The animal world is dependent on the plant world for the supply of food. A large number of animals are of the grazing habit and, naturally, such grazing means death and destruction for the plants. Plants are not capable of actively protecting themselves from the marauding animals. But, certain devices directly or indirectly help them in this struggle. Plants do not adapt these devices out of their own will but such devices are developed in plants in course of evolution as any plant thus protected will have a point in its favour in the *struggle for existence* (*cf.* Part V). Most conspicuous of such devices are the *armatures* which act as stationary weapons or shields to ward off the marauders. Besides the armatures there are also some other devices which help in this effort on the part of Nature to protect her weaker creation—the plants.

A. ARMATURE OR DEFENSIVE ORGANS

1. Thorns, spines and **prickles:** Although these are of different origin, they are similar and serve the same general purpose of warding off the marauders. *Thorns* are deep seated stem structures while *spines* are modifications of different organs. *Prickles* are very superficial structures which grow only from the epidermis. Such are the prickles on rose (Fig. 105) which are uprooted with the slightest pressure. Such superficial prickles are also common on the leaves and fruits of many plants like brinjal (*Solanum melongena*), *Solanum surattense*, etc. These thorny structures are often found in greater abundance on the lower (i.e., more mature) parts of the plant. Naturally, grazing animals do not like to disturb plants like the various *Cacti* (Fig. 242), *Argemone mexicana* (Fig. 243) or

FIG. 242

FIG. 243

SPINOUS PLANTS DREADED BY ANIMALS
FIG. 242. A cactus. FIG. 243. A twig of *Argemone mexicana*.

Amaranthus spinosus as plants like these present a uniform mass of spinous bodies which are detested and even dreaded by the former.

2. Stinging hairs: Certain plants have epidermal hairs, the tips of which are siliceous (therefore, glassy) and unicellular, containing a poisonous sap (Fig. 244). On contact with animal skin these needle-like hairs break and pierce the skin injecting (Fig. 244B) the poisonous substances. This causes irritation which may sometimes be very serious. These hairs are, therefore, called *stinging hairs*. Stinging hairs are commonly found in the nettles. There are various types of nettles belonging to the genera *Urtica*, *Fleurya*, *Laportea*, *Girardinia*, etc., of Urticaceae and *Tragia* of Euphorbiaceae. They are herbs and shrubs of various heights and of different degrees of poisonousness. Similar stinging hairs are found on the fruits of the extensive climber *Mucuna pruriens* (Papilionaceae). All these are naturally avoided by grazing animals.

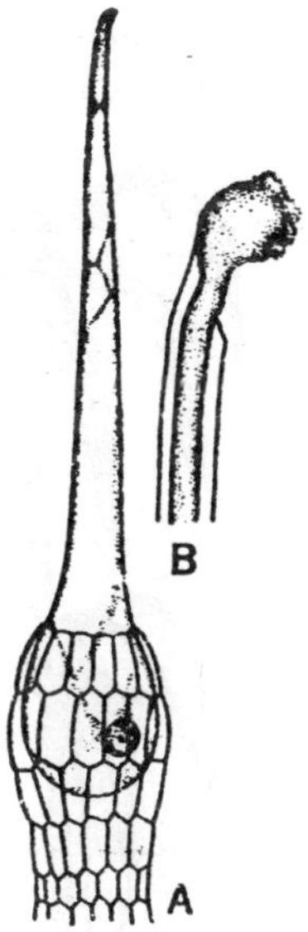

FIG. 244. Stinging hair of a nettle (*Urtica dioica*). A. An entire stinging hair. B. Broken apex of a stinging hair showing the sharp end and the irritating liquid coming out. (After Haberlandt).

3. Sticky glandular hairs: These are also epidermal outgrowths like the stinging hairs. They occur on leaves, stems, fruits, etc. These glandular hairs secrete some sticky substance so that the plant parts stick to the face of the animal causing much dislike on the part of the latter. They are, therefore, avoided by animals. Common examples are *Jatropha*, *Plumbago zeylanica*, *Boerhaavia*, tobacco, *Siegesbeckia*, etc.

4. A dense covering of hair on plants like *Calotropis* is always disliked by grazing animals so that they are saved from them.

5. Cork and **bark** act like a shield so that the maturer parts of the plants are avoided by the marauders.

B. OTHER DEFENSIVE DEVICES

1. Poisonous or repulsive substances: Many plants contain more or less poisonous substances so that they are avoided by animals either by instinct or by experience. These substances may be present in various forms:

(a) **Latex:** The laticiferous cells or vessels may contain some fluid which is repulsive to taste or is poisonous and is much disliked by animals as in *Calotropis*, *Opuntia*, *Thevetia*, *Nerium*, different species of *Ficus* and *Euphorbia*, etc.

(b) **Tannin,** present in barks and as cell-content, is disliked by most insects.

(c) **Alkaloids** are deadly poisons found in many plants, e.g., *nicotine* in tobacco, *morphine* in opium poppy, *quinine* in cinchona, *strychnine* in nux-vomica, *daturine* in datura, *caffeine* in coffee, *theine* in tea, *emetine* in ipecac, *atropine* in belladonna, etc. Although strong poisons, they are important medicines in mild doses.

(d) **Bitter taste** and **offensive smell** also are caused by chemical contents and ward off many animals. The bitter taste of margosa, *Momordica* and the smell of *Poederia*, *Blumea lacera* and many Labiatae are useful in this sense. The inflorescence of *Amorphophallus campanulatus* also emits a fetid smell.

(e) **Irritating substances:** The raphides and the sphaeraphides in various aroids are well-known sources of irritation and are effective in saving the plants.

(f) **Silica** on the epidermal cell walls of many grass leaves render them dangerously sharp for grazing animals and are invariably avoided.

2. Underground modified stems and some plants with **fleshy roots** have a peculiar life-history where a big part of their lives is spent under the soil out of the reach of grazing animals. They are, thus, rendered safe for this period of their life.

3. Mimicry

Certain plants strikingly resemble some dreaded animals so that they are carefully avoided by other animals. This is known as *mimicry* or copying other animals. Many

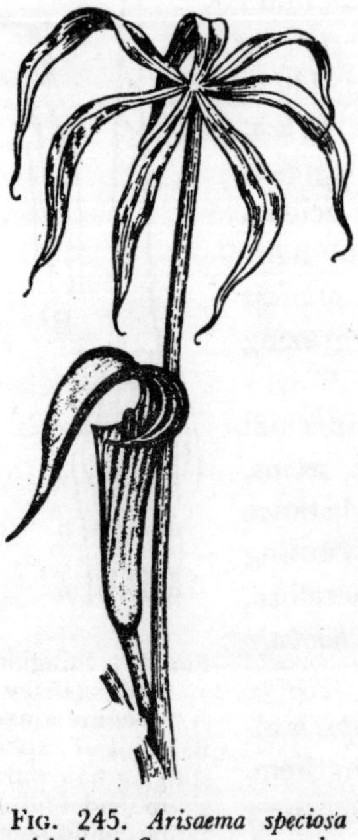

FIG. 245. *Arisaema speciosa* with the inflorescence spathe mimicking a snake hood.

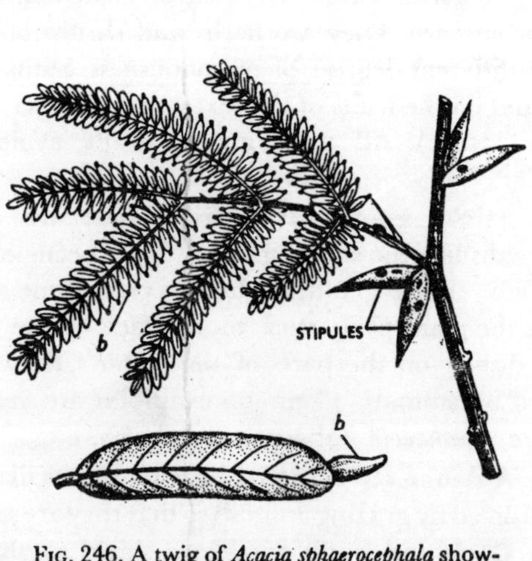

FIG. 246. A twig of *Acacia sphaerocephala* showing hollow stipules and Belt's corpuscles (*b*) serving as shelter and food for ants.

aroids resemble serpents and the most striking of them is *Arisaema speciosa* (Fig. 245) growing abundantly on the forest floors in Shillong and Darjeeling. The plant body is often spotted and the spathe of the inflorescence (green or greenish-purple), which is variously striped or spotted, is shaped just like the hood of a cobra. The plants look menacing from a distance and are carefully avoided by the marauders.

4. Myrmecophily

Another very interesting device is what is known as *myrmecophily* (literally, love of ants). These plants have something which is attractive to some war-like ants which then take their abode on such a plant and act as a troop of bodyguards against any marauding animal. This is generally the case in fruit trees like mango, litchi, *Syzygium jambos*, etc. In certain aroids with corms it is found that the underground corm is divided up into compartments which then serve as a castle for dangerous ants so that the corm cannot even be touched. The device is brought to its perfection in certain South American *Acacias*. In *Acacia sphaerocephala* (Fig. 246) the spiny stipules are hollow and serve as shelters for the ants. At the same time, certain food substances are secreted and accumulate as particles on the tips of leaflets. These are called *Belt's corpuscles*. So, the plant keeps ready both food and shelter for the troop of bodyguards which are always ready to accept them in lieu of the defence service undertaken.

CHAPTER VII

THE INFLORESCENCE AND THE BRACT

When the vegetative organs are fully developed, the plant becomes mature and the reproductive phase sets in. There are various factors controlling this initiation of the reproductive phase (*cf.* Part III). *Floral buds* are now formed which are either apical or axillary. In some annual plants the onset of flowering is the forerunner of death as the plants die as soon as seeds are formed. In perennial plants flowering and further growth goes on side by side.

Just like vegetative buds, axillary floral buds are usually borne at the axils of leaves. These leaves are specialised and called **bracts.** The floral buds may be borne singly when the resultant flower is **solitary,** or they may be borne in clusters on specialised branches. These specialised branches bearing groups of flowers are called *inflorescences.* Although the term **inflorescence** usually means a flower cluster, the solitary type of flower is also included within it. Flowers, whether solitary or not, usually have short stalks or **pedicels.** A flower without a stalk is called *sessile* and a flower with one is called *pedicellate.* The main axis (which may or may not branch) of an inflorescence bearing a group of flowers is called the **rachis** or **peduncle.** Sometimes, in a branched inflorescence, a distinction is made between the *peduncle* and the *rachis*—only the branches on which the flowers are borne are called *rachis* while the main axis (with any branch on which flowers are not directly borne) is called the *peduncle.* The flower stalk of a solitary flower is also sometimes called a *peduncle* when it cannot be distinguished from the *pedicel.* Sometimes the peduncle is flattened to form a fleshy cup-like structure called **receptacle.** The flowering shoot that comes out of *acaulescent* annual plants with radical leaves (e.g., radish, tube-rose, etc.) is called a **scape** (Fig. 85).

BRACTS AND BRACTEOLES

Bracts (*hypsophylls*—Chapter IV) are special leaves at whose axils the flowers are borne. They are not always present and, if a bract is found, the flower is called **bracteate** while, if it is absent as in the families Cruciferae and Boraginaceae, the flower is **ebracteate.** Bracts may drop off early (*deciduous*) or they may be *persistent.* When bracts are present, they serve for the protection of flower buds in the young stage. Morphologically, a bract may be formed by the entire phyllopodium or a part of it. Some shoots show transitional forms between normal leaves and bracts (Fig. 247). Additional bract-like small and thin structures are sometimes borne on the peduncle or the pedicel between the flower and the bract. These are called **bracteoles** (Fig. 248) and are of the nature of *prophylls* (Chapter IV).

Bracts and bracteoles may present various appearances some of which are described below:

(1) Leafy or **foliaceous bract:** Bracts resembling ordinary green foliage leaves may be seen in the china-rose. (*Hibiscus rosa-sinensis*—Fig. 283) and a number of other plants like *Acalypha indica* (Euphorbiaceae), *Adhatoda zeylanica* (Acanthaceae), *Gynandropsis* (Capparidaceae), etc.

(2) Petaloid bract: Bracts sometimes become brightly coloured simulating petals and perform the same function. These may easily be mistaken for petals. Beautiful examples are *Bougainvillea spectabilis* of Nyctaginaceae (Fig. 249) where three flowers cluster together and the large bracts are conspicuoisly bright red or orange, *Poinsettia*

(*Euphorbia*) *pulcherrima* of Euphorbiaceae (Fig. 250) with large leaf-shaped but brightly coloured red or orange bracts, and *Cornus* sp. (dogwood) of Cornaceae (Fig. 251) where the conspicuously large white bracts forming a whorl surrounding the cluster of small flowers in the centre, exactly look like a whorl of petals.

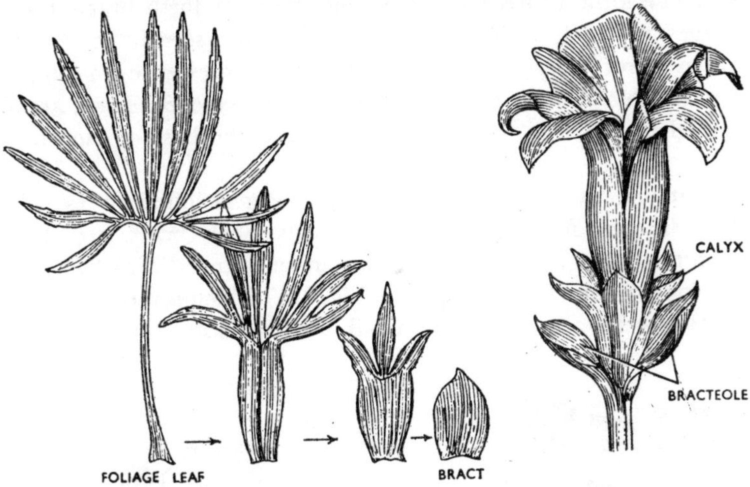

FIG. 247 FIG. 248

FIG. 247. Foliage leaf and bract showing transitional forms in *Helleborus foetidus*.
FIG. 248. Flower of *Gentiana acaulis* showing two bracteoles below calyx.

(3) Spathy bract: These are large, thick, boat-shaped bracts called *spathes* covering the whole or a part of a *spadix* inflorescence. These are often very large as seen in the inflorescence of *Amorphophallus titanum* (Fig. 266). In some palms the spathe is as much as twenty feet long. It is seen in many other plants like arum (Fig. 252), maize (Fig. 253), banana (Fig. 293), etc. Spathes also occur in some other types of inflorescences, e.g., the spike of *Polyanthes* (Fig. 265). It may be either green or coloured yellow, purple, etc. In

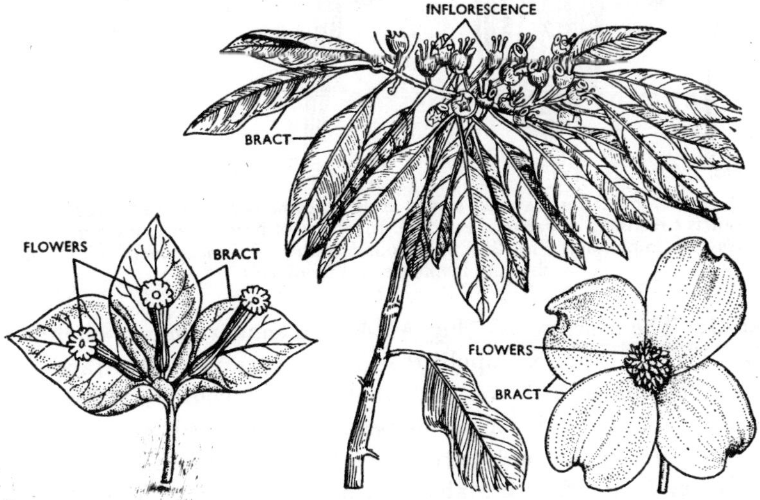

FIG. 249 FIG. 250 FIG. 251

PETALOID BRACTS in *Bougainvillea spectabilis* (FIG. 249), *Poinsettia* (*Euphorbia*) *pulcherrima* (FIG. 250) and *Cornus florida* (FIG. 251).

some palms the spadix is branched, the branches being covered by separate spathes which are sometimes given a special name, *spathellae*.

(4) Involucral bract: In an inflorescence of the *capitulum* type (Fig. 254) found in the family Compositae, whorls of green sepaloid bracts occur around the base of the inflorescence. These form an *involucre*. Sometimes bases of these bracts fuse forming a cup-shaped structure. Such an involucre of bracts is also seen in the family Umbelliferae where, as in carrot (*Daucus carota*—Fig. 255), the inflorescence is a compound umbel and there is an involucre at the base of the whole inflorescence. There are also smaller involucres (called *involucels* or *partial involucres*) round the base of each branch of the inflorescence.

(5) Scaly bract: Each individual floret of a *capitulum* often possesses its own *scaly bract* (Fig. 256) different from the involucral bracts. Scaly bracts are also seen interspersed between the florets of *catkin* and *cyathium* inflorescences (Figs. 270 & 290). In the *strobile* inflorescence of hop (*Humulus lupulus*—Fig. 271) which is used in the manufacture of beer, each flower is borne in the axil of a stiff, scaly, persistent bract.

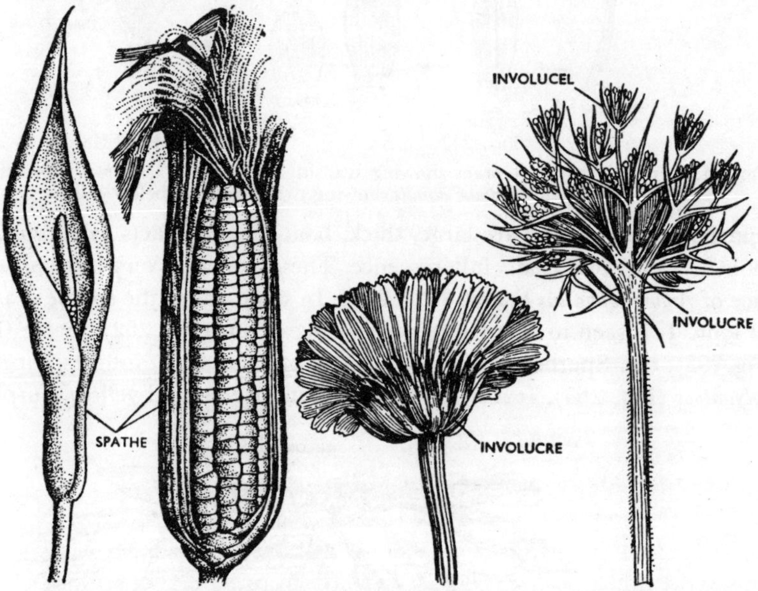

FIG. 252 FIG. 253 FIG. 254 FIG. 255

FIG. 252. Spathy bract on a spadix of the aroid *Colocasia esculenta*. FIG. 253. Multiple spathy bracts covering a maize cob. FIG. 254. Involucre of bracts surrounding a capitulum of *Calendula* (Compositae). FIG. 255. Compound umbel of *Daucus carota* (Umbelliferae) showing involucre and involucels.

(6) Cupule: The acorn of oak (Fig. 441A) and similar fruits of birch (*Betula*) and hazel (*Corylus*—Fig. 257), etc., of Cupulifereae possess a number of hard and woody bracts or bracteoles at the base of the flower somewhat like an involucre. As the fruit matures, these bracts fuse together forming a cup-shaped structure called *cupule* covering the base of the oak acorn and a similar husky structure is formed below the hazel nut.

(7) Epicalyx: In the family Malvaceae, as in china-rose (Fig. 283), a whorl of sepaloid green bracteoles occur just below the calyx and is called the *epicalyx*. Such an epicalyx is also seen below the calyx of the strawberry (*Fragaria* of Rosaceae—Fig. 258). An epicalyx is sometimes called an *involucre* because of the apparent resemblance.

(8) Glume: Each spikelet of the family Graminaceae (Fig. 259) bears as its base two scaly bracts called *glumes* or *empty glumes*.

(9) Lemma and **Palea:** Each flower in a spikelet of Gramineae (Fig. 259) bears a bract called *lemma* or the *flowering glume* which is sometimes *awned* at the tip, and a bracteole called *palea* which fits within the lemma.

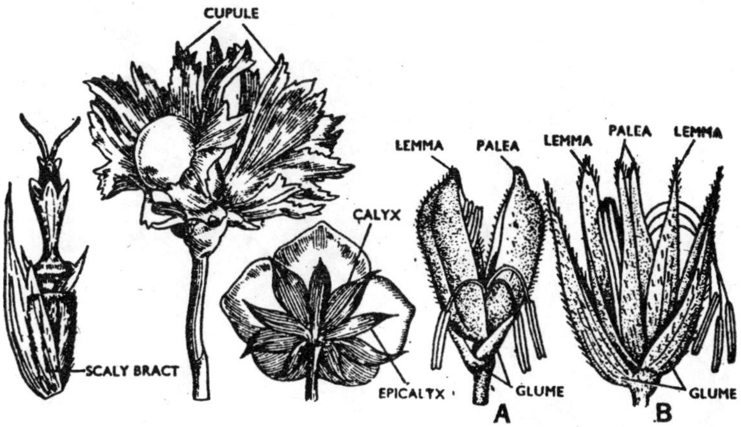

Fig. 256 Fig. 257 Fig. 258 Fig. 259

Fig. 256. Scaly bract of a disc floret of sunflower. Fig. 257. Cupule formed by husky bracts of hazel (*Corylus avellana*). Fig. 258. Epicalyx of strawberry (*Fragaria vesca*). Fig. 259. Spikelets of Gramineae showing glumes, lemma and palea. A. Single-flowered spikelet of rice. B. Two-flowered male spikelet of maize.

INFLORESCENCE

Like branching, inflorescence also may be **racemose** or **cymose.** It may also be of the **mixed** type when it is partly racemose and partly cymose. Within racemose and cymose again, if the rachis branches and the flowers are borne on the branches then the inflorescence is **compound.**

1. RACEMOSE OR INDEFINITE INFLORESCENCES

In this type the main axis of the inflorescence **(rachis)** grows indefinitely giving rise to younger and younger flowers in an *acropetal* order. The rachis never terminates in a flower, the apex being left free for further growth. Hence the other name *indefinite* for the racemose type. Sometimes the rachis, instead of developing into a long axis, is condensed to form a flat roundish structure called **receptacle.** On a receptacle the order of development of flowers is *centripetal*, i.e., towards the centre where the apex is actually located. Racemose inflorescences may be grouped and discussed as follows:

A. RACHIS ELONGATED

(*a*) *Flowers stalked*

1. Raceme: In a raceme the stalked flowers are borne acropetally on an elongated rachis. This type is quite common and may be seen in mustard (Fig. 260), radish, *Linaria* (Fig. 261), *Crotalaria*, *Cleome* and a large number of flowering plants.

Compound raceme or **Panicle:** When the axis of a raceme branches and the flowers are borne not directly on the axis but on its branches then the inflorescence is called a *panicle*. Such panicles are seen in *Yucca filamentosa* (Fig. 262), *Peltophorum ferrugineum* (Fig. 263), litchi (*Litchi chinensis*), gold mohur (*Delonix*), *Andrographis paniculata*, etc.

N. B. The term panicle is also used for a compound spikelet described on page 124.

(*b*) *Flowers sessile*

2. Spike: This is like a raceme but the flowers are without stalks. Among numerous common examples may be mentioned *Achyranthes* (Fig. 264), *Piper longum* (long

FIG. 260

FIG. 261

FIG. 262

FIG. 263

FIG. 260. Raceme of mustard. FIG. 261. Raceme of *Linaria striata*. FIG. 262. Panicle of *Yucca filamentosa* (after Schimper). FIG. 263. Panicle of *Peltophorum*.

pepper), tuberose (*Polyanthes*—Fig. 265) where each pair of flowers arises from the axil of a separate *spathe*, *Adhatoda vasica*, etc.

FIG. 264

FIG. 265

FIG. 267

FIG. 268

FIG. 264. Spike of *Achyranthes aspera*. FIG. 265. Spike of *Polyanthes* showing paired flowers at the axil of each spathe. FIG. 267. Spadix of *Colocasia esculenta*. FIG. 268. Compound spadix of coconut.

FIG. Spadix of *Amorphophallus titanum.* This Sumatran plant developed an 8 ft. 5 in. spadix with a spathe 4 ft. in diameter in eleven days out of a bud 3 ft. 6 in. high at the New York Botanical Garden. (Photo by courtesy Mr. H. W. Rickett, Prof. W. Robbins and the New York Botanical Garden.)

N.B. The inflorescence of *Polyanthes* (tuberose), although mentioned above, is not a typical spike as it is usually compound and shows variations.

Compound spike: If the axis is branched and sessile flowers are borne on these branches in the manner of a spike then the inflorescence is a *compound spike* as seen in *Amaranthus spinosus* and *Amaranthus viridis*.

N.B. The oat or wheat 'spike' is formed of *spikelets* described below.

3. Spadix: This is a variation of spike in which the rachis is thick and fleshy and the flowers are covered by one or more *spathy bracts*. A spadix may often be a huge structure as seen in the gigantic inflorescence of *Amorphophallus titanum* (Fig. 266). In arum (*Colocasia esculenta*—Fig. 267) the whole inflorescence is covered by a spathy bract. In this case the female flowers are borne towards the base of the axis. Above these are some sterile flowers, then the male flowers and, finally, the top of the axis is left bare as an *appendix*. The banana spadix (Fig. 293) is a mixed inflorescence and is considered later.

Compound spadix: This is formed when the axis of a spadix branches as in coconut (Fig. 268) and in many other palms. Usually the whole is covered by a stiff boat-shaped spathe but some palms may have smaller spathes or *spathellae* for separate segments of the spadix.

4. Catkin or **Amentum:** A catkin is a spike in which the axis is thin and weak. These are unisexual inflorescences which are usually deciduous, i.e., they drop off as a whole after fertilization or when mature. Examples are found in a number of families

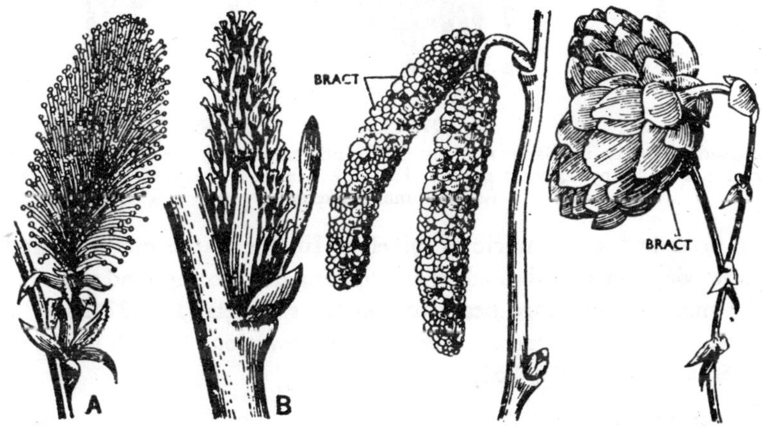

FIG. 269 FIG. 270 FIG. 271

FIG. 269. Catkins of willow (*Salix*). A. Male. B. Female. FIG. 270. Catkin of hazel (*Corylus*). FIG. 271. Strobile of hop (*Humulus*).

which are called *amentiferous*. These include oak (*Quercus*), willow (*Salix*—Fig. 269), birch (*Betula*), poplar (*Populus*), hazel (*Corylus*—Fig. 270), mulberry (*Morus*), *Trewia nudiflora*, *Croton*, *Acalypha*, etc. Catkins usually droop down forming a pendulous structure but sometimes they are erect as in the male and female catkins of *Salix*.

5. Strobile: This is a type of spike in which each flower is borne in the axil of a persistent membranous bract as found in the inflorescence of hop[1] (*Humulus lupulus*—Fig. 271).

6. Spikelet or **Locusta:** A spikelet is the ultimate component of a type of inflorescence found in the grasses (Gramineae) and the sedges (Cyperaceae). In a spikelet there are one to many flowers. The whole spikelet is subtended by two bracts or *glumes* at the base and, above it, every flower is borne in the axil of a bract called *lemma* and a bracteole called *palea*.

The ultimate inflorescences formed by the spikelets are complex and **compound.** **In rice,** each spikelet (Fig. 259A) is composed of one flower only and the whole inflorescence is branched like a panicle (Fig. 272) although different from a true panicle.

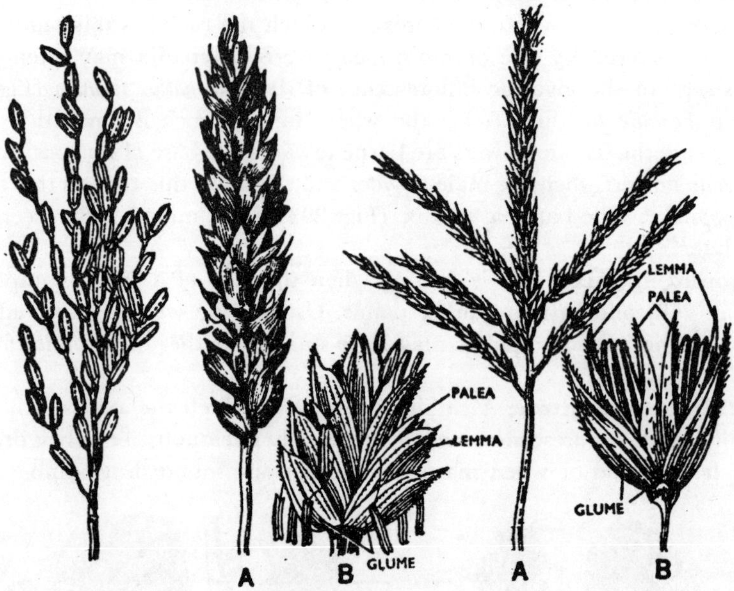

Fig. 272 Fig. 273 Fig. 274

Inflorescences formed by Spikelets. Fig. 272. Inflorescence of rice. Fig. 273. Wheat. A. Inflorescence. B. A many-flowered spikelet. Fig. 274. Maize. A. Male inflorescence. B. Two-flowered male spikelet.

This may be described as a panicle of spikelets. In wheat, the multiflowered spikelets (Fig. 273B) are borne on an unbranched axis looking like a compound spike (Fig. 273A). In maize the male inflorescence bears the paired spikelets (Fig. 274B) in a 'panicle' (Fig. 274A) while the paired female spikelets (the lower one of each pair being sterile) are arranged in a 'spadix' (Fig. 253).

B. RACHIS SHORTENED

7. Corymb: In a corymb the axis is not elongated to the same extent as in the previous inflorescences. The extent of shortening varies. Moreover, the stalks of the flowers are longer and longer as the flower is placed lower and lower on the axis so that all the flowers are placed almost at the same level. This is seen in different species of *Cassia* (Fig. 275). Young inflorescences of mustard look corymbose while they become racemes (Fig. 260) when mature. Similarly, the umbel inflorescence (Fig. 278) of cherry (*Prunus cerasus*) may become a corymb (Fig. 276) when mature.

Compound corymb: When the main axis branches and the flowers are arranged on the branches in a corymbose manner, the inflorescence is a *compound corymb* as seen in *Pyrus torminalis* (Fig. 277).

C. RACHIS SUPPRESSED

8. Umbel: The umbel differs from the corymb in that the axis is shortened further —almost to nil. The flowers (usually having stalks equal in length) then appear to form a cluster. It should be noted that the same inflorescence may change from umbel to corymb and from corymb to raceme due to the elongation of the rachis by growth. Umbels are found in cherry flowers (Fig. 278) and in *Centella asiatica* (Fig. 279) of Umbelliferae. In an umbel the central florets are younger and there are involucral bracts.

FIG. 275 FIG. 276 FIG. 277

FIG. 275. Corymb of *Cassia sophera*. FIG. 276. Corymb of cherry (*Prunus cerasus*) at maturity. FIG. 277. Compound corymb of *Pyrus torminalis*.

Compound umbel: When the axis of an umbel branches, one gets a compound umbel. Compound umbels are seen in the garden plant *Chaerophyllum* as also in the spices fennel (*Foeniculum*), coriander (*Coriandrum*), etc., and also in carrot (*Daucus carota*)—all belonging to the family Umbelliferae (Figs. 255 & 280). The involucres of the secondary umbels are called *involucels*.

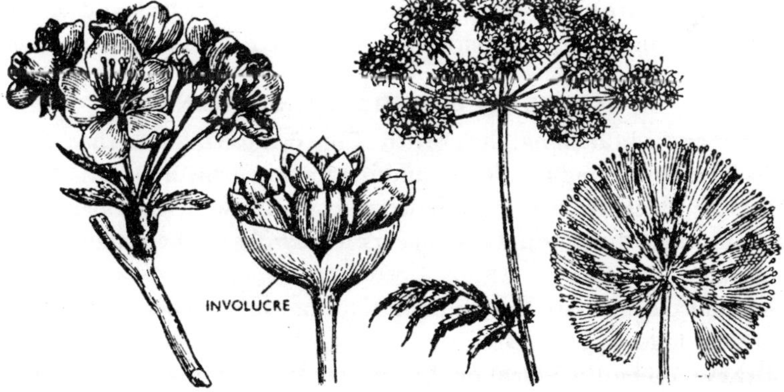

FIG 278 FIG 279 FIG 280 FIG 281

FIG. 278. Umbel of cherry when young. FIG. 279. Umbel of *Centella asiatica* with involucre of two bracts FIG. 280. Compound umbel of *Chaerophyllum temulum* (Umbelliferae). FIG. 281. Capitate inflorescence of *Albizzia lebbek*.

9. Capitate: When a large number of sessile flowers grow from a suppressed rachis giving rise to a more or less globose structure, it is termed a *capitate* inflorescence. This differs from *capitulum* (described below) in having no receptacle. This is seen in many leguminous plants like *Acacia, Mimosa, Albizzia* (Fig. 281), clover, etc.

D. RACHIS FLATTENED TO FORM RECEPTACLE

10. Capitulum or **Head** or **Anthodium:** This is the characteristic inflorescence of the family Compositae. The rachis forms a flattened, more or less convex *receptacle* on which the *florets* are arranged in a *centripetal* order. The whole capitulum is surrounded

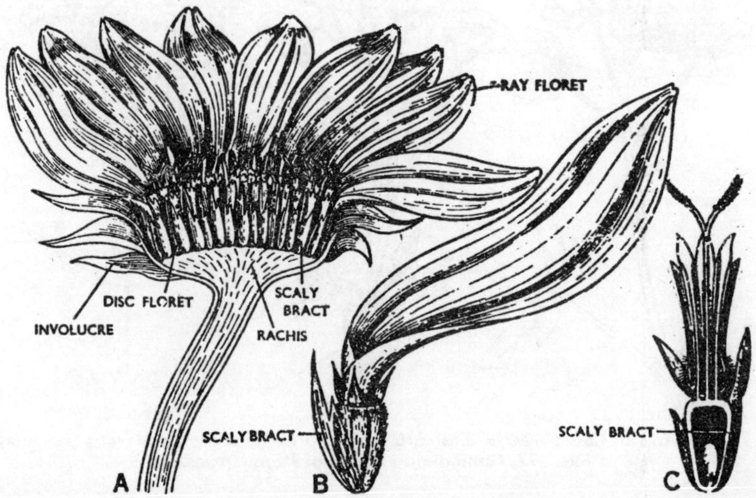

Fig. 282. A. Capitulum of sunflower. B. A ray floret. C. A disc floret (partly dissected).

by an *involucre* of *bracts* at the base and each floret usually has its individual *scaly bract.* In some capitula the peripheral or *ray florets* and the central or *disc florets* are different in appearance. Such difference is found in sunflower (Fig. 282) but there is no difference among chrysanthemum florets.

Compound capitulum: Here the main axis is branched and several capitula are included within the same involucre as in *Echinops echinata* of Compositae.

II. CYMOSE OR DEFINITE INFLORESCENCES

Just as in cymose branching, the growth of the cymose inflorescence is always checked vertically. The apical bud is a flower limiting further growth in that direction. As this apical bud is the oldest and younger flowers occur below it, the opening of flowers is *basipetal* (i.e., towards the base). This appears as *centrifugal* where the rachis is flattened to form a *receptacle.*

Cymose inflorescences are less common than racemose types. Inflorescences included herein are the following:

1. Solitary: When the apical or the axillary bud forms a single flower it does not form a real 'inflorescence' but this type is better included within the 'definite' group as further development is limited. In china-rose (Fig. 283) the stalk of the flower can be distinguished into a peduncle (axis) part and a pedicel (stalk) part with an articulation between the two.

2. Monochasial cyme or **Monochasium:** This is the same as the *uniparous* type of branching and may be *scorpioid* or *helicoid.* The main axis terminates in a flower and *one* lateral branch axis develops from its base also ending in a flower.

(a) In the **scorpioid** type the apex terminates in a flower and single lateral branch axes are given out alternately to the right and the left of it (Fig. 284A) as in scorpioid branching. Ultimate straightening out gives rise to an inflorescence as seen in *Heliotropium* of Boraginaceae or in *Ranunculus bulbosus* (Fig. 284B) of Ranunculaceae. Within scorpioid, again, two types are recognised—(i) in a **rhipidium** the lateral branches lie in the same plane as the main axis, while (ii) in a **cincinnus** the lateral branches lie in angular planes.

(b) In the **helicoid** type, each lateral branch develops on the same side forming a helix (Fig. 285A) as found in the Forget-me-not (*Myosotis palustris* of Boraginaceae—Fig. 285B), *Hamerocallis flava* of Liliaceae or *Hamelia patens* of Rubiaceae. Here again, there are two types—(i) **drepanium** with lateral branches on the same plane, and (ii) **bostryx** with lateral branches in different planes.

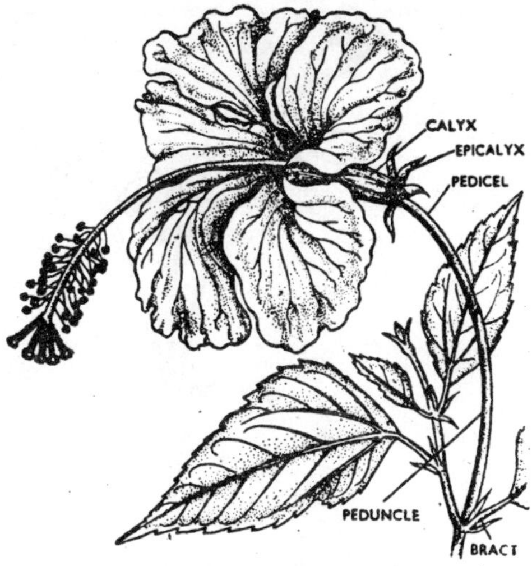

FIG. 283. Solitary flower of *Hibiscus rosa-sinensis*.

3. Dichasial cyme or **Dichasium:** This is the more normal cymose type where two lateral branches develop on the two sides of the terminal apical (oldest) flower. The

FIG. 284 FIG. 285

FIG. 284. Scorpioid cyme. A. General plan. B. Plant of *Ranunculus bulbosus* showing scorpioid arrangement of flowers. FIG. 285. Helicoid cyme. A. General plan. B. Inflorescence of Forget-me-not (*Myosotis palustris*) showing helicoid arrangement of ebracteate flowers.

lateral branches may again branch similarly after the manner of *biparous* cymose branching. Typical examples are found in jasmine (*Jasminum*), *Nyctanthes*, *Clerodendron* and pink (*Dianthus chinensis* of Caryophyllaceae—Fig. 286) flowers.

4. Polychasial cyme or **Polychasium:** These branch as in the multiparous cyme, there being more than two lateral branches from the base of the apical flower. An example may be found in *Calotropis* of Asclepiadaceae (Fig. 287) of Caprifoliaceae.

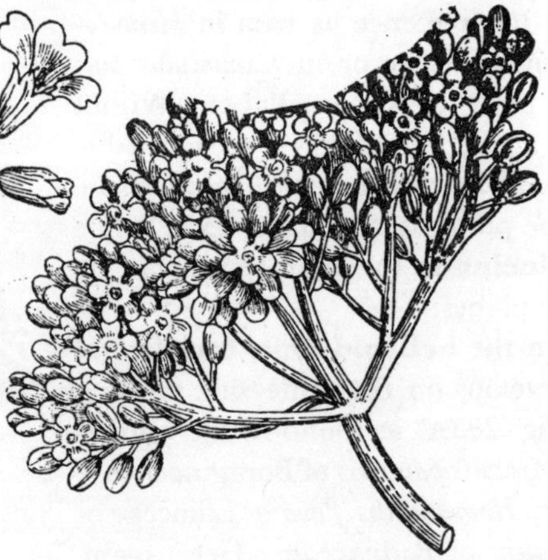

FIG. 286 FIG. 287

FIG. 286. Dichasium of *Dianthus chinensis*. FIG. 287. Polychasium of *Viburnum tinus*.

SPECIAL CYMOSE INFLORESCENCES

5. Hypanthodium. This is the typical inflorescence of *Ficus* or figs (Fig. 288). Here the cup-shaped *receptacle* is formed by the condensation of the rachis of three closely placed cymes. The receptacle forms an almost closed vessel with a small opening on the top. The flowers are arranged inside the receptacle in cymose groups.

6. Coenanthium: *Dorstenia*, belonging to the same family (Moraceae) as the figs, shows a different type of *receptacle*. Here the receptacle, instead of forming a closed structure, becomes saucer-shaped with the margins slightly curved up (Fig. 289).

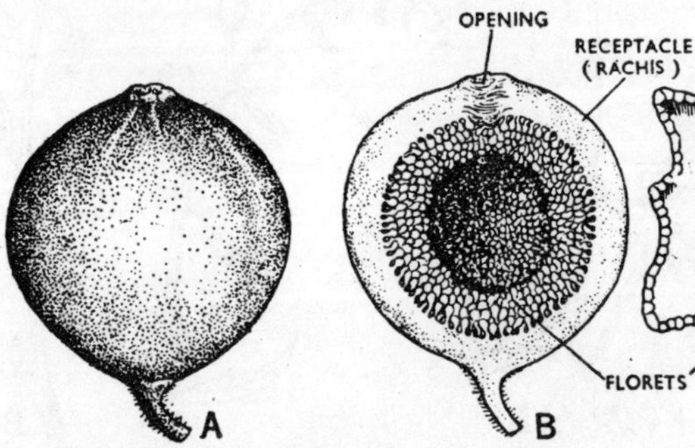

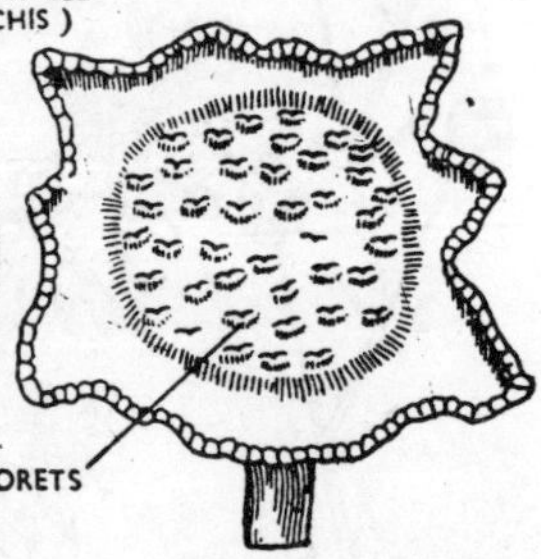

FIG. 288 FIG. 289

FIG. 288. Hypanthodium of fig (*Ficus cunia*). A. External view. B. L.s. showing florets inside receptacle formed by rachis. FIG. 289. Coenanthium of *Dorstenia cordifolia* showing florets on saucer-shaped receptacle.

7. Cyathium: The common garden *Poinsettia* (*Euphorbia*) *pulcherrima* (Fig. 250) shows this specialised cymose inflorescence which is covered by a cup-shaped green involucre formed by the union of bracts. The extremely reduced florets (Fig. 290) are placed on a convex receptacle. The central (i.e., locating the *top*) flower is the biggest and it is a female one showing only a single pistil. This is surrounded by a number of younger male flowers— each represented merely by a single stamen. Every floret has its individual stalk (pedicel) and is subtended in the axil of a thin and scaly bracteole. There are nectar glands on the involucre.

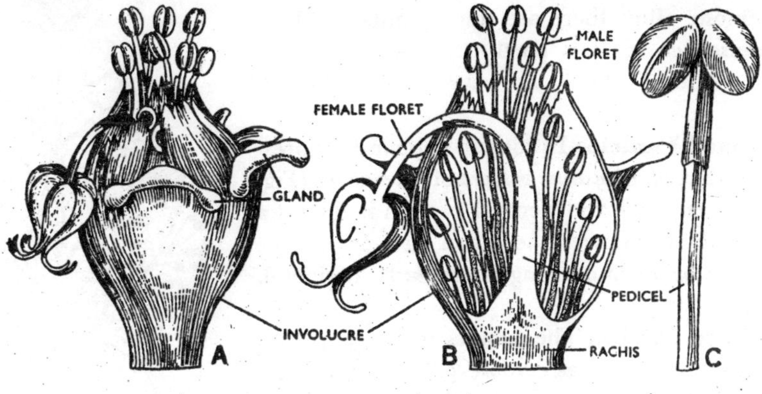

FIG. 290. Cyathium of *Poinsettia* (*Euphorbia*) *pulcherrima*. A. External view. B. L.s. showing male and female florets with bracteoles in between. C. A male floret formed of a single stamen.

8. Verticillaster: This is a complex inflorescence found in the Labiatae. There are opposite leaves in this family and one inflorescence develops from each of tne two opposite axils. Each inflorescence is a dichasial cyme which is reduced to two scorpioid cymes on two sides (Fig. 291A). As the sessile flowers cluster together, all of them (i.e., the two inflorescences together) appear to form a false whorl or *verticel* round the stem. Very common examples are *Leonurus sibiricus* (Fig. 291B) and *Leucas aspera*.

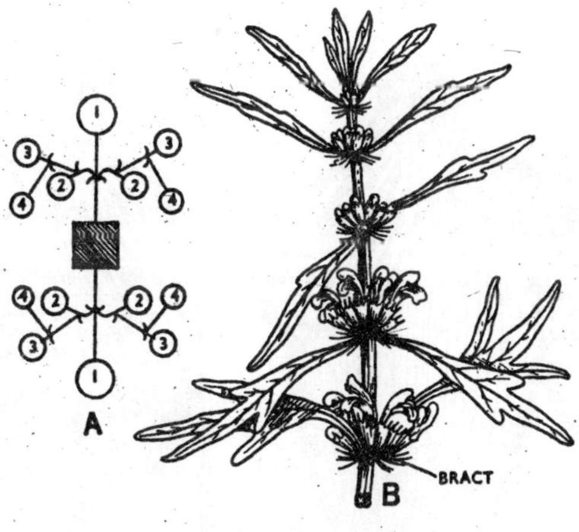

FIG. 291. Verticillaster. A. Ground plan.
B. Verticillasters in *Leonurus sibiricus*.

9. Cymose capitate: Clustering together of flowers on a suppressed rachis to form a capitate structure while the general plan is cymose is seen in plants like *Anthocephalus indica* of Rubiaceae.

III. MIXED INFLORESCENCES

There are quite a few instances when an inflorescence cannot be definitely assigned to either the racemose or the cymose group. Ordinarily, racemose inflorescences may show some cymose character. For example, the top apical inflorescence in flowering plants like dahlia, chrysanthemum, sunflower, etc., flower first and the lateral inflorescences lower down flower later. This is a cymose character in a plant where the inflorescences are otherwise racemose. But, there are some plants in which the inflorescence itself is partly cymose and partly racemose. Among these we may consider the following:

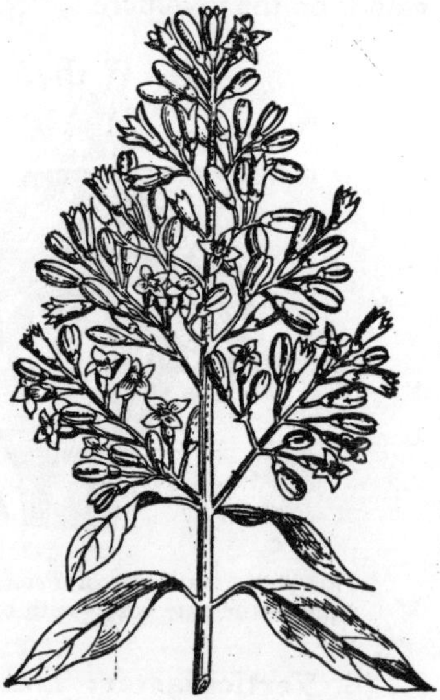

Fig. 292. Mixed panicle of *Ligustrum vulgare*.

1. Mixed Panicle: In some compound inflorescences like that of *Ligustrum vulgare* of Oleaceae (Fig. 292) it is seen that the general pattern is that of a panicle but the top flower opens first and in all the branches the order of opening is similarly basipetal, i.e., cymose.

2. Mixed Spadix: The banana *spadix* (Fig. 293) is a mixed inflorescence as cymose groups of flowers are arranged acropetally on the fleshy axis. Each cymose group is subtended by a spathe—an older spathe subtending the next younger. This spadix bends down by its own weight.

3. Cymose umbel: The inflorescence of *Allium cepa* (onion—Fig. 294) which looks like an umbel is found to be mixed on closer observation. The 'umbel' is formed by the grouping together of a number of smaller cymose inflorescences. This is also the case in *Calotropis*, *Crinum*, *Lantana*, etc.

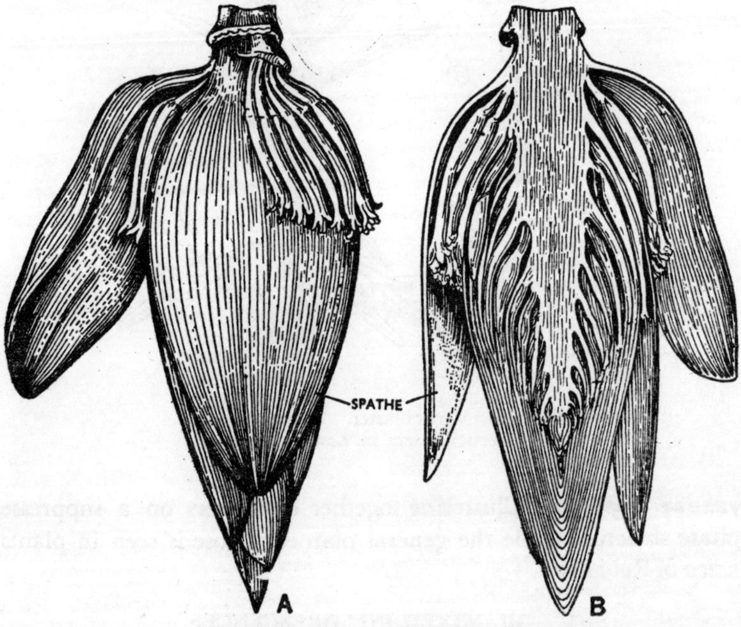

Fig. 293. Mixed spadix of banana. A. External view. B. L.s. of spadix.

4. Cymose corymb: This is seen in the inflorescences of *Ixora* (Fig. 295), *Alstonia scholaris, Holarrhena antidysentrica, Oldenlandia corymbosa,* etc., where a number of smaller

FIG. 294 FIG. 295 FIG. 296

FIG. 294. Mixed cymose umbel of onion. FIG. 295. Cymose corymb of *Ixora parvifolia*. FIG. 296. Thyrsus of grape.

cymose clusters are arranged in a corymbose manner on the axis bringing the clusters to the same level.

Fascicle is a special type of cymose corymb. In a polychasial inflorescence the successive axes are of equal lengths bringing the flowers to the same level in a corymbose manner. The flowers are very much clustered together. This is seen in some members of *Caryophyllaceae* and in *Garcinia* of Guttiferae.

5. Thyrsus: In this type a number of cymose clusters are borne acropetally on an unlimited axis in the manner of a raceme. This is seen in lilac and in grape vine (Fig. 296).

CHAPTER VIII

THE FLOWER

In the evolution of the land plants one finds a progressive specialisation of the reproductive structure until, in the flowering plants or *Phanerogams*, these reproductive structures or *sporophylls* are found to be borne in a specialised shoot called the **flower.** The flower is the most conspicuous structure in the Angiosperms and the evolution of the Angiosperms means, mainly, the evolution of this flower structure. The importance of the flower lies in its connection with reproduction which is a very manifest property of life.

PARTS OF A FLOWER

The different parts of a flowering shoot are best understood by referring to a typical flower (Fig. 297).

Like any shoot, there is an axis of the flower called the **thalamus.** On this thalamus the floral members are arranged as the leaves are arranged on any stem. Just like the phyllotaxy of leaves, the arrangement of the floral leaves on the thalamus may also be **spiral** or **cyclic.** A typical flower shows the *cyclic* or *whorled* type of **phyllotaxy.** Most common flowers show this *cyclic* arrangement in which the members of the consecutive whorls alternate with one another. Lotus, Magnolia, etc., show the *spiral* arrangement and these families are sometimes considered as more primitive. In some flowers like the rose, the floral phyllotaxy is partly spiral and partly cyclic when it is termed **spirocyclic** or **hemicyclic.**

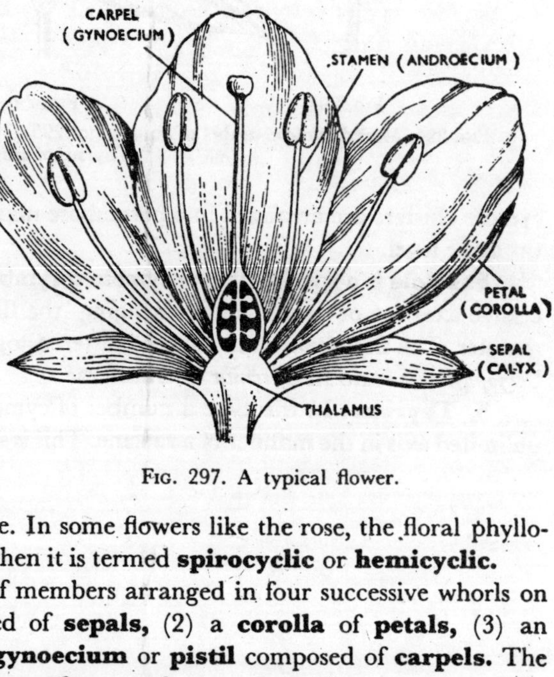

FIG. 297. A typical flower.

A typical flower shows four sets of members arranged in four successive whorls on the thalamus—(1) A **calyx** composed of **sepals,** (2) a **corolla** of **petals,** (3) an **androecium** of **stamens** and (4) a **gynoecium** or **pistil** composed of **carpels.** The two outermost whorls, *calyx* and *corolla*, serve for secondary purposes like protection of the flower and making it attractive. So, these two whorls are called the **accessory whorls.** On the other hand, the *stamens* are the **microsporophylls** and the *carpels* are the **megasporophylls** which are the male and the female sex organs essential for reproduction. Hence these two whorls, i.e., the *androecium* and the *gynoecium*, are the **essential whorls.**

In some flowers the *calyx* and the *corolla* cannot be distinguished, there being only one set of these *accessory* members which can neither be termed *calyx* nor *corolla*. Then this accessory outermost whorl is termed **perianth.**

While the above is the general plan of a flower there is a large number of flowers which do not conform to this scheme.

When all the four sets of members are present in a flower it is termed **complete;** if not, **incomplete.** When both the essential whorls, i.e., stamens and carpels, are present in a flower it is called **perfect** which is also **bisexual** or **hermaphrodite.** A flower missing either stamens or carpels is **imperfect** and also **unisexual** or **diclinous.** Unisexual flowers are either **staminate** (male) or **pistillate** (female). If male and female flowers develop on the same plant as in *Cucurbita maxima* or maize, the plant is called **monoecious.** Some Compositae bear hermaphrodite and female flowers on the same plant and these are called **gynomonoecious** while *Veratrum* of Liliaceae shows the **andromonoecious** condition bearing both hermaphrodite and male flowers on the same plant. If the male and the female plants are quite separate from one another, it is termed **dioecious** (also spelt *dioicous*) as seen in papaw (*Carica papaya*), fan palm (*Borassus flabellifer*), palwal (*Trichosanthes dioica*), etc. In thyme (*Thymus* of Labiatae) it is found that some plants are completely female while others bear hermaphrodite flowers. This condition is known as **gynodioecism.** Similarly, there are **androdioecious** species showing hermaphrodite plants and male plants. If staminate, pistillate and hermaphrodite flowers develop on the same plant (as in mango or litchi) it is called **polygamous.** On the other hand, plants like *Silene* (Caryophyllaceae) show separate plants of three kinds —male, female and hermaphrodite, so that they are **trioecious.**

There is a number of families with flowers showing neither sepals nor petals. Flowers are formed only by stamens and carpels—either both or only one (unisexual). Such flowers are called *naked* or **achlamydeous.** A flower having only one accessory whorl (calyx or corolla or perianth) is called **monochlamydeous** or **haplochlamydeous.** Ordinary flowers with both calyx and corolla are **dichlamydeous.** Pea and rose are dichlamydeous. Tuberose is monochlamydeous while plants belonging to the families Piperaceae, Salicaceae, etc., are achlamydeous. A flower having no functional stamen or carpel is **neuter** or *sterile* as are some ray florets of many Compositae flowers.

In the typical flower illustrated, the floral members are proportionally arranged round the thalamus so that the flower divides into two equal and symmetrical halves if one cuts the flower into two through any vertical plane passing through the axis. Such a flower as seen in rose, china-rose, etc., is termed **regular** or **actinomorphic.** On the other hand, there are some flowers where members of the different whorls are not arranged so uniformly. Here, the flower can be cut into two equal halves only through *one* particular vertical plane. Such flowers, as those of pea, bean, *Cassia*, *Ocimum*, etc., are **irregular** and **zygomorphic.** Some flowers are such that they cannot be cut into two symmetrical halves along any plane. Such a flower, as found in *Canna*, is termed **irregular** and **asymmetrical.**

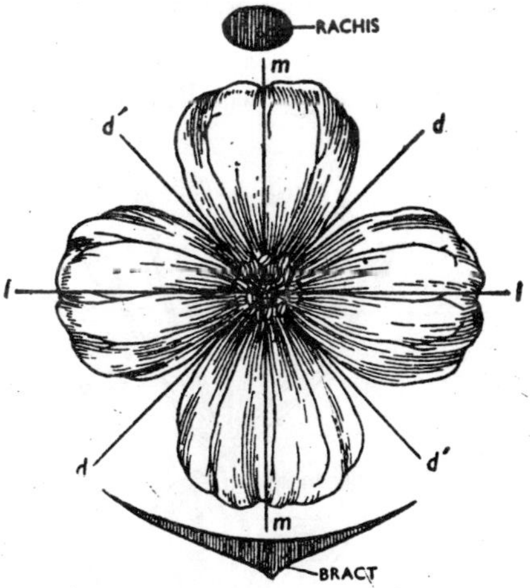

FIG. 298. Top-view of a flower showing its plan. *mm*=median plane; *ll*=lateral plane; *dd* & *d'd'* =diagonal planes.

In flowers, specially in irregular ones, it is seen that the flower is oriented on the stem in a definite way. The side of the flower which always faces the stem or the rachis is called the **posterior** or *back* while the side opposite, i.e., the side facing the subtending bract is the **anterior** or *front*. The plane passing through the axis of the stem, the back and the front of the flower and the bract is called *median* or the *antero-posterior plane* while the plane at right angles to it is the *lateral* or *transverse plane.* Similarly, there may be two *diagonal planes* (Fig. 298).

When the number of sepals, petals, stamens and carpels is uniform, i.e., the same or multiple of the same, the flower is termed **isomerous.** According to the number of floral members a flower may be **bimerous, trimerous, tetramerous** or **pentamerous** according as the number is a multiple of 2, 3, 4 or 5. When the number of members of different whorls is irregular, then the flower is termed **heteromerous.** *Trimery* is common among monocotyledons and *pentamery* among dicotyledons. It should be noted, however, that these numbers often vary. This variation is discussed later.

A typical flower has one whorl (cycle) of each of calyx, corolla, androecium and gynoecium. It is, therefore, **tetracyclic.** But, some flowers have two whorls of androecium. It is then called **pentacyclic.** Some dicotyledons have two whorls of sepals and are also pentacyclic. If the number of whorls increases further, then it becomes **polycyclic.** Thus wild rose and *Potentilla* with several whorls of androecium and gynoecium are polycyclic.

FLOWER AS A MODIFIED SHOOT

The *sporophylls* (i.e., stamens and carpels) are the essential parts in a flower so that the **flower** may be defined as a *shoot bearing sporophylls*. An ordinary shoot has an axial stem divided into nodes and internodes, the foliage leaves being borne at the nodes in an alternate or cyclic manner. The flower is exactly a similar shoot, only it has been modified for the special purpose of reproduction. Here the axial stem is the thalamus which is so much condensed that it is difficult to distinguish the nodes although they are there, and the specialised *floral leaves* (which again are modifications of leaves—Chapter IV), are borne on these nodes. The phyllotaxy is *spiral* in flowers like *Nymphaea* (Fig. 302) or *Magnolia* whereas it is *cyclic* in a vast number of flowers (Fig. 297). In the latter case, at the first node there is a whorl of sepals at the next node a whorl of petals, at the third node a whorl of stamens and at the apex a carpel or a whorl of them. The internodes are indeed small.

We may argue as follows to establish that *the flower is a modified shoot:*

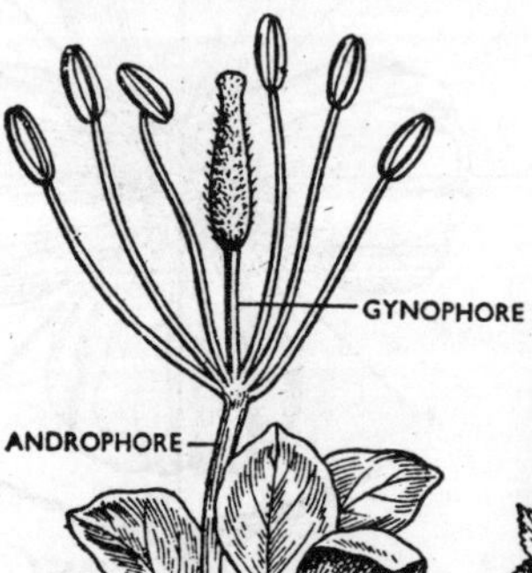

FIG. 299

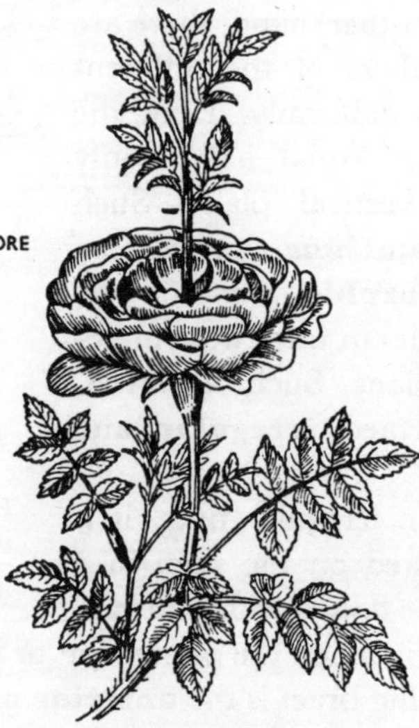

FIG. 300

FIG. 301

FIG. 299. Thalamus internodes (androphore and gynophore) in *Cleome gynandra*.
FIG. 300. Monstrous development in rose. FIG. 301. Monstrous development in pear.

1. Homology of the flower bud

(*a*) The flower bud has the same axillary or terminal position as that of a vegetative (shoot) bud and develops in the same way.

(*b*) Floral buds may sometimes be transformed into vegetative buds or bulbils as in *Agave* (Fig. 79), *Globba marantina* (Fig. 80), *Allium sativum*, etc.

2. Axis nature of thalamus

(*a*) The axis nature of the thalamus is not apparent as the internodes are so much condensed. But, in certain exceptional cases, the internodes do develop showing the stem character clearly. Such development of internodes may be seen in *Cleome gynandra* (Fig. 299) and *Capparis sepiaria* (Fig. 307) of Capparidaceae, *Passiflora suberosa* of Passifloraceae (Fig. 308), *Pterospermum acerifolium* of Sterculiaceae, etc.

FIG. 302

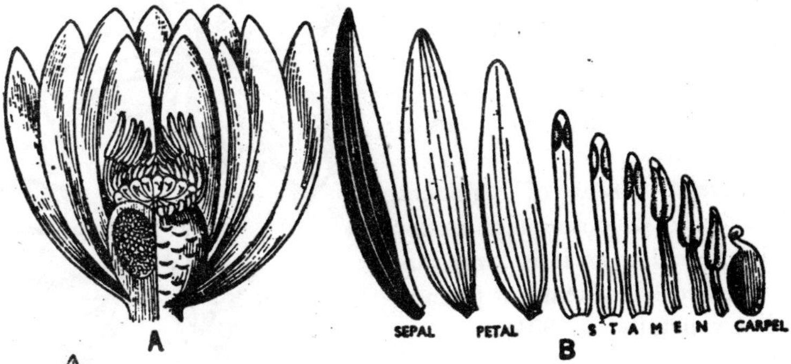

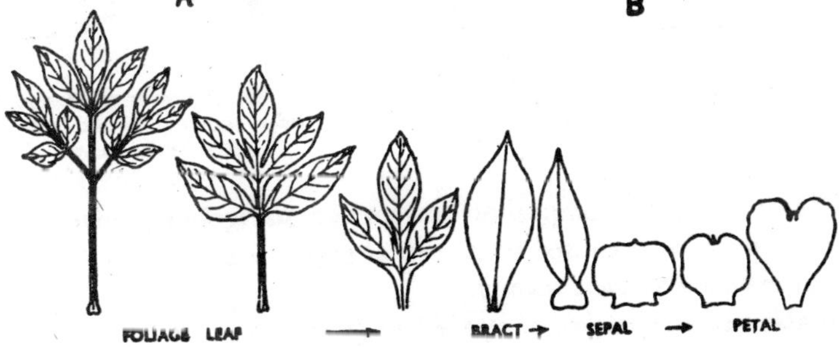

FIG. 303

FIG. 302. Transition of floral members in *Nymphaea alba*. A. Flower partially dissected. B. Dissected floral members showing gradual transformation. FIG. 303. Gradual transformation from foliage leaf to petal in *Paeonia officinalis*.

(*b*) The growth of the thalamus is limited by the carpels but sometimes the thalamus grows beyond the gynoecium and bears a leafy shoot or a flower above the first flower. Such *monstrous development* has been seen in rose (Fig. 300), pear (Fig. 301) and some other cases.

(*c*) In *Michelia champaca* of Magnoliaceae (Figs. 368 & 454A), *Polyalthia longifolia* of Annonaceae (Fig. 457B), etc., the region of the thalamus bearing carpels (arranged spirally) elongates like an ordinary stem giving rise to an aggregate fruit.

3. Leaf nature of floral members

The sepals, petals, stamens and carpels reveal their leaf nature in the following:

(*a*) The phyllotaxy of floral leaves on the thalamus (*cyclic* or *spiral*) strongly resembles that of foliage leaves on the stem. The floral leaves even show *ptyxis* and *prefoliation* (*aestivation*) in the same manner as foliage leaves.

(*b*) The four types of floral members (viz., sepals, petals, stamens and carpels) and foliage leaves often change from one to the other.

(i) Clear *gradual transition* of one type of floral member into another is seen in the flower of water-lily (*Nymphaea* sp.) where there is a large number of floral leaves arranged spirally on the thalamus (Fig. 302). The lowermost sepals are leafy green showing venation, these gradually merge into petals, petals gradually narrow down and merge into stamens developing anthers on the top. This need not signify that petals changed into stamens. More probably it is the stamens which became sterile and changed into petals. The carpels come last after the stamens.

(ii) In peony (*Paeonia* of Ranunculaceae) one can see transition from leaves to sepals and from sepals to petals showing the sameness of all these (Fig. 303).

(iii) In the green rose the sepals look like foliage leaves and even the petals are green and leafy.

(iv) In *Mussaenda frondosa* of Rubiaceae one of the sepals looks out of proportion in the calyx as it is large and vein-marked like a leaf but pigmented like a petal (Fig. 304).

FIG. 304. A flower of *Mussaenda frondosa* showing an odd leaf-like sepal in the calyx.

(v) In many flower species one finds 'single' and 'double' varieties. *Doubling* is usually the transformation of some stamens into petals as one sees in rose, china-rose, etc.

N.B. The doubling of Compositae flowers like marigold is, however, different. Doubling may also be caused by *chorisis* as explained later.

(vi) *Canna* flowers show ***petaloid stamens***.

(vii) Carpels also become ***petaloid*** or ***sepaloid*** in *Zinnia*, etc.

(viii) In certain flowers like the legumes (peas, etc.—Figs. 379 & 380), the carpel (specially the ovary part) clearly looks like a folded green leaf.

(*c*) While buds are developed in the axils of leaves, this does not ordinarily happen in the case of floral leaves. But, in some cases of ***monstrous development***, buds are developed at the axils of floral leaves (e.g., petal) and these buds may develop into secondary flowers so that one flower develops above another. This is called ***prolification***. In a similar way a carpel may develop at the axil of a carpel causing the development of one papaw fruit within another.

(*d*) In the lower groups like the Pteridophytes and the Gymnosperms, the sporophyll is clearly seen to be an ordinary leaf, modified or not, bearing sporangia.

THALAMUS

Thalamus (also called **torus**) is the axis of the floral shoot which is the direct prolongation of the *pedicel* and bears four sets of floral members. Usually, it is a slightly swollen knob-like structure but sometimes it may be somewhat elongated and conical bearing floral leaves spirally as in *Artabotrys* (Fig. 305), *Michelia* (Fig. 368), etc., or like an inverted cone with a spongy flat top as in lotus (Fig. 306), or it may be slightly convex as in brinjal, etc. In perigynous and epigynous flowers the thalamus becomes flat or cup-shaped as explained under leaf insertion.

The thalamus of most flowers has three internodes which are very much condensed. But, in certain flowers these internodes are elongated. The first internode (between calyx and corolla) is called the **anthophore** (*anthos*=flower), the second internode (between corolla and androecium) is the **androphore** (also called **gynandrophore**) and the third internode (between androecium and gynoecium) is called the **gynophore.** *Cleome gynandra* (Fig. 299) shows both androphore and gynophore (gynandrophore). *Capparis sepiaria* (Fig. 307) and *Pterospermum acerifolium* show gynophores only. Passion-flower (*Passiflora suberosa* of Passifloraceae) shows an androphore (Fig. 308).

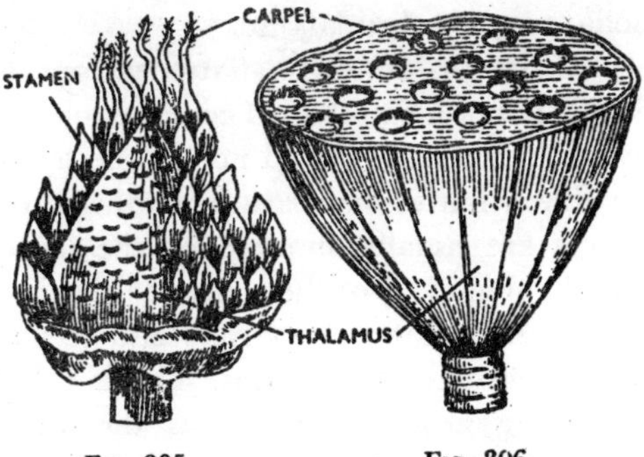

FIG. 305 FIG. 306

FIG. 305. Conical thalamus of *Artabotrys odoratissima* of Anonaceae. FIG. 306. Thalamus of lotus (*Nelumbium speciosum* of Nymphaeaceae).

The thalamus is terminated by the pistil except in cases of monstrous development. But, in the family Umbelliferae the thalamus normally projects into the ovary and the

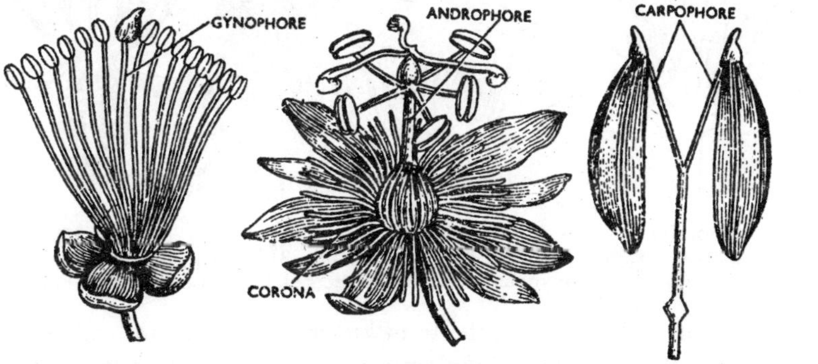

FIG. 307 FIG. 308 FIG. 309

FIG. 307. Gynophore in *Capparis sepiaria*. FIG. 308. Androphore in *Passiflora suberosa*. FIG. 309. Carpophore in *Foeniculum vulgare*.

carpels remain attached to it separating at maturity. This prolongation of the thalamus is called the **carpophore** (Fig. 309). This is seen in the fruits of anise, *Foeniculum*, *Coriandrum*, etc.

Any portion of the thalamus may be modified in certain flowers into characteristic structures called **discs.** The discs often bear nectar glands. Thus, the family Rutaceae (orange, lemon, etc.) has fleshy annular discs surrounding the bases of the pistils (**Fig.** 310). In grape vine (*Vitis* sp. of Vitaceae) the disc is formed of some scaly structures (**Fig.** 311). The disc may be *hypogynous*, *perigynous* or *epigynous* according as it is below, around or on the top of the ovary.

INSERTION OF FLORAL LEAVES ON THE THALAMUS

In a typical flower the thalamus is a slightly swollen (conical, convex, etc.) structure with the calyx, corolla, androecium and gynoecium developed successively on it. It is terminated by the pistil so that the gynoecium is on the top (Fig. 312). Such a flower is **hypogynous** (i.e., other members below gynoecium). The ovary here is *superior* and other members *inferior*. This condition may be seen in many common flowers like magnolia, mustard, brinjal, etc.

If the thalamus be flattened out to form a flat or a cup-shaped concave top, the gynoecium will be placed not on the *top* of the flower but at the *centre*. The other three whorls are usually inserted on the rim of the flat (Fig. 313) or the concave (Fig. 314) thalamus. Such an ovary may also be partially sunken in the thalamus. This is called the **perigynous** condition as the other members surround the gynoecium. Examples of the first (i.e., thalamus more or less flat) type may be found in the peas, strawberry, etc., while the second type (thalamus forming a concave *receptacle*) is found in rose, etc. Here also the gynoecium is usually described as *superior* although it is not actually

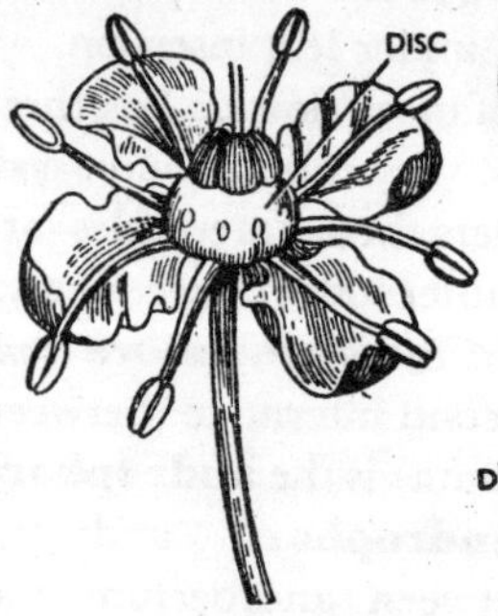

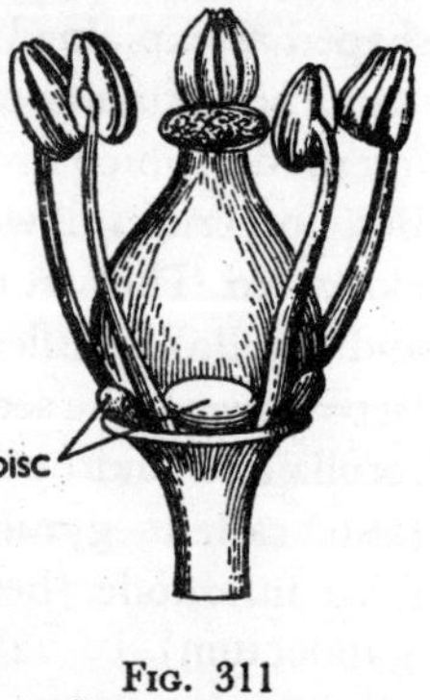

FIG. 310 FIG. 311
FIG. 310. Disc in *Ruta* of Rutaceae.
FIG. 311. Scaly disc in grape (*Vitis* of Vitaceae).

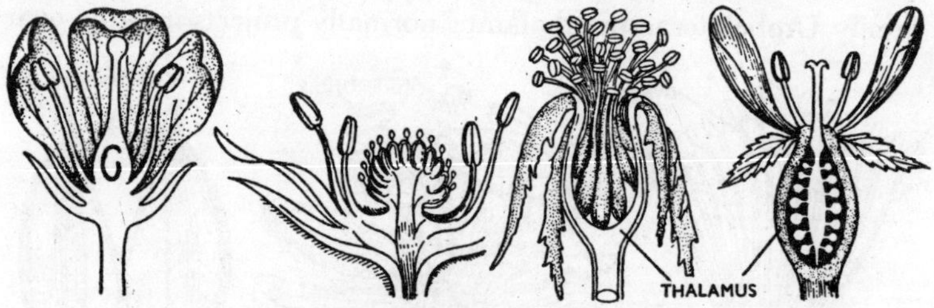

FIG. 312 FIG. 313 FIG. 314 FIG. 315
INSERTION OF FLORAL LEAVES. FIG. 312. Hypogynous. FIG. 313. Perigynous with flat receptacle in strawberry. FIG. 314. Perigynous with concave receptacle in rose. FIG. 315. Epigynous.

on the *top* of the flower. Sometimes, the term *half-inferior* is used instead of superior.

In a third type the cup-shaped thalamus as seen in Fig. 314 may fuse with the wall of the ovary giving the appearance as in Fig. 315. It now appears that the gynoecium is the lowest or the *inferior* whorl of the flower while all other members are *superior*, i.e., the condition is **epigynous.** Epigyny may be seen in the unisexual female flowers of cucurbits and in plants like banana, guava, sunflower, etc. It should be noted that in an epigynous flower the wall of the ovary has also got the thalamus fused with it. Thus, any fruit developed out of this ovary has this thalamus tissue on the outside. This statement is further strengthened by the fact that in some gooseberry one or two foliage leaves are sometimes seen to develop on the fruit wall. This is possible because of the stem nature of the thalamus.

CALYX

The lowermost axillary whorl is the **calyx.** An *epicalyx* (Figs. 248 & 283) sometimes develops below the calyx but it is formed of bracteoles and is not actually a part of the flower. **Sepals** forming a calyx are usually green, somewhat resembling abnormal leaves. When the flower is in the bud stage it is protected by the calyx and, to be able to perform this function better, it is sometimes quite thick and hardy although it may show veins and stomata like ordinary leaves.

Like the entire flower the calyx may be **regular** or **irregular.** The sepals may be quite free from one another when the calyx is termed **polysepalous** or **dialysepalous** as seen in mustard or water-lily flowers. In other flowers the sepals are wholly or partially united with one another. Such a calyx is **gamosepalous.** There may be different degrees of union. It may involve only the bases of the sepals (*partite*), or the fusion may extend up to the middle (*cleft*), or the union may extend almost up to the top when the sepal tops are seen as some teeth as in *Datura* and the condition is described as *toothed*. In the extreme case the fusion is complete (*entire*), the individual sepals being no longer distinguishable. In a united calyx the united portion is called the *tube* and the free portion the *limb*. Sometimes the number of teeth on the limb does not correspond to the number of sepals as the teeth may be further subdivided. Forms of **regular gamosepalous** calyx may be *tubular* (*Datura*), *bell-shaped* (china-rose), *urceolate* (*Hyoscyamus*—Fig. 316), etc., which are the same as explained in connection with corolla. **Irregular gamosepalous** calyx also may similarly be *bilabiate* (*Leonurus* and *Leucas*—Fig. 317), etc.

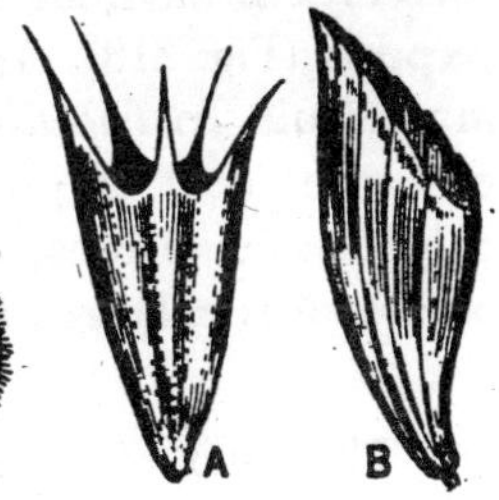

Fig. 316 Fig. 317
Gamosepalous calyx.
Fig. 316. Urceolate in *Hyoscyamus*.
Fig. 317. Bilabiate in A. *Leonurus*; B. *Leucas*.

Sepals are usually devoid of stalks (i.e., *sessile*) and with entire margins. Incised sepals may, however, be seen in rose (Fig. 314).

MODIFICATIONS

The calyx may sometimes be modified or have some appendages. The sepals may be **petaloid**, i.e., petal-like as seen in *Sterculia roxburghii* with a scarlet calyx and also in

Fig. 318 Fig. 319

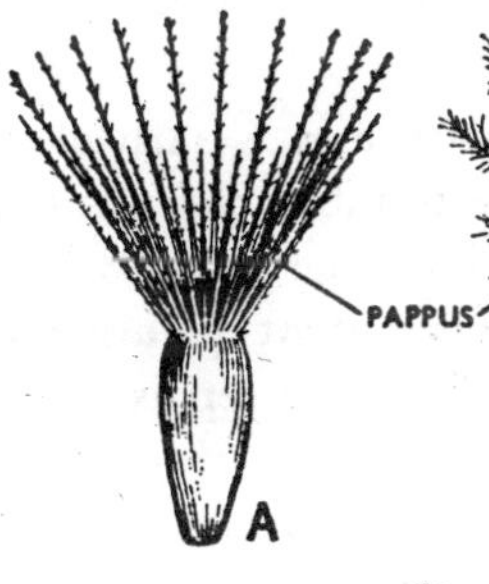

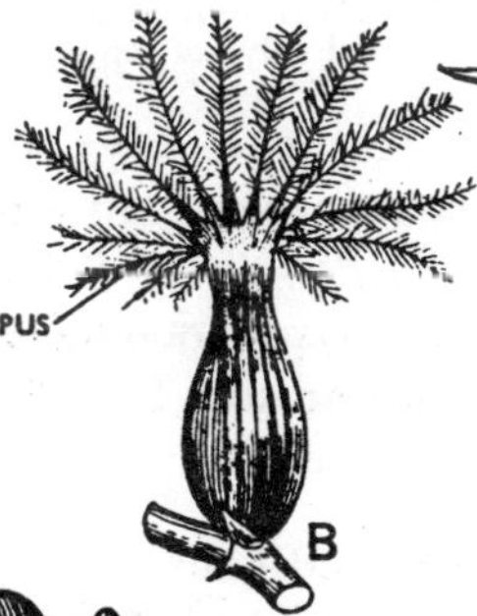

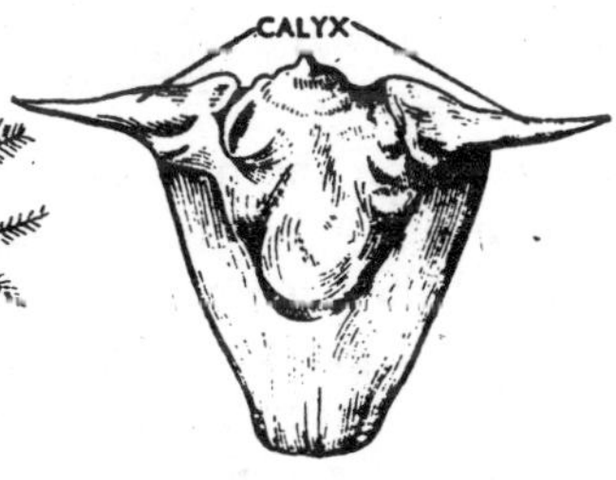

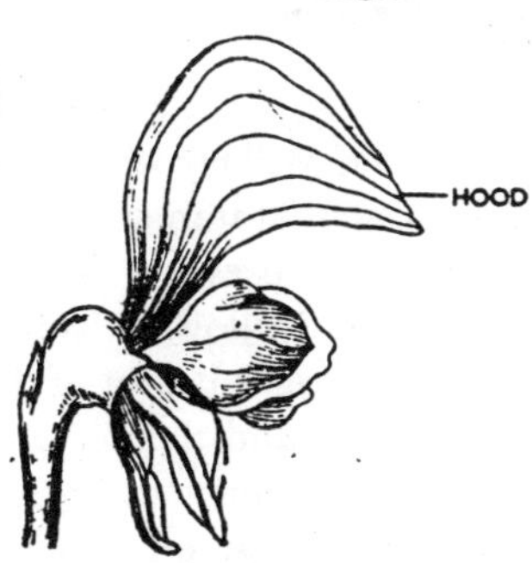

Fig. 320 Fig. 321

Fig. 318. Pappus in A. *Mikania scandens* (Compositae); B. *Valeriana* (Valerianaceae). Fig. 319. Spinous calyx in *Trapa bispinosa*. Fig. 320. Spurred calyx of *Impatiens balsamina*. Fig. 321. Hooded calyx of monkshood (*Aconitum napellus*).

Saraca asoca and *Caesalpinia pulcherrima* of Caesalpinieae, etc. In *Mussaenda* (Fig. 304), while four sepals are normal the fifth one is large and petaloid although showing vein marks. In the family Compositae and the allied families Dipsaceae and Valerianaceae, the calyx is modified into two or more hairy, scaly or feathery structures forming the **pappus** (Fig. 318). In *Trapa bispinosa* (water chestnut—Fig. 319) the calyx is **spinous** in the fruit. In *Impatiens balsamina* (Fig. 320), *Tropoeolum majus*, *Delphinium* (larkspur), etc., the sepal is prolonged downwards into a tubular process called *spur* and the calyx is said to be **spurred.** In *Aconitum* (monkshood of Ranunculaceae—Fig. 321) one sepal is enlarged forming a **hood** over the flower.

DURATION OF THE CALYX

The calyx of some flowers drops off as soon as the flower opens as in poppy (Fig. 322), *Argemone mexicana*, etc. This type of calyx is termed **caducous** or **fugacious.** In many flowers the calyx is, **deciduous,** i.e., it falls off along with the petals just after fertilization as in mustard, etc. In brinjal, *Datura, Ocimum, Trapa* (Fig. 319), etc., the calyx is **persistent** and remains attached to the ripe fruit. In inferior fruits like guava, the

FIG. 322

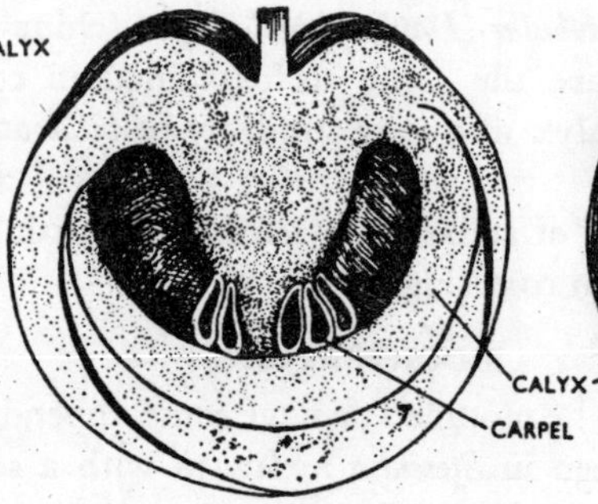

FIG. 323

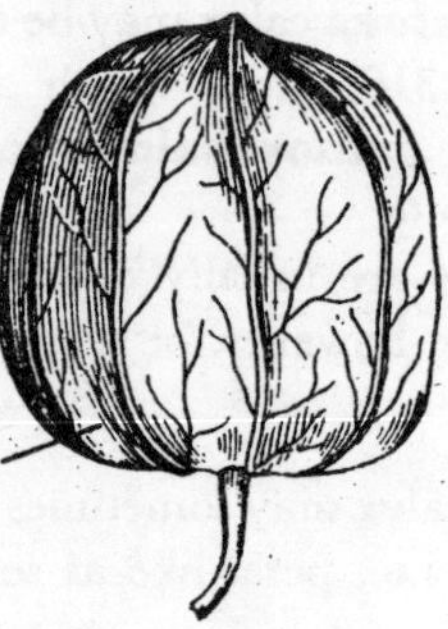

FIG. 324

FIG. 322. Caducous calyx of poppy. FIG. 323. Persistent edible calyx of *Dillenia indica*. FIG. 324. Persistent inflated calyx of *Physalis*.

calyx often persists on the fruit. If a persistent calyx assumes a shrivelled dried-up appearance as in guava, it is termed **marcescent.** In a number of fruits like those of *Dillenia indica* of Dilleniaceae (Fig. 323), *Physalis* of Solanaceae (Fig. 324), *Shorea* of Dipterocarpaceae (Fig. 443), etc., the calyx is not only persistent but also grows in size along with the fruit. Then it is called **accrescent.** In *Dillenia* the sepals are fleshy containing much acid and mucilage. These are edible.

COROLLA

Corolla is the second axillary whorl composed of **petals.** Like the calyx, the corolla also protects the more essential stamens and carpels within. The petals are usually brightly coloured because of the presence of water-soluble *anthocyanin* (red, orange, violet, blue, etc.) and *anthoxanthin* (yellow to ivory white) pigments or the *carotenoids* contained in the chromoplasts. The bright colour of the petals combined with the scent of essential oils present in some flowers and the nectar secretion of special glands make the flowers highly attractive to insects which act as agents for pollination.

Petals may sometimes be green or have some dull colour like the sepals when they are termed **sepaloid** as in *Annona, Polyalthia, Artabotrys,* etc., of Annonaceae or in the green rose. Petals are usually smooth but sometimes the surface may be hairy. Though usually thin, sometimes they may be thick fleshy structures. In a petal the lower part is usually narrow like the stalk of a leaf and is called the **claw** or *unguis*. When a claw is present the petal is called *clawed* or *unguisculate*. The expanded portion of the petal is the **limb** (Fig. 325). A clawless petal is *sessile*. A petal resembles the lamina not only in this but also in other characteristics. It may be shaped *linear*, *oblong*, etc., and its margin may be *entire*, *dentate*, *serrate*, etc., just as in the leaf lamina. In some flower petals, as in *Dianthus* (Fig. 331–2), the margin is deeply slashed or divided giving it a frilled appearance when it is called *fimbriated* or *fringed*.

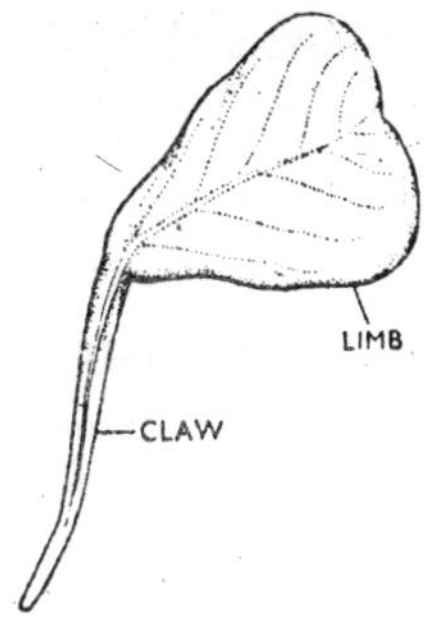

FIG. 325. A petal showing limb and claw.

The corolla may be **regular** or **irregular.** It is **polypetalous** or **dialypetalous** if the petals are free from one another and **gamopetalous** or **sympetalous** if there is any degree of cohesion. When united, the united portion is the **corolla tube** and the free portion above shows the **corolla lobes.** The junction of the tube and the lobe is called the **throat.**

Considering its duration, the corolla is **caducous** (e.g., grape vine) or, more commonly, **deciduous.** Very rarely it is **persistent,** as in heather, where it remains in a dry and shrivelled up *marcescent* form.

Like the calyx, the corolla also may have **appendages.** In *Antirrhinum* (snapdragon —Fig. 326) the tube of the corolla is slightly dilated on one side forming a pouch. This

FIG. 326

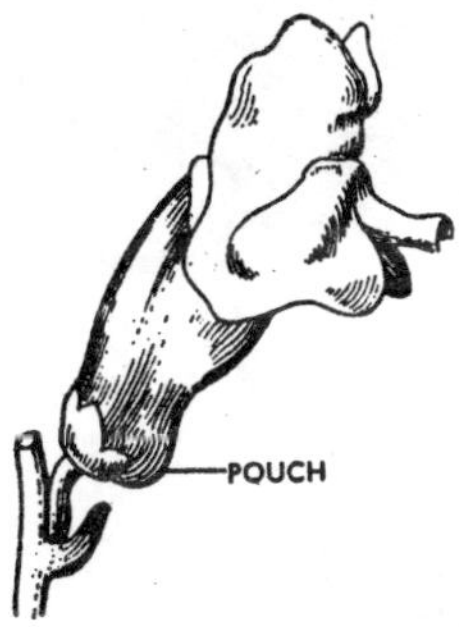

FIG. 327

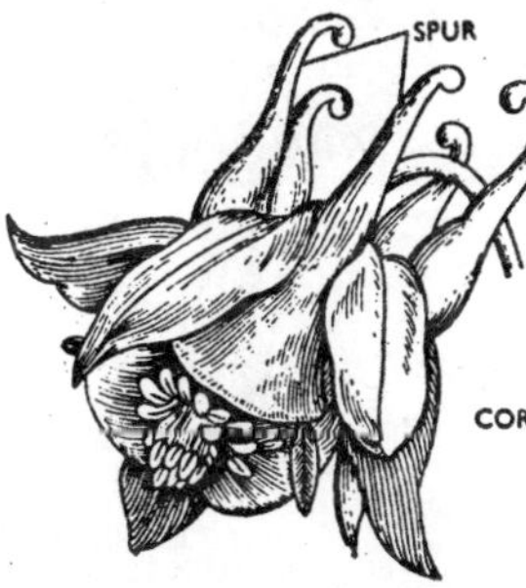

FIG. 328

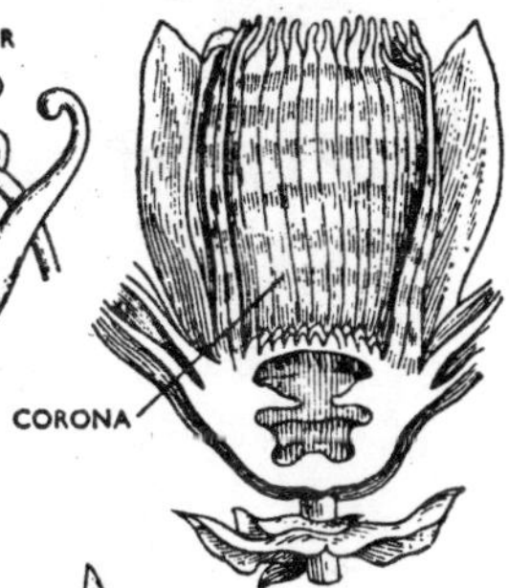

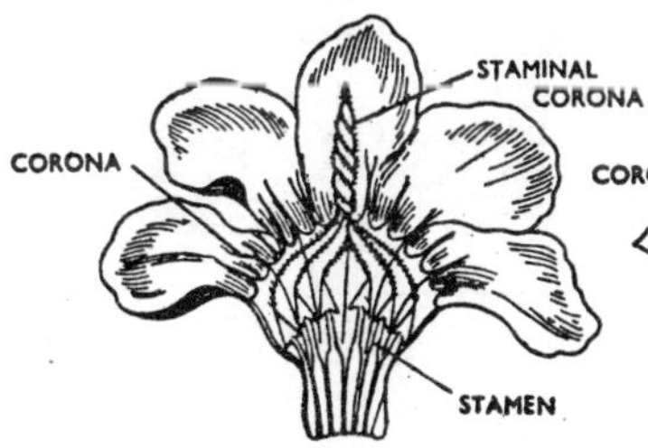

FIG. 329

FIG. 330

FIG. 326. Saccate corolla of *Antirrhinum*. FIG. 327. Spurred petals in *Aquilegia vulgaris*. FIG. 328. Corona developed on corolla of *Passiflora*. FIG. 329. *Nerium odorum* showing corona developed on corolla as well as staminal corona developed by staminal appendages. FIG. 330. Tubular corona developed on corolla of daffodil (*Narcissus pseudonarcissus*).

condition is termed, **saccate** or **gibbous.** In some cases one or more petals, or the tube itself is prolonged downwards, forming a *spur* which usually stores nectar. Such a flower is called **spurred.** There may be only one spur as in pansy or there may be several spurs as in *Aquilegia vulgaris* (Fig. 327) where each petal is spurred. Sometimes appendages of different kinds, such as scales, hairs, etc., develop from the inner wall of the throat. This gives rise to what is known as the **corona.** Beautiful coronas are seen in passion flower (Fig. 328), oleander (Fig. 329), etc. The corona of daffodil (Fig. 330) is a united tubular structure.

FORMS OF COROLLA

According to the nature of cohesion, shape, etc., corollas of the following types are usually met with:

A. POLYPETALOUS OR DIALYPETALOUS

(*a*) *Regular forms*

1. Cruciform: Four free clawed petals are arranged in the form of a cross as in the Cruciferae family, e.g., mustard (Fig. 331–1).

2. Caryophyllaceous: This is formed by five free clawed petals with limbs at right angles to the claws as in the family Caryophyllaceae, e.g., the pink flower (*Dianthus*—Fig. 331–2).

3. Rosaceous: There are five sessile (or with very short claws) petals with the limbs spreading outwards. The odd petal is anterior. This is found in the family Rosaceae (e.g., wild rose—Fig. 331–3) and also in tea (*Thea chinensis*). Most cultivated roses however show more petals because of reasons explained on page 158.

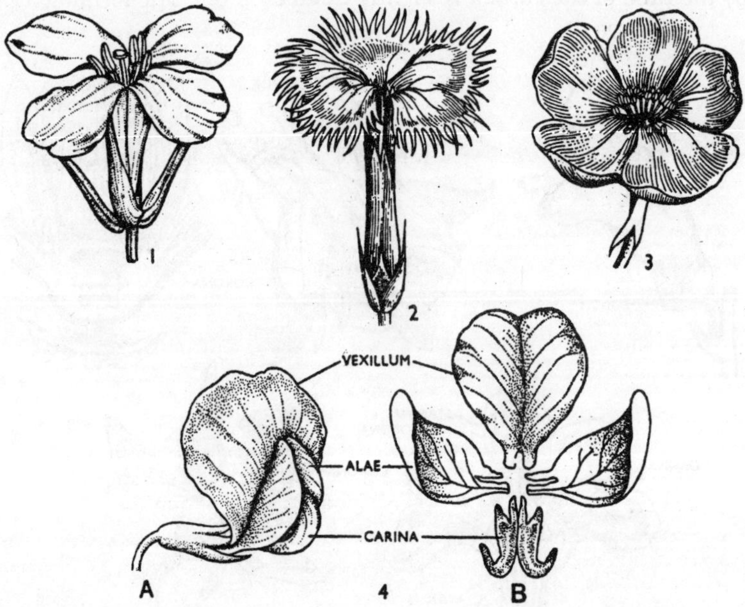

FIG. 331. POLYPETALOUS COROLLA FORMS. 1. Cruciform in mustard (*Brassica nigra*). 2. Caryophyllaceous in pink (*Dianthus chinensis*). 3. Rosaceous in wild rose. 4. Papilionaceous in pea. A. Complete flower. B. Corolla dissected to show arrangement and forms of petals.

(*b*) *Irregular forms*

4. Papilionaceous: Here five free petals resemble a butterfly (*papilion*=butterfly). The posterior superior petal is larger than the others and is termed the *vexillum* or *standard;* two anterior inferior ones are usually more or less united forming a boat-shaped structure and are called the *carina* or *keel.* Two lateral petals called *alae* or *wings* overlap the carina and are themselves overlapped by the vexillum. This is a characteristic of the subfamily Papilionaceae, e.g., the pea flower (Fig. 331–4).

B. GAMOPETALOUS OR SYMPETALOUS

(*a*) *Regular forms*

1. Tubular: The corolla tube is nearly cylindrical throughout and the limbs are not spreading. The central florets of most Compositae are tubular, e.g., sunflower disc florets (Fig. 332–1).

2. Campanulate or **bell-shaped:** Corolla tubes are rounded at base gradually widening upwards like a bell. This may be seen in the family Campanulaceae and in many cucurbits (Fig. 332–2), in *Physalis*, etc.

3. Infurdibuliform or **funnel-shaped:** The corolla resembles an inverted cone like a funnel as seen in *Datura* or many plants of the family Convolvulaceae like *Ipomoea pulchella* (Fig. 332–3).

4. Hypocrateriform or **salver-shaped:** The corolla tube is long and narrow with the limb placed at right angles to it as seen in *Vinca rosea* of Apocynaceae (Fig. 332–4).

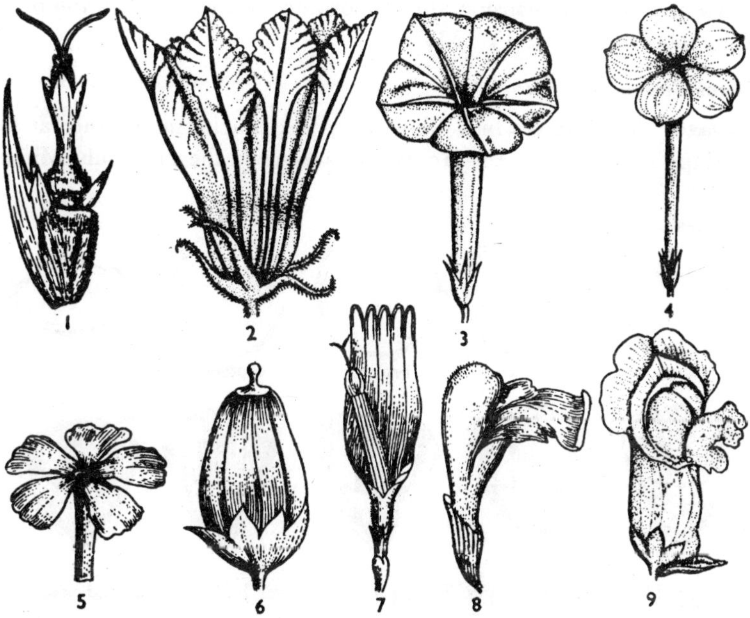

FIG. 332. GAMOPETALOUS COROLLA FORMS. 1. Tubular in sunflower disc floret. 2. Campanulate in *Cucurbita*. 3. Infundibuliform in *Ipomoea*. 4. Hypocrateriform in *Vinca*. 5. Rotate in *Nyctanthes*. 6. Urceolate in *Kalanchoe* . 7. Ligulate in marigold ray floret. Note 5 lobes signifying 5 petals. 8. Bilabiate in *Leucas*. 9. Personate in snapdragon.

5. Rotate or **wheel-shaped:** Here the tube is shorter than in hypocrateriform while the limb is at right angles to it as seen in brinjal (*Solanum melongena*), *Nyctanthes arbor-tristis* of Oleaceae (Fig. 332–5), etc.

6. Urceolate or **urn-shaped:** Corolla tube is swollen in the middle tapering towards both base and apex as in *Kalanchoe pinnata* of Crassulaceae (Fig. 332–6).

(*b*) *Irregular forms*

7. Ligulate or **strap-shaped:** Five petals unite to form a short tube at base which splits on one side and becomes flattened like a strap above as seen in many Compositae, e.g., the ray florets of marigold (Fig. 332–7) where one can see that the strap is formed by the union of five petals.

8. Bilabiate or **labiate** or **two-lipped:** This irregular corolla is united in such a way that the limb is divided into an upper posterior part (usually formed by the union of two petal lobes) and an unequal lower anterior part (usually formed by the union of three petal lobes) with the mouth gaping wide open. This is characteristic of the family Labiatae (e.g., *Leucas*—Fig. 332–8) and is also seen in some allied families (e.g., *Hygrophila* and *Adhatoda* of Acanthaceae).

9. Personate or **masked:** It resembles the bilabiate but the two lips are placed so close together that the mouth is closed. The projection of the lower lip closing the mouth is called the *palate.* This type of corolla may be seen in snapdragon (*Antirrhinum*—Fig. 332–9) or *Lindenbergia.*

AESTIVATION

Aestivation is the arrangement of either the sepals or the petals in a flower bud with relation to one another as *vernation proper* is the arrangement of young leaves in the vegetative bud. Aestivation is of considerable importance in the classification of plants. This may be of the following types (Fig. 333):

(*a*) *Floral leaves in one whorl*

1. Valvate: Floral leaves in a whorl may just touch one another at the margins without overlapping as seen in the family Anonaceae or the subfamily Mimoseae.

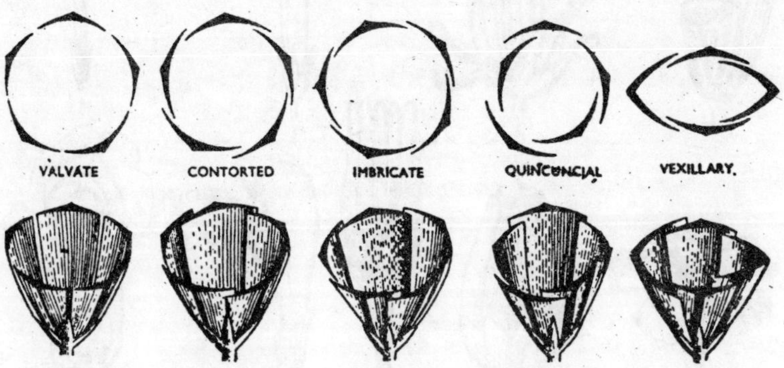

FIG. 333. Aestivation types. T.s. on top and lower half of flower bud below.

2. Contorted or **twisted:** When overlapping is regular in one direction so that one margin overlaps the next member on one side while its other margin is overlapped by the one before, giving a twisted appearance to the bud. This is seen in Malvaceae (china-rose, cotton, etc.) and Apocynaceae (*Nerium, Thevetia,* etc.).

(*b*) *Floral leaves not in one whorl*

3. Imbricate: When the margins overlap one another but not in any particular order as in the subfamily Caesalpinieae. In imbricate flowers the petals do not actually lie in a single whorl.

4. Quincuncial: The floral leaves are not in a whorl but spirally arranged as shown in Fig. 333. Leaves 1 and 2 are external, 3 partly external, 4 and 5 internal. This is seen in guava (*Psidium guyava* of Myrtaceae), etc.

5. Vexillary: This is the typical aestivation of the papilionaceous corolla, The posterior *vexillum* overlaps the two *alae* which again overlap the paired anterior *carina.*

PERIANTH

In most monocotyledons the calyx and the corolla are not differentiated and the general accessory whorls form the **perianth** as seen in *Polyanthes*, *Crinum*, etc. The perianth is often brightly coloured or **petaloid** as in *Gloriosa superba* (Liliaceae—Fig. 335), *Crinum asiaticum* (Amaryllidaceae), etc. The perianth in some members of Amarantaceae is **membranous** and **persistent.** In Graminaceae the perianth is represented by two **lodicules** as seen in *Festuca* (Fig. 334).

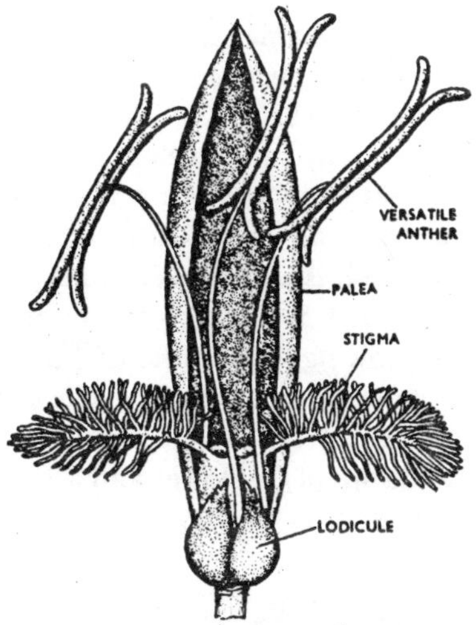

FIG. 334. Flower of *Festuca pratensis* with lemma removed showing lodicules, feathery stigmas and versatile anthers. (After Strasburger).

FIG. 335. Polyphyllous petaloid perianth in *Gloriosa superba* flower. Also note extrorse anthers and style placed at right angles to the ovary.

If the perianth members are free from one another as in *Gloriosa superba* (Fig. 335), the perianth is termed **polyphyllous.** When the perianth members are united as in *Polyanthes tuberosa* (Fig. 265), the perianth is **gamophyllous.**

SPOROPHYLLS

While calyx and corolla form the outer *accessory* whorls the two higher whorls are composed of the more important elements, the **sporophylls,** without which seed-setting is not possible. They are, therefore, the *essential whorls*.

ANDROECIUM

The **androecium** is the third set of floral organs composed of **stamens** or **microsporophylls.** Ordinarily, each stamen is composed of a slender stalk-like **filament** supporting a knob-like spore case or the **anther** (Fig. 336). Each anther consists of two lobes (**anther lobes**) connected by a **connective** which can be clearly seen on the dorsal side as an extension of the filament. Each anther lobe, again, has two **pollen sacs** or **pollen chambers** placed longitudinally. There are longitudinal grooves or *sutures* along the ventral face of the anther demarcating the pollen chambers. Each pollen chamber represents a **microsporangium** and contains innumerable **microspores** or **pollens.** The stamen, therefore, is a *microsporophyll bearing four microsporangia.* While this is the normal case, there are some flowers where the anther possesses only two pollen chambers (i.e., *bisporangiate*) and in Malvaceae even these two pollen chambers fuse developing a mature *unilocular anther.* A flower may sometimes be reduced to a single stamen as seen in the cyathium inflorescence (Fig. 290)

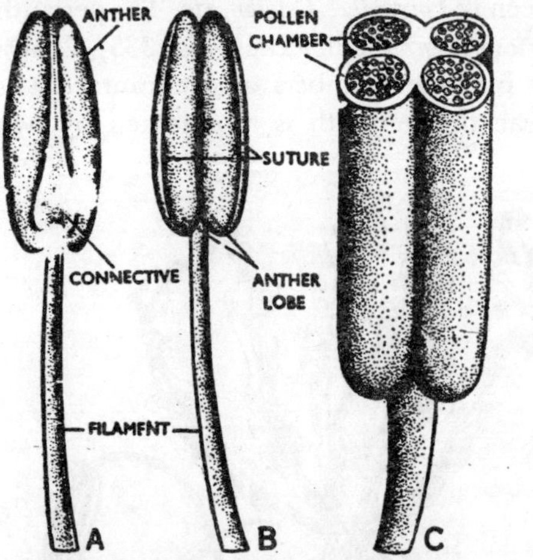

Fig. 336. Stamen showing parts. A. Dorsal view. B. Ventral view. C. Enlarged portion showing t.s. of anther.

FILAMENT

In rare cases a stamen may be devoid of a filament or *sessile* as seen in the stamens of *Arum maculatum* (Fig. 337). On the other extreme, a stamen may not develop any fertile anther when it is *sterile* and termed a **staminode** as seen in *Cassia* and *Canna* (Fig. 338). In the latter case it is also *petaloid.* The showy *labellum* (Figs. 338 & 339) of Scitaminae is formed of staminodes. In water-lily the filament is *flat* showing its transition from petals (Fig. 302). The filament may be white or coloured yellow, blue, black, etc., like petals. While the filament is ordinarily simple, in *Ricinus communis* (Fig. 340) it is found to be *branched.* When filaments are very long, stamens protrude out of the flower and are termed **exserted.** On the contrary, when stamens remain within the flower, they are termed **inserted.**

Fig. 337

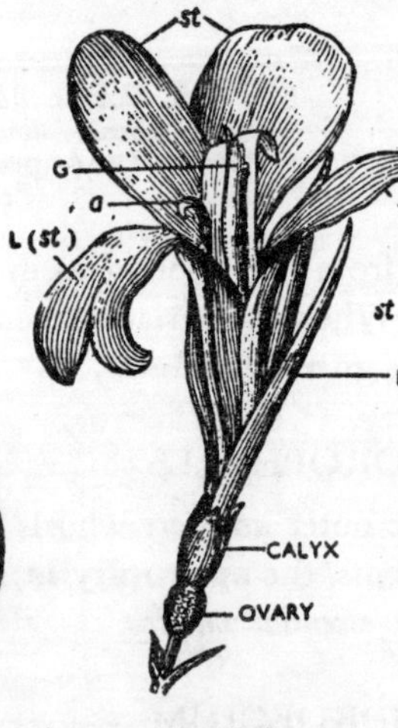

Fig. 338

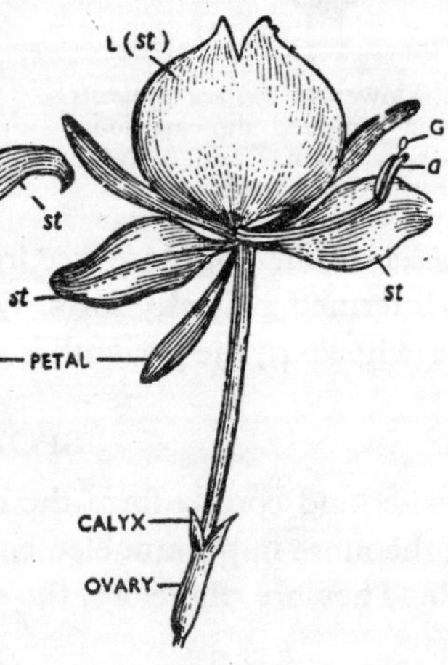

Fig. 339

Fig. 337. Sessile stamen of *Arum maculatum.* Fig. 338. Flower of *Canna indica* (Cannaceae of Scitaminae). Fig. 339. Flower of *Hedychium coronarium* (Zingiberaceae of Scitaminae). *st*=staminode; L=Labellum; *a*=anther; G=pistil.

Filaments sometimes bear *appendages.* Most characteristic of these is the **staminal corona** which is horny in *Calotropis* (Fig. 341) and cup-shaped in *Eucharis* (Fig. 342), *Pancratium* and some other flowers of Amaryllidaceae.

Fig. 340

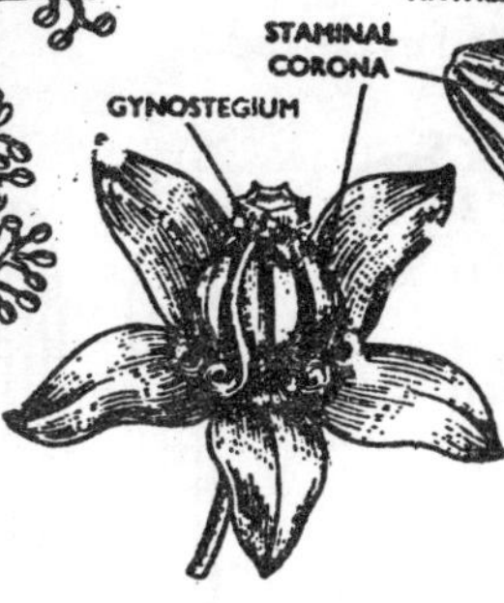

Fig. 341

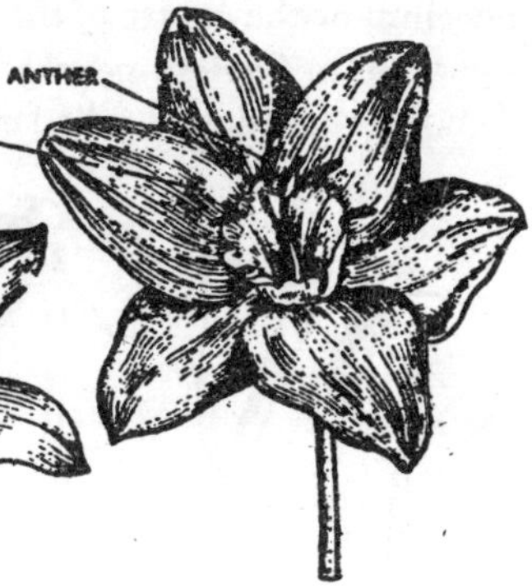

Fig. 342

Fig. 340. Branched stamen of castor (after Van Tieghem). Fig. 341. Staminal corona in *Calotropis procera* flower. Fig. 342. Staminal corona in *Eucharis amazonica.*

CONNECTIVE

Ordinarily, the *connective* is a patch of tissues connecting the two parallel anther lobes (Fig. 336). It is a prolongation of the filament and contains the conducting strands. The connective, however, may be (1) extremely small or altogether wanting as in some species of *Euphorbia* (Fig. 343) and in *Adhatoda zeylanica* (Acanthaceae) where the anther lobes are very close together. This condition is termed **discrete.** (2) In the lime tree (*Tilia*—Fig. 344) and in *Justicia gendarussa* (Acanthaceae) the connective is called **divaricate** as it develops in such a way that the two anther lobes are separated from one another. (3) In *Salvia* (Labiatae) a peculiar condition called **distractile** is noticed where the connective is a long stalk-like body (Fig. 345) placed crosswise on the filament separating the two anther lobes. One anther lobe is fertile while the other is abortive, usually represented merely by a deltoid plate.

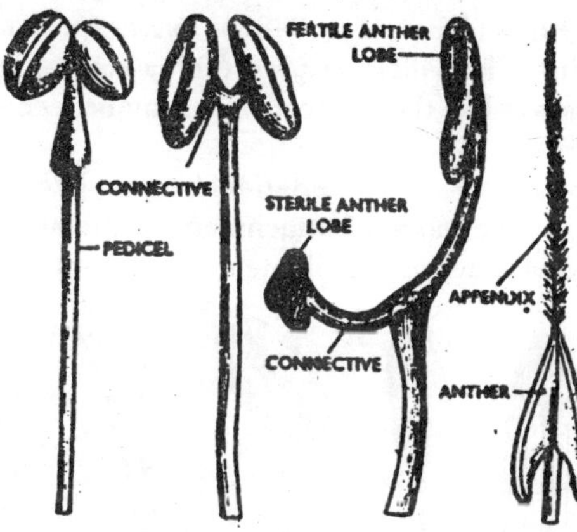

Fig. 343 Fig. 344 Fig. 345 Fig. 346

Connective. Fig. 343. Discrete in *Poinsettia* (*Euphorbia*) *pulcherrima.* Fig. 344. Divaricate in *Tilia.* Fig. 345. Distractile in *Salvia verbenaca.* Fig. 346. Appendiculate in *Nerium odorum.*

The connective also may bear appendages when it is called **appendiculate.** The connective is prolonged into a feathery *appendix* beyond the sagittate anther of the oleander (*Nerium odorum*—Fig. 346) and some other flowers of Apocynaceae. These appendices in *Nerium* unite to form a *staminal corona* (Fig. 329).

ANTHER

All Angiospermous anthers are bilobed and quadrilocular (i.e., formed of four *microsporangia*) at an early stage of development and this condition is seen in most mature stamens. Rarely, however, the anther becomes *unilocular* or *one-chambered* either by the abortion of one lobe and destruction of the partition wall between the two chambers or the destruction of the entire partition tissue separating the four chambers. This condition is seen in the family Malvaceae (Fig. 347). The grooved ventral side of an anther usually faces the gynoecium or the centre of the flower and this condition is known as **introrse;** but, in a few cases as in *Gloriosa superba* (Fig. 335), *Iris*, *Colchicum*, etc., the anther faces the petals when the condition is called **extrorse.**

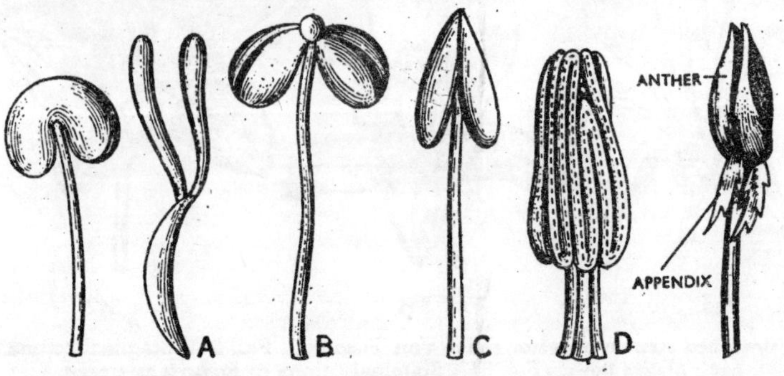

FIG. 347 FIG. 348 FIG. 349

ANTHER. FIG. 347. Unilocular and reniform in *Hibiscus rosa-sinensis*. FIG. 348. A. Linear in *Acalypha*. B. Rounded in *Mercurialis*. C. Sagittate in *Vinca*. D. Sinuous in *Cucurbita maxima*. FIG 349. Appendiculate in *Erica cinerea*.

Anthers may be *linear* (*Acalypha*—Fig. 348A), *rounded* (*Mercurialis*—Fig. 348B), *sagittate* (*Vinca*—Fig. 348C), *sinuous* (peculiar ഗ-shaped appearance as seen in the cucurbits—Fig. 348D), *reniform* (china-rose—Fig. 347), etc. The anther also may be **appendiculate** like the connective as may be seen in *Erica cinerea* of Ericaceae (Fig. 349).

ATTACHMENT OF THE ANTHER TO THF FILAMENT

The mode of attachment of the anther to the filament varies (Fig. 350). (1) It is **adnate** when the filament or its continuation, the connective, appears to be attached throughout the whole length of the back of the anther as seen in magnolia and water-lily. (2) In mustard, *Carex* and other members of Cyperaceae, etc., the filament ends just at the base of the anther, the latter being firmly fixed on the top of the former. This condition is called **basifixed** or **innate.** (3) The attachment is **dorsifixed** when the filament is firmly fixed to some position on the back of the anther as in passion-flower, *Sesbania*, etc. (4) In most grasses and in many lilies the attachment is **versatile** where the filament is attached merely at a point about the middle of the connective so that the anther can swing on it freely.

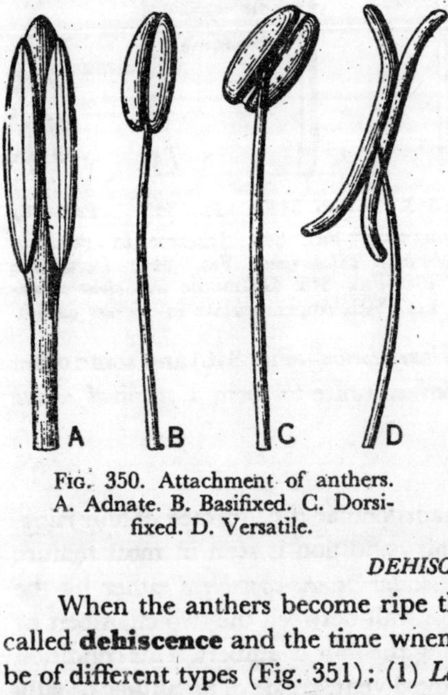

FIG. 350. Attachment of anthers. A. Adnate. B. Basifixed. C. Dorsifixed. D. Versatile.

DEHISCENCE OF ANTHERS

When the anthers become ripe they burst discharging the dry pollens. This act is called **dehiscence** and the time when this takes place is called *anthesis*. Dehiscence may be of different types (Fig. 351): (1) *Longitudinal*—this is the common type of dehiscence when the anther lobes burst along the longitudinal sutures (i.e., the lines of fusion of the two pollen chambers in the two anther lobes) as may be seen in *Datura*, etc.; (2) *Transverse*—seen in some unilocular anthers as those of Malvaceae (it appears to be transverse as the suture is placed that way); (3) *Porous* or *apical*—the discharge of pollens is through apical pores as seen in potato, brinjal, etc.; (4) *Valvular*—when the whole or portions of the wall of the anther open out like trap-doors releasing the pollens as seen in *Berberis*, *Laurus*, *Cinnamomum*, etc.

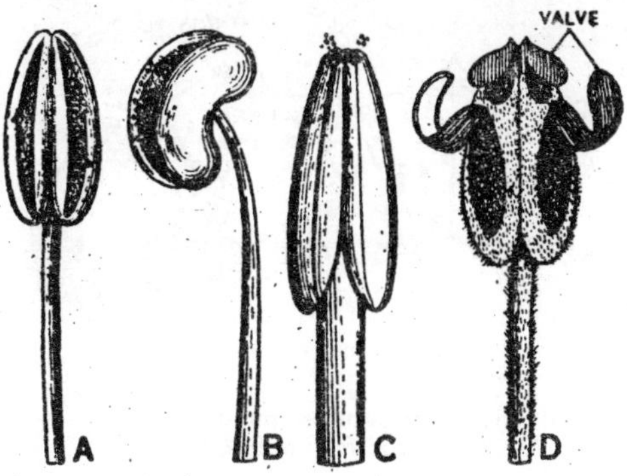

FIG. 351. Dehiscence of anthers. A. Longitudinal. B. Transverse. C. Porous. D. Valvular in *Laurus* sp.

NUMBER AND INSERTION OF STAMENS

A flower may be **monandrous** (*Poinsettia*), **diandrous** (Acanthaceae), **triandrous** (many monocots), **tetrandrous** (Labiatae), **pentandrous** (most dicots), **hexandrous** (rice, bamboo, etc.) or **polyandrous** (Rosaceae) according as the usual number of stamens in the flower is 1, 2, 3, 4, 5, 6 or many. The number of stamens, however, may sometimes vary as discussed later.

When the stamens form a single whorl and the number of stamens is the same as that of the sepals and petals, the flower is **isostemonous.** In such a flower the stamens alternate with the petals, i.e., they are *antisepalous.* Occasionally, however, such stamens may be *antipetalous* as found in different members of Rhamnaceae, Portulacaceae, etc. Sometimes there are two whorls of stamens, the first whorl alternating with petals (antisepalous) and the second whorl alternating with sepals (antipetalous). This type of flower is termed **diplostemonous.** A third condition is seen in some Rutaceae (Fig. 310) where there are two whorls of stamens of which the first whorl is antipetalous and the second whorl is antisepalous. This condition is described as **obdiplostemonous.**

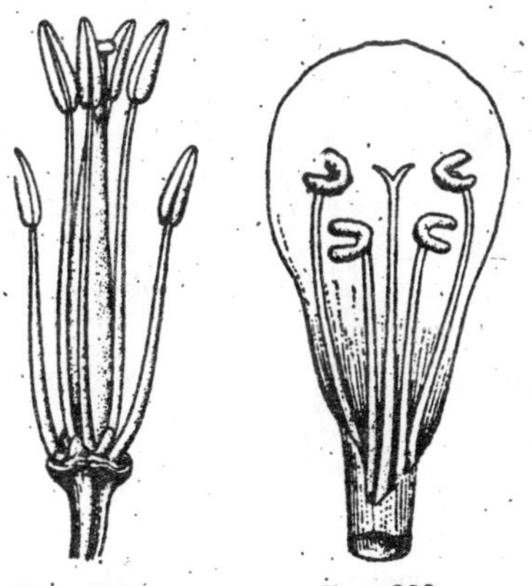
FIG. 352 FIG. 353

FIG. 352. Tetradynamous stamens in mustard. FIG. 353. Didynamous stamens in *Leonurus*.

Like other floral members, stamens also may be *epigynous*, *perigynous* or *hypogynous* in their insertion on the thalamus.

The stamens in an androecium may not be of the same length. Two conditions are rather common. (1) In the family Cruciferae there are six stamens of which the four in the inner whorl are taller than the two in the outer whorl. This condition is termed **tetradynamous** (Fig. 352). (2) Similarly, it is **didynamous** (Fig. 353) when out of four stamens present two are longer than the two others. This is found in Labiatae and the allied families Acanthaceae, Verbenaceae and Scrophulariaceae. Presence of stamens of different sizes in the same whorl, as often seen in *Cassia* flowers, is known as **heterostemony.**

UNION OF STAMENS

Union of stamens may involve **adhesion** (union with *other* members, viz., petals, perianth leaves or gynoecium) or **cohesion,** i.e., among the stamens themselves. When stamens adhere to petals they are termed **epipetalous**—a condition found in many flowers (Fig. 354). When the adherence is to perianth leaves, the condition is termed **epiphyllous** as seen in the tube-rose. Another interesting adhesion is between stamens and carpels (*gynandrous* condition) as seen in the **gynostegium** of Asclepiadaceae (Fig. 355) and the **gynostemium** of Orchidaceae (Fig. 399).

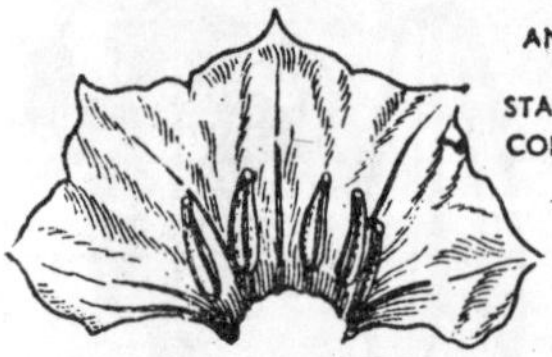

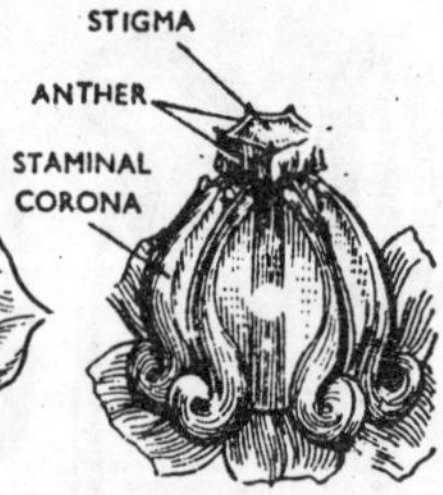

FIG. 354 FIG. 355

FIG. 354. Epipetalous stamens in brinjal.
FIG. 355. Gynostegium in *Calotropis*

Cohesion usually involves either only the filaments (**adelphy**) or only the anthers (**syngeny**). In adelphy, all the stamens may unite by their filaments forming one bundle of stamens with all the anthers free. This is the **monadelphous** condition. In the family Malvaceae and in many other flowers the united filaments form a *staminal tube* through which the long style of the pistil passes (Fig. 356). *Oxalis* (Oxalidaceae) also shows a similar staminal tube in which the few stamens are clearly unequal (Fig. 357). In unisexual female flowers of *Jatropha* (Euphorbiaceae), the filaments unite to form a central column. **Diadelphy** (two bundles) is very commonly seen in Papilionaceous flowers where nine stamens form one bundle and the tenth remains free as the second bundle (Fig. 358). In

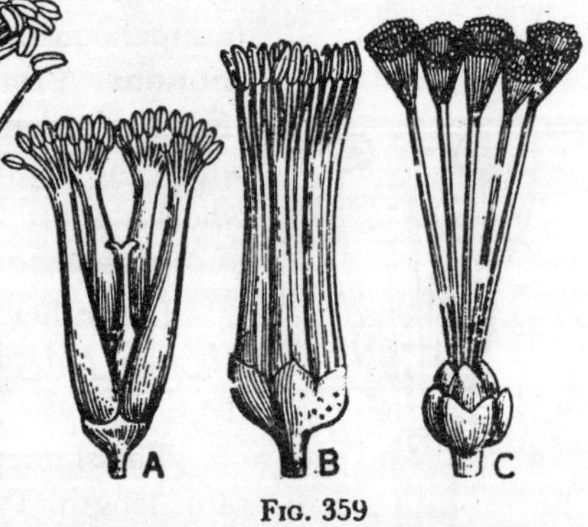

FIG. 356 FIG. 357 FIG. 358 FIG. 359

FIG. 356. Monadelphous stamens of *Hibiscus rosa-sinensis*. FIG. 357. Monadelphous and unequal stamens of *Oxalis*. FIG. 358. Diadelphous stamens of pea. FIG. 359. Polyadelphous stamens in A. *Bombax ceiba* ; B. Orange and C. *Melaleuca* (after Van Tieghem).

the silk-cotton tree (*Salmalia* or *Bombax ceiba*) the stamens form several **separate** groups with the filaments uniting to form several bundles or fascicles (Fig. 359A) giving rise to the **polyadelphous** condition. This is often seen in the families Guttiferae, Tiliaceae, Bombacaceae, Rutaceae (e.g., orange—Fig. 359B), Myrtaceae (e.g., *Melaleuca*—Fig. 359C), etc. When the stamens unite only by the anthers leaving the filaments free, the condition is termed **syngenesious.** This is characteristically shown by the family Compositae (Fig. 360). Here, the syngenesious anthers form a tube enclosing the style and the stigma. In Compositae the syngenesious stamens are also epipetalous (not shown in figure).

In the family Cucurbitaceae, of the five stamens four unite in pairs so that the androecium shows three bundles 2+2+1. Each composite structure of two stamens shows complete union of the filaments as well as the sinuous anthers (Fig. 361). This is called the **synandrous** condition.

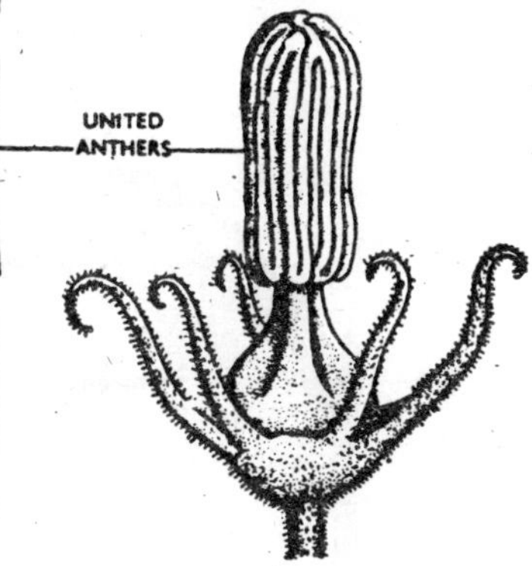

FIG. 360 FIG. 361

FIG. 360. Syngenesious stamens of Compositae.
FIG. 361. Synandrous stamens of *Cucurbita maxima*.

GYNOECIUM

The **gynoecium** (also spelt *gynaeceum*) or **pistil** is the central or the topmost whorl of the flower usually terminating the thalamus. It is composed of one or more **carpels** or **megasporophylls.** When there is a single carpel the pistil is called **simple** or **monocarpellary** which is not very common although it is a characteristic of the large families of Leguminosae and Gramineae (Fig. 334).

A typical carpel (Fig. 362) has three parts—**ovary, style** and **stigma.** The lowermost swollen part is the *ovary* containing one or more swollen bodies called **ovules** which are the rudiments of seeds. Above the ovary the carpel is protruded into a long or short *style* which ends in a somewhat rounded and usually sticky *stigma* on which the pollens are deposited during *pollination.* A *sterile* pistil devoid of fertile ovules is called a **pistillode.**

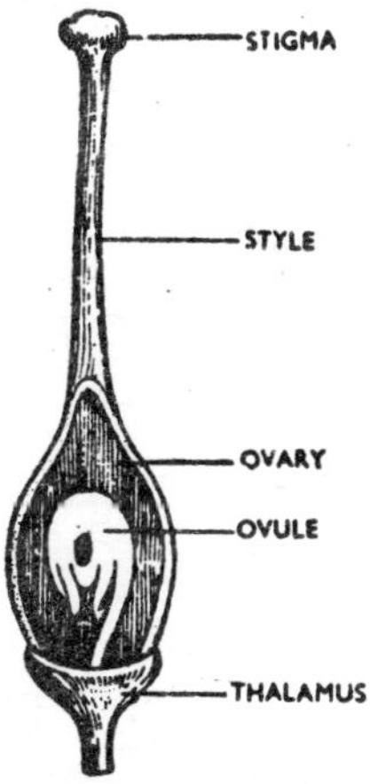

FIG. 362. A typical carpel.

Compound or **polycarpellary gynoeciums** are much more common than the *simple type.* In such a gynoecium, the different carpels may remain completely free from one another when it is termed **apocarpous** (apocarpous *multiple,* as opposed to *simple,* as there are multiple carpels) or the carpels may unite with each other, wholly or partially, forming **syncarpous** gynoeciums. *Apocarpous* gynoeciums may be seen in the families Annonaceae, Magnoliaceae, Ranunculaceae (Fig. 363), Nymphaeaceae, Rosaceae, etc. In Annonaceae (Fig. 305) and Magnoliaceae (Fig. 364) the carpels are arranged on a more or less elongated thalamus, in Nymphaeaceae (Fig. 306) they are sunk on the flat spongy top of the thalamus shaped like an inverted cone, while in Rosaceae (Fig. 314) they are arranged on the bottom of the cup-shaped thalamus. The *syncarpous* gynoecium is much more common and involves union of different degrees, e.g., in *Solanum* the carpels are completely united (Fig. 365); in china-rose (Fig. 366) the ovaries are united to form a five-locular compound ovary, the styles are united completely while the five stigmas are free; in *Linum usitatissimum* of Linaceae (Fig. 367) and in the pink flower (Fig. 368) the ovaries are united but the styles and stigmas are free. In Apocynaceae and Asclepiadaceae the two carpels have free ovaries and styles uniting only at the stigmas (Fig. 369) at the *gynostegium.* Rather interesting is the case of the union of the gynoeciums of two different flowers. In *Lonicera* of Caprifoliaceae (honey-suckle or woodbine—Fig. 370) the ovaries of two adjoining flowers may fuse while the other parts of the flowers remain free. This is termed **syngynia.** Ovaries of banana, etc., fuse abnormally in a like manner giving rise to paired fruits.

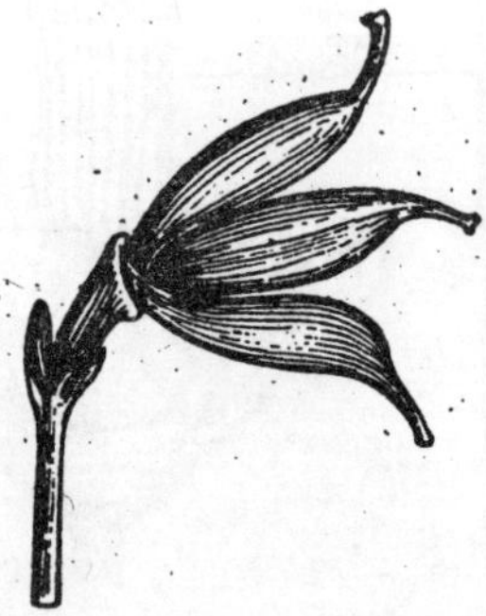

FIG. 363

FIG. 364

FIG. 365

FIG. 366

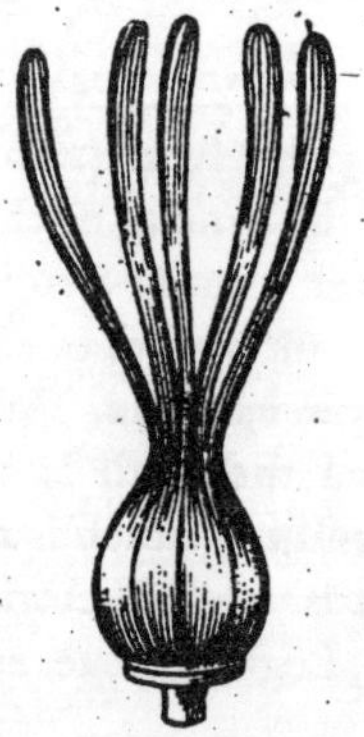

FIG. 367

FIG. 363. Apocarpous pistil of three carpels in *Aconitum* of Ranunculaceae. FIG. 364. Apocarpous pistil of many carpels in *Michelia champaca* of Magnoliaceae. Note elongated thalamus. FIG. 365. Completely united pistil of *Solanum melongena*. FIG. 366. Pistil of china-rose with free stigmas. FIG. 367. Pistil of *Linum usitatissimum* with free styles and stigmas.

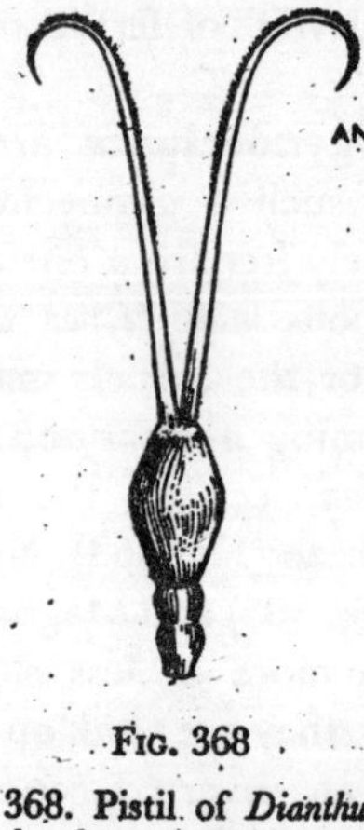

FIG. 368

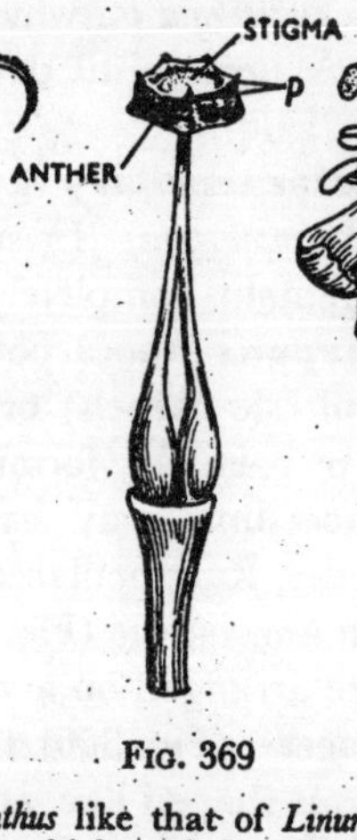

FIG. 369

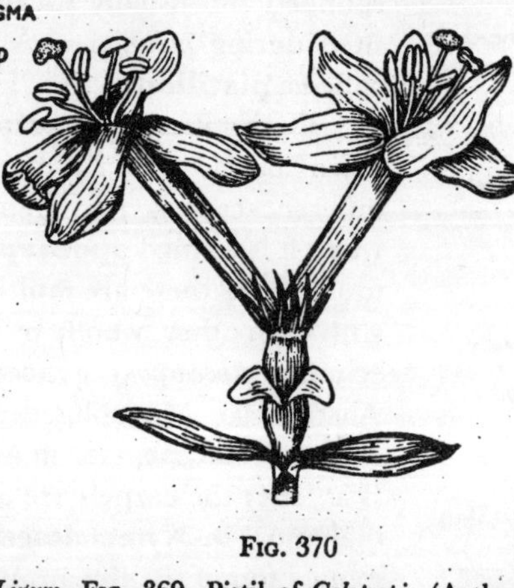

FIG. 370

FIG. 368. Pistil of *Dianthus* like that of *Linum*. FIG. 369. Pistil of *Calotropis* (Asclepiadaceae) united only at the stigma which again is united with anthers forming *gynostegium*. (Filaments and corona removed). *p* shows location of pollinia. FIG. 370. Syngynia in *Lonicera*.

STYLE AND STIGMA

The **style** connects the ovary with the stigma and usually arises from the top or the summit of the ovary, i.e., it is *apical*. In some cases, however, the ovary apex itself may be deflected so that the style may appear to originate from near the base **(basillar)** or from the side **(lateral)** as in *Alchemilla* and mango (Fig. 371). In plants belonging to the Labiatae the ovary is peculiarly four-lobed so that the ovary apex is depressed at the centre of the four lobes. As a result, the style appears to arise from the central base of the ovary (Fig. 372) and is termed **gynobasic.** In *Gloriosa superba* (Fig. 335) the style is placed at right angles to the ovary axis instead of being its direct prolongation.

The style is usually deciduous, dropping off after fertilization. But, in some cases, as in *Naravelia zeylanica*, *Clematis*, *Digitalis*, etc., it may be **persistent** (Figs. 455A & 464). The style of *Canna* (Fig. 338), *Iris*, etc., is **petaloid.** The base of the style in the family Umbelliferae is swollen forming what is known as the **stylopodium.**

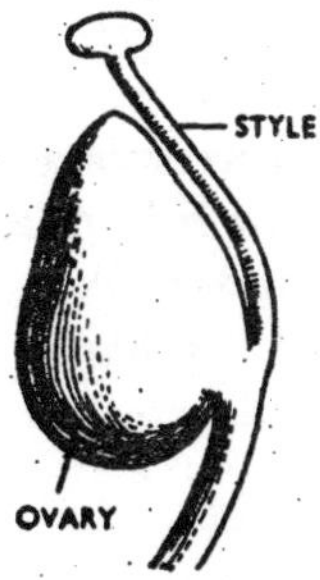

Fig. 371

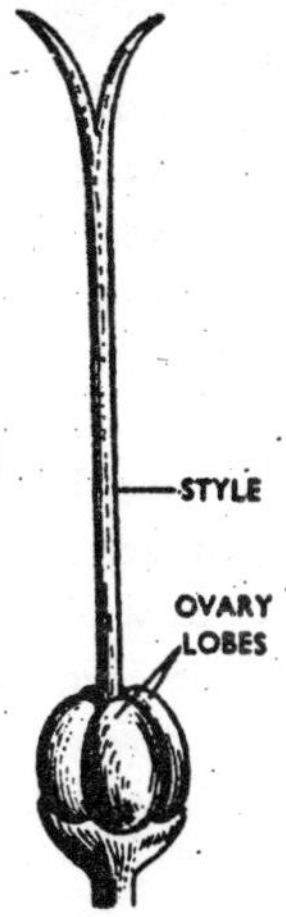

Fig. 372

Fig. 371. Lateral style.
Fig. 372. Gynobasic style.

Fig. 373. Branched styles. (After Green).

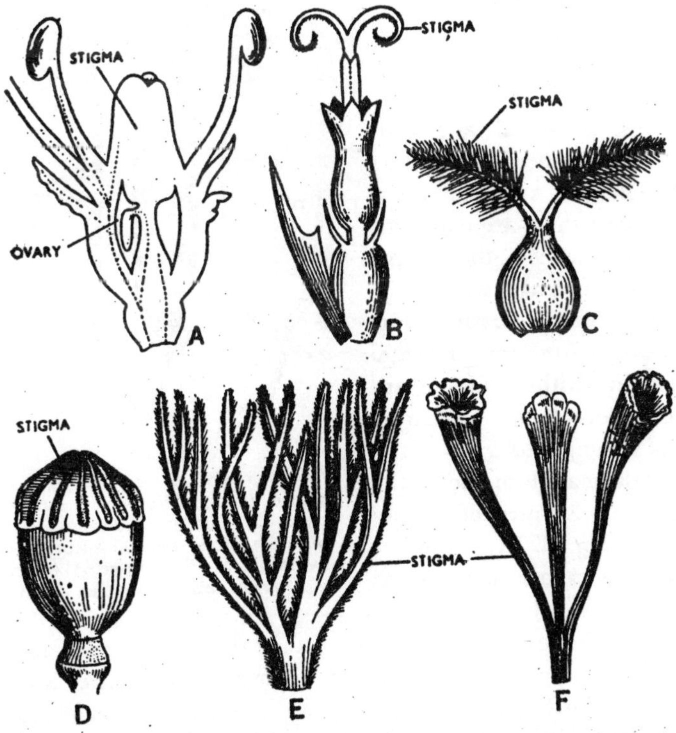

Fig. 374. Stigma. A. Sessile in *Sambucus nigra*. B. Bifid in sunflower. C. Bifid and feathery in rice. D, Striate and sessile in poppy. E. Highly branched in *Begonia*. F. Funnel-shaped in *Crocus sativus*.

The styles of a compound pistil may sometimes be completely united as in china-rose (Fig. 366). But, even when united, anatomical study shows separate vascular supplies for the different carpels. Thus, a cross-section of the china-rose style shows five vascular traces corresponding to the five carpels. When the styles are free, usually there is one style for each carpel but sometimes the style may be branched as in some Euphorbiaceae (Fig. 373).

The **stigma** is usually placed on the style. Sometimes, there may be no style, the stigma being placed on the top of the ovary as in *Sambucus* (Fig. 374A), *Berberis*, lotus, etc. Then it is termed **sessile.** The stigma top is usually rough, papillose or even hairy (Fig. 334) and somewhat sticky due to secretions. This shows a *receptive surface* where the pollens alight and germinate. In a syncarpous ovary there may be separate stigmas as in china-rose (Fig. 366) or the stigma may be **lobed** when it is described as *bifid* (e.g., Compositae—Fig. 374B), *trifid*, etc. Usually, the number of lobes correspond to the number of carpels but, monocarpellary flowers of Graminaceae show *bifid feathery stigmas* (Fig. 374C). The stigma of poppies (*Papaver*) is *sessile* as well as **striate** showing a star-like radiate appearance (Fig. 374D). The *Begonia* stigma (Fig. 374E) is highly branched. The three stigmas of *Crocus* (Fig. 374F) have peculiar funnel-shaped forms. These are the stigmas which, when dried, form the *saffron* (*zafran*) of commerce.

OVARY

Ovary is the most important part of the carpel as it contains the ovules which develop into seeds. A carpel without a functional ovary is sterile. The foliar origin of the ovary is rather clear in the simple ovary of pea (Fig. 375). It is even clearer in the inflated ovary of its relative *Colutea arborescens* (Fig. 376). One may easily realise that a leaf-like carpel, i.e., a *megasporophyll* as seen in the Gymnosperms, is folded about its midrib and forms a chamber by the fusion of the margins. There is a special tissue called **placenta** along the margin and the marginal line along which the carpel fuses is called the **ventral suture,** the midrib being the **dorsal suture** (Fig. 375). **Ovules** develop from this placental tissue and remain within the ovary chamber. When more than one carpel form the ovary of a compound gynoecium, the different carpels may fuse along the margins giving the appearance of a one-chambered ovary (Fig. 377A). Another way of joining of the carpels may be as shown in Fig. 377B. In this latter case each carpel first fuses along its own margin and then the three carpels fuse so that the ventral sutures all meet together at the axis of the ovary which latter is clearly divided into as many *loculi* or *chambers* as there are carpels. The ovaries of syncarpous pistils are frequently formed as in the second case. But, sometimes the number of chambers do not correspond to the number of carpels due to the development of false partition walls or

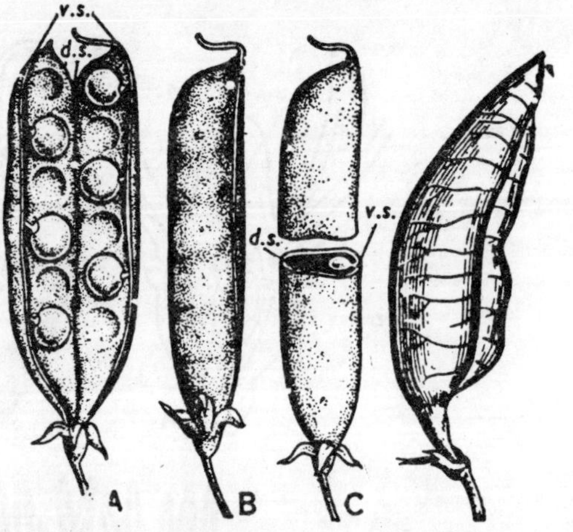

FIG. 375 FIG. 376

FIG. 375. Pod of pea. A. Split along ventral suture (*v.s.*); *d.s.*=dorsal suture. B. Whole pod. C. Cut transversely into two. Note placenta along ventral suture giving rise to ovules. FIG. 376. Inflated ovary of *Colutea arborescens* looking like a folded leaf.

septa within the ovary chamber. Thus, many bicarpellary flowers of Solanaceae (*Datura*—Fig. 380, brinjal, etc.) show four-chambered ovaries. Such false partition walls usually develop as outgrowths of the *placenta*.

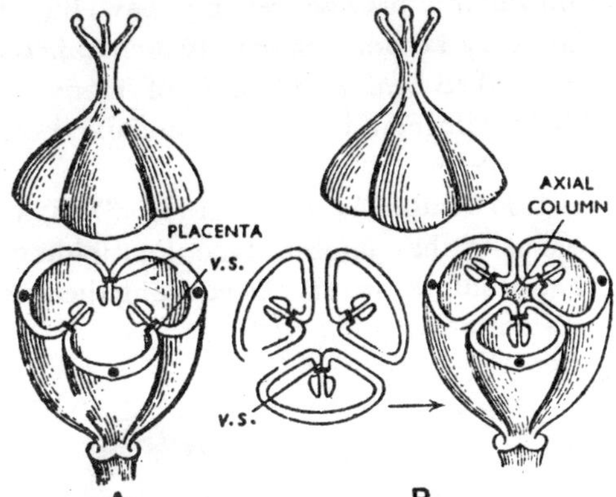

FIG. 377. Union of carpels. A. Fusion along margins (*v.s.*) only. B. Fusion along margins (*v.s.*) to form an axial column

PLACENTATION

It has already been seen that the placental tissue develops along the margin of the megasporophyll so that, when the latter closes to form a chamber, the *placenta* is located along the *ventral suture*. But, margin is not the only place where *placenta* develops. *Placenta* may also develop on a direct prolongation of the thalamus at the base of the carpel. As a result, we get different types of **placentation,** i.e., distribution of placenta, in different ovaries. As this placenta is the tissue on which the ovules or the future seeds develop, a study of *placentation* is of importance in the study of the fruit and the flower. The following types of placentation are usually met with:

(1) Marginal: This is the placentation of the simple ovary of pea (Leguminosae). The pea ovary is developed by the union of one megasporophyll along the *ventral suture*. The *placenta* forms a ridge along this suture and the ovules are borne on this forming two rows. The ovary is one-chambered and superior (Fig. 378).

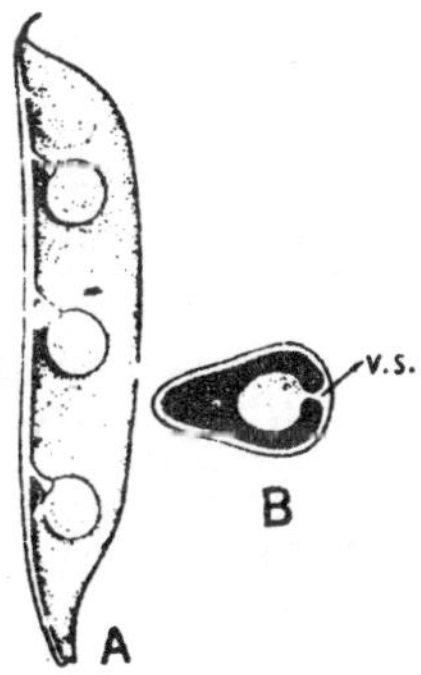

FIG. 378

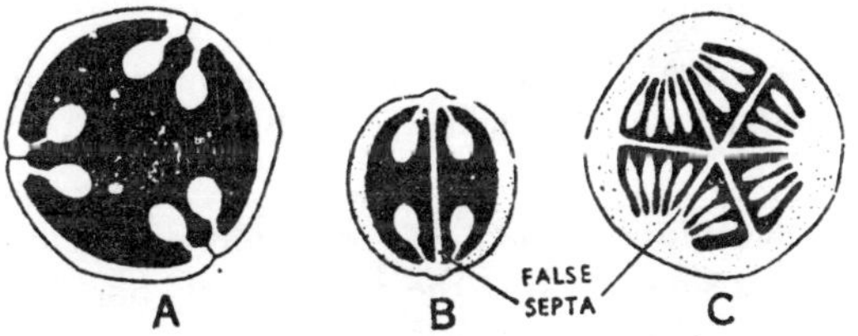

FIG. 379

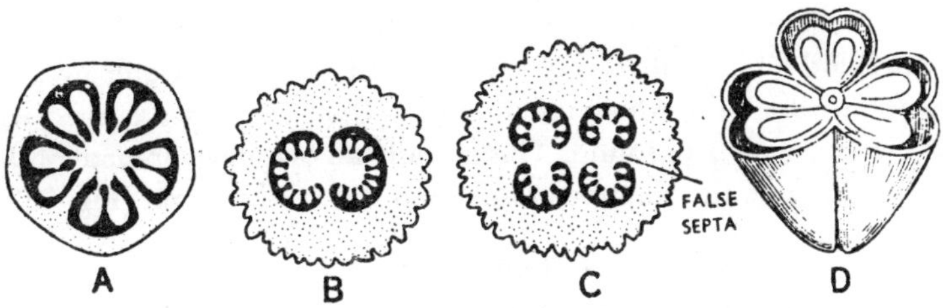

FIG. 380

PLACENTATION TYPES. FIG. 378. Marginal. A. Longitudinal half of pea pod. B. T.s. FIG. 379. Parietal. A. Typical. B. Mustard—Falsely bilocular. C. Cucumber—Falsely trilocular or hexalocular. FIG. 380. Axile. A. Typical. B. *Datura* showing normal bilocular ovary. C. *Datura* showing tetralocular ovary by formation of false septa. D. Axile in inferior ovary of lily.

(2) **Parietal:** When two or more carpels unite along the ventral sutures as shown in Fig. 377A, the ovules also are located on the placenta along these sutures. The ovary is unilocular as may be seen in papaw. But, in Cruciferae showing this type of placentation, the ovary becomes falsely two-chambered by the development of a false partition wall called *replum*. Ovaries of many Cucurbitaceous plants (also *parietal*) become falsely trilocular, or even hexalocular, because of later projections of the placental tissue (Fig. 379).

(3) **Axile:** As shown in Fig. 377B, a number of carpels independently form chambers as in the marginal type, i.e., folding inwards and then fusing together so that the ventral sutures are all placed along the axis. This is termed **axile** as all the placenta and, therefore, the ovules are axial. The ovary is divided into as many chambers as there are carpels. This is found in Solanaceae, Malvaceae, Rutaceae, Liliaceae, etc. But, here also the number of chambers may increase by the development of false partition walls (Fig. 380).

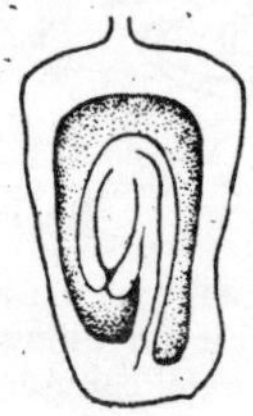

FIG. 381

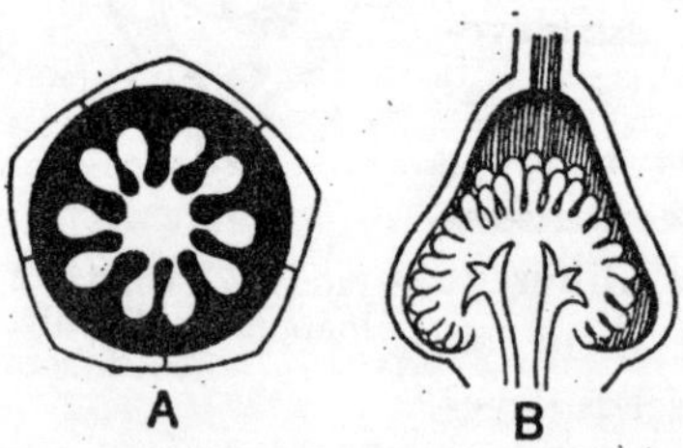

FIG. 382

FIG. 383

PLACENTATION TYPES. FIG. 381. Basal in Compositae. FIG. 382. Free-central. A. T.s. of ovary showing typical free-central placentation. B. Free-central in Primulaceae (l.s.) where the axial column of placenta seems to be a projection of the thalamus. FIG. 383. Superficial in water-lily.

(4) **Basal:** In Compositae, a single ovule is developed at the *base* of the unilocular simple ovary. The placenta is apparently placed on the tip of the thalamus at the floor of the ovary (Fig. 381).

(5) **Free-central:** In this type the ovules are found to develop on an axial column which is not connected with the ovary wall (Fig. 382A). Such an ovary may arise in two different ways:

(i) In a placentation which was originally *axile*, the partition walls may break down at a later stage making the ovary falsely one-chambered. This is the case in the family Caryophyllaceae (pink flower, etc.).

(ii) The thalamus may slightly extend into the ovary and placenta may develop on it as in the family Primulaceae (Fig. 382B).

(6) **Superficial:** The multicarpellary flower of *Nymphaea* (water-lily) develops multilocular ovaries in which the whole inner walls of the chambers are lined with placental tissue so that ovules develop all round. This is really an overdevelopment of the axile type, the placenta spreading to the chamber walls from the axial column (Fig. 383).

THE OVULE

The **ovule** is the **megasporangium** contained within the ovary. There may be one or more ovules inside an ovary and these are destined to be the seeds.

When fully formed, the tissues in a typical ovule (Fig. 384) are as follows: The ovule is attached to the **placenta** by the **funicle** or **funiculus** which meets the ovule at the

hilum. An ovule devoid of any funicle and directly attached to the placenta is termed *sessile.* **Raphe** is an extension of the funicle and may extend up to the **chalaza** which is the *base* of the ovule. (The ovule commonly remains reversed as shown in the figure). The general tissue of the ovule is called **nucellus** and the **embryosac** (the **female gametophyte** within the **megaspore**—its development is described in Chapter X) is in the *top* part of it. The ovule is enveloped by two **integuments** (*inner* and *outer*) which leave an opening called **micropyle** at the top. Abnormally, there may be a single integument as in some *Compositae* while in some parasites like *Loranthus*, *Viscum album*, etc., there may be no integument at all.

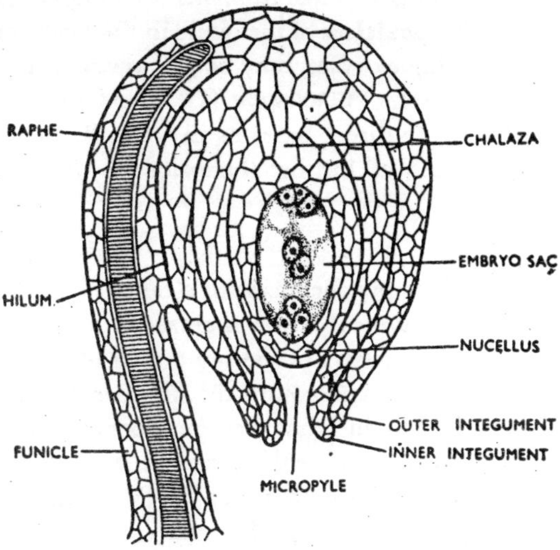

FIG. 384. Longitudinal section of a typical ovule (anatropous).

The ovule shown in Fig. 384 is the **anatropous** or reversed type which is the commonest and may be seen in pea, bean, castor, etc. When the ovule remains straight (Fig. 385A), as in Polygonaceae, Piperaceae, Urticaceae, etc., it is called **orthotropous** or **atropous.** It is **amphitropous** when placed transversely at right angles to the funicle (Fig. 385B). This

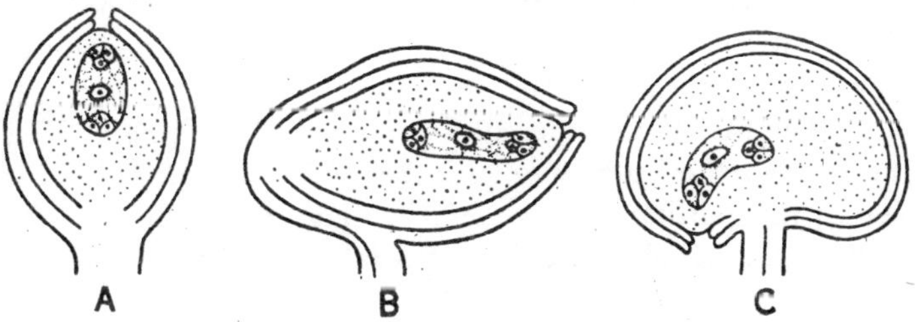

FIG. 385. OVULES. A. Orthotropous. B. Amphitropous. C. Campylotropous.

condition is not common but is met with in *Ranunculus*, *Lemna* and poppy. Sometimes, a transverse ovule may be bent like a horse-shoe so that the micropyle is brought nearer

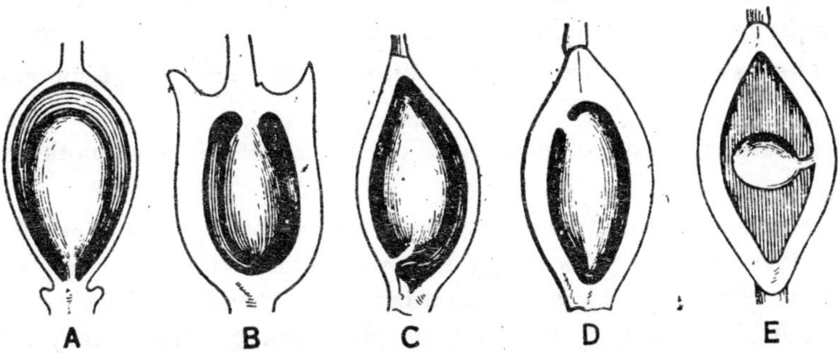

FIG. 386. Position of ovules. A. Erect. B. Pendulous. C. Ascending. D. Suspended. E. Horizontal.

to the chalaza (Fig. 385C). This is the **campylotropous** (*kampylos*=curved) condition found in many plants of Cruciferae, Chenopodiaceae, Caryophyllaceae, etc., and some other plants like *Mirabilis* (Nyctaginaceae).

The **position** of the ovule in the ovary chamber (Fig. 386), specially in ovaries containing single ovules, may be **(A) erect** as in Compositae, **(B) pendulous** (hanging from top) as in some Annonaceae, **(C) ascending** (rising obliquely from a side of the ovary wall and not the direct base) as in some Ranunculaceae, **(D) suspended** (hanging from a side of the wall but not the exact top) as in castor, or **(E) horizontal** as in *Podophyllum*.

VARIATIONS IN FLORAL MEMBERS

The chapter begins with a description of the typical flower but it may easily be seen that such a typical flower is extremely rare. Moreover, one is baffled to find variations in the floral members of the same type of flower. If one examines a number of tomato or brinjal flowers and counts the number of petals, sepals, chambers in the ovary, etc., one is surprised to find that these numbers are not constant. Such deviations may be the constant features of some flowers or they may occur accidentally in some specimens. Deviations usually met with are of the following types:

(1) Irregularity of form: A sepal or a petal may develop a spur or some such appendage so that the flower otherwise regular becomes irregular.

(2) Meristic variations: The fundamental number of members of different whorls may vary in related species or the same species or even within the same plant. This may involve increase or decrease as follows:

(a) *Pleiomery* or *augmentation:* Number of members of a whorl or even the number of whorls may increase. Thus, poppy flowers have developed several whorls of stamens.

(b) *Chorisis:* This may be considered as a special type of *pleiomery* causing increase in the number of members of a whorl. This usually happens by the splitting of some member (Fig. 387). 'Doubling' of flowers (i.e., increase in the number of petals) may be caused by *chorisis*.

FIG. 387. CHORISIS. Increase in number of stamens in *Vella* by splitting. (After Velenovsky).

(c) *Meiomery:* This involves the suppression of entire whorls or some members of a whorl. Thus, the number of sepals may be reduced from five to three in a *pentamerous* flower. *Meiomery* often involves stamens and carpels. *Abortion* also is a change of this type.

(3) Displacement of members: A floral member may be shifted from its natural position in the whorl. An opposite organ may become alternate.

(4) Cohesion: *Cohesion* between the members of the same whorl is found to be widespread.

(5) Adhesion: *Adhesion* of members of one whorl with those of another is also very common.

(6) Metamorphosis: Transformation of one type of organ into another has been found to be very interesting. 'Doubling' of flowers is often caused by the transformation of stamens into petals. *Phyllody* is a peculiar condition when floral leaves are found to be metamorphosed in foliage leaves. Metamorphosis of the thalamus may give different shapes to the flowers.

FLORAL FORMULA

It is sometimes found convenient to describe a flower by a simple and concise formula known as the **floral formula.** In this formula K represents calyx, C=corolla, P=perianth, A=androecium, G=Gynoecium. The number of members of a whorl is written after the symbol for a whorl. ∞ signifies a large and indefinite number. If the members are present in different whorls, the different whorls are written separately with a + sign between them. Thus, A_{2+2} means androecium of four stamens in two whorls of two each. When members are united (*cohesion*), the number is placed within first brackets. *Adhesion* is expressed by a line joining the two whorls. Superior or inferior position of gynoecium is denoted by a line below or above it. A male flower is signified by the symbol ♂, female flower by ♀ and hermaphrodite flower by ⚥ or ☿. The symmetry of the flower also may be signified by symbols like ⊕ for actinomorphic and ·|· or ↓ for zygomorphic flowers. Thus,

Mustard flower (Cruciferae) has the formula ⊕ ⚥ $K_{2+2}\ C_4\ A_{2+4}\ \underline{G_{(2)}}$

China-rose (Malvaceae) ,, ,, ,, ⊕ ⚥ $K_{(5)}\ \overline{C_5\ A_{(\infty)}}\ \underline{G_{(5)}}$

Datura (Solanaceae) ,, ,, ,, ⊕ ⚥ $K_{(5)}\ \overline{C_{(5)}\ A_5}\ \underline{G_{(2)}}$

Pea (Papilionaceae) ,, ,, ,, ·|· ⚥ $K_{(5)}\ C_5\ A_{(9)+1}\ \underline{G_1}$

Cucurbita (Cucurbitaceae) ,, ,, ,, ♂⊕ $K_{(5)}\ C_{(5)}\ A_{(5)}$ and ♀⊕ $K_{(5)}\ C_{(5)}\ \overline{G_{(3)}}$

Tuberose (Amaryllidaceae) ,, ,, ,, ⊕ ⚥ $\overline{P_{(3+3)}\ A_{3+3}}\ \overline{G_{(3)}}$

FLORAL DIAGRAM

It sometimes becomes very lucid if the ground plan of a flower be represented in the form of a **floral diagram.** In a floral diagram the position of the inflorescence axis or

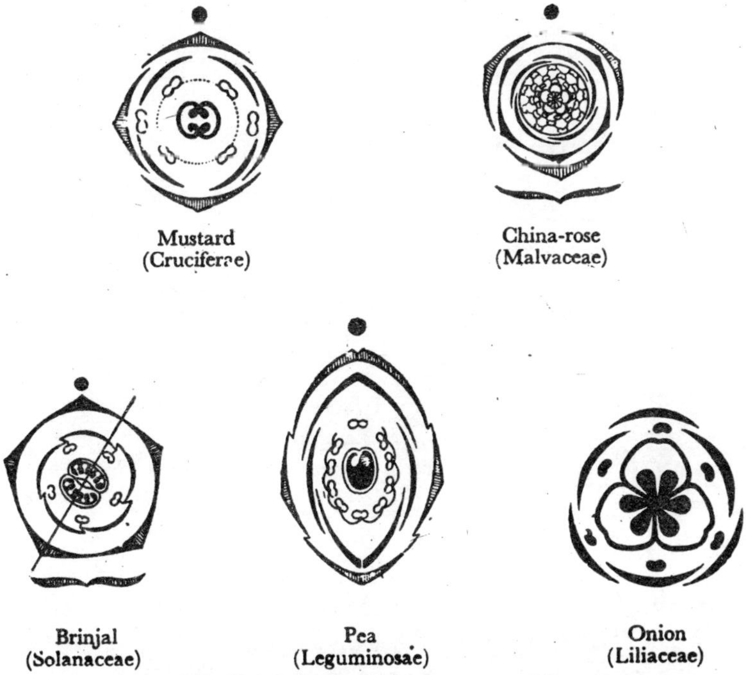

FIG. 388. Floral diagrams of some common flowers.

stem is shown by a dot or a small circle while the sepals, petals and stamens are put in concentric circles (or spirals when the floral phyllotaxy is spiral), the gynoecium being put at the centre. It has become customary to represent petals by thick first brackets (representing t.s. of petals), sepals by the same symbol but showing the midrib, stamens by anthers and carpels by placentation. The median (*mm*), lateral (*ll*) or diagonal planes (*cf.* Fig. 298) may be shown as straight lines. The bract may be shown by a second bracket. Thus, the orientation and symmetry of the flower and its different members may be shown very clearly. *Adhesion* and *cohesion* are shown by connecting lines.

The floral diagrams of some common flowers illustrating their constructions ar shown in Fig. 388.

CHAPTER IX

POLLINATION

In order that a seed may be set, the male and the female elements must come together. The male element is contained within the *microspore* (pollen) which develops as the *male gametophyte* while the female element is contained within the embryosac which is the *famale gametophyte* developed within the *megaspore* and is located within the ovule or the *megasporangium*. So, the next biological phase is **pollination** which means *the transference of the pollen from the anther to the receptive stigma whether of the same flower or of a different flower.* As the pollen is not capable of locomotion, this act either involves some agent for this transference or the anther must be placed in the flower right above the stigma so that pollens may drop directly on the stigma by sheer gravity.

If a stigma be pollinated by the pollen of the same flower, it is a case of **self-pollination** or **autogamy.** When the pollen of one flower pollinates the stigma of a different flower but on the same plant, it is called **geitonogamy.** But, from the biological point of view, there is little difference between *geitonogamy* and *autogamy* as all the flowers on a plant have the same genetic constitution. When the pollen of a flower pollinates the stigma of another flower located on a different plant, whether of the same kind or not, it is called **cross-pollination** or **allogamy.** Cross-pollination within a species (this may be inter-varietal) is called **xenogamy.** Cross-pollination *involving different strains* of plants yields **hybrids.** Self-pollination is possible only in *hermaphrodite* flowers while only cross-pollination or geitonogamy is possible in *unisexual* flowers. Self-pollination ends in **self-fertilization** and cross-pollination in **cross-fertilization.**

N.B. There is some confusion in the use of the terms 'self-pollination' and 'cross-pollination' in many books on Botany. Strictly speaking from the genetic point of view, 'cross-pollination' means only those cases where pollination involves flowers of different *genotypes* (*pure lines* or *strains*). This excludes not only 'geitonogamy', but even those cases where the flowers involved are from different plants but of the same *genotype*. In considering the 'contrivances for cross-pollination', which are of some classical importance, the term 'cross-pollination' is, however, used somewhat loosely, even in this book.

As to the merits and demerits of self-pollination and cross-pollination, the question assumed great importance when Darwin published his book on 'Cross- and Self-pollination'. Darwin was greatly impressed by the effect of cross-pollination and concluded that a species could not survive without it. Modern researches, however, have shown that, while the immediate effect of crossing is striking because of *hybrid vigour* and continuous self-pollination may sometimes be injurious because of the isolation of recessives, species may well survive without cross-pollination and *continuous* cross-pollination is not at all necessary. As a result, it is found that in nature there are not only flowers particularly adapted for cross-pollination as Darwin emphasized but there are also flowers particularly adapted for self-pollination.

CONTRIVANCES FOR CROSS-POLLINATION

It seems that Nature favours cross-pollination as opposed to self-pollination. It is a study of this partiality on the part of Nature that so greatly impressed Darwin. All unisexual flowers and a large majority of bisexual flowers are naturally cross-pollinated. Special contrivances ensuring cross-pollination as noted below are very conspicuous:

1. Dicliny: Cross-pollination is the rule among diclinous plants, i.e., those bearing unisexual flowers. In *dioecious* plants nothing else can take place. In *monoecious* plants the only alternative is geitonogamy which, however, has the same effect as self-pollination.

2. Self-sterility: This is the condition when a flower cannot be fertilized by the pollen of the same flower or, sometimes, from any flower of the same strain of plants. In some orchids, flowers wither away if pollinated by its own pollen. Many species of *Solanum* (potato, tobacco, etc.) and the tea plant are self-sterile because of genetic reasons. Cross-pollination is obligatory in such plants.

3. Dichogamy: When stamens and carpels of a bisexual flower mature at different times, pollination between them becomes ineffective. Sometimes, however, it is found that self-pollination may take place at a later stage if cross-pollination fails. Dichogamy may be of two types:

(a) **Protandry** : The anthers ripen first as in most Compositae (Fig. 399), many Umbelliferae, Malvaceae, etc. As a result, when the anther bursts, it pollinates stigmas of other flowers but not its own stigma which is not yet ripe.

(b) **Protogyny** : The carpel matures first as in many members of Annonaceae (e.g., *Annona, Polyalthia*) and Magnoliaceae (e.g., *Magnolia, Michelia*) as well as in *Arum maculatum*. When the stigma is receptive, its own pollen is not ripe so that it has to depend on foreign pollens.

4. Herkogamy (*herkos*=barrier): In some flowers there may be some physical barrier between the anther and the style so that pollination between them is rendered difficult or even impossible. In many Cruciferae and Caryophyllaceae, the stigma extends far beyond the stamens so that pollens from the latter are not likely to reach the former. The extrorse anthers of *Gloriosa* (Fig. 335) dehisce the anthers out of reach of its own stigmas. In *Calotropis* and orchids, where the pollens are aggregated in pollinia, the pollination (described later) is entirely at the mercy of insects.

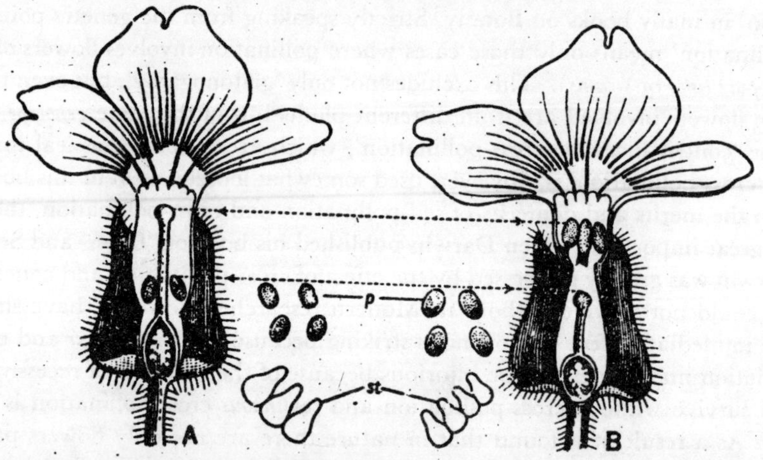

Fig. 389. Dimorphism in flowers of *Primula sinensis*. 'A' type flower has larger pistil with larger stigma papillae (*st*) and lower anthers with smaller pollens (*p*). 'B' type flower has shorter pistil with smaller stigma papillae and higher anthers with large pollens.

5. Heteromorphism: In certain plants there are flowers of two (**dimorphic**) or three (**trimorphic**) different forms with anthers and stigmas at different levels. This dimorphism or trimorphism usually involves **heterostyly** (styles of different lengths) and **heteroanthy** (i.e., different types of anthers). The primrose (*Primula sinensis* of Primulaceae—Fig. 389) shows an interesting case of *dimorphism*. In the first form, the anthers are placed deep down in the corolla tube and the stigma lies at the entrance. The pollens of this type are smaller and the stigma papillae larger. In the second type the anthers are placed at the entrance while the stigma is deep down. Moreover, in this case the pollens are larger and the stigma papillae smaller. Different types of insects moving about these flowers will naturally touch floral organs at the same level because of the difference in

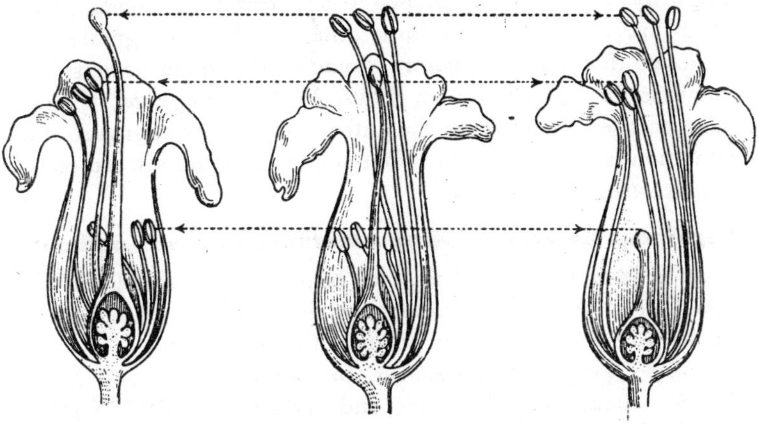

FIG. 390. Trimorphism in flowers of *Lythrum salicaria*. Three types of flowers show pistils and stamens of three different heights. Pollination usually takes place between organs of the same height.

the lengths of their organs (proboscis, legs, etc.), so that the short style will be cross-pollinated by pollens from low anthers and *vice versa*. This will involve pollens and stigmas of similar luxuriance and growth. Such *dimorphism* is also shown by jasmine and other sweet scented flowers of Oleaceae, by *Linum*, by *Fagopyrum* (buckwheat) of Polygonaceae, etc. Some species of *Oxalis*, *Linum* and *Lythrum* (Lythraceae) show *trimorphism* (Fig. 390). The three types of flowers show three positions of anthers and stigmas so that there is cross-pollination involving three heights. According to Tischler, dimorphic or trimorphic conditions may be altered by changed nutrition.

FIG. 391. Chasmocleistogamous *Commelina benghalensis*. *a*=normal *chasmogamous* flowers; *b*=underground *cleistogamous* flowers.

CONTRIVANCES FOR SELF-POLLINATION

Although cross-pollination seems to be favoured by Nature, there are cases where self-pollination is ensured.

1. Cleistogamy: In these cases the flowers never open (*kleistos*=closed) as opposed to most flowers which show **chasmogamy** (i.e., flowers open normally during *anthesis*). In *cleistogamous* flowers the pollens are shed within the closed flowers so that self-pollinaion is obligatory. *Cleistogamy* is seen in the underground flowers of *Commelina benghalensis* (Fig. 391) which are small and inconspicuous. This plant also bears normal *chasmogamous* blue flowers above. Such plants bearing normal as well as cleistogamous flowers are called **chasmocleistogamous.** Many rice varieties also are cleistogamous in the sense that the anthers in them shed their pollens and pollination is complete before the flowers open. *Cleistogamy* or *chasmocleistogamy* is also seen in balsam (*Impatiens balsamina*), pansy (*Viola tricolor*), *Oxalis, Portulaca,* etc.

2. Homogamy: *Homogamy*, as opposed to *dichogamy*, simply means that the stamens and carpels of a flower mature at the same time. So, there is a greater chance of self-pollination although that is not obligatory. Some homogamous flowers, however, show special mechanisms for self-pollination. Thus, in *Mirabilis jalapa*, when the stamens mature the filaments recoil and bring the anthers near to the stigma so that when they burst self-pollination is achieved. Somewhat similar adaptations are seen in *Argemone mexicana. Grewia subinaequalis,* etc.

3. Dichogamous flowers showing adaptations for self-pollination:

(a) Flowers of *Ixora, Gardenia, Vinca,* etc., have their anthers placed at the mouth of the corolla tube (*throat*). As the stigma elongates from below, it pushes out through these anthers which are already ripe and are pollinated in so doing.

(b) Compositae flowers are *protandrous*. In sunflowers the bilobed stigma is still young and hidden within the syngenesious tube of anthers when the latter ripen and shed the pollens within the tube. At this stage self-pollination is not possible as the receptive surfaces of the stigmas are not only immature but also hidden (Fig. 392A). In the second stage the bifid stigma grows through the anther tube pushing out pollens and opens out above (Fig. 392B). Cross-pollination usually takes place at this stage through the agency of insects. But, if cross-pollination fails, it is seen that the stigma lobes curl back so that the receptive surfaces brush against any pollen still sticking on its surface and are thereby self-pollinated (Fig. 392C).

It is seen in actual practice that a large number of agricultural crops are naturally self-pollinated. Among these are rice, wheat, barley, oats, potato, peas, beans, tobacco, linseed, tomato and most jute varieties (specially *Corchorus capsularis*).

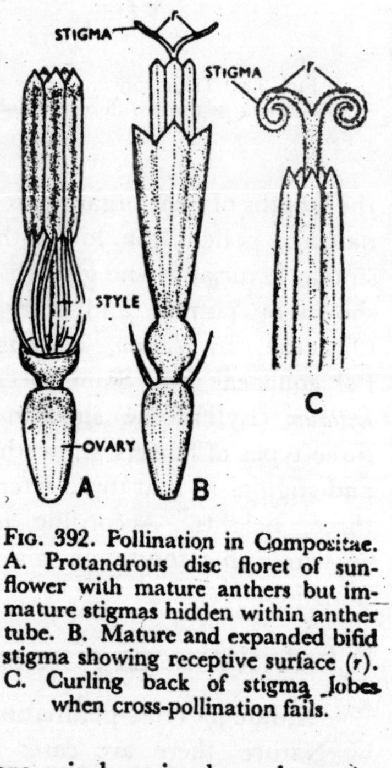

FIG. 392. Pollination in Compositae. A. Protandrous disc floret of sunflower with mature anthers but immature stigmas hidden within anther tube. B. Mature and expanded bifid stigma showing receptive surface (*r*). C. Curling back of stigma lobes when cross-pollination fails.

AGENTS FOR POLLINATION

It has already been stated that pollination, in a vast majority of cases, involves some agent. The usual agents are wind, animals and water. Accordingly, the types of pollination are **anemophily** or **anemogamy** (*anemos*=wind), **zoophily** or **zoidiophily** or **zoogamy** or **zoidiogamy** and **hydrophily** or **hydrogamy.** The terms *zoogamy* and *zoidiogamy* should better not be used as they are of ambiguous meanings. *Zoophily* is further subdivided according to the type of animal involved.

I. ANEMOPHILY

Anemophilous or wind-pollinated flowers are inconspicuous and not showy. They are also devoid of scent, nectar, etc. On the other hand, they produce a very large quantity of dusty pollens so that in great forests, the air is often charged with these. Such air may be healthy as in the proximity of pine forests where the abundant pollens are described as 'sulphur showers', or, the pollen-laden air may give rise to an epidemic of allergic hay fever. Another adaptation of anemophilous flowers is the branched bushy stigma (Figs. 374C, 374E & 393) capable of catching pollens from air easily as is seen in cereals. Such flowers are often unisexual and occur in bunches. The anthers are often versatile, swinging freely in air (Fig. 334) and the pollens are dry, light and smooth-walled. Some wind-pollinated flowers like *Urtica dioica* show an explosive mechanism by which the anthers burst and surcharge the air with pollens. In many anemophilous trees, flowers appear when they are bare of all leaves so that wind can play in full. In evergreen plants or herbs also the anemophilous flowers are well exposed to wind. Anemophily is found in most cereals and palms and in a number of Archichlamydeous families like Salicaceae (poplar, willow), Betulaceae (alder, hazel, birch), Fagaceae (oak, beech), Ulmaceae (elm), Urticaceae (*Urtica*), etc. According to some, anemophily is a primitive character in the evolution of plants.

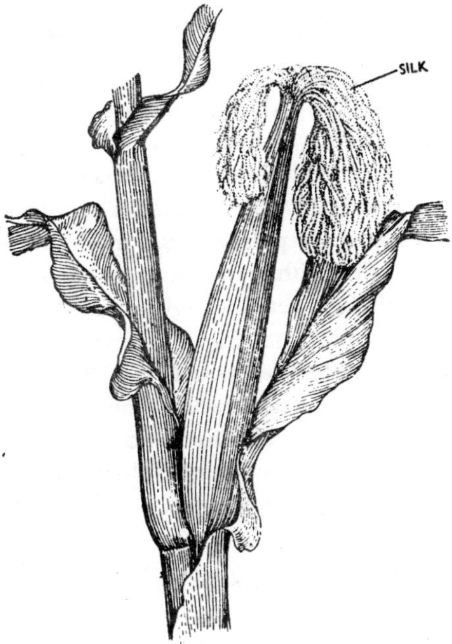

FIG. 393. A maize *cob* showing the bushy 'silk' formed of innumerable long and feathery styles and stigmas very suitable for catching wind-borne pollens.

II. ZOOPHILY

Zoophilous flowers are again subdivided into **(A) Entomophilous** (insect-pollinated), **(B) Ornithophilous** (bird-pollinated), **(C) Chiropteriphilous** (bat-pollinated) and **(D) Malacophilous** (slug- and snail-pollinated). Of these *entomophily* is by far the most important.

A. ENTOMOPHILY

A great majority of the flowers that we see about us today are insect-pollinated. It has been said that the evolution of flowering plants has gone on side by side with the evolution of insects. Insect-pollinated flowers are made attractive to insects in different ways and the pollens are sticky with a rough surface so that they may easily stick to insect limbs. The stigma also is similarly sticky to be able to receive the pollens more easily. To attract insects the following qualities have passed the test of evolution:

(1) *Conspicuousness*

A flower may be rendered conspicuous by the bright colour of its petals or perianth leaves. Sometimes bracts, sepals or even stamens may become petaloid and serve the same function. The conspicuousness may be due to the individual flower or to the aggregation of flowers forming such inflorescences as capitula. Müller's suggestion that certain insects show preference to certain colours has been found to be true. Thus, bees seem to prefer blue flowers, butterflies red and so on.

(2) *Scent*

Flies seem to be attracted by the nauseous odour (*indoloid* odour) emanating from certain flowers like *Rafflesia* and some aroids. But how far insects can distinguish scents is not known so that it cannot be said definitely whether the flies are attracted by the smell or by the fleshy appearance of the flowers. The actual role of odours in pollination is not yet clear. However, certain flowers, inodorous during day-time, emit a strong scent during night and are found to be pollinated by night-flying moths. This suggests a definite role of scent in pollination. Many essential oils are present in different flower petals being produced in internal glandular cavities.

(3) *Nectar*

Nectar glands are situated in different positions of flowers secreting a sugary fluid called nectar or honey much sought after by insects. As a matter of fact, nectar is the most valuable thing for which insects visit flowers. Bees collect nectar and elaborate it into commercial honey. Nectar glands may be located on the thalamus, on the outside of sepals, at the bases of sepals, petals, carpels and even in the septa within ovaries. The nectar is often concealed within corolla tubes or spurs so that insects have to do much searching for this and thereby get thoroughly dusted with pollens. Often there are spots and lines on the petals which converge towards the nectar glands and are called *nectar-guides* as they are supposed to guide insects towards the nectaries. Sometimes nectaries are present on extrafloral parts as on leaves, stipules, etc. These are supposed to divert unwanted insects.

(4) *Edible sap*

Some flowers (e.g., some orchids like *Orchis morio*) do not have nectaries but secrete an edible sap.

(5) *Edible pollen*

Pollens of very few flowers are sought after by insects. Nevertheless, bees collect a large amount of pollens to nourish their youngs and, at the breeding season, collect more pollen than nectar. Bees also utilise the wax on the pollens to build their combs.

(6) *Special mechanisms*

Certain entomophilous flowers show special mechanisms favouring pollen dissemination by insects:

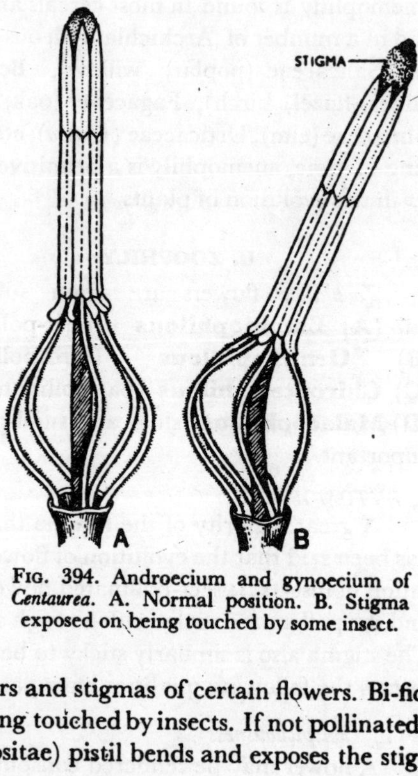

FIG. 394. Androecium and gynoecium of *Centaurea*. A. Normal position. B. Stigma exposed on being touched by some insect.

(i) *Irritability* is a feature shown by the anthers and stigmas of certain flowers. Bi-fid stigmas of *Bignonia*, etc., close the two flaps on being touched by insects. If not pollinated, they open again. Corn-flower (*Centaurea* of Compositae) pistil bends and exposes the stigma on being touched (Fig. 394).

(ii) *Explosive mechanism:* Some flower anthers burst and discharge the pollens on being touched.

(iii) *Ballistics:* Versatile stamens and other balancing features (viz., *Salvia* described below) often greatly help the dusting of insects with pollens.

According to the principal attraction in the flower, two groups may be recognised within entomophilous flowers:

I. Pollen Flowers

Flowers sought after mainly for pollens are usually actinomorphic with a large number of stamens. Flowers of *Argemone* and other poppies, *Anemone*, *Magnolia*, *Anona*, water-lily, potato and other Solanaceae, etc., fall under this group.

II. Nectar Flowers

A large majority of flowers fall under this group. The nectaries may be fully exposed, partly concealed, fully concealed or the flowers may be borne in compact groups (*social flowers*). The position of the nectary is the most important factor in determining which type of insect visits the flower.

The chief insects that visit flowers belong to the groups *Hymenoptera* (bees and wasps), *Lepidoptera* (butterflies and moths), *Diptera* (flies), *Coleoptera* (beetles), *Thysanoptera* (thrips), and *Hemiptera* (bugs). It is found that usually certain groups of flowers are visited by certain groups of insects. This adaptation is regulated by the advantage that the insect gains (pollen, nectar, etc.), the season and the hour when the plant flowers, habits of the insects and, above all, the accessibility of the nectary, etc., to the particular type of insects.

Of these insects, the *Lepidoptera* have the longest tongues, next come the *Hymenoptera* while the other groups have rather short tongues. Accordingly, flowers can be arranged in groups in respect of its adaptation to types of insects.

(i) Flowers adapted to short-tongued insects, i.e., *Coleoptera* (beetles), *Thysanoptera* (thrips) and *Hemiptera* (bugs). Naturally, these insects are limited to flowers which have either (a) the *nectary fully exposed* as in Rutaceae (nectaries in exposed disc—Fig. 310), Vitaceae (disc—Fig. 311), Umbelliferae, Euphorbiaceae, etc., or (b) *partially concealed nectaries* as in some Cruciferae, Ranunculaceae, Berberidaceae, Rosaceae, etc.

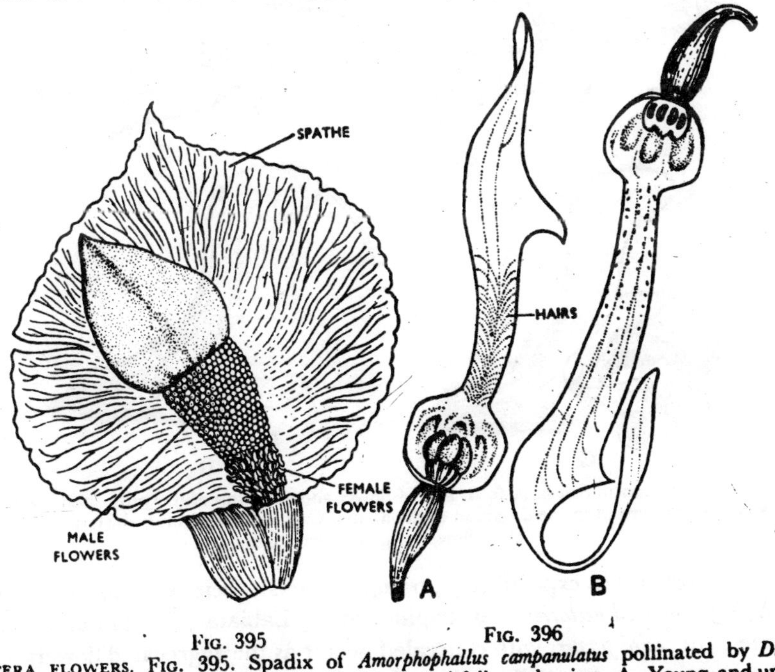

Fig. 395 Fig. 396

Diptera flowers. Fig. 395. Spadix of *Amorphophallus campanulatus* pollinated by *Diptera*. Fig. 396. Flowers of *Aristolochia clematitis* showing pitfall mechanism. A. Young and upright flower with mature stigmas. Flies can crawl down but cannot get out because of the hairs pointed down. B. Mature drooping flower with burst anthers and withered hairs allowing free movement of flies.

(ii) Diptera flowers

Diptera are small feeble flies which carry on pollination by crawling into flowers. These flies do not seem to gain much by their visits except, possibly, a shelter. Probably they are attracted by nauseous smells. Among these we may consider two instances:

(*a*) *Nauseous flowers:* These include the common aroids, e.g., *Arum maculatum* (Fig. 267), *Typhonium trilobatum*, *Amorphophallus campanulatus* (Fig. 395) *Alocasia indica, Colocasia esculentum*, etc., which emit strong fetid odour during night. On the spadix female flowers are located below and male flowers on top. Flies crawl up and down and thereby pollinate them. *Rafflesia arnoldi* and similar evil-smelling flowers also fall in this group. The odour or the fleshy appearance of such flowers attracts carrion flies.

(*b*) *Pitfall flowers:* These possess special traps for *Diptera* flies. *Aristolochia clematitis* flower (Fig. 396) serves as a good example. The flower is *protogynous*. In the young stage it is upright and flies can crawl down the corolla tube pushing the downward pointed hairs but they cannot come up until these hairs wither away. By that time the stigmas wither, the anthers mature bursting out the pollens and the flower droops down. So, the flies come out pollen-smeared. They can now enter a second younger flower, pollinate its stigma and again remain trapped until the anthers of that flower mature.

(iii) Hymenopterous flowers

Flowers of this group are pollinated by members of the *Hymenoptera* (bees, wasps, etc.). While a varied and large number of flowers are included here, red, blue or violet coloured zygomorphic flowers predominate. While these insects may visit flowers wherein the

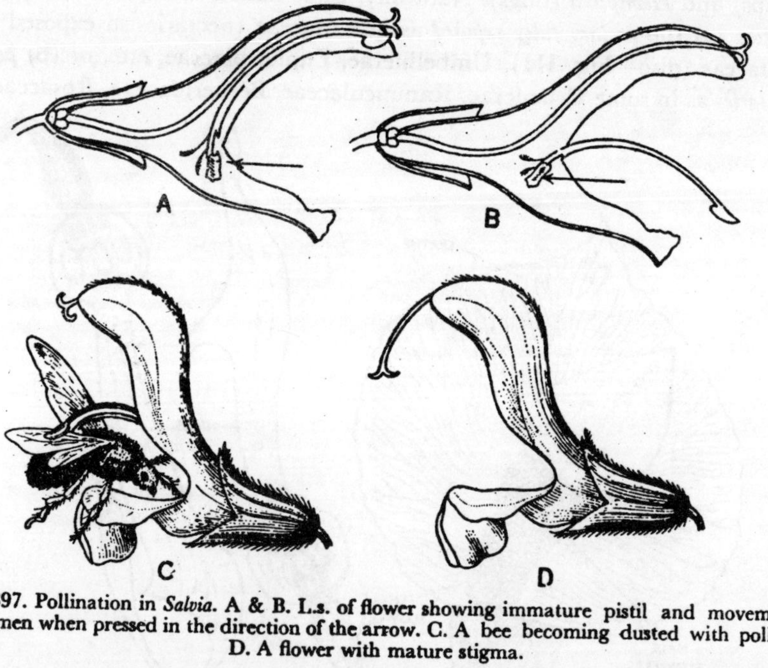

FIG. 397. Pollination in *Salvia*. A & B. L.s. of flower showing immature pistil and movement of stamen when pressed in the direction of the arrow. C. A bee becoming dusted with pollen. D. A flower with mature stigma.

nectaries are more or less exposed (e.g., orange flowers), their range also includes (a) flowers *with fully concealed nectaries* as in Papilionaceae, Labiatae, Scrophulariaceae, Orchidaceae, etc.; (b) *social flowers* with concealed nectar as the aggregated flowers of Compositae. These two groups are pollinated by *Hymenoptera* with shorter tongues (up to 6 mm). (c) Flowers with long tubes with the nectar hidden below so that they may be pollinated only by insects with proboscis 6 to 15 mm long. Within this group, again, one may distinguish some flowers favoured by special types of *Hymenoptera*. Some of these flowers are also visited by butterflies.

Within this group the interesting mode of pollination of a few flowers may be discussed:

(1) *Salvia* of Labiatae (The Sage flower)

Salvia has got a bilabiate corolla with two fertile epipetalous stamens. The stamens and pistil remain hidden under the upper lip (Fig. 397A). The flower is *protandrous* and the short epipetalous filament of each stamen is connected to the peculiar *distractile* connective (Fig. 345) which is long and lever-like, its two unequal arms separating the two anther lobes. The basal lobe of the anther is sterile while the upper lobe is fertile. A slight pressure on the lower anther lobe brings the upper lobe down (Fig. 397B). *Salvia* is bee-pollinated. Bees alight on the lower lip of the corolla and enter the flower to reach the nectary at the end of the corolla tube. In so doing, they push against the united lower anther lobes (Fig. 397C) thereby bringing down the fertile anther lobes which dust the bee's back with pollens. When the pistil of the flower becomes ripe, the stigmas protrude out of the upper lip (Fig. 397D) so that any bee entering the flower brushes against the stigma thereby pollinating it with pollen already on its back.

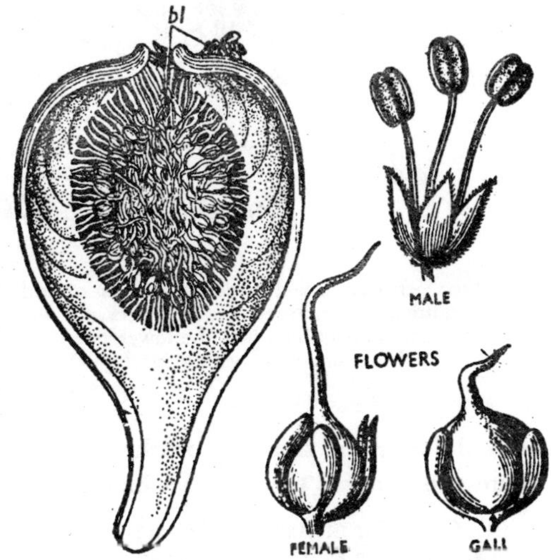

FIG. 398. L.s. *of Ficus carica* fig showing the orifice and *Blastophaga* (*bl*) wasps entering and coming out through it. Some *galls* are seen to have developed inside. Male flowers (from top), female flowers (from bottom) and gall flowers are shown separately.

(2) *Ficus carica* and other *Ficus* spp. of Moraceae

Flowers of *Ficus* plants are enclosed within the hollow pear-shaped hypanthodium inflorescences. There is a narrow orifice for entering the receptacle (Fig. 398) within which there are three types of flowers: male, female and gall. The male flowers are situated on the top near the orifice while lower down are the long-styled female flowers and the short-styled 'gall flowers' which also are female. Figs are pollinated by the gall wasp (*Blastophaga*) which crawls into the receptacle and lays eggs inside the ovules of the 'gall flowers' which it can easily reach by its ovipositor because of the shortness of the style. The eggs develop larvae which feed on the ovules and form *galls*. After passing the pupa stage the larvae develop into mature wasps and crawl out of the fig. In so doing they brush against the male flowers near the orifice carrying away pollens on their bodies. These pollen-laden insects then enter fresh figs where they pollinate the long-styled female flowers and lay eggs within the 'gall flowers' in their turn.

(3) Orchids of Orchidaceae

Orchids are peculiarly adapted for bee pollination with their pollens aggregated in pollinia (Fig. 399). The bee alights on the conspicuous *labellum* of the perianth and, being unable to enter the opening in front of the gynostemium, sends its tongue into the spur to reach the nectar below. In so doing, its head pushes against the stigmatic surface with the *rostellum* (which is also a part of stigma) which latter gives way so that the pollinia on the top of it come out and their sticky discs get stuck to the forehead of the bee. The pollinium at first remains erect but soon droops down on the bee head. When the bee goes to a

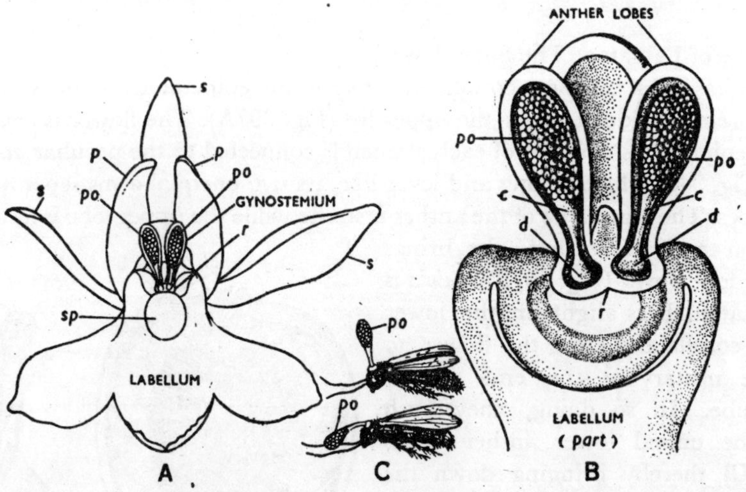

FIG. 399. Pollination mechanism in an orchid. A. A flower showing the different parts. B. The gynostemium enlarged. C. Two bees, the top one showing an erect pollinium on its forehead just after collection while the lower bee shows the later drooping of the pollinium facilitating pollination. *sp*=opening into spur; *r*=rostellum; *po*=pollinia; *s*=sepal; *p*=petal; *c*=caudicle; *d*=disc.

second flower and similarly brushes against a stigmatic surface, the bent and sticky pollinium gets stuck to the stigma which is thereby pollinated.

(4) *Calotropis* of Asclepiadaceae

Flowers of Asclepiadaceae are provided with *gynostegiums* formed by the union of the prismatic stigma with the androecium (Fig. 400). The pollens are in pollinia as in Orchidaceae. Two pollinia (from two different anthers) are attached to a glandular adhesive disc at the stigma angle (*translator* mechanism). These sticky discs get stuck to the

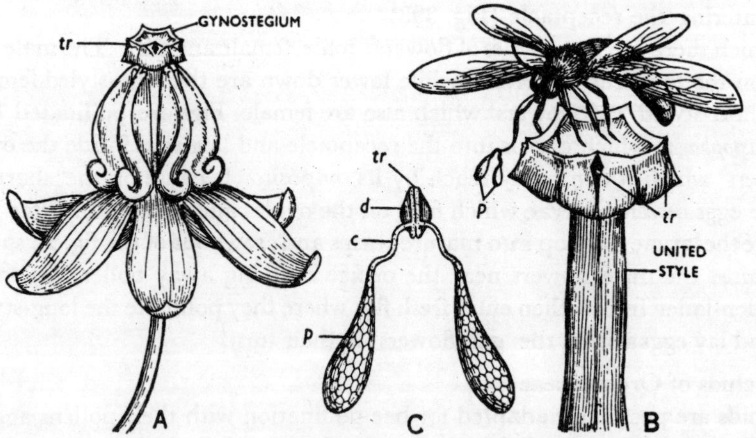

FIG. 400. Pollination mechanism in *Calotropis gigantia* of *Asclepiadaceae*. A. A flower. B. A bee removing a pair of pollinia from the gynostegium. C. A pair of pollinia. *tr*=translator locating paired pollinia; *d*=disc; *c*=caudicle; *p*=pollinia.

proboscis or legs of visiting bees so that the pollinia are pulled out when the bee moves. When the bee visits another flower, the latter gets pollinated with the sticky pollinia.

(iv) Lepidopterous flowers

These flowers are provided with the ***longest corolla tubes***, even longer than in the previous group so that the nectar can be reached only by the *Lepidoptera* (butterflies and moths) which have very long tongues (20 mm or longer). Many sweet scented flowers come under this group—lily, lavender, *Nyctanthes*, *Polyanthes*, jasmine, *Nicotiana*, pink, etc. Of these, the night-flowering ones are pollinated by moths and the day-flowering ones by butterflies. *Yucca* flowers (Fig. 401) are visited by the small moth *Pronuba yuccasella* for laying eggs within ovaries just in the same manner as figs are visited by *Blastophaga* wasps. While about 20% of ovules are destroyed in the process there is no seed-setting in Yucca plants without the visit of *Pronuba* moths.

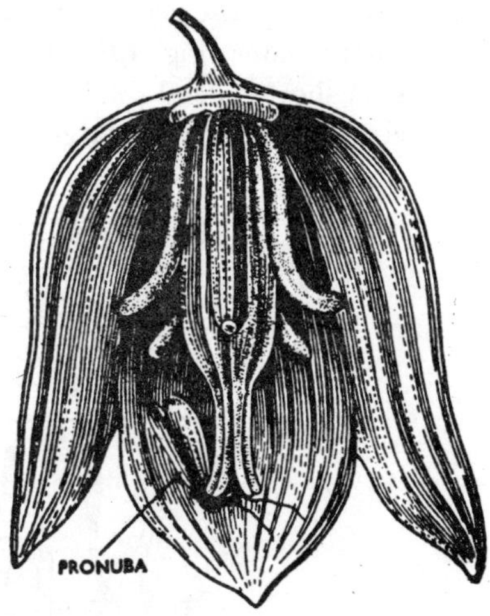

Fig. 401. A mature *Yucca* flower with a *Pronuba* moth in the act of pollination.

B. ORNITHOPHILY

Bird-pollinated flowers are not many in number. Tiny birds like the humming-birds and the honey-thrushes (birds only an inch or so long) feed on the nectar of flowers like *Bignonia capreolata* (Fig. 402) and thereby pollinate them. Large flowers of *Strelitzia* (Musaceae) are pollinated by a honey bird called *Nectarina afra*. In our country, silk-cotton (*Salmalia* or *Bombax*), *Erythrina* and a few other trees are visited by crows and mynas when in flower. These birds may play some part in the pollination.

C. CHIROPTERIPHILY

Bauhinia megalandra of Java, *Eperua falcata* (Leguminosae) and a few other trees are known to be pollinated by bats. Several trees in this country, like *Anthocephalus chinensis* (Rubiaceae) and silk-cotton, are also visited by bats which possibly help in the pollination although their role is not definitely known.

D. MALACOPHILY AND POLLINATION BY OTHER ANIMALS

Snails and slugs visit certain flowers and may have a role in their pollination. Many aroids, which are usually pollinated by *Diptera* flies are also visited by snails.

Fig. 402. A humming bird collecting nectar from *Bignonia capreolata* flowers, thereby pollinating them. (After Bates).

Squirrels visit a large number of flowering trees and may have some role in their pollination.

III. HYDROPHILY

Certain plants, as those of the families Naiadaceae, Ceratophyllaceae, Potamogetonaceae, Hydrocharitaceae, etc., are completely aquatic so that their pollination is adapted to such conditions. Pollination takes place completely under water (**hypohydrogamous**) in *Naias* and *Ceratophyllum* while it takes place on the water surface (**epihydrogamous**) in the common water weeds *Vallisneria, Hydrilla* and *Elodea* of Hydrocharitaceae. The dioecious plant *Vallisneria* having strap-shaped leaves grow in the mud at the bottom of stagnant water (Fig. 403). The male flowers are borne low down amongst the radical leaves on short-stalked spadix inflorescences out of which the individual flowers get detached and float freely in large numbers on the water surface. These flowers open on the

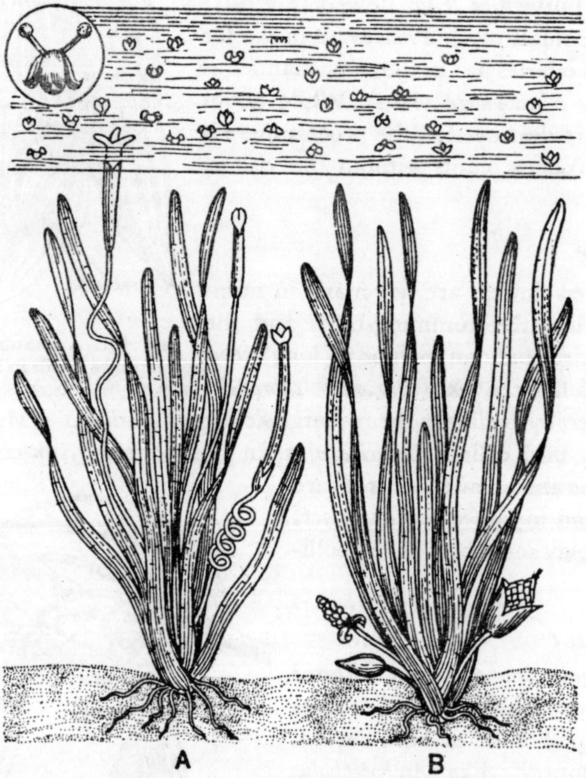

FIG. 403. Epihydrogamy in *Vallisneria*. A. A female plant. B. A male plant. A large number of male flowers and a female flower are floating on the surface. A male flower with the two erect stamens is shown enlarged.

surface and the three perianth leaves open widely exposing the two stamens vertically. The female flowers are borne singly on long wiry stalks which grow in such a manner that the flowers float on the water surface when mature. As the female flower is somewhat waxy, it causes a slight depression in the film of water because of surface tension. This depression, as well as wind, causes the detached male flowers to cluster around the floating female flower and, when the anthers burst, sticky pollens get attached to the stigma. Soon after pollination, the long stalk of the female flower begins to coil bringing the female flower again below water level until it reaches almost the tank base where the fruit matures.

CHAPTER X

SPOROGENESIS AND THE GAMETOPHYTES

Alternation of generations is a remarkable aspect of the life cycle of all higher plants. Angiosperm plants are *diploid* **sporophytes,** i.e., spore-bearing plants and the *haploid* spores are developed by reduction division or meiosis. The male spores or *microspores* (pollens) are developed by reduction division within the *microsporangium* (pollen sac) while the female spores or *megaspores* are similarly developed within the *megasporangium* (ovule). These, in their turns, develop the male and the female **gametophytes** which are *endosporous* (i.e., developing inside the spores) in the case of the seed-bearing plants (Spermaphytes). The process of development of the microspore (pollen) is termed **microsporogenesis** while that of megaspore development is **megasporogenesis.**

MICROSPOROGENESIS AND MICROSPORE

During the development of the microsporangium, the anther is seen at first as a homogeneous mass of meristematic cells, oblong in cross-section and surrounded by an epidermis (Fig. 404A). It then becomes more or less four-lobed and four longitudinal rows of **archesporial cells** are differentiated. The archesporial cells are marked off from the surrounding cells by their more deeply staining cytoplasm and conspicuous nuclei. There may be only one such archesporial cell in each of the four lobes (Fig. 404B) as in *Boerhaavia*, etc., or there may be more of them forming a plate (*Ophiopogon*, etc.). Longitudinally, also, there may be one to many of them. Each archesporial cell now cuts off a *primary parietal cell* towards the epidermis and a *primary sporogenous cell* on the inner side (Fig. 404C). The parietal cell now divides by periclinal and anticlinal walls giving rise to several layers of cells forming the wall of the anther while the sporogenous cell usually divides a few times giving rise to a number of **microspore** or **pollen mother cells** (Fig. 404D). The innermost layer of wall cells directly abutting on the sporogenous tissue forms the **tapetum** which is a nutritive tissue nourishing the developing microspores (Fig. 404E). The wall cells just below the epidermis form the **endothecium** which later loses the cell-contents, usually becomes fibrous, and forms the dry coat of the mature anther in which the epidermis becomes rather inconspicuous. Between the tapetum and the endothecium there are one to three *middle layers* of cells. The middle layers and the tapetum are usually crushed by the time actual meiosis occurs in the sporogenous cells. During **microsporogenesis** (i.e., development of *microspores* or pollens) the nucleus of each *microspore mother cell* undergoes *meiosis* or reduction division ultimately giving rise to four *haploid* (i.e., possessing **'*n*'** number of chromosomes) nuclei (Fig. 404—F & G). These four nuclei are arranged tetrahedrally (Fig. 404—H & I) and are soon invested with cell walls. Many variations are known of this typical pattern of meiosis, e.g., in maize a wall is formed across the dyad (2-nucleated condition). They are now the **microspores** or **pollens** which soon dry up and become powdery while the tapetum becomes absorbed. The anther now becomes a dry structure, the partition walls between the sporangia (i.e., loculi) are usually destroyed (Fig. 404J) and the microspores (pollens) are soon liberated by dehiscence of the anther.

The tapetal cells often become multinucleate and plays a great part in the nutrition of the pollens. Sometimes they develop a *plasmodium* after disintegration and play a part in the development of the *exine* of the pollen. Even a part of the sporogenous tissue may break down and serve for nutrition instead of developing spores.

While the pollens are dry and powdery in most flowers, peculiar conditions are often met with. In *Annona, Elodea, Typha*, etc., the four spores in a tetrad never separate but form

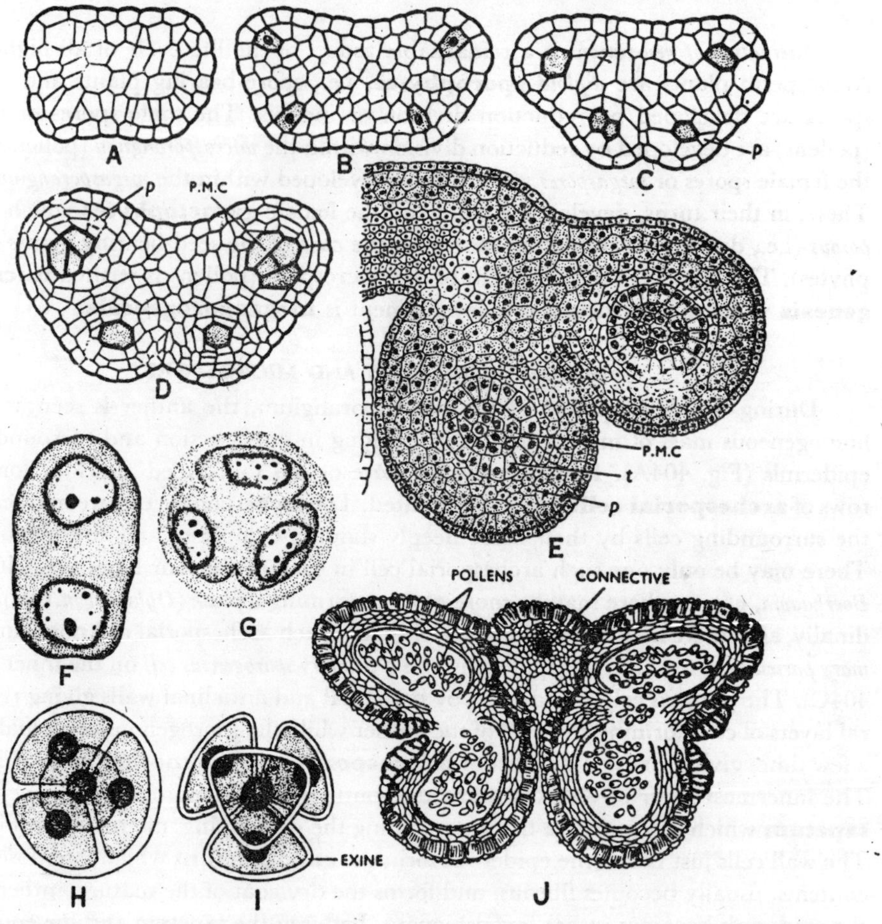

FIG. 404. Stages of anther development and microsporogenesis. A. T.s. of young anther. B. Differentiation of four rows of archesporial cells (shaded) in the four microsporangia. C. Primary sporogenous cells (shaded) and parietal cells (*p*). D. Microspore or pollen mother cells (P.M.C.). E. Microsporangia (pollen chambers) showing pollen mother cells (P.M.C.) and tapetum (*t*). F. Dyad stage of meiosis in P.M.C. G. Tetrad stage (the fourth nucleus is behind). H. & I. Tetrahedral arrangement and development of pollens. J. T.s. of mature anther.

compound pollen grains. In the Mimoseae 8 to 64 pollens often aggregate together while in the *gynostegium* of *Calotropis* and the *gynostemium* of orchids all the pollens of each anther lobe from a characteristic mass called **pollinium** (Fig. 405). Each pollinium is provided with a stalk called *caudicle* and a sticky base called *disc* or *corpusculum*.

The *pollen* or the *microspore* is a very minute structure (0·025 to 0·125 mm in diameter). It is unicellular and usually round although it may be oval, pyramidal, polyhedral, etc. It is provided with two coats—an inner, delicate cellulose layer called **intine** and an outer tough, cutinised layer called **exine** or **extine.** The exine is often sculptured or provided with spines, warts, etc. Occasionally, it is smooth. The exine may have a waxy coating to render the pollen more or less waterproof. Very often, there are some definitely thinner, circular spots or slits in the exine called **germ pores** or **slits.** These weak spots are utilised during the germination of the pollen. The pores are sometimes provided with lids (Fig. 409B) which open out like valves during germination. Very often, before the pollen is discharged from the anther it becomes *binucleate* (Fig. 406), the original single

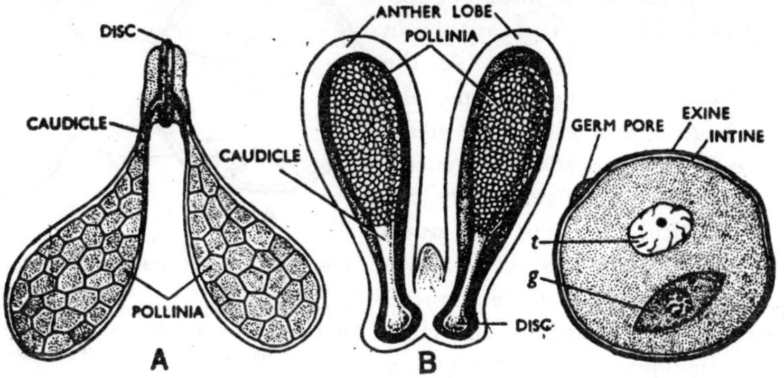

FIG. 405 FIG. 406
FIG. 405. Pollinia of A. *Calotropis* and B. Orchid. FIG. 406. A binucleate pollen showing tube nucleus (t) and generative cell (g).

nucleus dividing into a **tube nucleus** and a **generative nucleus.** The latter, with some cytoplasm surrounding it, becomes the **generative cell** (described later). Sometimes the pollen may even become *trinucleate* (Fig. 410), as in most cereal crops, by complete development of the male gametophyte even before it is shed. Fig. 407 shows some different forms of mature pollens. **Palynology** is the science involving the study of pollens.

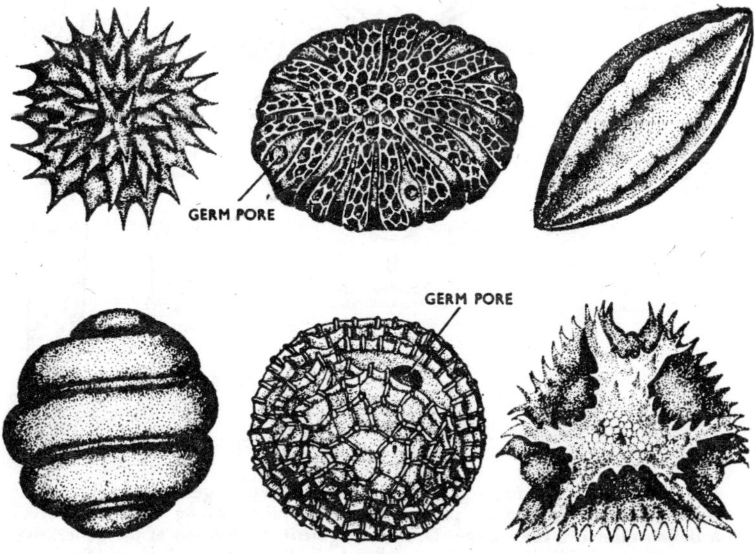

FIG. 407. Different pollens showing various types of sculpturing.

DEVELOPMENT OF THE MALE GAMETOPHYTE

The nucleus of the microspore begins to divide very soon after it is formed. Its division is usually quicker in the warmer climate than in the cooler regions. Such division may begin even before the microspores dissociate from the tetrad condition.

When the pollen is lodged on the stigma, usually its nucleus has already divided.

The microspore cell divides into two very unequal cells with the smaller cell attached to the wall (Fig. 408A). The latter soon loses contact with the wall (Fig. 408B) and becomes the lens-shaped **generative cell** floating freely in the cytoplasm of the large **vegetative** or **tube cell** (Fig. 409A). It is then in the *binucleate* (or two-celled) *stage*. As the pollen germinates and the pollen tube comes out through a *germ pore* (Fig. 409B) the vegetative nucleus precedes the generative cell (Fig. 409C). The generative cell soon divides mitotically to form two **male gametes** (Fig. 409D) and this act is known as **spermatogenesis** as the male gametes of Angiosperms are equivalent to the ciliated *sperms* of the lower groups of plants. The gamete nuclei are enveloped in cytoplasmic sheaths, thus forming *gamete cells*. At a later stage the two *male gamete cells* are seen to follow the *vegetative nucleus* (Fig. 409E).

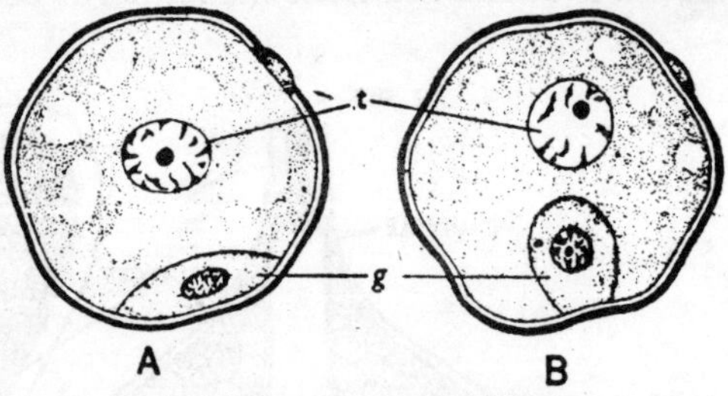

FIG. 408. Early stages of the development of the male gametophyte. A. The first division: t=tube nucleus, g=generative cell attached to wall. B. Later condition of the above: generative cell is detaching itself from the wall.

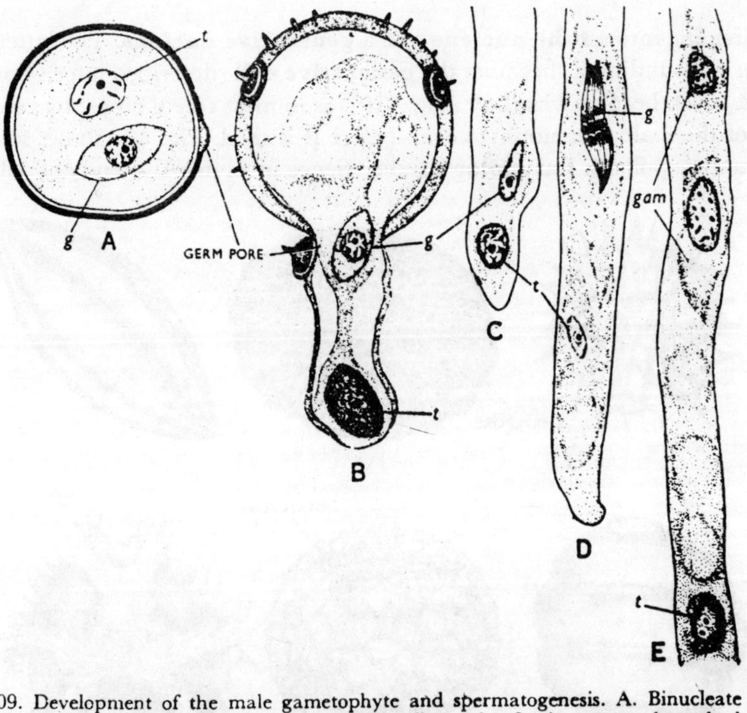

FIG. 409. Development of the male gametophyte and spermatogenesis. A. Binucleate stage. B. Intine coming out through a germ pore as a pollen tube forcing open the valvular lid. C. Tip of a pollen tube at a later stage. D. *Spermatogenesis* or division of the generative cell to form two male gametes. The tube nucleus is ahead. E. Pollen tube developed further showing two male gametes and tube nucleus. (A, C, D & E are stages in water hyacinth after Ganguee); t=tube or vegetative nucleus; g=generative cell; *gam*=male gamete.

Soon the *vegetative nucleus*, which seems to be of no importance being a mere vestigial organ, disappears. As the pollen is the *microspore*, the contents within the pollen and the pollen tube formed by the germination of the microspore is the **male gametophyte** of Angiosperms.

While the male gametophyte, thus, usually develops after the pollen has germinated, there are many instances where it develops within the pollen before the pollen tube is formed and, in some cases, even before pollination. **Trinucleate pollens** with tube nucleus and two gamete cells are observed in such cases (Fig. 410). Such trinucleate pollen formation is the rule in many plants (e.g., *Portulaca*), specially the cereal crops like wheat, rice, maize of Gramineae and sugarcane of the same family.

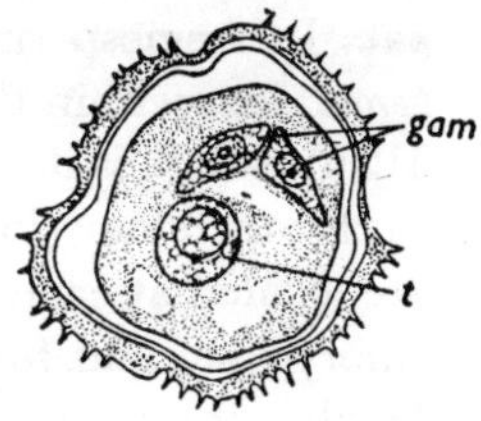

FIG. 410. Trinucleate pollen of *Portulaca* (after Cooper). *t*=tube nucleus, *gam*=male gamete cells.

MEGASPOROGENESIS AND DEVELOPMENT OF THE FEMALE GAMETOPHYTE

The *ovule* or the *megasporangium* develops as a small protuberance of the placental tissue. In the very young ovule a single hypodermal cell is differentiated as the **archesporium** (Fig. 411A). This archesporium cell may or may not cut off some *parietal cells* and then becomes the **megaspore mother cell** (Fig. 411B). The megaspore mother cell

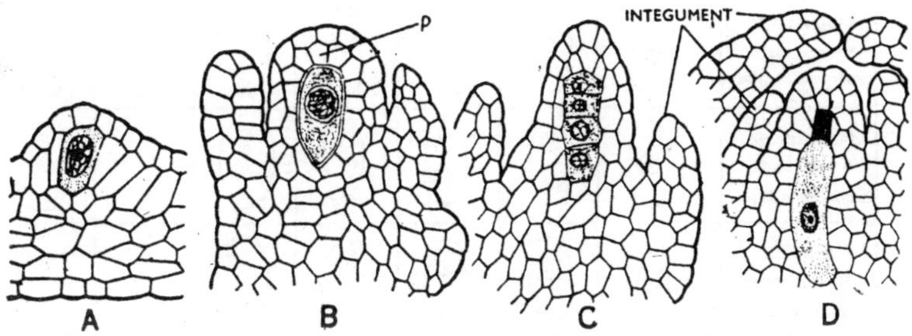

FIG. 411. Stages of ovule development and megasporogenesis. A. Protuberance of placental tissue and differentiation of archesporium (shaded). B. Parietal cells (*p*) and megaspore mother cell (shaded). One integument is developing. C. Linear tetrad of megaspores (shaded). D. Disintegration of the three upper megaspores and enlargement of the functional megaspore. Both integuments have developed.

now undergoes *meiosis* or reduction division, and, usually, a linear row of four haploid **megaspore cells ('linear tetrad')** is formed (Fig. 411C). Meanwhile, two integuments develop from the base of the ovule. Of the linear tetrad of megaspores, usually the lower-

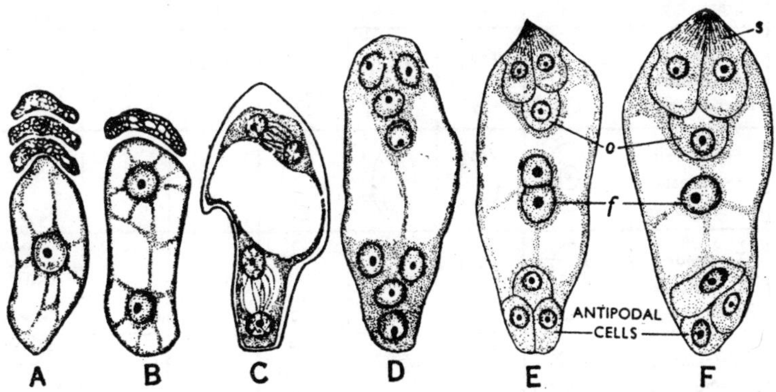

FIG. 412. Development of the female gametophyte or the embryo sac. A. Functional megaspore. Disorganising megaspores on top. B. 2-nucleate stage. C. 4-nucleate stage. D. 8-nucleate stage. E. Coming together of two nuclei at centre and organisation of embryo sac. F. Mature embryo sac showing egg apparatus with two synergids (*s*) and an egg cell (*o*) at micropylar end, fusion nucleus (*f*) at centre and antipodal cells at chalazal end.

most one enlarges and becomes the functional megaspore while the three on top disintegrate (Fig. 411D).

The *functional megaspore* now develops the **female gametophyte** or the **embryo sac.** In Angiosperms, the development of the female gametophyte is completely *endosporous*, i.e., within the megaspore. In a typical case, the nucleus of the embryo sac (Fig. 412A), which is the same as the functional megaspore (Fig. 411D), divides into two (Fig. 412B), then four (Fig. 412C) and finally, eight daughter nuclei (Fig. 412D) four of which are located at each pole. Then, one nucleus from each pole moves to the centre of the embryo sac and fuses there forming the **fusion** or **secondary nucleus** (Fig. 412E). Finally, the embryo sac or the female gametophyte becomes organised (Fig. 412F). Three nuclei at the base form the **antipodal cells.** The *secondary nucleus* remains at the centre. On the top, three cells form the **egg apparatus** which consists of two flask-shaped

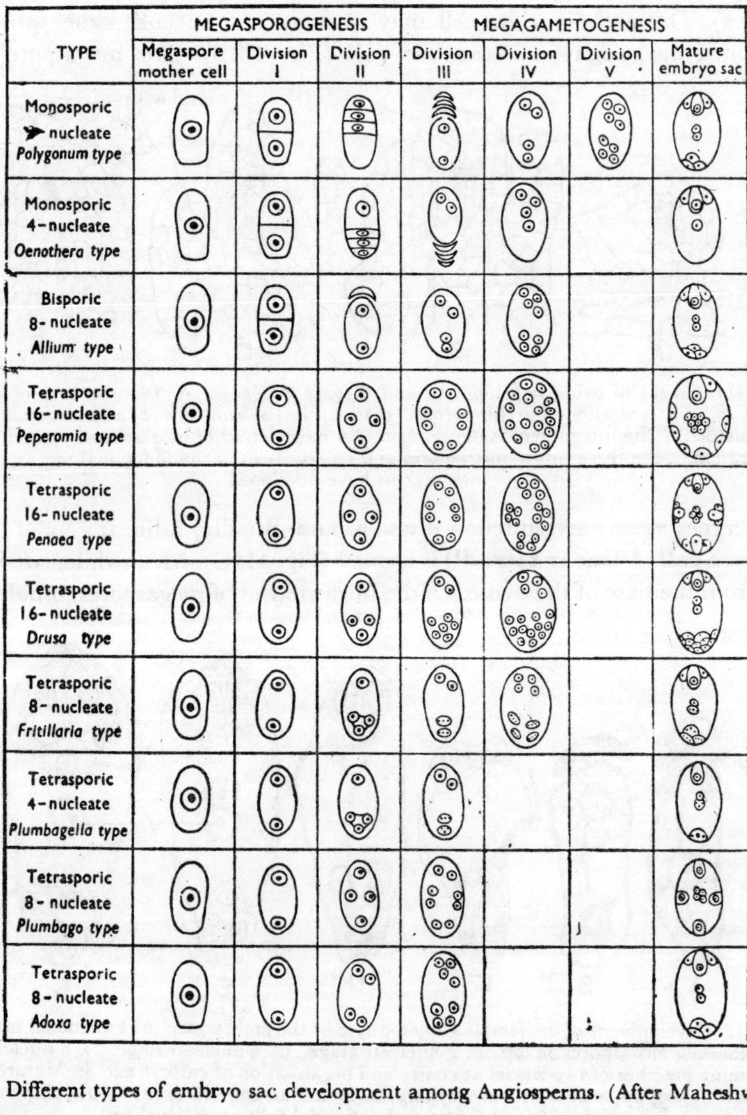

FIG. 413. Different types of embryo sac development among Angiosperms. (After Maheshwari).

synergids and a round **egg cell** (or **ovum** or **oosphere**) hanging between and below them. The synergids usually are somewhat notched by an indentation and they show striations at the tip ('filiform apparatus'). There is prominent vacuolation below the synergid nuclei and above the egg nucleus, showing accumulation of cytoplasm at different regions.

While the above method of gametophyte formation is usually described as the typical or *normal type* it has been found that many Angiosperms do not conform to this rule but show variations. The more intelligent students should be able to follow the different types of embryo sac development from the accompanying diagram (Fig. 413) prepared by Professor P. Maheshwari. First, there is a great variation in megasporogenesis itself. Often, the usual *four* separate megaspore cells do not develop. If four megaspores develop the female gametophyte (embryo sac) is **monosporic,** being developed from the usual *single* megaspore. Here we have the so-called *normal type* which is the *Polygonum type* in the diagram being shown only by a few plants like *Polygonum*. In *Oenothera*, the development is monosporic but only 4 nuclei (the antipodals are eliminated) develop instead of 8 (*Oenothera type*). In *Allium*, the megaspore mother cell divides only once giving rise to two haploid cells one of which develops a **bisporic** (as it contains the nuclei of *two* megaspores) embryo sac which is not normal although it develops the normal 8 nuclei (*Allium type*). In a number of cases no cell wall formation accompanies meiosis so that the normal separate megaspores are not formed. Here the embryo sac is **tetrasporic** as all the nuclei of the *four* megaspores are within the single cell wall. Reduction division (meiosis) of course takes place but only separate nuclei, and not cells, are formed. There are many variations of the tetrasporic embryo sac as shown in the diagram. In *Peperomia*, 16 nuclei are formed inside the megaspore mother cell (*Peperomia type*) and the diagram shows how these are arranged. In *Adoxa*, the megaspore mother cell develops 8 nuclei simulating the normal (*Polygonum*) type although it is tetrasporic (*Adoxa type*). There may be great variations in the appearance of the ultimate gametophyte, viz., the egg apparatus may show only one synergid (*Peperomia*), the fusion nucleus may involve only one (*Oenothera type*) or more than two (*Peperomia, Penaea, Plumbagella* and *Fritillaria types*) polar nuclei, the antipodals may vary from none in *Oenothera type* to a large number (*Peperomia, Penaea* and *Drusa* types; as many as 300 antipodal cells are reported in *Sasa paniculata* of *Bambusae*) and so on.

CHAPTER XI

FERTILIZATION, EMBRYOLOGY, SEED AND THE LIFE CYCLE

Pollination ends in a copious dusting of the stigma surface with pollen grains (Fig. 414). The next stage is the germination of the pollen to form a pollen tube wherein the male gametes are carried to the female gametophyte or the embryo sac whose development is already complete. **Fertilization** or **syngamy** or **fecundation** is the final phase of this biological process whereby the male and the female gametes fuse with one another giving rise to the **embryo.**

The pollen grain, after being liberated from the anther is *viable* for a short or long period. If it is transferred to the stigma during this period, it germinates. The **germination of the pollen** may be helped by the fluid of the stigma. Pollens may also be made to

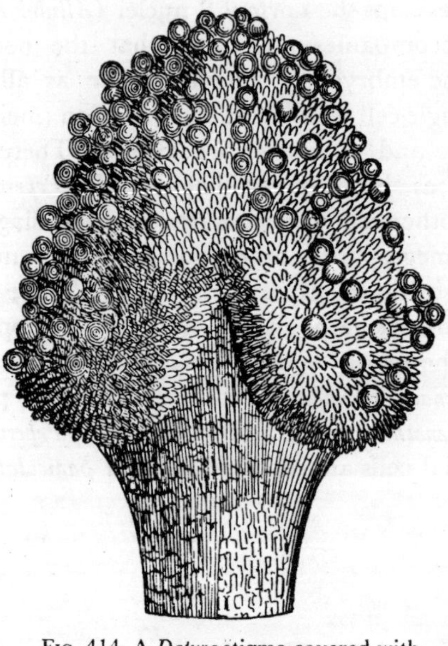

FIG. 414. A *Datura* stigma covered with pollens. (After Figuier).

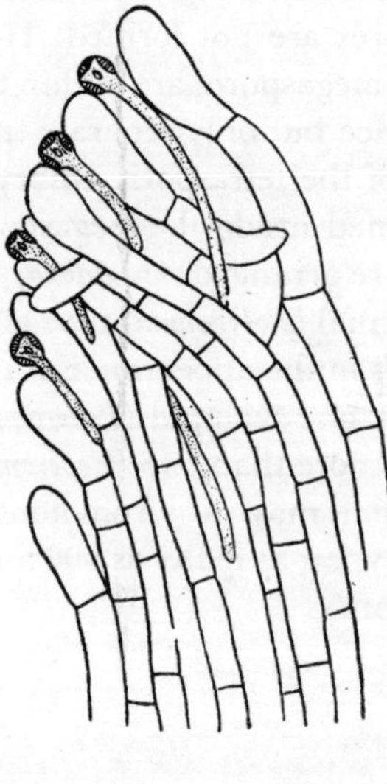

FIG. 415

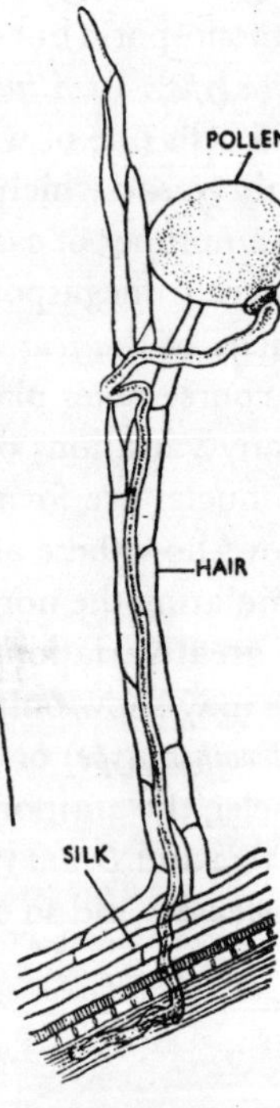

FIG. 416

FIG. 415. L.s. of a part of a stigma to show the growth of pollen tubes through stigma papillae.
FIG. 416. The long pollen tube of maize passing through the *silk*. (After Miller).

germinate in artificial culture media successfully. Germination may begin almost immediately after pollination as in sugarcane (*Saccharum*) *Sorghum* or it may take several hours or even days. The pollen absorbs liquid from the moist surface of the stigma and the intine comes out through one of the *germ pores* in the form of a single unbranched tube (Fig. 409). Abnormally, however, several pollen tubes may develop out of the same pollen as in some Malvaceae, Cucurbitaceae, etc., or, the single pollen tube may be branched. The pollen tube grows easily through the stigmatic papillae (Fig. 415) and then passes into the tissues of the style. The length of the stylar tissue which the pollen tube has to traverse may vary from nil in sessile stigmas to some 50 cm in the maize plants provided with silk (Fig. 416). Styles are solid or hollow. The centre may be filled with a longitudinal transmitting tissue or, it may be a canal full of mucilage. The pollen tube may grow through this or through the tissue of the stylar wall. Usually, only the distal portion of a long pollen tube contains living protoplasm and, as the nuclei within it passes forward, callose plugs are left in the empty portions behind them.

When the pollen tube (now with the male *gametes*) reaches the ovary its course is to find out an ovule and to reach the *embryo sac.* The pollen tube usually enters the ovule (orthotropous or anatropous) through the micropyle. This is known as **porogamy** (Fig. 417). But, in some cases the pollen tube is found to enter through the chalaza. This latter method, known as **chalazogamy** (Fig. 418), is found in Casuarinaceae, Betulaceae, Fagaceae and other members of Archichlamydeae. Formerly, much phylogenetic importance used to be attached to this character but, now it seems that the phenomenon is more of physiological importance as the same species may show both types of behaviour. In some cases (e.g., *Cucurbita*) the pollen tube enters the embryo sac piercing the integuments. This is termed **mesogamy.**

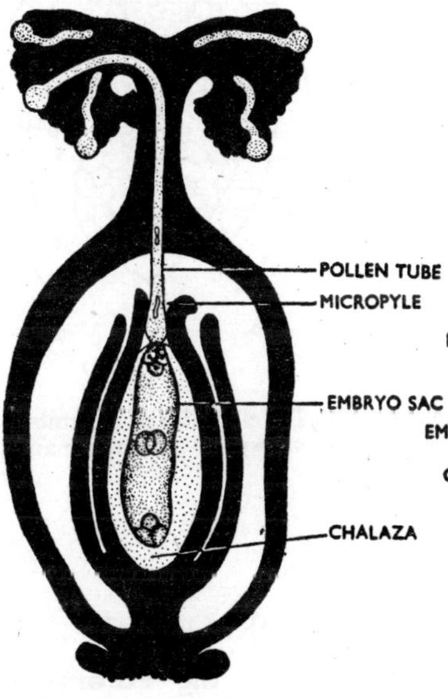

FIG. 417. Porogamy.

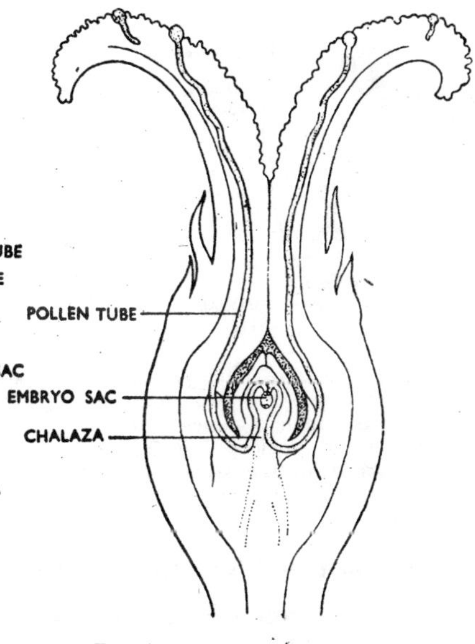

FIG. 418. Chalazogamy in *Juglans*. (After Strasburger).

In the passage of the pollen tube towards the micropyle, it has to pass through some empty space. Sometimes the pollen tube is directed in its growth through this distance by an **obturator** which forms a sort of a plug on the micropyle and through which the pollen tube may grow. The *obturator* is usually of placental origin.

FERTILIZATION

On penetrating the nucellus, whether through the micropyle, or, by a round about way through the chalaza, or, by piercing the integuments, the pollen tube penetrates the embryo sac and enters the *egg apparatus*. Ultimately, the tip of the pollen tube bursts and both the gametes are discharged. Probably the synergids play little or no role in this act. One of these gametes fuses with the *egg cell* of the egg apparatus. The other gamete fuses with the *secondary fusion nucleus*. This behaviour of the two male gametes is termed

double fertilization which was first observed by Nawaschin (1898) in *Lilium* and *Fritillaria* species. As a result of the first fertilization the **oospore** cell is formed which is the *mother cell of the embryo* and is a **diploid** cell containing 2*n* complement of chromosomes whereas the microspore and all nuclei of male and female gametophytes are **haploid** with *n* complement of chromosomes. The *secondary fusion nucleus* of the embryo sac, however, normally becomes 2*n* at the time of the fusion undergone by it. (Exceptions to this have already been mentioned in connection with female gametophyte development). So, the fusion of this nucleus with the second male gamete is **triple fusion** and the resultant nucleus is **triploid** or 3*n*. This is the *first nucleus of the endosperm.*

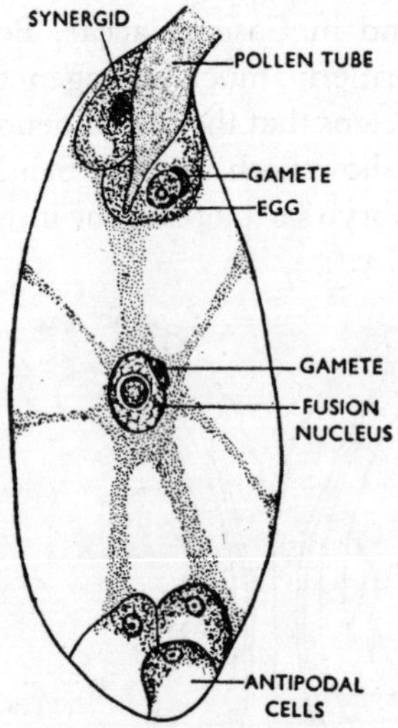

FIG. 419. Fertilization. The pollen tube has discharged the two male gametes. One male gamete is fertilizing the egg nucleus and the other is fusing with the secondary fusion nucleus.

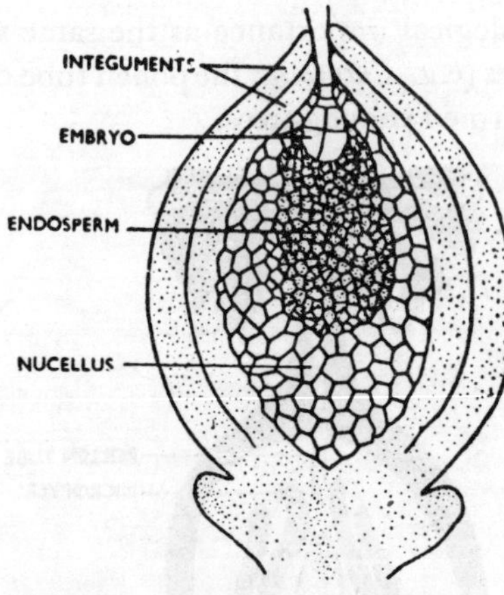

FIG. 420. Development of the embryo and the endosperm (diagrammatic).

EMBRYOGENY AND ENDOSPERM FORMATION

In the post-fertilizational changes within the ovule, the **embryo** and the **endosperm** are seen to develop simultaneously (Fig. 420). The *oospore* forms the embryo while the *fertilized fusion nucleus* (product of *triple fusion*) develops the endosperm. The other nuclei or cells within the embryo sac (*synergids*, *antipodal cells*) may behave differently but are destined to disorganise sooner or later.

DEVELOPMENT OF THE ENDOSPERM

The Angiospermic endosperm, except in special cases, is a *triploid* (3*n*) tissue as it is a product of triple fusion involving double fertilization. It is, thus, distinct from the endosperm of Gymnosperms and heterosporous Pteridophytes where the endosperm is a simple haploid (*n*) tissue of the gametophyte not involving any complication like polar fusion or fertilization. The real function of the endosperm is supplying nutrition to the growing embryo.

There are at least two families, Orchidaceae and Podostemonaceae, where the product of double fertilization soon disintegrates and endosperm development is completely suppressed. In other plants, three types of endosperm development, **nuclear, cellular** and **helobial,** are usual:

(1) *Nuclear type*

This is the common type of endosperm development. Here, the endosperm nucleus gives rise to a number of free nuclei (Figs. 421A, 424) which remain in the peripheral layer of embryo sac cytoplasm surrounding a large central vacuole. The nuclei, usually, become cut off by cell walls at a later stage if they be not already absorbed by the growing embryo. Such cell wall formation usually begins from the basal periphery. The cells soon organise into an endosperm tissue and goes on increasing further. Development of the nuclear endosperm in *Capsella bursa-pastoris* (Fig. 424) is described later. In some cases the central vacuole may not be filled up even in the mature seed. This is seen in the palms. In coconut, the central cavity full of coconut water is the original embryo sac vacuole while the nuclei around it form the peripheral endosperm kernel.

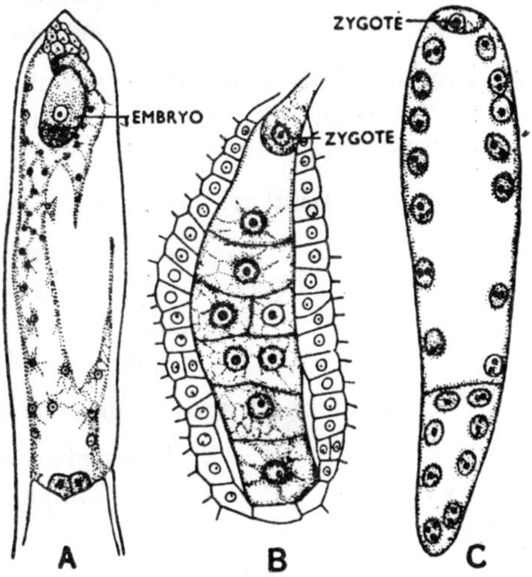

Fig. 421. Types of endosperm development. A. Nuclear. B. Cellular in *Villarsia* (after Stolt). C. Helobial in *Eremurus* (after Stenar).

(2) *Cellular type*

In a number of plants belonging to Annonaceae, Gentianaceae, etc., and in *Adoxa*, *Peperomia*, *Villarsia* (Fig. 421B), etc., division of the original endosperm nucleus is immediately followed by wall formation so that the endosperm is cellular from the beginning although some cells may enclose more than one nucleus.

(3) *Helobial type*

Among members of Helobiales (e.g., *Vallisneria*, *Eremurus*, *Limnophyton*, etc.) there is a type of endosperm development which is intermediate between the nuclear and the cellular types. Here, a partition wall develops between the two nuclei resulting from the first division of the endosperm nucleus so that the embryo sac is divided into two compartments. A large number of free nuclei is now developed in the upper chamber while the lower nucleus forms a few of them or may not divide at all (Fig. 421C).

DEVELOPMENT OF THE EMBRYO

Side by side with the development of the endosperm, the zygote or oospore is developing the *embryo*. Both dicotyledons and monocotyledons begin embryo development in the same way but, there is considerable difference in later differentiation. In all Angiosperms the zygote (Fig. 422A) divides to develop a two-celled **proembryo** (Fig. 422B). In most plants the first wall between the two cells is transverse while only in a few cases the first wall is more or less vertical (*Piperad type*, Fig. 423A$_1$). Of the two cells the one near the micropyle is termed the **basal cell** while the

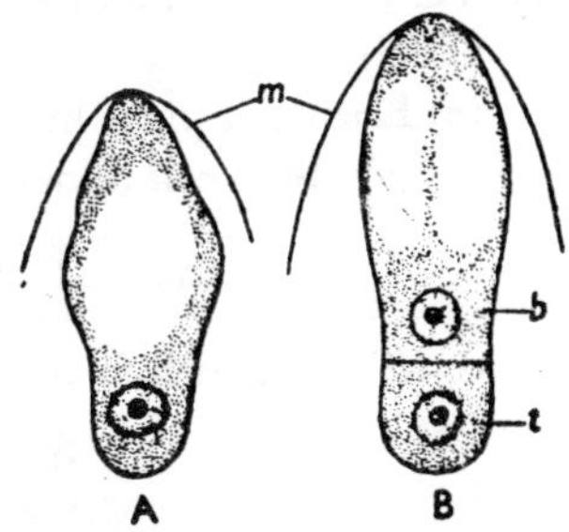

Fig. 422. A. Zygote, *m*=embryo sac membrane. B. 2-celled proembryo with transverse partition wall. *b*= basal cell (micropylar end), *t*= terminal cell.

one pointing towards the centre of the embryo sac is called the **terminal cell.** The basal cell usually forms the **suspensor** and may or may not contribute towards the future embryo in whose development the terminal cell takes the main part.

Johansen (1950) recognises six types of embryo development among the Angiosperms according to the manner of differentiation of the basal and terminal cells:

A. Zygote divides by a more or less vertical wall (Fig. 423A)

.. **1. Piperad type**
viz., *Peperomia, Balanophora*

B. Zygote divides by a transverse wall

I. Terminal cell divides by a vertical wall forming a ⊥-shaped proembryo (Figs. 423B, 424B, 425B)

(*a*) The basal cell plays little or no part in the formation of the embryo

.. **2. Onagrad** or **Crucifer type**
viz., Onagraceae, Cruciferae, Ranunculaceae, Rutaceae, Liliaceae, Juncaceae, etc., may be considered as the typical mode of embryo development.

(*b*) Both basal and terminal cells contribute to the formation of the embryo

.. **3. Asterad type**
viz., Compositae, *Polygonum*, etc.

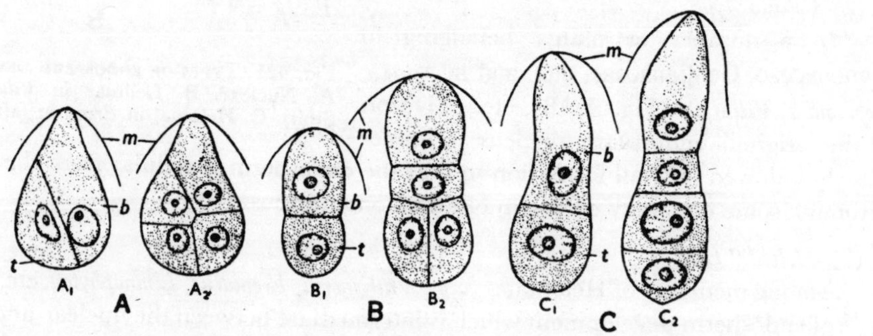

FIG. 423. TYPES OF EMBRYO DEVELOPMENT. A_1 & A_2. 2- and 4-celled proembryo of *Scabiosa* of *Piperad type* (after Souèges). B_1 & B_2. 2- and ⊥-shaped, 4-celled proembryo of *Onagrad type* (after Johansen). C_1 & C_2. 2- and straight 4-celled proembryo of *Nicotiana* of *Solanad type* (after Souèges).

II. Terminal cell divides by a transverse wall forming a straight proembryo (Fig. 423C)

(*a*) Basal cell plays little part in the future embryo

(*i*) Basal cell undergoes no further division but becomes a large suspensor cell

.. **4. Caryophyllad type**
viz., Caryophyllaceae, *Saxifraga, Medicago, Myriophyllum*, etc.

(*ii*) Basal cell forms suspensor of 2 or more cells

.. **5. Solanad type**
viz., Solanaceae, *Papaver, Linum*, etc.

(*b*) Both basal and terminal cells contribute to the formation of the embryo

.. **6. Chenopodiad type**
viz., Chenopodiaceae, *Myosotis*, etc.

Although there is no essential difference between the embryogeny of monocotyledons and that of dicotyledons in the early stages, later differentiation makes the two embryos so different that the development of two such embryos are described separately below.

(1) *Typical development of dicotyledonous embryo and endosperm*

The classical example of dicotyledonous embryo development is the plant *Capsella bursa-pastoris* of Cruciferae. The ovule is *campylotropous* so that the embryo sac and the

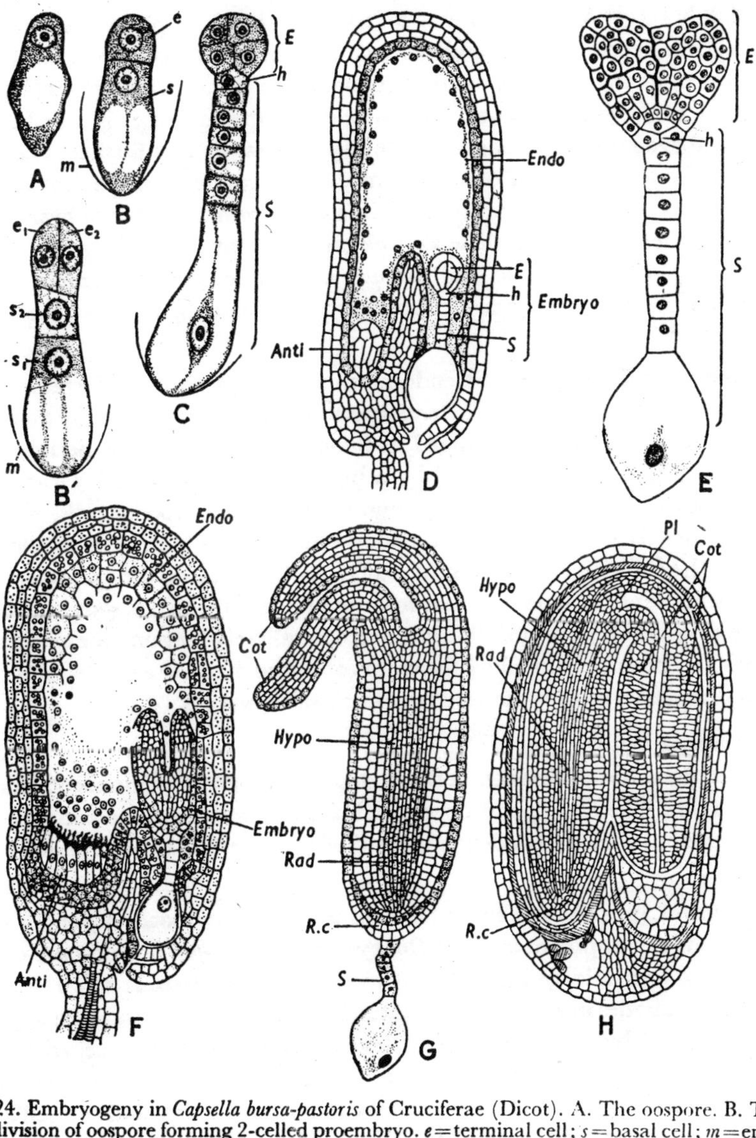

FIG. 424. Embryogeny in *Capsella bursa-pastoris* of Cruciferae (Dicot). A. The oospore. B. Transverse division of oospore forming 2-celled proembryo. *e*=terminal cell; *s*=basal cell; *m*=embryo sac membrane. B′. 4-celled I-shaped proembryo; e_1, e_2 are from terminal cell; s_1, s_2 are from basal cell. C. Further development of embryo. *S*=suspensor; *h*=hypophysis; *E*=embryonal mass. D. L.s. of ovule. *Endo*=endosperm in free-nuclear stage. *Anti*=antipodal tissue. *Embryo*=developing embryo. E. Embryo showing further development of embryonic octants and hypophysis. F. L.s. of ovule. Endosperm becoming cellular. G. Embryo. *Cot*=cotyledons; *Hypo*=hypocotyl; *Rad*=radicle; *R.c*=root-cap. H. Mature seed. *Pl*=plumule. Endosperm has been consumed almost completely.

later developed endosperm as well as embryo are horseshoe-shaped. Moreover the micropyle being pointed downwards, the embryo looks upside down. Embryo development here is of the typical *Onagrad* or *Crucifer type* described above The oospore (Fig. 424A) divides by a transverse wall forming a large **basal cell** and a smaller **terminal cell** (Fig. 424B). The basal cell now divides transversely while the terminal cell divides by a vertical wall developing a ⊥-shaped 4-celled proembryo (Fig. 424B′). Next, the second basal cell quickly divides by a number of transverse walls giving rise to a row of cells called the **suspensor** (Fig. 424C). The lowest (micropylar) cell of the suspensor remains disproportionately large. As the suspensor increases in length, it pushes down the terminal embryonal cells deeper into the embryo sac. The embryonal cell, meanwhile, divides by three walls at right angles to one another giving rise to eight cells or *octants* (Fig. 424C). This is called the **embryonal mass.** The lowest cell of the suspensor is called the **hypophysis.** The embryonal mass, along with the hypophysis, divides further (Fig. 424 D & E). Ultimately, the four terminal octants form the two cotyledons, the four micropylar octants form the hypocotyl and the core of the radicle while the hypophysis forms the cortex and the epidermis of the radicle as well as the root-cap (Fig. 424 F & G). In the final stage (Fig. 424H) we find the mature embryo with the plumule developed also out of the four terminal octants. The suspensor gradually withers as the radicle is developed.

As the embryo is developing, the endosperm also keeps pace. Endosperm development, in this case, is strictly *nuclear*. The endosperm mother nucleus rapidly gives rise to a number of free nuclei which are arranged in the cytoplasm lining the periphery of the embryo sac while the centre is occupied by a big vacuole (Fig. 424D). Gradually, cell wall formation begins from the periphery (Fig. 424F). The antipodal cells, in this case, form a tissue which is short-living. The endosperm supplies nutrition to the embryo mainly through the suspensor. As the embryo of *Capsella* increases, the endosperm presses upon and sucks in plenty of food material from the nucellus which is soon consumed. In the final stage (Fig. 424H) the nucellus has been completely consumed and most of the endosperm has also been consumed by the embryo. In the mature seed even this remnant of endosperm disappears so that the seed becomes exalbuminous.

(2) *Typical monocotyledonous embryo*

There is no essential difference between the embryogeny of monocotyledons and that of dicotyledons. The endosperm develops in the same way but, as a monocotyledonous embryo develops a single cotyledon instead of two, there is some difference in later differentiation.

Luzula forsteri of Juncaceae may be taken as a type for monocotyledonous embryogeny. Here also the development is of the *Onagrad* or *Crucifer type*. The oospore divides by a transverse wall to form a large **basal cell** *a* and a smaller **terminal cell** *b* (Fig. 425A). The terminal cell *b* divides into two by a longitudinal wall and the basal micropylar large cell *a* cuts off a cell *a′* by a transverse wall developing a ⊥-shaped 4-celled proembryo (Fig. 425B). The two cells *b* now divide into four quadrants by another wall at right angles to the first, the cell *a′* divides into two by a longitudinal wall and the cell *a* cuts off a second cell *a″* by a transverse wall (Fig. 425C). The cell *a* acts as the **suspensor**—while all the other cells take part in the formation of the embryo. The quadrants of *b* divide further giving rise to a tissue of which the lower half *b*(*l*) is destined to give rise to the lower part of the single cotyledon and the upper half *b′*(*l′*) to the upper part of the cotyledon, the hypocotyl with the core of the radicle and, later, the plumule. The cell *a′*

on the top of it, forms a tissue which develops the upper part of the radicle while a'' develops the root-cap tip (Fig. 425 D, E & F).

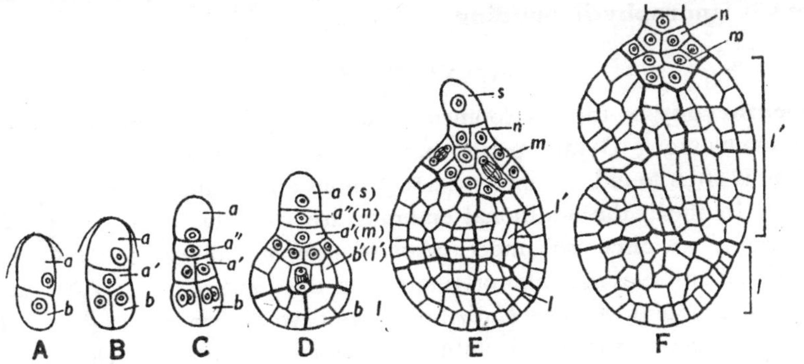

FIG. 425. Embryogeny in *Luzula forsteri* of Juncaceae (Monocot). A. Transverse division of oospore developing a 2-celled proembryo. a=basal cell; b=terminal cell. B, C & D. Stages of development (see text). E. Embryo showing initiation of different organs. s=suspensor=a; n=root-cap developed from a''; m=radicle and part of root-cap; l''=hypocotyl and part of cotyledon; l=lower part of cotyledon. F. Developed embryo showing the organs initiated in E further developed.

APOMIXIS

Apomixis is a term coined by Winkler to signify any asexual method of propagation not involving the normal production of embryo by fertilization. This definition includes even propagation by bulbils. But, following the usual practice, here a number of cases are being discussed which involve *seed* or *embryo formation without fertilization.*

Parthenogenesis means development of an embryo directly from an egg cell or a male gamete. **Apogamy** means the development of a sporophyte (i.e., embryo) out of *any* gametophytic cell without the union of gametes. **Apospory** means the formation of a gametophyte (i.e., embryo sac or pollen) on a sporophyte without any reduction division.

Parthenogenesis, therefore, is a type of *apogamy*. Parthenogenesis of a normal haploid egg into an embryo has been noticed in *Solanum nigrum*, etc. This embryo and the resultant plant is haploid. This is termed **haploid parthenogenesis.** Haploid plants are usually sterile. Cases are also known (as in different species of *Nicotiana* and *Crepis*) where such embryos actually developed from male gametes. A second type of parthenogenesis, called **diploid parthenogenesis,** has also been noticed. In these cases, the embryo sac develops from a diploid, i.e., sporophytic cell of the ovule *without any reduction division* (i.e., by *apospory*) so that the embryo sac and all cells within it are diploid. The diploid egg of this embryo sac develops a diploid embryo. This has been noticed in *Taraxacum albidum* and other plants.

Apogamy of other types also have been noticed. Any other haploid cell of a normal gametophyte (e.g., synergid or antipodal cell) may develop into an embryo. The resultant plant also is haploid as has been noticed in the orchid *Orchis*. On the other hand, if a synergid or an antipodal cell of a *Taraxacum* type embryo sac (i.e., diploid developed *aposporously*) develops into a diploid embryo, it will be a case of *diploid apogamy.*

Apospory, i.e., development of a gametophyte on the sporophyte without reduction division as seen in *Taraxacum albidum* has also been noticed in a few other plants like *Eupatorium glandulosum*, several species of *Hieracium*, etc.

An embryo may also develop directly from any diploid sporophytic cell (e.g., cells of nucellus, integument, etc.). This may be considered as vegetative growth of the category of bulbils although they are not always so simple. This is sometimes called **adventive embryony** or **sporophytic budding.**

POLYEMBRYONY

Polyembryony or the development of several embryos within the same ovule is quite common among Gymnosperms but rare among Angiosperms. This is often found in *Citrus* species (lemons and oranges —Fig. 426). In these, the additional embryos are, usually, of nucellar origin, as discussed under *adventive embryony* above. Such additional embryos may also originate from the *cleavage* or splitting of the zygote (*Cymbidium bicolor*) or proembryo as seen in *Erythronium americanum.* Additional embryos are known to develop by epidermal budding of the first embryo (*Nicotiana rustica*) or from a filamentous proembryo (*Nymphaea advena*). More commonly, two embryos are found in a seed of which one is normal and the other is *apomictic,* the latter being usually haploid. Such haploid-diploid twin embryos are known in many cultivated plants like rice, linseed, wheat, potato, chilli, cotton, etc.

N.B. Johansen is of opinion that the term *polyembryony* should be strictly applied only when all the embryos are developed from the zygote, the proembryo, or by budding of the first embryo. The haploid-diploid embryos or such apomictic embryos are, thus excluded.

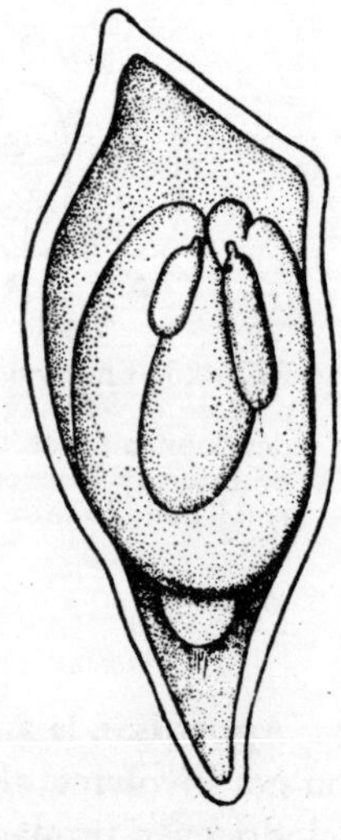

FIG. 426. Several embryos within a seed of orange. (After Penzig).

THE SEED

When the embryo is developed fully, the ovule or the megasporangium becomes the **seed.** With increase in the size of the ovule the integuments slowly dry up and form the protective seed coat, the outer integument forming the testa and the inner the tegmen. Simultaneously, the internal organs also lose moisture and all food substances are transformed into insoluble reserve forms. Physiological activities are reduced to a minimum until the whole structure becomes *dormant* and almost dry when we call it *seed.* The seed is, therefore, a *mature integumented megasporangium.*

It has been seen that the developing seed contains a growing embryo within a growing endosperm, and the whole is contained within the nucellus. As the embryo increases, the endosperm decreases and the growth of both decreases the nucellus. Finally, an equilibrium is reached when seed development is complete. According to the final equilibrium, it is possible to get three types of seeds among both dicotyledons and monocotyledons:

A. If the embryo completely fills up the ovule consuming both endosperm and nucellus, the seed is **exalbuminous,** there being no endosperm (Fig. 427). In such a seed food for the embryo is kept stored in the cotyledons.

B. If the endosperm be not completely consumed the embryo remains comparatively small and a large quantity of stored food (oily, mealy, horny, etc.) remains within the *endosperm* which the embryo could not consume but would consume in the germinating stage. Such a seed is **albuminous** (Fig. 428).

C. If even the nucellus is not consumed fully, it remains in the seed as a layer of **perisperm** (Fig. 429) over and above the endosperm and the embryo. This is the real nature of the *perisperm* as has been described in a number of monocotyledonous and dicotyledonous seeds (Chapter I).

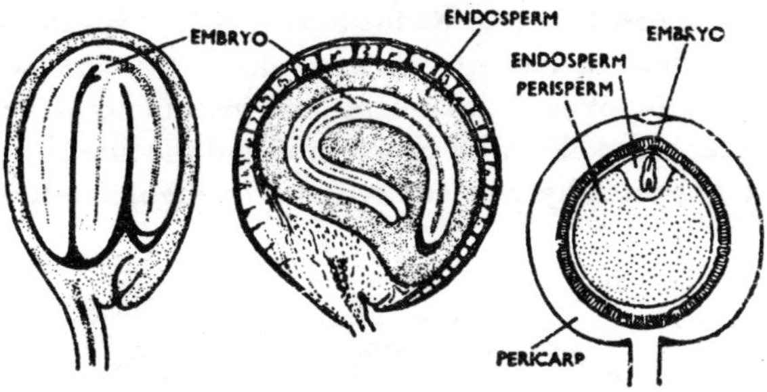

FIG. 427 FIG. 428 FIG. 429

FIG. 427. Exalbuminous seed of *Capsella*. FIG. 428. Albuminous seed of *Datura*. FIG. 429. Black pepper fruit with single seed showing endosperm and perisperm.

ARIL AND CARUNCLE

After fertilization, an outgrowth looking like a third integument is sometimes seen to develop from the funicular (i.e., hilar) end. In litchi (*Litchi chinensis*) such a juicy fleshy outgrowth develops within the pericarp. This (Fig. 430) is called the **aril.** In nutmeg (*Myristica fragrans* of Myristicaceae), such a red-coloured, dry, leathery and branched *aril* (Fig. 431) forms the valuable spice 'mace' of commerce. There is a pulpy *aril* in *Garcinia cowa* and *Garcinia mangostana*.

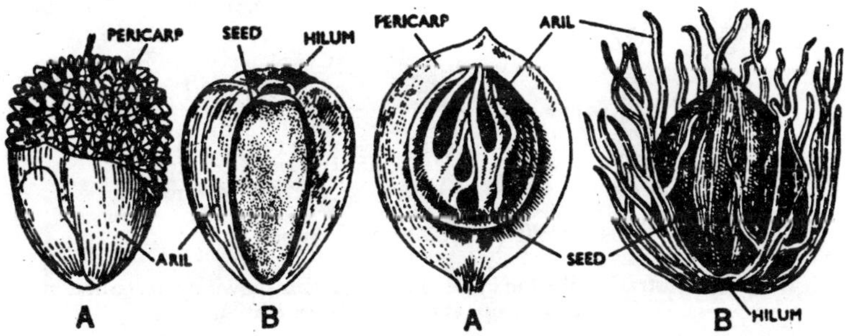

FIG. 430 FIG. 431

FIG. 430. (*Litchi chinensis*)(litchi). A. Fruit with part of pericarp removed to show aril. B. L.s. of fruit after removal of pericarp. FIG. 431. *Myristica fragrans* (nutmeg). A. L.s. of fruit showing aril. B. Aril ('mace') on seed.

In castor and some other Euphorbiaceae, a smaller outgrowth develops near the micropylar end. This is called the **caruncle** (Fig. 35). In pansy (*Viola*) such a *caruncle* develops at the hilum.

LIFE CYCLE OF ANGIOSPERMS

With the development of the seed, the life cycle of Angiosperms comes to a close. If we now trace the life-history (Fig. 432) we shall find a regular *alternation of generations* as in most groups of plants.

The plant is a **sporophyte** with 2*n* or *diploid* cells. The flower is the reproductive shoot or the *storobilus* with **microsporophylls** or stamens and **megasporophylls** or carpels. The microsporophyll contains **microsporangia** or pollen sacs and the megasporophyll contains **integumented megasporangia** or ovules. Within the microsporangium are many **microspore mother cells** and within the megasporangium there is a single **megaspore mother cell.** Reduction division or **meiosis** initiates the **gametophytic** or **haploid generation** with *n* cells resulting in innumerable **microspores** or pollens and four **megaspores.** Each microspore germinates developing a **micro-** or **male gametophyte,** i.e., pollen with pollen tube and the functional megaspore forms a **mega-** or **female gametophyte** or embryo sac. In the microgametophyte there is *no*

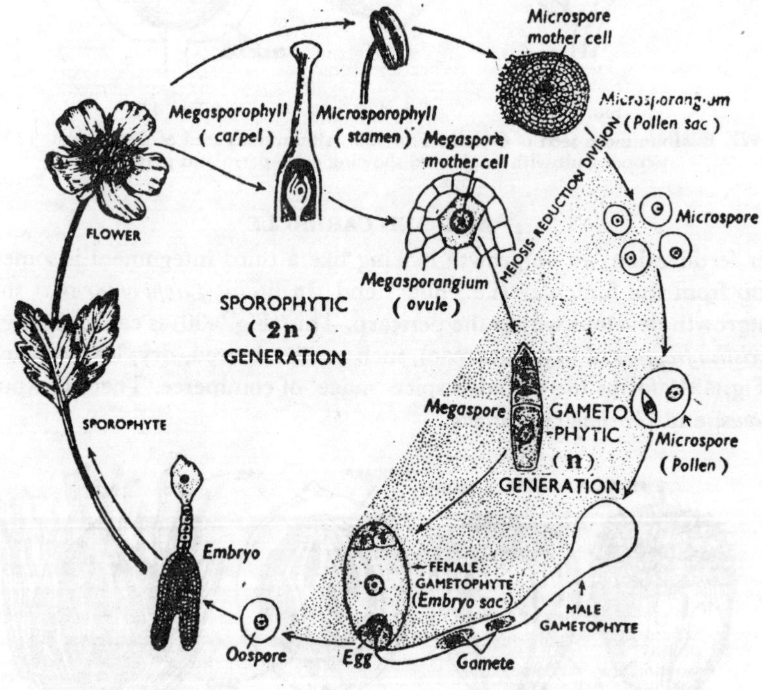

Fig. 432. Diagram illustrating the life cycle of Angiosperms showing alternation of sporophytic and gametophytic (shaded) generations.

prothallial cell and the generative cell with the vegetative nucleus represent a much reduced **antheridium.** In the megagametophyte the antipodal cells are **female prothallial cells** and the egg apparatus represents a reduced **archegonium.** The naked **sperm cell** or **gamete** now fertilizes the **egg cell** completing the *gametophytic generation* and reverting to the *sporophytic* or *diploid generation.* The *gametophytic generation,* therefore, is extremely brief and completely *endosporous* (i.e., confined to the walls of the microspore and the megaspore, if we remember that the pollen tube is an extension of the intine). The product of fertilization, the **oospore,** forms the **embryo** which, after a period of dormancy in the seed, germinates into the new sporophytic plant.

It should be pointed out again that the gametophytic generation in Angiosperms is reduced to the extreme. **As a result,** ***the seed represents three generations.*** The seed coats (integuments) and the funicle of the seed represent the first sporophytic generation while the embryo within it represents the second sporophytic, i.e., a third generation. The intervening second generation is represented by the short-living gametophyte (embryo sac) within which the embryo and the endosperm develop. The perisperm, if present, also represents the first sporophytic generation.

The endosperm of the Angiosperms is something new in the evolutionary history of plants. The endosperm of the Gymnosperms and the heterosporous Pteridophytes is formed of haploid gametophytic tissue. But, because of triple fusion, the Angiospermic endosperm is neither diploid nor haploid but is **triploid.** So, it belongs neither to the gametophytic nor to the sporophytic generation but forms a separate entity within the seed.

CHAPTER XII

THE FRUIT

The stimulus of fertilization not only develops ovules into seeds but also brings about other changes in the flower and even in the inflorescence. Usually, all the parts of the flower except the ovary wither away. The ovary begins to enlarge simultaneously with the development of the seed and ultimately becomes the **fruit.** These changes usually involve a copious development of succulent parenchyma within the ovary. The contents of these parenchyma cells also change. They become charged with vegetable acids, sugars and often different flavouring matters. These changes may also be associated with structural changes within the ovary, viz., a multilocular and multiovular ovary may change into a unilocular and one-seeded one by the dissolution of some partition walls and non-development of some ovules. On the other hand, a unilocular ovary may become multilocular by the development of false partition walls. If any other part of the inflorescence is incorporated within the fruit that also is enlarged along with the ovary. When ripe, some fruits remain *fleshy* and pulpy while others *dry* up becoming papery or stony. The biological need of the fruit is in the protection of the seed forming a *case* or a vessel round it. This is the main point where the Angiosperms (*angeion*=case or vessel) differ from the *naked* seeded Gymnosperms (*gymnos*=naked). The second importance of the fruit is in the dispersal of seeds. If a fruit is valued by other animals, naturally it will be carried by them and the seed will get more easily distributed.

The fruit is usually developed by the stimulus of fertilization but cases are know where a fruit may be formed without fertilization. This is called **parthenocarpy.** Sometimes, the stimulus of pollination (without fertilization) may be enough to cause fr formation and sometimes parthenocarpy may be induced by hormones, etc. Parthenocarpic fruits are, necessarily, seedless. Since seedless fruits are in great demand, parthenocarpy is of great practical value. World's best varieties of fruits like bananas, pineapples, grapes, etc., are seedless. Horticulturists are always trying to induce parthenocarpy in more and more varieties of fruits.

A **true fruit** is one which is developed out of a single ovary of a single flower and in the development of which no other part outside the ovary has taken a share. A fruit is *spurious* or *false* **(pseudocarp)** when the contrary takes place. A large number of fruits are *spurious*, in this sense. Thus, all *aggregate* and *multiple* fruits (described later) are false. Fruits resulting from inferior ovaries are false as the wall of such an ovary has some thalamus tissue fused with it. Fruits of apple and pear (Fig. 453) are false as the edible part is mainly fleshy thalamus. Brinjal is false as it has a persistent calyx and *Dillenia* (Fig. 433) fruits are false as the edible sour, fleshy sepals form the bulk of the fruit. Similarly, the persistent calyx makes the *Physalis* (Fig. 324) fruit false.

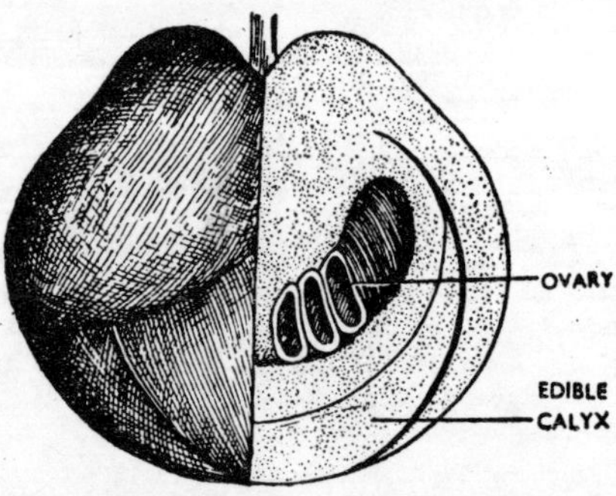

FIG. 433. Pseudocarp of *Dillenia indica.*

A fruit has got a wall or **pericarp** which was the wall of the ovary. When well developed, this shows three layers—an outer **epicarp,** a middle **mesocarp** and an inner **endocarp.** These are usually more evident in the simple fleshy fruits.

Fruits are usually classified into three groups—**simple, aggregate** and **multiple.**

I. SIMPLE FRUITS

When a single fruit develops from a single ovary of a single flower it is called a **simple fruit.** It is immaterial whether the ovary belongs to a monocarpellary simple gynoecium or to a polycarpellary *syncarpous* gynoecium. Such fruits, again may be **dehiscent** or **indehiscent.**

A. DEHISCENT FRUITS

Dehiscent fruits burst automatically on ripening and discharge their seeds. Such fruits generally become *dry* at maturity. The **dehiscence** may be of different types. If on dehiscence the fruit splits up into pieces or *valves*, the dehiscence is called *valvular*. If each broken piece encloses within it one or more seeds then the dehiscence is *schizocarpic*. Besides these types, a fruit may dehisce *transversely* or by development of *pores*. As most types of dehiscence occur among *capsules* these are discussed again under that type of fruit.

Dehiscent fruits may be:

1. Legume or **pod:** This is a characteristic of the Leguminosae family showing a superior monocarpellary one-chambered fruit. The fruit splits along both the sutures.

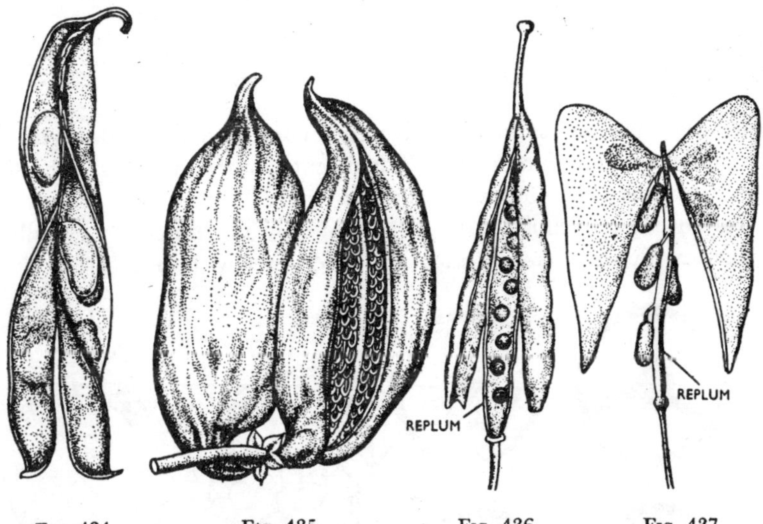

FIG. 434 FIG. 435 FIG. 436 FIG. 437

FIG. 434. Legume of a bean. FIG. 435. Paired follicle of *Calotropis*. FIG. 436. Siliqua of mustard. FIG. 437. Silicula of *Capsella*.

Numerous examples may be found among peas, beans, pulses, etc. (Fig. 434). The *lomentum* is a variation of the legume but since it is frequently *schizocarpic* it is considered under that group of fruits.

2. Follicle: The follicle resembles the legume in developing from a superior monocarpellary ovary but the fruit dehisces along one suture (usually the ventral) only. There is no common example of follicle among simple fruits. *Calotropis* (Fig. 435) of Asclepiadaceae shows a paired follicle which is actually an *aggregate fruit* from a bicarpellary ovary. Follicles are discussed under *aggregate fruits* (p. 199).

3. Siliqua: This is developed from a bicarpellary superior ovary which is one-chambered but appears two-chambered because of the development of a false partition wall called *replum*. The dehiscence is along both the sutures from base upwards while the seeds remain attached to the replum. This is a characteristic of the family Cruciferae (Fig. 436).

Silicula is a variation of the *siliqua* which contains fewer seeds and is much shorter (as broad as long) and flattened. This is seen in *Capsella bursa-pastoris* (Fig. 437) and a few other plants also belonging to the family Cruciferae.

4. Capsule: This develops from a superior syncarpous ovary and dehisces by valves. There may be capsules from inferior ovaries as well (e.g., *Iris* of Iridaceae and *Canna* of

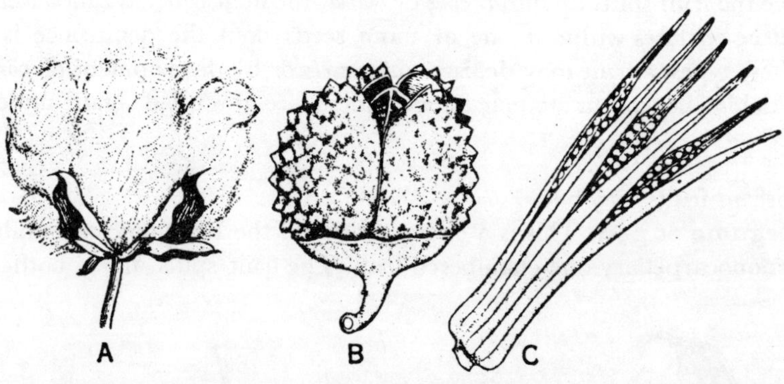

Fig. 438. Capsules of cotton (A), *Datura* (B) and lady's finger (C).

Cannaceae) but these are sometimes given a separate name, **diplotegia.** Capsule is, more commonly, a common name for all dehiscent syncarpous fruits with three or more carpels and examples (Fig. 438) are numerous.

The various types of capsules show different types of dehiscence. Usually, the dehiscence is *valvular*. Types of **dehiscence** (Fig. 439) met with are: (1) *Loculicidal*—when

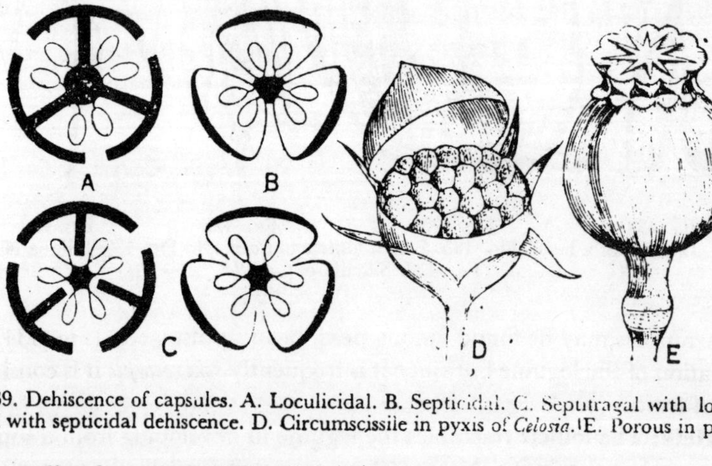

Fig. 439. Dehiscence of capsules. A. Loculicidal. B. Septicidal. C. Septifragal with loculicidal and with septicidal dehiscence. D. Circumscissile in pyxis of *Celosia*. E. Porous in poppy.

each loculus splits along the middle of its back (i.e., dorsal suture) as in the families Malvaceae (cotton, lady's finger, etc.) and Acanthaceae (*Andrographis*, *Adhatoda*, etc.). (2) *Septicidal*—when the splitting takes place along the septa thus separating each loculus as in linseed and castor. (These fruits actually show the schizocarpic type of dehiscence so that they are better considered under schizocarpic fruits). (3) *Septifragal*—when dehiscence is loculicidal or septicidal but the valves fall off leaving the seeds attached to the central azis as in *Datura*. (4) *Circumscissile*—when the dehiscence is *transverse* so that the top comes off as a lid as if exposing a box of seeds. This type of fruits is given a special name, **pyxis** (Fig. 439D), and may be seen in cock's comb (*Celosia cristata*), *Amaranthus gangeticus* and other plants of Amaranthaceae. (5) *Porous*—when dehiscence gives rise to a number of pores on the top of the ripe ovary through which granular seeds are liberated by *censer mechanism* (explained in Chapter XIII) as seen in the poppies (*Papaver*—Fig. 439E).

5. Schizocarps: This is a type of fruits which may be considered intermediate between dehiscent and indehiscent. The fruit breaks up into a number of indehiscent bits called *mericarps* each of which encloses one or more seeds which latter are liberated only on the rotting of the pericarp. Several types of such fruits are distinguished:

(a) Lomentum: This is a variation of legume. The fruit (Fig. 440) is constricted between the seeds and usually breaks up into bits containing one or more seeds. This is seen specially in the Mimoseae subfamily of Leguminosae as in *Acacia*, *Mimosa*, etc. The

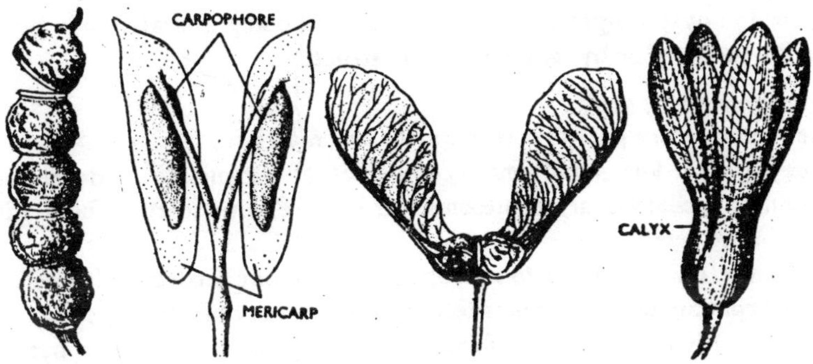

FIG. 440 FIG. 441 FIG. 442 FIG. 443
FIG. 440. Lomentum of *Acacia*, FIG. 441. Cremocarp of Umbelliferae. FIG. 442. Samara of maple. FIG. 443. Samaroid fruit of *Shorea*.

large lomentums (3 ft. or more long) of *Entada gigas* break in the forests with the sound of large crackers.

(b) Cremocarp: This is the two-seeded fruit (Fig. 441) developed from the inferior bicarpellary ovary of Umbelliferae. The two *mericarps* split along the *carpophore* (prolongation of thalamus) to which they (each enclosing a seed) are attached.

(c) Samara: This is a type of two or more chambered winged fruit developed out of a syncarpous ovary (Figs. 442 & 466). The wings are developed from the pericarp and the fruit breaks up into component parts—each part enclosing a seed. Examples are found in maple (*Acer*), *Dioscorea*, *Hiptage bengalhensis*, etc.

Fruits of *Shorea* (Fig. 443) and some other members of Dipterocarpaceae are also winged but here the wings are persistent sepals and such a fruit is termed **samaroid.**

(d) Regma: These are formed out of polycarpellary pistils and break up into as many component parts (called *cocci*) as there are carpels. Castor (Fig. 444A) dehisces into three *cocci* while the fifteen carpels of *Hura crepitans* (Fig. 444B) are seen to burst away from one another with great force. The *regma* of *Geranium* (Fig. 470B) breaks into five *cocci* as it is formed of five carpels.

(e) Carcerule: This is the fruit developed by the superior bicarpellary pistil of

Labiatac (Fig. 445) which is falsely four-chambered and develops a gynobasic style. The four chambers separate into four indehiscent parts enclosing one seed each.

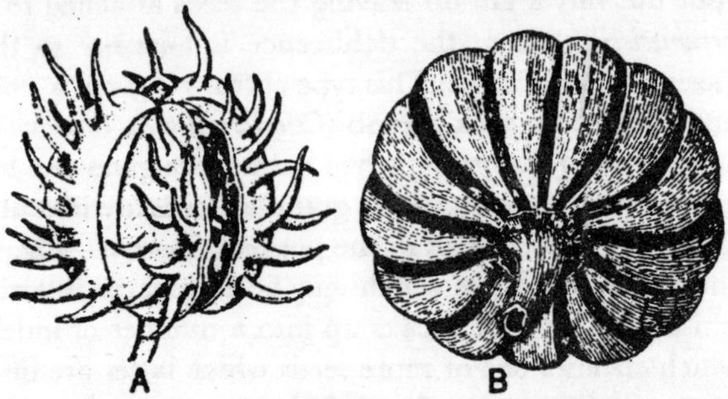

Fig. 444

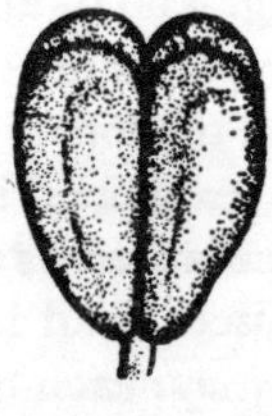

Fig. 445

Fig. 444. Regma of castor (A) and *Hura* (B). Fig. 445. Carcerule of Labiatae.

B. INDEHISCENT FRUITS

While dehiscent fruits are all dry, indehiscent fruits may be *dry* as well as *fleshy*. Indehiscent fruits do not split open at maturity, the seeds being liberated only by the decomposition of the pericarp or by its accidental destruction.

(A) *DRY INDEHISCENT FRUITS*

These present a dry pericarp at maturity and may be:

1. Caryopsis: The small one-seeded fruit of Gramineae developing from a superior monocarpellary ovary represents this type. The pericarp is fused with the seed coat as seen in the cereals.

2. Achene: This is similar to caryopsis in being one seeded and superior but pericarp and seed coat are free from one another. Achenes may be seen in *Mirabilis jalapa*, etc., but more commonly they are found forming *aggregate fruits* described on page 199.

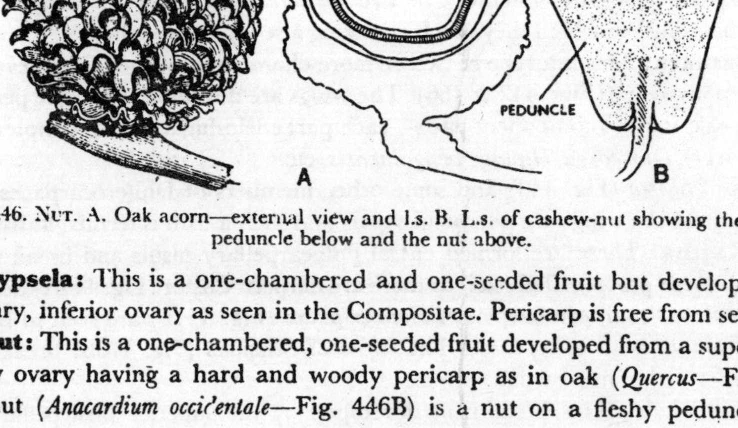

Fig. 446. Nut. A. Oak acorn—external view and l.s. B. L.s. of cashew-nut showing the fleshy peduncle below and the nut above.

3. Cypsela: This is a one-chambered and one-seeded fruit but developed from a bicarpellary, inferior ovary as seen in the Compositae. Pericarp is free from seed coat.

4. Nut: This is a one-chambered, one-seeded fruit developed from a superior polycarpellary ovary having a hard and woody pericarp as in oak (*Quercus*—Fig. 446A). Cashew-nut (*Anacardium occidentale*—Fig. 446B) is a nut on a fleshy peduncle. Litchi

(*Lichi chinensis* — Fig. 430) also is considered as a nut with a leathery fused epimesocarp, a papery endocarp and a fleshy aril which is not a normal component of a fruit.

(B) FLESHY FRUITS

Fleshy fruits usually remain succulent and juicy even when fully mature. These include:

1. Drupe: This is normally a one-seeded fruit with a skinny *epicarp*, a fibrous and usually juicy *mesocarp* and a stony hard *endocarp*. Drupes are called 'stone fruits' because of the stony hard endocarp. Herein are included mango (Fig. 447A), plum (*Prunus domestica*), peach (*Prunus persica*), apricot (*Prunus armeniaca*), Indian plum (*Zizyphus mauritiana*—Fig. 447B), etc. Sometimes there may be more than one seed in a drupe but, in that case, every seed has got a separate endocarp round it as seen in palmyra palm (*Borassus flabellifer*—Fig. 447C). Some fruits included within drupes may have the mesocarp dry

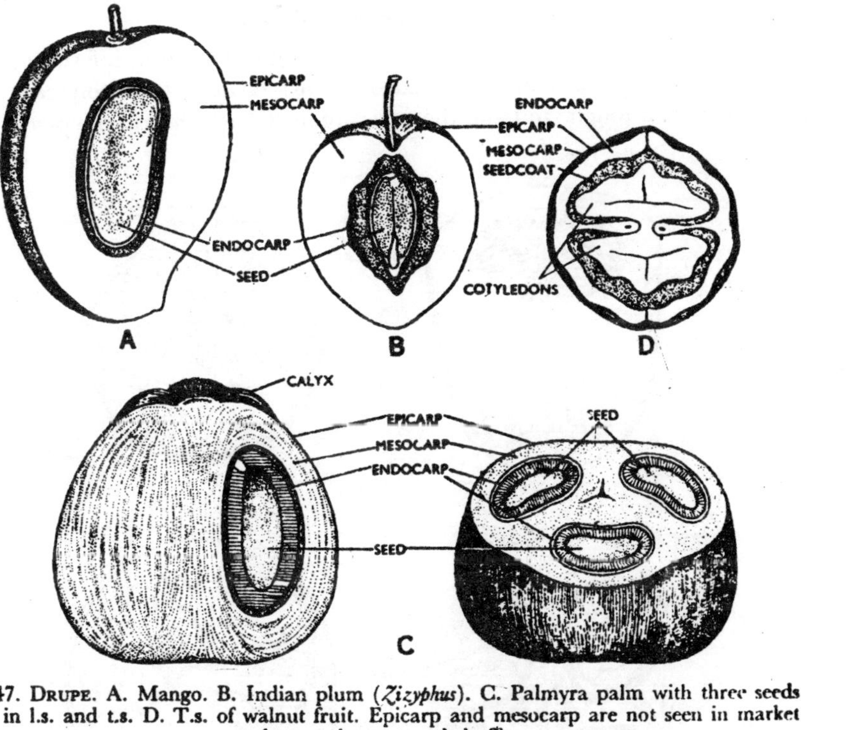

FIG. 447. DRUPE. A. Mango. B. Indian plum (*Zizyphus*). C. Palmyra palm with three seeds shown in l.s. and t.s. D. T.s. of walnut fruit. Epicarp and mesocarp are not seen in market walnut as they get peeled off.

and *fibrous* as in coconut (*Cocos nucifera*) or dry and tough as in almond (*Prunus dulcis*). Betel-nut may also be considered as a *fibrous drupe* like coconut although its endocarp is not so hard. The walnut (*Juglans regia*—Fig. 447D) also is a drupe from a syncarpous ovary but containing a single seed. Its thin epicarp and mesocarp peel off during ripening so that only the strong endocarp is seen in the market walnut. Ingrowths from the endocarp pass in between the cotyledons covered by a seed-coat giving them a ruminated appearance.

2. Berry or **Bacca:** This term is usually applied to any fleshy fruit in which there is no hard part excepting the seeds. One or the other of epicarp, mesocarp and endocarp may sometimes be differentiated or they may form a pulp in which the seeds are embedded. The seeds at first remain attached to the placenta but, later on, become detached and embedded in the pulp. Tomato, brinjal, papaw, banana, guava are common examples of berry. The *false* inferior fruits included here, specially guava, includes some thalamus tissue in the outer pulp. In these fruits the epicarp is often distinguishable while the mesocarp and endocarp form a common pulp embedding the hard seeds (Fig. 448). The fruit of date (*Phoenix sylvestris*) is sometimes considered as a one-seeded berry in which the endocarp is thin and papery. But, the endocarp is so distinct in this case that it is not wrong to consider date as a variety of *drupe* in which the endocarp has become papery instead of stony.

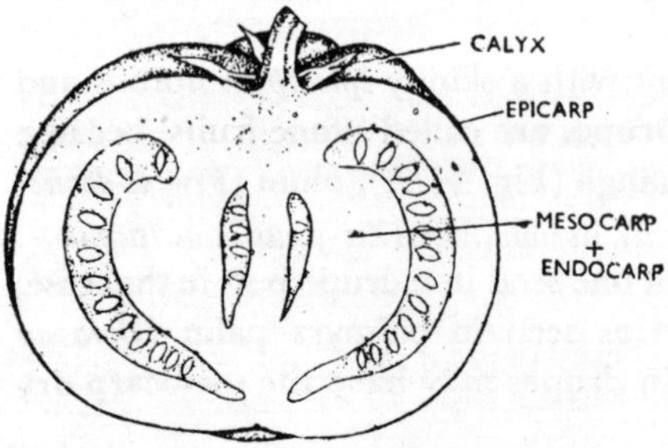

FIG. 448. L.s. of tomato berry.

There are a number of fruits of the *baccate* or *berry type* which show some variations from the normal type:

(a) Pepo: This is an inferior one-celled or spuriously three- or more-celled, many-seeded fruit as seen in the Cucurbitaceae (Fig. 449). The epicarp forms a tough rind and the seeds are firmly attached to the placenta.

FIG. 449

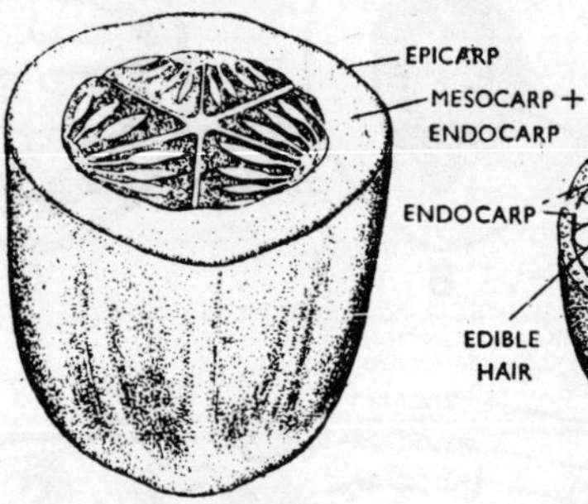

FIG. 450

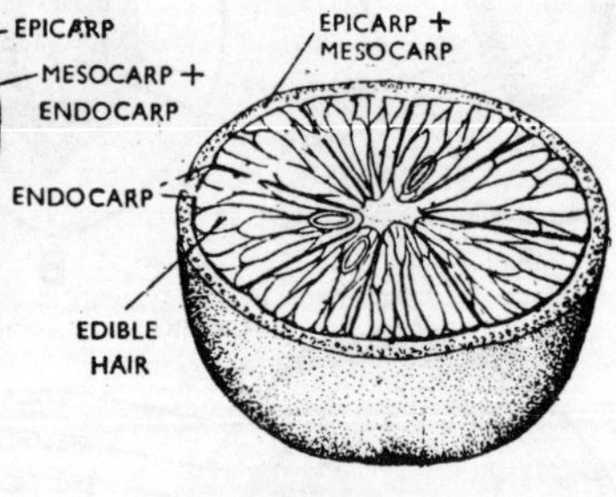

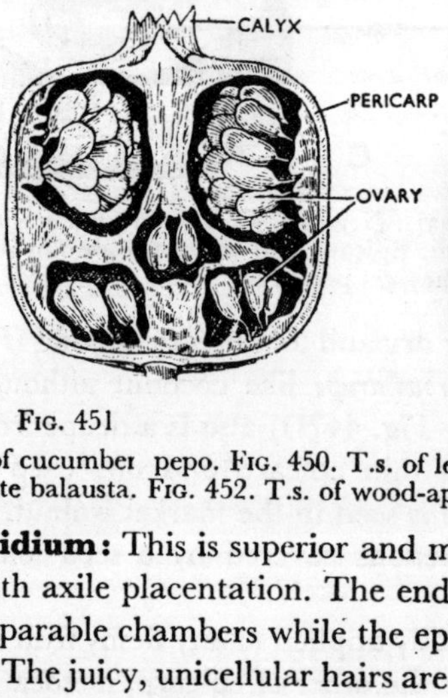

FIG. 451

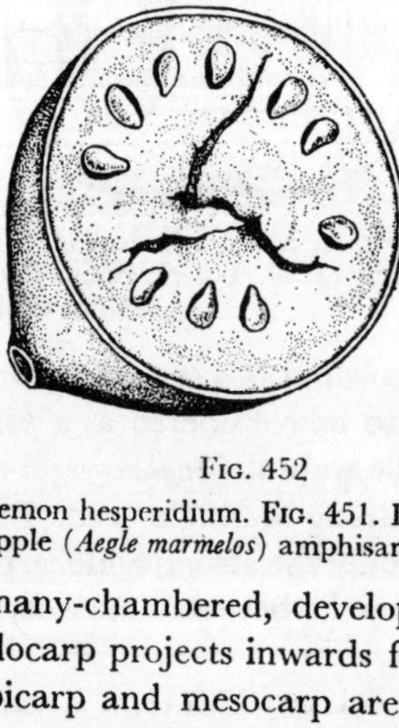

FIG. 452

FIG. 449. T.s. of cucumber pepo. FIG. 450. T.s. of lemon hesperidium. FIG. 451. L.s. of pomegranate balausta. FIG. 452. T.s. of wood-apple (*Aegle marmelos*) amphisarca.

(b) Hesperidium: This is superior and many-chambered, developing from syncarpous ovary with axile placentation. The endocarp projects inwards forming distinct and sometimes separable chambers while the epicarp and mesocarp are fused together forming the skin. The juicy, unicellular hairs are outgrowths from the endocarp. Oranges and lemons are fruits of this type (Fig. 450).

(c) Balausta: This is the special inferior, many-celled and many-seeded fruit of pomegranate (*Punica granatum* of Punicaceae). The pericarp is tough and two rows of carpels are placed one above the other. Seeds are attached irregularly and the calyx persists on the top. The succulent testa is edible (Fig. 451).

(d) Amphisarca: This is a superior many-chambered and many-seeded fruit with a woody pericarp. The placenta and, possibly, the inner layers of the pericarp become pulpy and edible. The testa is mucilaginous. Examples are *Aegle marmelos* (Fig. 452) and *Limonia acidissima* of Rutaceae.

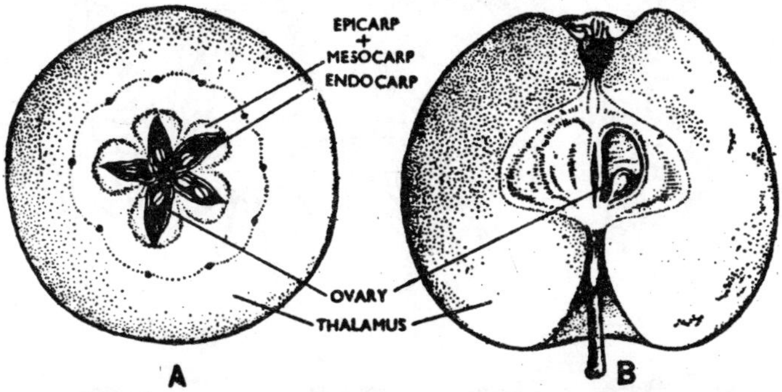

FIG. 453. Pome of apple. A. T.s. B. L.s. The real fruit with its pericarp is surrounded by a fleshy edible thalamus. Of the pericarp the epicarp and mesocarp are fleshy while the endocarp is cartilaginous.

3. Pome: This is an inferior frui developed out of an inferior syncarpous ovary. The fruit is *false* as it is surrounded by a fleshy thalamus which is edible while the actual fruit lies within (Fig. 453). Typical examples are found in apple (*Malus sylvestris*) and pear (*Pyrus communis*).

II. AGGREGATE FRUITS

Flowers with apocarpous multiple gynoecium are apt to give rise to a number of fruits as there are a number of free ovaries each capable of giving rise to one fruit. But as all these fruits develop from a single flower, they together form an **aggregate fruit.** Sometimes, these fruits (better called **fruitlets**) coalesce together giving the appearance of a single fruit while in many other cases the fruits remain free from one another forming a bunch or **etaerio** of fruitlets. An aggregate fruit is named according to the nature of the fruitlets:

1. Etaerio of follicles: Follicles rarely occur singly as they result from flowers with apocarpous multiple gynoecium. Follicles commonly develop from the apocarpous multicarpellary flowers of Ranunculaceae, Asclepiadaceae, Apocynaceae, etc. In *Aconitum* (Fig. 363) of Ranunculaceae three fruits develop from each flower while *Calotropis* (Fig. 435) of Asclepiadaceae and *Vinca* of Apocynaceae show paired fruits from the same flower. Other examples of such aggregate follicles are *Michelia champaca* (Fig. 454A) and *Magnolia grandiflora* (Fig. 454B) of Magnoliaceae which have infinite carpels in each flower. As the flower matures, the thalamus elongates so that the aggregate fruit looks like a bunch of fruits. The two follicles of *Asclepias* are united at the base and such a fruit has the special name of **conceptaculum.**

2. Etaerio of achenes: Achene type fruits are found very commonly as aggregate fruits. Aggregate of achenes may be seen in *Naravelia* (Fig. 455A) and *Clematis* (Fig. 464) of Ranunculaceae where the fruitlets are provided with feathery persistent styles. The hip of rose (Fig. 455B) shows a number of achenes within the cup-shaped thalamus. In strawberry (*Fragaria* of Rosaceae—Fig. 455C) the minute achenes are buried as small brown specks on a fleshy edible thalamus. The spongy thalamus of lotus (*Nelumbo*—Fig. 306) is also an aggregate of achenes, the fruitlets being sunk on the spongy thalamus.

3. Etaerio of drupes: The raspberry (*Rubus idaeus* of Rosaceae—Fig. 456) is an example of an aggregate of drupes. Here, the fruitlets are crowded pressing against one another on a raised thalamus.

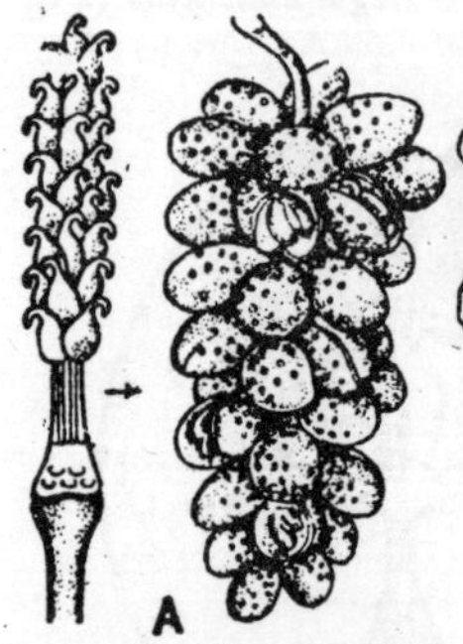

FIG. 454

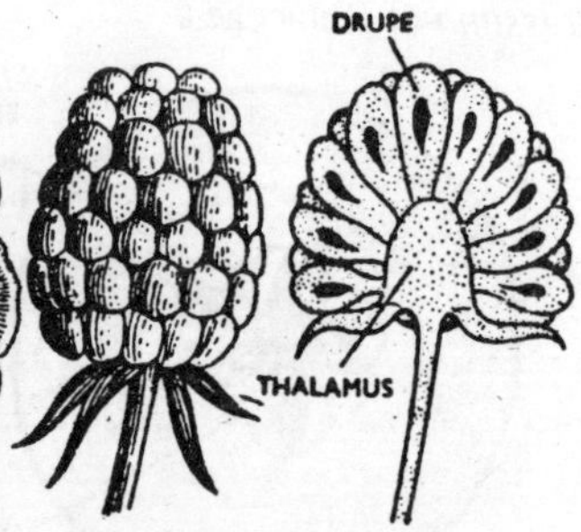

FIG. 456

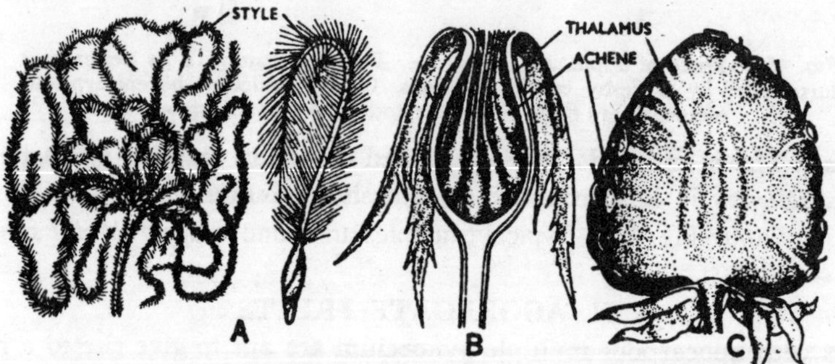

FIG. 455

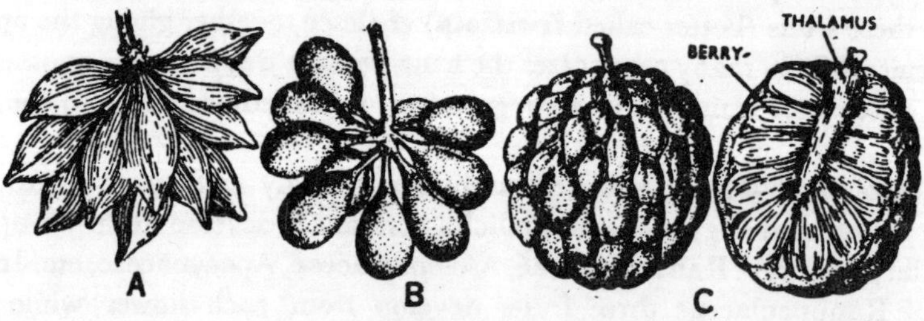

FIG. 457

AGGREGATE FRUITS. FIG. 454. Etaerio of follicles. A. *Michelia champaca*. The pistil on left develops into the aggregate fruit on right. B. *Magnolia grandiflora* fruit. The fruitless dehisce in the manner of follicles. FIG. 455. Etaerio of achenes. A. *Naravelia zeylanica*. An achene fruitlet is shown separately. B. Hip of rose. C. Strawberry. The achenes are on the surface of the fleshy thalamus. FIG. 456. Etaerio of drupes in raspberry. External appearance and l.s. FIG. 457. Etaerio of berries. A. Artabotrys hexapetalus. B. *Polyalthia longifolia*. C. *Anona squamosa* showing both external view and l.s.

4. Etaerio of berries: These are very common in the family Anonaceae. In *Artabotrys hexapetalus* (Fig. 457A) and *Polyalthia longifolia* (Fig. 457B) the berries occur in a bunch. In *Anona squamosa* (Fig. 457C) and *Anona reticulata* the berries become very fleshy and being crowded together on a thick thalamus form a complex single fruit. The apices of the berries fuse together forming something like a common rind.

III. MULTIPLE OR COMPOSITE FRUITS

While aggregate fruits are formed out of a number of free ovaries within a single flower, the **multiple** or **composite fruit** is formed by all the flowers of a whole inflorescence grouped together. Two types may be considered:

1. Sorosis: In this type the rachis and all the floral parts of a spadix type of inflorescence fuse together forming a composite fruit. In jack-fruit (*Artocarpus heterophyllus* of Moraceae—Fig. 458A), the thick club-shaped rachis has the flowers arranged on it. The fertile fruits have edible, juicy perianth leaves and the bracts form more or less juicy chaffs around them. The spines on the tough rind represent the stigmas of the carpels. Each

Fig 458

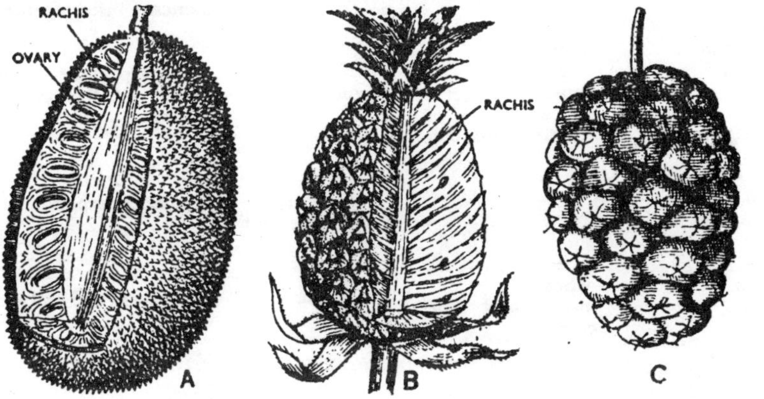

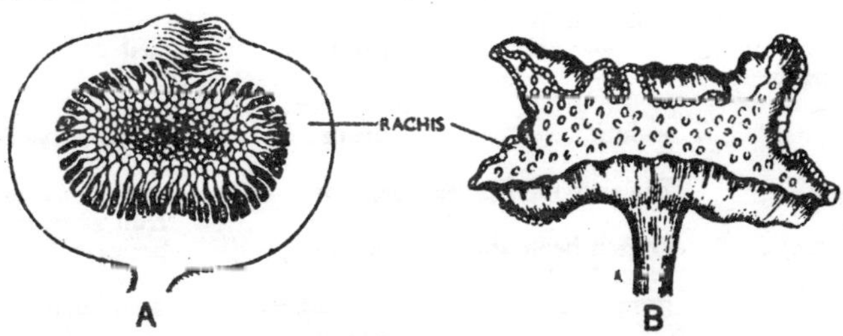

Fig. 459

Composite fruits. Fig. 458. Sorosis. A. Jack-fruit. B. Pineapple. C. Mulberry.
Fig. 459. Syconus. A. *Ficus* (Fig). B. *Dorstenia.*

seed, which is also edible, is covered by a membranous testa. *Pandanus* of Pandanaceae has got a multiple fruit somewhat similar to jack-fruit. In pineapple (*Ananas comosus* of Bromeliaceae—Fig. 458B) there is a similar structure but the ovaries are not so conspicuous, the edible portion being formed more by the rachis, perianth and bracts. Small seeds may be found. Each polygonal area on the surface represents a flower. The mulberry (*Morus indica* of Moraceae—Fig. 458C) has a similar fruit in which the fleshy perianths enclose dry achenes.

2. Syconus: This is the other type of multiple fruit which develops on the ripening of the *hypanthodium* or the *coenanthium* inflorescence. In *Ficus* plants or figs (Fig. 459A), the female flowers within the closed receptacle (rachis) of the inflorescence develop achenes giving rise to a multiple fruit of achenes. A similar fruit is formed by *Dorstenia* (Fig. 459B) where the *coenanthium* develops achenes on the open receptacle.

EDIBLE PARTS OF SOME COMMON INDIAN FRUITS

Family and Scientific name	Common English name (if any)	Type of fruit	Edible part
Anacardiaceae			
Anacardium occidentale	cashew-nut	nut	cotyledons (also fleshy peduncle)
Mangifera indica	mango	drupe	mesocarp
Anonaceae			
Anona squamosa	custard apple	etaerio of berries	pericarp (except apices)
Anona reticulata	—		
Bromeliaceae			
Ananas sativus	pineapple	sorosis	outer rachis, bract, perianth and pericarp fused together
Cucurbitaceae			
Cucurbita maxima	sweet gourd	pepo	mesocarp and endocarp; epicarp fused with thalamus edible when young
Cucumis sativus	cucumber		
Cucumis melo	musk-melon (cantaloup)		
Lagenaria vulgaris, etc.	bottle-gourd, etc.		
Dilleniaceae			
Dillenia indica	—	pseudocarp; aggregate fruit	accrescent sepals
Gramineae			
Oryza sativa	rice	caryopsis	endosperm and embryo
Triticum spp.	wheat		
Zea mays, etc.	maize and other cereals		
Guttiferae			
Garcinia mangostana	mangosteen	berry	aril
Garcinia cowa	—		
Garcinia xanthochymus	—		
Juglandaceae			
Juglans regia	walnut	drupe	lobed cotyledons
Leguminosae			
Pisum sativum, etc.	Pea and other pulses	legume	cotyledons and embryo (whole fruit when green)
Phaseolus vulgaris, etc.	French bean and other beans	legume	pericarp (young) and seeds
Arachis hypogea	ground-nut	constricted legume	seed (mainly cotyledons)
Tamarindus indica	tamarind	constricted legume or lomentum	mesocarp
Malvaceae			
Abelmoschus esculentus	lady's finger (okra)	capsule	whole fruit (pericarp and seeds) when young.
Moraceae			
Ficus spp.	figs	syconus	rachis when green; rachis with seeds in sweet figs.
Artocarpus heterophyllus	jack-fruit	sorosis	bract, perianth and seed
Morus indica	mulberry	sorosis	perianth
Musaceae			
Musa paradisiaca	banana	berry	mesocarp and endocarp
Myrtaceae			
Syzygium cumini	Indian black berry	berry	pericarp with thalamus fused
Syzygium jambos	rose-apple		
Syzygium malaccensis	star-apple		
Psidium guyava	guava		

Family and Scientific name	Common English name (if any)	Type of fruit	Edible part
Nymphaeaceae			
Nelumbo nucifera	lotus	etaerio of achenes.	thalamus and seeds
Palmaceae			
Cocos mucifera	coconut	fibrous drupe	endosperm and cotyledon
Borassus flabellifer	palmyra or fan palm	drupe	juice from mesocarp (also cotyledon and endosperm)
Phoenix sylvestris and *P. dactylifera*	date	berry (or drupe)	mesocarp
Punicaceae			
Punica granatum	pomegranate	balausta	succulent testa
Rhamnaceae			
Zizyphus mauritiana	Indian plum	drupe	epicarp and mesocarp
Rosaceae			
Prunus domestica	plum	drupe	epicarp and mesocarp
Prunus persica	peach	drupe	epicarp and mesocarp
Malus sylvestris	apple	pome	fleshy thalamus
Pyrus communis	pear	pome	fleshy thalamus
Fragaria spp.	strawberry	Etaerio of achenes	fleshy thalamus
Rutaceae			
Citrus spp.	orange, grape-fruit, lemon, shaddok, etc.	hesperidium	juicy, unicellular hairs from inner wall of endocarp
Aegle marmelos	wood-apple	amphisarca	succulent placenta and inner pericarp
Limonia acidissima	elephant-apple	amphisarca	succulent placenta and inner pericarp
Sapindaceae			
Lichi chinensis	litchi	nut	fleshy aril
Dimocarpus longana	—	nut	fleshy aril
Solanaceae			
Solanum melongena	brinjal	berry	pericarp and placenta
Lycopersicon lycopersicum	tomato	berry	pericarp and placenta
Physalis peruviana	Cape gooseberry	berry	pericarp
Vitaceae			
Vitis vinifera	grape	berry	pericarp and placenta

CHAPTER XIII

DISPERSAL OF FRUITS AND SEEDS

Plants, normally, produce a large number of fruits and seeds. A plant is put in a biologically advantageous position if these seeds get widely distributed in space. If, instead of such dispersal, all seeds of a plant get heaped under its own shade, there would be great struggle among the germinating plantlings for the limited resources available in that small space. As a result, most of the seedlings would not survive and that would be a severe check on the propagation and ultimate evolution of the species. On the other hand, if there be some direct or indirect means for a wide distribution of the seeds of a plant, there would be a greater chance for the survival and evolution of the species as more plants of the race are likely to survive in the wider area covered by it.

In dehiscent fruits, the seeds become exposed and it is these that are dispersed while in indehiscent fruits the whole fruits have to be distributed.

Just like the pollen, the seed or the fruit has no power of locomotion so that it has to depend on some *agent* (wind, water or animal) or some special mechanism for its dispersal. The different methods of dispersal may be considered accordingly:

I. DISPERSAL BY WIND

For easy dispersal by wind seeds have to be light so that their buoyancy may enable them to float on air over long distances. Seeds specially adapted for wind dispersal are characterised by the following:

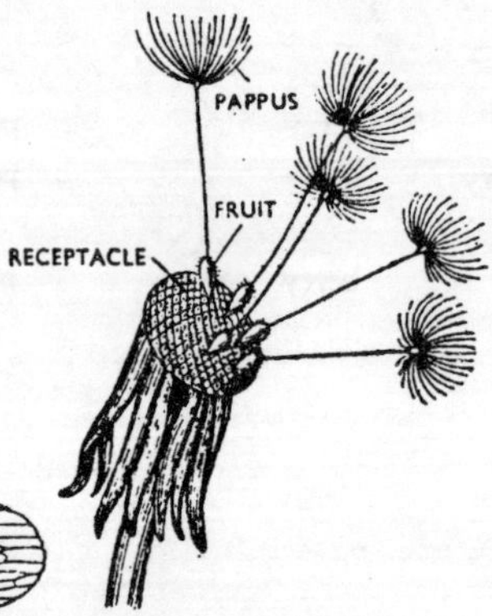

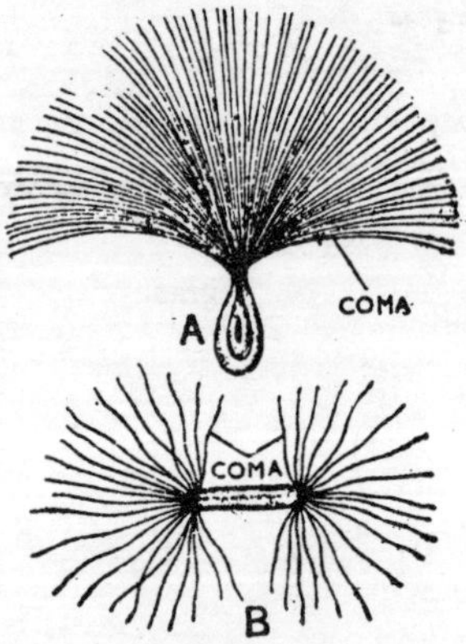

FIG. 460 FIG. 461 FIG. 462

FIG. 460. Minute seed of orchid. FIG. 461. Parachute-like pappus of dandelion (*Taraxacum officinale*). FIG. 462. Coma of *Calotropis* (A) and double coma of *Alstonia* (B).

1. Very small, dry and dusty seeds as those of orchids (weighing as little as 0·004 mg in some cases—Fig. 460) are carried by wind like pollens. Seeds of *Cinchona* are also extremely small and at the same time winged.

2. Parachute mechanism: Certain seeds and a few fruits are provided with appendages which act like parachutes in helping them to float in air. Among such appendages one may consider:

(i) Pappus (Figs. 318 & 461) formed by the modification of the persistent calyx in Compositae and the allied families Dipsaceae and Valerianaceae. This pappus is of different appearance in different species and in many cases it is large and light enough to keep the cypsela fruit floating in air for a long time.

(ii) Coma (Fig. 462) is a tuft of hair developed as a crown on the seeds of *Calotropis*, *Holarrhena*, *Alstonia* (two tufts) and most plants of Apocynaceae and Asclepiadaceae.

(iii) Hairy outgrowths on the testa (Fig. 463) completely cover cotton seeds. Such hair develops from the inner walls of the silk-cotton pod.

(iv) Persistent hairy styles (Fig. 464) of the achenes of *Clematis*, *Naravelia* (Fig. 455A), *Anemone*, etc., help them to float in air.

FIG. 463

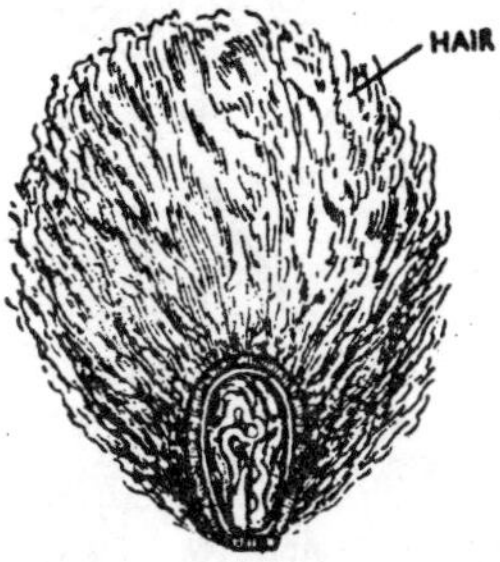

FIG. 464

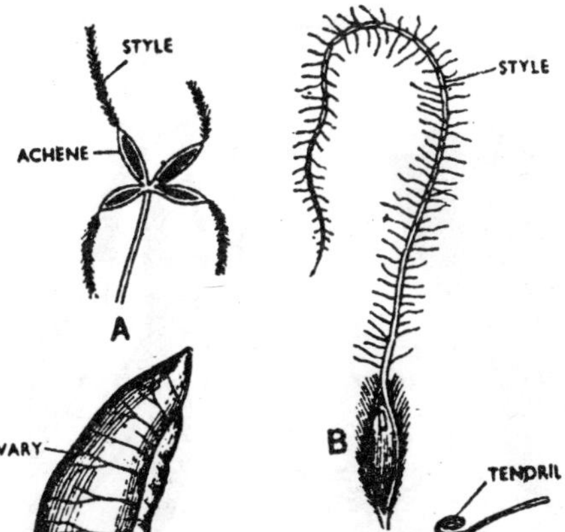

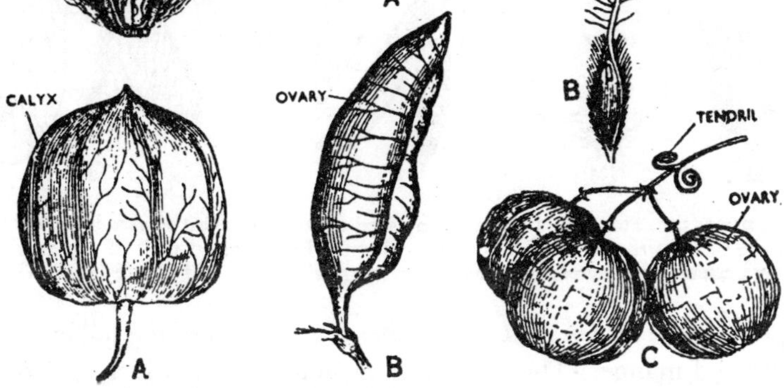

FIG. 465

FIG. 463. Hairy outgrowth on cotton seed. FIG. 464. Persistent hairy style of *Clematis*: A shows an etaerio of achenes while B shows an enlarged achene with the persistent hairy style. FIG. 465. Balloon type mechanism. A. Inflated persistent calyx of *Physalis*. B. Inflated ovary of the legume *Colutea*. C. Inflated ovary of *Cardiospermum* capsule. Note tendril developed by the modification of inflorescence axes.

(v) Balloon type inflated persistent calyx of *Physalis* enables the entire fruit to float on air for some time. The same purpose is served by the inflated ovaries of the pod of *Colutea arborescens* (Leguminoseae) and the capsules of the balloon vine (*Cardiospermum halicacabum*) and also a few other plants of Sapindaceae (Fig. 465).

3. Wings are developed on fruits and seeds of different types (Fig. 466) rendering them buoyant:

(i) Seeds of *Moringa oleifera*, *Oroxylum indicum*, *Lagerstroemia speciosa*, *Swietenia mahogoni*, *Cinchona*, etc., are provided with wings developed from the testa.

(ii) *Samaroid* fruits of *Shorea robusta* and some other members of Dipterocarpaceae have persistent sepals forming wings.

(iii) The *samara* fruit of many plants like *Dioscorea*, *Fraxinus* (ash), *Acer* (maple); *Hiptage*, etc., have wings developed by the fruit pericarp.

4. Censer mechanism: In certain fruits the seeds are discharged through small openings on the fruits. The outlets are so narrow that only a few seeds can escape at a time. Thus, in *Argemone mexicana* and the poppies (Fig. 467) of Papaveraceae, after the apertures are opened by porous dehiscence, as the capsule swings in air, the minute seeds are dispersed through the pores. A similar type of dehiscence may be seen in the peculiar

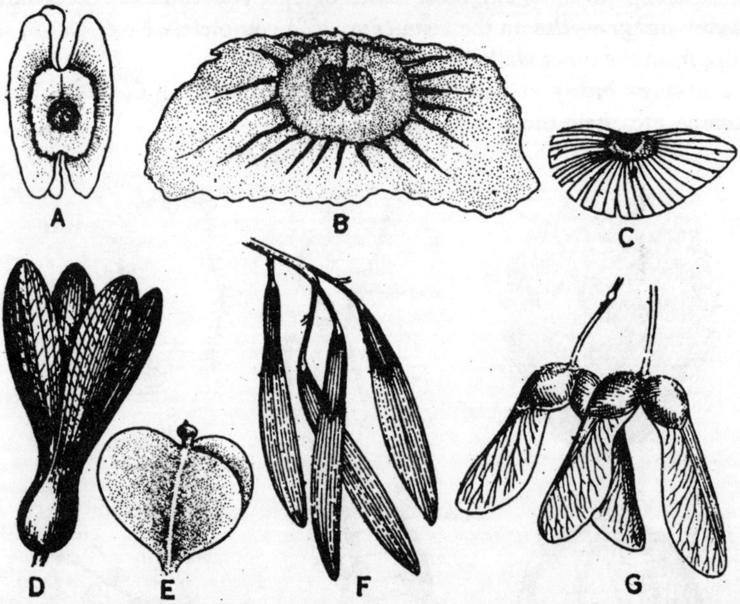

WINGS FOR DISPERSAL. FIG. 466. Developed by seed-coat in *Moringa oleifera* (A), *Oroxylum indicum* (B), *Lagerstroemia speciosa* (C); by calyx in the samaroid fruit of *Shorea robusta* (D); by pericarp in *Dioscorea* (E), ash (*Fraxinus excelsior*—F) and maple (*Acer*—G).

pelican-shaped flowers of *Aristolochia gigas*. Seeds escape from the follicles of *Aconitum* in the same restricted manner. The capitulum of sunflower, in which the seeds are heavy and not provided with pappus, shows a censer mechanism in the restricted escape of seeds when it is swung by wind.

5. A rather interesting type of wind dispersal is shown by certain weeds like *Salsola kali tenuifolia* which grow on sea coasts. When the plant matures, the branches become incurved giving the plant a spherical form. The main stem now breaks and the plant laden with ripe fruits and seeds is rolled like a ball miles after miles by the strong wind. Seeds are dropped all along this route.

II. DISPERSAL BY WATER

Dispersal by water is of great convenience in fruit trees growing by water-side. This is of special importance in seashore plants like the coconut which has migrated thousands of miles in the Pacific and Indian Oceans floating on the sea waves from island to island. Such fruits must be provided with a coat which is simultaneously waterproof, salt-resistant and buoyant. The dry fibrous mesocarp and the hard endocarp of coconuts render them specially suitable for this type of dispersal. Similar properties are possessed by

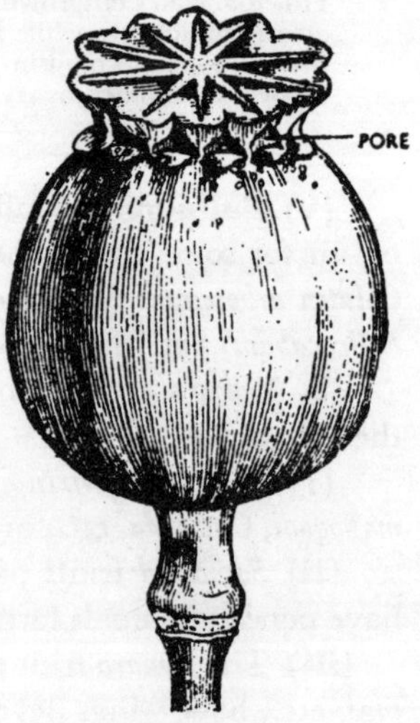

FIG. 467. Capsule of opium poppy (*Papaver somniferum*) showing censer mechanism.

the fruits of some other palms like *Areca catechu* and *Nypa fruticans* and those of *Cerbera odollam* of Apocynaceae (Fig. 468), *Pandanus* (Pandanaceae) and also of a few leguminous plants. Gulf streams often carry seeds and fruits of a large number of species.

Among fresh-water plants, the lilies disperse as their spongy thalamuses bearing fruitlets float about in water. Seeds are deposited at different places as the thalamus rots.

Seeds of many aquatic plants like water-lily, *Alisma*, *Sagittaria*, etc., are very light and waterproof so that they can float easily. Many of these seeds are provided with spongy arils rendering them more buoyant.

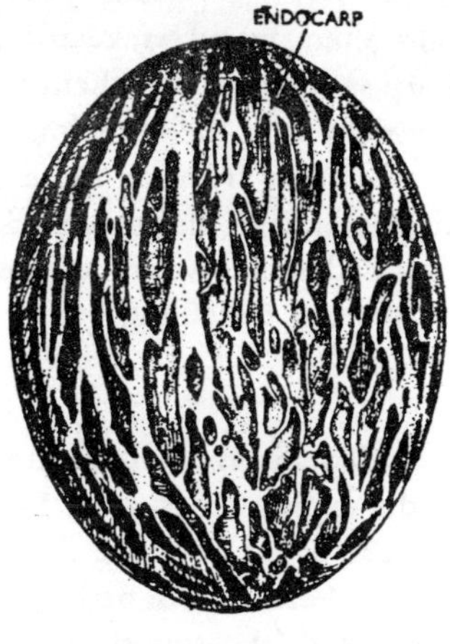

FIG. 468. *Cerbera odollam* fruit showing fibrous buoyant endocarp enabling the fruit to float easily. (After Schimper).

III. DISPERSAL BY ANIMALS

Edible fruits, specially those that are brightly coloured, are devoured by various animals and, in many cases, their seeds pass through animal alimentary canals without any injury. In this way, these seeds are dispersed through animal excreta often many miles from the place of their origin and often to very inaccessible places like tree tops and building cornices. Numerous herbivorous animals like birds, bats, squirrels, jackals, monkeys, etc., are thus responsible for such dispersal. Even ants play a role in dispersal by dragging seeds from one place to ano her. Human beings play no minor role in this mode of dispersal. Various economic plants are distributed all over the world by them. A large number of fruit trees, vegetable and ornamental plants, etc., have been brought to India from foreign countries by the mankind. Examples may be found in potato, tomato, tobacco, water hyacinth, etc. Even coconut is an exotic plant according to our mythology ('created', i.e., probably, introduced by Viswamitra).

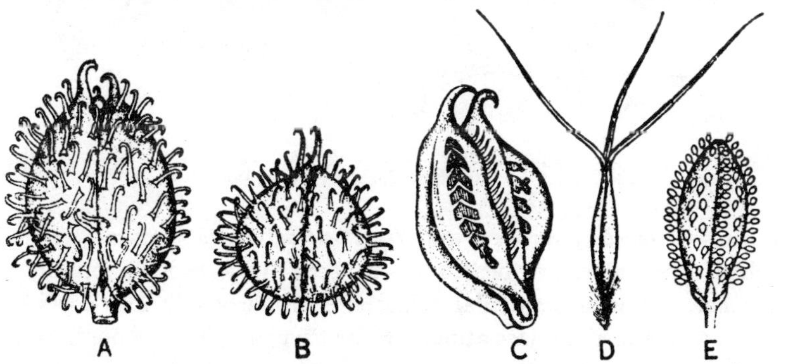

FIG. 469. Devices for animal dispersal. Hooks on fruits of *Xanthium strumarium* (A), *Urena lobata* (B), *Martynia annua* (C), Stiff hairs on the fruit of Aristida (D), Sticky glands on the fruit of *Boerhaavia repens* (E).

Sometimes, fruits and seeds are provided with appendages (Fig. 469) or sticky secretions which facilitate their mechanical dispersal by animals:

1. Fruits and seeds of many plants are provided with hooks, spines, barbs or stiff hairs (Fig. 469) so that if an animal grazes or brushes against them, these stick to the animal's body or clothing. Later on, these drop off somehow or other and are thereby dispersed. Thus, the fruits of *Xanthium strumarium* (cocklebur), *Urena lobata, Martynia annua* (tiger's claw), etc., are provided with hook-like structures which are effective in their dispersal. *Pupalia* fruits are provided with bunches of spreading, hooked bristles. Fruits of the lovegrass (*Andropogon aciculatus*), spear grass (*Aristida*), etc., are provided with stiff hairs often curved backwards so that they stick easily. Fruits of *Tribulus* are provided with sharp spines enabling them to stick to cattle hoofs. Fruits of *Achyranthes aspera* (Fig. 264) are sticky with their scarious bracts and perianths.

2. Some fruits as those of *Boerhaavia repens* (Fig. 469E), *Cleome viscosa*, etc., possess sticky glands so that they stick to the faces and bodies of grazing animals and are thereby dispersed. Many seeds, as those of *Aegle marmelos*, *Plumbago*, mistletoe, etc., are sticky and are thereby benefited.

IV. MECHANICAL DISPERSAL

All dehiscent fruits scatter the seeds when they burst. This dehiscence is accompanied by the expression of great force in many fruits so that seeds are jerked a considerable distance away from the mother plant. Such fruits are called **explosive fruits.** It has already been mentioned that the metre long lomentums of *Entada gigas* explode in the forests with the sound of crackers. Legumes of the mountain climber *Bauhinia vahlii* similarly explode with loud noise scattering the seeds yards away in all directions.

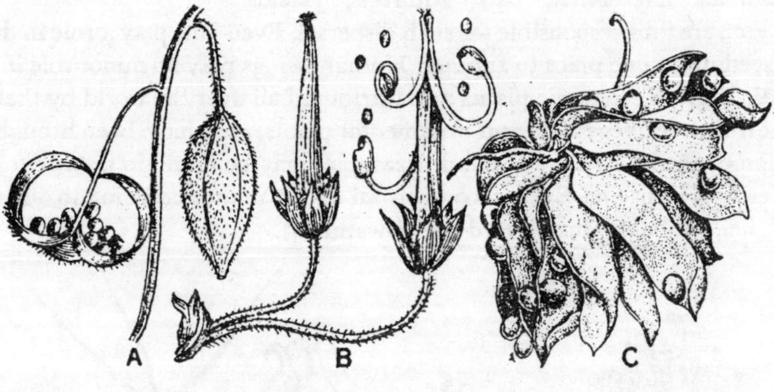

FIG. 470. Explosive dispersal of seed. A. *Impatiens balsamina*. B. *Geranium palustre*. C. A bunch of *Abrus precatorius* pods.

Ripe fruits of *Oxalis corniculata* and *Impatiens balsamina* (balsam) explode suddenly when touched. The valves are bent and the seeds are jerked off forcibly (Fig. 470A). Fruits of *Geranium* showing septicidal dehiscence similarly split into cocci scattering the seeds in this act. Here, the styles attaching the carpels to the carpophore curl upwards and outwards throwing out the seeds like slings (Fig. 470B). The ripe pods of *Abrus* (Fig. 470C), peas and beans (Fig. 434), *Clitoria*, etc., suddenly twist on bursting and thus scatter the seeds.

Fruits of a number of plants belonging to the family Acanthaceae (viz., *Ruellia tuberosa, Andrographis paniculata, Barleria prionitis*) suddenly burst into two valves ejecting the seeds (Fig. 471). In these pods the seeds are arranged in four rows and the flat lens-shaped seeds have peculiar hood-like projections, called *jaculators* or *retinacula*, from their funicles which straighten out during dehiscence so that the four rows of seeds are ejected crosswise in four different directions.

A very interesting type of seed ejection is seen in the mediterranean cucurbit *Ecballium elaterium* named the 'squirting cucumber' (Fig. 472). The internal content of the fruit remains under pressure and the stalk of the fruit acts like a stopper. When the fruit ripens it gets detached from the stalk, the internal pressure is released and some of the contents, including the seeds, are literally 'squirted' out. Seeds may be carried as much as 20 ft. away by this squirting act.

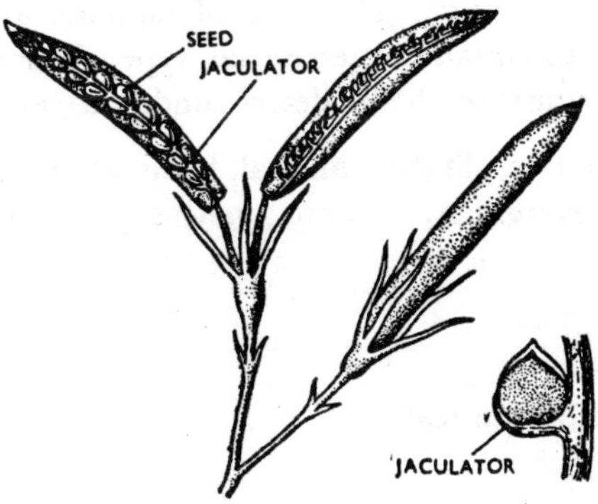

Fig. 471. Explosive mechanism in *Ruellia tuberosa*. Note *jaculators* or *retinacula*.

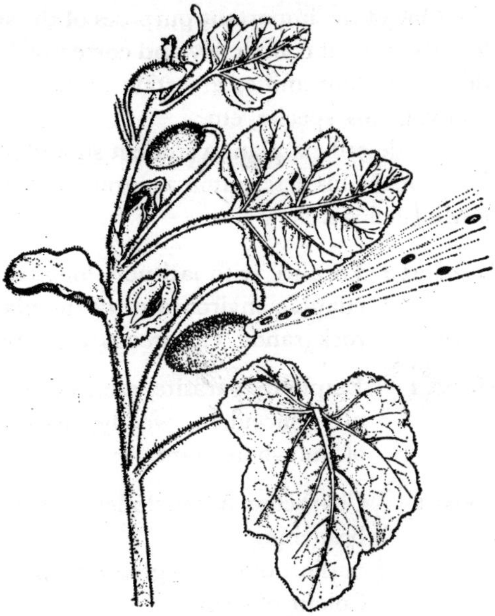

Fig. 472. A twig of the cucurbit *Ecballium elaterium* showing the 'squirting' of a fruit. (After Bocquillon).

XIV

SCHEME FOR THE COMPLETE DESCRIPTION OF A PLANT

One of the important purposes of the study of External Morphology is to be able to describe a plant completely and correctly. Without such a description it is not possible to identify a plant and to ascertain its place in the plant kingdom, i.e., to assign it to its family, genus, species, etc.

To describe a plant a student should follow the scheme given below. It will be seen that to make his description accurate, the student will have to exercise all his knowledge of Morphology.

Habitat : Locality with latitude, longitude and altitude. The natural climatic and edaphic environment of the plant, viz., aquatic or terrestrial, type of soil or rock (sandy, calcareous, etc.), bright sunshine, shade, desert conditions, etc.

Habit : Epiphyte, parasite, etc.; herb, shrub, tree, climber; annual, biennial, perennial; succulent, woody, deciduous, evergreen; size, etc., general characters of the whole plant.

Stem : Characters not included under habit.
Branching.
Surface of stem—glabrous, hairy, etc., as in leaf.
Colour of stem.
Shape—round, square, ribbed, etc.
Hollow or solid, jointed or not.
Any special modification—phylloclade, etc.

Leaf : Phyllotaxy.
Insertion—radical or cauline, petiolate or sessile, peltate, etc.
Simple or compound—type of compound.
Leaf-base—stipulate or exstipulate, kind of stipule, any speciality.
Petiole—any speciality.
Leaf lamina—shape, venation, margin, incision, apex, base, surface, texture, glands, colour, odour and taste.
Modifications and specialities (ligule, stipel, heterophylly, etc.)

Root : Tap or adventitious.
Modification.

Inflorescence : Type—racemose, cymose, etc.

Flower

General: Bracteate or ebracteate—type of bract.
Bracteole—if any.
Sessile or pedicellate.
Complete or incomplete—dichlamydeous, monochlamydeous, achlamydeous, etc.
Bisexual or unisexual—monoecious, dioecious, etc.
Symmetry—regular, zygomorphic or asymmetrical.
Floral phyllotaxy—cyclic, spiral, hemicyclic.

Insertion—hypogynous, perigynous, epigynous.
Erect, pendulous, etc.
Size.
Any speciality—disc, nectary, etc.

Thalamus: Any speciality not included under flower.

Calyx: Cohesion—polysepalous or gamosepalous.
Number of sepals or lobes.
Superior or inferior.
Aestivation.
Calyx shape—sepal shape.
Size (of calyx and sepals).
Colour.
Duration—caducous, deciduous, persistent.
Any speciality (modification, appendage, etc.)

Corolla: Cohesion—polypetalous, gamopetalous.
Adhesion.
Number of petals or lobes.
Superior, inferior.
Aestivation.
Corolla shape (regular or irregular)—petal shape.
Size (corolla and petals).
Colour, duration, texture, scent.

Perianth: (When calyx and corolla cannot be distinguished).
Sepaloid or petaloid.
Other characters as in calyx and corolla.

Androecium: Cohesion—in filament, in anther, in both.
Number—fertile and sterile ones as also separate whorls separately counted.
Adhesion—with corolla or gynoecium.
Superior or inferior.
Filament—any outgrowth, sessile or not, length.
Anther—attachment of filament, dehiscence, introrse or extrorse, number of pollen sacs, appendages.
Pollen—powdery, etc.; appearance; pollinia.
Colour.
Speciality—didynamous, etc.; obdiplostemonous, etc.; exserted or inserted; nectaries, etc.

Gynoecium: Cohesion—apocarpous, syncarpous, etc.
Number.
Adhesion—with androecium.
Ovary—shape, number of loculi, placentation, superior or inferior.
Style—number; free or united; terminal, lateral, gynobasic, etc.; shape; size; colour; any peculiarity.
Stigma—number; sessile or not; shape—simple, lobed, branched, etc.; smooth or papillose; any speciality.

Ovule—number; number in each loculus; ascending, horizontal, etc.; anatropous, campylotropous, etc.

Any speciality in the gynoecium.

Floral Diagram and Floral Formula.

Pollination: Type if observed.

Fruit : True or false; dry or fleshy; dehiscent or indehiscent.
Type of fruit, manner of dehiscence (if dehiscent).
Size.
Any speciality (edible, poisonous, etc.; which part edible).

Seed: Number of seeds in fruit.
Morphology—perisperm, endosperm (nature of food reserve), embryo.
Size and shape.
Any speciality (aril, etc.).
Manner of germination.
Method of dispersal.

Any description should be accompanied by neat and representative labelled sketches.

PART TWO

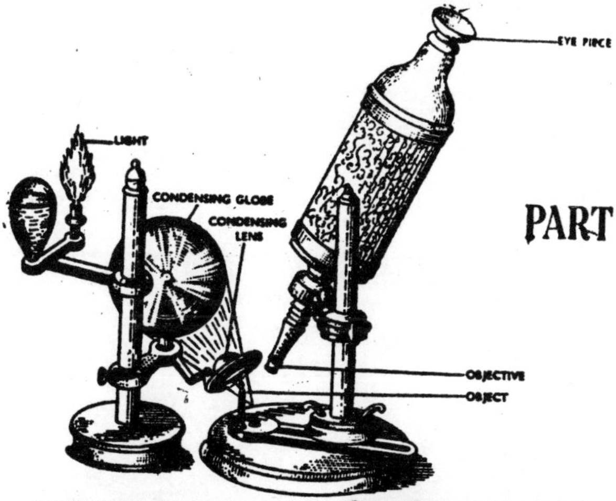

COMPOUND MICROSCOPE
USED BY
ROBERT HOOKE
IN THE SEVENTEENTH CENTURY

Plant Anatomy

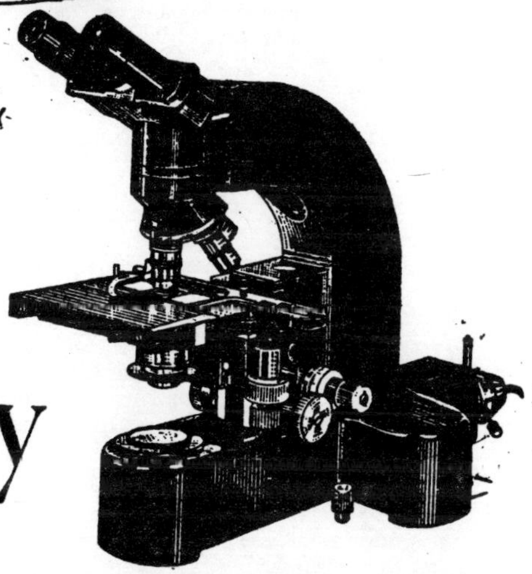

MODERN RESEARCH MICROSCOPE
OF
TWENTIETH CENTURY

CHAPTER XV

INTRODUCTION

The word **anatomy** has derived its origin from *ana*, meaning asunder, and *temnein*, meaning—to cut. So it is regarded as the branch of botany which considers the internal structures and organization of the plant members. It is also known as **internal morphology.** Unfortunately the term **histology** (*histos*=cobweb) is used as a synonym for anatomy. Histology deals with the minute internal structures like cellular arrangements, tissues and tissue systems; but gross internal structures like steles, vascular skeleton, leaf-trace and branch-trace systems do not come under histology. Specialised study of the cell belongs to **cytology** which is now treated as an independent branch of biology. Anatomy is really concerned with both minute and gross internal structures. Thus histology is covered by anatomy.

There is no denying of the fact that anatomy is rather a recent branch in comparison to morphology and taxonomy. Though really serious studies on anatomy started with the

Fig. 473. Aristotle (384-322 b.c.).

Fig. 474. Theophrastus (371–285 b.c.).

invention of the microscope in the later part of the sixteenth century, but this branch had also its beginning. Long before Christ and Greek Civilization ancient Hindus of India devoted their attention to the subject. They distinguished five regions in the body of the plants, viz. *tvac* (skin), *mansa* (soft tissues or bast), *asthi* (wood or bone), *majja* (pith), and *snayu* (fibres in the bast). *Mansa* was further divided into *valkala* (bark) and *sakara* (bast fibres). History of biological studies is usually traced from the time of great Aristotle (Fig. 473) of Greece who lived in the fourth century b.c. Theophrastus (Fig. 474), a pupil of Aristotle, is credited for his work on internal structure of plants. He said—"plants are made of *phloios* (bark), *zylon* (wood) and *metra* (pith), when pith is present". Thus the very commonly used terms xylem and phloem were coined by Theophrástus. He worked on the anatomical problems rather extensively, of course with the limited and meagre resources at his disposal. He is thus regarded as the father of botanical science as regards anatomy in particular. Since then for a pretty long time there was nothing noteworthy in anatomical work due to the want of a magnifying apparatus, though considerable advances were made in other branches like morphology and more so in applied sciences like medicine and agriculture.

Real interest re-awakened with the invention of the microscope. The first compound microscope was produced in about 1590 by Dutch spectacle makers, Jans and Zacharius Janssen. During the first part of the seventeenth century more improved models were designed. This wonderful instrument aroused curiosity in the workers and led to many significant discoveries. Robert Hooke (1635–1703) in England examined a piece of bottle cork under a microscope (front piece of this part) improved by himself which looks like a fire-extinguisher to twentieth century eyes, and noted the cells His observations were published in Micrographia in 1665. Curiously enough the credit of such an important discovery goes to Hooke, who was for some time a collaborator of Robert Boyle, curator and later Secretary of the Royal Society, but not a student of biological science. It is suggested in some quarters that Hooke probably used his tool more as a play-thing than as something for serious scientific pursuit. Nehemiah Grew (1641–1712, Fig. 475), an English physician and Marcello Malphigi (1628–'94, Fig. 476), an Italian physician and professor, independently carried on very useful works on internal structures, though the

FIG. 475. NEHEMIAH GREW (1641–1712).

FIG. 476. MARCELLO MALPHIGI (1628–'94).

approach of both of them; being physicians themselves, had been mainly from the point of view of medicine. Grew published his work in 1682 under the title—The Anatomy of Plants, where he discussed the gross structures of plants, and compared the vascular skeleton of the plants with that of the animals. The commonly used terms of anatomy like parenchyma, vessels, cortex, etc. were introduced by him. Malphigi's investigations were more or less in the same line. He discovered the spiral vessels and the stomata. His pioneering works on vascular plants came out in his 'Anatome Plantarum' in 1675. Both Grew and Malphigi made attempts to relate the anatomical structures with their physiological functions. So spectacular progress made in the seventeenth century practically waned in the next century. Again in the nineteenth century works on anatomy started with fresh energy by Charles Francois Mirbel (1776–1854), Kurt Sprengel (1766–1833) and Others. These scientists primarily worked on the origin of the cells. The method of formation of a vessel from a series of cells became more clear. Nucleus was first discovered by Robert Brown (1773–1858) in the skin of orchid leaves. Though some earlier workers noted this centrally located body in the cell, they failed to recognise its importance. Till this time no body could unravel the mystery of protoplasm. Suggestions had been put forward by many workers as regards the existence of the living contents of the cells, and different names like nourishing juice, 'sarcode', etc. were proposed, but the nature of the mysterious fluid remained practically unknown. Hugo von Mohl (1805–'72, Fig. 477) was the first man to recognise the importance of protoplasm and was instrumental for shifting the attention of the workers to the contents from the wall. The conception of a cell became clearer; and that it is essentially a mass of protoplasm with its outer secretory product, the cell wall or without it, holds good even now. Matthias Jacob Schleiden (1804–'81, Fig. 478), a botanist, and Theodore Schwann (1810–'82, Fig. 478), a zoologist, worked on plant and animal tissues respectively and postulated a theory known as Cell

Fig. 477. Hugo von Mohl (1805–'72).

Matthias Jacob Schleiden (1804–'81).

Theodor Schwann (1810–'82).

Fig. 478.

theory or Cell doctrine in 1838. The Cell theory demands that all living things are essentially cellular in nature. Thus they defined the basic unit of life. Hugo von Mohl, apart from putting the cell theory on a solid foundation, carried on extensive investigations on plant anatomy, viz., nature and formation of vessels, structure of epidermis, lenticel and cork and bark; and also traced the course of vascular bundles in stems and leaves. He carried his work with so much of accuracy and thoroughness that a new epoch was opened in plant anatomy. Rudolf Virchow, a great German physician, asserted that cells originate only from pre-existing cells. Carl von Nageli (1817–'91, Fig. 479) was another contemporary worker who made very valuable contribution to anatomical studies. Nageli's works were particularly confined to ontogenetical studies—on primary and secondary meristem, the types of vascular bundles—common, cauline and foliar bundles. Nageli was really the originator of the terms xylem and phloem, as we understand them today. It can be said that after Robert Hooke's discovery of cells, Grew and Malphigi laid the foundation of plant anatomy, and Hugo von Mohl and Carl von Nageli developed its superstructure; so that they may be called the founders of modern plant anatomy. Edward Strasburger could describe the chromosomes and their behaviour during cell division using Abbe microscopes, which were available then, and basic Aniline dyes. Spetacular advances were made in the studies of the structures and forms of plant tissues. Cytoplasmic and nuclear structures were naturally much better understood.

FIG. 479. CARL VON NAGELI (1817–'91).

In the modern period studies on the meristems and zonation of tissues at the root- and stem-apices gained prominence. Hanstein postulated the 'histogen theory', asserting that three histogens or tissue-builders occur at the tips, which in course of time develop into the three tissue-systems. Julian von Sachs (1832–'97) first introduced that there are three tissue-systems in plants. Increasingly more emphasis was being laid to the physiological aspect, rather than morphological, what early workers did, of anatomy. Simon Schwendener (1829–1919) and more particularly the great Gottileb Haberlandt were the exponents of this approach. Haberlandt published the results of his work in his classical book 'Physiological Plant Anatomy'. According to his own statement 'homologies of the tissues were of no interest to him' and that he was 'solely concerned with analogy'. He put the tissues into twelve systems, viz., dermal system, absorbing system, etc. purely on the basis of function. Without in any way underestimating Haberlandt's great contribution, it should be said that physiological anatomy is really of value to the physiologist, rather than to the anatomist, because origin, contiguity, topography and phylogeny of the tissues were rather ignored in these systems.

Anton de Bary (1831–'88), primarily a student of fungi, worked on plant anatomy and presented the same in his book 'Comparative Anatomy of Phanerogams and Ferns' in 1877. Edward van Tieghem (1839–1914) made further improvement and established the stelar theory. In collaboration with his students Tieghem worked on the various types of steles and explained them on the basis of comparative morphology and phylogeny. The contribution of Edward Charles Jeffrey towards knowledge of stele and stelar nature of vascular plants also merits special mention. Anatomical characters were being used in taxonomy more frequently than what found expression in the very important works—Engler and Prantl's 'Die Naturliche Pflanzenfamilien' and Solereder's 'Comparative Anatomy of the Dicotyledons'.

In the recent years developmental plant anatomy has come to the forefront. Several investigators in different parts of the world are now engaged in searches on the developmental history of the plants. Very valuable contributions on this aspect have been made by I. W. Bailey, A. S. Foster, K. Esaw, G. P. Majumder and Others. Due to their very important and useful works the conceptions about meristems, the zonation and structural development, histogenesis of the organs—shoot, root and flower are becoming increasingly more clear. A very useful instrument microtome was devised in late nineteenth century and gradually improved models were available. This tool has proved to be of inestimable value in studying developmental history of plant organs because serial sections of uniform thinness could be made. Fortunately modern workers have at their disposal more improved instruments and technical devices which would enable them to make spectacular advances in their search for truth.

Plant anatomy is an important basic science. For properly understanding the fascinating physiological processes going on inside the plant body and phylogenetic relationship existing between the various plant groups a thorough knowledge of anatomy is essential. Ecological adaptations to different environmental conditions can be more rationally interpreted in terms of their anatomical peculiarities.

CHAPTER XVI

AN OUTLINE OF THE PLANT-BODY

The plant itself originates from the morphologically unicellular zygote, which in course of time develops into the embryo and then to the mature plant. The higher plant is the result of evolutionary specialisation through a long period.

In spite of great diversities existing in the vascular plants as regards size, structure and form, there is a fundamental uniformity. The structural plan of a higher plant (Fig. 480 A) is such that each plant has an axis, the lower part of which is the subterranean

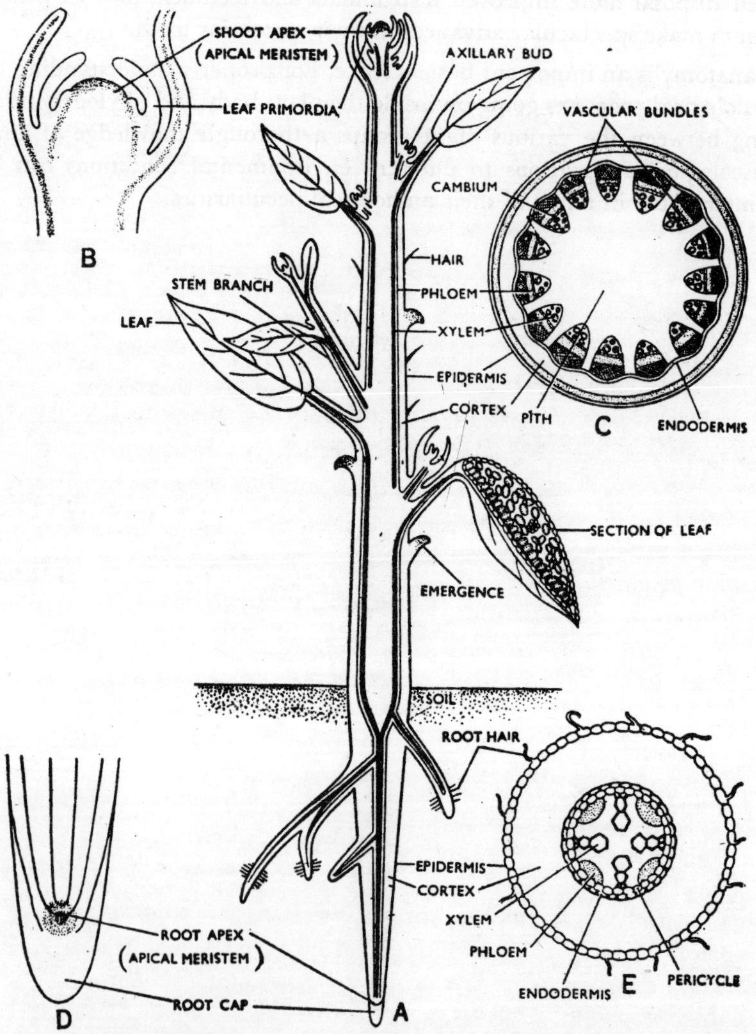

FIG. 480. The structural plan of a vascular plant. A. The plant showing the parts and appendages. B. *L.s.* of shoot-apex. C. *T.s.* of stem. D. *L.s.* of root-apex. E. *T.s.* of root.

root and the continuous upper aerial part is the **stem.** The root-stem axis bears appendages, which are of three types. The most important ones are those in which vascular strands pass from the axis. **Leaves** are such appendages. They are characteristics of the stem and are absent in the roots. Leaves have a definite order of development and are in close association with the stem. Thus they are regarded as lateral expansions of the stem. The root, stem and the leaves are the three main organs of a plant. These three organs are intimately related both from developmental and structural points of view. The flowers are reproductive organs which come out after the plants have attained maturity. It is an accepted view that the flower is homologous to a shoot and floral parts with the leaves. The second type of appendages, called **emergences,** develop from the external portions of the stem, the epidermis and cortex usually taking part in their formation. The **hairs** arising from the outermost layers of the stems and the leaves constitute the third type of appendage. These are superficial outgrowths. Emergences and hairs, unlike the leaves, have no definite order of development.

THE AXIS

The root and stem form a continuous axis of the higher plant. Three zones or regions may be usually located in the axis (Fig. 480 C & E). The **epidermis** or the skin forms the outermost zone. It is usually uniseriate due to anticlinal (at right angles to the axis) division of the mother cells, and is *protective* in function. The second zone internal to epidermis is the **cortex.** It is many layered and is composed of different types of tissues performing functions like support, storage, etc. The last layer of cortex is the **endodermis** which is also normally uniseriate. Next comes the central core, known as **stele** or central cylinder, which practically remains enveloped by the cortex. Stele includes the **vascular bundles,** consisting of complex tissues, *xylem* and *phloem*, and other tissues. Functions of vascular bundles are twofold—firstly, conduction, and, secondly, mechanical support. Xylem is meant for conduction of water with dissolved mineral matters absorbed from the soil and also for mechanical purpose, while the other tissue phloem is there for conduction of elaborated food matters in solution. A lateral meristem, called *cambium*, may be present between xylem and phloem, as in the stems of dicotyledons and gymnosperms. Xylem and phloem may remain radially, as in the roots, or they may occur on the same radius, as in other organs. According to the arrangement of the tissues the stele may be like a solid rod, a hollow cylinder, etc. In some cases they enclose a soft parenchymatous portion in the central region, which is known as **pith** or *medulla*. Continuations of parenchymatous pith passing in between the vascular bundles may look like rays radiating from the pith. They are called *pith rays* or **medullary rays.** Outside the vascular bundle occur one or more layers of non-conducting cells forming what is known as **pericycle.** It lies internal to endodermis and forms the outermost sheath of the stele.

PRIMARY BODY AND SECONDARY BODY

The axis with the appendages constitutes the fundamental part of the plant. All the tissues forming this body are the derivatives of the *apical meristem* present at the tips or apices of the axis (Fig. 480 B & D). These tissues are called **primary** tissues, and the body composed of primary tissues is the **primary body.** During the formation of this body, what is called primary growth, the axis grows in length, appendages develop and the branch systems are laid down. *Lateral meristems* like cambium occur between primary xylem and primary phloem in the vascular bundles of dicotyledonous and gymnospermic stems. Those cells divide and produce new tissues—secondary xylem and secondary phloem between primary xylem and primary phloem. After formation of these secondary tissues a new meristem, called *phellogen* or cork cambium, arises in the peripheral portion of the organ. This meristem which is also *lateral*, cuts of *phellem* or cork cells with suberised walls on the outer side and parenchyma cells called *phelloderm* or secondary cortex on the inner side. As a result, the axis grows in girth or thickness. These tissues formed later by the lateral meristems are known as **secondary** tissues; and the growth in thickness due to addition of secondary tissues is referred to as **secondary growth.**

Thus plant body is composed of *organs* like stem, root and leaves. These are distinct morphological structures, each having specific function to perform. Every organ, in its turn, is a collection of *tissue-systems* which carry on restricted functions. A tissue-system is made up of a number of contiguous *tissues*. A tissue is, in fact, a group of *cells* formed in response to a basic division of labour. So the ultimate units of the plants are the cells. Thus it is customary to begin studies on anatomy with the cell.

METHODS OF ANATOMICAL STUDIES

It goes without saying that the aid of a compound microscope is indispensable for the studies of the minute structures of plants. Very thin and uniform sections are taken for observation under the microscope. In case of cylindrical bodies like the axis sections are taken both at right angles to the axis (*transverse* or cross-section) and along or parallel to the axis (*longitudinal* section). Longitudinal planes are of two types, viz. that dividing the axis radially is known as *radial* plane; and that at right angles to the radial plane is the *tangential* plane (Fig. 481). For detailed studies sections along all the planes mentioned above are necessary. Normally preparations are made free-hand with the help of a sharp razor. For specialised studies an apparatus known as **microtome** is employed. For making sections by microtome the materials have got to be killed, fixed and ultimately embedded in paraffin, which involves long-drawn elaborate processes. Nevertheless microtome has positive advantage for two reasons, viz. first, the sections are always uniformly thin, and, secondly, serial sections may be obtained. In the studies of developmental anatomy microtome is of inestimable value.

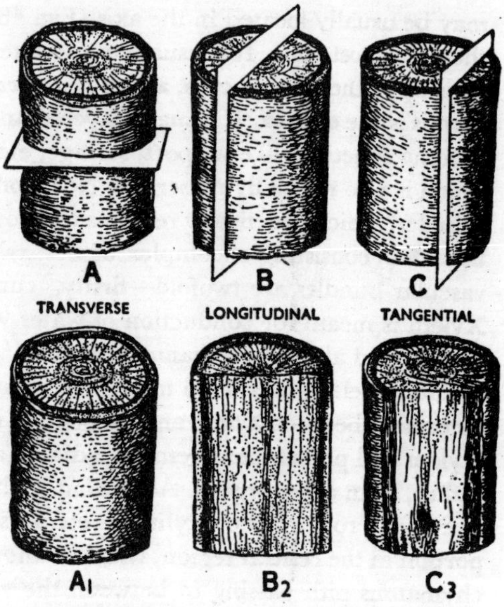

FIG. 481. Diagrammatic sketches of cylindrical axes to show transverse (A & A_1), longitudinal (B & B_2) and tangential (C & C_3) sections.

Another method, called **maceration,** is also employed for working out the detailed structure of the individual cells and elements. This process involves complete or partial dissolution of the 'cementing material'—the intercellular substance lying between the cells, by chemical treatment. Thus the cells become isolated and can be conveniently observed as three-dimensional bodies. The cells tend to create two-dimensional concept in sections. Moreover, many characters of the cells and particularly the nature of the wall with various types of localised thickenings become much more clear. Different techniques of maceration have been in use.

Other methods now employed are whole mounts, peel mounts and scrape mounts. Often materials are mounted entire without maceration or sectioning, though trimming of excess of tissues or reduction of the sizes of the materials may be necessary. This method is commonly employed for observation of lower plants like algae, fungi, thalloid bryophytes and prothalli of ferns. In higher plants it is used conveniently for the study of leaf-epidermis and stomata, hairs and trichomes, small flowers and pollen grains. In case of leaf-epidermis and stomata practically small portions are peeled off and mounted as usual. So it may be called peel method as well. Apart from cell-contents, scraping of portions reveals clear nature of tissues like cambium, tracheids and tracheae.

The value of the magnifying apparatus light microscope has already been mentioned, but it has its limitations. Transparency to visible light is a positive handicap in observing the cells in living state. Polarised light microscope has proved quite helpful particularly in case of structures like the cell wall with highly oriented molecules. Phase-contrast microscopes with monochromatic light source developed later were very useful in watching the process of cell division—undoubtedly a dramatic biological phenomenon. In modern times a new tool—the electron microscope with high resolving power—200 times higher than that of light microscope, has enabled us to understand the finer details of many structures, as well as in discovering many new structures hitherto unknown. Here a beam of high speed electrons is focussed by electromagnetic lenses in place of a beam of light. The unit of measurement in a light microscope is a micron, written as μm, one thousandth part of a mm, whereas the unit in case of an electron microscope is an Angstrom, written as Å=one ten thousandth part of a micron or ten millionth of a mm or 0·1 nm (nanometre). So one μm=10,000Å or 1,000 nm. Structures up to 15Å (1·5 nm) can be studied under electron microscopes compared to about 3,000Å (300 nm) visible under highest power light microscope.

CHAPTER XVII

THE CELL

Living things are composed of material units called **cells.** A cell is defined as a *unit of structure and function*. A typical plant cell consists of the essential living substance, **protoplasm** with the surrounding **cell wall.** The latter is a secretory product of the protoplasm itself. Structurally, plant cells are more distinct, in contrast to animal cells, due to the presence of rigid non-living cell wall (Fig. 482). Protoplasm, the mysterious fluid, has been aptly described as 'the physical basis of life'. The organised mass of protoplasm of a cell is known as **protoplast.** A plant cell may be considered separable into two parts—the protoplast and the cell wall. But this separation is, by no means, complete. Very fine fibrils or threads of protoplasm pass through the minute perforations of the cell wall and connect the protoplasm of adjoining cells. These threads are called ***plasmodesmata*** (Fig. 516). The organic continuity of protoplasm in a living body is thus established.

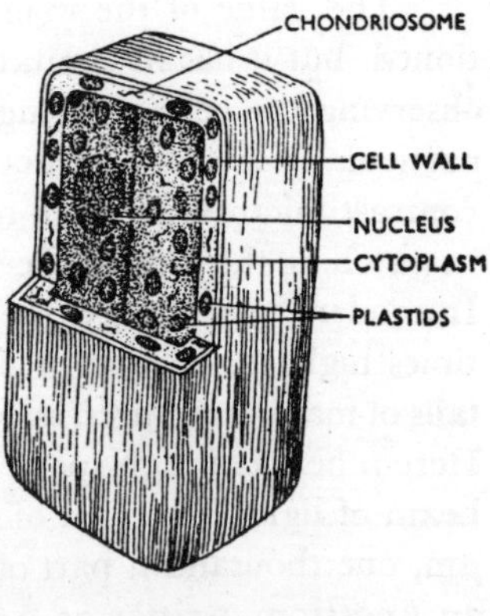

FIG. 482. Model of a plant cells, a portion removed to show central vacuole, nucleus, cytoplasm, plastids and chondriosomes, and the cell wall.

The use of the term cell has been associated with an interesting account of anatomical studies since a long time. It has been discussed in the introductory chapter that really serious attempts to explore the internal organisation of the plants and animals started with the invention of the compound microscope by Jansens in the later part of sixteenth century, though there are evidences that even early ancients in India and Greece had some knowledge about tissues and organs. In 1665 an Englishman Robert Hooke, observed a thin slice of bottle cork under a microscope improved by himself and saw some compartments. As each of them looked like a chamber of a honeycomb Hooke called it a cell (Latin *cellula*=a small apartment). His observations were published in his classical book 'Micrographia' in 1665 (Fig. 483). Obviously Hooke noted the more distinct cell wall. Though the mystery of living protoplasm remained unknown, he mentioned of some sort of 'nourishing juice' present in the cells. In the same century Italian professor Marcello Malphigi and English physician Nehemiah Grew carried on observations on the internal structures of the plants. Protoplasm escaped their notice as well. Corti (1772) and Fontana (1781) saw something like living sap in plants. Dutrochet (1824), a French biologist, also worked on cellular structures and laid emphasis on the cells as individuals. Sometime in 1835 Dujardin in France appreciated the importance of the living substance in the cells of lower animals, to which he applied the term 'sarcode'. Hugo von Mohl was the first man to recognise the importance of the living contents of the cell and the term *protoplasm*, meaning first-formed substance, was proposed by him. Robert Brown observed some round bodies at the central regions of

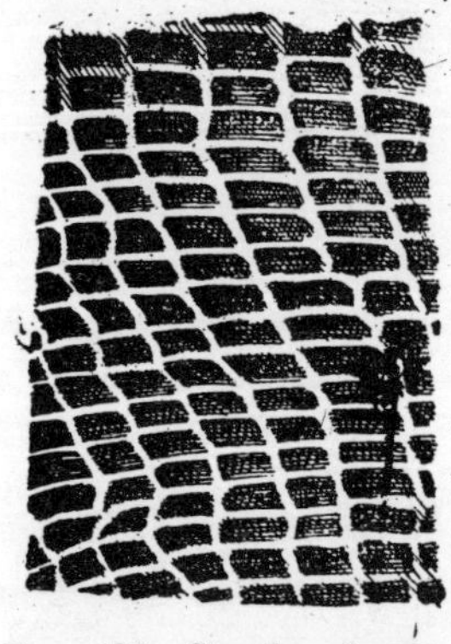

FIG. 483. Sketch of cells from bottle cork as appeared in the Micrographia in 1665.

the cells in the skin of orchid leaves. He suggested that each of these centrally located bodies be called a *nucleus*. After the discovery of the nucleus two biologists, Matthias J. Schleiden, a German botanist, and Theodor Schwann (Fig. 478), a German zoologist, postulated a theory known as the **cell theory** in 1838, which claimed that all living bodies are essentially cellular in nature, i.e. they are composed of cells. They recognised the cells as the units and thus 'primary agents of organisation'. Von Mohl's observations about the importance of protoplasm certainly put the cell theory on a strong foundation. The announcement of cell theory was an epoch-making incident in the history of biological studies. Schultze in 1861 put forward the *protoplasm doctrine*, stating that all masses of protoplasm are the units of organisation and that protoplasm in general, is similar in all organisms. The cell wall, on which the early workers put so much emphasis, practically shrank to the background and was considered of secondary importance. The cell was thought to be the mass of protoplasm with a nucleus, which would originate from a pre-existing cell, another generalisation made by Virchow To this organised mass of protoplasm Hanstein in 1880 applied the term *protoplast*. The protoplasm doctrine received support from many biologists. Huxley in his famous essay very aptly described protoplasm as 'the physical basis of life'.

In the mean time the concept of cell had undergone considerable modifications and some biologists like de Bary, Sachs and Others severely criticised cell theory. In fact, they proposed a new theory called **organismal theory.** The idea that a multicellular body is an aggregate of cells, did not appeal to the protagonists of the new theory. They were of opinion that a multicellular body was a continuous mass of protoplasm which was incompletely subdivided into small centres of activities—the cells, during the development of the body. Thus according to this theory the cells are not to be regarded as the ultimate units or seats of activities, but the whole is to be considered the 'primary agent of organisation'. Then growth is not due to multiplication of elementary units, but due to formation of cells by differentiation within the entire mass of protoplasm.

It is really difficult to explain the coenocytic bodies like the members of the green algae Siphonales in terms of cell theory, according to which a nucleus with the surrounding bit of cytoplasm is to be treated as a cell; but in the light of organismal theory the whole coenocytic body would be a single cell.

The term cell has been rather loosely used, some people included the wall and some excluded the outer cover. The present-day conception of a cell is that it is essentially a mass of protoplasm, the protoplast, with or without the cell wall. In modern terms cell may be defined as "a unit of biological activity delimited by a selectively permeable membrane and capable of self-reproduction in a medium free of other living organisms" —Loewy and Siekevitz. It has even been suggested that the term cell be abandoned and let it be replaced by protoplast. In plants cells like tracheids and sclerenchyma lose protoplasm with maturation and thus become dead. Here cell practically means the dead wall. The compartments observed by Robert Hooke from bottle cork were also nothing but cell wall. It is, at any rate, desirable to retain the term cell which would include the protoplast and the surrounding wall, particularly in view of the intimate relationship existing between the two and establishment of protoplasmic connections between adjacent cells through plasmodesmata. Though vast majority of the plant cells are walled, cells without walls are not altogether absent. The reproductive cells are as a rule naked bodies. In lower groups of plants very often naked masses of protoplasm are noticed.

The cell is regarded as the basic unit—the common denominator of the innumerable variety of living forms. Recent studies with electron microscope revealed two plans of cellular organisation—the simpler **procaryotic** and the more complex **eucaryotic.** The procaryotes including the bacteria and the blue-green algae are small ones (0·05 to 3 μm) usually have primitive nucleus without membrane and also lacking in organelles like chloroplasts, mitochondria, Golgi bodies and lysosomes. The genetic information here is located mostly in the single chromosome consisting of a double strand of DNA and that also without basic proteins—histones. Mitotic apparatus and nucleolus are also absent. The procaryotic cells are surrounded by cell wall made of carbohydrates and amino acids. The plasma membrane lying directly beneath the cell wall often forms intrusions into cytoplasm—the *mesosomes.* The eucaryotes are the cells of remaining groups from the algae to higher plants showing high degree of organization. They possess extensive membrane system—endoplasmic reticulum, Golgi bodies, well-formed nuclei, plastids, mitochondria and lysosomes. Nucleoli and more than one chromosome with DNA and histones are present.

The starting point in the life of every individual is a single cell. If it continues as such the organism remains **unicellular,** as in many lower plants like yeasts, diatoms, desminds, etc.; but in large majority of plants the cell goes on dividing, so that ultimately a distinct **multicellular** body is formed. In fact, the cellular complexity of an organism is based on the systematic position of the plants in the line of evolution.

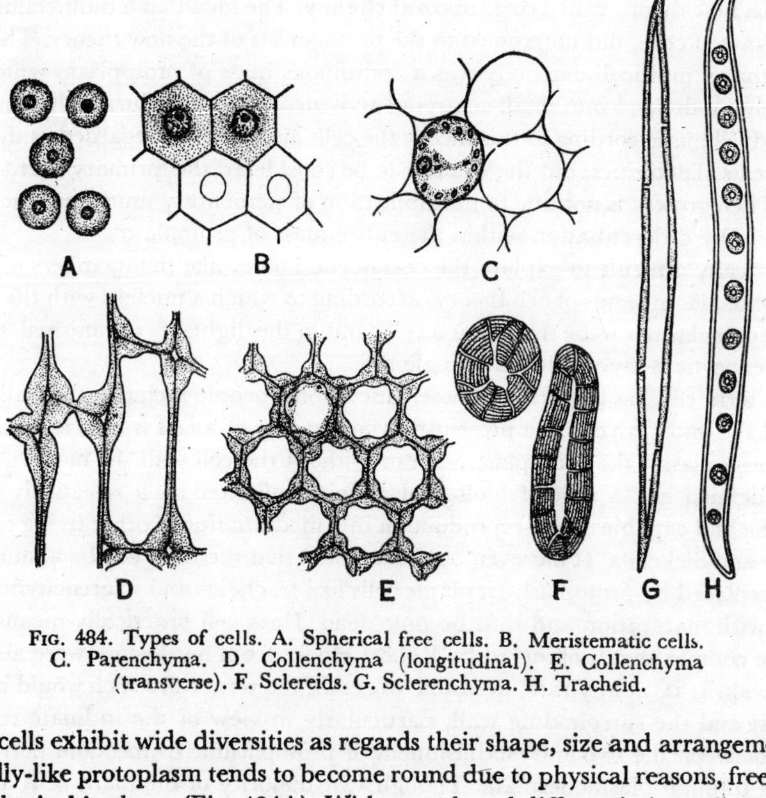

FIG. 484. Types of cells. A. Spherical free cells. B. Meristematic cells. C. Parenchyma. D. Collenchyma (longitudinal). E. Collenchyma (transverse). F. Sclereids. G. Sclerenchyma. H. Tracheid.

The cells exhibit wide diversities as regards their shape, size and arrangement. As a mass of jelly-like protoplasm tends to become round due to physical reasons, free cells are usually spherical in shape (Fig. 484A). With growth and differentiation cells attain different shapes. Two chief types, however, are *polyhedral* with diameters more or less same, and *elongate.* All transitional forms between the two extremes are possible (Fig. 484). In case of young cells of same size and age the shape becomes polyhedral due to mutual contact and compression. They may be fourteen-sided (Fig. 485) approaching in shape the mathematically determined form tetrakaidecahedron.

The size is closely related to shape. The average diameter of a common polyhedral cell ranges between 0·1 to 0·01 mm though there are cells much larger and much smaller than that, so much so that it may be even 1 mm in diameter. Very small cells are not common in higher plants. The fibres are often of unusual length. In some monocotyledons and dicotyledons of family Urticaceae the fibres may have length as much as 550 mm. On the average, the length of the fibres in higher plants ranges between 1 to 8 mm.

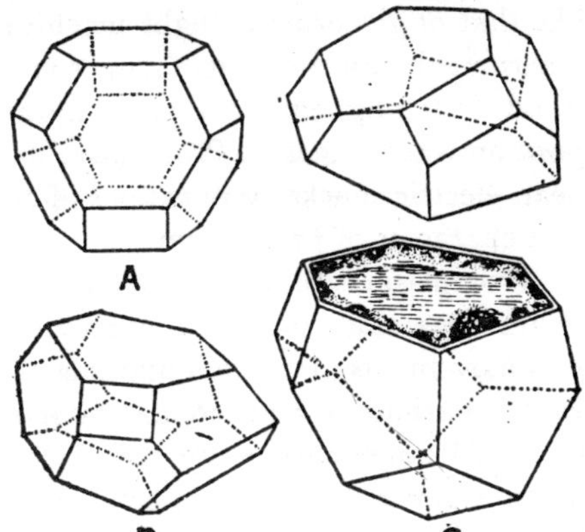

A B C

FIG. 485. The shape of cell. Diagrams of orthic tetrakaidecahedron with 14 faces. C. Same cut into two halves to show cytoplasm, nucleus and the central vacuole with cell-sap.

It has been stated at the outset that a plant cell has two parts—the living protoplast and the surrounding non-living cell wall. Protoplasm is the seat of all physiological activities of a cell. The wall is of secondary importance, as it is a secretory product of the protoplasm. Nevertheless, there exists an intimate relationship between the two. The presence of the wall not only makes the plant cells structurally more distinct, but have other advantages too. Apart from serving as a protective membrane against loss of water, intense heat, attacks of insects, fungi, bacteria, it is positively helpful in delimiting the physiological units and adding to the mechanical strength of the organs. The shapes of the cells are determined by the walls. At any rate, protoplast is much more important part of the cell, and it should be considered first.

PROTOPLASM

Protoplasm is the essence of life. It has been very fitly described by Huxley in the nineteenth century as 'the physical basis of life'. All the characteristics of the living things, such as metabolism, irritability, growth and reproduction are nothing but the external manifestations of the internal changes of this mysterious fluid protoplasm. Every plant or animal has its own type of protoplasm, but fundamentally it is same in all organisms. Protoplasm was first recognised by Hugo von Mohl in the middle of nineteenth century. Even today in spite of all the advances and attainments in science our knowledge about this mysterious fluid is rather vague and superficial. In recent years special techniques have been employed for unravelling the mystery. The investigators have taken the help of dark-field illumination, ultra-violet photography, polarised light, high-speed centrifuge, micromanipulator and the electron microscope for the purpose.

Protoplasm is composed of water and various substances like proteins, fats, carbohydrates and inorganic salts, but it is, by no means, a mere mixture of all these materials. It is continually undergoing constructive and destructive changes, a dual process of waste and repair. Thus it is a complex dynamic system of substances. Sharp has described it as 'a system in dynamic equilibrium'. Thomson has called it 'a marvellous form of matter in motion'.

Physical Properties. It appears under the microscope as a clear, optically homogeneous fluid saturated in a high percentage of water. The specific gravity of protoplasm is slightly above that of water. A large number of minute granules and globules of varying size and shape remain distributed in water. Thus it looks granular under the microscope. Due to occurrence of many vacuoles the protoplasm of plant cells has often foamy or alveolar appearance. It is a viscous and elastic fluid. The viscosity has been found to be like that of glycerine or light machine oil. The elastic nature of protoplasm has been determined with the help of a micromanipulator. It has been found that protoplasm can stand stretching and can be drawn out like a thread, which would return to its normal position when released. Protoplasm has the power of responding to external stimuli like heat, electric shocks, applications of chemicals, what is referred to as *irritability*, an inherent character of protoplasm.

Protoplasm exhibits *streaming movements* of different types which are commonly noticed in amoebas, slime moulds, and particularly in the plant cells with large vacuoles. Rotatory movements in the cells of leaves of aquatic plants like *Vallisneria* and *Elodea* of family Hydrocharitaceae and circulatory movements in the staminal hairs of *Rhoeo distolor* of family Commelinaceae are the glaring examples. In rotation protoplasm moves in a definite direction along the cell wall (Fig. 486); whereas in circulation the course of movement is irregular (Fig. 487). Besides the movements stated above, commonly called *cyclosis*, the streaming movements of slime moulds have been observed with the help of

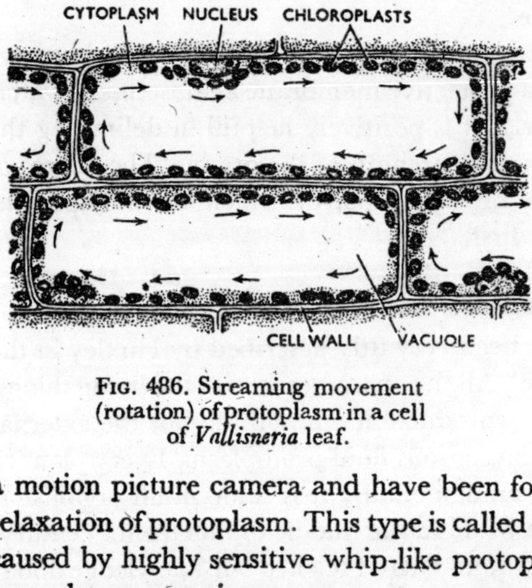

FIG. 486. Streaming movement (rotation) of protoplasm in a cell of *Vallisneria* leaf.

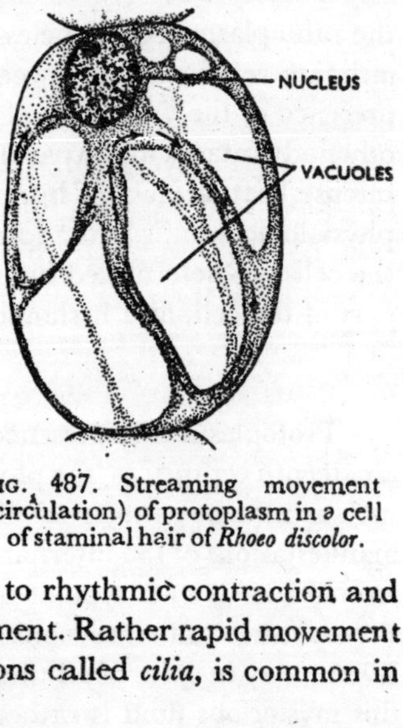

FIG. 487. Streaming movement (circulation) of protoplasm in a cell of staminal hair of *Rhoeo discolor*.

a motion picture camera and have been found to be due to rhythmic contraction and relaxation of protoplasm. This type is called *amoeboid* movement. Rather rapid movement caused by highly sensitive whip-like protoplasmic extensions called *cilia*, is common in many lower organisms.

Attempts to ascertain the nature of protoplasm had been going on since a long time and early workers proposed a number of theories, viz. (*i*) *reticular* theory, stating that it is something like a fine network, (*ii*) *fibrillar* theory, claiming that it consists of fine fibrils, (*iii*) *granular* theory, demanding that it is composed of innumerable minute granules, and (*iv*) *alveolar* or foam theory, stating that it is a collection of minute droplets of a liquid distributed in another liquid as an emulsion. The most important physical character of protoplasm, as known today, is that it is a complex *colloidal system*. The properties like viscosity, surface tension, alterations of physical state due to variations in external conditions and internal reactions are clear indications of a colloidal system. Though the tw e of colloidal structure is not clearly known, it is probably an *emulsion* with proteins, lipides and other substances suspended in the water medium.

Chemical Composition. The chemical composition of protoplasm, itself a dynamic system, cannot be determined without destroying it during the process of analysis. A general idea about its composition may be had after subjecting it to normal procedure of analysis. It is found that protoplasm is a very complex substance, which in normal active state consists of more than 75% of water and less than 25% of other materials. Of the dry matters nearly 90% are organic and the rest inorganic. The reactions are either slightly alkaline or neutral but seldom acidic.

As already stated water forms the most important constituent of protoplasm. In fact, life is inconceivable without water. In aquatic plants the proportion of water may go up to 95%. Low water content is noticed in the inactive and dormant organs like spores and seeds when it is something like 10 to 15%.

The organic matters present in protoplasm are proteins, fats and carbohydrates, of which the proteins are the most important ones. They are very complex matters, composed of carbon, hydrogen, oxygen, nitrogen, often sulphur and sometimes phosphorus. The nucleo-proteins occurring in the nucleus in particular are of special significance. The fatty substances entering into the composition of protoplasm are the true fats, composed of carbon, hydrogen and oxygen, and more commonly the phospholipides which are complex derivatives of fats containing nitrogen and phosphorus. The carbohydrates which directly enter into actual constitution of protoplasm are the pentose sugars. But other sugars, the hexoses, and their condensation products, the insoluble polysaccharides, have also importance in the activities of protoplasm. They usually combine with proteins and lipides. The inorganic salts are of different varieties; in plants they occur mainly as phosphates, sulphates, chlorides and carbonates of magnesium, potassium, sodium, calcium and iron. Many other elements may be present in minute traces.

It hardly needs mentioning that the account of the chemical composition given above does not reveal any information as regards the *organization* of protoplasm. It simply gives a general idea about the kinds and proportions of compounds present.

PROTOPLAST

Protoplast is the organized mass of protoplasm of each cell. It is the smallest part of the essential living substance. It possesses a highly organized more or less rounded body known as **nucleus**, what figuratively means the centre of activities. The remaining portion goes by the name **cytoplasm,** which may be looked upon as the ground substance

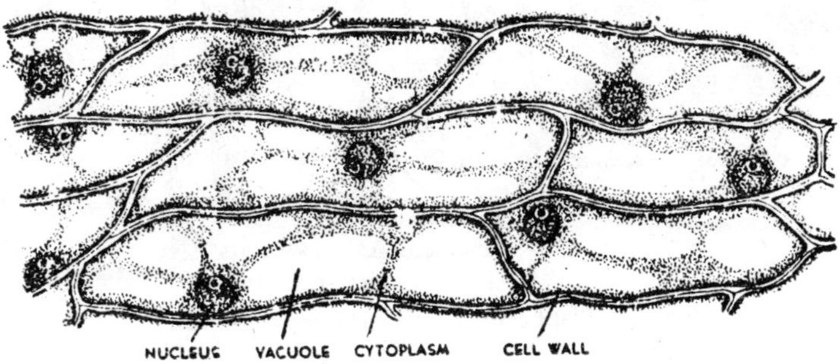

FIG. 488. A few cells from the epidermis of an onion (*Allium cepa*) scale.

of protoplasm (Fig. 488). The term cytoplasm was used by Kolliker in 1862 to designate the living material surrounding the nucleus.

CYTOPLASM

Cytoplasm is hyaline and jelly-like in nature and looks granular under the light microscope, due to suspension of many droplets and granules in the watery matrix. It is very complex, both physically and chemically. Various organic and inorganic substances remain distributed in the watery background either in true solution or in colloidal condition. The hyaline ground substance is also called *hyaloplasm.* The outermost layer of cytoplasm is, however, non-granular and firmer in consistency. It goes by the name **plasma membrane** or *ectoplasm.* The term ectoplasm is not very happy, as it has been used in a different sense in case of *Amoeba.* Plasma membrane is an extremely delicate layer of ultra-microscopic thinness, comparable to the surface film of a drop of water. It is *semipermeable* in nature. Though invisible under light microscope it can be seen as a single layered smooth and continuous membrane under electron microscope. This layer plays a great role in the metabolic activities. It regulates the entry and exit of fluid to and from the cells. The plasma membrane practically covers the protoplast, even including the plasmodesmata connections. It goes without saying that absorption of dissolved mineral matters from the soil would have been impossible, if the plasma membrane were perfectly semipermeable. But it allows certain substances to pass through it under one set of circumstances while under another set their entry is restricted. It appears that plasma membrane is formed by proteins and fatty substances particularly lipides. Electron microscopy revealed that plasma membrane is three-layered, made of two dense layers of protein, each roughly 20Å (2 nm) wide, with a lighter layer of lipid, 35Å (3·5 nm) thick, sandwiched between them (Robertson—1959; unit membrane theory). The remaining granular portion of cytoplasm is called **endoplasm.** The cytoplasmic portion of the protoplast is referred to as *cytosome.* A good number of technical expressions like popliopla-sm, have been suggested from time to time to designate the different parts of cytoplasm.

Liquid-filled cavities, called **vacuoles,** are of universal occurrence in the cytoplasm of plant cells. In the meristematic cells the vacuoles are small, they may be even at the limits of microscopic visibility. With gradual maturation of the cells the vacuoles become increasingly larger and prominent as cell enlargement is associated with a massive uptake of water. Frequently the major portion of an adult cell may be occupied by a large central vacuole (Fig. 489). The protoplast in that case is pushed towards the cell wall forming a lining layer, what is known as *primordial utricle.* The origin of vacuoles has been a

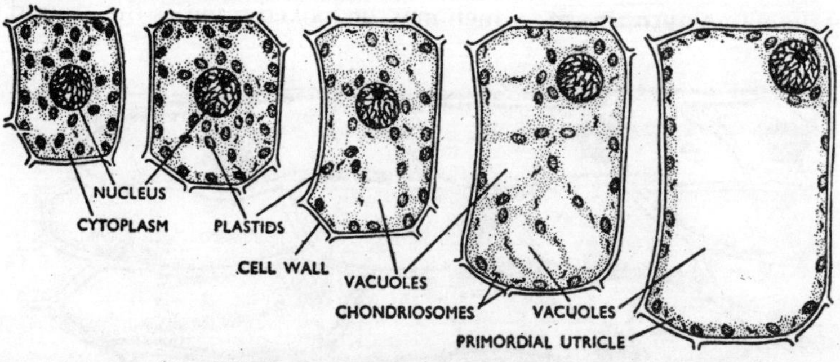

Fig. 489. Plant cells at different stage of growth.

subject of diverse opinions. It is, however, believed that vacuoles arise by hydration and collection of some cytoplasmic colloids. It is, at any rate, true that vacuoles are already present in all cells, even meristematic cells. So the old idea that young cells have no vacuoles and they 'become vacuolated' with the growth of the cell has been found to be incorrect. Some vacuoles, called *contractile* vacuoles, commonly present in lower animals and also plants exhibit a peculiar phenomenon of rhythmic filling and discharge of contents. They appear to be excretory passages. The layer of cytoplasm forming the vacuole membrane is known as **tonoplasm.** It is also a surface film of extreme fineness. Some people consider that tonoplasm is similar to plasma membrane in nature and composition both being relatively rich in lipoidal matter; whereas others are of view that the two differ both in structure and function. But it is admitted at any rate that tonoplasm is a distinct entity and that it is different from inner cytoplasm.

Finer details of the cell structure are revealed by the electron microscope. It appears that optically clear hyaloplasm is made of two components—viz., the *matrix* or ground substance and a membrane-enclosed vesicular system known as *endoplasmic reticulum* (Porter—1945). They usually occur in three forms—*cisternae, vesicles* and *tubules.* The cisternae are elongated, flattened and arranged in parallel lines, the vesicles are circular and the tubules are of various shapes. This system appearing as slender vesicles is continuous from the nuclear envelop throughout the cytoplasm to the surface membrane, even through the wall in meristematic cells, whereas in adult cells it appears as discontinuous vesicles. Numerous small particles, *ribosomes*, occur on the outer surfaces

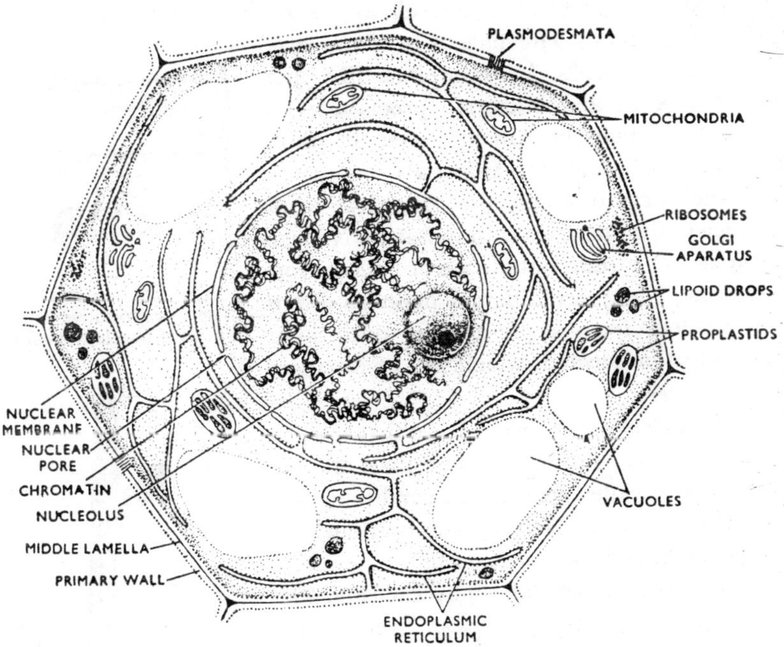

FIG. 489A. A meristematic cell (diagrammatic) under electron microscope.

of endoplasmic reticulum or freely in the matrix. They are often 150Å (15 nm) in diameter and rich in ribonucleoprotein. Ribonucleic acid—RNA and basic proteins play significant role in protein synthesis. The ribosomes with fragments of endoplasmic reticulum form which are called *microsomes*. Fig. 489A represents the diagram of a meristematic cell under electron microscope.

The watery content of the vacuole is called **cell-sap.** It is mainly water with various non-protoplasmic substances either in solution or in a colloidal condition. Inorganic salts, plant food matters like carbohydrates, amides; waste products like tannins, organic acids and mineral crystals are of common occurrence in the cell-sap of the vacuoles. Vacuoles are quite helpful in maintaining the turgor of the cells, which is necessary for the support of the herbaceous plants. Further, they serve as storehouse of water, food matters and various by-products of metabolic changes of protoplasm. The cell-sap is normally slightly alkaline or neutral (*p*H 7), but with gradual enlargement during differentiation of the cell it becomes acidic (*p*H 5 or even lower). Water-soluble pigments, called *anthocyanins*, are often present in the cell-sap of the vacuoles. Anthocyanin usually imparts red to blue colorations to the plant parts particularly to flowers and fruits. These are glycosidic compounds formed by chemical reaction between a sugar—generally a glucose and one or more non-sugars. With changes in the acidity and alkalinity the anthocyanins change their colour. *Cyanin*, the pigment of dahlia, rose, etc., is red when sap is acidic (*p*H 3), violet in slightly alkaline sap (*p*H 8, 5) and blue in highly alkaline sap (*p*H 11). Anthocyanins always occur in nature in form of chlorides. Besides attraction of insects for pollination, it is also supposed to serve as a screen against intense sunlight, particularly ultra-violet rays.

NUCLEUS

The nucleus is much denser than cytoplasm and is usually round or elliptical in shape. It is regarded as the 'brain' of the cell, as it is the controller or regulator of all the activities of the protoplast. The nucleus is of universal occurrence in all plant cells. In lower plants like blue-green algae (Fig. 490) and bacteria (procaryotes) a well-formed nucleus may be absent; even in those cases corresponding nuclear materials must be present. In the sieve-tubes of the complex tissue phloem the nucleus disintegrates with maturity. Nucleus never arises *de novo* or afresh, but always originates from a pre-existing one. There are different methods of formation of a nucleus from a pre-existing one. Some methods are quite complicated, which will be discussed in succeeding chapters.

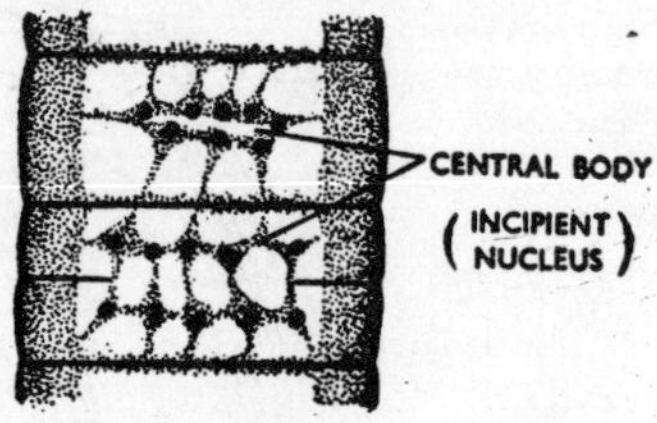

FIG. 490. Central body (incipient or primitive nucleus) in a cell of *Oscillatoria*—a blue-green alga.

Usually one nucleus occurs in a cell. In that case the cell is called *uninucleate*. Presence of more than one nucleus in a cell, what is called *multinucleate* condition, cannot be altogether ruled out. Non-septate multinucleate bodies, technically called *coenocytes*, occur in the order *Siphonales* of grass-green algae Chlorophyceae (Fig. 491) and some fungi of the group *Phycomycetes*. Even in some higher plants more than one nucleus may arise due to disease and other reasons. Formation of many nuclei in a cell may be due to (*i*) repeated division of the nucleus without corresponding cytoplasmic division; and (*ii*) absorption of walls between the cells.

The nucleus is usually round or elliptical in shape (Fig. 488). In highly elongated cells the nuclei are also correspondingly elongate. Peculiar shapes like crescent, dumb-bell, are found on rare occasions. It occupies comparatively a small space of the protoplast. In a

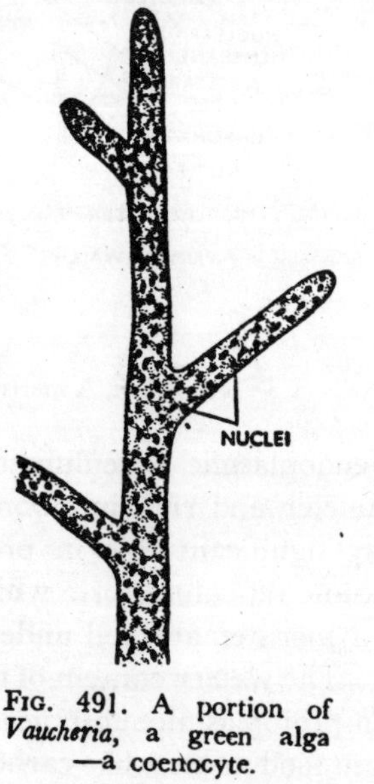

FIG. 491. A portion of *Vaucheria*, a green alga —a coenocyte.

meristematic cell it is quite prominent (Fig. 489). As the gradual enlargement of the cell is not accompanied by corresponding growth of the nucleus, the latter does not look so conspicuous in an adult cell. As regards size, it shows wide variations. In lower plants like *Mucor* the diameter of a nucleus is hardly more than 1 micron (micron, written as μ is equivalent to one thousandth part of an mm.). The egg cell of *Dioon*, a gymnosperm, may have a nucleus as large as 600μm in diameter. But consideration of a large number of cells reveals that average diameter of a nucleus varies from 10 to 15μm. The variations in the size and shape of the nucleus are not confined to different types of plants. Curiously enough considerable variations may be noticed in the different tissues of the same plant.

In meristematic cells with small vacuoles the nucleus is located more or less at the centre of the cell (Fig. 484B). In adult cells with a number of vacuoles the nucleus occupies more or less the same position; but when the vacuoles coalesce to form a large central vacuole, the nucleus with cytoplasm, etc. is pushed towards the cell wall (Fig. 489). Very peculiar position of nucleus is found in some algae like *Spirogyra*, where it is kept suspended in the midst of a large central vacuole by fine cytoplasmic threads (Fig. 493A). Whatever may be its position, the nucleus always remains associated with cytoplasm, never touching either the vacuole or the cell wall. The nucleo-cytoplasmic ratio usually ranges between 1 : 7 to 1 : 10. Hartwig (1960) suggested nucleo-cytoplasmic index—

$$NP = \frac{V_n}{V_c - V_n}$$

where NP is nucleo-cytoplasmic index, V_n—nucleus volume, V_c—cytoplasm volume.

The nucleus is much more viscous than cytoplasm. In fact, with the help of micromanipulator it may be pushed from one side to other in some cells without any visible injury. It has been observed that the nucleus may be stretched many times its normal size, which would come back to its original size on removal of the micromanipulator needles. In living condition it appears homogeneous. But it may be fixed and suitably stained for observation of detailed structure. A metabolic nucleus after suitable staining presents the following structure (Fig. 492):

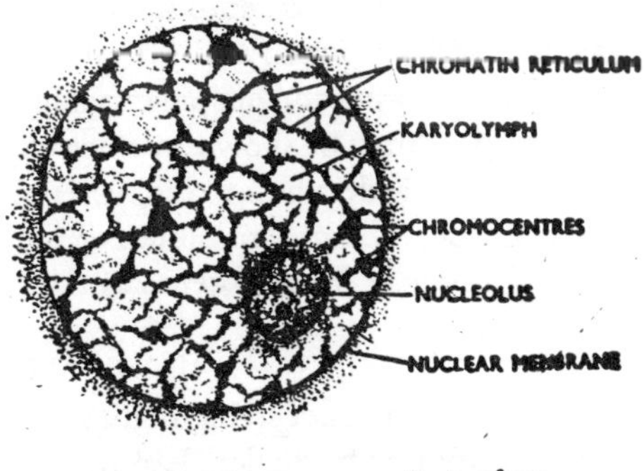

Fig. 492. The structure of a nucleus.

The main body of the nucleus is surrounded by a fine delicate membrane called **nuclear membrane.** It demarcates the main body of the nucleus from surrounding cytoplasm and has been found by electron microscopy to be double-layered with pores and projections into the cytoplasm continuous with the endoplasmic reticulum (Fig. 489A). Each layer is usually 80-100Å (8-10 nm) wide and the space between two layers —*perinuclear space*, is 100-300Å (10-30 nm). Chemically it is made of proteins and lipids. The membrane disappears at certain stage of nuclear division only to reappear at the end of the process. It is regarded as a structure formed by the **karyolymph** or jelly-like fluid filling up the main body of the nucleus. This fluid is also known as *nuclear sap* or *nucleoplasm*. A definite number of highly crooked threads, called **chromonemata,** remain suspended in the karyolymph. These threads anastomose and form a network which is called **chromatin reticulum.** The threads have strong affinity for taking up stains. The chromonemata are the most important bodies, because they represent the **chromosomes** at this stage. The most important constituent of the chromosome seems to be the nucleic acid—DNA, deoxyribose nucleic acid, which is the actual chromatic material causing staining of the chromosomes. Differential staining at different regions of chromosome means difference in its chemical nature. The nucleus contains the basic genetic information of the cell in form of long strands of this chemical material DNA, the molecules of which are believed to be in the form of double strands of intertwined helices in the chromosomes. At a certain stage of nuclear division another stainable substance, *matrix*, is associated with the chromonemata giving the chromosomes compact forms (Fig. 521). At that stage chromosomes lose water and acquire a fibrous coat of nucleic acid (stainable matrix). Densely stainable lumps are often found in the midst of chromonemata in some nuclei. These are called **chromocentres,** the regions where chromonemata are more closely packed and have more matrix. Besides these, there may be one or two highly stainable bodies inside the nucleus. This internuclear body is called **nucleolus** or plasmosome. Like nuclear membrane it also disappears at a certain stage of nuclear division and reappears at the close of the process. The number of nucleoli is variable. It may so happen that a number of them fuse at the end of cell division to form one nucleolus. Its function is not definitely known. According to one school, the nucleolus is nothing but accumulated materials, somehow related to metabolic activities of the nucleus. It is formed at a particular locus on one of the chromosomes and is mainly composed of ribonucleic acid—RNA.

Chemically, the nucleus practically resembles protoplasm, excepting that it contains nucleo-proteins made of nucleic acid and certain basic proteins. Though proportion of proteins is higher in cytoplasm, the nucleus is more rich in nucleic acids, DNA and RNA which are combinations of phosphoric acids with some carbohydrates and bases. DNA, as already reported, is the carrier of genetic information and RNA is concerned with protein synthesis. RNA is the prevailing type of nucleic acid in cytoplasm. It has been found that in an electric field the nucleus carries negative charge, whereas the cytoplasm is positive.

Functionally, the nucleus is of highest importance. It plays very prominent part in cell division, growth and reproduction. In fact, the physiological activities depend on the interaction of the nucleus and cytoplasm, the nucleus being the controlling centre. It is manifest in the formation of epidermal hairs in the roots, where the nucleus always moves on to the tip. It may be shown by microdissection that the dissected part of the cell without the nucleus is destined to die, whereas the other part continues quite well secreting a new cell wall. The chromosomes of the nucleus are the carriers of hereditary characters which occur in form of genes in the chromosomes.

PLASTIDS

Plastids are distinct living bodies or organs embedded in the cytoplasm of the plant cells. They are much smaller than the nuclei and usually a good number of them occur in a cell. Plastids are of universal occurrence in the plant cells with the exception of lower plants like bacteria, fungi, slime moulds and blue-green algae. Like nucleus plastids do not arise *de novo*, but originate from pre-existing ones. They have viscous bodies which may change shape in amoeboid fashion. Plastids are perfectly distinct from cytoplasm in which they remain embedded, possibly due to presence of a membrane resembling the plasma membrane on their outer margins. They have distinct protoplasmic basis and so are intimately connected with the metabolic activities of protoplasm. The body or ground substance of the plastid appears like colourless cytoplasm and is known as **stroma** (Fig. 493). Different pigments remain distributed on the stroma, and accordingly the plastids attain different colours. Broadly speaking, plastids are of two types, viz., the colourless plastids, called **leucoplasts,** and the coloured ones called **chromoplasts** or chromatophores. The term chromatophore, meaning colour-bearer, though suggestive is now used to refer to the plastids of algae; so the other term is preferred The coloured plastids or chromoplasts again may be green due to the pigment **chlorophyll,** or coloured other than green, the colour usually ranging between red to yellow. The green ones are called **chloroplasts** and others are referred to as chromoplasts. For the sake of convenience it is said that plastids are of three types, viz., chloroplasts, chromoplasts and leucoplasts.

Chloroplasts. Chloroplasts are the most important plastids, because they are responsible for the all important function, *photosynthesis*. For that wonderful and unique property chloroplasts occupy the most strategic position in the living world. They are green due to the presence of the pigment chlorophyll on the stroma. Chlorophyll is really a collection of four pigments, chlorophyll *a*, chlorophyll *b* and the yellow pigments, carotin and xanthophyll. Under light microscope chloroplasts look granular or homogeneous. Recent studies with electron microscope have revealed that chlorophyll remains distributed in the form of delicate granules or platelets called *grana* (Fig. 493) in the colourless proteinaceous ground substance stroma. Chloroplast is bounded by a two-layered membrane. An elaborate system of membranes called lamellae run parallel to one another throughout the stroma along the length of the chloroplast. They are differentiated into thin stroma lamellae and more dense and thick disc-like grana lamellae piled one above the other like a stack of coins. The latter form the grana. Minute granules, osmiophilic droplets and starch grains are present in the stroma. Fig. 493A shows a chloroplast under electron microscope. After extraction of chlorophyll grana have been found to consist chemically of protein and lipide. Chlorophyll has resemblance with haemoglobin, the efficient oxygen

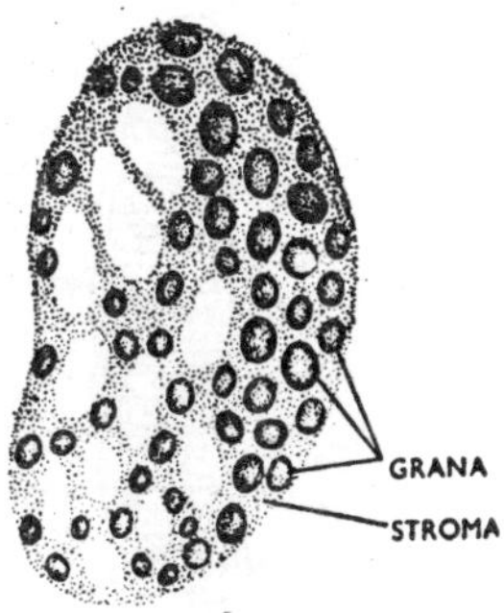

FIG. 493. A chloroplast (much enlarged to show stroma and grana).

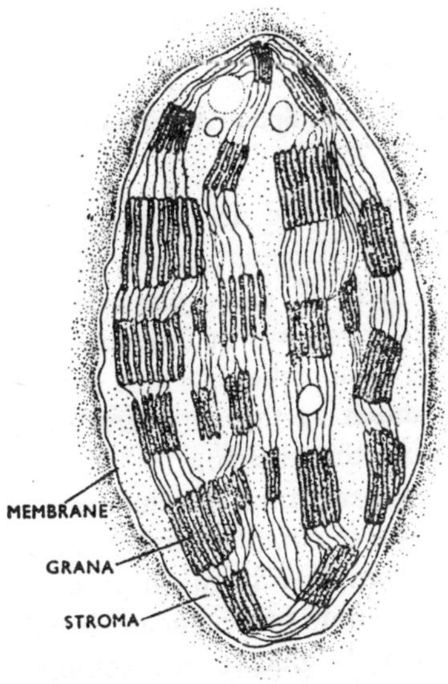

FIG. 493A. A chloroplast under electron microscope.

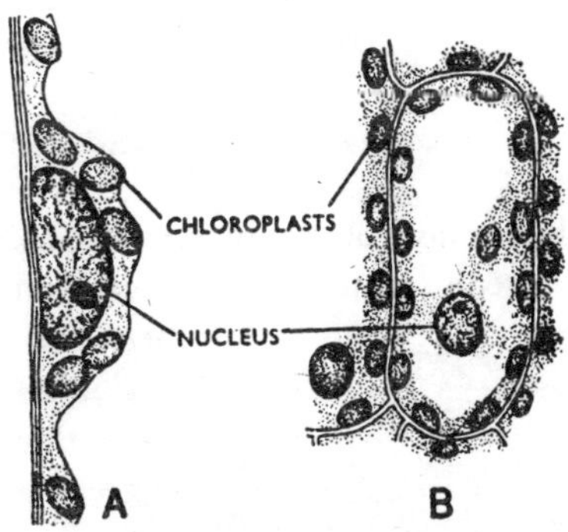

FIG. 494. Chloroplasts (magnified) from mesophyll cells of leaves. A. A part of the cell from an aquatic plant. B. From a land plant.

carrier of animal bodies—the place of magnesium in chlorophyll being occupied by iron in haemoglobin.

Chloroplasts occur only in those organs which are exposed to sunlight. They are abundant in the mesophyll cells of the leaves and young stems, and are normally absent in the underground organs and deep-seated regions of aerial organs where sun's rays do not penetrate. But they may be found in the parenchymatous cells of vascular tissues and in the embryo and endosperm of seeds. Chloroplasts have regularity of shape, usually being round, elliptical and discoidal (Fig. 494). Very peculiar chloroplasts are characteristics of lower plants, particularly grass green algae (Chlorophyceae). In *Spirogyra* chloroplasts occur in form of spiral bands (Fig. 495B); *Zygnema* (Fig. 495C) and desmids have stellate or star-shaped ones; those of *Ulothrix* are girdle-shaped (Fig. 495A); *Oedogonium* (Fig. 495D) has peculiarly reticulate green plastids.

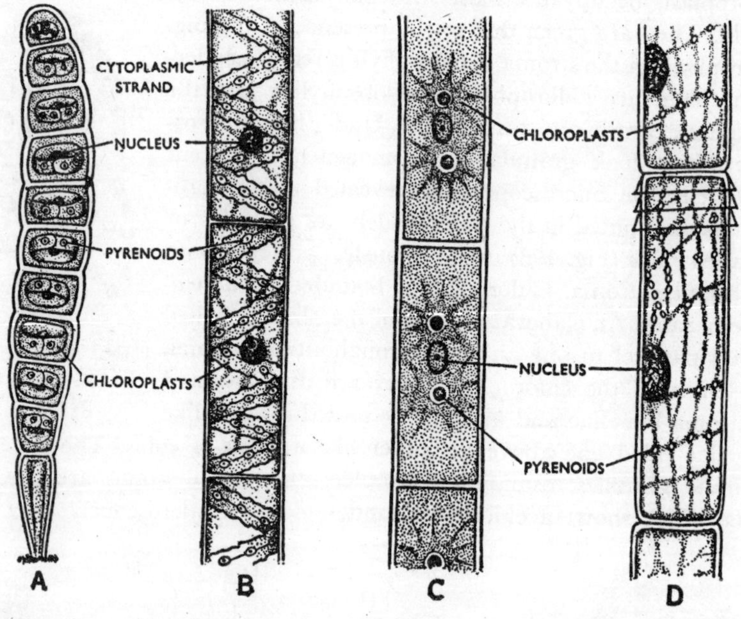

FIG. 495. Chloroplasts—peculiar shapes. A. Girdle-shaped in *Ulothrix*. B. Spiral in *Spirogyra*. C. Stellate in *Zygnema*. D. Reticulate in *Oedogonium*.

Chloroplasts of the liverwort *Anthoceros* (Fig. 496) are spindle shaped. In higher plants, as already stated, they have regular shapes and the average diameter usually ranges between 4 to 6 micra. The number of chloroplasts is also variable—it may be one, as in *Anthoceros*, two in *Zygnema*, and more than that in many plants. In most of the cells they usually aggregate near the nucleus, so much so that in many cases they form a lump almost covering the nucleus.

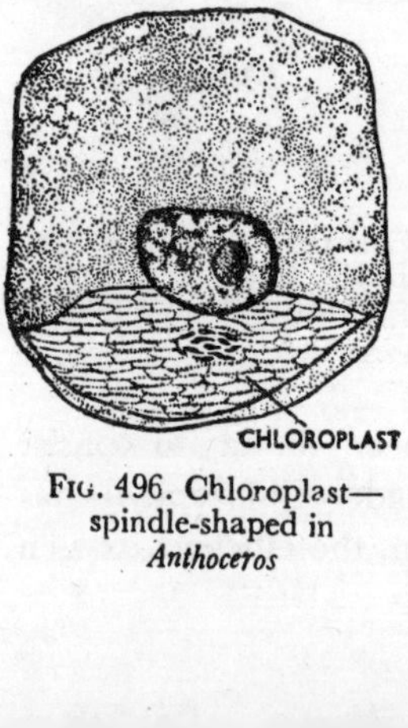

FIG. 496. Chloroplast—spindle-shaped in *Anthoceros*

The most important function performed by chloroplasts is *photosynthesis* involving elaboration of complex carbohydrate food matters out of water and carbon-dioxide gas in presence of sunlight. To start with, simple soluble sugar, *glucose*, is synthesized. They also convert the simple glucose into complex insoluble starch grains. These starch grains, called *assimilatory starch*, are re-converted into sugar with nightfall.

Chloroplasts of some plants bear special refractive bodies called **pyrenoids** (Figs. 495 & 496). These have protein centres usually surrounded by starchy layers. They are found in lower plants like algae and liverworts. In many members of grass-green algae (Chlorophyceae) one or more pyrenoids are present on the chloroplasts. In red algae (Rhodophyceae), pyrenoids are naked, where surrounding starchy layers are absent. These bodies may arise *de novo* or by division of pre-existing pyrenoids.

Chromoplasts. Chromoplasts are coloured other than green, the colours usually ranging between red and yellow. These colours are due to yellow pigments, carotin and xanthophyll. Chromoplasts have no definite shape. They may be angular, forked, rod-like, acicular (Fig. 497), etc. The carotenoids may remain diffused in the stroma, may occur in form of granules or may even appear to be crystalline. Chromoplasts are common in brightly coloured parts of the plants like the petals of the flowers and fruits. Attractive hues of all flowers are not due to chromoplasts alone. Water-soluble pigments, anthocyanin, present in the cell-sap of the vacuoles also impart colours to the plants. Tomato fruits, when ripe becomes red due to appearance of a pigment, lycopen, in the chloroplasts. The main function of chromoplasts is colour manifestation for attracting insects and other visitors which bring about pollination and dispersal of fruits and seeds. They are often present in the underground organs like carrots. Here the function is rather obscure.

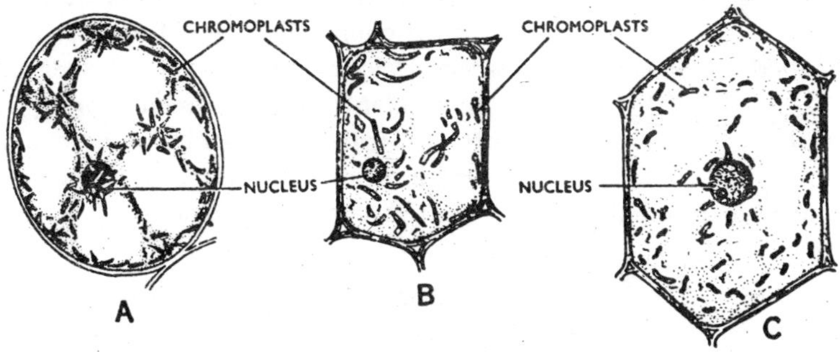

FIG. 497. Chromoplasts. A. A cell from the flesh of *Lycopersicum* (tomato). B. Same from the root of *Daucus* (carrot). C. Same from the petal of *Canna*.

Leucoplasts. These are colourless plastids. They do not bear any pigment. Leucoplasts occur only in those organs which are not exposed to sunlight. Thus they are found in the underground organs and deep-seated parts of the aerial organs. They may also occur in the epidermal cells. Leucoplasts are irregular in shape (Fig. 498)—may be small, granular and often larger. Small leucoplasts are converted into chloroplasts when exposed to sunlight. The larger ones, called **amyloplasts,** are responsible for conversion of simple sugars into insoluble starch grains for the purpose of storage. These starch grains are called **reserve** starch or storage starch, and function performed by the amyloplasts is referred to as *amyloplastic* function. Leucoplasts connected with the formation and storage of fatty substances are known as *elaioplasts* (Fig. 499) which

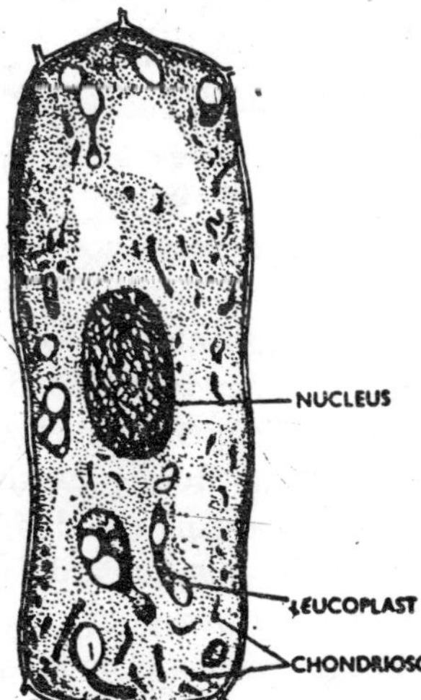

FIG. 498. Leucoplasts. A cell from the root of *Secale cereale* (rye) with leucoplasts and chondriosomes.

occur mainly in the liverworts and monocotyledons. *Aleuronoplasts* concerned with formation of proteinous matters have been reported in *Ricinus.* Leucoplasts are abundantly present in immature cells and usually aggregate near the nucleus.

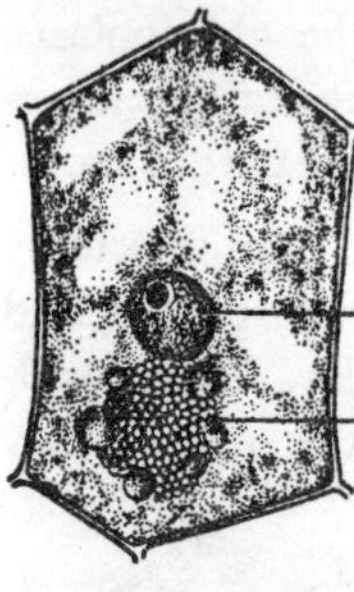

FIG. 499. Elaioplast. An epidermal cell from perianth of *Polyanthes tuberosa* (tube rose) showing elaioplasts with oil-droplets. (*After Politis.*)

The origin of plastids has been a subject of keen investigation since a long time. In meristematic cells they are usually very small and non-pigmented. These are **proplastids** or *plastid primordia* which may be called the precursors of all the three different types of plastids. With gradual maturity and elaboration of the organs the plastids multiply and grow into different types of mature plastids. Possibly they multiply by constriction. Division of the proplastid is not usually associated with the division of other parts of the cell. But in some plants like the green alga, *Zygnema* and the liverwort, *Anthoceros* the plastid divides regularly before or during the division of other parts of the cell. The entire plastid-complex of an organism has been referred to as *plastidome.*

CHONDRIOSOMES OR MITOCHONDRIA

These are very small living bodies, much smaller than plastids usually ranging in size from 0·2 to 3·0μm (Figs. 482, 489A & 498) universally present in the cytoplasm of plant and animal cells. They are more dense than cytoplasm and appear as small granules, rodlets or nets. Mitochondria occur more abundantly in the cells which are metabolically more active. Electron microscopy reveals that they have a double membrane, the outer limiting one and the inner one being convoluted into many plate-like folds called *cristae* projected into the lumen of the mitochondrion, which remains filled with a proteinous ground substance—matrix (Fig. 500). The outer and inner membranes are 40-75Å (4-7·5 nm) wide and the space between them being 20-60Å, (2-6 nm). Very small granules—rough and smooth are present on the outer and inner membranes. The nature and function of chondriosomes have been a subject of long-drawn controversy amongst cytologists. Some workers regard them as permanent 'organelles' of cytoplasm which originate from pre-existing ones and multiply in number. As they are often constantly appearing and disappearing, the suggestion that they may be formed anew has also been put forward. Chemically they are mainly composed of proteins and lipides, mainly phospholipids, and small amount of RNA and often DNA.

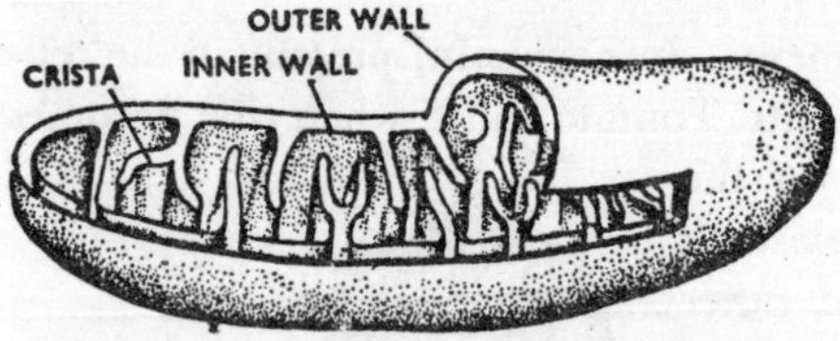

FIG. 500. A mitochondrion under electron microscope.

The function of chondriosomes has also been a vexing problem. The idea that they are proplastids or at least some of them develop into plastids has not received universal support. Due to occurrence of a large number of enzymes and co-enzymes they are regarded as the 'power-house' of the cell capable of breaking down complex molecules by *oxidation* and phosphorylation. Through oxidation carbohydrates, proteins and fats are broken down with consequent release of energy. The stored energy remains bound in *Adenosine triphosphate* or ATP. Thus mitochondria are connected with a variety of activities like respiration, secretion and enzyme action.

CENTROSOMES

These are minute cytoplasmic inclusions invariably present in animal cells. In some lower plants, algae and fungi, they occur near the nucleus. A centrosome (Fig. 500A) has a deeply-stained central granule, called **centriole,** which usually remains embedded in a mass of hyaline substance known as **centrosphere;** but both of them may occur alone. During cell division a number of radiating rays, called **astral rays,** collect round the centrosome. They usually divide before cell division and move to the two opposite poles of the bipolar spindle. Centrosomes may also be associated with the formation of cilia, the organs for locomotion, in some reproductive cells. In view of the fact that in many cases centrosomes are found to appear during cell division and disappear after the process, they are not regarded as permanent part of the cell.

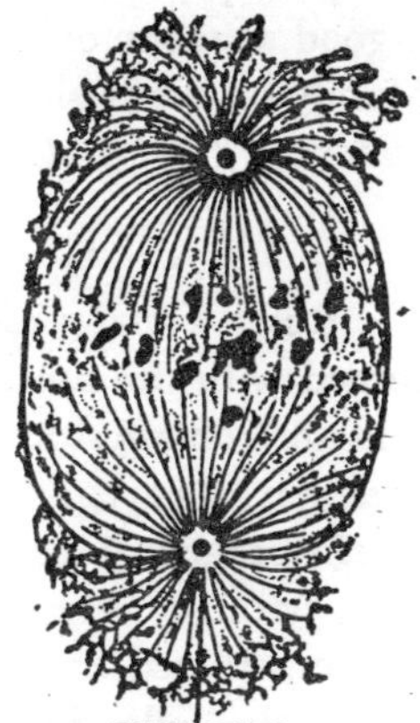

FIG. 500A. A centrosome in a red alga.

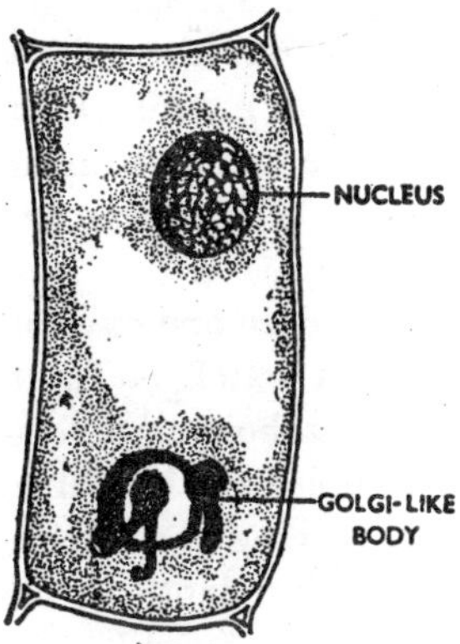

FIG. 501. A cell from the root-tip of lily showing Golgi-like body.

GOLGI BODIES

These cytoplasmic inclusions, named after the discoverer Golgi, have been definitely found in animal cells. Special techniques with osmic acid and silver salts have to be employed for demonstrating them. Using similar device certain cytoplasmic bodies with double membrane and vesicles, often dilated at the ends, have been reported in plants, particularly in the meristematic cells, which are supposed to be homologous to the Golgi bodies of animal cells (Fig. 489A, 501). They have been noticed with the aid of electron microscope in the root tip of *Allium* sp and root cells of *Pisum sativum.* Doubts exist in certain quarters that these are mere artifacts—non-living materials collected and rendered visible during staining. The role of Golgi bodies in secretion in animal cells has been accepted. It has also been established that the Golgi complex, also called *dictyosomes*, has positive role in enzyme synthesis and regeneration of membrane systems.

LYSOSOMES

These are small mitochondrion-like bodies having densely-packed granules with membraneous covering. Apart from their abundance in different animal cells they have been reported in yeasts. Due to pressure of hydrolytic enzymes lysosomes take active role in breakdown of different cell substances. The materials in the lysosome often form crystals. The function of lysosome membrane is that of separating the hydrolytic enzymes from other parts of the cell, thus avoiding self-digestion. On death of the cells enzymes are released which rapidly digest the remainder of the cell. They may be considered analogous to the digestive vacuoles in the unicellular organisms.

ERGASTIC MATTERS

All non-protoplasmic substances present in the cell constitute the **ergastic matters** (*ergon*=work). These are non-living inclusions of the cell produced during the metabolic activities of the protoplast. They are fairly large in number. Ergastic matters may remain in solution in the cell-sap of the vacuoles, dispersed in colloidal conditions, and also in form of insoluble granules or crystals. They may be present even in the cytoplasm itself. Some ergastic matters serve nutritive purpose, ultimately being used in the formation of new protoplasm; some of them are helpful indirectly in diverse biological functions, and a good number of them are absolutely useless by-products of protoplasmic activities. Mainly on the basis of utility ergastic matters are put into three groups, viz., (*i*) *Reserve* materials, (*ii*) *Secretory* materials and (*iii*) *Excretory* materials.

I. RESERVE MATERIALS

Reserve materials are the food matters of the plants, and so they are directly concerned with the nurition of the plants. These substances are synthesised by the living cells and remain stored up in different parts of plants. Though they may be present practically in all parts of the plants, the underground stems and roots, buds and seeds are the chief storage organs. Mineral matters absorbed in solution in water from the soil are often wrongly referred to as food. Really food matters are all organic in nature, which are constructed through complicated physiological processes. Reserve materials again are of the following types, according to their chemical nature—(*a*) *Carbohydrates*, (*b*) *Nitrogenous matters* and (*c*) *Fats* and *oils*.

Carbohydrates. These are chemically composed of carbon, hydrogen and oxygen, hydrogen and oxygen always occurring in the same proportion as in water. So in all carbohydrates hydrogen-oxygen ratio would be two : one. Of many carbohydrate food matters, some are simple and soluble in water, whereas others are complex and insoluble. On heating they lose water and a black mass is left behind, which is nothing but carbon. Carbohydrates are very important energy-giving food, present more abundantly in the storage regions. An account of the common carbohydrate food matters is given below:

1. *Sugars.* These are soluble carbohydrate food matters, usually sweet to taste. They occur as reserve materials in many monocotyledons and also in some dicotyledons like beet and carrot. Many of them are strong reducing agents. The common sugars of the plants have the formula, $C_6H_{12}O_6$. They are called *hexoses*, as six carbon atoms are present. *Pentoses* or sugars with five carbon atoms are not common in plants. Those having one basic sugar group are referred to as *monosaccharides*. Monosaccharides like glucose and fructose, particularly the former, are abundant in all plants. **Glucose,** $C_6H_{12}O_6$, is really the bssic material of metabolism, being the first stable product of photosynthesis. It is synthesised by the chloroplasts out of raw materials, water and carbon dioxide gas in presence of sunlight. So it is present in all green plants. **Fructose** or fruit sugar with the same formula, $C_6H_{12}0_6$, though differing in the structure of molecules, occurs in the fruits. Two other common hexose sugars of plants are **mannose** and **galactose.**

Sucrose or cane sugar is a *disaccharide*, having two of the sugar groups, with the formula $C_{12}H_{22}O_{11}$. It is our table-sugar. It is widely distributed in some plant tissues, often reaching very high concentration in storage organs. Sucrose is quite valuable commercially and is extracted mainly from sugar-cane stems and beet roots. On hydrolysis—breaking down by chemical addition of water, it yields one molecule of glucose and one molecule of fructose. **Maltose** is another disaccharide sugar commonly found in the germinating seeds during digestion of starch. It yields two molecules of glucose on hydrolysis.

The most widespread *trisaccharide* in plants is **raffinose,** which occurs in very small quantities. On hydrolysis it yields three molecules of hexoses—glucose, fructose, and galactose.

Tests: Sugar solution is taken in a test-tube. Equal quantities of Fehling's solutions

I and II are added and boiled. Orange-red precipitate of cuprous oxide indicates the presence of a reducing sugar. To sucrose solution add a little sulphuric acid and boil gently. It is hydrolysed to glucose and fructose. Now it would give Fehling's reaction.

2. *Inulin.* It is a polysaccharide carbohydrate which forms a powder-like compound and occurs in colloidal condition in the cell-sap of the vacuoles of plants like *Dahlia*, Artichoke—of the family Compositae. It is by no means of common occurrence in plants. Inulin may be precipitated by treatment with alcohol or glycerine. In fact, when blocks of root tubers of *Dahlia* are allowed to remain in alcohol or glycerine for some time and sections are examined under the microscope, beautiful fan-shaped crystals (Fig. 502) are noticed in the cells. On hydrolysis inulin yields only fructose. Inulin solution turns yellowish brown when treated with acid phloroglucin. On precipitation the crystals are easily recognised.

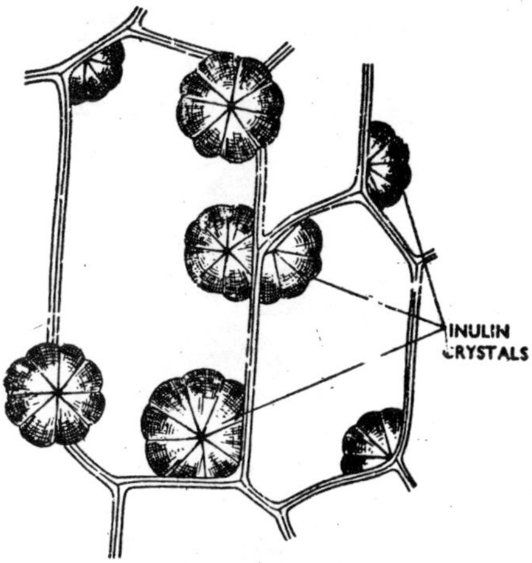

FIG. 502. Cells from root-tuber of *Dahlia* with inulin crystals.

3. *Starch grains.* Starch grains are the complex insoluble carbohydrates. They are the most important carbohydrate reserve materials of the plants and are of universal occurrence in the green plants with the exception of some algae. In some monocotyledons also the carbohydrate reserve is in form of sugars. Starch grains may occur in all parts of the plants, but they are more abundant in the storage organs, cereals, fruits and seeds. They are very complex chemically with the formula, $(C_6H_{10}O_5)_n$, where the value of n is not clearly known. It is formed out of unknown number of hexose groups and is thus a *polysaccharide.* For utilisation by the plants starch is ultimately converted into glucose by enzyme actions.

There are two types of starch grains: *assimilatory* starch and *reserve* starch. Assimilatory starch grains are temporary bodies formed by the chloroplasts in day-time during manufacture of glucose. As glucose is soluble in water the cell-sap becomes more and more concentrated with increasing manufacture of glucose, so much so that the saturation point is reached soon. As further manufacture would not be possible then due to physical reasons, the chloroplasts convert glucose into insoluble starch, which are called assimilatory starch. They are granular in appearance and temporary in duration, as they are reconverted into sugar with nightfall.

Surplus of sugar is converted into permanent starch grains to be stored up for future use by the colourless leucoplasts—the amyloplasts. These are reserve starch (Fig. 503). They occur profusely in storage organs like roots and underground stems and in seed and fruits. They are of diverse size and shape. Some of them are large enough to be seen even with the naked eye, whereas most of them are microscopic. Every starch grain has a shiny refractive point, known as *hilum*, which is the starting point. Starchy materials are deposited in layers around the hilum, ultimately giving the grain a distinct stratified appearance. The layers are referred to as *lines of stratifications*. The stratified nature of the grain is mainly due to the fact that the layers deposited in light and darkness differ in water content and density as a result of diurnal periodicity of the action of the plastids. The innermost layer is particularly rich in water which explains the cracks frequently noticed in desiccated starch. The alternate layerings become quite clear in the photographs taken in polarized light. Starch grains are regarded as sphaero-crystals consisting of layers of $C_6H_{10}O_5$ units arranged in a regular 'space-lattice'.

In a starch of potato tuber the hilum is located at one end of the grain, the deposition of layers being unequal. These are called *eccentric* grains, whereas in grains occurring in pulses, the hilum is more or less at the centre due to uniform deposition of the layers. These are *concentric* grains. A grain having one hilum and its surrounding lines of stratifications is known as *simple*. More than one grain may be adpressed together making it *compound*. In rice and other cereals, potatoes, etc., compound grains are of common occurrence. A peculiar type of grain, referred to as *semi-compound* or half-compound is also

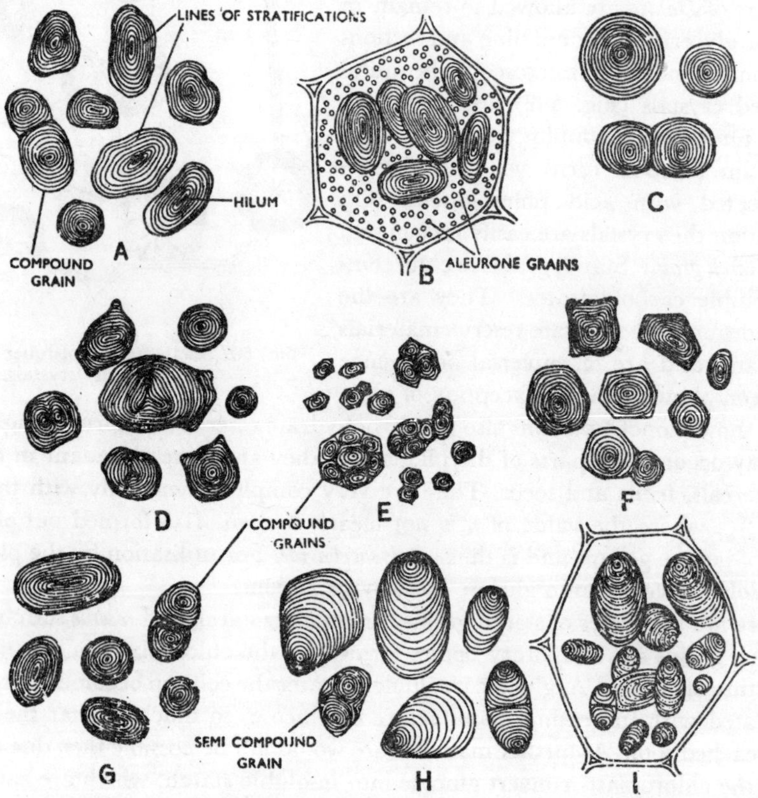

FIG. 503. Starch grains. A. From cotyledon of *Pisum* (pea) seed. B. Same *in situ*. C. From flesh of *Musa* (banana). D. From tuberous root of *Ipomoea batatas* (sweet potato). E. From endosperm of *Oryza* (rice) grain. F. From endosperm of *Zea* (maize) grain. G. From cotyledon of *Cicer* (gram) seed. H. From tuber of *Solanum tuberosum* (potato). I. Same *in situ*.

found. In that case there are two hila with their lines of stratifications, which are ultimately bounded by common lines of stratifications.

Starch has two components, viz., *amylose* and *amylopectin*, which differ in physical properties. Amylose is more water-soluble and turns deep-blue with iodine solution, whereas amylopectin is more waxy in nature and gives light blue-violet coloration with iodine solution. Iodine reaction is, at any rate, the most characteristic test of starch. Starch is synthesised by the plastids out of glucose; and on complete hydrolysis it yields glucose.

Starch grains vary considerably in size and shape. They are usually oval in potato varying from 70 to 100 micra in diameter, 30–45 μm in wheat, 12 to 18 μm in maize. But the shape is more or less constant in a particular type. That is how the plant source of starch in a given sample of flour can be determined. The percentage of starch is fairly high in storage organs, it is 15- 30% of dry weight in potato, even 50 to 70% of dry weight in the cereals.

4. *Glycogen*. It is another insoluble polysaccharide with the formula $(C_6H_{10}O_5)_n$. As it is abundantly present in the muscles and liver of the animals, it is also called animal starch, though the term is not a happy one. Glycogen occurs in blue-green algae, slime moulds, fungi and bacetria and serves as reserve food. It is ultimately broken down to glucose by enzyme action and utilised as such. It gives reddish coloration with iodine solution.

5. *Hemicellulose*. Cellulose $(C_6H_{10}O_5)_n$, is another polysaccharide carbohydrate which is much more complex than starch. It is particularly valuable, because the cell wall is primarily composed of this substance. Cellulose is not only insoluble, but it is also indigestible. So it is seldom used as food. In the seeds of plants like date palm, ivory nut palm another insuluble carbohydrate resembling cellulose is deposited as extra layers making the walls very thick and hard (Fig. 504). This is reserve cellulose or *hemicellulose*. Though insoluble in water, it readily dissolves in alkalies. Hemicellulose is utilised as food by proper enzyme action.

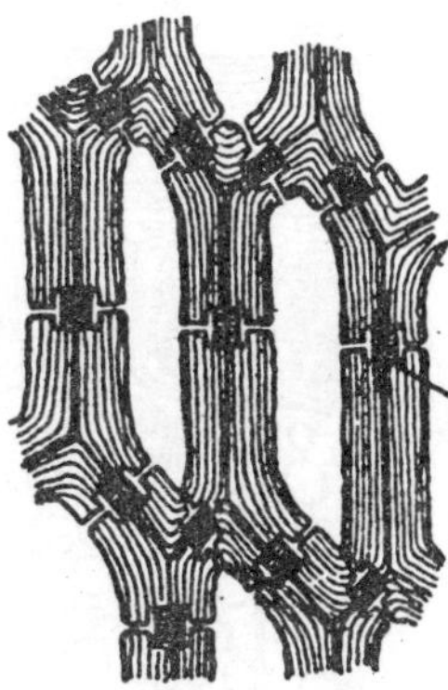

Fig. 504. Hemicellulose on the walls of cells from endosperm of *Phoenix* (date palm) having plasmodesmata connections.

6. *Pectins*. Pectins, also called fruit jellies, are present in many plants, particularly in the fruits. They are readily soluble in water. Apart from being used as food by the plants and animals, pectins can increase the water-holding capacity of the cells. The cementing material between two adjacently-lying cells, the middle lamella, is made of pectin compound. They are abundantly used in the preparation of jams and jellies.

7. *Gums* and *Mucilages*. Gums and mucilages are widely distributed in the plants. Though they do not dissolve in water, they form a slimy mass when moistened. Besides their utility as reserve food, they help the plants in diverse ways like holding water, checking rapid loss of water and in facilitating dispersal of seeds. To human beings they are particularly useful in industries and medicine.

Nitrogenous matters. Nitrogenous matters, commonly called *proteins*, are the most complex organic food matters of the plants. They are specially needed in the regions of active growth. Apart from their utility as very important tissue builders, they form an integral part of protoplasm itself, and so are present in every living cell. Chemically these extremely complex matters are composed of carbon, hydrogen, oxygen and nitrogen, and often sulphur and sometimes phosphorus. Chlorophyll, the most essential green pigment of plants, is a magnesium-containing protein and haemoglobin of blood, the most efficient oxygen-carrier of our body is an iron-containing protein.

The nitrogenous matters may be put into two categories: simple soluble ones, called **amino-acids**, and complex insoluble ones occurring in form of grains, known as **proteid grains.** It goes without saying that all proteid grains are formed out of amino-acids, which have been very aptly called 'building stones' of proteins, through complicated chemical processes. In fact, all the steps in protein synthesis are not clearly known. The amino-acids are often found in colloidal solutions in water. Complex-proteins are broken down to amino-acids by proper enzyme action. In that form they move to the different parts of the plants where they are needed and utilised. Commonly occurring amino-acids of plants are tyrosine , histidin, etc. **Amides** are salts of amino-acids, corresponding to inorganic salts; common examples are asparagine, glutamine.

The insoluble proteid grains are very complex. The formulae of a few common proteids would go to show how tremendously large the protein molecules are: *gliadin*, a protein of wheat: $C_{635}H_{1068}N_{196}O_{211}S_5$; *zein*, a protein of corn: $C_{736}H_{1161}N_{174}O_{208}S_3$. In the storage regions they occur as grains called **aleurone grains.** Pretty good number of plant proteins have been found occurring in crystalline or amorphous forms. On the average a plant protein is said to be composed of 51% carbon, 25% oxygen, 7% hydrogen, 16% nitrogen, 0·4% sulphur and 0·4% phosphorus.

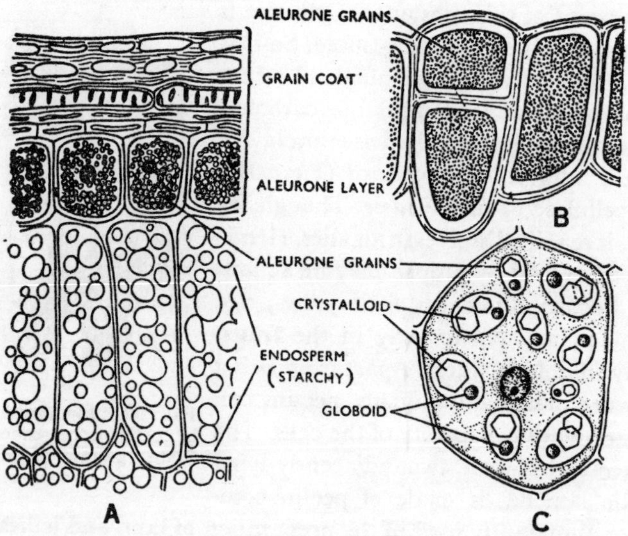

FIG. 505. Aleurone grains. A. In *Triticum* (wheat). B. In *Zea* (maize). C. In *Ricirus* (castor bean).

It has been said that most of the proteid grains stored as reserve food occur as small aleurone grains. In cereals a layer of cells just beneath the graincoat, composed of pericarp and testa, remains filled with aleurone grains (Fig. 505A). This layer is referred to as *aleurone layer*. They are also present in maize (Fig. 505B) and other cereals. In pulses like pea (Fig. 503B) they remain as small grains mixed with comparatively larger starch grains. More complex aleurone grains are found in the endosperm of castor-oil seed (Fig. 505C). Here the grain has a protein matrix in which occurs a large crystalline body, called *crystalloid*, and a small round body, known as *globoid*. Of the two, it is the crystalloid which is nitrogenous, the globoid, chemically, being a double phosphate of calcium and magnesium.

Test: Two common microchemical tests for proteins are used. (*i*) A section or suspension of tissue in water is treated with strong nitric acid. A yellow coloration is noticed. Now a few drops of strong ammonium hydroxide are added. The colour changes to orange. This is *xanthoprotein* test. (*ii*) One ml of 20% sodium lydroxide and one drop of

1% copper sulphate solution are added to the material. A violet colour is formed which indicates the presence of protein. This is *Biuret* test.

Fats and Oils. Fats and oils are also very important energy-giving reserve materials of plants. They are composed of carbon, hydrogen and oxygen, but not in the same proportions as in carbohydrates. They have much more of carbon and hydrogen and relatively very low oxygen. The formula of common fat, *olein*, present in olive oil—$C_{57}H_{104}O_6$, is suggestive. Due to occurrence of relatively low oxygen and high carbon and hydrogen fats and oils release much more energy than carbohydrates. Fats which are liquids at room temperature are called oils. They are lighter than water, the specific gravity ranging from 0·875 to 0·970, and are insoluble in water. They dissolve in substances like chloroform, ether and acetone. Fats leave greasy spots on a piece of paper. As oils they are probably present in all living cells, often in form of drops in the cytoplasm. These insoluble food matters are compounds (esters) of fatty acids and glycerine. Principal fatty acids are oleic acid, palmitic acid. They are abundantly present in storage organs, in seeds, embryos and meristematic tissues. Fats are possibly derived directly from carbohydrates. This inference is based on the facts that during the maturation of oily seeds there is a decrease of carbohydrates and increase in fats, and also that during germination the process is reversed, fats being converted into sugars. They are supposed to be formed directly by cytoplasm and also by special fat-forming plastids called elaioplasts. Fats usually occur in the plants as oils. These oils are non-volatile and, hence are called *fixed* oils. They are generally extracted by pressure. Some fatty oils are also used as medicines, e.g. chaulmoogra oil from *Hydnocarpus Kurzii* of family Flacourtiaceae in leprosy and croton oil from *Croton tiglium* of family Euphorbiaceae as one of the most strong purgatives.

Besides their utility as source of food, fats are also used in many industries. They saponify when treated with caustic soda or caustic potash. As a result of the said treatment soap is formed, which is nothing but a sodium or a potassium salt of a fatty acid.

Test: Fats and oils stain red with alkalin and Sudan IV. When treated with 1% solution of osmic acid in water they turn black in colour.

II. SECRETORY MATERIALS

These ergastic matters do not serve as food, and so are not directly concerned with the nutrition and growth of the plants. but as they have positive biological significance, their utility in the life of the plants cannot be questioned. They are often present in special sacs or glands. The following are the common secretory materials:

(a) Colouring matters. Plants possess different types of pigments which impart attractive colours to the parts. The green colour of the leaves and other aerial organs is due to the most essential pigment chlorophyll. Photosynthesis, the all-important process of manufacture of carbohydrate food, is solely dependent on the presence of chlorophyll. It is a combination of two green pigments, chlorophyll *a* and chlorophyll *b* and the yellow pigments, carotin and xanthophyll. The attractive colours, ranging from red to yellow, of the flowers and many fruits are due to carotenoids—carotin and xanthophyll. These are helpful in bringing about pollination, a necessary prerequisite to fertilisation, in case of flowers, and also in the dispersal of fruits. The blue, violet and pink colorations are imparted by water-soluble pigments, anthocyanin, present in the cell-sap of the vacuoles. Even lower plants like algae, fungi and bacteria possess some pigments.

(b) Enzymes. Enzymes are the ferments or digestive juices secreted by protoplasm. These are colloidal substances. Enzymes are the organic *catalysts*. They can catalise breaking down of complex food materials into their simpler soluble and diffusible forms. There are different types of enzymes for the purpose, viz., *diastase*, for breaking down starch ultimately into glucose; *protease* for nitrogenous matters; and *lipase* for fats and oils. Enzyme action is usually specific and often reversible. Thus the enzymes operating on starch have no action on nitrogenous matters or fats and oils. Furthermore, the enzymes breaking down starch into sugar are equally responsible for the formation of starch out of sugar, showing thereby the reversible action of the enzyme.

Hundreds of enzymes are present in the plants. Apart from their role in digestion as stated above, they regulate or control all physiological processes involving chemical reactions. They influence the metabolic processes like photosynthesis, respiration and fermentation.

(c) Nectar. This sweet substance is usually secreted by the floral parts of many plants. It is chiefly composed of sucrose together with glucose and fructose and is an excellent food. Nectar attracts a large number of insect visitors. The latter in course of their search for nectar carry pollen grains to the stigmas of flowers and bring about pollination. It is thus used as a device for pollination in entomogamous (insect-pollinated) flowers.

III. EXCRETORY MATERIALS

These ergastic matters are hardly of any use to the plants, so these by-products of metabolic activities are called *excretory* materials or waste products. Though they are useless to the plants, many of them, as will be subsequently found, are very useful to mankind. It is an admitted fact that the rate of metabolism in plants is less vigorous than that of animals; the waste products formed are also lesser in quantity. Moreover, the plants unlike animals, have no definite excretory systems or passages. Excretory matters may remain accumulated somewhere in the plants instead of passing off, usually away from the centre of protoplasmic activities. The plants, however, may ultimately get rid of the waste products to some extent through the leaves before seasonal fall and through fruits, seeds and barks of trees. Otherwise the accumulation of waste products would create toxicity and interfere with the activities of the living cells.

An account of the common excretory matters is given here:

Alkaloids. Alkaloids are nitrogenous waste products found in many plants. They are supposed to be decomposition products of proteins. Chemically, they are composed of carbon, hydrogen, oxygen and nitrogen. They occur in storage organs like roots and seeds; in bark, leaves, wood and other parts. Alkaloids are usually insoluble in water, but readily soluble in alcohol. They have bitter taste, and many of them are deadly poisonous. Alkaloids are the active principles of a large number of medicines derived from the plants. Quinine present in *Cinchona* bark, reserpine in the root of *Rauwolfia serpentina*, strychnine in nux vomica seeds, cocaine in coco, caffein in coffea seeds, ephedrine from *Ephedra*, nicotine in tobacco, are the common alkaloids.

Glucosides. These bear close resemblance with alkaloids in many porperties, but they are decomposition products of carbohydrates. Many glucosides are used in medicine, e.g. digitoxin, the active principle of the important heart stimulant digitalis—*Digitalis purpurea* of family Scorphulariaceae.

Tannins. These are usually bitter and astringent substances related to glucosides. They occur in the cell-sap of the vacuoles, in the cell wall, in bark and woody parts of plants. Tannins are abundantly present in many fruits like myrobalans, in leaves, seeds, bark and also in pathological galls. The percentage of tannin is usually high in unripe fruits, and it gradually decreases with ripening. Tannins are almost of universal occurrence in individual cells, in special sacs or forming connected system. They have been noticed even in the meristematic cells. Tannins are used in many industries. They can react with the proteins of animal skins and can produce a hard-substance. For this property they are used for tanning leather. Further, tannins turn blue-black in colour in contact with iron salts. Thus they are extensively used for manufacture of ink. Some of them have medicinal value as well.

Latex. Latex is usually a milky fluid, often watery or brownish, present in some families of angiosperms. It is really an emulsion with watery background and various substances like proteins, gums, resins, alkaloids, etc., suspended in water. Dumb-bell-shaped starch grains are frequently noticed in latex. Its utility to the plants is problematic, but it is, at any rate, helpful in healing of wounds. This milky fluid is very valuable as the chief commercial source of rubber. The majority of plants yielding rubber belong to the families Moraceae, Euphorbiaceae and Apocynaceae. *Hevea brazilensis* of family Euphorbiaceae needs special mention as a source of rubber. Latex of papaw contains papain, the enzyme helpful in digestion of protein food. The so-called 'Cow-tree' *Brosimum galactodendron* of family Moraceae of Venezuala yields the latex which is taken by the natives as a substitute of milk.

Essential oils. These are volatile oils found in many plant parts. They usually occur in special cells or glands. The fragrant odours of many flowers and leaves, of rinds of some fruits, bark and wood of some trees and even of some seeds are due to presence of essential oils. They are usually extracted by distillation. It goes without saying that these oils are of advantage to the plants in bringing about pollination and dispersal of fruits and seeds by attracting insect visitors. They are used in manufacture of perfumes, soaps, etc., and also in medicine, e.g. eucalyptus oil from *Eucalyptus globulus* of family Myrtaceae and valerian from garden heliotrope—*Valeriana officinalis* of family Valerianaceae. Essential oils may also have antiseptic value.

Resins. Resins are complex in composition and are probably oxidation products of essential oils. They occur in special glands or ducts either alone or in association with gums and essential oils. They are not soluble in water but dissolve in solvents like ether and alcohol. Resins are of three kinds from commercial point of view. Those which are secreted alone are usually solid brittle substances. They are called *hard resins* which readily dissolve in alcohol and are abundantly used in paint and varnish industry as water-proofing substance. Lacquer and shellac are the examples. Resins associated with essential oils are aromatic and more or less liquid in nature. They are known as *oleo-resins*, e.g. turpentines and Canada balsam. The third kind, called *gum resins*, are mixtures of resins and gums. They are usually milky in appearance and may be collected by injuring the plants. The common Asafoetida of Afganisthan, yielding a foul odour and acrid taste is the example. Many resins are used in medicine.

Gums. Gums are decomposition products of cellulose cell wall. They are soluble in water. They are exuded from the plants in natural course as mucilaginous masses or in response to injuries. They are of considerable commercial value as adhesives, and are also used in other industries. Gum arabic of commerce, an exudate from *Acacia senegal* of family Leguminosae is particularly used in medicine as an emulsifier.

Mineral crystals. Inorganic materials in form of crystals occur as waste products in many plants. They are mainly salts of calcium—as calcium oxalates and carbonates. Silicon crystals remain densely impregnated on the surface of plants belonging to grass family Graminaceae and horse-tails (*Equisetum*).

Calcium oxalate crystals are practically present in all higher plants. They may occur in any part, but are more abundant in storage organs, in pith and cortical tissues. Solitary crystals are of varying shapes—rod-like, prismatic, pyramidal, rhomboidal, needle-like. They can be demonstrated in large number in the papery tunic of onion (Fig. 506A) and in the cortical cells of *Leonurus sibiricus* (Fig. 506B) of family Labiatae, and many other plants. Needle-like or acicular crystals in form of bundles, called **raphides,** are common in the aroids and aquatic plants (Fig. 506 C, D & E) like *Pista* and water hyacinth. They often occur in special mucilage-containing cells called *idioblasts*. The needles are ejected when the mucilage swells causing bursting of the wall of the idioblast. A good number of crystals may remain conglomerated together to produce a roundish crystal-aggregate. They are called **sphaerapides** or *druses*. Druses may be seen in plants like

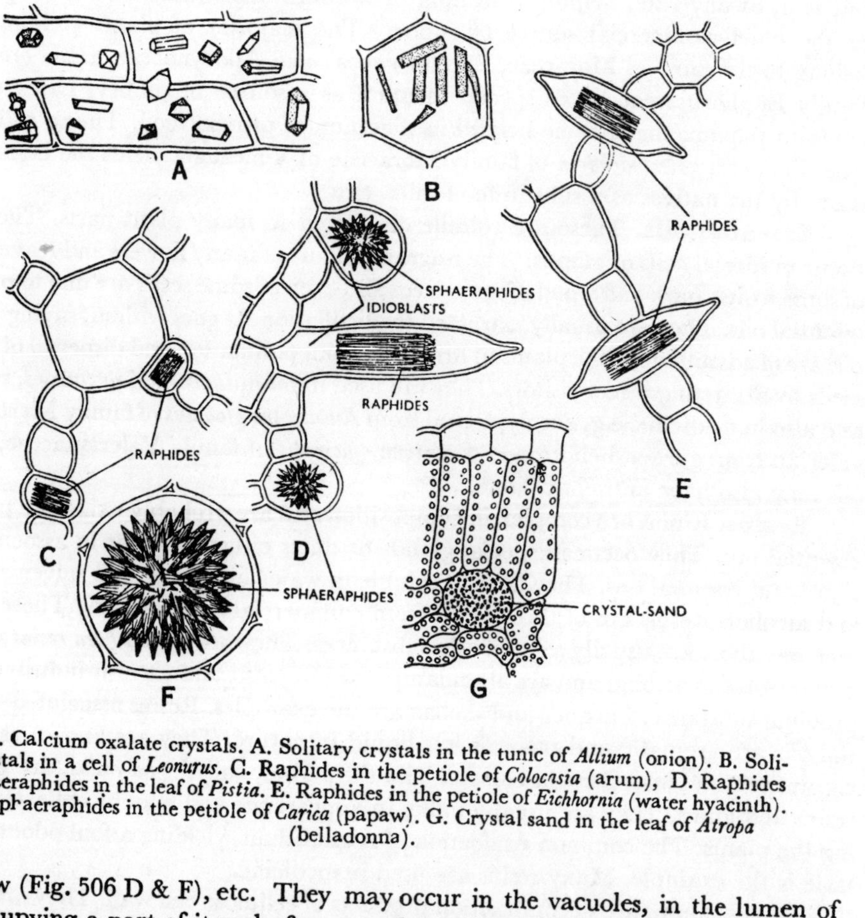

FIG. 506. Calcium oxalate crystals. A. Solitary crystals in the tunic of *Allium* (onion). B. Solitary crystals in a cell of *Leonurus*. C. Raphides in the petiole of *Colocasia* (arum), D. Raphides and sphaeraphides in the leaf of *Pistia*. E. Raphides in the petiole of *Eichhornia* (water hyacinth). F. Sphaeraphides in the petiole of *Carica* (papaw). G. Crystal sand in the leaf of *Atropa* (belladonna).

Pislia, papaw (Fig. 506 D & F), etc. They may occur in the vacuoles, in the lumen of the cell, occupying a part of it and often completely filling up the cell. Small solitary crystals often remain together forming what is called *crystal sand* (Fig. 506G).

Calcium carbonate crystals are less common, as they usually occur in the epidermis impregnating a portion of cellulose cell wall. In India rubber (*Ficus elastica*) or banyan these crystal-aggregates are found in the form of a bunch of grapes. In fact, the epidermis here is multiseriate and some cells of the innermost layer enlarge, and calcium carbonate crystals remain aggregated on a peg-like projection of the cellulose wall inside the enlargement (Fig. 507A). Such a crystal-aggregate is called **cystolith.** They also occur in the leaves of many plants of families Cucurbitaceae (Fig. 507B) and Acanthaceae (Fig. 507C).

Silicon crystals remain embedded on the parts of the plants belonging to grass family and also on the pteridophyte *Equisetum* (horse-tail). They impart roughness to these plants.

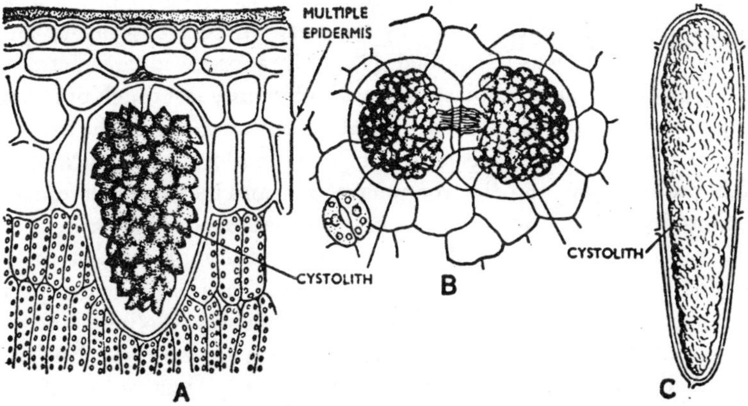

FIG. 507. Calcium carbonate crystals. A. In the leaf of *Ficus elastica* (India rubber plant). B. Double cystolith in *Momordica*. C. Elongated cystolith in *Ruellia* of Acanthaceae.

Organic acids. These are widely distributed in plants, more abundantly in the leaves and fruits. The common organic acids are *tartaric* acid in tamarind, *oxalic* acid in *Oxalis*, *critic* acid in citrus fruits, *malic* acid in apples. They may also occur as salts, particularly calcium oxalate, as already discussed. These acids are soluble in water. They make the cell-sap toxic when they reach high concentration. They may combine with calcium and other metals and are converted into insoluble crystals. Organic acids have roles to play in respiration and other metabolic processes.

CHAPTER XVIII

THE CELL WALL

Most of the plant cells are provided with tough rigid cell wall and this is taken as an outstanding point of difference between plant and animal cells. Naked protoplasts are noticed in lower groups of plants. The reproductive cells of higher plants as a rule lack walls. The cell wall was actually noticed by the early workers long before the more important matter protoplasm was recognised. Naturally it received considerable attention then. With the discovery of protoplasm the wall obviously lost much of its importance. In recent years the cell wall has again attracted serious attention of the investigators. In fact, spectacular advances in the technique of investigations, in application of X-rays, polarized light photography and electron microscope, and increasing commercial importance of cellulose have given new impetus in investigations on the wall.

The cell wall is usually looked upon as a secretory product of the protoplast. In that sense it ought to be a non-living membrane surrounding the protoplast. It protects the protoplast from adverse external influences, delimits the same, gives mechanical stability to the cells and also imparts shape to the cells. But the very fact that wall is secreted by protoplast and that an intimate relationship exists between the two have led some workers to believe that it may not be completely non-living.

Formation of the cell wall and the process thereof have got to be sought in the method of cell division. During mitosis two daughter nuclei are constructed after having

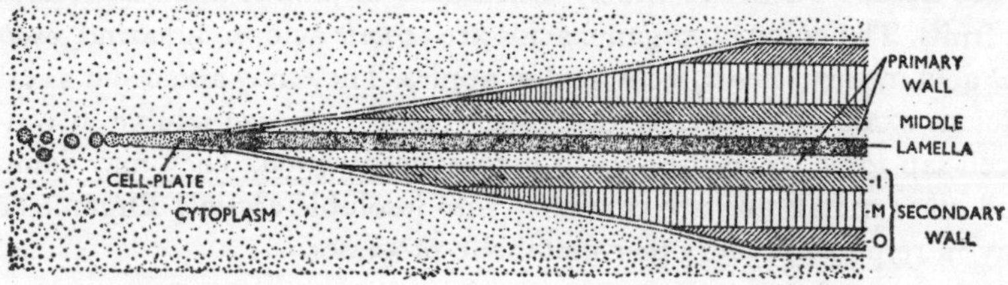

Fig. 508. Formation of cell wall (diagrammatic), showing origin of cell plate, its transformation into middle lamella and deposition of primary and secondary walls. I, M & O, indicating inner, middle and outer layers respectively of secondary wall. (*From Sharp*).

gone through a series of complicated changes. The nuclear division is then followed by cytoplasmic division (cytokinesis). In this process protoplasmic matters accumulate in the equatorial region in form of small droplets, which cohere in course of time and form a continuous plate called *cell plate* (Fig. 508). This plate undergoes physical and chemical changes and is ultimately converted into the intercellular substance, the *middle lamella*. In fact, middle lamella is the 'cement' between the two cells. By proper chemical treatment middle lamella may be dissolved, when the cells lose the cementing material and become isolated. Middle lamella is optically inactive and colloidal in nature. It is composed of pectates of calcium and often magnesium. The protoplast goes on secreting cell wall materials on the middle lamella, and ultimately a soft, delicate and plastic wall is formed. This is *primary* cell wall (Fig. 509). It consists mainly of cellulose and pectic compounds and may also contain other polysaccharides. This is really the first formed cell wall, which may persist in many cells as the only wall. The primary wall is optically active.

Primary cell wall gets more and more stretched during the growth of the cell. In course of time secondary cell wall materials are deposited on the primary wall. It may happen that primary walls of the adjacent cells become fused with the intercellular substance and form the common layer between the two cells. In that case middle lamella is not purely the cementing material. Some workers have attributed the term compound middle lamella to such layer. The secondary wall is mainly composed of cellulose which

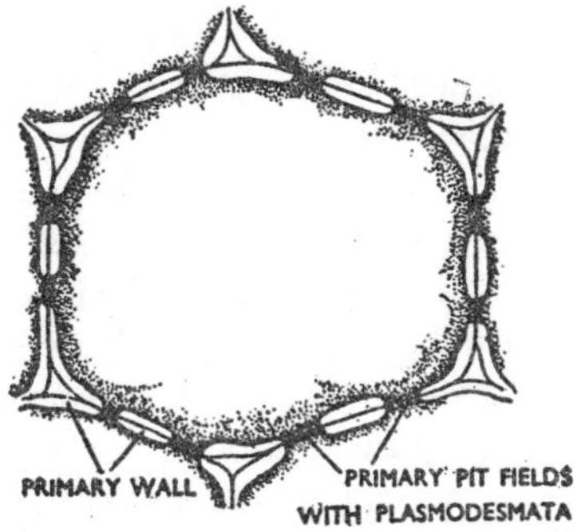

FIG. 509. Cell wall—a parenchyma cell with primary wall having primary pit fields with plasmodesmata.

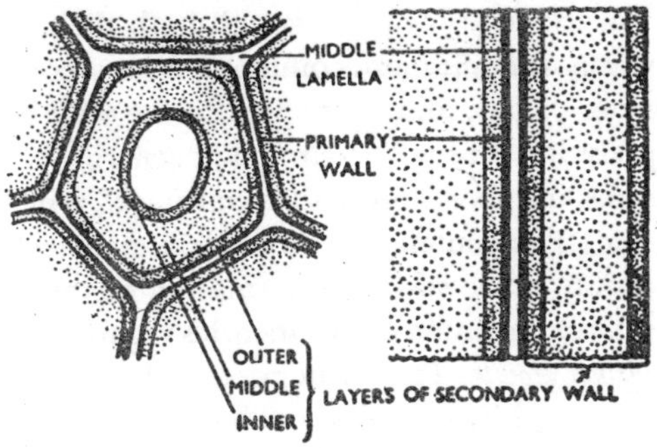

FIG. 510. Cell wall—diagram showing three layers of secondary wall in transverse (left) and longitudinal (right) views.

may undergo other modifications. In many cells it is usually three-layered, the layers differing in physical and chemical properties, of which the middle one is the thickest (Fig. 510). Some authors have suggested that another layer with a different chemical composition may be present in addition to the inner layer of secondary wall, and preferred to call this layer tertiary or terminal one (Frey Wyssling). While secondary materials are deposited on the primary wall small thin areas are left out. These areas are called *primary pit fields* (Fig. 509) through which fine fibrils of cytoplasm pass from one cell to another and thus establish the organic continuity of protoplasm. These fibrils are known as **plasmodesmata** (Fig. 516). As the secondary walls are laid down after the cell has ceased to enlarge, it does not grow like the primary wall. So it may be regarded as a collection of supplementary layers making the wall thick and massive and mainly giving mechanical support. Thus the cell wall is made of at least three layers—intercellular substance, primary and secondary wall.

The **submicroscopic** nature of the cellulose wall has attained prominence in recent years as a result of studies made with X-rays, polarized light and electron microscope. It is believed that the wall is made up of a porous matrix of cellulose. The most minute units are some extremely delicate thread-like bodies called *microfibrils* which are grouped together in larger bundles and anastomose to form a three dimensional network with an interfibrillar system of microcapillaries. The fibrils are the aggregates, referred to as *miscelles*, of a large number of long chain-like cellulose molecules arranged along the longer axis of the fibril. Colloidal pectic compounds and other non-cellulosic substances fill up the microcapillaries of the anastomosing system.

GROWTH OF THE CELL WALL

It has been stated that primary cell wall is soft and plastic. It is composed of cellulose and pectic substances. The pectic matters are hydrophilic colloids; they can absorb and hold water. Due to softness and plasticity the primary wall gets considerably stretched during the growth of the cell, when new cell wall materials are deposited on the primary wall

Two processes of the growth of cell wall have been described, viz., growth by *intussusception* and growth by *apposition.* The first method, though rather difficult to explain, involves wedging in or interpolation of new cell wall materials between the existing ones of the stretched primary wall. This process brings about growth in surface area. During growth by apposition new materials are laid down layer after layer on the inner side of the primary cell wall like the pages of a book. This method causes growth in thickness mainly on secondary walls. Though both the methods of growth are possible locally, growth by apposition is more common. Thus secondary walls are placed in layers, except in the region of pits, and consequently the wall becomes quite thick and massive. Growth by apposition is usually *centripetal,* occurring from outside. In unicellular bodies like pollen grains and spores, growth is *centrifugal,* which explains the prominently thick and spined exine. The contents of the degenerated cells surrounding the spores (tapetal cells) are possibly involved in the formation of the exine. Electron microscopic studies have led to the enunciation of two theories regarding the methods of growth, viz., *mosaic growth* and *multinet growth.* The former demands that new microfibrils are pushed into the thin areas penetrated by cytoplasm. The second one, a more recent concept, maintains that new lamellae with denser microfibrils are added centripetally on the previous ones. Thus the classical controversy as regards the growth of the wall by intussusception and apposition is still continuing.

In many plants cells the primary wall may persist up to the permanent stage. In that case wall remains comparatively thin and more or less uniform. In woody parts of the plants, particularly in the water-conducting portions (xylem) cells and vessels have massive and peculiarly thickened walls. Here secondary matters are localised to certain regions on the primary cell wall, instead of being deposited all over the same. Due to localised nature of thickening the following peculiar designs or patterns (Fig. 511) are produced:

(*i*) **Annular** or ring-like thickening is noticed in the protoxylem elements where secondary matters are placed centripetally in form of rings at regular intervals (Fig. 511 A & B).

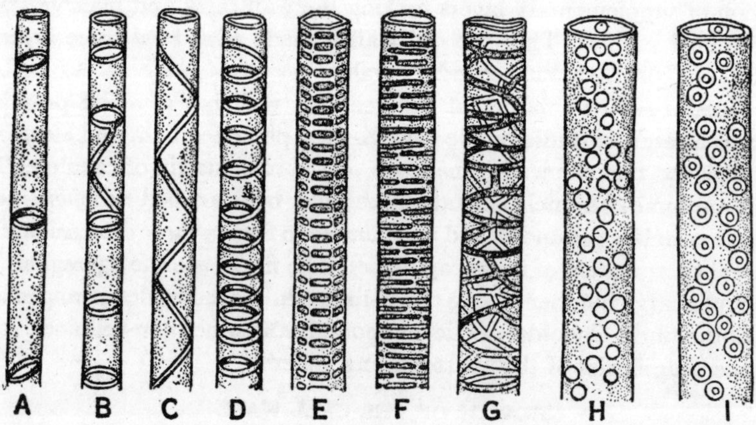

Fig. 511. Cell wall—different types of localised thickenings of the wall in longitudinal views. A & B. Annular. C & D. Spiral. E & F. Scalariform. G. Reticulate. H. Pitted (simple). I. Pitted (bordered).

(*ii*) **Spiral** thickening is also found in protoxylem elements, secondary wall being deposited in form of spiral (Fig. 511 C & D).

(*iii*) **Scalariform.** Secondary matters look like the rungs of a ladder here (Fig. 511 E & F).

(*iv*) **Reticulate.** The secondary matters here assume the form of a network (Fig. 511G)

(*v*) **Pitted.** In this case secondary cell wall materials are deposited practically all over the primary wall, only leaving some small thin areas here and there. These unthickened areas are the *pits* (Fig. 511 H & I). The pit fields present on the primary wall, may develop into pits with growth and maturation of the cell, or the pits may arise over primary wall parts which do not bear pit fields. Pits, in fact, are the areas through which diffusion of fluids takes place from cell to cell. In front view they look like circular, oval or pentagonal areas on the wall. Pits are usually formed in pairs, i.e., if a pit is present on one cell, a complementary pit will be formed in the adjacent cell. The picture becomes clear in sectional view (Fig. 512). So at the region of the pit the only wall between the two adjacently-lying cells is the primary cell wall together with the middle lamella. It is referred to as the *pit membrane* or closing membrane. There are two types of pits. The one just described is a simple pit. In cells with thick walls the simple pits may break into two or more cavities and look like branched canals. These are called **ramiform** pits (Fig. 513).

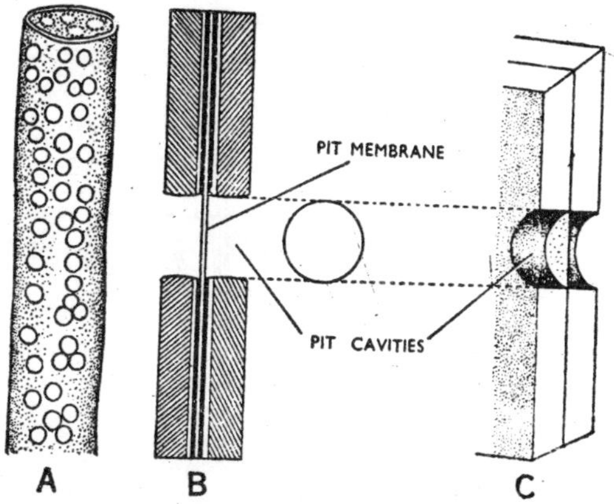

Fig. 512. Simple pits (diagrammatic). A. A vessel with simple pit in front view. B. Same in sectional view. C. Perspective diagram of the same.

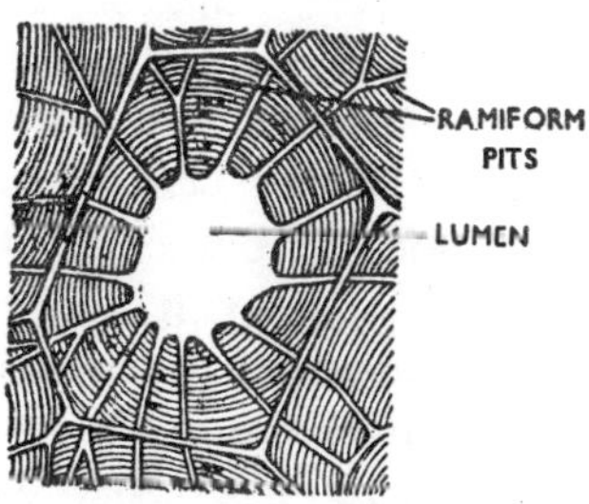

Fig. 513. Sclereids showing ramiform pits.

The second type, where the secondary wall arches or overhangs from both sides, is called a **bordered** pit. In this case due to the said overhanging of the materials a dome-shaped body is formed, which may be represented by two circles, the smaller one at the opening indicating the pit aperture (Fig. 514) and the larger one inside showing the pit cavity or pit chamber. In front view these pits exhibit two circles, the pit cavity forming a border round the pit aperture, and hence the name. In bordered pit the pit membrane is, as usual, composed of primary walls of two cells and the middle lamella. The central part of the pit membrane swells up. The swollen portion is called *torus*. The torus may shift its position. Due to changes of pressure it may swing and come to the mouth of the pit, and thus block the pit aperture (Fig. 514D). Presence of torus is characteristic in the bordered pits of the gymnosperms—Coniferales, *Ginkgo* and Gnetales. They occur rarely in other groups of vascular plants. Normally pits are of either simple or bordered types. In some cases it may be a combination of the two—a bordered pit being complemented by a simple pit (Fig. 515). This type of pit is called **half-bordered.** Here the torus is either poorly developed or lacking. A large pit on one side may be complemented by two or more small pits on the opposite side. It is known as unilateral compound pitting. Blind pit, a pit without the complementary one, is rather rare. It may occur, when, for example, a pit is placed opposite an intercellular space.

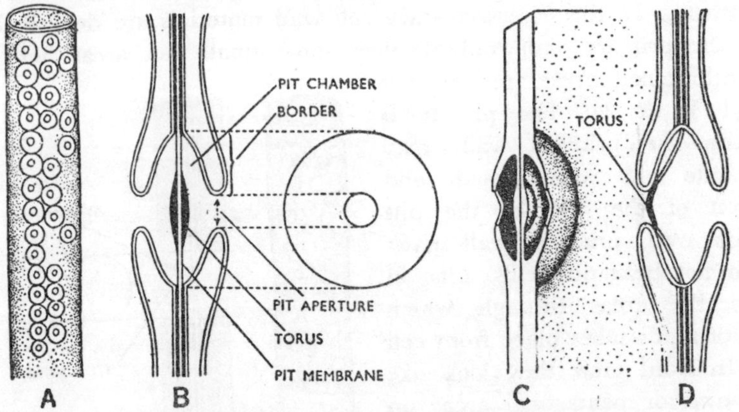

FIG. 514. Bordered pits (diagrammatic). A. A vessel with bordered pits in front view. B. Same in sectional view. C. Perspective diagram of the same. D. Sectional view of a bordered with changed position of torus.

In the tracheary elements of secondary wood of some families of dicotyledons like Leguminosae, Cruciferae, Myrtaceae and Caprifoliaceae minute outgrowths are formed from the free surface of the secondary walls of the bordered pits in form of sculpturings. These pits, which are thus rendered sieve-like in appearance, are called *vestured* pits. The projections forming the vesture are refractive and of diverse types. In half-bordered pits they occur only on the bordered side of the pair. They are found in phylogenetically more advanced xylem and are considered to be an advanced form of pit.

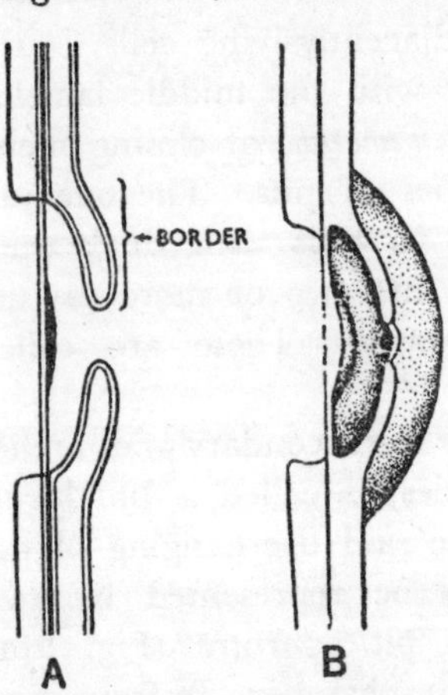

FIG. 515. Half-bordered pits (diagrammatic). A. Sectional view. B. Perspective diagram.

In the secondary wood of conifers rod-like or bar-like sculpturings are common in the lumen of the tracheids extending across from one tangential wall to the other. These are known as **trabeculae,** which usually occur in long radial series of cells.

In stained preparations of the tracheids of some gymnosperms bar-like or crescent-shaped thickenings of intercellular materials and primary walls are noticed on outer and lower margins of pit pairs partially encircling them. These are characteristic thickenings of the middle lamella and primary walls which are visible through the secondary walls in stained preparations. They are called **crassulae** (Fig. 538). Crassulae represent the borders of the primary pit fields of young cells from which they have developed.

The terms 'bars of Sanio' or 'rims of Sanio', after Sanio who extensively worked on wood of conifers in the nineteenth century, had been applied to the crassulae for a long time. But it was considered rather confusing, because these terms were also applied to trabeculae.

CHEMICAL NATURE OF THE CELL WALL

The cell wall is mainly composed of an insoluble polysaccharide, **cellulose,** with the empirical formula $(C_6H_{10}O_5)_n$. Pectic compounds, hemicelluloses and other polysaccharides may remain associated with cellulose. The middle lamella, as already reported, is mainly made of pectates of calcium, and often magnesium. The primary and secondary walls are fundamentally composed of cellulose. Other substances may be added to cellulose or the latter may be modified in various ways.

Cellulose is the most complex insoluble carbohydrate. It is the principal material in building up the framework of a plant. Normally cellulose wall is not digestible; excepting in case of some fungi and bacteria it is not used as food. Cellulose wall is permeable to water and solutes. Economically it is of great importance. Cotton fibres, commercially valuable fibres of flax and hemp and paper pulp are mainly cellulose materials. Rayons, cellophene, nitrocellulose and some plastics are also manufactured from cellulose.

It gives the following microchemical reactions: (*i*) *violet* colour with chlor-zinc-iodine solution, (*ii*) *blue* colour with iodine followed by 50% sulphuric acid.

Lignin. Lignin occurs in the walls of the woody tissues. It can give sufficient strength and rigidity to the plant members. Cellulose may be modified into lignin or may be associated with the same. As a result of infiltration of this complex substance the walls become quite thick and hard. With gradual lignification of the wall most cells lose protoplasm and ultimately become dead. It appears in the middle lamella and the primary wall and later may even be found in the secondary wall. Like cellulose lignin is also permeable to water and solutes. It gives requisite strength and rigidity to the plant members, so that they can stand against strains like tension and compression. Most of the mechanical tissues have lignified walls.

Lignified walls turn (*i*) *bright yellow* when treated with aniline sulphate solution; (*ii*) *red* with phloroglucin and hydrochloric acid and (*iii*) *yellow* with chlor-zinc-iodine solution.

Cutin. It is a waxy or fatty modification occurring on the outer walls of epidermal cells of aerial organs. Cutin remains associated with cellulose, and often with pectic materials. Walls impregnated with cutin are said to be *cutinised*. Cutin is impermeable to water and so cutinised walls can greatly check evaporation of water from plant surface. Though normally it is localised to outer walls of epidermal cells, often subepidermal cells are also cutinised. Impregnation of cutin associated with deposition of other waxy materials results in the formation of a strong waterproofing layer on the outer surface of aerial organs. This effective layer is called *cuticle*. Some fruits also possess strong cuticle for retaining water.

Cutin turns (*i*) *yellow*, if treated with caustic potash solution, (*ii*) *yellowish-brown* with iodine followed by sulphuric acid and (*iii*) *pinkish-red* with Sudan III.

Suberin. It is another fatty substance resembling cutin in many properties. It is an important constituent of the walls of cork cells. Walls made of suberin, *suberised* walls as they are called, are relatively more impermeable to water and hence can effectively check water loss from the surface. Though normally restricted to the peripheral layers, suberin may also occur in the internal cells of endodermis and exodermis.

Suberin turns (*i*) *yellowish-brown* with chlor-zinc-iodine solution, (*ii*) *brown* with caustic potash solution.

Mucilage and Gum. Mucilages and gums are substances related to carbohydrates of the cell wall. They can absorb water quite quickly and can retain the water. Mucilage is present in the seed coats and aquatic plants. The seeds of common 'topmari'—*Plantago* *ovata* of family Plantaginaceae is a nice example. In plants inhabiting dry regions like deserts mucilage is definitely helpful to germination, what is really the most serious problem there. In aquatic plants mucilage protects them from insects.

Mucilaginous hairs turn *violet*, when treated with iodine solution followed by sulphuric acid.

Mineral matters. Deposition of various inorganic salts are not uncommon on the cell wall. Silicon particles are infiltrated on the wall of the plants belonging to grass and sedge families and horse-tails. Oxalates of calcium are present on the walls of some plants. Calcium carbonate crystals of peculiar shapes occur in the leaves of plants of families Moraceae, Urticaceae, Cucurbitaceae, Acanthaceae, etc. *Cystoliths* are conglomerate crystals of calcium carbonate deposited on a stalked projection of the cellulose cell wall. Due to strong incrustation of lime they attain irregular shapes and often fill up the volume of enlarged cells (Fig. 507).

Besides these mineral matters, various organic substances like resins, tannins, fatty matters, aromatic oils, etc., may also be impregnated on the wall.

PLASMODESMATA

The organic continuity of protoplasm in a living body has been discussed in the preceding chapters. Delimitation of the protoplast in plant body due to presence of sharp wall is by no means complete. The primary wall is hardly uniform. Small breaks or meshes on the wall are traversed by extremely delicate and fine strands of cytoplasm to establish the connection between two adjacently-lying cells. These threads are called **plasmodesmata** (Fig. 516). They are characteristic of all living cells. Plasmodesmata have been found in red algae, liverworts, mosses, pteridophytes and spermatophytes. They appear to remain aggregated in the primary pit fields (Fig. 509). After formation of secondary walls they possibly remain restricted to the closing membranes of the pits. On account of their extreme delicacy plasmodesmata cannot be normally demonstrated. With some special techniques they may be readily seen in the endosperm of seeds with profuse stored food like date-palm—*Phoenix* (Fig. 504) and in the cotyledons. They are very thin in gymnosperms.

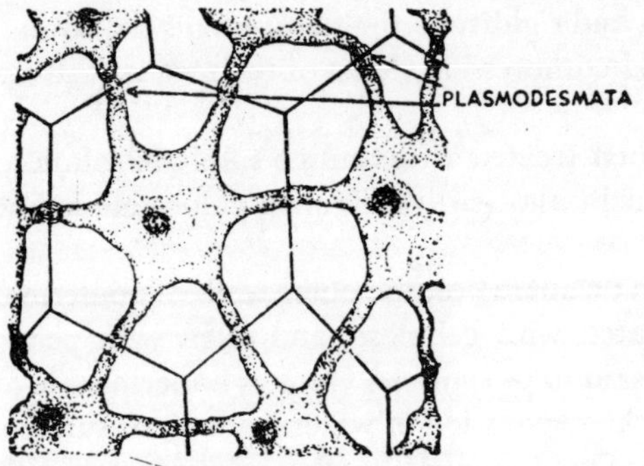

FIG. 516. Plasmodesmata in the cells of endosperm of seed of *Phoenix* (date-palm).

Though the presence of plasmodesmata definitely establishes the close relationship between the protoplast and the wall, their origin and development have been debatable. Some workers believe that at the early stage of growth the wall is penetrated by cytoplasm, which becomes gradually narrower as a result of accumulation of microfibrils and pectic matters and ultimately form thin plasmodesmata threads. Others are of opinion that plasmodesmata exist at the early cell-plate stage, and they also increase with the increase of the wall surface and often split. Various suggestions have been put forward from time to time about their functions. Though nothing has been definitely known, it has been suggested that (*i*) they may be instrumental in conduction of stimuli —external and internal, through plant tissues; (*ii*) may be mainly concerned with translocation of food, particularly in the regions like endosperm; and (*iii*) may serve as channels for movement of some viruses. Their occurrence in the haustoria-like bodies of parasites like *Cuscuta* and *Orobanchi* is suggestive of their roles in the movement of food and virus.

INTERCELLULAR SPACES

The cells usually remain compactly arranged in the meristematic condition. During gradual differentiation the close contact between the adjacently-lying cells may be partly broken down, so that some spaces appear between them. These are known as **intercellular spaces.** They usually remain filled up with air or water and constitute a continuous system for aeration or conduction. The spaces vary to a wide extent in shape, size and arrangement. They may be very small and, at the same time, some of them may be fairly large, so that the tissues become loose and spongy. The larger ones are usually called *chambers* and the elongated ones are known as *canals*.

The intercellular spaces are usually formed by separation of the walls along the areas of contact followed by contraction of separated parts. Spaces thus formed are called **schizogenous** (*schizo*=split), because these were formerly believed to be formed as a result of splitting of the middle lamella. But it has been conclusively proved by Martens in 1937–38 and others that the process is much more complex, particularly in case of dividing cells. According to them during the division of a cell the new middle lamella, formed by physical and chemical changes in the cell plate, between newly-formed cells at first comes in contact with the primary wall of the mother cell (Fig. 517) and not with

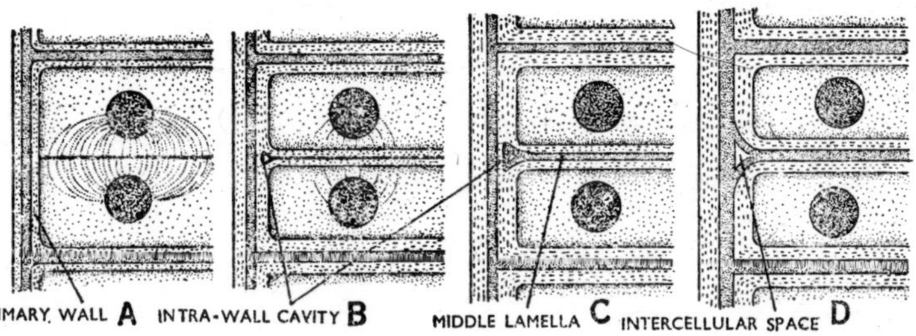

Fig. 517. Formation of the intercellular spaces—stages. (*After Martens.*)

the middle lamella of that cell. Thus new and old middle lamellae are separated by cellulose wall. A small cavity, referred to as *intra-wall* cavity, which is usually triangular in cross-section, appears at the point of contact. Now the parent cell wall dissolves opposite the cavity and the latter gradually reaches the outer middle lamella and thus the space is formed. A separate cavity is often formed within the old middle lamella, which is truly intercellular, either before the formation of the intra-wall cavity, or during its formation or even following its formation. The two cavities may fuse up giving rise to a large space.

Aquatic plants possess particularly large spaces which often extend like canals through the internodes. Schizogenously formed spaces often constitute specialised structures, the *secretory ducts* in conifers and some spermatophytes. The resin ducts of *Pinus* (Fig. 518A) and the secretory ducts of some members of families Compositae and Umbelliferae are the examples. Here the cells lining the duct cavity are secretory and the products are released into the canals. The secretory cells are called *epithelial cells*.

Another type of intercellular space may arise as a result of destruction of the cells formerly occupying the position of the space. These are known as **lysigenous** (*lysis*=loosening) spaces. Large air-spaces of some aquatic plants and of some monocotyledonous roots like maize, and the oil-cavities of *Citrus* fruits (Fig. 518B), *Eucalyptus*, etc., are formed by this method. In case of secretory cavities the cells that are destroyed release their

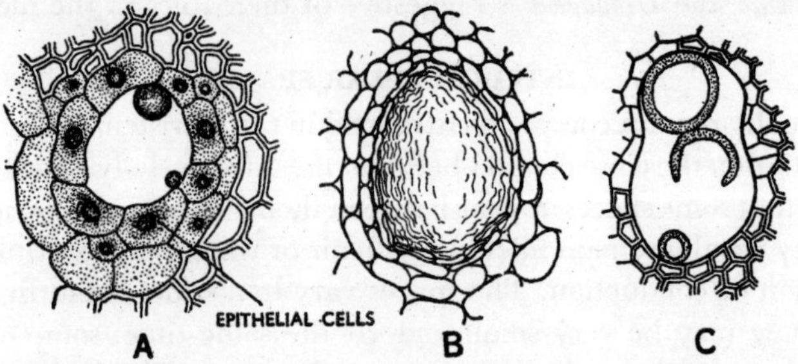

FIG. 518. Specialised intercellular spaces. A. Resin duct of *Pinus* (schizogenous). B. Oil-cavity of citrus fruit (lysigenous). C. Protoxylem lacuna of maize (schizo-lysigenous).

secretions in the space and they themselves remain in collapsed condition around the cavity.

Some cavities, though less common, are formed by the combination of both the methods. They are referred to as **schizo-lysigenous** ones. Protoxylem lacunae (Fig. 518C) of plants like maize are the examples.

CHAPTER XIX

CELL FORMATION

Growth is a fundamental feature of the living things. In unicellular organisms growth means simple cell enlargement, but in multicellular plants, which, in fact, form overwhelming majority, growth involves much more complications. The starting point in the life of a plant with sexual reproduction is the fertilised egg—the single-celled zygote. In course of time millions of cells are formed. Thus here growth involves formation of new cells from the pre-existing ones, instead of simple enlargement. Cell division is the most important method of formation of new cells. There are different methods of cell division, an account of which is being given here.

SOMATIC MITOSIS

Somatic mitosis is the most common method of cell division taking place in the vegetative parts which really constitute the body (*soma*) of the plants. It is a complicated process where the nucleus of the cell plays the prominent role. In fact, the nucleus divides first into two identically equal daughter nuclei. This division is referred to as **mitosis** or *karyokinesis.* Nuclear division is immediately followed by the division of the cytosome or cytoplasmic division. The second division is called **cytokinesis.**

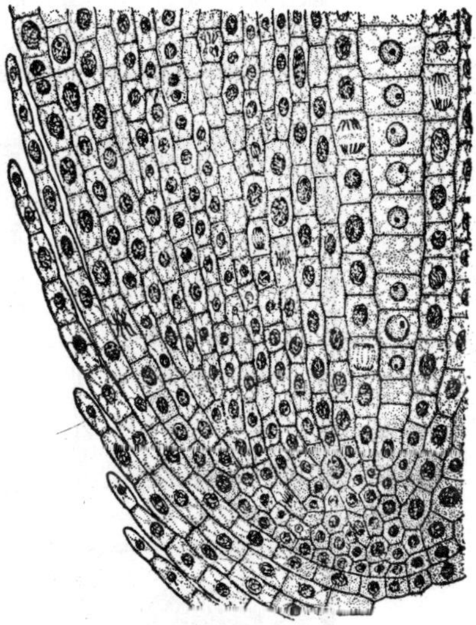

FIG. 519. Longitudinal section of the root-tip of onion showing different stages of mitosis in the cells.

As already stated, the method involves a series of complicated changes in the cell. Once it starts, it continues till the division is complete. So the process is definitely continuous. For the sake of convenience of study it is customary to divide the process into four parts or phases. Though purely man-made, the four phases have been recognised in the biological world. Somatic mitosis has been studied in different parts of the different plants. An outline of the method as found in the root-tips (Figs. 519 & 520) of plants is being discussed now.

Prophase. Prophase or early phase (Fig. 520 B, C & D) begins with the earliest recognisable changes in the nucleus. The nucleus before the onset of mitosis is said to be in *metabolic* condition (Fig. 520A). With the commencement of mitosis the chromatin reticulum unravels itself into slender and crooked threads, the *chromonemata,* which could not be traced as individuals in the earlier condition. Each chromonema, on close examination, reveals a longitudinally double structure. With advancement of the process the chromonemata tend to uncoil and become thick, so that double nature is more clear. These threads are called the **chromosomes.** In the mean time a second chromosomal element, *matrix*, appears; and the chromonemata, which again start coiling, remain embedded in the matrix. At this stage the chromosome distinctly shows a double structure, consisting of two daughter chromosomes or *chromatids.* Each chromatid now possesses matrix and a chromonema, which again splits lengthwise into two, remaining twisted about each other. So the chromosome now consists of two chromatids, each having two chromonemata or in other words, each chromosome breaks into four half-chromatids. Deeply stainable matrix renders the chromonemata rather invisible. The chromosomes are now shorter, thicker and denser bodies. Their number in a diploid nucleus is normally even and constant in a particular species of plant. The nucleolus and nuclear membrane gradually disappear in the mean time.

Metaphase. Now the nucleus passes to the metaphase (Fig. 520E). Probably due to some complicated changes in the karyolymph some fine fibrils make their appearance

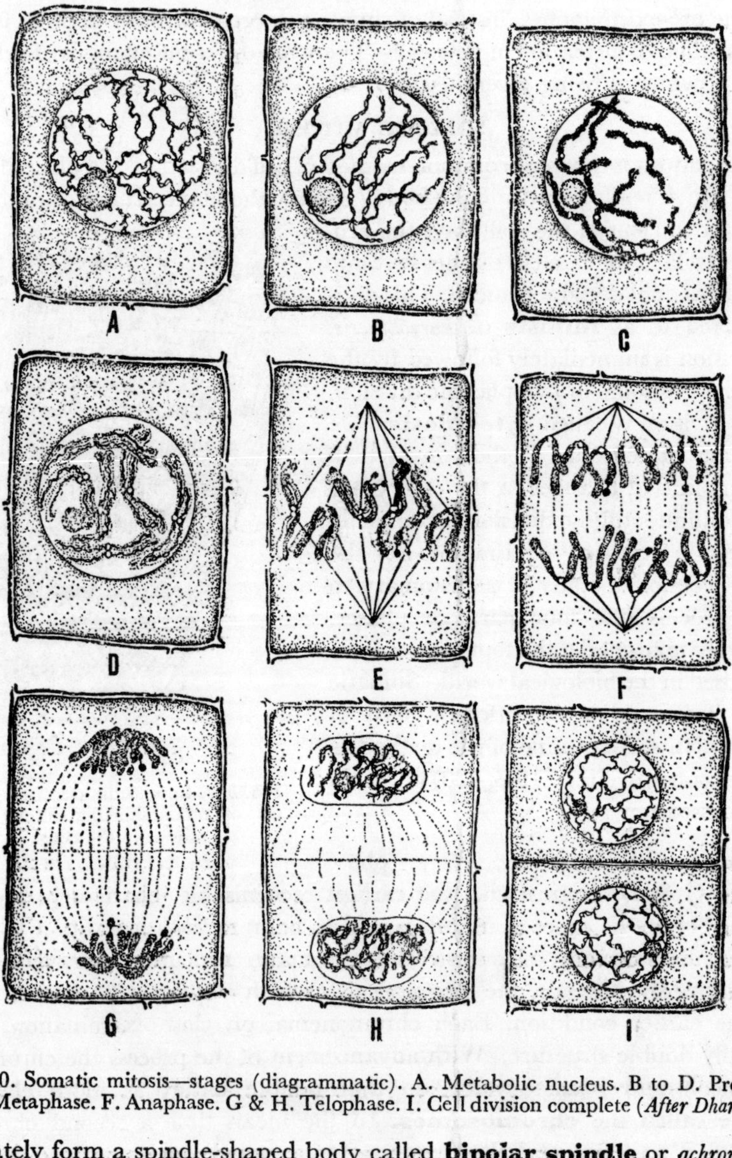

Fig. 520. Somatic mitosis—stages (diagrammatic). A. Metabolic nucleus. B to D. Prophase. E. Metaphase. F. Anaphase. G & H. Telophase. I. Cell division complete (*After Dharp*).

and ultimately form a spindle-shaped body called **bipolar spindle** or *achromatic figure*, so-called obviously for less affinity for stains. This appearance is probably caused by the arrangement of chemical molecules along the longitudinal axis of the spindle and a stratification of the material into firm and fluid layers. The two ends of the spindle are referred to as the poles and the middle plane as the equator. The double chromosomes or chromatids lying intimately associated to each other arrange themselves at the equator of the spindle. Each chromatid has a distinct region where it remains attached to the fibril. The attachment regions are called **centromeres** or *kinetochores* or primary constriction or points of attachment. These usually appear as unstained spots on the chromatids. The two sides, of equal or unequal lengths, of the chromatid are referred to as *arms*. The centromeres face opposite poles of the spindle, the arms may lie in any position. The fibrils directly attached to the bodies of the chromatids at the centromeres are called *tractile* fibres.

The nature of the tractile fibres is controversial. Suggestions have been made that it may be a local modification of spindle substance, an extension of the chromatid or something extruded from the chromatid. But no agreement could really be reached. Other fibrils running from pole to pole are known as *continuous* or *supporting* fibres. Studies by microdissection have revealed that the whole spindle with the chromosomes may be extracted from the cell, which goes to show that the spindle is a definite body and not a mere appearance.

Anaphase. At this stage (Fig. 520F), the chromatids become separated and start moving away from the equator, the two sets towards the two poles. The distribution of original chromosomes in the two sets are exactly equal, both qualitatively and quantitatively. It is the centromere which moves ahead along the tractile fibre, the arms trailing behind (Fig. 521). The chromosomes are quite distinct at this stage; they usually assume inverted V- or L-shaped appearance, depending on the length of the arms. The tractile fibres gradually disappear at the close of anaphase, and also the matrix, so that the chromonemata again come to view.

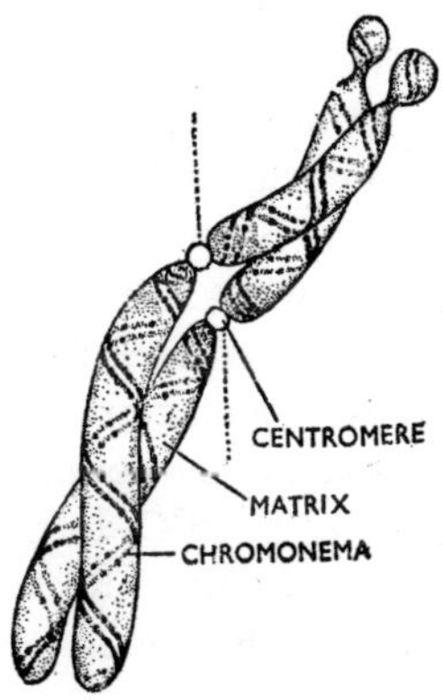

FIG. 521. A pair of chromosomes showing anaphasis split.

The cause of anaphasic movement of chromosomes has been a subject of keen investigation and has aroused good deal of controversy. Different views have been put forward, but no agreement could be reached on this point. It is not, however, unlikely that some sort of tractive force is exercised by the fibres, probably due to contraction of the protein molecules, of which they are made.

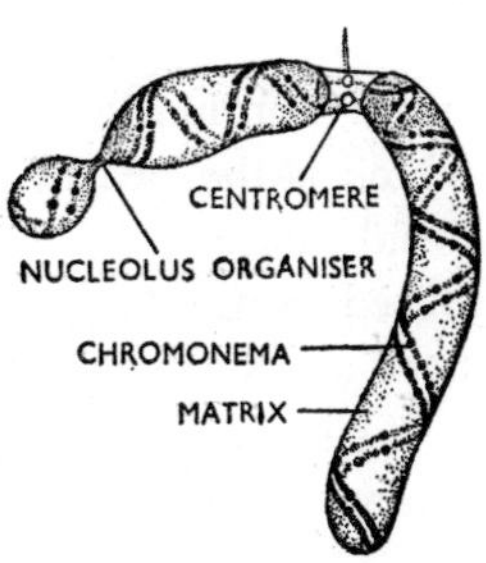

FIG. 522. A. sat-chromosome with nucleolus organiser. (*After Sharp*).

Telophase. During this phase (Fig. 520 G & H) the chromosomes reach the poles and two daughter nuclei are reconstructed. Having reached the poles the chromosomes become crowded together, their individuality being completely lost. The matrix loses its stainability and ultimately disappears. The crooked chromonemata become closely associated and anastomose to form the new reticulum. The nuclear membrane and the nucleolus reappear.

The relation of the nucleolus to the chromosomes is interesting. In the diploid nucleus there is at least one pair of peculiar chromosomes, having a distinct constriction at one end. Matrix does not collect round that point. The portion of the chromosome beyond the constriction, meaning the portion isolated from the main body due to pressure of constriction, is called *sattelite*; and the pair of such chromosomes are called *sat-chromosomes*. The constriction is known as the *nucleolus organiser* (Fig. 522), because it is connected with the formation of the nucleolus. Each sat-chromosome forms a nucleolus. The two nucleoli may fuse to form one, if the two happen to lie side by side, or they may remain separate.

CYTOKINESIS

The division of the nucleus, though the more important part of the process, is not the whole of it. It is accompanied by **cytokinesis** or division of the cytoplasm, ultimately ending in laying down of the cell wall between two protoplasts.

Cytokinesis usually takes place at the telophase stage when the daughter nuclei are reconstructed. The fibrils then extend outwards and touch the lateral walls, thus assuming barrel-shaped appearance (Fig. 520H). Protoplasmic materials now accumulate at the equatorial region of the spindle in form of droplets which coalesce in course of time and give rise to a plate, called **cell plate.** It soon undergoes physical and chemical changes and is transformed into the intercellular substance—middle lamella (Fig. 508). New cellulose wall is laid down by the protoplast on the middle lamella. In the mean time the fibrils become more and more faint and ultimately disappear.

In some cases cytokinesis also occurs by furrowing. Here instead of formation of cell plate, the plasma membrane undergoes furrowing inwards and the protoplast is ultimately separated into two parts.

DURATION AND PERIODICITY OF MITOSIS

Mitosis is confined to the cells of specialised regions called *meristems*, present in root-tip, stem-tip, leaf-primordia, etc. Once a nucleus has divided into two, the newly formed ones may have a resting stage or may again repeat the process in rapid succession. The stage between two mitosis is known as **interphase.**

Time required for mitotic process varies with different organisms, tissues and also with external factors like temperature. It has been found that in staminal hairs of *Rhoeo discolor* of family Commelinaceae mitosis takes 30 minutes at 45°C., 75 mins. at 25°C. and 135 mins. at 10°C. In the hair cells of the stigma of some grasses (*Arrhenatherum*) at 19°C. prophase occupies 36 to 40 mins., the metaphase 7 to 10 mins., the anaphase 15 to 20 mins. and the telophase 20 to 35 mins.; and the total 78 to 110 mins. excluding interphase.

The relative duration of the phase may be deduced from the data available. The prophase is the longest phase occupying something like 40% of the duration, next comes telophase with 32%, anaphase comes next with nearly 20%, and, lastly, the shortest phase, metaphase, with less than 10% of the total duration.

It has been stated that mitosis is the most common method of division in all vegetative or somatic cells. During this process, each chromosome, which is the bearer of hereditary units, is divided longitudinally into two exactly equal and identical halves; and the two sets consisting of the halved chromosomes, are distributed to the daughter nuclei. So the daugher nuclei formed are similar to each other, both qualitatively and quantitatively, and also to the nother nucleus. Somatic mitosis is thus **equational.** Since the nuclei are formed from pre-existing ones by mitosis and they have same chromosome complements, the somatic nuclei are all similar. The number of chromosomes thus remains constant in a given species of plant.

AMITOSIS

Amitosis or direct nuclear division is very simple, and at the same time, a rare process found in some lower organisms like bacteria, in the internodal cells of a green alga,

Chara. Here the nucleus undergoes constriction and ultimately divides into two parts, which are in all probability unequal (Fig. 523). There is nothing like chromosomes, their characteristic manoeuvres nor the spindle.

MEIOSIS

In all organisms with sexual method of reproduction the union of the two gametes results in the formation of the zygote. The zygote, which is the starting point of the new organism, has thus 'diploid' or '2*n*' chromosomes, the two sets being contributed by the two gametes, male and female. The two sets of chromosomes do not unite in the zygote. In fact, for each individual in the set of the male gamete, there is a corresponding one in the female gamete. The two matching chromosomes are usually not only alike in external characters, but also carry similar or matching hereditary units or *genes*. Such chromosomes are said to be *homologous*. The somatic cells develop from the zygote and thus possess '2*n*' or diploid chromosomes. But *haplosis* or *reduction* of the number of chromosomes does take place somewhere in the life cycle

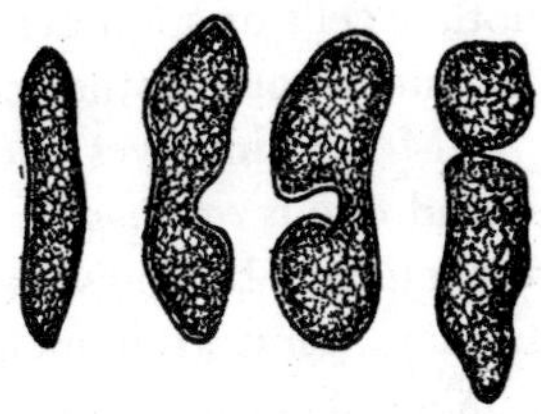

FIG. 523. Amitosis—stages in the internodal cells of *Chara*.

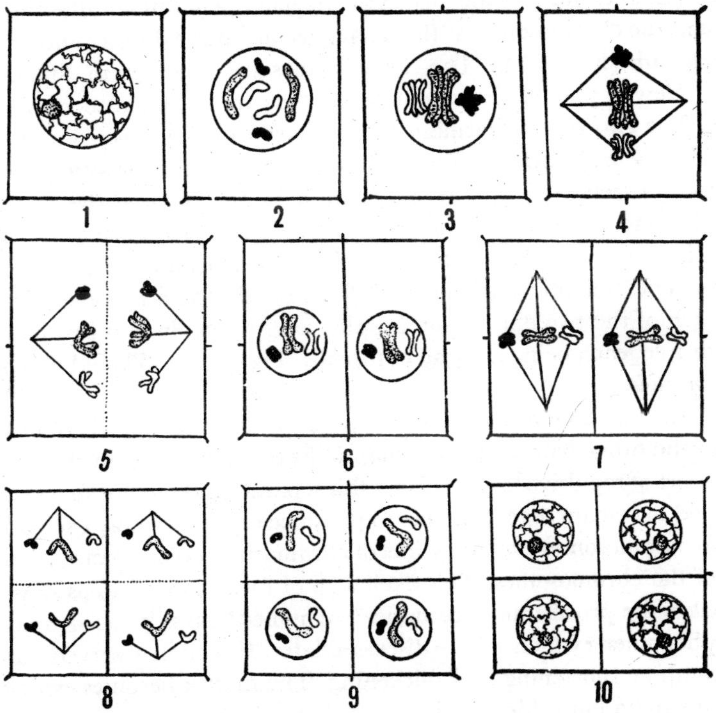

FIG. 524. Meiosis (diagrammatic). 1. Metabolic nucleus. 2. Prophase I. 3. Same, pairing of homologous chromosomes. 4. Metaphase I. 5. Anaphase I. 6. Daugheter nuclei reconstructed. 7 to 9. Division II. 10. Division complete with nuclei in daughter cells (spores).

when '2*n*' or diploid number is reduced to '*n*' or haploid number. In plants reduction or haplosis occurs before the formation of spores or gametes, so that the '2*n*' number may be restored during union of the gametes. Thus two *cardinal* points in the life cycle of plants with sexual method of reproduction are (*i*) gametic union, i.e., fertilisation or conjugation, when the chromosome number becomes diploid ($2n$); and (*ii*) haplosis or reduction of the diploid number into haploid (n) before formation of the spores.

Meiosis is the process of cell division in which the diploid number of chromosomes is reduced to haploid. It is a very complicated process restricted only to the reproductive cells. The cells in which meiosis is initiated are called *meiocytes.* In plants usually the spore mother cells or sporocytes are the meiocytes. In heterosporous plants the meiocytes are the microspore mother cells and megaspore mother cells.

Meiosis involves two divisions, of which the first division is *reductional,* whereas the second one is *equational.* The two divisions were formerly called *heterotypic* and *homotypic* respectively; but now they are referred to as first meiotic division and second meiotic division. Thus after meiosis four new cells are formed, each with reduced or haploid number of chromosomes (Fig. 524), in contrast to two cells with diploid chromosomes in mitosis. Like mitosis it also passes through four phases—prophase, metaphase, anaphase and, finally, telophase of the two divisions I and II.

Prophase I. The prophase as usual begins with the earliest recognisable changes in the nucleus. It is rather a lengthy phase involving a series of complicated changes. So it warrants further subdivisions into a number of stages. The changes occurring in the nucleus and the chromosomes are being discussed with reference to those stages.

Leptotene. The reticulum of the meiocytic nucleus opens out and the chromonemata, which represent the chromosomes at this stage, become more distinct and appear as long and slender threads (Fig. 525A). The chromosomes are obviously in diploid number.

Zygotene. Now a pair of chromosomes become intimately associated, but do not actually fuse. The two threads forming the pair are called *homologues* and the chromosomes are said to be in *bivalent* condition. This selective pairing or **synapsis** of homologous chromosomes seems to begin at one or more points, often at the centromeres and proceed side by side as if drawn by a zipper till it is complete, when they twist or coil round each other. Now the threads become gradually shorter and thicker (Fig. 525B).

Pachytene. The synaptic chromosomes forming the pair become more thick and short. Now each of the two chromosomes splits longitudinally into two halves. So they are now quadruple, each being a tetrad of threads (Fig. 525C). These half-chromosomes are called *chromatids.*

Diplotene. At this stage a tendency of separation is noticed in the four chromatids, one pair from the other pair, as if the synaptic force which brought them near is being replaced by a sort of repulsive force. During this separation there are one or more points where they tend to remain together. Due to presence of these points, known as *chiasmata,* where the two chromosomes exchange bits of chromatids, a disturbance is noticed in the dissociation of the chromosomes. This exchange of bits, referred to as *crossing-over,* undoubtedly makes the process more complicated; none-the-less it has significant bearing on genetical phenomena which will be discussed later. It is often observed that the paired chromatids continue shortening and thickening. The matrix becomes evident now with coiled chromonemata embedded in the same (Fig. 525D).

Diakinesis. Now the compact chromatid tetrads are well-distributed in the nucleus, their individuality becoming more clear due to abundance of matrix. This is an ideal stage for counting the chromosome number when they assume different shapes depending on the chiasmata (Fig. 525E). There prophase comes to an end.

Metaphase I. In metaphase the nuclear membrane and nucleolus disappear and the achromatic figure (spindle) is formed. The chromatids in tetrads arrange themselves at the equator, the centromeres facing the poles (Fig. 525F).

Anaphase I. Anaphasic movements begin now. Two chromatids of each tetrad, forming a diad, are now separated from the other two and move towards opposite poles. This separation is referred to as **disjunction.** The two chromatids forming pairs are called **diads.** Some resistance is offered at the chiasmata, but ultimately the diads become

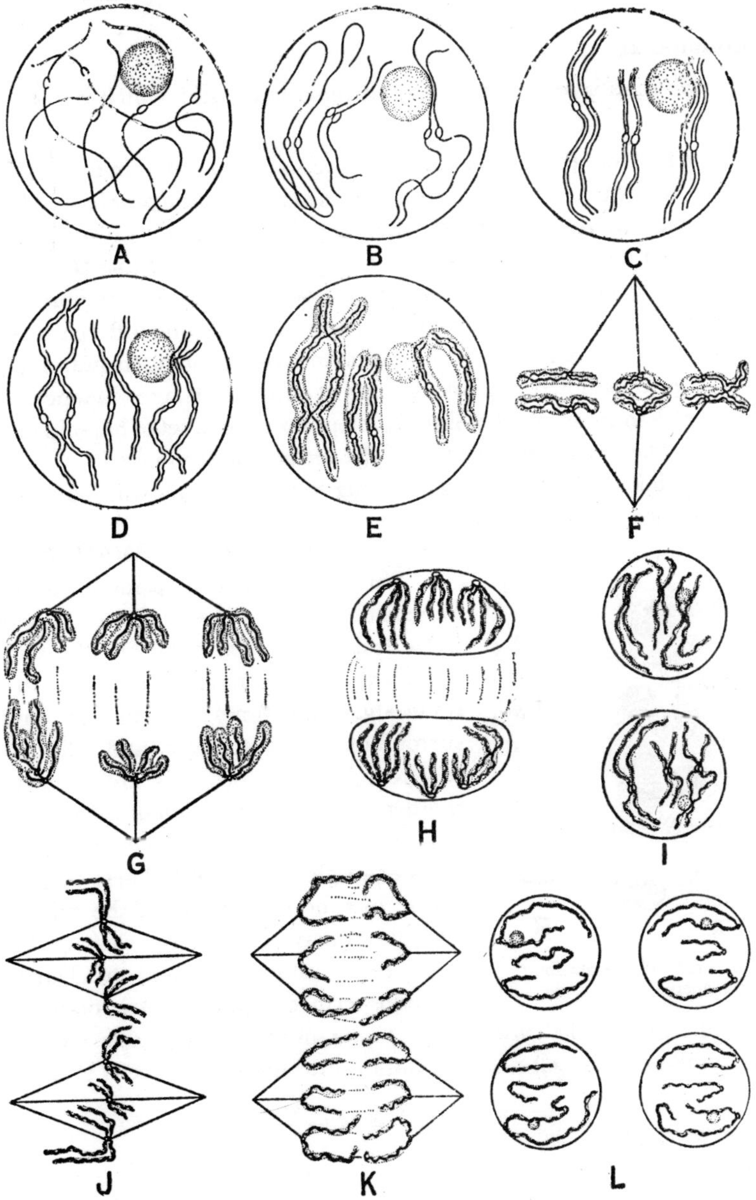

FIG. 525. Meiosis—stages (diagrammatic). A to E. Prophase I. A. Leptotene. B. Zygotene. C. Pachytene. D. Diplotene. E. Diakinesis. F. Metaphase I. G. Anaphase I. H. Telophase I. I. Prophase II. J. Metaphase II. K. Anaphase II. L. Telophase II. (*After Sharp.*)

free (Fig. 525G). This stage is significant, in view of the fact that the two homologues brought together in synapsis are separated now.

Telophase I. Now the chromosomes undergo telophasic changes resembling the same in mitosis (Fig. 525H). Cytokinesis does not always occur immediately. The two newly reorganised nuclei now pass through division II. The interval between the two divisions is called *interkinesis*, the duration of which is quite variable. The two nuclei are fairly large at this stage and the chromatids occur as slender threads.

Meiotic division II

The first division is followed by second meiotic division which is *equational* and *not reductional*.

Prophase II. Here the chromosomes still remain associated in diads (Fig. 525I), in contrast to the closely parallel split chromosomes of the somatic prophase. Matrix appears now. It is followed by **metaphase II** when achromatic figure is formed and the chromosomes arrange themselves at the equator (Fig. 525J). **Anaphase II** involves moving apart of the two chromatids of the *diad* towards two poles along the fibres (Fig. 525K). The centromeres, as usual, move ahead and chromatids assume different shapes. During **telophase II** the chromatids reach the poles and are reorganised into daughter nuclei (Fig. 525L). It is to be noted that four groups of chromatids, now independent chromosomes, reconstruct four new nuclei. Each nucleus possesses a set of haploid chromosomes; each set having one chromatid of each of the tetrads of prophase I.

Cytokinesis follows by furrowing, forming four cells, each with haploid or reduced number of chromosomes in the nucleus.

FREE CELL FORMATION

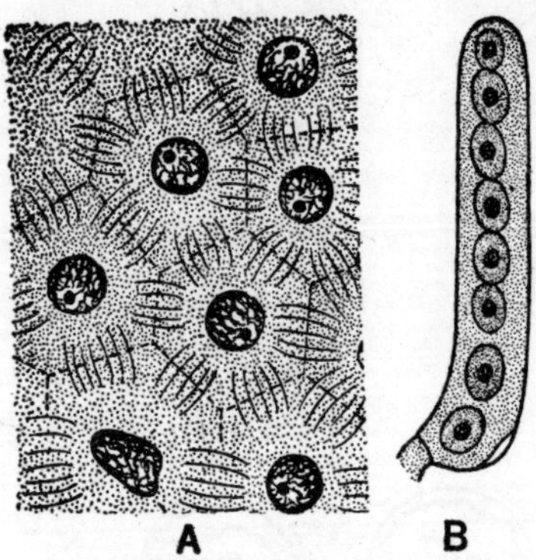

Fig. 526. Free cell formation A. In endosperm. B. In ascus.

This process is really a modification of mitosis. Here the nucleus of the mother cell divides into two by mitosis, those two again divide, and so on, till a good number of nuclei are formed in the cell. In some algae and fungi each nucleus thus formed is surrounded by cytoplasm and ultimately wall is secreted. Thus a few cells are formed within the mother cell. The formation of ascospores within the ascus in the group Ascomycetes of fungi may be cited as an example (Fig. 526B). These cells are set free in course of time by rupture of the original cell wall. Cytoplasm may also undergo cleavage forming a few naked cells within the original one. During formation of endosperm of many higher plants and embryo cells of Gymnosperms like cycad and *Ephedra* multinucleate mass (coenocytic stage) undergoes cytokinesis by cell plate formation into uninucleate cells (Fig. 526A).

CHAPTER XX

THE TISSUES

Lower plants are composed of either single cells or colonies of cells which are more or less similar in size, shape and origin. Utmost differentiation that is noticed is that some of them constitute vegetative parts and others reproductive. There exists a gradual ascent in the complexity of structure in the plant kingdom. In fact, higher plants are built up of diverse types of cells, varied in shape, size, origin, and function. A group of cells more or less alike in size and shape, having same origin, same methods of development and same function, constitute a **tissue.** Tissues are really formed in response to a basic division of labour. Every tissue has a particular function to perform and all the tissues work harmoniously in the life of the organism. This definition is rather loose, particularly so, in case of higher plants with cellular complexities. In fact, distinctly dissimilar cells are often found to be associated in the formation of a tissue, as for example, the complex tissue, xylem and phloem.

There are two viewpoints as regards the conception of a tissue. From the point of view of general anatomy cells constituting a tissue are contiguous and form a structural part of a plant. On the other hand, from physiological considerations even widely separated cells may form a tissue. Xylem consists of conducting elements, supporting elements and storage elements; all of them are contiguous and form the complex tissue. But from the point of view of physiological anatomy the conducting elements would form the 'wood', supporting cells would belong to mechanical tissue and the storage cells to a third tissue.

Tissues may be classified on the basis of different characters, viz., nature of cells, origin and methods of development, position in the plant body, etc. However, the tissues are put into two main groups—(*i*) **meristematic** and (ii) **permanent,** on the basis of stages of development.

MERISTEMATIC TISSUE

Meristematic tissue, commonly called *meristem*, is composed of cells which are immature, not fully differentiated ones, and which possess the power of cell division. The vascular plants exhibit an 'open system' of growth involving formation of new tissues and organs throughout its life. It is possible due to presence of certain organised regions where the cells are in perpetually immature condition and can form new cells throughout. These organised regions are the meristems. Primitive plants have cells all similar, carrying on different functions. With advance in the line of evolution growth has been restricted to certain parts of the plant body, the function of cell division being confined to the meristematic cells. The open system of growth in plants is in marked contrast to the development of an animal body.

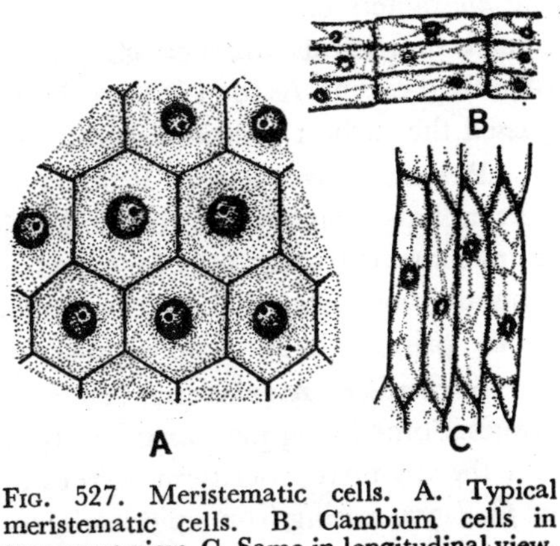

FIG. 527. Meristematic cells. A. Typical meristematic cells. B. Cambium cells in transverse view. C. Same in longitudinal view.

The meristematic cells have generally some distinctive features (Fig. 527). The cells are usually isodiametric in shape; they are compactly set without evident intercellular spaces; have dense cytoplasm with small vacuoles and large prominent nuclei; ergastic matters are lacking; the plastids are in proplastid stages; cell wall, made of cellulose, is thin and homogeneous. Though in general, meristematic cells possess the characters stated above, but departures cannot be ruled out. The meristematic cells of vascular cambium are fusiform in shape (Fig. 527 B & C) often with conspicuous vacuoles. Comparatively thick walls with primary pit fields are found in some meristematic cells. Ergastic matters like starch and tannins may also be present. Considering these departures some authors are in favour of using the term *eumeristem* or true meristem for those having the general characters stated above.

It has been stated that meristems are the formative regions where new cells are added to the body. Besides, the meristems perpetuate themselves. The cells which remain meristematic and thus continue cell division are called *initiating cells;* whereas the cells formed by them, the so-called *derivatives*, gradually change their shape, enlarge, lose the power of cell division and ultimately become mature cells with some definite characters and functions. These changes involving enlargement and specialisation represent the process which may be referred to as *differentiation.* The differentiation of a particular tissue has been characterised as "a progressive loss of embryonic features of meristematic cells and the progressive attainment of the state of maturity" by a well-known authority. The derivatives of the initiating cells gradually differentiate into mature cells and lose the power of cell division, at least temporarily. These completely differentiated cells are called **permanent** ones. Some permanent cells may get back the meristematic nature, a phenomenon which has been referred to as *dedifferentiation* by some workers. That shows that though cells have attained permanent form, but they have retained the power of cell division. The living parenchyma cells and the epidermal cells are common reversible permanent cells. In fact, all living cells retain the potentialities of division and growth, though they have become permanent. Strictly speaking, permanent cells are those which have completely lost the capacity of division, as for example, irreversibly specialised cells like the sieve tube elements and the dead elements like tracheids and cork cells.

Various systems of classification of meristems have been proposed, based on characters like origin and nature of initiating cells, stages of development, topography and function. No system is exclusive and rigid. The common important types of meristems according to their origin and development are the following:

Promeristem or Primordial meristem. Promeristem is the very foundation stage, the region of formation of new organs and tissues. It may be called the earliest embryonic condition consisting of young initials and their derivatives. All the cells possess the characters of true meristematic cells, viz., diameters alike, dense cytoplasm, large nuclei, proplastids, absence of ergastic matters and thin walls. Promeristem is definitely of limited extent. As soon as the cells begin to show tendency of differentiation, they have passed the earliest promeristematic condition.

Primary meristem. Primary meristems are composed of cells which are direct descendants of embryonic cells and which have all through retained meristematic nature. So primary meristem may be called a later developmental stage. Chief primary meristems are those located at the tips of stem, root, and appendages. Primary meristems build up the primary body of the plants.

Secondary meristem. Primary meristems gradually differentiate into permanent tissues. Some living permanent cells may regain the power of cell division. They constitute the secondary meristem, as they originate from permanent cells. The cork cambium or *phellogen*, arising from epidermis, cortical and other cells during secondary increase in thickness, is an example of secondary meristem.

It is to be noted that primary meristems are responsible for the building up of the primary body of the plant; and the secondary meristems, formed later, add new cells to the primary body with definite purposes like effective protection and repair. The definitions given above are not always accurate. In case of adventitious organs the apical meristem develops within permanent tissues secondarily though they are primary in structure and function.

According to their position in the plant body meristems are put into three groups, viz., **apical, intercalary** and **lateral** (Fig. 528).

Apical meristem. Apical meristems occur at the apices of the stems, roots, main and lateral, of the vascular plants. Growth in length of the axis is entirely due to their activities; so they are also called *growing points.* The initiating cells may be solitary or in groups. Solitary initiating cells go by the name *apical cells,* whereas those occurring in groups are called *apical initials.* Solitary apical cells (Figs. 530 & 533A) occur in many pteridophytes like ferns and horse-tails. Other vascular plants possess apical initials which may be terminal or terminal and subterminal (Figs. 532 & 533). The apical initials may occur in one or more tiers. In case of one tier all the cells of the plant body derive their origin from it; whereas if there are more than one tier different parts of the plant body originate from different tiers of initials.

The derivatives of the apical meristem differentiate in course of time into permanent tissues which together constitute the primary body of the plant.

Intercalary meristem. These are the portions of apical meristems which are separated from the apex during the growth of the axis and remain intercalated between permanent cells. The position is such that apical meristems go ahead during development and a portion of it is left behind which is inserted between permanent cells.

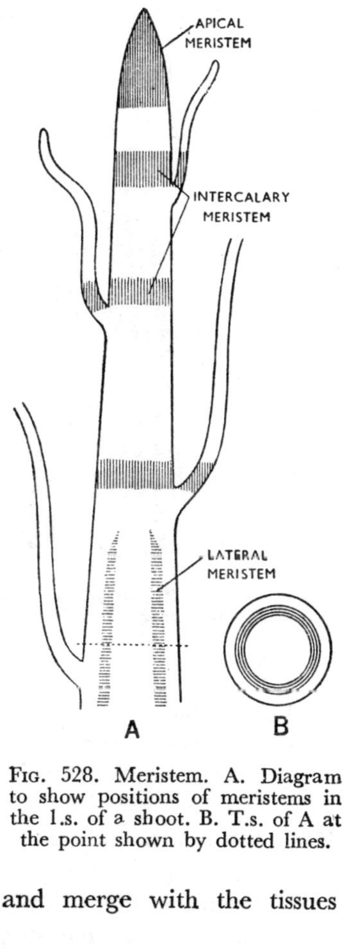

FIG. 528. Meristem. A. Diagram to show positions of meristems in the l.s. of a shoot. B. T.s. of A at the point shown by dotted lines.

Intercalary meristems are found in the stem and leaf sheaths of many monocotyledons, particularly grasses, and in the horse-tails. Here nodal regions are usually composed of permanent cells and so intercalary meristems are internodal. These meristems are short-lived, they very soon become permanent and merge with the tissues surrounding them.

Lateral meristem. These meristems occur laterally in the axis, parallel to the sides of the organs in dicotyledons and gymnosperms. They are composed of initials which divide periclinally. The derivatives gradually differentiate into permanent tissues called **secondary tissues.** These tissues are added to the existing ones and are responsible for increase in thickness. The growth in thickness thus secured due to addition of secondary tissues is referred to as secondary growth. The cambium of the vascular bundles and the phellogen or cork cambium are lateral meristems. It should be noted, however, that cambium is a primary meristems, whereas phellogen or cork cambium is secondary in origin.

A classification of meristems chiefly based on function was followed in physiological anatomy. Haberlandt in 1890 suggested that primary meristem at the apex of the stem and root is segregated into three tissues, viz., **protoderm, procambium** and **ground meristem** (Fig. 529). The protoderm develops into epidermis, the procambium into

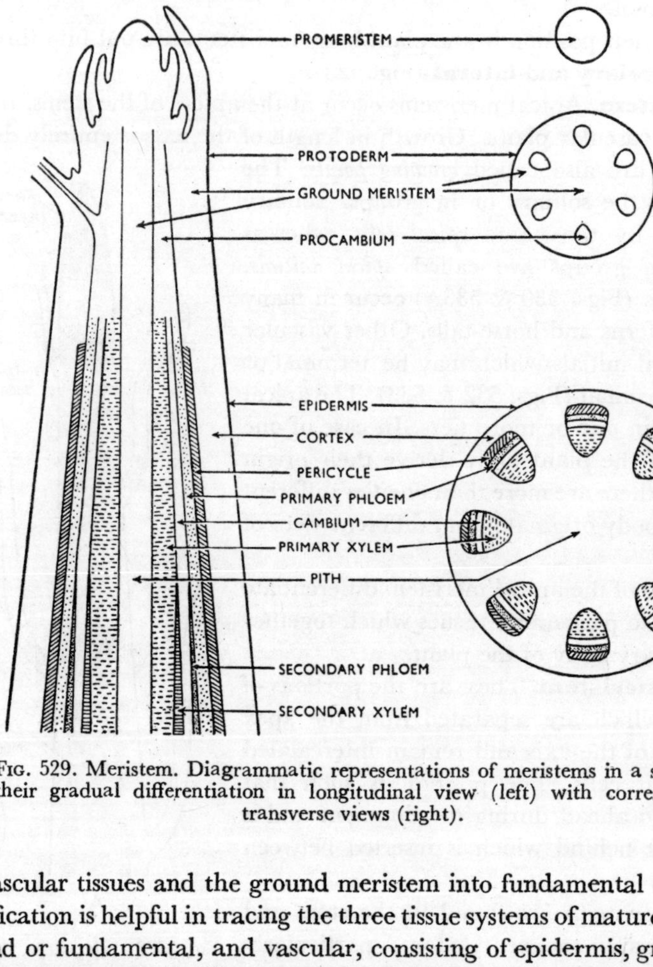

FIG. 529. Meristem. Diagrammatic representations of meristems in a stem and their gradual differentiation in longitudinal view (left) with corresponding transverse views (right).

primary vascular tissues and the ground meristem into fundamental or ground tissues. This classification is helpful in tracing the three tissue systems of mature regions—epidermal, ground or fundamental, and vascular, consisting of epidermis, ground tissues and vascular tissues respectively. The advantage of this set of terminology from topographical point of view is undeniable, though the interpretation of development of shoot and root from these zones is a quite complex problem.

Meristems on the basis of *planes of divisions* are of three types, viz., *mass* meristem, *rib* meristem, and *plate* meristem. Mass meristems grow by dividing in all planes, so that the bodies formed are either isodiametric or have no definite shape. This pattern of growth is noticed in endosperm, young embryo and also in the formation of spores and sperms. The rib meristems, also called *file* meristems, divide anticlinally to the long axis and give rise to longitudinal files or rows of cells. This growth pattern is clear in the development of cortex and pith. The plate meristems divide chiefly anticlinally in two planes, so that new cells are formed but number of layers does not increase. The uniseriate epidermis and multiseriate flat blade of the leaf are illustrations of growth forms by plate meristems.

THEORIES OF STRUCTURAL DEVELOPMENT AND DIFFERENTIATION

The classical concept of the apices of shoot and root had been that they represent the earliest self-perpetuating promeristems consisting of homogeneous eumeristematic or truly meristematic cells. Modern workers are of opinion that a distinct zonation is exhibited by the immediate derivatives of the promeristem. By zonation is meant the existence of distinct regions differing from one another by characters like nature of cells, plane of cell division, position of the initiating cells and rate of maturation of cells. The growth and development of the apical meristems of shoot, root, and flower and their differentiation have been most notable problems of plant anatomy. Intensive studies on those problems have been going on since the nineteenth century and a few theories have been proposed from time to time. A brief review of those theories is being given here.

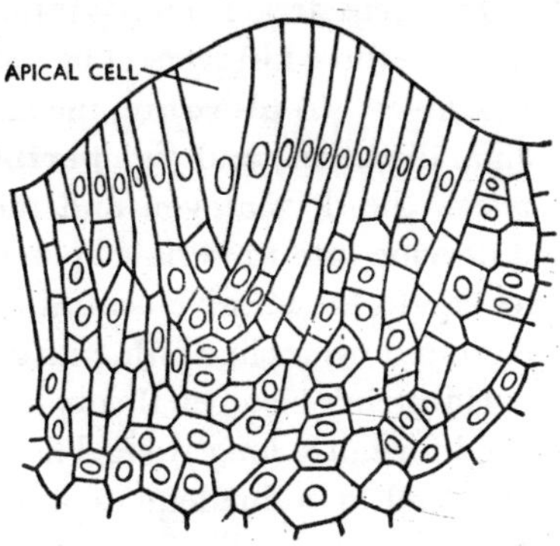

FIG. 530. Meristem. Apical cell in the apex of rhizome of *Pteridium* (longitudinal).

Apical cell theory. The solitary apical cells (Figs. 530 & 533A) were discovered in the cryptogams—in algae, bryophytes and pteridophytes. This led early workers to believe that a solitary apical cell was a constant unit of apical meristem governing the whole process of growth. They assumed that the same condition prevails in all higher plants. On that assumption **apical cell theory** was advanced by Hofmeister in 1857 and supported by Nageli (1878) in the nineteenth century. Subsequent works revealed that complex apices of gymnosperms could not be interpreted in the light of this theory. So it was not applicable to seed plants.

Histogen theory. The older theory was replaced by **histogen theory** as an attempt towards interpretation of the growing points of seed plants. Hanstein proposed

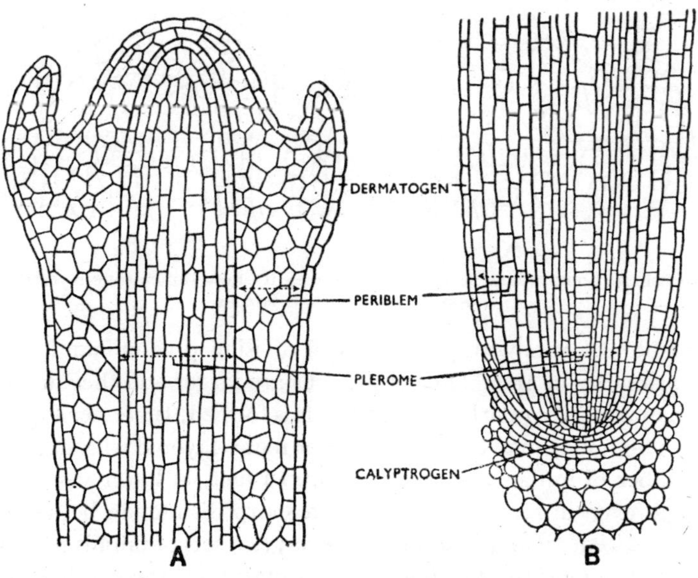

FIG. 531. Meristem. Histogens in the shoot apex (A) and root apex (B) [longitudinal sections].

this theory in the nineteenth century and received support from Strasburger (1868). According to this theory plant body develops from groups of initials forming a mass of meristem of considerable depth, as opposed to superficial apical cells of former theory, and that three distinct zones or strata can be recognised at the growing points of stem and root. Every zone consisting of a group of initials was called a *histogen* or tissue builder. The histogens arise from separate sets of initials and have different courses of development. The three histogens (Fig. 531) suggested were (*i*) **dermatogen,** the outermost uniseriate layer; (*ii*) **plerome,** the massive central core consisting of cells extended in longitudinal direction; and (*iii*) **periblem,** the region composed of isodiametric cells lying between the dermatogen and the plerome. The dermatogen cells divide anticlinally and develop into uniseriate epidermis. The periblem forms the cortex; and the plerome gives rise to the massive central cylinder or **stele,** consisting of primary vascular tissues and ground tissues like pith, pith rays and pericycle.

Hanstein's histogen theory as a basis of interpretation of the shoot and root apex held ground for fairly long time. Though it is followed even now in case of root apex, it has been found to be inadequate as regards shoot apex of angiosperms mainly for two reasons, viz. (*i*) sharp distinction between dermatogen and periblem is absent; and (*ii*) origin of different regions from the sharply defined histogens cannot be demonstrated. Moreover in many gymnosperms and angiosperms shoot apices hardly show any distinction between periblem and plerome.

Tunica-corpus theory. As an interpretation of the apical growth in the shoot apex the third theory—**tunica-corpus theory**—was proposed by Schmidt and supported by Foster and others in the early part of the twentieth century. According to this theory, two tissue zones occur at the apex. They are the **tunica** or cover, consisting of one or more layers of cells forming the outer enveloping region, and the **corpus,** the central core, a mass of cells surrounded by tunica (Fig. 532). The two regions differ in structure and appearance due to varying rates and methods of growth. The cells of tunica

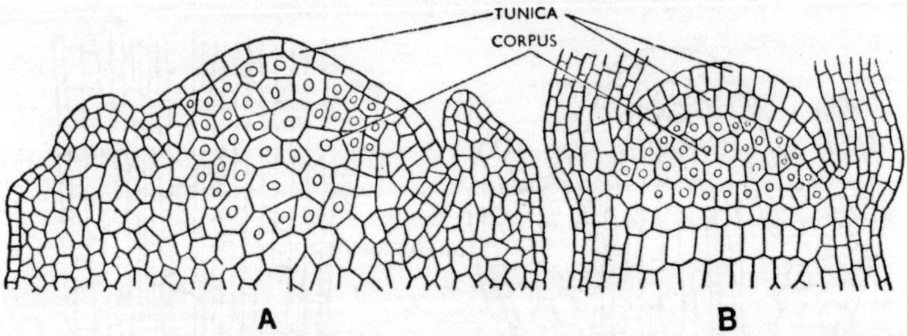

FIG. 532. Meristem. Longitudinal section of shoot apex to show tunica (one-layered in A and two-layered in B) and corpus.

are smaller and lie in layers with planes of cell division predominantly anticlinal. So growth here is primarily in area. The corpus cells are larger and they divide in various planes, so that a mass of irregularly arranged cells is formed. Increase of the mass here is thus in volume. The tunica may be one-layered or many-layered with massive or slender corpus. The number of initials are also variable. In fact, in lower vascular plants like pteridophytes and even in some gymnosperms a sharp distinction between tunica and corpus cannot be traced. In angiosperms, however, two sets of initials may be definitely located, the two sets giving rise to tunica and corpus. Thus the two have independent origin.

In spite of the fact that many variations and fluctuations occur, the tunica-corpus theory has given a renewed impetus in the interpretation of the apical growth in the stems of angiosperms. As a result, the stages in the development of the primary body from the initiating cells is better understood.

THE SHOOT APEX

The terminal part of the shoot with the leaf primordia is the shoot apex where primary organisation of the shoot is initiated. In size and shape it varies enormously, but in a general way it can be said that the shoot apex is more or less convex in longitudinal section. The apical meristem widens considerably before the initiation of the leaf and it again becomes narrow after the appearance of the leaf primordium, thus exhibiting a rhythmic phenomenon. The term *plastochron* has been used for the period between the initiation of two successive leaves. The shoot apex is usually small in angiosperms, but they are much wider in gymnosperms like *Gingko* and *Cycas*.

In Pteridophyta (Fig. 532A—1 & 2) one (apical cell) or more (apical initials) are easily distinguishable from others and they give rise to all the cells of the apex. The apical cell, usually pyramidal in shape, as found in Psilotales, *Equisetum* and leptosporangiate ferns divide to produce new cells on all sides, whereas the eusporangiate ferns possess two to four apical initials. It is believed that ferns with apical cells are more advanced from evolutionary point of view than those with apical initials.

In Gymnosperms (Fig. 532A—3 & 4) the cells in the surface of the apex divide both anticlinally and periclinally. As a result an initiating zone of the entire apex, known as *surface meristem*, is formed which gives rise to the epidermis and other meristems. Due to periclinal divisions a subsurface zone of *central mother cells* develop where the cells have large nuclei, prominent vacuoles and often thick wall. The rate of cell division is more pronounced on the peripheral side of this zone, and the derivatives together with those formed by anticlinal divisions of the apical initials constitute the peripheral zone or *flank meristem*. This zone produces the cortex, the procambium and the leaf primordia. The basal part of the subsurface region develops into a zone of *rib meristem* consisting of vertical series of cells dividing in all planes. Pith originates from this zone. Three types of gymnospermic apices (Popham, 1952) have been recognised which are nothing but differences in detail in an organisation that is basically uniform. In one of the types—the Gingko type another meristematic zone, a narrow transitional one has been found to

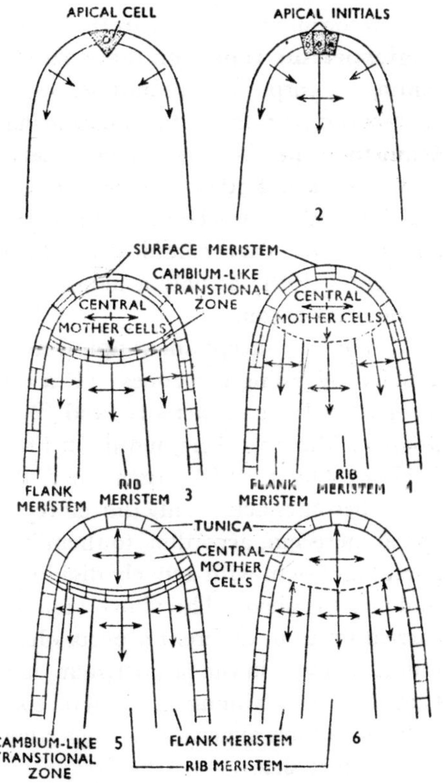

Fig. 532A. Shoot apics (diagrammatic). 1. A pteridaphyta with a single apical cell; 2. Same with apical initials; 3. *Gingko* type; 4. *Cryptomeria-Abies* types; 5. *Opuntia* type; 6. Usual angiospermic type. (*After Popham*).

occur below the central mother cells. This cup-shaped portion called *cambium-like transitional zone* consists of cells which undergo rapid periclinal division. Unlike other gymnosperms the Gnetales exhibit a definite separation into a surface layer and inner core derived from the initials, which correspond to the tunica and corpus of the angiosperms, and thus suggesting the most advanced type of apical organisation amongst gymnosperms.

As already reported the shoot apex of **angiosperms** has been the subject of intensive investigation for pretty long time, and a number of theories (discussed) have been enunciated regarding the growth and differentiation of the same.

The generally accepted concept, tunica-corpus theory, demands that the tunica with one or few layers of cells forms an envelop round the centrally-located initials of the corpus. The number of layers of tunica may vary (one to nine) in the same family, genus and even in the same plant at different stages of growth. Two types of cells have been recognised in the tunica. One or few initials forming a central apical zone having larger size, more prominent nuclei and conspicuous vacuoles. The cells of this zone take light strain. The second zone consisting of comparatively smaller and deeply staining cells occur on the sides of the apex between the initials and the leaf primordium.

According to internal arrangement, the corpus may be of two types—(1) the **usual angiospermic type** with three main zones, viz., (*a*) a zone of *central mother cells* representing the corpus initials, just below the apical portion of the tunica, (*b*) *rib meristem*, and (*c*) *flank meristem*, the last two appearing as continuations of central mother cells; (2) the **Opuntia type,** in which a cup-shaped cambium-like transitional zone between the mother cells and other meristems have been recognised (Fig. 532A—5 & 6) This zone which is always associated with leaf primordium differs from the rest in that the height and diameter vary considerably during the plastochron. It is considered by some (Philipson, 1954) as only a temporary feature, because it disappears towards the end of the plastochron.

Another concept about apical meristem developed on the basis of the rate of mitotic activity suggesting two zones—(1) a central apical zone with tunica and corpus initials having less frequent activity, and (2) a peripheral zone with pronounced mitotic cell division. Many workers, mostly in France and hence called 'French school' (Plantefol, 1947-48; Buvat, 1952) maintained that an inactive zone exists at the shoot apex, which becomes active only during reproductive stage. The vacuolated distal part of the apical meristem, according to them is of no significance in vegetative growth, the peripheral meristem with actively dividing cell and the rib meristem play the main roles in building the shoot. The above view was seriously challenged and investigations were carried on from different aspects—histological (Popham, 1958) asserting considerable mitotic activity in the promeristem, in colcichine-induced cytochimeras (Dermen, 1951; Clowes, 1958) claiming ontogenetic continuity between promeristem and mature tissues, in tissue culture experiments suggesting the necessity of promeristem for the restoration of the entire shoot.

THE ROOT APEX

In comparison to the stem apex the apical meristem of the root is simpler, because of the absence of nodes and internodes and lateral appendages. But it has a protective cap, which acts as the buffer between the root-tip and the soil particles. As cap occupies the terminal position, the apical meristem is subterminal here. Curiously enough, growth in the root-tip proceeds in two directions opposite to each other—towards the tip in the cap and away from the tip in the root proper.

Though histogen theory has been practically discarded in case of stem apex, it is followed in interpreting the structure and growth of root apex. The apical meristem here is rather short. Considerable variations exist in the relation between the cap and the tip. In fact, root apices are of a few types depending on the mode of origin of cap and relations between histogens and primary tissue regions of the root proper.

In vascular cryptogams a solitary cell (Fig. 533A) occurs at the root apex. In course of time the root proper and the cap are formed from the apical cell, though the cap and the root become structurally quite distinct.

Two groups of initials occur (Fig. 533B) at the apex of roots of many gymnosperms. Of the two the inner group forms the plerome, and the outer forms the periblem and the cap. Unlike other groups dermatogen does not occur at the apex. It is formed from the periblem a bit away from the apex as a 'distal proliferation'. The outstanding features here are the formation of the cap from the cortex and at the same time outer layer of cortex becoming the epidermis—thus both cap and epidermis from the cortex.

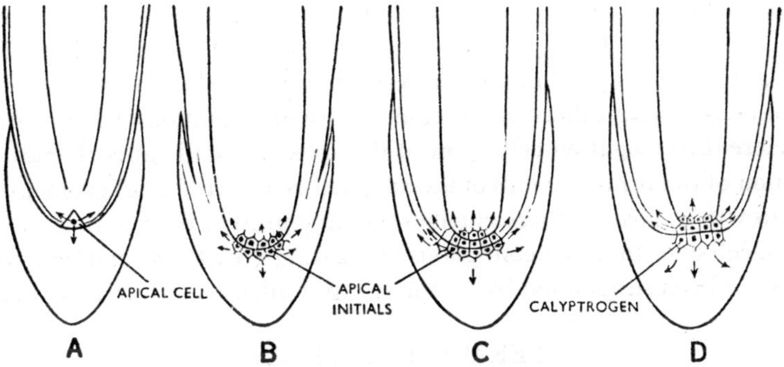

Fig. 533. Root apices (diagrammatic)—A. Type found in pteridophytes with solitary apical cell. B. As found in gymnosperms. C. As found in dicotyledons. D. As found in monocotyledons.

In angiosperms three groups of initials usually occur at the root apex. In dicotyledons (Fig. 533C), of the three groups, the terminal one forms the dermatogen and the cap; the median one forms the periblem; and the innermost one gives rise to the plerome. So here the cap and dermatogen have common origin. Thus cap may be considered a specialised development of the epidermis. In monocotyledons (Fig. 533D), the outermost group of initials produces the cap; the medium one, the dermatogen and periblem; and the innermost one, the plerome. The group of initials forming dermatogen and periblem is usually one-layered, but it may be two or three layers in thickness. Independent origin of the cap is a notable feature. The histogen concerned with the formation of the cap is referred to as **calyptrogen.**

The lateral and adventitious roots show same type of organisation.

In recent years many investigators (Jensen, Clowes, and others) have claimed that a zone of low mitotic activity composed of the cells of the central part of the promeristem occur. It has been called *quiscent centre*. A theory—*Körper-Kappe theory*, more or less similar to the tunica-corpus of the shoot apex, was enunciated on the basis of planes of cell division. According to this theory there are two regions, outer (Kappe) and inner (Körper), and the cells divide in a pattern known as T-division. Kappe cells first divide horizontally and the derivatives divide at right angles to the plane of first division—the planes of two thus forming a T. In the inner region Körper T is inverted as the second division takes place in the upper daugher cells.

GROWTH PATTERNS

Differentiation involves some striking changes in the characteristics of the individual cells, viz., enlargement accompanied by increase of vacuolar sap, collection of ergastic matters, development of plastids with colours from the proplastids, disappearance of protoplasts in extreme cases, and growth and chemical changes of the cell wall. The increase in the size of the cell may be more or less uniform in all directions or due to unequal rate of enlargement in different directions the cells may appear distinctly different from the meristematic cells from which they have derived their origin.

The cellular arrangement of a tissue is really determined by the growth pattern of its meristem. The increase and changes in the shape of the cells in a differentiating tissue bring about some adjustments in their relation on each other. The formation of intercellular spaces, either small ones at the corners or fairly large ones changing the very appearance of the tissue is quite common. In expanding organs mutual adjustment occurs due to growth of the entire walls of the cells in a co-ordinated manner without involving separation of the walls. This type, called *symplastic* growth, is noticed in the formation of the derivatives of the apical meristem. A second type, called *intrusive* growth, involving intrusion of cells among others is common in case of formation of fibres, tracheids and laticifers. Here the elongating cells grow at the apex, normally at both ends, resulting in the separation of the primary walls of the adjacent cells. Early workers mentioned *gliding* or sliding growth as basis of these changes. Gliding growth used to mean the slipping of the wall of a cell over that of a contiguous one during growth, so that new contacts were made. This has been supplanted by symplastic and intrusive growth mentioned above.

PERMANENT TISSUES

It has been stated in the preceding pages that meristematic cells gradually differentiate and become mature or **permanent.** Permanent tissue is thus composed of cells in which growth has stopped. Of course, all living cells retain the potentialities of further growth. Classification of permanent tissue has been a perplexing problem. Mainly on the basis of constituent cells, permanent tissues are put into two groups—**simple** and **complex.** Simple tissues are composed of similar cells, and so they are *homogeneous* in nature. The complex tissues are *heterogeneous*, being made up of different types of cell elements. There is a third group of permanent tissue, called **secretory** or special tissues. Though this classification is widely followed, it has some demerits. In higher plants, tissues are very seldom homogeneous from morphological or physiological point of view.

Main simple permanent tissues of plants are *parenchyma*, *collenchyma* and *sclerenchyma*.

PARENCHYMA

Parenchyma (Fig. 534) is the most common simple tissue of the plants with relatively little specialisation. It is composed of cells which are usually isodiametric in shape with intercellular spaces. The cells have active protoplast. This tissue is universally distributed in all the plants, the softer portions like epidermis, cortex, pith, pericycle, whole or part, of stems and roots, mesophyll of leaves, pulp of the fleshy fruits, embryo and endosperm of the seeds being composed of parenchyma cells. It is called the fundamental tissue of the plant, because it really constitutes the ground substance where other tissues remain embedded. Bodies of lower plants are made of parenchyma cells. The meristematic cells are also parenchymatous in nature. Thus parenchyma is the precursor of all other tissues. So, it is considered to be the most primitive tissue, both phylogenetically and ontogenetically.

Parenchyma occurs as homogeneous mass in many portions, but it may also be associated with other elements in complex tissues like xylem and phloem. Normally parenchyma cells are polyhedral in shape with profuse intercellular spaces. Unspecialised cells may approach 14-sided tetrakaidecahedron in shape (Fig. 485). In the mesophyll of the leaves they are slightly elongated. Irregular shapes as a result of folding (Fig. 534D), lobation, etc., are also not uncommon.

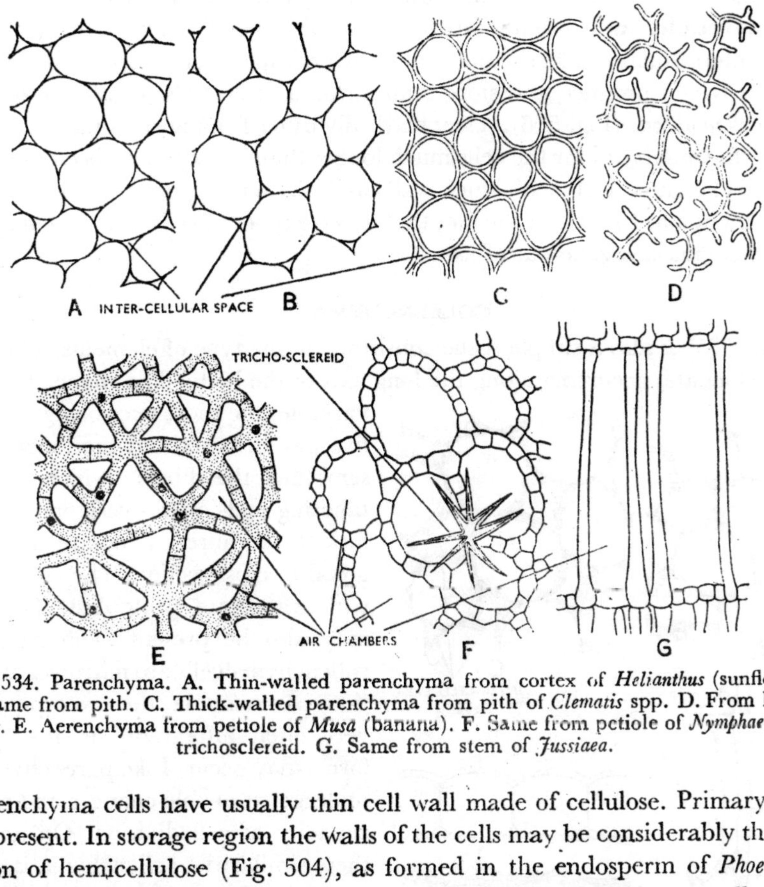

Fig. 534. Parenchyma. A. Thin-walled parenchyma from cortex of *Helianthus* (sunflower). B. Same from pith. C. Thick-walled parenchyma from pith of *Clematis* spp. D. From leaf of *Pinus*. E. Aerenchyma from petiole of *Musa* (banana). F. Same from petiole of *Nymphaea* with trichosclereid. G. Same from stem of *Jussiaea*.

Parenchyma cells have usually thin cell wall made of cellulose. Primary pit fields may be present. In storage region the walls of the cells may be considerably thick due to deposition of hemicellulose (Fig. 504), as formed in the endosperm of *Phoenix* (date-palm). Often thick and lignified walls are present in the parenchyma cells of xylem, particularly secondary xylem.

From the point of view of function it is a very important tissue. Due to presence of active protoplast this tissue is the seat of all essential vegetative functions like photosynthesis, storage of food matters, secretion and excretion. Parenchyma cells occurring in xylem and phloem help in the conduction of water and food matters in solution. Parenchyma with thin cellulose wall can also serve as a supporting tissue due to turgid condition of the cells, what is particularly evident in herbaceous plants. Epidermal cells with cutinised outer walls have protective function. Parenchyma cells retain the potentialities of cell division. Secondary meristems usually originate in this type of cells. Thus they are concerned in increase in thickness and also in healing of wounds and formation of adventitious roots and buds. In some xerophytic plants they are specially adapted for storage of water. Water-storing parenchyma cells are large with prominent vacuoles, scanty chloroplasts and thin cell wall. Mucilaginous matters are present in the cell sap which increase water-holding capacity.

The parenchymatous cells of leaves and sometimes other organs like stem contain enough chlorophyll. These cells with that all-important function photosynthesis are also called **chlorenchyma.** Fairly large air cavities may be present in the parenchyma cells of aquatic plants in particular, where the volume of cavities is often greater than that of the cells. As a result they often taken up star-like or stellate or armed appearance (Fig. 534 E, F & G). The air spaces give buoyancy to the plants in addition to normal aëration. The term **aerenchyma** is usually attributed to this type of parenchyma. Parenchyma with small intercellular spaces are noticed in the endosperm cells. Specialised parenchymatous cells which produce and store tannins, oil and crystals of calcium oxalate are referred to as *idioblasts* (Fig. 506). They markedly differ from normal parenchyma cells.

The term *prosenchyma* for the cells much longer than breadth had been used by early authors. All elongated cells with thick walls and having supporting function were put under prosenchyma. In view of the fact that diverse types of cells may be elongated, the term prosenchyma has gone out of use.

COLLENCHYMA

It (Fig. 535) is also a simple tissue consisting of one type of elements. The cells are somewhat elongate, occurring along the long axis of the body. The shape of the cells is variable, the short ones being more or less like parenchyma and the longer ones resembling the fibres, which may have tapering ends often overlapped and interlocked like fibres. They are usually polygonal in cross-section. The cells are living with vacuolate protoplast. Chloroplasts may also be present. Though normally collenchyma cells are narrower and longer than parenchyma, the two types of tissues very often intergrade; even transitional forms may occur. Like parenchyma it can undergo reversible changes and retain the capacity of cell division. On account of these similarities collenchyma is considered a type of parenchyma specially adapted for supporting function.

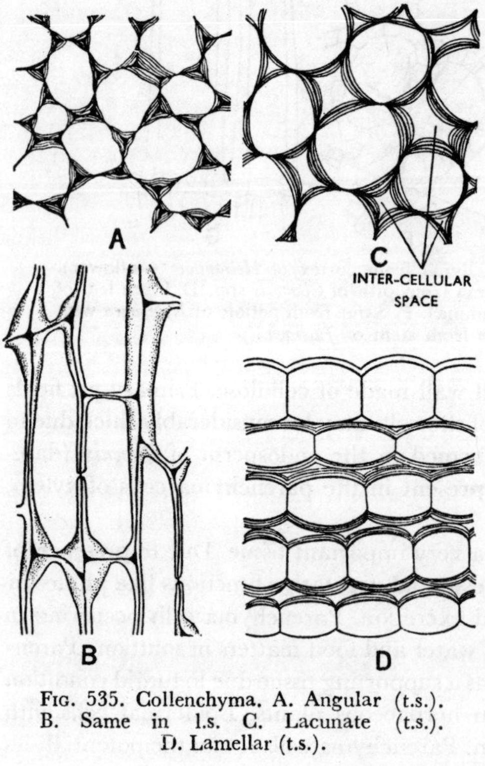

FIG. 535. Collenchyma. A. Angular (t.s.). B. Same in l.s. C. Lacunate (t.s.). D. Lamellar (t.s.).

The most distinctive character of the collenchyma cells is the cell wall which becomes unevenly thickened. There are different methods of deposition, but commonly, the thickenings are confined to the corners of the cells. Often the degrees of deposition may be so much pronounced that cells look circular in cross-section. On the basis of thickening of the cell wall and arrangement of cells three forms of collenchyma have been recognised. They are: (*i*) **Angular** collenchyma (Fig. 535 A & B), the most common type, where deposition is localised to the junctions between the cells. This typical collenchyma is a compact tissue consisting of irregularly arranged cells without intercellular spaces. Due to continued thickening of the cell wall the lumen may look circular. A term *annular* collenchyma has been used by some workers for this type which has lost the angular appearance. (*ii*) **Lacunate** or **tubular** collenchyma is the second type in which intercellular spaces are present and thickenings are restricted to the walls of the regions bordering on spaces (Fig. 535C). (*iii*) **Plate** or **lamellar** collenchyma consists of compactly arranged cells with vigorously thickened tangential walls (Fig. 535D). As a result the cells appear like plates or bands. Though these are the three types of collenchyma recognised, it should be noted that sharp distinction between them hardly exists. In fact, all the three types may occur in the same strand or one type may merge with another.

The wall though considerably thickened is primary in nature. It is composed of cellulose and pectic materials with high percentage of water. In some plants cellulose-rich and pectin-rich layers alternate on the wall. In some cases the wall may undergo sclerification. Simple pits may be present both on the thin and thick portions of the wall.

Collenchyma occurs in the peripheral portions of rapidly elongating organs like young stems, petioles of leaves, floral stalks and the leaves. They are noticed most commonly as homogeneous bands beneath the epidermis, or they may occur as discontinuous patches. In ribbed organs like the stem and petiole of *Cucurbita* and square stems of members of family Labiatae, collenchyma occurs as patches in the ribs and at the corners of square stems. In leaves they may be present on both sides of the veins or along the margin. This tissue is normally absent in underground organs, though it may rarely occur in roots, particularly when they are exposed to light. It is usually absent in the monocotyledons, both stems and leaves.

Collenchyma is an effective mechanical tissue of the growing organs, where it can give considerable strength and elasticity. The closely packed cells with thick walls have the capacity of increasing in surface and in thickness when the organ is still growing. In growing organs it provides sufficient tensile strength till more effective mechanical tissues like sclerenchyma are differentiated. Here it serves temporary supporting function and may be crushed afterwards. In herbaceous stems collenchyma usually continues to function permanently, because secondary increase in thickness is poor in these organs. In leaves they give support occurring on both sides of the bundles or as isolated patches. Though chloroplasts may often be present, this tissue probably has no photosynthetic function.

SCLERENCHYMA

Sclerenchyma is another simple tissue nicely adapted for mechanical function. The component cells are usually 'prosenchymatous', a term once used to designate the cells much longer than breadth. The walls are considerably thickened, often heavily lignified with simple pits. The presence of hard elastic secondary wall with low water-content distinguishes sclerenchyma from collenchyma possessing plastic primary walls with high percentage of water.

The shape and size of the cells constituting this tissue are variable. They may be broadly placed into two groups: very much elongate cells, called **sclerenchyma fibres,** and short cells, isodiametric or irregular in shape, known as **sclereids** or **sclerotic cells.** Though these are the two forms, intermediate types are not uncommon. Moreover, both the forms may remain intermixed in the same strand to carry on the same function.

Fibres. Fibres (Fig. 536) are very much elongate sclerenchyma cells usually with pointed needle-like ends. The fibres of ramie, *Boehmeria nivea* of family Urticaceae may measure as much as 55 cm. in length. These are really the longest cells in the higher plants. Though typical fibres have acute ends, blunt and even branched ends may also be noticed. At maturity these cells lose protoplast and become dead. It has been worked out that protoplast becomes multinucleate during differentiation and ultimately disappears. Here is another anatomical feature by which sclerenchyma fibres may be distinguished from parenchyma and collenchyma. Of course some fibres may retain protoplast even up to the permanent stage. In cross-section sclerenchyma cells look angular.

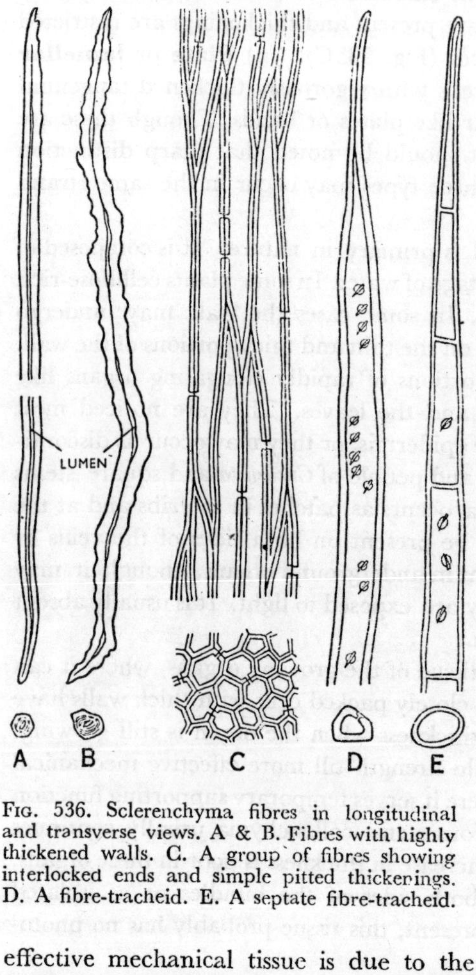

Fig. 536. Sclerenchyma fibres in longitudinal and transverse views. A & B. Fibres with highly thickened walls. C. A group of fibres showing interlocked ends and simple pitted thickenings. D. A fibre-tracheid. E. A septate fibre-tracheid.

The wall is usually hard, uniformly thickened and lignified. Small pits, round or slit-like in appearance, are frequently present. Often the secondary wall is so much thickened that the central lumen is almost obliterated (Fig. 536B). Some fibres have non-lignified walls. In fact, highly thickened secondary walls of fibres of flax, *Linum usitatissimum* of family Linaceae are made of pure cellulose. Their walls exhibit lamellations. Some fibres may have gelatinous walls which may swell and fill up the entire lumen. They have been called *gelatinous* or *mucilaginous* fibres.

Fibres usually occur as groups or sheets along the longer axis of the body in different parts of the plants. They have peculiarly overlapped or interlocked ends (Fig. 536C). The value of sclerenchyma as the most effective mechanical tissue is due to the interlocking of the ends and considerably thickened walls. They may also occur singly as idioblasts as in the leaflets of *Cycas*.

The distribution of sclerenchyma in the plant members has direct relation to the strains and stresses to which they are subjected. They are abundantly present in cortex, pericycle, xylem and phloem. As regards classification of fibres a number of systems has been in use. Often fibres are put in two groups, viz., *intraxyllary* fibres or wood fibres, and *extraxyllary* fibres, also called bast fibres. Fibres associated with xylem differ from other fibres particularly in the occurrence of bordered pits. These fibres are put into two groups: *libiriform* fibres and *fibre-tracheids* (Fig. 536D). Libiriform fibres actually resemble other fibres, whereas fibre-tracheids are regarded as reduced tracheids due to presence of bordered pits. These are really something intermediate between the tracheids and fibres. Some xylem and phloem fibres develop partition walls later, e.g., *Vitis*, which are called *septate* fibres (Fig. 536E). They are living and have been found to contain starch, oils, resin and crystals and are thus thought to have storage function. Fibres occurring outside xylem—the so-called extraxyllary fibres are often called *bast* fibres. This term is misleading, because phloem is also commonly known as bast. It is possibly desirable to designate extraxyllary fibres on the basis of their topography as cortical fibres, pericyclic fibres and phloem fibres. All these fibres are typically elongate bodies with simple pits. They commonly occur as continuous bands or isolated strands in the cortex, in the pericycle, as caps or sheaths on and around the vascular bundles and as patches in the leaves of monocotyledons in particular.

As already stated sclerenchyma is the most effective mechanical tissue of the plants. It can very successfully withstand strains and stresses like flexion, compression and longitudinal pull in cylindrical bodies and shearing stresses in leaves. The fact that some fibres, even fibre-tracheids retain protoplasts may suggest other physiological functions apart from support.

The fibres have great commercial value. Jute, hemp, flax, ramie, etc., are common sclerenchyma fibres.

SCLEREIDS

Sclereids or *sclerotic cells* (Fig. 537) are non-prosenchymatous cells, usually isodiametric or irregular in shape (Fig. 537C). That is a marked point of distinction between

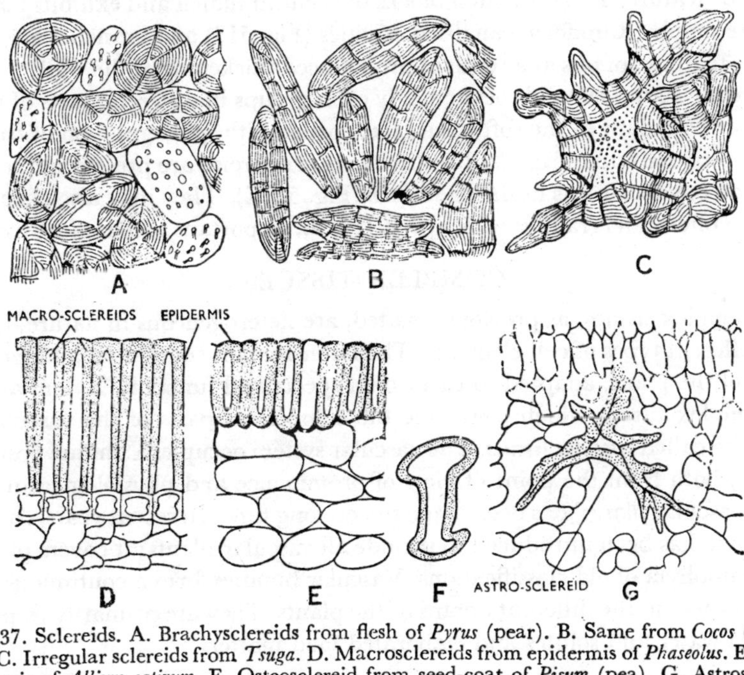

FIG. 537. Sclereids. A. Brachysclereids from flesh of *Pyrus* (pear). B. Same from *Cocos* (cocoanut). C. Irregular sclereids from *Tsuga*. D. Macrosclereids from epidermis of *Phaseolus*. E. From epidermis of *Allium sativum*. F. Osteosclereid from seed-coat of *Pisum* (pea). G. Astrosclereid from a leaf.

sclereid and sclerenchyma; but it has to be taken into account that sclereids originate from parenchyma cells by continued thickening and lignification of the wall and sclerenchyma fibres from meristematic cells. But some of them may be slightly elongated (Fig. 537 B & D), specially those occurring in seed and fruit coats. Normally sclereids become dead with maturity. Cases have been recorded where they may retain protoplasts. They may occur singly or in groups or patches in different parts of plants, more commonly in the cortex and phloem. As they vary considerably in size, shape and nature of the cell wall, sclereids may be put broadly into a few groups:

Brachysclereids, also called *stone cells* (Fig. 537A), are usually isodiametric ones like parenchyma cells. They are abundantly present in the soft parts of the plants like cortex, pith, phloem and flesh of fruits. The gritty texture of pears is due to presence of this type of sclereid. **Macrosclereids** are rather rod-shaped elongated cells which form a palisade-like epidermal layer on the seed-coats in *Phaseolus* (Fig. 537D), *Pisum* and other members of pulse family. **Osteosclereids** are bone-like sclereids consisting of columnar cells, more or less dilated at the ends (Fig. 537F). They occur in the leaves of some dicotyledons and seed-coats. **Astrosclereids** have irregularly branched arms and thus attain star-like or stellate appearance (Fig. 537G). This type is common in many dicotyledonous leaves. This classification does not really cover all the types. Very often sclereids develop into idioblasts, which are distinctly different from surrounding parenchyma cells. In aquatic plants like *Nymphaea* armed idioblastic sclereids are found to project into the intercellular spaces (Fig. 534F). These are also called 'internal hairs' or *trichoblasts* or **trichosclereids.** Some parenchyma cells have thick lignified secondary walls. They are called sclerotic parenchyma.

The wall is typically hard and lignified. Strong secondary wall with distinct pits is the characteristic feature. The wall often blocks the central lumen and exhibits lamellations. The pits are simple. Ramiform canal-like cavities (Fig. 513) and branches are frequently formed due to fusion of pits as a result of pronounced thickening of the wall.

As already stated sclereids occur singly or in groups in different parts of the plants, but are most abundant in the soft tissues like cortex, pith, phloem, flesh of fruits, leaves and also in seed-coats and fruit wall. Solid layers of sclereids are present in the epidermis of some protective scales, as in *Allium sativum* (Fig. 537E). They may also occur in xylem and phloem often intergrading with fibres. It is an important mechanical tissue.

COMPLEX TISSUES

The complex tissues, as previously stated, are heterogeneous in nature, being composed of different types of cell elements. The latter remain contiguous and form a structural part of the plant, adapted to carry on a specialised function. Xylem and phloem are the complex tissues which constitute the component parts of the vascular bundle. They are also called *vascular tissues*. The vascular system occupies a unique position in the plant body, both from the point of view of prominence and physiological importance. The term 'vascular plants' has been in use since a long time. In recent years a new phylum Tracheophyta has been introduced to include all vascular plants; it covers pteridophyta and spermatophyta of old classifications. Vascular bundles form a continuous and interconnected system in the different organs of the plants. They are primarily responsible for transport of water and solutes and elaborated food matters.

XYLEM

Xylem is a complex tissue forming a part of the vascular bundle. It is primarily instrumental for conduction of water and solutes, and also for mechanical support. Primary xylem originates from the procambium of apical meristem, and secondary xylem from the vascular cambium. As a complex tissue it consists of different types of cells and elements, living and non-living. The tissues composing xylem are **tracheids, tracheae** or **vessels, fibres,** called xylem fibres or wood fibres, and **parenchyma,** referred to as xylem or wood parenchyma. Of the above mentioned elements only the parenchyma cells are living and the rest are dead. A term *hadrome* was once used for xylem. It included the elements excepting the fibres.

Tracheids. A tracheid is a very much elongate cell (Fig. 538) occurring along the long axis of the organ. The cells are devoid of protoplast, and hence dead. A tracheid has a fairly large cavity or lumen without any contents and tapering blunt or chisel-like ends. The end walls usually do not uniformly taper in all planes. Tracheids are round or polyhedral in cross-section. They are really the most primitive and fundamental cell-types in xylem from phylogenetic point of view. The wood of ancient vascular plants were exclusively made of tracheids. This is the only type of element found in the fossils of seed-plants. In modern plants they practically occur in all groups including the angiosperms, though they predominate in lower vascular plants, the pteridophytes and gymnosperms. More effective conducting elements, tracheae or vessels, have evolved from the tracheids.

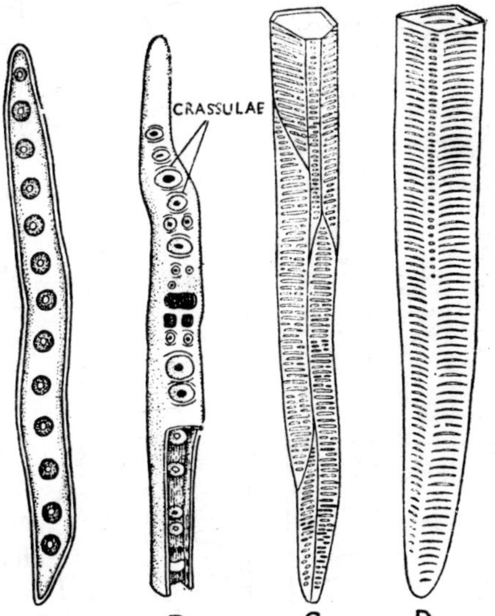

Fig. 538. Tracheids. A. A tracheid with bordered pits. B. A part showing bordered pits with crassulae. C & D. Parts with scalariform thickening.

The wall is hard, moderately thick and usually lignified. Secondary walls are deposited in different manners, so that the tracheids may be annular, spiral, reticulate, scalariform or pitted. But pits of the bordered type are most abundant. Through these pits they establish communication with adjoining tracheids and also with other cells, living or non-living. The nature of the pits on the walls of the tracheids is variable; in lower vascular plants the pits are elongated giving them scalariform appearance (Fig. 538 C & D), those of gymnosperms and angiosperms have round pits with well-developed borders (Fig. 538 A & B). Tracheids occur both in primary and secondary xylem.

Due to the presence of central lumen and hard lignified wall tracheids are nicely adapted for transport of water and solutes. They also serve as supporting tissue.

A typical fibre differs from a tracheid in more pronounced thickening of the wall and correspondingly much smaller lumen, as well as in reduction of the size of the pits. An intermediate type of cell element, called fibre-tracheid (Fig. 536D), is found in some plants. They have smaller pits with reduced or vestigial borders. In some cases protoplast persists up to the mature stage, and may even divide, so that transverse partition walls are noticed within the original wall. These are called *septate* fibre-tracheids (Fig. 536E).

Tracheae or Vessels. These are long tube-like bodies ideally suited for the conduction of water and solutes. A trachea or vessel is formed from a row of cylindrical cells arranged in longitudinal series where the partition walls become perforated, so that the whole thing serves like a tube. In tracheids the only openings are the pit-pairs, whereas the vessels are distinct 'perforate' bodies. Thus translocation of solutes becomes more easy in a vessel, as it proceeds more or less in a straight line; but the line of conduction is rather indirect in a group of tracheids. Perforations are commonly confined to the end-walls, but they may occur on the lateral walls as well. The walls undergoing perforations are referred to as *perforation plates*, which are mainly of two types: *multiple* plates and *simple* ones. In primitive plants it has been found that the end-walls between the cells do not completely dissolve, but the openings or perforations remain either in more or less parallel series like bars called *scalariform* perforation (Fig. 539A) or in form of a network known as *reticulate* perforation, or even may form a group of circular holes (*foraminate* perforation). In advanced types of plants the dissolution of the end-wall is more or less complete, and the perforation occurs in form of a single large circle. This is referred to as *simple* perforation (Fig. 539B). There is anatomical evidence in support of the fact that the single large circular or oval perforation has been formed by gradual disappearance of the transverse bars of scalariform and other types. The vessels are considerably long bodies; in ash plant, *Fraxinus excelsior* of family Oleaceae vessels have been reported to be as long as 10 ft. Like tracheids these elements are devoid of protoplast and have hard and lignified cell-wall with different types of localised thickenings. Some forms intermediate between typical tracheids and vessels have been noticed. These elements, analogous to fibre-tracheids, are called *vessel-tracheids*.

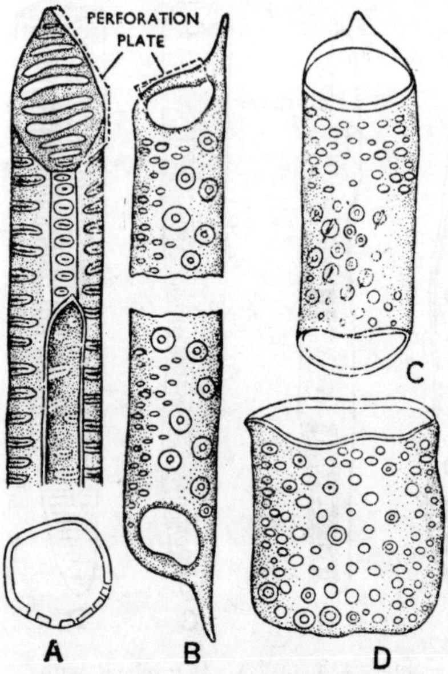

FIG. 539. Tracheae—different types. A. With multiple perforation plate in l.s. and t.s. B, C & D. With simple perforation plates.

Ontogeny of a vessel. A vessel or a trachea originates from a row of meristematic cells of procambium or vascular cambium which remain attached end on end in longitudinal series (Fig. 540). As usual the cells grow and secondary walls are laid down, only the primary walls where perforations will take place remain uncovered. The secondary walls undergo lignification and other changes. The protoplast in the mean time becomes progressively more and more vacuolated and

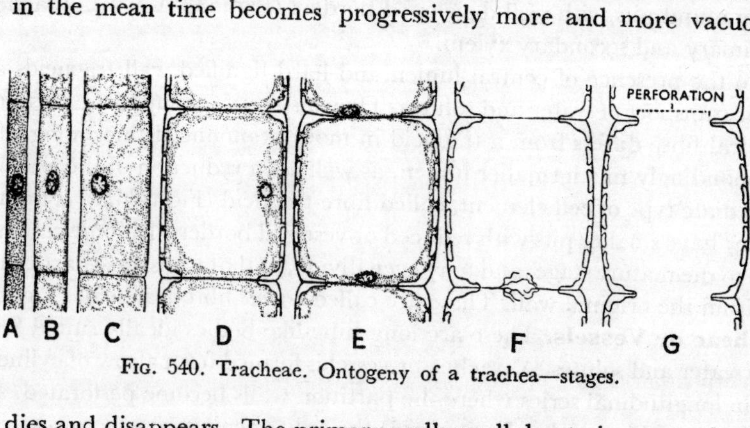

FIG. 540. Tracheae. Ontogeny of a trachea—stages.

ultimately dies and disappears. The primary walls swell due to increase of pectic intercellular substance and break down, thus forming the continuous vessel. It should be noted that a vessel or trachea arises from a group of cells, unlike a tracheid, which is an elongate 'imperforate' single cell. The individual cells taking part in the formation of the vessel are called vessel elements.

The walls of the vessels are thick, hard and lignified. The secondary walls are deposited in different patterns, so that the thickenings may be ring-like, spiral, scalariform, reticulate or pitted (Fig. 511). The pits are mostly of bordered types.

Tracheids are more primitive than the vessels. In fact, in the primitive types of vessels the form of a tracheid is maintained, but with advance in evolutionary line the diameter of a vessel may so much increase that it may become drum-shaped (Fig. 539 C & D) in appearance. Vessels have originated phylogenetically from the tracheids; and occur in the pteridophytes *Pteridium* and *Selaginella*, in the highest gymnosperms, Gnetales, and in the dicotyledons and monocotyledons. Comparative studies on the dicotyledons have revealed that evolution of vessel members have proceeded from the long narrow elements with tapering ends to short ones with wider cavities having transverse or inclined end-walls which ultimately dissolved. Suggestions about independent development of vessels by parallel evolution has also been put forward (Cheadle, 1953). Vessels first appeared in the secondary xylem and then proceeded towards primary xylem. In some dicotyledons belonging to the families Winteraceae, Trochodendraceae and Tetracentraceae and others of the lowest taxonomic group, curiously the vessels are absent (Bailey and others). They do not occur in some xerophytes, parasites and aquatic plants. These have been interpreted as cases of reduction of xylem tissues involving evolutionary loss. In monocotyledons vessels are not present in secondary xylem (which tissue is lacking in many monocotyledons). Here vessels first appeared in the roots and then extended to the aerial organs (Cheadle, '53; Fann. '54).

These are the most important elements of xylem. They are primarily adapted for easy transport of water and solutes, and, secondarily, for mechanical support.

Xylem fibres. Some fibres remain associated with other elements in the complex tissue, xylem, and they mainly give mechanical support. As previously stated, fibres are very much elongated, usually dead cells with lignified walls. Xylem fibres or wood fibres are mainly of two types: **fibre-tracheids** (Fig. 536 D & E) and **libiriform** fibres (Fig. 536 A & B) which usually intergrade, so much so that it is difficult to draw a line of demarcation between them. Fibre-tracheids, as already reported, are intermediate forms between typical fibres and tracheids; they possess bordered pits, though the borders are not well-developed. Libiriform fibres are narrow ones with highly thickened secondary wall. The central lumen is almost obliterated and pits are simple. They resemble the phloem fibres, and hence the name. They occur abundantly in many woody dicotyledons.

Phylogeny of tracheary elements. The tracheary elements have developed during the evolution of land plants (Bailey, '53). In the lower vascular plants the function of *conduction* and *support* were combined in the tracheids. With increasing specialisation woods evolved with conducting elements—the vessel members being more efficient in conduction than in providing mechanical support. On the other hand fibres evolved as principal supporting tissue. Thus from the primitive tracheids *two lines* of specialisation diverged—one toward the *vessel* and the other toward the *fibre*.

The following structural features may be taken as the basis in support of the evolution of the tracheary elements from primitive tracheids which are usually long imperforate cells with small diameter, angular in cross-section, having lignified scalariformly pitted walls.

(*i*) The primitive vessels are also elongate bodies like the tracheids with rather small diameter and tapering ends. Similar condition is still noticed in lower dicotyledons. With evolutionary advance they gradually become shorter and wider, often becoming drum-shaped in appearance.

(*ii*) The wall of the primitive tracheid is rather thin, more or less of equal thickness, and it is angular in cross-section. Same condition prevails in primitive vessels. With progressive advance considerable thickening appeared and the vessels became circular or nearly so in cross-section.

(*iii*) In the primitive vessels the perforation plates are *multiple*, usually scalariform with numerous bars, and *oblique* end-walls. Progressive increase in specialisation led to gradual decrease in the number of bars and their ultimate disappearance, so that the perforation plates become *simple* with *transverse* end-walls. These are positively advanced characters.

(*iv*) The pitting of the vessel wall also changed from early scalariform arrangement, characteristic of tracheids, to small bordered pit pairs, first in opposite (arranged in transverse rows) and ultimately in *alternate* (arranged spirally or irregularly) pattern. Moreover the pit pairs between vessels and parenchyma changed from bordered to half-bordered and then to simple.

In the specialisation of the xylem fibres adapted for more efficient support there has been steady increase in thickness of the wall leading to decrease in cell-lumen. The pits changed from elongate to circular, the borders becoming reduced and functionless, and ultimately disappeared. Thus the evolutionary sequence was from *tracheids*, through *fibre-tracheids* to *libiriform* fibres.

Xylem parenchyma. Living parenchyma is a constituent of xylem of most plants. In primary xylem they remain associated with other elements and derive their origin from the same meristem. In secondary xylem parenchyma occurs in two forms: *xylem parenchyma* (Fig. 541A) are somewhat elongate cells and lie in vertical series attached end on end; *ray parenchyma* (Fig. 541B) cells occur in radial transverse series in many woody plants. Parenchyma is abundant in the secondary xylem of most of the plants, excepting a few conifers like *Pinus*, *Taxus* and *Araucaria*. These are the only living cells in xylem.

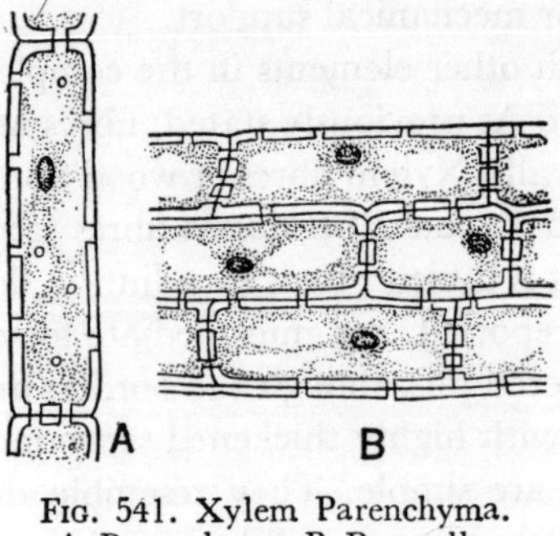

FIG. 541. Xylem Parenchyma. A. Parenchyma. B. Ray cells.

The cells may be thin-walled or thick-walled. If lignified secondary wall is present, the pit-pairs between the cells and the adjacent xylem element may be bordered, half-bordered or simple. Between two parenchyma cells the pit is obviously simple.

These cells are particularly meant for storage of starch and fatty food; other matters like tannins, crystals, etc., may also be present. As a constituent part of xylem they are possibly involved in conduction of water and solutes and mechanical support.

PHLOEM

The other specialised complex tissue forming a part of the vascular bundle is *phloem* It is composed of **sieve elements, companion cells, parenchyma** and some **fibres.** Sclerotic cells may also be present. Phloem is chiefly instrumental for transloca-tion of organic solutes—the elaborated food materials in solution. The elements of phloem originate from the procambium of apical meristem or the vascular cambium. Two terms, *bast* and *leptome*, have been used for phloem, though they are not exactly synonymous with it. Bast, derived from the word 'bind', was introduced before the discovery of sieve elements; it mainly meant the fibres. The soft-walled parts of phloem, obviously excluding the fibres, were referred to as leptome.

Sieve elements. The most important constituents of phloem are the sieve elements, the **sieve tubes** and **sieve cells.** From ontogenetic point of view a sieve tube resembles a vessel and a sieve cell a tracheid.

Sieve tubes (Fig. 542) are long tube-like bodies formed from a row of cells arranged in longitudinal series where the end-walls are perforated in a sieve-like manner. The perforated end-walls are called the **sieve plates,** through which cytoplasmic connections are established between adjacent cells. The perforations or *sieve areas,* as they are called, may be compared to the pit fields of the primary wall with plasmodesmata connections. But the sieve areas are more prominent than pit fields and the connecting strands are more wide and conspicuous. It may be that a number of plasmodesmata fuse to form a connecting strand. Moreover, an insoluble substance, called *callose,* (probably a carbohydrate of unknown chemical composition, is impregnated into cellulose or replaces cellulose forming a case round each connecting strand which passes through the sieve area (Fig. 543A). A sieve area in surface view looks like a depression on the wall having a pretty good number of dots. Each dot represents a connecting strand in cross-section and remains surrounded by a case of callose (Fig. 543). In sectional view sieve areas appear like thin places on the wall through which the connecting strands pass from one cell to another (Fig. 543). The sieve plate or the perforated end-wall is really the primary walls of two cells with the middle lamella in between them. The end-walls may be obliquely inclined or transverse. A sieve plate is called *simple* (Figs. 542 & 543), if it has only one sieve area, whereas the plate may be *compound* (Fig. 544) with several sieve areas arranged in scalariform, reticulate or other manners. Though rare, the sieve areas may occur on the side walls as well. From evolutionary point of view simple sieve plates on transverse end-walls are more advanced characters than compound plates on oblique walls. The cylindrical cells which take part in the formation of the sieve tube are called sieve tube elements. Like vessel elements the sieve tubes have also undergone decrease in length with evolutionary advance.

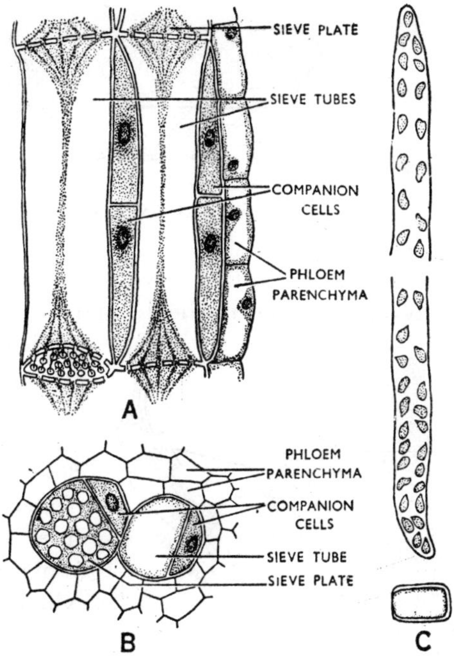

FIG. 542. Sieve elements. A. Sieve tubes in l.s. B. Same in t.s. C. Sieve cell in l.s. and t.s.

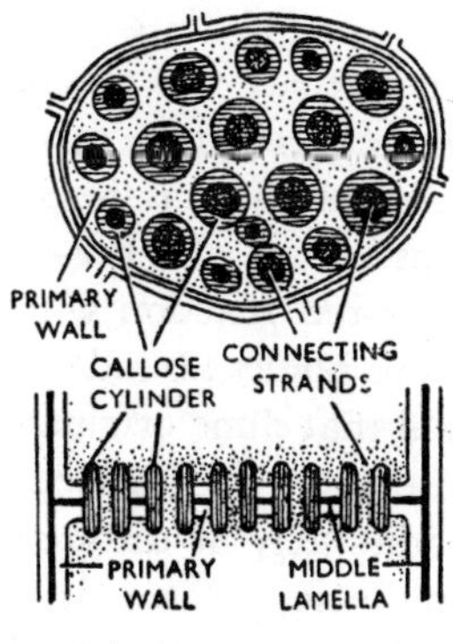

FIG. 543. Sieve tube. Structure of sieve area of an angiosperm in surface (upper) and sectional (lower) views (diagrammatic).

Sieve cells (Fig. 542C), which may be compared to the tracheids, are narrow elongated cells without conspicuous sieve areas. They usually have greatly inclined walls, which overlap in the tissue, sieve areas being more numerous in the ends. Sieve cells are more primitive than the sieve tubes. They occur in lower vascular plants and

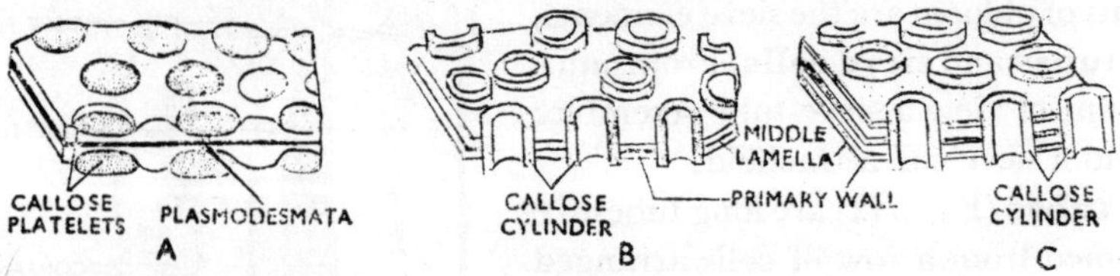

FIG. 543A. A sieve plate—showing stages of development and maturation.

gymnosperms. In fact, sieve tubes have evolved from the sieve cells, as vessels have evolved from the tracheids, and so sieve tubes occur in all angiosperms. In monocotyledons, unlike the xylem elements, sieve tubes first appeared in the aerial organs, the course being from the leaves to the stem and, lastly, to the roots.

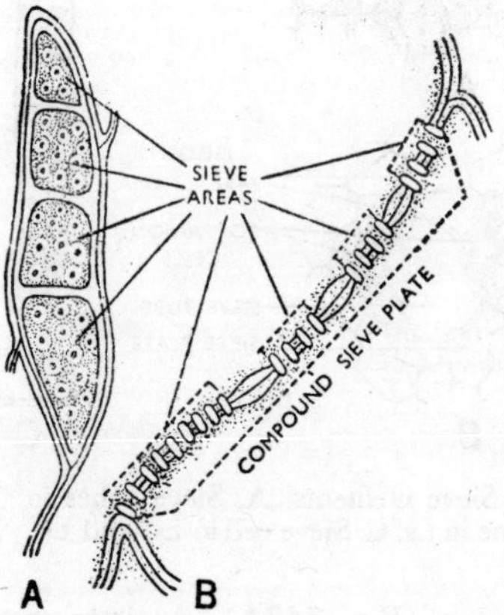

FIG. 544. Sieve tube. Compound sieve plate. A. Surface view. B. Sectional view.

Ontogeny of the sieve elements. In spite of close ontogenetic resemblance between tracheary elements of xylem and sieve elements of phloem, the latter unlike the former, are living. They originate from the *mother cells* (Fig. 545) which are usually short cylindrical or elongate ones. The mother cell divides longitudinally into two daughter cells, one of which serves as the sieve element and the other one becomes the companion cell, of course in those cases where companion cells occur. The sieve element undergoes gradual differentiation. It grows in length, cytoplasm gets more and more vacuolated, so that it may have a lining layer of cytoplasm round a large central vacuole. The most outstanding character is the disintegration of the nucleus with the maturity of the sieve elements. In fact, a distinct nucleus is present in every cell at the meristematic stage. During differentiation the nucleus disorganises (Fig. 545F). It is the only living functioning element without a nucleus. Small colourless plastids are also present in the protoplast. They contain carbohydrates which give wine-red reaction with iodine and are interpreted as starch grains. Slimy proteinaceous bodies abundantly occur in the sieve tubes, what is commonly called *slime*. It is said that slime originates in the cytoplasm as small discrete bodies, which eventually fuse and get dispersed in the vacuoles. In fixed preparations funnel-shaped slime bodies may be distinctly seen in form of plates referred to as *slime plugs* (Fig. 545H), on the sieve plates. In this connection a very interesting statement has come from a well-known authority, Prof. K. Esau, to the effect that in some plants the nucleolus is extruded from the nucleus before it finally disorganises and that the nucleolus persists in the tube. Slime bodies have not been observed in pteridophytes, gymnosperms and monocotyledons. Sieve areas develop from the primary pit fields and the connecting strands originating from one or a group of plasmodesmata become more conspicuous which remain surrounded by callose cylinders. Another theory demands that pores are formed by dissolution of cell wall and no plasmodesmata occur at the pore sites. The connecting strands were thought to be entirely cytoplasmic in nature; but it is argued that they may contain vacuolar substances and thus establish connections between vacuoles of neighbouring elements.

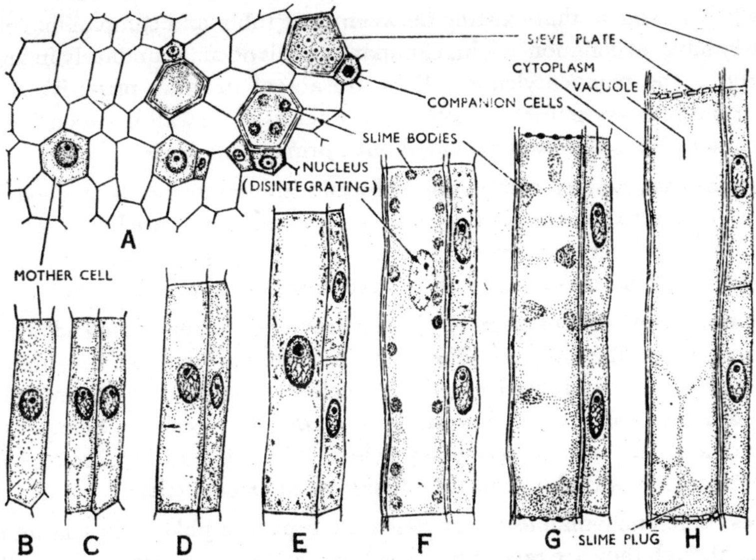

FIG. 545. Sieve tube. Differentiation of sieve tube in an angiosperm. A. T.s. through phloem showing stages. B to H. L.s. through phloem showing stages (*After Esau.*)

The wall of sieve elements is primary and chiefly composed of cellulose. Thick walls are found only in exceptional cases. The tubes often cannot withstand the pressure from adjoining cells and ultimately get crushed.

It has been stated that protoplasmic strands pass through the pores of the sieve areas and that the strands remain surrounded by callose. With the differentiation of the tube the amount of callose increases and finally forms something like a pad on the sieve plate. This pad is referred to as **callus pad.** Due to its formation the cell to cell communication is considerably cut down or entirely prevented. The callus pad is usually formed with the approach of resting or inactive season; and it disappears when the active season (spring) sets in. This type is known as *seasonal* or *dormancy* callus. In old functionless sieve tubes callus becomes permanent, what is called *definitive* callus. Though the term definitive callus is often used to designate the former type, it is desirable to confine it to permanent callus of old and functionless tubes.

Companion cells. Companion cells (Figs. 542 & 545) remain associated with the sieve tubes of angiosperms, both ontogenetically and physiologically. These are smaller elongate cells, having dense cytoplasm and prominent nuclei. Starch grains are never present. They occur along the lateral walls of the sieve tubes. A companion cell may be equal in length to the accompanying sieve tube element or the mother cell may be divided transversely forming a series of companion cells (Fig. 545). A sieve tube element and a companion cell originate from the same mother cell. Their functional association is evident from the fact that companion cells continue so long the sieve tubes function, and die when the tubes are disorganised. The companion cells are so firmly attached to the sieve tubes that they cannot be normally separated by maceration. In transverse section it appears as a small triangular, rectangular or polyhedral cell with dense protoplast (Figs. 542 & 545).

In pteridophytes and gymnosperms some small parenchymatous cells remain associated with sieve cells, which are known as *albuminous cells*. They die in natural course when the sieve cells become functionless. Thus the relation between sieve cells and albuminous cells is similar to that existing between sieve tubes and companion cells, excepting that they have no common origin. Companion cells occur abundantly in angiosperms, particularly in the monocotyledons. They are absent in some primitive dicotyledons and also in the primary phloem of some angiosperms. The wall between the sieve tube and companion cell is thin and provided with primary pit fields.

Parenchyma. Besides companion cells and albuminous cells, a good number of parenchyma cells remain associated with sieve elements. These are living cells with cellulose walls having primary pit fields. They are mainly concerned with storage of organic food matters. Tannins, crystals and other materials may also be present.

The parenchyma cells of primary phloem are somewhat elongate and occur with the sieve elements along the long axis (Fig. 542). In secondary phloem they may be of two types. Those which occur in vertical series are called *phloem parenchyma;* and others occurring in horizontal planes are known as *ray cells*, the position being just like the parenchyma and ray cells of secondary xylem. The cell wall is primary, composed of cellulose. Parenchyma is absent in the phloem of monocotyledons.

Fibres. Sclerenchymatous fibres constitute a part of phloem in a large number of seed plants, though they are rare in pteridophytes and some spermatophytes. They occur both in primary and secondary phloem. These are typical elongated cells having interlocked ends, lignified walls with simple pits. The fibres of primary phloem are essentially similar to those occurring in cortex and secondary phloem. They are of considerable commercial importance, as these fibres are abundantly used for the manufacture of ropes and cords. The flax fibres, unlike others, have non-lignified walls. Sclerotic cells are often present in primary phloem. They probably develop from parenchyma with the age of the tissue. So it is a case of 'secondary sclerosis'.

SECRETORY TISSUES

The cells directly concerned with secretions like resins, essential oils, mucilage, latex and similar substances together constitute **secretory** or **special tissues.** These cells have neither common origin nor morphological continuity. They may occur as isolated patches in any part of the plant, or may form well-organized structures.

Two types of cells are usually found in this tissue. In some of them the materials formed are exuded from the cells. Nectaries (Fig. 547A) exuding sugary fluids are the examples. These are also called *excretory cells*. In others the materials formed remain stored up in the cells to be released only after the breakdown of the cells. The cells belonging to the first type have rich cytoplasm and prominent nuclei, whereas those of the second type are large cells with well-developed cavity where the secretion remains stored up (Fig. 518B). The ducts containing essential oils and mucilage are the examples.

GLANDS

Glands are well-organized secretory structures, composed of diverse types of cells. The secretory materials are produced and liberated by the protoplasts of the cells. The substances may be at once exuded from the seats of formation as in nectaries, or they may remain stored in some cavity inside. The glands may be *external* or superficial, naturally formed on the epidermis; or they may be *internal* where the cavities correspond to the intercellular spaces formed either schizogenously or lysigenously. Glandular hairs and trichomes, common in many plants (Fig. 244) are superficial in origin. The common glands are those which secrete digestive enzymes, known as *digestive glands*, nectar-secreting ones or *nectaries*, and, similarly, *resin ducts*, *oil ducts*, latex-secreting glands called *laticiferous ducts*, and water-secreting ones called *hydathodes*.

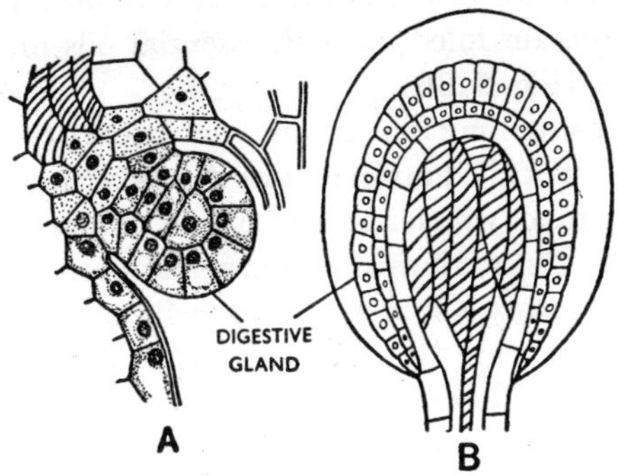

FIG. 546. Digestive gland in sectional views. A. Of *Nepenthes*. B. Of *Drosera*. (*After Haberlandt*.)

Digestive glands. It is an established fact that plants in general have intracellular digestion. Here the living cells secrete enzymes, no specialised structure being present for the purpose. The insectivorous plants possess special digestive glands which secrete proteolytic (protein-digesting) enzymes and thus part of the nitrogen requirement is obtained from the bodies of the insects they catch. So it is a case of extracellular digestion. In *Drosera* or sundew the gland (Fig. 546B) is located at the tip of the tentacle; in *Nepenthes*, the pitcher plant, glands (Fig. 546A) occur along the wall of the pitcher and secrete enzyme in the liquid present inside the pitcher. In *Dionoea* (Fig. 23) the glands are normally inactive, but contact with the insect-body stimulates them.

Nectaries. These are special glands usually located on the floral parts. They secrete the sugary substance, nectar or honey, and thus attract the pollinating insects. These glands are superficial, usually consisting of epidermal cells. In some cases the cells forming the glands are columnar or papillose ones (Fig. 547A), with dense protoplast, whereas in others the secretory cells may be more or less like normal epidermal cells, but without cuticle. Nectar is exuded through the walls and exposed on the outer surface of the gland. Nectaries may also occur on purely vegetative parts, in that case they are called *extrafloral* nectaries. Those present on the rim of the pitcher of *Nepenthes* which really lure the poor insects, and those occurring on the stipules of *Vicia* spp. (Fig. 547B) of family Leguminosae, are nice examples of extrafloral nectaries.

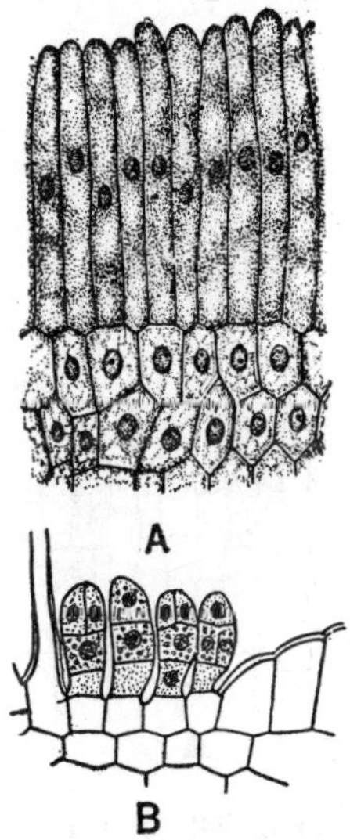

FIG. 547. Nectary in sectional view. A. Of *Euphorbia pulcherrima*. B. Extrafloral nectary on the stipule of *Vicia* spp.

Resin ducts and Oil ducts. Substances like resins, oils, gums are copiously secreted in gymnosperms in general and also in many angiosperms. The materials are secreted and conducted through special glands, which are also called *ducts*. In gymnosperms like pine these ducts form extensive canals occurring in vertical and horizontal position. But ducts present in the fruits of umbellifers are localised in certain regions. The resin ducts are formed schizogenously and appear as tube-like bodies having a lining of small parenchyma cells with dense protoplast (Fig. 518A). The latter are called *epithelial cells* which excrete resin. The oil ducts of umbellifers, though localised and limited in extent, are also formed in the same manner.

The characteristic oil glands present on the rinds of citrus fruits like lemons and oranges are formed lysigenously. The cavities remain filled up with essential oils and other substances due to disorganization of tissues (Fig. 518B).

Laticiferous ducts. Long tube-like bodies containing the viscous fluid, latex, occur in a large number of angiospermic families. These are called **laticiferous ducts** or tubes. Latex is usually a milky, often yellowish or watery fluid, which is readily exuded when the plants containing it are injured. It is an emulsion of various substances like proteins, sugars, enzymes, rubber, etc., in a watery matrix. Latex is secreted by the cells and conducted through them often forming extensive system inside the plant body. The commercial value of latex has already been reported. Latex of *Hevea* of family Euphorbiaceae and some other allied plants is the chief commercial source of rubber, and that of papaw contains the very useful enzyme papain, and *Papaver*, a useful alkaloid.

From ontogenetic point of view laticiferous ducts are of two types, viz., **non-articulate** latex ducts or *latex cells* and **articulate** latex ducts or *latex vessels*. They are also called *simple* and *compound* laticifers respectively. There is no difference between the two, so far as the contents are concerned.

Non-articulate latex ducts or latex cells (Fig. 548A), are single cells. They arise as small meristematic cells at the very embryonic stage. With the growth of the plants these cells also elongate and make their way through other tissues. The growth of the tip is mainly apical and often branches are given out. Thus as much elongate tube-like bodies growing continuously they penetrate into all the tissues of the plant, even invading the newly-formed organs like leaves, buds and lateral roots. The latex cells are coenocytic (multinucleate) bodies with scanty cytoplasm. They do not form a network. This type of latex duct is common in members of families Asclepiadaceae, Apocynaceae, Euphorbiaceae, etc.

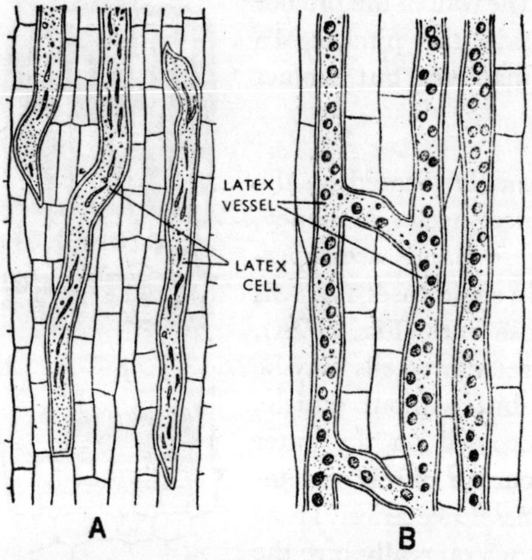

Fig. 548. Laticiferous ducts in sectional view. A. Non-articulate duct from *Euphorbia pilulifera*. B. Articulate duct from *Carica papaya* (papaw).

In some plants like *Vinca* of family Apocynaceae and some members of Urticaceae the latex cells remain unbranched. The cells here originate not in the embryo, but in meristem of developing shoot. But instead of extending like ones previously discussed, these cells remain restricted to one internode and to the attached leaf and branch.

Articulate latex ducts or latex vessels (Fig. 548B) originate from a row of cells in continuous series by partial or total dissolution of the end-walls. So from ontogenetic point of view it is a compound structure resembling a xylem vessel. But the latex vessel is living and coenocytic. The rows of cells may be even irregular. The ducts send out branches frequently, which extensively penetrate through other tissues and eventually form a complex system of network. Like non-articulate ones this type also originates in the early embryonic stage where fusion of cells has been observed. Latex vessels occur in poppy of family Papaveraceae, papaw of Caricaceae, banana of Musaceae and the rubber-yielding plant *Hevea* of family Euphorbiaceae. It is interesting to note that the nature of duct does not bear any relation with the systematic position of plants. Most of the plants of family Euphorbiaceae possess latex cells, but *Hevea*, the most important rubber-yielding plant of the same family has latex vessels.

Hydathodes. These are specialised structures present in many angiosperms, through which exudation of water takes place. So they are also called *water-stomata* or water-pores. Exudation or guttation is the process of escape of water in liquid form. It occurs in humid climates when there is rapid absorption of water by roots, but transpiration rates are reduced. On cool nights following sultry days water is found at the tips and margins of leaves of many plants in form of dew-drops. Every drop marks the location of a hydathode through which water has come out. In many plants like *Tropaeolum* (garden nasturtium), tomato, hydathode is not a simple pore. Here the ends of vascular elements, the tracheids, are in contact with some loosely arranged cells, called *epithem*, which possess scanty chlorophyll. External to epithem there is a cavity and, finally, the pore. The whole structure is referred to as hydathode (Figs. 549 & 550). These pores or stomata differ from ordinary stomata of the leaves in a number of points, viz., they are larger in size, are located at the vein-endings, and particularly in the fact that they always remain

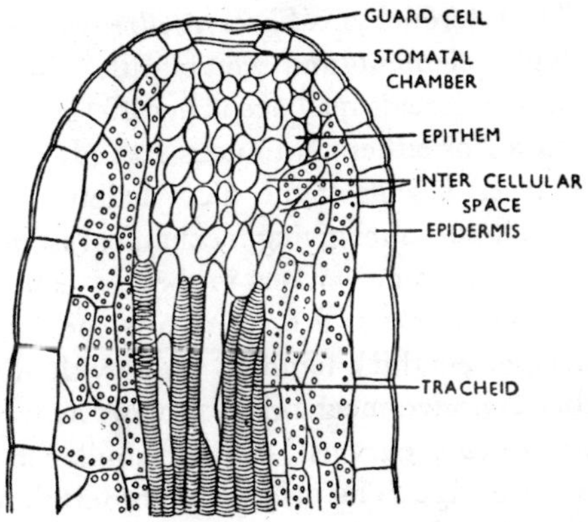

FIG. 549. Hydathode from *Primula* spp. (sectional view). (*After Haberlandt.*)

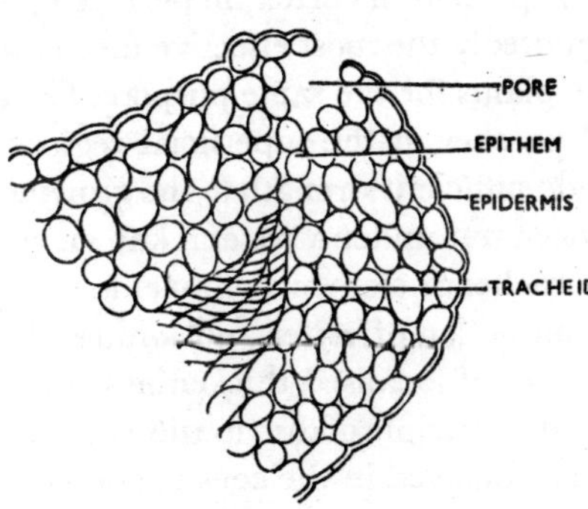

FIG. 550. Hydathode, at the margin of tomato leaf—sectional view.

PRIMARY PERMANENT TISSUES

- **Simple**
  1. Parenchyma
  2. Collenchyma
  3. Sclerenchyma
     - fibres
     - sclereids
- **Complex**
  1. Xylem
     - Tracheids
     - Tracheae
     - Xylem parenchyma
     - Xylem fibres
  2. Phloem
     - Sieve elements
       - Sieve cells
       - Sieve tubes
     - Companion cells
     - Phloem parenchyma
     - Phloem fibres
- **Secretory**
  1. Secretory cells
  2. Glands
     - Digestive glands
     - Resin ducts
     - Oil ducts
     - Laticiferous ducts
       - Latex cells
       - Latex vessels
     - Hydathodes

open, the characteristic mechanism of opening and closing of ordinary stomata being absent. In some plants the epithem may be lacking, where water moves to the pore through the mesophyll cells. Morphologically hydathodes may be compared to the stomata. These are the passages for exudation of water, often with some dissolved salts, particularly occurring in the plants of humid climate

A summary of the primary permanent tissues is given in a tubular form in page 293.

MECHANICAL TISSUES AND THEIR DISTRIBUTION

It has been discussed in the preceding pages that some tissues are specially meant for providing mechanical strength or support to the plant members which are frequently subjected to various kinds of strains and stresses. The problem is not so acute in case of small herbaceous land plants and plants adapted to grow in water. The cell wall and the turgidity of the cells may be adequate to keep them in proper position. But comparatively larger plants, and, in fact, their main organs have to encounter considerable, often vigorous, strains operating from outside. They possess conspicuous mechanical tissues for withstanding them.

The most important mechanical tissues are *sclerenchyma fibres* with highly lignified walls and peculiarly interlocked ends, *sclereids,* with massive lignified walls, and *collenchyma* with unevenly thickened cellulose walls. The sclerenchyma fibres, as already stated, may be present in cortex, in pericycle, with vascular elements and even in pith. They are undoubtedly the most effective mechanical tissues. Sclereids may occur in different parts of the plants for the same purpose. Collenchyma forms either continuous strands or isolated patches in the superficial regions of the aerial organs of the dicotyledons They provide sufficient strength to the growing organs. These tissues giving mechanical strength had been put under a system, known as **stereome,** by some workers in the last century. The tracheary elements of xylem—the *tracheids* and *tracheae*, are primarily meant for the conduction of water and solutes. But as they possess thick lignified walls with different types of localised thickenings, they possibly can give mechanical support as well.

The principle of distribution of mechanical tissues is, curiously enough, similar to the practice followed in the construction of a house or bridge. The primary consideration of an engineer or architect is to secure maximum strength with minimum expenditure of materials. Thus the guiding principle is to 'economise materials', as Haberlandt has suggested in his classical book, without in any way impairing the strength.

In the later part of the nineteenth century Schwendener made comprehensive studies on the distribution of mechanical tissues in plants and asserted that the principle was in conformity with the practice followed by the engineers in constructions. Without in any way underestimating the classical importance of his work it should be said that engineering practices have considerably advanced in the meantime, so that one demerit of his contention cannot be ignored now. He treated sclerenchyma as an isolated system and disregarded the soft tissues in which they remain embedded. An analogy with ferroconcrete structure will be helpful in understanding that the soft tissues are to the sclerenchyma what the concrete is to the iron framework. So the soft tissues support the framework of sclerenchyma and take up a part of the strain.

An account of the distribution of mechanical tissues in accordance with the principle stated above, is being given :

Inflexibility. If a load is put at the middle portion of a straight girder supported at the ends, the result would be a curvature when the upper surface would be shortened and the lower lengthened. That shows that the upper surface is subjected to *compression* and the lower to *tension*, while at the middle portion tension will come to zero. So requisite materials should be concentrated on the two surfaces, which are regions of greatest tension. The typical girders are constructed accordingly, so that they appear as I in cross-section (Fig. 551). Here the top and bottom portions, called *flanges*, are really stronger and the upright middle portion, known as *web*, connecting the two flanges may be made of lighter materials. In fact, web represents the neutral line or null-line where no strain operates.

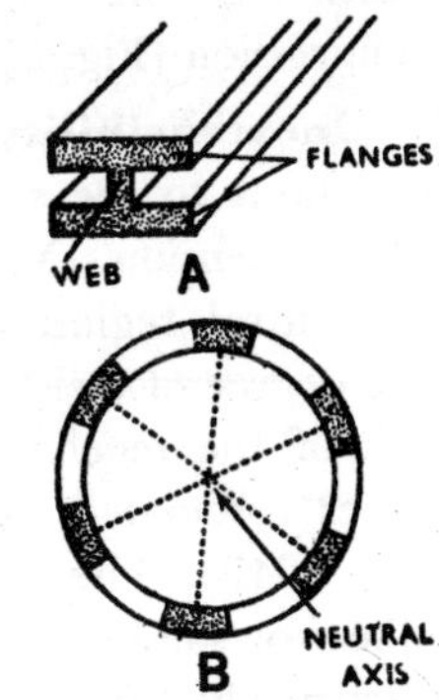

FIG. 551. I-girders (diagrammatic). A. Single girder. B. A system of girders.

Many plant organs, particularly the cylindrical ones like stems, etc., are frequently subjected to *bending stress* or *flexion* involving compression on one side and tension on the other. The mechanical tissues are distributed at the peripheral region here, so that they approach the I-girders in construction, the web being made of parenchyma or other tissues. Typical dicotyledons like sunflower possess mechanical tissues in hypodermis, in the region of pericycle and in association with vascular tissues (Fig. 552A) where the large pith at the central region serves as the web. Thus it resembles a group of composite I-girders (Fig. 551B). In the square stems of members of family Labiatae like *Leonurus*, extra patches of collenchyma occur at the four corners, so that they appear like diagonally-placed I-girders (Fig. 552B). In a typical monocotyledon

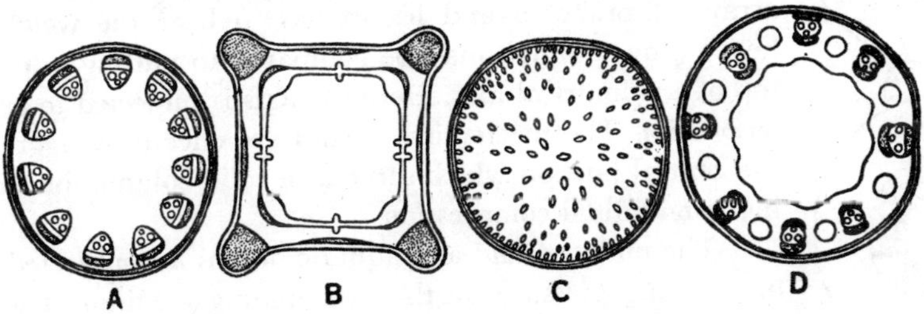

FIG. 552. Mechanical tissues in stem (diagrammatic) in sectional views A. *Helianthus* (sunflower) stem. B. Square stem of *Leonurus*. C. *Zea* (maize) stem. D. *Cyperus* (sedge) stem.

like maize sclerenchyma is present as a band in hypodermal region and the bundles with sclerenchymatous sheaths remain more crowded towards the periphery (Fig. 552C). Some members of sedge family, Cyperaceae possess patches of mechanical tissues just internal to epidermis and corresponding semilunar patches of the same on the lower side of the bundles, the two patches constituting the flanges of a girder (Fig. 552D). The central portion being hollow (web) the vascular bundles with associated mechanical tissues constitute composite I-girders. In some monocotyledons, as in the onion family, Liliaceae, the vascular bundles remain completely embedded in a peripherally located cylinder of mechanical tissues.

Bilaterally symmetrical organs like the foliage leaves have mechanical tissues arranged in form of I-girders parallel to one another and at right angles to the surface. In the leaves of many grasses and sedges subepidermal I-girders extend from one surface to another (Fig. 553A), the web being composed of the bundle and parenchyma. In case of long leaves the upper surface is subjected to more vigorous tension, and so patches of subepidermal mechanical tissues occur at the outer surface to prevent tension, and the small I-girders present at the lower surface are meant for withstanding compression (Fig. 553B).

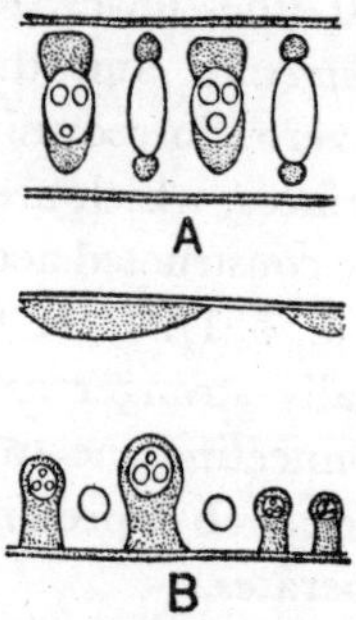

Fig. 553. Mechanical tissues in leaves (diagrammatic); in sectional views.

Inextensibility. The roots and other organs which attach the plants to the soil or other substratum suffer from *longitudinal pull* or tension. Mechanical tissues are advantageously put in the central region in form of a compact mass in these organs. The degree of resistance of course depends on the cross-sectional area of the mechanical elements. Thus the roots have mechanical tissues associated with the vascular elements inside the stele (Fig. 554B). The underground rhizome has also centrally located hollow or solid mechanical strand (Fig. 554A). The aquatic plants, particularly those which are adapted to grow in rapidly running water, e.g., *Potamogeton* of family Potamogetonaceae are inextensible and naturally possess mechanical tissues at the central region. The same principle is applicable to the inflorescence axis and pendulous fruit-stalks. The stilt roots of the members of grass family, e.g., maize, have to encounter both flexion and longitudinal pull. So here in addition to aggregation of mechanical tissues at the central region forming a massive stele, a peripheral band is also present (Fig. 554C).

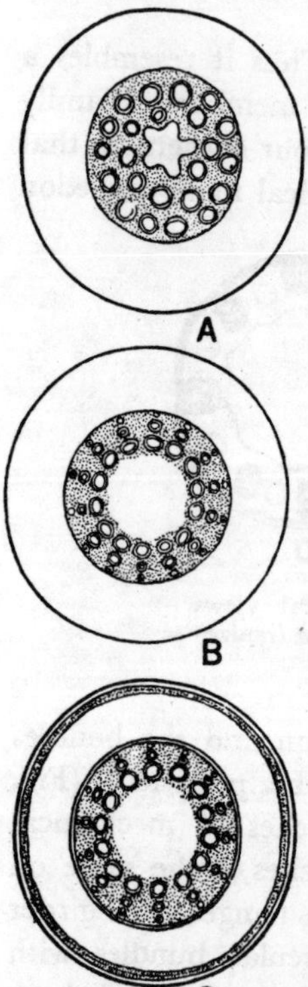

Fig. 554. Mechanical tissues in underground organs in sectional view. A. In a rhizome. B. In monocot root. C. In a stilt root.

Incompressibility. The axis of a spreading tree with its arrays of branches and leaves has to bear the weight of the heavy crown, which may be compared to putting a load at the top of a cylindrical axis. Here the axis is subjected to *longitudinal compression*. The mechanical tissues are effectively aggregated at the central portion which serve as a solid column for withstanding longitudinal compression.

The underground and aquatic organs are subjected to *radial* or crushing pressure by the surrounding medium. The aquatic plants have loosely-arranged cortical cells which give requisite protection. In plants like *Sagittaria*, *Alisma*, etc., the epidermis also becomes associated with the outer layers of cortex for the purpose. The roots of grasses in particular develop tubular sheaths of cells, often with suberised walls.

Shearing stresses. The flat organs like the leaves are often subjected to violent *shearing stresses* due to movement of surrounding air or water. The wind currents work at right angles to the surface of the leaves and cause considerable laceration. To stand against this stress the I-girders present for securing inflexibility are firmly held together by a large number of cross-ties in form of veins which often form a network. The I-girder arrangement is more pronounced in monocotyledonous leaves, usually having parallel venation (Fig. 553). The margins of the leaves are particularly exposed to shearing stresses. They have special arrangement for protection by increased thickness of the epidermis, and frequent occurrence of thick-walled collenchyma in the subepidermal region.

CHAPTER XXI

TISSUE SYSTEMS

Tissues are large in number and also varied in nature and function. They are aggregated into **tissue systems** on the basis of topographical continuity or physiological similarity or both. In fact, attempts to put tissues into tissue systems have been made since a long time and two systems from two entirely different points of view have been proposed.

Sachs in 1875 proposed a classification mainly on the basis of topography of the tissues, according to which three tissue systems were recognised. They are **epidermal** tissue system, **ground** or **fundamental** tissue system and **vascular** tissue system. The epidermal tissue system is composed of the epidermis in a primary body which forms the outer cover. It is thus protective in function. The vascular tissue system consists of the vascular bundles, which are made of the complex tissues, xylem and phloem, the principal conducting tissues of plants. The ground or fundamental tissue system includes the rest, i.e., all the tissues other than epidermal and vascular. Apart from topography, this classification is also helpful in interpreting the meristems and tracing the course of their development into permanent region. The cells and their derivatives of the apical meristem, as already stated, exhibit three zones—the protoderm, the procambium and the ground meristem. These meristematic tissues may be called precursors of the three tissue systems—the epidermal tissue system developing from the protoderm, the vascular tissue system from the procambium and the ground tissue system from the ground meristem (Fig. 529).

	Meristematic region (*apical*)	**Permanent region**	**Tissue system**
Promeristem	→Protoderm →	Epidermis →	Epidermal tissue system
	→Ground meristem →	Cortex Stele: { Pericycle, Medullary rays, Pith }	→ Ground tissue system
	→Procambium →	Vascular bundles →	Vascular tissue system

THE EPIDERMAL TISSUE SYSTEM

This system solely consists of the outermost skin or **epidermis** of all the plant organs beginning from the underground roots to the fruits and seeds. This layer represents the point of contact between the plants and the outer environment and, as such exhibits diversities in structure. It is primarily a *protective* tissue, which protects the internal tissues against excessive loss of water by transpiration and mechanical injury. Subsidiary functions like storage of water, mucilage, secretion and, though rarely, even photosynthesis, may also be carried on. But for stomatal and lenticular openings the epidermis is a continuous layer. Normally it is uniseriate—typically consisting of one layer of cells. It derives its origin from the protoderm of the meristematic region.

The protoderm cells divide anticlinally and in course of time uniseriate epidermis is formed. Many-layered or multiseriate epidermis, usually called **multiple** epidermis, is found in some organs like roots of orchids, leaves of *Ficus* spp. (Fig. 507A), *Nerium, Peperomia*, etc. Normally it may be assumed that these layers have originated from the protoderm by periclinal divisions The outermost layer of multiple epidermis is similar to ordinary uniseriate one. The inner layers are different from other tissues in absence of chlorophyll. There is room for doubts if all these layers belong to epidermis from ontogenetic point of view. They may be outer layers of cortex originating from the ground meristem, but resemble the epidermis both in structure and function.

The epidermal cells are living with lining layer of protoplast around large central vacuole. The plastids are normally small and colourless. Chloroplasts are present only in the guard cells of the stomata in case of organs exposed to sunshine, but they occur in the epidermal cells of aquatic plants and plants growing in moist and shady situations. Mucilage, tannins and crystals may occasionally be present. Anthocyanins may occur in the cell-sap of the vacuoles. Epidermal cells retain the potentiality of cell division. During normal course of development or due to external stimuli they may divide and produce new cells.

Epidermal cells exhibit wide diversities as regards their size, shape and arrangement. But they may be said to be essentially tabular in shape (Fig. 555 A & B). compactly set,

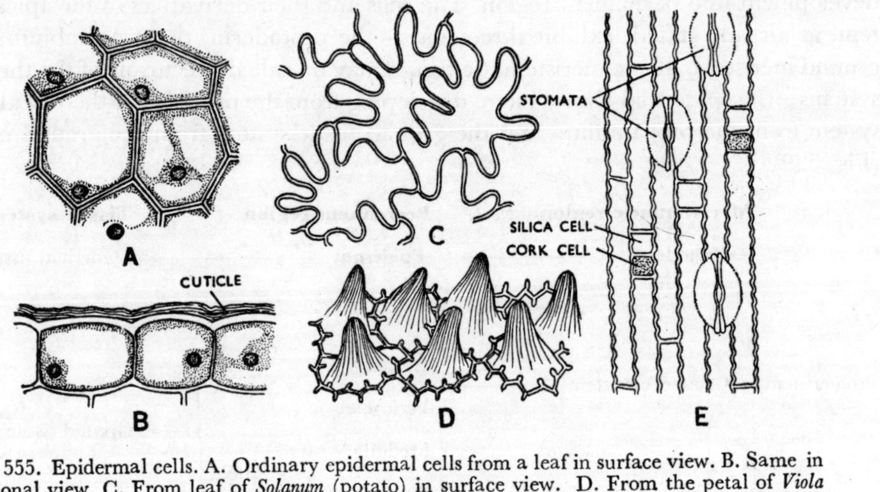

Fig. 555. Epidermal cells. A. Ordinary epidermal cells from a leaf in surface view. B. Same in sectional view. C. From leaf of *Solanum* (potato) in surface view. D. From the petal of *Viola* showing ridge-like infoldings of lateral walls and protruding papillae. (*After Strasburger*). E. Of *Saccharum* (sugar-cane) with silica cells and cork cells.

so that a continuous layer without intercellular spaces is formed. Only in the petals of some flowers intercellular spaces are found, but they remain covered by outer cuticle. In surface view they are more or less isodiametric in shape. In the leaves and petals of flowers they may have irregular shapes, often with teeth and flanges (Fig. 555 C & D) which remain peculiarly interlocked with one another. In monocotyledonous stems and leaves with parallel venation the epidermal cells are rather elongated in the direction of the long axis (Fig. 555E), so much so that in extreme cases they may be fibre-like in appearance.

Epidermal cells have unevenly thickened walls, the outer and radial walls being much more thick than the inner walls. In some cases they may be so massive that the central lumen is almost obliterated. The walls are strongly cutinised, what is very important for protection against mechanical injuries and prevention of loss of water. The fatty substance cutin is found in the wall—in interfibriller and intermicellar spaces of the cellulose and forms the *cuticle* occurring all over the outer wall of the epidermal cells (Fig. 556 A & B). It remains as a separate layer and in some cases it may be removed as a whole. The cuticle is often found to project into the radial walls as peg-like bodies (Fig. 556C). Cuticle is absent only in the epidermis of roots and some submerged aquatic plants. The thickness of the outer walls of the epidermal cells depend on the environmental conditions of the plants. It is quite thin in plants with adequate water supply, and it is unusually thick in plants growing in dry situations. The surface of the cuticle may be smooth or may possess ridges and cracks. The cutinised portion of the walls, the portion lying beneath the cuticle, has been found to consist of alternating layers of cutin and pectic materials. Waxy matters are often deposited on the cuticle in form of rods and granules (Fig. 556E). The so-called 'bloom' of many fruits and glaucous characters of many stems

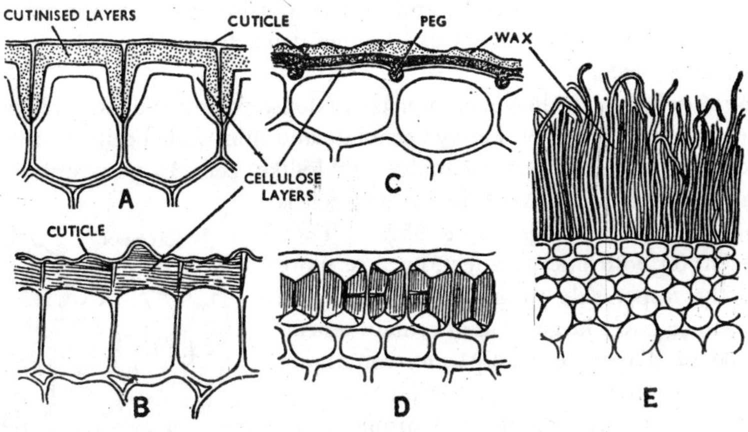

FIG. 556. Epidermal cells. A. From *Aloe*. B. From *Allium* (onion). C. From *Musa* (banana). D. Lignified cells in *Pinus* (Pine). E. Coating of waxy rods in *Saccharum* (sugarcane).

and leaves are due to these deposits. Lignification is rather rare in epidermal cells. It occurs in the pine needles (Fig. 556D), in cycad, in grass leaves outside the sclerenchyma patches and in a few dicotyledons. Deposition of silica is common in the epidermal cells of horse-tails (*Equisetum*) and grasses. It is really interesting to find long epidermal cells having corrugated margin (Fig. 555E) associated with two kinds of short cells—the *silica* cells and *cork* cells in grasses. The silica cells contain silicon oxide and cork cells with suberised walls contain organic materials. In some dicotyledonous families like Malvaceae, Rutaceae, etc., the epidermal cells individually or in groups undergo mucilaginous changes, particularly in the seeds. Special sac-like cells remain scattered in the epidermis of some members of family Cruciferae. These are idioblastic cells resembling the laticifers, but they contain an enzyme, *myrosin*, and so they are called myrosin cells. It has been stated in a preceding chapter that many dicotyledonous families like Urticaceae, Moraceae, possess cystoliths. The cystolith-containing cells of epidermis are referred to a *lithocysts*. The epidermis is often made up of a layer of sclereids, as found in the seed-coats of *Pisum* and *Phaseolus* of family Leguminosae (Fig. 537D) and in the scales of garlic—*Allium sativum* of family Liliaceae (Fig. 537E). Radial and inner walls of epidermal cells possess pit-fields. Plasmodesmata have also been reported, those on the outer walls of leaves have also been called *ectodesmata*.

The origin of the shoot epidermis may be traced from the apical meristem. It arises from the outer layers of tunica, according to tunica-corpus theory, or from the dermatogen of Hanstein or protoderm, as suggested by Haberlandt, which may be called primordial epidermis. But the root epidermis fundamentally differs from that of shoot in origin, structure as well as in function. It has been pointed out in the previous chapter that epidermis of root is related to the root-cap or the cortex from the developmental point of view. The cells are tabular, lack in cutinisation of wall and their function is mainly absorption of water and solutes. So the terms *epiblema*, *piliferous layer* or *rhizodermis* have been applied to it.

Epidermis, as a rule, persists as uniseriate layer throughout its life in the organs where distinct secondary growth does not take place. In some monocotyledons, though secondary increase is absent, a kind of periderm is formed, and thus the epidermis is destroyed. In organs with distinct secondary growth in thickness epidermis continues till cork cells are formed. In leaves, flowers and fruits, it persists as long as the organs do. In roots the epidermis with a part of cortex becomes dead, lignified or suberised after the root hairs are destroyed.

Bulliform cells. In the leaves of monocotyledons, excepting a few families, a peculiar type of comparatively larger, highly vacuolate and thin-walled cells occur in the epidermis. These are called *bulliform* (meaning, bubble-like) cells. In transverse section they appear as a fan-like band because the median cell is usually the largest in size (Figs. 557 & 557A). They may be present on both sides of a leaf, but are more common on the upper side running parallel to the veins. They either cover large areas or remain restricted to the grooves. These are mainly water-containing cells with no chlorophyll. The walls are usually thin, but the outer walls may be thick and cutinised like other epidermal cells, often filled with silica.

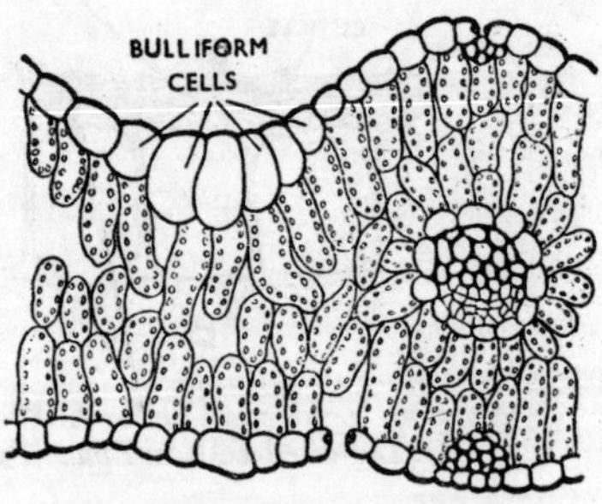

FIG. 557. Bulliform cells in a section of a leaf of *Triticum* (wheat).

There are three views as regards the functions of bulliform cells. According to the first view they are concerned with the unrolling of the developing leaves. It is suggested that these cells undergo sudden and rapid expansion at a certain stage of leaf development and consequently bring about unfolding of the leaves. The second view is that they have a role to play in the hygroscopic opening and closing movements of mature leaves, due to changes in turgor. They

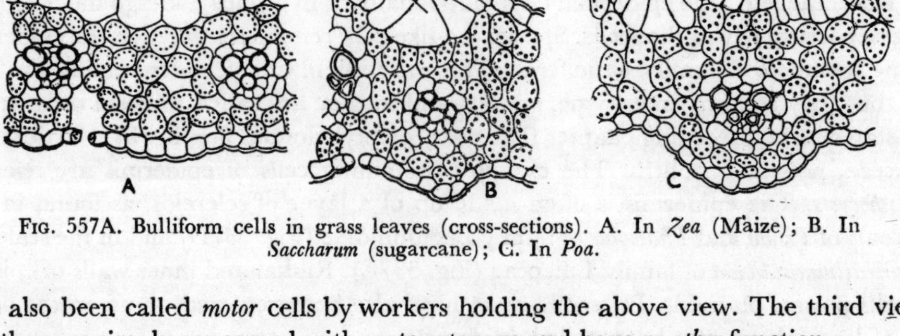

FIG. 557A. Bulliform cells in grass leaves (cross-sections). A. In *Zea* (Maize); B. In *Saccharum* (sugarcane); C. In *Poa*.

have also been called *motor* cells by workers holding the above view. The third view is that they are simply concerned with water-storage and have *no other* function.

Stomata. The continuity of the epidermis of aerial organs is interrupted by the presence of some minute pores or openings on it. These pores are called the **stomata,** through which exchange of gases takes place between the internal tissues and the outer atmosphere. A stoma has a small slit or pore and two specialised epidermal cells, called **guard cells,** on the two sides. Often other epidermal cells adjacent to the stoma undergo modifications. They differ from other epidermal cells and become associated with the stoma functionally. These are referred to as **subsidiary** or *accessory* cells (Figs. 559 & 561). Though gaseous interchange actually occurs through the pore, called **stomatal aperture** or opening, the term stoma includes the whole thing, the pore, guard cells and subsidiary cells, when present. In surface view the guard cells look cresent or kidney-shaped in appearance, being attached to each other at the margin of the concave side with the aperture lying in between them (Fig. 558A). They may be easily distinguished from ordinary epidermal cells, because they possess dense cytoplasm, prominent nuclei, chloroplasts, and even starch grains. A cavity is present just beneath the stoma, what is called **sub-stomatal chamber** or cavity (Fig. 558B). It is in communication with the intercellular space system of the internal tissues.

The walls of the guard cells are unevenly thickened, the wall along the aperture being strongly built and that away from the aperture being thin and extensible. The guard cells have cutinised outer walls with a layer of cuticle which extends through the aperture and joins the inner wall. Due to strong cutinisation often ledges of wall materials are noticed on the upper and lower sides of the ventral wall, so that in sectional view they appear like horns or beaks. The ledges project above and below and overarch the two chambers, referred to as the *front cavity* and *back cavity,* which communicate with each other through the pore (Figs. 558B & 560A).

The guard cells, due to uneven thickening of the wall, what is really an outstanding character, can regulate the opening and closing of the stomatal aperture. Normally stomata remain open in daytime and close up with nightfall. The opening is influenced by the changes in the turgor of the guard cells. With increase of turgor the thinner walls of the guard cells get stretched and the thicker walls become more concave, thus the gap becomes wide. In the grass and sedge families the guard cells of the stomata are peculiarly dumb-bell-shaped where the middle portion is straight and strongly thickened and the two ends are swollen or bulbuous (Fig. 559) and thin-walled. Here increase in turgor causes further swelling of the bulbuous ends and, as a result, the straight median portions get separated from each other. Decrease in turgor brings about reverse changes. The

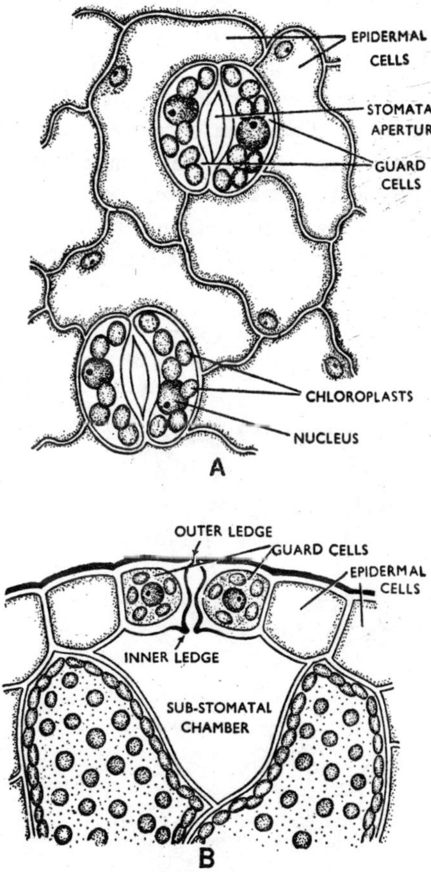

FIG. 558. Stomata. A. In surface view. B. In sectional view.

physiological factors influencing detailed mechanism of the opening and closing of stomatal aperture will be taken up in the portion on plant physiology.

Stomata occur in all aerial parts of the plants, most abundantly in the foliage leaves. Those present on the floral parts and in the aquatic plants are normally functionless. In leaves they may occur on both upper and lower surfaces. In woody plants with dorsiventral leaves they are located on the lower epidermis. In herbaceous plants with isobilateral or centric leaves they occur on both the surfaces. Even in that case stomata are more abundant on the lower side than on the upper. In floating leaves they occur only on the upper epidermis. In an individual leaf stomata are more numerous near the apex and minimum near the base, the middle portion having a distribution, which is an average of the apex and base. In leaves with parallel venation, as in the monocotyledons, and the needles of conifers stomata remain arranged in parallel rows (Figs. 555E & 559), whereas in reticulately-veined leaves they lie scattered (Fig. 563). The number of stomata occurring on the epidermis of leaves is fairly large, which may range between a few thousand to over a hundred thousand per square cm. It has been estimated that a maize plant may have more than two hundred million stomata. So one can hardly estimate the number in a large tree.

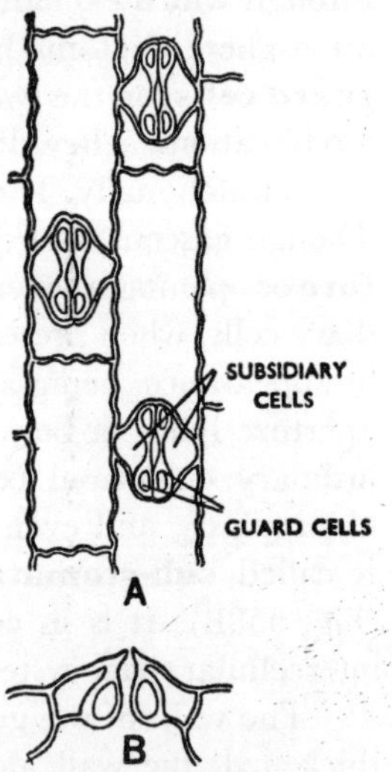

FIG. 559. Stomata from leaf of *Zea* (maize). A. Surface view. B. Sectional view.

The guard cells may be at the same level with adjacent epidermal cells or they may be placed above or lie sunken below the surface of the epidermis. Sunken stomata (Fig. 560) are characteristic of the plants of dry situations, where they often appear to be located at the bottom of a cup-shaped depression, which is called the external cavity or outer chamber. This is an effective mechanism for reducing transpiration. The subsidiary cells are highly thickened here. In the leaves of *Nerium* a groove or depression is formed, what is called stomatal pit (Fig. 560B) and stomata remain very much sunken. Stomata raised above the surface of epidermis (Fig. 560D) are found in the peduncle of *Cucurbita* where they appear to be placed at the summit of a conical papilla. Stomata also

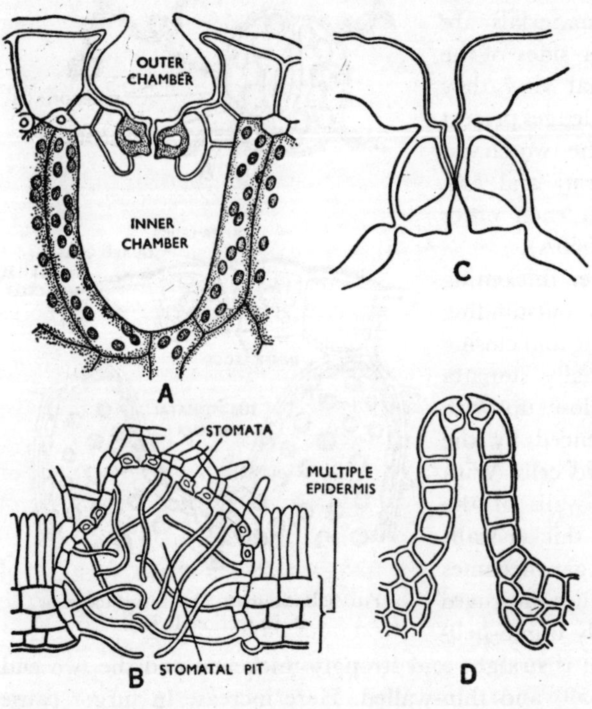

FIG. 560. Stomata in sectional view. A. Sunken stomata in *Hakea*. B. In *Nerium*. C. In *Allium* (onion). D. Raised stoma in peduncle of *Cucurbita*.

occur on the sporophytes of bryophytes like *Anthoceros* and mosses. The stomata of mosses representing really the simplest types, show departure from other types in the nature of thickening of the wall—ventral walls being thin and dorsal thick (Fig. 561) and in the mechanism of opening and closing of the aperture.

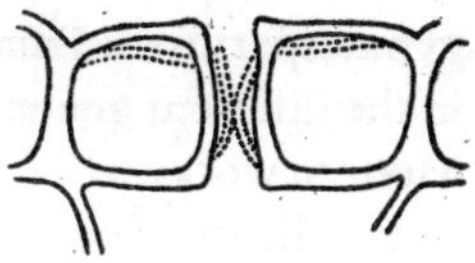

FIG. 561. Stomata of moss sporophyte.

Ontogeny of the stomata. Stomata arise from the protoderm cells. Normally a protoderm cell undergoes anticlinal division, one of them serves as the stoma mother cell. It eventually divides into two cells leaving a small slit between them (Fig. 562). The two cells develop into two kidney-shaped guard cells and the slit into the stomatal aperture. In many families the protoderm undergoes several divisions before the stoma mother cell differentiates. Commonly subsidiary cells arise from protoderm cells lying adjacent to the stoma mother cell. They may be sister cells of the mother cell or may arise by division of the cells lying adjacent to the mother cells.

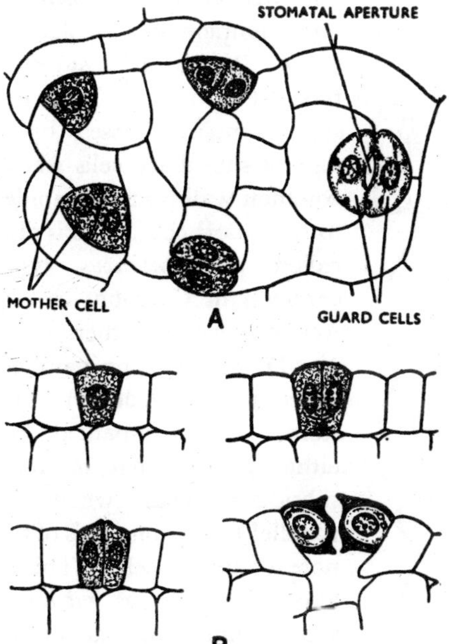

FIG. 562. Stomata—development A. Surface view. B. Sectional view.

A number of types of stomata has been recognised on the basis of their modes of development, relation with neighbouring cells and occurrence and number of subsidiary cells. Without going into detail the following types may be cited as common ones:

In *Allium*, *Iris*, etc., the protoderm cell divides anticlinally into two unequal cells; the smaller one serves as the stoma mother cell which gives rise to the stoma. The subsidiary cells are absent. It is a very common type of stoma. In *Zea*, bamboos and other members of grass and sedge families the guard cells are peculiarly dumb-bell-shaped in appearance. Two subsidiary cells arise by division of the protoderm cells lying adjacent to stoma mother cell and they occur on two sides of the guard cells. This is referred to as *Zea* type. In *Tradescantia* four subsidiary cells are formed which originate from four protoderm cells surrounding the stoma mother cell. In *Bryophyllum* the protoderm cells have been found to produce a series of spirally arranged subsidiary cells, and finally they give rise to the guard cells. The stomata occurring in bryophytes as found in the sporophyte of *Mnium*, are the simplest where wall ledges are absent and, unlike other types, the ventral wall is thin and the dorsal wall thickened.

In recent years intensive investigations have been in progress regarding the mode of development of stomata, their relation to the neighbouring cells and the occurrence of the subsidiary cells. In fact, these characters have been used in problems of classification and phylogeny. The stomata on the basis of investigations particularly in the gymnosperms (Florin and others) have been put into two types: viz., (1) *Haplocheilic*, where the guard cells originate by a single division of the stomatal initial, and some of the neighbouring cells become modified into subsidiary cells. (2) *Syndetocheilic* type, when the guard cells and subsidiary cells originate from the same mother cell. It was thought that haplocheilic type is more primitive than the syndetocheilic one, but actual studies on a large number of plants do not support that contention. Both the types have been noticed in gymnosperms and many families of angiosperms. In fact, different types have been found in the different genera of the same family, and even in different species of the same genus. Modern workers (Cf. Metcalfe and Chalk) have suggested the following types of stomata in the dicotyledons on the basis of the characters stated above.

A. *Anomocytic* or irregular-celled type (Fig. 563A): Stoma remains surrounded by a limited number of cells which cannot be distinguished from other epidermal cells. Thus the subsidiary cells are absent. This is also called *ranunculous* type, common in the families Ranunculaceae, Capparidaceae and others.

B. *Anisocytic* or unequal-celled type (Fig. 563B): Here the stoma remains surrounded by three subsidiary cells of which one is distinctly smaller than the other two. It is otherwise known as *cruciferous* type common in Cruciferae.

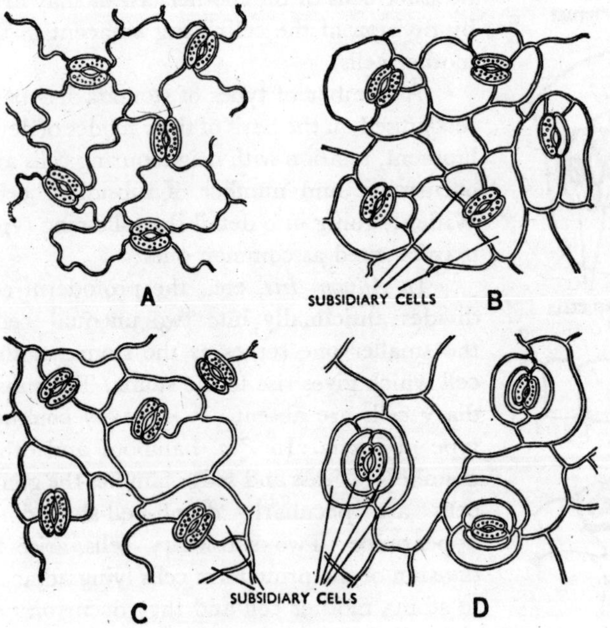

FIG. 563. Stomata—types. A. *Anomocytic* or irregular-celled type in *Clematis*. B. *Anisocytic* or unequal-celled type in *Iberis*. C. *Diacytic* or cross-celled type in *Dianthus*. D. *Paracytic* or parallel-celled type in *Gardenia*.

C. *Diacytic* or cross-celled type (Fig. 563C): Here the stoma remain enclosed by a pair of subsidiary cells whose common wall is at right angles to the guard cells. This is also called *caryophyllaceous* type, common in Caryophyllaceae, Acanthaceae and others.

D. *Paracytic* or parallel-celled type (Fig. 563D): The stoma is accompanied on either side by one or more subsidiary cells which lie parallel to the long axis of the pore of guard cells. This is also referred to as *rubiaceous* type common in Rubiaceae, Magnoliaceae and others.

In view of the fact that diversities occur as regards the nature of the stomata the terms *ranunculous*, etc., are rather confusing, and *anomocytic*, etc., suggested by Metcalffe and Chalk appear to be more appropriate.

Another classification on the basis of development was devised (Pant, 1965), and stomata have been put in three categories:

(1) *Mesogenous* type—guard cells and subsidiary cells derived by consecutive division of a mother cell, e.g., Rubiaceae, Cruciferae.

(2) *Mesoperigenous* type—where the surrounding cells are of dual origin, some from the mother cell and some from the neighbouring cell, e.g, Ranunculaceae, Caryophyllaceae.

(3) *Perigenous* type—all neighbouring and subsidiary cells having independent origin, e.g., Cucurbitaceae, Nympheaceae.

In the monocotyledons the most common one is the *graminaceous* or grass type (Fig. 559). Here the two guard cells are dumb-bell-shaped having a narrow middle portion and bulbuous ends. Two distinct subsidiary cells lie parallel to the long axis of the pore. In recent years intensive investigations have revealed a few other types as well (Fig. 563A). In Orchidaceae, Amaryllidaceae and others the guard cells are not associated with any subsidiary cells. Guard cells surrounded by four to six subsidiary cells have been noticed in many species of Araceae, Commelinaceae, Musaceae and others. In Palmae, Pandaceae guard cells have four subsidiary cells—two of them are lateral and two polar ones. The latter ones are smaller in size and round in shape. The term *tetracytic* has been used for this type.

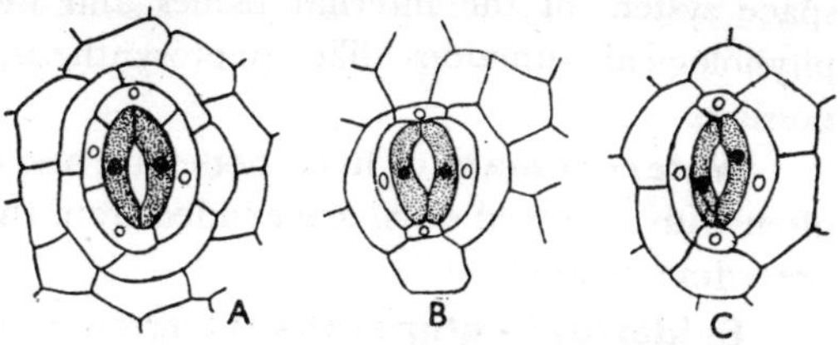

FIG. 563A. Stomata types in monocotyledons. A. In *Strelitzia*; B. In *Commelina*; C. In *Pandanus*.

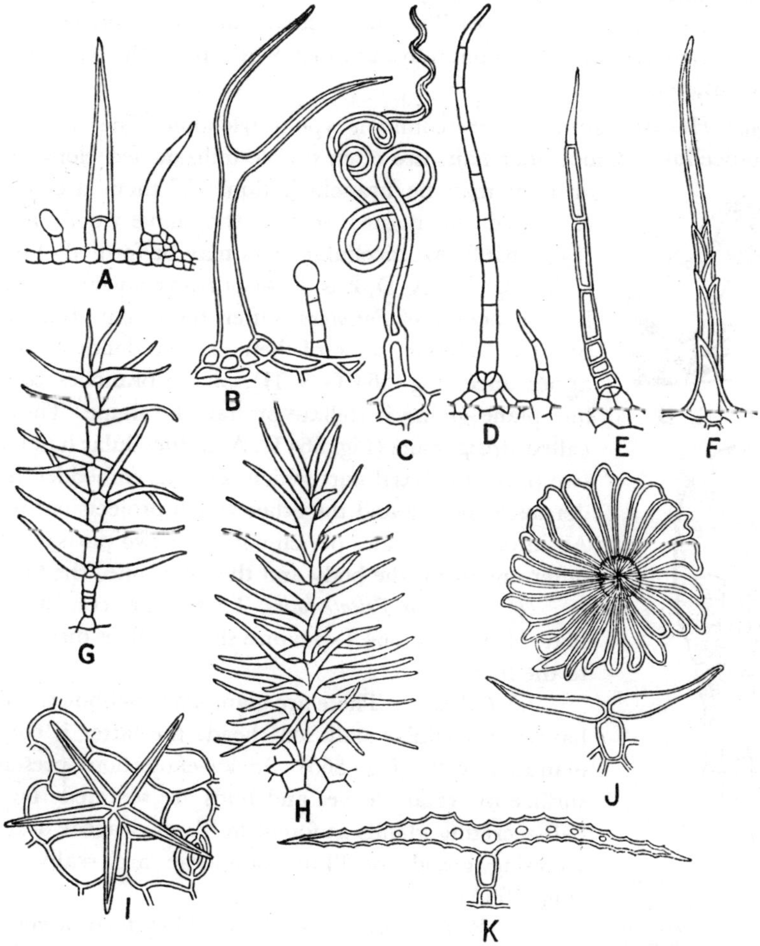

FIG. 564. Epidermal hairs—different types. A. Sharp hairs of *Lantana*. B. Lobed hair of *Amarantus*. C. Wooly hair of *Banksia*. D. Of *Lycopersicum* (tomato). E. Of *Helianthus* (sunflower). F. Of *Mimosa*. G. Dendroid hair of *Platanus*. H. Same of *Mimosa* spp. I. Stellate hair of *Althaea* (Malvaceae). J. Peltate hair in surface and side view of *Olea*. K. T-shaped hair of *Dicrasus* (Amarantaceae).

It has been suggested that stomata with many subsidiary cells are primitive, and those with few or no subsidiary cells have been derived by reduction.

The stomata are very important from physiological point of view. It is through them that interchange of gases takes place between the intercellular space system of the internal tissues and the outer atmosphere and thus important physiological functions like photosynthesis, respiration and transpiration become possible.

Water-stomata or hydathodes are also epidermal openings through which liquids often with dissolved salts, are exuded from the plants. They have been discussed in the preceding chapter.

Epidermal outgrowths. Outgrowths of diverse forms, structures and functions develop from the epidermis. All these appendages which are epidermal in origin, are referred to as **trichomes.** Thus they are different from the *emergences* like the prickles of roses, as the latter are formed by epidermis and a part of cortex. Trichomes may occur on all parts of the plant body. Some of them persist throughout the life of the organs, whereas many of them are ephemeral bodies. They may remain alive or become dead and continue as such. Trichomes have been put into a number of groups on the basis of their morphological characters.

(*a*) *Hairs.* Hairs constitute a very common type of trichome. They may be unicellular or multicellular. Unicellular hairs are often simple unbranched elongated bodies or they may be branched. Some of them are very much elongated and twisted, so that they have woolly appearance (Fig. 564C). Multicellular hairs may be formed of one row of cells (Fig. 564 A, D, E & F) or of many layers as found in the base of petiole of *Portulaca.* Often these hairs branch in very peculiar fashions; some of them assume dendroid or tree-like appearance (Fig. 564 G & H), or the branches come out in one plane giving it stellate or star-like shape. They are also called *stellate* hairs (Fig. 564 I). A multicellular hair has usually two parts, the basal part which remains embedded in the epidermis is the *foot* and the other which projects out is the *body.* An initial cell divides periclinally into two parts, of which the outer one forms the body and the inner one, the foot.

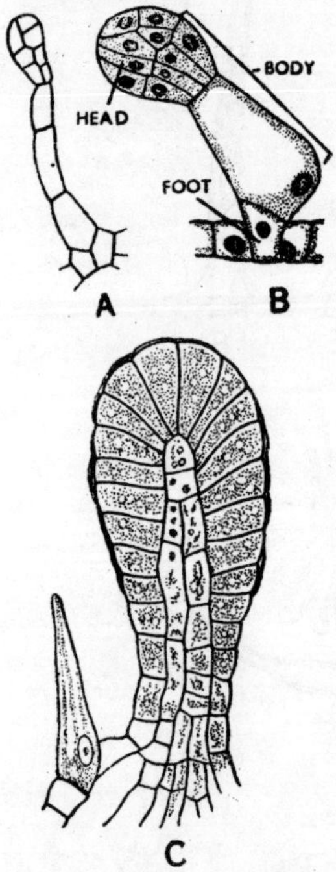

FIG. 565. Colleters (glandular trichomes). A. Of *Nicotiana* (tobacco). B. Same enlarged C. From stipule of *Viola.*

(*b*) *Scales* or *Peltate hairs.* These hairs consist of disc-like plate of cell (Fig. 564 J) put on a short stalk or directly attached to the foot.

(*c*) *Colleters.* These are glandular trichomes. Some hairs have multicellular stalk and head, the latter is composed of glandular cells (Fig. 565). Sticky exudations present on the surface of certain leaves and buds are secreted by colleters. Salt-secreting glands as found in Tamaricaceae and calcium-secreting glands of Plumbaginaceae are really interesting (Fig. 565A).

(*d*) *Water vesicles* or *bladders.* They form a very interesting type of trichome where some epidermal cells become greatly distended and serve as water reservoir. They occur in the so-called 'ice-plant' (*Mesembryanthemum crystallinum* of family Aizoaceae) where the surface of the leaves and young stems appear to be covered by ice-beads

(Fig. 566). Those occurring in *Artiplex*, also called *vesiculate* hairs, dry up with maturity and persist as a white layer on the leaf surface (Fig. 565A).

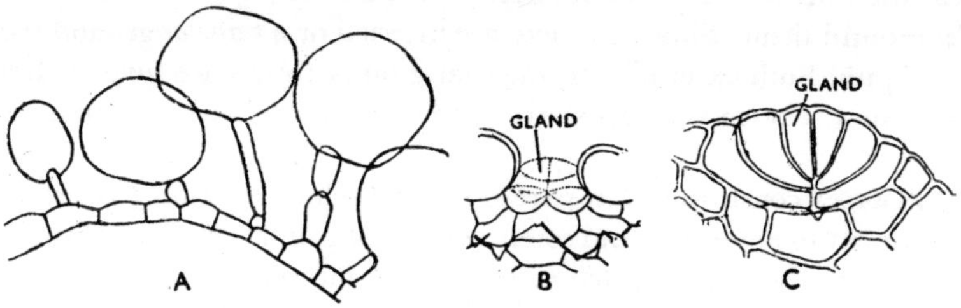

FIG. 565A. Trichomes. A. Vesiculate hairs of *Artiplex*. B. Salt gland of *Tamarix*. C. Calcium-secreting gland of *Plumbago*.

The walls of trichomes are commonly of cellulose covered by cuticle. They sometimes remain impregnated with silica and calcium carbonate. Trichomes other than glandular ones have highly vacuolated protoplast. The cotton fibres, which are really hairy outgrowths from the seeds, have secondary walls of almost pure cellulose. The stinging hairs of nettle (*Urtica dioica*) possess a peculiar type of wall structure for releasing the contents of the gland. The hair (Fig. 244) resembles of fine capillary tube with silicified upper end and calcified lower end. The base remains embedded in the epidermal cells. Coming in contact with the skin the tip breaks at a predetermined point and the sharp edge penetrates into the skin when the contents (histamine and acetycholine) are injected, so to say, to the wound.

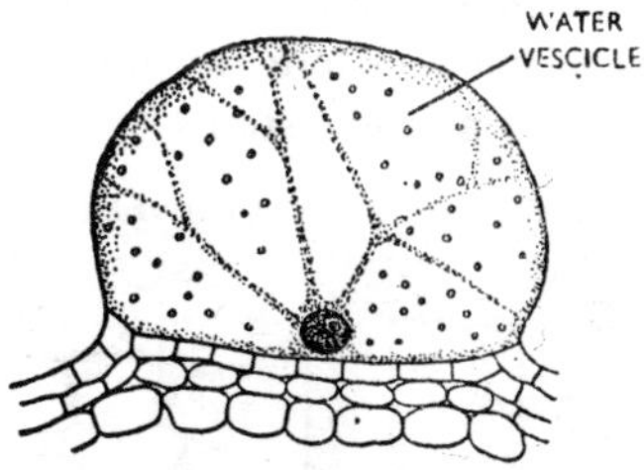

FIG. 566. Water vesicle of *Mesembryanthemum crystallinum*. (*After Haberlandt.*)

Root-hairs. As already reported the root epidermis fundamentally differs from shoot epidermis in origin and in absence of cuticle and stomata. But it bears hairs at a particular zone. Unlike the hairs and trichomes discussed above, the root-hairs are not outgrowths or appendages, but they are prolongations of the epidermal cells. During the formation of root-hairs, growth in length of the epidermal cells is checked. It has been found in some plants that root epidermis possesses two types of cells, *short* cells and *long* cells due to unequal division, and the hairs are formed from the short ones (Fig. 567) which are called *trichoblasts*. It comes out as a protuberance, continues elongation and thus the hair is formed. It has vacuolate protoplast and the nucleus moves on to the tip. The wall is thin, composed of cellulose and pectic materials. Root-hairs are short-lived bodies. During the growth of the root, old hairs are destroyed and replaced by new ones. They are responsible for the absorption of water and mineral solutes from the soil.

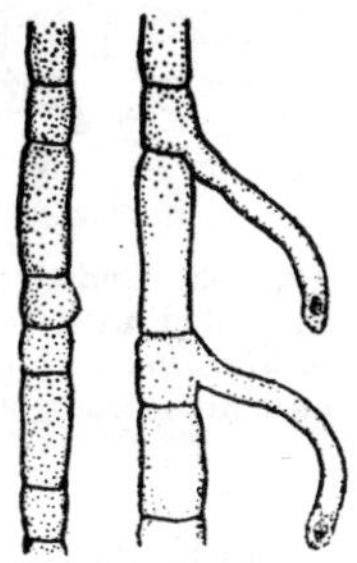

FIG. 567. Root-hairs from short cells in *Cyperus*.

THE GROUND OR FUNDAMENTAL TISSUE SYSTEM

This system includes all the tissues excepting the epidermis and the vascular bundles. Thus it is the largest or the most exhaustive system, which begins from the layer next to epidermis and continues right up to the centre of the organs in cylindrical bodies. Ground tissue system is heterogeneous in nature, consisting of diverse types of cell elements adapted to carry on different functions. In the axis of higher plants the vascular bundles occupy a restricted position inside the *stele* or central cylinder. The ground tissues occurring outside the stele, and, in fact, surrounding it, form the **cortex,** what may be called external or *extrastelar* ground tissue. Similarly there are internal or *intrastelar* ground tissues inside the stele, e.g., pith. Both external and internal ground tissues are further differentiated to specialised zones. These tissues are derived from the ground meristem of the embryonic region.

Cortex. Cortex constitutes that part of the axis which envelopes the stele. It begins from the layer next to epidermis and solely consists of primary tissues. During growth in thickness secondary tissues are added to the cortex, forming what is known as *secondary* cortex. The cortex may be rather thin, consisting of only a few layers of cells, or it may be quite massive. It is essentially composed of parenchyma which is the most predominant cell type. Collenchyma very often occurs in the superficial regions of herbaceous stems and other rapidly elongating bodies like leaf stalks and floral stalks. It may be present as discrete patches at the ridges in the bodies with wavy outline (Fig. 584), at the corners in the angular stems (Fig. 586), or as continuous bands consisting of a few layers of cells beneath the epidermis (Fig. 582). Sclerenchyma fibres are also common as continuous sheets or in patches. In some organs sclerenchyma forms a band of supporting tissue just beneath the epidermis (Fig. 590). The term **hypodermis** is applied to that zone of protecting or supporting tissues composed of collenchyma or sclerenchyma. The parenchyma cells of the cortex are loosely-arranged isodiametric ones usually containing chloroplasts. Sclereids, resin ducts, oil-cavities and laticifers may also occur. Starch, tannins and crystals are the common inclusions. The cortex of root (Fig. 601) is more homogeneous, as it consists of only parenchyma cells.

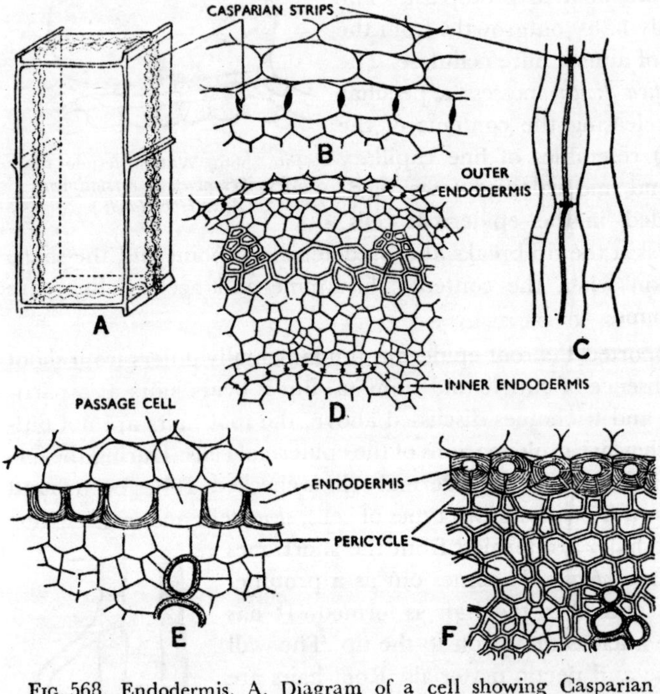

FIG. 568. Endodermis. A. Diagram of a cell showing Casparian strip. B. Endodermal cells in t.s. C. Same in l.s. D.T.s. of rhizome of *Marsilea*, with outer and inner endodermis. E. Thick-walled endodermis in root of *Zea* (maize) with passage cell. F. Same in root of *Smilax*.

Though cortex is often composed of various types of cells, naturally with different functions, but it is primarily a protective layer. Other functions like support, photosynthesis, storage, etc., are secondary.

Endodermis. It is a uniseriate zone composed of modified parenchyma cells which are compactly set, so that intercellular spaces are absent. Endodermis is considered to be the innermost layer of cortex; but some anatomists are of opinion that it is the limiting layer of stele. Though ontogenetic studies are not conclusive on this point, its intimate association with the vascular system is undeniable. The endodermal cells are elongate ones with long axes parallel to the vascular tissues. In cross-section the cells are essentially tabular in shape (Fig. 568).

A very distinctive feature of the endodermal cells in the presence of a waxy substance like suberin in form of bands or strips on the radial and transverse walls. These are known as *Casparian strips* or bands (Fig. 568 A, B & C), so named after Caspary who first noted them in 1865. The bands also show lignin reactions. In some cases the radial and inner, and often, all walls (Fig. 568 E & F) become very much thickened, so much so that the lumen of the cells may be obliterated. These walls are also strongly suberised. In this type of endodermis, what is common in monocotyledonous roots, thin-walled cells occasionally occur here and there. These cells, called *passage* cells or *transfusion* cells (Fig. 568E), are supposed to facilitate flow of water and solutes. In roots passage cells are located just opposite to the protoxylem vessels. The terms *primary* and *secondary* have also been attributed to the thin-walled and thick-walled endodermis respectively. The first type is common in pteridophytes and most dicotyledons and the second type in the monocotyledons. In an endodermal cell the cytoplasm usually remains firmly attached to the Casparian strips (Fig. 568C), so much so that they do not easily separate even on plasmolysis. Some worker believe that it is due to strong penetration of cytoplasm into the band, while others are of opinion that it is for higher viscocity of cytoplasm of endodermal cells. This phenomenon is of considerable importance in interpreting the functions of endodermis.

The endodermis lies internal to the cortex (recognised as the innermost layer of cortex by some authors) and surrounds the vascular tissues. Another layer, known as **pericycle,** is often placed between the two—the endodermis and vascular tissues. At any rate, endodermis delimits the stele from the cortex. In some stems, as in *Equisetum*, *Marsilea*, etc., another layer of endodermis occurs, which limits the vascular tissues from the internal ground tissue, pith. Such a layer is known as *inner endodermis* (Fig. 568D). In some ferns and angiosperms individual bundles remain surrounded by endodermis. It is of universal occurrence in all roots and in all the organs of most of the pteridophytes. In many gymnosperms it is absent in the stems, though it occurs in the leaves. Endodermis is present in the stems of majority of herbaceous angiosperms, in aquatic plants, in creepers and rhizomes. It is absent in the woody stems of dicotyledons and gymnosperms, and leaves of angiosperms.

In young stems of some angiosperms the innermost layer of cortex contains abundant starch grains. This layer is referred to as **starch sheath** (Fig. 583). In view of the fact that it occupies the position of the endodermis, starch sheath is considered homologous with the endodermis, the distinct wall character, the Casparian strips, are of course absent. Moreover, there are cases when the starch sheath in the young stem may lose starch and become thickened at a later stage. The starch grains present in the root of some plants are often large, and they move to the lowest side of the cell. They have been called **statoliths** by some authors, who have further suggested that they have a function in connection with geotropism, being receptors of gravitational impulse.

The function of endodermis has been a subject of long-drawn controversy, but no agreement could really be reached. Mainly on the basis of its relation to water and vascular tissues, quite a good number of functions has been ascribed to it. It is often considered as a 'watertight' layer between vascular system and surrounding tissues. Due to absence of intercellular spaces and occurrence of highly thickened walls it is said to be 'a sort of a water-dam'. Endodermis has also been called a protective layer, something like an accessory epidermis; its connection with root-pressure has been suggested, it has been called an 'air-dam' meant for preventing the conducting elements being clogged with air. The cells of thin-walled endodermis retain the power of cell division. The lateral roots and adventitious buds often originate from this layer.

Pericycle. The pericycle is the region consisting of one or more layers of cells which surround the vascular tissues. It is located between the endodermis and the vascular bundles. Pericycle is regarded as the limiting layer of stele; thus it forms the outermost part of intrastelar ground tissues, the other portions being the pith and medullary rays. Though occurring inside the stele it is considered to be distinctly separate from the vascular tissues. It typically consists of parenchyma (Fig. 601), as found in all roots and pteridophytes. Heterogeneous pericycle or pericycle consisting of parenchyma and sclerenchyma or sclerenchyma in form of discrete bands have been recognised since a long time. But modern anatomists have shown that most of the so-called pericyclic fibres of stems are really phloem fibres. The economically important fibres of hemp, *Cannabis sativa* of family Urticaceae, and flax, *Linum usitatissimum* of family Linaceae, definitely belong to phloem. So it is absent in many angiospermic stems. Pericycle, however, is of universal occurrence in the roots of higher plants, only some aquatic plants and parasites excepted, and also in the stems and roots of pteridophytes.

Pericycle is the seat of origin of lateral roots. During secondary growth in thickness phellogen or cork-cambium arises in this layer. In case of anomalous steles it forms secondary cambium. The term 'pericambium' was formerly used for the pericycle of roots because of its property of becoming meristematic and giving rise to new tissues. In monocotyledons, which usually have no secondary increase, the pericycle generally becomes sclerified, either partly or wholly. The parenchyma cells of pericycle serve for storage. They may contain laticiferous tissues and secretory ducts.

It is usually one-layered in the roots of angiosperms (Fig. 601). But in quite a good number of monocotyledons like *Smilax, Agave* of family Liliaceae, palms and some grasses and in some dicotyledons like *Morus* of family Moraceae, it may be multilayered (Fig. 568F). The pericycle of gymnosperms is typically multiseriate.

Pith. The pith is the main internal ground tissue present at the central portion of the organs limited by the vascular bundles. Often it is quite massive, consisting of parenchyma cells with profuse intercellular spaces. The cells are isodiametric in shape and sometimes remain arranged in longitudinal series due to their development from rib meristems. At the mature stage the cells of pith lack chloroplasts, when colourless leucoplasts are usually present. The walls are thin and made of cellulose. The pith cells serve as organs for storage of starch, fatty substances and often of crystals and tannins. Secretory tissues are frequently present. Fibres and sclereids may also occur, though rarely. In some cases the outer part of the pith is composed of smaller cells with thicker walls and dense protoplasm and thus become morphologically distinct from the rest. This distinct outer region has been called *perimedullary zone* or *medullary sheath*. The cells may be parenchymatous, as found in many members of families Euphorbiaceae, Boraginaceae; sclerenchymatous, as in Compositae, Umbelliferae; or both parenchyma and sclerenchyma may be present. In many herbaceous plants the pith is destroyed during rapid elongation accompanied by radial expansion of the stem. *Hollow pith,* as it is called, with broken walls lining the cavity may be seen in the family Cucurbitaceae (Fig. 584) and many grasses (Fig. 593). Pith is absent in many dicotyledonous roots. When present it resembles that of the stem, but is more homogeneous and normally is not destroyed.

Medullary rays. Parenchymatous ground tissues passing in between the vascular bundles, thus occupying the interfascicular areas, constitute the medullary rays (**Figs.** 582 & 589). These cells, often slightly elongate, link the parenchyma of pith, also called *medulla*, with those of cortex. So they often look like radiating portion of pith. Due to their origin from early meristems, they are referred to as primary medullary rays. During secondary increase in thickness some of these cells become meristematic and produce secondary tissues.

GROUND TISSUES OF THE LEAF

Ground tissue system of the foliage leaves markedly differs from that of stem and root. In leaves the ground tissue is called **mesophyll** tissue which is mainly composed of thin-walled parenchyma cells (chlorenchyma) with profuse chloroplasts, so that photosynthesis may be effectively carried on. In many monocotyledonous leaves with parallel venation mesophyll consists of more or less isodiametric parenchyma cells with intercellular spaces (Fig. 618). But in most of the dicotyledons mesophyll is differentiated into two types of cells. Beneath the adaxial or upper epidermis mesophyll cells are elongated or columnar ones arranged more or less at right angles to the epidermis. These are known as **palisade** parenchyma (Fig. 613). Those occurring towards abaxial or lower epidermis are isodiametric or irregular in shape with more intercellular spaces. They are called **spongy** parenchyma (Fig. 613). In some leaves, as for example *Eucalyptus*, palisade cells may be present on both the sides. Chloroplasts are abundant in mesophyll cells, which is, in fact, the region of most active photosynthesis. Secretory cells and sclereids may often be present.

THE VASCULAR TISSUE SYSTEM

The vascular tissue system consists of the complex tissues, **xylem** and **phloem,** which constitute discrete conducting strands called **vascular bundles.** These are usually primary in nature. The elements of xylem and phloem have already been discussed in the preceding chapters. In spite of the occurrence of supporting and other cells the function of the vascular bundles is primarily *conduction*, xylem for the conduction of water with dissolved mineral matters, and phloem for the conduction of elaborated food matters in solution. The term *fibrovascular* bundle had been in use in the past mainly because of association of sheaths of fibrous tissues with the bundles. But in view of the fact that the fibrous sheaths do not always form a part of the bundle, the term fibrovascular bundle has been discarded and replaced by simply vascular bundle.

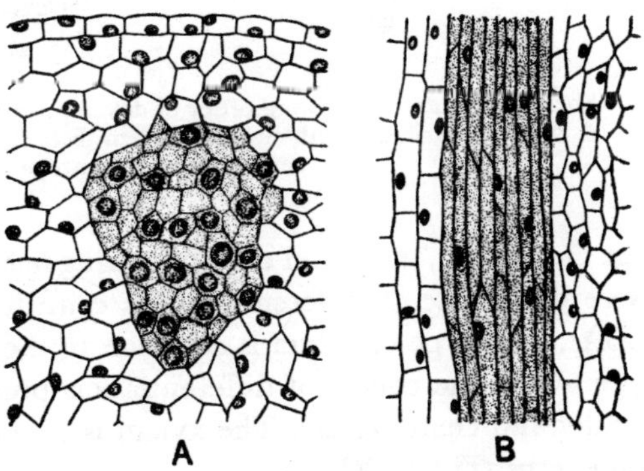

Fig. 569. Procambium. A. Transverse view. B. Longitudinal view.

The vascular bundles originate from the *procambium* of the apical meristem. In the promeristem stage normally all the cells are isodiametric. During the longitudinal divisions they undergo, some cells are set apart as patches or strands. These cells are smaller and elongate ones and possess dense cytoplasm. This meristematic tissue (Fig. 569) is called *procambium*, which is destined to produce the elements of vascular bundles. In recent years the terms 'provascular tissue' or 'provascular meristem' have also been used to designate this tissue. Procambium appears early near the apex of the axis and gradually differentiates out backwards, so that the course of development is acropetal. The procambium shows early differentiation into two parts. The part destined to give rise to phloem takes dense stain and shows different planes of division than the other part which would eventually produce xylem. So the terms *phloic* procambium and *xyloic* procambium have been used for the two parts. This fact really justifies the use of the term provascular meristem.

The phloem elements mature earlier than the xylem elements; the order of maturation is always acropetal. But that of xylem elements may be both acropetal or basipetal. Thus the progressive development of the vascular elements from the procambium strands may be both centripetal and centrifugal. The order of differentiation in phloem is probably always centripetal, i.e., towards the centre of the axis. But in case of xylem three different conditions are possible as regards the order of differentiation of elements. In the first type the initial xylem elements are located furthest from the axis, where the course of development is obviously centripetal or towards the axis. This type of xylem is said to be *exarch*, what is characteristic of the roots (Fig. 570B). In the second type, the condition is

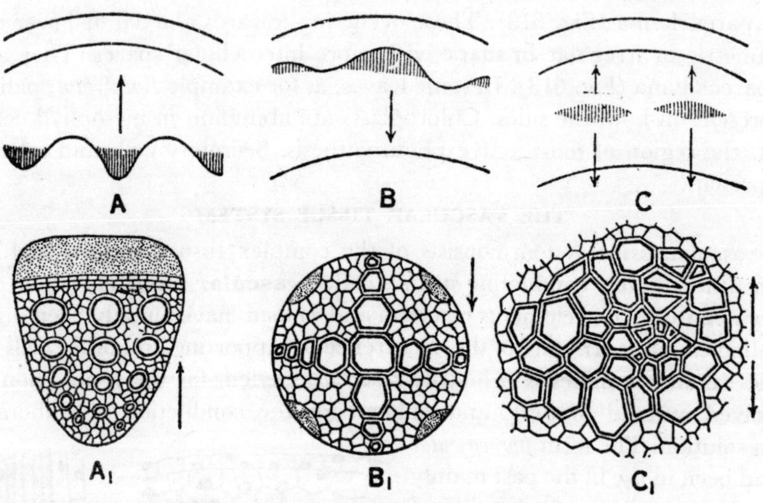

Fig. 570. Diagrams showing order of development of primray vascular tissue in transverse views. A & A_1. Centrifugal with endarch xylem. B & B_1. Centripetal with exarch xylem. C & C_1. Both centrifugal and centripetal with mesarch xylem.

just the reverse, the initial ones occurring nearest the axis and the latest furthest from it. Here the course of differentiation is centrifugal and the xylem is called *endarch* (Fig. 570A). This type of xylem is common in the stems of spermatophytes. There is a third type where the course of differentiation proceeds in two directions, i.e., both centripetally and centrifugally. The xylem is that case is known as *mesarch*, what is found in some ferns (Fig. 570C).

Primary xylem. The first-formed xylem is called **protoxylem** and the lately-formed one is known as **metaxylem.** Protoxylem differentiates from the procambium when the organ continues elongation and is often subjected to considerable stretching. It consists of tracheids, vessels and parenchyma, the fibres being usually absent. The cells and elements are elongate and slender bodies with cellulose cell walls, reinforced by lignified secondary walls. The secondary wall layers are deposited in form of rings and spirals (Figs. 583B & 585B) which keeps the thin and plastic walls of empty water-conducting elements in proper position and prevent collapse during the elongation of the organs. Annular and spiral thickenings are thus characteristic of the protoxylem elements. Scalariform or ladder-like thickening may also occur. Of all the above-mentioned types the spiral elements are much more abundant. The water-conducting elements of protoxylem are tracheids in pteridophytes, gymnosperm and some angiosperms and vessels in many angiosperms. The annular cells or vessels are often subjected to so much stress during elongation that the primary wall is destroyed and secondary wall distorted, so that even a canal-like body, called protoxylem *lacuna*, may be formed (Fig. 518C).

The lately-formed xylem or metaxylem elements are not subjected to stretching as they mature only when organs have completed growth in length. They have more extensive secondary walls in form of network (reticulate) or pits. Scalariform thickening may also be present. The two parts of the xylem often intergrade. Metaxylem is the main water-conducting portion in plants which have no secondary increase in thickness. It is more complex than protoxylem and possesses more tracheary elements. The type of primary xylem, whether exarch, endarch or mesarch, is determined by the position of protoxylem in relation to metaxylem (Fig. 570). In exarch condition protoxylem occurs towards the circumference and metaxylem towards centre; in endarch the position is just the reverse, i.e., protoxylem towards centre and metaxylem towards circumference; in mesarch protoxylem is flanked on two sides or remains surrounded by metaxylem.

Primary phloem. The first-formed elements of phloem are called **protophloem.** The cells of protophloem differ considerably from the lately-formed ones, known as **metaphloem** elements, in size and shape. The protophloem elements are slender and elongate bodies with cellulose cell wall. They are subjected to considerable stretching during the rapid growth in length of the organs. The metaphloem elements mature after the completion of growth in the surrounding tissues and so they are not subjected to stretching. Primary phloem consists of protophloem and metaphloem. In secondary phloem protophloem is absent, because secondary tissues are formed when growth in length has ceased.

The protophloem constitutes the first vascular elements to mature from the procambium. It is composed of sieve elements—sieve tubes in angiosperms and sieve cells in gymnosperms and pteridophytes. These are small cells more or less similar to the procambium ones. In keeping with the rapid elongation of the organs the cells are elongate and slender. Companion cells are scarce or lacking. Parenchyma and fibres are formed later from the procambium. Metaphloem is rather complex, consisting of all the elements —sieve tubes or cells, companion cells, parenchyma and fibres.

The primary phloem persists throughout the life of the organs and carries on its physiological functions, where secondary phloem is not formed. This condition prevails in the monocotyledons and in some dicotyledons like *Cucurbita*. But in plants having secondary growth primary phloem is of short duration, as the considerable amount of secondary phloem formed later takes over the physiological function, and the primary phloem often gets crushed. However, in some plants with secondary growth, like *Solanum* of family Solanaceae, *Aster* of family Compositae the amount of secondary phloem is small and primary phloem persists all through.

Normally phloem occurs on the outerside of the xylem in the vascular bundles of stems and on the abaxial side in the leaves and leaf-like organs. So phloem is said to be *external* with reference to xylem. But in a number of dicotyledonous families like Solanaceae, Cucurbitaceae, Compositae, Apocynaceae, Asclepiadaceae, Convolvulaceae and Compositae a part of phloem may be present on the internal side as well. This is known as *internal* phloem, as opposed to normal external one. It usually occurs as large or small strands, in close association with primary xylem, as in *Cucurbita* (Fig. 585), or it may often form independent strands in the outer part of the pith, as found in potato (Fig. 571), *Calotropis* (Fig. 588) and others. Early workers called it *intraxylary* phloem, but that term has been abandoned now. The external and internal phloem are practically similar in composition, structure and arrangement of cells, but internal phloem develops later than the external one. It should not, at any rate, be confused with another type called 'interxylary' or *included* phloem found in some dicotyledonous families like Combretaceae, Loganiaceae, Acanthaceae. Interxylary phloem is really secondary phloem formed due to peculiar behaviour of the cambium cells and it ultimately gets embedded in secondary xylem.

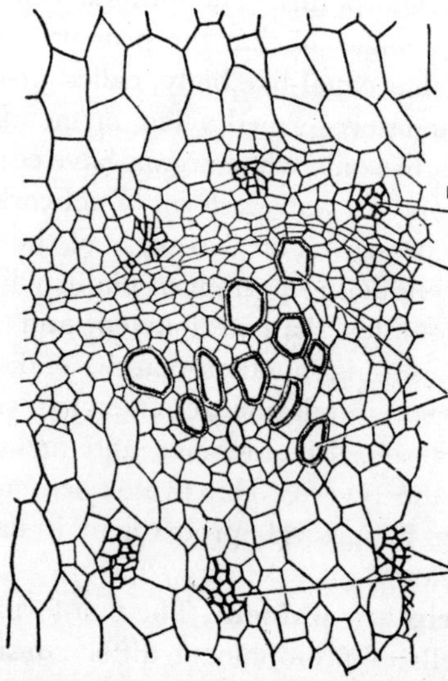

FIG. 571. Internal phloem in a young stem of *Solanum* (potato) in transverse section.

Cambium. As already stated the procambium cells differentiate and mature into xylem and phloem elements. In plants having no secondary growth in thickness, as in lower vascular plants and monocotyledons, all the procambium cells ultimately mature into vascular tissues. But in the stems of dicotyledons and gymnosperms, which grow in thickness, a part of the procambium remains meristematic. It is called **cambium** (Fig. 583). It is a *lateral* meristem occurring parallel to the axis. Cambium produces secondary tissues and is thus responsible for growth in thickness of the organs. The cambium cells have vacuolate protoplast and thin cell wall composed of cellulose, often with primary pit-fields. The cells divide periclinally and produce secondary tissues.

The vascular cambium is composed of two types of cells, viz., elongated cells with tapering ends, called *fusiform* initials, and small, more or less isodiametric cells known as *ray* initials. The fusiform initials by cell division give rise to the secondary tissues—secondary xylem and secondary phloem which remain arranged along the long axis of the organ. The ray initials produce the ray cells of xylem and phloem, which occur in transverse or horizontal series.

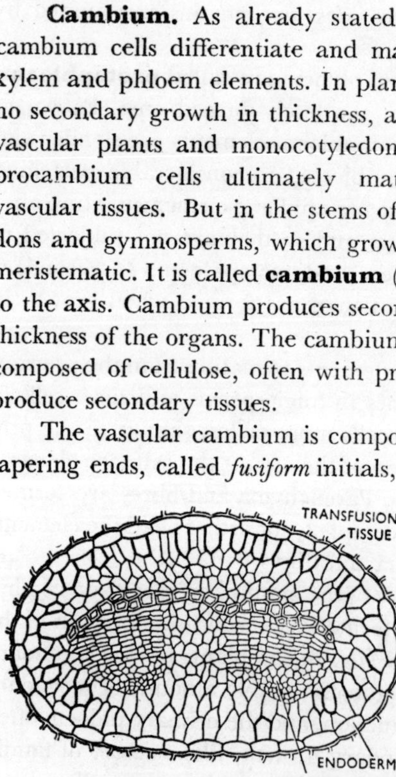

FIG. 572. Transfusion tissue in a leaf of *Pinus* (pine) in transverse view.

Transfusion tissue. The leaves of gymnosperms like pine possess a peculiar type of conducting tissue in addition to normal vascular tissues. This tissue, called **transfusion tissue** (Fig. 572), consists of rather short tracheid-like cells. They have thin cellulose walls with characteristic thickenings of the tracheids, viz., bordered-pitted, scalariform or reticulate ones. The cells are devoid of protoplasts. Barring those two characters—localised thickenings of the walls and absence of protoplast, they resemble elongate parenchyma cells. They occur adjacent to xylem in the bundle, partly or wholly surrounding the latter. Some workers are of opinion that transfusion tissue derives its origin from centripetal xylem, while others consider it to be transformed parenchyma cells outside the vascular tissues. Their function is uncertain. But as they connect the veins with the mesophyll of the leaves, taking the position of the vein-lets, they may be regarded as modified vascular tissues. According to some workers it is a water-storage tissue.

Types of vascular bundles. The complex tissues, xylem and phloem, are usually associated in the formation of the vascular bundle. They show three common types of arrangements, viz., (*i*) the two complex tissues occur side by side, (*ii*) one tissue remains surrounded by the other, (*iii*) the two tissues are separated from each other. According to mode of occurrence of the elements, the vascular bundles are of the following types:

1. **Collateral** (Fig. 573 A & B). This is the most common type of vascular bundle in the stems and leaves of angiosperms and gymnosperms. Here xylem and phloem remain side by side arranged on the same radius, phloem on the outer side, i.e., external, and xylem towards the pith, i.e., internal. In the stems of most dicotyledons and gymnosperms, a strip of lateral meristem, the cambium, occurs between xylem and phloem (Fig. 573A). The bundles in those cases are called *open*, whereas those without cambium, e.g., monocotyledons, are said to be *closed* (Fig. 573B).

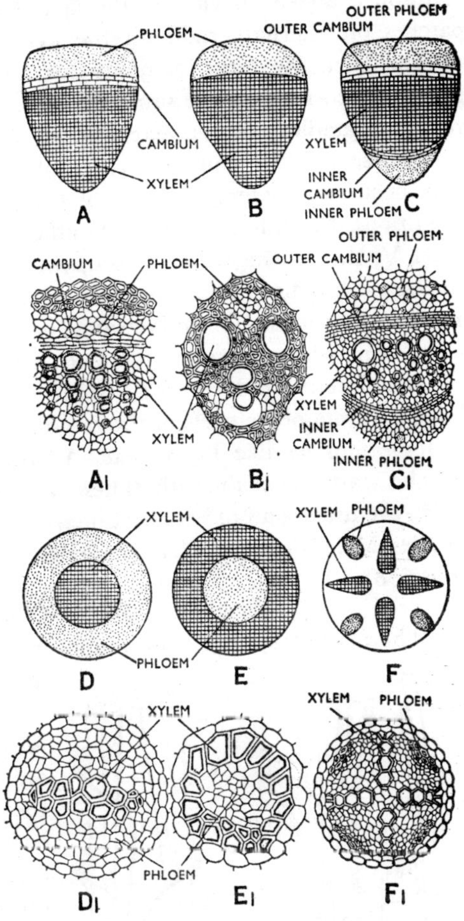

FIG. 573. Vascular bundles—types. A. Model of collateral open bundle. A_1. Same in t.s. B. Collateral closed (model). B_1. Same in t.s. C. Bicollateral (model). C_1. Same in t.s. D. Concentric-amphicribral (model). D_1. Same in t.s. E. Concentric—amphivasal (model). E_1. Same in t.s. F. Radial (model). F_1. Same in t.s.

2. **Bicollateral** (Fig. 573C). These are collateral bundles, where, in addition to the external phloem, another patch of phloem occurs on the inner side, what may be called *internal* phloem. Though rather uncommon, this type occurs in the family Cucurbitaceae. Two strips of cambium and two patches of phloem are present on the outer and inner sides of xylem. Naturally the sequence is outer phloem, outer cambium, xylem, inner cambium and inner phloem. Obviously the bundles are always open.

3. **Concentric** (Fig. 573 D & E). Here one kind of vascular tissue completely surrounds the other. Thus concentric bundles are of two types—xylem surrounding phloem, called **amphivasal** or *leptocentric* bundles (Fig. 573E) or, phloem surrounding xylem called **amphicribral** or *hadrocentric* bundles (Fig. 573D). Though less common, the amphivasal bundles occur in some monocotyledons, particularly in the nodal regions

and in some rhizomes. In dicotyledons the medullary bundles are amphivasal. Amphicribral ones are frequently found in the ferns. The small bundles of flowers, fruits and some leaves of dicotyledons are of this type. The concentric vascular bundles are always closed.

4. **Radial** (Fig. 573F). These bundles are fundamentally different from the types mentioned above, in view of the fact that here xylem and phloem occur in separate patches on alternate radii on the axis intervened by non-conducting tissues. Early workers used the term 'conjoint' for all the types discussed above with xylem and phloem occurring on the same radius, as opposed to the radial arrangement. Radial vascular bundles are characteristic of the roots.

These are the common types of vascular bundles. But variations in structure may occur and intergradations of the types are also possible, even with transitional conditions. In some grasses the bundles are collateral, where xylem occurs in form of letter V, the two metaxylem occupying the flanks and phloem located between them. The same bundle may differ in the arrangement of the elements along its course, so much so that it may be collateral at one level, amphivasal at another and even transitional somewhere between the two levels.

The internal phloem in the stems of some plants, e.g., family Solanaceae, may be detached from other parts of the bundle and occur as independent patches in the pith. In some families like Piperaceae, Amarantaceae, etc., even all the vascular bundles may be scattered in the pith. These are referred to as **medullary** bundles (Fig. 649), which are usually amphivasal. Similarly bundles may occur in the cortex in members of families Melastomaceae, Cactaceae, Oleaceae, etc. They are called **cortical** bundles (Fig. 658).

THE VASCULAR SKELETON

The vascular system is continuous in the two parts of the axis, the stem and the root, and is also connected with the lateral expansions, the leaves. The vascular bundles, in fact, form a skeleton comparable to the skeleton of the animal bodies. The skeleton (Fig. 574) differs in the various plant organs, and this difference is constant and characteristic. Every species has its own plan and arrangement what is different from other species. This vascular skeleton becomes increasingly more complex in the plant kingdom from pteridophytes to the spermatophytes.

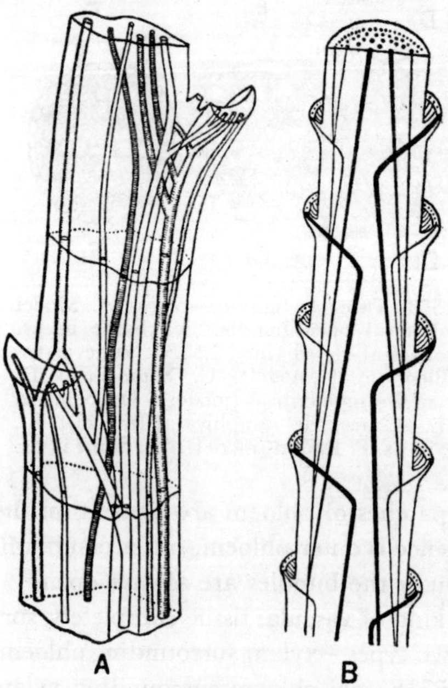

FIG. 574. The vascular skeleton. A. In a dicotyledon (stem of potato, *after Artschwager*). B. In a monocotyledon (palm type).

The two parts of the axis, stem and root, possess different types of vascular bundles—collateral bundles in the stem with endarch xylem and radial bundles in the root with exarch xylem. Nonetheless, the continuity of the vascular system in the axis is maintained. Some orientations take place in the region between the root and the stem, usually the hypocotyl. That region is referred to as *transition region*. There are diversities as regards the methods of changes taking place in transition regions (to be discussed in detail in succeeding chapters); but the xylem usually splits and swings laterally by 180 degrees and ultimately joins up with phloem. Thus the radial bundles of the root with exarch xylem become continuous with the collateral bundles of the stem having endarch xylem.

Nodal Anatomy

At each node of the stem the vascular bundle runs into the leaf; it is called a **leaf trace** or foliar trace (Fig. 575). A trace is nothing but an extension of the vascular tissues of the stem into the leaf. The terminal part of the trace bundle is made of xylem alone, and the basal part of xylem and phloem. The traces supplying the leaves, forming what is known as *leaf supply*, vary from one to many; but the number is constant for a particular species, and even for a family. The traces have characteristic forms. The bundles may remain separate in the stem; but if a trace is followed along its descent towards stem it is found to join ultimately with another bundle which has entered from a lower leaf. Questions actually arose on the point as to how much of the vascular supply belongs to stem and how much to the leaf. In fact, bundles had been said to be of three types, viz., *leaf trace* bundles, *cauline* bundles and *common* bundles. Leaf trace bundles, as already reported, are connected with the leaves. Cauline bundles (*caulis*=stem) are those which form the vascular skeleton of the stem and do not enter the leaves. Thus they belong to the stem proper. Common bundles run through the stem in unbranched condition for some distance and finally terminate as leaf traces. Thus they are common both to the stem and the leaves. In lower vascular plants like *Lycopodium* and *Selaginella* the leaves are very small and simple. The traces are naturally small and superficially connected with the strongly developed vascular skeleton. But higher vascular plants—ferns, gymnosperms and angiosperms, have larger leaf traces. So here the vascular system of the stem is closely associated with the leaves, in fact, forming a vascular skeleton with interconnected leaf traces.

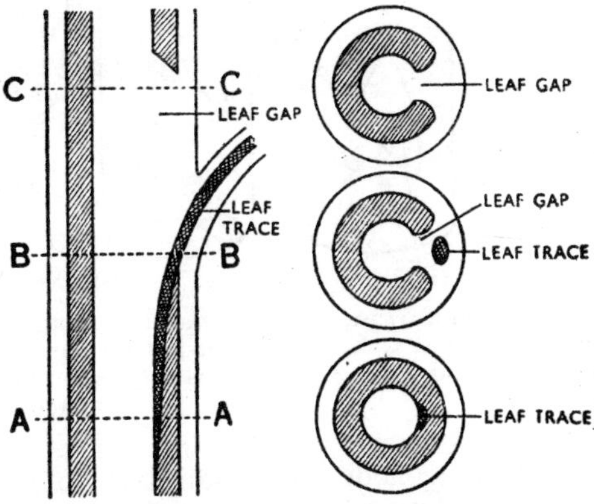

FIG. 575. Leaf trace and gap (diagrammatic). *Left*. Longitudinal section through a node showing leaf trace and gap. *Right*. Transverse sections at the regions A A, B B, and C C shown by dotted lines.

Just above the trace parenchymatous cells, instead of vascular tissues, differentiate up to a limited distance. These parenchymatous regions are called **leaf gaps** (Fig. 575). It should be noted that gaps are not breaks in the vascular system, but they are the areas or openings where cortex and pith become continuous. Lateral connections are established below and above the gaps, so that the continuity of the system remains undisturbed. Leaf gaps are absent in lower vascular plants like *Lycopodium*, *Equisetum*, etc., but they are constant in the ferns, gymnosperms and angiosperms.

The number of traces and gaps is variable (Fig. 575A). It is usually one in pteridophytes; one or two in gymnosperms; and one, three, five or many in angiosperms. Accordingly the common types of nodes in the dicotyledons are said to be *unilacunar* with one gap and one trace; *trilacunar* with three gaps and three traces to a leaf; and *multilacunar*, with many gaps and traces gaps being also known as *lacuna*. Anatomists believe that trilacunar condition is more primitive in the dicotyledons and the other two types have evolved either by reduction or amplification in the number of traces (Sinnot, 1934). This assumption has been refuted (Bailey and others) by many workers. In fact volume of recent researches on nodal anatomy (Gunkel & Wetmore, 1946; Marsden & Bailey, '56 & others) lead to the reputation of Sinnot's assumption regarding the primitiveness

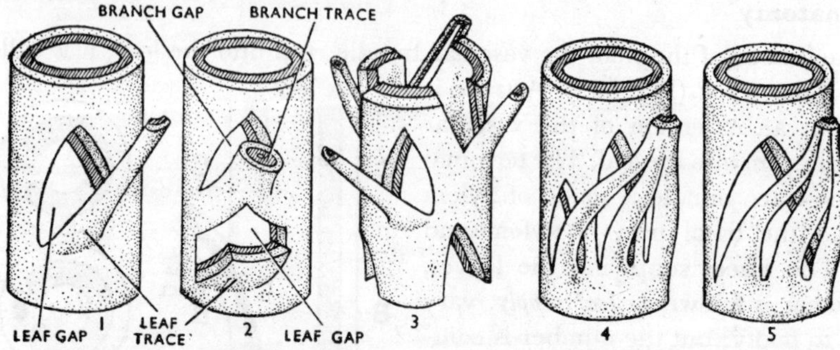

FIG. 575A. Vascular skeleton (diagrammatic) with different types of traces and gaps. 1. Unilacunar node with one trace. 2. Unilacunar node with branch traces and gaps. 3. Overlapping gaps forming a network. 4. Trilacunar node with three traces. 5. Unilacunar node with three traces.

of trilacunar node. Four main types of dicotyledonous nodes are now recognised. They are—(1) unilacunar *two-trace*, in which the two traces are connected to opposite halves of the eustele; (2) unilacunar with a *single trace*, (3) trilacunar with traces from three gaps; and (4) multilacunar, in which more than three traces and three gaps per node are present. The trilacunar type remains a form from which the unilacunar one trace and multilacunar types have been derived. The discovery of the fourth type of nodal anatomy (unilacunar—two traces) by Marsden & Bailey led to revision of concepts of nodal evolution. It now appears to be basic in angiosperms. Bailey, Canright ('55) & others have put foward the following reasons in support of their contention:—(a) this

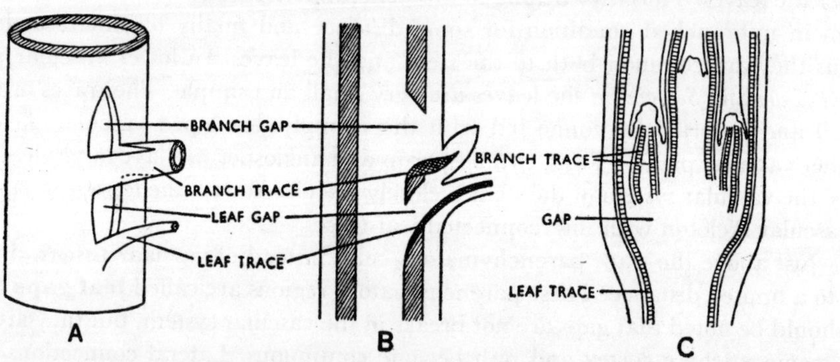

FIG. 576. Branch trace and gap (diagrammatic). A. A vascular cylinder showing departure of leaf and branch traces with associated gaps. B. Longitudinal section through a node showing the same. C. L.s. through a node with opposite leaf arrangement with branch and leaf traces and gaps.

condition seems to be wide-spread and basic in vascular plants other than angiosperms, what is expected in case of 'pteropoid' origin of angiosperms; (b) a large percentage of cotyledonary nodes appear to have two traces and retain that condition; (c) dicotyledons with many other primitive features exhibit unilacular two-traces or some derivative form, e.g. *Austrobaileya* showing this condition throughout the entire plant. Apart from many other families, occurrence of this condition has been curiously noted in advanced family like Verbenace. So the sequence followed in the evolution of nodal anatomy would be (1) two-traces unilacunar, trilacunar, and multilacunar; or (2) two-trace unilacunar, one-trace unilacunar, trilacunar, and multilacunar. The one-trace unilacunar could have been derived from the trilacunar as well.

In the monocotyledons the vascular bundles follow a different course. The common type of vascular skeleton here is known as 'palm type' which occurs in the palms and other monocotyledons (Fig. 574B). The traces supplying a leaf here are numerous. They are usually of two kinds—small and large. The small ones which pass into a stem from the leaf base are located peripherally in the stem practically encircling it. The larger traces penetrate up to the centre of the stem in the upper parts and move towards the periphery in the lower, where they fuse with others. Since the penetration of the traces in the stem is not uniform, the bundles appear scattered in cross-section.

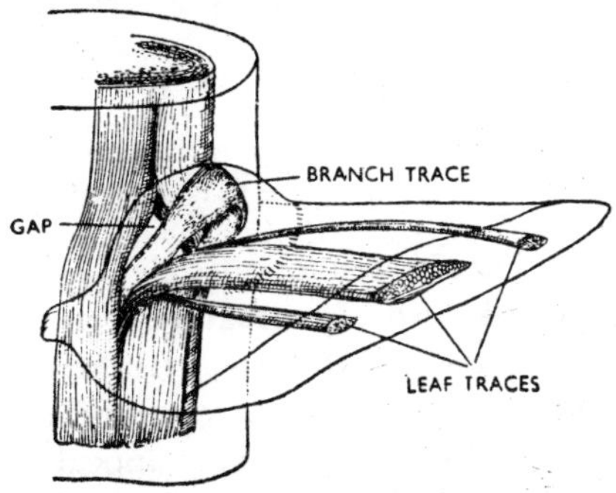

FIG. 577. Nodal region of *Phlox* (diagrammatic) showing relations of leaf and branch traces to the stele of stem. (*After Miller and Wetmore.*).

The continuations of the vascular system into the lateral branches constitute the **branch traces** or *ramular* traces (Figs. 576 & 577). Like leaf traces they also prolong into the axis and ultimately merge with the vascular system. In dicotyledons and gymnosperms there are usually two branch traces, in some plants there may be one, and in others they may be more than two. Gaps, known as **branch gaps,** are also present here accompanying branch traces. They occur in all vascular plants having pith. These are larger and more extensive than leaf gaps.

Thus the longitudinal course of the vascular bundles forming a discrete skeleton, is evident from the continuity of the root-stem axis, and occurrence of leaf traces and branch traces which tie up all the parts of the axis and the appendages. The anatomy of the node is being studied intensively now, particularly in view of its importance in taxonomy and comparative morphology of the organs concerned.

THE STELE

The central core of the axis is called **stele.** It includes the vascular tissues and the ground tissues like pericycle and pith, when present. These are referred to as intrastelar ground tissues. The stele remains surrounded by the cortex, what constitutes extrastelar ground tissue, the endodermis being the innermost layer. Both stem and root possess stele enveloped by the cortex. In view of this fundamental similarity in the two organs of the axis in anatomical nature, the *stelar theory* was proposed in the later part of the nineteenth century. The theory was readily accepted and it profoundly influenced investigations on comparative anatomy and proved to be immensely helpful in the interpretation of stem anatomy, particularly of the lower vascular plants. The proponents of stelar theory considered endodermis as the innermost layer of cortex and pericycle as the outermost portion of stele. In recent years some anatomists have suggested that the boundary between the stele and cortex is still doubtful, and so the stelar theory needs a thorough re-examination. But that at any rate does not minimise the classical importance of the stelar theory, which has been 'of unmistakable value in emphasising the unity of the structure of vascular system'—as stated by Prof. Esau. Different types of steles (Fig. 578) were recognised, a brief review of which is given here.

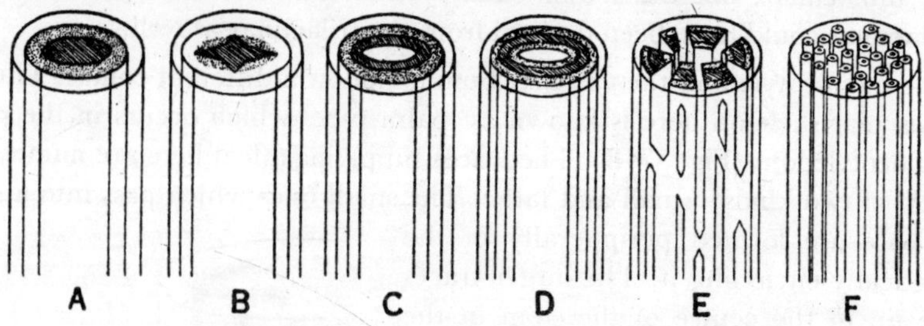

FIG. 578. Stele—type (diagrammatic). A. Protostele. B. Radial stele. C. Siphonostele—ectophloic. D. Siphonostele—amphiphloic. E. Dictyostele. F. Atactostele.

The simplest type of stele consists of a solid column of vascular tissues having no pith. This is known as **protostele** (Fig. 578A). This is the most primitive one from phylogenetic point of view, from which other types have evolved. In the simplest condition in a protostele xylem forms the core and remains completely surrounded by phloem. In cross-section the protostele may appear as a column—circular, angular, stellate or even irregularly lobed in outline. Accordingly a few terms have been in use. Protostele with smooth core of xylem is called **haplostele,** which is considered most primitive (Fig. 579A). That with xylem in form of radiating ribs is known as **actinostele** (Fig. 579B). The most advanced type of protostele is one where xylem and phloem intermingle and xylem occurs as separate plates usually lying parallel to one another. This is known as **plectostele** (Fig. 579C). The stele of the root of dicotyledons with pithless central column is also regarded as protostele. Some authors called it **radial** stele (Fig. 578B), in view of radial arrangement of vascular tissues where xylem radiates like arms towards pericycle, phloem patches lying alternatingly. Protostele is common in the lower vascular plants. It also occurs in the earliest parts of shoot of ferns and in some aquatic plants of the angiosperms. Haplostele is found in *Lycopodium cernuum*, *Selaginella kraussiana;* actinostele in *Isoetes coromandeliana*, *Psilotum triquetrum;* and plectostele in *Lycopodium clavatum.*

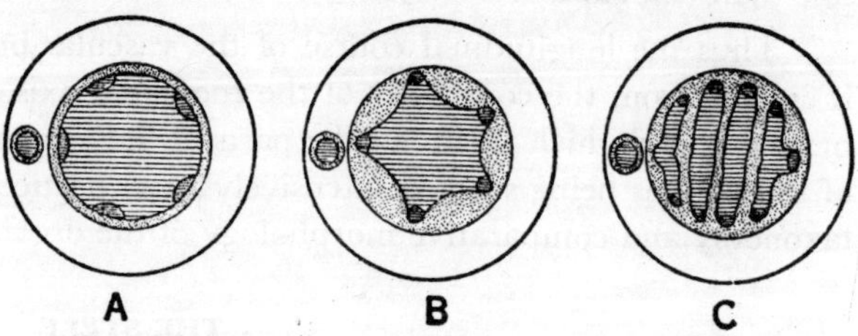

FIG. 579. Protostele—types (diagrammatic, sectional views). A. Haplostele. B. Actinostele. C. Plectostele.

The next type is known as **siphonostele** or tubular stele (Fig. 578 C to F), which is characterised by the differentiation of a pith in the central region. It is considered to be derived phylogenetically from the protostele and thus represents an advance from the point of view of evolution. This is undoubtedly the most prevailing type in ferns, gymnosperms and angiosperms. In cross-section siphonostele also shows various outlines. It is of two types, according to distribution of the vascular tissues, viz., **ectophloic** siphonostele (Figs. 578C & 580A), when phloem occurs on the outer side of xylem, and **amphiphloic** siphonostele (Figs. 578D & 580B), when the phloem is present both on the outer and the inner side (internal) of xylem. In simplest cases, as in lower vascular plants, siphonostele has no leaf gaps; in some others the gaps are very small and thus not overlapping, so that a section through the internode shows a continuous ring of vascular bundles. Such a siphonostele without overlapping gaps is also called **solenestele.** In ferns leaf gaps are fairly large and overlapping. As a result the whole stelar system is dissected into a net-like structure. This type is known as **dictyostele** or dissected siphonostele (Figs. 578E & 580C), and the intervening strands of the vascular tissues, each resembling a miniature protostele and occurring laterally to two overlapping gaps are called **meristeles.** As the term dictyostele was used by some early authors in a different sense, modern workers have preferred to use **eustele,** meaning true stele, for dissected siphonostele, what is very common in dicotyledons and gymnosperms. In majority of the monocotyledons the vascular bundles remain scattered in the ground tissues, so that the semblance of a stele is lost. This complex type of stele (Figs. 578F & 580D) consisting of dispersed strands has been called **atactostele,** (*atactos*, meaning, without any order).

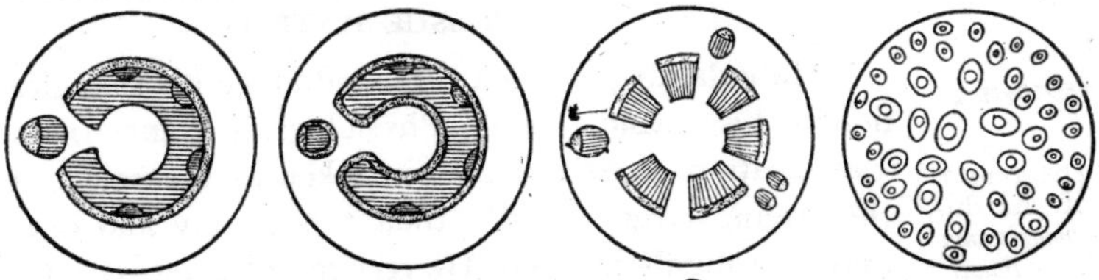

FIG. 580. Siphonostele—types (diagrammatic sectional view). A. Ectophloic siphonostele. B. Amphiphloic siphonostele. C. Dictyostele. D. Atactostele.

A very complex type of stelar construction is noticed in some pteridophytes. These are called **polycyclic** steles—having two or more concentric rings of vascular tissues. Those in the inner cylinder usually form a siphonostele and the outer one is either a solenestele (Fig. 581A), as in *Matonia pectinata;* or a dictyostele (Fig. 581B), as found in *Pteridium latiusculum.*

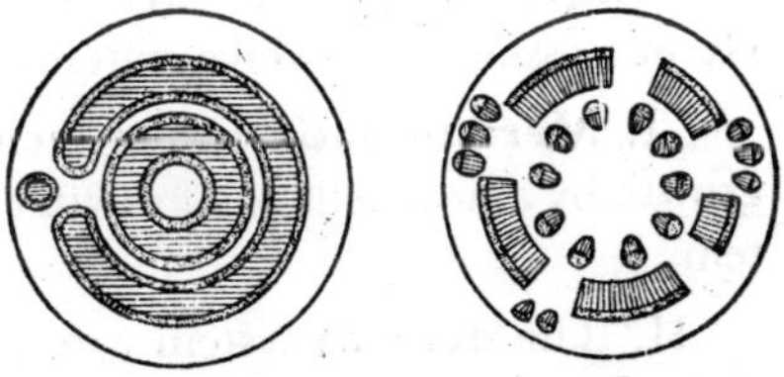

FIG. 581. Polycyclic stele—types (diagrammatic. sectional views). A. Polycyclic solenestele. B. Polycyclic dictyostele.

The terms *monostele* and *polystele* had been used by early workers. Monostele used to mean one stele with vascular tissues forming a unit structure. Polystele was applied to the strands of dissected siphonostele, where each strand, particularly one with amphicribral bundle, appeared like a protostele in cross-section. Thus protostele and unbroken siphonostele had been called monostele, and some dissected siphonosteles were termed polystele. These terms have been discarded now. True polysteles do not occur in living plants.

It is clear from the above discussion that two principal types of steles are protostele and siphonostele. It is generally accepted that siphonostele has evolved from protostele. The methods of evolutionary changes have been controversial. In fact, two theories had been proposed in this connection. The first theory holds that the central part of the stele remains unspecialised during the process of evolution and ultimately becomes pith. Thus according to proponents of this theory the origin of pith is *intraxylic*—it morphologically belongs to vascular tissues. They have suggested that by metamorphosis of vascular tissues parenchymatous pith has been formed. This is known as *expansion theory*—rather an unfortunate expression, because expansion of cortex to pith is not

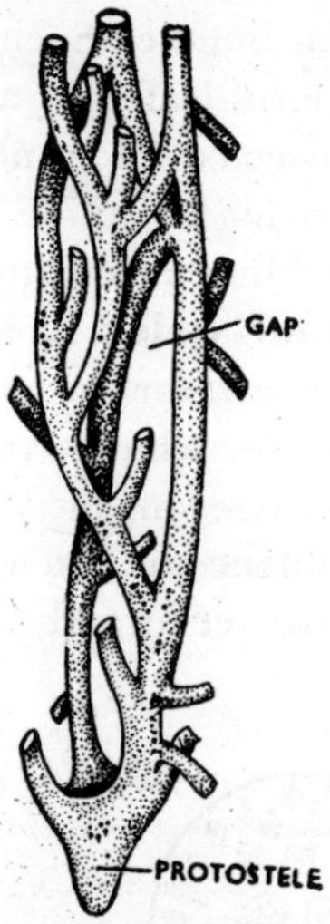

FIG. 581A. Diagram of the stele of the rhizome of *Ophioglossum lusitanicum*. (*After Fann*).

established here. In some primitive plants tracheary elements have been found scattered in parenchymatous pith. The stele in the rhizome of *Ophioglossum lusitanicum* (Gewirtz & Fann, 1960) is peculiarly protostelic at the base and siphonostelic at the upper portion (Fig. 581A) with parenchymatous pith mixed with tracheidal elements. These are strong points in support of above theory. The second one, known as *invasion theory*, demands that cortex has invaded the central cylinder during phylogenetic advance in the vascular plants, the leaf gaps and branch gaps being the channels of invasion. According to this theory pith is cortical in origin and thus does not belong to stele. So pith is extrastelar in nature from morphological point of view.

ANATOMICO-PHYSIOLOGICAL CLASSIFICATION OF TISSUE SYSTEMS

Haberlandt proposed the 'anatomico-physiological' classification in his classical book 'Physiological Plant Anatomy'. In his classification, *function* has been taken as the sole criterion while the origin, nature of cells, their topography and other characters were thoroughly ignored. He recognised a few systems like dermal system, absorbing system, and so on. In spite of all thoroughness and advantages, this system has been confusing, mainly due to utter disregard of the origin of the tissues. Epidermis and cork cells have different origin, but they have been put in the dermal system because of the same protective function. The guard cells of the stomata and the root hairs originate from protoderm cells; here guard cells have been put in the ventilating system and root hairs in the absorbing system. To Haberlandt 'homologies of the tissues were of no interest', his concern was 'solely with analogy'. He put tissues into twelve systems, an outline of which is being given.

I. **Meristematic** tissues included *primordial* meristem-marginal and apical cells, growth by apical cells and growth by apical initials, *primary* meristem and *secondary* meristem.

II. The **dermal** system consisted of *epidermis* including its nature, wall characters and contents, the *multiple epidermis;* the epidermal appendages; the secondary meristem *phellogen* and secondary protective tissues *cork* and *bark*. The epidermal openings were excluded.

III. The **mechanical** system considered the *bast fibres*, *wood fibres*, *collenchyma*, *sclerenchyma* cells with *sclereids;* strength and elasticity of mechanical cells; the *principles* governing the distribution of mechanical tissues in plants and the influence of external factors on the development of the mechanical system (Ref. Chapter XX).

IV. The **absorbing** system included *root hairs*, *rhizoids*, *velamen*, the absorbing tissue of aerial roots; and water-absorbing hairs of foliage leaves, as regards absorption of inorganic compounds; and systems of absorbing organic food materials among seedlings, saprophytes, parasites and carnivorous plants.

V. The **photosynthetic** system included chloroplasts, chlorophyll containing parenchyma—*palisade* and *spongy* cells of leaves, and influence of light upon the distribution and organisation of the photosynthetic system.

VI. The **vascular** or **conducting** system consisted of the *vascular bundles*, tracheary and sieve elements, xylem parenchyma, and phloem parenchyma; laticiferous tissues; and the structure of the vascular bundles.

VII. The **storage** system consisting of *water tissues*, *storage tissues* of seeds, tubers and bulbs.

VIII. The **aerating** or **ventilating** system was concerned with intercellular spaces, stomata and lenticels; and pneumatophores of breathing roots.

IX. The **secretory** and **excretory** systems included hydathodes, digestive glands, nectaries, oil-, resin-, mucilage- and gum-secreting glands.

X. The **motor** system considered the flying hairs and tissues, hygroscopic tissues, vital motor tissues, etc.

XI. The **sensory** system covered papillae, hairs, bristles, statolith organs, tactile bristles of carnivorous plants.

XII. The **stimulus-transmitting** system considered the protoplasmic connections, stimulus-transmitting tissue of *Mimosa pudica*.

CHAPTER XXII

THE PRIMARY BODY : THE STEM

The embryo develops into a plant with root-stem axis and the appendages. In fact, three important organs of a plant are the stem, the leaves and the root. The continuity of the tissues, and particularly the vascular system, has been discussed in the preceding chapter. Apical meristems occurring at the tips of the axis, the stem tip and root tip, divide and produce new cells. In course of time these tissues become permanent and the fundamental body of the plant is laid down. The tissues deriving their origin from the apical meristems are known as primary permanent tissues and body made of primary permanent tissues is designated as **primary body.**

THE STEM

In lowest vascular plants, as in Psilotales, a sharp demarcation between the stem and leaves is hardly possible. The boundary between the stem and leaves in the seed plants is, in fact, uncertain. Both of them derive their origin from the same apical meristem—the shoot apex, and they are to a great extent interdependent in the process of growth and differentiation. For these reasons some workers prefer to put the stem and its foliar appendages (leaves) under the broader concept of *shoot*. However, it is a matter of practice to take up the three important vegetative organs separately.

An outline of the internal structure of the stems is given here. In view of the fact that wide diversities exist as regards the nature of the plants, a few common dicotyledons and monocotyledons have been selected for the study of anatomical structures. The anomalous types will be taken up in a separate chapter.

DICOTYLEDONS

1. *YOUNG STEM OF SUNFLOWER*

Transverse and longitudinal sections through the internode of a young sunflower stem (*Helianthus annuus* of family Compositae) should be taken and stained suitably for the study of internal structures. It would exhibit the following plan of arrangement of tissues from the periphery to the centre of the organ (Figs. 582 & 583).

I. **Epidermis.** It is the outermost uniseriate zone, consisting of tabular cells attached end on end without leaving intercellular spaces. The cells are living with vacuolate protoplast. The chloroplasts are usually absent. The outer walls of the epidermal cells are cuticularised for checking loss of water. Stomata may be present here and there. A large number of multicellular hairy outgrowths develop on the epidermis.

II. **Cortex.** It occurs next to epidermis and represents the extrastelar ground tissues. In sunflower stem cortex is differentiated into three zones. Just internal to epidermis there are a few layers of collenchyma, usually angular ones, forming a continuous band. This collenchymatous band meant for giving mechanical support to the growing stem, is called **hypodermis.** Next to hypodermis a few layers of thin-walled parenchyma occur which have conspicuous intercellular spaces. A few glands, each having a hollow cavity surrounded by small densely protoplasmic epithelial cells, are present here and there in the parenchymatous portion. Some authors used the term 'general cortex' for this part (parenchymatous part) of cortex; but that is rather misleading and hence has been abandoned. The last layer of cortex is a wavy band made of compactly-set barrel-shaped parenchyma cells. Starch grains are abundantly present in these cells. This band is referred to as **starch sheath.** It is morphologically homologous to the endodermis commonly found in the roots. The starch sheath is the limiting layer of the extrastelar ground tissue, cortex.

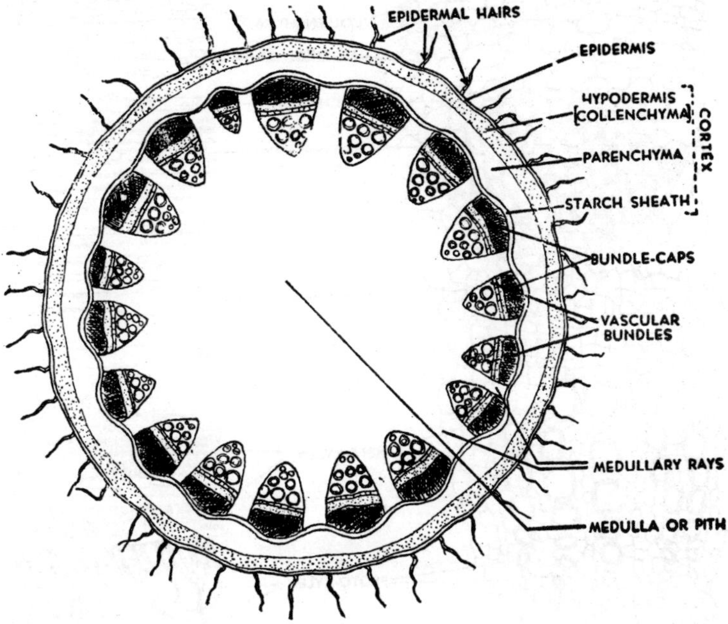

FIG. 582. Transverse section of young stem of *Helianthus annuus* (sunflower—diagrammatic).

III. **Stele.** All the tissues occurring internal to the starch sheath constitute the stele. It is a dissected siphonostele, consisting of vascular bundles arranged in form of a ring, and intrastelar ground tissues.

Vascular bundles: The vascular bundles are typically **collateral** and **open** ones, with xylem and phloem on the same radius, xylem being internal and phloem external. A strip of lateral meristem, **cambium,** is present between xylem and phloem.

The xylem is characteristically *endarch,* showing centrifugal mode of differentiation from the procambium. The protoxylem elements, tracheids and tracheae, with smaller cavities and annular and spiral thickenings occur towards the centre; and the metaxylem elements with pitted and other types of thickenings are present towards the circumference. Xylem parenchyma cells are smaller than other parenchyma cells of the stele. The phloem is composed of sieve tubes, companion cells and phloem parenchyma. The cambium of the vascular bundle, what is called *fascicular* (*fascicle*=bundle), appears to be composed of two or three layers of fusiform cells, which look rectangular in cross-section. This tissue is responsible for growth in thickness.

Intrastelar ground tissues: Against every vascular bundle there is a patch of sclerenchyma forming something like a cap. This has been called **bundle cap.** The patches of sclerenchyma and intervening parenchyma cells were considered to constitute the pericycle, the outermost portion of the stele. But modern workers are of opinion that the fibres associated with the vascular bundle in sunflower stem, and as a matter of fact, in many other plants, belong to phloem. So the term *hard bast* has also been attributed to this patch. A complete section (Fig. 582) shows that the bundles are located towards the periphery and the large central portion remains occupied by parenchyma cells with profuse intercellular spaces. This part is known as **pith** or *medulla*. Continuations of the pith

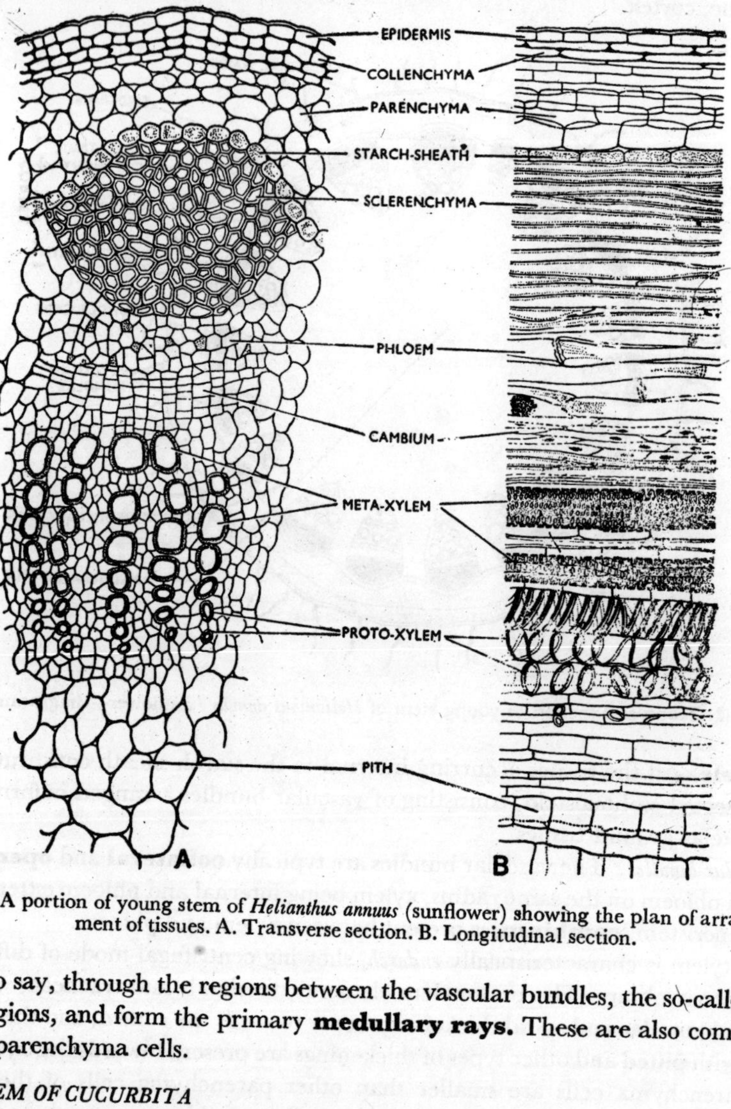

FIG. 583. A portion of young stem of *Helianthus annuus* (sunflower) showing the plan of arrangement of tissues. A. Transverse section. B. Longitudinal section.

radiate, so to say, through the regions between the vascular bundles, the so-called interfascicular regions, and form the primary **medullary rays.** These are also composed of thin-walled parenchyma cells.

2. *YOUNG STEM OF CUCURBITA*

It (*Cucurbita maxima* of family Cucurbitaceae) is a weak plant climbing with the help of the tendrils. Transverse and longitudinal sections of the stem are taken, stained suitably and studied under the microscope. It is wavy in outline and thus shows distinct ridges and furrows (Fig. 584). The vascular bundles occur in two rings. One ring usually consisting of five smaller vascular bundles is present against the ridges. These are leaf-trace bundles. Another ring of same number of bundles occurs against the furrows. The central part of the stem is hollow due to disintegration of pith. The *Cucurbita* stem shows the following plan of arrangement of tissues (Fig. 585):

I. **Epidermis.** It is as usual the single-layered zone made of compactly-set tabular cells. The cells are living with vacuolate protoplast and cuticularised outer walls. A large number of multicellular hairs develop from the epidermis. Stomata may be present in the young stem.

II. **Cortex.** It is differentiated into three zones. The **hypodermis** is made of collenchyma cells forming conspicuous patches particularly at the ridges. Collenchyma cells do not form a continuous band, the number of layers become reduced towards the furrows and are interrupted by some parenchyma cells containing abundant chloroplasts. These photosynthetic cells actually occur just beneath the epidermis where stomata are present. Chloroplasts may be present in the collenchyma cells as well. Next to hypodermis there are a few layers of parenchyma, larger in size and having lesser number of chloroplasts. The last layer of cortex is the **starch sheath,** a layer of barrel-shaped cells, larger than other cells of cortex. These cells contain abundant starch grains. The cortex naturally is more wide at the regions of ridges than at the furrows.

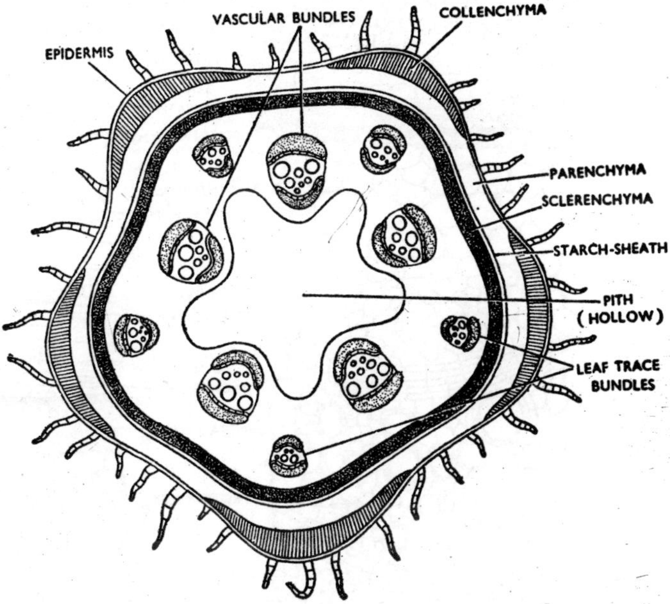

FIG. 584. Transverse section of a young stem of *Cucurbita* (diagrammatic).

III. **Stele.** The central cylinder or stele containing the vascular bundles and intrastelar ground tissues remains enveloped by the cortex.

A few layers of sclerenchyma occur on the inner side of the starch sheath in form of a continuous band. Some parenchyma cells are usually present between the sclerenchymatous band and the vascular bundles. These tissues together form the **pericycle.** Pith, as already stated, disintegrates rather early in *Cucurbita* stem. The cells occurring in between the vascular bundles are referred to as internal parenchyma.

The vascular bundles are typically **bicollateral,** having two patches of phloem and two strips of cambium on either side of xylem. So the sequence of tissues in the vascular bundle is outer phloem, outer cambium, xylem, inner cambium and inner phloem. The bundles, particularly those occurring against the furrows are fairly large and thus are quite suitable for study of detailed structure. The outer phloem is more massive than the inner one and is composed of sieve tubes, companion cells and phloem parenchyma. The sieve tubes are quite conspicuous and sieve plates are noticed here and there. The xylem has the characteristic elements. The metaxylem vessels are of very large size, what is characteristic of most climbers. They have pitted thickenings. A good number of protoxylem vessels with comparatively narrower cavities occur towards the centre. They have usually spiral and annular thickenings. Tracheids and fibres are rather few or lacking, whereas xylem parenchyma cells are abundant. The vessels do not remain arranged in radial rows. The cambium occurs on either side of xylem. The outer cambium is many-layered and is made of cells more or less rectangular in cross-section. The inner cambium is much smaller and is usually curved.

As formation of secondary tissues is confined to individual bundles in climbing plants like *Cucurbita*, the vascular bundles may contain some secondary tissues as well, in addition to the primary tissues. In that case outer phloem may be made of primary phloem occurring towards pericycle and secondary phloem against the cambium. The

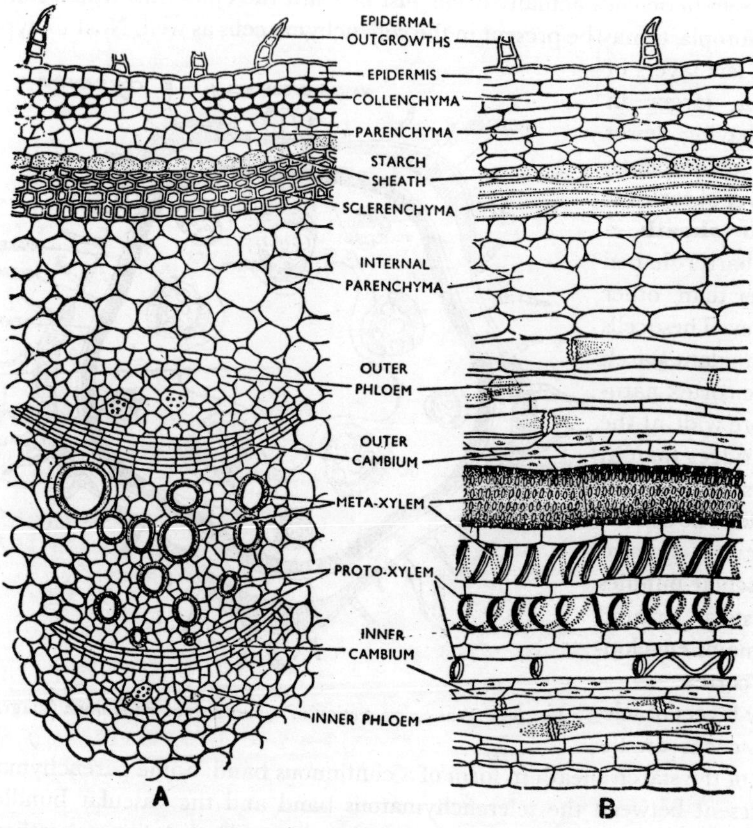

FIG. 585. A portion of young *Cucurbita* stem showing the plan of arrangement of tissues. A. Transverse section: B. Longitudinal section.

elements are practically same, so that the two are hardly distinguishable, only the secondary phloem has larger elements. The internal phloem is solely primary. Secondary xylem similarly may occur internal to outer cambium. The elements, particularly the vessels, are much larger.

3. *YOUNG STEM OF LEONURUS*

A very young stem of *Leonurus sibiricus* of family Labiatae should be selected, because secondary growth commences unusually early in this plant. Transverse sections are taken and stained suitably for the internal structure. The stem is square in cross-section. It shows the following plan of arrangement of tissues (Figs. 586 & 587):

I. **Epidermis.** It is a uniseriate zone composed of tabular cells attached end on end, without intercellular spaces. The cells are living with vacuolate protoplasts. The outer walls are strongly cuticularised. Many multicellular hairs develop on the epidermis. Stomata may be present here and there.

II. Cortex. Like sunflower stem the cortex is differentiated into three zones, though it is less massive. Next to epidermis occurs **hypodermis** composed of collenchyma cells.

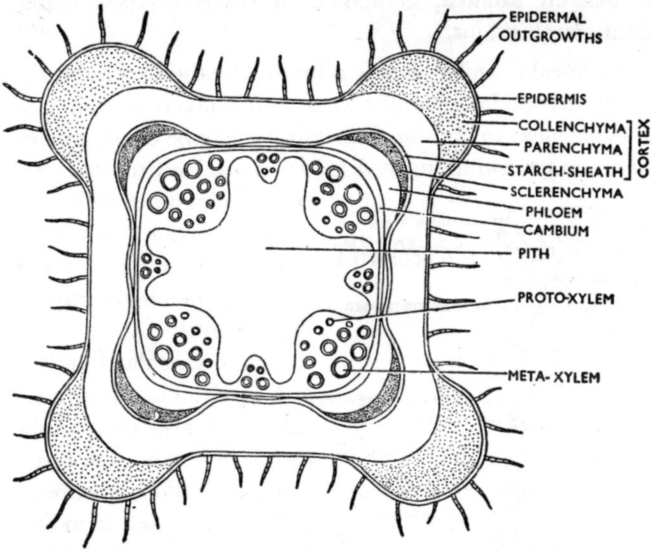

Fig. 586. Transverse section of young *Leonurus* stem (diagrammatic).

These cells aggregate densely at the four corners of the stem, so that these patches serve as diagonally placed I-girders for withstanding flexion. Collenchyma cells extend beyond

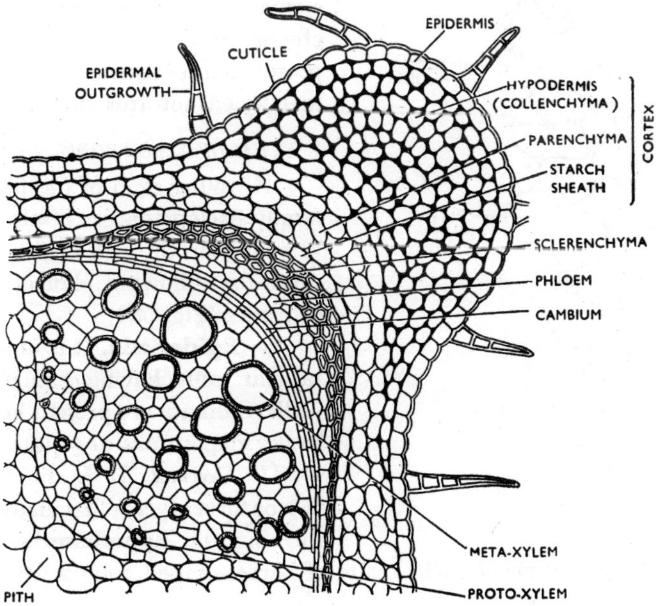

Fig. 587. A portion of young *Leonurus* stem showing the plan of arrangement of tissues (transverse section).

the corners, but do not form continuous bands. There are a few layers of thin-walled parenchyma just internal to collenchymatous hypodermis. These cells contain abundant chloroplasts (chlorenchyma) and appear as a green belt under microscope. At the portions where collenchyma is absent, chlorenchyma cells occur next to the epidermis. These are usually the regions where stomata are present on the epidermis. The last layer of cortex is the **starch sheath,** composed of barrel-shaped compactly-arranged cells with abundant starch grains.

III. **Stele.** The central cylinder or stele is limited by the starch sheath, and is made of vascular strands and intrastelar ground tissues. A few layers of sclerenchyma forming a continuous band occur next to the starch sheath. It may be called **pericycle** or perivascular tissue. The vascular bundles are disposed more towards the periphery; a large parenchymatous pith is present at the central portion. As secondary growth in thickness is initiated unusually early, the primary medullary rays very soon lose their identity.

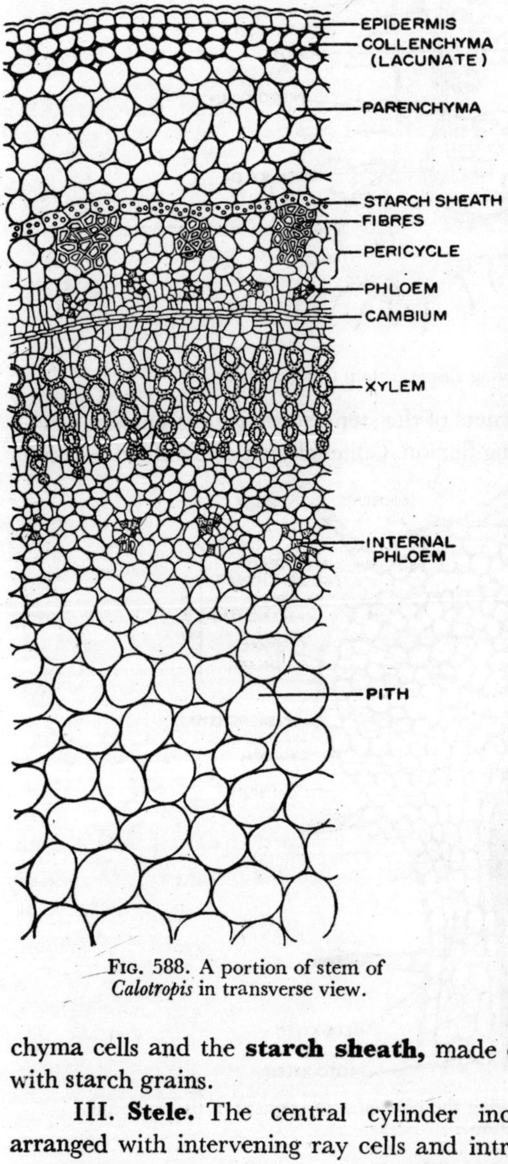

Fig. 588. A portion of stem of *Calotropis* in transverse view.

The vascular bundles are **collateral** and **open.** The phloem is external, i.e., it occurs towards the periphery and has the usual elements—sieve tubes, companion cells, and phloem parenchyma. Next to phloem there is a strip of cambium, made of a few layers of fusiform cells appearing more or less rectangular in outline. Cambium ring is formed rather early. Xylem occurs next. It has the usual tracheary elements—tracheids and tracheae, parenchyma and fibres. Xylem is endarch—protoxylem occurring towards centre and metaxylem towards circumference.

4. *STEM OF CALOTROPIS*

A young stem of *Calotropis procera* of family Asclepiadaceae is selected. A transverse section through the internode would show the following plan of arrangement of tissues (Fig. 588):

I. **Epidermis.** It is uniseriate—with tabular cells having cuticularised outer walls. Stomata are present here and there. Waxy depositions are quite common.

II. **Cortex.** It is differentiated into two or three layers of lacunate collenchyma forming the **hypodermis,** large parenchyma cells and the **starch sheath,** made of barrel-shaped compactly-arranged cells with starch grains.

III. **Stele.** The central cylinder includes the vascular bundles—compactly arranged with intervening ray cells and intrastelar ground tissues.

The pericycle is quite massive, made of parenchyma (a few layers) having patches of cellulosic fibres here and there. The central portion is occupied by a large parenchymatous pith.

The vascular bundles are made of internal xylem with endarch arrangement, external phloem and cambium ring in between the two. Presence of patches of internal or *intraxylary* phloem is a distinctive character.

5. *STEM OF AN AQUATIC DICOTYLEDON*

Enhydra fluctuans of family Compositae is selected. It is an aquatic plant, and naturally the stem is soft with abundant air chambers and scanty mechanical tissues. A transverse section through the internode would show the following plan of arrangement of tissue (Fig. 589).

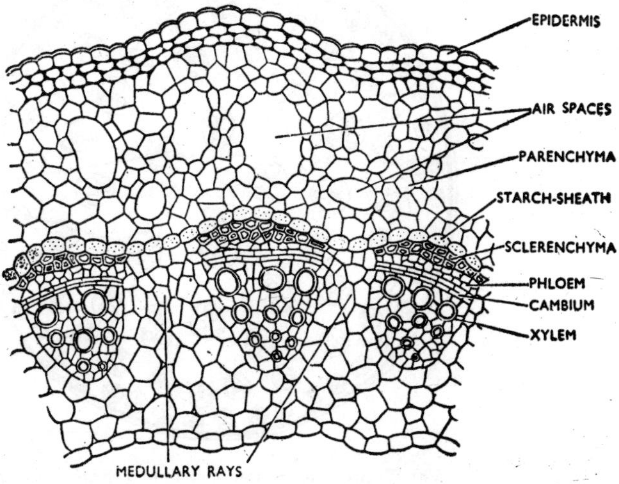

FIG. 589. A portion of *Enhydra* stem showing the plan of arrangement of tissues (transverse section).

I. **Epidermis.** It is a uniseriate layer consisting of comparatively small thin-walled cells. The outer walls have feeble cuticularisation.

II. **Cortex.** It is less massive here than in other terrestrial dicotyledonous stems. The differentiation of the cortex into three zones is noticed. Next to epidermis, just a few layers of parenchyma with comparatively thicker walls occur as a continuous band forming the **hypodermis.** The absence of peripheral collenchyma is thus a notable feature. The next zone consists of quite a few layers of parenchyma, larger in size and having abundant spaces and conspicuous air chambers. These are helpful in giving buoyancy to the plants, apart from normal aeration. Chloroplasts are present in the parenchyma cells. The last layer of cortex is as usual the **starch sheath**—a wavy layer composed of barrel-shaped compactly-arranged cells with starch grains.

III. **Stele.** The central cylinder or stele includes the vascular bundles arranged in a ring and intrastelar ground tissues.

Against every vascular bundle there is a patch of sclerenchyma, like sunflower stem, but much smaller here, forming the so-called hard bast or bundle cap. The sclerenchyma patch together with intervening parenchyma has been called pericycle, though doubts prevail as regards the origin of the fibres. The pith disintegrates leaving a large cavity at the central region, so it may be called hollow pith. Some parenchyma cells occur in the interfascicular regions.

The vascular bundles remain arranged in a ring. They are **collateral** and **open.** Phloem is rather small and has the usual elements—sieve tubes, companion cells and phloem parenchyma. A strip of cambium occurs between xylem and phloem. Xylem is also small and is made of a few tracheary elements as protoxylem and metaxylem showing endarch arrangement, and a few parenchyma cells.

MONOCOTYLEDONS

A few typical monocotyledonous plants are selected for the study of internal structures. Grasses are undoubtedly the most suitable materials for the purpose. Transverse and longitudinal sections are taken, stained suitably and plan of arrangement of tissues is noted.

1. *STEM OF MAIZE*

It (*Zea mays* of family Graminaceae) also shows the three tissue systems, but the

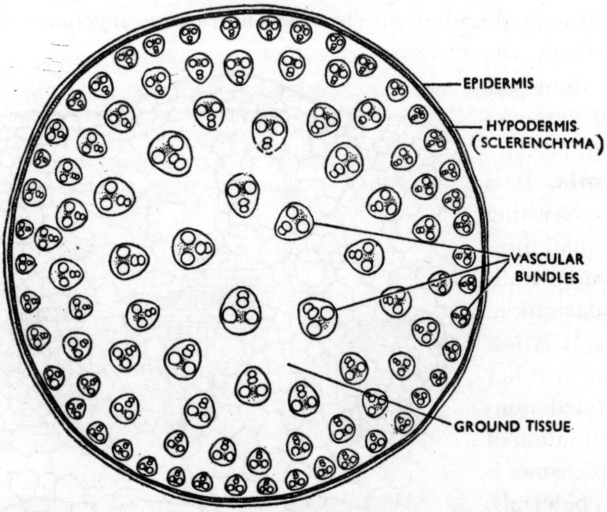

FIG. 590. Transverse section of stem of *Zea mays* (maize) [diagrammatic].

number of vascular bundles is much larger and they remain irregularly scattered in the ground tissue (Figs. 590 & 591). The bundles occurring towards periphery are smaller in

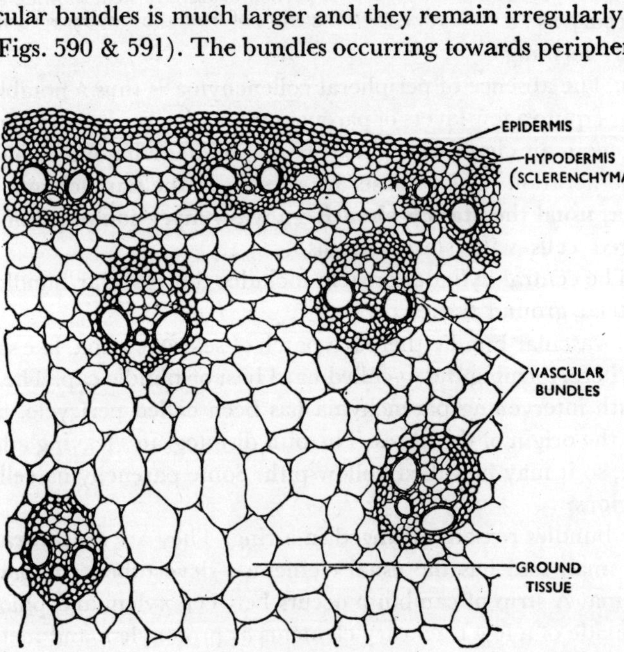

FIG. 591. A portion of stem of *Zea mays* (maize) showing the plan of arrangement of tissues (transverse section).

size and more crowded, whereas those at the central region are larger in size and more spaced. All the bundles are common to the stem and leaves, the central ones from the median veins of the leaf blade and small peripheral ones form the marginal bundles. Unlike most of the grasses maize stem is not hollow in the internodal region.

I. **Epidermis.** It is a uniseriate zone, composed of small tightly-set cells. The outer walls are cuticularised. The hairy outgrowths, so characteristic of dicotyledons, are usually absent.

II. **Cortex.** The cortex and, in fact, the ground tissue system is not well-differentiated here. A few layers of sclerenchyma, rather parenchyma undergoing sclerosis, occur next to epidermis. This band usually continuous, but may be interrupted here and there by internal parenchyma, is referred to as **hypodermis.** All the ground tissues lying internal to hypodermis are composed by thin-walled parenchyma cells with profuse intercellular spaces. In fact, the bundles remain embedded in parenchymatous ground tissue.

III. **Stele.** Due to irregular distribution of the bundles in the ground tissue the semblance of a stele is lost. As already stated, this type is called *atactostele.* The vascular bundles (Fig. 592) are **collateral** and **closed.** Xylem occurs in form of letter Y—the two

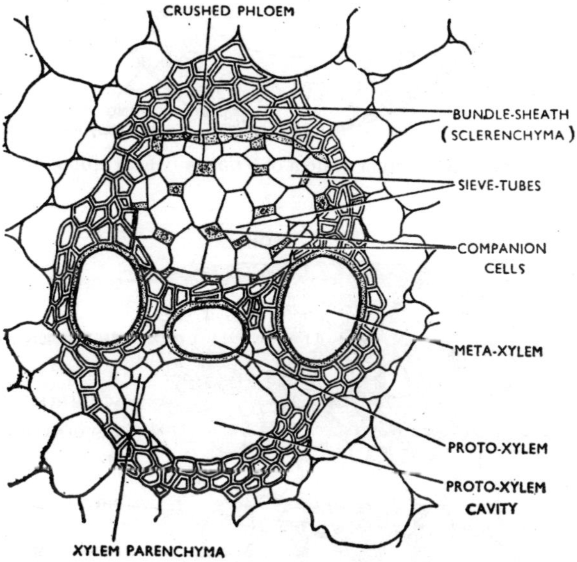

FIG. 592. A vascular bundle of stem of *Zea mays* (maize) [magnified].

metaxylem vessels with wider cavities and pitted thickening at the two arms, and protoxylem vessels usually one or two, with narrow cavities and spiral or annular thickening at the base. A few tracheids occur near the protoxylem. In a mature bundle the xylem elements undergo more and more lignification and the lowest protoxylem disintegrates forming a lacuna or cavity known as **protoxylem cavity.** Parenchyma cells also remain associated with other elements in xylem. Phloem is rather small. It is composed of only sieve tubes and companion cells, phloem parenchyma being absent. In a mature bundle the protophloem often gets crushed due to pressure from internal tissues. The whole vascular bundle remains surrounded by sclerenchyma cells, forming what is known as **bundle sheath.** The sheaths of small peripherally located bundles and the hypodermal sclerified cells often coalesce, so that those bundles appear to be embedded in the sheath.

2. *STEM OF WHEAT*

The wheat stem (*Triticum vulgare* of family Graminaceae) and, in fact, many grasses are hollow at the internodes with vascular bundles arranged in two rings. A transverse

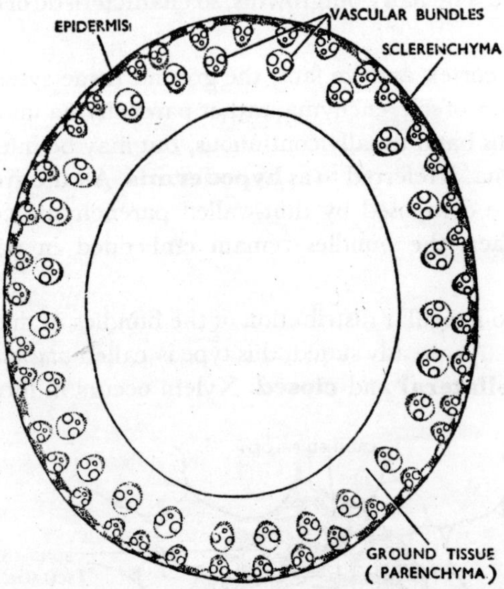

FIG. 593. Transverse section of stem of *Triticum* (wheat) [diagrammatic].

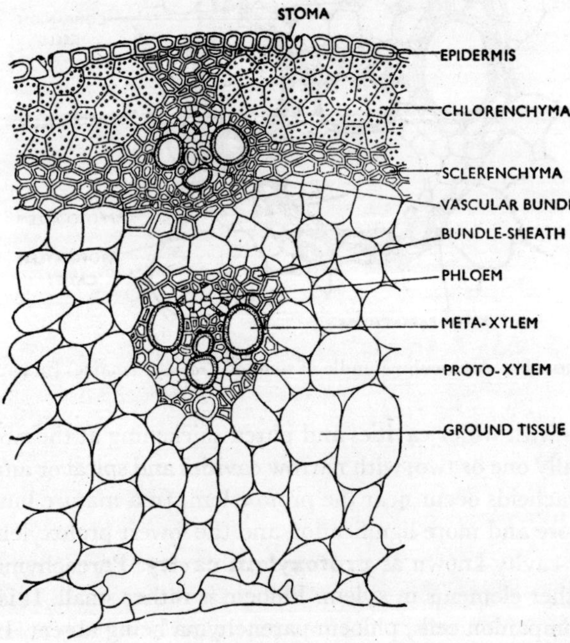

FIG. 594. A portion of stem of *Triticum* (wheat) in transverse section.

section (Figs. 593 & 594) taken through the internode shows the following plan of arrangement of tissues:

I. **Epidermis.** It is as usual single-layered, composed of compactly-arranged tabular cells with cuticularised outer walls. Stomata may be present here and there.

II. **Ground tissues.** A few sclerenchyma cells occur as patches in the subepidermal region. They are not arranged in a continuous band, but are interrupted by chloroplast-containing parenchyma cells here and there. The latter occur at the regions where stomata are present on the epidermis. The remaining portion of the ground tissue is parenchymatous, the cells having conspicuous intercellular spaces. The central portion is hollow.

III. **Vascular bundles.** Vascular bundles occur in two series. One ring of small bundles remain associated with subepidermal sclerenchyma, so that they appear to be embedded in sclerenchymatous patches. The second ring placed lower is composed of larger bundles. Like those of maize stem the bundles are **collateral** and **closed** ones. They usually remain surrounded by sclerenchymatous bundle sheath. The sheath of bundles of outer circle actually reaches the epidermis.

3. *STEM OF AN AQUATIC GRASS*

A transverse section of the stem of an aquatic grass (*Leptochloa chinensis* of family Graminaceae) is taken and stained suitably. It would show the following plan of arrangement of the tissues (Fig. 595).

I. **Epidermis.** It is as usual uniseriate, composed of a row of compactly-set cells with cuticularised outer walls.

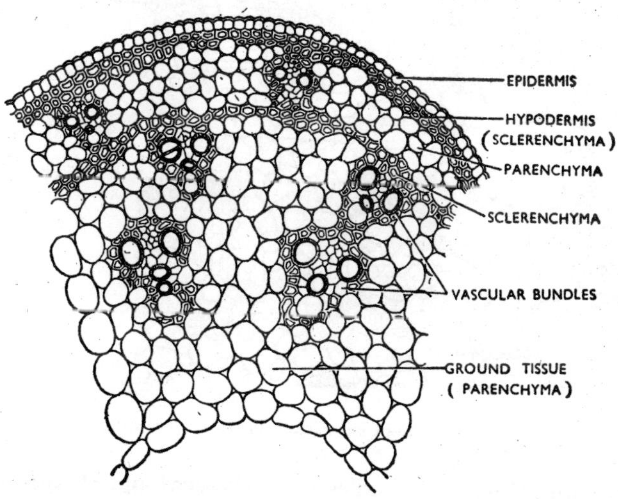

FIG. 595. A portion of stem of aquatic grass (*Leptochloa chinensis*) in transverse section.

II. **Ground tissues.** Internal to epidermis there are two or three layers of sclerenchyma forming the **hypodermis.** Parenchyma cells with lots of intercellular spaces occur next to hypodermis. Another band of sclerenchyma is present—usually one- or two-layered in the ground tissue. The central portion is hollow.

III. **Vascular bundles.** These are fairly large in number and remain scattered in the ground tissue. Those occurring towards periphery are smaller in size. The bundles resemble those of typical monocotyledon, maize. They are **collateral** and **closed** ones. Xylem consists of two metaxylem vessels with pitted thickenings and one or two protoxylem vessels with spiral or annular thickening. Protoxylem cavity is present. Phloem is smaller, composed of sieve tubes and companion cells; phloem parenchyma is absent. The bundle remains surrounded by sclerenchymatous sheath.

4. *SCAPE (FLORAL AXIS) OF CANNA*

It (*Canna indica* of subfamily Cannaceae) also shows monocotyledonous structure (Fig. 596), though a bit different from the maize and wheat stems.

I. **Epidermis.** It is the uniseriate protective zone with cuticularised outer walls.

II. **Ground tissues.** A few layers of parenchyma occur next to epidermis forming a small cortex. It is immediately followed by a band of chloroplast-containing cells. This band is referred to as *chlorophyllous tissue*. Patches of sclerenchyma remain attached to the chlorophyllous tissue here and there. The remaining portion is parenchymatous ground tissue enclosing the scattered vascular bundles.

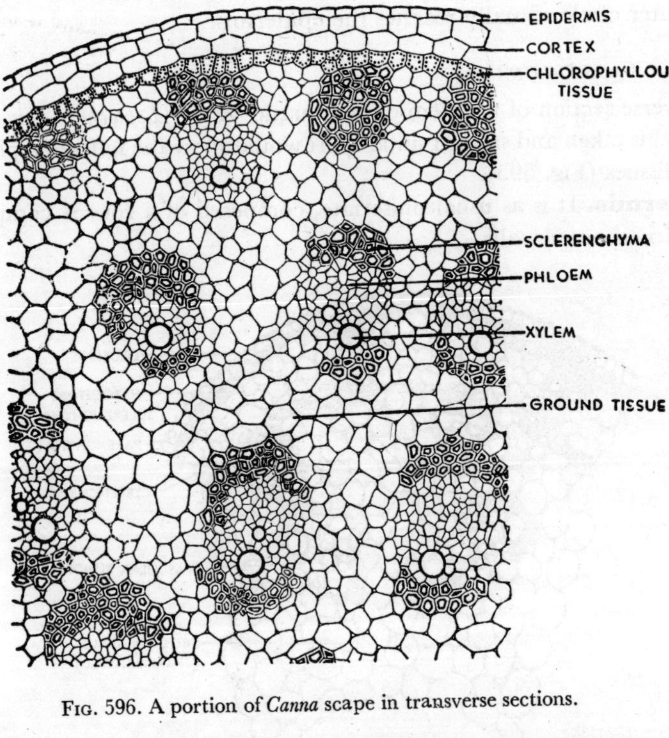

FIG. 596. A portion of *Canna* scape in transverse sections.

III. **Vascular bundles.** The bundles are **collateral** and **closed.** Each of them has small amount of xylem and phloem, the latter with only sieve tubes and companion cells. Sclerenchyma cells do not form a sheath surrounding the whole bundle, as in maize and wheat stems, but remain in two patches on the outer and inner sides of the bundles. The patch on the outer side is larger forming something like a cap and that on the inner side is proportionately much smaller.

1. *STEM OF ASPARAGUS*

A transverse section through the stem of *Asparagus* (*A. racemosus* of family Liliaceae) would show the following plan of arrangement of tissues (Figs. 597–599):

I. **Epidermis.** It is single-layered composed of roundish cells with cuticularised outer walls.

II. **Ground tissues.** Next to the epidermis there occur a few layers of parenchyma cells containing abundant chloroplasts, so that they together form something like a green

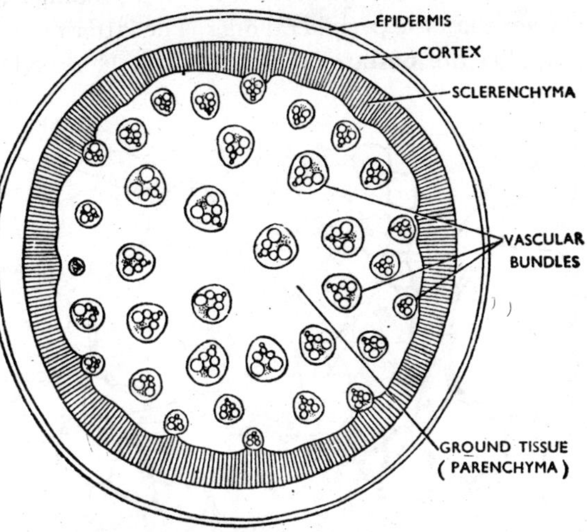

FIG. 597. *Asparagus* stem in transverse section (diagrammatic).

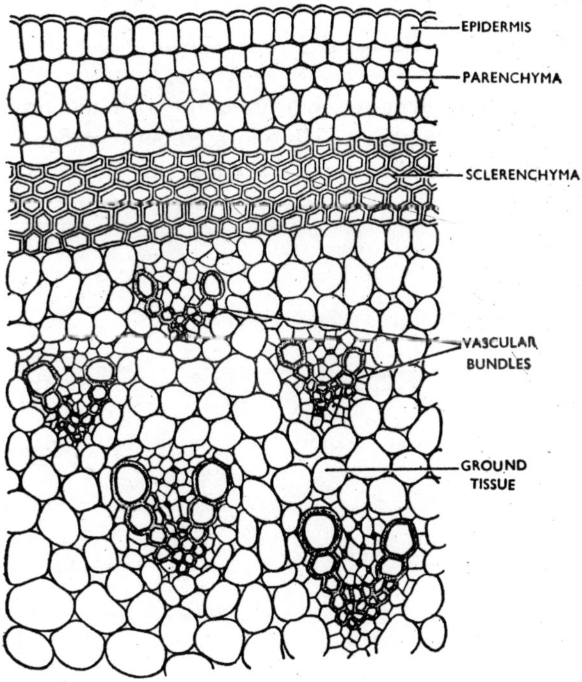

FIG. 598. A portion of *Asparagus* stem in transverse section.

belt. This band made of parenchyma may be called **cortex.** The layer internal to chlorophyll-containing parenchyma consists of a single row of compactly-set cells, forming the **starch sheath.** Next occur quite a few layers of sclerenchyma in form of a band. The central part of the stem is composed of thin-walled parenchyma with distinct intercellular spaces. Vascular bundles remain embedded in this central parenchymatous portion.

III. **Vascular bundles.** As in other monocotyledons, bundles remain scattered; the central bundles are larger than the peripheral ones. The latter may touch the sclerenchymatous band. The bundles are **collateral** and **closed.** Here xylem occurs in form of

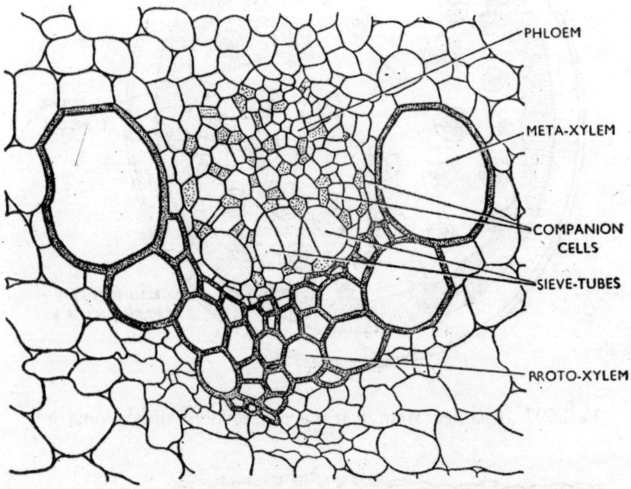

FIG. 599. A vascular bundle of *Asparagus* stem (magnified).

letter V (Fig. 599), the metaxylem vessels forming the arm and protoxylem, the base. Phloem with sieve tubes and companion cells is located between two arms of V. Sclerenchyma cells surrounding the bundle—the so-called bundle sheath is absent.

CHAPTER XXIII

THE PRIMARY BODY : THE ROOT

The root develops from the radicle of the embryo. Due to the fact that the extreme tip of the root remains covered by a cap, the apical meristem here is subterminal, as opposed to the terminal apical meristem of the stem. However, the primary body consists of three tissues systems; and, in fact, boundaries between the tissue systems are more precise here. The outstanding characters by which root differs from the stem are the following: the epidermis in roots is usually uniseriate, composed of thin-walled cells. Cutinisation of outer walls and cuticle are absent. Hence the terms *epiblema*, *piliferous layer* and *rhizodermis* have been proposed for the epidermis of root. The stomata, so characteristic in aerial organs, are absent. Formation of root hairs, which are prolongations of epidermal cells themselves, is confined to a particular zone of the root. The cortex is comparatively more simple and homogeneous. It is made of mainly parenchymatous cells and is often

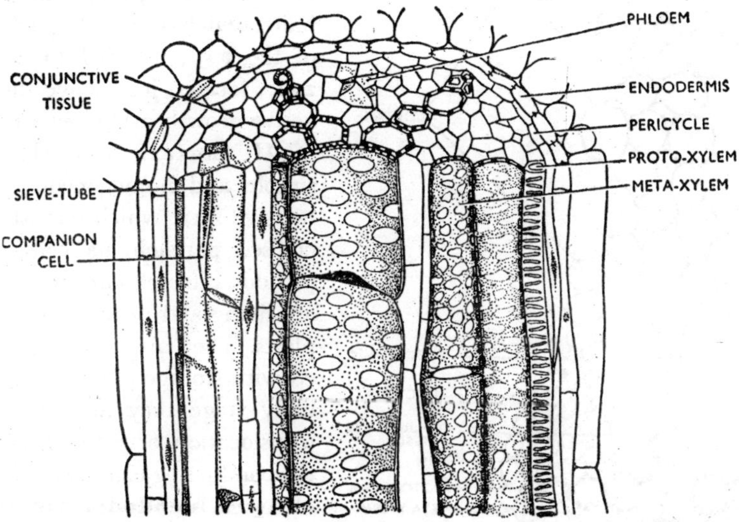

FIG. 600. Stele of a typical dicot. root in transverse and longitudinal section (diagrammatic).

massive for the purpose of storage. The endodermis, the last layer of cortex with characteristic Casparian thickenings, is of universal occurrence. Pericycle, normally uniseriate and rarely multiseriate, invariably occurs next to the endodermis. The vascular cylinder is more compact, firstly due to absence of gaps, and secondly, due to presence of endodermis and pericycle. The complex tissues, xylem and phloem, occur as separate patches showing radial arrangement. Xylem is always exarch, due to centripetal mode of differentiation from the procambium, so protoxylem occurs towards circumference and metaxylem towards the centre. According to the number of protoxylem groups roots may be *monarch*, as in *Trapa natans*, *diarch*, as in *Lycopersicon*, *Nicotiana; triarch*, as in *Pisum; tetrach*, as in *Vicia*, *Cicer* ; *pentarch*, as in *Ranunculus*. *Polyarch* condition with many xylem groups is characters of monocotyledons. In dicotyledonous roots xylem plates usually join at the centre forming a solid core Hence the stele is regarded as a protostele (Fig. 600). Endogenous development of the branch roots from the pericycle, as opposed to exogenous formation of a branch from the growing point of a stem, is a marked feature.

A few typical roots are selected for the study of internal structures.

DICOTYLEDONOUS ROOTS

1. *ROOT OF GRAM*

A transverse section of the root of gram (*Cicer arietinum* of subfamily Papilionaceae) is taken and stained suitably for the study of internal structure. It shows the following plan of arrangement of tissues from the epidermis to the centre of the organ (Fig. 601):

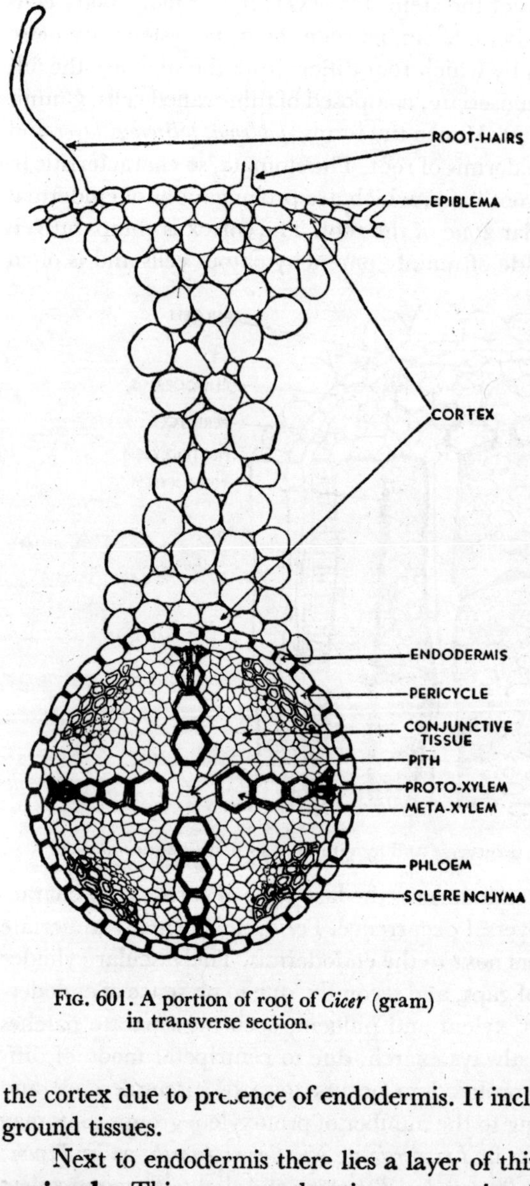

FIG. 601. A portion of root of *Cicer* (gram) in transverse section.

I. **Epidermis.** Epidermis, also known as *epiblema* or *piliferous layer*, is typically uniseriate outermost zone consisting of tabular living cells. Cuticle on the outer walls and stomata are absent. Some epidermal cells prolong to form the typically unicellular root hairs, which occur at a particular zone of the root, referred to as root hair zone, located just above the region of active growth and elongation.

II. **Cortex.** It is relatively more simple and homogeneous, forming a massive zone which consists of unspecialised parenchyma cells with conspicuous intercellular spaces. The cells are living and possess abundant leucoplasts. They are particularly concerned with storage of food, though at the early stage they are responsible for translocation of water and solutes to the conducting elements. The last layer of cortex is **endodermis.** It is of universal occurrence in roots and consists of compactly arranged barrel-shaped cells forming a distinct zone surrounding the stele. The endodermal cells possess Casparian thickenings on the radial walls.

III. **Stele.** The stele or central cylinder is precisely demarcated from the cortex due to presence of endodermis. It includes the vascular tissues and intrastelar ground tissues.

Next to endodermis there lies a layer of thin-walled parenchyma cells forming the **pericycle.** This non-vascular tissue occurring inside the stele is the seat of origin of branch roots.

The vascular bundles are **radial.** Xylem and phloem occur in separate patches arranged on alternate radii, intervened by small parenchyma cells. The latter form the **conjunctive tissue.** The bundle is *tetrarch* here, because four patches of xylem alternate with equal number of patches of phloem. Protoxylem vessels occur towards periphery and metaxylem towards centre, thus showing centripetal mode of differentiation from procambium. This is the typical *exarch* xylem of roots. The central portion is usually occupied by a metaxylem vessel, so that all the plates of xylem are joined forming a solid core. Hence it is regarded as a protostele. Phloem patches are rather small and consists of sieve tubes, companion cells and phloem parenchyma. The outer part of this tissue lying next to pericycle is protophloem and the inner is the metaphloem; of course they are not readily distinguishable like protoxylem and metaxylem. A few sclerenchyma cells occur against every phloem patch. Pith is normally absent in dicotyledonous roots. At early stages a few parenchyma cells may be located at the central portion, which is very soon obliterated by development of metaxylem.

2. *ROOT OF PEA*

A transverse section of the root of pea (*Pisum sativum* of subfamily Papilionaceae) is taken and stained suitably for the study of internal structure. It shows the same plan of arrangement of tissues as found in gram root from the periphery to the centre (Fig. 602):

I. **Epipermis.** It is the uniseriate outermost layer with typical root hairs.

II. **Cortex.** Cortex is parenchymatous with intercellular spaces. The last layer is **endodermis** with distinct Casparian thickenings.

III. **Stele.** It includes the vascular elements and intrastelar ground tissues. The bundles are radial. Here it is *triarch*, three patches of xylem alternating with three patches of phloem. Xylem is typically *exarch* having metaxylem towards centre and protoxylem towards periphery. Phloem is rather small with constituent elements—sieve tubes, companion cells and phloem parenchyma. A few fibres occur against every phloem group. **Pericycle** lying next to endodermis is single-layered and parenchymatous. Small parenchymatous conjunctive tissues occur between xylem and phloem groups. A very small pith is noticed at early stage which is obliterated later.

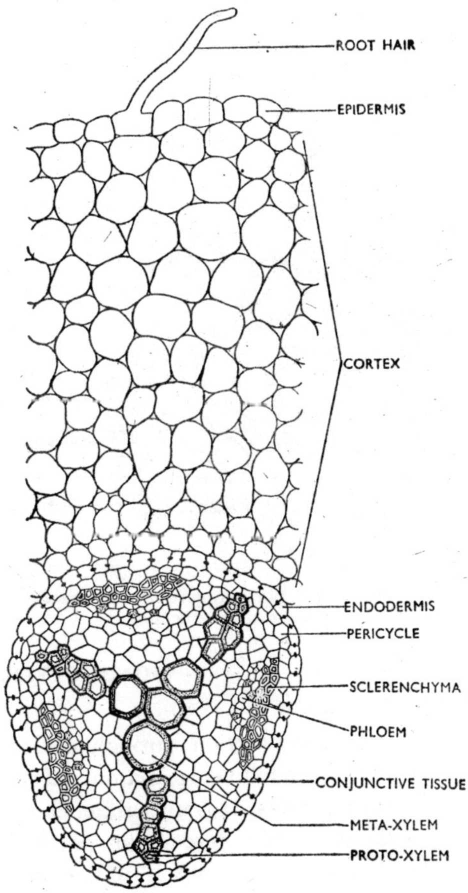

FIG. 602. A portion of root of *Pisum* (pea) in transverse section.

3. *ROOT OF BUTTERCUP*

A transverse section through the root of buttercup (*Ranunculus* sp. of family Ranunculaceae) would exhibit the following plan of arrangement of tissues (Fig. 603):

I. **Epidermis.** It is as usual a uniseriate zone, composed of a row of living tabular cells. It remains more or less in a collapsed condition and disorganised protoplast may be noticed in some cells.

II. **Cortex.** It is quite massive, as in all roots, enveloping the stele. A narrow zone, called **exodermis,** corresponding to the hypodermis of stems, occurs next to the epidermis. It is composed of comparatively smaller cells rather compactly arranged with very scanty intercellular spaces. Exodermis may be called the outer zone of cortex. So-called inner zone occurs internal to exodermis. It consists of quite a good number of larger parenchyma cells with conspicuous intercellular spaces. The cortical parenchyma cells contain abundant starch grains. The limiting layer of cortex is known as **endodermis.** It is made of a row of barrel-shaped cells with Casparian thickenings. Secondary depositions on the endodermal walls are also noticed. In that case some thin-walled cells, in which secondary deposition has not taken place, often occur, particularly against the protoxylem groups. These are called the *passage cells.*

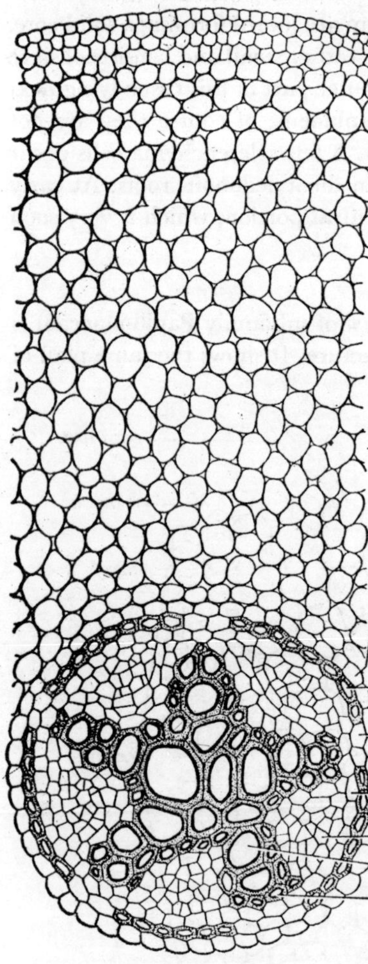

FIG. 603. A portion of root of *Ranunculus* in transverse section.

III. **Stele.** The outermost portion of the stele is the uniseriate parenchymatous **pericycle.** Vascular bundles are radially arranged. Four or five strands of xylem alternate with equal number of patches of phloem. Xylem is typically *exarch.* Phloem forms small patches. Parenchymatous conjunctive tissues occur between xylem and phloem. Pith is absent.

MONOCOTYLEDONOUS ROOTS

1. *ROOT OF ARUM*

A transverse section through the root of arum (*Colocasia* sp. of family Araceae) would reveal the following anatomical structure (Fig. 604):

I. **Epidermis.** It is uniseriate, composed of a row of tabular cells attached end on end without having intercellular spaces.

II. **Cortex.** The cortex is quite massive, as in other roots, and mainly consists of unspecialised parenchyma with profuse schizogenously formed spaces. In a slightly old root a few layers of cortex next to epiblema undergo chemical changes—suberisation, and thus give rise to a zone meant for protecting the internal tissues. This band is known as **exodermis.** Formation of exodermis may be initiated before the epiblema loses its function, but once epiblema is decayed exodermis takes over the function of protection. The last layer of cortex is the **endodermis.** It is composed of barrel-shaped compactly-set cells with conspicuous Casparian strips. Due to secondary thickening the endodermal cells may have considerably thick radial and inner walls. In that case some thin-walled cells usually occur against the protoxylem groups; obviously they are meant for ready diffusion of fluids. These are known as *passage cells* or transfusion cells.

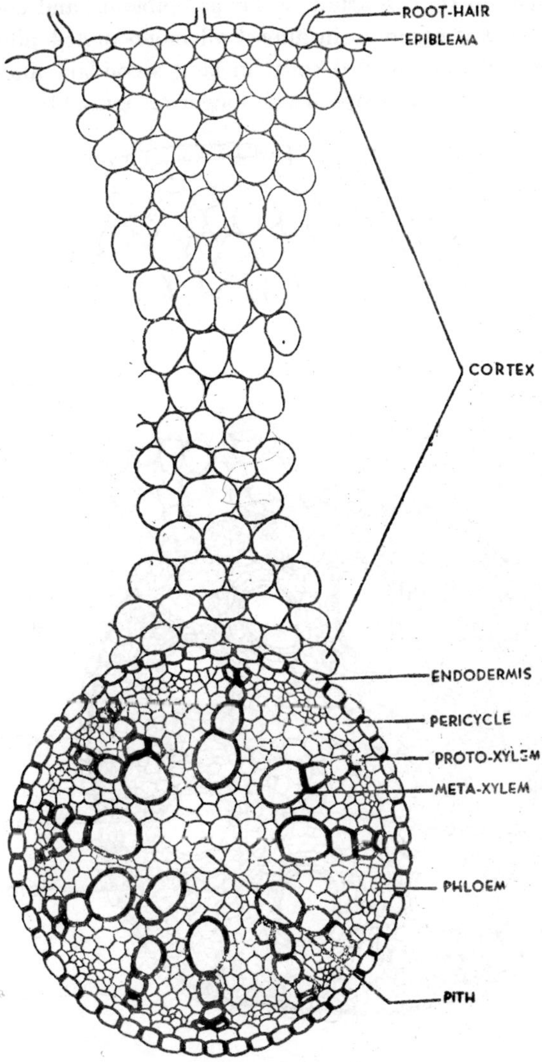

FIG. 604. A portion of root of *Colocasia* (arum) in transverse section.

III. **Stele.** The central cylinder consists of radially arranged vascular strands and intrastelar ground tissues. Uniseriate **pericycle,** made of thin-walled parenchyma cells, occurs next to endodermis. Xylem and phloem remain arranged alternately as separate patches, the xylem being typically *exarch*. As a good number of vascular strands are present, as opposed to the limited number of dicotyledonous roots (triarch, tetrarch, etc.), this is referred to as *polyarch*. Phloem is composed of sieve tubes, companion cells and parenchyma. Though protophloem occurs on the outer side and metaphloem on the inner, the two can hardly be distinguished. Small conjunctive parenchyma cells are present between xylem and phloem patches. The central part of the stele is occupied by a fairly large pith.

2. *ROOT OF MAIZE*

The internal structure (Fig. 605) of the maize root (*Zea mays* of family Graminaceae) is more or less similar to that of arum, so far as epiblema and cortex are concerned. Formation of exodermis is common in almost all monocotyledonous roots. The **endodermis** is composed of thick-walled cells; in fact, secondary and tertiary layers are deposited, so that the Casparian strips are no longer recognisable.

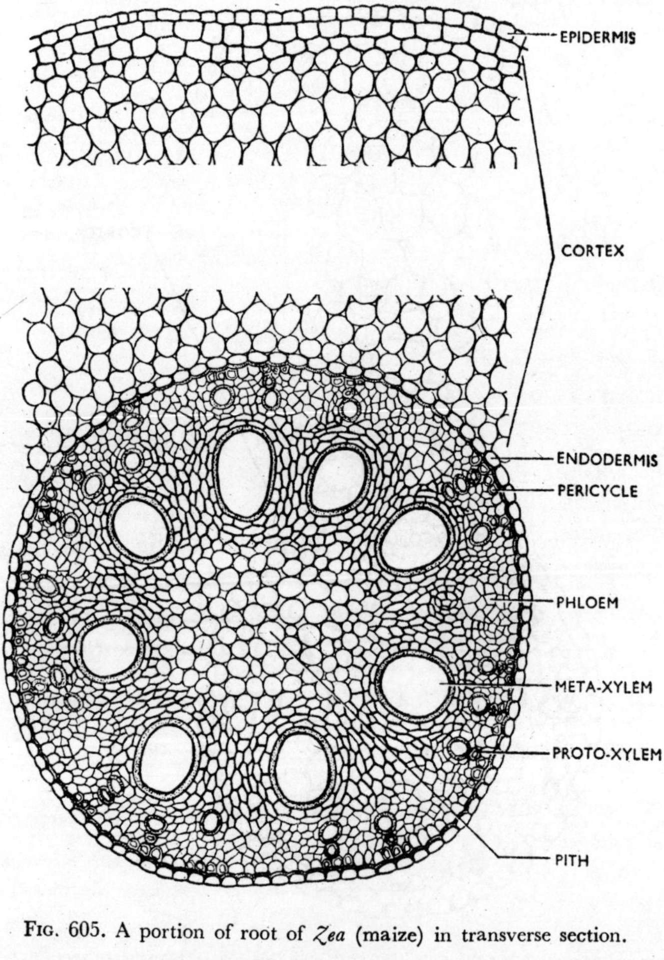

FIG. 605. A portion of root of *Zea* (maize) in transverse section.

The **pericycle** is uniseriate, but unlike that of the previous one, it is partly sclerenchymatous here. Vascular bundles are as usual *polyarch* with a pretty good number of xylem and phloem strands. Parenchyma cells associated with xylem undergo sclerosis and thus become thick-walled. The central portion is occupied by a large pith—made of loosely-arranged parenchyma cells containing abundant starch grains.

3. *ROOT OF SMILAX*

It (*Smilax* sp. of family Liliaceae) is a herbaceous monocotyledon, the roots of which are quite suitable for the studies of internal structure. Though in general it (Fig. 606) shows the same plan as found in other monocotyledonous roots, but some distinctive features are to be noted.

Epidermis is as usual uniseriate—made of parenchyma cells with rounded outer walls. A single row of heavily thick-walled cells occur just internal to epidermis, forming the **exodermis.** The rest of the **cortex** is composed of thin-walled parenchyma cells with distinct intercellular spaces. Starch grains are abundantly present in the cortical cells. **Endodermis** is of thick-walled type, where radial and inner walls in particular undergo considerable secondary thickening.

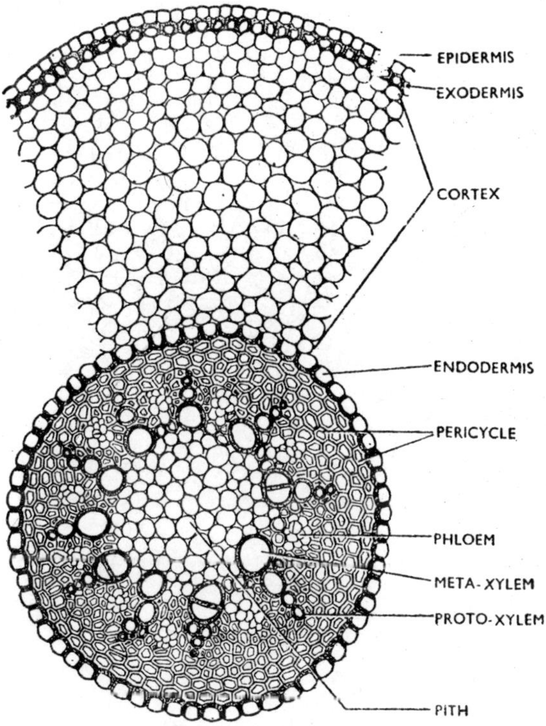

FIG. 606. A portion of root of *Smilax* in transverse section.

The **stele** consists of a large number of xylem and phloem strands arranged alternately, so it is also *polyarch.* The **pericycle,** unlike that of other roots, is multiseriate and consists of a few layers of thick-walled sclerenchyma cells. Protophloem occurring on the outerside are smaller than the metaphloem elements. Pith occupying the central portion of the stele is fairly large, and is made of thick-walled parenchyma.

4. *ROOT OF ORCHID*

The roots of epiphytic orchids (*Vanda* spp. of family Orchidaceae) possess a spongy outer tissue for absorbing moisture from the atmosphere. This tissue is known as **velamen** (Fig. 607). It consists of a few layers of compactly-set dead cells, which often form a silvery outer coat. The walls are usually porous, so that the cells work like a sponge. These empty cells have walls variously thickened by spirally or reticulately arranged fibres which take up the form of supporting ribs (Fig. 607B). During dry weather the cells remain filled with air, and during rains they quickly absorb water. Special structures, called *pneumatodes*, consisting of groups of cells with dense spiral wall thickening are present. They are helpful in gaseous exchange when the roots are saturated with water. Velamen is derivative of protoderm, and hence may be interpreted as a typical **multiseriate epidermis,** specially adapted to serve as an absorbing tissue. The outermost layer of velamen is known as limiting layer.

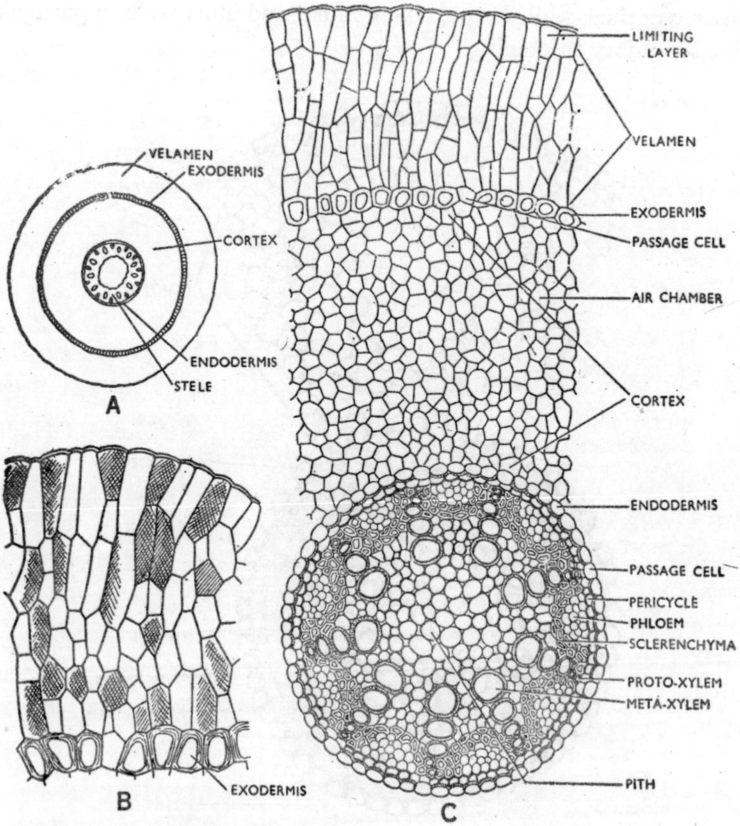

FIG. 607. Root of orchid. A. Complete transverse section (diagrammatic) ; B. Velamen with exodermis (enlarged). C. A portion of the root in transverse section.

Cortex. The outermost layer of cortex consists of a row of thick-walled cells forming what is known as **exodermis.** The thickenings due to deposition of suberin are more pronounced on the outer and lateral walls of the cells. Unthickened ones, called *passage cells,* occur here and there, which may serve as channels for flow of water absorbed by the velamen. The main bulk of the cortex occurs internal to exodermis. It is composed of a few layers of parenchyma cells with intercellular spaces. Chloroplasts are present in these cells; this fact explains the greenness of the roots, particularly when wet. The last layer of cortex is the **endodermis** with suberised radial and inner walls. Passage cells occur in the endodermis, usually opposite the protoxylem vessels.

Stele. Pericycle is uniseriate, made of thick-walled cells; only the cells just lying internal to passage cells of the endodermis are thin-walled. A good number of xylem and phloem groups occur alternating in the stele. Conjunctive tissues surrounding the phloem groups are sclerenchymatous. Central portion is occupied normally by the parenchymatous pith, but these cells may undergo sclerosis.

BREATHING ROOTS

(*PNEUMATOPHORES*)

These roots are found in the plants growing in situations with scanty oxygen. Unlike the normal roots, they come vertically upwards—thus becoming negatively geotropic and negatively hydrotropic. They absorb oxygen from the outer atmosphere through specially located lenticels at the tips. Thus strictly primary condition can hardly be noted. The anatomical structure resembles that of a stem in the nature and disposition of the vascular system in particular. A transverse section would show the following arrangement of tissues (Fig. 608):

I. **Cork.** A few layers of cork cells occur at the outermost portion with small **lenticels** which are really instrumental for absorption of oxygen.

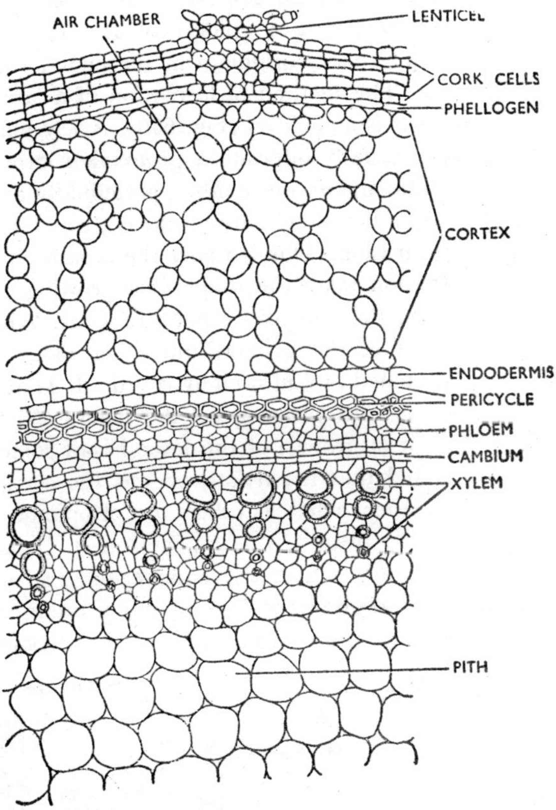

Fig. 608. A portion of a breathing root in transverse section.

II. **Cortex.** It is quite massive, composed of a few layers of more or less rounded parenchyma cells with well-developed intercellular space system. The last layer of cortex is as usual the **endodermis**—a uniseriate zone made of small barrel-shaped cells.

III. **Stele.** Just internal to endodermis occurs pericycle made of parenchyma and sclerenchyma. Usually the outer portion is parenchymatous and the inner sclerenchymatous. The vascular bundles actually resemble those of the dicotyledonous stems. They are **collateral** and **open.** In fact, the bundles form a continuous cylinder with xylem and phloem, the cambium occurring between them. The xylem is endarch. A large pith composed of parenchyma cells occurs at the central portion.

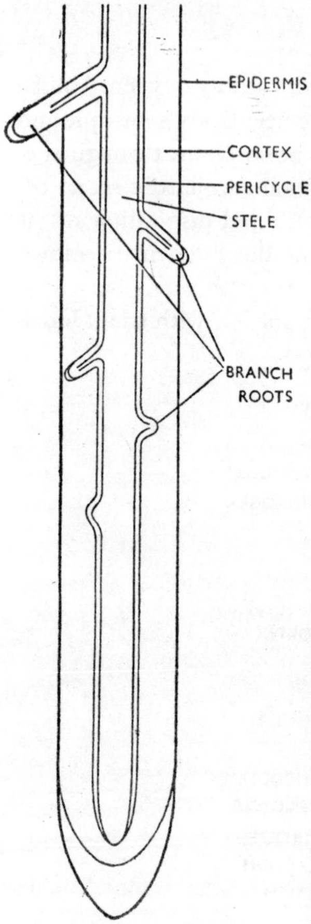

FIG. 609. Diagram showing origin of branch roots.

ORIGIN OF LATERAL ROOTS

The lateral roots are *endogenous* in origin. They develop from mature cells, some distance away from the apical meristem (Fig. 609) usually behind the root hair zone. The apical meristem of the root does not lay down any appendage. It is a marked point of contrast with the shoot where the primordia of the leaves and branches develop in the apical meristem.

In higher plants, angiosperms and gymnosperms, the development of lateral root is commonly initiated in the **pericycle**—an intrastelar ground tissue of the parent root, and the lateral root ultimately makes its way by piercing the cortex and epidermis. In lower vascular plants the branch roots originate in the endodermis.

The method of formation of a lateral root in a higher plant is as follows (Fig. 610). A few mature cells of pericycle, usually opposite a protoxylem group, become meristematic and go on dividing periclinally and anticlinally. Thus a number of cells are produced, which form something like a protrusion. This is really the *primordium* of the lateral root. It soon takes the shape of a growing point with its initial cells, the cap and other histogens. With gradual development of the primordium other tissues surrounding it get stretched and ultimately ruptured. That is how it eventually comes out piercing all the tissues. As regards the mechanism of growth of the lateral root some workers have suggested that it partially digests the cortical tissue during its advance; whereas others are of opinion that it is entirely a matter of mechanical penetration. It has been reported in some cases that the endodermis also undergoes anticlinal division and forms a layer surrounding the lateral root primordium; and by further periclinal division it may be even more than one layer in thickness. But at any rate those cells die and are shed when the lateral root comes out.

The lateral roots develop from the mature tissues in acropetal order, though there is no regularity in the order of development with reference to each other. In

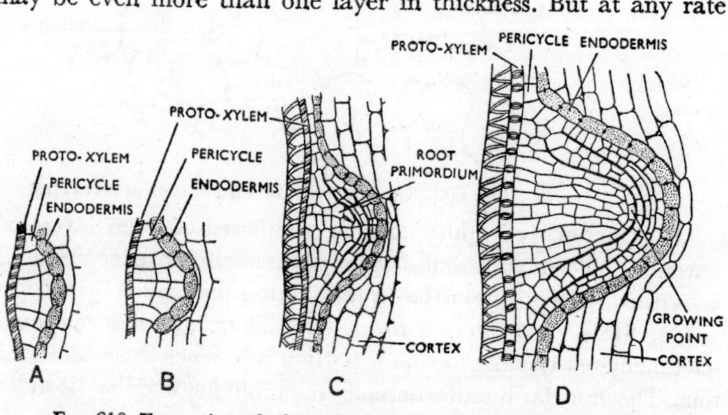

FIG. 610. Formation of a lateral root—stages in longitudinal views.

roots having more than two xylem strands the lateral roots commonly originate against the protoxylem groups, or less commonly opposite the phloem groups. Thus they come out in vertical rows, the number being equal to that of the xylem strands present. But in diarch roots having two xylem strands, the primordia develop at each side of the phloem group, so that the number of lateral roots formed is double that of xylem strands. They exhibit all the characters of the primary root, having four distinct regions. Thus an extensive root system is formed which ramifies through the soil particles. All the lateral roots do not grow equally vigorously. In fact, many of them continue normal growth and form the root system, whereas some remain undeveloped, or may even be lost. In some fleshy roots like those of carrot additional lateral roots may arise at the bases of original ones when the latter perish.

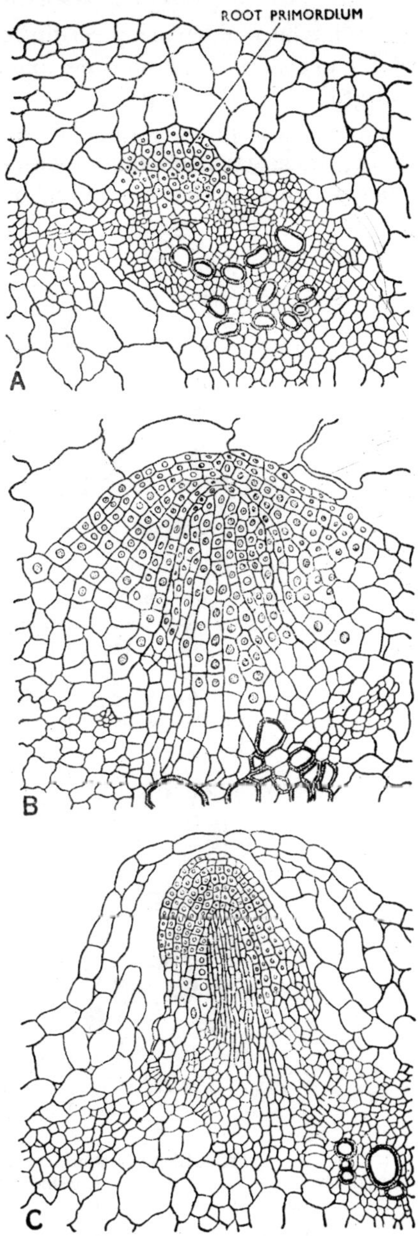

FIG. 611. Formation of adventitious root from the stem of *Lycopersicum* (tomato) —stages.

FORMATION OF ADVENTITIOUS ROOTS

Adventitious roots are of diverse types. They may arise either in association with buds or independently; either in the young organs or in the comparatively mature tissues which have retained the potentialities of cell division. They usually originate *endogenously* from the primordium already formed and lying in dormant condition, or they may form new primordium. Exogenous development cannot be ruled out. Primordia of adventitious roots may be formed from epidermis with cortical tissues, different internal regions, even from the tissues of leaf margins and petioles, e.g. *Begonia*, *Kalanchoe*.

Adventitious roots from the stems are the most common ones (Fig. 611). It has been established that initiation of such roots takes place near about the differentiating vascular tissues of the organ—by a group of cells forming the primordium near the periphery of vascular system in case of young stem, and near the vascular cambium in case of mature stem. The cells forming the primordium are derived from interfascicular parenchyma, medullary rays, or from vascular rays. It is often stated that the pericycle is the seat of origin of adventitious roots, but the very existence of pericycle in a stem has been rather doubtful.

Because of the origin of adventitious root from the tissues stated above, the developing root lies close to the xylem and phloem of the stem; and thus vascular connection between the two organs is easily established. It is similar to a lateral root as regards the organisation of the growing point, formation of cortex, etc., and mechanism of growth.

ROOT-STEM TRANSITION

It has been stated in a preceding chapter that the vascular skeleton in a plant is formed due to continuity of root-stem axis and the lateral appendages. The epidermal tissues and ground tissues are directly continuous in the two organs, stem and root. But the arrangement of primary vascular tissues is distinctly different in the two organs, roots having radial vascular bundles with exarch xylem and stems usually having collateral bundles with endarch xylem. A region actually exists where changes and adjustments take

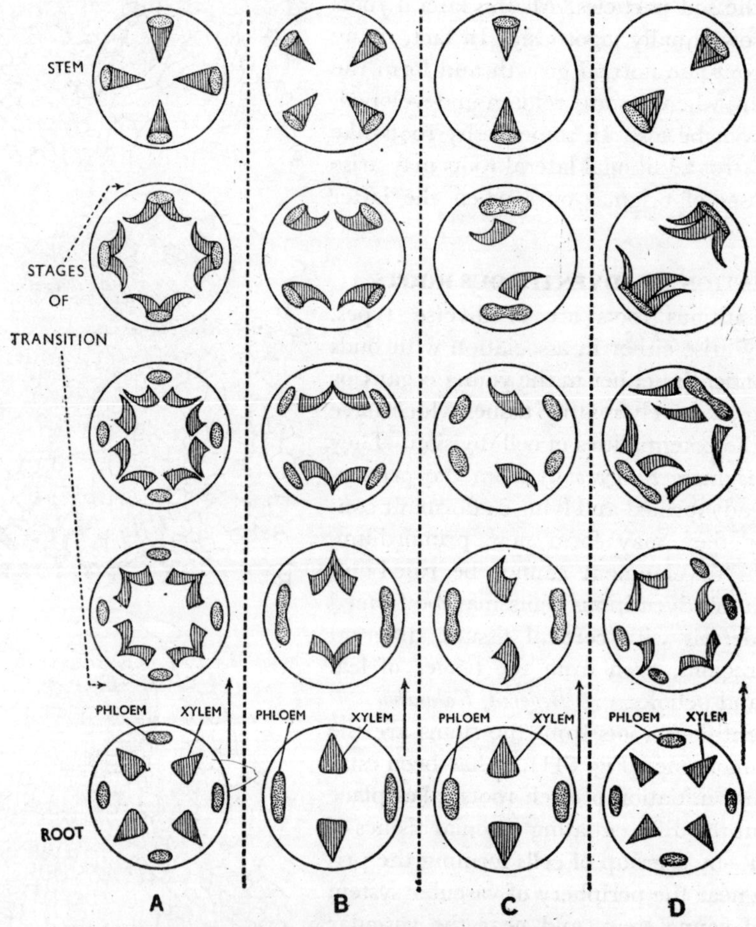

FIG. 612. Root-stem transition. Diagrams of four types A, B, C & D (modified *after Eames and MacDaniels*).

place, so that the two markedly different types of vascular tissues ultimately become continuous. The change involving inversion or twisting of xylem strands from one type of structure to another is referred to as **vascular transition,** and the region of the axis where changes occur is called **transition region.** This region is usually short, changes may take place gradually or rather abruptly at the top of the radicle and more commonly in hypocotyl—at its base, centre or upper part. The structure of the region often becomes more complex due to origin and departure of cotyledonary traces. The stele may enlarge in diameter in the transition region. The changes occur according to some plans. An account of a few types that have been studied is given here (Fig. 612).

I. In this type xylem strands fork by radial division and the two branches formed swing laterally, one to the right and one to the left by 180°, and join the phloem strands. The latter have remained unchanged all through, and run as straight strands from the root to the stem. That is how the radial bundles with exarch xylem become collateral ones with xylem endarch. The number of primary bundles in the stem here is equal to the phloem groups present in the root. This type of vascular transition has been noted in *Mirabilis* of family Nyctaginaceae, *Fumaria* of family Fumariaceae, etc.

II. The second type, what is really more common than the first one, involves forking in both xylem and phloem strands. The phloem branches remain in same position, whereas the branches of xylem strands swing laterally, as in the first type, and ultimately join up with the phloem strands. Thus the number of bundles in the stem is twice that of phloem groups present in the root. This type has been noticed in *Cucurbita* of family Cucurbitaceae, *Acer* and *Phaseolus* of family Leguminosae, *Tropaeolum* of family Geraniaceae, etc.

III. In this type xylem strands do not fork, but while passing upwards they swing laterally by 180°. The phloem strands divide; the branches swing and eventually join up with xylem strands. So the number of bundles in the stem is equal to that of the phloem groups of the root. This type occurs in *Lathyrus*, *Medicago* of family Leguminosae and *Phoenix* of family Palmae.

IV. Here half of the xylem strands fork and swing, whereas the other half do not divide, but become inverted. The phloem strands do not undergo any division, but simply fuse. In the meantime the triple xylem strands—the branches of one strand and an unbranched inverted one, join up with the phloem strands which have united in pairs. Thus, the bundle in the stem is the product of fusion of five strands and the number of bundles is half that of the phloem groups present in the root. This type, though of rare occurrence, has been found in some monocotyledons like *Anemarrhena* of family Liliaceae.

Some authors are, however, of opinion that this approach towards interpretation of transition region between the stem and the root is not happy. According to them "transition region represents connection not between two axial organs with somewhat different arrangement of tissues but between an organ with an axial vascular cylinder and one whose vascular system develops in relation to leaves".—*Esau*. In that case transition region should really explain the relations between the roots and the traces of the first-formed foliar organs of plants. It has been worked out in some cases that inversion of the xylem strands does not occur. In carrot the cotyledon has three traces—of which the median one consists of exarch xylem flanked by two lateral phloem groups, whereas the two lateral traces are collateral with outer phloem and inner endarch xylem. That shows the continuity between the vascular system of the cotyledon and root without inversion of the xylem strand.

CHAPTER XXIV

THE PRIMARY BODY : THE LEAF

The foliage leaves are characterised by green colour, thinness and flatness. They develop as protrusions from the shoot apex and are organs of limited growth. Leaves are very important vegetative organs, as they are chiefly concerned with the physiological process, photosynthesis and transpiration. Like other organs they also exhibit three tissue systems (Fig. 613). Epidermal tissue system consists of the epidermal layers occurring on the adaxial (upper) and abaxial (lower) sides. Occurrence of stomata and outgrowths are distinctive features. The ground tissue system, as already reported in a preceding chapter, is known as **mesophyll** tissue. It is often differentiated into columnar **palisade** parenchyma on the adaxial side and irregular or isodiametric **spongy** parenchyma on

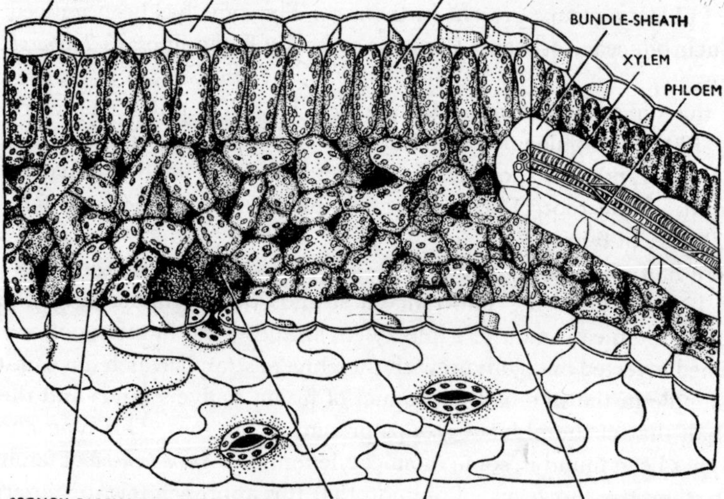

FIG. 613. Diagram of a dorsiventral leaf showing the tissues.

the abaxial side. A leaf showing this differentiation in mesophyll is referred to as *dorsiventral*, what is very common in dicotyledons. One with undifferentiated mesophyll, as commonly found in the monocotyledons, is known as an *isobilateral* leaf. Presence of conspicuous air spaces in the mesophyll is another marked feature. The gaseous exchange between the internal photosynthetic tissues and outside atmosphere thus becomes easy. The vascular tissue system is composed of vascular bundles which are usually **collateral** and **closed.** But the bundles entering the leaf occupy such a position that xylem occurs on the upper side and phloem on the lower. The vascular tissues, in fact, form the skeleton of the leaf, on which other tissues—the ground tissues, remain inserted. The petiole may continue into the midrib which bears branches and sub-branches ultimately ramifying in the leaf lamina in both reticulate and parallel type of venation. The ultimate branches are very small and terminate in what are known as bundle ends. Often these ends bend into minute specialised photosynthetic areas known as *vein islets* or they may just extend into the mesophyll. Bundle-ends vary considerably in the leaves, but commonly it consists of a single tracheid with a single sieve element or specialised parenchyma representing xylem and phloem respectively, surrounded by a parenchymatous bundle sheath (Fig. 613A). In extreme cases the phloem may be absent and the veinlet may be made of a single spiral tracheid. Thus the size of the bundle depends on the position one prefers to take while making a section. Very commonly vascular bundles remain surrounded by a row of cells, which may or may not contain chloroplasts. This row made of parenchyma cells is referred to as **bundle sheath** or border parenchyma. The leaves of monocotyledons often have two bundle sheaths—outer parenchymatous one usually with chloroplasts and an inner thick-walled one without chloroplasts. In some dicotyledons the bundle sheath extends up to the epidermis, either on one or on both sides of the leaf, and are termed bundle sheath extensions. Their contact with conducting elements on one side and mesophyll on the other and often the extension up to epidermis are suggestive of positive physiological functions. The morphology of bundle sheath was considered to be uncertain, but it is now regarded as an endodermis (Fann), where Casparian strips in some cases have been observed. Often parenchyma cells of the sheath contain starch, then it may be called a 'starch sheath'. The leaves of mesophytes possess highly-thickened epidermis and patches of mechanical tissues, either as isolated patches or in association with vascular tissues, as they have to withstand shearing stresses in particular.

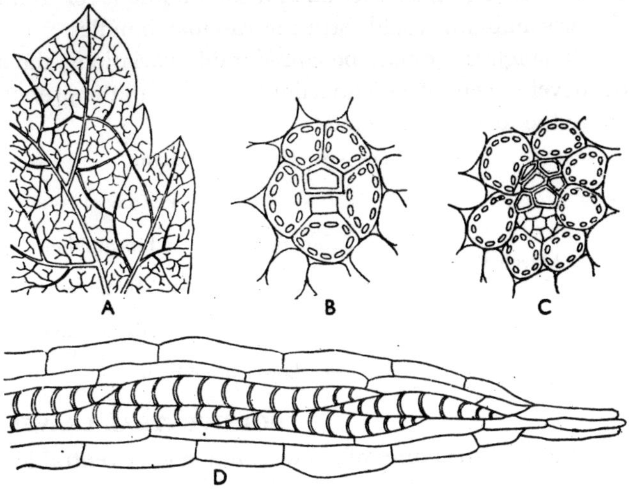

FIG. 613A. Vascular skeleton of leaves. A. a leaf of dicotyledon with vein islets; B. bundle end with one tracheid and one parenchyma cell; C. same with xylem and phloem; D. bundle end in longitudinal section-tracheids surrounded by parenchyma.

ONTOGENY OF THE LEAF

Leaves originate from the promeristem of the shoot apex. The outer layers of apical meristem just below the apex undergo anticlinal and periclinal divisions and give rise to a small rounded protrusion, what is the *leaf primordium*. The meristematic cells occurring at the tip of the primordium constitutes the apical meristem of the leaf. The apical growth takes place at the early stage. It stops soon, and further expansion giving the leaf proper form is due to rapid enlargement by intercalary growth. During the development of the blade the basal part of the protrusion remains unexpanded which ultimately forms a meristem that gives rise to the petiole. At the early stage of the development of the blade two strips of meristems, referred to as *marginal meristem*, occur along the two surfaces of the leaf axis. These are composed of files of initials known as *marginal initials*. They form the protoderm by anticlinal division. The cells occurring beneath the marginal initials, known as *submarginal initials*, divide in all planes and produce the internal tissues of the leaf. They are often differentiated into three layers—adaxial, abaxial and middle layers. A median procambium develops from the procambial strands of the shoot apex. After cessation of marginal growth further growth of the lamina is due to anticlinal division of the cells, so that surface area of the leaf is increased but in depth number of cell layers remain unchanged. The two protoderm layers produce upper and lower epidermis; the adaxial layer of submarginal initials gives rise to palisade parenchyma, the abaxial layer to lower spongy parenchyma and the middle layer to middle spongy parenchyma; and the procambium would form the vascular bundles.

Though there may be considerable variations, the above is the general account of the development of a dicotyledonous leaf from the primordium, which may be represented by the following scheme :

- Marginal initials
  - → Upper protoderm → Upper epidermis
  - Submarginal initials
    - → Adaxial layer → Palisade parenchyma
    - · Procambium → Vascular bundles
    - → Middle layer → Middle spongy parenchyma
    - → Abaxial layer → Spongy parenchyma
  - → Lower protoderm → Lower epidermis

DORSIVENTRAL LEAVES

Internal structures of a few common dorsiventral leaves are being described below:

1. *LEAF OF MANGO*

A thin section through the lamina of a leaf of mango (*Mangifera indica* of family Anacardiaceae) will show the following plan of arrangement of tissues (Fig. 614):

I. **Epidermis.** There are two epidermal layers on adaxial and abaxial surfaces of the leaf. Each is uniseriate, composed of a row of compactly-set tabular cells. The outer

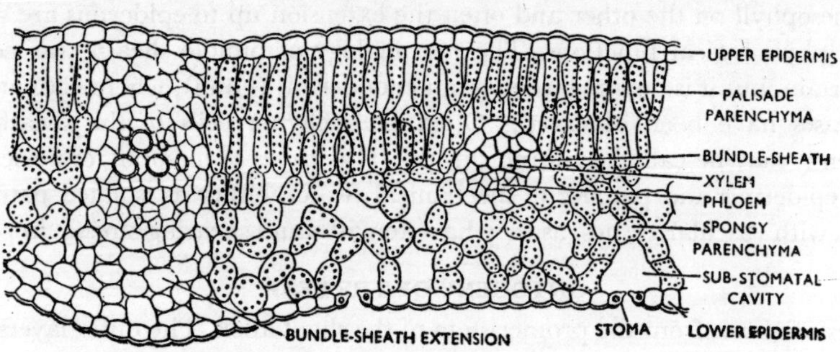

FIG. 614. A portion of leaf of *Mangifera* (mango) in transverse section.

walls are cutinised and possess thin cuticle, the thickness being more pronounced in the cells of the upper epidermis than those on the lower side. Stomata occur on the lower epidermis.

II. **Mesophyll.** The ground tissue forming the **mesophyll** is differentiated into palisade and spongy cells. The **palisade** cells occur towards upper epidermis. They are columnar cells with scanty intercellular spaces and remain arranged more or less at right angles to the upper epidermis. Chloroplasts are abundantly present, which particularly occur along the radial walls of the cells. There are two layers of palisade cells. The **spongy** cells occur towards the lower epidermis. They are quite loosely arranged with conspicuous intercellular spaces. The number of chloroplasts is naturally much smaller here, which explains the pale green colour of the lower surface of the leaf.

III. **Vascular bundles.** Bundles are **collateral** and **closed.** They are located in the mesophyll. The size of the bundle depends on the position one chooses to take in making a section. A bigger bundle is composed of xylem and phloem, the former occurring towards upper epidermis and the latter towards the lower side. The xylem is made of tracheary elements, and the phloem of sieve tubes and companion cells. The bundle remains surrounded by a row of colourless parenchyma cells. This band is referred to as **bundle sheath** or border parenchyma. Thus the bundle is not in direct contact with the mesophyll cells. Parenchyma and often collenchyma cells are present on the outer and inner sides of the bundle which may reach up to the two epidermal layers. These cells constitute what is known as *bundle sheath extension.*

2. *LEAF OF BANYAN*

A transverse section through the leaf of Banyan (*Ficus benghalensis* of family Moraceae) would reveal the anatomical characters (Fig. 615) more or less similar to the previous one.

I. **Epidermis.** The upper epidermis is multiseriate, being made of a few layer of cells. Lithocysts are frequently present and well-developed calcium carbonate crystals, the cystoliths, occur here and there. The lower epidermis is uniseriate. The outer layer of upper multiple epidermis and the lower epidermis as a whole are made of compactly-arranged tabular cells with cutinised outer walls having cuticle. The degree of cutinisation is more pronounced on the upper side. Stomata occur on the lower epidermis.

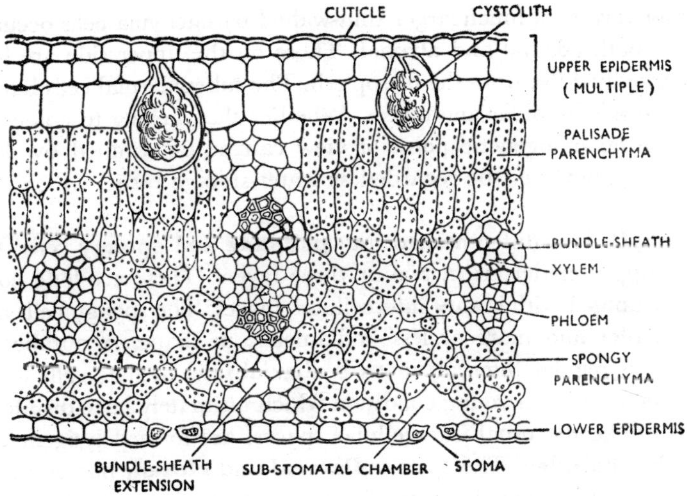

FIG. 615. A portion of leaf of *Ficus benghalensis* (banyan) in transverse section.

II. **Mesophyll.** It is differentiated into **palisade** and **spongy** cells. Two or three layers of columnar cells with abundant chloroplasts remain arranged more or less at right angles to the upper epidermis. These are palisade cells. This is the principal photosynthetic tissue. The spongy cells occurring towards lower epidermis are isodiametric, and often irregular in shape, and have profuse intercellular spaces. The number of chloroplasts is naturally much smaller here in comparison to palisade cells.

III. **Vascular bundles.** The bundles are as usual **collateral** and **closed** ones, with xylem lying on the upper and phloem on the lower sides. They remain surrounded by parenchymatous bundle sheath. In case of bigger bundles bundle sheath extensions are present.

3. *LEAF OF BANANA*

Though a monocotyledon, the leaf of banana (*Musa sapientum* of family Musaceae) is dorsiventral. It exhibits the following internal structure (Fig. 616):

I. **Epidermis.** The upper and lower epidermal layers are composed of compactly-arranged roundish cells with cuticularised outer walls. Slightly sunken stomata are present on the lower side.

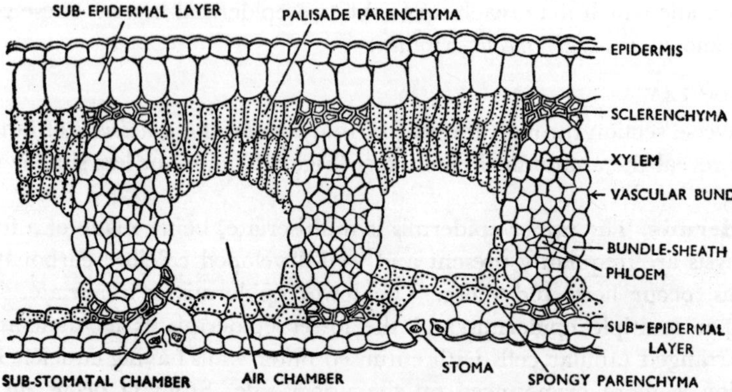

FIG. 616. A portion of leaf of *Musa* (banana) in transverse section.

One or more layers of much larger thin-walled parenchyma cells occur next to epidermal layers—both adaxial and abaxial. Those on the upper side are larger in size. These cells markedly differ from the mesophyll. The subepidermal cells have been designated as hypodermis by older anatomists. Whether they belong to epidermis or to the ground tissues from ontogenetic point of view can be determined by developmental studies. They may, however, be regarded as water-storage tissues which do not possess chlorophyll.

II. **Mesophyll.** It is differentiated into **palisade** and **spongy** cells. Palisade cells occur towards upper epidermis in two or three layers. These are long columnar cells rather tightly arranged with scanty intercellular spaces. The spongy cells are comparatively much smaller and more or less isodiametric in shape. A number of large air chambers occur at regular intervals towards abaxial side. Stellate cells were present in the regions of the cavities in young condition which ultimately have disintegrated. Thus these cavities, unlike the usual intercellular spaces, are formed lysigenously.

III. **Vascular bundles.** These are **collateral** and **closed,** but bundles are poorly developed with scanty xylem. Patches of sclerenchyma occur on the upper and lower sides of the bundles. Obviously they are meant for providing mechanical strength against shearing stresses.

4. *LEAF OF OLEANDER*

The leaf of oleander (*Nerium oleander* of family Apocynaceae) shows some distinct departures so far as the arrangement of tissues is concerned, from the ones described previously, exhibiting clear xerophytic adaptations. A transverse section would reveal the following structure (Fig. 617):

I. **Epidermis.** Both the adaxial and abaxial epidermal layers are multiseriate. They are composed of a few (usually three) layers of compactly-arranged isodiametric cells. The outer walls of the cells forming outermost layers are very strongly cuticularised. Epidermal outgrowths are present. The stomata are present on the lower epidermis, but they are sunken stomata located in a depression known as stomatal pit. A good number of trichomes develop from the cells bordering the depression. It is an effective mechanism for reducing transpiration.

II. **Mesophyll.** It consists of **palisade** and **spongy** cells. But unlike other dorsiventral leaves, palisade cells occur here both towards upper and lower epidermis and spongy cells are located in between them. On the adaxial side palisade cells occur in three or more layers forming a compact zone of columnar cells with little intercellular spaces. Chloroplasts are abundantly present, usually along the radial walls of the cells. Inner palisade, i.e., palisade occurring on the abaxial side, is usually one-layered and consists of rather small cells. The spongy cells located between the two palisade layers are isodiametric, much smaller in size and much loosely arranged with profuse intercellular spaces. Calcium oxalate crystals are present here and there.

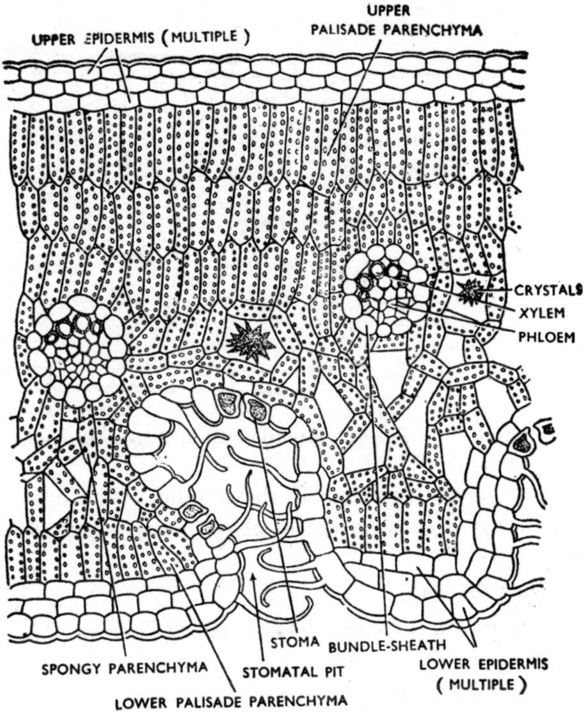

FIG. 617. A portion of leaf of *Nerium* (oleander) in transverse section.

III. **Vascular bundles.** The bundles are **collateral** and **closed** ones with xylem on the upper and phloem on the lower sides. They remain surrounded by parenchymatous bundle sheaths.

ISOBILATERAL LEAVES

These leaves are common in the monocotyledons. Here stomata occur on both the epidermal layers, though they are more abundant on the abaxial side. Mesophyll hardly shows differentiation between palisade and spongy cells. A few common isobilateral leaves have been selected for the study of internal structures.

1. *LEAF OF TUBEROSE*

A section through the leaf of tuberose (*Polianthes tuberosa* of family Amaryllidaceae), would show the following anatomical structure (Fig. 618):

I. **Epidermis.** Both the epidermal layers are uniseriate, composed of compactly-arranged rectangular cells with rounded cuticularised outer walls. Stomata are present on both the epidermal layers.

II. **Mesophyll.** It as usual forms the main bulk of the leaf, and is composed of isodiametric cells with intercellular spaces. Chloroplasts are abundantly present. Thus the differentiation of mesophyll into palisade and spongy cells is absent; all the cells are of spongy type.

III. **Vascular bundles.** These are present at regular intervals. The bundles are **collateral** and **closed** ones with xylem and phloem. A few sclerenchyma cells are

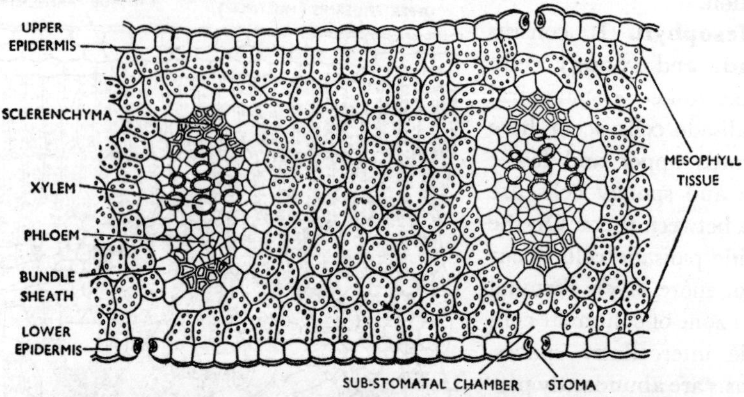

FIG. 618. A portion of leaf of *Polianthes* (tuberose) in transverse section.

present at the two ends of the bundles. The bundles remain surrounded by a row of parenchyma cells devoid of chloroplasts, which forms the *bundle sheath.*

2. *LEAF OF MAIZE*

A section through a leaf of maize (*Zea mays* of family Graminaceae) shows the following structure (Fig. 619):

I. **Epidermis.** Both upper and lower epidermal layers are uniseriate and composed of more or less oval cells with cuticularised outer walls. Upper epidermis may be

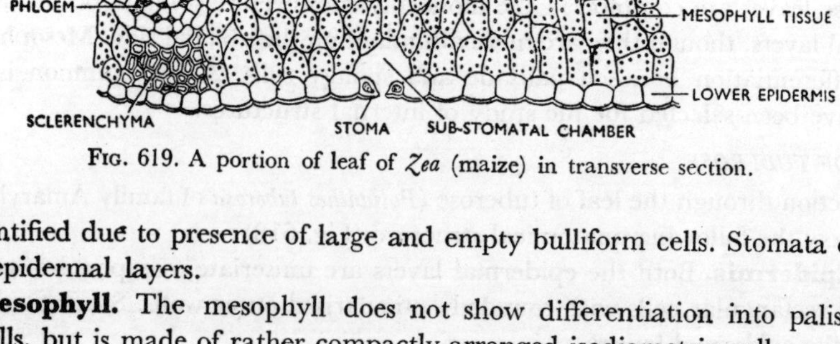

FIG. 619. A portion of leaf of *Zea* (maize) in transverse section.

easily identified due to presence of large and empty bulliform cells. Stomata occur on both the epidermal layers.

II. **Mesophyll.** The mesophyll does not show differentiation into palisade and spongy cells, but is made of rather compactly-arranged isodiametric cells.

III. **Vascular bundles.** The bundles are **collateral** and **closed** ones which remain arranged in parallel series. Majority of the bundles are small, but fairly large bundles occur at regular intervals. Small bundles have xylem on the upper and phloem on the lower sides surrounded by large parenchyma cells forming the *bundle sheath.* The cells of the sheath contain plastids, often with starch grains. It is assumed that this layer serves as a temporary storage tissue, apart from conducting the products of photosynthesis to the phloem. Xylem, as usual, consists of tracheary elements, and phloem of sieve tubes and companion cells. A large vascular bundle practically resembles that of a stem. Sclerenchyma cells occur in patches on both edges of the bundles, obviously for giving mechanical strength.

3. *LEAF OF BAMBOO*

A section through the leaf of bamboo (*Bambusa* spp. of family Graminaceae) would reveal the following internal structure (Fig. 620):

I. **Epidermis.** As usual there are two epidermal layers. The upper epidermis possesses a number of conspicuous bulliform cells. The lower one bears stomata and remain

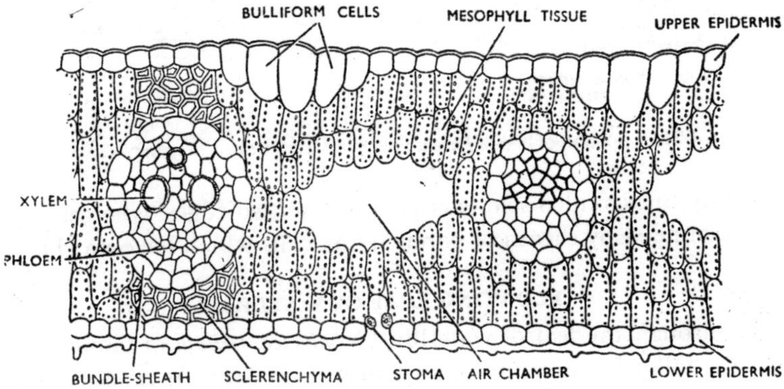

FIG. 620. A portion of leaf of *Bambusa* (bamboo) in transverse section.

covered with strong cuticle. Considerable deposition of silicon is a distinctive character. Stiff sharply pointed hairs are also present.

II. **Mesophyll.** It is composed of compactly-arranged cells, without showing any differentiation into palisade and spongy cells. The cells are of rather palisade type, though not much elongate. Some distinct cavities are present here and there.

III. **Vascular bundles.** These are **collateral** and **closed** ones which remain arranged in parallel series. Xylem occurs on the adaxial and phloem on the abaxial sides. Larger bundles have more distinct xylem and phloem surrounded by a *bundle sheath*, and have patches of sclerenchyma cells on the two sides.

4. *LEAF OF DATEPALM*

A section through a leaf of datepalm (*Phoenix sylvestris* of family Palmae) would show the following structure (Fig. 621):

I. **Epidermis.** The two epidermal layers are composed of compactly-arranged tabular cells with cuticularised outer walls. Stomata are present here and there. A layer of parenchyma cells with scanty chlorophyll occurs just internal to both the epidermal layers. These distinctly differ from the mesophyll. These subepidermal layers may be called special hypodermis.

II. **Mesophyll.** It is composed of more or less isodiametric cells with small intercellular spaces, thus the differentiation into palisade and spongy cells is absent. Patches of sclerenchyma occur more or less in parallel series towards both the upper and lower epidermis, as I-girders for withstanding shearing stress.

III. **Vascular bundles.** The bundles are **collateral** and **closed** ones. They remain arranged in parallel series. Each bundle has xylem on the upper and phloem on the lower side. The bundles are of two types, viz., large and small ones. The large bundles have

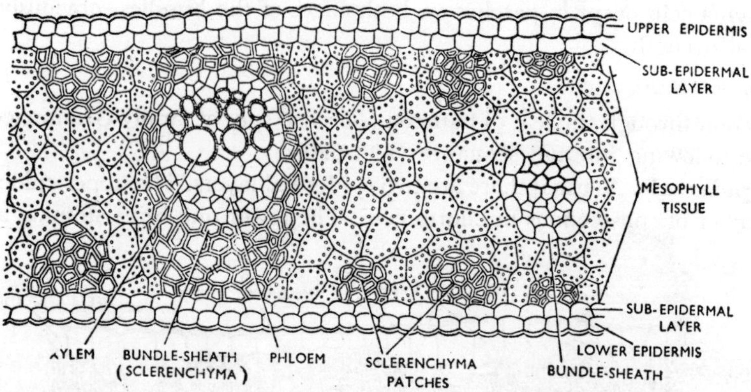

FIG. 621. A portion of leaf of *Phoenix sylvestris* (datepalm) in transverse section.

patches of heavily thick-walled sclerenchyma on the two edges, whereas the small bundles remain surrounded by sheath of parenchyma cells which have no chlorophyll.

FLOATING LEAF OF WATERLILY

A section through the leaf of waterlily (*Nymphaea stellata* of family Nyphaeaceae) would reveal the following anatomical structure (Fig. 622). As an aquatic plant it has extremely reduced vascular and supporting tissues and well-formed air chambers.

I. **Epidermis.** Epidermal layers are uniseriate both on the adaxial and abaxial sides. They are composed of closely-set cells. Stomata occur on the upper side.

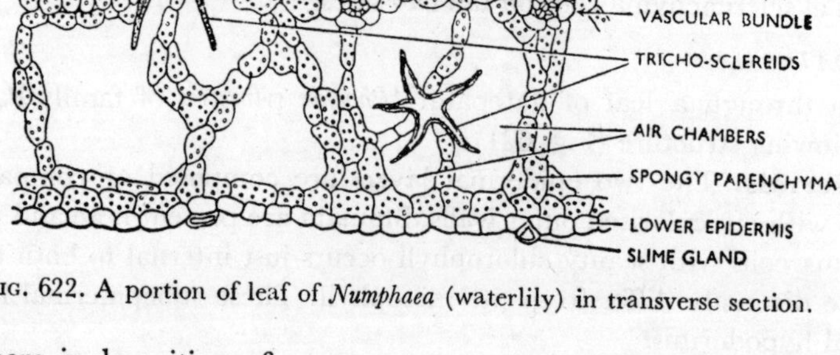

FIG. 622. A portion of leaf of *Numphaea* (waterlily) in transverse section.

Moreover, there is deposition of waxy matters which prevents wetting and clogging of the stomata.

II. **Mesophyll.** It is differentiated into **palisade** and **spongy** cells. A few layers of columnar cells occur towards the adaxial side forming the palisade. The spongy cells present towards lower epidermis and irregular in outline. Large air chambers are present in the mesophyll. Elongated sclerotic cells—the trichosclereids commonly called 'internal hairs', often with branched ends are frequently present.

III. **Vascular bundles.** These are very much reduced. As usual they are composed of xylem and phloem, and remain surrounded by parenchymatous bundle sheath.

STRUCTURE OF A SUCCULENT LEAF

A section through a succulent leaf with halophytic adaptations reveals the following structure (Fig. 622A):

The epidermis is uniseriate made of compactly-arranged tabular cells with cuticularised outer walls. A row of palisade parenchyma occurs next to epidermis forming

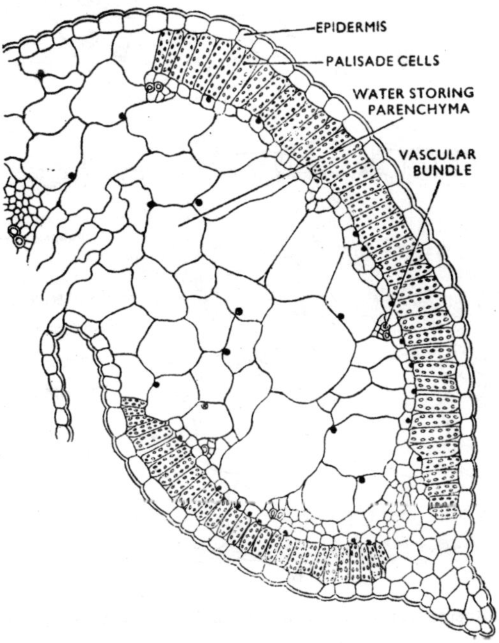

FIG. 622A. Succulent leaf of *Solsola*—half of lamina in cross-section.

more or less a compact band. The central part of the leaf is made of well-developed water-storage tissue, consisting of large parenchymatous cells with conspicuous vacuoles and intercellular spaces. Vascular bundles are very much reduced with very scanty mechanical elements and occur next to the palisade layer.

CENTRIC LEAF OF ONION

These leaves are acicular in shape and are more or less rounded in cross-section. Thus they do not possess two epidermal layers. The differentiation of tissues into epidermal, ground and vascular is clear. A transverse section through a leaf of onion (*Allium cepa* of family Liliaceae) would show the following internal structure (Fig. 623):

I. **Epidermis.** It is uniseriate and continuous one, made of small round cells with strongly cuticularised outer walls. Slightly-depressed stomata are present.

II. **Mesophyll.** The mesophyll consists of two types of cells. Just internal to epidermis there are usually two layers of columnar cells with abundant chloroplasts. These are palisade cells. They actually form something like a green belt. Next to this band occur a few layers of isodiametric parenchyma—spongy cells, with scanty chloroplasts. Conspicuous air spaces are present in the mesophyll. The central part of the leaf is hollow. The cells bordering the cavity are devoid of chloroplasts.

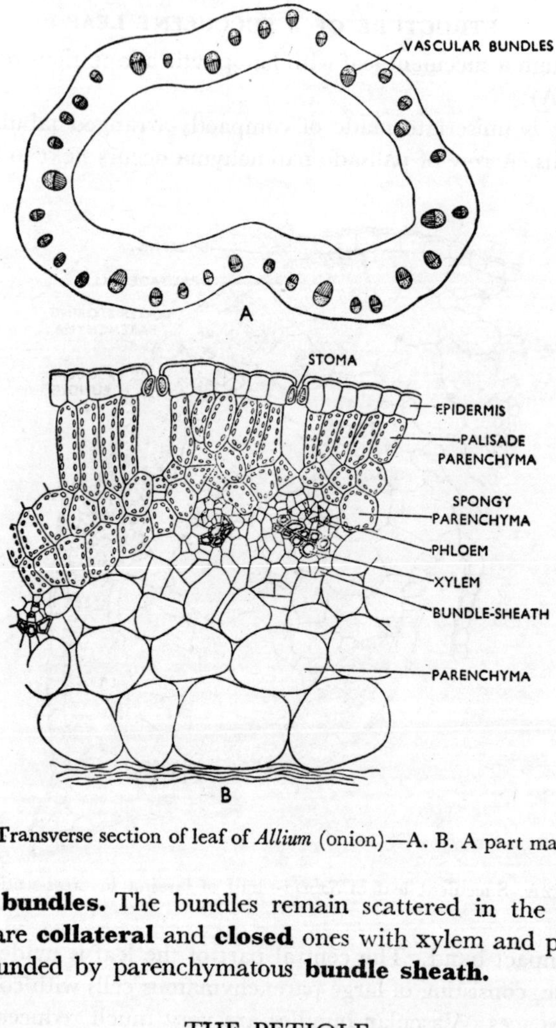

FIG. 623. Transverse section of leaf of *Allium* (onion)—A. B. A part magnified.

III. **Vascular bundles.** The bundles remain scattered in the lower part of the mesophyll. They are **collateral** and **closed** ones with xylem and phloem. Each bundle remains surrounded by parenchymatous **bundle sheath.**

THE PETIOLE

The petiole or the leaf stalk connects the lamina to the base, and thus supports it on the axis and exposes it to proper amount of light and air. As already stated, it is formed from an intercalary meristem located at the base of the finger-like protrusion of the shoot apex. The arrangement of tissues in the petiole more or less resembles that of the stem.

1. *PETIOLE OF CUCURBITA*

A transverse section through the petiole of *Cucurbita* would show the following structure (Fig. 624). It is wavy in outline having distinct ridges and furrows and a large hollow cavity in the central regions.

I. **Epidermis.** It is as usual uniseriate, made of a row of tabular cells with cuticularised outer walls. A large number of multicellular outgrowths are present.

II. **Ground tissues.** Patches of collenchyma cells with thickened corners occur at the ridges. The rest of the ground tissue is composed of thin-walled parenchyma with distinct intercellular spaces. A big cavity is present in the central region.

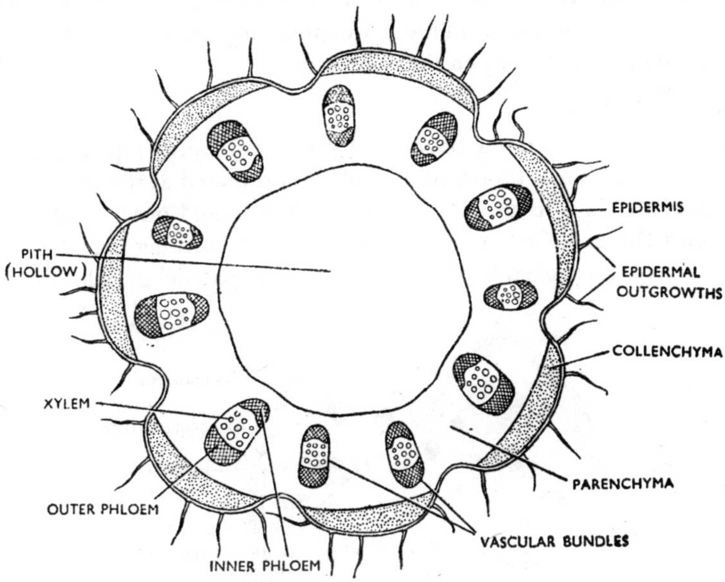

FIG. 624. Transverse section of petiole of *Cucurbita* (diagrammatic).

III. **Vascular bundles.** The bundles remain arranged in a ring. Individual bundles are **bicollateral** ones.

2. *PETIOLE OF SUNFLOWER*

Here (*Helianthus annus* of family Compositae) it is grooved at one side and rounded at the other (Fig. 625).

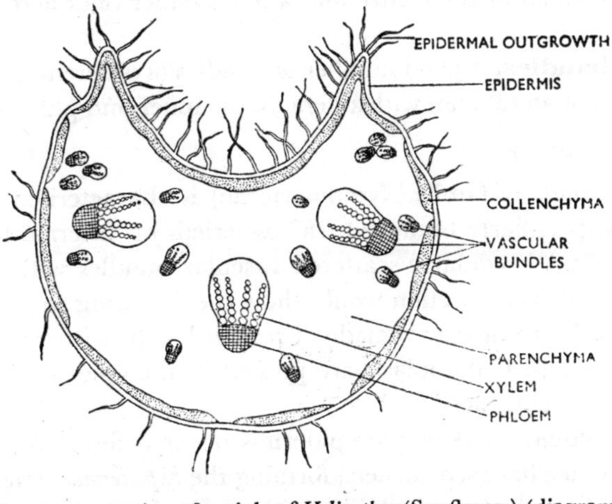

FIG. 625. Transverse section of petiole of *Helianthus* (Sunflower) (diagrammatic).

A transverse section would show the same plan of arrangement of tissues:

I. **Epidermis.** It is uniseriate with tabular cells and a large number of epidermal outgrowths.

II. **Ground tissues.** A band of collenchyma, intercepted by chlorenchyma here and there, occurs next to epidermis forming the hypodermis. The rest of ground tissues is made of isodiametric parenchyma cells with distinct intercellular spaces.

III. **Vascular bundles.** A number of vascular bundles occur in the form of a band; some bundles are small and some of them are quite large. Every bundle is collateral with xylem on the inner and phloem on the outer side.

3. *PETIOLE OF PIPER BETLE*

The section (Fig. 626) of the petiole of *Piper betle* of family Piperaceae is more or less heart-shaped with a distinct groove at one side and rounded at the other.

I. **Epidermis.** It is as usual uniseriate with cutinised outer walls.

II. **Ground tissues.** Collenchyma cells occur next forming hypodermis. The continuity of collenchyma is broken here and there by small patches of chlorophyll-containing

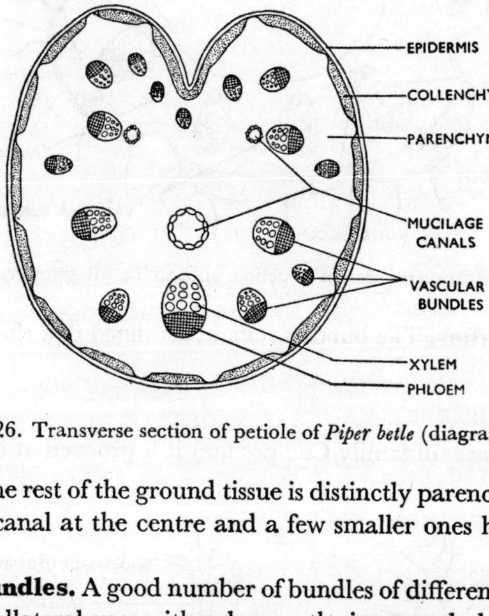

FIG. 626. Transverse section of petiole of *Piper betle* (diagrammatic).

parenchyma cells. The rest of the ground tissue is distinctly parenchymatous. Occurrence of a large mucilage canal at the centre and a few smaller ones here and there is a distinctive character.

III. **Vascular bundles.** A good number of bundles of different sizes occur in the form of an arc. They are collateral ones with xylem on the inner and phloem on the outer side.

4. *PETIOLE OF WATERLILY*

It (*Nymphaea stellata* of family Nymphaeaceae) is characterised by the presence of large air chambers, peculiarly branched trichosclerieds or 'internal hairs' with calcium oxalate deposition, and irregularly scattered vascular bundles with extremely reduced xylem elements. A transverse section would show the following structure (Fig. 627):

I. **Epidermis.** It is uniseriate, made of roundish cells with very scanty cuticularisation on the outer walls. Chloroplasts are present in the epidermal cells. Unbranched hairs are of frequent occurrence.

II. **Ground tissues.** Next to the epidermis occur a few layers, usually three, of collenchyma cells with thickened corners, forming the *hypodermis*. The rest of the ground tissue is made of thin-walled parenchyma with pretty large number of air chambers.

Trichosclereids of peculiar shapes, often with deposition of calcium oxalate crystals are abundantly present.

III. **Vascular bundles.** The bundles are very poorly developed and remain scattered in the ground tissues. The xylem vessels usually break down and form lacunae

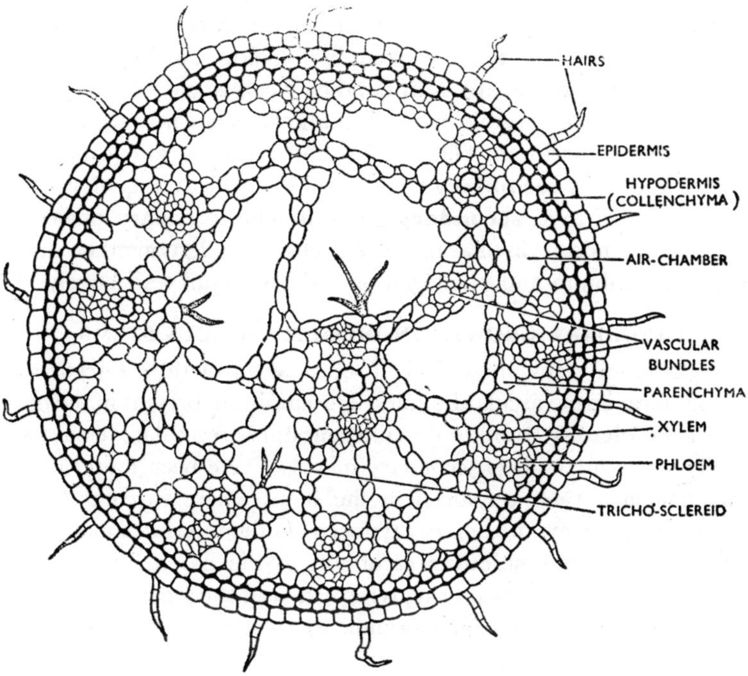

FIG. 627. Transverse section of petiole of *Nymphaea* (waterlily).

which resemble the air chambers. The phloem elements are normally developed. There are two types of bundles—small ones with one patch of phloem, and comparatively larger ones with two patches of phloem.

CHAPTER XXV

SECONDARY GROWTH IN THICKNESS

It has been stated in the preceding chapters that the apical meristem is responsible for the formation of the primary body of the plant. In fact, the primary permanent tissues derived from the apical meristems lay down the fundamental parts of the plant and thus cause growth in length and to some extent in thickness. Further increase in thickness is noticed quite distant from the apices in the dicotyledons and gymnosperms. The said growth in thickness becomes possible due to formation of some new tissues entirely by the activities of the lateral meristems like *cambium* and *phellogen*, also called *cork cambium*. These tissues are known as **secondary** tissues, and the growth in girth or thickness thus accomplished is referred to as **secondary growth.** The secondary tissues involved in the process are the secondary vascular tissues and periderm deriving their origin from the lateral meristems, cambium and phellogen or cork cambium respectively.

SECONDARY GROWTH IN DICOTYLEDONOUS STEMS

The dicotyledonous stems show distinct secondary growth in thickness, as they usually possess collateral open vascular bundles. The cambium present in the bundle, otherwise known as a **fascicular** cambium, mainly produces secondary tissues at the initial stage. Normally a few methods of secondary increase have been found in the dicotyledonous stems. The first type is noticed in the stems where vascular bundles remain arranged in rings and narrow medullary rays occur in the interfascicular regions. Due to the differentiation of **interfascicular** cambium in the parenchymatous cells of the medullary rays, a complete cambium ring is formed, and the secondary tissues are added on either side of the cambium ring. The secondary vascular tissues thus have the form of a continuous cylinder. This method is common in erect stems like those of members of sunflower family and, in fact, in many dicotyledonous stems. In some dicotyledons the primary vascular bundles remain arranged in a continuous cylinder, and the secondary vascular tissues are also added in the same form. This type is found in *Tilia*, *Nicotiana* (tobacco), etc. The other method prevails in climbers like *Aristolochia* where broad medullary rays are present between the vascular bundles in the primary condition. During secondary growth in this case the interfascicular cambium, a secondary meristem, goes on producing only parenchyma cells, and as a result a continuous ring of secondary tissues is not formed. Thus the secondary tissues also occur in strands. In climbers like *Cucurbita* secondary growth is confined to the individual bundles and does not extend to the interfascicular regions. Besides these, anomalous secondary growth occurs in a good number of families, which will be discussed in a separate chapter.

Here is an account of the process of secondary growth in thickness in a typical dicotyledonous stem. It starts in the intrastelar region and then extends to the extrastelar portion.

Cambium. In dicotyledons and gymnosperms a portion of the procambium remains undifferentiated during the progressive maturation of the tissues leading to the formation of the primary body. This undifferentiated part is the **cambium** of the vascular bundle which occurs between xylem and phloem. It is called **fascicular** cambium (*fascicle*=bundle). A few living parenchymatous cells of the medullary rays, which retained the potentialities of cells division, now resume meristematic nature, and thus new strips of meristems are formed in a line with the fascicular cambium. They are known as

interfascicular cambium, which may be formed either simultaneously with or after laying down of the primary body. The interfascicular cambium which is secondary in origin joins up with the fascicular cambium and thus a distinct continuous **cambium ring** or cylinder is formed.

The vascular cambium consists of two types of cells—spindle-shaped elongated ones, known as *fusiform initials*, and comparatively smaller and isodiametric ones, called *ray initials* (Fig. 628). The fusiform initials produce the secondary tissues which remain arranged parallel to the long axis of the organ, in fact forming the *vertical* or longitudinal system. The tracheary elements of xylem, fibres, parenchyma, and the sieve tubes, companion cells of phloem belong to this system. The ray initials give rise to the ray cells—the xylem and phloem rays which remain arranged horizontally and thus form the transverse or *horizontal* system. Thus secondary vascular tissues are composed of two systems—the vertical and horizontal, originating from fusiform and ray initials respectively of the cambium.

The cambium, strictly speaking, is one layer in thickness. But during active growth the cambium cells and their immediate derivatives form a zone, referred to as *cambium zone*, where the actual initials can hardly be distinguished from their immediate derivatives. The cells possess conspicuous vacuoles; primary pit-fields with plasmodesmata are present. On the basis of the arrangement of the cells as found in tangential views, the cambium is of two types (Fig. 628). In one type the fusiform initials occur in horizontal tiers, where ends of short fusiform cells of one tier remain more or less at the same level with another. This type of meristem is known as **storied** or *stratified* (Fig. 628A)

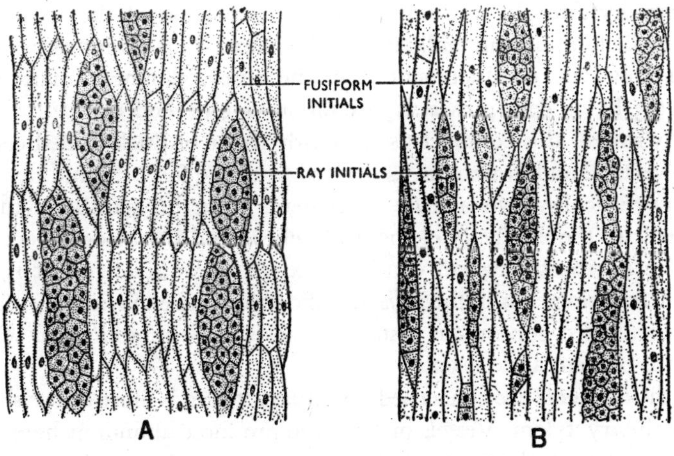

FIG. 628. Cambium. Fusiform initials and ray initials in longitudinal views. A. Storied cambium. B. Non-storied cambium.

cambium. In the second case comparatively longer fusiform initials do not occur in horizontal tiers but overlap at the ends. This is called **non-storied** or *non-stratified* (Fig. 628B) cambium. In all highly developed dicotyledons the cambium is of storied type.

The cambium cells divide tangentially, and secondary tissues, **secondary xylem** and **secondary phloem** are added on the internal and external sides respectively. A cambium cell divides into two daughter cells, one of the derivatives becomes either a xylem mother cell or a phloem mother cell depending on its position, while the other one remains meristematic (Fig. 629). The xylem or phloem mother cell by further differentiation and perhaps on further division ultimately develops into secondary xylem or secondary phloem element. The cambium cells continue to divide in this fashion producing secondary tissues on either sides. That is how the cambium perpetuates itself and thus remains active in plants for pretty long time. Usually the cambium produces more xylem than phloem, so that the cambium ring with all the tissues in front of it, is pushed more and more towards the periphery. Due to formation of secondary tissues the primary xylem and primary phloem which were lying side by side in the primary condition, are gradually pushed apart from each other. Usually in the plants growing in tropical regions cambium is active throughout the year. In temperate regions cambial activity may cease with the onset of unfavourable period and enter into a dormant state till favourable conditions are available.

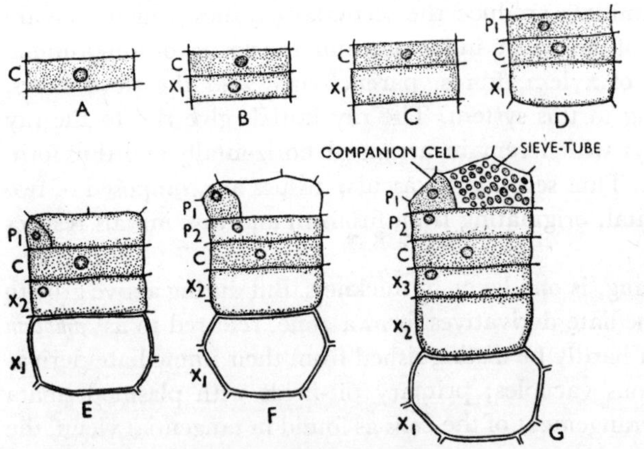

FIG. 629. Formation of secondary xylem and secondary phloem by the cambium—stages.

Secondary xylem. It constitutes the major portion of the secondary vascular tissues, and, in fact, the main bulk in woody plant. This tissue serves a number of important functions, viz., conduction of water and dissolved mineral matters, mechanical support; and the living cells provide space for storage of food.

As already stated, this compact tissue mostly made of thick-walled cells, consists of two systems—vertical and horizontal as derived from the fusiform and ray initials of the cambium. The vertical system is made of elements occurring parallel to the long axis of the organ. They are the tracheary elements, tracheae and tracheids, xylem parenchyma and fibres. The horizontal system consists of xylem rays arranged at right angles to the long axis. The living cells of the rays and the xylem parenchyma of vertical system form an interconnected continuous system which is in communication with the living cells of the pith, phloem and cortex.

So far as the elements are concerned, secondary xylem is more or less similar to those occurring in primary xylem. Vessels or tracheae are most abundant here and they are usually shorter and much larger than those of primary xylem and are mostly of pitted types. Annular and spiral thickenings of the vessel members are absent. The xylem parenchyma cells may be long fusiform ones, or more commonly, they are short cells. These living cells are particularly meant for storage of food—starch or fat. Tannins, crystals, etc., are also frequently present in these cells. Secondary wall may be absent; if present, the pits between the parenchyma and tracheary elements are bordered, half-bordered or even simple. Xylem parenchyma may occur abundantly in some species, may be comparatively smaller in amount or entirely absent. Much importance has been attached to this tissue in recent years in taxonomic considerations. They may remain either in association with the vessels or independent of them. In the former case they are called *paratracheal*, and in the latter, *apotracheal*. On the basis of distribution paratracheal parenchyma has been found to be of different types. Often they form entire sheaths, appearing circular or elliptical in cross-section, around the vessels. This type is termed *vasicentric* parenchyma. Apotracheal parenchyma may also be of different types. It is thought that apotracheal type is primitive and paratracheal advanced, highest development being reached in vasicentric type. The fibres of secondary xylem have thick walls and reduced bordered pits. The two main types of fibres are the *fibre tracheids*, and the *libriform* fibres. The fibre tracheids have bordered pits, but the borders are poorly developed, whereas the libriform fibres resembling the phloem fibres have simple pits. These are the important supporting elements of xylem.

The xylem rays forming the horizontal system remain arranged at right angles to the long axis of the organ. They originate from the ray initials and run radially as sheets of tissue and continue as band to the secondary phloem through the cambium, thus forming a continuous system of conducting tissue (Fig. 630). The expression medullary rays has often been used for this band, obviously on the basis of their similarity or supposed homology with the pith rays of herbaceous dicotyledonous stems. But, strictly speaking, they should be called **vascular rays,** as they are rays of vascular tissues, partly of xylem and partly of phloem, formed by the cambium.

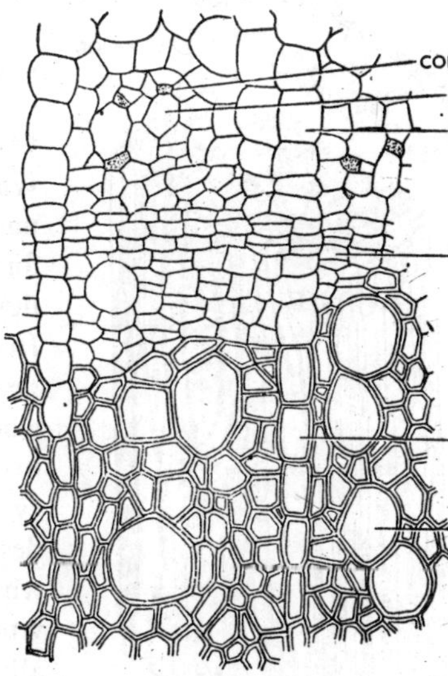

FIG. 630. Secondary tissues showing continuous xylem and phloem rays through the cambium.

The xylem rays establish transverse communication in the living cells of the vascular tissues. They make the gaseous interchange with the outer atmosphere easy through the intercellular spaces. Apart from providing space for storage of food, the ray cells are also responsible for the movement of water from the xylem to the cambium and phloem, and that of food from the phloem to the cambium and living cells of xylem.

The ray cells in the dicotyledons are usually parenchymatous. They vary in width —may be one-layered or uniseriate or two- to many-layered or multiseriate. In the angiosperms usually rows of ray cells occur at the margins which are different from the rest both in character and functions. They are always living and shorter than other ray cells. Special secretions are often present in these cells. Ray parenchyma cells may be upright or procumbent. Cells with their long axes oriented radially are *procumbent;* and those with long axes oriented vertically are the *upright* ray cells. If both the types combine, the tissue is called heterogeneous; whereas if it consists of only type of cell, it is known as homogeneous. Both uniseriate and multiseriate ray cells may be homogeneous or heterogeneous. Uniseriate type comprising vertically elongated cells and multiseriate heterogeneous rays are considered to be primitive and homogeneous rays to be advanced. In nearly all gymnosperms the rays are uniseriate. The cells of the xylem rays are usually living with rather thick walls. In gymnosperms the rays may be composed of only living parenchyma cells and tracheids. The terms *homocellular* and *heterocellular* rays have been used for the former and latter respectively. The tracheids, called *ray tracheids*, are dead, having lignified wall with bordered pits. They exhibit tendency of elongation.

ANNUAL RINGS. The activity of the cambium is often periodical, and as a result, distinct growth layers are formed in the xylem. In transverse views these growth layers appear as rings, and hence are referred to as *growth rings*. In most of the cases the periodicity in the activities of the cambium is seasonal, so the secondary xylem formed in a year constitutes the growth ring. Hence it is also an **annual ring.** Formation of annual rings is characteristic in plants growing in temperate climates with pronounced seasonal variations. In a quite good number of plants the annual or growth rings are found in tropical and subtropical countries as well.

The immediate reason for the visibility of the annual ring is the formation of different types of wood or secondary xylem in different parts of the growing season (Figs. 631 & 632). The wood formed in spring, which is the most favourable season for plant growth, is less dense and is made of vessels with wide diameters. The vigorous rate of assimilation in spring necessitates huge quantity of water, and so large vessels preponderate in the wood formed in that season. This is called **early wood.** In summer the rate of assimilation gradually decreases, and thus the secondary xylem formed is more dense and made of compact lignified elements. This part is called **late wood.** In temperate countries in particular the terms, *spring wood* and *summer wood*, are used for early wood and late wood respectively. But because the two parts are not strictly seasonal in some plants even in temperate climates, it is better to designate them as early wood and late wood. The winter is the resting season when cambial activities remain suspended. The early wood gradually merges with the late wood of the season. But there exists a sharp line of demarcation between the late wood of one season and the early wood of the next season. That explains the formation of concentric annual rings, each of which consists of two parts—inner early wood and outer late wood.

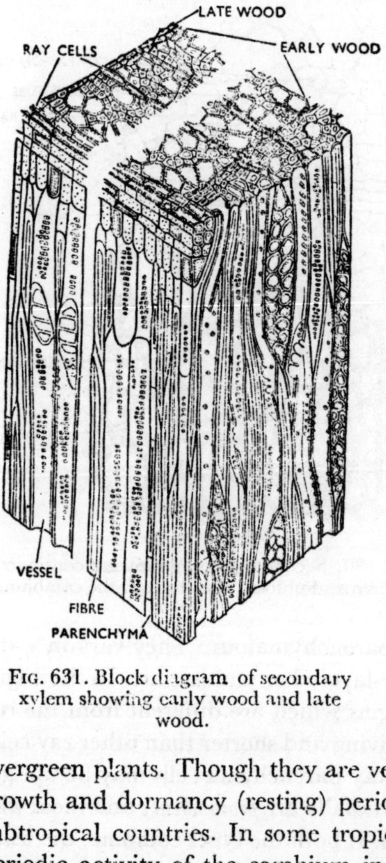

FIG. 631. Block diagram of secondary xylem showing early wood and late wood.

Annual rings are found in the deciduous and evergreen plants. Though they are very common in temperate countries with distinct growth and dormancy (resting) periods, they may also occur in plants of tropical and subtropical countries. In some tropical trees distinct annual rings are formed due to periodic activity of the cambium in alternatingly wet and dry seasons. Well-marked rings are found in the important timber-yielding plants—*Tectona grandis* (teak), *Dalbergia sissoo* (sissoo), *Albizzia lebbeck* (sirish), *Salmalia malabarica* (silk cotton), etc.

The seasonal activity of the cambium may be distributed by external influences, viz., adverse climatic conditions, drought, defoliation and disease. In that case normal development of the wood is checked, and growth is resumed later in the same season. As a result, additional layers, called *false annual rings*, are formed. Thus two or more false ring may be formed in a year, all of them constituting what is known as *multiple annual ring*.

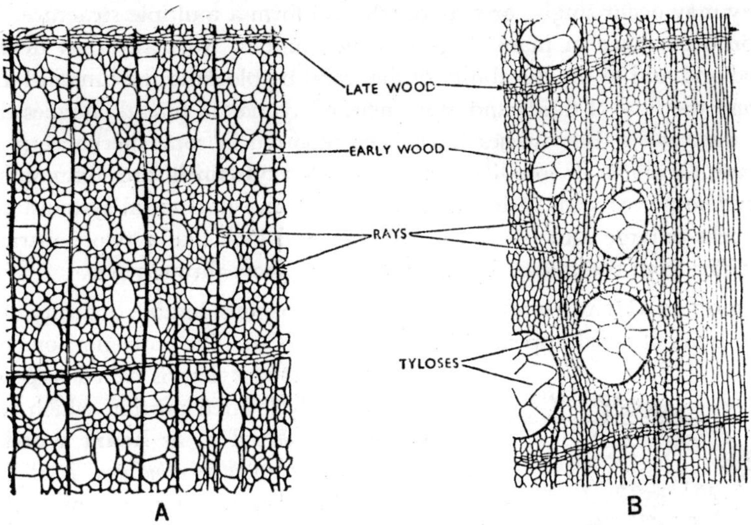

FIG. 632. Secondary xylem in t.s. showing diffuse porous (A) and ring porous (B) wood.

Normally large vessels occur in the early wood, making it more conspicuous than the late wood. Here the vessels are of unequal diameters, the largest ones being restricted to the early wood. As they exhibit a ring-like arrangement in transverse section, this type of wood is called **ring porous** (Fig. 632B). But in some plants the vessels are found to be of more or less equal diameters and they remain rather uniformly distributed throughout the wood or when there is only a gradual change in size and distribution throughout the growth ring. This type of wood is called **diffuse porous** (Fig. 632A). Intermediate forms may occur between the two types. From phylogenetic point of view ring-porous wood is considered more advanced than the diffused-porous one.

TYLOSES. In many plants the xylem parenchyma and ray parenchyma cells develop balloon-like protrusions into the tracheary elements. Such ingrowths are called **tyloses** (Figs. 632 & 633). They are characteristics of secondary xylem, though they may also develop in the primary xylem. Tyloses are formed by the enlargement of the pit membranes of the half-bordered pits occurring between a living parenchyma cell and a tracheary element—a vessel or a tracheid. They are often pretty large, so much so that the lumen of the elements is almost blocked. The nucleus with a part of cytoplasm flows from the living cells to the tylose. In fact, the delicate membrane grows and forms the large balloon-like protrusion. Starch, resin and other matters may be present in the tyloses. The walls of the tyloses are initially thin, but they may undergo lignification

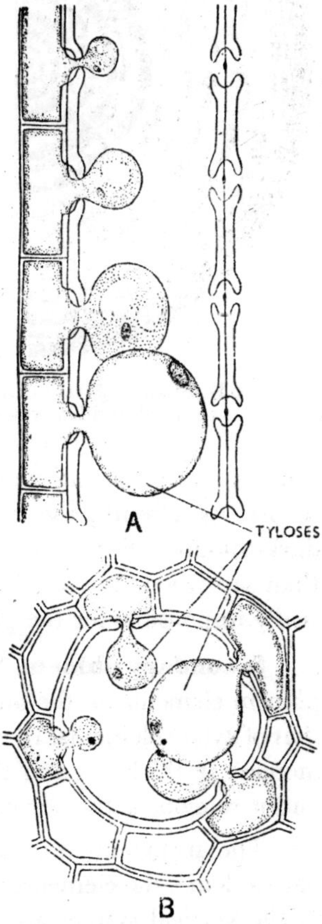

FIG. 633. Tyloses in longitudinal (A) and transverse (B) views.

later. They may occur singly, or may divide and form a multiple structure. Tyloses are of common occurrence in many angiospermic families. Normally they develop in the heart wood and add to the durability of the wood by blocking the lumen. Furthermore, they prevent the flow of fluids and movement of hyphae of fungal enemies through the lumen of the vessels. But tyloses have been reported in some herbaceous plants like *Coleus*, *Canna*, *Cucurbita*. Tylose-like intrusions called **tylosoids** are common in the wood of some gymnosperms with resin ducts in vertical and horizontal systems. In fact, the epithelial cells enlarge and block a resin duct and look like a tylose intrusion. They basically differ from the tyloses in that they never grow through pits.

Sap wood and Heart wood. After formation of considerable quantity of secondary xylem, two types of wood appear in the stem (Fig. 634). The outer portion consisting of recently formed xylem is called **sap wood,** and the centrally located portion which was formed earlier is referred to as **heart wood.** The sap wood, also known as *alburnum*, is of light colour and contains some living cells, in addition to the tracheary elements and fibres. This part actually carries on the physiological functions—primarily conduction of water and solutes, support and storage of food. The hard central part, the heart wood, also known as *duramen*, mainly consists of dead elements and is usually of dark colour. It simply serves as a solid mechanical column. The sap wood is gradually transformed into heart wood. During this transformation a number of changes occur in the wood—the living cells lose protoplasts, water contents are reduced, food matters are removed from the cells, tyloses are frequently formed and thus the lumens of the elements are blocked, the walls of the parenchyma cells undergo considerable lignification, and substances like gums, resins, oils and particularly tannins and colouring matters are infiltrated both on the walls and lumen of the elements. As a result, the heart wood becomes more compact, strong and durable and dark-coloured. From commercial point of view heart wood is obviously more valuable than sap wood; but the latter is more important from biological point of view, being really responsible for the physiological functions.

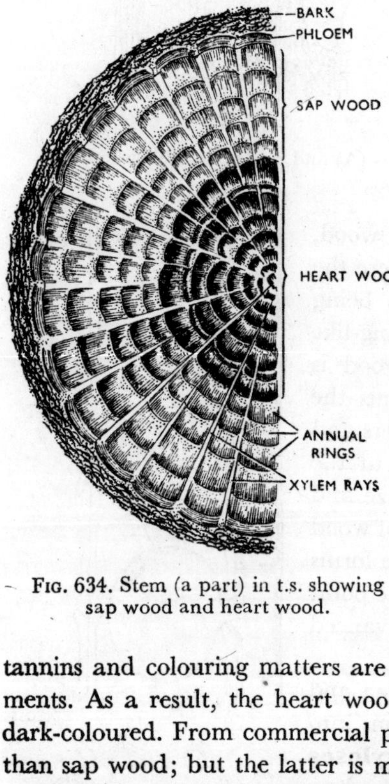

Fig. 634. Stem (a part) in t.s. showing sap wood and heart wood.

Secondary phloem. The cambium cells by tangential divisions produce secondary phloem elements on the outer side. Usually the number of phloem elements is less than that of xylem elements. In most of the dicotyledons the primary phloem gets crushed with increase in thickness and the secondary phloem takes over the physiological functions, and it remains active for considerable length of time.

The arrangement of cells in secondary phloem is similar to what prevails in secondary xylem—the elements being arranged in two systems, vertical and horizontal. Those in the vertical system are the sieve tubes, companion cells, phloem parenchyma and fibres. They derive their origin from the fusiform initials of the cambium, and are chiefly responsible for vertical conduction of elaborated food matters. The horizontal system consists of the ray parenchyma cells formed by the ray initials. Like secondary xylem storied and non-storied arrangement is noticed.

There is no fundamental difference in the primary and secondary phloem as regards the elements, but the latter is a more complex tissue (Fig. 635). All the cells here remain arranged in radial rows; the sieve tubes are usually shorter and larger in number and the fibres are more abundant. The complex tissue as a whole has a longer functioning life. The sieve tubes usually have simple sieve plates, but compound plates may also be present in some plants. They do not show any regularity of distribution in the tissue. The companion cells accompany the sieve tubes, as they develop from the same mother cell. There may be one companion cell extending along the entire length of the sieve tube or there may be a few ones. They are often quite distinct to be readily recognised in transverse views. In gymnosperms vertical system consists of sieve cells, parenchyma and albuminous cells. Parenchyma cells are abundantly present in the secondary phloem of the dicotyledons, but with gradual increase and maturity their number is decreased. The nature of the cells is variable, they may be elongated with pointed ends like the cambium cells from which they have originated, or may be short rectangular or slightly elongated cylindrical ones. Apart from their role in vertical conduction of elaborated food matters, the parenchyma cells also serve in storage of starch, crystals and other materials. Sclerenchyma is characteristic of secondary phloem. They occur abundantly either as discrete tangential bands, or as isolated patches, or even singly. In certain dicotyledons the fibres occur in patches regularly alternating with the bands composed of sieve tubes, companion cells and parenchyma. Sclereids are also of frequent occurrence, either independently or in association with the fibres. Some phloem elements become functionless in course of time. In non-functioning phloem the number of sclereids increases, mainly because the parenchyma cells undergo sclerosis there. Elongate fibre-like elements formed by sclerosis of parenchyma cells have been called *fibre-sclereids*. The fibres together with the sclereids provide mechanical strength. In fact, many fibres of commercial importance belong to this tissue.

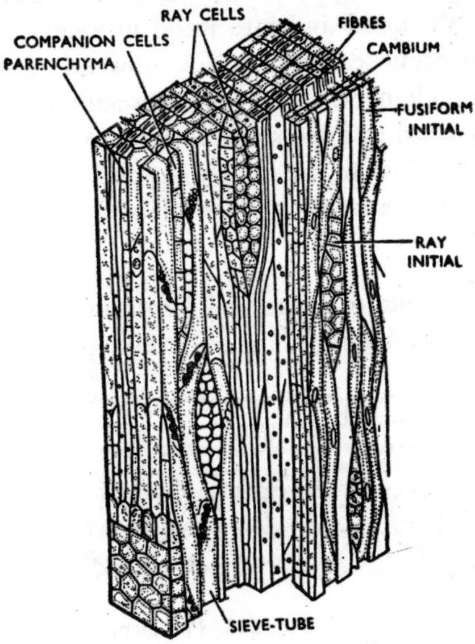

FIG. 635. Block diagram of secondary phloem with the cambium showing the elements.

PHLOEM RAYS. The phloem rays (Figs. 630 & 635) originate from the ray initials of the cambium and constitute the horizontal system of the tissue. With xylem rays occurring on the inner side these cells (phloem rays) form the vascular rays which establish communication in the living cells of the vascular tissues through the cambium, and are responsible for horizontal conduction to and from the xylem and cambium. Near the cambium xylem rays and phloem rays are of equal size, with maturity of the plant phloem rays have increased width. Like xylem rays they may be uniseriate with only parenchyma cells, as in conifers, or multiseriate, and homogeneous, composed of either upright or procumbent cells, or heterogeneous, made of both upright and procumbent cells. Phloem rays are smaller than their counterpart in xylem, because of the facts that the vascular cambium as a rule produces more xylem than phloem and that the outer part of phloem is periodically sloughed off.

Growth rings which are so characteristic of secondary xylem are practically absent in phloem. The elements of secondary phloem, as already stated, remain arranged in regular tangential series and thus exhibit something like concentric rings. But unlike the early wood and late wood of xylem, phloem formed in the early and late part of the growing season does not show any distinction. So a line can hardly be drawn between the elements formed in consecutive seasons. Something like rings have, however, been noticed in some species of dicotyledons which are formed due to differences in the cells produced at the beginning and end of the season. Tangential bands of fibres have also been observed in some plants. They at any rate do not indicate the age of secondary phloem.

PERIDERM

Due to continued formation of secondary tissues by the cambium cylinder in the intrastelar region, considerable pressure is exerted on the epidermis and other extrastelar tissues. The epidermis gets more and more stretched and ultimately tends to rupture. The protective function discharged by them in the primary condition is now taken over by a new secondary tissue, called **periderm,** formed in the extrastelar region.

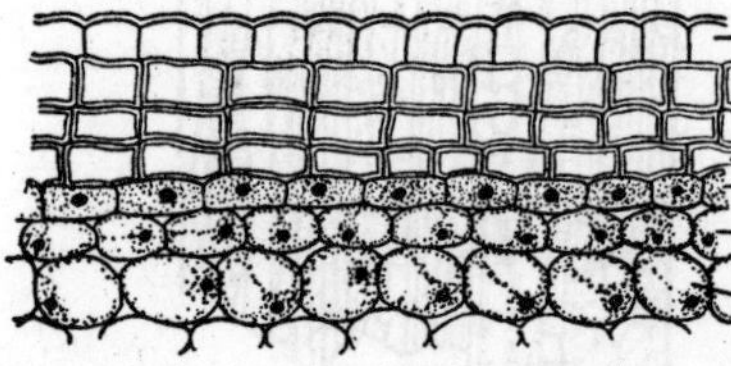

FIG. 636. Periderm—formation.

The periderm consists of three tissues, viz., (*i*) a meristem known as **phellogen** or *cork cambium*, (*ii*) cells cut off by the phellogen on the outer side, called **phellem** or cork cells, and (*iii*) cells formed by the phellogen on the inner side, called **phelloderm.**

As stated previously, when considerable pressure comes from the internal portion and the epidermis tends to rupture, the phellogen or cork cambium is usually formed in the superficial layers of cortex. It is a secondary meristem arising from the living permanent cells which retained the potentialities of cell division. It originates as a single layer of initiating cells either in the subepidermal portion, in the epidermis itself, in the deep-seated layers of cortex or may even extend up to the phloem. Phellogen is rather simple in comparison to the cambium, as it consists of only one type of initials. In transverse view the cells appear rectangular in shape and remain radially flattened. They are compactly arranged without having intercellular spaces. The cells have vacuolate protoplasts. This lateral meristem behaves like the cambium cells. They divide tangentially producing new tissues both centrifugally and centripetally (Figs. 636 & 637). The derivatives usually remain arranged in radial rows.

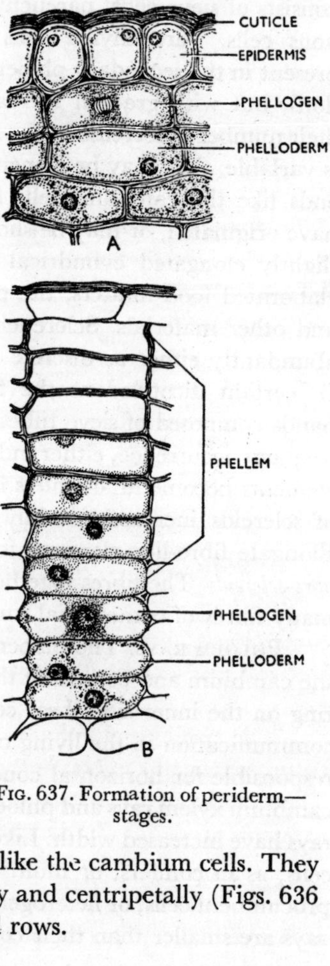

FIG. 637. Formation of periderm—stages.

Normally the phellogen produces much more phellem on the outer side than phelloderm on the inner. The cells constituting phellem, otherwise called cork cells, are uniformly formed ones. They remain arranged in distinct radial rows due to their origin from a tangentially dividing meristem. The cork cells are compactly set and have no intercellular spaces. They lose protoplasts after differentiation and become dead. The cells remain filled with air and coloured organic matters. The wall characters of the cork cells are most significant. The primary walls are mostly of cellulose or partly suberised or even lignified. A thick layer of suberin is deposited next to the primary walls. So the cork cells become impervious to water and gases. In some cases the cork cells may be thick-walled, where the cavities of the cells remain filled up with resinous and other materials as in *Eucalyptus*. But commonly they are rather thin-walled, radially elongate empty cells. Two types of phellem may occur in the same plant in alternating layers, as in *Betula*, where

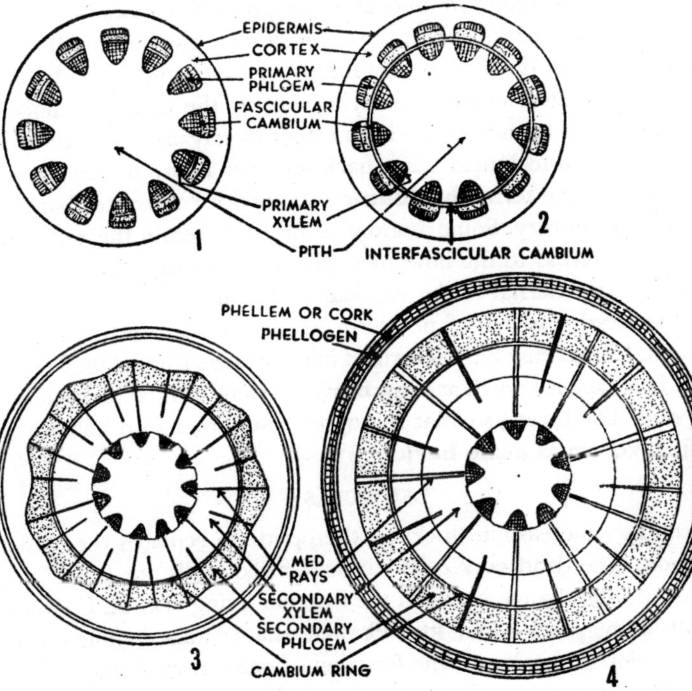

FIG. 638. Secondary growth in a dicot stem up to two years (diagrammatic)—stages in transverse views.

cork is peeled off like sheets of paper. Periderm can serve as an effective secondary protective tissue against desiccation and mechanical injuries mainly due to strong suberisation of the cork cells and their compact arrangement.

The phellogen produces phelloderm on the inner side. The cells constituting this tissue practically resemble those of cortex in their nature, wall structure and contents. These are living, more or less isodiametric cells with intercellular spaces. But they remain arranged in definite radial rows. These cells may carry on photosynthesis and serve in storage of food. Some authors use the expression secondary cortex for the phelloderm. But it is unhappy and confusing, because the same term has been applied to the whole of periderm by others.

Thus it is evident from the process stated above that secondary growth in a dicotyledonous stem is initiated in the intrastelar region with the activities of the cambium, fascicular and interfascicular. Secondary vascular tissues are formed by the cambium ring. Lastly, for withstanding the pressure from the inside periderm arises in the extrastelar region (Fig. 638).

BARK

It has been stated that phellogen first originates in the superficial portion—either in the epidermis itself or in the subepidermal cells. The phellogen cells go on dividing and produce cork cells and phelloderm on the outer and inner sides respectively. But the formation of secondary tissues continues in the intrastelar region. The periderm formed at the superficial region of the organ may not be adequate to withstand the pressure coming from inside. In that case additional layers of periderm are formed in the progressively deeper regions of the stem. So new phellogen layers may arise in the deeper regions of cortex, pericycle and even phloem. Once phellogen arises in the deeper region and it produces phellem or cork cells on the outer side, all the living cells present on the outer side of the phellogen do not get any supply of water and food. Eventually all of them become dead. All these dead tissues lying outside the active phellogen constitute the **bark** of the trees. This term bark has been rather loosely used and has created a good deal of confusion. In non-technical sense bark includes many tissues occurring at the outer part of the stem. The term **rhytidome** was used for the outer bark in the latter part of the nineteenth century. Some anatomists prefer to retain the term rhytidome to cover all the tissues external to the innermost phellogen (outer bark); the term bark to be applied to all tissues external to the vascular cambium.

Commonly successive layers of periderm are formed in the progressively deeper regions of the stem. In that case bark is formed in concentric rings (Fig. 634) surrounding the entire stem. This type is known as **ring bark.** In some plants the periderm is formed as overlapping scale-like layers, so that the outer tissues break up and are sloughed off in patches. This is known as **scale bark.**

LENTICELS

During secondary increase in thickness the periderm replaces the epidermis in the protective function. The dead cork cells with suberised walls are partly impervious to gases. Thus gaseous interchange between the internal living cells and outer atmosphere becomes difficult. Some pores which look like lens-shaped raised spots on the surface of the stem, now develop and take up the function of gaseous exchange. These pores are known as **lenticels.** Only a few plants, mostly climbers, do not possess lenticels though periderm is formed.

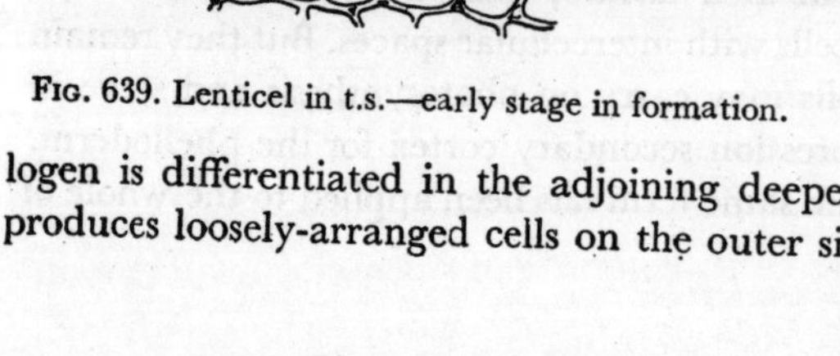

FIG. 639. Lenticel in t.s.—early stage in formation.

Lenticels are first formed just beneath the stomata. They usually originate either just before, or simultaneously with or after the initiation of the periderm. The parenchyma cells near about the substomatal chamber lose chlorophyll and divide in various planes, forming a mass of colourless loose cells. Sooner or later normal phellogen is differentiated in the adjoining deeper region (Fig. 639). The phellogen also produces loosely-arranged cells on the outer side, instead of giving rise to normal cork cells. All these cells—those formed by the division of the parenchyma in the substomatal region and those formed by phellogen on the outer side, are together known as **complementary** cells. With increase in the number of complementary cells the epidermis is ruptured, and the former protrude above and thus appear like so many raised spots (Fig. 640). The outermost cells often die due to exposure to outer atmosphere and are replaced by the cells cut off by the phellogen. Often massess of more dense and compact cells, known as **closing** cells, alternate with loosely-arranged complementary cells. The closing cells together form a layer referred to as *closing layer* which is helpful in keeping the loosely-arranged complementary cells in position. Continued formation of the complementary cells may cause rupture of the closing layer. This condition actually prevails in the growing season, whereas at the end of the same the formation of closing layer is more pronounced.

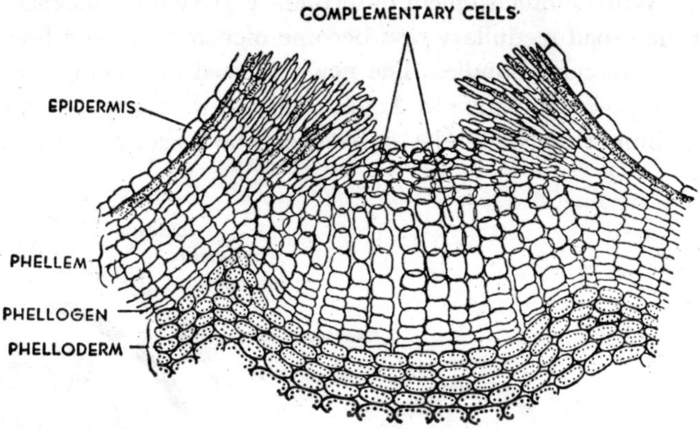

FIG. 640. Lenticel in t.s.—well-formed.

These complementary cells are characterised by thin walls and absence of suberin and loose arrangement due to presence of profuse intercellular spaces. The latter establish communication with the spaces of the internal tissues. Like the stomata the lenticels are thus concerned with gaseous exchange between the outer atmosphere and the internal tissues.

In some cases lenticels are formed independent of the stomata. Even it may happen that phellogen has produced cork cells for a while then it started forming loose complementary cells which would ultimately break through the cork and give rise to a lenticel. The number of lenticels formed in stem is variable. They may remain scattered or arranged in vertical or longitudinal rows. Rows of lenticels may occur opposite to the multiseriate rays, suggesting free interchange of gases.

SECONDARY GROWTH IN ARISTOLOCHIA STEM

It has been stated that this stem which is a liane, differs from the normal ones in the process of secondary growth. The most striking points of difference are the formation of only parenchymatous medullary rays by the interfascicular cambium, and consequent occurrence of secondary tissues in strands (Fig. 641).

In the primary condition the *Aristolochia* stem has the following structure. Epidermis is uniseriate with cuticularised outer walls. Cortex is differentiated into collenchymatous hypodermis, parenchymatous portion and the starch sheath. Chloroplasts are present in both collenchyma and parenchyma cells. A continuous band of sclerenchyma with strongly thickened walls occur. They may be called *perivascular fibres*, or may be said to form pericycle together with the adjoining parenchyma cells. Broad medullary rays composed of parenchyma cells occur between the vascular bundles. The central portion of the stem is occupied by a large parenchymatous pith. The vascular bundles remain arranged in a ring. The bundles are distinctly collateral and open ones, with xylem and phloem on the inner and outer sides and having a strip of cambium in between the two.

With commencement of secondary growth in thickness a few parenchymatous cells of the broad medullary rays become meristematic in a line with the *fascicular* cambium of the vascular bundles. The newly-formed meristem, a secondary meristem, is known as *interfascicular* cambium. It joins up with the fascicular cambium and thus a continuous cambium ring (Fig. 641A) is formed. The fascicular cambium, in fact, the cambial zone

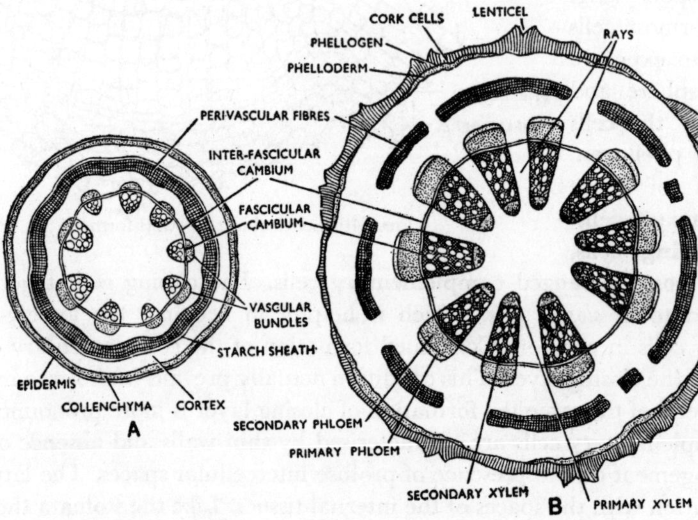

FIG. 641. *Aristolochia* stem in t.s. (diagrammatic). A. One-year old. B. Two-year old.

goes on dividing tangentially and produces secondary xylem and secondary phloem on the inner and outer sides respectively. Thus the primary xylem and primary phloem are pushed apart from each other. The secondary xylem has the usual elements arranged in vertical and horizontal systems. Metaxylem elements are fairly large in size. Annual rings with early wood and late wood are formed as a result of seasonal activities of the cambium. The secondary phloem pushes the primary one on the outer side, and the latter usually gets crushed due to the pressure. Bands of sieve tubes and associated cells alternate with bands of parenchyma in the secondary phloem; fibres are absent. On the whole the vascular bundles increase enormously in size due to continued activity of the fascicular cambium.

The interfascicular cambium simply produces parenchyma cells on the outer and inner sides. Thus the medullary rays become increasingly more broad and long. The cells remain arranged in more or less radial rows.

The formation of the secondary tissues brings about profound changes in other portions of the stem. The central pith gets more and more reduced in extent. Distinct disruption in the continuous cylinder of sclerenchyma is caused by the increase, so that the band is ruptured here and there, commonly in front of the medullary rays. The adjoining parenchyma cells fill up the gaps thus formed. These cells gradually undergo sclerosis and are ultimately transformed into sclereids. They, in fact, repair, so to say, the gaps caused by the onrush of the internal tissues. The band of hypodermal collenchyma also suffers from pressure and breaks down frequently. The parenchyma cells of the cortex make their way into the breaks and thus occur as strips amongst hypodermal collenchyma.

The epidermis gets stretched and ruptures. Periderm develops in the subepidermal layers. Phellogen is formed in patches. They divide and produce a thick layer of cork cells on the outer side and considerable phelloderm on the inner. Lenticels are formed.

SECONDARY GROWTH IN CUCURBITA STEM

The arrangement of the primary tissues in the stem of *Cucurbita* has been discussed in a preceding chapter. The process of secondary increase in thickness is more or less similar to that of *Aristolochia*. But as this plant is not so woody, the cambial activities remain confined to the individual vascular bundles (Fig. 642). Moreover, the perivascular fibrous band is not ruptured here.

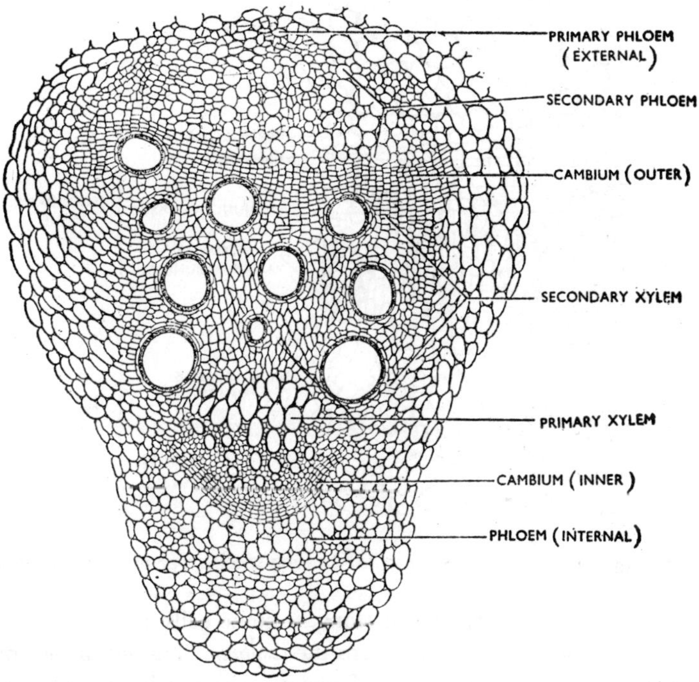

FIG. 642. A vascular bundle of *Cucurbita* stem in t.s. showing formation of secondary tissues.

The outer cambium continues tangential division and produces secondary xylem elements on the inner side. The vessels are usually of pitted types and are enormously large in size. So the characteristic radial arrangement and position of other tissues are considerably disturbed. The secondary phloem elements formed by the outer cambium on the outer side is hardly distinguishable from the primary phloem. The tissue consists of wide sieve tubes, rather narrow companion cells and parenchyma cells of intermediate size. Fibres and ray cells are absent. The sieve plates are simple and have large pores and connecting strands. Of the two strips of cambium, it is only the external one which divides and thus causes growth in thickness.

SECONDARY GROWTH IN MONOCOTYLEDONOUS STEM

Normally the vascular bundles of the monocotyledonous stems are closed ones. Thus due to absence of the cambium they lack secondary growth in thickness and the vascular system is wholly composed of primary tissues. The bundles remain irregularly scattered in the ground tissues, forming an atactostele, where the limits of cortex and other ground tissues can be hardly discerned.

Some monocotyledons belonging to the family Liliaceae, mainly the arborescent ones like *Dracaena* (Fig. 643), *Yucca*, *Cordyline*, *Agave*, *Aloe* and others exhibit a peculiar type of secondary increase in thickness, an account of which is given here.

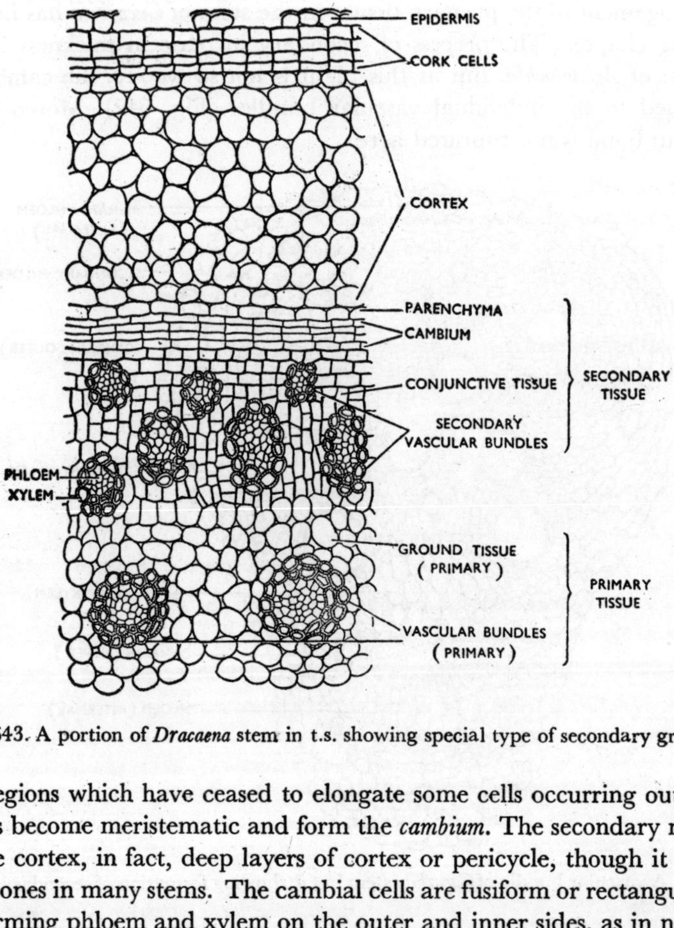

FIG. 643. A portion of *Dracaena* stem in t.s. showing special type of secondary growth.

In the regions which have ceased to elongate some cells occurring outside the vascular bundles become meristematic and form the *cambium*. The secondary meristem originates in the cortex, in fact, deep layers of cortex or pericycle, though it is difficult to locate those zones in many stems. The cambial cells are fusiform or rectangular in shape. Instead of forming phloem and xylem on the outer and inner sides, as in normal condition, the cambial cells go on dividing and producing secondary tissues on the inner side first, and later small amount of new tissues are cut off on the outer side as well.

Those formed on the inner side differentiate into oval-shaped vascular bundles and radially arranged parenchyma cells. These parenchyma cells in which the vascular bundles remain embedded are said to constitute the *conjunctive tissue*. The radial arrangement of the parenchyma cells of conjunctive tissue is due to their origin by tangential divisions of the cambial cells. So they may be easily distinguished from the irregularly arranged parenchyma of the primary ground tissues. They may be thin-walled or thick-walled.

A few derivatives of the cambium divide longitudinally, initially anticlinally, then periclinally and even haphazardly to form xylem and phloem elements of the secondary vascular bundles. These differ from the primary ones in presence of small amount of phloem and in absence of annular and spiral protoxylem elements. The secondary bundles are mostly *amphivasal*, some of them may be collateral as well. The small amount of phloem consists of short sieve tubes, companion cells and parenchyma. The xylem is made of only tracheids, usually with scalariform thickening and small amount of xylem parenchyma which have lignified walls. The tissues cut off by the cambial cells on the outer side are scanty in amount and are parenchymatous in nature. The primary bundles are comparatively larger. They are also mostly amphivasal or rather rarely collateral ones.

In Aloe, cocoanut and royal palm periderm is formed in the same manner as in the dicotyledons. A special type of protective tissue is produced in the majority of the monocotyledons with secondary increase in thickness. Parenchyma cells in the deeper layers of cortex divide periclinally several times and produce radially arranged layers of cells, suggesting something like a storied meristem. The series of cells thus formed undergo suberisation and differentiate into cork cells. This tissue is called *storied cork*, mainly because of its storied appearance.

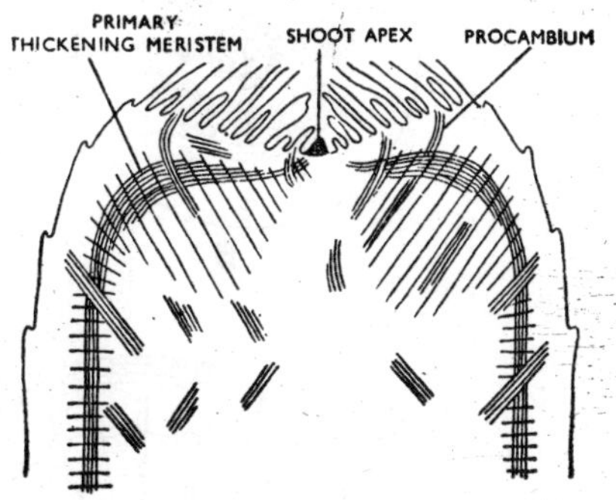

FIG. 644. Apical portion of shoot of palm showing meristems concerned with its growth. (*After Esau*)

Considerable thickening which is often noticed at the bases of many palms (Fig. 644) is not due to the activity of a cambial layer, but it is really a case of protracted or long-continuing primary growth. The apical meristem or shoot apex produces only a part of the primary body —a central column of parenchyma and vascular strands. Another meristem occurs just beneath the leaf primordia forming a mantle-like tissue region. This is known as *primary thickening meristem*. The cells of the same divide periclinally and form rows of cells which differentiate into ground parenchyma traversed by procambial strands. The latter ultimately develop into vascular bundles. There are further enlargement and divisions of the ground parenchyma, causing increase in thickness. Thus the primary thickening meristem produces the main bulk of the stem tissues, viz., ground parenchyma and most of the vascular bundles. The thickening is due to the activities of the two meristems stated above.

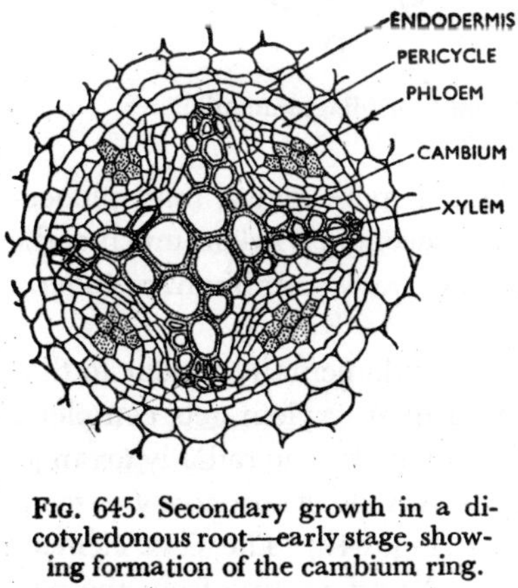

FIG. 645. Secondary growth in a dicotyledonous root—early stage, showing formation of the cambium ring.

SECONDARY GROWTH IN DICOTYLEDONOUS ROOT

The roots of some herbaceous dicotyledons and of all gymnosperms and woody dicotyledons show secondary increase in thickness, whereas most of the monocotyledonous roots, like those stems, are entirely primary. The secondary tissues formed in the dicotyledonous roots are fundamentally similar to those of the stem, but the process is initiated in a different way (Figs. 645 to 647).

The dicotyledonous roots have limited number of radially arranged vascular bundles with exarch xylem. Pith is usually absent. A few parenchyma cells beneath each phloem group become meristematic and thus form strips of *cambium*, the number of strips being equal to the number of phloem groups present.

Cambial cells go on dividing and produce secondary tissues. The cells of the uniseriate pericycle against the protoxylem group now divide and form a few layers. The first-formed cambium now extends both ways and reaches the innermost derivatives of the pericycle, and thus a continuous cylinder of the cambium is formed (Fig. 645). The cambium cylinder is wavy in outline, as it passes internal to phloem and external to

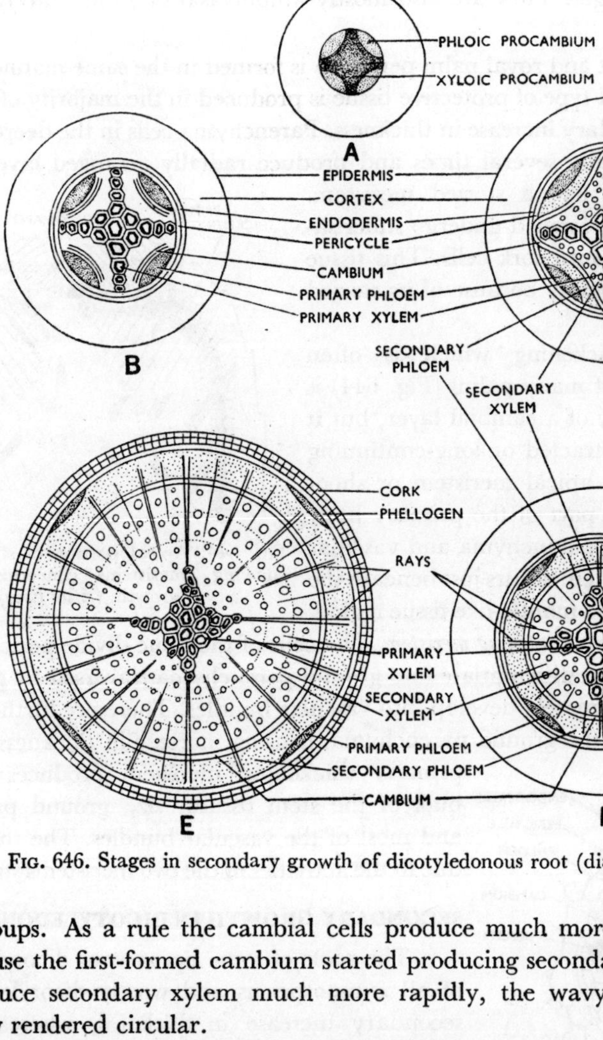

FIG. 646. Stages in secondary growth of dicotyledonous root (diagrammatic).

xylem groups. As a rule the cambial cells produce much more xylem than phloem; and because the first-formed cambium started producing secondary xylem much earlier and produce secondary xylem much more rapidly, the wavy cambium cylinder is ultimately rendered circular.

As already stated, the secondary vascular tissues are fundamentally similar to those of the stem. They form a continuous cylinder and the primary xylem gets completely embedded in it. At this stage the root structure is revealed only by the radially arranged exarch primary xylem located at the central region, the strands of secondary vascular tissues being collaterally arranged like those of the stem (Fig. 647). The sieve elements of the primary phloem often get crushed. The cambial cells originating from the pericycle against protoxylem groups function as *ray initials* and produce broad bands of vascular rays. These rays running between xylem and phloem through the cambium are characteristic of the roots. They are also called main medullary rays.

Periderm is formed in the outer region. Phellogen arises in the outer cells of the pericycle (Figs. 646 & 647). It produces phellem or cork cells on the outer side, and probably some phelloderm on the inner. The pressure caused by formation of secondary tissues

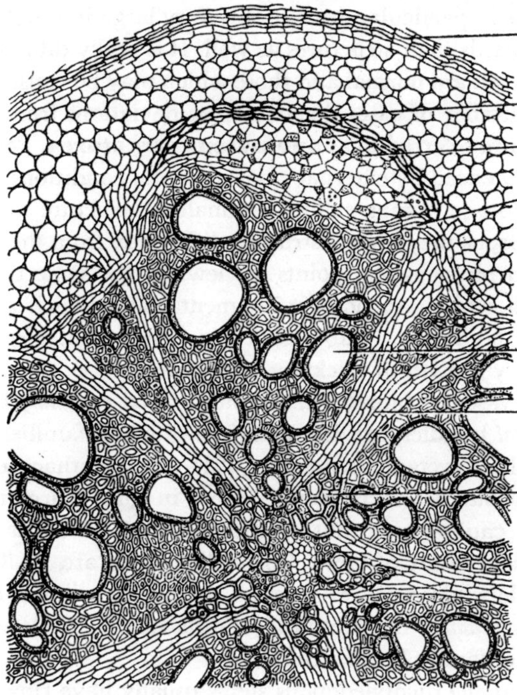

FIG. 647. Secondary growth in root of *Tinospora*—a portion in transverse section.

inside ruptures the cortex with endodermis, which is ultimately sloughed off. Lenticels may be formed. They usually occur in pairs as transversely elongated rough areas, one on each side of a lateral root.

CHAPTER XXVI

ANOMALOUS STRUCTURES

An account of the normal types of internal structures has been given in the preceding chapters. But the angiosperms in particular are not only very large in number, but they exhibit infinite varieties as regards their structures as well. It is really difficult to draw a line between the structures which are normal and those which may be considered abnormal. But in a pretty good number of cases departures or deviations from the normal structures are noticed. These are referred to as **anomalous structures.** Peculiarities in structures may be directly influenced by external environmental conditions. The woody climbers or lianes and the storage organs exhibit anomalous structure and abnormal secondary increase in thickness, as they are constructed differently from the normal ones both from morphological and physiological points of view. At the same time there are cases of anomalous condition which are non-environmental and which appear to be merely cases of variations of the design. Haberlandt has called them *adaptive* and *non-adaptive* anomalies of secondary growth in thickness respectively.

Occurrence of accessory vascular bundles in the pith, called *medullary* bundles, and in the cortex, known as *cortical* bundles, are characteristic of many families. Medullary bundles are common in families Nyctaginaceae (Fig. 651), Amaranthaceae (Fig. 653), Piperaceae, etc., where many bundles occur in pith either in rings or irregularly. They may be leaf-trace bundles or cauline ones, but even the latter form a part of the leaf-trace system from physiological point of view. Cortical bundles are really leaf-trace bundles which pursue an independent course through the cortex without entering the ring. They are common in *Calycanthus* of family Calycanthaceae, in *Nyctanthes* of family Oleaceae (Fig. 655), *Begonia* of family Begoniaceae.

It is an admitted fact that the dicotyledonous stems usually have ring-like disposition of the vascular bundles, whereas those in the monocotyledons remain irregularly scattered in the ground tissue. A number of cases showing dicotyledons with scattered bundles and monocotyledons with ring-like arrangement of bundles may be cited. Scattered bundles are found in *Thalictrum* of family Ranunculaceae, in *Podophyllum* of family Berberidaceae, *Papaver* of family Papaveraceae. Many bundles occur in Nymphaeaceae, Primulaceae and some others, where the connected bundles form something like a network, and they have no secondary increase in thickness. The tubular stems of monocotyledons, as in wheat (Fig. 593) and many grasses with hollow internodes, have vascular bundles arranged in rings. *Tamus communis* of family Dioscoreaceae, is a glaring example of a monocotyledon with bundles arranged in ring-like fashion towards the periphery round a large central pith.

Peculiarly inverted bundles are also noticed in plants like *Rumex crispus* and *Rheum officinale* of family Polygonaceae where phloem, instead of xylem, occurs towards pith in the primary bundles. In fact, a secondary meristem arises on the outer edge of each phloem strand, which produces xylem elements on the outer side. The inner bundles actually exhibit an inverse disposition of tissues.

A few cases have been cited above showing anomalous structures which are not directly influenced by environmental factors. But in good many cases anomalies particularly in the process of secondary increase in thickness are environmental. As already stated these structures are quite common in the climbing lianes and storage roots which are different from the normal ones from the points of view of mechanical construction and physiological functions. Though the methods leading to the formation of anomalous structures are quite variable, they may, however, be put into two categories: (*i*) where the cambium is of normal type, but due to irregular or peculiar activity the vascular tissues, xylem and phloem elements formed exhibit unusual arrangement and proportion; (*ii*) where the cambium itself is abnormal in disposition, and the secondary tissues consequently are unusual in structure and arrangement; or, the original cambium is replaced by other cambium layers formed later which are unusual both in arrangement and behaviour. The lianes are constructed more or less after the pattern of a twisted

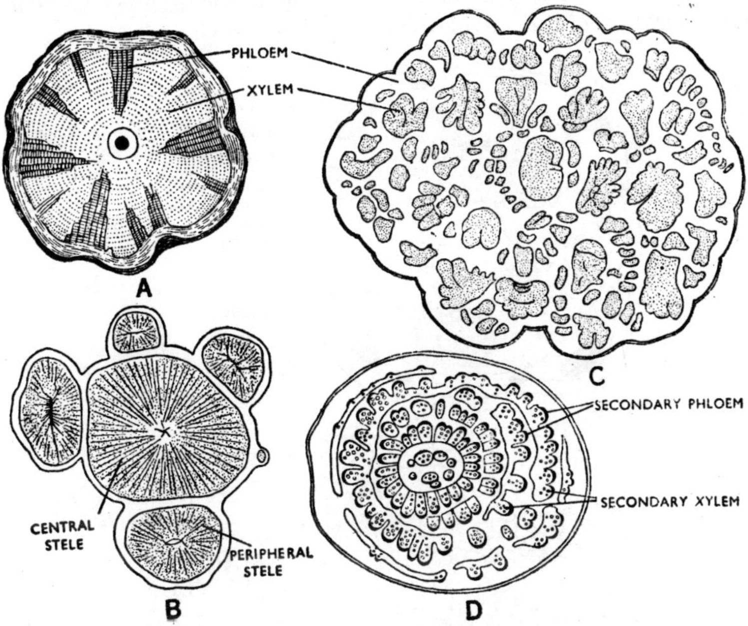

Fig. 648. Anomalous stem structures. A. In *Bignonia*. B. In *Serjania*. C. In *Bauhinia*. D. In *Boerhaavia*.

rope, and need high degree of tensile strength with pliancy. Unlike a solid mass of woody cylinder of an erect stem, here the cylinder splits up into separate strands with interpolated softer tissues, so that the strands can slip past one another.

A brief review of a few types is given here.

In some liane type of stem as in *Bignonia* and other members of family Bignoniaceae, the cambium is normal in disposition and activity to begin with, but it soon cuts off different proportions of xylem and phloem in different points. At four points arranged in form of a cross, formation of secondary xylem is reduced and that of secondary phloem correspondingly increased. As a result the woody cylinder appears to have four longitudinal grooves which become increasingly deeper with secondary growth. The cambium breaks up into a number of strips, widest ones occurring opposite the four projecting ridges of wood and the narrow ones at the bases of the grooves. As a result peculiar structure with ridged and furrowed xylem cylinder (Figs. 648A & 649) is formed. Deeply ridged vascular cylinder is formed in some plants due to the fact that the cambium produces only ray parenchyma cells at some points.

A very peculiar structure is noticed in *Serjania*, a climber of family Sapindaceae, where the cambium from the very beginning occurs in separate strips, each strip having surrounding portion of primary tissues. It thus appears like a deeply-lobed compound structure made of several stems fused together (Figs. 648B & 650). With age and consequent development of periderm the compound structure becomes more marked. The development of interfascicular cambium and its union with fascicular ones result in the formation of separate cambium rings. In fact, there is a large central stele, surrounded by a few peripheral ones. The cambium rings of the steles independently form secondary xylem, composed of vessels and tracheids, on the inner side, and a few layers of phloem on the outer. A more or less similar structure is formed in *Bauhinia* of family Leguminosae, where the original cambium breaks into strips and proliferation of xylem parenchyma brings about separation of the vascular cylinder (Fig. 648C).

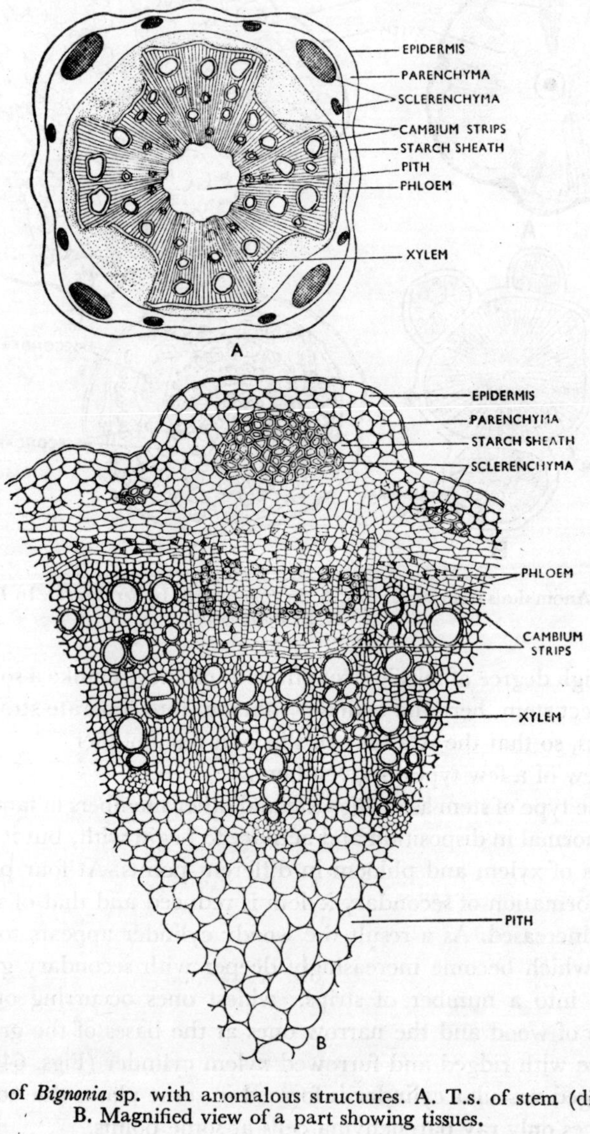

Fig. 649. Stem of *Bignonia* sp. with anomalous structures. A. T.s. of stem (diagrammatic). B. Magnified view of a part showing tissues.

A peculiar anomalous thickening due to occurrence of additional cambial layers—total or partial, outside the primary cambium is characteristic of some members of

Convolvulaceae, Menispermaceae and others. Here each central layer cuts off secondary xylem and secondary phloem for some time; its activities stop soon and a new layer of secondary meristem appears further out either in the primary cortex or even in secondary phloem. This process may be repeated several times. Another type of anomaly is noticed in *Tecoma*, where a secondary cambium arises on the inner side of the normal

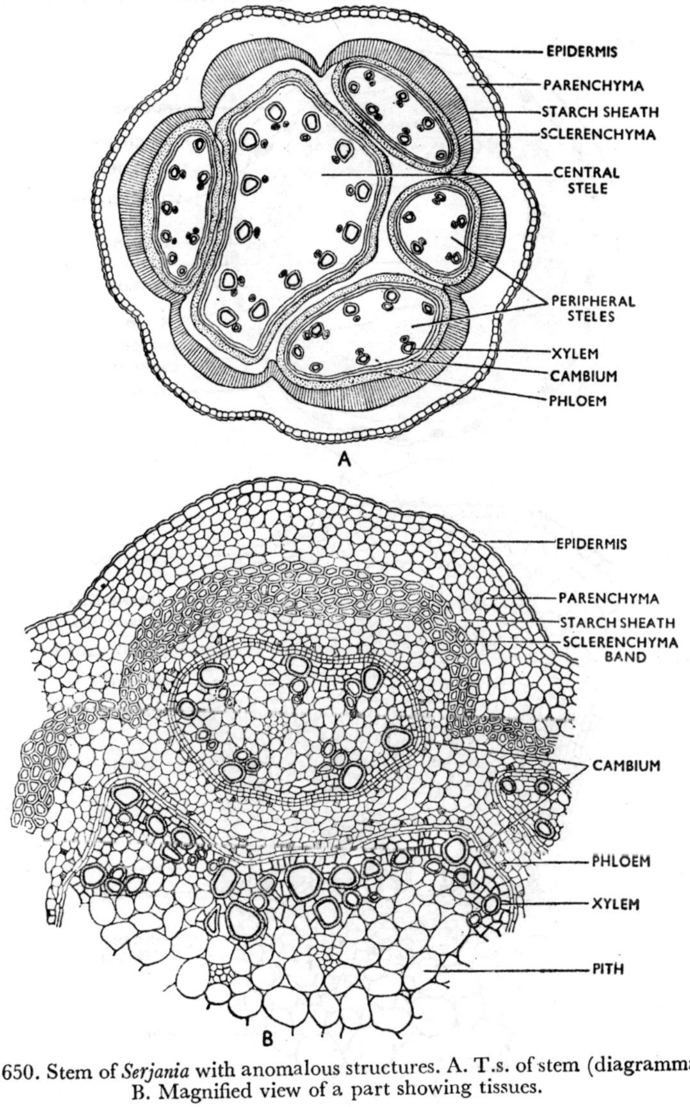

FIG. 650. Stem of *Serjania* with anomalous structures. A. T.s. of stem (diagrammatic). B. Magnified view of a part showing tissues.

woody cylinder. This internal cambium cuts off both xylem and phloem, but in reverse order. So phloem here is intraxylary and secondary in origin (Fig. 651).

In some members of family Combretaceae—*Combretum*, *Entada*, and *Strychnos* of family Loganiaceae anomalous structures result from occurrence of unusual type of phloem which remains embedded in secondary xylem. This type is known as *interxylary* or *included* phloem. Here the cambium cells for some time produce secondary phloem on the inner side instead of forming normal secondary xylem. Very soon the cambium takes to normal activities, i.e., it goes on cutting off secondary xylem. The secondary phloem is thus buried in the secondary xylem, and is termed included or interxylary phloem. It should not be at any rate confused with internal or intraxylary phloem which is normally primary in origin, but secondary in *Tecoma*. Often unusual or anomalous structures are formed due to development of secondary cambial zones which usually arise in the pericycle and function like the normal cambium. The secondary cambium originates either

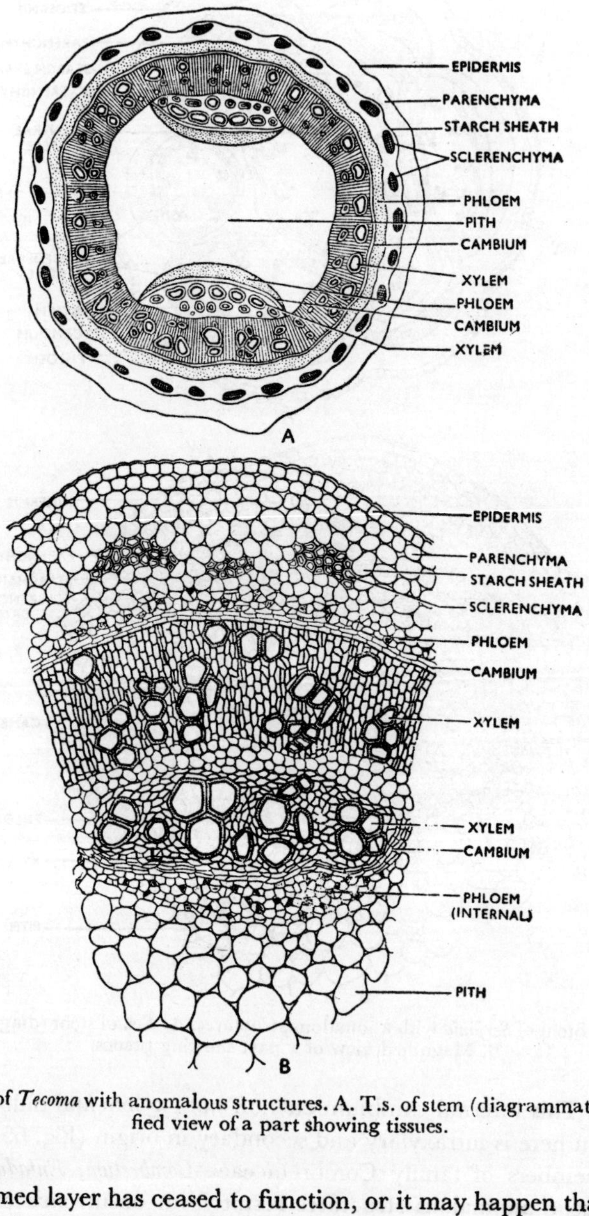

FIG. 651. Stem of *Tecoma* with anomalous structures. A. T.s. of stem (diagrammatic). B. Magnified view of a part showing tissues.

when the first formed layer has ceased to function, or it may happen that the additional layers arise successively and cease to function. In that case cylinders with alternate concentric layers of xylem and phloem are formed, as found in some climbers of family Polygalaceae and Nyctaginaceae (Fig. 648D). Variations in the nature and activities of the secondary layers lead to the formation of extremely unusual types of stems which are common in many lianes having special habits and mechanical requirements.

In the families Nyctaginaceae, Amaranthaceae, Chenopodiaceae, etc., anomalous structures are formed in rather different ways. The vascular bundles normally appear either in a ring or they remain irregularly scattered in the ground tissues (medullary bundles). They are as usual collateral ones with fascicular cambium. Cambial activities continue in the individual bundles for some time and soon cease. So secondary growth is rather limited. Secondary cambium arises on the outer side of the bundles. This cambium cuts off secondary bundles on the inner side, which remain embedded in a non-vascular tissue, known as *conjunctive tissue*. It consists of either thin-walled parenchyma or thick-walled lignified elongate cells. The parenchyma cells undergo gradual lignification and thus become thick-walled. The cambium produces very little tissue on the outer side. The secondary bundles may be irregularly scattered or may remain arranged in concentric rings.

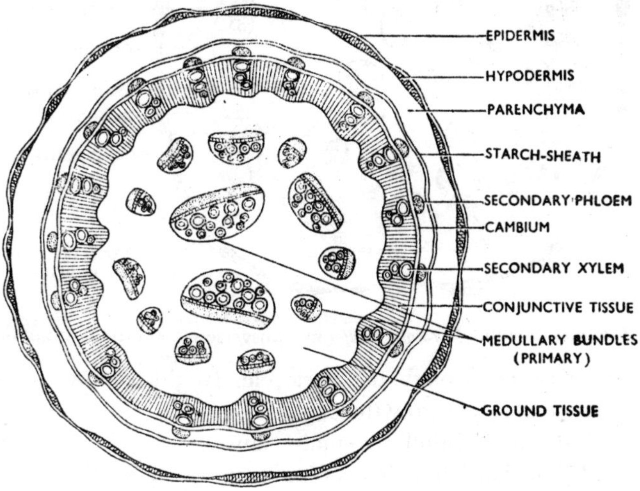

FIG. 652. Transverse section of stem of *Boerhaavia* sp. (diagrammatic).

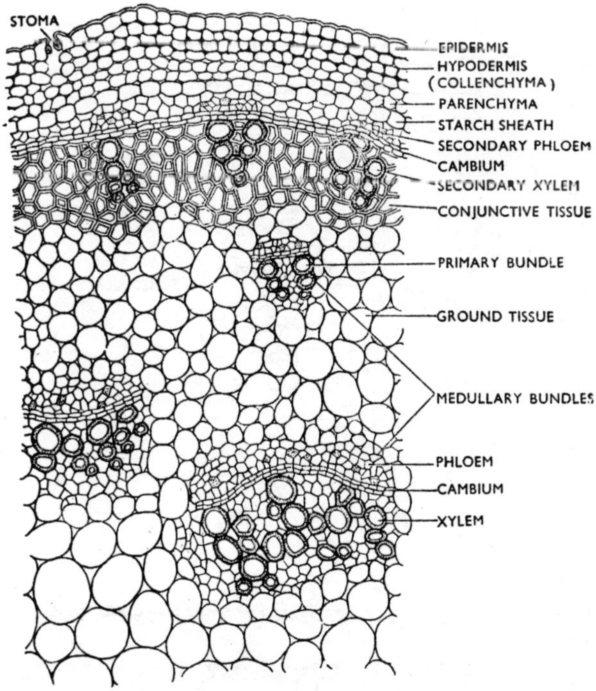

FIG. 653. A portion of stem of *Boerhaavia* in transverse section.

A transverse section through the stem of *Boerhaavia* sp., a common weed, of the family Nyctaginaceae would show the following characters (Figs. 652 & 653). The section is more or less circular in outline with uniseriate epidermis having stomata. A few layers of collenchyma in patches with intervening parenchyma cells form the hypodermis followed by parenchyma with distinct intercellular spaces. The stele, apart from ground tissues, includes vascular bundles in three rings. Two fairly large bundles occur at the central region surrounded by quite a few comparatively much smaller ones forming a loose ring. The outermost ring consists of pretty large number of much smaller bundles occurring just beneath the pericycle. The central bundles are collateral open ones; but cambial activity remains confined to individual bundles.

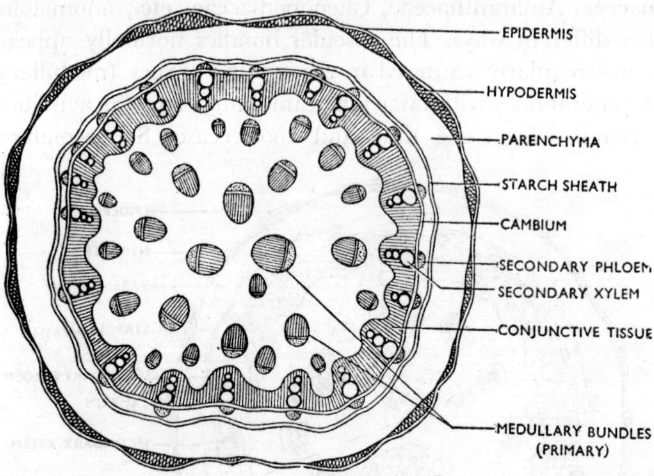

FIG. 654. Stem of *Mirabilis jalapa* in transverse section (diagrammatic).

Secondary increase in thickness is due firstly to the formation of a continuous cambium cylinder in the outer ring. These cells form secondary xylem and secondary phloem in the fascicular region and lignified conjunctive tissue in the interfascicular region. The activities of this cambium decline soon and a new cambium arises in the parenchyma outside the secondary phloem, in fact, passing through the pericycle. In this manner other cambia may arise outside producing a few growth rings. The newly-formed bundles remain embedded in hard lignified conjunctive tissue. Scanty secondary phloem is formed outside opposite to the xylem vessels.

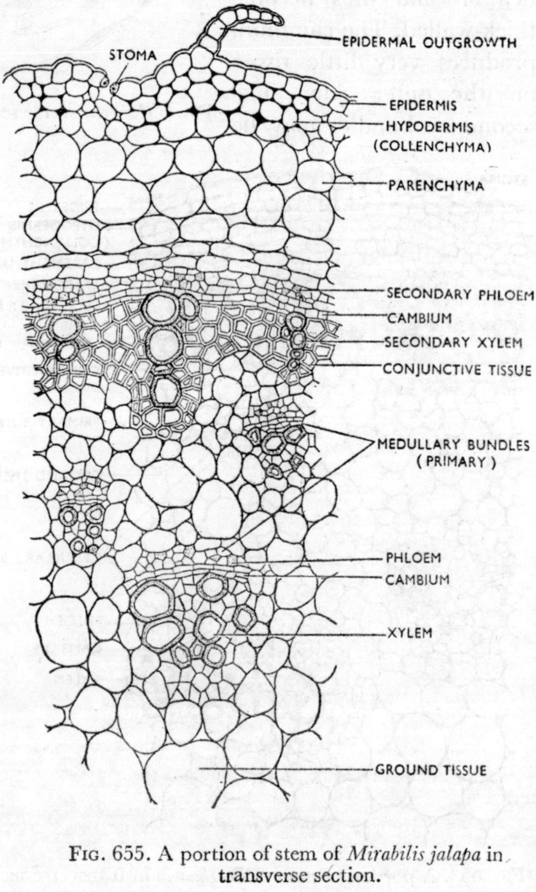

FIG. 655. A portion of stem of *Mirabilis jalapa* in transverse section.

The stem of *Mirabilis jalapa* (Figs. 654 & 655) of the same family is rather quadrangular in outline. It more or less resembles that of *Boerhaavia*, but the medullary bundles are more numerous. Those occurring towards periphery are smaller in size and more crowded, whereas those at the central region are larger and more spaced out (Fig. 654). Secondary growth is initiated by the formation of secondary cambium originating in the same manner as in *Boerhaavia*. This cambium cuts off secondary tissues, usually secondary xylem elements on the inner side which remain embedded in the conjunctive tissue. Secondary phloem elements are occasionally formed. The cambium produces very little secondary tissues on the outer side.

The stem of *Amaranthus* of family

Amaranthaceae also shows anomalous secondary growth (Fig. 656). The transverse section is more or less circular in outline with uniseriate epidermis. The ground tissues are well-differentiated. Collenchyma cells occur in the hypodermal region interrupted by chlorophyll-containing parenchyma cells here and there. The vascular bundles are medullary ones. They are large in number and remain scattered in the pith. The bundles are collateral and open. Cambial activity is confined to the individual bundles,

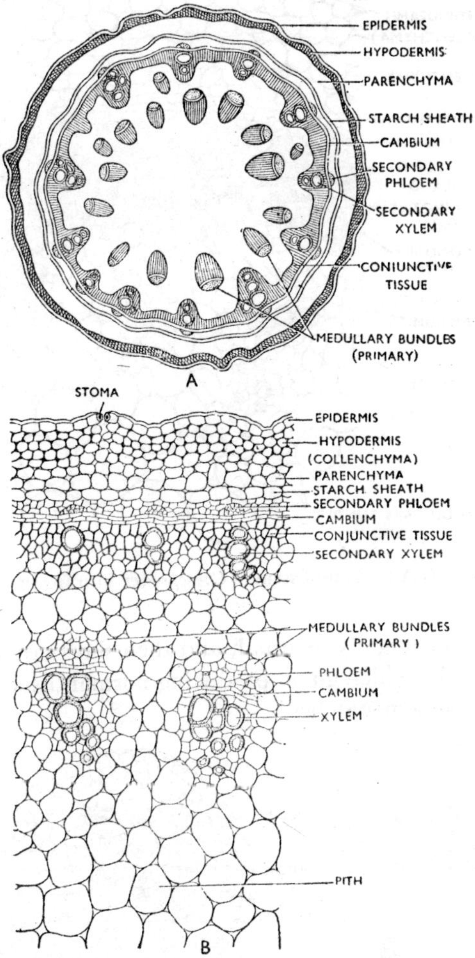

FIG. 656. Stem of *Amaranthus* with anomalous structure. A. T.s. of stem (diagrammatic). B. Magnified view of a part showing the tissues.

and it ceases soon. Secondary growth occurs due to development of a new meristem—the secondary cambium outside the stele. The cambium cuts off similar bundles with xylem on the inner side and phloem on the outer side. The secondary bundles remain embedded in thin-walled conjunctive tissue, which is wavy in outline on the inner side.

In the stem of *Chenopodium album* (Fig. 657) of family Chenopodiaceae there is a concentric ring of primary bundles (medullary) which are collateral open ones. They grow in thickness to some extent and then a peculiar anomalous secondary increase ensues when concentric zones of collateral vascular bundles arise from successive rings or arcs of secondary cambium originating in the pericycle or phloem. Conjunctive tissues with lignified walls are also formed by the secondary cambium in which the bundles remain embedded. A very interesting feature is the occurrence of isolated strands of phloem, called *phloem islands*, buried in the secondary xylem. It becomes possible because

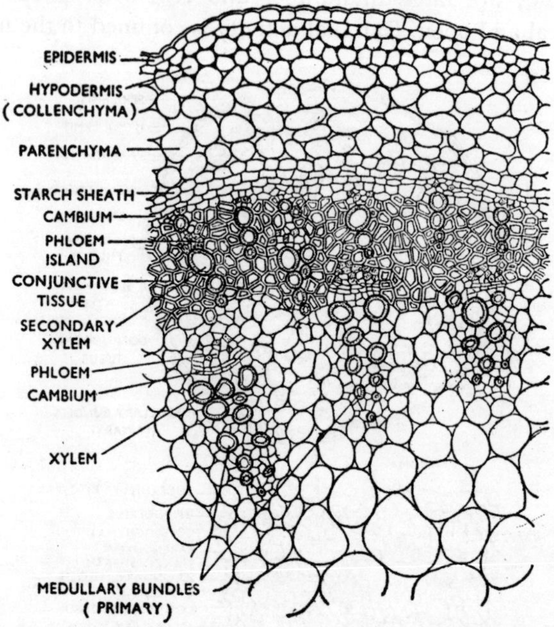

FIG. 657. A portion of stem of *Chenopodium* in transverse section.

phloem patches arise centrifugally as usual, and then short arcs of secondary meristem develop on their outer side, which go on producing normal tissues, resulting in the complete enclosure of the phloem patches.

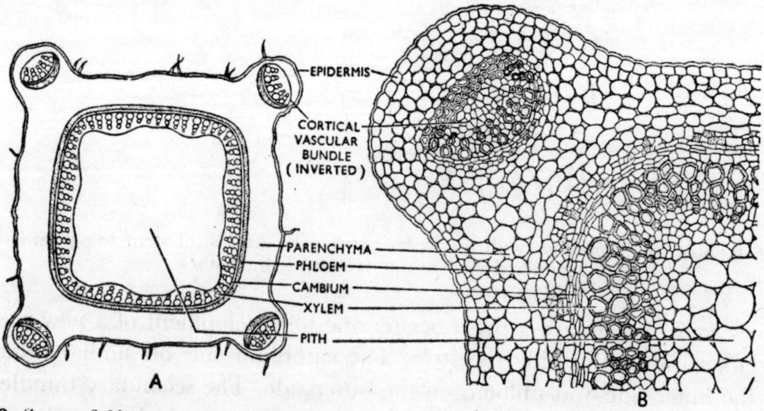

FIG. 658. Stem of *Nyctanthes* in transverse views. A. Complete section. B. A portion magnified.

In the stem of *Nyctanthes* of the family Oleaceae (Fig. 658), apart from normal vascular bundles occurring in more or less a ring, there are four cortical bundles at the ridged portions of the stem. These bundles are *inverted*. They are obviously leaf trace bundles.

The central vascular bundles are rather compact collateral open ones with intervening patches of parenchyma in form of rays.

In many dicotyledons the roots become fleshy and serve as organs of storage. Anomalies are often noticed in these organs which develop considerable storage parenchyma. Moreover, due to anomalous cambial activities the conducting elements are multiplied and suitably arranged for movement of food materials. Massive development of storage parenchyma is noticed in many cases in the region of cortex or secondary phloem. Many fleshy roots exhibit a notable deviation from the normal structure in the comparatively poor development of woody tissues. In the fusiform roots of radish (*Raphanus sativus* of family Cruciferae), which are formed by the association of the hypocotyl with the base of the tap root, the storage tissue mainly originates by proliferation of parenchyma in the pith, if pith is present and from the secondary xylem—the xylem parenchyma and xylem rays developing massive storage parenchyma. The vascular elements with some fibres remain arranged in concentric rings here. The fleshy nature of the conical root of carrot (*Daucus carota* of family Umbelliferae) is also due to massive development of storage parenchyma in the phloem and xylem.

The napiform root of sugar beet (*Beta vulgaris*) of family Chenopodiaceae exhibit a striking anomalous form of secondary growth, where several cambial layers are formed in succession (Fig. 659). The first cambium forms a ring of vascular bundles near the

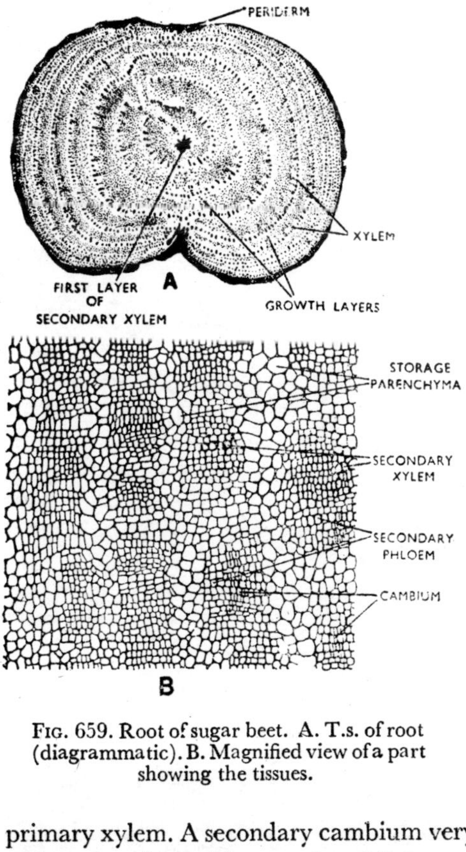

FIG. 659. Root of sugar beet. A. T.s. of root (diagrammatic). B. Magnified view of a part showing the tissues.

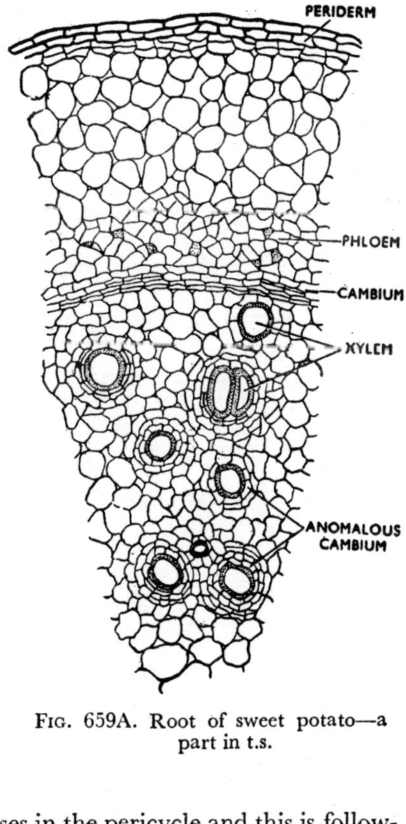

FIG. 659A. Root of sweet potato—a part in t.s.

primary xylem. A secondary cambium very soon arises in the pericycle and this is followed by others in rapid succession. All the cambial layers continue to function, but the rate becomes slower after the early period of activity. The cambium, though appearing on continuous bands, produces separate bundles with surrounding conjunctive tissue. The newly-formed cambia divide and multiply very rapidly to build up parenchymatous layers. As a result alternate layers of proliferated pericycle and vascular elements are formed. The bundles are largely parenchymatous with very scanty lignified elements. Continued growth takes place in the bundles by cambial activity and by proliferation of parenchyma of xylem and phloem. Concentric rings are thus formed. A complex system is often formed due to the fact that the layers do not produce complete cylinders and that irregular association occurs with other layers. The major bulk of the storage parenchyma, however, belongs to phloem. Sugar occurs as storage product in the parenchyma cells specially in those associated with vascular elements.

The fleshy adventitious root of sweet potato (*Ipomoea batatas*) of family Convolvulaceae exhibits another complex type of anomalous growth (Fig. 659A). Cambium arises in the normal position separating the xylem from the phloem. Both primary and secondary xylem elements develop in the usual manner, but contain a large proportion of parenchyma. With further development many secondary cambia arise in the parenchyma around individual vessel or vessel groups. These cambia produce a few tracheary elements towards the vessels, a few sieve tubes and laticifers away from the vessels, but a fairly large number of parenchyma in both the directions. At a still later stage tertiary cambia may be formed in the parenchyma. Periderm of pericyclic origin occurs as normal in the periphery.

CHAPTER XXVII

ABSCISSION AND HEALING OF WOUNDS

The plants commonly lose various parts periodically. Lower vascular plants usually do not shed their parts, which remain attached to the plants and ultimately get decayed. But in the higher vascular plants the fall of the parts—particularly of the leaves, floral members and fruits—is mainly caused by the formation of a distinct zone where usually the part gets detached from the mother plant. The process is referred to as **abscission,** and the zone as the *absciss zone*.

Periodic defoliation is a notable feature of the gymnosperms and most of the woody dicotyledons. The deciduous plants shed their leaves at the approach of winter. The evergreen plants have no fixed season for fall of leaves; but they may shed their leaves in any season. Some structural modifications actually take place at the base of the leaves, and finally they are abscissed. The formation of absciss zone is usually initiated before the leaf-fall. In fact, the leaves now become mature, turn yellowish in colour due to disintegration of chlorophyll and preponderance of carotinoid pigments. Moreover, the waste products are transferred to the leaves, so that the plants may get rid of them with the leaf-fall. It is really a vital phenomenon involving special changes in the cells occurring at the base of the petiole, what is evident from the fact that the leaves of the dead branches are not shed, but they get dried up and decayed in course of time.

The absciss zone is usually formed at the base of the petiole in case of simple leaves. In compound leaves the zone appears at the base of the rachis and also at the base of individual leaflets. It is a narrow zone composed of a few layers of cells (Fig. 660). Before

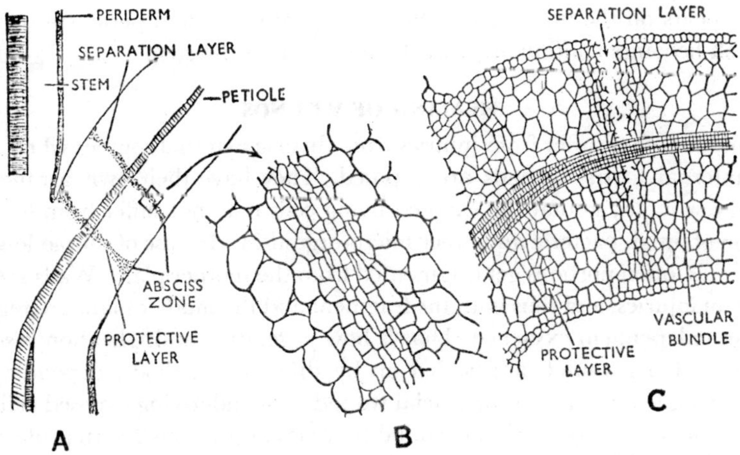

Fig. 660. Abscission of leaf. A. Petiole showing absciss zone. B. Magnified view of separation layer. C. Petiole magnified to show separation layer and protective layer.

the fall of leaf a distinct *separation layer* is formed in the absciss zone which may be considered the immediate structural reason for the fall of leaf. Besides this, there develops a *protective layer* within the absciss zone which protects the exposed surface after leaf-fall from loss of water and attacks of micro-organisms. Periderm is formed later and provides effective protection.

The absciss zone is the weakest spot on the petiole of the leaf. In some cases it becomes slightly constricted and thus rendered externally visible. The vascular bundles become reduced here and the strengthening tissues are minimum. Collenchyma cells are absent and the sclerenchyma, if present, are weakly developed. The separation layer consists of at least two rows of cells which are different from those occurring above and below the absciss zone in cell size and cell contents. The cells of this layer are smaller in size, have denser protoplasm and abundant starch grains. The vessels usually become clogged by gums and tyloses below the zone. Distinct changes take place in all the parenchyma cells of the separation layer. The middle lamella and the outer walls of the cells swell, become gelatinous and eventually break down. Calcium pectate of the middle lamella is converted into pectic acid and that into water-soluble pectin. Thus the cells now become separate from one another either along the middle lamella or the walls actually break down. Now that the separation layer is thus formed, the leaf remains attached to the stem only by the vascular elements. It breaks in time either by the mechanical weight of the leaf or slight swaying due to wind. During rains, the leaf gains in weight and gelatinisation of the cell wall is hastened, resulting in quicker abscission. The separation actually starts at the periphery of the petiole and extends inwards. It continues in the living parenchyma cells of the vascular bundles, but the reduced vascular elements break by mechanical weight. It has been suggested that middle lamella of the cells of the vascular tissues also disintegrate.

After the fall of leaf the surface is exposed which has to be protected from desiccation and outside injuries. A few cells in the absciss zone just beneath the separation layer now constitutes the protective layer which remains as a cover to the scar caused by leaf-fall. The walls of the cells of this layer undergo suberisation, and possibly partial lignification, and thus protect the exposed surface. That the substance defined as lignin is really wound gum giving reactions similar to lignin, is the opinion of some workers. Now phellogen appears on the inner side of the protective layer, and periderm is ultimately formed which becomes continuous with the periderm of the stem.

HEALING OF WOUNDS

Plant organs often suffer from injuries from diverse external sources when the living cells lying internal to the wounds are exposed. They have their own arrangement of healing of the wounds, so that the exposed living cells may not suffer from loss of water and infection of fungi, bacteria and other micro-organisms. In case of simple lower plants the thallophytes, often portions get fragmented from the mother plant. Walls are secreted at the points of injuries, and thus both the fragment and the mother plant are regenerated and carry on independent existence. That is how vegetative multiplication is secured.

In higher plants wounds are healed up by different methods, depending on the nature of the wounds. In case of superficial wounds the underlying exposed cells die and dry up. The layer of cells beneath the wound becomes impregnated with protective substances like suberin. Plants with laticifers heal up the simple wounds by secretion of latex, which coagulates and serves as a protective cover. In some cases the living cells beneath the injury take up meristematic nature and serve as a secondary cork cambium. Those cells cut off cork cells on the outer sides and thus the living internal portions are protected.

In case of deep wounds the uninjured living cells beneath the wound become meristematic and go on producing a mass of undifferentiated parenchyma cells known as **callus.** The callus actually fills up the wound. Very soon a cork cambium arises in the peripheral layers of callus, which produces a layer of cork, referred to as *wound cork*, on the outer surface and thus the wound is healed up. Very often callus overgrows the wound, and as a result, characteristic swellings or *knots* are noticed in the trees (Fig. 661). Knots are also frequently found in the wood of the trees which have considerable increase in thickness. When a branch is either cut off or broken down, the wound surface is covered by the callus. But with secondary increase in thickness, the secondary xylem elements gradually extend over the stump of the branch, so that it eventually gets embedded in the wood. In the mean time all the living cells of the base of the branch die, shrivel up, and the stump remains as a cone embedded in the woody portion. They appear as knots in the timber.

The wound may be very deep so as to injure the cambium of the vascular bundle. In that case callus is formed as usual and the wound is filled up. Then a few parenchymatous cells of the callus become meristematic to form a new cambium, so to say, in a line with the uninjured cambium of the organ. The new cambium joins up with the

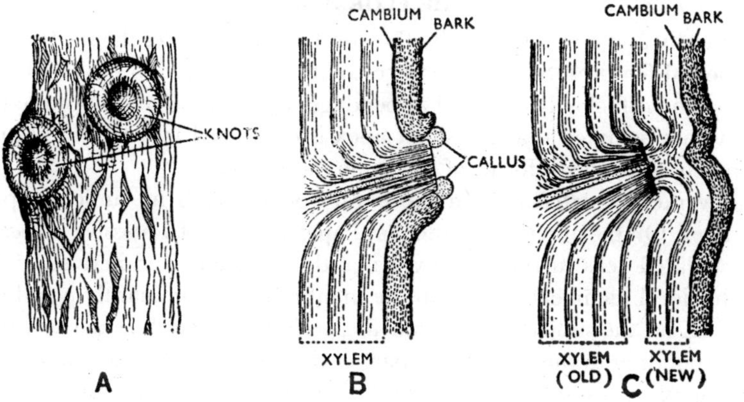

Fig. 661. Knots in a stem. A. Surface view. B. Longitudinal section view showing formation of callus. C. Same with old wood and new wood.

existing uninjured one, and thus continuity is established, so that the formation of secondary tissues is not disturbed. The horticulturists take advantage of this process in budding and grafting, where the cambia of the two shoots, referred to as *stock* and *scion*, brought together during grafting, unite, provided of course the two are of the same or closely-related species.

CHAPTER XXVIII

THE FLOWER

Though the nature of the flower from morphological and anatomical points of view and its phylogenetic relation to other parts have been subjects of long drawn controversy, majority of the botanists consider the flower as something equivalent to a shoot—a modified shoot, and the floral parts homologous to the leaves. The arguments in favour of the above statement have been discussed in detail in the chapter on flower in Part I. Like a vegetative shoot the flower has an axis—the *thalamus*, and the lateral appendages—the floral leaves or parts. The latter may be put into two categories—the fertile ones which take active part in reproduction, the *stamens* and *carpels*; and the sterile ones—the *sepals* and *petals* which only help the process and hence are called helping or accessory whorls.

THE FLORAL APEX

Structurally and developmentally the floral apex resembles the shoot apex in almost all fundamental features. Slight differences that exist are due to the fact that flower is a *determinate* stem with closely crowded appendages and very much suppressed internodes. It is at any rate assumed that the shoot apex is organized into the floral apex and the two are merely the growth forms of the same meristem.

Some workers asserted that in the floral apex there is no tunica-corpus zonation like shoot apex. On the other hand, the floral apex consists of a massive *central core* of thin-walled vacuolated cells surrounded by a mantle of smaller and actively dividing cells; and that the floral members arise from the mantle by predominantly periclinal division. Further works on the subject reveal that the peripheral mantle is derived from the cell layers corresponding to the tunica and part of the corpus of the shoot apex. The elongation of the apex ceases and growth activity is confined to the peripheral region due to determinate nature of the flower. Periclinal division at any depth of the floral apex is associated with the broadening of the thalamus and origin of the appendages. Thus floral apex is regarded as the ontogenetic modification of the shoot apex. The shoot is now considered a single unit composed of the stem and leaves; and in that light it is assumed that development of the flower is "parallel to that of the vegetative shoot, and not as being derived from it".

VASCULAR ANATOMY

The vascular organization of flower may in a general way be compared to that of a vegetative shoot (Fig. 662).

The axis exhibits three zones—epidermis, ground tissues and vascular system. The latter is really a complex of traces which run to the floral parts—often branch and again combine with each other in the axis in an irregular manner.

The accessory parts—sepals and petals resemble the leaves in internal structure. The ground tissues hardly show any differentiation into palisade and spongy cells. The epidermis is normally made of papillose cells, often with intercellular spaces covered by cuticle. Trichomes and stomata may occasionally occur. Special cells containing volatile oils may be present in the epidermis of petals imparting fragrance to the flowers. In spite of some variations it may be said that a sepal has as many traces as the leaf of that plant and it resembles a bract anatomically. A petal of dicotyledon normally has one trace, whereas a perianth member (tepal) of a monocotyledon has one to many traces.

Like those of foliage leaves the vascular bundles form complex systems in the sepals and petals and the veinlets are usually dichotomously branched.

The stamen is a simple structure with the filament surmounted by the anther. A single vascular bundle runs through the filament and ends bluntly in the connective which is located between two anther lobes. In some families three traces are present, what is considered a primitive condition. The bundle of the filament is usually amphicribral and remains surrounded by vacuolated parenchyma cells without conspicuous

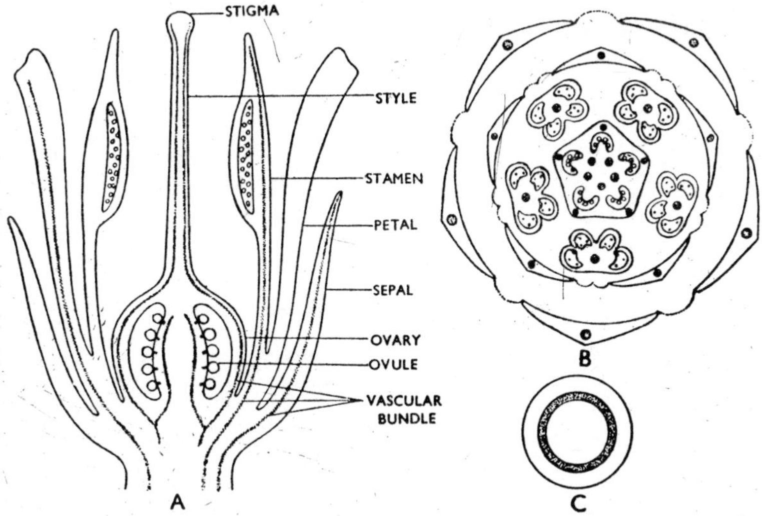

FIG. 662. A hypogynous flower showing vascular supply. A. Longitudinal. B. Transverse. C. T.s. of pedicel showing zones.

intercellular spaces. The carpel, either a simple or an apocarpous one, is a conduplicately folded structure, at the basal part forming the ovary and the upper sterile part—the style, which ends in the receptive part, the stigma. This is an advanced condition in the phylogenetic line. The ovules occur on the ovarian wall attached to the parenchymatous outgrowths—the placenta, normally at the regions where the margins of the folded carpels meet. Commonly a carpel has three veins—one dorsal and two ventral. The carpellary bundles run through the style. The ovules get the vascular supply from the ventral bundles or from placental branches of these bundles. Normally a single trace enters through the base and reaches the chalaza; branches may extend to the integuments but never to the nucellar region. The ovules are made largely of parenchymatous tissue with primary vascular system which may be functioning during maturation of the seed. The ovary and style are simple structures from histological points of view being composed of cuticularised epidermis, parenchymatous ground tissues and vascular bundles.

Abscission. Abscission of floral parts, the petals in particular, is similar to what happens in case of leaves. The petals fall off either with the complete opening of the flower or may remain attached temporarily or permanently till the fruit stage. They often undergo constriction at the zone of abscission. The separation layer is poorly formed of small closely packed vacuolated cells. As in leaves the separation is due to softening of the middle lamella, but formation of cork or deposition of suberin for protecting the scar is lacking. Other members—the sepals, stamens, and style may also absciss in the same manner. Male flowers normally fall off after shedding of pollen grains and both female and bisexual flowers may also drop in case fertilisation fails.

PART THREE

Plant Physiology

CHAPTER XXIX

INTRODUCTION

Understanding of the processes by which plants keep alive and grow is the aim of plant physiology. Thus plant physiology deals with the life processes of the vegetable organism. Any chemical or physical change occurring within a living plant cell or any interaction between the organism and its environment, would be considered as a physiological process. Such classifications of the plant and its activity into morphology, histology, physiology, ecology, etc., followed in text-books are rather arbitrary and are adopted for mere convenience. In fact, all of them are interrelated. Thus plant physiology cannot be studied separately without consideration of the other aspects of living organism as a whole.

The study of the life process of a plant evidently pre-supposes a thorough understanding of the structures of the organisms, the environment in which it lives, its hereditary make-up, etc. Structures and functions are so intimately related that it is impossible to separate one from the other.

A plant grows—so the processes concerned in its growth are certainly dynamic. The problems in plant physiology are concerned with *how*'s and *why*'s and *what*'s. *What* is the mechanism of absorption of water, gases and solutes by the plant from its surroundings? *Why* does a plant flower only at certain periods in its life cycle or under certain environmental conditions? *How* do the plants prepare their own food? The explanations of these and many such questions and interrelationship of one with the other give us an insight into the co-ordinated growth of the plant as a whole.

From the earliest times to the eighteenth and nineteenth centuries, and even up to comparatively recent times opinion was almost universal and unanimous that living organisms possess and operate through some intangible and unknown forces which distinguished them from the non-living. There are, however, few advocates of such 'vitalistic' theories at the present time. The most widely held assumption now is that the living organisms, whether plants or animals, also operate following the same physico-chemical principles that are true in the non-living world.

The behaviour of the living organism does not differ in many respects from the non-living bodies. Weight, rigidity, elasticity, conductivity of heat and electricity, etc., are certainly the properties of a solid non-living body and the plants and animals in spite of large amounts of water in them, are as a rule solid and show the same properties of a solid body.

What then distinguishes the living from the non-living and can we attempt at a definition of life? Is it the capacity for reproducing its kind accompanied by relatively slow and small transformations of energy which is the biochemical criterion of life? A candle flame may be said to be 'living'—it has got a shape or form, it can reproduce its kind—you can get one flame from the other. The transformation of energy in such a flame is, however, too rapid and if such quick transformations of energy occurred in the living body, the whole thing would go into a flame just like the candle. The virus nucleoprotein which is responsible for many animal and plant diseases, shows one of the fundamental properties of the living body in that it can be made to reproduce its kind when it thrives in the host body (virus restricts the normal activity of the host cells and compels those cells to produce more of virus particles—virus DNA or RNA nucleoprotein is incapable of dividing by itself outside a host cell). But the virus particles do not show even the minutest trace of transformation of energy to maintain their existence and self-duplication (they do not respire) which is an important manifestation of life from a purely biochemical point of view. Moreover, virus protein can be isolated and crystallised in the pure state; so where does virus protein take its position in the 'ladder of life'—from the inanimate to the animate world? Certainly virus protein and the other nucleoproteins in the chromosomes of the nucleus, e.g., genes, are the missing links in the biochemical evolution of life from the inorganic to the living organic world. Can we then not attempt at a definition of life in the light of the great advancement of knowledge in physical sciences during the last half of a century? We prefer the definition of life as given by John Perret as consistent and quite in keeping with the trend of development of modern science. Perret's definition of life as "a potentially self-perpetuating open system of linked organic reactions, catalysed stepwise and almost isothermally by complex and specific organic catalysts which are themselves produced by the system" certainly satisfies to a large extent the phenomenon and characteristics of a living organism. It can be further clarified if we learn to recognise that the laws of non-living matter merge continuously and perhaps smoothly in those governing the living. Potentially self-perpetuating—it is always capable of reproducing its kind—it may not be doing so all the time and thus perpetuate itself. The most important parts in the definition are the linked system of reactions and the specific enzymes. The linked system as such obeys the laws of chemistry, but it is probable that the system of reactions having a dynamic stability may adjust themselves to small but limited changes. The reactions are catalysed step by step by enzymes almost *isothermally*, i.e., without any large absorption or liberation of energy as heat; thus the temperature of the medium or the body in which the reactions take place does not show any large or rapid fluctuations in temperature as long as the reactions are taking place, i.e., as long as the organism is said to be living. Once the catalysts or enzymes have been produced by the system of reactions itself, the chemical laws governing the activity of enzymes take on the aspect of a biological law of adaptation, i.e., the living organism can adapt itself to a change in the surrounding environment in which it lives.

It is also true that the enzymes, to be effective agents in the linked system of reactions, must have physical characteristics such as size and coherence between the molecules due to opposite electric charges, which is only possible in very large or macro-molecules of polymers (compounds formed from polymerisation of smaller molecules). Among such polymers, the large and complex molecules of proteins formed by peptide linkage of comparatively smaller molecular units of amino acids occupy a pre-eminent position. We know, however, that many organic molecules exist in both left-handed and right-handed forms (*d* and *l* forms), yet most of these molecules *are found in only one* of these forms in living organisms. The proteins certainly represent, among others, any easily built-up set of molecules having just the properties needed to open the way for an evolving metabolism in contrast to a metabolism of a low order of efficiency, being merely repetitive. No wonder then that the viruses and genes which occupy a hypothetical position between living and non-living, are largely proteinaceous.

The first organic molecule to achieve characteristics of living must have been a protein. And what can be said with assurance is that there is actually a unique and nearly ubiquitous compound, with the empirical formula, $H_{2960}O_{1480}C_{1480}N_{16}P_{1.8}S$ called living matter!! The moment the protein molecule was formed, the chain of reactions leading to the evolution of life, as we know, was surely and steadily moving towards its ultimate goal. The presence of mostly one form of organic molecules, either *d* or *l* and the remarkable basic similarity among all forms of life, however, support a strong presumption that only *once* did a living molecule originate from non-living matter. Again, all amino acids, the alphabet of protein structure which is ultimately responsible for the specificity of variability of living matter derived from proteins, are α-amino acids (except glycine which is not optically active). This fact strongly suggests the common origin of all living matter on the earth. And for life to originate, an atmosphere without oxygen is necessary. Further spontaneous regeneration of life seems impossible once the green plants evolved which put oxygen into the atmosphere.

It is true that the complexity of many of the physiological processes taking place in the living world sometimes eludes our grasp but that should not be the reason to endow living organisms with some intangible and incomprehensible varieties of energy. Attempts should be made to explain them by the easily and truly recognisable physico-chemical laws based on experimental evidence and logical deductions. A knowledge of fundamental principles of physics and chemistry is, therefore, absolutely essential for a clear understanding of the physiological processes and this cannot be overstressed. It is impossible to learn even the rudiments of plant physiology without being familiar with the underlying principles of physical sciences and organic chemistry.

With unimportant exceptions, it is indisputable that the basis of all life on the earth is photosynthesis by green plants, a process that involves both physics (in the fixation of solar energy) and chemistry (in the union of CO_2 and water to form carbohydrates and more complex biochemical compounds).

The relation of plant physiology to agricultural sciences is intimate and profound. The plant world not only supplies us with all our food requirements but also supplies the raw materials for many of our basic industries. An industrial civilisation such as ours not only requires a wide variety of plant products but also naturally demands and expects certain standards of quality. In order to maintain or to improve this standard, cultivation of plants has become a skilled occupation requiring a thorough understanding of the physiological processes taking place in plants. The application of the principles of plant physiology is certainly helping to solve the various problems with which the ordinary farmers, the fruit grower, the horticulturist, the cotton and tea planters, the foresters and all others who cultivate plants are confronted daily.

The foundations of plant physiology may be said to have been laid with the publication in 1727 of Stephen Hales' 'Vegetable Staticks'. Long before that, however, many books on agricultural practices employed have been published and there was an extensive literature on agriculture in Roman times. Man's conscious interest in plants must have begun long before any recorded history and agriculture, that is the cultivation of economically important plants for food and clothing was a highly developed art ever since man began living in social groups and started his long progressive march towards the control of the environment around him.

The books and learned treatises published on agricultural practices mostly of course contained speculations based on observations through ages by men close to the soil but without any critical experimental evidence. Some of these ingenuous speculations have stood the test of time while others have been proved to be entirely erroneous. Thus Palissy's speculation in 1563 on the agricultural practices employed in Italy at that time about the importance of burning unused wheat straws in the field when growing this crop for second year in succession was fundamentally correct and must have inspired other investigators who came later in the field on the realisation of the importance of manures to crop yield. In the theories propounded by Van Helmont (1577) and great

Francis Bacon water was considered as the sole nutrient for plants; the purpose of growing them in soil was only to keep them upright and protect them from excessive cold or heat! This, of course, as we all know, is largely false. Plant physiology as a science made very little progress between the time of the publication of 'Vegetable Staticks' in 1727 and the announcement of discovery of photosynthesis by Joseph Priestley in 1781. What was the reason of so little interest in plant physiology shown by the naturalists between 1727–81, a period of about 50 years? It was certainly to a large extent due to the pioneering work on classification of plants by the great Linnaeus. Naturalists of that period were inspired by the work done by Linnaeus (his rather pompous statement, "God created plants and Linnaeus classified them" was illuminating showing the spirit of the man and of the age in which he lived) and almost all of them were busy in identifying, checking and classifying plants according to the new system of classification based on binomial nomenclature.

The advent of plant physiology as a science certainly coincided with the great advance the pneumochemists made in the latter half of the eighteenth century—rightly called the 'age of enlightenment'—in the discovery of various gases and in the determination of the composition of air. The names of Joseph Priestley who discovered oxygen and photosynthesis in green plants, Jan Ingen-Housz, Senebier, de Saussure, Boissingault, Liebig among others must be honoured as the founders of modern plant physiology. By careful experiments, precise reasoning and an innate capacity of grasping the truth, these pioneer plant physiologists have left a blazing trail for future investigators in this field, as in all others, to expand and enrich the realm of human knowledge.

CHAPTER XXX

SOLUTIONS AND THEIR PROPERTIES

Water is the most abundant compound present in all living plant or animal cells. Water which occurs in a liquid state invariably contains many substances dissolved in it and, naturally, also contains dispersed particles which are not in true solution. When the particles dispersed in water are within a range of sizes, the system of water and the particles falls into the category of a physical system known as colloids.

True solutions are systems in which one component falling below a certain range of sizes (the solute) is dispersed throughout the other (solvent). Thus when NaCl is dissolved in water, water is the solvent and sodium chloride is the solute, the particles of which are in the form of molecules of NaCl or ions of Na^+ and Cl^-.

Now, what are ions? Some aqueous solutions readily conduct an electric current; others do not. The former are called electrolytes; the latter non-electrolytes. The solutions of acids, bases and salts are electrolytes whereas solutions of such organic compounds as sugars, alcohols and ethers are non-electrolytes.

The passage of an electric current through an electrolyte results in its dissociation or it can also be said that conduction of an electric current by an electrolyte is due to the presence of ions. An ion is electrically charged atom or a group of atoms. An atom is, however, electrically neutral because the positive charges of the protons in the nucleus of the atom completely balance the negative charges of the surrounding electrons in their orbits. Now when an electrolyte such as NaCl is dissolved in water, some of the molecules, not all, dissociate or ionise into two kinds of particles, one positively charged Na^+ and the other negatively charged Cl^-. Na^+ becomes positively charged because it loses an electron to chlorine, and thus the number of positive charges in Na^+ becomes more than the negative charges and so it becomes electropositive. Similarly chlorine, which on ionisation gains extra electron from sodium which loses it, becomes electronegative. Two, three, four or even more ions may be formed from a single molecule, e.g., Na_2SO_4 dissociates in solution into Na^+, Na^+ and $SO_4^=$.

The positively charged ions which on electrolysis migrate towards the negative pole or cathode of the electrode, are called *cations* whereas the negatively charged ions which migrate towards the positive pole or anode are called *anions*. A cation may carry one, two, three or more positive charges; an anion may also carry one to several negative charges. In an ionised solution, the number of positive charges carried by the cations is always equal to the number of the negative charges of the anions.

Complete ionisation of all solute molecules in an electrolyte is never possible. Some of the molecules are continually dissociating into ions while free ions in solutions are continuously reuniting to form molecules and when the two processes proceed at an equal rate, a dynamic condition of equilibrium prevails. It is also apparent that solutions in which a large proportion of the molecules of the solutes is in a dissociated condition will conduct electric current better than solutions in which a smaller proportion of the molecules is ionised. It must be clearly understood that the more dilute a given electrolytic solution is, the larger must be the proportion of its dissociated molecules. It follows that in extremely dilute solutions, theoretically ionisation will be practically complete.

An atom, we know, is electrically neutral because the number of positive charges in the nucleus exactly balances the number of negative charges carried by the electrons in the outer orbit. The nucleus is mostly built up of positively charged particles called protons and electrically neutral particles called neutrons and also other particles. The number of protons or unit positive charges in the nucleus defines the element and this is also the atomic number of the element. Atomic weight of an element is very nearly equal to the number of protons *plus* the number of neutrons in the nucleus. Atoms of the same element, i.e., having the same atomic numbers in the nucleus which have different number of neutrons in the nucleus, i.e., also different masses, are said to be *isotopes* of the element. The isotopes of an element though they differ in atomic weight because of the difference in the number of neutrons, have the same chemical properties because electrons in the outermost orbit remain unchanged.

Most of the atoms of the elements existing in nature are more or less stable. The nuclei of some of the heavy metals are, however, unstable and tend to revert to a stable state by spontaneous disintegration losing energy in the form of ionising radiations such as the α-particles, the positively charged helium nuclei, the β-particles and also γ-rays. Such atoms with unstable nuclei which tend to revert to a stable state are said to be radioactive and the elements which have such atoms are known as *radioelements*, and the process of disintegration is known as radioactivity. Radium and radon are examples of such spontaneous radioactive elements which are ultimately converted in a particular time into lead.

But there are elements, whose atoms are normally stable but when they are bombarded by particles such as neutrons from a cyclotron, the atom splitting apparatus. They gain excess energy and become unstable and tend to revert to a stable state by disintegration. Isotopes can thus be artificially made by increasing or decreasing the number of protons and neutrons in the nuclei. Stable ^{14}C atoms can be converted into isotopic heavy ^{14}C. Similarly ^{16}O, ^{14}N and many others which are ordinarily stable can be converted into isotopic heavy ^{18}O or ^{15}N by addition of new neutrons to the neutrons already present in the nuclei. Isotopic ^{11}C, ^{13}C or ^{14}C have identical properties of normal ^{12}C but they differ only in atomic weight and they are artificially prepared (they also occur in very small quantities naturally). The same is also true for ^{18}O, ^{15}N and others.

The production of these many kinds of radioactive isotopes has opened new fields of research in medicine and biology, particularly plant physiology, and it would be no exaggeration to say that working with carbon, oxygen, nitrogen and other isotopes, investigators in plant physiological problems have revolutionised the outlook towards the functioning of the living cell to such an extent that some of the fundamental aspects which were eluding our grasp for more than a century are being known correctly for the first time and many of our strongly held and cherished conceptions are being altered radically. Thus the nuclear physicists have given a magnificent tool in the hands of plant physiologists who are using this new implement successfully to throw new light on many plant physiological problems which had remained baffling or only partially understood so long.

Radioactive isotopes are used as *tracers* in plant physiological research and also in some diagnostic procedure and as radio-therapeutic agents in medical research. The use of isotopes as tracers is dependent on mainly two assumptions: (1) the living cell does not distinguish between the isotopes of the same chemical element and (2) the radiations emitted do not affect the process to be studied. Using them as *tracers* in metabolic processes in the living cell is more or less similar to tracing a marked coin as it is allowed to move and circulate. You can lose sight of it sometime but as it is marked (i.e., emitting radiations) you are sure to find it again because it retains its identity from others and it is not difficult to trace its course in its passage through a certain period and through a certain region.

Isotopes are extensively used nowadays in medicinal and biological research but because of their radioactivity, proper precaution must be taken in their transport, storage and handling; for an unnecessary exposure to emitting ionising radiations might involve the serious risk of later carcinogenesis. Isotopes, in recent times, are supplied to Indian investigators in this exciting line of research, from Trombay's Bhaba Atomic Research Centre and may soon also be available from Calcutta's Variable Energy Cyclotrone (VEC —also a part of the Bhaba Atomic Research Centre).

pH—HYDROGEN-ION CONCENTRATION

An acid may be defined as a substance which, when dissolved in water, produces hydrogen ions (H^+). Similarly a base is a substance which produces hydroxyl ions (OH^-) when dissolved in water. The most common inorganic acids like HCl, HNO_3, H_2SO_4 play comparatively unimportant roles in the living cells of plants or animals compared to the large group of more abundant but complex and much weaker acids known as organic acids.

A normal solution of any acid will exactly neutralise an equal volume of a normal solution of any base, no matter how strong or weak may be the characteristics of the two reactants. During neutralisation, equal numbers of hydrogen ions of acid and hydroxyl ions of base unite to form water.

Why then should certain acid or base though of the same normality be designated as strong or weak? The strength of an acid depends on its degree of ionisation, the greater the proportion of hydrogen ions and acid produces in a solution of a given concentration, the stronger it is. The same is true for alkalis producing hydroxyl ions. Weak acids (e.g. organic acids, such as, oxalic, acetic, malic, citric, etc.) or alkalis, e.g., $Ca(OH)_2$ are those, which are only slightly dissociated in solution producing proportionately much less hydrogen or hydroxyl ions compared to strong acids and alkalis.

At room temperature at 0·1N concentration, the dissociation of HCl molecules is about 91% whereas in acetic acid it is only about 1·35%. It follows then a solution of 0·1N or (N/10) HCl is approximately 67 times (91/1·35) stronger than (N/10) acetic acid. Dissociation, however, increases with dilution. In a solution of (N/1,000) HCl ionisation is about 98% whereas in acetic acid it is now about 14%. So at this particular dilution, HCl is only 7 times (98/14) stronger than acetic acid.

Total acidity or alkalinity is customarily expressed in terms of normality. It is also possible to use the normal system for expressing the concentration of hydrogen ions. In normal solution of HCl, about 78% of HCl molecules are dissociated. The term normal used so long referred to the total acidity, i.e., to all the ionisable hydrogen whether present as ions or in the form of HCl molecules. In terms of actual number of hydrogen ions present, however, such a solution is only 0·78N. We may thus speak of a normal solution of hydrogen ions as well as a normal solution of an acid.

A normal solution contains one gram molecular weight of a dissolved substance divided by the number of replaceable hydrogen atoms or hydrogen equivalent per litre of solution. A normal solution of HCl contains approximately 36 grams of HCl per litre whereas a normal solution of H_2SO_4 contains 49 grams of H_2SO_4 because H_2SO_4 has two replaceable hydrogen atoms. Both those solutions contain, however, the same amount of hydrogen ions or 1 gram per litre. Similarly a normal solution of an alkali must contain 17 grams of hydroxyl (OH^-) ions per litre equivalent to 1 gram of replaceable hydrogen ions of an acid.

Now if we have 0·1 gram of hydrogen ions present in a litre of a solution, that solution is N/10 with respect to the hydrogen-ion concentration, which is actually an extremely strong solution. Much more frequently, the biologists have to deal with such hydrogen-ion concentrations as N/1,000, N/10,000, N/100,000, and even weaker solutions up to N/100,000,000. This method of expression is obviously very awkward and cumbersome, and to say the least, requires too much space to write down. The difficulties can be obviated to some extent by writing it as a power of 10. Thus we may express an N/1,000 solution as 10^{-3}, N/100,000 as 10^{-5}, N/100,000,000 as 10^{-8} and so on.

The chemist Sorensen was the first to devise a method which is now universally accepted and followed all over the world to express the acidity or alkalinity of a medium with reference to hydrogen-ion concentration. According to this method, all but the exponent of the above is omitted, the sign is changed and the symbol pH is substituted for the term hydrogen-ion concentration. Thus N/100,000 or N 10^{-5} becomes pH=5·0.

Actually pH is the common logarithm of the reciprocal of the hydrogen-ion concentration, or in other words the negative logarithm of the concentration of hydrogen ions in terms of normality. Stated mathematically, the relation can be written thus $pH = \log 10 \frac{1}{(H^+)}$. For example, if the hydrogen-ion concentration of a solution is equivalent to, say, N/250, the reciprocal is, of course, 250 and the logarithm of this reciprocal is 2·4. Thus the pH of such a solution is 2·4.

The relation between pH and the hydrogen-ion concentration being a logarithmic one, each succeeding whole pH unit in the descending scale from 14 to 0 represents a tenfold increase in the hydrogen-ion concentration. Thus a solution of pH=5 is ten times as concentrated as regard the hydrogen ions as another at pH=6, pH=4 has ten times the concentration of hydrogen ions as has pH=5 or one hundred times the concentration at pH=6 and so on.

The hydroxyl-ion concentration of solutions is naturally the reciprocal of the hydrogen-ion concentration. For example, at pH=6, the pOH is =8; at pH=2, the pOH is =12, the pH values automatically also defining pOH values.

The basis of this reciprocal relationship can be made clear if we consider ionisation of pure water which according to the principle of mass action can be written as follows:

$$\frac{H^+ \times OH^-}{H.OH} = \text{equilibrium constant of water, } K_W = 10^{-14}$$

i.e., the product of hydrogen and hydroxyl ions divided by the concentration of undissociated water molecules is equal to a constant and which later has been very accurately determined and found to equal 10^{-14} gram molecules per litre. Since water is present *almost* entirely in the undissociated form and for purpose of calculation here we are concerned with 1 litre of water at 25°C the expression may be simplified for all practical purposes by omitting the denominator. Thus since the product of hydrogen and hydroxyl ions has a value of 10^{-14} at 25°C, in a neutral solution where the concentrations of H and OH are exactly equal, the concentration of $H=10^{-7}$ or a neutral solution has a pH=7 at 25°C. A pH value below 7 indicates that the hydrogen-ion concentration is greater than the hydroxyl-ion concentration; this is the acid range;

above pH=7, the reverse is true; it is the alkaline range. Since dissociation is affected by temperature, pH is also affected by temperature. At 30°C more of water dissociates, the actual amount of H^+ is $10^{-13.8}$ grams per litre of pure water; the pH of this water is thus, 6·9 and not 7·0.

Solutions having pH values of less than 0 or greater than 14 are theoretically possible but biologists are seldom concerned in the practical field with such values. For the plant physiologist, the most useful range seems to lie between pH=3 and pH=9.

$$\text{pH } 0 \xleftarrow{\text{increasing acidity}} \text{pH } 7 \xrightarrow{\text{increasing alkalinity}} \text{pH } 14$$

An accurate knowledge of hydrogen-ion relationship is essential for the proper and adequate interpretation of the data with which investigators are concerned in many of the plant physiological problems, such as growth of plants, cell metabolism, opening and closing of stomata, infection of plants by pathogenic organisms, enzyme action, absorption of ions, changes in the permeability of the plasma membrane, availability of inorganic nutrients in the soil, etc.

If 1 ml of N HCl be added to 10 ml of N NaCl it will be found that the pH value of the solution will drop abruptly to a value of about 1. If, however, 1 ml of N HCl be added to 10 ml of sodium acetate solution, the change in the pH value will be small. It is evident that sodium acetate solution in some way retards changes in pH value by tying up free hydrogen ions produced by the formation of compounds (e.g., acetic acid) which are only slightly dissociated. Solutions of such compounds, such as sodium acetate, which are relatively resistant to changes in pH due to addition or loss of hydrogen or hydroxyl ions are known as *buffer sloutions* and solutions, such as sodium chloride, are called unbuffered solutions. If, however, 1 ml of N NaOH is added to 10 ml of N sodium acetate solution, the pH value will increase markedly, i.e., although sodium acetate is buffered against acid, it is not buffered against alkali producing OH ions. If a solution of N acetic acid is, however, substituted for sodium acetate, it is found that the acetic acid is buffered against the alkali. A mixture of equal volumes of N sodium acetate and N acetic acid is thus buffered against both acid and base over a considerable range of pH value.

The pH value of an unknown solution is determined with a fair degree of accuracy by matching the colour developed in the solution to be tested with buffer solutions of known H-ion concentration containing the same indicator. Modern electrical methods are, however, much more widely used because the determinations made by these methods are quick for routine work and the limits of accuracy are in the order of 0·01 pH unit. The instrument, the potentiometer, is generally referred to in terms of the type of electrode used in the circuit—glass electrode, quinhydrone electrode, etc.

The pH is one of the essential factors in determining the distribution of plants. It is also true that different species of plants show different requirements with regard to optimum pH. Among the cereals oats and rye, as also turnips and potatoes yield economically the most profitable crops when grown in a medium whose pH is slightly acidic, within a range of pH 5–6 while wheat, barley and particularly rice, etc., thrive best in a medium which is neutral or only slightly alkaline, pH 7–7·5.

CHAPTER XXXI

THE COLLOIDAL SYSTEMS

The complex nature of the living substance in the cell, the protoplasm, is due to its physico-chemical properties. Protoplasm is predominantly composed of substances in the colloidal state and certainly it is to these colloidal systems that it owes much of its characteristic properties—variation in viscosity from values not greater than that of pure water to those of jellies, pronounced capacity for imbibition of water, though itself high in water content sometimes behaving as if immiscible with water, properties of irreversible coagulation at very high or low temperatures or at high concentration of salts, etc.

Almost all the physiological processes and chemical reactions which occur in the living cells for the maintenance of life take place under the influence of organic catalysts or enzymes which are active only when they are in colloidal condition, owing many of their properties to that fact. So it is desirable and very important that in order to properly comprehend the properties of protoplasm we should have a clear understanding of the fundamental principles regarding the colloidal systems.

The word colloid is derived from the Greek word *kolla* meaning glue and *eidos*, appearance, i.e., glue-like appearance.

If a little sucrose of NaCl is dissolved in water the solid disappears and the resulting system will be a solution. The solute particles are dispersed throughout the solvent (water) in the form of molecules or ions or both and such particles are so small that it is impossible to detect them in the system.

If instead of sucrose or NaCl, some fine river bottom silt is stirred in water, the result is a different sort of system. The silt particles are simply dispersed throughout water and not actually dissolved, for the particles of silt are of such a size that they can readily be identified under the microscope. These systems are called *suspensions* for the particles of silt gradually settle down and the system becomes separated into its original components, silt and water, in a short period of time; these systems are not stable as true solutions. Similar systems can also be prepared by vigorously shaking together two immiscible liquids such as oil and water. Such systems are known as *emulsions*. Like suspensions, they also are not stable as oil soon separates out from water. These emulsions can, however, be stabilised by the addition of a third component called emulsifier.

A third type of system can be prepared if a trace of sulphur be dissolved in a small volume of alcohol which then can be poured into a beaker containing a larger volume of water. A cloudy, opalescent liquid will result which is composed of sulphur particles dispersed in water, not in the form of molecules or ions as in true solutions, but as in suspensions and emulsions in the form of aggregates of atoms or molecules, invisible under a microscope showing that the molecular aggregates here are smaller in size than those in suspensions and emulsions. They can, however, be detected under an ultramicroscope unlike true solutions. These systems are relatively stable as the particles may remain dispersed throughout the liquid indefinitely. A simple example of stable colloidal system is given by the above sulphur in water system. Other good examples are milk and blood, which are predominantly colloidal suspensions of protein in water.

Colloidal systems are thus composed of two phases like true solutions, but unlike true solutions, the particles or molecules of the dispersed phase are either much larger as in certain dyes, proteins, etc., or as has been stated before in the form of molecular aggregates composed of hundreds or even thousands of molecules lumped together.

If the molecular aggregates are very large, the particles readily settle out of the system as in suspensions and emulsions and stability is not attained; this stability is the essential attribute of any colloidal system. In general, if the dispersed particles fall within a range 1–100 mμ in diameter, the system is colloidal; if smaller than this it is a true solution, if larger, unstable suspensions or emulsions.

The individual molecules of some proteins and dyes are so large as to bring them within the colloidal range of dimensions and molecular dispersion. Such substances form true solutions and colloidal systems at the same time.

Thus colloids consist of two or even more substances intimately mixed only sometimes molecularly, but never ionically, dispersed as in true solutions but forming physical systems containing a continuous or dispersion phase and a discontinuous or dispersed phase.

Another property of the dispersed particles in a colloidal system is that though the particles of the dispersed phase are invisible to the naked eye, the particles are able to diffract or scatter light giving the so-called *Tyndall effect*. If a strong beam of light is passed through a colloidal solution and then observed at right angles, the path of the light is clearly seen through the medium. The visible bright spots are due to scattering of light by the large dispersed particles or molecular aggregates although the size and form of the particles in the system remain invisible.

The particles of the dispersed phase in a colloidal system are incapable of passing through a parchment membrane as was first pointed out by Graham. Graham's original classification of substances into crystalloids (substances forming crystals and in solution readily diffusing through parchment membrane) and colloids, is no longer valid because we know now that by application of proper physical methods, theoretically at least, any substance can be brought into the colloidal state.

The particles of dispersed phase show *Brownian movement*, particularly in a suspension, which is an erratic, zigzag movement performed by those particles owing to their being bombarded by the moving molecules of the continuous or dispersion phase. This is partly responsible for keeping the particles of the dispersed phase from settling down and precipitating.

The particles of the dispersed phase are electrically charged with respect to their surroundings and generally all possess the same electric charge of the same sign. Colloids have been classified as positive colloids and negative colloids in accordance with the nature of electric charge the dispersed particles in a colloidal system carry. They behave differently, however, from electrolytes in that they migrate as a whole to one pole or another depending on the electric charge they carry and do not separate as oppositely charged ions as in electrolytic dissociation. This migration of the particles of the dispersed phase as a whole is known as *cataphoresis* or *electrophoresis*.

Since particles in the dispersed phase are similarly charged, they keep repelling each other as they move about in the dispersion phase. By varying the pH of a colloidal system the electrical charge on the particles of the dispersed phase can be neutralised. When this is attained, the particles no longer migrate towards the poles in an electrical field and the colloidal sol is said to be at its *isoelectric point*. At this isoelectric point the colloidal sols are naturally and most easily precipitated by the removal of the electric charges of the dispersed particles, as stability of a colloidal solution, as we know, depends upon these charges. Such neutralisation or removal can be effected in the laboratory by mixing colloidal sols having opposite charges—a positive colloid such as ferric hydroxide is added to a negative colloid, e.g., fine suspension of gold—in proper proportion. The charges soon neutralise each other and the particles of the dispersed phases of both systems are precipitated. This is known as *mutual precipitation*. As milk, which is a colloid is treated with a few drops of an acid, there is a precipitation of protein (casein) particles. The addition of any electrolyte—a dissociable salt solution will also cause identical precipitation. In this case the ion having a charge opposite to that of the particles of the dispersed phase is responsible for the precipitation of the colloidal solution.

Types of colloidal systems. Unlike true solutions, colloidal systems are, as we have learnt, not homogeneous, but consist of at least two distinct phases. There are eight possible types of colloidal system of which two are very uncommon, e.g., the gas-in-solid (some minerals, pumice stone) and gas-in-liquid (some foams, whipped cream). Gases always exist as separate molecules which are distant from one another either in the dispersion phase or in the dispersed; thus it is impossible to have a gas-in-gas colloid. The other six types are as follows: liquid-in-gas (clouds, fog, mists, etc.), liquid-in-liquid (all emulsions; the naturally occurring important ones are milk, butter, latex, etc.), liquid-in-solid (pearls and certain other minerals), solid-in-gas (smoke, fine dust clouds, certain fumes, etc.), solid-in-liquid (all suspensions) and solid-in-solid (ice cream, certain types of coloured glass, alloys, precious stones, e.g., black diamond).

Only two of several types of colloidal systems, enumerated above, are of special interest to the biologist. These are emulsions and suspensions and though the suspensions are not commonly found in the living cells, they are present in the environment especially in the soils in which plants grow.

Colloidal systems consisting of proteins, starch, etc., when mixed with water are classed as emulsions even though the individual materials when pure are solids. The solids, however, adsorb large amounts of water and as a result both phases have the properties of liquids and thus can only be regarded as emulsions. Emulsions are commonly met with in the living cells of plants and animals and are generally believed to be essential component of the protoplasm.

Emulsions fall naturally into two groups, generally known as oil-in-water and water-in-oil emulsions. Both types occur in the living cells but the oil-in-water type is more common. When observed under high-power microscope protoplasm often appears as an emulsion of fat or fat-like substances dispersed through a liquid phase. Some common examples of oil-in-water emulsions are milk, cream (droplets of fats dispersed in water), emulsions of olive oil-in-water, latex, etc. One familiar example of water-in-oil is butter.

Emulsions, we know, lack stability unless stabilised by the presence of another component, *emulsifier*. In absence of the emulsifier, the components readily separate out, the oil being lighter rising to the top. Emulsifiers are a group of heterogeneous substances. Some of the best-known emulsifiers are soaps, gelatine, fine suspensions of sulphur, carbon, silica, etc. Emulsions commonly occurring in nature are stabilised by proteins. Protoplasmic emulsions are also stabilised by proteins although metals or soaps may also act as emulsifier in the living cell.

A colloidal system exhibiting properties of a fluid is known as *sol*. Such systems can be poured more or less easily from one vessel to another. They may appear to be true solutions to the naked eye but critical examination reveals their colloidal nature. Many *sols* set forming solid or nearly so, *gels*. Gelatine, custard, ordinary household jellies are familiar example of *gels*. They are fluid at first but soon set hard.

Sols may also be classified as *lyophobic* (solvent-hating) and *lyophilic* (solvent-loving).

In the former, no attraction exists between the particles of the dispersed phase and the dispersion medium whereas in the latter such affinity exists appreciably. When the dispersion or the continuous medium is water, as in most biological colloidal systems, the corresponding terms are hydrophobic and hydrophilic. Lyophilic or hydrophilic sols are thus easy to prepare but difficult to precipitate whereas the converse is true of lyophobic or hydrophobic sols. Most colloidal systems composed of metallic substances dispersed in water are familiar examples of lyophobic sols whereas gelatine, agar, starch, protein sols are typically lyophilic.

ADSORPTION

Adsorption is the condensation in the form of a film of molecules of a gas or of a solute or suspended particles upon the surface of a solid or liquid.

Specially treated charcoal is one of the best adsorbents known for both gases and solutes. Gas masks which are used as a protection against poisonous gases in warfare owe their efficiency to the adsorptive capacity of charcoal present in the gas masks. Soluble and coloured impurities are often removed from liquids by treating them with charcoal which adsorb them. The forces involved in adsorption are partly forces of cohesion, i.e., attraction between similar particles and adhesion, attraction between dissimilar particles, surface tension and partly also the electric charges on the surface of the adsorbent and on the particles it is adsorbing.

The phenomena of adsorption play significant and manifold role in practically all the activities of the living cells. Protoplasm and many of the constituents of the living cells are essentially colloidal in nature and adsorption phenomena are strongly exhibited by all colloidal systems. Imbibition of water and the action of enzymes which are of tremendous importance in the water relations and metabolism of plant cells both involve to a large extent the adsorption of water. Differential adsorption of certain dyes and colouring matter by various constituents of cells, has given us an insight into the structure of living cells. We are all familiar with the characteristic blue colour developed by starch grains when treated with iodine solution. This specific test for starch is entirely due to adsorption of iodine in potassium iodide solution on the surface of the starch grains. This is known as chemical adsorption.

The chromosomes are so named because they strongly adsorb certain stains while rejecting others and the entire technique of staining of biological matter depends on the differential adsorption of different dyes by different constituents of a living tissue.

IMBIBITION

We have seen before that colloidal systems invariably take up a large amount of water from their surrounding medium and as a result swell up considerably. Many other substances such as proteins, gums, starch, cellulose, agar, gelatine, etc., show such hydrophilic properties when placed in contact with water or lose water and shrink when devoid of water. This taking up of water and consequent swelling of colloidal materials and other substances is commonly called imbibition. This imbibition of water by dispersed particles in a colloidal system is responsible for many of the characteristic properties shown by the functioning colloidal system and is also responsible for a number of physiological processes continually taking place on the surface of protoplasm which is, as we well know, mostly in a colloidal state.

If a few seeds with seed coats permeable to water, are soaked in water, these seeds will swell up visibly within a few hours. They will also swell similarly if placed in other solutions. The amount of water or solution which would enter the seeds by imbibition is often very large in proportion to the dry weight of the seeds. Some seeds can take up as much as 15 times its own weight of water by imbibition.

Water is not only taken up as liquid but also as a vapour. The swelling of doors, windows and other wood-works during rains is an everyday example of the operation of the imbibitional phenomena.

The imbibitional force due to the entrance of water and solutions from medium in dry seeds can be easily demonstrated by tightly packing a bottle with dry seeds and a little water, whereupon the bottle would burst after some time due to the swelling force or pressure of the imbibing seeds. It has been estimated that this pressure of imbibing seeds may be as high as 1000 atmospheres or even more.

Imbibition is usually considered to be basically a process of diffusion (as osmosis) but capillary phenomena and surface tension are probably also involved to a considerable extent. Imbibing substances are, as a rule, permeated with minute capillaries and it is certain that a portion of water or liquid enters the imbibant (material which is imbibing) by capillary forces through minute pores. Over and above that, there seems to be some specific forces of attraction between the molecules of the imbibant and the water (or the liquid) it may be imbibing. It is well known that whereas dry seeds imbibe water and swell up considerably when put into water or certain other solutions, such as sugar solution, the same seeds do not swell up at all when immersed in ether or certain other organic liquids. A piece of rubber, however, which certainly does not imbibe water, readily imbibes appreciable amounts of ether and certain other liquids. These observations certainly point to the occurrence of specific forces of attraction mentioned before.

It is probable that the force or pressure, causing the swelling of colloids and other imbibants is partly comparable to the force causing the increase in volume within an osmotic membrane in a living cell, the expansion in volume being brought about by the inward movement of water from the surrounding medium. Difference in the diffusion pressure between the liquid in the external medium and the liquid already present in the cell (imbibant) must be there, as in osmosis, if the imbibition phenomenon is to be operative. As long as the diffusion pressure or concentration of the liquid in the imbibant is less than the diffusion pressure or concentration of the liquid in the surrounding medium, a net movement of liquid inward into the imbibing cell will continue. A dynamic equilibrium will be established, as in osmosis, only when the diffusion pressures of liquid in the imbibant and in the external medium have attained the same value.

Now, how does imbibition then differ from osmosis? Well, for one thing, there is no necessity of the presence of a differentially permeable membrane separating the imbibant molecules from the surrounding external liquid in imbibition phenomena whereas osmosis, as we know, is essentially a process of diffusion of water through a membrane which is semipermeable. Furthermore, imbibition differs from osmosis in that when colloids and other imbibants take up liquids and swell up, large amounts of solutions and *not just water* as in an osmotic system, may be readily absorbed. A block of gelatine or agar may swell nearly as much in a solution of 0·5 M sucrose as in pure water.

When an imbibant is taking up liquid from external medium, it always results in the release of energy in the form of heat. This can be easily demonstrated if some dry seeds are immersed in water in a vessel containing a calorimeter and noting the change in temperature. The imbibing molecules of water or liquid lose a large part of their kinetic energy as they are adsorbed, and this released energy reappears in the system, which is measurable by the calorimeter.

EXPERIMENTS ON COLLOIDS AND COLLOIDAL SYSTEMS

I. Preparation of suspensoids or suspensions

(*a*) Precipitate a solution of $BaCl_2$ with H_2SO_4. Shake well the fine precipitate of $BaSO_4$ formed. Note the gradual settling of the fine ppt.

(*b*) Grind clean charcoal in a clean mortar to a very fine powder. Weigh about 0·2 g (three samples). Shake each sample with 10 ml each of the following : (*i*) distilled water, (*ii*) 1% NH_4OH and (*iii*) 0·5% acetic acid solution. Allow the suspensions formed to stand and observe the rate of settling of the fine particles of charcoal.

II. Preparation of emulsoids or emulsions

(*a*) Take a drop of olive or castor oil in a test-tube; half fill the tube with absolute alcohol. Shake well and pour in a beaker of water. A fine white emulsion of oil-in-water is formed from which particles of oil (dispersed very finely) will not separate—the oil is obtained in such small drops that a degree of stability is attained.

(*b*) Take about equal quantities of olive oil in two test-tubes and add an equal quantity of water to each To one tube, add a drop or two of 10% NaOH. Shake well. An emulsion is formed in both but in the tube without the alkali, the oil separates out from water on standing. In the other, the emulsion is stabilised due to the fact that the fatty acid components of oil form soap with alkali, the soap then acting as an emulsifier, renders it permanent.

III. Preparation of suspensoid sols

Take 5 ml of 1% solution of $AgNO_3$; add dilute NH_4OH until the ppt. formed disappears. Dilute with 100 ml of water. Mix equal volume of this solution with tannic acid (0·5%). A colloidal solution of silver is formed.

IV. Preparation of emulsoid sols

Take 2 g of dry starch, mix well with cold distilled water. Pour the paste formed in 100 ml of boiling water in a beaker and boil further for a few minutes, stirring well. A colloidal solution of starch-in-water is obtained which remains unchanged when cooled.

V. Preparation of gels

Add 1 g of agar to 50 ml of water and boil for half an hour. The agar gives a thick opalescent solution (*sol*) which sets to a *gel* on cooling. This change is reversible. Take a little warm solution in two test-tubes and show that the agar-mucilage is precipitated by (*a*) alcohol and (*b*) saturation with solid $(NH_4)_2SO_4$.

VI. Precipitation or 'salting out' of suspensoid sols with electrolytes

The particles of metallic Ag (expt. III) carry negative electric charge; hence they are readily precipitated by di- or tri-valent cations, e.g., barium, aluminium, etc. Add a few drops of $BaCl_2$ to the $AgNO_3$ solution—the dispersed silver particles are precipitated. Very little electrolyte is necessary for this 'salting out' of suspensoid sols as these salts are very sensitive to traces of electrolytes.

VII. Precipitation of an emulsoid with electrolytes

Emulsions are less sensitive to electrolytes than suspensoid sols—large quantities of electrolytes are, therefore, necessary for 'salting out' of an emulsion.

(*a*) Take 2·5 g of wheat flour in 25 ml of distilled water; stir well; allow it to stand for some time and then filter. The extract contains a colloidal emulsion of protein-in-water. Shake again and note that it froths. Saturate the system with solid $(NH_4)_2SO_4$ (as in expt. V) and note that the protein is precipitated. The ppt. now readily dissolves in water.

(*b*) Perform similar salting out of a starch emulsion (expt. IV).

VIII. Demonstration of Brownian movement of dispersed particles in an emulsion

(*a*) A drop of milky latex from a fresh sample of *Euphorbia* is spread out on a slide. Cover with a cover slip and make the borders of the cover slip air tight with vaseline. Observe under microscope the erratic, zigzag movements of the particles in the colloidal state.

(*b*) 1 ml of olive oil is thoroughly shaken in a test-tube (about a minute) with 1 ml of 1% KOH solution and the Brownian movement is observed quickly under the microscope.

CHAPTER XXXII

DIFFUSION

The movement of various substances into a plant, usually from the soil, out of which the green plant synthesises the numerous complex organic compounds, is accomplished, principally through the agency of the process known generally as diffusion. In some cases, however, the operation of the diffusion phenomena is complicated by other factors. We know that in higher green plants, some substances enter the living cells through the aerial organs—the diffusion of CO_2 and O_2 from the atmosphere into the plants is principally through stomata. From soil, water and ions of simple inorganic salts pass into the plants through the root cells by a process which is basically diffusion, though greatly modified by other factors. Similarly, the loss of large quantities of water as vapour from leaves and other aerial organs into the atmosphere is also accomplished by diffusion. O_2 evolved in photosynthesis and the respiratory CO_2 also diffuse out of the plant into the atmosphere. The exhaled CO_2 in respiration of the root cells diffuses out into the soil.

The movement of substance within a plant, i.e., from one part of a plant to another, is mainly by diffusion. The gases move through the intercellular spaces and the water and solutes through the dead and living vessels and also from cell to cell by diffusion, each following its own independent diffusion pressure or concentration gradient.

There are, very few, if any, physiological processes occurring in plants which do not directly or indirectly involve the diffusion phenomena. What is diffusion then? According to the kinetic theory, the molecules or ions of all substances (gases, liquids, solutes or solids) are in constant motion in all directions, which is due to their individual kinetic activity. This is called diffusion. In a mixture of several substances separated by a membrane, the molecules or ions of each substance will diffuse in that direction which is from a region of its own higher concentration or pressure to its own lower concentration or pressure or in other words from a region of its own higher activity to one of its own lesser activity of the particular substance concerned. And the direction of this diffusion is independent of the direction of diffusions of ions or molecules of other substances present. From plants, the outward diffusing water vapour molecules in transpiration or oxygen evolved in photosynthesis do not carry respiratory CO_2 with them, but each gas diffuses independently of the other, depending only on the difference of its own concentration or pressure, inside and outside the plant.

Diffusion of gases

It is well known that when a gas is allowed entry in a room, the molecules of the gas, owing to their kinetic energy, become evenly distributed throughout the room. If a bottle of chloroform or ether is opened indoors, the distinctive odour can be perceived in all parts of the room within a short time. This is diffusion of gases and the rate of diffusion is dependent upon the concentration of the particular gas, the temperature, the density and also to some extent on the presence of other gases.

Diffusion of dissolved substances (solutes)

If a blue crystal of $CuSO_4$ is dropped in a beaker containing water, the diffusion of the dissociated ions of Cu^{++} and $SO_4^{=}$ or entire molecules of $CuSO_4$ can be observed by the slow change in the colour of water. This rate of diffusion of solute molecules or ions through the solvent is extremely slow and may principally be due to the fact that the densely packed particles of water (solvent) considerably impede the diffusion of particles of the solute $CuSO_4$. The direction of diffusion of a particular solute ion or molecule follows the same general pattern, i.e., from higher concentration or pressure to its lower concentration or pressure, regardless of diffusion, of other solute particles, if any, in the same system. The rates of diffusion are also generally controlled by principles which are essentially similar to those governing diffusion of gases.

DIFFUSION THROUGH A MEMBRANE

Diffusion takes place also through membranes. Three types of membranes are possible, *permeable*, *impermeable* and *semipermeable*. Permeable membranes are those which allow diffusion of both solvent and solute molecules or ions through them. Ordinary cellulose cell walls of young non-vacuolated cells are examples of completely permeable membranes. Lignified cell walls are also quite freely permeable to water and when wet, to solute ions and molecules as well. Impermeable membranes are those which prohibit the entry of both solvent and solute particles. Suberised or heavily cutinised cell walls in plants are practically impermeable as regards the passage of water and dissolved substances. Semipermeable or differentially permeable membranes are those which allow the diffusion only of the *solvent particles* through them while restricting the entrance of the solute particles, i.e., the membrane is permeable to one substance but impermeable to the other. In this connection, the only solvent we need to consider is water, for water is the only important liquid in living organisms which moves from cell to cell and hence all our further discussions of this phenomenon will naturally be in terms of water and aqueous solutions.

If a U tube (Fig. 663A) is divided by a permeable membrane, filled on one arm with pure water and the other with a solution of sucrose, the molecules of sucrose will diffuse through the permeable membrane towards water because the diffusion pressure

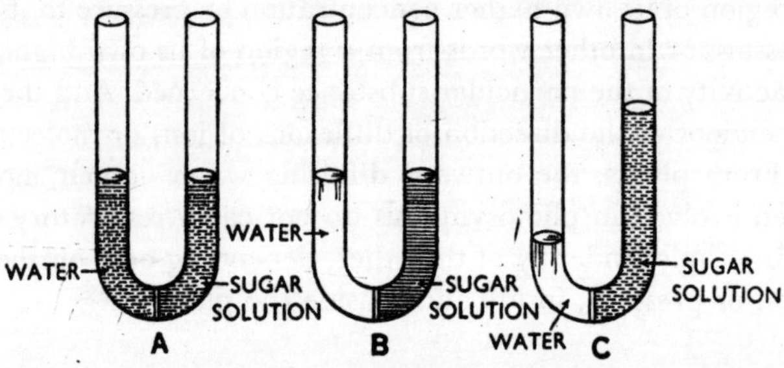

FIG. 663. Diffusion of solvent and solutes through different types of membrane. A. Permeable membrane. B. Impermeable membrane. C. Semipermeable membrane.

and concentration of sucrose molecules are higher on the right-hand than on the left-hand side and the water molecules will move in the opposite direction for its diffusion pressure or concentration is higher in the left portion than in the right. Thus there will be a thorough mixing of both sucrose and water molecules in the two arms of the tube. An equilibrium is eventually established between the two arms of the tube when equal quantities of sucrose and water molecules are present on the two sides of the partition. If the membrane is more rapidly penetrated by one of the two substances, that substance will at first pass more quickly than the other but ultimately equalisation of the concentration and the pressure of the solutions in the two arms must result.

The case will be very different, however, when the partition, separating the two liquids is semipermeable, i.e., permeable only to the molecules of water while molecules of sucrose are quite unable to pass through it (Fig. 663C). The result will be that owing to greater diffusion pressure (or concentration) of water in the left arm of the U tube, water molecules will pass through the membrane towards the right arm. Volume for volume, the number of molecules of water is certainly higher in the left arm than in the right arm since in the solution of sucrose, a large proportion of molecules of water is replaced by sucrose molecules. Thus there will be a greater accumulation of water molecules on the side of the membrane containing sucrose solution and the level of solution in the right arm will consequently increase considerably. This diffusion of water molecules through a semipermeable or differentially permeable membrane from the region of higher concentration and higher diffusion pressure of water to a region where it is lesser is known as *osmosis* and the pressure developed as a result of this is known as *osmotic pressure*. It must be pointed out here, however, that semipermeable membranes which completely restrict the movement of solute particles or will completely restrict the movement of water molecules from the right arm to the left arm of the U tube, are only theoretically possible. In practice, however, most semipermeable membranes are really only differentially permeable and allow passage to some sucrose molecules or even to some solvent (water) molecules from the right to the left of the tube. In U tube B, where the membrane is impermeable, there is theoretically no net diffusion of either the solvent or the solute molecules.

Obscurities have sometimes resulted from the application of the term osmosis to the movement of solutes through the membranes. The use of the term in this sense has not only led to much misunderstanding but the concepts of many students have remained vague as regards the process of absorption of water and absorption of solutes which two terms, they are apt to consider synonymous. The movement of solutes through a membrane, though certainly primarily a process of diffusion, may even be simple diffusion, but complicated enormously by a number of physico-chemical factors which do not play any part in the process of diffusion we know of.

According to the principles of diffusion, the molecules or ions of all substances in solution, gases or even dispersed particles in a colloidal system, tend to attain an equal distribution in terms of concentration through all parts of a system. The osmotic movement of water is in response to the same general principles; equal distribution is, however, seldom attained in living organisms.

Semipermeable membranes and the process of osmosis play the primary role in the absorption of water by vegetable cells. In studying the phenomena and the laws governing them, consideration of natural vegetable cells may be deferred for a while, and the study of the operation of the osmotic phenomena and the effects produced in an artificial cell with an artificially prepared synthetic membrane containing an osmotically active substance, might give us a clearer understanding of the process that actually takes place in a living cell.

An artificial osmotic chamber with a true semipermeable membrane can be prepared easily by taking a clay cell of galvanic elements coated on the inside with copper ferrocyanide. Copper ferrocyanide is prepared by mixing $CuSO_4$ with potassium ferrocyanide. Another type of artificial semipermeable membrane can be obtained by mixing a solution of gelatine with tannic acid

The naturally occurring semipermeable membranes include bladders of fishes and animals, the thin white membrane in eggs when the hard outer calcium coatings are removed by dissolving in acids and the most important of all, the *plasma membranes* of living vacuolated mature cells of the vegetable tissue. The plasma membrane is the semipermeable membrane in plant cells.

It is the presence of this differentially permeable membrane which gives to osmosis its distinctive aspect as compared with other diffusion processes.

How can you distinguish between a living and a non-living cell? Sometimes, the streaming movements of the protoplasm shows us that a cell is living. Another method is by staining a cell with a dye such as neutral red or methylene blue. These dyes only stain the cell contents in the vacuole and may accumulate there. Non-living cells, or living cells when killed never show this type of staining. A third characteristic of the living cells is that their cell contents can be shrunk or plasmolysed away from the cell wall when placed in a solution whose concentration is greater than that of the cell contents.

The cell contents thus can be separated from its surrounding cell wall. Obviously then, even the living plant cell consists of a non-living part (cell wall) and a living part (protoplasm). The non-living cell wall, however, has pronounced physiological significance since it markedly affects the protoplasm, particularly its water content and growth.

The most important part of a cell is this living protoplasm for here are synthesised and broken down all the organic chemical substances from which life itself is regenerated. Energy is absorbed in synthesis and released in the breakdown.

We are concerned here with the activities of living cells which certainly depend upon the properties of the living protoplasm. The nature of the protoplasm can, however, be only properly understood if the chemistry and the physics of the protoplasm are known to us. A chemical analysis of protoplasm obtained from some lower plants such as slime molds, shows that as percentage of the fresh weight, water content varies from 80–90%, proteins 7–10%, lipid substances 1–2%, other organic and inorganic materials, just over 1%. The analysis is gross, and it certainly does not mean that new protoplasm could be formed just by mixing the above substances in the proportion indicated above. It only gives us some insight into the essential physical and chemical properties of protoplasm.

Among the many kinds of proteins in the protoplasm are the nucleoproteins. The nucleoproteins are perhaps the most fundamental substance of life as indicated by the fact that the smallest living entity (?) virus, consists almost entirely of nucleoproteins which are the bearers of heredity in all organisms—plants or animals.

The phospholipids (e.g., lecithin) are the main fatty substances of the protoplasm being mostly confined to the structural components of protoplasm, e.g., plastids, mitochondria, etc. They are also associated with the cell membrane.

All these structural components are suspended in the ground substance of the protoplasm which is made up predominantly of proteinaceous sol or gel. The ground substance is usually considerably viscous than the cell sap of the vacuoles and since viscosity and gel formations as we know are characteristics of a colloidal system, the protoplasm is obviously a colloid. The non-living part of the cell wall also shows colloidal properties. The vacuolar cell sap also possess enough colloidal materials, e.g., tannins which hold dyes such as methylene blue. Optically, the cell sap, however, appears as homogeneous. The cell walls are made of colloidal gels of cellulose and other organic substances, e.g., pectin.

All the components of a living cell, the cell wall, vacuole (both non-living) and protoplasm contain large amounts of water which pass readily from the living to the non-living parts and *vice versa*. The planes separating the three components of a living cell—the protoplasm, cell wall and the vacuole are known as *interfaces*. At each interface, special, physical forces such as surface tension, operate freely. The substances near the interfaces tend to be concentrated sometimes producing the phenomenon of adsorption.

Water movement across the cell membrane has usually been expressed in terms of diffusion pressures and their differences and osmotic pressure of solutions. However, it is more reasonable to explain this phenomenon in thermodynamic terms, since movement takes place only when energy is available in some form.

All molecules at any temperature above absolute zero have a free energy with which it has a tendency to move about in the surrounding space. As the number of molecules per unit space increases, the total free energy also increases; the free energy of Avogadro's number ($6{\cdot}03 \times 10^{23}$) of a chemical species present in 1 gram mole is termed the *chemical potential* of that species—whether in the molecular or in the ionic form. In solutions both solute and solvent particles have chemical potentials. When we talk of chemical potential of water, we refer to it simply as *water potential*, usually designated by the Greek letter ψ (psi). When two solutions are separated by a permeable membrane both solute and solvent molecules diffuse across the membrane until the chemical potentials of both on two sides of the membrane are equal. Movement is always from a higher to a lower potential. There is diffusion even after equilibrium is attained, but the rate of diffusion in both directions is equal. When the membrane is semi-permeable, movement of only solute or solvent particles is permitted by the membrane. In the case of plant membranes only the solvent water molecules are permitted to pass.

Water molecules in pure water have the maximum chemical potential; however, when solute molecules are dissolved in it, this potential is decreased, the greater the number of such solute molecules, that is greater the concentration, less is the water potential. As in the case of the Celsius (centigrade) scale of temperature measurement, where the freezing point of water is arbitrarily fixed to be zero, water potential of pure water is arbitrarily taken to be zero. Consequently the water potential of any solution containing solute particles is negative. The water potential of a 0·1 M solution is higher than a 0·2 M solution, because it is more dilute and its water potential is less negative. Water potential is usually expressed in terms of pressure, i.e., bars on atmospheres. Increase in temperature or pressure increases water potential. Water potential is also affected by substances in the environment which absorbs water.

Movement of water across membranes practically impermeable to solute particles is called *osmosis*. Osmotic pressure of a solution is a function of its *solute* concentration at a particular temperature according to the equation

$$\pi \text{ (pi)} = CRT,$$

where π is the osmotic pressure, C the *molar* concentration (since pressure is the result of the total activity of the number of molecules), R is the Gas constant (0·082 litre atmospheres/mole degree or 0·0357 litre cal. per degree centigrade) and T is the absolute temperature. It is usually expressed in litre atmospheres. Although osmotic pressure is a positive quantity, *osmotic potential* is always negative, since it is the value by which water potential is lowered when the solute is dissolved in pure water.

Two major components of water potential of a solution are pressure potential and osmotic potential. At a particular temperature water potential is the sum of pressure potential (ψ_p) and osmotic potential (ψ_π). Pressure potential is the effect of atmospheric pressure or any other pressure applied. Thus, pressure potential is the differnece of water potential and osmotic potential. It may be positive or negative accordingly as pressure is applied or the cells are in a state of tension. Plant cells also contain many hydrophilic, colloidal gel-like and porous surfaces, e.g., the cell wall, which absorb and retain water with some tenacity; the water potential of this matrix is referred to as *matrix potential*, ψ_m. If the colloidal matrix and the cell solution are two distinct phases in equilibrium with each other, then osmotic potential of the cell solution may well be equal and either value added to the pressure potential should constitute the water potential of the system. However, several workers consider them as separate entities and the total effect is believed to be additive according to the equation

$$\psi(\text{psi}) = \psi_p + \psi_\pi + \psi_m.$$

Water movement in plant cells is along a water potential gradient. Although it is primarily a diffusion process, the rates observed are much higher than what may be provided by a pure diffusion process. There is evidence that there is a bulk movement of water. Presumably, the gradient in the transit region of the membrane surface is quite steep, so that the rate is considerably accelerated and water molecules are pulled in a bulk, somewhat as in the case of a siphon.

In most earlier text books water potential was considered as *diffusion pressure* of water and the difference in the water potentials of the external solution and cell solution as the *diffusion pressure deficit* or the *suction pressure*. We shall now explain the process of osmosis particularly with reference to an osmometer using some of this terminology to familiarise the student with these concepts.

If an osmometer, as the artificial cell mentioned before is called (or simply osmotic chamber), containing a strong solution of sucrose is immersed in a vessel containing pure water and a mercury manometer is attached to it as shown in Fig. 668, the rise of column of mercury in the manometer tube due to inward diffusion of water will denote the measure of the pressure of the solution inside the cell. The artificial membrane is here rigid and as this membrane is permeable only to water, the inwardly directed diffusion of water from outside into the cell will cause a pressure to develop inside the membrane. The maximum pressure thus developed is quantitatively equal to the pressure on the solution to prevent any increase in its volume due to entrance of water. The inflow of water from outside (endosmosis) due to higher water potential there, causes a hydrostatic pressure to develop inside the cell. This hydrostatic pressure is directed against the cell wall and this actual pressure developed within the cell as a result of osmosis, is known as *turgour pressure*. Actual pressure or turgour pressure developed during osmosis is seldom equal to osmotic pressure. If a solution, having an osmotic pressure of 10 atm. be immersed in a solution, having an osmotic pressure of 6 atm., water will diffuse inwards until at equilibrium, i.e., when the water potentials of the external and internal solutions are equal, the actual pressure developed, can at its maximum be only 4 atm. Even if the external liquid be pure water, the actual pressure or the turgour pressure developed in the internal solution would still not be quite equal to its original osmotic pressure unless the membrane is completely inelastic. Thus in an osmometer, the maximum turgour pressure developed can only equal the osmotic pressure but can never exceed it.

When a solution is confined within a semipermeable membrane and the membrane is immersed in water, the passage of water through the membrane results in the development of turgour pressure on the solution side of the membrane. But the maximum potential osmotic pressure which can develop in the solution must be equal to the excess of diffusion pressure of pure solvent over the diffusion pressure of water in the solution, but only when the solution is not under any pressure except, of course, atmospheric pressure. In plant cells, other pressures must be taken into account in order to determine the *suction pressure* or *diffusion pressure deficit* (DPD) of the cell sap. *Suction pressure* is water absorbing capacity of the solution in the membrane. For example, if a solution has an osmotic pressure of 10 atmospheres, it certainly means that the diffusion pressure of water in the solution is less than the diffusion pressure of pure water by 10 atm. or in other words, the suction pressure of such a solution is 10 atm. Hence if a cell with cell sap solution of 10 atm. (suction pressure of also 10 atm.) is immersed in a solution of 4 atm. (suction pressure also 4 atm.) and is separated by a membrane permeable only to water, water will diffuse towards the region of its lesser diffusion pressure or greater diffusion pressure deficit (suction pressure), i.e., towards the solution whose osmotic pressure and suction pressure are both 10 atm. If we disregard dilution for the present and the volume changes, the solution inside the membrane will exert a turgour pressure of 6 atm. and this will be the actual hydrostatic pressure developed within the cell. As each action has an equal and opposite reaction, imposition of a pressure on the wall of the membrane will result in an outwardly directed *wall pressure* of 6 atm. The wall pressure evidently is always equal in magnitude to turgour pressure but acts in the opposite direction.

The wall pressure will necessarily increase the diffusion pressure of the enclosed solution of 10 atm. by 6 atm. Since the initial diffusion pressure deficit of the enclosed solution was also 10 atm., the development of a wall pressure of 6 atm., raises the diffusion pressure of water in the internal solution by 6 atm. while at the same time *reduces* the diffusion pressure deficit of water in the internal solution to 4 atm. (10−6=4) which is the suction pressure or diffusion pressure deficit in the external solution.

Osmotic pressure is proportional to the *number* of particles of solutes per given volume of water under standard conditions of temperature and atmospheric pressure and irrespective of kind, weight or size of the particles. If one membrane contains a solution with twice as many particles—molecules or ions—as that in the second, the first, evidently will have twice as much osmotic pressure as the second. A solution, such as found inside the vacuoles of a mature plant cell containing a mixture of many solute molecules (or ions) will have the same osmotic pressure as a solution with a single salt if the ratio of the total number of dissolved molecules and ions to the number of water molecules is the same in each case. In order to determine this proportional number of particles it is necessary to know the proportional weights of the particles. Since molecular weights give the proportional weights of the molecules, it is customary to use molecular solutions. *Molar solutions* are solutions in which as many grams of solutes as its molecular weight are added to water to a final volume of 1,000 ml. Sucrose with an empirical formula of $C_{12}H_{22}O_{11}$ has a molecular weight of 342 g; the molecular weight of glucose is 180 g, etc. These amounts dissolved in 1,000 ml of water will give molar solutions (M) of sucrose and glucose respectively. A 0·1 M solution of glucose will develop the same osmotic pressure as 0·1 M solution of sucrose as these two solutions will contain the same number of particles of molecules. Equimolecular solutions of non-electrolytes are thus *iso-osmotic* or *isotonic*. A colloidal system should also possess an osmotic pressure, for osmotic pressure is determined by the proportion of the solute particles to solvent molecules and is independent of the kind and the size of particles. The particles of the dispersed phase of a colloidal system theoretically then will have the same effect on the magnitude of the osmotic pressure as smaller molecules or ions. Large molecules or molecular aggregates theoretically have the same effect as smaller molecules or ions. Thus osmotic pressure, in a colloidal system, will depend upon the concentration of the particles of the dispersed phase. As the total concentrations of the particles of the dispersed phase are relatively much smaller compared to particles in a true solution, all colloidal systems have very low osmotic pressure values, seldom exceeding a fraction of an atmosphere.

Sugars and complex organic substances are, as we know, non-electrolytes whereas the inorganic salts are electrically active, i.e., certain percentage of inorganic salts are dissociated when the salts are in solution. This increases the number of particles in a particular volume of water. At a concentration 0·1 M NaCl, 80% of NaCl molecules are dissociated and therefore for every 100 NaCl molecules, 20 molecules will remain undissociated whereas 80 molecules are ionised into equal number of Na^+ and Cl^- ions. There will, therefore, be 180 particles either as molecules or ions, instead of 100 particles and the osmotic pressure would be 1·8 times that of a substance of the same molar concentration with only undissociated molecules. A 0·1 M solution of NaCl, therefore, will have an osmotic pressure approximately equivalent to $0·1 \times 1·8$ or 0·18 M solution of cane sugar or glucose or any other non-electrolyte. The number which expresses by how much the osmotic concentration of an electrolyte is greater than an equimolecular solution of a non-electrolyte, is termed *isotonic coefficient* for that particular concentration; for 0·1 M NaCl, as we have seen, it is 1·8.

The conditions affecting osmotic pressure are very similar to those which affect gas pressure and within certain limits are the same. General laws of vapour pressure also hold good for osmosis. According to Avogadro's hypothesis: equal volumes of all gases under identical condition of temperature and pressure, contain same number of molecules or particles. One molar weight of any gas at standard conditions (0°C. and 1 atm.) occupies 22·4 litres. If this gas be compressed to a volume of 1 litre, it will certainly exert a pressure of 22·4 atm. at 0°C. (From Boyle's law) The osmotic pressure of a solution varies directly with concentration of the solution and inversely with the volume. This theoretical value of 22·4 atm. sometimes slightly differs from actually measured pressure due to various reasons—24·83 atm. instead of 22·4. The theoretical value 22·4 atm. can be corrected for any desired temperature and pressure. (At 25°C., the osmotic pressure of 1 M solution is aproximately 27 atm.)

We shall again consider the case of an artificial rigid osmotic chamber with a membrane, permeable only to water (semipermeable membrane) and a long vertical glass tube of very small diameter attached at the open end as in Fig. 664. Theoretically such a membrane, either inorganic or organic, should allow only the solvent molecules to pass through it, but as a matter of fact all artificial inorganic membranes—the same is true for organic as well—allow certain solute molecules to pass through them when certain conditions are fulfilled. The osmotic chamber containing a 0·45 M solution of sucrose is immersed in pure water as in Fig. 664(A). Net movement of water inwards will take place by osmosis resulting in raising the column of water in the tube. At equilibrium, the suction pressure of water in the chamber will be zero, because of the imposition of a hydrostatic pressure (turgour pressure) of 10 atm. on the internal solution. When equilibrium will be reached between osmotic chamber and the surrounding water, the level of solution in the vertical tube will remain stationary. What will be the maximum height of solution attained in the vertical tube at that condition? At this stage, the osmotic pressure of the solution in the chamber is equal to the actual pressure developed due to incoming water, i.e., turgour, which is also equal to 10 atm. at its maximum. 1 atmospheric pressure is equivalent to 76 cm of mercury and the specific gravity of mercury is 13·6. Thus the maximum height attained by the solution in the vertical tube attached to the osmotic chamber will be $76 \times 10 \times 13{\cdot}6$ cm or approximately 103 m. (We disregard here the dilution of the solution in the chamber and the changes in the volume.)

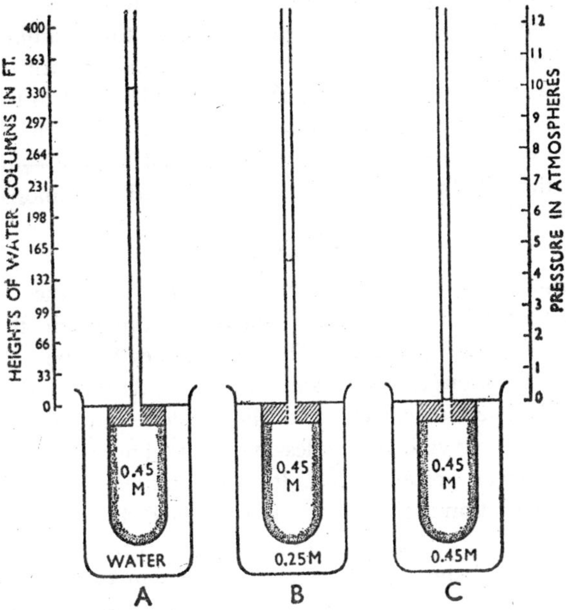

FIG. 664. Artificial osmotic chambers with attached vertical tubes in water and in solutions of 0·25 M and 0·15 M sucrose.

If the osmotic chamber is now taken out of water and then immersed in a solution of 0·25 M concentration as in Fig. 664(B), the level of column of water in the tube will certainly drop until the actual pressure developed on the membrane and the wall of the chamber is determined by the difference in concentration of the solutions inside and outside the chamber, i.e., approximately 4·48 atm. ($22{\cdot}4 \times 0{\cdot}2 = 4{\cdot}48$ atm.). The actual pressure developed (turgour) within the chamber can thus only attain a value of 4·48 atm. The suction pressure of the water in the solution inside the chamber (which was zero in pure water) will increase for it will now be subjected to a pressure of only 4·48 atm. Imposition of a turgour pressure of only 4·48 atm. now will make the suction pressure equal to $10 - 4{\cdot}48$, i.e., 5·52 atm., equivalent to 0·25 M, which is exactly the concentration of the external solution. Thus the water level in the tube will be stationary at a height of $4{\cdot}48 \times 76 \times 13{\cdot}6 = 46$ m (this height is due to the actual hydrostatic pressure or turgour developed within the chamber, keeping the water column at that height).

If the chamber with its attached tube is again removed and transferred to an external solution of 0·15 M concentration, water will again start entering the chamber by endosmosis and the turgour pressure will further rise and at its maximum will now be equal to 0·30 M (or 6·72 atm.). The level of water in the tube will show a further rise until it reaches $6{\cdot}72 \times 76 \times 13{\cdot}6 = 69$ m. Imposition of a turgour pressure of 6·72 atm. will reduce the suction pressure of the solution inside to 3·28 atm. which is exactly the suction pressure of water in the external solution (0·15 M=3·28 atm.).

If the chamber is now finally transferred to an external solution, having the same concentration as that in the chamber (0·45 M), water will come out again by exosmosis until the levels of water column in the tube and in the external solution are exactly the same. The turgour pressure will be progressively reduced to zero, and all the hydrostatic pressure developed on the walls of the chamber is thus released due to water exit. And as the suction pressure of water in the solution inside the chamber is progressively released from the imposed turgour pressure, the suction pressure value progressively increases until it attains a maximum of 10 atm. (0·45 M) equal to the suction pressure of the external solution. There is no net movement of water into the chamber when the chamber is thus brought into equilibrium with external solution—movement of water molecules certainly there is, but the rates of exosmosis and endosmosis are equal.

DIALYSIS

Dialysis is essentially a sort of filtration in which the pore size of the membrane permits the diffusion of smaller particles from inside to outside and *vice versa*, but larger molecules are retained. Thus coenzymes can be separated from apo-enzymes by dialysis. If the enzyme solution is taken inside a dialysis bag (collodion or other similar membranous structures) and the contents equilibrated with a buffer solution or water; the coenzyme molecules and metal ions which are small, pass out of the long but the larger protein molecules are retained within the bag. Several changes of the outside solution are required, so that the diffusion of the smaller particles is complete or almost complete. If the outside solution is not changed, after sometime equilibrium is attained and the small particles are present both inside and outside the dialysis bag or tubing. Kidney of man and animals is an efficient dialyser.

CHAPTER XXXIII

OSMOTIC RELATIONS IN PLANTS CELLS

Most living plant cells can act like artificial osmotic chambers previously described and as such osmosis or diffusion of water results when a weak solute solution or pure water is separated from a relatively stronger solution by a semipermeable membrane—a membrane showing characteristic differential permeability allowing water molecules to pass through it readily while restricting to a great extent, the diffusion of solute ions or molecules. In plant cells, semipermeable membrane, responsible for the osmotic phenomena, consists, as we know, of the non-living cell wall and the thin layer of protoplasm lining the walls of all living mature vacuolated plant cells. This membrane is, no doubt then, living with its characteristic features in contrast to inorganic artificial membranes. Naturally occurring semipermeable membranes also include bladders of many fishes and animals, the thin white membrane in eggs, when the hard white coating is removed by treating with mineral acids.

PLASMOLYSIS

If a plant cell is immersed in a solution having a higher osmotic concentration (*hypertonic*) than that of the cell sap, exosmosis will result; water will come out of the cell and the turgour pressure or the hydrostatic pressure upon the distended cell wall will be gradually released. If such a cell is now observed under a microscope, characteristic shrinkage of the volume of the cell (can actually be measured accurately) could be detected. The lining layer of protoplasm will gradually be seen to separate from the

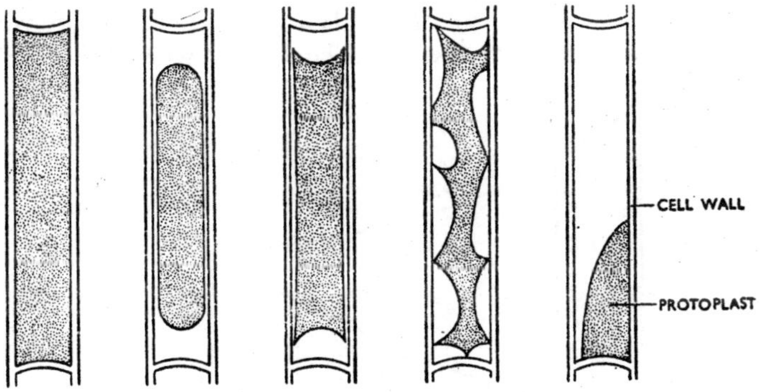

FIG. 665. Plasmolysis patterns followed by shrinking protoplasm in plant cells.

cell wall and if the external solution is strong enough, this separation soon becomes much pronounced. In many plant cells, the protoplasm may shrink completely away from the cell wall and collect as a spherical mass in the centre or in one corner of the cell (Fig. 665). This phenomenon is called *plasmolysis.* When this contraction or withdrawal of protoplasm away from the cell wall has just commenced, i.e., when the turgour or the hydrostatic pressure on the cell wall has just been released (turgour is zero), the cell is then said to be at *incipient plasmolysis* or *threshold plasmolysis.*

The pattern followed and set up by shrinking protoplasm seems to be typical for each kind of cell (Fig. 665) although it can be modified to certain extent by the physiological condition of cell at that time. The space between the cell wall and the contracting protoplasm will be filled up by the external solution as the cellulose cell wall without the adhering protoplasmic layer, is now permeable to both the solvent and solute alike and not differentially permeable.

In such a plasmolysed plant cell the release of turgour from the cell wall increases evidently the suction pressure of the cell, and the suction pressure being now freed from any other imposed pressure upon it, becomes equal to the osmotic pressure of the cell sap. Evidently, such a cell will potentially have the maximum water absorbing capacity for both the osmotic and suction pressures are now equal in magnitude. If such a plasmolysed cell is transferred from hypertonic solution and to pure water, it will slowly recover by the net inward movement of water (if *deplasmolysis*, as it is sometimes called or inward movement of water, is too rapid, it may result in the death of the cell due to sudden stretching of protoplasm) subjecting the cell wall to a progressively imposed turgour pressure and in consequence, reducing the suction pressure of the cell. When the turgour pressure becomes equal to the osmotic pressure, the suction pressure of such a cell is zero and the cell has taken up the maximum amount of water from the surrounding medium. Such a cell is known as *turgid* and the phenomenon associated with it, *turgidity*. Plasmolysis can be conveniently demonstrated in plant cells under microscope containing coloured (blue or red anthocyanins) cell sap since the contracting protoplasm can now be readily distinguished in the cell matrix.

OSMOTIC QUANTITIES OF PLANT CELLS

The osmotic pressure, turgour pressure (hydrostatic pressure actually developed on the cell wall) and the suction pressure (diffusion pressure deficit, DPD) are collectively known as *osmotic quantities of a plant cell.* It is the turgour pressure which imparts to plant cells, their usual rigid and distended condition, when the supply of water is

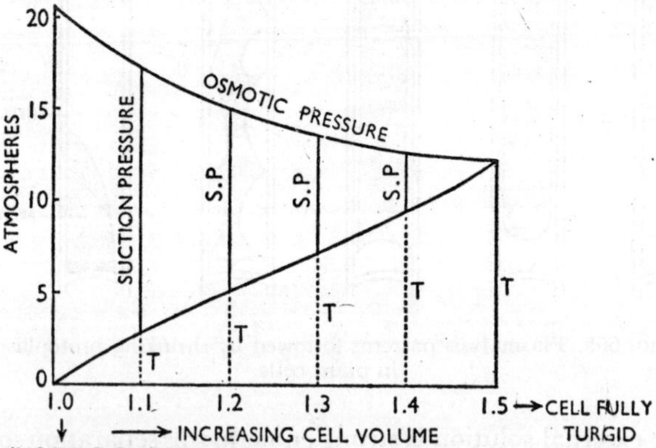

FIG. 666. Interrelationship of osmotic quantities of plant cells. S. P. Suction pressure. T. Turgour pressure.

abundant. This distended condition of the cell is variously termed *turgour*, *turgidity* or *turgescence*. Cells, low or entirely lacking in turgour, are sometimes referred to as *flaccid*.

The interrelationship of osmotic pressure, turgour pressure and suction pressure of a plant cell could be further clarified by a study of Fig. 666 in which the importance of changes in the volume of the cell has also been taken into consideration; we have disregarded the influence of changes in the volume of the cell on the osmotic quantities in our all previous discussions. Changes in the volume of the cell certainly bring about changes in the osmotic pressure values due to dilution of the cell sap. An increase in the volume of the cell decreases the osmotic pressure while progressively increasing the turgour pressure on the cell wall due to influx of water.

In all our foregoing discussions, it has been assumed that the membranes are inelastic. In living organisms, however, the plasma membranes of cells are elastic within limits. In appreciably elastic membranes, a complication actually arises—for changes in volumes of the cells take place with corresponding increase or decrease in the osmotic pressure values of the internal solution due to entrance or exit of water.

Let us assume that a plasmolysed plant cell with elastic membrane (osmotic pressure 10 atm.) is capable of permitting a fifty per cent increase in the volume of the cell before dynamic equilibrium is attained with the surrounding water medium. As the cell increases in volume, the internal solution becomes more dilute, i.e., less concentrated. We know that osmotic pressure of a solution decreases with diminution of its concentration. Now if we assume a proportionate decrease in osmotic pressure with decrease in concentration, the final osmotic pressure of the cell sap would be 6·66 atm. as shown below:

$$10 : X = 150 : 100$$
$$\text{or } 150X = 10 \times 100$$
$$\therefore X = \frac{10 \times 100}{150}$$
$$= 6{\cdot}66 \text{ atm.}$$

If the external solution instead of being water is a solution with an osmotic pressure of 5 atm. a hydrostatic pressure (turgour pressure) of 1·66 atm. would be sufficient to bring the cell in equilibrium with the external medium. If the cell wall were completely inelastic, the turgour pressure volume would certainly have been (10—5) or 5 atm. at equilibrium.

When a plant cell is completely limp and flaccid (i.e., at incipient plasmolysis), the relative cell volume evidently is the smallest and in Fig. 666 taken as 1·0, under such conditions the cell wall is subjected to no turgour pressure while the suction pressure of the cell is, therefore, equal to the osmotic pressure of the solution in the cell sap, and is certainly at its maximum. As this cell is immersed in pure water, the volume of the cell increases, progressively with the progressive increase in turgour pressure; the osmotic pressure also falls due to dilution of the cell sap associated with progressive increase in the volume of the cell. The suction pressure imposed upon by a progressively increasing turgour pressure, shows a corresponding rapid progressive decrease. When the cell ultimately attains a condition of full turgidity (it has taken up as much water as it could) and the cell volume is at its maximum, 1·5, the turgour pressure, the actual pressure developed within the cell becomes equal to osmotic pressure and the suction pressure, balanced by the imposed turgour pressure falls to zero.

The following fundamental relation among the three osmotic quantities of a plant cell (evident also from our previous discussion with artificial osmotic membranes) can be expressed in simple equation: S=suction pressure; P=osmotic pressure; T=turgour pressure;

$$S = P - T$$
$$\text{or } T = P - S$$
$$\text{or } P = T + S$$

When the osmotic pressure is determined for a plant cell at its incipient plasmolysis, i.e., when turgour pressure is zero and is compared with the values obtained by purely physical methods, e.g., *cryoscopic*, it is frequently markedly different instead of being identical. Sometimes positive values, as high as 7 atm., have been observed for some plant tissues. According to the classical view which we have discussed so far, there should have been no difference at all. Generally the osmotic pressure values found by plasmolytic method exceeded those found by purely physical methods.

This excess pressure suggests that the intake of water by plant cells may result from two entirely different types of force, possibly acting independently—one osmotic and, therefore, passive while the other non-osmotic or active.

The non-osmotic uptake of water by plant cells under certain conditions has been examined critically by determining the effects of electrolytes and non-electrolytes on the volume of the cell. When epidermal cells of onion are plasmolysed in a hypertonic solution of potassium chloride and then transferred immediately into an isotonic (the same strength as the cell sap) solution of sucrose, increase in the volume of the cell is observed. This observation is certainly against the classical view of osmotic phenomena—according to the classical view, the cell volume should have remained the same. A probable explanation offered is that in sucrose solution, sucrose molecules being non-electrolytes, both the passive osmotic and the non-osmotic active forces were operating freely. The active forces were drawing the extra water into the cell, thereby giving higher values for osmotic pressure determinations by plasmolytic method. In a solution of an electrolyte such as potassium chloride, the non-osmotic active absorption forces are in some way eliminated and the water is drawn into the cell only osmotically. The non-osmotic active forces can be of considerable magnitude and are sometimes referred to as *secretary forces* of the protoplasm and the absorption of water and solutes thus come under the general metabolic control of the tissues.

DETERMINATION OF OSMOTIC QUANTITIES OF PLANT CELLS

OSMOTIC PRESSURE

The osmotic pressure of a solution can be determined directly by placing the solution in an osmotic chamber with a semipermeable membrane and immersing the chamber in pure water. The pressure developed can be easily measured by a mercury manometer attached to the open end of the chamber (Fig. 668). It is, however, an impracticable method for determination of osmotic pressure of living plant tissues with more or less elastic cell walls.

The plasmolytic method and the purely physical cryoscopic method are in general use for determination of the osmotic pressure of plant cells. Both the methods are subject to many sources of error and neither of them is perfect.

Plasmolytic method. For determining the osmotic pressure of a given cell or tissue, the cell or tissue is placed in a graded series of solutions of known osmotic concentrations—preferably in volume molar concentration of mannitol (KNO_3 has been used previously as a plasmolytic agent but its use has now been discarded owing to the fact of excessive penetration of K^+ and NO_3^- ions into the cells which may bring about considerable errors).

Let us assume that the concentration of our known solutions fall within a range of 10 dilutions from 0·5 to 0·4 M of mannitol and pieces of vegetable tissues of more or less same size and thickness are immersed in these solutions and kept until an osmotic equilibrium is attained. After this, strips or sections of tissues are observed under microscope. Time required for the equilibrium between the tissue and the external solutions generally varies from 20 min. to 2 hours. The tissue or the individual cell may show strong plasmolysis, say, at 0·5 M mannitol indicating that the concentration of the external solution is higher than the cell sap. By routine testing with the series of solutions, it is possible to find one solution, slightly stronger than the cell sap (indicated by slight plasmolysis), say, at a concentration 0·46 M and another solution, a little weaker, which just fails to plasmolyse the cell, say, at 0·44 M. Average of the two concentrations gives us a value which is approximately equivalent to the concentration of the cell sap, i.e., at 0·45 M at incipient plasmolysis. At this stage, there is no pressure of the cell contents against the cell wall, i.e., turgour pressure is zero. Since molar solutions of undissociated compounds like mannitol theoretically have an osmotic pressure of 22·4 atm. at 0°C., the osmotic pressure of the cell sap would be $0{\cdot}45 \times 22{\cdot}4$ or 10 atm. approximately. Corrections could easily be made for the desired temperature at which the experimental readings are taken.

Advantage of the plasmolytic method of determination of the osmotic pressure of plant cells is that the measurements of single cells are possible by this method.

Instead of working with single cell, it is better to work with an integrated tissue or groups of cells. In the series of graded solutions mentioned before, there will be a particular concentration at which about half the cells are more or less completely plasmolysed and the other half not so. The average osmotic pressure of the cells comprising the tissue may be more or less correctly taken as equal to the osmotic pressure of the external solution.

Cryoscopic method. It is well known that there is a direct proportionality between osmotic pressure and the lowering of vapour pressure, the elevation of boiling point and the depression of freezing point of solutions. Thus osmotic pressure can be calculated from the observed results of any one of these physical quantities; most frequently determinations are made of the depression of freezing point of a solution and the osmotic pressure is calculated from this. The presence of a solute in a solution lowers the freezing point of the solvent water and the amount of lowering is directly proportional to the number of particles in the solution, i.e., the osmotic pressure. A 1 M solution of a non-electrolyte freezes at a temperature of −1·86°C. and its theoretical osmotic pressure, as we know, is 22·4 atm. If an unknown solution was found to freeze, say, at −0·46°C, the solution would have a molar concentration of 0·25 M which is equivalent to 5·6 atm.

Thus an equation relating to freezing point depression and osmotic pressure, reasonably applicable over a wide range of concentrations, can easily be derived. If we represent by A, the freezing point depression of an unknown solution (whose osmotic pressure is to be determined), its osmotic pressure, π may be calculated as follows:

$$\pi : 22{\cdot}4 = A : 1{\cdot}86$$

$$\text{or } \pi \times 1{\cdot}86 = 22{\cdot}4 \times A$$

$$\therefore \pi = \frac{22{\cdot}4 \times A}{1{\cdot}86}$$

If we know the freezing point of an unknown solution, the osmotic pressure of such a solution is easily found. If A is −0·23°C., the osmotic pressure is $\frac{22{\cdot}4 \times 0{\cdot}23}{1{\cdot}86} = 2{\cdot}8$ atm.

For actual determination of the depression of freezing point of the call sap, the cell sap at first must be extracted by killing the cells by freezing so as to destroy the protoplasmic membrane which might otherwise interfere with extraction process itself. The vessel containing the experimental tissue is surrounded with a mixture of ice and salt and the frozen tissue is then allowed to thaw or melt. The sap can now be easily extracted with a hand press. It is not even necessary to separate the sap from the other insoluble residues for the freezing point of the macerated material may be determined with little error for subsequent computation of osmotic pressure values.

This cryoscopic method of determination of osmotic pressures has been extensively used in long routine analysis in the laboratory or in the field. For all practical purposes when large number of determinations are to be made, the cryoscopic method is immensely superior to the plasmolytic method.

The relation of direct proportionality between elevation of boiling point and osmotic pressure of a solution has also sometimes been taken use of, although very rarely, for osmotic pressure determinations in routine analysis. The boling point of water is raised by 0·52°C. by a solution of a non-electrolyte compared to pure water. For example, if a certain solution was found to raise the boiling point by 0·13°C., the solution would have an osmotic concentration equivalent to 0·25 M (5·6 atm.) (*Ebulliscopic method*).

SUCTION PRESSURE (DIFFUSION PRESSURE DEFICIT, DPD)

Most methods of measuring the suction pressure of plant cells are based on the same principles, previously discussed, i.e., if a plant cell with inelastic walls or a tissue is immersed in a solution with an osmotic pressure equal to the suction force of the cell, an equilibrium is readily attained and *there is no change in the volume of the cell or tissue.*

For single cells or groups of cells, they are immersed in series of solutions of known molar concentrations as in the determination of osmotic pressure. When the equilibrium is attained, say, in about 20 minutes, the cells are measured under the microscope. The osmotic pressure of the external solution in which the cells show *no change in dimension*, compared to the initial value, is taken as equal to the suction pressure of the cell or tissue. A better method is to use thin, narrow strips of tissue cut from such structures as thin leaves or petals. Very good results are obtained with cylinders of tissue of nearly equal sizes cut from potato tuber, beet root and other tuberous organs. The equilibrium point can be determined by measuring changes either in weight or length or volume of the strips of tissues or cylinders of tissues. The solution in which the tissue neither gains nor loses weight (or in length or volume) is considered to have an osmotic pressure equal to the average suction force of the cells.

Determination of suction pressure by method described before should not be confused with plasmolytic determination of osmotic pressure of plant cells. In the determination of osmotic pressure, the critical measurement evidently is the osmotic pressure of the external solution *with which cells or tissues come to equilibrium at incipient plasmolysis, and when the cell volume is at its relative minimum* due to elimination of turgour brought about by exosmosis of water.

In suction pressure determination, it must be borne in mind, that the critical measurement is the osmotic pressure of the external solution with which cells come to equilibrium *without any change in their original volume* of the cells, i.e., without any change in their turgour pressure. The suction pressure values can be identical with osmotic pressure values only if the cells are initially at incipient plasmolysis.

TURGOUR PRESSURE

The turgour pressure or actual hydrostatic pressure against the cell wall due to presence of vacuolar cell sap evidently cannot be determined by direct methods. However, if osmotic pressure of a cell or tissue as well as its suction pressure are known, the turgour pressure can be calculated from the well-known equation, $S=P-T$ or $T=P-S$.

IMPORTANCE OF THE OSMOTIC PHENOMENA IN THE LIFE OF A PLANT

The absorption of water by root hairs from the soil solution and the movement of this water from cell to cell within the plant are in general controlled by osmosis. This movement of water from the non-living xylem vessels into living cells is also an osmotic phenomenon.

A plant is dependent on osmosis to give its cells and tissues rigidity, which is certainly due to the hydrostatic pressure of water or turgour on the cell walls. All young stems and tissues before they become woody due to development of lignified elements, depend largely on the turgour of cells resulting from osmosis to keep them erect, standing and extended.

As far as it is known the forces concerned in the movements of the guard cells in the opening and closing of stomata and the force used by the growing tissues are osmotic in nature. The force exerted by a young root in its passage through the soil is probably chiefly an osmotic force. The forces concerned in explosive bursting of many fruits and sporangia are largely osmotic in nature. The autonomous movements of the lateral leaflets of the Telegraph plant, the induced seismonastic movements of the leaflets of *Mimosa* are no doubt also controlled in part by the osmotic phenomenon.

Turgidity of the cells is essential for the growth of the plant. Normal functions of protoplasm are impeded when the cells lose their turgidity.

In the cell-enlargement phase of growth, cells produced by an apical meristem enlarge principally in a direction parallel to the axis of the stem due to the turgour pressure produced inside the cells. The cell wall being elastic is stretched by the pressure of incoming water molecules and while in this stretched condition, material substances of the wall—the cellulose and pectin substances—are deposited as additional molecules on the existing cell walls. The stretched condition of the cell is thus rendered permanent and irreversible, even when the pressure is removed from the cell wall by withdrawal of water from the vacuoles.

An increase in osmotic concentration of the cell sap most probably increases the resistance of the tissues to freezing temperatures, and also to drying due to excessive loss of water.

EXPERIMENTS ON OSMOTIC PHENOMENA AND IMBIBITION

A. IMBIBITION

I. Weigh out three lots of 10 or more seeds (maize, gram, etc.) and cover them with water and with 7·5% and 30% NaCl and also 5% sucrose solutions for 24–72 hours. Determine the percentage of imbibition of liquid by different seeds by weighing.

II. Place about 2 g of dry starch in a small beaker. Add water gently and stir carefully with a thermometer. Note that there is a rise in temperature due to imbibition of water by starch—imbibition means performance of work and release of energy.

III. Tightly pack gram or any other dry seed in a thin glass bottle of a test-tube. Add a little water. Seal with paraffin and leave it overnight. The bottle is seen to burst due to the imbibitional pressure of the seeds on the glass wall.

IV. A simple apparatus to demonstrate imbibitional phenomenon in seeds is shown here (Fig. 667). Due to pressure exerted by the swollen seeds on the flat disc which just covers them up, the pointer attached at right angles to the vertical rod connecting the disc, moves up and down along a graduated curved scale. At the beginning of the experiment, the pointer may be fixed at zero point on the scale, and any movement of the pointer may be read off from the graduated scale, indicating a measure of imbibitional force of the swollen seeds.

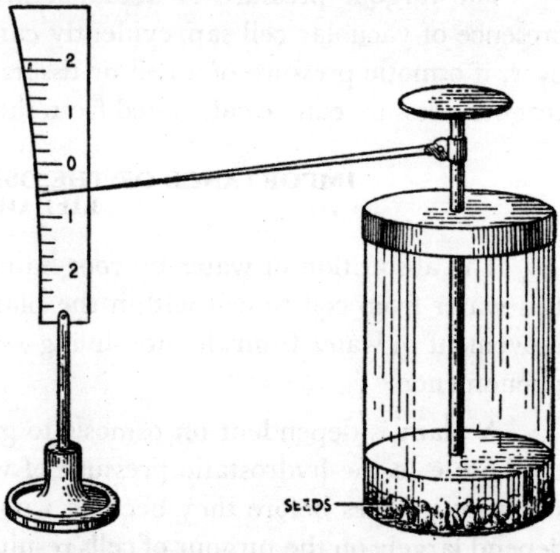

FIG. 667. An apparatus to measure imbibitional pressure in seeds.

B. OSMOSIS AND OSMOTIC PHENOMENA

I. Osmosis as a physical process—Experiment with artificial membrane

An artificial osmotic chamber whose walls are made of either collodion (solution of cellulose nitrate in a mixture of ether and alcohol) or parchment containing a strong solution of sucrose is immersed in a vessel containing pure water and a mercury manometer is attached to it as shown in Fig. 668. All connections are made air-tight and the

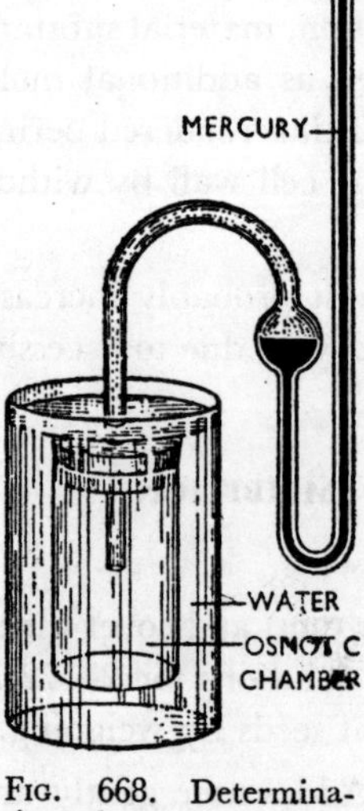

FIG. 668. Determination of osmotic pressure in an artificial osmotic chamber.

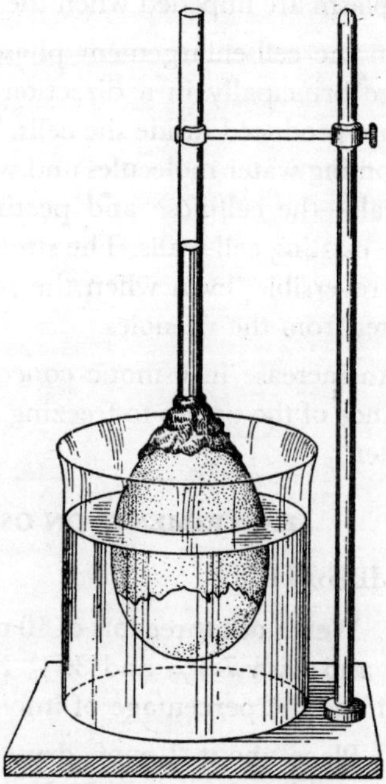
FIG. 669. Egg osmoscope.

level of mercury in the vertical manometer tube is noted. After a few hours, the rise of level of mercury in the manometer tube due to inward diffusion of water into the chamber indicates the magnitude of osmotic pressure of the sucrose solution contained in the chamber.

II. Experiment with semipermeable membrane of an egg—Egg osmoscope

A small hole is made in one end of an egg and the inner contents are taken out. The semipermeable membrane is then obtained by dissolving about a third of white calcareous shell of the egg very carefully with concentrated HCl and bringing out the soft inner membrane. Proper precautions and care must be taken so that the inner membrane (which is semipermeable) is not in any way injured. Wash the egg and the exposed membrane thoroughly in water and then attach and seal with sealing was a narrow tube through the hole. The tube should go some distance inside the now empty egg but must not touch the membrane. Now, through the tube, the egg with its membranous sac is filled with strong sucrose solution and also up to a certain height in the glass tube which is noted. The whole thing is then put into a beaker of water and clamped to a vertical stand (Fig. 669). After some time, the level of liquid in the glass tube is seen to rise due to endosmosis of water through the semipermeable egg membrane.

III. Osmosis in living plant cells—Potato tuber osmoscope

A large potato tuber is taken and its skin is removed. Slice off portions from all sides to make it a flat rectangle. Now the central flesh is scooped out carefully with a knife or a scalpel to obtain a clean deep hollow cavity, taking care not to pierce completely through the sides and bottom, as shown in Fig. 670. The cup-like cavity is now half-filled with strong sucrose solution and the level of the solution inside is marked with a pin. The cup is now put into a Petri-dish containing water, whose level reaches nearly to the rim of the osmoscope. Within a short time, the level of solution in the cavity is seen to rise due to increase in volume. This is certainly due to diffusion of water from the Petridish towards the solution inside the cup by cell-to-cell osmosis—through multicellular, semipermeable membranes.

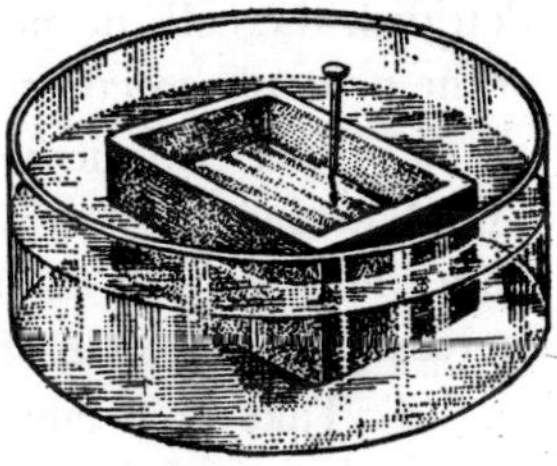

FIG. 670. Potato tuber osmoscope.

IV. Plasmolysis

(*a*) Peel thin strips of tissue from the lower, coloured epidermis of *Rheo* leaf or any other leaf of flower petals with coloured cell sap. Examine under the microscope and note that the epidermal cells are filled with pink-coloured cell sap. The thin strips are now covered with a cover glass and are irrigated with strong sugar solution, about 0·5–1 M. Note the effect after a few minutes on the coloured cell sap and observe the plasmolysis patterns. The patterns developed by shrinking protoplasm appear to be specific for each kind of cell (Fig. 665).

(*b*) Treat some *Spirogyra* filaments with 50–70% of methyl alcohol. Observe the bursting of the filament cell walls and the gradual emergence of the protoplast together with the chloroplastids.

V. Demonstration of the living nature of the protoplasmic membrane

Semipermeable nature of the living protoplasmic membrane is destroyed by (*a*) *high temperature*, (*b*) *high concentration of solutes* and by (*c*) *toxic substances*, e.g., $CuSO_4$, $HgCl_2$, alcohol, etc.

(*a*) Cut a number of pieces of beet root about 2 cm/1 cm/0·25 cm of about the same size. Wash them thoroughly in tap water to remove all traces of red colour from the injured cut surfaces (injury destroys the semipermeable characteristics of the plasma membrane).

(*i*) Take several test-tubes containing water at temperatures of 25°, 30°, 35°, 40°, 45°, 50°, 55°, 60°, 70°C. and put a slice of beet root in each test-tube. Stir the slices thoroughly in their tubes and note any excretion of colour from the slices into the water of the test-tubes. After about 15 min. note the temperature at which the colour comes into the water. This is the lethal temperature for protoplasm (*thermal death point*)—indicating that the semipermeable nature of the cell membrane is destroyed at that temperature.

(*ii*) Mount a few filaments of *Spirogyra* in water on a slide and heat one end of the slide carefully over a Bunsen flame. Examine under microscope and note any change in the appearance of the cell. Heating kills the living plasma membrane and with the loss of water from the cell, there is drastic plasmolysis. Dip the filaments again in water and examine under microscope. There is no deplasmolysis and the cells do not revive, showing irreversible changes in the protoplasm brought about by high temperature, again proving the living nature of the membrane.

(*b*) Keep some *Spirogyra* filaments in a very strong solution of sucrose, say about 2 M for about 15 minutes. Examine under microscope and observe pronounced plasmolysis of the cells. If these plasmolysed cells are now taken out and put in pure water in Petridish the cells do not revive (there is no deplasmolysis) showing irreversible injury to the protoplasm constituting the plasma membrane.

(*c*) Take two slices of beet root and dip them in two test-tubes containing (*i*) 0·01 N $CuSO_4$ and (*ii*) 2 M ethyl alcohol. After about an hour, note the excretion of red colour from the slices into the test-tubes. The permeability of the protoplasmic membrane is increased to a point that is injurious to it, leading to its ultimate destruction.

(*d*) Take epidermal peelings from leaves of any suitable plant. Put the peelings into a solution of dilute methylene blue (or neutral red) and observe under the microscope the diffusion of the dye into the cell sap, staining it blue or red. Living cells, when killed or non-living cells never show this type of staining.

VI. Determination of osmotic pressure of integrated plant tissues—by Plasmolytic method

Soak epidermal peelings from the coloured undersurface of leaves of *Rheo* or any other tissue with coloured cell sap in Petridishes containing solutions of sucrose of different concentrations, ranging from 0·05 M to about 0·30 M. Make about six to seven gradations of solutions within the desired range and after about an hour examine the strips under microscope. Count the *number* of cells plasmolysed in the field (degree of plasmolysis need not be taken into account) at different concentrations used and plot the percentage of cells plasmolysed in different concentrations against sugar concentration on a graph paper. Estimate the concentration of sugar corresponding to the 50% of the cells plasmolysed. This concentration of sugar can be approximately taken as, equivalent to the osmotic concentration of the cell sap. The corresponding osmotic pressure can be easily obtained (1 M=22·4 atm. at N.T.P.). Desired corrections could be made for the experimental temperatures.

If osmotic pressure determination is desired for individual cells, it might be done simply by observing the point of incipient plasmolysis and the corresponding concentration of external solution. In practice, however, it is very difficult to know exactly the point at which the shrinkage of protoplasm has just started at the corners. It may be better if a concentration is found at which there is just plasmolysis and one at which there is none. As for example, if a cell shows plasmolysis say at a conc. of 0·26 M and no plasmolysis whatsoever at 0·24 M, the concentration at incipient plasmolysis then must lie between 0·26 and 0·24 M, i.e., 0·25 M.

VII. Determination of mean suction pressure (diffusion pressure deficit, DPD) of cells

Divide a beet root or potato transversely into slices 2·5 to 3 cm thick. With a cork borer cut vertical borings from both outer and central zones (2–3 cm in diameter) and divide into 5 similar samples of about 5 g each. Weigh the samples quickly and immerse one sample in each of several gradations of sucrose solutions from 0·1 to 0·5 M in labelled specimen tubes. Leave for about an hour, blot, dry and reweigh.

Plot the percentage change in weight, either increase or decrease against sugar concentration and find out the concentration of sucrose corresponding to zero change. Calculate the suction pressure of the tissue from molar concentration-osmotic pressure standard table and make temperature corrections.

It may be mentioned here that if cylinders of tissues of either beet root or potato are worked with the same determinations could be made if instead of weight, percentage increase or decrease in length or volume are taken into consideration. The critical point of determination is the concentration of the external solution with which the tissue comes into equilibrium without any change in the volume of the cell.

VIII. Determination of temperature coefficient of water absorption by plant tissues

Take three similar samples of about 5 g each of discs (about 2·5 cm to 3 cm thick and 2–3 cm in diameter) of potato tuber, cut out from both outer and central zones. After correctly noting their fresh weights, put them in three beakers of water (with a thermometer in each) maintained at different temperatures (*a*) at room temperature, (*b*) at 10°C. or so above room temperature and (*c*) at approximately 10°C. below room temperature. The temperatures are maintained at the required levels by adding hot or cold water as needed. After about an hour, take the weights of the disc-samples at different temperature after carefully blotting them to remove all adhering water. The difference between the first and second weighing is the measure of the amount of water absorbed by the tissue samples. Calculate the results as percentages of original weight and determine the temperature coefficient (Q_{10}) of water absorption.

CHAPTER XXXIV

THE ABSORPTION OF WATER BY ROOTS

Land plants obtain their water requirements from the soil in which the roots grow. Now what is actually meant by the term soil and how is the soil formed?

THE FORMATION OF SOIL

The upper portion of the earth's crust in which the roots of plants grow is known as soil and is formed by the weathering of various parent rocks. The soil is disintegrated to smaller and smaller particles by weathering agencies such as climate, by the activity of micro-organisms such as bacteria, by earthworm and also actually by the plant roots themselves. Rain falling on the earth's surface takes the more soluble components from the upper layers and carries them below where they may again be redeposited. In the final form, the soil contains in addition to a fragment of the parent rock, much larger quantities of entirely new minerals. A soil is thus a mosaic of rock particles, minerals, dead and living roots and micro-organisms and the spaces between these various particles are either filled by air or water. Of the numerous components of soils, the two major ones essential for plant growth are the soil solutions and the oxygen carrying air.

A soil is generally composed of particles of wide range of sizes—from the macroscopic, easily seen with naked eye to microscopic and even smaller, too minute to see even under a microscope. A soil may be characterised in part by the proportion of particles in each size category which it contains and for convenience individual names have been applied to particles of particular size. The very large particles are generally referred to as *gravel*, somewhat finer as *sand* and still smaller particles as *silt* and the minutest particles as *clay*. The distinction of particles among this wide range of sizes has been taken as the basis for the naming of the soil types. Thus a soil in which the greater proportion of the particles are of very large size are referred to as gravel whereas one in which the minutest-sized particles predominate is known as clay.

To be capable of supporting good growth of plants a soil must contain a readily available supply of all the nutrient elements essential for plant growth in addition to large amounts of easily available water. The predominant minerals of the soil, originally formed from rocks are generally oxides or other derivatives of silicon, which are scarcely utilised in plant nutrition. The elements calcium, magnesium and potassium are also in the soil as components of the parts of rocks in small proportions but in those forms they are not available to the plants; they are only made available for plant growth by being adsorbed on the surface of clay particles from which they may pass into solution. The element phosphorus is present in the soil in a variety of chemical forms, the majority of which are insoluble and hence are unable to provide the plants with the much sought for phosphate ions essentially required for plant nutrition. So the water-soluble phosphate ions, essential for normal plant growth must be continually formed from less available, somewhat insoluble forms and the phosphate economy of soil is based on a balance between the ability to supply the plant and ability of the insoluble forms which resist washing out by rain. Owing to this delicate balance, phosphorus deficiency is of widespread occurrence under agricultural conditions. The addition of phosphate fertiliser is thus a recognised agricultural practice.

Nitrogen, unlike the nutrients discussed above, is not contained in the primeval rocks from which soils are derived and all the nitrogen content of the soil can, in the last analysis, be traced to the molecular form in the atmosphere which is biologically fixed by micro-organisms in forms generally available to the higher plants.

The red or yellow colours of soil are due to insoluble iron salts, particularly to the more complex iron silicates.

Other nutrients, both the major and the trace elements may be present in the soil either in the soluble form or are precipitated in unavailable forms.

The constant removal of nutrients by plant from the soil by the roots and translocation and elaboration of these nutrients to form various organic plant products constitute a continued drain on the mineral reserve of the soil. The fact that mineral elements are removed from the soil by crop plants was first formulated more than hundred years ago by Liebig in 1840 who also suggested that the soil fertility can be re-established and maintained by the artificial addition of manure in the form of inorganic mineral nutrients.

In a vegetation where no crop is removed from the field, several essential elements are continuously cycled, that is they are taken up by the plants, utilised and elaborated inside the plant into various products and with the death of the plants are again returned to the soil ready for further absorption by other plants. But in our modern agricultural practice, the crop once ready and removed for consumption, is not returned to the soil and thus if the soil has a small reserve of one nutritive element, it constantly requires replenishment. The nutrients which are in most demand as replenishments in the soil, are nitrogen, phosphorus and potassium. The crop plants require them most for normal growth and development and as a result these are also taken up from the soil in large quantities and a deficiency of nitrogen, phosphorus and potassium frequently occurs in the soil. Thus most of the modern artificial fertilisers which are added to the soil to maintain the fertility and ensure rich harvests from year to year contain the three elements in different proportions of N, P and K.

As a plant completes its life span and dies, all its tissues normally return to the soil either directly as plant remains or indirectly in the form of dead animal remains and excreta. Thus the soil is constantly kept supplied with organic material and all soils generally contain organic matter in addition to purely inorganic minerals. These plant and animal remains do not long retain their original state but are rapidly attacked and disintegrated by the soil microflora consisting of many forms of bacteria and fungi. A portion of the organic matter is utilised by the bacteria and fungi themselves as substrate for their metabolic activities particularly respiration but the bulk of the organic matter is liberated into the atmosphere as CO_2, the end product of microfloral respiration and this may again be available for photosynthesis of green plants.

In the process of decomposition of plant and animal remains, the simple sugar and amino acids are naturally preferentially attacked whereas more resistant constituents such as lignified woody parts are attacked much more slowly. Thus as the decomposition of the organic material in the soil progresses, the lignified woody parts of the plant remains become more and more abundant in the soil. The brownish or black colours of the soil which are rich in organic matter or humus consist mostly of lignified, woody plant remains and the relatively slowly decomposed organic materials.

The mineral and water resources of the soil are taken by the plants through a ramified system of roots and root hairs which bring the plant into intimate contact with the soil particles surrounding it. The search for water of the soil by roots is achieved not only by repeated branching of the root as it goes downwards but also by production of unicellular outgrowths from the epidermal cells of roots—the *root hairs*—which develop in enormous number. These hairs prodigiously increase the area over which the root is in actual contact with the soil and thus provide the predominant portion of the actual absorbing area of the root system. The root hairs are certainly the main point of entrance for water into the plant but the solutes and dissolved minerals are supposed to enter by the actual growing regions of roots themselves. No root hairs are found on the older portion of roots. The diameters of the root hairs seldom exceed 10μ and in length generally less than 1 mm. Largest root hairs are found in some aquatics such as *Trianea* (8 mm), *Elodea* (4 mm) and among the land plants *Tradescantia* (3 mm), *Avena* (2·5 mm) and *Pisum* (2·5 mm) have quite long root hairs.

The extent to which the roots penetrate and ramify in the soil can be gained from the fact that a single wheat plant has been found to possess as much as 64–80 Km of roots if added together while a single rye plant growing in containers (12″ ×44″) produced about 640 Km of total roots in 4 months of its existence! In general, roots of herbaceous plants grow more rapidly and profusely than those of shrubs and trees. With most of the cultivated plants in agricultural lands, the roots are mainly concentrated in the upper 4–7 ft. of the soil.

The soil water reservoir may be considered to consist of a vast number of interconnecting spaces, the spaces between the soil particles; some of these spaces are quite large as between larger particles and granules while others between individual clay particles are minute. After a shower of rain the soil is at first saturated with water in the larger spaces, after which water drains out and goes downwards obeying the laws of gravity and the empty spaces draw in air from the atmosphere above the soil. Some water is, however, retained by surface tension as films around individual particles and also in the smaller spaces or capillaries in which the capillary forces are large enough to counteract the forces of gravity.

The root hairs come into contact with the surrounding soil particles whose surfaces are covered with thin films of water held by various forces including gravity. The plant absorbs from the soil far greater quantities of water than of any other single substance. A determination of the total water content of a particular soil does not however give us any information about the amount which is actually available for plant growth for most of that water may be retained by the soil particles with so much force that it is practically unavailable to the plants.

Soil air plays an important role in root respiration and maintenance of microbial activity. Roots are strongly aerobic and active respiration is required particularly for uptake of solutes as growth of the root system. When the soil gets waterlogged, dissolved O_2 is the only source of root respiration. Such waterlogged soils gradually become a micro-aerophilic or almost anaerobic environment. While rice plants may thrive well ever under such conditions, this does not hold true for most other plants. The main object of tilling the soil, is to make the soil porous and allow fresh air to enter the upper strata of the soil, permitting the accumulated CO_2 to escape. Development of anaerobic conditions help the growth of anaerobic micro-organisms some of which are harmful. Denitrification resulting in the loss of N_2 is also favoured by anaerobic conditions. Microbial activity in the soil is important not only for the fertility of the soil but also to keep the soil porous due to activities resulting in gas exchange.

TYPES OR STATES OF WATER IN THE SOIL AND THEIR AVAILABILITY AND UTILISATION BY PLANTS

We have seen the soil is porous medium which admits water, allows some to pass through while retaining the remainder. Soil water and ground water are closely related but whereas soil water is always biologically important, the importance of ground water may range from being trivial to being all that matters. The soil is a kind of buffer, between rainfall and ground water. Water deficit of the soil can be completely restored by rain before there can be any surplus water available to move down to ground water. The classes of water in the soil can be roughly divided into the following 4 types:

Gravitational water or Ground water. After a heavy rain or on application of irrigation water, the surface layer of a soil is temporarily saturated. This water obeys the laws of gravity and thus descends rapidly through the dry layers, leaving a moist zone in its path. If there is sufficient water this wet layer penetrates deeply. The rate of this downward percolation of this gravitational water is determined by the number, size and continuity of non-capillary or larger pores. The gravitational water is available to plants only when rain showers follow one another in rapid succession; otherwise it percolates below the reach of the roots within a few days and thus remains unavailable.

Capillary water. As ground water drains out of the upper layer of the soil, it leaves behind considerable moisture in the form of films, coating smaller soil particles and as droplets suspended in the angles of larger pores or completely filling the smaller pores. This water is retained by the forces of surface tension and this capillary water does not respond to the gravitational pull. The forces are, however, small—seldom more than a fraction of an atmosphere depending on the diameter of the capillaries. This is capillary water. It is primarily this capillary water which is readily available to the plant and this is the source of practically all the water a plant extracts from the soil.

Hygroscopic water. Owing to evaporation from the soil surface and absorption by roots, the capillary water held by the soil is gradually depleted. As depletion progresses, the forces of molecular attraction or adsorption between the soil particles and the thin films of water held against their surfaces increase until finally the remaining water passes into a state where it is no longer in the liquid condition and thus ceases to be chemically or biologically active. The forces of molecular attraction increase rapidly as the water film surrounding soil particle grows thinner until a point is reached when the films are only a few molecules thick and the forces of attraction may reach values as high as 1000 atm. Evidently plants can absorb only a relatively small amount of this hygroscopic water. The hygroscopic water cannot be entirely evaporated from a soil under ordinary atmospheric conditions, but it can be done by heating soil to a constant weight in an oven at approximately 150°C.

The above three types of soil are not sharply defined but form a continuous series from water which is not retained by the soil, to water which is held with great force.

Chemically combined water. The water chemically combined in the structure of soil minerals is known as combined water. After the elimination of hygroscopic water by heating soil to about 150°C., the only water that remains is in the hydrated oxides of aluminium, iron, silicon, etc. This water is absolutely unavailable to the plants and can only be driven off from the soil by resorting to very high temperature but not before bringing about irreversible changes in the physical and chemical composition of the soil itself.

ABSORPTION OF WATER BY ROOTS

Now for the actual absorption of water by the roots—it can be shown that when a root hair is in contact with a number of cortical cells of the root and finally a trachea or a xylem vessel, water will enter the root hair, pass from there into the cortical cells and finally into the xylem if *there be a gradient of water potential from the root hairs to the xylem vessels.*

It is the gradient of water potential from the root hair to the xylem vessels that is essential for the absorption of water by the roots.

If the root hair A has a high osmotic pressure and is exposed to the surrounding water in the soil and the cells B to J and the xylem vessel K have progressively lower pressures, water will pass from A⟶K by diffusion from the soil to the roots following the same osmotic relations we have discussed so far. Thus A, the root hair will take up water from the surrounding medium as water tends to move by diffusion from regions of higher water potential to regions of lower water potential. The first essential condition for the diffusion of water to take place from outside to inside root hairs is that the root hair cell sap must have higher salt concentration or in other words lower water potential than the water in the soil. Thus A, the root hair will take up water from the soil and it will ultimately become turgid and as a result its water potential will fall below B. B will now take up water from A and will in turn become turgid and now its absorbing capacity will fall below B_1; B_1 will now draw water from B and so the process will continue until the xylem vessel K is reached. The force with which water will

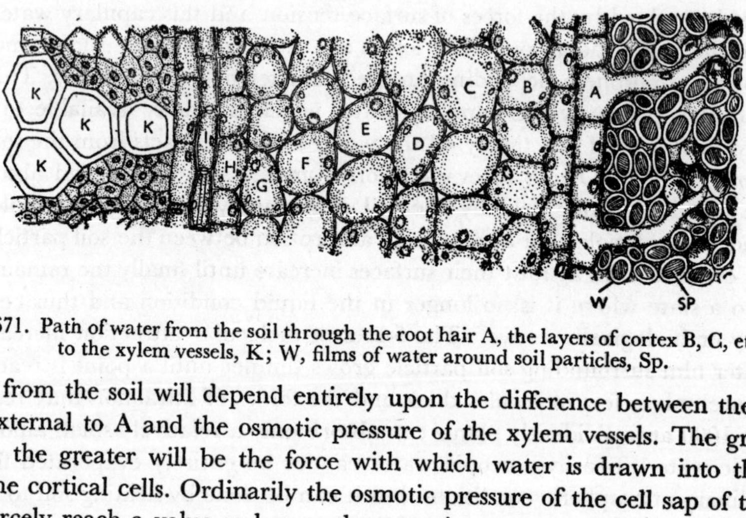

FIG. 671. Path of water from the soil through the root hair A, the layers of cortex B, C, etc., to the xylem vessels, K; W, films of water around soil particles, Sp.

be drawn from the soil will depend entirely upon the difference between the osmotic pressure external to A and the osmotic pressure of the xylem vessels. The greater the difference, the greater will be the force with which water is drawn into the vessels through the cortical cells. Ordinarily the osmotic pressure of the cell sap of the xylem K can scarcely reach a value as low as the osmotic pressure of the surrounding soil solutions which (vessels) form a continuous pipe line from the roots to the leaves, for once the water from the soil reaches the main transpiration current in the xylem vessels, water is taken upwards to the leaves for utilisation and for ultimate escape of the excess water through stomata. As a result there generally is a higher osmotic pressure in the sap of the xylem vessels than the water outside the root hair in the soil.

The osmotic pressure of the root hair cells generally varies from 3 to 5 atmospheres. Thus whenever water potential of such soil exceeds that of young root cells and root hairs, water will move from soil into the root and since the osmotic pressure of the soil solution in inert soils is only a fraction of an atmosphere, the absorbing capacity or suction need not be very great before water will enter them.

The absorption process which has been described above undoubtedly accounts for the intake of most of the water which the roots absorb, but it is certainly not the only mechanism of absorption known to operate in plants. This mechanism of absorption of water is sometimes referred to as passive absorption because the entry of water into the roots is brought about probably by conditions which originate in the top of the plant and the root cells apparently play a passive or subsidiary role. 'Originate in the top of plant' means that the osmotic concentration of the root cells is due to the presence of soluble metabolic products which are synthesised in the aerial parts of the plants and translocated downwards to all the tissues including the root cells.

In many plants, an internal pressure known as root pressure often develops in the xylem. This can be demonstrated in some plants which are growing vigorously, particularly in spring. If the root system of a suddenly decapitated plant is immersed in a potometer, absorption of water can be shown clearly by the movement of the bubble in the horizontal arm of the potometer. Thus it is seen that water can be absorbed by root cells by forces which develop there. In this type of absorption, the mechanism involved is localised in the root system and is often called *active absorption*. This active absorption apparently takes place in plants when transpiration rate is relatively low and the soil contains water in abundance. This active absorption of water can only be in very small amounts since water would tend to leak out so rapidly by diffusion that enormous amount of respiratory energy would be required to maintain the gradient.

EFFECT OF EXTERNAL CONDITIONS ON WATER ABSORPTION BY ROOTS

Any factor which influences water potential in the young root cells or in the soil will undoubtedly influence the rate of absorption of water. When the water at the superficial layers of soil becomes limiting, absorption of water by roots can only occur if the growth of the root continues downwards. Thus the factors which influence the rate of growth of roots also have ultimately important effects on the amount of water which can be absorbed. It is also true that although there is no direct connection between the amount of water absorbed by roots and amount of water lost from leaves by transpiration, the absorption of water is greatly influenced by the rate of transpiration, and so any factor influencing rate of transpiration, will also indirectly affect the rate of water absorption by roots.

The more important soil factors which influence the rate of absorption of water by roots are the following:

Concentration of the external soil solution

Absorption of water, in general, depend upon the difference between the osmotic concentration in the cell sap and the external soil solution. Thus an increase in the concentration of the external water medium, must reduce the difference between the osmotic concentration in the external solution and the root hair cell sap at least for some time, thereby reducing the rate of water absorption. Such reductions are actually observed in the plants growing in normal water culture solutions, where an addition of 1% NaCl or some other salt immediately brings about a decrease in the rate of water absorption. This reduction may only be temporary as the plants soon adapt themselves by increasing the osmotic concentration of the cell sap. This adaptation or increasing the osmotic concentration of the cell sap may be due to absorption and diffusion of dissolved substances into the cell from the surrounding medium or it may be due to the production inside the cells of osmotically active substances. This adaptation is, however, limited. Most species of plants develop normally only when the osmotic concentration of the soil solution does not exceed a few atmospheres. Exceptions are provided by the halophytes which are native to the saline soils of high osmotic pressure. Halophytes adapt themselves to such soils by increasing proportionately the osmotic concentration of the root cells.

Soil temperature

In general, the rate of water absorption by roots increases with rise in temperature of external medium, but at 35°C. or above in most species of plants the water intake is greatly reduced and soon replaced by withdrawal. The withdrawal of water from the cell sap may be due to the effect of high temperature on the cell organisation, particularly the plasma membrane of the cell which most probably loses its semipermeable nature.

Aeration of the soil

In general, absorption of water by root systems of most species of plants certainly proceeds much more rapidly in well-aerated soil than those which are poorly aerated. Poor aeration actually results in a reduced water absorption most probably due to a deficiency of oxygen in the neighbourhood of roots or due to an accumulation of carbon dioxide which results from anaerobic respiration of the root cells under such conditions. The growth of hydrophytes in water-saturated soils with scarcely any oxygen, however, can take place only because these plants have well developed and continuous air passages from the leaves down to the roots which store large amounts of oxygen in their intercellular spaces which is available for the normal respiration of root cells.

Some substances present in the soil

Some substances, occurring in the soil, affect the absorption of water on account of their chemical nature quite apart from their osmotic properties. One of them is *humus acid*, which retards water absorption in bog soils, owing to the chemical nature of the compound which probably is toxic to the plasma membrane of the root cells.

ROOT PRESSURE

If a well-aerated plant growing vigorously in spring is cut off slightly above the ground, water is seen to exude from the cut end of the stump through the xylem. This positive pressure occurs in the xylem either in deciduous trees when the leaves are shed in early winter or when the plant is in a saturated or near saturated condition. This exudation is particularly marked and prompt in conditions where transpiration from the leaf surface is very low, and the conditions favour rapid absorption of water from the soil. A pressure is thereby developed in the xylem. The exudation of the xylem fluid under these conditions, although slow, can take place against considerable pressure. The magnitude of this pressure can be measured by attaching a closed manometer to the cut end of the stump. This pressure, which is a push from below and not due to any tension developed in the xylem vessels due to transpiration and which seems to originate in roots has been termed *root pressure*. The quantities of water moving upwards through the xylem by root pressure are, however, very small compared to the amounts moved upwards during active transpiration. Under soil conditions favouring rapid water absorption combined with slow transpiration, water is also excreted from the leaves in *Nasturtium*, *Colocasia*, grasses, etc. in liquid form, particularly early in the morning when the soil is warm and moist and the atmosphere nearly saturated. This is sometimes referred to as *guttation* or *bleeding*. Guttation or bleeding and root pressure are now considered to be merely different aspects of the same phenomenon.

The development of this root pressure in the dilute sap of the xylem vessels originating certainly in the root cells is not quite fully understood as yet. Some investigators believe that this is due to a development of higher osmotic concentrations in the xylem vessels than in the external soil solution. The magnitude of the hydrostatic pressure developed due to this difference, is a measure of root pressure. Osmotic movement takes place from the soil solution to the xylem through a multicellular semipermeable membrane—the intervening cortical cells of the root being fully turgid with a consistently decreasing gradient of water potential, allow flow of soil solution through them passively. This osmotic explanation of root pressure can, at best, be only partial for the rate of exudation from a cut stump is usually too rapid to be explained only in terms of simple osmosis. It is more than probable that other mechanisms, as yet not fully understood, are involved.

The magnitude of the root pressure rarely exceeds 2 atmospheres (pressure as high as 7 atm. has been reported in tomato roots), and is generally much less and thus the magnitude of the pressure developed is seldom sufficient to force water to the top of any but small herbaceous plants. The fundamental objection to the idea that root pressure plays any prominent part in the ascent of sap in plants lies in the fact that root pressure is usually negligible in plants of temperate regions during the summer where, in general, conditions of the soil rarely favours excessive absorption for any considerable length of time.

During the periods of rapid transpiration particularly in midsummer, it has sometimes been observed that the cut end of a stump absorbs water, instead of exuding, if water is supplied at the cut surface. This has sometimes been referred to, very inappropriately, negative root pressure. This should really be called *root suction.*

It must be understood here that root pressure is a hydrostatic pressure developed in the sap of the xylem vessels. But how is this root pressure maintained in the non-living xylem vessels? The non-living xylem cells certainly do not possess any semipermeable membrane like the living cells of the plant body. Active absorption of solutes from the soil solution and secretion of solutes by the vessels from the adjacent living cells of the root must accompany rapid absorption of water in order to maintain root pressure in the xylem vessels. The leakage of the solutes from the vessels into the neighbouring living cells and ultimately out of the roots into the soil is presumably prevented by the narrow, usually one-row thick, hollow cylinder strip of cells—the endodermis—with its cell walls peculiarly thickened with fatty substances—the *casparian strip*. Soil conditions favouring rapid water absorption by root such as low concentration of the solutes, higher temperature, soil aeration, etc., accompanied simultaneously by conditions which greatly reduce transpiration rates from the leaves make the development of the root pressure possible.

EXPERIMENTS ON THE DEVELOPMENT OF ROOT PRESSURE IN PLANTS

Cut across the stem of a vigorously growing healthy potted plant, a few inches above the ground level, preferably in the morning in spring. Fix by means of strong, thick rubber tubing, a mercury manometer to the decapitated stump as shown in Fig. 672.

Make all connections air tight with vaseline or paraffin and freely water the soil. The level of mercury in the vertical arm of the manometer is noted. After a few hours, the level of mercury in the manometer is seen to rise and difference in the level is a quantitative measure of the magnitude of the root pressure developed in the xylem pushing out water from the cut end of the xylem vessels into the tube.

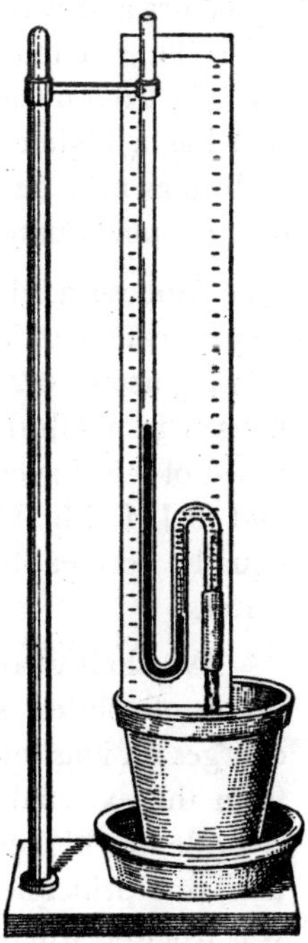

FIG. 672. Experiment to demonstrate root pressure in plants.

CHAPTER XXXV

THE ABSORPTION OF SOLUTES BY ROOTS

It will be observed from the discussion in the previous chapter on osmotic phenomena in plants that we have not said anything as yet on the absorption of minerals from the soil solution. Elementary text-books on biology define osmosis to include the diffusion of solutes through the membrane and implication is that they enter the root hairs by a process of simple diffusion. Thus the statement certainly means that the diffusion of a solute molecule or ion from outside into cells takes place as a result of its greater concentration or greater diffusion pressure in the external solution than in the cells. However, plenty of data have accumulated in recent years to indicate clearly that the absorption of a solute, particularly an electrolyte, by plant cells is under many conditions a much more complicated process than could be explained by simple diffusion.

We know that a perfectly true semipermeable membrane allows diffusion of only the solvent molecules and not the solutes. It is clear then that if the plasma membrane of the root cells were a true semipermeable membrane, nothing but water could have entered from outside into the root cells and the continued existence of the plant and ultimately of animals would have been impossible. But because the living plasma membrane, in contrast to the artificial semipermeable membrane is not perfect, it also allows some solutes to diffuse through it. It must be clear, however, that in the diffusion of solute molecules through the plasma membrane, even if we assume that the absorption of solutes follows simple laws of diffusion, each one of the solute molecules will follow its own individual diffusion pressure gradient independently of diffusion of other solute molecules and water, of course, if the membrane is permeable to the solutes and solvent alike. Thus it may happen, that at any particular time water may be diffusing into the cells from outside following its own concentration gradient, i.e., if the external solution has a higher water potential than the cell sap and at the same time the solute molecules may be coming out of the cell or getting into the cells each according to its own chemical potential gradient, i.e., from higher concentration to lower concentration.

Suppose a plant cell with a cell sap osmotic concentration of 1M (composed of 0·4M sucrose, 0·1M glucose, 0·2M KNO_3, 0·3M $MgSO_4$) is immersed in a solution of 0·8M concentration (composed of 0·5M sucrose, 0·1M fructose, 0·1M KNO_3 and 0·1M $MgSO_4$) separated by a membrane equally permeable to solutes and the solvent, water.

We shall disregard other factors such as ionisation of solutes, kind and amount of the electric charges on ions, etc. Water will enter into the cells from the external medium as the water potential is evidently higher outside than inside. Following the same principle sucrose molecules will go in, so will fructose whereas glucose, KNO_3 and $MgSO_4$ will come out of the cell into the external medium.

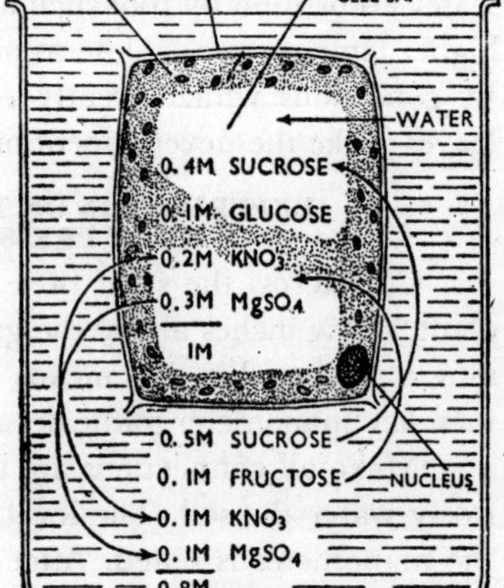

FIG. 673. Theoretical diffusion of water and solutes between a cell and external medium.

It has been frequently observed that plasma

membrane of the living cells allows certain solute molecules to pass through it very slowly and at the same time other solute molecules comparatively rapidly though the membrane is apparently equally permeable to both the solute molecules, even if the concentration gradient is in the right direction. Under another set of circumstances the reverse may be true. Frequently it is observed that the membrane is permeable to larger molecules of solutes but at the same time impermeable or only partially permeable to much smaller molecules (or ions). This is what is known as *selective absorption*. Various theories have been proposed to explain the differential permeability of the plasma membrane. For convenience, most of them are grouped under the following heads:

The sieve or ultrafiltration theories. These theories assume that the plasma membrane acts as a molecular sieve with intermolecular spaces which are very minute and only molecules below a certain size are supposed to diffuse through the membrane, while those exceeding the size limit are prevented from entering the cell. According to later advocates of these theories, the spaces or pores are not fixed but are capable of constant change and adjustment.

The diameter of the pores are supposed to enlarge by aggregation of the particles of protoplasm into larger ones when large molecules of solutes are to enter the cell. Under another set of circumstances, the segregation of the same protoplasmic particles is supposed to make the sieve pores smaller, thus restricting the entry of large molecules while at the same time freely admitting the relatively smaller solute molecules.

The most serious objection to this theory seems to be that it is difficult to conceive this proposed sieve-like structure of the plasma membrane. Again it has been frequently observed that in certain plant cells, very large molecules of certain alkaloids enter through the plasma membrane easily and very rapidly while at the same time the entry of relatively smaller amino acid molecules are significantly restricted.

The solution theories. According to these theories the plasma membrane is supposed to have definite solvent properties. Membranes are supposed to be permeable to only such substances or molecules which will dissolve in them while impermeable to others. This also is known as *lipid theory*. The plasma membrane is supposed to be constituted chiefly of proteins and lipids and only those solutes, which readily dissolve in lipid matrix, will pass through the plasma membrane.

The serious objection to this theory is that all the inorganic solutes which for the most part are insoluble in lipid substances, enter through the plasma membrane freely and in large amounts. The later advocates of this theory have attributed to the plasma membrane a mosaic structure consisting of an aqueous phase which permits the passage of water-soluble substances and a lipid-containing phase which facilitates the entry of fat-soluble materials.

The absorption and chemical reaction theories. The advocates of these theories proposed that particles of the protoplasm of the plasma membranes adsorb the molecules of solutes to which the membrane is permeable. The stronger the adsorptive forces between the membrane particles and the solute molecules, the more readily the solute molecules will pass through the membrane.

It is also possible that chemical reactions are involved in the transport of certain solute substances across the cell membrane. According to the chemical reaction theory molecules are supposed to combine chemically with certain carrier substances on the outer boundary of the membrane forming an intermediate compound and it is in this bound form that the solute-molecule moves across the width of the membrane itself.

Once the inner boundary of the membrane that is lining the vacuolar cell sap is reached, a reverse chemical reaction is supposed to occur which releases the molecule from the membrane to the cell sap.

It is possible that once the entering solute molecules are dissolved, adsorbed or chemically combined with particles in the lining layer of the plasma membrane the penetration of the solutes across the membrane is facilitated by movement or kinetic activity of the membrane particles; each particle might be assumed to carry a load (as if on their back) of solute molecules across the membrane and once it is past the membrane, the load is freed from its carrier into the cell sap. A more probable hypothesis of course will be that the dissolved, adsorbed or chemically bound penetrating solute molecules are not carried all the way across the membrane by the activity of *one* single particle. Rather, the loads of solute molecules are transferred from one carrier particle to the other, as if in a 'relay race' as impacts occur between them, on the inward journey of the particle-solute combination across the membrane.

The colloidal theories. These theories are based on the concept of protoplasm as an emulsion of oil and water and the differential distribution of water between the two liquid phases by phase inversion or a reversal of the continuous and discontinuous phase.

According to these theories, when the continuous phase of the emulsion constituting the plasma membrane consists of oil, the membrane becomes more permeable to penetrating fat-soluble molecules. With phase inversion taking place, the continuous phase becomes water; the membrane is now more readily and easily accessible to penetrating inorganic solute molecules. It must be admitted, however, that experimental evidences are as yet lacking to support this phase reversal idea of the protoplasmic emulsion in the cell membrane.

ION EXCHANGE AND ACTIVE ABSORPTION

In the foregoing paragraphs we have assumed that both the electrolytes and non-electrolytes enter through the cell membranes as molecules. It is true that an electrolyte when dissolved in water is largely or entirely dissociated into ions but there is evidence which indicates that such dissociation is inhibited by the lipid constitution of the plasma membrane. Other investigators, however, believe that electrolytes penetrate the cell membrane only in the form of ions.

If penetration of solutes through the cell membrane is in the form of ions it is evident that either an equivalent number of oppositely charged ions must enter the cell at the same time or there must be an exchange of ion of like charge with the cell if the electric potential across the membrane is to remain undisturbed. Let us consider the case of a vegetative cell immersed in KNO_3 solution. There are several possibilities concerning the movement of dissociated K^+ and NO_3^- ions; K^+ and NO_3^-, K^+ and OH^- (H^+ and OH^- obtained from dissociation of water) or an ionic exchange between the cell sap and the external solution might occur, an ion bearing a negative charge say HCO_3^- moving out of the cell as penetration of NO_3^- occurs or in the same manner, K^+ ions move into the cells from the external solution as Na^+ ions of the cell sap move out.

It has been observed that a rapidly penetrating anion seems to facilitate the entrance into the cell of a cation. If two cations are entering a cell, the entry of one is retarded by the other. Na^+ ions always decrease the rate of entry of K^+ ions. This is also true for anions; a fast moving NO_3^- ion always retards the intake of anions such as

$SO_4^{=}$. In general, it is found that a rapidly penetrating ion retards the rate of absorption of another ion of the same charge while increasing the rate of absorption of an ion of the opposite charge.

In all our previous discussions we have assumed that diffusion of inorganic solutes in the form of molecules or ions into cells takes place as a result of their greater concentration in the external medium than in cell sap, i.e., following the right concentration gradient. All this have been termed *passive absorption.*

The result of a number of investigations indicate, however, that the absorption of ions by plant cells is under many conditions a much more complicated process than simple diffusion. It has been observed that the cells of fresh-water alga, *Nitella* accumulate K^+ ions to a concentration of more than 1000 times greater than the concentration of K^+ ion in the surrounding medium and in fact concentration of every ion analysed was far greater in the cell sap than in the water in which the alga grows. The total concentration of electrolytes in the cell sap was about twenty-five times as great as in the external medium. In the marine alga *Valonia* similar, but not as striking, accumulation of ions takes place. Thus the entrance of ions by plant cells is characterised by the feature that the living cell can accumulate ionic material or in other words they can continue to take up an ion even if the concentration level of this ion inside the cell sap in many times higher than the concentration of same ions in the external medium. Apparently a very different mechanism from simple diffusion must be at work. This is *active absorption* since the penetration takes place against the gradient. This must not be confused with salt absorption in which no equilibrium is attained with the external solution because the penetrating solute is removed from the cell sap through utilisation in growth or by absorption or by precipitation. While large amounts of salts may be held by a plant cell in this manner, an equilibrium is never attained in such cases between the cell sap and the external medium and they are not accumulated against the gradient.

The passive absorption of ions is, we know, essentially an exchange adsorption, i.e., absorption taking place mainly by diffusion on the surface of the cells whereas in active absorption, ion-building compounds and carriers are most probably involved. The passive absorption is primarily non-linear with time with equilibrium approaching within a very short time, even less than an hour. In active absorption, equilibrium is never really attained and the curve is more or less linear with time. In active absorption, ions are essentially non-exchangeable whereas in osmotic uptake ions are readily exchangeable. The active absorption of ions or group of ions seems highly selective, hence highly specific with regard to competition for site and entry in contrast to osmotic uptake where there may be considerable indiscriminate competition among various ions with respect to entry.

Active absorption cannot be explained completely satisfactorily. The penetration of ions into the cells from the external medium or retention of the ions in the cell sap in greater concentration than in the external solution must require a continuous expenditure of energy and this energy is most possibly supplied by respiration. Though the complex relationship between respiration and salt accumulation by plant cells is not as yet fully understood, there is no doubt, however, that the energy liberated in respiration is utilised in the accumulation and retention of ions.

The phenomenon of *salt respiration* or *anion respiration* and its relation with accumulation of ions in plant cells against a concentration gradient have been studied brilliantly by Lundegardh. The increased respiration rates (*salt respiration*) in many tissues upon addition of salts over its normal respiration in distilled water (*ground respiration*) have been related to the uptake of anions only by plant cells, rather than the absorption of cations. Thus Lundegardh bases his theory on the primary assumption that the absorption of anions and absorption of cations are independent of each other to such an extent, that different mechanisms may be responsible for each, which assumption is, however, disputed by other workers, who prefer to attribute salt respiration phenomenon specifically to the anion uptake.

The ratio of the moles of the extra oxygen consumed when tissues are immersed in salt solutions over ground respiration to that of moles of salts absorbed appears to approach *four* as a limit. Since each molecule of oxygen accepts four electrons and is equivalent to oxidation of four molecules of ferrocytochrome c to ferricytochrome, evidences are strongly in favour of migration of cytochrome c in its two valence forms and the consequent entrance of anions of a salt into a plant cell from external medium. If one anion is accumulated for each electron transferred by the cytochrome system, then apparently for each molecule of extra oxygen utilised in salt respiration, *four* anions of a salt should accumulate, assuming 100% efficiency. Actually with carrot tissue values close to four have been obtained.

It has actually been observed that cells which show markedly high respiration rates, such as those in the growing meristematic regions of roots and other similar tissues, also show considerable ion accumulating ability. It has been demonstrated in excised roots and tissues of potatoes that salt accumulation takes place only in presence of molecular O_2.

The respiration of the root cells is thus one of the essential factors for vigorous absorption of ions. This explains the favourable influence of the aeration of the soils and bubbling of air through water culture solutions for optimal salt absorption. The digging of the soil in the neighbourhood of roots from time to time is a normal agricultural practice for supplying air to the root system for the normal respiration of the root cells and which certainly contribute largely to the maximum uptake of minerals from the soil. A probable connection between respiration and absorption of ions into the cells from soil may be that CO_2 released in cellular respiration combines with water of the cell sap to form H_2CO_3. This H_2CO_3 is dissociated into H^+ and HCO_3^- ions. This H^+ ions and HCO_3^- ions in the cell sap can be exchanged with other cations and anions of the external soil solution, thus facilitating rapid absorption of K^+, Ca^{++}, Mg^{++} and anions such as NO_3^-, PO_4^{---}, etc., which are essential for normal growth of plants. Recent researches seem to indicate, however, that ions which are actively absorbed enter essentially irreversibly into the vacuoles of the root cells and these ions are not, therefore, free to move to the aerial parts. This would seem to indicate that only osmotically absorbed ions are available for movement through the plant and these ions occur only in the cytoplasmic outer space of the root cells.

ANTAGONISM OF SALTS OR IONS

Antagonism means enmity and so the term antagonism of salts suggests that there is enmity or better active opposition between salts or to be more precise, between different ions of the same electric charge.

The first observation of antagonism of salts was made with eggs of *Fundulus* (a marine organism). It was observed that the development of the eggs was greatly inhibited if immediately after fertilisation, they were placed in a solution of NaCl having the same concentration as this salt has in sea water. If, however, small quantities of the salt of a divalent metal were added—salts of calcium, strontium or magnesium—development proceeded normally. It was the first indication that solutions of single salts were much more toxic in their action than solutions composed of two or more salts. Thus a number of salts when present together in solution appears to be able to antagonise the toxic effects of each other; such solutions are called *balanced solutions.*

If a slice of freshly cut beet root is placed in a solution of NaCl of a particular conc. after thoroughly washing it to remove all the colour from the cut surface, the red colour gradually diffuses into the NaCl solution. If the slice of beet root is properly washed in pure water, no diffusion of colour from the tissue to the outside is observed. It shows that the effect of NaCl on the beet root tissue is to increase the permeability of the cell walls to the red pigment of the vacuole which diffuses out into the solution. If, however, $CaCl_2$ is added to the solution of NaCl in proper proportion, no outward diffusion of colour could be detected. Evidently the presence of Ca ions in someway counteracts the usual effects of Na ions on the permeability.

It was later seen that the cause of diffusion of red colour from beet root tissue out wards is that the single salt solution of NaCl increases the permeability of the plasma membrane to a point that is injurious to it whereas in a solution of $CaCl_2$ only, the reverse is true—the permeability of the plasma membrane is reduced to the point of injury. If, however, these two salts, NaCl and $CaCl_2$ are mixed in proper proportion, their combined effect on the plasma membrane will neutralise the toxic effects of each other and the permeability of the membrane may remain unaltered.

Other examples of antagonism of salts can be cited. In an experiment with roots of lupins, the roots elongated only about 3·5 mm per day in a very dilute solution of $CuCl_2$. When, however, proportionately sufficient $CaCl_2$ was mixed with $CuCl_2$ and the roots allowed to grow in the mixture, the increase in length per day was about 11 mm. The antagonism between Cu and Ca ions was sufficient to greatly reduce the toxicity of Cu ions alone. In another experiment with wheat roots the rate of growth was about six times or more when the roots were allowed to grow in a mixture of different salt solutions like NaCl, $MgCl_2$ and KCl combined in proper proportions compared to their growth in individual single salt solutions. However, these effects are complicated by the relative beneficial effects of Ca^{++}, Mg^{++} and K^+ ions for nutrition and the growth effects observed may not be entirely due to antagonism.

We know that in mixture of salt solutions, antagonism disappears for the toxic effect of one salt or ion is antagonised by the presence of other. Such solutions where the antagonism has been neutralised are, as stated before, balanced solutions. Sea water is typically balanced for in addition to NaCl, the sea water also contains small amounts of $CaCl_2$ and other salts. Ordinary soil water and artificial sand and water culture solutions are really balanced solutions. The various salts present are in themselves toxic to the wellbeing of the plant, but when mixed together in a solution, they are harmless and appear to antagonise each other's toxic effects. The mineral solution which is certainly a mixture of various salts in the soil in which the roots grow must also be in a balanced state for the minerals in the soil are readily available to the roots.

It has been observed that the greatest amount of growth in plants takes place in a balanced solution of a mixture of different salts, when the concentration of one of the salts is slightly in excess of *half* the total concentration of the mixture.

Several explanations have been offered to account for the phenomenon of salt antagonism. It is known that antagonism is only effective as long as the plasma membrane remains intact; so the seat of antagonism must be there.

It is apparent that any theory which accounts for the antagonism of salts must explain: (1) why both NaCl and $CaCl_2$ are toxic individually? (2) why then when mixed in proper proportions their toxicity is diminished? (3) why NaCl causes an increase in the permeability of the plasma membrane and $CaCl_2$, a decrease? (4) why the reduction in permeability caused by $CaCl_2$ is later followed by an increase and (5) why toxicity disappears in sea water?

One explanation of salt antagonism is provided by the prevalent concept of protoplasm of the plasma membrane being a system of emulsion of water and oil. This system is reversible, but for the normal functioning of protoplasm the system must be at the *transition stage*, that is neither an emulsion of oil-in-water nor water-in-oil but somewhat intermediate between the two. When a single salt solution, e.g., NaCl is present in the medium, it produces a pure protoplasmic emulsion of oil-in-water while in $CaCl_2$, the system is reversed, producing an emulsion of water-in-oil. In both the cases, the normal functioning of the plasma membrane is hampered. If, however, both NaCl and $CaCl_2$ are present in proper proportions, a condition of equilibrium is obtained for the protoplasm of the plasma membrane, where the effect due to one is counterbalanced by the opposite influence due to other, producing the desired system at the transition stage essential for the normal functioning of the protoplasm. As a result, the permeability of the plasma membrane is supposed to remain unaltered in a mixture of two or more salts.

Another hypothesis bases its idea on several primary assumptions such as: (1) the normal charge upon the particles comprising the plasma membrane is negative; (2) salts may be classified in respect to their action upon the plasma membrane, as positive or negative. *Negative salts*, such as NaCl (or alkalis) cause the particles on the plasma membrane to separate and thus increasing permeability while *positive salts*, such as $CaCl_2$ (or acids) have the opposite effect during the first few minutes of exposure, because it causes the particles on the plasma membrane to approach each other and hence the permeability is decreased; (3) antagonism exists between salts which produce dissimilar electric effects upon the protoplasm.

If to a given protoplasmic surface of the plasma membrane a molecule of NaCl is added, the positive charge of sodium will neutralise a negative protoplasmic charge on the particles comprising it. The substituted Cl^- unit, being smaller than the protoplasmic negative charge on the particles, has, it is supposed, a more concentrated surface charge and as a result there will be a readjustment of the relative positions of the particles on the protoplasm at a little greater distance than before. The total effect is loosening up of the mesh or a significant increase in the permeability of the plasma membrane. When all the negative charges on the particles of protoplasm are neutralised and substituted by negative Cl^- units, the permeability is at a maximum.

When a salt like $CaCl_2$ is allowed to come in contact with the protoplasmic surface of the plasma membrane *alone*, its first effect (for the first few minutes) will be to cause the particles on the protoplasmic surface to approach each other and the permeability is decreased. As the negative charges on the protoplasm are later replaced by the negative charges of Cl^- units, the negative charges again repel each other and the permeability is increased considerably as in the case of NaCl—only difference is that, increase in permeability with $CaCl_2$ occurs *later*, and not immediately, as is true for NaCl.

In sea water or in other balanced solutions, toxicity disappears because of the balance existing between the various ions which compose it.

It is interesting to note that of the six normal plant elements taken in quantity from the soil solution by the roots, three occur in the cation with five positive charges (K^+, Ca^{++}, Mg^{++}) while three occur in the anion with six negative charges (NO_3^-, SO_4^{--} and PO_4^{---}. This is really too nearly balanced to be a mere coincidence.

SYNERGISM

Combined activity of agencies, e.g., chemicals, hormones, which separately influence a certain process in the same direction, such that the effect produced is greater than the sum of the effects of each agency acting alone, is referred to as synergism.

The term may also be used for combined activity such that effect is either the sum of the separate effects (*summation*), or greater than the sum of the separate effects (*potentiation*), it does not matter which, i.e., they (agencies) are not antagonistic to each other.

CHAPTER XXXVI

TRANSPIRATION—THE LOSS OF WATER FROM PLANTS

Of all the materials required by a plant for its existence and normal development that is taken up in the largest amount, is water. Water is absolutely essential to animal and plant life. Plant and animal cells, deprived of water to any considerable extent soon die. The bulk of this water absorbed, however, takes no permanent part in its development or in metabolic processes, but evaporates into the air from the leaves and other aerial parts. This loss of water in the form of vapour from living plants, particularly from the aerial parts, is known as *transpiration.* The process is in principle one of evaporation and diffusion.

Loss of water vapour may occur from any part of the plant which is exposed to the air. All aerial parts lose water by transpiration, although in some tissues due to the presence on some organs, of superficial layers which are impervious to water, e.g., cork cells, the rate of water loss is almost insignificant compared to the water lost from leaves through stomata. Most of the water vapour lost from leaves occurs through the stomata, this is termed *stomatal transpiration;* smaller amounts of water vapour are also lost by direct evaporation from the epidermal cells, when the stomata are closed, through the cuticle; this is usually termed *cuticular transpiration.* As a general rule transpiration rates are higher when the stomata are open, lower when they are closed. In other words, stomatal transpiration is usually many times higher than cuticular transpiration, commonly about 4 times as high during a warm sunny day. But there are all gradations from one extreme to the other. In the case of succulents like cacti, cuticular transpiration is usually nil. On the other hand, in some plants, cuticular transpiration can be as high as stomatal transpiration.

The general need of water for plants can be discussed under the following heads:

It is best to point out here that water for its capacity to form hydrogen bonds is largely responsible for the maintenance of the three dimensional structures of such macromolecules as nucleic acids and proteins and of the various subcellular organelles —essential for their biological function. Water scarcity or any other condition which destroys or disrupts their physical structures impairs the activity of these macromolecules and organelles, which ultimately may have disastrous consequences for plant life. Among all natural compounds, water also gains and loses heat slower than any other known substance. Because plants contain so much water, the rapid changes of temperature that may take place in the surrounding atmosphere have less effect on the plant than they would, if water gained or lost heat quickly. We know that chemical reactions are speeded up by heating and slowed down by cooling. The abundance of water in the plant maintains a fairly constant temperature so that chemical reactions inside the plant are not speeded up or slowed down so rapidly between the extremes of temperature to which a plant cell may often be subjected. In nearly all its physical properties, water is either unique or at the extreme end of the range of a property. From these physical and chemical characteristic of water, the biological importance of water is realised.

Among its thermal properties, water has the greatest specific heat known among liquids; the same is true of water's latent heat of vaporisation. Finally, with the exception of mercury, water has the greatest thermal conductivity of all liquids.

As constituent of the protoplasm. It is by far the most important integral part of the protoplasm. Large amounts of water are necessary for the proper functioning of the protoplasm. The younger, the more vigorous, the more rapidly growing protoplasm is, the more water it contains. In the actively growing meristematic regions of plants more than 90% of water is directly associated with protoplasm. Water deficiency during the conjugating phase of meiosis (reduction division) greatly decreases the number of the terminal junctions and chiasmata. If the normal proportion of cell-water is removed from the protoplasm, it becomes less and less active until a point (below 30–50%) is reached, when death of the protoplasm by desiccation ensues.

In the vacuoles of the mature cells. As has been discussed previously a certain amount of water is absolutely necessary to maintain the normal condition of turgour essential for the growth of the cells also for giving rigidity and erectness to the young plant.

For translocation of minerals and synthesised food from one part of the plant to the other. All inorganic minerals which move upwards through xylem and the movement of foods synthesised in the leaves downwards through the phloem and all from cell to cell are translocated in the form of solution in water.

For chemical combination. Many of the various complex organic substances synthesised in plant tissues are formed by chemical combination of water with certain inorganic materials that enter the plant from air and soil. Taking the fresh weight of a maize plant of average height (about 3–4 m) as for example, growing vigorously in well-watered and aerated soil, to be about 3,000 g, the total dry wt. was found to be about 850 g. The water content is thus about 2,150 g, i.e., the amount of water the plant needed for the protoplasm, for maintenance of turgour of the cells and used in the translocation of materials through the plants. The amount of water necessary for all the chemical combinations in which water is known to take part would not certainly exceed 250 g. Thus, about 2,150+250=2,400 g (less than $2\frac{1}{2}$ litres) of water a plant would need during its entire growing season which will be sufficient for all its essential requirements. In practice, however, under normal conditions much greater quantity of water is taken up by the plants from soil—about 250 litres or approximately 100 times more than is needed. This large excess of water absorbed by the plants from soil is eliminated by the plant in the form of vapour into the atmosphere. How do plants achieve this?

TRANSPIRATION AND LEAF ANATOMY

The occurrence of transpiration seems to be a natural consequence of the basic facts of leaf anatomy. The leaves consist of water-filled mesophyll cells, the water of which certainly also saturates the protoplasm lining the vacuoles and the cell walls. This water is supplied to the mesophyll cells of the leaves through the xylem bundles which form the network of veins of the leaves. The wet cell walls of these cells of the leaves are in intimate contact with the intercellular spaces which cover from 5–75% of the total area of the leaf. Hence evaporation of water takes place by direct absorption of radiant energy by leaves from the wet cell walls into the internal atmosphere of the intercellular spaces just as it will occur from any wet surface into the surrounding air. Recent evidences indicate that the mesophyll cell walls in the leaves are hydrophobic in nature, i.e., they are not rapidly wetted. Thus, it would seem that water would remain deep within the walls, leaving the wall surfaces relatively dry. The intercellular spaces form a connected system extending throughout the leaf and in turn lead through the stomata to the outside atmosphere. Water evaporating from the wet mesophyll cells is conducted through the continuous intercellular spaces and if the stomata are open, to the external air. If the stomata are closed, however, the only effect of evaporation from the cells walls will be gradual saturation of the entire internal atmosphere of the intercellular space with water vapour.

It must be clearly understood that the structure of leaf is adapted for absorption of gases that it needs and for excretion of those, which arise as end products in plant metabolism. It is certainly very poorly adapted for retention of water within the plant so that continuous threat of loss of water is a constant menace to the life of plant. Under certain conditions, the intercellular spaces can become filled with liquid water (guttation); normally they are occupied by air.

Transpiration and its relation with radiant energy. Leaves exposed to direct sunlight absorb large quantities (as much as 80%) of radiant energy and unless the energy absorbed is dissipated, it will be converted to heat energy raising the temperature of the leaf to such an extent as to be lethal to the protoplasm of most plants in a very short period of time. Only a very small proportion of the total absorbed energy (0·3–5%) is utilised under natural conditions (under best laboratory conditions, however, the efficiency of photosynthetic apparatus in converting absorbed light energy is at least 35%) in photosynthesis and stored in carbohydrate molecules. But actually, the leaf temperature seldom exceeds atmospheric temperature by more than 2–5°C. Part of the absorbed energy after being changed into heat is used in transpiration, i.e., in changing water into water vapour.

Since transpiration like evaporation of water is an energy consuming process, it has been assumed that in the evaporation of water from the leaves most of the energy absorbed by them is dissipated. A simple calculation, however, shows that even at maximal transpiration rates, transpiration can at the most account for the dissipation of only a part of the radiant energy absorbed by leaves in sunlight. Even if transpiration could account for the dissipation of most of the absorbed solar radiation, in doing so it apparently plays no essential role since the absorbed energy could as well, if not better, be dissipated by purely physical means, just as an object heated above its environmental temperature, does so by conduction, convection and radiation. Thus the function of transpiration seems to be redundant in protecting the leaves from a thermal death.

Transpiration during day and night: Amount of water lost. Transpiration occurs both during day and night but in general by far the greater amount of water, about 95%, is lost during the daylight hours.

In general more water is transpired during the afternoon than during the forenoon. The maximum rate of water loss takes place between 11 a.m.–3 p.m.

Amounts of water lost by plants per day undoubtedly depends upon the prevailing climatic conditions. A sunflower plant, under normal climatic conditions, may lose as much as 4 litres per day on the average.

THE SIGNIFICANCE OF TRANSPIRATION

Although transpiration has been studied extensively, very little is known concerning the significance of transpiration in the life of the plant. There is even now some controversy whether transpiration is a necessary and unavoidable evil or whether it is in some way essential for the welfare of the plant. In the first category, opinions prevail that transpiration brings nothing but harm to the plant, and the harmful effects ascribed to transpiration far outweigh its supposed beneficial effects, while in the second, there are opinions which consider transpiration as important as photosynthesis and respiration. From the former point of view the primary function of stomata is to permit exchange of CO_2 and O_2 between the plant and the atmosphere in the process of photosynthesis and respiration and water is lost through the open stomata, because there is no stopping the passage of water vapour, while permitting CO_2 and O_2 exchange. Thus the primary function of stomata is not to be thought of as one permitting water loss any more than the function of a main door in a busy office building is to let heat escape from the building in winter. Just as loss of heat cannot be avoided if the door must be kept open to continuous streams of persons who enter and leave the building during business hours, similarly loss of water vapour from the leaf is inevitable when stomata are open to permit exchange of CO_2 and O_2 between the plant and the atmosphere.

From the other point of view transpiration fulfils perhaps, only partly, these main functions:

Transpiration promotes absorption and translocation of solutes. It has often been assumed that the more rapid the rate of transpiration, the greater is the rate of absorption of solutes from the soil. This view certainly implies that the dissolved mineral salts are swept into the plant along with water. This implication, however, completely ignores much evidence that the mechanism of water absorption and the absorption of mineral salts are very different and possibly independent of each other. The results of certain experiments, however, do indicate that somewhat larger quantities of mineral salts accumulate in plants under conditions favouring high transpiration compared to the similar plants growing under conditions where transpiration rate is low. But as there is certainly no direct relationship between the volume of water absorbed and the volume of water lost by transpiration, it is evident that at the most only a slight, if any, correlation exists between the rate of transpiration and the rate of absorption of solutes from the soil. Since, however, plants grown even under conditions favouring low transpiration rates obtain their usual and normal requirements of various mineral salts, provided they are present in the soil, it is difficult to see how this effect of larger intake of salts is of any great particular advantage to the plant.

It is also generally considered that transpiration plays a significant role in the translocation of solute solution in the xylem vessels from the roots to the leaves. This factor has been the subject of the most controversy. That mineral elements absorbed from the soil are translocated mainly through the xylem vessels and rise upward with the transpiration stream, is indisputable. The question is, however, sometimes raised whether or not the rate of translocation of solutes in the vessels depends on the rate of transpiration. A critical examination, however, shows that if the solutes from the soil enter plants by active absorption, the amount of solute reaching the leaves per unit time will depend not on the rate or speed of the transpiration current through the xylem vessels but on the rate of absorption of mineral salts from the soil. There is also considerable evidence that at least a part of mineral salts absorbed from soil may move through the phloem and not through the xylem and it is evident that transpiration can have little or no effect upon such movement of salts.

Transpiration reduces leaf temperature. It has frequently been stated that transpiration cools the leaves and prevents their death or injury by high temperatures. The change of water from the liquid stage to vapour does involve absorption of a considerable amount of heat energy and thus vaporisation would tend to lower the temperature of the leaf. It has been shown that the actual cooling effect of transpiration is not so great as some investigators have believed; it is generally not more than 3–5°C. compared with the temperature of the atmosphere prevailing at the time—a difference, which so far as our knowledge of protoplasm goes, would scarcely be of any marked benefit in preventing injurious effect of high temperature. According to some investigators, however, a matter of even a very few degrees may mean at its worst, difference between life and death or between efficient and inefficient functioning of the cells. In a few experiments it has been claimed that leaves exposed to sun becomes much hotter than air temperature, when transpiration is stopped artificially or slowed down by wilting. The temperature of leaf undergoing rapid transpiration may remain, in exceptional cases, as much as 20° C. below the temperature of the surrounding air.

STOMATA

The stomata constitute the main route for the escape of water vapour from the leaves of land plants to the surrounding atmosphere.

Stomata are very minute openings or pores surrounded by two semilunar guard cells. They occur only in the superficial layer, in the epidermis of all the plant organs except in the roots, being much more numerous in the leaves. They are formed by the ordinary cell division of a young elliptical epidermal cell forming two daughter cells. By the dissolution of the middle lamella between the two cells, a pore or intercellular space is formed. The two cells resulting from the division are known as the *guard cells* and the pore between them is the stomatal pore or aperture. The guard cells differ from epidermal cells not only in shape but also in possession of rich protoplasmic material, chloroplastids and peculiar localised thickening of their cells walls. The chlorophyll of the guard cells can be formed either in light or in darkness in contrast to the chlorophyll of the mesophyll cells which requires light for its formation. If a leaf is kept in dark the chlorophyll of the mesophyll is decomposed whereas chlorophyll of the guard cells is not. The parts of the guard cell wall which abut on the stomatal pore are much thicker than the remainder of the wall which are in contact with ordinary epidermal cell walls. Because of this uneven thickening of the walls of the guard cells, increased turgour leads to opening, and decreased turgour leads to closure of stomata. The epidermal cells which border the guard cells are also somewhat different from the ordinary epidermal cells and are sometimes called *subsidiary cells* or *accessory cells*.

Distribution, number and size of stomata. The number of stomata per unit area of the leaf is characteristic for each species of plant. In most dorsiventral leaves, the distribution of stomata is usually restricted only on the lower epidermis of the leaves. In such leaves, approximately 95–97% of total gas exchange occurs through the lower epidermis and only about 3–5% occurs directly through the cuticle of the upper epidermis. In isobilateral leaves of most monocotyledons, stomata are more or less evenly distributed on both the upper and lower epidermis. In wheat leaves, actually more stomata are found in the upper epidermis. In floating leaves such as those of waterlily, stomata are restricted only on the upper epidermis, the lower epidermis being in contact with water of the medium. Stomata usually occur with a frequency of 5,000 to 30,000 per sq. cm. of the leaf area, although up to 130,000 per sq. cm have been frequently reported.

The area of the stomatal aperture or pore, when fully open, as in maize, can be as high as $100\mu^2$. There is no conclusive evidence as yet whether environmental conditions influence the number of stomata formed per unit area of the leaf. Species, in which the stomata are relatively smaller in size, have more per unit area than species in which stomata are relatively large. The aggregate area of stomata, when the pore is fully open, is approximately 1–5% of the total leaf surface.

The number of stomata per unit area varies from leaf to leaf of the same plant and even in different parts of the same leaf. In any particular leaf, the greatest number of stomata per unit area is at the top, the lowest towards the base and the middle frequency is in the centre of the leaf. As a rule, the younger leaves in the tops of the plant have greater number of stomata per unit area than those situated below.

In general, if the stomata are open, no correlation direct or indirect is found between the rates of transpiration and the size, number and distribution of stomata per unit area, other factors, particularly environmental conditions, being much more important.

Capacity of stomata for diffusion of gases. Practically all the interchange of gases between the interior of the leaf and the outside atmosphere, takes place through stomata. Although stomata occupy only about 1–5% of the total area of the epidermal surface, nevertheless, they have a carrying capacity for gases greatly in excess of that needed by plants for all their metabolic activities. Consequently the stomatal pore can close to a considerable degree and yet provide for all necessary gaseous exchange between the plant and the atmosphere.

The rapid rate of diffusion of gases and water vapour into and out of the leaf is understandable on the basis of fundamental laws of diffusion : the diffusion constant of gases is high, about 10,000 times that of liquid water; the area across which diffusion occurs, is also large. The difference in vapour pressure between intercellular spaces and outside atmosphere is large, and the distance between the internal air surface and outside atmosphere is small. But the area can only be considered large, if we consider the whole leaf area. Although the area of stomata only seldom exceeds 1% of the total leaf area, yet a leaf may lose almost as much water vapour as a free water surface of the same total area. This certainly means that the rate of water loss per unit stomatal area is of the order of about 100 times that from a unit area of free water surface.

It has been shown that the loss of water from a free water surface is practically unaltered when it is covered by a perforated sheet, even though the area across which diffusion occurs is reduced to only a small fraction of total area.

It has also been experimentally determined that if the openings or pores of the perforated sheet are as close together as 10–20 times their diameter, maximum diffusion occurs. Doubling the diameter of these openings only, have no effect on the rate of diffusion of water vapour. The same fact largely holds good for stomatal pore on the surface of the leaves.

Regulation or control of water loss by stomata. The general view came to be held that the stomata by opening and closing control or regulate the transpirational water loss from the plant.

The earlier idea was that stomata could close in anticipation of wilting and thus conserve the water supply of the leaf and prevent it from wilting. When later it was found that the stomata do not close until the leaf has wilted to some extent, it was then realised that the regulation, if any, was passive since under such conditions, the stomatal closure is not due to any special activity of the guard cells but to a decrease in their turgour caused by loss of water from the leaf cells.

The rate of transpiration is generally higher than the absorption of water by roots from soil during the day and the reverse is true at night. The water content of a leaf may vary over a considerable range during a 24-hour period without any visible signs of wilting of the aerial organs.

We have discussed before that no correlation generally exists between the number of stomata per unit area of the leaf surface, and the rate of transpiration.

The general view now held by most investigations is that when the stomata are fully open or nearly so, the transpiration rate is primarily determined by the same physical factors which control evaporation from a free water surface. Transpiration may actually increase when stomata begin to close and may decrease when they open wider. This is sometimes due to the fact, as we have seen before, that when the diameters of the stomatal pores are about 1/10th to 1/20th the distance between the two pores, no further opening of the pore can affect the rate of diffusion of water vapour through them. As the stomata gradually close, influence of physical factors of evaporation is lessened until at 50% opening of stomatal pore, variable atmospheric conditions become the primary factors in controlling transpiration, physiological regulation by the stomata, of the water loss, taking only a subsidiary role. Only when the pores are almost closed, the guard cells begin to exert a controlling influence on the water loss *regardless* of evaporation and climatic factors prevailing at the time.

MECHANISM OF OPENING AND CLOSING OF STOMATA

Because of the uneven and peculiar thickening of the walls of the guard cells (guard cell walls are thick only on the side bordering the pore, the rest of the wall being thin), increased turgour of the cell sap of the guard cells leads to opening, loss of turgour leads to closure. Since, we know that turgour is a hydrostatic pressure due to entry of water into the cells, stomatal movement is controlled evidently by water exchange and any factor that alters the water content of the guard cells will in turn affect their turgour and therefore the stomatal opening. The turgour pressure, as we very well know, at its maximum can equalise the osmotic pressure of the cell sap. Any increase in osmotic pressure of the cell sap due to an increase in solutes will therefore favour opening of the stomata; any decrease will favour its closure. It must be remembered, however, that any

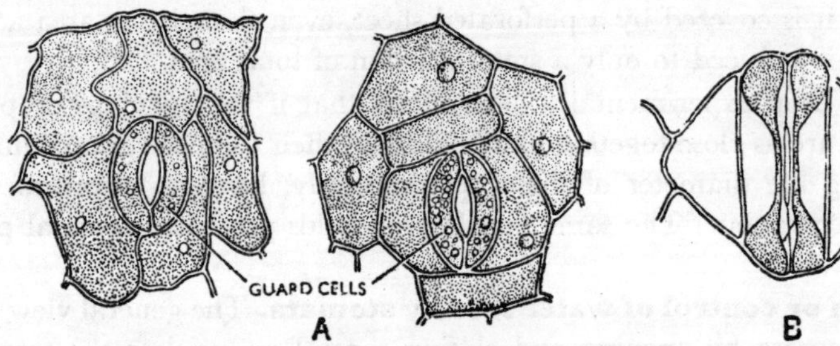

FIG. 674. Stomata with guard cells.
A. Stomata open. B. Stomata closed.

increase in solute content of the cell sap can affect only the osmotic pressure directly and the actual change in turgour pressure can occur only as a result of movement of water into or out of the guard cells. Any real change in the turgour pressure consequently is accompanied by a change in the degree of opening of the pore if the elastic cell wall is capable of a further stretch or shrinking. The other epidermal cells, particularly the subsidiary cells in the neighbourhood of the guard cells if they are turgid, exert, so to say, a back pressure on the guard cells and if all the epidermal cells were to increase equally in osmotic pressure and turgour pressure, no stomatal opening would immediately result. Similarly, if all the epidermal cells decreased equally in osmotic and turgour pressures, no closure would occur. Under such conditions stomatal opening may actually result from decreased turgour in the adjacent epidermal cells (subsidiary cells), even in darkness. In fact, a simultaneous water loss from all the epidermal cells may result in the opening of the stomata if the guard cells lose less water than the other epidermal cells, because the guard cells are the last of the epidermal cells to become flaccid and to reach zero turgour. The importance of the osmotic pressure of the guard cells in stomatal movement is shown by the high values—20–100 atm. that have been recorded for open stomata and the corresponding low values, 5–10 atm. in the case of closed ones.

The classical idea about changes brought about in the guard cells, was thought to be due to *three* possible causes of such change: (*a*) accumulation of sugars due to photosynthesis in the guard cells; (*b*) starch$\rightleftharpoons$sugar transformation (starch$\longrightarrow$sugar leading to opening and sugar$\longrightarrow$starch, to closure) and (*c*) active absorption of solutes by the guard cells from surrounding epidermal cells.

The following classical theory as to the mechanism of stomatal opening and closing in most dicotyledons is based on perhaps not much experimental evidence, but the theory, for a long time, certainly looked good enough on paper: (1) when the plant is exposed to light, photosynthesis occurs in the green cells of the leaf, including the guard cells; (2) as a result, the CO_2 content of the leaf is reduced due to its utilisation in photosynthesis; (3) the pH of the guard cells rises, i.e., become alkaline; (4) starch is converted into sugar in the guard cells; (5) the permeability of the plasma membrane of the walls of the guard cells increases in light; (6) osmotic pressure of the guard cells rises; (7) water enters the guard cells from the surrounding epidermal accessory cells; (8) turgour pressures increase in the guard cells; (9) the guard cells are forced apart due to their uneven cell wall-thickening, opening the stomatal pore.

The starch$\rightleftharpoons$sugar conversion is an enzymatically controlled chemical reaction and it is well known that enzymes are usually most active at some pH and inactive at others. The enzyme responsible for the reversible conversion of starch$\rightleftharpoons$sugar—*phosphorylase*—is present in the guard cells. It has been experimentally shown that in the presence of this enzyme at pH 5, the ratio of starch to sugar is about 4 times than it is at pH 7. More recent investigations indicate, however, that starch to sugar transformation is absolutely insufficient to explain the speed with which stomata open or close in most plants.

As a result, the starch$\rightleftharpoons$sugar balance theory of stomatal movement in light and darkness, given above, has always been suspect, to say the least.

The recently discovered fact that when the CO_2 content of the substomatal spaces is reduced artificially from 0·03% (normal percentage present in the air) to 0·01%, the stomata open. The closing in the dark is supposed to be caused by the increase in the CO_2–content in the intercellular spaces produced by dark respiration. It has also been shown that sufficient photosynthesis can occur in the guard cells from internally derived respiratory CO_2– and not dependent on an external atmospheric CO_2–supply—to permit usual rates of stomatal opening in light. On the other hand, however, all the behaviour of the stomata is not fully explained by this theory. Thus the night opening of the stomata, when it takes place in many plants cannot be related either to photosynthesis or to a reduction of CO_2–content as there is no photosynthesis during the hours of darkness.

Transpiration as a metabolic process

In the light of recent much detailed and critical work in this long neglected field of research, it is now accepted by most investigators that transpiration is actually a metabolic process, like photosynthesis and respiration and like them needs considerable usage of ATP molecules, photosynthetically phosphorylated by the guard cell-chloroplasts.

It has never been shown that guard cells completely devoid of chloroplasts are capable of opening in light. It has been seen that the development of the stomatal light response, in etiolated wheat and onion leaves, closely followed the increase in chlorophyll-*a* content of the tissue.

It has been shown recently that the behaviour of epidermal strips of *Senecio* provides strong evidence that, photosynthesis in the guard cells themselves, is responsible for the maintenance of stomatal opening in light. When these epidermal strips are treated with DCMU (dichlorophenyl dimethyl urea) even at 10^{-5} M dilution, there was complete closure of stomata in light and it is well known that DCMU is a particularly strong inhibitor of photosystem II in photosynthesis as also non-cyclic photophosphorylation.

Naturally all these observations raised the inevitable question of whether photosynthesis of the guard cells was actually needed for the production of carbon compounds, as was supposed before or is primarily required for the production of ATP molecules only and that too non-cyclically.

A number of lines of evidence—and certainly there are contradictory evidences—indicated that glycolate ($CH_2OH.COOH$) metabolism plays an important part in the opening of stomata in light—either to re-oxidise NADPH, facilitating non-cyclic photophosphorylation or glycolate may possibly be an intermediate in the synthesis of carbohydrates (there seems to be no doubt that sugars and other soluble carbon compounds, produced in photosynthesis may contribute to the increase in the osmotic value of the guard cells, especially at the later stages of stomatal opening).

There is, however, the question of stomatal opening *in the dark*. Allway & Mansfield (1967) found that stomata on leaves in which photophosphorylation was strongly inhibited, could open widely in response to a low CO_2-concentration, which suggests that ATP formed *directly* from photophosphorylation is not *essential* in stomatal opening, everywhere and in all cases. Evidences also seem to suggest the participation of an active transport mechanism, using ATP derived from oxidative phosphorylation of aerobic respiration. Evidences have also come to light, indicating that when epidermal strips are immersed in a solution containing K^+ ions, these accumulate in the guard cells, using respiratory ATP, as stomata open in darkness. (Fujino, 1967; Fischer, 1968). Guard cells do have an unusually high concentration of K^+ when they are open in the light and relatively small quantities when they close in the dark. The influx of K^+ ions is accompanied by an efflux of H^+ ions mediated by a H^+–K^+ exchange pump located in the plasma membranes of the guard cells. Electrical neutrality is maintained probably by anions of malate which are synthesized actively in the guard cells by photosynthetic fixation or respiration in light (or dark CO_2 fixation (?) in Crassulacean plants where stomata open at night), at the same time that K^+ influx occurs.

It is now known that some of the plant hormones may play important roles in the regulation of transpiration. The most remarkable effect is shown by abscisic acid (ABA). When water stress induces stomatal closure, the stomata remain closed even after the stress is relieved. It is now known that water stress results in a rapid movement of ABA from mesophyll tissue to epidermal cells and the ionic and metabolic status of guard cells is altered. Natural or applied ABA accumulates in guard cells. The effect of ABA slowly disappears. ABA probably regulates stomatal opening and closure by affecting the water level in the guard cells. It has been suggested that ABA may in-influence the permeability of cells.

The temperature of the air surrounding the leaf affect the speed with which the stomatal pore opens. Within limits at which protoplasm functions (up to 35–45°C.) the length of the time for opening is reduced to *one half* for every 10°C. rise in temperature. The explanation is most probably osmotic and also to some extent chemical. Increase in temperature increases the permeability of the unthickened part of the cell wall of the guard cells, thereby facilitating rapid entry of water into the guard cells. The enzymatic transformation of starch to sugar is also quickened by an increase in temperature. Rates of photosynthesis increase as also the rate of utilisation of CO_2 of the intercellular spaces due to rise of temperature and all contribute to a rapid opening of the stomata due to increased temperature.

Daily movement of stomata. In general, the stomata tend to show a diurnal periodicity closing at night and opening during the day. This periodicity is clearly related to light. There is great variation in the behaviour of the stomatal pore of the different leaves of the same plant and even on the different parts of the same leaf. The stomata on the upper surface may behave differently from those on the lower.

Some stomata, however, open at night, others close at noon when the light intensity is at a maximum. Wilting may cause most of stomata to close regardless of the light factor. There are several types of diurnal movement of the stomata:

(1) Stomata of this type under favourable conditions are open all day and closed all night. Under less favourable moisture conditions the stomata close partially for a time during the middle of the day, when the rate of entry of water into the guard cells is incapable of keeping pace with the high rates of transpiration. The period of partial midday closure increases to complete closure as the moisture conditions in the guard cells become less and less favourable. With the appearance of midday closure, night opening develops which progressively increases with the increase in the day closure, until finally there may be a partial opening of the stomata all night and a closure all day. This may happen under extreme scarcity of moisture in the leaves. This type of opening and closing is observed mostly in thin leaved mesophytes, e.g., peas, beans, etc.

(2) Movement of stomata of this type is markedly different from the previous type under the same favourable conditions. The stomata of potato are open continuously during day and night except for about 3 hours following sunset. Midday closure is seldom observed and that only under extreme moisture deficiency conditions. Potato, cabbage, onion, banana, etc., afford examples of this type of stomatal movement.

(3) In the third type, which is typical of all cereals, there is absolutely no opening during the hours of darkness, no matter how slight and how short the day opening may be. There are many closed stomata even during the daytime.

STRUCTURAL FEATURES OF PLANT WHICH INFLUENCE THE RATE OF TRANSPIRATION

A number of different species of plants growing under identical environmental conditions differ some times very greatly in the rates of water loss. Physiological factors, such as cell sap concentration, stomatal behaviour, etc., are no doubt partially responsible for the fact that all plants even under same edaphic and climatic conditions do not lose water vapour at the same rate. Structural differences in leaves also certainly account in part for different rates of transpiration. Many of the supposed effects of structural differences in plants influencing transpiration rates are not, however, verified by experimental evidence. Our discussion will be limited to the following structural features of plants supposed to influence transpiration rates:

Cutinised and waxy thickening of the epidermal cells. Water certainly evaporates more readily from uncutinised epidermal cell walls than from these which are coated with a layer of cutin. It has been observed, that cell walls which are heavily cutinised or covered with wax, show considerably less water loss from them. Cutin as also suberin are certainly very effective in checking loss of water from leaf surface. The deposition of cutin on the epidermal cell walls, however, has one serious disadvantage; it tends to bring about a closure of stomata, thus severely restricting exchange of gases like CO_2 and O_2 between the plant and the atmosphere—a function absolutely essential for normal metabolic activity of the green plants. A heavily cutinised leaf surface is usually shiny and reflects much of the solar radiation falling on it. Theoretically, because most of the radiations falling on the cutinised leaf surface is reflected, less is available for absorption. The total effect will be to reduce transpiration rates through reduced heat absorption.

It must also be understood clearly that magnitude of the thickness of cuticle is not a limiting factor in retarding rates of transpiration. After a particular thickness of deposition of cutin on the epidermal cell walls is attained, any further increase in thickness has no appreciable influence on the rates of water loss.

Epidermal hairs. Living hairs on the epidermal cells most probably increase the rate of cuticular transpiration by increasing the surface area of the leaves. Dead hairs, on the other hand, act as a mechanical barrier to the effect of light and wind on stomatal transpiration. It should be pointed out clearly, however, that many actual experiments have indicated that the efficiency of a coating of hairs dead or living in lowering transpiration rates may be negligible.

Ratio of internal/external surface of leaves. We know that most of the water vapour lost from leaf surface evaporates from the walls of the mesophyll cells which are bound by intercellular spaces and which constitute the internal evaporating surface of the leaves. This proportion area of internal evaporating surface to external total leaf surface varies greatly not only in leaves of one species compared to those of the other, but may also vary in the leaves of same species if they have developed under different environmental condition. In general, the larger the ratio of the internal evaporating surface, i.e., the total area of the intercellular spaces to the total leaf surface, the greater is the rate of transpiration.

Sunken stomata. Stomata of many species of plants, adapted to conditions of extreme soil and atmospheric drought, are sunken in pits below the level of the epidermal surface and likewise are well below the level of atmospheric disturbances in the vicinity of the leaf. Plants with sunken stomata have generally been found capable of significant reduction in transpiration. Diffusion through a small microscopic opening such as stomata is certainly slower if the water vapour must pass through a relatively longer tube (depth of the pit) before reaching the external air than diffusion through a shorter tube (thickness of the stomatal pore only).

Structure and distribution of the root system. Under some environmental conditions, a particular plant may develop a greater total leaf surface in proportion of the extent of its root system. This structural adaptation will certainly favour maintenance of higher transpiration rates. The distribution of root system in the soil frequently influences the water balance in plants. In a habitat where deep-rooted and shallow-rooted plants are growing side by side, the former may transpire more rapidly during dry periods than the latter. In the former the roots penetrate to a soil horizon which may still contain available water while the roots of the latter are in a moisture-deficient surface soil due to drying of superficial layers of soil if caused by atmospheric and soil drought.

ENVIRONMENTAL FACTORS AND TRANSPIRATION

The rate of transpiration of a plant varies from day to day, from hour to hour and frequently still more rapidly. The difference in the rapidity with which water is lost by plants when the stomatal pores are even half open is primarily due to the effects of environmental conditions. When the stomatal pores are fully open, the diffusion of water vapour from the intercellular spaces of the leaves to the external air is primarily controlled by the laws of diffusion. In other words, the larger the difference between the vapour pressure or concentration within the intercellular spaces of the leaf and that of the surrounding external air, the greater is the transpiration rate. So any environmental factor which will tend to steepen this gradient will facilitate exit of water vapour from the plants to the air outside.

The important environmental factors, which help in the development of a difference between the vapour pressure inside and outside the plant and thus facilitating rapid diffusion of water vapour out of the plants, are given below. In this discussion we assume that the supply of water is abundant in the leaves and absorption keeps pace with transpiration and the stomatal pores are not less than 1/3 open:

Light. Though light occupies a position of prime importance among the environmental conditions, its effect on the development of a difference in vapour pressure within the plant and the outside air is negligible except in so far as its indirect undoubted effect in increasing the temperature of the leaf. The principal effect of light in transpiration, as we have seen in our previous discussions, is predominantly through its influence on the opening and closing of stomata. In the complete absence of light, in most species of plants under normal environmental conditions, the stomatal pores are usually closed.

Humidity. In general, the greater the vapour pressure of the atmospheric air, the slower is the rate of transpiration, other factors remaining unchanged. We know that the air in the intercellular spaces of the leaf is maintained nearly always at the saturation point if the supply of water to the leaf is uninterrupted and abundant while in the outside atmosphere, conditions are rarely favourable for attainment of saturation vapour pressure. It is well known that plants show very high transpiration rates on dry days compared to moist ones and as a result wilting of the aerial parts is a common phenomenon on hot, dry days if the supply of water to the leaf is not fast enough to keep pace with rapid diffusion of water vapour from the leaf to the external air. On rare occasions, when the vapour pressure of the air outside approaches saturation point and in extreme cases when the diffusion pressure difference between inside and outside becomes zero, no transpiration occurs even when the stomata are fully open.

Temperature. Before active transpiration starts leaves exposed to direct sunlight usually have temperatures from 2–5°C. in excess of that of the atmosphere.

When the temperature of the leaf and the surrounding atmosphere is raised by direct sunlight, unless the leaf is markedly deficient in water, the vapour pressure of internal atmosphere, i.e., the intercellular spaces of the leaf maintains essentially a saturation vapour pressure for the prevailing temperature. As the internal volume is constant in size and is small, the evaporation of water from wet mesophyll cells is sufficient to saturate the comparatively small volume of air inside the leaf. In the surrounding atmosphere, however, vapour pressure conditions are very different—on clear days, there is frequently little change in the vapour pressure—evaporation into the atmosphere is insufficient to permit a rapid building up of the vapour pressure towards the value for a saturation atmosphere. Increase in temperature certainly brings about an increase in the movement of the water molecules and if the volume of the external atmosphere remains constant a small increase in vapour pressure would result. But this again is nullified by an expansion of atmospheric volume brought about by an increase in temperature.

Thus, within a certain range of temperature, at which protoplasm remains fully active, the effect of rise in temperature always results in an increase in the difference between vapour pressure within the plant and the outside atmosphere and hence an increase in the rate of transpiration.

Wind. If all other external conditions are favourable, usually a moderate increase in the velocity of wind, results in an increase in the rate of transpiration. It may be due to the effect of wind in removing the near saturated layer of air in contact with the leaf surface. The result of this removal of saturated atmosphere from the immediate vicinity of the leaf will be a steepening of the vapour pressure gradient, that is in a reduction of the distance across which the vapour pressure difference exists. This will, however, primarily affect the cuticular transpiration and only relatively slightly, the stomatal transpiration. This is due to the fact, that the removal of saturated layer of air in contact with leaf, will alter only the distance across which vapour pressure difference exists in stomatal transpiration compared to the cuticular transpiration, only a little, where the distance is simply from one side of the cuticle to the other. In stomatal diffusion of water vapour the said distance is from the wet mesophyll cell walls across the intercellular spaces and the stomatal pore to the external surface of the leaf and this distance is only slightly affected by removal of saturated layer from the vicinity of the leaf surface.

The effect of wind in removing the saturated air from the immediate vicinity of the leaf surface is probably of less importance than its effect in causing swaying of branches and shoots and the bending, twisting and fluttering of the leaves. It has been shown experimentally that immobile leaves usually transpire much less than leaves which are fluttering or bending in the wind, when both are exposed to winds of equal velocity. Such bending may increase the rate of water vapour loss from leaves in part, by sudden compressing of the intercellular spaces which actually shakes the water vapour out of the intercellular spaces through the stomata into the external atmosphere.

Winds of high velocity may sometimes actually reduce transpiration rate by causing closure of stomata. If the wind produce a cooling effect on the leaves due to increased heat emission, this may conceivably bring about a reduction in the rate of water loss.

Atmospheric pressure. The effect of atmospheric pressure on the transpiration rates is in general only of theoretical importance. It can be shown experimentally that a reduction in atmospheric pressure results in an increase in the rates of water vapour loss from leaves. The theoretical explanation will be that a reduction in the density of the atmosphere would permit the diffusion of water vapour to occur into it more rapidly. However, for all practical purposes, in any given locality, the variation in atmospheric pressure is too slight to have any significant effect upon rates of transpiration. The high altitude flora shows a general trend of high transpiration rates compared to the flora of the plains, if other soil and atmospheric factors are not limiting.

Antitranspirants. There are certain compounds which act as inhibitors of transpiration rate through their highly specific action on guard cells. The most reliable of such inhibitors found, thus far, is *phenyl mercuric acetate.* This compound closes stomata in many plants (such as *Vinca*, tobacco, maize, etc.) when sprayed on leaves, at even as low as concentration as 10^{-4} M. From critical experimentation, this compound was found to inhibit primarily the stomatal diffusive resistance—a single treatment may close the stomata for about 6 weeks—the growth rate of the plants, however, remaining unaffected. Several growth retardants also reduce transpiration rates.

Other compounds have also been successfully tried, such as *monomethyl ester* of *nonenyl succinic acid* and *decenyl succinic acid* (10^{-3} M), CO_2 etc.

EXUDATION OF WATER IN LIQUID FORM

Guttation. We have already touched on the phenomenon of guttation in our previous discussions of root pressure and active water absorption. By far the commonest and conspicuous manifestation of root pressure is the exudation of drops of liquid from the edges and tips of leaves. This process of escape of water in the liquid form from uninjured plants, particularly from leaves, is called guttation. Guttation has been reported from more than 300 genera of flowering plants, although there are many species in which it has not been observed.

Guttation occurs more frequently and abundantly under conditions which favour rapid absorption of water by roots combined with reduced transpiration rates. Thus guttation is most common in plants growing in moist warm soil with their aerial parts surrounded by humid air. Only when the rate of absorption is sufficiently in excess of transpiration so as to cause development of positive root pressure in the xylem vessels, guttation takes place. Conditions which hinder intake of water by roots such as a cold, dry soil, a high concentration of solutes in the external medium, and also conditions favouring high transpiration rates, prevent guttation. Guttation can be stopped by reducing the root pressure, for example, simply by watering the soil with a dilute solution of sugar or KNO_3, thus increasing osmotic pressure of the soil solution, which may greatly reduce water absorption by roots.

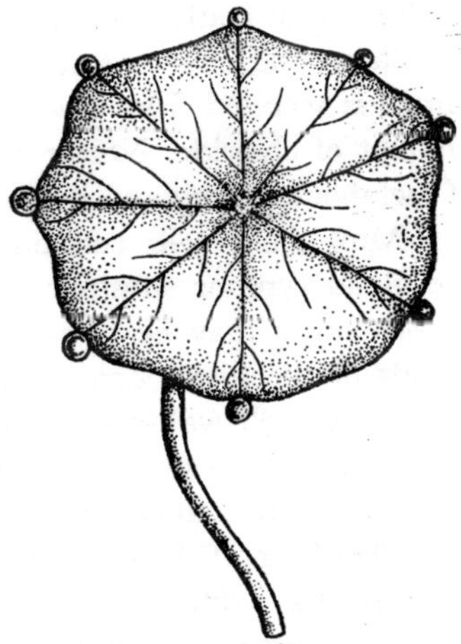

FIG. 675. Guttation. Exudation of drops of liquid from *Nasturtium* leaf.

Guttation is very commonly and frequently observed from the tips and edges of leaves of grasses, *Nasturtium*, *Colocasia*, tomato, etc., early in the morning, after a moist and warm night.

Guttation occurs through specialised structures known as *hydathodes* or *water stomata* or water pores. The term water stomata is very inappropriate, for hydathodes have no osmotically active guard cells but consist of a large opening, below which there is a rather large air chamber, bordered by a mass of thin-walled, loosely arranged parenchymatous cells (*epithem*). The xylem of a small vein terminates below each air chamber.

The positive pressure, originating in the root and developed in the xylem vessels forces water in the liquid form out of the xylem into the intercellular spaces. The intercellular spaces are thus injected with liquid water (under all normal conditions they are occupied by air, always at near saturation point, by evaporation of water from wet mesophyll cells) which floods the intercellular spaces of the epithem, ultimately causing an overflow through the pore of the hydathodes to the exterior of the leaf.

Guttation may also occur through ordinary stomata as in some grasses and legumes. While most plants exude only a few drops of water during an entire night, young leaves of species of *Colocasia* have been observed to lose as much as 10–100 ml of liquid water in a single night by guttation. Although *Colocasia antiquorum* exudes almost pure water, guttation liquid usually contains small quantities of soluble sugars and salts. Instances of injury to the leaves in the nature of *tip-burn* have been reported in some species of plants where large concentrations of salt are left on the surface of the leaves in the region of apices and edges, when evaporation removes the guttation water. The white incrustation which is sometimes seen on the surface of the leaves, is also due to evaporation of guttation water, leaving the dissolved salts as a thin layer on the surface of the leaves.

It is probable that guttation results in a thorough soaking of the leaves, which may conceivably make such leaves more susceptible to the attacks of pathogenic bacteria and fungi.

Glands. Glands are found on almost all parts of the plant. Many of them are nothing but modified epidermal hairs or cells. Nuclei of gland cells are usually relatively large compared to the total cell volume. Certain types of glands secrete a very dilute solution, mainly of sugars and salts. Sometimes nectar, resins, volatile oils and enzymes are also secreted by glands. This exudation of sap is commonly called *secretion* and is apparently caused by forces which develop within the gland and not by the hydrostatic pressure developed in the xylem vessels. Glands certainly are not closely connected with xylem elements as hydathodes are. The mechanism of the development of the forces causing secretion by glands, is not clear.

The outermost cell layer of the glands located in the lower third of *Nepenthes* pitchers is an epithelium with cylindrically shaped typical gland cells, having a dense cytoplasm and characteristic cell wall incrustations. The digestive fluid in the pitcher lumen is secreted from this inner pitcher wall; pitcher fluid contains an appreciable amount of chloride. Nectar secretion, in leaves and flowers, requires the expenditure of metabolic energy and it is also hormonally controlled.

Most gland cells contain a very large number of mitochondria, irrespective of the gland type or substance transported. In the excretion of salt, the role of these mitochondria may be to supply energy for the working of ion pumps. The *hydropote* glands of the lower epidermis of floating water-lily (*Nymphaea*) leaves contain large nuclei; the cell wall-protuberances and the space left in glands are filled with mitochondria. A high proportion of activity of the transport of at least SO_4^{--} seems to be associated with mitochondria.

Bleeding. If the stem of a herbaceous or woody plant be cut or broken, especially in the spring, a slow exudation of sap often occurs from the cut stump. This phenomenon is called bleeding and under certain favourable conditions large quantities of sap solution, as much as 5–6 litres, may be lost in a single day.

This process differs notably from the phenomenon of bleeding from a wound in animals by the fact that whereas the composition of blood of one species of animal is approximately constant, that of the bleeding sap of a plant may vary within wide limits, according to the season of the year and the place of the wound. Bleeding, however, is not a natural phenomenon.

Some plants yield a bleeding sap that contains a high percentage of sugar, up to about 18–20%. They are nearly all monocotyledonous palms.

It is interesting to note that in plants that exude a bleeding sap containing a high percent of sucrose, the bleeding sap originates in the phloem, whereas in many herbaceous plants (e.g., tomato) the exuding bleeding sap is from the xylem. In the xylem-bleeding sap, the sugars are largely absent whereas the concentration of inorganic ions, especially calcium, is very high.

In some species of plants, bleeding appears to be due to the development of a positive pressure in the xylem vessels—root pressure; in others to pressures developed in the sieve tubes in phloem instead of xylem, and in still others it may be due to locally developed pressure in the neighbourhood of injury.

WILTING CO-EFFICIENT

(ALSO CALLED WILTING PERCENTAGE OR WILTING POINT)

Wilting co-efficient is a physiological measure of the water relations of the soils. It is defined as the percentage of water content (expressed as % dry wt.) of a soil when the plant or plants growing in it has just reached the condition of permanent wilting. A permanently wilted plant is usually considered to be one which will not recover its turgidity unless water is supplied to the soil. Permanent wilting is not necessarily an accurate indicator of the wilting co-efficient. Thus in the seedlings of bean, a relatively large proportion of water content of the plantis stored in the lower part of the hypocotyl. During loss of water, transpiring leaves draw upon this reserve water in the stem when the supply of water from the soil is cut off. Thus the critical point of soil moisture may be reached previous to permanent wilting.

In order to determine the wilting co-efficient of a sample of soil, the sample is enclosed in a water-proof vessel and the 'test' plant is generally allowed to develop from seed in the soil until it has attained a small size. The soil surface is then sealed over so that all loss of water from the soil occurs through the plant. Since no water is added to the soil the plant eventually shows signs of permanent wilting due to loss of water from leaves by transpiration. As soon as this occurs, a sample of soil is removed and its water content determined by drying in an oven.

Before extensive determinations of wilting percentage for plants growing under different soil and climatic conditions, it was supposed that plants differed markedly in their capacity to reduce the water content of the soil. It was assumed that xerophytes, which endure drought conditions could deplete the moisture content of a soil to lower value before showing permanent wilting than those species growing in normal supply of water in the soil. Extensive investigations have shown, however, that hydrophytes, xerophytes, mesophytes—all reduce the water content of a given type of soil to about the same value, before showing permanent wilting.

While the wilting point of a given soil shows no appreciable variation when measured by means of different plants growing in it, this value varies greatly with the type of the soil. The percentage of water remaining in a soil when permanent wilting of plants growing in it occurs, ranges from 5-10% for sandy loam, 10-15% for silty loam and 15-20% for clay soil. Thus we see that different soils have different water-holding capacities. The wilting point at which plants can no longer remove water from the soil whatever the texture of the soil may be, sand, silt or clay corresponds to a force of about 15 atmospheres. No root cells can possibly have osmotic pressure higher than 15 atmospheres and water held in the soil at a force of more than 15 atmospheres is thus completely unavailable to the plant. Thus, the amount of water available to the roots of most plants is held in the soil by forces between 0·3 atm. (i.e., at field capacity) and a maximum of 15 atm. The quantity of available water certainly varies with the type of the soil; it is relatively large in fine textural clay and relatively small in a coarser sandy soil. The force with which water is held in the soil increases with decreasing water content and hence at permanent wilting point it is independent of the species of plant growing in a soil. In general, wilting point seems to be controlled almost entirely by soil conditions and type of soil—it is only slightly influenced by the species of plant growing in it or by climatic conditions to which the plant is exposed.

The significance of wilting point lies in the fact that it is essentially a measure of that fraction of soil water which is unavailable to the plants and is in turn a measure of the hygroscopic and chemically bound water present in the soil. The determination of wilting percentage under field conditions, presents many difficulties compared to the more or less easy manipulation which could be obtained in the laboratory. Large discrepancies are sometimes found between the values determined by laboratory methods and those determined actually in the field, for the same type of soil. In the laboratory, small water-proof containers are used and effective penetration by roots in the small soil sample can be taken for granted and a fairly uniform reduction of water content of the small soil sample can be assured. In field, conditions are very different, however, soil is less effectively penetrated by roots, some portion of the soil in the field may be less completely drained of water than others, soil at the immediate vicinity of the roots may be at wilting point while more remote portions of the soil may still be at field capacity, etc. Other difficulties which stand in the way of correct determination of wilting point under field conditions are the direct evaporation of water from the soil reducing the water content of the soil, local variation in the soil texture, i.e., some parts of the field may have more clay while others, more silt or sand, lack of uniformity of root development, etc.

To surmount the difficulties encountered in the determination of wilting co-efficient of the soil under field conditions and from our knowledge that wilting co-efficient is practically the same for a given soil for any plant under all climatic conditions, several indirect purely physical methods have been proposed for the determination of wilting percentage under field conditions. These methods are based on the relationship of the wilting point to the moisture retentiveness of the soil as measured by certain physical standards. By determining in the laboratory, some factors on the moisture retentiveness of a sample of soil from a field in which the plants are growing, wilting co-efficients can be calculated quite accurately for all types of soil. In these determinations of wilting co-efficients, the part played by plants is altogether eliminated.

Large scale determinations of wilting co-efficients have been one of the most important routine work in ecological and agronomic ones in recent times. By determining the wilting points of different crop fields, we can obtain some idea about the texture of the soil—suppose, the wilting point of a field is about 15%, we can guess that the soil is mostly clay, whereas a value of about 5-10% will show that the soil is mostly sandy. Consequently if the soil is clay, more water is needed for normal growth of the plants and water must be added to the soil by irrigation when there is no rainfall than when the soil is sandy where less irrigation-water will be needed. Of the various physical methods employed for the determination of wilting percentage of soils, the two most commonly used are following:

(1) *Relation of the wilting percentage to the moisture equivalent of the soil*. The moisture

equivalent of a soil is defined as the percentage of water that a soil can retain in opposition to a centrifugal force 1000 times that of gravity.

Wilting co-efficient=moisture equivalent/(1·84±0·013).

(2) This method depends upon the determination of percentage of coarse, medium and fine particles in several samples of a soil from a field and then multiplying the three with three constant factors and finally adding them up.

Wilting co-efficient=percentage of sand (coarse) particles ×0·001+percentage of silt (medium sized) particles ×0·12+percentage of clay (very fine) particles ×0·57.

WATER BALANCE PROBLEMS IN PLANTS

Primitive plants, not to speak of life itself, are believed to have originated in the seas where the control of such phenomena such as transpiration, wilting and drought did not arise at all. In the aqueous medium, protoplast evolved to a high degree of complexity which adapted fully to a water-saturated condition. Later, when plant life began to encroach upon the land, the most significant advance was represented by the translation from an aquatic to terrestrial habitat. In this translation two main obstacles had to be surmounted: overcoming gravity and mechanical forces of air movement and overcoming drought. Both the obstacles were most successfully overcome by plants coming to the land from water which possessed or developed a vascular system and especially those in which the soma was divided into a root for water absorption and an aerial shoot with stem and leaves. The problem of keeping the water content of the protoplasm of the cells above a certain high minimum, necessary for its normal and continued existence (in the terrestrial environment the risk of almost continual loss of water from the aerial parts to the atmosphere was great) was solved and the landward migration became possible with the gradual development of more and more efficient adaptations to cope with the entirely new conditions in which the migrated plants found themselves.

Adaptations permitting the maintenance of a satisfactory balance between the absorption and loss of water evolved mainly in two different directions. On the one hand, there were developed nearly water-impervious coverings of cutinised and suberised tissue which greatly helped in checking excessive water loss by transpiration. On the other hand, roots and rhizoids, structures highly efficient in extracting moisture from the deeper layers of soil, were evolved. In the water habitat, absorption was no problem for water flows freely and abundantly to aquatic plants.

Cutin and suberin are, to some extent, effective agents in regarding the loss of water from a plant surface. However, at the same time they have the serious disadvantage of restricting the exchange of gases between the plant and the atmosphere—a function which is absolutely necessary for both photosynthesis and normal respiration. This difficulty was surmounted by pulling apart of certain epidermal cells so as to leave intercellular gaps or pores (stomata) in the superficial layer which are encrusted with cutin or suberin. Such a solution to the problem was at its best only an apparent compromise for even though all but a very small fraction of the total exposed surface is protected by water-proof coverings, the stomata and lenticels which constitute the small area, permit the loss of great quantities of water vapour. Exchange of gases like CO_2 and O_2 and water vapour between the plant and the atmosphere takes place by diffusion that is always from higher concentration or pressure to lower concentration or pressure. Furthermore air outside the leaf contains only 0·03% of CO_2 and 21% of O_2, while the internal atmosphere of the mesophyll of the leaves is nearly always saturated with water vapour under conditions of normal absorption by roots. Thus transpiration is much more ıpid than inward diffusion of either CO_2 or O_2.

Owing to the fact that the average vascular plant absorbs water through its roots at the same time that it loses water vapour to the air from the aerial parts, the ratio between these two physiological processes determines the amount of water contained in the tissues. *The ratio between the water gain and water loss is called the water balance of the plant.* The true significance of water to plants cannot safely be understood in terms either of water intake or of water loss but only as water balance.

In as much as absorption and transpiration are partly controlled by the environment and partly by structural organisation of the plant there are evidently both internal and external aspects of water balance. The external aspects consist of: (1) the amount of soil water which is available to the root system and (2) intensity of transpiration-favouring factors.

The internal aspect of water balance depends largely upon those structural and accompanying functional characteristics which tend to minimise or increase the natural limitations of the habitat.

In many terrestrial plants, absorbing and conducting systems are relatively inefficient in supplying water to meet the demands of high rates of transpiration particularly in summer, even at times when available water is plentiful.

As a result of this inefficient conduction of water from the absorbing to the transpiring surface, most plants, even if growing in moist soil, lose more water during daylight hours than can be fully compensated and replaced by absorption.

At night, however, transpiration rates decline sometimes to no more than a small fraction of the diurnal rate and this permits a reverse in the direction of water balance trend, for absorption by roots at night is more or less unaffected, so that the deficit due to transpiration during daylight hours is gradually changed to a surplus. *It has been truly said that if just once, night failed to follow the day, many of the plants on earth would perish.*

Susceptibility to an unfavourable water balance varies continuously throughout the life cycle of the average plant. Some seeds and spores are often capable of enduring extreme drought.

The seedlings of cereals retain a high degree of drought resistance until several days old when the first leaves unfold when they become very sensitive to desiccation.

In comparison with plants grown under optimal moisture conditions, those grown under an unfavourable water balance have all the xerophytic physiological characteristics. By xerophytic here, is meant non-inherited structural characteristics, which clearly results from an exposure of plants during tissue differentiation to an unfavourable water balance.

EXPERIMENTS ON STOMATA AND TRANSPIRATION

I. Determination of stomatal frequency or number of stomata per unit area of a leaf

Standardise an ocular micrometer with the help of a stage micrometer and calculate the value of 1o cular division. 1 mm=1000 μm; therefore 1 stage division is equivalent to 10μ. Then, if x stage divisions are equal to y ocular divisions, 1 ocular division is equal to $\frac{10x}{y}\mu$m. Standardise in both high and low powers of the microscope.

Now, find out the diameter of the field of vision of the microscope by the ocular scale. The area of the field of vision is easily obtained. (The area of a circle is πr^2 where r is the radius of circle, in this case half of the diameter of the field of vision.)

Count the number of stomata in this field. Take several readings by moving the epidermal peelings of leaves (e.g., *Crinum*) through the microscopic field of vision. The average of these readings divided by the area of the field is a measure of the *stomatal frequency* of the leaf, which is generally specific for each species of plant.

II. State of opening of stomata

(a) *Lloyds method.* Peel off epidermal tissue from *Rheo* and other leaves and quickly put them into hot alcohol. The alcohol fixes the stomata, preventing any further movement of the guard cells. Examine under microscope and measure: (1) the entire area of stomatal apparatus—guard cells and the aperture and (2) actual area of the stomatal opening with the help of standardised ocular. (If both those structures are taken as roughly ellipsoidal in nature, the areas could be very approximately taken as equivalent to $\pi(a \times b)$ where 'a' and 'b', represent half of the two axes of the ellipse.)

(*b*) *Infiltration method.* Place a drop of absolute alcohol on the leaf and observe the rate of penetration, i.e., the rate of formation of transparent patches due to penetration of alcohol through the stomatal openings and its accumulation in the intercellular spaces.

(c) *Durofix method.* Thinly smear both the surface of a suitable leaf while it is still attached to the plant carefully with *Durofix* adhesive. Allow the durofix to dry up quickly into a thin papery film. The film is now stripped off and the impressions of the stomata in the durofix-film can be observed under microscope.

III. Use of Darwins porometer

Darwins porometer (Fig. 676) is a useful apparatus for following the changes in stomatal apertures, i.e., degree of opening of stomata. It essentially consists of a vertical tube, one end of which is dipped in a beaker of water. The other end is fixed to a T-tube, into one arm of which is attached a rubber tubing provided with screw-cock and a small glass chamber (porometer cup) is attached with rubber tubing to the other arm. The porometer cup is affixed to the undersurface of a dorsiventral leaf (stomata restricted only on the lower epidermis) and water is sucked up to a certain fixed level in the vertical tube and the screw-cock is closed and all connections made air tight. The rate of fall of water in the vertical tube observed, provides a measure of the rate with which air passes out under suction through the stomatal openings of the leaf. If the cup is affixed to upper surface of such a leaf where there may not be any stomata, the water level in the vertical tube is maintained, but no fall in level is observed. This certainly demonstrates that air cannot pass through the cuticle. The rate of fall of the water level

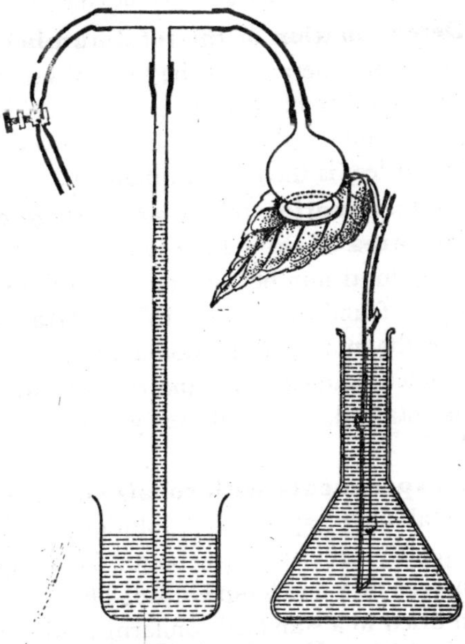

FIG. 676. Darwins porometer.

and hence that of passage air is thus roughly indicative of the size of stomatal openings. If the rate of fall diminishes or becomes extremely slow, it evidently indicates that the stomata are closed or about to close.

IV. Measurement of the leaf-area

(*a*) The vast majority of leaves of oval, ovate and obovate shapes, which are the commonest in a tropical rain forest, have ratios very close to 2/3 of the overall rectangle. The area of a blade can thus be, more or less, correctly and at the same time quickly obtained, by multiplying the recorded rectangular area (length × breadth) by 2/3, over a wide range of leaf-blade forms.

(*b*) *By graph paper.* Sketch the outlines of leaves on a graph paper and determine the area by counting the number of squares.

(*c*) *By weighing.* Trace the outline of a leaf on a thick uniform cardboard. The leaf outline is cut out by scissors and weighed in a balance. Then, weigh a known area, say, about 4 sq. cm of the same cardboard. The area of the leaf is thus easily found. The tracing of the outline of the leaf on cardboard paper is done in order to ensure uniform thickness of the area which is not obtained if actual leaves are used in weighing.

(*d*) *By planimeter.* This is a simple gadget for determination of areas of unspecified surfaces. For routine work, this is a very useful apparatus as areas could be read out quickly from readings in a small vernier attached to the instrument.

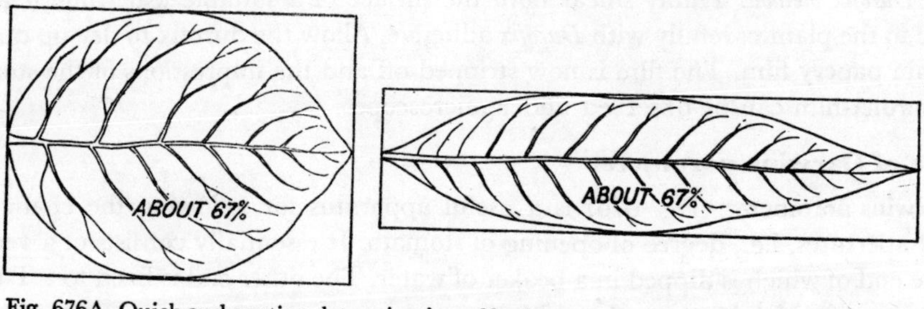

Fig. 676A. Quick and routine determination of leaf area; two-thirds of rectangular area.

V. Determination of the total number of stomata in a leaf

Find out the area of the field of vision of microscope under both powers and the average number of stomata present in this area. Area of the entire leaf is then determined by any of the four methods given above. The total number of stomata in the particular leaf is then easily obtained.

VI. Determination of the percentage of the total stomatal area in relation to the area of the whole leaf

The total number of stomata in the experimental leaf as also the area of a single stoma are found as in expts. IV and II(a) respectively. The product of these two values gives us the total area the stomata occupy in the leaf. The area of the whole leaf is then determined as in expt. IV and the percentage the stomatal area in relation to the entire leaf is obtained. Stomatal area generally is only between 1 and 5% of the total area of the leaf.

VII. Experiments with cobalt chloride paper

Soak filter paper in 5% solution of cobalt chloride and hang it up to dry. When dry, quickly cut into strips about 2·5 cm × 1·25 cm and dry these strips again thoroughly in an oven at a low temperature. The original colour of the wet paper is pale pink, but when dried to a standard uniform shade, the colour becomes an intense blue. The uniform-sized dry strips are stored in a desiccator.

The treated blue dry paper is thus a moisture detector, turning pink when left in the air (the air containing moisture) or when placed in contact or near an evaporating surface. The time taken for a blue colour to disappear and attain a standard pink colour (determined by a stop watch) is a measure of the amount of evaporation as also the rates of water loss from the surface of the leaf to which it is attached.

The following tests can be conveniently performed with these dry cobalt chloride paper strips:

(*a*) Select a suitable leaf attached to a potted plant. Lay a piece of strip on each surface of a leaf and immediately press them gently down between two slides, held one above another with the help of clips (Fig. 677). Compare the respective rates of water loss from both upper and lower surfaces of a leaf and also from young and old leaves.

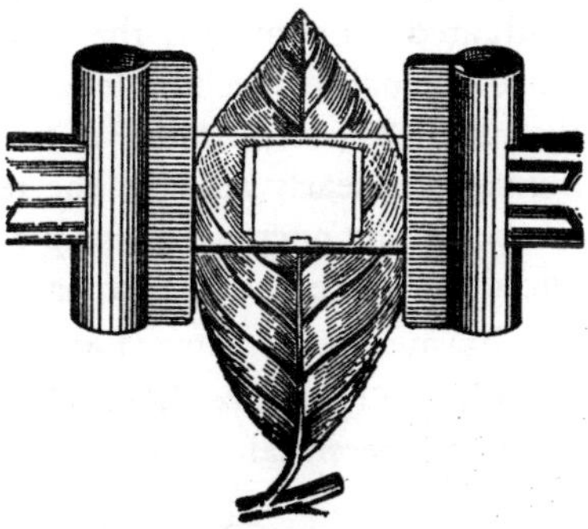

FIG. 677. Fixing of cobalt chloride paper strips on a leaf.

(*b*) Compare the respective rates of one leaf kept in darkness for some hours (stomata closed) and the other kept in the open.

(*c*) Compare the respective rates of water loss from a leaf and also from an open water-evaporating surface (physical evaporation). For open evaporating surface a petri-dish containing water can be taken and the blue paper-strip is held by suitable arrangement as close to the water surface as possible, without actually touching it.

The time in seconds for a standard colour change of cobalt chloride paper over a free water-evaporating surface (S) divided by the time for the same colour change on the leaf (E) is a measure of what is sometimes defined as *transpiration index*. It is convenient to multiply this index by 100 to obtain the water loss from the leaf surface as a percentage of the evaporation from a free water surface.

Transpiration index $S/E \times 100$.

VIII. Measurement of rates of transpiration

The rate of transpiration can be conveniently measured with the help of standard *potometers*, such as, Ganong's and Farmer's (Figs. 678 and 679). Ganong's potometer

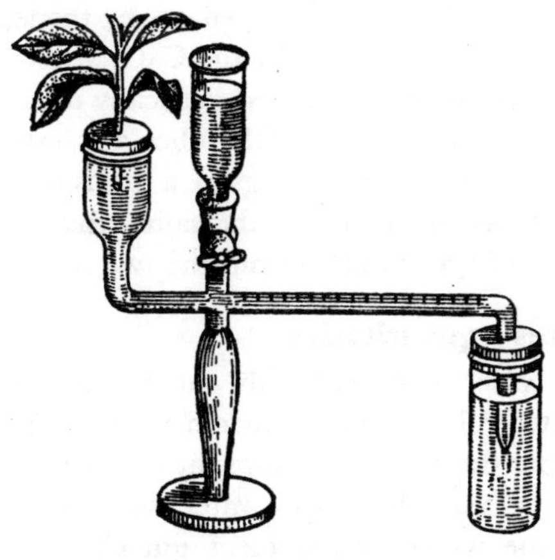

FIG. 678. Ganong's potometer.

FIG. 679. Farmer's potometer.

essentially consists of a long narrow graduated tube held horizontally with two vertical short tubes, one of which takes a leafy twig (cut under water) through a cork while the other fitted with a stop-cock, acts as a reservoir. The distal end of the horizontal tube is dipped into water in a beaker which may be coloured with eosin. The whole apparatus is filled with water while keeping the stop-cock of the reservoir tube closed and all connections are made air tight with paraffin. As water is lost by transpiration from the leafy twig, the coloured water is seen to enter horizontal arm from the beaker. The end of the tube dipping into water in the beaker is then momentarily removed from water to allow an air-bubble to enter it. The tube is dipped again in water and as transpiration proceeds, the air-bubble is seen to travel with the transpiration current through the graduated arm towards the transpiring object. The time taken by the air-bubble to move through a specified length in the graduated tube may be noted. If the volume of the tube is known, the rate of transpiration can be approximately calculated. When the bubble has nearly reached the reservoir tube, it can be pushed back into the water of the beaker by opening the stop-cock of the reservoir. A new bubble can again be introduced as before and the experiment repeated.

Farmer's potometer (Fig. 679) is essentially a modification of Ganong's.

Garreau's potometer (Fig. 680) has been conveniently used for quantitative measurement of differential rates of transpiration from both upper and lower surfaces of a leaf. It consists of two small bell jars placed one above another in between which a leaf is placed while still attached to the potted plant as shown in Fig. 680. At the narrow ends of the two bell jars, weighed amount of anhydrous $CaCl_2$ is placed in two very small tubes. The broad rims of the bell jars in contact with the leaf surface are made air tight with vaseline. At the two ends of the two bell jars, there are attached two oil manometers which ensure the maintenance of constant vapour pressure within the bell jars. A change in the vapour pressure within the bell jars shown by the manometers is indicative of the fact that either the connections are not air tight or that all the vapour given out by the leaf surface is not being absorbed by the $CaCl_2$-tubes. The whole arrangement is clamped vertically on a stand. After a few hours, the $CaCl_2$-tubes are taken out and weighed again. The difference between the two weighings is a measure of the amount of water lost from the leaf surfaces. As the areas are the same, this is also a quantitative measure of the differential rates of transpiration from the two surfaces.

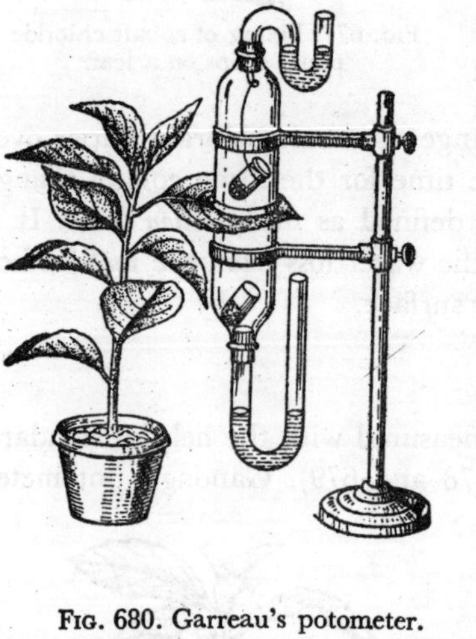

FIG. 680. Garreau's potometer.

IX. Measurement of transpiration rates—quantitative

Nearly fill a 500-ml conical flask with water. Insert a freshly cut (cut under water in order to maintain the continuity of water columns in the xylem vessels) sunflower or any other suitable leaf into the flask so that the petiole-end dips into the water. Pour some olive oil carefully into the flask to seal the open water surface, thereby preventing evaporation of water. Weigh the whole arrangement quickly and allow it to stand in the open air or in a well-ventilated room. After a few hours, weigh the whole arrangement again. Difference between the first and the second weights gives the amount of water transpired by the leaf. The water level of the flask is lowered, showing amount of water absorbed by the leaf. Find out the area of the leaf by any of the methods given before and express your results as the amount of water lost per unit area of the leaf surface. Divided by the number of hours the experiment has continued, the rate of transpiration, per unit area per unit time, is obtained.

This simple arrangement of conical flask with water and oil on the surface and with the experimental leaf can be conveniently utilised for obtaining quantitative measurements of various correlation behaviours of transpiration rates such as with: (a) *physical evaporation*, (b) *stomatal frequency*, (c) *the degree of opening of stomatal aperture*, (d) *the total area of the cross-section of xylem vessels in the petiole*, (e) *the number of xylem vessels in the petiole, etc. A quantitative idea of differential rates of transpiration from the two surfaces of the leaves can also be found out.* (Three flasks are taken: (*i*) with untreated leaf, (*ii*) with leaf whose upper surface is thinly smeared with vaseline and (*iii*) lower surface smeared. The rates of transpiration in all three leaves are determined and differential rates from either surface are easily obtained by subtraction. This quantitative measurement can also be obtained, as indicated before, by Garreau's potometer; a qualitative, as also roughly quantitative aspect, is given by cobalt chloride paper method.) If a dorsiventral leaf with no stomata on the upper surface is selected, the rates of *cuticular transpiration* can also be determined by this method. In each case the area of the evaporating surface must be obtained and the results expressed as g per unit area per unit time.

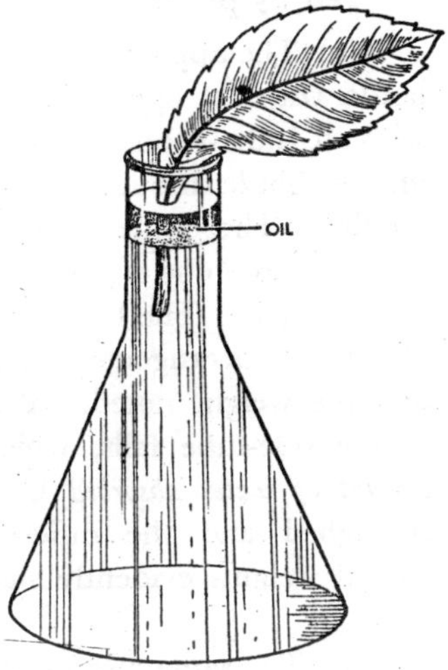

FIG. 681. Conical flask with water and leaf. Open and covered with a layer of oil.

If a quantitative estimate of amount of water lost per stoma in a leaf is desired, the total number of stomata in both upper and lower surfaces is determined, as also the total amount of water lost. Amount of water lost from the transpiring surface divided by the number of stomata gives us an approximate measure of the amount of water given out per stoma. If comparative data are wanted for different plants, the transpiration rates per unit area divided by the number of stomata in that area, indicate a measure of rates of water loss per stoma.

X. Determination of the effect of environmental conditions on transpiration rates in plants

A few experiments can be devised to show that environmental conditions greatly influence transpiration rates in plants.

(*a*) **Due to changes in the atmospheric humidity.** The same conical flask-oil-water-leaf arrangement can be conveniently utilised. The weighed control flask is kept in the ordinary air (relative low humidity). The second weighed flask with leaf is kept under a bell jar. Keep a petri-dish with water inside the bell jar and the inside walls of the bell jar are covered with soaked blotting paper which ensures maintenance of almost saturated vapour pressure conditions (high humidity) inside the bell jar compared with the ordinary air outside. Determine the areas of the two leaves and reweigh the conical flasks after the experimental period. Estimate the difference in the transpiration rates due to changes in the atmospheric humidity.

(*b*) **Due to increase in the velocity of wind.** The control flask is kept in relatively still air in a well-ventilated room. The second flask is kept at a short distance from an electric table fan going at a moderate rate. After the experimental period, the transpiration rates are calculated due to increase in the wind velocities.

XI. Simultaneous determination of the amount of water absorbed and the amount of water transpired by plants

(*a*) The apparatus (Fig. 682) consists of a wide-mouthed bottle with an open-end graduated side-tube attached to it with an India-rubber cork. A small-rooted plant is introduced into the bottle through a split cork and all connections are made air tight. The bottle is filled with water and the evaporation from the side-tube is prevented by pouring oil on the exposed water surface. The level of water in the side-tube is recorded. The whole arrangement is then weighed in a pan-balance and the weight noted. After some time, the level of water in the side-tube falls, which is certainly a measure of the *amount of water absorbed* by the plant. The arrangement is reweighed and the difference between the first and the second weights evidently shows the *amount of water transpired* from the aerial leaf surface.

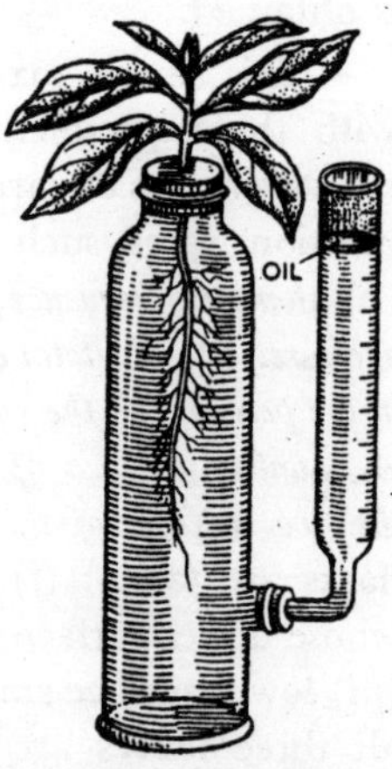

FIG. 682. Simultaneous determination of the amount of water lost and water absorbed.

(*b*) The same relation can be experimentally determined by the previous conical flask-oil-water-leaf arrangement. Let the weight of the flask+water+oil (on the free water surface to prevent evaporation) $=W_1$.

Flask+water+oil+leaf (cut under water) $=W_2$.

After a desired number of hours, take again the following weights:

Flask+water+oil+leaf $=W_3$

Flask+water+oil only $=W_4$.

Evidently W_1-W_4 =amount of water absorbed by the leaf

W_2-W_3 =amount of water transpired by the leaf.

XII. Experiment to demonstrate suction or pulling force developed due to transpiration

A glass-tube is taken and a leaf-shoot of *Catharanthus* (cut under water) is inserted into it with the help of rubber tubing and the connection made air tight with paraffin. The tube is now completely filled with water and holding it with a finger, the lower end is carefully dipped below the surface of mercury in a small vessel. The whole arrangement is clamped vertically to a stand and is kept in bright sunlight in open air. Within a short time, the mercury i seen to rise in the tube which indicates the suction or the transpirational pull exerted on the water in the xylem vessels transmitted to the water column in the tube and ultimately to the mercury level in the beaker causing the mercury to rise in the tube (Fig. 683).

The weight of this transpirational pull can be easily calculated if the diameter of the tube is known and the height attained by the mercury column in the tube within a specified time is noted. The volume of mercury raised is evidently $\pi r^2 h$ and specific gravity of mercury is 13·6. Thus the weight of the pulling force is $\pi r^2 h \times 13{\cdot}6$ g. If comparative data for different species of plants are required then the total area of the transpiring surfaces (i.e., the leaves) of each must be determined by graph paper. The results then can be expressed as so many g per unit area of the aerial surface.

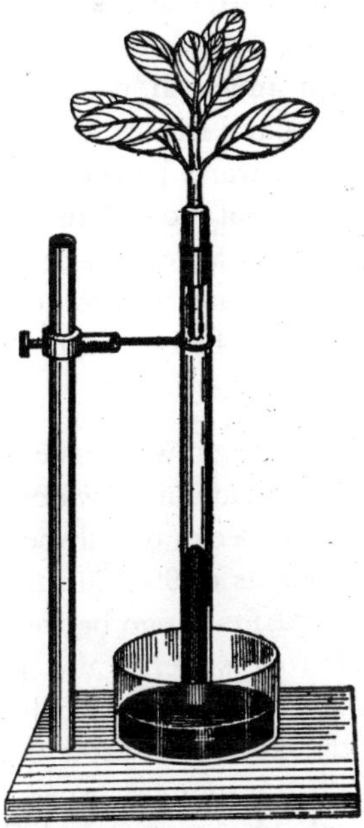

FIG. 683. Arrangement to demonstrate suction developed due to transpiration.

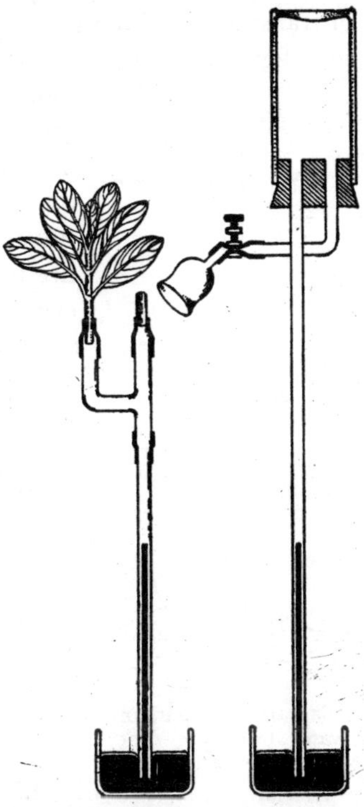

FIG. 684. Comparative demonstration of suction developed due to water loss from leaves and from a porous evaporating surface.

Experimental systems for demonstrating that both transpiration and the evaporation of water from a porous pot set up pulling forces which cause liquids such as water, mercury, etc., to ascend in the vertical glass-tubes, are shown in Fig. 684. In the right-hand figure, instead of a leafy twig, there is a pot whose surface is dotted with minute pores. The pulling force or the tension developed due to both the processes is transmitted throughout the system, causing them to behave in a similar manner.

CHAPTER XXXVII

THE MOVEMENT OF WATER IN THE PLANT

Most land plants obtain the necessary water for normal growth and development from the soil. By far, the largest proportion of absorbed water is lost as vapour in the process of transpiration from the aerial parts. Much smaller quantities are utilised for growth and for various other metabolic processes of plants. Water, therefore, must move continuously through the intervening tissues and organs from the absorbing regions of roots to the tissues in which it is utilised or from which, it escapes as vapour, the leaves. In small herbs and shrubs, the distance to be traversed by water on its onward journey from the root tips to the leaves, is usually not more than a few feet. But even in such plants, they may have such deep-seated root system, that the ascending water has to cover distances as great as 6–10 metres(m) before even reaching the surface of the soil.

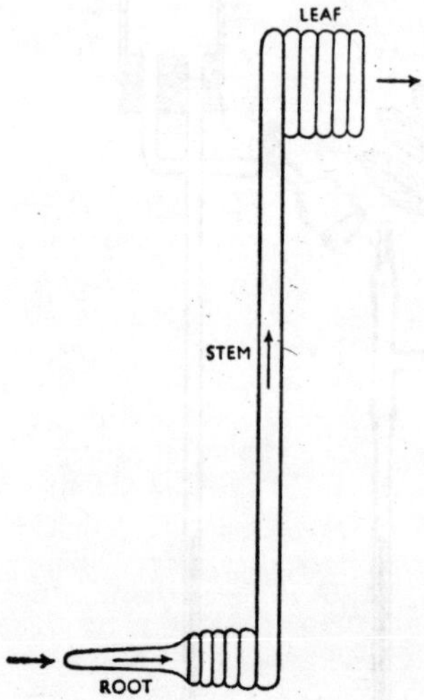

Fig. 685. Diagrammatic path of absorbed water from the root to the leaf and to external air.

It is in tall trees, however, in which the most striking illustrations of upward conduction of water occur. The tallest trees in record are a species of redwood in U.S.A., which attain immense heights of 90–120 m. The heights of our own eucalyptus and firs range between 60–90 m. Since in all trees, the root systems also penetrate at least a few feet into the ground, the actual vertical distance through which the absorbed water must be conducted in order to reach the topmost leaves, may easily be 120 m or even more! A simple calculation reveals that the force required to lift water to the top of tallest trees must be enormous indeed. For example, in case of a 90 m tree, a pressure of 10 atm. is just sufficient to support the column of water and according to all available experimental evidence, another 10 atm. is needed to overcome resistance due to friction. So altogether a minimum of 20 atm. is necessary if the water is actually moving up through the xylem vessels. The mechanism by which this feat is accomplished against the force of gravity has been the subject of much speculation and it has been one of the most fascinating problems of plant physiology which has intrigued generations of investigators, not to speak of laymen. This has been termed, *ascent of sap*.

We know that water enters the land plant entirely through the root hairs and after crossing the cells of the cortex, the peculiarly thick-walled endodermis and pericycle, finally reach the xylem vessels or tracheids of the roots. At this point, its upward movement begins. We know that xylem vessels form a continuous conducting system, just like a pipeline, from the root tips through the main tissues of the roots and through the stems, the petioles of the leaves and ultimately through veins and veinlets, ending up in the surrounding mesophyll tissue of the leaves. In the mesophyll cells, water moves onward from cell to cell and eventually the mesophyll cell walls, losing most of it by evaporation into the intercellular spaces, from where it escapes into the external air through stomata.

It was generally agreed, that xylem vessels are the main pathway of the upward conduction of water. A cell-to-cell lateral movement of water in a radial direction undoubtedly occurs along the vascular rays in the stems of most species of plants. That xylem is the water-conducting channel in plants has been recognised as early as 1671 when Malpighi did his famous '*ringing*' *experiments*. For over a hundred years now it has been definitely recognised that water is carried through the xylem, and that the other tissues, like pith, cortex, cambium, phloem are not directly concerned in the mechanism of water conduction in plants. Among the reasons for this recognition of primary functions of xylem are the following:

(1) the anatomy of xylem obviously and clearly fits it for conducting purposes and it is also true that among the vascular tissues, xylem vessels have the right cross-sectional area to allow the upward transport of the large amounts of water required by plants;

(2) in the classical ringing experiments, it is observed that the removal of a ring of tissue external to xylem from the stem does not interfere significantly with upward movement of water to the organs situated above the ring whereas the removal of a cylinder of xylem from the stem certainly seriously disturbs or even completely stops conduction of water. The net result in many cases is almost complete wilting of leaves attached to the stem above the ring;

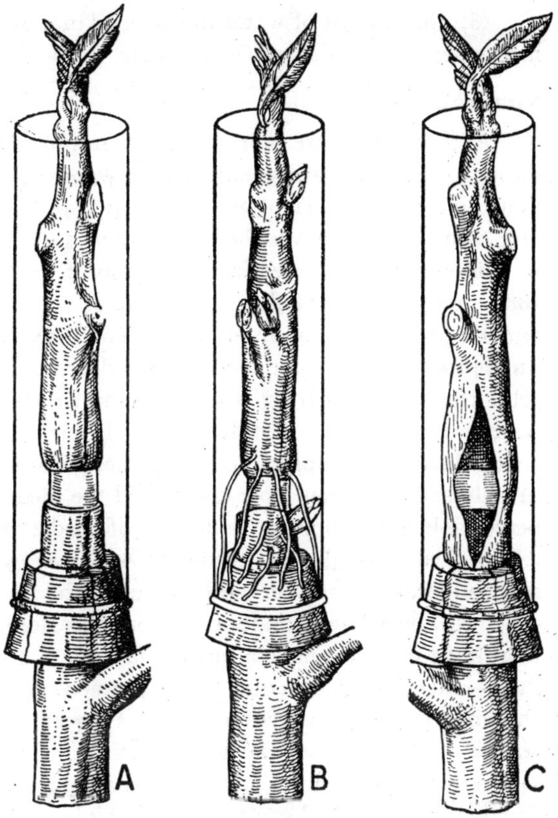

FIG. 686. **Ringing** experiments; A & B, all tissues external to xylem removed; C, only xylem removed.

(3) upward movement of water continues for some time even in shoots cut from plant, when placed in water. The rate of water movement (that is the length of the water column which will move past a given point) through xylem vessels varies greatly from almost imperceptibly slow, to speeds as high as 75 cm per min.

ASCENT OF SAP

The mechanism of water moving vertically in plants against the force of gravity is still a classical problem of plant physiology. Since diffusion is much too slow to account for the rates that commonly occur in plants, mass movement of water must be envisaged; the whole column must move simultaneously instead of molecular movement.

A number of different theories of the mechanism by which ascent of water is brought about in plants has been suggested and the present state of our knowledge justifies a discussion of only three possible mechanisms:

(1) that the ascent of water is caused by the activity of the living cells, particularly of the stem—*vital theories;*

(2) that upward movement occurs as a result of *root pressure;*

(3) that ascent of water has its explanations in a set of purely physical principles, supplemented by *cohesion of water theory*, proposed by Dixon.

Vital theories

As living cells are more or less in intimate contact with the dead xylem elements through which upward conduction of water occurs; suggestions have often been made that the motive power of ascent of sap is provided by the *vital* activity of living cells. Our own J. C. Bose was the most ardent advocate of the vital theory of ascent of sap.

Bose's idea was essentially an elaboration of the theory proposed by Godlewski in 1884. According to Godlewski, ascent of water resulted from periodic changes in the osmotic pressure of the living cells—wood-ray cells in the xylem. It visualised a repetitive process of alternate increase and decrease in the osmotic pressure of the wood-ray cells, resulting in driving water out and into the xylem vessel. Thus the movement of water was supposed to be due to an alternate contraction and expansion of wood-ray cells, which also supplied necessary energy for the mechanism, perhaps energy released in cellular respiration. Bose elaborated the idea still further. He claimed to have found perceptible galvanometric deflection of needle when an electric probe was very delicately pushed through stem tissue of *Desmodium gyrans*. The deflection of the needle was most noticeable when the probe reached the innermost layer of cortex, i.e., the row of cells just above the endodermis. From this Bose concluded that the cells of the innermost layer of cortex have a sort of pulsating movements, just like the heart beats of animals—alternate contraction and expansion. According to Bose, these living cells with rhythmic pulsatory activities, acted as a sort of system of relay pumps. As a cell contracted, it pumped up sap to the next higher cell, which (second cell) receiving the sap from the cell below (foot cell) expanded. The expanded cell (second cell) on contraction again pushed sap to the next higher cell (third cell), and so on.

Molisch reported confirmation of certain of Bose's results, observing that moisture movement was closely linked to a characteristic rhythmic activity of the plant, which was amenable to control by drugs. Molisch, like Bose, concluded that the mechanism of the ascent of sap was mainly physiological rather than physical.

The concepts of Bose and Molisch have been destructively criticised by a number of investigators, including Smith, Benedict, MacDougal, and others.

Elaborate experiments carried out by these investigators, however, showed conclusively that water would continue to ascend for some time through the plant after all the living cells present in a woody stem were killed either by soaking the cut end into boiling water or by dipping it into a strong solution of picric acid. However, it must be stated here that in many experiments the leaves at the top of such a treated stem, where all the living cells have been killed, sooner or later, in some cases after several days, showed definite signs of wilting. The supporters of vital theories have accepted this phenomena of wilting of leaves as sure evidence that the living cells of the stem are essential for the conduction of water through it. They argued that if the water-ascent was only through non-living xylem vessels in the stem, the leaves at the top of the treated twig would ever have remained fresh as long as the supply of water was there. The exponents of vital theories, however, forget that the retardation of the conduction through the stem may have been due to much simpler causes than due to the direct killing of the living cells.

The plugging of the vessels by formation of toxic substances in the treated region of the stem or the transport of the same toxic substances into the leaves may well be the explanation of the wilting of the leaves in such experiments. Even then, in all fairness, we admit the possibility that the living cells in the region of the xylem vessels may, in some way, contribute to the rise of water through the stem.

Root pressure

It has been claimed by some investigators that the positive hydrostatic pressure developed in the xylem vessels—root pressure—under conditions of excess absorption and low transpiration may be adequate to push the water to the tops of small herbs or shrubs. While it is true that root pressure does, in some plants under certain conditions, account for the transport of water in the upward direction through plants, the process seems to be wholly insufficient to account for the rate at which water is known to travel through plants and also for the amount of water actually translocated. In the second place, the magnitude of the root pressure developed is seldom adequate to raise water to the top of any, except perhaps trees of heights between 15-20 m. at the most. For, the maximum observed root pressure rarely exceeds 2 atm. (i.e., a rise of 20 m.) and often much less. Furthermore, many of the tallest trees, particularly of the temperate regions, such as conifers, have no demonstrable root pressure.

The most conclusive evidence that the ascent of sap is not commonly due to root pressure is the fact that at time of rapid water movement in plants, "negative pressure", (or tension) instead of positive root pressures, usually exists in the xylem vessels. This "negative force" has sometimes been called root tension. This root tension is easily demonstrated in rapidly transpiring plants when a cut is made to remove the aerial parts and water is poured on the cut end of the stump. The water (sometimes as much as a gallon) is drawn back into the root immediately.

Physical theories: Cohesion of water theory of ascent of sap

In the application of physical principles for an explanation of the mechanism of ascent of sap in plants, atmospheric pressure naturally comes first to mind. But it is at once apparent that atmospheric pressure can only account for a rise of about 10 m. Suggestions were then put forward that water rose by imbibitional forces through the thick walls of the xylem vessels. Physicists have shown that the forces of imbibition are very great, ranging from 100–1000 atm. and this at first sight would seem more than adequate for carrying water to any required height. However, the rate of movement of water through imbibing colloids are extremely slow compared with known rates of water conduction in the xylem. Moreover, it was soon shown that water actually moves not through the cell walls, which would be required for an imbibitional transport, but through the lumen or cavities of the xylem vessels.

We have seen that negative pressure (or tension) generally exists in the xylem during rapid transpiration from leaves. This can be shown by the use of a very sensitive instrument known as *dendrograph* which measures diameters of the tree trunk and such measurements definitely show, at times at least, a contraction of trunk-diameter during the day (caused by tension or pull due to transpiration) and expansion at night (tension less, due to stopping of transpiration).

Thus the ascent of sap is usually associated with a pull from above rather than a push from below (positive root pressure). Forces developing in aerial parts of the plants, especially in the leaves, cause the rise of water through the plant. It is easy to calculate mathematically whether this can be explained by a capillary pull in the vessels.

A vessel diameter of 0·1 mm would account for a rise of only 30 cm. But it must be understood clearly that it is not in the xylem vessels themselves that the main capillary pull occurs. Since this is clearly a surface tension phenomenon, the pull must occur at the water surface due to transpiration. We know that water system of a plant is continuous from the lowest root hairs to the leaves at the topmost parts of a plant. The upper surface of the water column, therefore, is in the leaves, actually at the outer surface of mesophyll cells that are in contact with intercellular spaces, that is in the *microcapillaries* of mesophyll cell walls. Since the microcapillaries are so fine, their diameters may be as small as or even less than 0·1μ ($1\mu = \frac{1}{1000}$ mm). Taking the radius as 0·05μ, a simple calculation reveals that as transpiration from the surface of microcapillaries on the cell walls takes place, they are capable of developing sufficient pull to support a water column about 300 m. high—nearly three times the height of the tallest trees!

The microcapillary pull, exerted on the mesophyll cell walls results in movement of water from the protoplasm into the cell walls and this in turn resulting in the movement of water from the vacuolar cell sap into the protoplasm lining the cell walls. If the particular mesophyll cell is in direct contact with one of the branches of xylem cavities which spread like a network in the lamina of the leaf, water will be pulled out from the xylem vessel into the cell by the tension or suction developed in the microcapillaries of the cell wall. This results in the development of tension in the water column terminating in this particular xylem element.

Since the column of water moves as a whole this means transpirational pull is transmitted from the microcapillaries in the mesophyll cell walls to the protoplasm and vacuolar cell sap, to the adjacent cells and to the nearest vessel. It is then transmitted down the vessels through the leaf blade, petiole, stem, and all the way to the roots. From the base of the vessel in the roots, the pull is transmitted through the adjacent cells of pericycle, endodermis, cortex to the epiblema cells and even to the medium surrounding the cells, i.e., external solution or soil particles. Thus the transpiration pull, it seems, is responsible not only for the movement of water within the plant, but also of absorption of water from the external root medium, i.e., soil solution.

The following three stages in the process then may be distinguished: (1) transpiration from the surface of microcapillaries, (2) capillary rise of fine threads of water due to the force of adhesion between water and cell wall, and (3) the whole column of water moves all together and resists breaking because of the force of cohesion between the water molecules—similar particles always have tendency to stick to each other. Of course, these three stages occur simultaneously. This concept is known as Dixon's transpiration-cohesion-tension theory of ascent of sap in plants. However, a column of water 90 m. high is subjected to a downward gravitational pull of 10 atm. all the time, tending to break the water column. This powerful gravitational force is counteracted by an equal (10 atm.) upward surface tension pull due to transpiration. The most pertinent question that could be asked is that whether liquid such as water has sufficient tensile strength to resist the pull without rupture or whether forces of adhesion between the water and the vessel walls are sufficient to prevent a separation of the column of water from the vessel wall.

Physical chemists have experimentally shown that pure water molecules have a cohesive force of theoretically about 1000 atm. This is about 20 times more than the necessary cohesive force even for a 120 m. tall tree. The tensile strength of water can be exhibited when the water is confined in very narrow tubes whose walls are rigid and incollapsible. Actually, values reported from plant cells for cohesive force of water range from 200–350 atm.; most striking values obtained are from the dead, thick-walled annulus cells of fern sporangia (300–350 atm.) at the time of violent bursting open of the sporangia for dispersal of spores. But the water in the vessels is not pure. Besides solids, gases are present in solution. The reduction in pressure due to the capillary pull may, therefore, reduce the solubility of the gases until they separate from liquid and expand, causing rupture in the column of water. Actually, vessels have, in fact, been found to become gas-filled. But since there are many columns of vessels side by side, it is not necessary for all of them to be continuous. The wet cell walls of vessels seem to be adequate in effectively preventing gas bubbles in one vessel from spreading into other units. When the tension is relieved at night (due to stoppage of transpiration) or by rain, the gases in the vessels go back into solution and the columns of water may become continuous again.

Tensions in the vessels of up to 100 atm. have been reported. This points to the need of thick-walled xylem vessels in order to prevent a collapse of the vessels and stopping the flow of water.

Thus we see that in ascent of sap the water column is pulled upward *en masse* from the roots very much as a solid wire or rope may be pulled through a tube, only difference being that the water, since it is fluid, completely fills the cavity within the vessels. Ascent of sap is like lemonade sucked through a straw from a bottle on a hot summer day!

Although the aim of plant physiology is to explain all living processes in terms of known laws of physics and chemistry, the purely physical explanations of a living process seem, in most cases, incompatible with physiological evidence. As a result it frequently becomes necessary to give a number of possible explanations of a particular phenomenon and to attempt to evaluate the merit of each one on the basis of accepted physico-chemical theories. Thus, in the case of ascent of sap, it has been considered quite logical to discard (?) all other hypotheses proposed as explanations of the mechanism of ascent of sap from time to time except one, Dixon's *Cohesion of water theory*, as it is not desirable in an elementary treatise such as this to discuss all of them critically. The transpiration-cohesion-tension theory is generally accepted, perhaps with a pinch of salt, not because it looks good, but it reasonably seem to agree with some known facts.

During recent years, the supporters of classical cohesion theory had also to face drastic criticisms from several investigators in England and elsewhere. Handley, from his investigations on maple seedlings, came to the conclusion that living cells only are involved in the ascent of sap as his test plants showed pronounced wilting of leaves where the temperature of the seedlings were lowered below 2°C. There was a complete recovery of wilted leaves when the temperature was increased above the so-called critical 2°C. As sufficiently low temperature had not been employed to freeze the water in the vessels, according to Handley, there should not have been any curtailment of water movement in the vessels. But actually there was pronounced wilting of the aerial parts. Handley was strongly supported by Preston, who from a study of Handley's findings, came to the same conclusion.

Handley's observations, together with others discussed by Preston, have prompted the latter to conclude that only *living* xylem is involved in water transport, and both of them called for the complete discarding of the classical cohesion theory. Where are we now, then?

According to these investigators, a chain of living cells continuous from roots to leaves was involved in the ascent of sap in plants and that the vessels and the supporting elements are not active in the actual water conduction but merely serve as reservoirs, a view point strikingly reminiscent of the classical vitalistic school led by Bose.

Lundegardh also proposed his idea about water movement in trees. The most salient points among his suggestions were that only a very small proportion of total water in a stem is mobile and as vessels and larger tracheids in large woody dicots are air-filled, they can scarcely play any part in the conduction of water, the movement of water being confined almost exclusively in the narrower tracheids in which there may be continuous water columns from roots to leaves. The activity of the living cell in the wood was supposed to be a strong contributory factor.

Lundegardh's suggestion that vessels play only a minor role in water conduction is not, however, corroborated by the experience of several other investigators who produced strong circumstantial evidence of the importance of vessels in water conduction in woody trees. Moreover Lundegardh's suggestion that the narrower, medium-sized tracheids are the main channels of water transport in trees is certainly not keeping with marked higher rates of water transport in dicotyledonous trees as opposed to those in conifers.

EXPERIMENTS ON CONDUCTION OF WATER IN PLANTS

I. Immerse a *Peperomia* plant and a few stalks of tuberose (*Rajanigandha*) in water coloured with eosin or acid fuchsin. After some time observe narrow streaks of pink in the transparent stem and pink veins of *Peperomia* leaves showing the movement of coloured water through inner conducting xylem strands. In tuberose the white flowers take a beautiful pink shade after about 24 hours, again showing upward movement of coloured solution through the stalks to the flowers themselves. The colouring of the white flowers involves also lateral diffusion of the dye.

II. Immerse leafy twigs of *Vinca*, *Trabernaemontana coronaria* (*Tagar*), etc., in red-coloured solutions. After some time the detection of the coloured soln. at various heights of the stem can be made by visual observation and also by making sections and looking through a microscope. It becomes at once evident that the red colourings are primarily confined to the xylem vessels, proving again that the xylem strands are the main path of the upward movement of the solutes in plants.

III. **Ringing experiment.** Take a forked twig of *Vinca* or any other suitable plant and leave its main axis intact. From the left-hand fork remove all the cortical portions from a narrow zone. Sever the central woody strand very carefully from the right-hand fork leaving the cortical tissue undamaged. (See Fig. 686.) Immerse the lower end of the main axis in water coloured with eosin in a beaker. Observe after a few hours. It is seen that leaves of the right-hand fork have considerably wilted and no eosin could be detected above the incision, proving that severance of vascular strands stops upward translocation of liquids, causing the leaves to wilt. For the same reason, the leaves of the left-hand fork are more or less unaffected and eosin can be easily detected above the stripped zone—removal of cortical tissue does not seriously interfere with water conduction if the xylem remains intact.

IV. Two leafy *Vinca* or *Tagar* twigs are taken, the lower end of one is dipped in boiling water for some time in one case and in a strong solution of $CuSO_4$ or any other toxic substance in the other. After such drastic treatment which kills all the living cells in the treated regions, the twigs are put in a beaker containing eosin-coloured water. After about 24 hours, eosin can be detected in the xylem strands throughout the stem and leaf of both the twigs (there is very little wilting of the leaves) indicating that living cells take a very small part, if any, in the upward conduction of liquid through the plant.

CHAPTER XXXVIII

MINERAL NUTRITION

THE CHEMICAL COMPOSITION OF THE PLANT

The nutrition of any living organism includes all the substances that must be supplied to it from outside. Those organisms which require both organic and inorganic substances from outside are called heterotrophic; those that need supply of only the inorganic substances are called autotrophic. Animals and non-green plants such as fungi and most bacteria are heterotrophic, whereas green plants are autotrophic for they can synthesise all their biochemical requirements, provided they are supplied with all the essential inorganic elements and grown under normal environmental conditions. The nutrition of green plants is, therefore, solely inorganic. It is in fact commonly called mineral nutrition and the elements, except carbon, hydrogen and oxygen, absorbed by the roots are essential mineral elements, since they are obtained directly or indirectly from the minerals present in the soil.

The conception of elements as plant nutrients was originated by Liebig in 1840 when he put forward his revolutionary theory of mineral nutrition. Revolutionary indeed, for formerly it had been supposed that plants drew their food somewhat mysteriously or even mystically from humus, the dead and decaying organic material in the soil. The soil was thought to be meant only for supporting the crops growing in it and not as a reservoir for minerals essential for plant life.

The mineral nutrition of plants embraces all the elements, essential to the existence of plants with the exception of C, H and O. Nitrogen, strictly speaking, is not a mineral element but is included with these since it is obtained by the plant from the soil.

The first and, perhaps, the most fundamental problem of mineral nutrition is to determine which elements are essential for the normal growth and development of plants. As a start, one might chemically analyse the plant to find out which are present, for any clear understanding of the complex metabolic changes taking place in the plant body undoubtedly requires a knowledge of the chemical composition of the plant.

Every one knows that plants contain a large proportion of water and a relatively small proportion of dry substances. By means of simple weighing it is easy to show how large is the proportion of water in the total weight of the plant. By drying plant tissues at a temperature slightly over 100°C., it is easy to ascertain the percentage of water and dry substances in the plant tissues. The proportion of water is very considerable; in woody parts some 40–50%, in herbs about 70–80%, and in succulent plants and fruits it may be as high as 85–95%. In aquatic plants, particularly algae, 75–98% of the weight of the plant consists of water. So we see that the percentage of dry substances may vary between 60% and 2%.

While we can thus distinguish by drying between the water and the dry substance of the plant, we are only able by burning the dry substance to distinguish between the combustible or organic material and the incombustible inorganic substance or ash. The fact that the plant leaves an ash is evident in burning of wood or other plant tissues or in the smoking of a cigar.

When the amount of ash contained in various organs of diverse plants is determined by burning, the percentage which this forms of the total dry substance shows a wide range—from about 17% (tobacco leaves) to as low as 1% in cotton fibres. The separate organs of the same plant differ widely in this respect, the straws of cereals, such as rye or wheat, contain more ash than the seeds. Leaves in general tend to have a larger proportion of ash than stems. It must be clearly understood here that ash is nothing but oxides of elements which are formed during burning or incineration. In the living cells, the metallic elements are mostly present as salts, especially of organic acids.

The constituents of ash vary according to the nature and character of the soil and other external influences. On the other hand, distinct species may accumulate different quantities of minerals even when exposed to the same external conditions.

While the majority of the more common elements occurring on the earth have been traced in the ash of plants, only a few elements are present in sufficient amount in ash to be quantitatively estimated by chemical methods. It must be understood, however, that the number of elements found in any particular case depends on the plant, the medium in which it is grown and the accuracy and the completeness of the analysis. But the analysis of ash certainly merely tells us that the elements not found in plants are not essential in any appreciable quantities. The mere presence of an element in a plant does not prove its essentiality. Actually, some 60–70 elements have already been found in plants, and undoubtedly all the rest of the 100 and odd known elements could be made to be traced in the ash by supplying them to the roots of plants. Even gold (*Equisetum arvense*), silver (*Eriogonum ovalifolium*), and mercury are known to be absorbed and accumulated by plants. Actually pure samples of these metals extracted from some species of plants, growing in the hills of Eastern Europe, have been obtained.

The organic substance consists in the first place of the elements C, H, O and N. On complete combustion (at about 400°C.) the organic substance is transformed into volatile compounds—CO_2, water vapour, NH_3 or free molecular nitrogen. In addition, the organic substances may contain some of the elements found in the ash. Chemical analysis is not required to show that the plant contains carbon in a combined form. Every burning of log or match stick shows by its charring that it contains carbon. The examination of a piece of charcoal in which the finest structure of the wood is retained shows further how uniformly the carbon is distributed in plant and how largely the substance of the plant consists of this element. Accurate weighing has shown that in general carbon accounts for about one-half (44%) of the total dry weight of the plant and oxygen accounts for about 45%. The presence of nitrogen in the organic substance is easily shown by collecting and chemically analysing the volatile ammonia and nitrogen gas produced during incineration. The presence of nitrogen can also be simply shown by heating some plant tissue in a test-tube with soda-lime. The faint smell of ammonia can soon be detected. The following fourteen elements have been found to occur in quite appreciable quantities in the tissues of most plants:
metals—K, Na, Ca, Mg, Fe and Al; and **non-metals**—C, H, O, N, S, P, Cl and Si. When the plant is growing, their amounts continually increase in the plant and consequently they must be absorbed all the time from without, from soil, and air.

The elements, C, H, O, N, P and S, are the building blocks of organic compounds of the cells. They can truly be called *fundamental biological elements*. Most carbohydrates and lipids contain C, H and O while in addition, all proteins contain N and sometimes S, and nucleic acids, N and P. Thus these *six* elements are the main constituents of the living machine. The ability of different permutations and combinations of these elements to produce the molecular diversity of the cell, makes them unique in their fitness to support life.

We know that, as a rule, only gases and liquids can enter the plant. Except carbon and oxygen which are invariably taken in by the green plants from the atmosphere in the form of CO_2 gas and O_2, all other elements generally enter the plant through the roots and root hairs in solution from the soil where the elements are present in very dilute concentration of soluble salts. Though the atmosphere contains about 78% of nitrogen, the atmospheric nitrogen is completely unavailable for higher green plants. So the green plants depend wholly on the nitrogen fixed by certain micro-organisms. Hydrogen is taken as water and so also is oxygen, in addition to the atmospheric supply.

It is not, however, correct to speak of these definite elements as the food materials of the plant and it is certainly not sufficient to supply the plant with elements found in the ash. Thus metallic K or pure S is of no use. The plants require the elements in the form of particular inorganic salts since these in part dissociate in water and the elements are absorbed as ions. Necessary cations are K^+, Ca^{++}, Mg^{++}, Fe^{++} or Fe^{+++}, Al^{+++}; Cl^-, PO_4^{---} and NO_3^- are necessary anions. All the metals thus generally enter the plants in the form of cations while the non-metals enter as anions. Nitrogen, however, could be absorbed by plants both as cation, NH_4^+ or as anion, NO_3^-, with almost equal facility.

We have stated before that presence of an element in a plant does not prove its essentiality. A more direct approach to the problem of essentiality is to grow or culture plants in complete absence of a given element. If the plant grows normally in its absence, the element is evidently not essential; if it fails to grow normally, the element is usually considered essential. In order to ascertain that any particular element is truly essential, it is necessary to show not only: (1) that a deficiency of the element makes it impossible for a plant to complete its whole life cycle—from the vegetative to the reproductive stage, but also (2) that it cannot be replaced by any other element, and (3) that the effect is not due to any interaction between the particular element and other elements or between the element and the plant. These three requirements have been called the *criteria of essentiality*. It has recently been shown, however, that some of the long-accepted essential elements can at least be partially replaced by others, as for example, Mg by Mn, K by Rb, etc. In some bacteria, strontium can substitute completely for calcium.

Chemical analysis ascertains all the elements present in the plant. It does not, however, show whether these substances are utilised by the plant nor whether the substances are necessary for nutrition and are thus to be regarded as food materials. How do we know that these elements are essential for plants and why? Such questions can only be decided experimentally. The search for essential elements which actually enter into the composition of various plant foods and ultimately into the composition of the plant tissue itself has thus required the development of techniques for growing plants in pure salt solutions in the absence of impurities and elimination of the natural root medium, the soil. In its place, a pure sand, gravel or best of all simple salt solutions have been used. The latter, the so-called *water* or *solution cultures*, however, offer the greatest difficulty since they fail to supply the normal support of soil for the roots and the normal amounts of oxygen. This requires development of adequate artificial support and aeration by stirring the solutions. Nowadays, the *sand* or *gravel culture* of plants are preferred to the classical water culture methods, as the sand or gravel cultures are easier to manipulate and the problem of aeration of the medium is also eliminated. This system of growing plants in absence of soil has been used to some extent on a practical scale for growing vegetables particularly tomatoes, gourds, cucumbers, etc., and has been named *hydroponics*.

In these artificial methods of plant culture, it is only necessary to add to the water the elements found in ash in various combinations of their salts and so to ascertain which elements are necessary and which superfluous. H and O, of course, are taken up by the roots in the form of water from the culture medium. Oxygen also enters the plant from the atmosphere through stomata for the respiration of the living cells and without atmospheric oxygen, higher plant life would have been impossible as it is with animals. Various types of water culture solutions have been used. Of the classical ones the following can be cited:

STANDARD NUTRIENT CULTURE TYPES FROM SACHS TO CRONE.
GRAMS PER POT.

Sachs' solution (1860)		Knop's solution (1865)		Pfeffer's solution (1900)		Crone's solution (1902)	
KNO_3	1·00	$Ca(NO_3)_2$	0·8	$Ca(NO_3)_2$	0·8	KNO_3	1·00
$Ca_3(PO_4)_2$	0·50	KNO_3	0·2	KNO_3	0·2	$Ca_3(PO_4)_2$	0·25
$MgSO_4$	0·50	KH_2PO_4	0·2	$MgSO_4$	0·2	$MgSO_4$	0·25
$CaSO_4$	0·50	$MgSO_4$	0·2	KH_2PO_4	0·2	$CaSO_4$	0·25
NaCl	0·25	$FePO_4$	a trace	KCl	0·2	$FePO_4$	0·25
$FeSO_4$	a trace			$FeCl_3$	small amount		
H_2O	1000 ml	H_2O	1000 ml	H_2O	1000 ml	H_2O	1000 ml

In a mixture of normal proportions of salts in each one of these types of solutions, plants growing in them form roots, shoots, flowers and fruits and increase its dry weight perhaps a thousand fold just as if they were growing in the soil.

From such water cultures, using only the first criterion of essentiality (i.e., no normal growth in the absence of any particular element) it was long ago found that the following eight elements—H, N, P, K, Ca, Mg, S and Fe in addition to C and O (C as CO_2 and O_2 from the atmosphere) are essential for normal growth and development of a typical green land plant. It was recognised long ago, that these ten elements must be regarded as indispensable plant foods which enter into the composition of the plant body itself, perhaps with the exception of Fe. These are termed *macroelements* or *macronutrients*.

For many years, chlorine was considered essential for higher animals but not for higher plants. Broyer established, however, the essentiality of chlorine for tomato and subsequent work has conducively demonstrated that it is essential for all higher plants.

Sodium is accumulated in large quantities by many plants particularly halophytes, but it seems to be essential only for blue-green algae, Cyanophyta.

It was soon discovered that plants can grow normally in four-salt-solutions containing KNO_3, KH_2PO_4, $CaCl_2$ and $MgSO_4$ (or other salts supplying the six elements) with a small amount of $FeCl_3$ (or other iron salts) provided the atmosphere contained CO_2 and O_2. Even three-salt-solutions like KH_2PO_4, $Ca(NO_3)_2$, $MgSO_4$ *plus* some iron were soon found adequate. Many formulae of such solutions have been adopted and extensively used. A simple precaution about making these solutions is that before putting the solution into the pots, the pH is adjusted to a value of about 5–6, for most plants can thrive best in a slightly acid or almost neutral medium.

A novel method of systematic study of the effects due to variation in the composition of the nutrient solution is to take three salts, e.g., KH_2PO_4, $MgSO_4$, $Ca(NO_3)_2$

containing among them all the essential elements and to arrange all possible combinations of the proportions of the three salts along the three axes of an equilateral triangle as shown in the Fig. 687.

In a triangle like this the sum of perpendicular distance of any point to the three sides inside the triangle is always constant. Thus any point along the axis, $Ca(NO_3)_2$ will have none of $Ca(NO_3)_2$ but only varying proportions of KH_2PO_4 and $MgSO_4$. At points Y and Z, it will be solely KH_2PO_4 and solely $MgSO_4$ respectively while at X, the nutrient solution will have only $Ca(NO_3)_2$. The nutrient solution represented by the exact centre of the triangle consists of all the three stock solutions in equal proportions. Thus a multitude of different solutions could be prepared by combining the three stock solutions in mathematical proportions increasing the proportions of each one of the stock solution along a line parallel to each of the axis of the triangle. The proportion of $Ca(NO_3)_2$ is measured along the line parallel to YZ, KH_2PO_4 along XZ and $MgSO_4$ along XY.

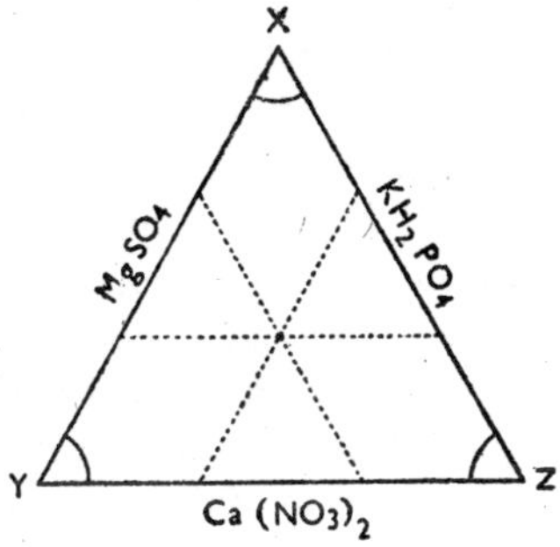

FIG. 687

Plants are grown with nutrient solution of these different compositions and the nutrient solution which produces optimal growth can be selected. At the same time deficiency symptoms due to individual elements can be observed in plants grown in incomplete mediums. It can be shown by these water culture methods that the complete absence of any one of these essential elements will, sooner or later, stop all growth. But under normal conditions in the soil, the elements are never completely absent. If, however, they are present in less than optimum quantities, growth will occur, but abnormalities known as deficiency diseases, will develop. Some of the more recent culture solutions do not eliminate the particular element completely in order to study the morphological deficiency symptoms and also the accompanying physiological disturbance in the cell metabolism. A normal culture solution of this type particularly suitable for cereals like wheat, rice, barley, etc., can be given as follows: In grams: $NaNO_3$, 9·11; Na_2HPO_4, 3·00; $CaCl_2$, 0·37; K_2SO_4, 1·85; $MgSO_4$, 1·25; added once (e.g., barley) or more, as required and a trace of $FeCl_3$ and $MnSO_4$ dissolved in 100 ml of distilled water for 3 plants per pot of sand culture.

In order to study the deficiency of potassium, the amount of K_2SO_4 present in the normal solution is reduced to 1/9th the level in one series of solution and in the next series reducing the amount of K_2SO_4 to 1/81 of the normal. Thus we may have three series of nutrient solutions, the normal where all the elements are present at the required normal dose for optional growth, 1/9 potassium, leaving all others unchanged and the third 1/81 potassium—arranged in geometric progression. Similarly, in order to study phosphorus deficiency the Na_2HPO_4 can be similarly reduced to 1/9 and 1/81 that of normal. We can call the three types of culture solutions as normal or control, medium-deficient series and extreme-deficient series. Deficiency levels can be made still further progressive, e.g., 1, 1/3, 1/9, 1/27 and 1/81. It must be noticed here that reducing Na_2HPO_4 to lower levels, also reduces the amount of Na at the same time. This is remedied by adding the equivalent quantity of sodium in the form of Na_2SO_4. If it is desired to study the effects of deficiency of nitrogen, the normal amount of $NaNO_3$ can be reduced to 1/9 in one and 1/81 at the other, at the same time supplementing the two nitrogen-deficient solutions with corresponding equivalent amounts of Na_2SO_4, so as to keep the cation concentration constant in all the solutions, disregarding the variation in the sulphur concentration due to addition of SO_4.

MICRONUTRIENTS AND PLANT GROWTH

The early culture solutions were made with salts which were not pure, i.e., the salts contained a lot of traces of other elements as impurities. It was found that as better-purified salts were used for culture solutions, the growth of the plants instead of getting better, definitely got poorer. This was followed up by direct evidence that other elements are needed for the normal growth of the plant but only in minute amounts. They are just as essential for the life and the growth of the plants as the macroelements and therefore no normal growth was possible in their complete absence. But in view of the fantastically small amounts needed, they are usually grouped separately from the macronutrients and called *trace* or *micronutrient elements*. Iron seems to occupy an intermediate position between the macro- and micronutrients. It does not enter into the composition of plant food, or in the composition of the plant itself. The amount of iron necessary for normal growth of the plants is very small compared to the other six macronutrients, yet it is hardly so small as to rank as a micronutrient.

Trace elements can be conveniently divided into *four* groups: (*a*) the *essential*—so far the following *six* have been conclusively proved to be essential for normal plant growth—B, Zn, Cu, Mn, Mo and Co; (*b*) the *probably essential*—elements like selenium, barium, etc.; (*c*) the *toxic*—all essential macro- and micronutrients in high dosages and (*d*) *physiologically inactive elements*—arsenic, etc.

In 1914, Maze, a French scientist, using very highly purified chemicals in water culture solutions found that his very pure salt solutions did not support plant growth satisfactorily. Thus, although this observation was at the time given very little attention, it really laid the foundation of the importance of micronutrients. Almost 50 years before that, however, Raulin (1869) when studying the role of zinc in the nutrition of *Aspergillus niger* concluded, with a remarkable insight for his day, that the microelements were required by plants in minute quantity and were present as impurities in the external medium.

The ratio between the amounts of micronutrients and the amounts of macronutrients will be roughly between 1 : 1000 and 1 : 10,000. In soils, however, quantities are somewhat larger—rate of application in soil, when there is a known micronutrient deficiency, are usually at the rate of a few pounds per acre.

Actual amounts of the trace elements or micronutrients range from as little as 2–3 parts in 100 million parts up to 10–100 parts in 100 million. These amounts are so fantastically small that sceptics may be tempted to argue that the whole subject is an exaggeration of science.

Since such small amounts of trace elements are required for plant growth in the soil, it might seem unlikely that soils should ever be unable to provide enough for all crops, yet it is a fact that in the past 15–20 years, increasing number of serious micronutrient deficiencies have been recognised. In many parts of the world, economic cropping would have probably ceased but for diagnosis of microelemental deficiency and subsequent remedial treatment.

Microelement shortage is not always induced by a real absence of the particular element in the top soil. A large supply may be present but it may be locked up as a result of soil condition and thus unable to enter the soil solution and become available to the roots. Severe Zn deficiency effects on fruit trees were experienced in many countries though the soil contained enough Zn for the whole orchard's need for hundred years!

Even the addition of Zn salts in many cases to the soil could not correct the deficiency for the added Zn was also quickly made unavailable. The remedy was found in the spraying of a very dilute solution of $ZnSO_4$ on the foliage of the trees. Another approach has been the injection of solid salt containing a trace element into the trunk of the tree. Like Zn, manganese deficiency is generally due to non-availability rather than to actual absence of the element from the soil. On the other hand, boron deficiency is usually due to absence of boron in the soil for borates tend to be easily washed out of the soils, particularly in sandy soil. Here, the remedy of applying boron to the soil is reasonably effective.

Micronutrient elements, although they are as essential as the macronutrients in minute amounts, soon become toxic to the plants if the beneficial rate is exceeded.

Most artificial fertilisers always contain trace elements in any case; for instance, chilean nitrate contains boron and basic slag contains manganese; superphosphates contain slight amounts of Cu and Zn, derived from the commercial H_2SO_4 used in the industrial manufacture of superphosphates.

Why are such minute quantities of some elements needed by plants? What indispensable roles do these elements play in the metabolism of the living cells?

We know that these elements are not direct plant foods. It was supposed also that they may act as catalysts or activators for certain chemical reactions in the plants and for that reason, it was understandable that their requirement will be much smaller, compared to the macronutrients, which directly enter into the composition of plant foods or living cells themselves.

Recent researches have given us the answers to the questions about the functions of the microelement which seemed so perplexing a few years before. There is no doubt whatsoever now, that the essentiality of heavy metals like Cu, Zn, Mn and Fe in minute traces for the normal growth of the plants is due to their forming constituents of the essential enzymes. The enzyme, *tyrosinase*, which transfers hydrogen from the amino acid, tyrosine, is active only in the presence of a single atom of copper as a coenzyme (prosthetic group). Other copper-containing enzymes are also known such as *Ascorbic acid oxidase* which destroys vitamin C. The non-protein prosthetic groups or coenzymes in these cases, consist only of a single metal atom of copper and nothing else. Terminal cytochrome oxidases (Cy a and Cy a_3) whose functions are to transfer electrons or hydrogen ions to molecular oxygen in the respiratory chain of oxidative phosphorylation Vitamin C, always contain either one or two atoms of copper per molecule of heme. Plastocyanin discovered by Katoh (1960) is a natural one-electron-transfer copper protein which occurs in chloroplastids in concentration of 1 : 500 chlorophyll. It may be a normal electron carrier between the two photochemical reaction. Thus it is only too true to suggest that copper serves a direct functional role in photosynthesis.

Atomic Zn is an activator component for a host of enzymes, e.g., *carbonic anhydrase* (catalyses the reaction $CO_2+H_2O \longrightarrow H^+ + HCO_3^-$), *alcohol dehydrogenase, lactic dehydrogenase, glutamic dehydrogenase, triosephosphate dehydrogenase, aldolase* (can be replaced by Co or Fe), etc. It has been known for quite a long time that Zn is essential for the formation of the most important plant hormone, indole acetic acid. More recently it was shown that Zn in minute traces is indispensable for the formation of the amino acid, tryptophan, the generally accepted precursor of indole acetic acid and is not directly concerned with the synthesis of the auxin. The catabolic breakdown of tryptophan to the indole nucleus of the auxin is accomplished by the enzyme *tryptophanase*. At the present moment, however, it is clear that the essentiality of Zn for the formation and breakdown of tryptophan is certainly due to its acting as an activator component of enzime *tryptophanase*.

Manganese forms activator components of several enzymes, such as some *dehydrogenases*, *decarboxylases*, *kinases*, *oxidases* and *peroxidases*, specifically and also non-specifically by other divalent cation-activated enzymes. Manganese is also reportedly required by *nitrite and hydroxylamine reductases* in lower non-green plants but not in higher plants. One of the fundamental discoveries of the last few years is that a manganese enzyme is specifically required for photosynthetic evolution of oxygen.

Iron is an activator constituent of many enzymes such as *peroxidase*, *catalase* and of *cytochromes*. In the case of iron, however, the iron atom does not form the coenzyme by itself, but forms a more complex molecule with a porphyrin (tetrapyrrole structure). And since the iron-porphyrin proteins are the primary catalysts in respiration of living cells, it is not surprising that minute traces of iron should be essential for the maintenance of life in animals as well as in plants.

Iron is also a constituent of non-heme iron proteins, such as, ferredoxin, which is involved in photosynthesis as the primary electron acceptor, as well as in nitrogen fixation and, also of respiratory-linked flavoprotein *dehydrogenases*.

It is now known definitely that molybdenum plays an important role in nitrate assimilation in plants and the fixation of atmospheric nitrogen by micro-organisms can occur only in the presence of minute traces of Mo. But in spite of great amount of work for almost 50 years on boron, its exact functions are as yet unknown. Boron may be necessary in order that the plant can obtain calcium in an efficient way. According to the more recent prevalent idea, boron forms a complex with sugar in the plant cells that can penetrate through the living cell walls, more rapidly than free sugars and is, therefore, more readily translocated to the growing meristematic cells where the carbohydrates are most needed.

Cobalt and iodine are essential elements for animal nutrition, but all attempts to prove that these elements are essential micronutrients for plants have so far failed though iodine is accumulated in sea-weeds in large quantities. Very recently, however, the importance of cobalt in promoting auxin formation in plants is being recognised—cobalt ion is known to depress the specific oxidase enzymes which destroy auxins. Likewise boron has been found to be essential to plants but there has been no definite proof that it has any function in animal nutrition.

This might very well be our ignorance; just as the gaps in Mendeleef's periodic table were for a long time no more than evidences of our inability to isolate the missing elements. So this strange anomaly in what seems to be an otherwise admirably planned natural arrangement (most of the element needed for plant nutrition as Ca, P, Fe, Mg, K and S are also essential for animal nutrition) will, it is hoped, in near future be satisfactorily explained.

Some investigators have produced evidence for the essentiality of a few other macro-elements, not mentioned before, such as aluminium, silicon and selenium, at least for certain plants, but other workers have failed to confirm these results. It is undoubtedly true for silicon (SiO_2) which occurs in large quantities in *Equisetum* and grasses, but silicon does not seem to be indispensable to such plants—*Equisetum* and grasses can get on quite as well without silicon. As a matter of fact all soil-grown plants contain silicon and in many *Gramineae* (in rice it is about 10–15%) in addition to grasses and also other monocots, considerable silica is deposited in the form of *opal*, hydrated amorphous silica ($SiO_2.nH_2O$). These silica deposits (natural crystalline silica SiO_2) have also been called *opal phytoliths* and their distribution in several species of *Gramineae* is known, without any indication, whatsoever, about the significance of their presence in these plants. On the other hand, silicon seems to be essential for diatoms, the cell walls of which are almost entirely composed of silicic acid and which gives permanence and rigidity to the diatom cells. Aluminium, while found in small quantities in almost all higher plants, is accumulated in any considerable amounts in only a few plants, such as *Lycopodium*. An interesting relation between large accumulation of aluminium in the cells and the development of blue-coloured flowers has been established recently in many blue-flowered or blue-fruited plants. Aluminium may also be more necessary to the water plants than to the land plants.

Elements such as aluminium and silicon have been called *ballast elements* by investigators who refuse to attribute any useful metabolic function to these elements because they are usually present in large amounts in plant tissues though the plant can be grown perfectly normally without them. As ballast (heavy material) in a ship's hold gives stability to the ship and maintains its equilibrium, so can the large quantities of aluminium and silicon in the living cells be used for maintaining the stability of ionic potential and thus preventing the cell system from 'capsizing'. Some plants are *accumulators*, i.e., they concentrate in their cells large quantities of certain elements, as for example, iodine in sea-weeds, silicon in grasses, sodium in halophytes, selenium by some species of *Astragalus* (about 5%), molybdenum by clover, etc.

An interesting point: Geological distribution of *Selenium* points to an accumulation of the element by plants, possibly as long ago as the paleozoic era (Rosenfield & Beath, 1964). It has even been suggested that during the mesozoic, herbivorous dinosaurs and other prehistoric giants ate themselves into oblivion with massive ingestion of seleniferous vegetation. The distribution of the species of *Astragali* was perhaps global at that time and much more numerous, which could accumulate selenium to such an extent as to render the herbage extremely lethal to the herbivores. Selenium replaces sulphur in several organic molecules and may become toxic.

Deficiency symptoms due to macro- and micronutrients

We know that abnormalities in structural and physiological make-up of plants are brought about by deficiencies of essential elements inside the plants. The symptoms of such deficiency are more or less specific for each element, but it is sometimes very difficult to distinguish the differences. Furthermore, different plants may show somewhat different symptoms. In some cases, the deficiency symptoms occur first on the oldest leaves (as for example, N, K, P and Mg) while in others the first symptoms are visible in the youngest leaves (such as S, Ca, Fe, Mn, B, Cu and Zn). The symptoms associated with deficiency of particular elements occur in the oldest leaves for the first group of elements because the elements of the first group are *mobile* and are transferred from the older inactive leaves to the younger leaves during the first stage of deficiency. The younger leaves can get a supply of element as long as there is any in the older leaves and as a result the older leaves show the symptoms before the newly formed young leaves. The elements of the second group are *immobile*—they remain in the old leaves and hence the younger leaves are deprived of supply of these elements when deficiency of a particular element occurs in the plants. Hence the younger leaves are starved and show the deficiency symptoms, though the older leaves do not.

The commonest type of deficiency symptoms are *chlorosis* and *necrosis*, reduction in the amount of green pigment chlorophyll in the leaf. But chlorosis can occur as a result of deficiency of any one of the several elements such as Fe, Mg, N, Mn and even S.

However, chlorosis, due to deficiency of each one of them, is different from the other. In case of nitrogen deficiency, the chlorosis of the leaf blade is uniform whereas in case of iron deficiency, chlorosis though uniform is more pronounced between the veins. Calcium deficiency results in increased chromosome fragility.

In many plants excesses of several metals including Mn, Cr, Cu, Zn, Co, Ni, Cd, may induce symptoms apparently identical with those of iron deficiency.

There are also deficiency diseases associated with microelements. Dry rot of sugarbeet and internal browning of cauliflower due to boron deficiency, little leaf of fruit trees due to zinc deficiency, yellow spots of pea seeds, etc., are usual symptoms of trace-element deficiency.

The deficiency of a particular element found in the plant may not correspond with actual deficiency in the soil. Thus definite iron deficiency symptoms of chlorosis may actually be due to non-availability of iron to a plant due to high pH conditions and not really to an actual absence; addition of iron in such soils does not improve the position to a great extent because the added iron is also soon made unavailable by precipitation. A recent method which has been successfully used in citrus orchards to overcome this difficulty is to apply iron combined with an organic compound to the soil in such a way that it can ionise and therefore cannot be precipitated. Sometimes the ratio of the quantities of elements in the soil may be more important in determining deficiency of particular elements in the soil than the absolute amounts (e.g., Fe : Mn and Ca : B).

THE PHYSIOLOGICAL ROLE OF MACROELEMENTS IN PLANTS

We have seen that sometimes how difficult it is to prove the essentiality of a particular nutrient element but it is still more difficult to find out why it is essential. If the element is a constituent of a substance whose function in the plant is known, this, of course, explains why the element is indispensable. It must always be remembered that a single element may play fundamentally different roles in the metabolic activities of plants:

Nitrogen—It is the most important constituent of proteins. As protoplasm or the living substance is predominantly proteinaceous, life is impossible without nitrogen. It is also a constituent of many vitamins, auxin and chlorophyll, all of which play fundamental roles in the metabolism of the living cells. Nitrogen also occurs in many substances, such as plant bases, e.g., purine and pyrimidine, the significance of their presence is fully realised now.

Phosphorus—Like nitrogen, it is a necessary constituent of many important substances, such as nucleoproteins, the phospholipids, high-energy phosphates, the coenzymes, etc. As high-energy phosphates of sugars they are the main agencies which drive the metabolic cycles of photosynthesis and respiration.

Sulphur—It is a component of many proteins, the amino acids, *cysteine*, *cystine*, *methionine* as also of some vitamins (vitamin B_1 and biotin). It is also a constituent of lipoic acid, coenzyme A and other compounds such as the antibiotic penicillin. It is a constituent of glutathione, the respiratory pigment as well as of mustard oil glycosides, sinigrin and glucobrassicin. The pungent odour of onion and garlic is also due to the presence of a sulphur compound, allyl sulphide.

Magnesium—It is a constituent of chlorophyll molecule; but as it is also essential for non-green plants, this cannot be its only function. We now know that magnesium is one of the constituents of the enzyme *pyruvic carboxylase* and also a host of other enzymes, particularly kinases. It also helps in the maintenance of the physical structures of subcellular organelles and chromosomes.

Calcium—Calcium occurs as calcium pectate in the middle lamella of adjacent cells. Unlike magnesium, however, it is not indispensable for some fungi and bacteria. But the role of calcium is obviously much more fundamental than as an element of cells. There is also enough evidence to indicate that protoplasm cannot maintain its living entity in the absence of certain ions and calcium is undoubtedly one of those ions; potassium is another. Calcium is essential for growth including the growth of pollen tubes. It can replace magnesium for the activity of several enzymes.

Potassium—It is the only univalent cation that is generally indispensable for all living organisms. The only concrete hypothesis of the role of potassium in plants is that like calcium the K ions are indispensable for the existence of protoplasm, the living substance. Although K^+ is needed in such large quantities by all plants, so far as it is known, it is not a constituent of any known essential organic substance. An important fact is that K^+ occurs in plant cells only in the ionic form. Its chief role seems to be to provide the necessary ionic atmosphere for the protoplasm, though it may also conceivably act as an activator of enzymes. K^+ is perhaps a necessary cation for photosynthesis. K^+ is necessary as an activator of enzyme *ALA-dehydrase* in the biosynthesis of chlorophyll. Rubidium has chemical properties similar to K^+ and it has been reported that rubidium could substitute for K^+ in the biological processes of a cell. Active absorption of potassium ions by the guard cells of stomata perhaps determine their opening and closing. Apical dominance in several plants appears to be lacking or very weak under potassium deficient conditions of the cells.

Application of Fertilizers

Although soil contains all essential minerals—both macro- and micro-intensive cultivation of crops results in a removal of such nutrients every year and the reserves of each element get depleted. This has to be replenished by adequate doses of fertilizers—organic or inorganic. The element required to be added in the largest quantities is N, followed by P and K. Nitrogen is usually applied in the form of $(NH_4)_2SO_4$, or urea, $CO(NH_2)_2$. Usually NH_4^+ ions are taken up more avidly than $SO_4^=$ ions, resulting in the development of acidity in the soil; $SO_4^=$ ions have a tendency to pick up H^+ from water and produce H_2SO_4. Soil pH is decreased and this affects the capacity of the roots to take up solutes from the soil solution. Urea is much less harmful in this respect, and that is why it is usually preferred to ammonium sulphate. It can be taken up directly; another advantage of using urea is that it encourages microbial activity. NH_4NO_3 and NH_4HCO_3 or liquid ammonia are also used in some countries as nitrogenous fertilizers. Acidity is usually checked by liming the soil which increases the pH. Alkaline conditions are also harmful, since several minerals (phosphates, for example) are precipitated at alkaline pH and are thus rendered unavailable to the plant. Phosphate is usually added as bone meal or superphosphate. In addition to its important role in plant nutrition, it helps in ion uptake; $(NH_4)_2SO_4$ as a fertilizer is more useful when applied along with phosphate than alone. Our soils usually have considerable S and thus application of S is not essential usually. K, however, is required in large quantities and this is usually applied as muriate of potash. 10 to 30 Kg of P and K per hectare is usually applied depending on the crop grown. Nitrogen is usually applied at 20 to 40 Kg per hectare but 60 or 100 Kg are applied for higher yields. The high yielding cultivations of wheat and rice require the application of high doses of fertilizers for production of high yields. Balanced fertilizers are now commercially available.

Fertilizers are usually applied in split doses, before sowing, at tillering stage of rice and wheat and just before ear emergence. Adequate irrigation is also important. Nutrients which are absorbed may also be applied in the form of a foliar spray. Urea is absorbed easily. Some micronutrients may also be applied in the form of a foliar spray, provided they are absorbed. Tracer studies have revealed that all inorganic cations are not translocated equally from the leaves, when applied as a foliar spray; some do not move at all.

Applications of organic manures have several advantages. They are less harmful and do not affect much the pH of the soil. Leaf compost, cow-dung, oil cakes, etc. are usually used by our farmers. They encourage microbial activity in soil and are decomposed; organic and amino acids are taken up by roots directly. However, the yield produced is usually much less than that obtained with synthetic fertilizers.

The practice of rotation of crops, in which a legume and a non-legume are grown alternately on the same soil, maintains the nitrogen content of the soil at particular level which may support the growth of the non-legume. After harvesting, the plants are ploughed into the soil for decomposition by micro-organisms. Legumes do not need the application of nitrogenous fertilizers, but P and K have to be applied. They also need Fe and Mo in a little larger quantity than other micronutrients.

Saline soils do not favour the growth of plants, unless such plants are salt tolerant. The halophytes in the mangrove vegetation are adapted to such conditions. Salinity-resistant varieties are now being developed for cultivation in saline conditions. Saline soils can be reclaimed by appropriate chemical treatment but this is an expensive proposition.

Soil is usually eroded by river and sea water as also by non-judicious application of fertilizers. Plants which can grow on sea shore and have extensive root systems are usually used. *Ipomea pescaprae* and *I. bilobata* are used as sand-binders. Appropriate chemical treatments are necessary to check the bad effects of the application of excessive fertilizers or industrial chemical wastes.

EXPERIMENTS ON WATER CONTENTS AND MINERALS IN PLANTS

I. Determination of dry weight, water content and ash content of plant tissues

(*a*) Weigh out carefully 5–10 g of different plant materials (root, stem, leaf, flowers, fruits, etc.; also succulents, aquatics, etc.). Particular care should be taken with root samples to ensure complete removal of sand and soil particles adhering to the roots (should be done in running water and the adhering water removed by carefully pressing with blotting paper). Put the samples in an oven at about 90°C. overnight. Cool in a desiccator and weigh again. Calculate the percentage of dry weights and water contents as % of fresh weights and water contents also as % of dry weights.

(*b*) Remove accurately weighed dried samples of tissues carefully into crucibles and ignite over a Bunsen flame or an electric muffle furnace at low red-heat (300–400°C.) for about 3–4 hours until the samples are completely free of black or grey colour (all combustible materials are removed). The soft, grey-white powdery stuff now contains only oxides of elements (formed during incineration) present in the plant tissues. When the incineration is complete remove the samples from the furnace, cool in a desiccator and weigh again (repeated weighing is necessary to obtain correct reading of as near

±0·1 mg). Express the amounts of ash present in different tissues as percentages of both fresh and dry weights.

II. Demonstration of the presence of volatile elements in plant tissues

(*a*) Place one or two crushed dry seeds (e.g., wheat, oat, pea, barley, rice, etc.) in a test-tube and heat carefully over a Bunsen flame. Observe the moisture which condenses on the sides of the tube.

(*b*) Heat a few crushed seeds of wheat or pea mixed with soda-lime. The evolution of NH_3, tested by smell and also by its action on litmus paper, proves the presence of nitrogen which escapes as ammonia.

(*c*) In a hard glass or metal test tube, held horizontally by a clamp, put some wheat or pea seeds and the open end of the tube is fitted with a cork, carrying a delivery tube which can be conveniently dipped in different jars, containing different reagents. The horizontal tube is strongly heated for some time and the delivery tube is put in turn into jars containing:

(i) a little anhydrous, whitish $CuSO_4$; turns blue (formation of $CuSO_4$, $5H_2O$) showing escape of water.

(ii) filter paper strips soaked in 5% cobalt chloride solution and dried. The treated dry, blue paper turns pink, showing again escape of water from burnt plant samples.

(iii) a solution of 10% lead acetate; black ppt. of PbS indicates the presence of *sulphur* (as H_2S).

(iv) Nessler's reagent; turns brown proving evolution of NH_3 (presence of nitrogen).

(v) baryta water (or lime water); turns milky-white showing evolution of CO_2.

III. Demonstration of the presence of nitrates in plant tissues

Nitrates give a deep blue colour with *diphenylamine sulphate* (0·5% in conc. H_2SO_4) solution. The coloured compound is soluble in water and thus fades on standing. Place carefully made thickish sections of leguminous plants (roots, stems, leaves, etc.), on a dry slide in a few drops of diphenylamine reagent. Observe quickly under microscope and look for the blue-coloured localised areas, especially abundant in or around the veins.

IV. Demonstration of the elements present in plant ash—qualitative ash analysis

To demonstrate the presence of various elements in plant ash, perform the following qualitative tests on ash sample (e.g., ash obtained preferably from mature tobacco leaves):

A. Dissolve about 0·5–1 g. of ash in 10 ml of warm 20% (by volume) HCl (conc. HCl is approximately 37%; one part conc. HCl and four parts distilled water, i.e., 20 ml of conc. HCl in 100 ml of water). Make up to 100 ml or stronger with water. Filter and perform the following tests with the filtrate:

Sulphur—Add a few drops of 5% $BaCl_2$ solution to a 10-ml portion of the filtrate. A white very fine crystalline ppt. of $BaSO_4$ proves the presence of sulphur.

Calcium—Take a second portion of 20 ml of the filtrate and make it slightly alkaline with dilute NH_4OH (equal portions of conc. NH_4OH and water). Filter and add a few drops of saturated ammonium oxalate solution. A white ppt. of calcium oxalate proves the presence of calcium. Save the solution.

Magnesium—Add an excess of ammonium oxalate to precipitate all calcium of the calcium-test solution. Filter and evaporate the filtrate to a volume of about 5 ml; this is important as conc. extracts give good results. Add about 1 ml of saturated disodium hydrogen phosphate. Cool and allow to stand. Crystals of ammonium-magnesium-phosphate—$(NH_4)_3Mg.PO_4$ prove the presence of magnesium. Precipitation may be hastened by rubbing the inside of the test-tube with a glass rod.

Iron—(*a*) Add a few drops of 2% solution of potassium ferrocyanide to about 10 ml aliquot of ash solution. A dark-blue colour—prussian blue—due to ferrocyanide being converted into ferricyanide proves the presence of iron.

(*b*) A few drops of 10% solution of potassium or ammonium thiocyanate (sulphocyanide) are added to about 10 ml of ash solution. A blood-red colouration due to the formation of iron thiocyanate—$Fe(CNS)_6$—proves the presence of iron.

Sodium—Test a loopful of ash solution, concentrated, if necessary, in a clear flame on sodium-free wire. A yellow-colour flame shows the presence of sodium.

B. Dissolve a second sample of ash in 10% (by volume) of HNO_3 and make it up to 100 ml (stronger solutions may be prepared, if desired). Filter.

Chlorine—To a small portion of the filtrate, add a few drops of $AgNO_3$ solution. Observe heavy, white precipitate of silver chloride, indicating the presence of chlorine.

Phosphorus—To another 10 ml aliquot of the filtrate in a test-tube, add ammonium molybdate solution. Heat on a steam-bath for few minutes and cool. A profuse yellow crystalline ppt. of ammonium-phospho-molybdate—$(NH_4)_3PO_4(MoO_3)_{12}$—proves the presence of phosphates.

Sodium—To another portion of the filtrate, add uranyl acetate and zinc acetate in acetic acid. Pale yellow ppt. confirms the presence of sodium.

CHAPTER XXXIX

THE METABOLISM OF NITROGEN IN THE SOIL

NITROGEN CYCLE

We have seen in our previous discussion that though nitrogen constitutes about 78% of the atmospheric air, the higher green plants are unable to utilise this gaseous nitrogen. The source of carbon for green plants, we know, is CO_2 gas of the atmosphere which is readily available to all green plants, even though CO_2 constitutes only about 0·03% of the atmospheric air. It is a perplexing thought why CO_2 from atmosphere readily diffuses into the green plants through stomata, and the much more abundant nitrogen is unable to do so. It will be more correct to say that nitrogen from the atmosphere certainly enters the plant but before it can be trapped and converted into combined forms of NO_3 or NH_4 by the plant, nitrogen again diffuses out into the atmosphere, whereas CO_2, however low its quantity may be, is fixed or converted into sugar in presence of light. Thus, the atmospheric nitrogen being completely unavailable to most higher green plants, the only source of nitrogen for the rooted plants must be the soil. Without nitrogen, life would have been impossible on this earth for protoplasm is almost completely proteinaceous. But, how did the soil get its nitrogen supply, for none of the primeval rocks—the rocks of the first age of the earth—contains nitrogen as a constituent? Thus the earth's whole quantity of nitrogen must originally have been present exclusively in the atmosphere. The primitive earth atmosphere might have contained some ammonia, and oxides of nitrogen formed by electric discharges during thunderstorms were (and are still) carried down to the earth by the rains. We also know now that the inherent ability of certain micro-organisms in converting atmospheric molecular nitrogen into combined forms such as NH_4^+—the process of biological nitrogen fixation—has made the continued existence of life possible in this earth. These micro-organisms possess certain biological mechanism completely lacking in higher green plants by which the gaseous nitrogen can be trapped and converted into combined form. This ability to assimilate elementary nitrogen by a comparatively few species of micro-organism is not only of great biological interest but in importance for maintenance of life on this earth, is second only to photosynthesis. The exclusive source of nitrogen of higher green plants is the soil or water which must be replenished all the time with the nitrogen, biologically fixed by micro-organisms, inhabiting soil and roots or leaves of higher green plants. Yet in the nitrogen cycle, this link is the weakest, severely counteracted by *denitrification* (the reverse process of nitrogen fixation; the reduction of NO_3^- in the soil into molecular nitrogen) which has exactly the opposite effect.

The economic importance of biological nitrogen fixation thus cannot also be underestimated and it is true, that there are many aspects of nitrogen fixation still to be explored. The growth of the plants, the leaching (washing away) effects of rain water and the activities of denitrifying soil micro-organisms, all tend to remove from the soil the nitrogenous compounds fixed in the soil, which are essential for the maintenance of most forms of plant life, as a matter of fact all animal life too. It is true that this loss is, in part, restored naturally by the decomposition of plant and animal remains, and artificially by the application of inorganic or organic fertilisers to the soil. A little amount of atmospheric nitrogen is also fixed by the electric discharges in the atmosphere, ionising radiations, meteor trails, etc., which may momentarily provide the high-energy needed for nitrogen to combine with hydrogen or oxygen, which ultimately is brought down to the soil by rainfall. The total amount of fixed nitrogen thus delivered to the earth by rainfall is small, not more than 25 million metric tons per year. Delwiche (1970), however, thinks that more than 70% of this total is previously fixed nitrogen, cycling through the biosphere, and that only about 30% is fixed by lightning and other atmospheric phenomena. Fixed nitrogen compounds produced commercially (artificial fertilisers) probably account for not more than 15% of the total nitrogen returned annually to the soil. Thus, by far the greatest proportion of annual replenishment of the soil nitrogen is due to nitrogen fixation by biological agents. There is evidence that bacteria associated with a few genera of insects possess capacity for fixing atmospheric nitrogen and it is also true for certain types of micro-organisms living in symbiotic association in the lumen (caecum) of certain herbivorous farm animals. But the amount of nitrogen fixed must be very small. It has been estimated that free-living nitrogen-fixing soil or water organisms and the symbiotic legume-bacteria system are responsible respectively for returning to the soil 4.37×10^6 and 5.46×10^6 tons of atmospheric nitrogen per year.

As regards the protein source for man, it ultimately depends upon the nitrogen fixed by micro-organisms. But oh! how long and laborious the method employed is and even then running at a loss—atmospheric nitrogen fixed in the soil or converted into artificial fertilisers by a great many intricate processes, taken up by the higher plants and built up into plant proteins—this again decomposed first by animal and then built up again into animal proteins—which ultimately serve as food for man. But it is a thought for the nutrition experts of the world, whose problem must be to produce more food for additional tens of millions new mouths every year, particularly in the tropics, how to cut out the intermediate links in the nitrogen cycle and to make available the much sought for proteins as indispensable food for human beings, without so many perhaps unnecessary conversions and reconversions.

The nitrogen metabolism in soil covers the sequence of chemical changes undergone by the nitrogen molecule in biological processes taking place in the soil, from its initial fixation in the soil, until it goes back into the atmosphere.

The main processes comprising this cycle of changes are as follows:

I. *The transformation of atmospheric gaseous nitrogen by micro-organisms and certain photosynthetic bacteria and blue-green algae into the combined form of* NH_4^+ *nourishing the bacteria and the green plant.*

This process is generally known as biological nitrogen fixation and is accomplished largely by three sets of organisms, the first one consisting of micro-organism living in endo- or ectosymbiotic association with a number of plants, the second consisting of organisms living in the soil independent of host plants (*Azotobacter* and *Clostridium*, the frequently chemo-synthetic *Desulphovibrio*, etc.) and the last consisting of a group photosynthetic bacteria and the blue-green algae.

II. *The transformation in soil of organic nitrogenous compounds arising from decomposition and autolysis of all forms of biological materials, or from excreta of animals or from products of metabolism of living soil organisms into* NH_4^+ *ions.*

The process is accomplished by a great variety of micro-organisms which may be anaerobes or aerobes, and which break down the organic nitrogenous compounds into NH_4^+ ions as a part of their metabolism or the formation of NH_4^+ ions may be brought about by the action of hydrolytic and oxidation enzymes present in the micro-organisms.

III. *The conversion of ammonium cations to nitrite and nitrate anions.*

The process is generally referred to as *soil nitrification* and is accomplished largely by two groups of organisms—*Nitrosomonas* (also *Nitrosocystic* and *Nitrosospira*) which form nitrites from NH_4 ions and *Nitrobacter* which converts nitrites to nitrates.

IV. *The reduction of nitrates into nitrites and finally into NH_4^+ ions or into free molecular nitrogen which goes back to the atmosphere.*

Reduction of NO_3^- into NO_2^- and NH_3 and to gaseous nitrogen is accomplished by various micro-organism. Several bacteria can reduce nitrate to NH_4^+ for assimilation purposes. However, there are a group of bacteria which reduce nitrate to the level of gaseous nitrogen, which usually escapes into the atmosphere. This process is known as denitrification. Examples of denitrifying bacteria include *Bacillus denitrificans*, *Pseudomonas stutzeri*, *Ps. denitrificans* and *Thiobacillus denitrificans.*

I. BIOLOGICAL NITROGEN FIXATION

(A) SYMBIOTIC—ROOT NODULE-LEGUME-BACTERIA SYSTEM

For several centuries it has been a matter of common observation and agricultural practice that soils impoverished by the growth of the cereal crops can be revitalised by growing leguminous plants. Yet, not until the end of nineteenth century were adequate explanations forthcoming as to why these procedures caused such beneficial effects.

It was in 1837, that the first observation was made that the fixation of atmospheric nitrogen takes place during the growth of legumes, such as clover, peas, lucerne, etc., whereas no such fixation occurs, during the growth of cereals, like wheat or oats. In 1857, it was demonstrated that plots of lands cropped to non-legumes without addition of artificial manures, gave rise to only low yields while plots cropped to legumes maintained relatively high yields even without any manurial treatment. Moreover if a non-legume followed a legume in a rotation [crop rotation such as wheat (1st year)⟶ potato, beet, etc., (2nd year)⟶barley, oat, etc., (3rd year)⟶legume (4th year) ⟶wheat (again)] the yield was as high as if the field has been previously left uncropped (fallow). And this occurred in spite of the fact that large quantities of nitrogenous compounds were removed by the preceding leguminous crop. By 1866, it was definitely established that certain bacteria in soil infect legumes forming nodules which enabled the nodule-bacteria system to fix atmospheric nitrogen; these bacteria have no infecting power on cereals. The supply of available nitrogen is so commonly a limiting factor in plant growth that the almost unique ability of leguminous plants to fix and to draw on atmospheric nitrogen (hence to be largely, if not entirely independent of soil nitrogen supply) is of the greatest importance in the development of economic cropping systems in the rotation of crops.

The organisms responsible are known as root nodule bacteria. They are species of the genus *Rhizobium*, once collectively known as *Bacillius radicicola*. They were first isolated in the pure state by Beijerinck.

The exact mechanism of nitrogen fixation by the root nodule bacteria is not well understood; on the other hand great deal is now known of the complex interrelations between *Rhizobium* and its host. It has been established that fixation of N_2 occurs in the nodules and that its amount is proportional to the volume and duration of life of the bacterial cells, so that it appears that the bacteria are, in fact, the actual agents of fixation. When bacteria and the host plant are in close symbiotic association nitrogen fixation can take place actually. *Rhizobium* cells growing some distance away from cultures of even non-legumes have been reported to be capable of fixing nitrogen. In 1975 it was demonstrated independently in five different laboratories that *Rhizobium*, particularly slow growing rhizobia belonging to the soybean or cow pea group, can fix nitrogen in laboratory culture media supplemented with pentozes like xylose, and arabinose, galactose, succinate and a small amount of glutamine.

The site of nitrogen fixation within the root nodule is the bacteroid. These bacteroids are usually X- or Y-shaped and are not found in laboratory cultures unless treated with some alkaloids or malic acid. There are some biochemical differences between *Rhizobium* and its bacteroid form. A bacteroid transferred to laboratory cultures produces cells having the usual morphology of *Rhizobium*. Bacteroids are stable only within the root nodule. The genetic basis or the bacteroid form is not understood. Nitrogen fixation is controlled by the *nif* (nigrogen fixation) gene responsible for the synthesis and function of the enzyme—nitrogenase which has two components—one of which contains iron alone and the other iron and molybdenum. Both iron and molybdenum are essential for nitrogen fixation.

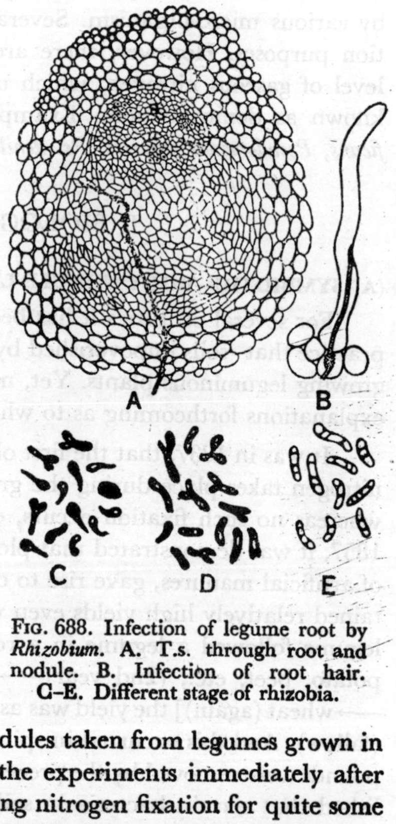

FIG. 688. Infection of legume root by *Rhizobium*. A. T.s. through root and nodule. B. Infection of root hair. C–E. Different stage of rhizobia.

For vigorous fixation, it is essential that the host plant should be intact and healthy; nodules excised from the roots, for example, rapidly lose their powers of fixation, although in recent years it has been conclusively demonstrated by subjecting only soluble nitrogenous compounds of the nodules to isotopic analysis with ^{15}N, that excised nodules taken from legumes grown in the fields (not in the greenhouse) and performing the experiments immediately after detachment from the roots, are capable of continuing nitrogen fixation for quite some time. It has been found that under certain conditions, nitrogenous products formed in the nodules may be excreted from the roots and become available to other plants and it is true, that it is sometimes possible to grow successfully mixed cultures of legumes and non-nigrogen-fixing plants such as oats without supplying any combined nitrogen.

Nodulated legumes require illumination for the host for nitrogen fixation and *nodule slices* showed enhanced fixation when supplied with sucrose, fructose and glucose in that order, which are well known to be prominent photosynthetic products. Interest with regard to the symbiotic system is centred around the question why association with leguminous plants should endow one or the other of symbionts with the ability of nitrogen fixation. When the seed of a legume develops in a soil containing *Rhizobium leguminosarium*, the latter are attracted to the region of developing root hairs. The result of the presence of the bacteria in the near neighbourhood of the root hairs is to produce a 'curling' or deformation of the hairs. Specific chemical substance, possibly the growth-promoting auxin, indoleacetic acid is responsible for this curling for bacterial extracts in absence of rhizobia produce the same effect. (For the curling of the root hairs, a conc. of 0·01 p.p.m. of indoleacetic acid is sufficient.) The response, however, is biologically non-specific, i.e., may be induced by extracts of bacteria other than rhizobia. At the site of deformation of the root hair, rhizobia invade the root tissue and proliferate within the hair in the form of a thread directed towards the cells of the roots (Fig. 688) and a sheath is laid down by the cells of the host separating the infected tissue from the rest of the plant. Cell division is stimulated, the newly formed tissues are invaded by more bacteria and thus a nodule is formed. Bacterial cells are included in a thread-like structure known as the infection thread. Many cells in the vicinity of the infection thread are triploid and some are tetraploid. Direct vascular connection of nodules formed with the host plant is maintained as long as fixation takes place. The local stimulation of host cells to divide is also due to the specific hormonal substances secreted by the bacteria. Both the bacteria and the host cells fail to proliferate as the nodule becomes older, and finally the whole nodular tissue becomes necrotic. The nodule softens, its interior is digested and finally falls off. The *Rhizobia* return to the soil.

The exact *site* of fixation inside leguminous nodules remains obscure. Some investigators believe that fixation occurred on a *membrane* surrounding single bacterium or groups of them in the nodule.

Rhizobium induces nodule formation in a restricted number of leguminous plants. There is a number of different species of *Rhizobium*, and each can function satisfactorily on certain legumes only. The species of Rhizobium concerned in N_2 fixation are: *R. meliloti*, *R. trifolii*, *R. leguminosarium*, *R. phaseoli*, *R. japonicum* and *R. lupini*. For convenience, leguminous crops are divided into cross inoculation groups, each group capable of being infected by the specific strain of bacteria belonging to that group only. Eight cross inoculation groups have been identified: (1) alfa alfa group (alfalfa, lucern), (2) clover group (clover, berseem), (3) soybean group (soybean), (4) pea group (pea, lentil, cicer etc.), (5) cowpea group (cowpea, sunhemp), (6) lupin group, (7) bean group and (8) only specific strain group. Thus if only cowpea group bacteria are present in the soil, effective nodules are formed in all the plants belonging to that group leading to a successful crop of plants of that group. The cropping of legumes of other groups, as for example clover or soybean, will not, however, be profitable because the particular strain either does not form any nodule on clover or soybean, plants of different groups, or even if the nodules are formed they are ineffective, for the quantity of nitrogen fixed by ineffective nodules per unit volume of the nodular tissue compared to those produced by effective strains of bacteria is insignificant.

A noticeable feature of the association of the bacteria with the legume is the presence of a red haemoglobin-like pigment (*leghaemoglobin*), similar to the vertebrate blood pigments, in the effective, healthy and mature nodules only. This leghaemoglobin (LHb) shows the characteristic absorption spectrum (in red), identical with myoglobin, a variety of haemoglobin specifically found in muscle fibres of animals. This haemoprotein was first isolated and crystallised from soybean root nodules as two proteins each having *one haeme* (about 0·34% Fe) group but differing in molecular weight—15,400 and 16,800—and amino acid composition. LHb is localised *outside* of the bacteroids. LHb thus can conclusively be eliminated as an integral part of *nitrogenase* enzyme, as it (enzyme) is only found *inside* the bacteroids. There also seems to be a complete lack of requirement for added LHb for the actual nitrogen fixation process. This haemoglobin is never formed in either partner when alone. It can easily be seen that mature nodules on a healthy pea plant are pink, whereas those on plants which are unhealthy or which have been kept in the dark for some days, are white or green. The infecting rhizobium apparently not only induces growth and multiplication of the root cells of the legume but also supplies the proliferating cells, directly or indirectly with a factor or factors necessary for the synthesis of this haemoglobin. The fact at first pointed naturally to the possibility that in the root nodule, haemoglobin itself or the mechanism connected with its synthesis may be directly linked with nitrogen fixation. A more probable explanation was there all the time, that haemoglobin may act only *indirectly* by securing in the root nodule necessary optimal conditions for oxidative processes with which nitrogen fixation may be associated by regulation of oxygen diffusion. The demonstration of inhibition of nitrogen fixation by nodules and its bacteroid suspensions by 2-4 dinitriphenol (DNP) at a conc. normally used for uncoupling phosphorylation, gives strong indications that oxidative phosphorylation is the major source of ATP for nitrogen fixation in the nodulated system.

The intensity of nitrogen fixation is proportional to the amount of haemoglobin present. The pigment is only formed after the nodules have been properly established and concurrent with the change of the red colour of the pigment to a green colour, fixation of N_2 by the nodular bacteria declines.

The rhizobia certainly need a continuous supply of oxygen to enable them to fix nitrogen efficiently. *Rhizobium* is a strict aerobe. It has been demonstrated that rhizobia in the root nodules possess a high rate of respiration, much more intense than that of the other parts of host plant. It might, at first sight, seem that the O_2 evolved in photosynthesis of green host plant would possibly more than suffice for the purpose. It is, however, not possible for this to occur because photosynthesis occurs in the host plant only in cells containing chlorophyll, that is, in the aerial parts and O_2 evolved must be transported from these parts to the nodular bacterial tissue and also there is no photosynthesis at night. The transport of O_2 from the exterior to the nodular bacteria, as also between the nodule and host plant is greatly hampered in many nodules due to the presence of common endodermis characterised by the presence of complete suberin lamella in the outer tissue of the nodule, in addition to individual endodermal sheath around the vascular strands of the nodule. This considerable reduction in the direct communication between the nodule tissue and the exterior results especially in a greatly reduced oxygen supply to the tightly packed rhizobia inhabiting nodular tissue. Thus, it is evident that coupled with high rates of respiration of the bacteria in the nodules and the difficulty of supplying oxygen from exterior or from the host, due to the common endodermal sheath, the aeration of the nodules is greatly restricted and the nodular bacteria are in a state of partial anaerobiosis. This is, however, useful for *Rhizobium*, since the enzyme nitrogenase is highly sensitive to O_2.

In order that N_2 fixation in the root nodules may continue at a normally high rate and the supply of O_2 does not become a limiting factor in the fixation mechanism, the nodules certainly need an efficient means of absorbing and transporting O_2 to the growing bacterial cells. In the course of biological evolution, the problem of efficient oxygen absorption arose early and because of its urgency it was solved first by certain worms, such as earthworm by synthesising haemoglobin, the most efficient oxygen carrier of all.

Similarly in order that nitrogen fixation can proceed uninhibited, some device must be adopted by legume host plant which can enhance the supply of oxygen to the deep-seated bacteria. The net result apparently was believed to be the formation of haemoglobin in the nodules from the porphyrin precursors, apparently always available in plants as in animals. It has been proved that within the nodules the leghaemoglobin displays the usual property of reversible oxygenation, characteristic of vertebrate blood. Some investigators believed that the rhizobia in the root nodules of legumes may meet at least a part of their oxygen requirement by the same biochemical mechanism as used by all vertebrates, including man himself, although it has been definitely proved that haemoglobin-bound oxygen in the nodules would suffice for only less than 4 minutes of respiration of an average-sized nodule. Thus there is really very little evidence that the red pigment in the nodule can function as an oxygen store for the nodular tissue.

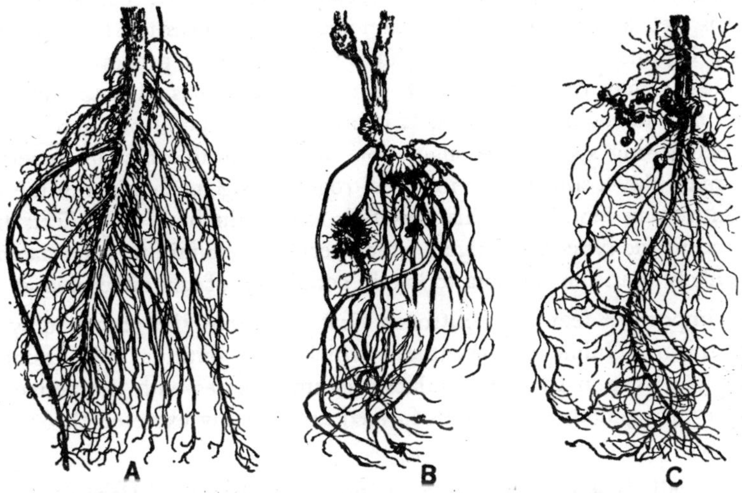

FIG. 689. Root system of legumes containing nodules with *Rhizobia*.
A. Clover. B. Sweet Pea. C. Soybean.

The role of LHb which now appears to be most probable (1971) is that proposed by Yocum (1964). Scholander (1960) more than a decade ago, demonstrated that the diffusion of O_2 through a membrane which was saturated with a solution of a haemoprotein—both haemoglobin and myoglobin—was much faster than the diffusion of N_2 in the same system. It has been proposed that this process of facilitated diffusion was an important way in which O_2 could be transported through tissues which had a high oxygen consumption. The LHb seems to be suited for the role in facilitated O_2-diffusion, under low conc. of O_2 which, as we know, obtains in the nodules. No doubt, in legume root nodules then O_2 is required for bacteroid respiration and consequent provision of ATP for nitrogen fixation. Without LHb, it would appear that the diffusion of O_2 through the dense nodular tissue would be completely inadequate to meet the necessary ATP-requirement. (Bergersen, 1971)

It is interesting to note that formation of haemoglobin in the plant kingdom, may be only under special circumstances and identified as yet only in the root nodules, solves one more discrepancy and supplies us with one more evidence that the world of living organisms, whether of animal or plant, is one and there is a splendid natural arrangement governing the both.

The presence of combined nitrogen, nitrates or ammonium salts, in the soil in which the host legume is growing, apparently makes the legume resistant to attack by rhizobia; the combined nitrogen impede the development of nodules, fewer root hairs are deformed and as a result fewer nodules are formed. The net effect is a great reduction in the quantity of nitrogen fixed and under certain circumstances this quantity is almost nil. Organic nitrogenous compounds such as proteins which have to be broken down before they can be assimilated, inhibit nitrogen fixation to a much lesser extent. Molybdenum appears to be of special significance in all organisms which fix atmospheric nitrogen. However, whilst molybdenum undoubtedly influences the growth of all nitrogen-fixing organisms there is now direct evidence that it is specifically concerned with the fixation mechanism itself. Recent work with the nitrogen-fixing blue-green algae *Anabaena*, indicates that Mo may also be concerned in phosphate metabolism. There is a unique requirement for cobalt for the nitrogen-fixing system; copper may also be essential. It is well known that carbon monoxide gas reduces the fixation of nitrogen in nodulated legume. This effect of CO on the fixation process in legumes and on vertebrate animal respiration may thus be analogous, due to the presence of Cytochromes or haemoglobin or both. It is interesting to point out here that recently there has been strong speculative suggestion that the antipernicious anaemia factor, vitamin B_{12} may play a similar role in the nitrogen-fixing mechanism in the nodulated legume, as in human blood. But this view fails to explain how CO affects fixation in *Azotobacter* and in the blue-green algae, which certainly contain no such haemoproteins and are less likely than the rhizobia to live in conditions of any anaerobiosis.

A most important feature of nitrogen fixation is that the process only occurs in growing organisms. Unlike photosynthesis which can continue after cessation of all growth, nitrogen fixation by the nodulated legumes ceases with the active growth of rhizobia. The nitrogenous compounds formed are possibly excreted through the roots of the host plant.

Taken overall, biological nitrogen fixation must be an endergonic reaction and, therefore, to put it in terms which the biologist understands, energy released, as indicated before, in the form of ATP, during respiration must be supplied to make it work.

All nitrogen-fixing organisms, including the rhizobia in the root nodules, generally grow best in neutral or slightly alkaline medium and there appears to be an inhibition of nitrogen fixation under acid conditions.

Ammonia is the most obvious candidate for the position of key intermediate in the mechanism of biological nitrogen fixation for it is often excreted by nitrogen-fixing organisms and its formation from elementary nitrogen by addition of hydrogen is a reaction of apparent simplicity. In experiments with isotopic ^{15}N with various nitrogen-fixing organisms, it is always found that the heavy ^{15}N appears in the greatest extent in glutamic acid just as it does if ^{14}N is supplied. Glutamic acid is well known to be the main substance through which NH_3 enters into the general metabolism of the cells and thus this would strongly suggest that NH_3 is the key intermediate in nitrogen fixation by various micro-organisms. Ammonia perhaps is produced from elementary nitrogen in several stages, one of which stage may conceivably be represented by hydroxylamine, NH_2OH, even now believed by some investigators, as the key intermediate in the nitrogen-fixing mechanism.

The leguminous genera *Cassia*, *Cercis*, *Gleditschia*, *Gymnocladus*, etc., contain species, not known to form any nodules in the roots. As a matter of fact it seems that effective root nodules are only formed in species belonging to *Papilionaceae;* species of the two other subfamilies of *Leguminosae—Caesalpinieae* and *Mimosae*—are rarely known to form symbiotic association with bacteria. *Gleditschia* belongs to a small group of legumes known from Cretaceous and early tertiary strata and this probably suggests that the association between the legume and rhizobia began later than that time. *Rhizobium* may form nodules on the roots of some non-leguminous plants also, e.g., *Trema cannabina* of Ulmaceae, which grows in New Guinea. The nodules are pink in colour, but the pigment is not leghaemoglobin.

(B) A few micro-organisms other than *Rhizobium* also form nodules on roots and leaves of non-leguminous plants. Some of these associations are mentioned below:

(i) SYMBIOTIC—ROOT NODULE—MYRICA—ACTINOMYCES SYSTEM. (Frankia)

(ii) SYMBIOTIC—ROOT NODULE—ALNUS—ACTINOMYCES SYSTEM.

(iii) SYMBIOTIC—ROOT NODULE—CASUARINA—BACTERIA SYSTEM.

(iv) LEAF NODULE—PAVETTA, DIOSCOREA, etc.,—KLEBSIELLA SYSTEM. (Fixation doubtful)

The agricultural importance of legumes has resulted in a concentration of attention upon their relations with *Rhizobium* and a neglect of other nitrogen-fixing symbiotic associations. Such symbioses are perhaps not so highly developed as the legume-bacteria system but their contribution in maintaining the fertility of the earth is by no means negligible.

Azospirillum (=*Spirillum lipoferum*) also forms endosymbiotic association with maize roots. Döbereiner in Brazil has claimed that the maize plants benefit substantially from this association in so far as its nitrogen nutrition is concerned. Some strains of *Bacillus* sp. also fix nitrogen in association with certain genotypes of wheat. *Azospirillum*, *Azotobacter* and *Beijerinckia* have been detected in the rhizosphere of several plants, particularly grasses including sugarcane. They utilize root exudates.

Several nitrogen fixing bacteria grow in ectosymbiotic association with the leaves of plants, particularly in the tropics. The organisms include *Klebsiella, Beijerinckia* and *Azotobacter*. Some active strains have been found to improve the growth of crop plants, when sprayed on their foliage. These organisms derive the energy for N_2 fixation and for their life activities from the organic matter leached out from the leaves when they come in contact with water, as for example, when it rains or when the leaves collect dew drops.

The bog plants, *Myrica gale* and *Myrica cerifera* (Fam. *Myricaceae*) have root nodules produced by a species of *Actinomycetes* (filamentous organisms classified with bacteria; occupies a position somewhere between bacteria and fungi. Many antibiotics, such as streptomycin, chloromycin, terramycin are produced by species belonging to this group of soil micro-organisms) and thus infected, the plant is able to fix atmospheric nitrogen. Free-living *Actinomycetes* alone are incapable of fixing nitrogen, so are the *Myrica*, if uninfected by *Actinomycetes*. Root nodule *Actinomycetes* of *Myrica gale* can fix nitrogen when the reaction of the soil or water in which they live is distinctly acid. Except for a few species of *Azotobacter* which can fix nitrogen in acid medium, all the nitrogen-fixing mechanisms seem to be inhibited by acid conditions. It has been shown that *Myrica* has got a molybdenum requirement for N_2 fixation. Nodulation occurred with or without added molybdenum, but N_2 fixation was increased tenfold by a trace of Mo. An interesting feature of the root nodules of *Myrica* is the development of negatively geotropic pneumatophores from the nodules which come above the soil water. The production of pneumatophores is certainly a biological adaptation for obtaining oxygen for the *Actinomycetes* inhabiting the nodular tissue and suffering like rhizobia, from an acute shortage of oxygen. The same difficulty of suboptimal oxygen conditions of the *actinomycetes* is effectively surmounted by the production of pneumatophores in the root nodules from *Myrica* from the upper surface of the nodules. The root nodules of *Myrica* are somewhat brownish in colour; the pigment is most probably a flavone glycoside.

The alder, *Alnus*, also develops root nodules as a result of infection by an endophytic *actinomycete*; the alder is then able to grow in the absence of a supply of combined nitrogen. Fixation rate is high and it is probably extracellular to the endophyte.

Trees of the subtropical genus *Casuarina* (and also *Podocarpus*, *Eleagnus*, etc.) are also evidently able to fix nitrogen by virtue of symbioses with a root-nodule-forming as yet unidentified bacterium.

In some cases, micro-organisms which alone can fix nitrogen may enter into symbioses with others which cannot. These organisms appear to be able to fix nitrogen when isolated and grown in culture. It is not clear how much benefit is derived by the micro-organism partner from this type of association. If it is symbiosis, it is evident that it is neither so highly developed nor so successful as the legume bacteria system. Evidences of such associations are afforded by the nodules in the *leaves* of some tropical shrubs, such as *Pavetta* belonging to the family *Rubiaceae* (also *Chomelia* and *Psychotria*) and also by the monocotyledon, *Dioscorea*. Altogether, evidences have accumulated during recent years that more than 400 spp. of non-leguminous plants harbour bacteria in the leaf nodules. *Psychotria* exhibits leaf nodules which are the seats of nitrogen fixation. The endophyte has been isolated and it has been established that it belongs to the genus *Klebsiella*. These bacteria isolated from the leaf-knots do not fix nitrogen. The leaf nodules of *Pavetta* are economically important; they are widely utilised as 'green manures' in the tropics, particularly in Ceylon.

Species of sensitive aquatic *Mimosaean* genera, *Neptunia*, are known to form beautiful nodules harbouring rhizobia, in their floating stems which root at the nodes.

(C) NON-SYMBIOTIC NITROGEN FIXATION

(i) BY FREE-LIVING NITROGEN-FIXING BACTERIA OF AZOTOBACTER AND CLOSTRIDIUM GROUPS. Winogradsky discovered an anaerobic soil micro-organism, *Clostridium pasteurianum* which will fix free molecular nitrogen, when supplied with carbohydrate, the amount of nitrogen fixed, being roughly proportional to the amount of carbohydrate broken down. About 10 years after, Beijerinck isolated from soil two aerobic organisms capable of fixing atmospheric nitrogen. They were *Azotobacter chroococcum* (the common non-motile type) and *Azotobacter agilis* (the motile variety). These nitrogen-fixing bacteria were found to differ strikingly in that, *Clostridium* (rod-shaped) is obligately anaerobic, i.e., able to grow only in the absence of oxygen whereas *Azotobacter* species are aerobes, growing only under condition of good aeration (obligate aerobes).

Azotobacter species have a world-wide distribution although it seems they are absent from some arctic soils. Species of *Clostridium* are more widespread for they can thrive well where anaerobic conditions prevail and in an examination of 15 different species of the genus, only three have been found unable to fix nitrogen. As far as it is known, in the sea, biological nitrogen fixation does not occur to any appreciable extent. *Azotobacter* and *Clostridium* species have been isolated from marine habitats but there is no evidence that these or any other nitrogen-fixing organisms occur in anything but low numbers in the open waters of the ocean. Compared with legume bacteria, they are relatively unimportant sources of nitrogen for higher plants, fixing perhaps not more than about 15% of the total nitrogen fixed in the soil or aquatic surface per year.

Optimal condition of the soil necessary for maximum intensity of nitrogen fixation by *Azotobacter* and *Clostridium* as well as by most other known nitrogen-fixing organisms is, as we know, neutral or only slightly alkaline. *Clostridium* under certain circumstances can tolerate a pH of approximately 5, but the fixation of nitorgen is greatly reduced. The growth of *Azotobacter* completely ceases at a pH less than 6. Thus it is evident if the soil is distinctly acidic in nature, all types of fixation of nitrogen come to a halt sooner or later except perhaps, as we have discussed before, in root nodule-*Myrica-actinomycete* system which can tolerate distinctly acid conditions of bog soils. It appears that the actual

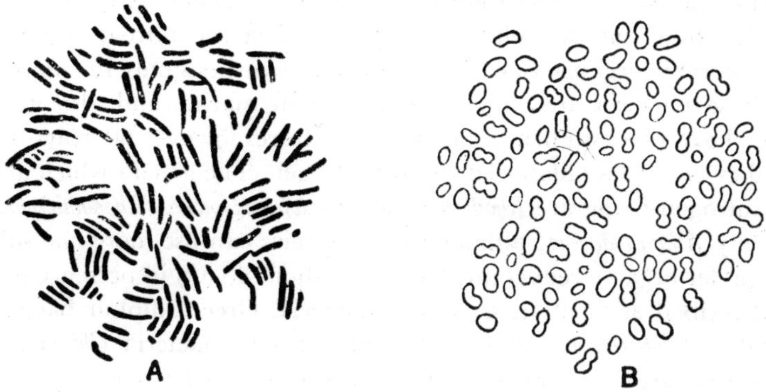

FIG. 690. Nitrogen-fixing bacteria. A. *Clostridium pasteurianum*. B. *Azotabacter chroococcum*.

fixation process is specifically inhibited under acid conditions for *Azotobacter* supplied with NO_3 may still grow under slightly acid conditions that totally inhibit growth of the bacteria in presence of molecular nitrogen only. Recently, however, several species of nitrogen-fixing bacteria closely resembling *Azotobacter*, except in details of cell form, have been grouped under a newly created genus *Beijerinckia*. Like the nodulated *Myrica*, they are very tolerant of distinctly acid conditions of the medium of the soil or water in which they live. Is there, then, a physiological difference between the nitrogen-fixing mechanisms of acid tolerant organisms and the vast majority of the nitrogen-fixers, which certainly prefer a neutral or alkaline medium? It does not necessarily imply that, for it is quite possible that alkaline conditions may be maintained in certain parts of the cell interior, even though the environment and the bulk of the protoplasm are acid in reaction.

The most important single factor influencing nitrogen fixation in soils is the presence of NO_3^- or NH_4^+ salts. With both *Azotobacter* and *Clostridium*, the presence of utilisable nitrogenous compounds diminishes the rate of nitrogen fixation, NH_4^+ or NO_3^- salts being most effective in this way. Inhibition of fixation by *Azotobacter* is complete in the presence of NH_4-nitrogen at a conc. of 0.5 mg nitrogen per 100 ml. The same is true for root nodule bacteria also, as we have seen before. Thus, when excess combined nitrogen is available in the soil, little or no fixation of atmospheric nitrogen takes place. This inhibition, however, may be partly counteracted by an increase in the carbohydrate concentration of the medium.

There is no doubt now that NH_3 is the key intermediate product in the nitrogen-fixing mechanism of all micro-organisms including *Azotobacter* and *Clostridium*. The enzyme system responsible for nitrogen fixation has been identified and the whole enzyme system has once been termed *azotose*, the individual enzyme, first reacting with elementary nitrogen being called *nitrogenase*. *Azotobacter* also possesses a *hydrogenase*. Certain chemicals have been found to have a specific inhibitory effect on the nitrogen-fixing enzymes. We have seen that carbon monoxide is one of them and the other is gaseous hydrogen. However, hydrogen has no marked inhibitory effect on nitrogen fixation by *Clostridium* or by photosynthetic bacteria (*e.g.*, *Rhodospirillum*). Excretion of organic nitrogenous compounds by roots, has been reported before from nodulated legumes; excretion of basically similar nature has also been observed in *Clostridium*. The key intermediate, NH_4^+ must represent the end of fixation reaction and the start of assimilation of the fixed nitrogen into the organic molecules of the organism.

(ii) BY FREE-LIVING COLOURED PHOTOSYNTHETIC BACTERIA. A most interesting discovery has been that coloured photosynthetic bacteria are nitrogen-fixing. All these bacteria are anaerobic organisms, carrying out photosynthesis only in complete absence of oxygen. These bacteria are widely distributed in marine and also in fresh water habitats. They have been studied in laboratory cultures for over 60 years before it was noticed that they could assimilate elementary nitrogen. The genera which are known to be definitely nitrogen-fixing are *Rhodospirillum, Rhodopseudomonas, Rhodomicrobium*, belonging to the family *Athiorhodaceae;* although these species can use inorganic sulphur compounds, they prefer organic reductants for assimilation of CO_2 in presence of light; they can best be described as facultative sulphur bacteria. Green sulphur bacteria, such as *Chlorobium* and *Chlorobacterium*, and the purple sulphur bacteria *Chromatium* are also N_2-fixers. In *Rhodospirillum*, nitrogen fixation appears to be closely associated with photosynthesis, for although this bacterium can grow in the dark if provided with suitable carbohydrate supply, assimilation of nitrogen is then really very slight.

Although the contribution of photosynthetic bacteria to the total amount of nitrogen fixed in soil and water may not be considerable, it cannot be denied that in the geological eras, their nitrogen-fixing activities must have been much more significant. The photosynthetic bacterium *Rhodopseudomonas capsulatus* abundantly found in tropical paddy fields, can certainly fix considerable quantities of nitrogen, thereby increasing the fertility of such soil.

(iii) BY FREE-LIVING COLOURLESS SULPHUR BACTERIA. These are one of the most interesting groups of recently discovered sulphate-reducing bacteria. These are obligate anaerobes, i.e., quite unable to grow in air. Instead of using oxygen to oxidise their food (as in normal aerobic respiration), they use sulphate and as a by-product, the sulphide is formed and energy liberated. This exergonic energy could be utilised for the formation of the cell material of the bacteria in presence of carbonates which are available in the water in which the bacteria live. If carbohydrates are not available for oxidation by sulphate, *Desulphovibrio desulphuricans* can utilise hydrogen equally, if not with more facility. It is only very recently established beyond all doubt that *Desulphovibrio* is also capable of fixing elementary nitrogen. It may be pointed out here that extreme thermophilic *Desulphovibrio* must be occupying a unique position in the hierarchy of living organisms—for it can survive and evolve on a planet completely devoid of oxygen and also of sunlight! And life perhaps will not disappear from this planet of ours when sunlight and oxygen be in short supply and insufficient for highly specialised forms, such as ourselves (in another 3,000 million years?). A comforting thought, indeed!

(iv) BY FREE-LIVING YEAST CELLS. There have been many reports that certain yeasts and also other fungi are able to fix atmospheric nitrogen. But the reported gains in nitrogen have generally been so small, as to make one suspect that errors of technique, e.g., absorption of NH_3 or oxides of N_2 from air, have occurred rather than fixation of the element. Recently, however, conclusive evidence has been obtained by means of Kjeldahl and other methods and also by use of isotopic techniques with ^{15}N, of fixation of N_2 by a variety of yeast, isolated from *heath soils* (waste flat land, usually covered with shrubs).

It seems quite possible, also, that the capacity of nitrogen fixation is present in some fungi on first isolation, but this capacity is rather easily and rapidly lost if cultured in artificial media for any length of time.

Some free-living nitrogen-fixing bacteria isolated from soil are *Pseudomonas radiobacter* and *Flavobacterium fulvum* in mixed culture.

(D) NITROGEN FIXATION BY BLUE-GREEN ALGAE (MYXOPHYCEAE)

Towards the end of last century, even before the isolation of *Clostridium* by Winogradsky, it was claimed that some of the blue-green algae are able to fix atmospheric nitrogen. However, the purity of such cultures was doubted by most of the investigators for the gelatinous sheath with which most members of *Cyanophyceae* (=Cyanobacteria) are invested, provides a natural medium for the development of bacteria and renders the removal of these bacteria by ordinary culture methods extremely difficult. Plausibility was lent to the prevalent idea by the actual isolation of free-living nitrogen-fixing bacteria from the algal sheath. So the idea—that blue-green algae are unable to fix nitrogen and where they were observed apparently to grow in absence of nitrates or ammonium salts, they were in close association with nitrogen-fixing bacteria—prevailed until about 1928, when the fixation of nitrogen was definitely shown to take place in absolutely pure cultures of the algae (pure cultures are nowadays obtained easily by using ultra-violet light of suitable irradiation, which would kill the bacteria but not the algae). Recently, however, conclusive evidence of considerable nitrogen fixation by members of blue-green algae has been given by the use of heavy isotopic ^{15}N. Of some 40 species of algae, many of them belonging to the family *Nostocaceae* tested so far, for nitrogen-fixing capacity, more than half have been found to possess it.

Fixation of nitrogen, perhaps not quite as efficient as some species of *Nostocaceae*, has also been demonstrated in species of family other than *Nostocaceae*, e.g., *Tolypothrix* sp. (*Scytonemataceae*), *Calothrix* (*Rivulariaceae*), *Mastigocladus laminosus* (*Stigonemataceae*), etc. Some of the fixed nitrogen, not all, is liberated into the medium. Blue-green algae are common in soil and fresh water and have a world-wide distribution, being particularly abundant in the tropics. Blue-green algae may be the most important nitrogen fixers in certain types of fresh water and where anaerobic conditions prevail, *Clostridium* or the photosynthetic bacteria may also fix appreciable quantities. These algae are common in alkaline or near neutral soil but are rarely present under acid conditions. They are certainly more abundant in the surface layers of soil exposed to light but the demonstration that some species are capable of heterotrophic nutrition in presence of carbohydrate suggests that they are not necessarily confined to such regions only. Like all other nitrogen-fixing organisms, blue-green algae have a high nitrogen content, about 8% of the total dry wt. Nitrogen fixation by the blue-green algae occurs only under certain well-defined conditions, such as: (1) fixation has never been found to take place in resting material; active growth is essential, (2) as with the case of all other nitrogen-fixing bacteria, assimilation of free nitrogen does not take place in presence of readily available combined N_2, NO_3^- or NH_4^+ salts, (3) traces of molybdenum are necessary, like all other N_2-fixing organisms, for fixation of nitrogen by the algae, (4) as in the case of nodulated legume and *Azotobacter*, a slightly alkaline medium is the most favourable for the growth of the algae—no fixation takes place below a pH of approximately 5.7 and (5) like those of *Azotobacter* and the symbiotic system, the enzyme system of the blue-green algae is specifically inhibited by CO and O_2.

As the cyanophycean algae are photosynthetic organisms capable of synthesising carbohydrate, they are able to flourish in situations such as bare rock faces, open sea, etc., where nitrogen-fixing bacteria cannot thrive due to lack of carbohydrate. On the other hand, environments into which light cannot penetrate, such as all but superficial layers of soils and water, although suitable for bacteria, would appear unsuitable for the algae.

In combining the nitrogen-fixing and photosynthetic modes of nutrition, blue-green algae resemble root nodule-bacteria symbiotic system. However, in the latter, the seat of photosynthesis (in the green aerial parts) and nitrogen fixation (in the root nodules) are separated in space if we regard the whole system as a composite organism. In the algae, nitrogen and carbon assimilation may proceed side by side *in the same cell*, indicating the close relation between the two mechanisms. It may very well be that there are common intermediate compounds. (We know that in higher green plants, the same intermediate compound of photosynthesis, the phosphoglycerates may be tapped and a part may be preferentially elaborated to proteins in young growing cells.)

Recent evidences in studies on cell-free extracts on *Anabaena cylindrica* support the conclusion that N_2 fixation is localised together with photosynthesis on the chromatoplasmic lamellae.

The blue-green algae as well as the photosynthetic bacteria must have undergone their greatest development at a stage when the shortage of combined nitrogen on the earth was really becoming acute. Higher plants and animals appear to have evolved from non-nitrogen-fixing stocks but the frequency with which symbioses have arisen between higher photosynthetic plants and nitrogen-fixing organism, photosynthetic and heterotrophic, certainly to a large extent, testifies to the decided biological advantage of the combination (both nitrogen-fixing and photosynthetic at the same time) found in the blue-green algae and photosynthetic bacteria.

The nitrogen fixed by blue-green algae, like all others, eventually passes into general circulation and is assimilated by plants, unable to fix it for themselves. Two possible ways, which are true also for other nitrogen-fixing flora, in which (i) the nitrogen compounds formed in the cell released are excreted into the surrounding medium during the life of the algae and (*ii*) they may be released by death and subsequent decomposition of the algae.

While blue-green algae are of general occurrence in soils of agricultural value in temperate regions, it may be doubted whether they are all nitrogen-fixing, contributing to the fertility of soil, or that they occur in appreciable number.

In certain tropical soils, blue-green algae appear to play an important part in the maintenance of the fertility of the soil. In India, rice may be grown on the same land for many years without addition of fertilisers to the soil. During the season of growth of rice (the monsoon), the rice fields are flooded by incessant rain and luxuriant growth of blue-green algae, including many species that have been shown to be nitrogen-fixing, occurs in the waterlogged fields. The principal nitrogen-fixing blue-green algae in Indian rice fields are *Aulosira fertilissima* (*Microchaetaceae*). The chroococcalean, non-heterocystous algae, *Chlorogloea fritschii* which occur abundantly in paddy fields, where rice is cultivated, also fix nitrogen. The nitrogen fixed by the algae becomes available to the rice plants when the algae decay and there is good evidence that the algae are the main agents responsible for maintaining the level of fertility of rice fields (it is true, that the level, at least in India is not very high). It has also been shown that the growth of the blue-green algae on waterlogged soil substantially increases the content of organic matter and combined nitrogen, either organic or inorganic of the soil and when the rice plants and the blue-green algae are grown together, each seems to do better than if grown separately. This *quasi symbiosis* between rice and *Aulosira* or *Anabaena* is certainly interesting from physiological point of view. The claim that uninfected rice plants can fix-N_2, has not been substantiated. There also appear to be considerable potentialities for increased use of the blue-green algae in the tropics in land reclamation.

As we have seen before, blue-green algae are abundant in fresh water and copious growth of planktonic forms of microscopic plants, including blue-green algae in fresh water produces a phenomenon which is known as 'water bloom'. The growth of the algae is sometimes so luxuriant that fishing there becomes hazardous.

Some members of the blue-green algae produce nodules on roots or stems. Root nodules are found in several gymnosperms and stem nodules are formed on the angiosperm *Gunnera*. Probably the most remarkable symbiotic association between a blue-green alga and a green plant is the *Azolla—Anabaena azollae* association. *A. azollae* fixes N_2 inside the leaf cavities of the water fern *Azolla*. When the dead remains of *Azolla* decay, the soil accumulates the nitrogen which is utilized by crop plants. In Viet Nam, China and Indonesia *Azolla* is grown extensively on standing water in rice fields. Rice plants gain substantial quantities of nitrogen in this way. The technology is simple and its judicial use may enhance crop yield considerably. Use of *Azolla* is now recommended for adoption in this country on a large scale.

It may be interesting to point out here that all the species of blue-green algae (e.g., *Anabaena*, *Nostoc*, *Cylindrospermum*, *Aulosira*, etc.) known definitely to be nitrogen-fixing, possess the enigmatic structure called *heterocysts* (except perhaps *Chlorogloea fritschi* and perhaps also one or two *Nostoc* species) whereas those as yet known to be quite unable to assimilate free nitrogen are for the most part non-heterocystous. We do not as yet know the significance of the presence of heterocysts in blue-green algae. It is possible that heterocysts may in some way or other be functional in the mechanism of nitrogen fixation in these algae. The thick wall of heterocysts provide the anaerobic environment required for nitrogenase; the heterocyst and the adjacent cells are believed to be functional in nitrogen fixation. However, a few non-heterocystous unicellular forms, e.g., *Gloeocapsa* and *Aphanothece* also fix nitrogen; thus, the presence of heterocysts is not essential for nitrogen fixation by the blue-green algae.

Although higher plants, such as rice and barley uninfected by bacteria, aphids and man himself have been reported from time to time to be capable of assimilating elementary nitrogen, critical experiments have always failed to substantiate these assertions. Thus it seems highly probable, inspite of incompleteness of our knowledge, that the property of nitrogen fixation is confined to microorganisms only. And again among microorganisms, a capacity for nitrogen fixation apparently occurs quite independently of other nutritional characteristics. Thus blue-green algae are photosynthetic and *Desulphovibrio* is frequently a chemautotroph, while *Azotobacter* requires a source of organic carbon. Thus, given a suitable source of energy, these and all other nitrogen-fixing organisms are able to synthesise all their complex organic requirements from inorganic substances. Nitrogen fixation has never been demonstrated in *holozoic* organisms, as for example in parasites, saprophytes or insectivorous plants and in fact there appears to be definite incompatibility between morphological specialisation and nitrogen fixation.

There is now more or less definite evidence that the mechanism of nitrogen fixation is basically the same in different organisms, be it *Azotobacter* or nodulated legume, blue-green algae or other nitrogen-fixing organisms. In heterotrophs the electrons required for reduction of nitrogen are provided by organic compounds. In *Clostridium pasteurianum* where N_2-fixation has been demonstrated in cell-free systems, pyruvate is oxidised to acetyl phosphate releasing $2H^+$ and 2e. The acetyl phosphate in presence of ADP is then cleaved to acetate with the concomitant formation of ATP, the energy being provided by the hydrolysis of the carboxyl phosphate bond (a high energy bond) of acetyl phosphate. This ATP then is used by nitrogenase to split the triple bonds connecting the two N atoms in the N_2 molecule. The electrons of pyruvate are accepted by ferredoxin (in some systems flavodoxin is used), the redox potential of which is very close to that of the hydrogen electrode; the enzyme hydrogenase may also provide electrons from molecular hydrogen. Reduced ferredoxin then transfers the electrons to the enzyme which finally reduces N_2 in several steps to NH_3. The postulated intermediates include HN=NH (diimide), hydrazine (H_2N-NH_2) and NH_2OH (hydroxyl amine). $^{15}N_2$ supplied to nitrogen-fixing cells or tissues is recovered in NH_3 as the first stable intermediate and in amino acids, particularly glutamic acid. (Fig. 690a).

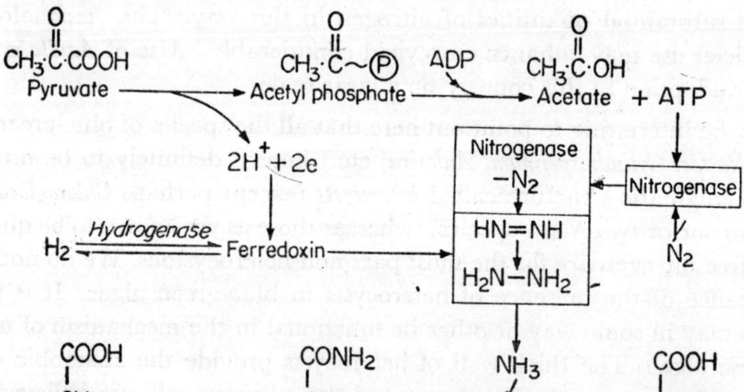

FIG. 690A. Schematic representation of some major reaction sequences in biological N_2 fixation. Pyruvate acts as an electron donor in *Clostridium pasteurianum*.

Although glutamic acid is the key intermediate in amino acid metabolism glutamic dehydrogenase which catalyses the reaction $NH_3 + \alpha-$oxoglutarate $+ NADPH + H^+ \rightleftharpoons$ Glutamate $+ NADP^+ + H_2O$ does not play a major role here, since it acts only in presence of relatively high concentrations of ammonia. The enzyme which is involved here is glutamine synthetase which converts glutamic acid to glutamine; glutamine then donates the $-NH_2$ group to α-oxoglutarate forming two molecules of glutamate as follows:

$$\text{Glutamate} + NH_3 + ATP \xrightarrow[\text{Synthetase}]{\text{Glutamine}} \text{Glutamine} + ADP + Pi$$

$$\text{Glutamine} + \alpha\text{-oxoglutarate} + NADPH + H^+ \xrightarrow[\text{Synthetase}]{\text{Glutamate}} 2\ \text{Glutamate} + NADP^+$$

$$NH_3 + ATP + \alpha\text{-oxoglutarate} + NADPH + H^+ \rightarrow \text{Glutamate} + ADP + Pi + NADP^+$$

The amino group of glutamate is then transferred to various keto acids, by trans aminases, forming a variety of amino acids which are then utilized for synthesis of proteins or other cellular metabolites.

The *nif* gene which codes for nitrogenase is plasmid (extra chromosomal DNA) carried in several nitrogen-fixing microorganisms. Attempts are now being made in many laboratories round the world to introduce the *nif* gene into eukaryotic cells. If this is achieved, this will be a major break-through since it will no longer be necessary to apply nitrogenous fertilizers to such plants. However, there are many obtstacles which have to be crossed before them.

One more interesting point before we conclude: Isn't it rather misleading to speak of 'fixing nitrogen'? Elementary nitrogen is chemically inert and undergoes appreciable reaction with other substances only at high temperature or pressure, which is quite outside the range to be found in the living cells. We have discussed before that nitrogen fixation is an endothermal or endergonic process, in which energy is absorbed and this energy must be supplied from outside. Thus, it is evident that the fixation of nitrogen requires an investment of energy. Before nitrogen can be fixed, it must be 'activated', which means that each molecular nitrogen must be split into two atoms of free nitrogen. This step requires at least 160,000 cal for each mole of nitrogen (equivalent to 28 grs). The actual fixation primary step, in which two atoms of nitrogen combine with 3 molecules of hydrogen, to form two molecules of ammonia is: $\begin{matrix} N_2 \longrightarrow N + N \\ 2N + 3H_2 \longrightarrow 2HN_3 \end{matrix}$; this, however, releases 13,000 calories. Thus the fixation of a molecule of nitrogen requires a net energy supply or input of at least 147,000 calories (equivalent to about 18 ATP molecules). For photosynthetic bacteria and blue-green algae, this energy requirement can be met directly from sunlight which is a photo phosphorylation; for *Rhizobium*, from host leguminous plant, and for free-living *Azotobacter* and *Clostridium*, from cellular respiration.

DEMONSTRATION OF THE PRESENCE OF ROOT NODULE BACTERIA (RHIZOBIA) IN LEGUMES

MATERIAL: Roots of gram, bean, pea, etc. Wash the dug-up roots carefully and note that the roots are studded with small pink nodules. Cut through a large root nodule with a blade or razor and put the cut surface on the centre of a perfectly clean slide and press. Spread the pink juice which comes out and allow the smear to dry and then heat it gently over a Bunsen flame or pass it thrice through the flame. The slide should feel hot when placed on the back of the hand but not too hot to bear. The heating fixes the bacteria to the slide, but it is imperative not to overheat. When the slide is cool, stain the film with methylene blue for 2-3 min; wash the stain off gently with water and blot the slide dry by careful pressure with a filter paper. When quite dry, mount in canada balsam. Bacteria, like minute rods, are distributed singly or in groups—very variable in length, sometimes mere dots; some branched Y-shaped forms are also present (see Fig. 688 C, D & E).

II. AMMONIFICATION

Nitrogen compounds in the plant and animal residues are decomposed in the soil to form NH_3 so long as the ratio of carbon to nitrogen in the organic matter does not greatly exceed 10 : 1.

Proteins and other organic nitrogenous compounds are broken down in soil by a variety of soil micro-organisms, the end product being NH_3. Whether NH_3 appears or not, depends on the rate of growth of organisms in the soil themselves requiring NH_3 for their own anabolic operations. If ample utilisable carbohydrates are present, NH_3 will not appear, as it is entirely used for building up fresh bacterial or fungal cells.

The ammonification of a molecule of glycine in presence of oxygen to ammonia, carbon dioxide and water releases 176,000 calories (about 22 ATP).

Many soil micro-organisms are involved in NH_3 formation, the mechanism of which varies from one organism to other. Proteins are broken down to peptides and then amino acids and the amino acids may yield NH_3, by the action of a host of enzyme systems, *e.g.*, *tyrosinase*.

III. NITRIFICATION

The metabolic process whereby NH_3 is finally converted to NO_3 is called *nitrification*. *Nitrosomonas* and *Nitrobacter* are the two most important genera of strictly aerobic chemosynthetic autotrophs, which respectively obtain their energy by oxidation of NH_3 to nitrite (NO_2) and nitrite to nitrate (NO_3) (*see* later).

These nitrifiers prefer a slightly alkaline environment; — *Nitrosomonas* is unable to grow in organic media; growth of *Nitrobacter* is stimulated in certain cases by a small amount of peptone.

$$2NH_3+3O_2 \xrightleftharpoons{\textit{Nitrosomonas}} 2HNO_2+2H_2O+130{,}000 \text{ cal}$$

$$2HNO_2+O_2 \xrightleftharpoons{\textit{Nitrobacter}} 2HNO_3+35{,}000 \text{ cal}$$

Nitrifying bacteria generally grow on the surface of the soil where NH_3 or NH_4 ions may be held (adsorbed). It has also been found that the adsorbed NH_4^+ ions are preferentially nitrified by micro-organisms on the surface of the soils.

$KClO_3$ in minute traces inhibits the growth of *Nitrobacter* but not *Nitrosomonas*. Chloromycetin, an antibiotic (specific against typhoid and typhus pathogens) containing a nitro group is especially active against soil organisms like *Nitrobacter*, oxidising NO_2 to NO_3.

IV. DENITRIFICATION

In contrast to the limited number of organisms capable of oxidising NH_3 and NO_2, several species accomplish the reverse process, namely, the reduction of nitrate and nitrite to N_2O and molecular N_2. While NH_3 may be retained in the organism or in the medium, gaseous products such as N_2O and N_2 pass readily into the atmosphere with the result that the overall nitrogen content of the soil is decreased. This biological reduction of NO_3 and NO_2 is often termed *denitrification*. Examples of organisms known to reduce NO_3, include *Pseudomonas fluorescence*, *Pseudomonas denitrificans*, *Pseudomonas stutzeri*, *Bacillus* sp. *Thiobacillus denitrificans*, etc.

Denitrification is generally, but not always, encouraged by poor drainage and lack of aeration due to waterlogging as well as by a plentiful supply of organic matter. The presence of O_2 tends to suppress the reduction of NO_3.

Under anaerobic conditions, the chemosynthetic autotroph, *Thiobacillus denitrificans* can obtain energy only by the oxidation of elementary sulphur or other sulphur compounds at the expense of reducing NO_3, i.e., using NO_3 in place of free O_2.

When denitrification takes place in absence of oxygen, and with the help of nitrate ion in presence of sulphur and chalk, the equation can be written thus:

$$5S + 6KNO_3 + 2CaCO_3 \longrightarrow 3K_2SO_4 + 2CaSO_4 + 2CO_2 + 3N_2 + 660{,}000 \text{ cal.}$$

A scheme for the probable pathways of denitrification by micro-organisms is given below though direct evidences about some of the intermediates shown in the scheme are lacking. The liberation of N_2 may also be achieved without the participation of enzymes, since an amino compound coming in contact with NO_2^- may liberate N_2.

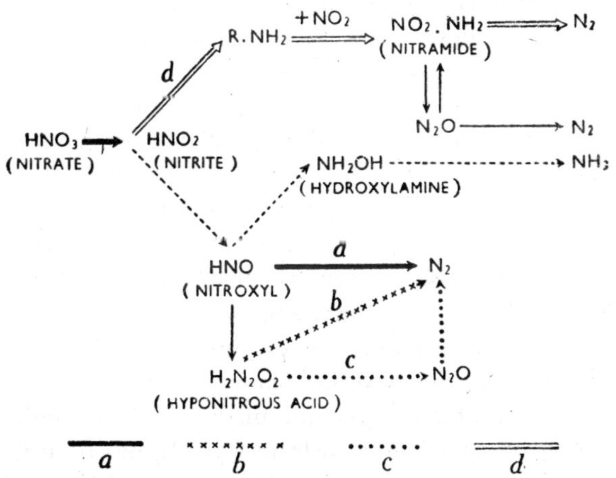

FIG. 691. Pathways of denitrification. Routes *a* & *b* operative in *Pseudomonas denitrificans*; route *c* in *Bacillus subtilis*; route *d* in *Pseudomonas stutzeri*.

The gaseous products such as N_2O, N_2 or NO_2 formed by denitrifying organisms are responsible for much troublesome odour and consequent loss in certain industries. NO_3^- and NO_2^- are commonly used for curing and preserving meat products and their decomposition is often due to unwelcome activities of denitrifying bacteria. It is possible that denitrifiers can convert appreciable quantities of fertilisers, such as $(NH_4)_2SO_4$ and KNO_3, added to the soil, into gaseous N_2 or oxides of N_2 which escape into atmosphere, thereby significantly decreasing nitrogen available for plant growth.

ECOLOGICAL ASPECTS OF N_2-FIXING ORGANISMS AND NITROGEN ECONOMY OF NATURE

In arctic and subarctic conditions, the bacterial activity would necessarily be slow and consequently the soil has relatively small supply of nitrates. The plants in these areas must either be adapted to a low nitrogen intake or have some means of utilising atmospheric nitrogen. Species of *Alnus* which are prominent in such areas are definitely known to form root nodules infected by actinomycetes and this symbiotic relationship may help *Alnus* to fix N_2 and thrive in those freezing cold areas. Legumes seem very poorly represented. In extreme arctic conditions lack of carbohydrate due to very short growing season appears more critical than nitrogen shortage and as a result legumes are completely absent from those areas. In very sandy leached soils, however, a characteristic flora develops in which legumes and species of *Casuarina* are prominent. It is interesting to note that spp. of *Casuarina* are also known to form nitrogen-fixing nodules.

Blue-green algae and the photosynthetic bacteria are commonly the first plants to colonise bared areas of rocks and soils. The best known example of this is seen in the recolonisation of Krakatoa group of islands in the Pacific where a volcanic explosion in 1883 denuded the island of all visible plant life. It is probable of course that photosynthetic and the thermophilic bacteria, *Desulphovibrio*, must have survived this drastic treatment for, at least in the case of some varieties of *Desulphovibrio*, it is known to thrive in a temp. of 80°C.! Blue-green algae and most probably photosynthetic bacteria were the first plants to appear in quantity on the pumice and volcanic ash a few years after the eruption—photosynthetic bacteria which use inorganic sulphur compounds in photosynthesis can be imagined as relishing such an environment more than any other organism, except perhaps anaerobic *Desulphovibrio*. In part, the success of these organisms as colonisers is, no doubt, due to a marked ability to withstand adverse conditions such as desiccation, high temperature and high concentration of salts, particularly SO_4^-, but a capacity for nitrogen fixation cannot be discounted as a possible factor determining their presence on such unpromising environment. Members of the blue-green algae are equally important pioneers under less impressive circumstances: (*i*) a community of blue-green algae forms the initial stage in plant succession on certain eroded soil and (*ii*) they play a conspicuous part in colonisation and stabilisation of salt marshes. It is perhaps significant that majority of species occurring in the list of algae as pioneer plant communities belong to the family *Nostocaceae*.

Besides free-living forms there are many blue-green algae that occur in more or less intimate association with other plants, for example, with liverworts, with water-fern *Azolla*, with Cycads and with the flowering plant *Gunnera* (Fam. *Haloragidaceae*). *Azolla* contains an *Anabaena* and is able to grow in a medium free of combined nitrogen, NO_3 or NH_4 salts. A species of *Nostoc*, isolated from Cycads, has been found to be scarcely capable of autotrophic nutrition and needs to be supplied with carbohydrate for growth and nitrogen fixation. The host Cycad can grow quite well without the algal partner and it is not clear how much benefit is derived from the N_2 fixed by the algae or how successful the symbiotic association is. Blue-green algae form symbiotic associations with fungi; the algal partner in many lichens is often a blue-green alga. In the lichen *Collema*, the alga *Nostoc* has definitely been shown to be nitrogen-fixing. Nitrogen fixation by the algal partner may well be the most important factor in enabling these composite organisms to live on bare rocks and other similar inhospitable situations. It has been found that some lichens contain the nitrogen-fixing bacteria *Azotobacter*. It seems, therefore, that many lichens are *triple alliance of organisms* and are capable of assimilating elemen-

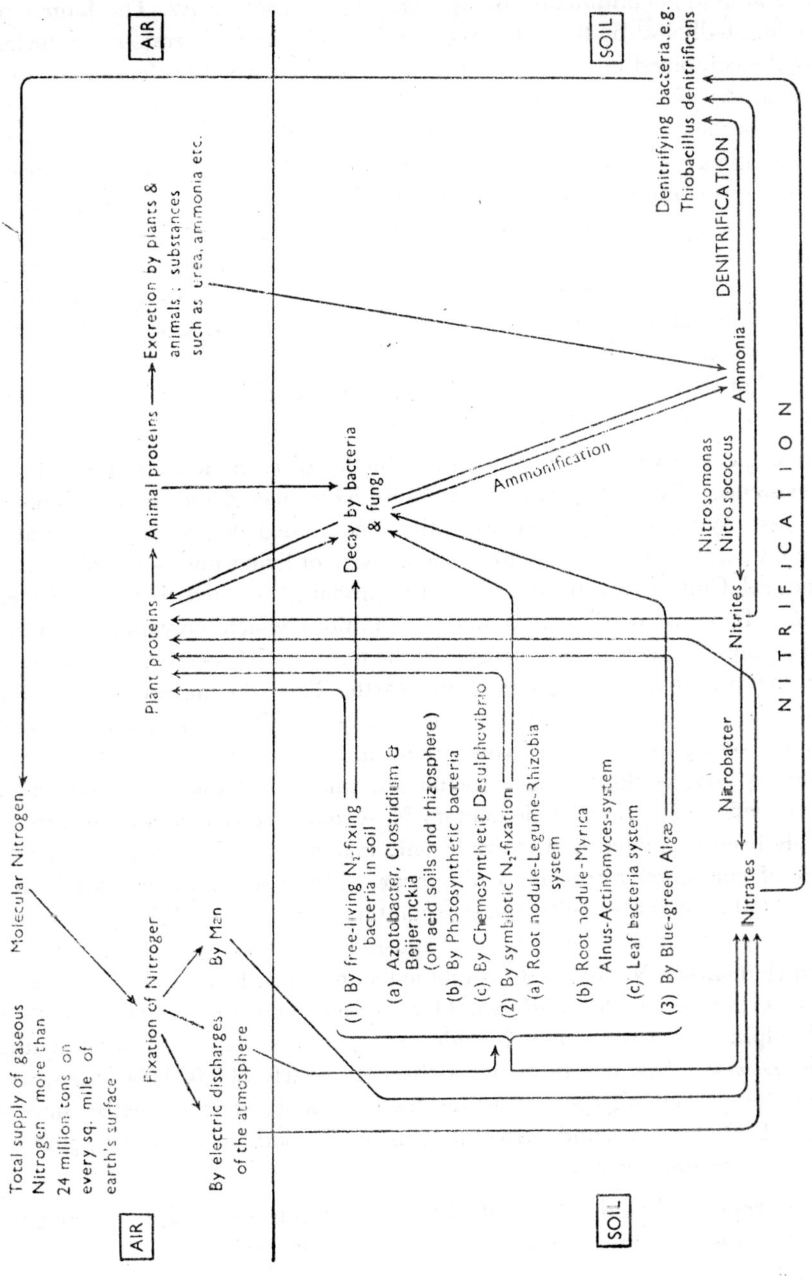

FIG. 692. The nitrogen cycle.

tary N_2 even if the blue-green algal partner fails in its duty and even if the algal partner is not a blue-green alga at all!

Agricultural land is continually losing nitrogen in various ways. The largest proportion is undoubtedly lost by the removal of crops and by animals grazing on the land. Very little of the combined nitrogen contained in these is returned to the soil due to the modern method of unforeseeing sewage disposal. As a refuse, it ultimately finds its way to the sea and becomes largely unavailable to man. Comparatively large amounts are also lost by being carried away to the sea as a result of leaching or washing away of soluble nitrogenous compounds from the soil and also by the removal of soil itself by erosion. Again by the activity of denitrifying bacteria, flourishing in a plentiful supply of organic matter and bad aeration of the soil as a result of waterlogging, the combined nitrogen is being continually reduced to elementary form, which escapes into atmosphere. That there must be a compensatory process for the maintenance of life on earth is evident from the observation that soils from which crops have been taken out are not totally depleted of combined N even if no manures are applied to the soil, since lands left without cropping rapidly gain in nitrogen.

Thus biological nitrogen fixation is undoubtedly of extreme importance for the continued existence of life on earth and it must have played an equally important part in the past. With certain minor exceptions, combined N does not occur at the present time on earth except as a result of the activity of living micro-organisms in the soil. The natural Chilean nitrate deposits have probably resulted from the bacterial action in past. It seems possible that under conditions which prevailed upon earth when life first appeared, combined nitrogen in the form of NH_3 occurred in substantial amounts in the primitive atmosphere of the earth. Thus the necessity for nitrogen fixation did not arise for hundreds of million years. This must have provided the only source of N_2 for the first living organisms. However, a stock of combined nitrogen such as NH_3 could not last indefinitely, since oxidation to nitrate and the subsequent reduction of nitrates to elementary N_2 would soon reduce the stock to zero. It has actually been estimated that it would only take about 60,000 years for the total earth's stock of combined nitrogen to disappear without any compensatory scheme of bringing back the elementary nitrogen into combined form again. If life first appeared on the earth, say, about 3000 million years ago it is clear that organisms, capable of transforming elementary N_2 back into combined state, must have appeared at a very early stage of evolution of life or life would have soon become extinct. Actually the ancestry of sulphur bacteria which include nitrogen fixers, such as photosynthetic bacteria and *Desulphovibrio*, has been traced back to nearly 800,000,000 years, which indicate that the gaseous nitrogen-fixing mechanism has been in existence since—at least that time! How young and arrogant man seems with his ancestry going back perhaps to a meagre 100,000 years.

A scheme representing the cycle of stages of nitrogen metabolism in soil and in air—**nitrogen cycle**—is shown on the previous page (Fig. 692).

CHAPTER XL

ENZYMES AND DIGESTION

In order to understand fully the true functions of raw materials which are taken in by the growing plant, it is necessary to know the mechanism by which the raw materials are elaborated into complex organic substances. Carbohydrates are synthesised in photosynthesis. From the carbohydrates produced, the plant is capable of manufacturing a host of other organic compounds of diverse nature known to be present in the plant, directly or indirectly. There, in the tissues of plant, synthetic processes are continually being carried on and at the same time sugar and other substances are continuously being broken down into simpler constituents. The chemical reactions resulting in the synthesis and degradation of complex organic substances are known as plant *metabolism.* The synthesis of complex organic substances from raw materials is usually called *anabolism,* the breakdown of those substances to simpler, organic or inorganic substances, *catabolism.* The anabolic reactions are generally energy absorbing, i.e., endothermal or *endergonic* and catabolic, energy releasing, i.e., exothermic or *exergonic.* (If only the heat of the reaction is taken into account the terms endothermal and exothermal are used—in some cases heat of reaction and energy change are almost equal.)

From the point of view of the second law of thermodynamics, exergonic reactions are spontaneous. The endergonic or endothermal reactions, on the other hand, cannot take place unless free energy is supplied from some external source. It must be clear, however, that many chemical reactions are taking place in the plant at any one time. Therefore, it is natural to assume that most of the endergonic or energy-absorbing processes are taking place at the expense of released free energy by the exergonic processes.

In general then, the anabolism of plants in which energy is absorbed is possible only if catabolism is simultaneously taking place. It is, however, true that free energy needed to drive some endothermal reaction may be obtained from sources other than exothermal reactions. The most fundamental endothermal biochemical reaction taking place in plants, i.e., photosynthesis is undoubtedly driven by the free energy obtained from sunlight. Much of the food produced or accumulated in various parts of plant body is stored in the form of large organic complex molecules.

The conversion of these complex substances, usually insoluble, into simpler soluble forms is known as **digestion.** Some of the commonest of these temporary storage products are starch, fats and proteins. The most familiar digestive reaction in plants is the conversion of starch into sugar which is a hydrolytic decomposition or *hydrolysis,* i.e., breaking down into simpler forms by combination with water. The significance of digestion lies in the fact that the simple products of digestion are usually more soluble, more diffusible, are more readily transported from cell to cell and are also more directly available for utilisation by cells. Thus, starch which is not directly available for utilisation and assimilation by the living cell is changed in all these respects by the process of digestion. Many other storage compounds such as a host of complex carbohydrates, proteins, fats and other organic compounds, temporarily stored in more specialised storage organs of seeds, roots, leaves, stems (underground stems), etc., become more directly available to the growing cell through digestion.

This digestion is brought about by specific substances or organic catalysts that are commonly called *enzymes.* Again, gradual and orderly release of energy by the exergonic reactions, such as respiration at the right time and in the right place enables the plant to perform all its energy-consuming processes, such as active absorption of solutes, protoplasmic streaming, cell growth, etc. And none of these energy-consuming processes would have been possible in the absence of enzymes. Because the enzymes are highly active, they easily reveal their presence. It can be easily demonstrated that a substance which will not break down of itself, may readily do so when mixed with a plant juice. This indicates that the plant juice contains an enzyme that can catalyse the breakdown of the particular substance.

Thus a complex substance may be broken down step by step to simpler and simpler substances and at each step a packet or certain amount of energy is released which may easily be used to drive an endergonic or energy absorbing reaction.

The term enzyme was given to these substances by Kuhne in 1878. In 1860, Pasteur was the first to explain the phenomena of alcoholic and lactic acid fermentations and to think that they must be due to the vital activities of living cells. It was later found that it was not the living cells themselves which brought about the decomposition of sugar but actually the breakdown was accomplished by certain chemical substances secreted by the living cells. This was later fully confirmed when it was shown that a chemical substance extracted from yeast converted large quantities of sugars into alcohol in the total absence of living cells.

Because the enzymes are very similar, although more complex in their activity to the inorganic catalysts, it is assumed that they are required only in extremely small quantities. It has been estimated that a single molecule of the enzyme *catalase* can effect the decomposition of approximately 5,000,000 molecules of hydrogen peroxide per minute, even at 0°C. *Invertase*, the enzyme which catalyses sucrose, can cause the breakdown of sucrose, 1,000,000 times that of its own weight without exhibiting appreciable diminution in its activity.

Enzymes which operate within the cells in which they are produced are sometimes called *endoenzymes*. In rare cases the enzymes are excreted outside the cell and accomplish their effects in external medium. These are called *exoenzymes*. Metabolism of many bacteria and fungi grown in culture medium containing starch is due to the excretion of enzymes into the medium shown by gradual digestion of the medium.

PROPERTIES OF ENZYMES

The number of known enzymes runs into hundreds and is increasingly daily. About 100 have been sufficiently purified and concentrated to obtain in crystalline state. All such enzymes have proved to be proteins or protein complexes of high molecular weights, ranging from 12,000 (Ribonuclease) to over 4,000,000 (urease).

Solubility. Enzymes are soluble in water, dilute glycerin, alcohol, saline and buffer solutions.

Specificity. Each enzyme may be pictured as having a definite shape into which only certain other molecules will fit like pieces in a jigsaw puzzle. The enzyme is able to piece together or to split up only those molecules which do fit it precisely. Each individual enzyme is thus restricted in its catalytic activity to one particular reaction or one group of related chemical reactions. Proteolytic enzymes will thus act only on proteins; *zymase* will only ferment certain monosaccharides, not all. Examples of extreme specificity are afforded by the enzyme, *invertase* which will act only on sucrose and not even on any other disaccharide; the enzyme *urease* only acts on urea, hydrolysing it into CO_2 and NH_3. In some cases an enzyme may show specificity for a particular chemical linkage, as for example, the enzyme *lipase* which catalyses the cleavage of ester linkages in fat, without any particular regard for the nature of fatty acid involved in a fat.

SENSITIVITY. Enzymes are very unstable in nature. Due to their protein nature, enzymes are heat labile. They are readily inactivated or destroyed by heat (60–100°C.) or by long exposure to light (particularly to ultra-violet or lights of still shorter wave-lengths). Enzymes are inactive at temperatures near freezing point but this low temperature inactivation is usually reversible. For each enzyme there is one specific temperature at which its activity is at a maximum—between 25–30°C. in most plant cells.

The enzymes are very sensitive to variations in pH and also to the presence of certain inorganic salts. *Trypsin*, a proteolytic enzyme, is completely inactive in an acid pH, whereas *pepsin* requires an acidic medium. The pH inactivation is, however, reversible as in the case of low temperature inactivation if the pH change is not too extreme. Most enzymes act best in a neutral or slightly alkaline medium. The activity of *trypsin* is at its maximum in a solution of 1% Na_2CO_3.

CATALYTIC ACTIVITY. By the agency of enzymes chemical reactions readily occur in the living cells which otherwise can only be brought about, as in hydrolysis, by boiling with reagents like strong acid. A living cell cannot heat substances up in a test-tube in order to get them to break up or to combine as we do in the laboratory. Thus the enzymes join simple molecules together to form complex molecules or break down the latter into simpler molecules at room temperature and only at atmospheric pressure.

Enzymes are very similar to inorganic catalysts accelerating the rate of chemical reactions but are themselves found to be chemically unchanged at the end of the reaction. The concentration of the enzymes affects the speed of the reaction but not the position of the equilibrium and the final quantity of product is unchanged. As we have seen before, an enzyme transforms many times its own weight of the substrate (the compound or compounds on which the enzyme acts), as molecular ratio of one to one million will easily be the normal order. During the time the reaction is proceeding, an enzyme gradually disappears, reappearing at the end of the reaction, apparently unchanged in activity or in amount.

ENZYME INHIBITORS

Inorganic poisons may inactivate many enzymes when present in very small quantities; for example, cyanide (0.0001 M) inactivates some respiration enzymes. Other known inhibitors are sulphides, azides, malonate, antimycin, amytal, fluorides, iodoacetates, etc. These inhibitors are usually specific, frequently inhibiting only one or few enzymes.

Zymogens

Cells lacking a specific enzyme in its active form are frequently found to contain chemical precursor of the enzymes, which could be converted, under certain conditions, into the enzymes proper. Such substances were termed *zymogens*. The prosthetic groups combine with different proteins to constitute different enzymes.

ENZYME-COENZYME-ACTIVATOR RELATIONSHIPS

Certain of the enzymes appear to consist solely of proteins. This is true, for example, of certain proteolytic enzymes which break down proteins to amino acids and also of *amylases* which break down starch to glucose. But it has been known for a long time that certain enzymes require the presence of some other compound to be enzymatically active.

Thus, it appears that though some enzymes are active without any other additional component, most enzymes are enzymatically active only in presence of certain non-protein, thermostable component in addition to the protein (terms, *holoenzymes* and *apoenzymes* have also been used to denote complete enzymes and protein components respectively). The non-protein component may be firmly attached to the protein when it is called a *prosthetic group* or it may be readily removed by *dialysis* in which case it is called a *coenzyme*. In contrast to the protein portion, the coenzyme or prosthetic group is heat stable, being unaffected by boiling.

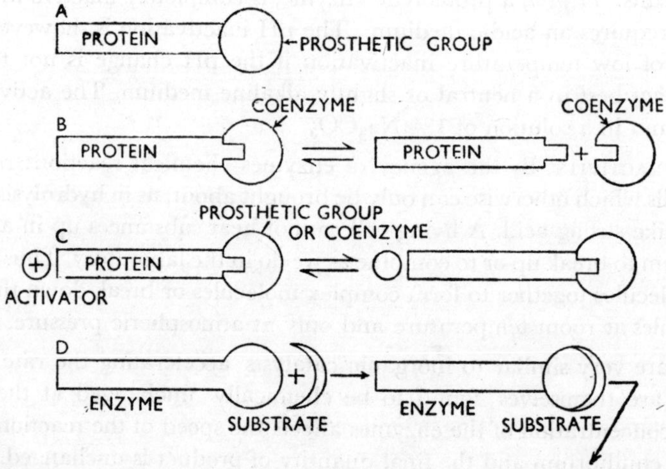

FIG. 693. Diagrammatic representation of enzyme systems: A. Protein *plus* non-removable component; B. Protein *plus* removable component; C. Protein *plus* one or two removable components and D, enzyme-substrate reaction. Substrate shown cresent-shaped.

If the non-protein component is a metal ion it is sometimes called *activator* since the protein is inactive in absence of the metal ion. Perhaps the simplest case of this kind are enzymes in which the prosthetic group consists of a single metal ion. The enzyme *tyrosinase* which attacks the amino acid tyrosine is made up of a protein portion combined with an atom of copper. If the copper atom is removed, the protein at once becomes inert. The presence of copper is essential to the functioning of these enzymes in catalysing oxidation of tyrosine.

The chemical compound on which the active enzyme acts is known as the *substrate*. In general, it is usually the prosthetic group or coenzyme which reacts with the substrate leading to its (substrate) breakdown. The thiol or sulphydryl (SH) group of the protein component itself may also combine with the *substrate*, the starting material of a reaction. The specificity of an enzyme, however, depends on the protein component and not on the non-protein part. It is apparent that this specificity depends ultimately upon the variety of amino acids present in the protein chain, the sequence in which they are arranged and the way in which the chain forming protein is 'folded'. Thus several enzymes may contain the same coenzyme, but because of the difference in the protein component, each will act specifically on one substance. *Tyrosinase* and *ascorbic acid* (vitamin C) *oxidase* both have copper as their coenzymes yet the former acts on tyrosine (and several similar substances) and not on vitamin C; the latter acts on ascorbic acid but not on tyrosine.

There are enzymes which contain metals as activators other than copper; zinc, manganese, magnesium and iron are all known to function as non-protein activator components of various enzymes. In case of iron the metal is not usually bound directly to the protein but is incorporated into a more complex molecule, an ironporphyrin which then becomes the prosthetic group proper. Such ironporphyrin enzymes (*cytochrome oxidase*) include several enzymes which are essential in the chemical reactions of respiration. The iron-complex is an activator also of *peroxidase* and *catalase*.

The perplexing problem of the importance of micronutrients in plant growth in fantastically small amounts is getting clear now. The unquestionable importance of them in minute amounts for normal growth and development of plants is certainly because they form constituents of many essential enzymes. The names of some specific microelement activators of enzymes have been cited before.

Many enzymes possess relatively complex organic compounds as their coenzymes and in addition may also contain an activator metal ion. Thus it would appear that many enzymes consist of three components—*protein*, *coenzyme* and *activator*. In absence of either the coenzyme or the metal ion, the protein component is inactive, i.e., the whole enzyme is inert. Similarly, the coenzyme or the activator is inactive in absence of the protein component. It is a striking fact of advance of science, that several vitamins first discovered as essential microfood elements for living cells, particularly in animals, are now known to function as coenzyme components of many enzymes. This may well prove to be their sole function. In the enzymes of higher plants also, the vitamins of the B-complex group such as vitamin B_1, vitamin B_2, vitamin B_3, vitamin B_6, vitamin B_{12}, niacin, adenine, pantothenic acid, choline (an organic plant base, constituents of some essential fats, such as egg-yolk), etc., all constitute coenzyme groups. Ascorbic acid (vitamin C) is a coenzyme of the enzyme of *myrosinase* which is active on a number of glucosides, particularly the mustard oil glucosides, *sinigrin* and *glucobrassicin*. The enzyme *pyruvic carboxylase*, which catalyses the removal of hydrogen from the carboxyl group of pyruvic acid is a protein, diphosphothiamine magnesium (DPT), where thiamine (vitamin B_1) phosphate is the coenzyme group and magnesium is the activator. Vitamin B_1 (thiamine or aneurine) is structurally a thiazopyrimidine compound. So the coenzyme can alternatively be called pyrimidine thiazolediphosphoric ester; the ring thus contains the essential element sulphur in addition to carbon and nitrogen. The protein-coenzyme-activator ratio here is about 75,000 gm protein : 1 gm mol thiamine phosphate : 1 gm atom of Mg! The coenzymes of *dehydrogenase* group of enzyme include among others nicotinamide adenine dinucleotide, NAD (previously known as coenzyme I and until recently, DPN) and nicotinamide adenine dinucleotide phosphate, NADP (coenzyme II, TPN) in which both niacin and adenine appear together. The nucleotide contains a pentose sugar, ribose. The NAD and NADP can be combined with any of the several proteins to yield a whole series of different enzymes all of which, however, remove hydrogen atoms from different oxidisable substrates.

We have seen before that when the non-protein prosthetic group of an enzyme is removable by dialysis it is called a coenzyme. The separated coenzyme is inert so also is the protein component; united again, the active enzyme is reconstituted. The protein-coenzyme relation is supposed to act like a lock and key arrangement. The protein component of an enzyme, the lock, is inactive without the coenzyme, the key, just as a lock cannot be opened without its particular key and the key is useless without the lock. As a key fits a lock and opens the lock so does a coenzyme unite with the protein to initiate the activity of the whole enzyme. As a particular lock has only one key which fits and opens the lock so has one particular protein component only one coenzyme for the initiation of the activity of this enzyme as a whole. Again just as one master key can open a number of locks so may one coenzyme convert a number of proteins into a series of highly active different enzymes by combining with the different proteins.

NATURE OF ENZYME ACTION

We have discussed before that exothermal (exergonic) reactions are thermodynamically spontaneous, i.e., they are losing reactions as far as energy is concerned. But this certainly does not mean that the exergonic breakdown can take place without any outside help. A substance, S at energy level E, can therefore be converted to substances, X and Y, spontaneously at the lower energy level, EL. Then why S does not fall of itself to energy EL, at the same time splitting into substances X and Y? The reason is that there is an energy barrier and the substance S may be too stable at energy E to roll down slowly as it were to energy level, EL. In order that the substance S can go over the energy barrier it must first be activated, that is, it must be given some extra energy as for instance, such as supplied by increasing the temperature, enabling S to go uphill over the peak and for this reason the rate of reaction, i.e., S breaking into X and Y increases with rise of temperature. But this reaction can also be made to take place without any rise in temperature, in the presence of highly reactive substances acting in extremely small quantities and known as catalyst—inorganic or organic. If the reaction involves two substances, the enzyme may accelerate it by adsorbing these substances on its surface, bringing them closer, and thereby reducing the energy barrier. When a gas stove is turned on, the gas does not burn until and unless the activation energy is supplied by a lighted match stick.

The enzyme certainly forms an unstable intermediate compound with the substance, S. The energy barrier may thus be lowered to such a value that the energy of the moving molecules in the solution can now easily overcome the barrier and the substance S, falling to the lower energy level EL, breaks down into substances X and Y with the release of exergonic energy:

$$\text{S} + \text{enzyme} \longrightarrow \text{enzyme.S}$$
$$\text{enzyme.S} \longrightarrow \text{X} + \text{Y} + \text{enzyme} + \text{energy.}$$

The net result is, therefore, that the catalyst has permitted the reaction to take place by removing the so-called energy gap. It is evident that the catalyst may be active in minute quantities because it is regenerated after each reaction.

CLASSIFICATION OF ENZYMES

The classification of enzymes is generally based on the nature of the reactions they catalyse rather on the nature of the enzyme itself, since in so many cases, the exact chemical nature is still largely unknown. The terminology of the enzymes is often arbitrary; the modern tendency is to name them after substrate or the nature of reaction, adding the Greek suffix *-ase*. Certain classical names such as ***pepsin***, ***trypsin***, ***bromelin***, are, however, still retained.

Enzymes are classified into six classes: (1) Oxidoreductases; (2) Transferases; (3) Hydrolases; (4) Lyases; (5) Isomerases; and (6) Ligases. They are further subdivided according to the groups within the substrate, which they attack, modify or transfer, the bonds broken down or formed, the nature of acceptors etc. The International Union of Biochemistry has given each enzyme a number which gives a general idea regarding the nature of its action. Thus, alcohol dehydrase which catalyses the reaction.

$CH_3CH_2OH + NAD \rightleftharpoons CH_3CHO + NADH + H^+$ is given the number 1.1.1.1 because it is a dehydrogenase (Class 1), it acts on the CHOH group (Subclass 1), it was NAD as the acceptor (Sub-Subclass 1) and it is the first enzyme in the Sub-Subclass.

We shall now discuss very briefly some major types of enzyme reactions with a few illustrations.

(A) Hydrolytic and condensation reactions (Hydrolases)

In most cases hydrolysis results in a large release of energy (exothermal reactions) while in condensation, energy is absorbed (endothermal reactions). Hydrolysis always involves use of at least one molecule of water. These reactions can be represented as follows:

$$A.B. + H_2O \underset{\text{condensation}}{\overset{\text{hydrolysis}}{\rightleftharpoons}} A.H. + B.OH$$

A.B is a substance which splits into A.H. and B.OH in presence of the enzyme. Enzyme usually accelerates the hydrolysis. If, however, the substance A.B. can be removed as rapidly as it is formed, the enzyme can also accelerate the condensation reaction, at least in some cases. These are called hydrolases.

Most of the complex organic substances in plants are stable in aqueous solutions but are readily split into their constituents if the necessary enzyme is present. The hydrolytic enzymes are subdivided into *four* groups. The presence of water is essential for the chemical reactions to take place:

(1) *Esterases*

$$\underset{\text{ester}}{R_1COOR_2} + H_2O \rightleftharpoons \underset{\text{acid}}{R_1COOH} + \underset{\text{alcohol}}{R_2OH}$$

The most important esterase is *lipase* which controls the hydrolysis of fats into fatty acids and glycerol. Other esterases are ***phosphatases, chlorophyllases,*** etc.

(2) *Carbohydrase*

The reaction is:

$$\underset{\substack{\text{sucrose with}\\ \text{glycosidic bond}}}{R_1{-}O{-}R_2} + H_2O \rightleftharpoons \underset{\text{glucose}}{R_1OH} + \underset{\text{fructose}}{R_2OH}$$

or

$$C_{12}H_{22}O_{11} + H_2O \overset{\textit{invertase}}{\rightleftharpoons} C_6H_{12}O_6 + C_6H_{12}O_6$$

The *carbohydrases* catalyse the hydrolysis of the glycosidic bonds in carbohydrates to set free the constituent sugars. *Invertase* controls the hydrolytic splitting of sucrose (cane sugar) into glucose and fructose. Other carbohydrates and glycosides (compounds of carbohydrates with one or more non-sugar components), are also hydrolysed by these enzymes, e.g., *amylase (diastase), cellulase, maltase, myrosinase,* etc. In every case, the carbohydrates are ultimately split into the simplest hexoses or pentoses.

(3) *Proteolytic enzymes*

These enzymes accelerate the splitting of complex protein molecules into proteoses, peptones, polypeptides and ultimately into amino acids. The most important proteolytic enzymes are *trypsin, pepsin, papain, bromelin*, etc. *Papain*, we know now, consists of 202 amino acids, the sequence and the order of which is now known almost completely. Papain is a *sulphydril protease* of approx. molecular weight, 23,000.

(4) *Amidases*

Asparaginase splits the amide asparagine into asparagine and ammonia.

(*B*) *Transfer reactions* (*Transferases*)

In these reactions enzymes catalyse the transfer of radicals from one compound to the other:

$$A.B + C \rightleftharpoons C.B. + A$$

These reactions usually result in small energy changes and hence are readily reversible. Enzymes of this group accelerate both synthesis and breakdown reactions. These are called transferases.

Many chemical groups, e.g., CH_5, $CH_2OH\text{-}CHO$, glycosyl, nucleotidyl, etc. can be transferred by these enzymes from one compound to another. *Transaminases* transfer amino groups; *transphosphatases* transfer the phosphate group (transfers PO_4 group from H_3PO_4 to starch, etc.).

(*C*) *Oxidation-reduction reactions* (Oxidoreductases)

In this type of reactions enzymes themselves or some other compounds act as hydrogen acceptors in the breakdown of one of the substances concerned, getting reduced in the process:

AH_2	+	B	$\rightleftharpoons$	A	+	BH_2
hydrogen donor		hydrogen acceptor		oxidised donor		reduced acceptor

The substance is oxidised by the removal of hydrogen which (hydrogen) is taken up by the enzyme (coenzyme), getting reduced in the process. The reduced coenzyme may in turn be oxidised by another hydrogen acceptor and (e.g., another enzyme) thus freed again to take part in the reaction:

$$BH_2 + \text{enzyme} \rightleftharpoons B + \text{enzyme}.H_2$$

These are the most heterogeneous of all the three classes. *Zymase*, the first known of all the enzymes, falls into this class. Zymase is now known to be a complex of several enzymes, all of which catalyse either oxidation or reduction reactions in plant and animal cells. *Carboxylase*, one of the previously known components of zymase complex which splits off CO_2 from carboxyl groups, can be compared to a *dehydrogenase* for the reaction is really a dehydrogenation of the carboxyl group into acetaldehyde. As we have seen before, the most important enzymes of this class are those controlling the transfer of hydrogen atoms from one compound to another, when one compound is oxidised and the other reduced. These are known as *dehydrogenases*. The enzyme, actually the coenzyme, may itself act as a hydrogen acceptor; for example:

ethyl alcohol		coenzyme *dehydrogenase*				
CH_3CH_2OH	+	NAD	$\rightleftharpoons$	CH_3CHO	+	$NADH^- + H^+$
hydrogen donor		hydrogen acceptor		acetaldehyde oxidised donor		reduced coenzyme

Other well-known *dehydrogenase* coenzymes are FMN (flavin mononucleotide) and FAD (flavin adenine dinucleotide); FAD acts specifically as electron or hydrogen ion carrier between the nicotinamide coenzymes (NAD or NADP) and the *ubiquinone, coenzyme Q*.

In the above reaction alcohol is oxidised, i.e., dehydrogenated by the removal of hydrogen atom to aldehyde and the NAD is reduced. The reduced NADH can in turn be oxidised by another hydrogen acceptor. If the NADH or any other similarly reduced enzyme is reoxidised not by removal of hydrogen but by molecular oxygen, it is called *oxidase*.

$$AH_2 + \text{enzyme} \longrightarrow A + \text{enzyme}.H_2$$
$$\text{enzyme}.H_2 + \tfrac{1}{2}O_2 \longrightarrow \text{enzyme} + H_2O$$

One of the most important of these oxidase enzyme-complexes are the various *cytochromes* and *cytochrome oxidases*.

Two other well-known oxidation-reduction enzymes are *peroxidases* and *catalases;* both of these control the decomposition of H_2O_2 or organic peroxide. When a *peroxidase* acts on peroxide, inorganic or organic, the molecular oxygen thus released at once oxidises another compound. When *catalase* acts on a peroxide, free molecular oxygen is released. The presence of *catalase* in plant tissue can easily be demonstrated if a block of potato tissue is dropped in a test-tube containing H_2O_2; the evolution of oxygen can at once be seen. The metabolic significance of the universal presence of *catalase* in plant or animal cells is not clear. It is possible that *catalase* acts as a safety-valve in preventing accumulation of peroxides, produced as a result of cell metabolism.

(D) *Carboxylase* is a lyase type of enzyme which catalyses reversible reactions.

$$CH_3CO.COOH \xrightleftharpoons{\textit{pyruvic de-carboxylase}} CH_3CHO + CO_2$$

The *lyases* snap C—C, C—O, C—N and C—S bonds, without the agency of water. *Isocitrate lyase* converts isocitrate into succinate and glyoxalate; *phhospophyruvate carboxylase* degrades the substrate oxalacetate in presence of inorganic phosphate to phosphoenol pyruvate, CO_2 or H_2O.

The *isomerases* isomerise the same or different molecules, e.g., alanine recemase, ribulose-phosphate 3-epinaerase, Triosephosphate isomerase, etc.

The *ligases* form C—O, C—S or C—N bonds. They are responsible for syntheses of many aminoacyl-tRNA, acyl coenzyme AS, asparagine, glutamine, nucleic acids and proteins.

Distribution of enzymes in the cell

From all observations as yet available it appears that the enzymes which the plant cells contain are usually localised in the protoplasm. One of the most amazing facts about enzymes is the huge number of different kinds that are certainly located in the extremely small space occupied by protoplasm of a single mature cell.

Some enzymes are present as the soluble component of the protoplasm whereas many other enzymes are bound firmly into the structure of various particles which are always found in protoplasm. Numerous enzymes, as for example *phosphorylase*, are found to be functionally associated with protoplasm. Other enzymes are found attached to the microscopic particles of protoplasm such as *mitochondria* (minute, 1–$\tfrac{1}{4}\mu$, semi-solid bodies found in the protoplasm of all cells except bacteria and blue-green algae), or the plastids.

The number of enzymes in a cell is large. The orderly way in which the series of metabolic reactions take place in a cell, points to the distinct possibility that the organisation of the enzymes in a cell is according to a pattern evolved for the functioning of the cell as a living entity.

EXPERIMENTS ON ENZYMES AND ENZYME ACTION

A. *HYDROLYTIC ENZYMES*

I. Amylase (or diastase)

(*a*) Make thin starch paste by shaking up a little ordinary starch or wheat flour with 50 ml boiling water and standing it to cool. Get some germinating pea seedlings, 2–3 g. (in which the radicle has grown out at least an inch); remove the seed coats, grind the cotyledons with water in a mortar and filter through a funnel. Pour the starch paste into three white saucers and label these. Leave saucer 1 as it is; into 2 pour a few drops of iodine soln. (it turns blue) and into 3 pour some of the watery extract from the cotyledons. Set the three saucers in a fairly warm place (about 35-40°C.) and from time to time test 1 and 3 with iodine soln. Note that 3 soon gives only a reddish-brown colour with iodine soln. and ultimately remains uncoloured with iodine, showing that starch has disappeared—starch has been converted into sugar by the enzyme amylase which has been extracted from the cotyledons.

(*b*) Squeeze the milky contents of about half-a-dozen sprouting wheat grains into a test-tube with a little water and stir well. Filter and add a few drops of mixture of Fehling A and B to the clear filtrate and heat. A brick-red ppt. of Cu_2O indicates the presence of simple reducing sugars produced by the enzyme diastase in the germinating seeds.

II. Invertase

Mix about 10 ml of yeast extract (1 g.) with 10 ml of a 2% soln. of pure sucrose (test it before to see whether it is non-reducing with Fehling soln.). Put the flask or beaker at about 38°C. After 30 mins. to 1 hr. add Fehling soln. Observe considerable reduction of Fehling due to the conversion of sucrose into reducing glucose and fructose. This may also be done with extracts of germinating seeds.

III. Lipases: Fat- or oil-splitting enzymes or esterases

(*a*) Take some linseed oil (olive oil will also do) and test for acidity, if any. Neutralise, if necessary. Take some germinating *Ricinus* seeds which have just begun to germinate, and crush or chop them up. Add them to the oil mixture, shaking vigorously. Test again with litmus paper after a few hrs. The reaction is now markedly acidic due to the hydrolysis of fats into fatty acids and glycerol.

(*b*) (i) Take a small portion of the liquid in a test tube. Then add a little $CuSO_4$ soln. and a few drops of NaOH soln. A deep blue colouration is produced, but there is no ppt. of $Cu(OH)_2$ which is kept in solution by glycerol.

(ii) To another portion of the liquid add a little solid $KHSO_4$ and heat strongly; a pungent odour (acrylic aldehyde) confirms the formation of glycerol.

IV. **Proteolytic enzymes**

Trypsin: Pound up some fresh tissue or seeds of peas and beans, carrot and cabbage with 100 ml water containing a little toluene as preservative. Filter and then add a small quantity of peptone and shake up thoroughly. Keep in a warm place (about 35°C.) for 2-3 days, then test for the amino acid, tryptophan. Filter the extract and then acidify it with drops of acetic acid (dil). Add bromide water carefully, drop by drop, until a purple colour develops, indicating the presence of tryptophan. Then shake up with a little amyl alcohol; on standing the alcohol will rise to the top carrying with it any of the purple colour formed.

B. *OXIDATION-REDUCTION ENZYMES*

V. **Direct oxidases**

Cells containing oxidase systems become discoloured after injury due to the presence of these enzymes. The browning or discolouration of many tissues (e.g., potatoes) and fruits when cut (as in apples, pears, etc.) is due to the activity of these oxidases.

Apply a drop or two of an alcoholic solution (freshly made) of *gum guiacum* (inner portion of a lump of gum may also do) on the cut open surface of a potato tuber. Dry the surface gently with a filter paper. A deep-blue oxidation product (guiacum blue) is formed. Guiacum soln. is one of the most sensitive reagents used for oxidation as even the minutest trace of oxygen may be detected by this reagent.

VI. **Indirect oxidases or peroxidases**

Cells of many plants cannot directly bring about aerobic oxidation of gum guiacum. The peroxidases cannot turn guiacum blue until hydrogen peroxide is added. Cells containing these enzymes do not turn brown when injured but if both H_2O_2 and gum guiacum are applied to the cut surfaces of horse radish root, deep-blue colour is obtained. Pyragallol, catechol and quinone may also be used instead of guiacum— (with pyragallol, the coloured product is red).

It has been shown that the presence of peroxidases in plant tissues may intensify the colour changes brought about by oxidases, when H_2O_2 is added.

VII. **Dehydrogenases**

These enzymes bring about simultaneous oxidation-reduction reactions when one compound is oxidised by removal of hydrogen and the other reduced (acceptor of hydrogen).

(*a*) In a test tube take 1 ml of a 15% solution of glucose and 1 ml of either living yeast cells or Baker's yeast (zymin). Fill up the tube with a very dilute solution of methylene blue. Cork carefully and excluding all contact with external air place the tube for a short time in a warm place. The dye methylene blue is seen to decolourise owing to the formation of leucomethylene blue, i.e., methylene blue is reduced by hydrogen derived from hydrogen donor, glucose, which is oxidised.

(*b*) If the tube containing colourless leucomethylene blue is uncorked and contents shaken in air, the blue colour reappears. This is because leucomethylene blue is autooxidisable; it can act as a hydrogen donor, giving up hydrogen readily to atmospheric oxygen which acts as a hydrogen acceptor, forming water.

VIII. **Catalase**

Catalase is usually present in plant tissues but it does not bring about oxidation as in the case of other oxidation-reduction enzymes. The oxygen released by catalase is in the active form. Catalase prevents accumulation of H_2O_2 or organic peroxides in plant tissues. It is one of the most active enzymes that has yet been studied.

It is present in yeast cells (or in Zymase extract) and is thermolabile.

(*a*) In a test tube, which is half-filled with a mixture of equal volumes of H_2O_2 and water, add a few slices of freshly cut potato tuber. Evolution of oxygen bubbles can be seen immediately.

(*b*) As in previous expt., add about 1 gm. of potato tuber (ground up with alcohol and filtered dry with a filter pump) in a test tube containing equal volumes (10 ml) of a mixture of water and H_2O_2. Invert the open mouth of the tube (holding it by thumb) below the surface of water in a basin and hold it vertically by means of clamp arrangement. A rapid evolution of oxygen takes place which accumulates by displacement of solution in the tube.

(*c*) If the experiment is performed with vigorously boiled potato tuber tissue, no evolution of oxygen is seen, as heating destroys the enzyme.

(*d*) A suspension of Baker's yeast (0·5—1·0 g.) added to a dilute solution of H_2O_2, liberates oxygen almost instantaneously.

CHAPTER XLI

PHOTOSYNTHESIS

Every day, every hour, every minute millions of tones of organic matter are oxidised into its ultimate constituents, CO_2 and H_2O, which go back to the atmosphere or return to water. At this rate of destruction, all the organic matter on the earth will be consumed in a matter of mere 20 or 30 years. But, during the same time an equal amount of organic matter is synthesised by the green plants to compensate for the loss. O_2 is driven out from its stable union with CO_2 and water, and thus freed, the C and H atoms are knitted into patterns which spell the secret of organic life.

Oxidation is, however, not merely a disaster, it is also the prime-mover of life. Life really thrives on death, not only because the material for organic synthesis, CO_2 and H_2O come from the destruction of the organic matter, but also because oxidation or respiration is the main source of energy of life. Every movement of the living body, every chemical activity of the digestive system, every flash of thought in the brain consume energy and the main source of the energy is respiration. Thus it is evident that only a certain part of the chemical energy contained in the organic matter is utilised for the maintenance and manifestation of life, while the larger proportion sooner or later is lost in death and decay of the living bodies, fallen leaves and branches and excretion. All living matter reverts into CO_2, H_2O and N_2 and the energy it once contained becomes converted into heat. Only a small fraction of the organic matter and its contained energy escape rapid heat-destruction and may be preserved, perhaps, for millions of years underground, in the half decomposed forms of coal, peat or mineral oil under protective layers of rock.

For continued existence of life on the earth, it requires that its chemical elements dissipated by the decomposition of organic matter be brought back into combination and the energy converted into heat, be replaced. We can see that the regeneration of organic matter can occur by a cyclic process—the same CO_2, H_2O and N_2 which are liberated by destruction of organic matter, can be used all over again for synthesis. Laws of thermodynamics postulate that the energy liberated in the destruction of organic matter, i.e., oxidation, is dissipated in the vastness of atmosphere and is lost beyond recovery. The liberated energy, for all practical purposes, becomes unavailable for the reverse conversion into chemical energy. Thus the energy required for organic synthesis must come from some external source and the only external source which is continuously available on the surface of the earth is sunlight. Only about tenth of one per cent of the energy received from the sun by the earth is fixed in photosynthesis although most investigators now agree that up to about 35% of the available light energy could be theoretically transformed into the chemical energy of the sugars produced. Theoretically, no thermodynamic restrictions stand in the way of complete conversion of light energy into any other form of energy but it certainly requires a quick, efficient mechanism able to intervene immediately after the light strikes the absorbing surface (leaf) and to prevent the energy of this impact from being converted into heat (if it is converted into heat, it is lost beyond recovery). Man has not yet solved this apparently simple problem, but because the chlorophyll of the green plants could solve this lit engineering problem of converting light energy into the chemical energy of the carbohydrates, the evolution and existence of man have been possible on this earth and an increasing fraction of that energy, we know, is being diverted to the support of one living species, man, himself. The process by which the green plants synthesise organic compounds from CO_2 and H_2O with the energy absorbed from sunlight is, beyond all doubt, the most fundamental of all biochemical reactions.

It is roughly calculated that about 15 million cubic kilometers of water on the earth are split by photosynthesis and reconstituted by respiration, once every 2 million years or so. Photosynthetically generated oxygen temporarily enters the atmospheric 'bank' whence it is itself reconstituted, perhaps once every 2000 years at the current rates.

Of course, organic synthesis of one kind or another is carried on by all living organisms but the green plants alone start from a scratch; all other plants and animals use already prepared organic material which they may transform in thousands of ways, the chemical reaction being sometimes heat absorbing or sometimes heat liberating. We know that sugar is converted into fat in certain fat-storing seeds and also, unfortunately for some of us, when our bodies convert sugars into fats. This certainly looks like that this transformation leads to a higher accumulation of energy (in fats) than the carbohydrates, from which they are produced. But a critical examination shows that actually this accumulation of chemical energy (in fats) is achieved only at the expense of the cost of degradation of another quantity of sugar. So the moment one part of carbohydrate is promoted to an energy-rich compound, e.g., fat, another part is degraded to a compound, CO_2, of lower energy content than carbohydrate.

It has been approximately estimated that the total yield of photosynthesis (as glucose) on the earth per year is about three hundred thousand million tons of which, though it seems utterly unbelievable, about nine-tenths is contributed by the green planktonic microflora of the oceans and seas, all the green land plants contributing only about one-tenth of the total amount!

The liberation of gas in the form of bubbles from free water surfaces in which green plants grow has been a very ordinary observation of intelligent men. But the identity of the gas bubbles, and how and why they were produced, had to wait till about the early part of the 18th century.

From the time of Aristotle until the 18th century it was generally supposed that plants derived all their nourishment from the soil. This view was questioned by Stephen Hales, an illustrious contemporary of Isaac Newton in 1727, who first thought that the green plants may get some part of the nourishment through their leaves and sunlight may have something to do with it. The clarification had to wait for almost another 50 years (1772) when oxygen was discovered by Joseph Priestly by an accident. Priestly grew some green plants (mint) in what was then called impure air, air made impure by burning candles and which would extinguish a candle. We know now that the impure air was really CO_2. After some time Priestly found that the impure air was no longer impure (it would not extinguish a candle and it was not at all inconvenient to a mouse!) if a green plant was introduced and the green plant in some mysterious way seemed to have purified the foul air. The momentous discovery of photosynthesis was thus made!

$$CO_2 \xrightarrow{\text{green plant}} O_2$$

Another name associated with the discovery of the overall mechanism of photosynthesis is Ingen-Housz, a contemporary of Joseph Priestly, who was the first man to realise and demonstrate the importance of sunlight for the purification of impure air (CO_2).

$$CO_2 + \text{light} \xrightarrow{\text{green plant}} O_2$$

Ingen-Housz also correctly diagnosed the fact of plant respiration. Jean Senabier, the Swiss, was the first man (1782) to suggest that CO_2 came from the atmosphere and was not taken up by the roots from the soil as envisaged by Ingen-Housz.

De Saussure (1804) made the comparison of the volume of CO_2 absorbed to the volume of O_2 liberated and he proved beyond all doubt that the increase in weight of the plant caused by the assimilation of certain quantity of CO_2 was considerably greater than the weight of carbon contained in it. Since the equivalent of all the oxygen in the decomposed CO_2 is evolved in the air, the source of weight increase could come only from the participation of water in the actual mechanism of photosynthesis.

$$CO_2 + H_2O + \text{light} \xrightarrow{\text{green plant}} \text{or ganic matter} + O_2$$

In 1845, Robert Mayer announced the laws of conservation of energy and in 1848 he proclaimed his idea of organic synthesis and energy transformation, which remains qualitatively complete and true today:

$$CO_2 + H_2O + \text{light energy} \xrightarrow{\text{green plant}} \text{organic matter} + O_2 + \text{chemical energy.}$$

This is the concluding chapter in the history of discovery of photosynthesis. We had to wait for another hundred years or so to get an insight into the quantitative aspect and chemical mechanism of photosynthesis. But we cannot say even now, that all the aspects of the most fundamental biochemical process are completely known. Only that a fair, clear picture is available now and although different workers hold somewhat different opinions about the precise steps involved, the broad outlines are no linger in question.

CARBOHYDRATES

As carbohydrates are the visible products of photosynthesis, an understanding of their chemical nature, functions and interrelations are essential before we can thoroughly grasp the metabolism of living cells.

It appears certain that the great majority, perhaps all, of the many different chemical compounds in plant are manufactured in one way or another from the sugar produced in photosynthesis. Many of the materials which go to make up the bulk of the plant—starch and cellulose—are chemically closely related to the simple sugars produced in photosynthesis. The simple sugars, and the hexose sugar in particular, are of functional importance not only as the principal materials utilised in respiration but also as the raw materials from which the plant synthesises hosts of other organic compounds.

The simple *hexose* sugars belong to the general class of chemical compounds known as carbohydrates. As the name indicates, these contain the elements C, H and O, the latter two are usually, but not invariably, present in the same ratio in which they are found in the water molecule, i.e., two hydrogen atoms for every oxygen atom.

The *monosaccharides* are the group of carbohydrates from which more complex carbohydrates can be produced by condensation. They are classified according to the number of carbon atoms which they contain. Of the possible seven groups of monosaccharides, perhaps only two, 5-carbon atom (*pentoses*) and 6-carbon atom (*hexoses*), are of importance as stable plant products. The 7-carbon monosaccharide *heptose* has recently been identified in green plants as an intermediate compound in the continual production of 5-carbon *ribulose diphosphate*, presumed to be the primary CO_2 acceptor in photosynthesis. Like the pentose phosphate ester this heptose compound is also a ketose (heptulose). In the pentose sugar, 5C atoms, 10H atoms and 5O atoms are united to form $C_5H_{10}O_5$ whereas in the hexose sugar, 6C, 12H and 6O atoms are involved—$C_6H_{12}O_6$. The molecules of monosaccharides—hexoses and pentoses—may be united to form more complex derivatives such as *disaccharides* (two molecules of hexoses or simple sugars linked together) or *trisaccharides* (3 molecules), etc. Among the disaccharides of higher plants, the most important is sucrose or cane sugar which is perhaps the principal sugar of translocation. *Polysaccharides*, such as starch and cellulose, are built up from condensation of a large number of monosaccharides and possess both metabolic and structural functions.

The hexose sugar (sometimes called *dextrose*) possesses a chemical structure in which 5 of the carbon atoms are linked to OH groups. The remaining carbon atom is in an aldehydic or ketonic group:

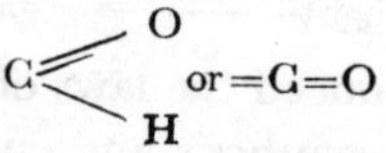

These aldehyde or ketonic groups are readily oxidised by reagents [such as Fehling A and B; $CuSO_4$ and NaOH when mixed together gives $Cu(OH)_2$], which (reagents) are themselves reduced in the process. The oxidation of glucose or fructose, etc., to the corresponding acid takes place when hexoses or pentoses are boiled with a solution of $Cu(OH)_2$, $Cu(OH)_2$ being simultaneously reduced to red cuprous oxide which precipitates out. This is the standard test for reducing sugar.

The four saccharides (mono, di, tri and tetra) are collectively called sugars—all of them possess the property of sweetness and all of them are white, more or less crystalline compounds and are soluble in water.

Most of the sugars are like many other organic compounds, optically active when in solution. When the plane of polarised light is rotated clockwise by a sugar solution, the sugar is *dextrorotatory;* those sugars which rotate it to the left, counter clockwise, are *laevorotatory.*

An important point regarding sugars is that many isomers may exist with the same molecular formula. Thus *glucose, fructose, mannose, galactose* all have the same empirical formula of $C_6H_{12}O_6$, but differ from each other due to differences in the atomic grouping within the molecule. Another type of isomerism—*stereoisomerism*—is possible in case of these sugars. The stereoisomeric forms may exist for each of these sugars. Here exactly the same atomic groupings are present but one is arranged as the mirror image of the other. Thus *d-glucose* and *l-glucose* are exactly similar, not only in molecular formula ($C_6H_{12}O_6$) but also in atomic configuration but they differ in that the right hand atomic grouping in one becomes the left hand grouping of the other. Sugars are not always designated *d* and *l*, however, because of the direction of their rotating power, but actually because of their structural relationship to *d*- and *l*-glucose respectively; *d-fructose*, for example, is not dextrorotatory but strongly laevorotatory.

The four hexose sugars—*glucose, fructose, mannose* and *galactose*—appear to be present in every living plant. Of these only *d-glucose* and *d-fructose* ordinarily occur dissolved in molecular form in the protoplasm or in the cell sap of the vacuoles. Galactose is one of the most common sugar components of fungi.

There are four pentoses or 5C sugars—*arabinose, xylose, ribose, deoxyribose* which appear to be important components of plant materials. None of them, however, is found dissolved as such within the plant, but all occur as derivatives, bound in larger polysaccharide molecules. *Xylose* and *arabinose* occur in plants as constituents of the cell wall, the polysaccharides, *xylan* and *araban* respectively. Neither of these sugars seem to be utilisable in respiration; they seem to be purely structural materials. The pentose sugar, *ribose* [$CH_2OH(CHOH)_3CHO$] is found in the plant as a component of certain respiratory enzymes, in high-energy phosphates, e.g., ATP and also a constituent of certain nucleic acids (components of genes, the unit of inherited material). The sugar-like compound, *deoxyribose* [$CH_2OH.CH_2(CHOH)_2CHO$], closely related to ribose but lacking in one oxygen atom, is a universal constituent of living matter (forming DNA, the main constituent of genes and chromosomes). It is now known that a ketoribose phosphate ester (*ribulose diphosphate*) acts as a preliminary acceptor of CO_2 in the photosynthetic reaction of C_3 plants.

The open chain structure of sugars has been depicted in our discussion so long. Sugars actually occur in a ring or cyclic structure, the exact nature of the ring depending on the sugar involved. In *glucose*, for example, a 6-membered ring may be formed by the formation of an oxygen bridge.

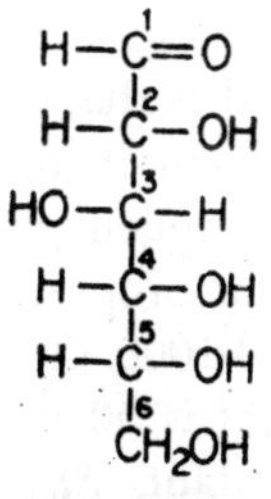

Open chain structure of α-β-glucose

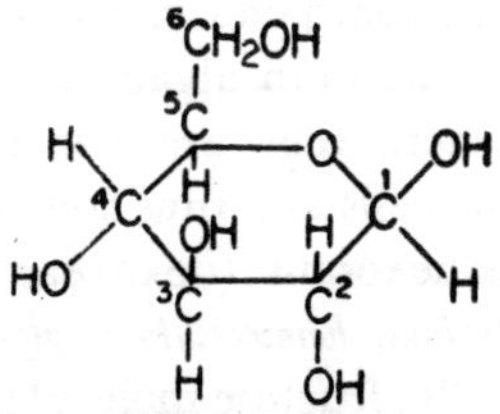

Ring structure of β-D-glucose

Fig. 694. Molecular structure of glucose.

The principal disaccharide of higher plants is sucrose or cane sugar. In *sucrose*, as in all disaccharides, two monosaccharide molecules are bound together through a linkage formed by the elimination of one molecule of water. Thus sucrose has the empirical formula $C_{12}H_{22}O_{11}$. On treatment with warm dilute acid or with enzyme invertase, one molecule of water is taken up, and one molecule each of glucose and fructose are produced. Sucrose is not a reducing sugar, so that it does not, for example, reduce alkaline copper sulphate solution [$Cu(OH)_2$]. It is evident that the free aldehyde and ketone groups of its constituents, glucose and fructose, have been rendered ineffective for it is precisely these two groups, CHO and C=O, which are involved in glucose-fructose linkage in sucrose. *Maltose*, which breaks down into 2 molecules of glucose on hydrolysis, though a disaccharide is, however, a reducing sugar; during linkage, the bond which unites the two glucose molecules, leaves one aldehyde group free. It is interesting to note that in the cyclic structure of sucrose, though the glucose unit occurs as a 6-membered ring, the fructose occurs as a 5-membered one.

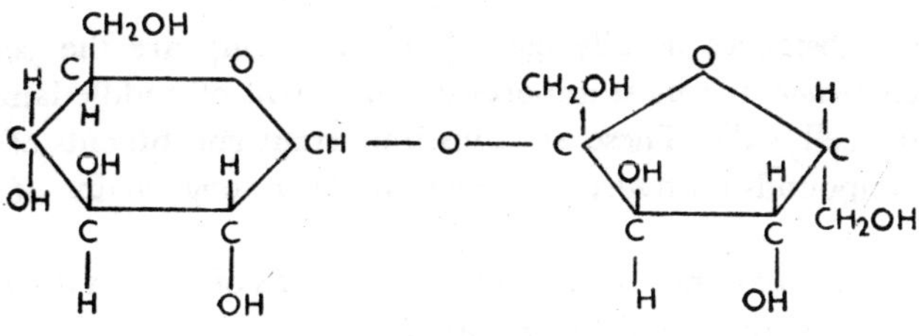

Fig. 695. Molecular ring structure of sucrose.

Other disaccharides in plants are *gentiobiose*, *melibiose* and a glucose containing disaccharide, *trehalose* (produced by some fungi). These are all reducing sugars. The disaccharide, *rutinose* is widely distributed in plants as a constituent of scarlet and crimson anthocyanins in flower petals. The rutinose on hydrolysis breaks into two monosaccharides, a hexose; glucose and a pentose, rhamnose.

Raffinose is the most wide-spread *trisaccharide* in plant, though present only in small quantities. Other trisaccharides in plants are *gentianose* and *melicitose*. On complete hydrolysis, raffinose yields three molecules of hexoses, *glucose, fructose* and *galactose*. It is a non-reducing sugar, the linkage connecting the constituent hexoses apparently involves all the three aldehyde and ketonic groups.

The polysaccharides are carbohydrates formed by the condensation of large number of monosaccharide molecules. Polysaccharides are not sweet like sugar, they are non-reducing and they are also insoluble in water. They are divided into two great classes, the *pentosans* of empirical formula $(C_5H_8O_4)n$ and *hexosans* $(C_6H_{10}O_5)n$, *n* denoting the number of such pentose or hexose units.

The principal pentosans in plants are *araban* and *xylan* and perhaps, *fucosan* (fucose methyl pentose; from algae, e.g., *Phaeophyta*) which mostly occur in the cell of water plants. In cacti, they are also important constituents of the mucilaginous materials present in the cells and thus contribute largely to the hydrophilic properties of mucilage.

Of the hexosans (dextrosans) *starch* (or *amylose* of 200-1000 glucose residues), *mannans, cellulose, hemicellulose, glycogen, inulin* (molecular weight is 5000—on hydrolysis gives essentially, fructose molecules), etc., are all synthesised in plants. *Starch* is the almost universal form of storage carbohydrate in higher plants, while *cellulose* forms the principal constituent of cell walls of these plant cells. Starch and cellulose are insoluble in water but on complete hydrolysis break up, through dextrin and maltose, finally into glucose. *Glycogen* (sometimes called animal starch) is also a hexosan which serves as a storage carbohydrate in animal tissues, particularly in muscle and liver. It is unknown in green plants but found in many fungi (abundant in yeast). Hemicelluloses are found in fungi.

Starch molecules are built up by linear condensation of molecules of glucose, each glucose being connected to the next, by an *oxygen bridge*.

In certain species of plants, such as members of the family *Compositae* (dahlia, Jerusalem artichoke, etc.) plants do not produce starch as reserve material but accumulate instead, the polysaccharide, *inulin*. Inulin is seldom, if ever, found in the aerial parts of the plant but may constitute as much as 10 to 15% of some tubers or fleshy roots.

Compound carbohydrates

These are compounds which have been derived from carbohydrates and non-carbohydrate groups, the latter, however, often being in close chemical relation with the carbohydrate.

Among the substances usually classed in this group are the pectic compounds, especially *calcium pectate*, the most important constituent of middle lamella between the adjacent primary cell walls. These are thus important constituents of the cell walls of plants. Pectic compounds on hydrolysis generally form long chains of basically galactose molecules.

Chitin (acetyl glucosamine) is found in the cell walls of most of the true fungi, except *Oomycetes* which are characterised by cellulose.

The *gums* are complex carbohydrates which upon hydrolysis yield hexoses or pentoses or both and complex organic acids. The *mucilages* are also complex carbohydrates, perhaps similar to gums and are abundantly present in the cells of cacti and other succulents. They are also found on the outside walls of many aquatics in which they prevent too rapid diffusion of water through cell wall. *Agar agar* is a mucilage from seaweeds (*Rhodophyta*) and appears to contain H_2SO_4 in the molecule, probably as an ester.

Gums usually exhibit acid character. In nature, they are present as salts of Ca, Mg and K.

The *glycosides* are compounds formed by a chemical reaction between a sugar (generally a glucose—*glucoside*) and one or more compounds which are non-sugars. They are optically active and soluble in either water or alcohol. They may be found in almost any part of the plant. All glycosides can be hydrolysed by the enzyme *emulsinases* or dilute mineral acids.

Among the various *glycosides* identified, *anthocyanins* (which will be discussed in detail next) *amygdalin* and *sinigrin* are well known. *Amygdalin* occurs in the seeds and leaves of apples, cherries and peaches. Upon hydrolysis it breaks up into glucose, hydrocyanic acid and benzaldehyde:

$$\underset{\text{amygdalin}}{C_{20}H_{27}NO_{11}} + 2H_2O \xrightarrow[\textit{myrosinase}]{\textit{emulsinase or}} \underset{\text{glucose}}{2C_6H_{12}O_6} + \underset{\substack{\text{hydrocyanic}\\\text{acid}}}{HCN} + \underset{\text{benzaldehyde}}{C_6H_5CHO}$$

Sinigrin is called the mustard oil glucoside. It is found in the black mustard (*Brassica nigra*) and on enzymatic by *myrosinase* hydrolysis produces the commercial mustard oil. *Myrosinase* contains vit. C as its coenzyme and perhaps this is the first clearly demonstrated physiological function of vit. C:

$$\underset{\text{sinigrin}}{C_{10}H_{16}O_9NS_2K} + H_2O \xrightarrow{\textit{myrosinase}} \underset{\substack{\text{allyl}\\\text{isothiocyanate}\\\text{or mustard oil}}}{C_3H_5CNS} + \underset{\text{glucose}}{C_6H_{12}O_6} + \underset{\substack{\text{potassium}\\\text{hydrogen}\\\text{sulphate}}}{KHSO_4}$$

Tannins which are of common occurrence in plants can only be classed with glycosides for their possible remote relation with them. On hydrolysis, they produce certain complex organic acids and nearly always a sugar (usually glucose) and sometimes also other substances. The proportion of sugar in the hydrolytic products of tannin is relatively very small and tannins seem to have distinctive properties.

Tannins are sometimes soluble in water and are thus found dissolved in cell sap. More frequently, however, they accumulate in large quantities in the cell walls or in dead tissues such as barks. Unripe fruits of plum, guava, etc., contain relatively large quantities of tannins. The most spectacular accumulation of tannin is seen in the leaves of tea—about 15–20% of the dry weight, though in oak barks, tannin may even constitute half of the total dry weight.

Anthocyanins and other flavonoid glycosides

The most important glycosidic compound in the plant is the water-soluble pigment of the cell sap of the vacuoles—the *anthocyanins, flavonols* and *flavones*—which impart red, pink, blue, violet and some yellow colour chiefly to flowers and fruits. They are one of the most important of all the secondary plant products.

The anthocyanins are glycosides, the sugar-free components *aglycons*, of which are called *anthocyanidins*. The various anthocyanidins (six have actually been isolated from plants) are hydroxy-derivatives of flavone-type, heterocyclic 2-phenyl-benzopyrillium chloride. Various anthocyanidins, such as *pelargonidin, cyanidin, delphinidin*, etc., have been distinguished by substituting hydroxyl or methyl groups or both in different positions of the parent molecule. The corresponding anthocyanins are *pelargonin*, *cyanin*, *delphinin*, etc. The anthocyanidins, which are insoluble in water, occur in living cells as glycosides. The combination with sugar is important in providing sap solubility and stability to light and to enzymic attack. As a rule, they occur in nature attached to one or more molecules of sugars, as anthocyanins. The sugar component is generally a monosaccharide, *glucose*, *rhamnose*, *xylose*, *arabinose* (*Theobroma* and *Rhododendron*) or *galactose* or it may also be a disaccharide (*genitiobiose* and *rutinose*). It may even be a trisaccharide, e.g., *gentionose*. Usually only one sugar molecule is present but as in the anthocyanin, *cyanin* (the red pigment of dahlia, rose, cornflower, etc.) two glucose molecules may also be attached to the anthocyanidin nucleus. If there is only one sugar molecule, it is usually attached to 3-hydroxyl group. If there are more than one, the second molecule is either linked with the first or may also be attached to the 5-OH group, the so-called *diglycosides* which are quite common; however one 3,7-diglycoside anthocyanin is known to occur in nature.

Anthocyanins are found in all orders of flowering plants, with one notable exception, the *Centrospermae*. In this order there is a number of colouring matters, such as *betanin* (*betacyanins*) which are similar in colour but quite different in structure. The best known betanins are represented in this order by the red colour of beet root, by the purple bracts of *Bougainvillea* as well as colour of flowers of many cacti. They always have a nitrogen content.

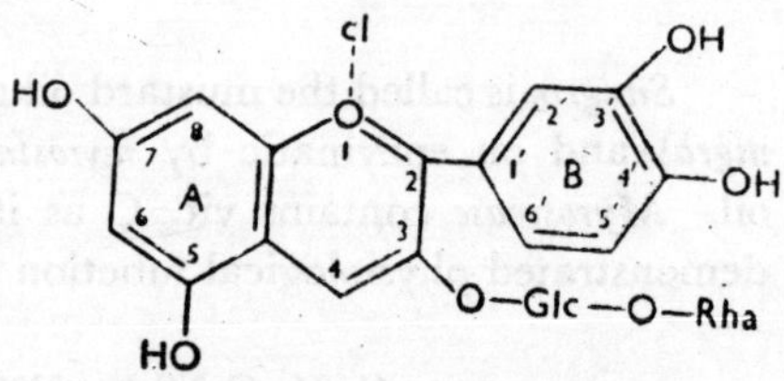

Fig. 696. Anthocyanin, *cyanin*, as chloride. (Glc=glucose; Rha=rhamnose.)

As will be evident from the structure of anthocyanin, the oxygen is basic and tetravalent and forms salts with HCl (*oxonium salts*). Anthocyanins generally occur in nature in the form of chlorides. The anthocyanins may change their colour due to change in acidity or alkalinity of the cell sap. The anthocyanin, *cyanin* of rose petals is red in pH 3 or less, violet in pH 8·5 (Cl is removed; oxygen is usual and divalent) and blue at pH 11 (formation of sodium salt of the colour base). It must be pointed out here, however, that the role of the much discussed '*acidity-factor*', has been to some extent, over-emphasised. It is true, that though it is easy to produce a wide range of colours between red and blue by variation of the acidity in the laboratory, the colour which is displayed by the petals in nature, often does not depend solely on the acidity of the cell sap. *Cyanin* is blue in acid cell sap (pH about 4·8) of corn flowers whereas the same anthocyanin is responsible for the colour of the red rose with approximately the same pH value of the cells. A pH value of above 7 is of very rare occurrence in plant cells and most of the deep blue or violet colours are not definitely dependent on the alkaline reaction of the cell sap. An examination of more than 200 species of plants containing anthocyanin in flowers and fruits reveals that the pH value was always acid, below 7, irrespective of whether the flowers were red, blue or violet in colour.

Some early investigators have made use of this property of anthocyanins utilising them as indicators. Because of the change of colour of anthocyanins with change in pH, anthocyanins have been called 'vegetable chameleon' (chameleon is an animal which frequently and rapidly changes the colour of its skin).

Anthocyanins are optically active (absorb very strongly in the investigated range between 200–600 mμ and an absorption maximum is present in the visible range) water-soluble pigments and are widely distributed in nature. Apart from its presence in flowers and fruits, it is also known to be present in mosses and ferns and even in conifers. The scarlet, crimson and purple anthocyanins in flowers and fruits—the *cyanidin-3-rutinosides*—are the most important group of colouring matter in higher plants, apart from chlorophylls. In some families of flowering plants such as *Crassulaceae*, *Saxifragaceae*, *Compositae*, etc., red root tips are sometimes observed. A rather striking presence of anthocyanin is in the root tips of *Eichhornia*, the blue pigment being exclusively limited to the cellular walls of the epidermis and subepidermal cortical cells.

The artificial synthesis of almost all anthocyanin pigments has already been achieved and a general method for synthesis of anthocyanins occurring in nature and those not isolated yet, have been successfully developed and perfected. The structural set-up of many anthocyanins have also been established beyond all doubt.

Light is believed to be essential for the formation of anthocyanin in most flowers and fruits and the idea is that both a photochemical and a dark chemical reaction are involved in the synthesis of anthocyanin in nature. However, anthocyanin formation has been frequently observed in the dark.

There seems to be a close correlation between the accumulation of aluminium in the cell sap and anthocyanin formation. A majority of the plants with blue flowers or fruits have also been found to show striking aluminium accumulation.

In *Hydrangea*, the blue colour of the flowers is due to an acid-stable dephinidin-aluminium complex. The cell sap is not alkaline, but acid. It is presumed that the blue colour in many flowers and fruits is due to the formation of metal complexes with various ions, particularly with aluminium; these are stable and remain blue even in acid cell sap. The formation of blue colour of corn flowers, even in high cell sap acidity, is perhaps, due to the presence of a colloidal substance in the cell sap, which seems to stabilise the blue colour.

In the biosynthesis of anthocyanidins and other flavonoids, most investigators believe that they are formed from small structural elements such as C_6-C_3-C_6-framework and not from flavone or similar compounds.

Other flavonoid pigments (*anthoxanthins*): They are usually yellow in colour. The aglycon of these pigments are either *flavonols* or *flavone* itslelf. Both of them have got double bond ketonic oxygen at 4-carbon position, but true flavones lack 3-hydroxyl group present in flavonols and anthocyanidins. Except this, the basic hydroxylation is similar in all aglycon (pigment-free part) groups, and they differ only in the number of hydroxyl or methoxyl groups in the 'B' ring.

As regards the physiological significance of the presence of flavonoid pigments in plant organs, it is difficult to give a satisfactory explanation except, of course, for brightly coloured flowers which certainly help to attract birds and bees to bring about pollination. They may have some favourable influence on the enzymatic reaction in plant tissues owing to their optical properties. It is a common phenomenon that young leaves of many plants are red or pink in colour. The anthocyanin here may very well act as a light screen or filter, protecting the young tissues from excessive light. But it is difficult to explain its formation in the foliage which sometimes gives a bright red colour to the leaves in autumn—the so-called autumnal colouration—before the leaves are shed. The active participation of anthocyanins or other flavones and also tannins (owing to their optical properties) in photosynthesis can readily be dismissed for, as we have seen, these pigments are usually found in vacuolar cell sap or are in precipitated form and are at such a distance from the chloroplastids, either in separate cells or at least isolated by the tonoplast (inner plasma membrane bordering the vacuole), so as to have little chance of passing absorbed energy into the photosynthetic system. It is possible that the formation of anthocyanins, except in flowers and fruits, is nothing but a symptom of some physiological disturbance in plants or they may only be metabolic waste products of not quite normal activities of the protoplasm of cells. These waste products may be insoluble in water and they are only glycosided in order to make them soluble, facilitating their transfer to the cell sap.

The formation of different colours of anthocyanin in brightly coloured flowers—red, blue and violet—seems to have a biological significance. Birds, for example, can perceive red better than any other colour and as a result it has actually been found that brightly coloured red flowers appear especially in those plants which are pollinated by birds. Bees can perceive only blue-violet and yellow colours and actually blue-violet flowers predominate in plants pollinated by bees.

Flowers are rather late acquisition in the process of plant evolution. If flowers are considered to be derived from vegetative shoots and homologous to it, questions may be asked whether the metabolic activities of flower parts are only specialised functions of normal leaves or whether the colours result from specific activities of calyx, corolla, etc.

The production and accumulation of pigments other than chlorophyll are not essential functions of flowers as such. The primitive types of flowers, produced early in the course of evolution, had certainly not yet developed and relationships with insects and other animals with light-sensitive organs such as eyes. It was only a further step in evolution which resulted in flowers, attracting insects, birds and other animals for pollination. Colours, however, are not exclusively developed for serving the purpose of attracting animals. The bright-yellow anthers in *Graminaceae* and red stigma in many cases, are conspicuous examples of what appear to be purposeless colour. Colour development is not thus necessarily connected with flower formation. Colour of organisms can be produced in two different ways—by refraction of light on physical structures and surfaces or by absorption of definite wave-lengths of light. In the plant kingdom, the latter type predominates. White colour of flowers certainly originates from total light reflection.

In the foliage leaves yellow or red colour occurs at two different stages of leaf development, red in the very early phase of leaf elongation and red and yellow, in senescent leaves. Is there any close relation then between the very early and very late phases of leaf development (when normally colour appears in them) and flower colouration? The probability is high that there is such an intimate relation. Although, generally similar classes of pigments occur in flower parts and normal leaves, we certainly find a greater variety of anthocyanins and other pigments in flowers than in leaves, just as the range of morphological variations of flower parts is much wider than the modifications of green leaves.

The origin and development of colour patterns such as speckled, spotted or striped and variegated colour on petals are much more complicated than those of natural colour tones of anthocyanins.

Various flowers, normally characterised by differently coloured spots and stripes, lose or change their pattern when grown under abnormal light or temperature conditions. As for example, in many *Petunias*, violet pigmented areas or spots on the petal, increase if the temperature is lowered. On the other hand, if the light intensity is increased (only effective if the increased light intensity is given during the formation of flower buds), the petal becomes less spotted, i.e., the spotted areas are now fewer. Thus in the development of flower colour by anthocyanin, light and temperature seem to influence the amount of anthocyanin formed almost in opposite direction. Thus low temperature combined with high light intensity during the flower-bud formation reduces the coloured areas. But if the treatment is continued during further development of flower petals, it results in high pigment concentration, giving deep blue spots, although the coloured areas are now *fewer* in number.

The genetic interpretation of variegated flowers, either with striped or spotted petals, is that the phenomenon is either due to patterns manifested by certain genes or is caused by mutable genes with or without the co-operation of cytoplasm.

PLANT STRUCTURE IN RELATION TO PHOTOSYNTHESIS

In photosynthesis carbon dioxide, water and energy are consumed, oxygen gas is released and carbohydrate formed. The chemical reactants are CO_2 and water, the source of energy is sunlight and the place where synthesis occurs is primarily the leaf.

Something more may be said at this very beginning of the energy changes involved. The potential chemical energy of compounds of carbon depends on their state of reduction. Unreduced CO_2 on the left-hand side of the equation has no potential chemical

$$CO_2+H_2O \longrightarrow CH_2+O_2$$

energy whereas the reduced carbon compound CH_2O (carbohydrate) produced in photosynthesis has a considerable amount. Since the total potential chemical energy of the substances on the right hand of the equation exceeds that on the left, energy must be supplied from outside if the reaction is to take place. In photosynthesis this energy is received as light by chlorophyll but essentially the same process of CO_2 reduction can occur, using chemical energy, as in the case of certain bacteria which can grow in the dark if they are supplied with CO_2 and inorganic compounds of high potential chemical energy from which they may derive energy by oxidation (*chemosynthesis*).

Since only the visible light is appreciably active in photosynthesis, the photochemical reaction must involve the absorption of visible light for energy requirements and must therefore be localised in the green intracellular chloroplast bodies of green plants. Light reflected or transmitted by a substance cannot produce chemical change in it. Hence it is among the pigments of the leaf that we must look for the absorbing substance since other materials are white or colourless because they reflect or transmit all the light which falls on them.

The pigments of the leaf are mostly confined to the mesophyll cells. Thus in the mesophyll cells the protoplasm and its coloured inclusions, chloroplastids, must be regarded as the seat of photosynthesis. The two inorganic reactants of the photochemical reaction, brought about by the light absorbed by the chlorophyll, must be taken to this seat of photosynthesis, i.e., to the chloroplastids of the assimilatory cells.

We have seen before that the water absorbed by the roots from the soil is taken up by the transpiration current and ultimately to the chloroplastids of the mesophyll cells. The other reactant, CO_2 gas, diffuses from the air outside the leaf up to and through the stomata, across the intercellular spaces and then through the wet mesophyll cell walls to the chloroplastids, where the photosynthetic reaction occurs by the light or energy trapped by the green pigment in the chloroplastids. For synthesis of one gram of hexose, the land plant consumes CO_2 equal to that in 2,500 litres of air and at the best, the plant utilises only two-thirds of the CO_2 of air. Thus the acutal volume of air required is greater than 2,500 litres. The concentration of CO_2 in the atmosphere is small compared to oxygen (0·03% against 21%) yet the rate of photosynthesis of a leaf under natural conditions can be 10 times as great as the rate of respiration. The rate of diffusion of CO_2 is proportional to the difference in concentration of CO_2 outside the leaf and the leaf intercellular spaces, to the area of the diffusion path and inversely proportional to the length of path. Organisms such as the unicellular algae possess the simplest diffusion path—CO_2 entering through the whole surface of the cell. The leaf of the higher green plants shows the development of a more complex diffusion path; the external surface is generally covered by a cuticular layer and this certainly offers considerable resistance to gaseous diffusion. But we know that this layer is broken at intervals by stomata which lead to a continuous series of air spaces. The resulting structure allows CO_2 diffusion up to the wall of the assimilating cell itself (mesophyll). Thus the main pathway of diffusion of CO_2 and other gases from the outside is *via* the stomata and the rate of diffusion is controlled by the concentration of CO_2 in the intercellular spaces of the leaf, in terms of external concentration. The internal concentration will be determined by the diffusion path and the net rate of CO_2-consumption. CO_2 which diffuses in the gaseous state up to the cell walls, subsequently must diffuse in an aqueous phase through the mesophyll cell wall and to the front where reaction occurs (i.e., the chloroplastids). Some control of the diffusion of CO_2 into and out of the leaf is perhaps achieved by the changes in the aperture of the stomatal pore. But even when the stomata are completely closed, there still remains some diffusion of CO_2 through the cuticle.

Because the initial step in photosynthesis involves the entrance of CO_2 into the leaf, a close relationship between the degree of stomatal opening and photosynthesis has often been inferred, but the degree of stomatal opening was never seriously thought to be a limiting factor. Recently, however, it was found that if the stomata are closed when the leaves are illuminated, no photosynthesis occurred until the stomata began to open. So it is apparent that only the cuticular diffusion of CO_2 may not keep pace with the CO_2-requirement in normal photosynthesis.

CHLOROPHYLLS

In the mesophyll cells, the cytoplasm and its inclusions, chloroplastids must be regarded as the seat of photosynthesis. The chloroplastids are clearly defined structures within the cytoplasm and the pigment chlorophyll is confined to them.

In gross composition, the chloroplastids are 50% water, 25% protein, 15% lipides and 10% pigments.

Chloroplastids of the mesophyll cells are ellipsoidal or have an unsymmetrical lens shape. They vary in size, the diameter often being about 5–10μ. The number of chloroplastids (usually more numerous in the palisade cells than in the spongy) varies from several hundred down to only a few (average about 50) in individual cells. In algae frequently each cell contains a single chloroplast, the shape of which varies (stellate in *Zygnema*, cup-shape in *Chlorella*, long spiral in *Spirogyra*, etc.). The chloroplastids show many points of similarity with mitochondria but it is not absolutely certain that these two protoplasmic inclusions are analogous.

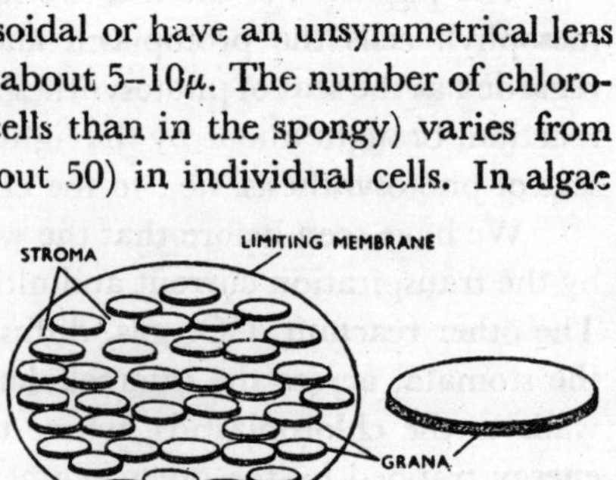

Fig. 697. Diagrammatic structure of a chloroplastid.

Chloroplastids usually appear to consist of small discrete green bodies, the dark *grana* which are embedded in a lighter coloured *stroma*, the whole being surrounded by a double membrane with semi-permeable properties, thereby giving the chloroplastids a lamellate or granular structure. It is certain that the photosynthetic pigments, the green chlorophyll and the yellow carotenoids are largely confined to the *grana* and are very firmly bound to the grana protein, rather lipoprotein. Actually the proteins of the chloroplastids are contained principally in the grana and very little, if any, in the stroma. The grana contain many materials in addition to proteins and the pigments; substances such as lipides and also small amounts of inorganic elements such as calcium, iron, copper and zinc are also present. If the grana do not serve merely as energy traps but they are also at the same time the loci of CO_2-assimilation, it is evident that they must contain proteins in addition to pigments. Many individual enzymatic activities (e.g., catalase) have also been found to occur in the grana. The actual protein content of the grana is only about one-half of the total weight of the grana.

The number of grana in a single chloroplast of higher plants may vary from about ten to several hundred. In algae, in general, the chloroplast shows no sharp differentiation into individual grana, though consisting of continuous layers or *lamellae* like the sheets of a book parallel with each other; thus in algae, the chloroplast itself is equivalent to a single granum. All chloroplasts, which have been examined so far, show this lamellate form and are mostly circular of about 1μm in diameter. Thus, as mentioned before, sometimes sheets of lamellae run through the whole chloroplast (algae) but in others, there are a number of more or less separate piles of sheets (granular chloroplasts of higher plants). Each lamella is known to consist of two membranes, joined at the ends and each membrane is about 100–200 Å (10–20 nm) thick and the space between them is about the same thickness; this is approximately the thickness of a small protein molecule. Following Menke (1960) the double lamellae became known as *thylakoids*. Lamellar system appears to be made of uniform thylakoids, 5–10 μm in diameter or of smaller thylakoids of about 1μm in diameter, the bigger ones passing through several stacks. In algae, it appears that only the bigger thylakoids are present. It is likely that the membranes consist of films of protein with a lipide layer between them. It has been suggested that the chlorophyll molecules form a film between the protein and the fatty substances. The structure of photosynthetic apparatus may be different in different plants, but lamellae definitely occur in all. This strongly suggests that the presence of lamellae is a prerequisite for photosynthesis.

In the blue-green algae and photosynthetic bacteria no true chloroplasts are visible. But it is known that the pigments are present in the cytoplasm as lamellae (*chromatoplasm*) in the blue-green algae, too small to be visible without electron microscope. In the photosynthetic bacteria the particles appear as disc-like lamellate structures which are most likely free grana.

Evidence as to whether the locus of the whole of the photosynthetic mechanism is only confined to grana is still not complete. Investigations with radio-active carbon suggests that this is not the case. In dark, carbon dioxide fixation occurs almost exclusively in cytoplasm whereas during illumination almost all the activity was found in chloroplasts, more properly in the granum. The process involving CO_2, in photosynthesis may, therefore, be associated with cytoplasm and the photolytic decomposition of water, exclusively with the granum. This would certainly involve diffusion process; diffusion is thus an important step of overall photosynthetic mechanism. Hydrogen is made available from photolysis of water by some carrier system from the granum to the stroma cytoplasm.

It is apparent from our previous discussion that the stroma is also in lamellate form. The stroma lamellate contains most probably only PSI-activity, whereas the grana regions contain both PS_I and PS_{II} (see later).

Pigments of chloroplastids

The pigments of the chloroplast belong to two chemical classes: *tetrapyrrolic* compounds to which class chlorophyll belongs and the *hydrocarbon* carotenoids which include the orange carotenes and the yellow xanthophylls. In the blue-green and also in red algae, additional pigments which often mask the green colour of the chlorophyll are always present, intimately associated with the chlorophyll. These are chemically related to chlorophylls and are *open chain tetrapyrrolic* compounds in contrast to chlorophylls which are cyclic in nature. The cyclic tetrapyrrolic structures are widely distributed in nature and are termed generally as *porphyrins*. The iron-containing red haeme pigments of mammalian blood and the coenzyme components of oxidation-reduction haeme enzymes, essential for biological oxidation, such as various cytochromes, cytochrome oxidases, peroxidases and catalases, are also cyclic tetrapyrrolic compounds.

CHLOROPHYLLS

The chlorophylls are readily extracted from fresh green tissues of plants by acetone and from powdered dry leaves by both acetone and alcohol.

The chlorophylls are methyl phytyl esters of the parent dicarboxylic acid (containing two COOH groups), the *chlorophyllins*. *Phytol* ($C_{20}H_{39}OH$) is a long (about one-third of the chlorophyll molecule by weight) chain alcohol which is related to the carotenoids (carotenoids also contain phytol) and can be regarded as derivable from vitamin A by hydrogenation. The enzyme *chlorophyllase* hydrolyses the phytol group from chlorophyll and thus in an alcoholic solution of chlorophyll, the product is an *ethyl* or *methyl chlorophyllide*, in which phytyl group is replaced by ethyl or methyl groups.

The chlorophylls are remarkable in being a group of natural substances in which magnesium is present in a non-ionic form. Treatment of chlorophyll solution in alcohol with dilute acids results in the formation of a brown colour — magnesium-free compound, the *phaeophytin*. This phaeophytin readily reacts with copper, zinc or iron salts in contrast to its relative inertness towards magnesium, once the magnesium atom is removed. In organic solvents, containing water under neutral conditions, chlorophyll cannot thus be reformed, even with excess of magnesium salts; it is, however, slowly formed in absence of water by heating Mg-free phaeophytin with a modified *Grignard reagent* (ethoxy-magnesium bromide). Ordinary Grignard reagent reacts immediately at ordinary temperatures to give a magnesium derivative, but not chlorophyll because in this treatment, the side chains are also attacked. The Cu-derivative is green, stable to light and not affected by acids; the Zn-derivative is like chlorophyll so that treatment with acid readily removes zinc; and the iron-derivative is dark-green and shows likeness to iron-porphyrin compounds.

When both the phytol structure and the central magnesium atom are removed from the chlorophyll molecule by treating an alcoholic solution of chlorophyll with dilute HCl, the product formed is known as a *phaeophorbide*.

The atom of magnesium is situated in the centre of the chlorophyll molecule and is surrounded by four nitrogen-containing stable pyrrole nuclei linked by CH bridges and thus chlorophyll has a similar structure to haeme, the red pigment of blood, the latter, however, containing a central iron atom. Magnesium is the only metallic constituent of the chlorophyll. Pure chlorophyll leaves a residue composed solely of magnesium oxide. Chlorophyll contains approximately 2·7% of magnesium.

In addition to the cyclically linked four pyrrole units, there is an additional cyclopentane ring in the chlorophyll molecule which is absent from the haeme pigments. In chlorophyll there are two extra hydrogen atoms in position 7 and 8 in pyrrole ring IV. The hydrogenated nucleus IV is thus readily susceptible to oxidation and reduction. Due to the presence of the extra hydrogen atoms nucleus IV, the term magnesium-dihydroporphyrin for chlorophyll also seems appropriate.

The approximate dimension of chlorophyll porphyrin ring is 15Å (1·5 nm) square. The phytol tail is 20Å (2·0 nm) in length.

A pigment isolated from etiolated seedlings (the yellow colour of the etiolated seedling is due to the preponderance of carotenoid pigments) is kown as *protochlorophyll*, which seems like an oxidation product of chlorophyll, i.e., without the two extra hydrogen atoms in pyrrole ring IV. The formation of chlorophyll proper from protochlorophyll or protochlorophyllide has been envisaged as a process of *photoreduction*, i.e., insertion of two hydrogen atoms in nucleus IV in light.

Chlorophyll a and *chlorophyll b*, the two classical components of chlorophyll, are functionally and chemically identical, the only difference being that the chlorophyll *a* molecule contains two atoms more of hydrogen and one atom less of oxygen than a molecule of chlorophyll *b*. Actually in the atomic configuration of the chlorophyll, the only difference being that in chlorophyll *b* in pyrrole nucleus II, in the position 3, the side-chain CH_3 is replaced by a formyl, CHO grouping. Whether chlorophyll *b* could actually arise from chlorophyll *a* by such a simple substitution was a matter of conjucture; it was found impossible to convert chlorophyll *a* to chlorophyll *b* in the laboratory and it seemed highly improbable, at one time, that it could happen in nature even though chlorophyll *b* certainly looked like an oxidation product of chlorophyll *a*.

The present concept of biosynthesis of chlorophyll *b* from chlorophyll *a* is based on specific radio-activities of both *a* and *b* and it definitely supports the conclusion that the formation of *b* from *a* is a dark reaction, even for exogenous chlorophyll *a*. This exogenous *a* seems to compete in the reaction with an endogenous one (Shlyk, 1971). On the

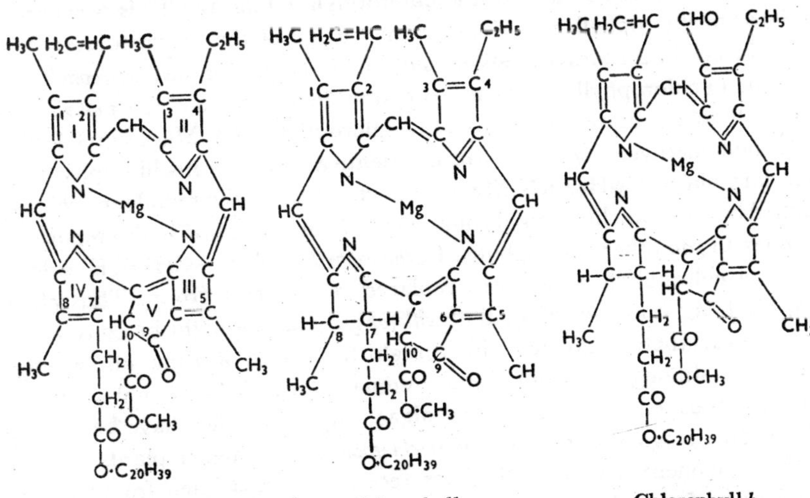

Protochlorophyll Chlorophyll *a* Chlorophyll *b*

FIG. 698. Structures of protochlorophyll and chlorophylls *a* & *b*.

strength of all the evidence, the verdict now reached is that there is a *sequential scheme* of chlorophyll *b* from chlorophyll *a* formation (a special pool of more labile, young chlorophyll *a* molecules, as the selective precursor). These experiments showed that the transformation of chlorophyll *a* to chlorophyll *b* should and could be performed in the laboratory.

As a matter of fact, pigments whenever they occur in pair, such as carotene-xanthophyll, phycocyanin-phycoerythrin, bilirubin-biliverdin and bacteriochlorophyll-chlorobium chlorophyll, one, at one time seemed to be, the oxidised product of the other. Perhaps this idea will have to be modified as there seems to be considerable structural dissimilarity between bacteriochlorophyll and chlorobium chlorophyll in the side-chains while structurally the known biliproteins in plants, cyanobilin and erythrobilin are similar.

It may be interesting to point out here that there seems to be a certain incompatibility between starch formation and chlorophyll *b*-deficiency. It is well known that all *b*-deficient coloured algae never form true starch as storage carbohydrate. This may also be true for flowering plants—monocotyledons as a class seldom store carbohydrates in their vegetative organs in the form of starch and monocotyledons relatively are also found to be deficient in chlorophyll *b*. These observations have prompted some investigators to suggest that chlorophyll *b* may be the specific sensitiser for starch formation in green plants. In *Vaucheria*, which once was put into green algae group *Chlorophyta*, foods are always stored in the form of fat and oil globules and it alone, among the green algae, is chlorophyll *b*-deficient.

Chlorophyll *b* has never been found independently of chlorophyll *a* and it seems certain that chlorophyll *b* is incapable of photosynthesising by itself in absence of chlorophyll *a*. Recent evidences indicate, however, that efficient energy transfer takes place from chlorophyll *b* to chlorophyll *a*, wherever they are present together in green cells.

Chlorophyll *a* has got a molecular formula of $C_{55}H_{72}O_5N_4Mg$ (mol. wt. 893); chlorophyll *b*, $C_{55}H_{70}O_6N_4Mg$ (mol. wt. 907) while protochlorophyll is $C_{55}H_{70}O_5N_4Mg$ (mol. wt. 891).

In brown algae, diatoms and some flagellates, a third chlorophyll component has been isolated and termed *chlorophyll c*. The proportion of chlorophyll *c* probably does not exceed 10% of the total quantity of the chlorophyll. Chlorophyll *c* is perhaps not very closely related to chlorophylls proper but rather to one of the intermediates (Mg-pheoporphyrin) formed in the biosynthesis of chlorophyll. It may be that *chlorophyll c* is not in the direct line to chlorophyll but is most probably in a side-shoot. *Chlorophyll c* does not contain the phytol structure. Most probably chlorophyll *c* is a Mg-2-vinyl-4-ethylpheoporphyrin monomethyl ester. This can be identical with protochlorophyllide (?); its molecular formula is $C_{35}H_{32}O_5N_4Mg$ (mol. wt. 712). In the red algae, *chlorophyll a* is accompanied by a fourth chlorophyll component, *d* (an isomeric *d'* is also reported) which is very similar to the other chlorophylls. In *chlorophyll d*, vinyl $-CH{=}CH_2$ side-chain at C_2 position of chlorophyll *a* has been replaced by $-CHO$ group. Thus chlorophyll *d* is $C_{54}H_{70}O_6N_4Mg$ (mol. wt. 895). Both *chlorophyll c* and *d* might be considered as examples of photosynthetic pigments which might have been of prime importance in the distant past and although they only retain remnants of such functional activity only in certain algal groups, e.g., brown and red algae, physically, they have been completely replaced, by physiologically more efficient *chlorophyll a*, in higher plants.

A fifth component, chlorophyll *e*, has recently been isolated from two species of algal *Xanthophyta*, from *Vaucheria* zoospores and *Tribonema* species. The chemistry of chlorophyll *e* is not yet fully known.

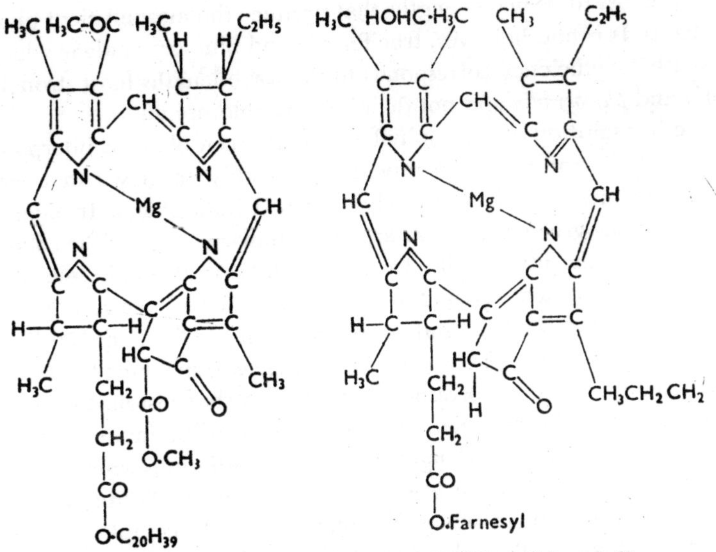

Bacteriochlorophyll   Chlorobium Chlorophyll

FIG. 699. Structures of bacteriochlorophyll and chlorobium chlorophyll.

Both chlorophyll and the phytol chain have been artificially synthesised in the laboratory since 1960 by Woodward and also independently by Strell *et al.*

Bacteriochlorophyll is the active pigment of the primitive anaerobic purple sulphur and non-sulphur bacteria. It is very similar to *chlorophyll a* of the higher green plants and algae but has the vinyl group $CH{:}CH_2$ in nucleus I (position 2) oxidised to an acetyl group, $CO.CH_3$, and an additional hydrogenation of nucleus II; so *bacteriochlorophyll* has *two* hydrogenated pyrrole rings and not one as in chlorophyll. In *chlorobium chlorophyll* (previously known as *bacterioviridin*), the pigment of green sulphur bacteria which was supposed to be an oxidised product of bacteriochlorophyll, the vinyl side-chain at C_2 position has been replaced by hydroxyethyl group, CH_3CHOH. There is probably a normal propyl side-chain at C_5 position in place of CH_3 and the carboxymethyl group at C_{10} is completely eliminated. The propionic acid side-chain at C_7 is esterified by another long-chain terpene alcohol, *farnesol* ($C_{15}H_{24}OH$). The empirical formula of *bacteriochlorophyll* is $C_{55}H_{74}O_6N_4Mg$ (mol. wt. 911) but that of *chlorobium chlorophyll* is not yet exactly known.

Five different types of bacteriochlorophyll Bchl *a* have been distinguished recently. They have been designated Bchl 800 nm, Bchl 820 nm, Bchl 850 nm, Bchl 870 nm, and Bchl 890 nm according to their maximum light absorption in the infra-red region.

A few species of *Rhodopseudomonas* do not contain the above five types, but a different one, Bchl *b* in contrast to five previous ones of Bchl *a*. The absorption maxima of this pigment in acetone lies between 1017-1020 nm as well as a weak band between 830-835 nm. (Sauer, 1968; Drews, 1965; Olson and others 1964-1967.) Chlorobium chlorophyll occurs in two forms—distinguished by their maximum light absorption region—*chlorobium chlorophyll* 650 nm and *chlorobium chlorophyll* 660 nm.

Properties of chlorophyll in solution

In green algae and higher green plants the photochemical reaction involves principally, if not solely, chlorophyll since those wave-lengths of light most strongly absorbed by chlorophyll are just those wave-lengths that produce the most efficient photosynthesis in these organisms. It is not, however, true for red and blue-green algae where the optimum photosynthetic efficiency corresponds to the wave-lengths most strongly absorbed by *phycoerythrin* and *phycocyanin* and not definitely by chlorophyll.

When an ether solution of chlorophyll of higher green plants is interposed between a source of white light and a spectroscope, it can readily be shown that certain wave-lengths of light are much more completely absorbed than others. If pigments are extracted by grinding the green cells in water, absorption spectrum of the colloidal extract is the same as that of the intact cells. In a clear ether or alcoholic solution, however, absorption spectra are shifted as compared with those of the spectrum in nature. This certainly points to the fact that the state of pigments in the living cells differs from that in solution in organic solvents. The wave-length-range within which light can be utilised for photosynthesis, happens to correspond roughly to that visible to the human eye, except perhaps in some photosynthetic bacteria which utilise invisible infra-red radiations for photosynthesis. The wave-lengths of the sun's radiations which reach the earth's surface range from 300 nm (in the ultra-violet) to about 2,600 nm (in the infra-red). The regions of the spectrum in which complete or nearly complete absorption takes place appear as dark bands. Both the *chlorophylls, a* and *b* of higher green plants, exhibit maximum absorption in the blue-violet region (400-510 nm) and a secondary maximum in the short red (651-688 nm) and these are certainly the wave-lengths (or the absorbed light) most effective in photosynthesis. At the red end of the spectrum, the maximum of *chlorophyll a* absorption lies at a longer wave-length than that of *chlorophyll b*

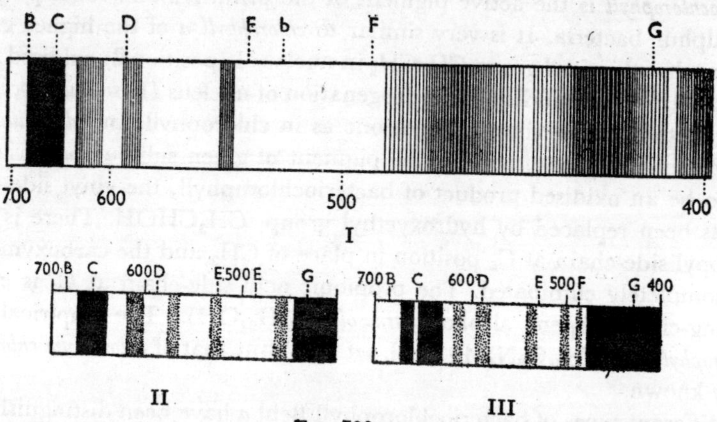

FIG. 700.

I. Spectra of chlorophyll in ether solution. II. Chlorophyll *a*. III. Chlorophyll *b*. B, C, D, E, b, F & G are Fraunhofer lines and numbers indicate wave-lengths in nm. The black and shaded portions (in red and blue-violet) are those where the light is strongly absorbed.

and conversely at the violet end. *Chlorophylls c, d* and *e* show absorption bands very similar to the chlorophylls of higher plants. The absorption bands of *protochlorophyll* are also similar to the other chlorophylls. Of the two bacterial green pigments, *chlorobium chlorophyll* shows similar bands to chlorophylls but *bacteriochlorophyll* is remarkable in that it shows strongest absorption in the infra-red (800-900 nm) and again at 1017-1020 nm. So invisible light may be effective in photosynthesis and this was proved by the demonstration that purple bacteria could thrive in complete darkness if they were allowed only infra-red radiations (lights of longer wave-length than visible red).

Fluorescence

Fluorescent substances have two properties: that of absorbing radiant energy and that of re-irradiating this energy, usually at longer wave-lengths. This re-irradiation takes place with extreme rapidity, often in less than 10^{-7} second.

When radiant energy is absorbed by chlorophyll, the excited chlorophyll molecules may release their excitation energy in three ways: as heat, fluorescence and chemical energy. The absorbed energy may merely add to the already existing thermal energy of chlorophyll molecules (depends on the movement of whole molecules). Light absorbed in this way is not effective in photosynthesis. When wave-lengths of light in the blue-violet and the red region of the spectrum are absorbed by chlorophyll, the most important effect occurs within the atoms of which chlorophyll is composed. An electron from an atom constituting the chlorophyll molecule, may be displaced outwards and it may now travel in an orbit further from the central nucleus and the potential energy of the atom rises accordingly. If the light energy absorbed is sufficiently high, the electrons may be displaced to such an extent that they are lost from the atom altogether. An electron displaced to a higher energy level—but not altogether lost—sooner or later reverts to its normal position and the energy may be released in various ways, as pointed out before. Chlorophyll in plants is fluorescence-activated and if this were not the case, the excitation energy due to absorption of light by chlorophyll would be lost to heat internally, long before it would be used in photosynthesis. The energy may also be used to produce chemical changes, more truly photochemical reactions and this is the fundamental one in photosynthesis, or this energy may be re-emitted as light. If the last occurs almost instantaneously it is, as we know, called *fluorescence*, whereas if there is a time lag, it is called *phosphorescence*. Chlorophyll-fluorescence, however, is much weaker in nature than in solution. In organic solvents, both chlorophylls show characteristic fluorescence-spectrum consisting of a strong band in the red and a weaker one in the far-red. The intensity of fluorescence of an acetone or ether solution represents about 10% of the light energy absorbed, while in the living plant the fluorescence yield is still smaller, being less than 1%. If a solution of chlorophyll is prepared of the same concentration as occurs in the chloroplast (0·2M), all fluorescence ceases. A colloidal dispersion of chlorophyll in water is non-fluorescent; why and how then, chlorophyll in the aqueous medium of cell is capable of fluorescence at all?

Chlorophyll does not appear to possess *phosphorescence*. In both algae and higher plants, light is sometimes given out by plants after illumination had ceased for a period, which may be as long as two minutes. Heat and high concentration of CO_2 destroy this ability of luminescence but the depression in case of CO_2-concentration can be reversed. Chloroplast suspensions show similar effects. It is, however, not phosphorescence, for the emitted light seems to originate from a reversal of some steps of the photosynthetic mechanism. This can be called *chemiluminescence* (*bioluminescence*). The emission spectrum is that of chlorophyll but it is of very low intensity. It suggests that a part of the light energy, originally imparted to chlorophyll for bringing about chemical reactions, can be returned to chlorophyll *via* some pick-up substance in the chlorophyll. The extent of fluorescence provides some guide to the amount of energy being consumed in photochemical reaction. If the photochemical reaction is slowed down or altogether stopped, fluorescence increases accordingly. An actively photosynthesising leaf may show very little fluorescence for almost all the energy absorbed is consumed in bringing about the primary mechanism of photosynthesis and very little is left for re-emission as light again. It has been observed, during recent years, that an *outburst of fluorescence*—the so-called *Katusky effect*—occurs in the chlorophyllous tissue during the first few moments of illumination. The fluorescence-burst seems to be related to the induction period of photosynthesis. Is it, that immediately after light is absorbed by chlorophyll, most of it is re-emitted as light and is not available for photosynthesis? Does it also suggest that photosynthesis attains its maximum efficiency only *after a time lag*?

Distribution of Chlorophylls

The contents of chlorophyll in green leaves vary between 0·05 to 0·20% of fresh weight, generally consisting of about 2·5 to 3·5 parts of chlorophyll *a* to one part of chlorophyll *b*, and this ratio is remarkably constant. Leaves of yellow-green variety may have far greater relative amount of *chlorophyll a*, as also the plants growing in high altitudes (alpine plants). The so-called *sun plants*, plants adapted to strong light, can have an average ratio of about 5·5. Green algae, however, behave as extreme *shade plants* with an exceptionally large proportion of chlorophyll *b*, the average ratio, *a* : *b*, being 1·4 : 1.

Coloured algae, living in much deeper water than the green algae, could thus be expected to contain proportionately even more chlorophyll *b*. But surprisingly, all coloured algae such as the brown algae, diatoms, the blue-green and the red algae contain no chlorophyll *b*. Perhaps this may partly be due to the fact that sunlight becomes bluish-green after its passage through several feet of water and the deep-living marine algae thus are deprived of most of the utilisable light (blue-violet and red) that can be profitably absorbed by chlorophylls and carotenes. The preponderance of chlorophyll *b* in shade plants is perhaps due to the fact, that chlorophyll *b* is better adapted for maximum utilisation of the wave-lengths of light, comparatively more abundant in the blue-green (450-489 nm) shade of the overhanging foliage (*chromatic adaptation*). *Fucoxanthol*, in the brown algae which absorbs at least some bluish-green light (up to 500 nm) and the red *phycoerythrin* (the presence of chlorophyll *b* in the coloured algae would have been superfluous) seem to be eminently suitable for absorption of such light (green), for *phycoerythrin* shows maximum absorption efficiency in the green part of the spectrum (500-550 nm).

The well-known fresh-water green algae *Vaucheria*, we know, contains no *chlorophyll b* and this evidence has prompted many investigators to remove *Vaucheria* from *Chlorophyta*, and at first put it in *Heterokontae*, and now in *Xanthophyta*.

In the class of algae, e.g., *Flagellatae*, some species (e.g., *Volvox* and *Chlamydomonas*)

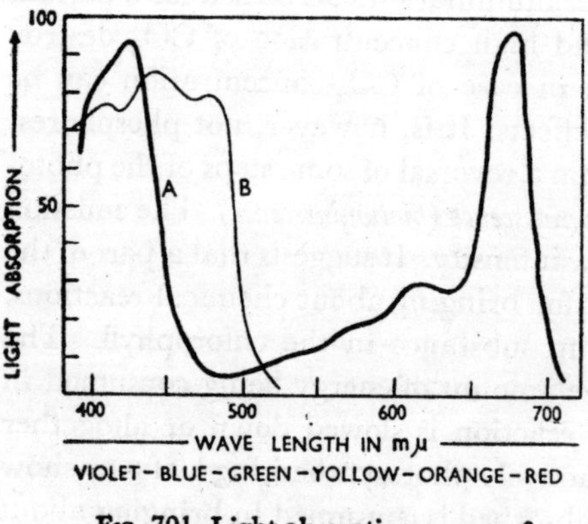

Fig. 701. Light absorption curves of chlorophyll *a* (A) and carotene (B).

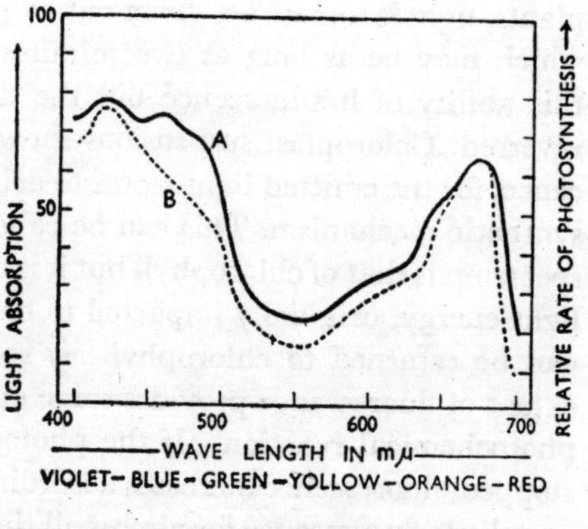

Fig. 702. Spectral absorption curve of a green alga (A) compared with its photosynthetic efficiency (B).

(For mμ read nm)

contain components *a* and *b* in normal proportions, while others contain only a very small amount of *b* (*Euglena*) or none at all (*Peridinium*). Thus plants can be classified into two large groups, '*a*+*b*' plants comprising all higher green plants and the algae, *Chlorophyta* including *Euglenophyta*, and '*a*' plants comprising all the rest of the coloured organisms, excluding the green and purple bacteria, which contain none of either *a* or *b*. The occurrence of identical plastidial pigments in *Chlorophyta* and the higher green plants certainly suggest a close phylogenetic relationship between them and is quite in accord with the views of the morphologists. Similarly, all the *b*-deficient coloured algae may be phylogenetically related to and evolved from a common ancestral source such as the flagellate, *Peridinium*, with no chlorophyll *b*.

Biosynthesis of Chlorophylls

The pyrrole rings of the tetrapyrrolic molecule (porphyrins) are thought to have originated from the simple amino acid *glycine* (CH_2NH_2COOH) and a four-carbon compound such as *succinic acid* ($COOH.CH_2 : CH_2.COOH$). The amino acid and the succinate are supposed to be building units or blocks from which ultimately the pyrrole ring is derived. The biosynthetic chain starts with *glycine* and *acetate*. This has already been demonstrated by the use of labelled ^{14}C atom. The acetate can easily be converted into *succinyl CoA* through the series of reactions of the TCA cycle. The succinyl coenzyme A then reacts with glycine in presence of *aminolevulinic acid synthetase* (*ALA-ase*) and two molecules of aminolevulinic acid (ALA) are formed. Later on, the two molecules of ALA condense to form porphobilinogen (PBG) in presence of *ALA-dehydrase enzyme*, K^+ ion being the specific activator for this reaction. Four molecules of PBG are then united for the first time to form the first cyclic tetrapyrrole of the *Uro*-type. The biosynthesis of both chlorophyll and haem compounds (haemoglobin, cytochrome, catalase, peroxidase, etc.) till now follows a common path until iron or magnesium is inserted into the centre of the protoporphyrin ring. If it is iron, the protoporphyrin becomes a haem; if magnesium,

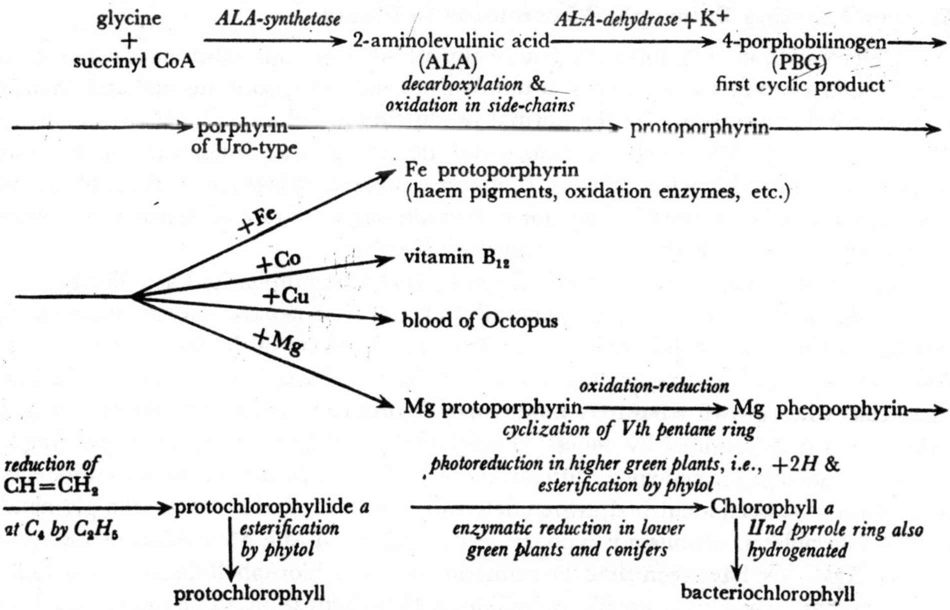

FIG. 703. Probable biosynthetic pathway for chlorophyll formation.

it is a magnesium protoporphyrin. Oxidation or reduction in side-chains of the last compound followed by the formation of the additional fifth cyclopentane ring leads to the formation of *protochlorophyllide a* via magnesium-vinyl-pheoporphyrin (an oxidation product of chlorophyll *a* isolated from etiolated seedlings). In case of higher green plants, excluding conifers, *protochlorophyllide a* itself absorbs light and is photoreduced to chlorophyll, i.e., addition of two extra H at C_7 and C_8, and subsequent esterification by phytol chain, gives chlorophyll *a*. Esterification by phyto is a dark reaction (*see* Fig. 698). In higher green plants no slow dark chemical reaction is involved in this transformation of protochlorophyllide to chlorophyll. In lower green plants, including bryophytes, and conifers, where light is not necessary for the conversion of *protochlorophyllide a* to *chlorophyll a*, the reduction, i.e., the addition of two hydrogen atoms at C_7 and C_8 is thought to be brought about by a dark enzymatic reaction.

This photochemical (or thermochemical) reduction of protochlorophyllide to chlorophyll is of especial interest since, as we have seen before, it brings about the addition of two hydrogen atoms to protochlorophyll, a change, which certainly involves a storage of energy in chlorophyll molecule formed. The questions arise as to whether all the energy involved in this reduction is supplied by the photons absorbed by protochlorophyll or whether light is required merely as a 'trigger mechanism' to permit some enzymatic reductive process to act. In view of the formation of chlorophyll in dark in lower green plants, the latter view seems to be more probable for if the absorbed photons supplied all the energy for the photoreduction process, this would certainly be akin to the process of photosynthesis. But no evolution of O_2 has yet been detected during reduction of protochlorophyllide to chlorophyll. Thus this reaction might be considered as a kind of *protosynthesis*, the energy of the absorbed sunlight being used for enzymatic reduction with no simultaneous evolution of O_2.

This scheme representing chlorophyll formation in plants seems to be in agreement with experimental evidence obtained so far but is certainly oversimplified (FIG. 703).

Factors Affecting Chlorophyll Formation in Plants

Chlorophyll, in common with practically all other organic substances which occur in plants, is a product of synthesis; a number of conditions, both internal and environmental, are thus necessary for the normal production of chlorophyll. If any one of the factors is limiting, chlorophyll synthesis is inhibited to a greater or lesser extent, resulting in a condition often known as *chlorosis*. In mature leaves there is evidence that chlorophyll is being constantly renewed by new formation although the rate of destruction is extremely small compared with the rate of photosynthesis.

The factors influencing chlorophyll synthesis can be summarised as follows:

1. *Genetic factor*. Chloroplasts generally do not arise *de novo* but are independent self-duplicating entities, inherited generally from cytoplasm of the female reproductive cell. In *Euglena*, cells are formed without chloroplasts or leucoplasts when grown in dark; these cells when grown separately never form chloroplasts again. The development of chloroplasts is determined by genes (Mendelian factors) and by factors depending on the functional and physiological differentiation of tissues. In maize, numerous independent genes have been identified, which determine albinism and influence the pigmentation of chloroplasts. Albinism is a characteristic inherited as a Mendelian recessive.

2. *Light*. We have seen that the reduction of protochlorophyllide to chlorophyll is due to light absorbed by protochlorophyllide itself. So light seems to be indispensable for chlorophyll formation in angiospermic flowering plants though exceptions are provided

by particular species, which can form chlorophyll in complete absence of light. In algae, bryophytes, pteridophytes and even conifers, however, light does not seem to be essential for chlorophyll synthesis and true protochlorophyll has not yet been isolated from these green plants. It may be that the mechanism of chlorophyll synthesis is slightly modified in cryptogams and conifers with the possible elimination of the photochemical stage so that light energy absorbed by protochlorophyllide in these plants is not really necessary for reduction to chlorophyll (reduction can, perhaps, be brought about by enzymes in the absence of light).

3. *Oxygen.* It is observed from the scheme presented here for the biosynthesis of chlorophyll that oxidation of the side-chains is necessary (as also the oxidase enzymes) for the formation of protoporphyrin from the *Uro*-type porphyrin precursor and also for various other processes (oxidation reactions to form the side-chains and carboxylation). Hence a supply of oxygen is one of the essential conditions for chlorophyll formation.

4. A supply of carbohydrates or other photosynthates must be continually provided at the seat of chlorophyll formation. Glycine and succinate ultimately must be derived from carbohydrates, perhaps in the respiratory breakdown of pyruvates in the Krebs cycle.

5. Nitrogen becomes a part of the chlorophyll molecule itself from the initiation of its biosynthesis, for the amino acid glycine containing the NH_2 grouping, is ultimately incorporated into the chlorophyll molecule proper. It is apparent that without N_2 there cannot be any synthesis of the porphyrin structure in the living cells essential for the maintenance of life. As protein and constituents of nicotinamide nucleotides, it is involved in all the enzymatic reactions for the synthesis of chlorophyll.

6. *Magnesium.* Magnesium itself enters into the composition of the chlorophyll molecule. Traces of Mg also act as activator component of the prosthetic front (co-enzyme) of the *carboxylase,* involved in the synthesis of chlorophyll.

7. *Iron.* Though iron does not enter into the composition of chlorophyll molecule, minute quantities of iron are essential for synthesis of chlorophyll. We know that iron is in the prosthetic group of all the oxidation and reduction enzymes, including cytochrome oxidases. It is evident that presence of minute traces of iron must be an essential condition for various enzymatic reactions involved in chlorophyll synthesis.

8. *Temperature.* The maximum rate at which chlorophyll is synthesised seems to lie between 25-30°C. The photochemical (reduction of protochlorophyllide to chlorophyll) and thermochemical reactions leading to the synthesis of chlorophyll appear to operate in some kind of integrated fashion, during greening. They cannot as yet be separated from the phenomena involved in growth and development of chloroplasts themselves.

9. *Supply of water.* Desiccation of leaf tissue not only inhibits synthesis of chlorophyll but also accelerates the disintegration of the chlorophyll already present. The effect of a plentiful supply of water possibly acts through the activation of the enzyme system in various ways such as: (*i*) increase in amount of soluble enzymes—both oxidative and hydrolytic, (*ii*) soluble substrates are brought into the aqueous phase and (*iii*) release of oxidative enzymes due to hydrolysis of cellular tissue.

CAROTENOIDS

These are visible, short wave blue-violet absorbing, tetraterpene (rubber is a polyterpene) accessory pigments in plants.

The orange-yellow carotenoids occur in all green plants in the chloroplastids generally in association with chlorophyll. Some of the carotenoids are probably photosynthetically inactive but others are definitely accessory. Members of this group of pigments are also found in the non-green parts of plants, such as the perianth and brightly coloured flowers and fruits and in orange roots of carrots (the name carotene is derived from this), in separate oil globules, which topographically exclude them from participation in photosynthesis, for they are too far away from the main centres of photosynthetic activity (mesophyll cells). An energy transfer under such circumstances, when carotenoids are found alone, seems to be out of question. Carotenoids are often called *lipochromes* or coloured fats because they are soluble in fat solvents and usually found closely associated with fatty substances.

Carotenoids are also widely distributed in all classes of algae and they have also been observed in fungi and bacteria including the photosynthetic bacteria; carotenoids present in photosynthetic bacteria seem to be specific for the group and are not found in higher plants. The carotenoids are also of wide occurrence in the animal kingdom—one of much importance in photochemical reactions in animals—the visual purple of the eye contains the carotene, *retinene*. The photochemical reactions in the visual purple play an essential part in vision. Moreover, the red light sensitive eye spots of certain unicellular flagellates contain carotenoids. Chromatographic analysis has revealed the presence in the higher plants of a wide variety of members of the class xanthophyll; in some cases the relative amount, of even some minor constituent of which being shown to be controlled by a genetic factor. In algae, taxonomic groups may be defined by the presence of a particular carotenoid pigment, e.g., among the green algae *Chlorophyta*, the old class *Siphonales*, to which *Vaucheria* belonged, contain the asymmetrical α-carotene as the main and only constituent carotenoid pigment. The higher green plants and all the coloured algae have carotene β as their main constituent. Insignificant quantities of α-carotene are, however, present in higher green plants and red algae. It appears that the old class *Siphonales* should form a taxonomic group by itself and be even separated from the *Xanthophyta*. The physiological evidence of absence of *chlorophyll b* (and presence of chlorophyll *e*) in some members of *Siponales* and the almost exclusive presence of carotene α (in contrast to all other green plants) strongly support the idea. Carotenoid pigments have been traced in mitochondria of both plants and animals.

The molecular ratio of carotenoids to chlorophylls present in leaves is very variable. In normal green leaves there are many more molecules of chlorophyll than carotenoid. In shade leaves the ratio of chlorophyll to carotenoids approaches 5 : 1. In etiolated leaves the carotenoids are present before greening has occurred.

The original identification of two carotenoid compounds in green leaves, *carotene* and *xanthophyll*, has now been extended and a number of different isomeric forms of carotenes and variously oxidised alcohols, xanthophylls (*carotenol*) have been recognised. All carotenoids contain the long-chain alcohol, *phytol*. Carotenes are hydrocarbons with unsaturated straight carbon chains with the general formula $C_{40}H_{56}$. It crystallises easily in steel-blue crystals. In most organic solvents, carotenoids form a yellow solution but in carbon bisulphide they form an orange solution. In common with all other carotenoids, carotene gives deep-blue colour with concentrated H_2SO_4. *Four* isomeric forms of carotene are now recognised, α, β, γ and δ, of which β *is the only symmetrical one*, the other three being asymmetrical. This is interesting from a biochemical standpoint since vitamin A ($C_{20}H_{29}OH$), which is one-half of a carotene molecule and is formed from enzymatic hydrolytic cleavage of carotene β in the intestinal walls of animals, plays an essential part in vision. Thus carotene has aptly been called *provitamin A*.

$$C_{40}H_{56}+2H_2O \xrightarrow{\textit{carotenase}} 2C_{20}H_{29}OH$$

The *xanthophylls* or *xanthols* (*Carotenols*) are the oxidised pigments (alcohols) containing oxygen ($C_{40}H_{56}O_2$). The carotene-xanthophyll ratio in normal green leaves is about 1 : 2, i.e., proportionately there are more xanthophyll than carotene. The most common xanthophyll in green leaves is, *luteol* ($C_{40}H_{56}O_2$). *Fucoxanthol*, the brown pigment masking the chloroplast of the brown algae, is the most strongly oxidised carotenol ($C_{40}H_{60}O_6$) and along with *peridinol* is abundantly found in brown algae, diatoms and also in one red alga species of *Polysiphonia* as also in chlorophytan, *Zygnema pectinata*, perhaps compensating for the absence of chlorophyll *b*, in these species of algae. The absence of chlorophyll *b* and the presence of fucoxanthol and peridinol in diatoms and brown algae do certainly subscribe to the prevailing idea of a phylogenetic relationship between these algae.

Lycopene, a red pigment found in the fruits of tomato, red pepper, etc., is isomeric with carotene and all carotenoids are thought to be derivatives of this red *lycopene* pigment. The lycopene is not, however, found in all carotene-synthesising cells.

Orange-yellow autumnal colouration of the leaves of many plants, seen just before they are shed, is due to changes which occur in the carotenoid constituents of the leaf. *Zeaxanthol* (a β-carotene derivative and an isomer of *luteol*) becomes the most abundant constituent in such a yellow leaf.

The carotenoids overlap chlorophyll absorption in the violet and blue end of the spectrum and thus strongly absorb light in the violet end, actually in the blue-green, of the visible spectrum, showing two or three separate bands. The maximum absorption ranges between wave-lengths 425-490 nm. Carotenoids of bacteria absorb farther to the red than those of higher plants.

In green plants, one conception is that the unsaturated phytyl alcohol acts as a precursor of carotenoids. Phytol is apparently fundamental to the synthesis of both chlorophylls and carotenoids. The phytol group can be synthesised in absence of light. Once the phytol group is synthesised, it can esterify the propionic acid side-chain of an intermediate (Mg.-vinyl pheoporphyrin) in the chlorophyll synthesis and thus can form chlorophyll or alternatively be further elaborated, in a different line, to carotenoids.

Carotenoids and Photosynthesis

As regards the function of carotenoids, there seems to be little doubt now that light absorbed by carotenoids can be fully or partially utilised in photosynthesis. A comparative examination of the two curves (FIGS. 702 and 704) is that at wave-length 480-500 nm, where most of the absorption of light is by carotenoids, photosynthesis occurs to a greater extent than would be expected if the light absorbed by chlorophyll alone were effective. This certainly seems to show that the light absorbed by carotenoids is also used in photosynthesis. As we shall see later, this is also doubly true for phycobilin pigments of red and blue-green algae; actually chlorophyll, as a light absorbing pigment in these groups of algae is considerably less effective than phycobilin pigments. Since, however, it is certain that photosynthesis cannot occur when the carotenoids or other pigments, e.g., phycobilins *alone* are present, it must be supposed that excitation energy absorbed by carotenoids and phycobilins are transferred to chlorophyll *a*. The efficiency of this carotenoid-chlorophyll transfer is, however, relatively low in higher green plants, about 20-40%. It is also known that such energy transfer between different pigments is possible. If photosynthesis could be shown to occur in complete absence of chlorophyll such as non-chlorophyllous carotenoid-coloured organs, we have to postulate an entirely separate, parallel system of enzymes, able to carry on photosynthesis independently of chlorophyll. This seems highly improbable in absence of any evidence of photosynthesis with light absorbed by accessory pigments alone. Thus, it seems to be true that accessory pigments channel their abosrbed energy entirely into chlorophyll *a*. In higher green plants the contribution of the carotenoids to the total light absorption must only be negligible and this function of carotenoids which can, at its best, only be subsidiary is unlikely to provide an adequate explanation of their invariable presence in the photosynthesising cells. Thus despite the obviously intimate connection between carotenoids and the photochemical processes in some living organisms, there are reasons for believing that in higher green plants and green algae, the photochemical reaction involves principally, if not solely, chlorophyll *a* since the wave-lengths of light, most thoroughly absorbed by chlorophyll, are just those wave-lengths that produce the most efficient photosynthesis in higher plants and green algae. In the blue end of the spectrum absorption of light is at least half due to carotenoids and the probability is that some of the carotenoids in many higher green plants absorb light effectively. It is true, as we have said before, that the absorption of light by carotenoids is not quite as effective as chlorpohyll, because the photosynthetic efficiency in the blue (absorbed primarily by carotenoid) is only about 85% as high as in the red (absorbed by chlorophyll *a*). In other cases, photosynthesis is appreciably lower in the blue region of the spectrum, strongly suggesting inactive absorption of light by carotenoids in that region.

In some green algae (*Chlorella*), the fall in the photosynthetic efficiency in the blue or in the blue-green (480-490 nm) compared to the efficiency in the red end of the spectrum (absorbed by chlorophyll only), is perhaps due to inactive carotenoid absorption. But such absorption even in the blue-green region could hardly be entirely inefficient, for carotenoids here absorb about 60-70% of the incident light, yet photosynthetic efficiency is down in most cases to only about 25% of the chlorophyll efficiency in the red. The situation is still better at 420 nm where carotenoids absorb only about 50% of the light, yet the efficiency is only about 10% less than that in red. Some orange algae, such as *Trentepohlia* look orange due to presence of large amounts of β-carotene dissolved in numerous oil globules. *Trentepohlia* and other orange algae seem to inhabit regions of considerable drought such as mediums with excessive dissolved sodium chloride or spray-swept trees; the high carotene content in these algae may be a response to desiccating conditions. In any case, these algae scarcely use blue light at all (absorbed primarily by carotenoids); photosynthesis in these orange algae seems to be restricted to the red end of the spectrum (absorbed by chlorophyll only). This probably is a simple shading effect; blue light is absorbed by the abundant lipochromic carotene in non-chlorophyllous separate structures before it can reach chlorophyll.

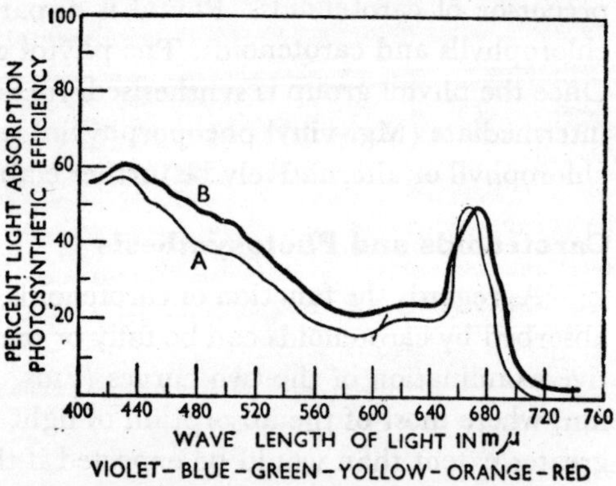

FIG. 704. B represents absorption of light by thallus of a brown alga. A. relative photosynthetic efficiency. (For mμ read nm)

In contrast to variable and sometimes completely inactive role of carotenoids in the green algae, the brown algae and the diatoms supply us with clear cut examples of effective chlorophyll-carotenoid collaboration in photosynthesis. Compared to green algae, many of the brown algae show enhanced photosynthesis in the blue end of the spectrum (light absorbed mostly by carotenoids). Indeed the rate is often higher than in the red (light absorbed by chlorophyll). The inevitable conclusion is that carotenoids greatly assist photosynthesis in brown algae, and because of its high concentration, *fucoxanthol* must be the responsible carotenoid. More recently it was shown that in diatoms, photosynthetic efficiency was nearly equal throughout the visible spectrum—except around 480 nm, the so-called already familiar *depressed zone*—where photosynthetic efficiency in most groups of algae always seem to be very low due to some reasons yet unknown. In brown algae and diatoms, energy absorbed by fucoxanthol enters the chain of chemical dark processes *via* chlorophyll—the efficiency of this transfer of the excitation energy approaching almost 100%. The brown algae and diatoms certainly absorb fairly completely through the spectrum and utilise most wave-lengths of absorbed light and it is perhaps only natural, that diatoms in number and brown algae in size, are among the most successful of all marine algae. The classical name *Melanophyceae* for brown algae, now seems to be more than justified, for although individual cells of the brown algae can transmit yellow, orange or green light, increased thickness of the thallus in many cases enhances absorption of almost all wave-lengths of visible light, making some of them look almost black !

Because of their unsaturated nature, even though carotene and xanthophyll may not act as an oxidation-reduction catalyst, they certainly serve as transmitters of energy and also as oxygen carriers. An interesting function has been ascribed to carotenoids —that the presence of carotenoid pigments was essential in the chloroplasts for a photochemical light-catalysed reversible reaction using light energy of wave-length 468-509 nm, which was at one time thought to be specifically involved in the photolytic decomposition of water. The absorption of light in this region of spectrum is primarily by carotenoids, and in analogy with reactions of visual pigment of the eye, this reaction was supposed to be mediated by a carotenoid pigment. The carotenoid pigments would then be specifically required in all green photosynthesising cells as a natural hydrogen acceptor in the decomposition of water which, however, seems extremely unlikely. The carotenoids were then regarded as a hydrogen transfer agent to the nicotinamide nucleotides (e.g., NADP), by the reversible release of hydrogen from carotenoid, and the reduced NADP utilised for reduction of CO_2-fixation product to the carbohydrate level in normal photosynthesis. A similar function was also ascribed to vitamin K, which like chlorophylls and carotenoids, contains the long-chain phytyl alcohol. Vitamin K (or the related compound plastoquinone) now is thought to be closely connected with cyclic photosynthetic phosphorylation, leading to the formation of high-energy phosphates, ATP (*see* the biochemical mechanism of photosynthesis), for ATP formation is, perhaps, the fundamental and the earliest event in photosynthesis and is intimately related to it.

One of the most interesting function ascribed to the carotenoids is the absorption of blue light by carotene to produce phototropic curvatures in oat coleoptile, in *Phycomyces* (e.g., *Mucor* sporangiophore) and also in certain blue-green algae. These positive phototropic curvatures towards incident light have been supposed to be due to unequal distribution of auxins on the two sides of the coleoptile. The farther side (away from the light) has more auxin than the lighted side due to light induced inactivation or destruction of auxins in the lighted side. But a difficulty soon arose; a solution of auxin (indoleacetic acid) was found to be perfectly stable in light. The effect of light cannot,

TABLE OF PHOTOSYNTHETICALLY ACTIVE PIGMENTS

Pigment Chlorophylls		Distribution in the plant kingdom	Empirical formula and molecular weight	Chemical structure	Maximum absorption of light	Photosynthetic efficiency	State of pigment in the plastid
Chlorophyll	*a*	All green plants	$C_{55}H_{72}O_5N_4Mg$; mol. wt. 893.	Cyclic tetrapyrrole with phytol. Ring V with 5C atoms. Ring IV with two extra hydrogen; Mg atom.	Red and blue-violet; 651-688 nm. 410-500 nm.	Primary photosynthetic pigment in all green organisms except red algae.	Pigment in chloroplastids as lamellate grana in higher plants and as plastidial lamellate structure in algae except blue-green.
,,	*b*	Absent from all coloured algae, e.g., brown, red, diatoms and blue-green.	$C_{55}H_{70}O_6N_4Mg$; mol. wt. 907.	In chlorophyll *b*, the CH_3 group in position 3, replaced by CHO.		When present, as active as chlorophyll *a*.	Probably loosely-held protein-pigment complex.
,,	*c*	Only in brown algae and diatoms.	$C_{35}H_{32}O_5N_4Mg$; mol. wt. 712.	Ring IV without 2H. No phytol.		May have been active photosynthetically in the distant past.	
,,	*d*	Only in red algae.	$C_{54}H_{70}O_6N_4Mg$; mol. wt. 895.	Cyclic tetrapyrrole in which—$CH{=}CH_2$ at C_2 is replaced by —CHO.	Farther in the red; 696 nm.	A rather poor co-partner for chlorophyll *a* in red algae.	
,,	*e*	From *Xanthophyta*	Not exactly known.				
Protochlorophyll		Etiolated plants.	$C_{55}H_{70}O_5N_4Mg$; mol. wt. 891.	Ring IV without 2H.	Near red and blue-violet.		Part hydrophilic tetrapyrrole nucleus; part hydrophobic (phytol or farnesol). Soluble in organic solvents.
Bacteriochlorophyll 800, 820, 850, 870, 890 nm	*a*	Purple sulphur bacteria.	$C_{55}H_{74}O_6N_4Mg$; mol. wt. 911.	Both rings IV and II with 2H. Ring I position 2, the vinyl ($CH:CH_2$) radicle of chlorophyll is oxidised to acetyl group (CO.CH.).	Infra-red, 800-890 nm and above, up to 1020 nm; also lue-violet.	Photosynthetically active pigments of green or purple autotrophic bacteria.	
Bacteriochlorophyll above 1000 nm	*b*	The same.					
Chlorobium chlorophyll 650, 660 nm		Green sulphur bacteria.	Not quite known.	Only Ring IV with 2H; C_2-position, $CH{=}CH_2$ replaced by CH_3CHOH. Phytol replaced by farnesol; at C_5-position, CH_3 replaced by *n isopropyl* side-chain, $CH_3CH_2CH_2$; carboxymethyl group at position C_{10} eliminated.	Red and blue-violet.		No true chloroplastids. Pigment in free-lamellate grana.

TABLE OF PHOTOSYNTHETICALLY ACTIVE PIGMENTS (CONTD.)

Pigment Carotenoids		Distribution in the Plant Kingdom	Empirical Formula and Molecular Weight	Chemical Structure	Maximum Absorption of Light	Photosynthetic Efficiency	State of Pigment in the Plastid
Carotene	α	Primarily in algae group, *Siphonales* (*Chlorophyta*). Very small amounts in higher green plants and red algae.	$C_{40}H_{56}$; mol. wt. 536.	Long-chain unsaturated hydrocarbons. Phytol. β-carotene symmetrical; α, γ and δ, asymmetric.	Blue and blue-green; 425-475 nm.	In higher green plants, photosynthetic efficiency, insignificant. In algae, contribution of light energy absorbed may sometimes be considerable. Absorbed light energy utilised *via* chlorophyll.	Pigment in the plastids; in the grana in higher plants and as lamellate structures in algae.
,,	β	All green plants and algae except *Siphonales*.	The same.				
,,	γ	Very uncommon.	The same.				
,,	δ	Only traces in green plants and algae.					
Zanthophylls (*Carotenols*)		Higher plants, ferns, mosses, green algae and also in red algae.	$C_{40}H_{56}O_2$; mol. wt. 568.	Long-chain hydrocarbons with 1 or more oxygen atoms (alcohols). Phytol.	Blue and blue-green; 450-490 nm.	The same.	Fat soluble lipochromes. Always associated with lipides; soluble in fat-solvents.
Luteol		All plants.					
Zeaxanthol		All higher plants and green algae.	The same.				
Fucoxanthol		Mostly in brown algae and diatoms; also in one red alga sp., *Polysiphonia*.	$C_{40}H_{60}O_6$; mol. wt. 636.			In brown algae and diatoms considerable photosynthetic efficiency acting through and in collaboration with chlorophyll.	
Spirilloxathin		Only in photosynthetic bacteria.	Very similar to other carotenols.				No plastid; pigment in free grana.

TABLE OF PHOTOSYNTHETICALLY ACTIVE PIGMENTS (CONTD.)

Phycobilins or bilichromoproteins		Distribution in the plant kingdom	Empirical formula and molecular weight	Chemical structure	Maximum absorption of light	Photosynthetic efficiency	State of pigment in the plastid
Phycocyanin	C-	Characteristic pigment of blue-green algae; in a few red algae and also in *Cryptomonad* algae.	Protein-free pigment —cyanobilin (mesobiliviolin).		Orange; 615 nm.	Photosynthetic efficiency as great as chlorophyll *a*, if not greater. Absorbed light energy is utilised probably *via* chlorophyll *a*.	No chloroplastid; pigment in cytoplasmic lamellae; (blue-green algae) water-soluble chromoproteins.
	R-	Only in red algae. Traces in *Cryptomonad* algae.			Orange: 615 nm; A very small absorption peak in green; 500 nm.		
	P-	(*Allophycocyanin*) in both blue-green and red algae.			Further into orange; 650 nm.		
Phycoerythrin	C-	In blue-green algae; also in *Cryptomonad* algae.	Protein-free pigment —erythrobilin (mesobilierythrin).	Linear open chain tetrapyrroles. No metal or phytol; water-soluble.	Green; 550 nm.	Primary photosynthetic pigment in red algae—role of chlorophyll *a* in light absorption is minor and subsidiary, most of it probably being in inactive state; phycoerythrin-absorbed light energy is utilised by transfer mechanism *via* small active chlorophyll *a* portion.	Pigment in plastidial lamellae, intimately associated with protein; water-soluble chromoprotein (red algae).
	R-	Only in red algae.			Green and yellow; 540 nm and 565 nm.		
	B-	From primitive members of red algae (*Bangiales*; *Porphyra* sp.)			Green and yellow; 550 nm and 565 nm.		
	a and *b*	Found as yet *only* in *Cryptomonad* algae.					

therefore, be a direct one but must be mediated by light absorbed by some other pigment, e.g., *riboflavin* (vitamin B_2) which occurs in many plant organs and are also found in oat coleoptile. Riboflavin has been found recently to be involved in some way or other in the photodecomposition of auxins, i.e., the light absorbed by riboflavin brings about the photochemical reaction, leading to the destruction of auxin. But another difficulty is that, why does not riboflavin-absorbed light destroy auxins on both sides of the coleoptile in equal amounts as the coleoptile sheath is only about 2 mm thick? The answer is most probably that carotenoids which are found in the coleoptile apparently prevent the destruction of auxin by light. Without carotenoids in the coleoptile, the light would reach both sides of the coleoptiles in equal amounts and in such circumstances, the phototropic curvature would be impossible. With carotene present in the coleoptile, it may provide a filter which may cause a suitable drop in light intensity across the coleoptile, leading to a gradient of increasing auxin concentration from the lighted side to the darker side. In the photosynthetic purple bacteria, the phototactic movements shown by them seem to be due to light absorbed solely by specific carotenoids.

A remarkable observation that pollens generally carried by insects for pollination contain carotene, whereas carotenoid pigments are rarely detected in the pollens of wind-pollinated flowers.

PHYCOBILINS OR BILICHROMOPROTEINS OR BILIPROTEINS

As their name implies, these accessory pigments are found only in the algae and are associated with the photosynthetic mechanism of the red and blue-green algae. Phycobilins have not yet been identified in flowering plants and in true green algae but it is probable that minute traces of their presence in higher plants may have escaped our notice. They were so named by Lemberg who showed their chemical affinity to the bile pigments of the animals. They differ completely from the chlorophylls and carotenoids being soluble in water but though water-soluble, the phycobilins are found in the plastids of the red algae and probably in the cytoplasmic lamellae of the blue-green algae.

In some red alga (*Porphyridium*) electron microphotographs show small phycobilin-containing granules, approximately 35 nm in diameter, attached to the chloroplast lamellae. The name *phycobilisomes* (Gnatt and Conti, 1966) has been suggested for them.

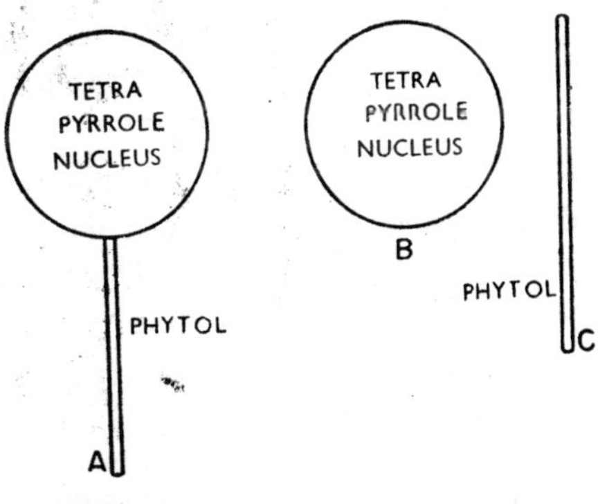

FIG. 705. Diagrammatic representation of chlorophyll: A; chromoproteins, e.g., haemoglobin and phycobilins: B; and lipochromic carotenoids: C. The 'head' (tetrapyrrole) is hydrophilic and the 'tail' (phytol), hydrophobic.

Phycobilins are chromoproteins or coloured proteins like haemoglobin, the chromophores or the pigment-prosthetic group being combined with proteins of the *globulin* class and not globin as in haemoglobin. Both chlorophyll and carotenoids are now generally assumed to be conjugated with proteins in the grana of the plastid. But this chlorophyll-carotenoid-protein connection is relatively feeble and is easily broken by gentle warming or by organic solvents like methyl alcohol. The phycobilins, on the contrary, are remarkably stable chromoproteins, water-soluble and crystallisable (crystallise independently into brilliant red and blue crystals of phycoerythrin and phycocyanin respectively) and of definite molecular weights. They were indeed among the earliest proteins to be studied physico-chemically because of their stability and colour. The molecular weights of *phycoerythrin* and *phycocyanin*, the two phycobilins from the blue-green and the red algae vary from 134,000 (*allophycocyanin*, previously known as P-*phycocyanin*; the allophycocyanins show striking similarity with phytochrome pigment, as regards their light absorption maxima. Recent findings also indicate that phytochromes are chemically closely related to phycobilins, e.g., allophycocyanin) to 290,000 (R-*phycoerythrin*). *Phycoerythrin* and *phycocyanin* are to be regarded as two classes of chromoproteins. Differences which depend on the plant source are known to occur and are due to the protein part (as in enzyme and coenzyme relation) rather than to the prosthetic group. Thus R-*phycoerythrin* and C-*phycoerythrin*, regarded as two different phycoerythrin pigments, have been isolated from the red, the blue-green and also from *Cryptomonad* algae respectively and the difference between these two pigments is due to the difference in the protein component rather than to the pigment, which part is exactly the same in the two phycoerythrins. The amino acid composition of the protein component of these pigments resembles the general cytoplasmic proteins and the general picture of amino acid composition of these algal biliproteins is that they are richer in dicarboxylic (asparatic and glutamic acids) than in basic amino acids and they have a high content of amino acids, having hydrophobic side-chains. The amino acids are also rich in sulphur, e.g., methionine and theronine. A possible phosphate attachment has also been suggested.

As pointed out before, the protein-free pigment components, *cyanobilin* (*mesobiliviolin*) and *erythrobilin* (*mesobilierythrin*) from *Cyanophyta* and *Rhodophyta* show very close similarity with the bile pigments such as bilirubin, biliverdin and urobilin. The pigment groups are very firmly bound to proteins and require strong acid-hydrolysis to separate them. Chemically both the pigments are tetrapyrrolic derivatives with the pyrrole rings in open chain, and not cyclic, as in chlorophyll and haemoglobin and

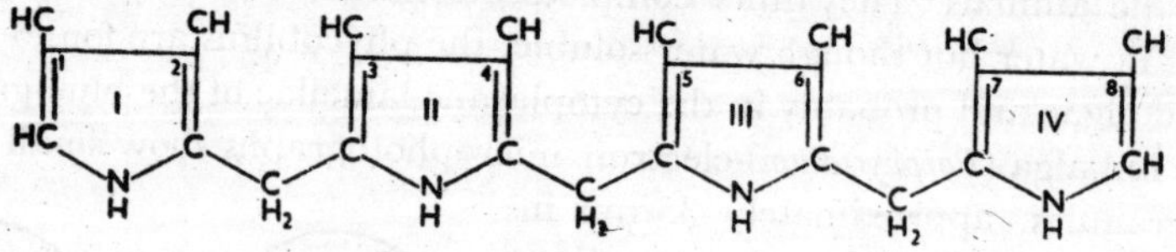

Bilan: $C_{19}H_{20}N_4$.

FIG. 706. Molecular structure of the parent linear tetrapyrrole nucleus, *bilan*.

without any metal or phytol. The two pigments which are readily interconvertible are derived from the parent linear tetrapyrrolic structure, *bilan*, from which the animal bile pigments are also derived. The bile pigments are *open chain tetrapyrroles* as mentioned before (Fig. 706).

Each molecule of phycobilin chromoprotein contains 16 molecules of pigment and the ratio (16 erythrobilin per 290,000) of pigment to protein weight is more or less similar to that in haemoglobin (4 haemes per 68,000), in nodular haemoglobin, leg-haemoglobin (1 haeme per 16,000), myoglobin (1 haeme per 17,000) and in cytochrome *c* (1 haeme per 14,000).

In analogy with the actual conversion of haeme pigments to biliverdin by opening up the closed ring by oxidation and the removal of iron by acids, evidences are strongly in favour of a similar conversion of chlorophyll to phycobilin occurring in the algae.

Informations available as yet seem to show that the occurrences of phycobilins are restricted mainly to the two classes of algae, *Rhodophyta* and *Cyanophyta* although reports about their presence in unicellular, *Cryptomonad* flagellates have been completely substantiated during recent years. These pigments may be present in amounts equal to or greater than those of chlorophyll *a*.

Optical properties

The lavender-red phycoerythrin absorbs light in the visible range, mainly in the *green region* of the spectrum and shows three maxima between 495-565 nm. Phycocyanin absorbs somewhat farther, in the orange (615 nm). Both the pigments show strong fluorescence; orange in phycoerythrin and red in phycocyanin.

Phycoerythrins are almost completely complementary in colour to chlorophyll, closing the gap in the middle of the spectrum. R-phycoerythrin (isolated from red algae) has three absorption maxima at 495 nm, 540 nm, 565 nm—most of it evidently in the green region. Schizophycean phycoerythrin, C-phycoerythrin (mostly from the blue-green algae) has but one absorption peak in the green, close to 550 nm while a newly isolated phycoerythrin, B-phycoerythrin (from primitive red algae, *Bangiales*) rather resembling it, but with one additional peak in the yellow at 565 nm. All of them transmit well in the blue end of the spectrum and almost completely in the red end. There is really very little overlap with chlorophyll and carotenoids.

The preponderance of C-phycoerythrin over all other pigments in *Oscillatoria erythraea*, a blue-green alga, paradoxically, is the reason for the name of Red Sea. These algae floating along the water surface often present a distinct reddish aspect.

The remaining spectral region where none of the other preceding pigments show strong absorption is the orange-red (600-640 nm). This gap is, however, filled competently by C-*phycocyanin* with absorption maxima at 615 nm. This is the characteristic pigment of the blue-green algae, *Cyanophyta*, but is also found in small quantities in red algae, *Porphyra*. Another type, R-*phycocyanin*, prevails in most red algae and in addition to its principal maximum at 615 nm, it has a very small absorption peak at 550 nm (in green). A new phycobilin, having a much smaller molecular weight, less than half compared to other phycobilins, P-*phycocyanin* (*allophycocyanin*), has been, as stated before, isolated (absorption peak around 650 nm, in the red) from a number of blue-green and red algae.

It has recently been surmised that primitive plants of the type of blue-green algae have later developed in two directions: one in which the phycobilins have gradually disappeared, and the other in which one particular form of phycobilin, the phycoerythrin has largely accumulated. Plants of the first type have taken overland and the surface layers of the ocean while plants of the second kind have mostly concentrated in the depths of oceans. This division, as we know, is certainly not strictly sharp, with some species of one kind invading the domain of the other (Rabinowitch and Govindjee, 1970).

Phycobilins and Photosynthesis

As far as it is known, phycobilins were thought to be restricted to two classes of algae (the blue-green *Cyanophyta*, regarded as one of the oldest forms of life upon the earth and the highly evolved, *Rhodophyta*). This long-known distribution certainly looks queer and perplexing.

Since 1960 both phycocyanin and phycoerythrin have been traced in certain unicellular, flagellate *Cryptomonad* algae.

In the red algae, we encounter one of the most remarkable pigment situations known. These red algae have abundant chlorophyll *a*, the closely related chlorophyll *d*, (as yet unique to them) in rather low concentration and at best, two carotenoids, common with higher angiosperms, β-carotene and the xanthol, luteol. Absorption of light by all these pigments is often as great as in green algae of comparable thickness and this would seem more than sufficient for all their photosynthetic requiremens. Yet these algae invariably contain relatively high concentration of the phycobilins, effectively closing the gap in the chlorophyll-carotenoid absorption spectra and being almost complementary to them. Absorption of visible light is more nearly complete here, therefore, than in any other photosynthetic organism. But the question is, how effective is this total absorption? The answer seems to be as remarkable and as consistent as the colour of the bright red thalli: chlorophyll was strikingly less active than phycoerythrin.

As early as 1883, Engelmann working with the red algae, *Ceramium*, found that photosynthetic activity as measured by O_2 evolution showed a sharp peak, strikingly sudden, at 575 nm, in the green. There was a barely perceptible foot hill in the red and nothing but a valley in the blue. Clear and exclusive participation of phycoerythrin was indeed indicated and the chlorophyll-carotenoid activity was so low as to be startling, perhaps even to cast doubt upon the method. It seems to have taken us nearly 75 years to concede that Engelmann was really right! Actually more than a score of investigators working with over 15 genera of the red algae came roughly or precisely to Engelmann's conclusion—all found considerably enhanced photosynthesis in the middle of the spectrum (green) compared to red or blue; just the reverse is found in green and brown algae. Most of these blue-green and red algae showed 200 or 300% more O_2 evolution in the middle of the spectrum (in the green, 530-540 nm) compared to either at 430 nm (absorbed by carotenoids) or at 670 nm (absorbed by chlorophyll). Since the total light absorption by phycoerythrin is not much greater than by chlorophyll in these algae, a considerable lower activity (1/2 to 1/3 that of phycoerythrin) of chlorophyll is clearly indicated. The combined chlorophyll-carotenoid activity is even less than this. In some other red algae, the photosynthetic activity was found to be nearly 3 times as great in the green (545 nm, light absorbed by phycoerythrin only) as in the very short red (620-660 nm, light absorbed mostly by chlorophyll) and 6 times as great as in the blue (435 nm, light absorbed mostly by carotenoids).

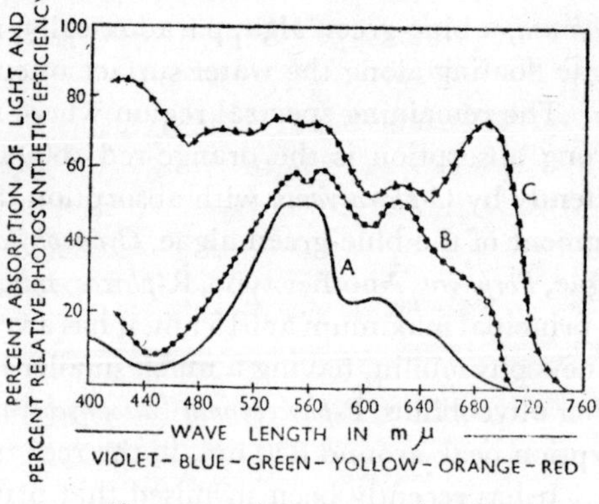

[for mμ read nanometre (nm)]

FIG. 707. Action and absorption spectra for red alga. A: absorption of light by phycoerythrin only; C: absorption of light by the whole thallus and B: photosynthetic efficiency. Note that photosynthetic efficiency is least in the blue (light absorbed by carotenoids), blue-violet and red (absorbed by chlorophylls). Photosynthetic efficiency is greatest in the green (light specifically absorbed by phycoerythrin).

In *Porphyra*, which contains B-phycoerythrin with its two absorption maxima in the green and yellow (550 and 565 nm), the photosynthetic activity almost literally and faithfully records this peculiarity with a two-peaked photosynthetic curve ! It also shows an increased rate of photosynthesis at 620 nm corresponding to an increased phycocyanin content found in this species of red algae. In some unicellular red algae which contain a single-peaked C-phycoerythrin (550 nm), photosynthesis is maximal at this peak. In all these cases, photosynthesis seems to be minimal or rapidly falling towards a minimum at the very point where chlorophyll absorption begins in the red (675 nm) and it is even lower, almost insignificant, in the blue where carotenoids are also absorbing with chlorophyll (around 450-470 nm). It seems apparent and without the least doubt that when light is available for absorption by phycoerythrin, phycoerythrin monopolises the photosynthetic mechanism almost exclusively to itself thereby making chlorophyll and carotenoids look on helplessly !

Chlorophyll activity is indeed depressed in the red algae. Whereas green algae might require about 14 quanta to release a molecule of O_2 and the brown algae only about 10, the red algae require 25-50 quanta in the regions of chlorophyll and carotenoid absorption (red and blue-violet). Only in the middle green, where the absorption is only by phycoerythrin, the quantum requirement is 15, comparing favourably with that of green and brown algae.

It seems evident then that red algae in utilising phycoerythrin to increase effective absorption in the medium wave-lengths (green and yellow) have lost efficiency correspondingly at the two ends of the spectrum—red and blue. Chlorophyll efficiency is down to a third of that in green or brown algae and carotenoid efficiency is even lower. This certainly seems too high a price to pay for utilisation of green light. Why is it that about 66% of the red algal chlorophyll *a* is apparently non-fluorescent and inactive and certainly cannot receive any energy from phycoerythrin? There seems to be no known answer to it yet although it is true that red algae seem to yield their chlorophyll *a* very slowly to acetone or other organic solvents. Is much of it then bound to different proteins so closely as to be unavailable to the photosynthetic enzyme system? Is the other chlorophyll component, chlorophyll *d* so very inadequate and indifferent partner for chlorophyll *a*? Another possibility is that the greater part of chlorophyll *a* is so located topographically (the establishment of definite spatial arrangement required for the effective functioning of most of chlorophyll *a* may have been absent) in the plastid that it cannot transfer the energy absorbed to the small active fraction, although phycoerythrin apparently can.

In the blue-green alga, *Oscillatoria*, Engelmann's results strongly indicated the participation of phycocyanin in photosynthesis. The photosynthetic efficiency was maximum in the orange, at 620 nm, very close to the absorption peak of phycocyanin (615 nm). Chlorophyll was only 80% as effective at its maximum absorption range (670 nm) while the photosynthetic efficiency was down to only 40% of phycocyanin in the short end of the spectrum (475–480 nm, absorbed both by chlorophyll and carotenoids). These investigations were later confirmed in unicellular blue-green alga, *Chroococcus*, where the light absorbed by phycocyanin is used almost as efficiently as that absorbed by chlorophyll, while carotenoid absorption was strikingly ineffective. In marine *Oscillatoria* and *Anabaena* with C-phycocyanin, the photosynthetic efficiency was maximum only at 620 nm (corresponding almost exactly to the absorption peak of C-phycocyanin) falling away rapidly towards both chlorophyll and carotenoid absorption regions. *Phormidium*, a blue-green alga with abundant C-phycoerythrin in addition to the normal C-phycocyanin content, shows a shift of the photosynthetic maximum, understandably to the green. Thus in the blue-green algae, the situation is somewhat different from the red algae, for although the phycobilin is active, so also is chlorophyll, may be to a lesser extent. The carotenoids, however, seem singularly inactive with perhaps only 20% efficiency of phycobilins and chlorophylls.

There is no evidence, however, as yet that photosynthesis can proceed with the light absorbed by the accessory pigments when they *alone* are present in complete absence of chlorophyll *a* (or bacteriochlorophyll). So it must be supposed that the accessory pigments are able in some way to transfer the energy which they absorb entirely to chlorophyll or alternatively the phycobilins may have an entirely separate parallel system of enzymes, able to carry on photosynthesis independently of chlorophyll. Most evidence points to the first alternative of a common pathway. The most conclusive evidence for such a common pathway *via* chlorophyll is that of fluorescence. It has been demonstrated that in living diatoms, fluorescence was just as intense in blue-green light (absorbed by fucoxanthol) as in red light (absorbed by chlorophyll). Since photosynthetic activity was also high in such light, the conclusion is inevitable that the accessory pigment. fucoxanthol, passed its absorbed energy to chlorophyll efficiently. In the red algae, green light absorbed by phycoerythrin causes the fluorescence of chlorophyll much more efficiently than does blue light (absorbed by chlorophyll itself). Phycoerythrin thus excites chlorophyll-fluorescence more effectively than does chlorophyll itself; on the other hand, absorption of light by the chlorophyll and carotenoids does not seem to excite the fluorescence of phycoerythrin at all. So the energy transfer must be a one-way traffic, from accessory pigments to chlorophyll and never in the opposite direction. Such energy transfer mechanism is also definitely known to be operative in the photosynthetic purple bacteria between different bacteriochlorophyll types but always from pigment types absorbing light of relatively shorter wave-lengths, B800 nm, B820 nm and B850 nm to pigment types, B890 nm and B1020 nm-B1030 nm approx., absorbing light of relatively longer wave-lengths and never in the opposite direction. It has also been shown that absorption of light by phycoerythrin can excite a longer wave-length fluorescence characteristic of phycocyanin (red, 650 nm). The energy is apparently transferred from phycoerythrin to phycocyanin and probably from the latter to chlorophyll. If phycocyanin is the necessary mediator in the energy transfer mechanism, it is easily understandable why chlorophyll cannot excite phycoerythrin fluorescence while phycoerythrin apparently excites chlorophyll. Phycocyanin absorption lies at a longer wave-length than that of phycoerythrin. An indication that chlorophyll can be excited by accessory pigments as a necessary and normal part of photosynthesis is that in green light (absorbed by phycoerythrin, 550 nm), the *Katusky burst* of fluorescence, during the first moments of illumination, is restricted only to chlorophyll-fluorescence, the phycoerythrin-fluorescence is unchanged and steady. Since this fluorescence burst is accepted as closely related to the *induction period of photosynthesis* it clearly points to the active participation of chlorophyll in photosynthesis even when excited only indirectly *via* another pigment, e.g., phycoerythrin.

BIOCHEMICAL MECHANISM OF PHOTOSYNTHESIS

Ever since plant physiologists began to study the effects of various conditions on photosynthesis it became clear that the production of carbohydrate from water and CO_2 cannot be a simple process but must take place in a series of steps. For example, it was found by Emerson and Arnold (1932) that if light is supplied in short flashes separated by dark periods, a higher photosynthetic yield for a given amount of light is possible than if light of the same intensity is given in one uninterrupted period of illumination. This certainly suggests that photosynthesis consists of at least two separate types of reactions, one for which light is necessary and another which can proceed in the dark. When light is given continuously the rate of photosynthesis is limited by the dark reaction. In intermittent light the intervening dark periods can keep pace with the faster light reaction and thus a greater amount of photosynthesis is possible.

Emerson also showed that light reaction mechanism was saturated by brief flash of light (10^{-5} sec) and that up to 100 m/sec of dark period was then required to completely use up the products of the light reaction.

Thus we see that photosynthesis is at least a two-stage process involving two distinct types of reactions, one photochemical and the other chemical or dark reaction. What is the nature of these two processes?

The primary light reactions are exceedingly fast and are completed within 10^{-15} to 10^{-9} sec. This is essentially a physical process. The reactions which follow are consummated within 10^{-9} to 10^{-4} sec. The dark reactions which are chemical in nature proceed after this time. Energy is transmitted by inductive resonance or hole migration and singlet-triplet transition. The two electrons which are shared in every chemical bond spin in opposite directions (as the earth spinning round itself moves round the sun). Every electron makes in an orbit around the nucleus of the atom spinning round itself and may be considered to have three energy levels $-\frac{1}{2}$, 0 or $+\frac{1}{2}$. As light energy is absorbed and chlorophyll gets excited an enegetic electron is ejected. This can be detected in an ESR (electron spin resonance) instrument in the form of a signal; the molecule soon returns to its ground state and an electron is accepted from the surrounding environment. As an electron is ejected, a hole is developed in that part of the molecule, which is then filled by an electron coming from another part of the molecule, which then develops a hole. As this process continues the holes seem to migrate from one point to another due to movement of electrons. Chlorophyll is a resonating molecule when many single bonds alternate with double bonds and ejection of one electron from a single bond results in development of a deficit; the molecule tries to adjust with this situation by forming a double bond in this region resulting in a series of progressive shifts of double bonds and single bonds. This resonance may be induced within the same molecule or may involve adjacent molecules. The nature or the primary reactions, however, is not well-understood and more experimental evidence is required for a clear understanding.

PHOTOCHEMICAL REACTIONS

Purely photochemical reactions are influenced by the intensity of light and are independent of temperature effects. When the intensity of light is low and there is abundant supply of CO_2, the temperature coefficient of photosynthesis is 1, i.e., rate of photosynthesis is influenced only by the intensity of light. Purely photochemical reactions have also temperature coefficient of 1. This again proves that photochemical reactions are involved in the photosynthetic mechanism.

QUANTASOME. By dividing the number of chlorophyll molecules in a particular photosynthetic system by the number of CO_2 molecules fixed or reduced during the dark period the number of chlorophyll molecules engaged for the fixation of one molecule of CO_2 was calculated to be approximately 2500. This was termed *photosynthetic unit* (Emerson and Arnold).

Since it is generally accepted, that approximately 8-10 quanta of light energy are required for the reduction of one molecule of CO_2 in higher green plants, Kok (1956) assumed that ten (?) or eight "flashes" of light will normally be required for the same. Thus each individual "flash-light" unit would involve 2500/10, i.e., 250 chlorophyll molecules.

The morphological expression of the photosynthetic unit (is it a functional unit or just an architectural accident?) has been seen through electron microscope, on the inner surface of chloroplast lamellae, as small, oblate spheres, arranged in a linear, repeating pattern. These are the *quantasomes*, 20 nm in diameter and 10 nm thick. They are the elementary particles of photosynthesis (comparison with oxysomes of mitochondrial respiration). From actual experimental work, the detailed distribution of chemical substances in a quantasome was later obtained (Park and Biggins, 1964). It has, roughly, a molecular weight of about two million which in addition to the pigments, also contain proteins, glycerides, sulpholipids, sterols, and Cu, Fe (both as cytochromes and ferredoxin), manganese, etc.

The number of chlorophyll molecules actually found -230- surprisingly close to the number envisaged by Kok (1957), 250, certainly speaks for the correctness of the method employed. Of the 230 chlorophyll molecules, 160 were chlorophyll *a* and 70, chlorophyll *b*. There were 48 carotenoids.

Purple photosynthetic bacteria, however, contain only about 50 bacteriochlorophyll molecules in each of its chromatoplasmic lamellar quantasome.

Photosynthesis is now believed to depend on *two* separate photochemical reactions driven by two different sets of pigments. The existence of these two separate light reactions is deduced from the striking increase in photosynthesis obtained when two pigments both absorb light as contrasted with illumination of either pigment alone. The idea is now strongly favoured that the *photosystem* I which powered by light energy absorbed by *chlorophyll a* 683, C*a* 695 and the energy sink P700 is specifically concerned with the reduction of ferredoxin and ultimately causing electrons to flow, in the *photophosphorylation* of ATP molecules. The *photosystem* II which may be driven by accessory pigments, C*a* 673, C*b*, C*c*, C*d*, carotenoids and phycobilins is mainly concerned in inducing photolysis of water, i.e., oxygen evolution. In the anaerobic photosynthetic bacteria and green and also other algae adapted to molecular H_2, instead of H_2O, there is no photolysis of water (no O_2 evolution) and consequently, *photosystem* I is generally concerned here with the production of the reducing substance, NADPH, the electrons or hydrogen being obtained from readily oxidisable compounds (*photoxidation*) other than water.

The formulation of photosynthesis which is currently accepted by the majority of research workers in this area is a *series one*, in which the photosystems co-operate in *sequential manner*. In bacterial photosynthesis they may be connected in a parallel manner.

At present, it is true that reduced ferredoxin FdH is considered the first stable, isolable and recognisable chemical reductant formed in the energy conversion process. Very little is known as yet concerning the mechanism of electron transfer between excited (photo-activated) chlorophyll molecules in the grana and ferredoxin.

The resultant photosynthesis in higher green plants using specifically water as a reductant, is the evolution of O_2 and the assimilation of CO_2 into carbohydrate and then into cell materials at the expense of light energy. Under normal physiological conditions in nature, neither process is observed in the absence of the other.

Thus the major function served by photosynthesis is now generally accepted as being the production of ATP molecules through both non-cyclic and cyclic electron transfer, a process known as *photosynthetic phosphorylation*. The importance and the universality of photophosphorylation among the many types of photosynthetic species is now fully recognised. Some investigators feel that *photophosphorylation is the major and perhaps the only direct chemical consequence of light absorption.*

Formation of ATP in photosynthesis

As we have discussed before, photosynthetic phosphorylation demonstrated almost simultaneously by Arnon and Frenkel in chloroplasts and chromatophores respectively may be of two types—*non-cyclic* and *cyclic*. Of the two, there is no doubt, that cyclic phosphorylation is more primitive in an evolutionary sense. Cyclic phosphorylation is now known to occur specifically in all organisms where water is not utilised as a donor of electrons or hydrogen ions, i.e., in bacterial photosynthesis as also in photoreduction exhibited by many algae.

The array of pigments present in photosystem (or pigment system) *I* transfers the energy absorbed finally to P700 (a dimer of chlorophyll associated with water like molecules), which then gets excited and emits a highly emergetic electron, the potential of which is raised by about 1·5 volts, strongly increasing its reducing power (Fig. 708),

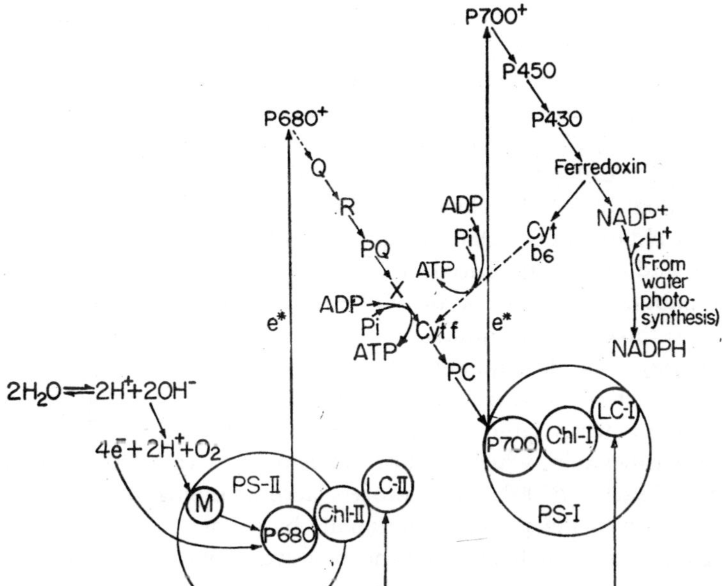

FIG. 708. A modified form of the Z scheme of Hill and Bendoll (based on Govindjee, 1980). PS, photosystem; Q and R, unidentified quinones; PQ, plastoquinones; PC, plastocyanin; P430, P450 and P700 form D chlorophyll -absorbing maximally at 430, 450 and 700 nm; X, unidentified Fe-centre; LC I & II, light-harvesting complexes; Cyt, cytochrome; Chl I & II, Chlorophylls; M, Mn-containing O_2-ending enzyme.

the electron moves down hill through P430 and P450 to ferredoxin, the redox potential of which is close to that of the hydrogen electrode (—0·42 volts at pH 7·0). Receiving this electron ferredoxin becomes a strong reducing agent and with the help of ferredoxin—NADP reductase reduces NADP+ which is then converted to NADPH (52 Kcal) using a H^+. The electron of reduced ferredoxin instead of being donated to NADP+ may also be passed on to such cytochromes like cytochrome b_6, cytochrome b559 and cytochrome f; this electron transport through the cytochrome system involves a drop in potential, which at one or two points is enough to form ATP from ADP and inorganic phosphate, storing about 7 Kcal. in the pyrophosphate bonds formed. The electron is then passed on to plastocyanin, which then transfers it to P700 which gets back the electron lost and thus its electrical neutrality is maintained. As the electron loses most of its energy in its down hill movement it is almost at its ground state, when it is received by P700. P700 again may be excited absorbing light energy and an electron may again be ejected and returned as before. This electron thus moves in a cycle producing ATP in each turn. Since ATP is formed without any net loss of electron, the P/e ratio of this process, known as *cyclic phosphorylation* is α. This type of phosphorylation is common in photosynthetic bacteria but exists in higher plants also.

In green plants another method of phosphorylation—*non-cyclic phosphrylation*—i.e., in which the electron supplies energy for ATP synthesis without returning to P700, is operative. Here the loss of electron from P700 is met by water through photosystem II. In photosystem II in addition to P680 (a monomer or chlorophyll bound to protein) and the light harvesting a complex including accessory pigments is associated with a few other participants in the photochemical reaction, e.g., the Mn-containing hypothetical O_2 evolving enzyme which also acts as a charge accumulator, and some compound having the properties of a quinone (the quinone and Mn may also be bound to a protein). Like P700, P680 absorbing light energy transmitted to it, emits an energetic electron, making P680 positively charged. This positive charge is neutralised by electrons liberated during the photolysis of water. The considerable energy required to split the O—H bond in the OH^- group is provided by light and from two molecules of water $4H^+$ and $4e^-$ are produced. The quinone present in photosystem II probably helps in the photolysis of water as the Hill reaction. The energetic electron shot from the P680 passes through three quinones—Q, R and plastoquinone as it moves down hill, losing energy; the plastoquinone passes on the electron to an iron centre. (* in Fig. 708) which then transfers it to cytochrome f. It is in this step that the potential drop is equivalent to an energy large enough to permit ATP synthesis. The electron is then passed on to plastocyanin, which in turn hands over to an electron donor bound to it before it finally reaches $P700^+$. The process accompanying phosphorylation here is non-cyclic, electrons being supplied continually by water.

Green plants may have both cyclic and non-cyclic phosphorylation. In photosynthetic bacteria instead of water electrons are donated by H_2S and organic acids like malate and succinate. Since water is not used as a source of electrons no O_2 is evolved. Here the *C* type of cytochromes play a major role and P890 is the important pigment involved.

In photophosphorylation 4 light quanta—two for each photosystem—provide the energy for the formation of one molecule of NADPH or ATP in non-cyclic phosphorylation. The energy stored in these high energy compounds is 52 Kcal and 7 Kcal respectively. If we consider quanta of red light (one einstein, i.e., Avogadro's number of quanta in the red region is 41 Kcal), then the energy provided is 164 Kcal. The energy efficiency works out to 36%. During CO_2 fixation at least 8 quanta provide the energy. Since 112 Kcal are consumed for fixation of one gram mole of CO_2, the efficiency will be of the order of 30%. However the number of sites of phosphorylation has not been conclusively proved. If more than one molecule of ATP is formed or two electrons transferred, then the efficiency would be higher.

Photophosphorylation can also be explained by the *Chemiosmosis* Theory of Mitchell. In this case ionization of water and the oxidation of $2OH^-$ to $2H^+$ and O_2 by light help in the removal of water. However in this case H^+ accumulates on the inside of the membranes of grana—just opposite to what happens in mitochondrial electron transport.

The final products of light reaction, ATP and NADPH, are utilised for the reduction of CO_2-fixation product (PGA) to triosephosphate sugars (G3P and DHAP), as will be described later.

Photolysis of water

As we have seen before, the electrons ejected from illuminated, excited chlorophyll molecule split up water. The end result of this photolysis of water is, in some ways, similar to the end result of the electrolysis of water for in both processes O_2 is released as gas. Chloroplast photolysis differs, however, from normally observed electrolysis in that hydrogen does not appear simultaneously with oxygen; in photolysis, the hydrogen, instead of being released as gas, is used to reduce a series of compounds present in the chloroplast until finally it is taken over by NADP which gets reduced in the process.

The OH of water, *a*, is split into highly reactive components. The reduced NADP, is of high potential chemical energy and is used for reduction of CO_2-fixation product in the subsequent dark reactions. Many compounds, such as cyanides previously considered to inhibit the photochemical reaction of photosynthesis was also found to inhibit the production of hydrogen donor by chloroplasts in light. These facts suggest that the hydrogen donor formation in the reaction of chloroplasts may be identical with the primary photochemical reaction of photosynthesis, the difference being that in the latter case, the hydrogen donor formed ultimately reduces the CO_2-fixation product to carbohydrates. As no direct reduction of CO_2 can result in the photolysis of water by isolated chloroplasts, this reaction can only represent one-half of photosynthesis and only one aspect of it, i.e., evolution of O_2.

$$2H_2O + 2X \xrightarrow[\text{chlorophyll}]{\text{light energy}} 2H_2X + O_2$$

This formulation suggests that photolysis by illuminated chloroplasts will go on as long as there is a supply of an electron or hydrogen acceptor X; when all the X initially present in the chloroplast has been reduced, photolysis would stop. This agrees with experimental observation; oxygen evolution from illuminated, isolated chloroplasts soon stops but can be restarted if some compound capable of accepting hydrogen is added to the chloroplast suspension. Ferric iron has been successfully used as such an electron acceptor. Robin Hill (1937) found that ferrocyanide added to an illuminated chloroplast suspension was reduced to ferrocyanide while a corresponding amount of oxygen was released (*Hill reaction*). Many other compounds, e.g., acetaldehyde, benzaldehyde, quinone, etc., have been successfully used as oxidants in the Hill reaction, as acceptors of hydrogen in the photolytic decomposition of water, getting reduced in the process.

In 1954, Arnon developing the basic idea of Hill reaction, was able to demonstrate that isolated chloroplasts, not only carried out Hill reaction, but were also capable of ATP synthesis, from inorganic phosphate supplied externally, as for example:

$$2Fe^{+++} + H_2O + ADP + Pi \longrightarrow 2Fe^{++} + ATP + O + 2H^+$$

Later on it was found that as attention was given more and more towards obtaining pure chloroplast extracts, such very pure extracts when suspended in water failed to produce any oxygen on addition of ferricyanides or other Hill oxidants. The oxygen evolution, however, could be restarted if salts of potassium, e.g., KCl were added to the chloroplast suspensions. This certainly points to the fact that chloride forms an essential constituent of the photosynthetic apparatus in the grana, which (chloride) being water soluble gets washed out during the process of purification (general practice of purification of chloroplast preparations nowadays consists in washing the extracts with salt solution so that the isolated chloroplast may retain their full Hill-reaction activity).

That all the oxygen evolved in photosynthesis came from photolytic decomposition of water and perhaps none from CO_2 was proved by Reuben, Randall, Kamen and Hyde (1941) working with heavy oxygen isotope ^{18}O.

Further work in this line, however, seemed to support conclusion to the extent that at least a *major part* (at least 85%) and perhaps not all of oxygen evolved in photosynthesis is derived from water. A conclusion about all the oxygen cannot be made because of uncertainty about the extent of exchange reactions involving ^{18}O, for it is known that this rate of exchange increases with decrease in pH, i.e., with increasing acidity. At pH=6, the rate of exchange between oxygen from water and the oxygen from CO_2 is fast enough to invalidate the claim that none of the oxygen evolved in photosynthesis comes from CO_2. And the cell sap conceivably may have pH values below 6.

Any how, we are still inclined to believe, that is, normal photosynthesis in higher green plants with water as the specific hydrogen donor, in illuminated cells, occasions the reduction of CO_2 according to the overall equation:

$$CO_2+2H_2O \longrightarrow CH_2O+H_2O+O_2$$

The insertion in this equation of a second molecule of water apparently introduced *ad hoc*, marks the turning point between the classical conception of photosynthesis as a unique process of CO_2 decomposition by light and the modern idea that it should be looked upon as one of a group of similar oxido-reduction processes all following the same general formula:

$$CO_2+2H_2A \longrightarrow CH_2O+H_2O+2A$$

and only when this A is oxygen, we have the normal green plant photosynthesis. This A can, however, be S or any other sulphur-containing inorganic radicle or it may even be an organic radicle. Implicit in the view expressed above is that oxygen liberated from green cells in normal photosynthesis originates, at least, primarily from water consumed in the process and that the partial reaction in which CO_2 is fixed is distinct from that in which oxygen is liberated. Before considering the transfer of hydrogen and the reduction of CO_2, something must be said about how oxygen is actually formed in the oxidation of water.

The hypothesis which emerges from the previous discussions is that: (*a*) the main light reaction in photosynthesis is the production of reduced ferrodoxin and oxidised OH^-, (*b*) oxygen is liberated from the oxidised fragment OH^- or from a derivative of it, and (*c*) CO_2 gas from the atmosphere enters the green mesophyll cells of the leaves and combines with an acceptor substance, already present in the green cells, in a dark reaction and which can proceed continually only in the presence of a reducing agent NADPH and ATP which are derived from the reduced fragment of the water molecule.

Evolution of O_2 in photosynthesis

We are completely ignorant concerning the enzymatic events associated with the process of O_2 evolution from OH^- ions released from water—one of the least understood portions of photosynthesis concerns these reactions leading to the formation of O_2 by green illuminated cells adapted to water as a reductant.

It is true that hydroxyl ion, OH^-, at one time was assumed to combine to form inorganic peroxide (H_2O_2) or an organic peroxide. This peroxide in presence of haeme enzyme, *catalase*, was supposed to break up in one or more dark reactions into water with oxygen evolution.

$$4(OH) \dashrightarrow 2H_2O_2 \xrightarrow{\textit{Fe-catalase}} 2H_2O + O_2$$

But this was one of the earliest assumptions which has *completely negatived by the observation of cyanide-insensitivity of O_2 evolution in isolated chloroplasts.*

The possible involvement of *manganese* as an activator of an unidentified enzyme (photoenzyme) as yet and a requirement for chloride are, however, no longer in doubt. An FMN-mediated reaction may also be functional.

Hydrogen transfer mechanism

All evidences thus strongly indicate that nicotinamide nucleotides such as NADP act as the ultimate hydrogen donor for the reduction of CO_2-fixation product. The reduced ferredoxin which subsequently transfers hydrogen to NADP and thereby provides a pathway for the formation of high-energy reduced NADPH, one of the two most active agents produced in the photochemical reaction. The energy level of these reduced coenzymes is not, however, sufficient to accomplish the reduction of CO_2 to carbohydrate level in photosynthesis and as we know, additional energy required must come from high-energy phosphate, ATP. Thus the ultimate products of the photochemical reaction are the active agents, e.g., NADPH and high-energy phosphate, ATP available for the reduction of carbon-dioxide-fixation product and also for supplying the fuel or energy requirements of photosynthesis. The continuous regeneration of NADPH and ATP are essential for the completion of the photosynthetic cycle.

THE REDUCTION OF CARBON DIOXIDE

Fixation of CO_2. Experiments in which algae, mostly *Chlorella*, were permitted to photosynthesise for short periods in the presence of isotopic $^{14}CO_2$ established beyond all doubt that the three-carbon *phosphoglyceric acid* is the first product of photosynthesis. Within two seconds, almost 90% of radio-active ^{14}C could be traced in the three-carbon acid. This is undoubtedly the fundamental carboxylation reaction in the green cells. The term 'first product' here means that the earliest, stable, tangible organic compound to contain a carbon atom derived from CO_2 is phosphoglyceric acid. A pentose-phosphate ester—*ribulose* (a ketonic derivative of pentose sugar, ribose) *diphosphate*—has been established definitely as the primary acceptor of CO_2 in photosynthetic dark reaction.

$$\underset{\text{(RuDP)}}{\text{ribulose diphosphate}} + CO_2 \xrightarrow[\text{Mg}^{++}]{\textit{RuDP-carboxylase}} \underset{\text{3-PGA}}{\text{2 molecules of phosphoglyceric acid}}$$

FORMATION OF CARBOHYDRATES IN PHOTOSYNTHESIS

The phosphoglyceric acid formed in the dark reaction when ribulose dipohsphate combines with CO_2, reacts with NADPH and ATP, the products of the photochemical reaction of the decomposition of water to form the *triose-sugar phosphates*, phosphoglyceric aldehyde and also its ketoisomer, dihydroxyacetone phosphate, the ratio of their formation being 60 : 40. This is dependent on the light reaction for there must be a continuous supply of NADPH and ATP for the reduction of phosphoglyceric acid to phosphoglyceric aldehyde (the triose-sugar) *via* diphosphoglyceric acid (ATP is specifically required for the formation of diphosphoglyceric acid from phosphoglyceric acid).

$$\text{ribulose diphosphate} + CO_2 \underset{}{\overset{\textit{RuDP-carboxylase (carboxydismutase)}}{\rightleftharpoons}} \underset{\text{3-PGA}}{2 \text{ phosphoglyceric acid}} \quad (i)$$

$$2 \text{ phosphoglyceric acid} + 2ATP \overset{\textit{phosphoglycerokinase}}{\rightleftharpoons} \underset{\text{1, 3-DGPA}}{2 \text{ diphosphoglyceric acid}} + 2ADP \quad (ii)$$

$$2 \text{ diphosphoglyceric acid} + 2NADPH \overset{\textit{triose-phosphate dehydrogenase}}{\rightleftharpoons} \underset{(\textit{triose-sugar} \text{ phosphates})}{2\text{glyceraldehyde-3-phosphate(G3P)}} + 2NADP + 2H_3PO \quad (iii)$$

It is evident that the photosynthetic cycle continually needs hydrogen from the photolytic breakdown of water to keep the cycle going because the cycle is essentially a reductive one and hydrogen as well as CO_2 must continuously be introduced into it.

It is also evident that in the reduction of diphosphoglyceric acid to phosphoglyceric aldehyde, hydrogen or electrons ultimately derived from a suitable reductant (NADPH) is transferred to carbon, derived from CO_2.

Strong evidences support this conclusion: if photosynthetic period, using $^{14}CO_2$ is extended to one and half minutes, most of the radio-active carbon can be traced occupying the third and fourth carbon positions of a hexose molecule formed.

The sequence of changes for the continual cyclic regeneration of ribulose diphosphate and the synthesis of carbohydrates as also proteins and fats have been diagrammatically represented in Fig. 709.

Photosynthetic carbon reduction cycle (*PSCR cycle*)

The photosynthetic carbon reduction cycle is essentially a *reductive* pentose-phosphate cycle. The discovery of pentose-phosphate pathway (direct oxidation pathway) of aerobic respiration in plant tissue during the last decade has given ample support to the suggested series of reactions, culminating in the formation of RuDP. The validity of the PSCR cycle as proposed by Calvin and Bassham seems no longer in doubt but another alternative pathway of carbon traffic is operative in certain tropical and subtropical plants.

Calvin and Bassham's cycle as shown in Fig. 709 proposes linking of the cycle with all metabolic synthesis, leading ultimately to synthesis of carbohydrates, proteins, fats and other products.

In PSCR cycle, as will be seen from the accompanying diagram, the first stable carboxylation product is 3-phosphoglyceric acid. The remaining stable intermediates are all sugar phosphates and diphosphates with carbon skeletons from 3-7 carbon atoms in length. In the simplest terms, the PSCR cycle consists mainly of three stages: (*i*) RuDP

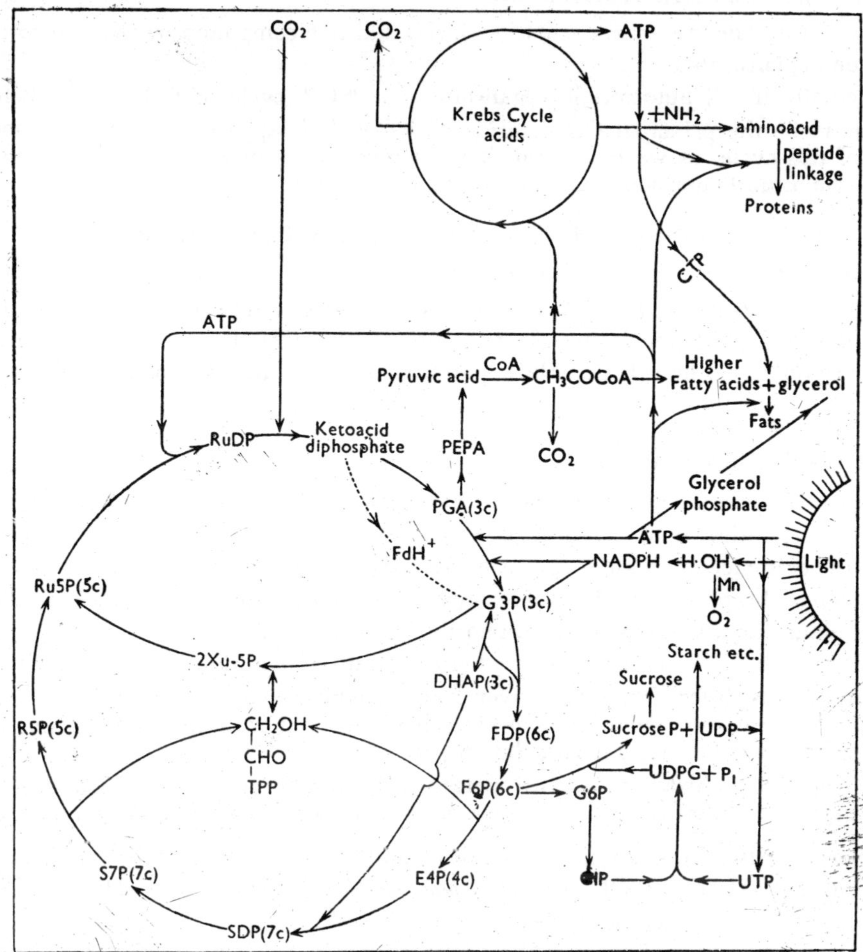

FIG. 709. The Calvin Benson photosynthetic carbon reduction cycle (*PSCR Cycle*). Abbreviations include: PGA, phosphoglyceric acid; G3P, glyceraldehyde-3-phosphate; DHAP, dihydroxyacetone phosphate, FDP, fructose-1, 6-diphosphate; F6P, fructose-6-monophosphate; G6P, glucose-6-monophosphate; G1P, glucose-1-monophosphate; TPP, thiamine pyrophosphate and $CH_2OH.CHO$. TPP, glycolaldehyde thiamine pyrophosphate additive compound; E4P, erythrose-4-phosphate; SDP, sedoheptulose-1, 7-diphosphate; S7P, sedopeptulose-7-monophosphate; R5P, ribose-5-phosphate; RuDP, ribulose-1, 5-diphosphate; Ru5P, ribulose-5-monophosphate, Xu-5P, xylulose-5-phosphate; FdH, reduced ferredoxin; PEPA, phosphoenolpyruvic acid; CH_3COCoA, acetyl coenzyme A; NADPH, reduced nicotinamide adenine dinucleotide phosphate; UDP and UTP, uridine diphosphate and triphosphate; UDPG, uridine diphosphate glucose. Several reactions are reversible and the rates of the reactions may not be equal, leading to accumulation of some of the products in varying degrees.

is carboxylated, as we have seen before, by the enzyme, *carboxydismutase* or *RuDP carboxylase* to give two molecules of PGA; (*ii*) PGA is reduced through light-reaction products, ATP and NADPH to triosephosphates (G3P and DHAP) and (*iii*) the pentose phosphate molecules are phosphorylated by ATP to give 1,5-RuDP. Carbon which enters the cycle as CO_2 from the atmosphere is later drained off as reduced carbon in the form of sugar phosphates. These pathways may easily lead to the synthesis of cell constituents.

The part played by thiamine pyrophosphate (TPP) glycolaldehyde (2 carbon) additive compound formed during the *transketolase* reactions of the cycle is accepted beyond doubt. The enzymatic oxidation of this additive compound by electron carriers has been demonstrated.

Actually in all autotrophic organisms, the PSCR cycle of Calvin and Benson appears to be the principal CO_2-fixation pathway. It is apparent that for the operaation of this cycle, as we have seen, both energy and reducing power are required. Hence the overall equation for photosynthesis would be:

$$6CO_2+18ATP+12NADPH+12H^+ \longrightarrow C_6H_{12}O_6+18ADP+18Pi+12NADP+6H_2O$$

Thus the Calvin cycle requires 3ATP molecules for the reduction of one molecule of CO_2, *one* to form RuDP from Ru5P and *two* molecules of ATP to convert the 2 PGA molecules into two triosephosphates, 60% G3P+40% DHAP, one also can easily be converted to the other. Hatch-Slack pathway requires an additional ATP, i.e., 4 molecules.

Hatch-Slack-Kortschak pathway

C_4-DICARBOXYLIC ACID PATHWAY. It has been realised for a long time that some major uncertainties about Calvin-Benson cycle were still unsolved and thus the possibility of the existence of an alternative carboxylation reaction (Zelitch, 1965) has always been there. The proposals, that there might be *more than one path* of carbon traffic in higher plant photosynthesis, may be unorthodox, but not unexpected.

Hatch and Slack first demonstrated (1966) that the photosynthetic carbon-fixation in many plants differs from the conventional Calvin cycle. As a result, a new modification of the classic Calvin-Benson cycle, has emerged with the development of 4-carbon dicarboxylic acid, β-carboxylation pathway.

The HSK-pathway is said to operate in a number of monocotyledonous tropical grasses of *Graminae* and *Cyperaceae*, and also spectacularly in sugarcane leaves; it is operative also in certain dicotyledonous plants, e.g., *Amarantaceae* (*Amaranthus edulis*) (Johnson and Hatch, 1968), certain *Chenopodiaceae* (Johnson and Hatch, 1968), *Portulacaceae* (Hatch and Slack, 1971), etc.

Another interesting feature of these HSK-plants is that these species are extremely efficient in photosynthesis and *photorespiration* is difficult to detect in this group of plants. These plants are referred to as *low* compensation species in light, because of their capacity to deplete the ambient or surrounding atmosphere to very low CO_2 concentrations.

The essentials of the sequence of this pathway is that the primary carboxylation reaction is catalysed by *phosphoenolpyruvate carboxylase*, so that the first product would be oxalacetate (Hatch, 1967):

$$PEP+CO_2+H_2O \xrightarrow{\textit{PEP-carboxylase}} \text{oxalacetate}+\text{orthophosphate}$$

The oxalacetate formed may easily give rise to malate or aspartate and may also serve as a substrate for transcarboxylation, donating energy-bound CO_2, to an appropriate receptor, such as RuDP. The overall reaction may be written as follows:

$$\underset{\text{(or malate)}}{RuDP+\text{oxalacetate}}+H_2O \longrightarrow 2\ 3\text{-}PGA+\text{Pyruvate}$$

PEP might then be regenerated from pyruvate as follows:

$$\text{pyruvate}+\text{orthophosphate}+ATP \longrightarrow PEP+P-P+AMP$$

As it will be evident, that for Hatch-Slack-pathway there must be provision for an **extra ATP, i.e., 4 in all, compared to classical cycle requiring, as we have seen, only 3.** Work with $^{14}CO_2$ has proved the validity of the HSK-pathway about the intermediates formed, the 4-carbon dicarboxylic acids.

Thus, the overall equation for the synthesis of one mole of hexose sugar, following HSK-pathway, may be written as:

$$6CO_2+24ATP+12NADH+12H^+ \longrightarrow C_6H_{12}O_6+24ADP+24Pi+12NAD+6H_2O$$

It has also been found that HSK-plants contain rather low levels of *RuDP-carboxylase* but high levels of *PEP-carboxylase*.

The primary evidence of the operation of HSK-pathway was obtained by Kortschak (1965) and Hatch (1967) from experiments with sugarcane leaves, exposed to $^{14}CO_2$ after obtaining a steady state of photosynthesis in ordinary $^{12}CO_2$. The products, after approximately 1 sec., were principally aspartate (93%), malate and oxalacetate rather than phosphoglycerates and sugar-phosphates.

Evidence is also available, which showed that radio-activity, incorporated into these C_4-dicarboxylic acid pool is transferred to sugars *via* 3-PGA. The Calvin cycle intermediates appeared, however, only when the exposure time was prolonged. After longer periods, large proportions of radio-activity appeared in 3-PGA, hexose monophosphates and sucrose; the labelling pattern in hexoses were consistent with their formation from 3-PGA. The reaction giving rise to C_4-dicarboxylic acid appears to be the only quantitatively significant carboxylation reaction (Hatch and Slack, 1970).

There are important anatomical differences between the two types of plants. Low compensation *Graminae*-HSK-plants have dense, well-developed bundle sheath cells which are almost entirely occupied by large specialised chloroplasts—these cells at maturity contain very few grana or none at all and the thylakoids are essentially parallel with little tendency for adhesion. These chloroplasts contain large amounts of starch, whereas the dimorphic form of smaller, less numerous chloroplasts in the surrounding mesophyll, contain distinct grana but *very little starch.* These features are in contrast to Benson-Calvin plants (high compensation species) and other monocotyledonous, e.g., wheat plants, in which bundle sheath is poorly defined and the mesophyll cells contain **starch. Low compensation HSK-species of the Dicots, e.g., *Amaranthus* and *Atriplex*, have leaf anatomy and cell structure similar to low compensation *Graminae* although the large chloroplasts of bundle sheath *cells do contain grana.***

Mitochondria of the bundle sheath cells in HSK-plants are larger than those of the surrounding palisade cells (Downton, 1969).

It must be pointed out here that in the archtype Calvin-plant, like *Chlorella*, on illumination in air containing $^{14}CO_2$, malate, citrate and aspartate are found to be prominent among the radioactive fixation products.

How such di- and tri-carboxylic acids are actually produced in C_3 plants has not been adequately explained. The rate of CO_2 fixation via PEP carboxylase in a C_4 plant is much faster than that mediated by RuDP carboxylase in a C_3 plant. The anatomical differentiation further increases the efficiency of the process.

The similarity between the early CO_2 fixation reactions of a C_4 plant and the dark CO_2 fixation reactions discussed later particularly in relation to CAM plants is also noteworthy. In Crassulacean plants the stomata are usually closed **during the day and open at night. The CO_2 is fixed into malate (via oxalacetate) in dark, which is broken down in light to pyruvate and CO_2 which is then refixed via**

carboxylation of RuDP producing phosphoglyceric acid. Many non-Crassulacean plants also fix CO_2 in darkness and the major products of fixation are malate, citrate and aspartate, produced from oxalacetate. PEP formation may be accelerated in presence of a higher level of ATP arising via photophosphorylation. But the PEP carboxylase of non-Crassulacean plants may not be a chloroplastic enzyme.

The C_2 fragment carried by thiamine prophosphate from transketolase mediated reaction may be **converted** to glycollate which may be metabolised to glycine. Two molecules of glycine may give rise to serine. These reactions take place on the peroxisomes, minute particles present in the cytoplasm, concerned with photorespiration.

Synthesis of sucrose

The possible pathway of formation of sucrose and polysaccharides, as also proteins and fats, from the intermediates of PSCR cycle has been schematically represented in the diagram (FIG. 709). Uridine triphosphate (UTP), formed from ATP, is thought to react with GIP (*Cori ester*) obtained from G6P (*Robinson ester*) to form UDPG (uridine diphosphate glucose). UDPG in presence of *sucrose phosphorylase* reacts with F6P to form sucrose-phosphate and UDP. The sucrose-phosphate is then hydrolysed to sucrose and inorganic phosphate by *phosphatase* action.

The biosynthesis of sucrose may also occur by a slightly different pathway: by transfer of glucose from UDP-glucose to fructose by *sucrose synthetase* enzyme action:

$$\text{UDP-glucose} + \text{fructose} \xrightarrow{\textit{sucrose synthetase}} \text{sucrose} + \text{UDP}$$

For starch and cellulose synthesis adenosine diphosphoglucose (ADPG) may also be used.

The probable pathways of amino acid (protein) and fat synthesis can also be seen from the diagram.

FACTORS AFFECTING PHOTOSYNTHESIS

The process of photosynthesis is conditioned by a number of different factors, some external others internal. Four environmental factors are of primary importance in influencing the rate of photosynthesis: (1) light, (2) concentration of CO_2 in the atmosphere, (3) temperature and (4) supply of water. Other environmental factors which influence the rate of the process include: (*a*) oxygen, (*b*) various chemicals, (*c*) wounding or injury and (*d*) supply of mineral salts. The latter factors do not greatly influence the plant under natural conditions but they can be shown to be effective in laboratory experiments.

A number of internal conditions which also affect the process can be cited: (*i*) chlorophyll content of the assimilating surface, (*ii*) hydration of the protoplasm, (*iii*) **anatomy of the leaf, (*iv*) photosynthetic enzymes, coenzymes (e.g., NADPH, NADH, etc.) and ATP and (*v*) accumulation of the products of photosynthesis within the cell.**

(1) *Light.* Photosynthesis means synthesis in presence of light. So without light there cannot be extensive synthesis of carbon compounds in plants except in chemosynthetic bacteria, even if all other environmental conditions are favourable, i.e., not limiting. Under suitable conditions of temperature and CO_2 supply, the amount of photosynthesis, in a given time is proportional to the total amount of incident light or in other words, the total photosynthesis is proportional to the product of light intensity and the time of illumination. In intermittent light when brief light and dark

periods alternate with each other—though the total amount of light reaching the green cells is reduced, rate of photosynthesis is not correspondingly reduced and the efficiency of the process, measured as the amount of total photosynthesis by a given amount of light, actually increases. This must point to the only conclusion, as we have seen before, that the photochemical or photoactivation process is far more rapid than the subsequent temperature-controlled chemical reactions. If intermittent light is only supplied, the short period of light is sufficient for gathering enough energy or accumulation of sufficient reducing power to keep the dark chemical reaction going for an appreciable period. The chemical reaction will, of course, diminish with time for when the light is off, the supply of the product of the photochemical reaction will ultimately show depletion. Under conditions of continuous illumination, the product of the photochemical reaction simply builds up to a steady concentration and the rate of total photosynthesis is determined by the rate of the dark or chemical reaction. Light supplied in excess of the amount needed to produce this steady concentration is, if absorbed, uselessly dissipated as heat. These considerations on rate-limiting factors leave no doubt that photosynthesis is at least a two-stage process, involving two distinct types of reactions.

(2) *Concentration of CO_2*. The curve, expressing the relation between the rate of photosynthesis and the concentration of CO_2 when all other external factors are kept constant, seems to consist of 3 distinct parts (see Fig. 710), the ascending part of the curve showing proportionate increase in the rates of photosynthesis with increasing concentrations of CO_2. The relationship is linear. In the second part, the line merges with or without a sharp break into a part approximately parallel to the axis of CO_2 concentration when other factors such as light, rate of stomatal diffusion of CO_2, etc., are limiting the process; and in the third part, the curve may pass over into a descending part (not shown in Fig. 710), due to increasing CO_2-concentration, by its toxic effect, lowering the rate of photosynthesis.

(3) *Temperature.* In the previous chapter, we have seen that some temperature-controlled reactions are operative in the photosynthetic mechanism. All chemical reactions are, as is well known, independent of the intensity of light and influenced only by temperature. It can take place either in light or in darkness and following *Vant Hoff's rule* every 10°C. rise in temperature brings about a two-fold increase in the velocity of a chemical reaction, i.e., $Q_{10}=2$. Over a range of about 10–25°C., if light intensity and CO_2-concentration are relatively high, temperature coefficient of photosynthesis actually lies between 2 and 3. Above 25°C., however, this relation does not hold good, probably due to partial inactivation of enzymes. So, provided that no other factor is limiting, the rate of photosynthesis increases with an increase in temperature up to 25°C. This point was first clearly elucidated by F. F. Blackman—so the chemical reaction is sometimes also called *Blackman reaction.*

(4) *Supply of water.* One of the reactants of photosynthesis is water—the photolysis of which sets free the reducing hydrogen atoms essential for the reduction of CO_2 to carbohydrates and storage of solar energy. But the amount of water needed for a supply of hydrogen atoms must be very small indeed and seldom limiting for subsequent reduction under normal circumstances. The secondary effect of scarcity of water may influence photosynthesis by its effect on the closing of stomata, thereby reducing the rate of diffusion of CO_2 in the green mesophyll cells. Indeed, if stomata completely close due to scarcity of water in the leaf, it can conceivably limit the photosynthetic process.

THE PRINCIPLE OF LIMITING FACTORS

If we consider the environmental factors, there seems to be a *minimum*, *optimum* and *maximum* for each relative to photosynthesis.

The first important step in the clarification of the influence of various factors upon photosynthesis was taken when Blackman (1905) enunciated the *principle of limiting factors*. This is essentially an elaboration of the law of minimum which states that the rate of a physiological process is limited by the factor which is present in the relative minimum. The quantitatively smallest factor does not necessarily condition the rate of the process, since it may be necessary only in traces, while larger amounts of some other factor or a greater intensity may be necessary. To illustrate this, suppose we assume that 10 units of *a*, 2 units of *b*, and 1 unit of *c* are necessary for the formation of 1 unit of *d*. If only 5 units of *a* are available, none of *d* can be formed regardless of the quantities of *b* and *c* available. Although *c* is in absolute minimum here, *a* acts as a limiting factor controlling the formation of *d*.

Blackman's law of limiting factor, as enunciated by its author states 'when a process is conditioned as to its rapidity by a number of separate factors, the rate of the process is limited by the pace of the slowest factor'.

The explanation of this principle can be presented in terms of an illustration given by Blackman (Fig. 710):

If we assume that the intensity of light, under a particular circumstance, is just high enough to permit a leaf to utilise 5 ml of CO_2 in an hour, the rate of photosynthesis is limited by the concentration of CO_2, i.e., the supply of CO_2. As the supply of CO_2 is increased, the rate of photosynthesis is also increased until 5 ml of CO_2 enter the leaf per hour. Supply of CO_2 has been the limiting factor till now and the rate of photosynthesis can only be increased by increasing the supply of CO_2. But once the supply of CO_2 reaches 5 ml per hour, any further increase will have no influence on the rate of photosynthesis for the light energy available is insufficient for utilisation of more than 5 ml of CO_2. Light has now become the limiting factor and further increase in the rate of photosynthesis can only be brought about by an increase in the intensity of light. Thus we see that a progressive increase in the rate of the process takes place with a quantitative increase in one of the factors (CO_2) until the point is reached at which some other factor becomes limiting, such as light here. At this point, the rate of increase stops abruptly, and the rate of photosynthesis becomes constant, parallel with the increasing CO_2-concentration axis.

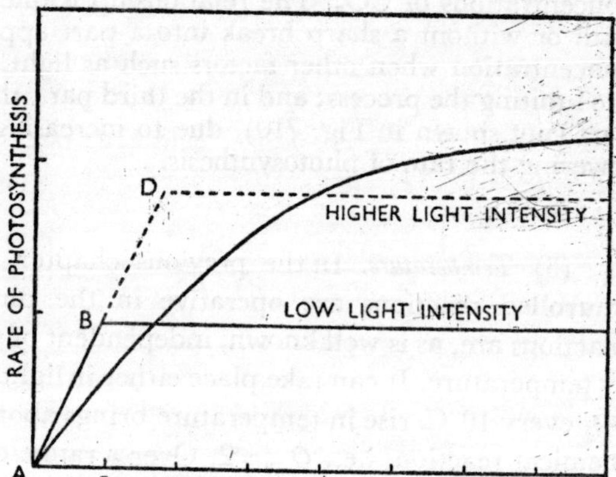

FIG. 710. Curves to illustrate Blackman's conception of the operation of law of limiting factors. The black smooth transition curves are those obtained by some later investigators (Harder and others).

If the light intensity is sufficient to permit the leaf to utilise 10 ml of CO_2 per hour, the rate will again increase with increase in CO_2 supply (CO_2 limiting factor again) up to a value twice as great as that at which the maximum rate of photosynthesis was attained at the low light intensity. Light, then, again becomes the limiting factor.

Light and CO_2-concentrations are not the only pair of factors which can be limiting in the photosynthetic process. Theoretically any one of the factors, which influences this process, can under certain conditions, become limiting.

If the law of limiting factors is a correct proposition, it is imperative that there must be a *sharp break in the curve* (see Fig. 710) as soon as one factor (CO_2 supply) ceases to be limiting (at points B and D in the curve) and the other of the pair, becomes limiting (light). With aquatics, some investigators obtained typical limiting factor curves as in the figure shown here but others failed to find any sharp break in the curve. The later investigators found *smooth curves*, without a sharp break, and that certainly speaks against the operation of the law of limiting factors, as there was a considerable range during which the rate of photosynthesis was neither directly proportional to nor independent of the CO_2-concentration. If the curves are actually smooth, that certainly points to the conclusion that the two factors, light intensity and CO_2-concentration are influencing the photosynthetic process simultaneously and the law of limiting factors cannot be strictly operative or hold good, in the Blackman sense.

The explanation of the discrepancy between Blackman's results and others, who found smooth curves, is not very clear. In Blackman's experiments, the range of the curve, where the rate of photosynthesis was directly proportional to CO_2-concentration, i.e., limited by a single factor (Fig. 710, AB and BD), it is more logical to refer to these regions as limited by a limiting process and not by a limiting factor. In the regions BC and DE where the rate is independent of CO_2-concentration, the rate is determined and limited, not only by light but also by all the factors affecting the rate of consumption of CO_2, e.g., temperature and possibly the chlorophyll content of the leaf.

The present position of the theory of the limiting factors is that it should not be taken as an unalterable law and that Blackman himself only put forward the theory as an *approximation* and never intended it to be rigidly followed. In the absence of any other theory, which can satisfactorily explain the reasons associated with smooth curves, actually obtained in many investigations, the general principle of limiting factors perhaps cannot yet be replaced altogether.

Chlorophyll content of the leaf

The effect of concentration of chlorophyll on the rate of photosynthesis of the green plant was investigated by Willstatter and Stoll. They showed that as a result of photosynthesis, there was no significant change in the concentration of chlorophyll in the plant. They also attempted to determine the maximal rate of photosynthesis in mg CO_2 per hour/mg chlorophyll—a quantity they called *assimilation number*. A considerable range in the assimilation number values was found among different species of plants. This indicated that photosynthetic rate was not proportional, to the amount of chlorophyll present in the tissue. Yellow varieties of leaves were found to have higher assimilation numbers than leaves of the corresponding green varieties. In green leaves, during the course of a season (say from June to October) it was found that as the leaf aged, the chlorophyll concentration increased but the increase of the rate of photosynthesis was of much lesser extent. It appears, therefore, that chlorophyll content is seldom the limiting factor of the photosynthetic process. Some recent researches, however, indicate that in *weak light* when photosynthesis is proportional to the light intensity, increasing chlorophyll concentration may lead to an increase in the rate of photosynthesis up to a point; any further increase in chlorophyll content actually depresses the rate. Is there then a critical concentration of chlorophyll, which may vary from plant to plant?

BACTERIAL PHOTOSYNTHESIS AND CHEMOSYNTHESIS

Until the end of the 19th century, there appeared to be a sharp distinction between green plants on the one hand which were able to develop in the light on a completely inorganic medium—the *autotrophs*, and all other plants together with animals, which required a supply of already prepared organic substance for growth—the *heterotrophs*. Prior to the discovery of the existence of photosynthetic bacteria, the synthesis of organic matter by the green plants appeared to be a unique process, unrelated to all other biochemical reactions in living organisms. The simple distinction between autotrophs and heterotrophs thus indicated, has become less useful as a result of the discovery of the wide range of biochemical activity exhibited by bacteria.

In 1887, Winogradsky discovered a group of organisms which could assimilate CO_2 in the dark. This they could do only in presence of H_2S and as a result of the reaction, elementary sulphur in the form of globules was deposited *within* the bacterial cells; in the absence of further supply of sulphide, the deposited sulphur disappeared and sulphate was excreted into the medium and eventually the bacteria died. A few years later, Winogradsky discovered a different group of organisms which grew only in the presence of CO_2 and NH_3 or nitrite. Both groups of organisms, were colourless and although they grew in the dark free of organic matter, were autotrophic. The energy necessary for the reduction of CO_2 was obtained from simultaneous oxidation of simple inorganic chemical compounds; such as H_2S or NH_3 or nitrite, a type of metabolism described as *chemosynthetic* or *chemautotrophic* or *chemolithotrophic*. Prior to Winogradsky's discovery of colourless chemosynthetic bacteria utilising sulphur compounds, Engelmann in 1883 had described pigmented bacteria which assimilated CO_2 in the presence of sulphide or other simple inorganic sulphur compounds, but which apparently needed light. In subsequent work with these bacteria of Engelmann, several difficulties arose, since it was difficult to understand why, if the bacteria were photosynthetic, they should also require a supply of sulphide; furthermore, in the light no oxygen production could be detected and indeed the cells grew only under strictly anaerobic conditions—the bacteria were obligate anaerobes. Another striking point was that it was found that some of these bacteria apparently could grow and develop in complete darkness if they were exposed to infra-red radiations. The difficulty was soon settled for the pigment (bacteriochlorophyll) isolated from the bacteria was found to be very similar to chlorophyll *a* but with its maximum absorption of light in the infra-red (800-890 nm and beyond). For the first time, organisms were found to be capable of utilising invisible light for photosynthesis. Since these bacteria required sulphide for growth there was naturally a tendency to suppose that they were related to the sulphur-utilising organisms of Winogradsky but differed in the presence of pigment, imparting a purple colouration, and in the necessity of light, visible or invisible, for growth and development. Later, yet other purple non-sulphur bacteria were found which required both light energy and a supply of organic substances for growth. In these organisms, light is apparently necessary for assimilation of the organic substance; and these bacteria, requiring organic nutrients but capable of assimilating them *only* in light, were rightly terned *photoheterotrophs*.

The difficulties encountered were partly due to the fact that all these organisms at first could not be obtained in pure culture. Later when this was achieved, Van Niel was able to make clear the interrelationship between the pigmented autotrophic and the colourless chemautotrophic, mostly aerobic bacteria.

The following *five* physiological groups of bacteria thus could be distinguished:

I. Vast majority of bacteria which can utilise and grow only in organic nutrient medium—completely dependent upon a supply of organic compounds for assimilation, either in light or in darkness. They may include both aerobic and anaerobic types. These are the true *heterotrophs*.

II. The chemosynthetic or chemolithotrophic colourless bacteria, mostly oxidising various inorganic compounds, with simultaneous reduction of CO_2, in absence of light; they may include a few obligate anaerobic types such as *Thiobacillus denitrificans* and *Desulphovibrio*. *Thiobacillus denitrificans* is distinguished by its ability to use nitrates in place of free oxygen for the oxidation of sulphur and its derivatives. *Desulphovibrio* is one of the most interesting species of the bacterial flora. They are anaerobic, i.e., not needing oxygen; another striking characteristic of *Desulphovibrio* is that they are also nitrogen-fixers. They are actually *sulphate-reducing bacteria*—transforming SO_4^{2-} to H_2S. Instead of using oxygen to oxidise their food and obtain energy, they use SO_4^{2-}—as an acceptor of electrons, i.e., for oxidative purposes, releasing energy for the assimilation of CO_2. The chemosynthetic sulphur bacteria. *Thiobacillus thiooxidans* may be cited as an example of the transition form, from a heterotroph to an autotroph (*chemautotroph*); these bacteria require for normal growth, presence of minute trace of some organic substances—growth factors (usually compounds of vitamin B group), in addition to the inorganic sulphur compounds.

III. Photosynthetic bacteria (*Thiorhodaceae*) which utilise various inorganic sulphur compounds, e.g., H_2S, $Na_2S_2O_3$, elementary sulphur, etc., non-sulphur compounds, elementary selenium and at times, even molecular H_2. They are coloured, purple and green, due to the presence of pigments, *bacteriochlorophyll* and *chlorobium chlorophyll* respectively which are structurally essentially similar to chlorophylls. These bacteria are obligately anaerobic and perhaps all of them are nitrogen-fixers:

(*a*) Green sulphur bacteria (*Chlorobacteriaceae*) which assimilate CO_2 in light in presence of sulphide and certainly cannot utilise organic compounds. The pigment is *chlorobium chlorophyll*;

(*b*) Purple sulphur bacteria which utilise sulphide in light but can in addition, utilise some organic compounds. Some species can utilise molecular H_2 as a reductant. The pigment is *bacteriochlorophyll*.

IV. The photosynthetic purple non-sulphur bacteria, *Athiorhodaceae* which require minute traces of growth factors, e.g., vitamin B_1, and in general, use only organic compounds as reductant for assimilation of CO_2, *in light*. They have been truly called—*photoheterotrophs*. The pigment is *bacteriochlorophyll*.

V. *Rhodomicrobium* occupies an intermediate position between III and IV, since it requires no growth factors, but utilises organic compounds for assimilation of CO_2, in light. *Rhodomicrobium* as well as *Rhodospirillum rubrum*, members of this group of non-sulphur pigmented bacteria, are also important nitrogen-fixers.

CHEMOSYNTHETIC OR CHEMOLITHOTROPHIC BACTERIA

The chemosynthetic bacteria use inorganic compounds for oxidation and release energy (mostly in the form of ATP) for assimilation of CO_2. Because, they work with chemical energy instead of light energy for CO_2-assimilation, they can grow and develop in complete darkness in organic matter-free medium, and naturally do not need any light-absorbing pigment. Reduction of CO_2 by this method is known as chemosynthesis, which as a means of maintaining life may be of minor importance, compared to photosynthesis.

A wide range of organisms of this type is known which utilise the energy of oxidation of a great variety of substances for the assimilation of CO_2. In addition to the sulphur-oxidising bacteria and nitrifiers, organisms are known which oxidise iron salts, hydrogen and even carbon monoxide and methane. Several soil bacteria are *facultative heterotrophs*, living usually heterotrophically in the soil but capable of growth in purely inorganic medium when there is a supply of CO_2. It is of utmost interest to note that in these chemosynthetic organisms, the energy produced by the oxidation of one molecule of oxidisable inorganic substance is much less than required for the reduction of one molecule of CO_2 and presumably only a part of energy, thus released during the oxidation reactions, becomes available in the form of energy-rich phosphate group (ATP). Thus, it is evident that energy liberated from the substrate must be in some way ***pooled***.

Bioenergetics of Nitrobacter chemosynthesis

It has been shown that the electron transport from nitrite to molecular oxygen, catalysed by *Nitrobacter* (see chemosynthetic nitrifiers) yields inorganic phosphate/oxygen ratio, (P/O) of 1·0. The reduction of NAD by nitrate requires approximately 5 moles of ATP, per mole of NAD reduced (Shaw, 1964). In order to synthesise *one sixth* of a hexose molecule involving Calvin cycle, 3ATP and 2NADH molecules, we know, will be required; the process then can only be driven by the energy of oxidation of at least, as is evident, 13 moles of nitrite (for a complete hexose molecule formation, would thus require the oxidation of 78 molecules of nitrite to nitrate).

Assuming that the enzyme ***nitrate reductase is also coupled to energy generation***, a further investment of 5ATP would be needed to reduce nitrite to the ammonia level for the biosynthesis of amino acids, and an additional requirement of 2ATP is essential for the incorporation of these amino acids into cellular proteins. The reduction of NADP by NADH consumes another ATP per molecule of nucleotide reduced. Thus, the completion of chemosynthetic reaction, i.e., production of more of itself, *Nitrobacter* would require, the availability of at least, $13+5+2+1=21$ moles of nitrite for ready oxidation to nitrate.

The assimilation of CO_2 requires energy and it appears very probable that different types of organisms do not necessarily differ as to the mechanism of assimilation of CO_2 but rather in the mechanism by which the energy is supplied. The photosynthetic organisms supply energy from the sun *via* some light-absorbing pigment, the chemosynthetic, from the oxidation of inorganic substances and the heterotrophs from the oxidation or catabolic breakdown or organic compounds.

The assimilation of CO_2 by the photosynthetic bacteria, using H_2S requires 17·4 Cal.; using H_2 gas only 2·7 Cal.; whereas higher green plants, using water, requires 112 Cal. per mole of CO_2 assimilated values. It has been found experimentally that the number of light quanta absorbed per mole of CO_2 reduced, is only 1 in the photosynthetic bacteria and in the higher green plant, about 8-10 quanta.

TYPES OF CHEMOSYNTHETIC BACTERIA

1. Colourless sulphur bacteria

They can be classified mainly into *three* groups:

(a) Sulphur-oxidising bacteria.
(b) Sulphide-oxidising bacteria.
(c) Sulphate-reducing bacteria.

(a) **Sulphur-oxidising bacteria.** *Thiobacillus thiooxidans*, *Thiobacillus thioparus* and various other species are closely related organisms, able to grow with sulphur, thiosulphates or thiocyanates. They are obligate aerobes, obligate autotrophs and are largely distinguished, according to their acid tolerance.

Thiobacillus (the word is a hybrid of Greek and Latin)—sulphur-oxidising rods, like most bacteria, live in water.

$$2S+3O_2+2H_2O \longrightarrow 2H_2SO_4+\text{energy}$$
$$Na_2S_2O_3+2O_2+H_2O \longrightarrow Na_2SO_4+H_2SO_4+\text{energy}$$

These reactions perform functions similar to that of photosynthesis in green plants; they provide energy which the bacteria can utilise to convert inorganic carbon (as CO_3^{2-} or CO_2) to cell material; they cannot utilise organic compounds and their growth is sometimes actually inhibited by them.

The ability of *Thiobacillus* and other related species to form H_2SO_4 has several economic significance. The bacteria are able to survive in surprisingly strong acid solutions; during its oxidation of sulphur, it can produce and survive in solutions stronger than normal sulphuric acid (5% H_2SO_4)! This toleration of H_2SO_4 by bacteria has been used to cure infertile alkaline soils; if such soils are liberally dressed with sulphur, Thiobacilli rapidly develop and neutralise the alkalinity of the soil by forming H_2SO_4. But most of the activities of Thiobacilli are harmful. Perhaps the most unpleasant compounds, which pollute the atmosphere over industrial towns, are H_2S, oxides of sulphur, and sulphur particles. They are deposited on the stone and concrete cement which the various Thiobacilli find very congenial for their activities and readily oxidise these sulphur compounds to H_2SO_4, which may corrode stone or concrete.

Thiobacilli are by no means the only sulphur-oxidising bacteria. Some of the sulphide-oxidising bacteria can also oxidise sulphur and certain blue-green algae can also do this. But Thiobacilli are remarkable for the amounts of acid they can form and tolerate. Sulphur is extremely insoluble in water and yet the bacteria conduct this oxidation in a watery medium. The question, of how the sulphur gets into the bacteria to be oxidised, is completely unsolved.

Thiobacilus concretivorus is particularly active in corroding concrete and can utilise sulphur as well as sulphides. The most interesting Thiobacillus, which is remarkably active in the process of denitrification in the soil is *Thiobacillus denitrificans*. It is a facultative aerobe, i.e., it usually survives in presence of O_2; some varieties can also grow in absense of O_2, even air. *Thiobacillus denitrificans* is distinguished by its ability to use *nitrate* in place of free oxygen for the oxidation of sulphur and release of energy:

$$6KNO_3+5S+2H_2O \longrightarrow K_2SO_4+4KHSO_4+3N_2+\text{energy}$$

a-Keto dehydrogenase and *NADH* (or *NADPH*) *oxidase* have not yet been traced in some obligate chemautotrophs, as for example *Nitrobacter* and some *Thiobacillus denitrificans*. Smith et al (1967) are of opinion that in the above two organisms, Krebs cycle appears to be blocked at α-oxoglutarate level (see later). Neither the catabolic nor the anabolic role of Kerbs' cycle enzymes, have been established conclusively for any of the chemosynthetic bacteria. Smith (1967) proposed, therefore, that the biochemical basis of obligate chemautotrophy might be the lack of these two fundamental enzymes.

Thiobacillus ferro-oxidans, oxidises iron sulphides forming H_2SO_4—acid water in mines. These microbes are responsible for considerble damage to the pumping machinery of the mines. Finally and perhaps, the most remarkable of all, is *Thiobacillus thiocyanoxidans* which is able to grow on thiocyanates in addition to thiosulphate. Thiocyanates contain only one atom each C, N and S; from CNS in aerated water, these bacteria can synthesise all the C, N, H, O and S containing compounds that constitute its cell. A prodigious effort, indeed.

(b) **Sulphide-oxidising bacteria.** Several of the Thiobacilli can oxidise sulphides to sulphur, but too strong a solution of sulphide is toxic to them and they can therefore only deal with small amounts at a time. Sulphide-oxidising bacteria, on the other hand, are entirely dependent for their development on the oxidation of sulphides; they obtain energy from reactions which may be written:

$$2Na_2S+O_2+2H_2O \longrightarrow 2S+4NaOH+\text{energy}$$
$$2H_2S+O_2 \longrightarrow 2S+2H_2O+\text{energy}$$

Na_2S is oxidised in aerated water to form caustic soda and sulphur—the sulphur is deposited in the form of granules *inside* the cells. This reaction tends to make the medium alkaline, but there is always sufficient CO_2 in solution in the watery medium and thus the end products are generally carbonates and bicarbonates. *Beggiatoa* (now regarded as colourless, hence non-photosynthetic *Cyanophytan* algae), an example of this type of micro-organism consists of motile filaments which can be as long as 1 cm, certainly gigantic in size for a bacterium! If the sulphide is in short supply, they are able to oxidise the stored sulphur to sulphate.

(c) **Sulphate-reducing bacteria.** Very large amounts of sulphides are formed in nature by a single group of bacteria called the *sulphate-reducing bacteria*. They are obligate anaerobes—quite unable to grow in air. Instead of using oxygen to oxidise their food and obtain energy, they use sulphates; their sulphate-reducing process thus corresponds to the respiration of normal organisms. And as a by-product, the sulphide is formed and some energy liberated. Some sulphate-reducing bacteria are also efficient molecular N_2-fixers.

They also contain a specific cytochrome pigment. *Desulphovibrio* can grow without sulphate, if supplied with certain organic compounds, e.g., pyruvic acid—showing then a fermentative type of metabolism, such as seen in yeast.

These bacteria, which are comma-shaped structures, about 2μm in length with a long tail, swimming about rapidly and smelling horribly, are extremely thermophilic being knwon to survive at a temperature of about 80°C.!

Desulphovibrio, as seen before, is almost always heterotrophic, depending upon oxidation of organic carbon compounds by sulphates. But in one unique circumstance it can behave like an autotroph—it can utilise hydrogen gas instead of organic carbon compounds. If incubated with H_2, sulphate and carbonate and without any organic food, they are able to multiply and grow. Undoubtedly then, they can also be classified as autotrophs—*chemautotrophs*—because they may grow without any organic food:

$$4H_2+H_2SO_4 \longrightarrow H_2S+4H_2O+\text{energy}$$

The sulphate-reducing bacteria have several effects of economic importance—some harmful and some beneficial. Among the harmful effects, the most striking is their role in water pollution. If natural water, in a river, sea, or lake or stored water in a tank becomes contaminated with organic material, all sorts of bacteria grow and rapidly exhaust the dissolved air. But there is always sulphate in natural waters and as the supply of dissolved air is exhausted, sulphate reducers take over and become very active. Having a positive dislike for oxygen, they flourish, living on the organic matter and sulphate, and forming amounts of sulphides. This escapes as H_2S, causing part of the extremely unpleasent smell of polluted water. The phenomenon of black sand has a similar cause. Sand, particularly at the mouths of estuaries where much organic matter is deposited, is frequently blackened because the sulphide, these bacteria form, reacts with iron salt in the sand, to form black iron sulphide.

To make up for these harmful acts, *Desulphovibrio* is probably primarily responsible for the world's supply of sulphur. The major deposits of sulphur of the world are not volcanic in origin but are associated with dried up seas from long ago. The calcium sulphate of drying sea water is reduced by these bacteria to sulphide and sulphides are perhaps oxidised by the sulphide-oxidising bacteria or by the atmospheric air, spontaneously to sulphur.

Sulphur deposits are commonly associated with oil deposits and it has long been suspected that the *Desulphovibrio* may have something to do with oil formation.

2. Nitrifiers

These bacteria oxidise NH_3 into NO_2^- and NO_3^-. They belong to two groups:

(*i*) *Nitrosomonas*, also *Nitrosocystis* and *Nitrosospira* which oxidise NH_3 into NO_2

$$2NH_3+3O_2 \longrightarrow 2HNO_2+2H_2O+133{,}000 \text{ cal.}$$

(*ii*) *Nitrobacter*, which oxidise NO_2^- to NO_3^-

$$2KNO_2+O_2 \longrightarrow 2KNO_3+35{,}000 \text{ cal.}$$

The activities of these nitrifiers have already been discussed in connection with their importance in the mechanism of soil nitrification—a primary link in the nitrogen cycle in nature. If the soil contains only the first group of nitrifiers, e.g., *Nitrosomonas*, the transformation of NH_3 stops at the NO_2^- stage. Only in presence of both the groups of nitrifiers, the NH_4^+ ions are ultimately converted into nitrates.

The *cell-free extracts* of *Nitrobacter* and also extracts from *Thiobacillus dentrificans* can, however, oxidise NADH under *anaerobic* conditions with *nitrate as the terminal electron acceptor* and this process has been shown to be coupled to phosphorylation (Aleem, 1968).

The bioenergy requirements of these nitrifiers for assimilation of carbon and other cell-substances have already been discussed.

3. The iron bacteria

Gallionella, *Leptothrix ochrucea*, *Spirophyllum ferrugineum*, etc., oxidise ferrous salts to ferric salts:

$$2Fe(HCO_3)_2+\tfrac{1}{2}O_2+H_2O \longrightarrow 2Fe(OH)_3+4CO_2+\text{energy}$$

4. The hydrogen bacteria

Baccillus pycnoticus and others oxidise H_2 gas to form water:

$$2H_2+O_2\longrightarrow 2H_2O+\text{energy}$$

The hydrogen oxidising, *Hydrogenomonas*, case utilise the emergent hydrogen (obtained from photolysis of water formed) and exergonic energy thus released, can be directly used as a source of free energy, as well as an agent for CO_2-reduction in chemosynthesis of *Hydrogenomonas*.

5. The carbon bacteria

These bacteria oxidise CO to CO_2:

$$O_2+2CO\longrightarrow 2CO_2+\text{energy}$$

Bacillus oligocarbophilus is an example of the carbon monoxide oxidisers.

6. Methane oxidisers

The bacteria, *Methanomonas*, under this group bring about an oxidation of methane, CH_4 to CO_2:

$$CH_4+2O_2\longrightarrow CO_2+2H_2O+\text{energy}$$

BACTERIAL PHOTOSYNTHESIS

The Photosynthetic Bacteria

(a) **The green sulphur bacteria** (*Chlorobacteriaceae*) have a pigment, *chlorobium chlorophyll*, similar, but not identical, in structure to chlorophylls. They occur in hydrogen sulphide medium and their photosynthetic activity is restricted to photoreduction of CO_2 with inorganic sulphides. The overall reaction can be represented thus:

$$2H_2S+CO_2\longrightarrow \text{photosynthetic product } (CH_2O)+2S+H_2O+\text{energy}$$

This reaction, as seen from the above equation, is weakly exergonic in contrast to the primary endergonic reaction of photosynthesis in higher green plants. The three most widely distributed species are *Chlorobium*, *Chloropseudomonas* and *Chlorobacterium. Chlorobacterium*, as we have seen before, is also able to fix atmospheric nitrogen and as such these green sulphur bacteria must be regarded as one of the most completely autotrophic organisms known—they can thrive and grow in situations where all other forms of life (except perhaps certain cyano-bacteria (=blue-green algae), living anaerobically) seem impossible, for these bacteria are also anaerobic.

The green *Chlorobium* is a very small organism, unable to move, which deposits sulphur *outside* the cell and for many years it was believed to be incapable of oxidising this sulphur further. However, it has recently been shown that they can in fact utilise this sulphur; they can oxidise sulphur right through to sulphate without depositing any sulphur. The economic importance of *Chlorobium* in preventing the spread of pollution in lake waters has only been understood recently. Lake mud is rich in bacteria, able to produce sulphides, which are toxic to fish and which are among the most disagreeable components of water pollution. An analysis of some lake water showed, that though the deep waters were rich in sulphides, layers at some 25 feet from the surface were rich in *Chlorobium*, above which no sulphide was detectable. Clearly the *Chlorobium* layer was oxidising the sulphides and converting them to innocuous elementary sulphur.

Some cyano-bacteria—the most promitive among green plants—though able to grow in air, can also grow without it, and then they behave very much like *Chlorobium*.

(b) **The purple sulphur bacteria** (*Thiorhodaceae*) have the characteristic pigment, *bacteriochlorophyll*. They can utilise a variety of sulphur compounds, in addition to sulphides, and also molecular H_2 and sometimes selenium compounds or selenium itself, for simultaneous assimilation of CO_2. In absense of the inorganic sulphur compounds, they can live heterotrophically utilising organic carbon compounds:

$$2CO_2 + 5H_2O + Na_2S_2O_3 \longrightarrow 2CH_2O + 2H_2O + 2NaHSO_4 + \text{energy}$$
$$3CO_2 + 8H_2O + 2Se \longrightarrow 3CH_2O + 3H_2O + 2H_2SeO_4 + \text{energy}$$

It is well known that the metabolism of sulphur and selenium is very similar. So selenium or selenium compounds can be oxidised by this group of bacteria as readily as sulphur compounds.

The most important member of this group is the purple, *Chromatium*. Sometimes waters of the lakes polluted by the formation of hydrogen sulphide turns a brilliant pink or purple colour (the phenomenon of 'bloody waters') owing to luxuriant growth of *Chromatium*. When this happens, the stinking nuisance caused by H_2S, escaping from water is much reduced, because these bacteria can convert it to harmless sulphur. Clearly a deliberate control of water pollution by biological methods is a practical proposition. In place of black, smelly and stagnant pools, one may expect the beautiful, pink or green lagoons when biological methods of water purification are economically employed in future.

(c) **The purple non-sulphur bacteria** (*Athiorhodaceae*). These bacteria are unable to use sulphur compounds (with the exception of some strains of *Rhodopseudomonas*) but grow in the presence of simple organic substances, such as organic acids, alcohols, etc. All of them require traces of growth factors such as vitamin B complex group for normal growth, thus differing from *Thiorhodaceae*, which do not require such growth factors. These bacteria oxidise organic compounds with simultaneous reduction of CO_2, in contrast to other photosynthetic bacteria, but this assimilation of CO_2 is only possible with the help of light absorbed by their pigment. So, they can be regarded as both autotrophs and heterotrophs—*photoheterotrophs*. If isopropyl alcohol is the organic substrate available, these bacteria quantitatively convert it into acetone and at the same time an equivalent quantity of CO_2 is consumed and elaborated into carbohydrate, energy required for the reaction coming from sunlight (only short infra-red) absorbed by the purple pigment, *bacteriochlorophyll*.

$$2CH_3.CHOH.CH_3 + CO_2 + \text{light energy} \longrightarrow CH_2O + 2CH_3COCH_3 + H_2O$$

The number of light quanta absorbed per mol. of CO_2 reduced is approximately the same or in the same order, in all photosynthetic organisms, be they pigmented bacteria or higher green plants. This certainly suggests that *in all these organisms, the primary photochemical reaction is the same* and consists in the removal of hydrogen from a suitable hydrogen donor with the formation of NADPH, which ultimately donates its hydrogen for the reduction of CO_2-fixation product.

In photosynthetic bacteria such as ***Chlorobium*** and ***Rhodospirillium***, the CO_2-fixation in photosynthesis, operates via Calvin cycle. CO_2 fixation by reversal of Krebs' TCA cycle may also take place in these organisms. An alternative pathway has also been suggested.

THE EVOLUTION OF AUTOTROPHISM

The most of evolution of the photosynthetic process and the chemosynthesis of autotrophic bacteria concern the roles which these processes may have played in the development of the earth. It may be supposed that in remote antiquity, under very different environmental conditions which existed at that time, organic molecules arose spontaneously from inorganic molecules. Eventually, a self-perpetuating unit was formed which was capable of utilising the energy derived from the oxidation of some inorganic compounds for further production of organic compounds. It may be terrifically interesting to speculate about the time when the chemosynthetic mechanism first evolved. From studies of biological isotopic ratio of the sulphur deposits of the earth, it is possible now to get an idea about the age of particular rocks which are in many cases of microbial origin. The determination of biological isotopic ratio provides added evidence for or against the bacterial origin of the sulphur deposits (sulphur subjected to biological transformation finishes with greater proportion of lighter isotope and lesser proportion of heavier isotope. *Desulphovibrio* can partially separate the two isotopes). Sulphur compounds, with correct biological isotopic ratio, have been obtained from rocks up to 800,000,000 years old which indicated that this chemosynthetic sulphur bacteria have probably been in existence, since at least that time! The first fossil records begin from 500,000,000 years and man seems so young with his utmost 100,000 years. Undoubtedly the sulphur bacteria and perhaps other autotrophic micro-organisms represent very primitive forms of life and it is not surprising that some of them are anaerobes, since at the time life originated, there may have been very little or no oxygen in the atmosphere of this planet. At that time, our earth was less settled in its chemical ways than now, and not only H_2S but also free H_2, might have been available in the atmosphere. From chemosynthetic bacteria which oxidise inorganic compounds for energy requirements, the development might have progressed to pigmented bacteria and ultimately to green plants. The development of the pigment system allowed the utilisation of external solar energy but, in the first place, the cells were still dependent on the supply of suitable hydrogen donors. Photosynthetic bacteria, utilising hydrogenated sulphur compounds as reductants for the reduction of CO_2, might rightly be regarded, as Rabinowitch puts it, the missing link between the world of green plant photosynthesis and the chemosynthetic way of life. Finally, in the green plants, a mechanism was evolved in which water was the sole extraneous hydrogen donor for the reduction of CO_2; a further stage in independence of the external environment was thus achieved. Thus, step by step, complete emancipation was achieved by the green plants by the acquisition of a more complex mechanism and attaining at each stage a greater independence of the external reducing substance. The transition from bacteria to algae which liberated the plants from dependence on uncertain, dwindling supplies of unstable sulphides and other inorganic compounds, has allowed life to spread over the whole surface of the globe. The capacity of certain green algae, e.g., *Scenedesmus*, for the adaptation to molecular H_2 for CO_2-assimilation, may be reminiscent of their genetic relationship to the autotrophic bacteria.

We know that energy becomes available in the biological system as the direct or ultimate result of oxidation reactions. From elementary chemistry, we know that oxidation of one substance must necessarily be accompanied by reduction of another and biological oxido-reduction reactions involve the transfer of hydrogen atoms or electrons. The complete oxidation of any one substance proceeds by one or more simple steps, each catalysed by the appropriate enzyme and in all knwon reactions, the transfer of hydrogen atoms or, to be more precise, electrons, to the ultimate H-acceptor, is affected by one or more intermediate groups—as in photosynthetic transfer of hydrogen. In aerobic organisms, molecular O_2 serves as the H-acceptor and according to the enzymes concerned, the end product is water. Organisms which by chance or by necessity, live in an anaerobic enrivonment, must use a substance other than oxygen for this purpose. Such a substance may be derived from the environment or it may be a product of the organism's own catabolism, as in lactic acid bacteria, where pyruvate (the intermediate product) is reduced to lactic acid. Among the autotrophic organisms, the highest evolved green plants reduce CO_2 by means of water in light; the photosynthetic bacteria reduce CO_2, also in light, but inorganic sulphur compounds like H_2S or even organic compounds are used as reductants or electron donors, in place of water. Colourless sulphur bacteria reduce CO_2 by means of energy derived from oxidation of inorganic sulphur compounds, *without light.* This comparison shows the existence of a hierarchy of autotrophic organisms and encourages speculation as to the genetic relationship between them. This comparison may well start from the heterotrophs for the original distinction between a heterotroph and an autotroph on two counts, viz., (*i*) that heterotrophs are unable to incorporate the carbon of CO_2 into organic molecules and (*ii*) that autotrophs live exclusively and entirely at the expense of inorganic substances is no longer valid, and requires drastic modification. The growth of heterotrophs is in fact dependent on the presence of CO_2 and they are also known to possess enzyme systems accomplishing its fixation (dark fixation of CO_2 or heterotrophic CO_2-fixation) when CO_2 is metabolically fixed or enter into biochemical reactions, (CO_2 added on to an already existing organic molecule) apart from those of the photosynthetic process and this *dark fixation* is thought to be of universal occurrence in all living organisms—heterotrophs or autotrophs. The CO_2 is incorporated into the carboxyl group (COOH) and the end product is an acid, which can be further elaborated into more complex organic molecules of sugars or proteins by a reversal of some of the steps of glycolysis. So it is a truism to say that carbon of the CO_2 molecule can be incorporated into the pre-existing organic molecules of heterotrophs. This fixation of CO_2 by heterotrophic or autotrophic organisms was discovered by Wood and Werkman working with the propionic acid bacteria. The first step, the incorporation of CO_2 into pyruvic acid to form oxalacetic acid, is often referred to as *Wood and Werkman reaction.*

Crassulacean acid metabolism (CAM)

This discovery by Wood and Werkman that, CO_2 can be metabolically fixed in the synthesis of organic acids by certain bacteria, quickly led to the opening of new fields of experimentation to explain, the increase during the night or when they are placed in the dark, in the titrable acidity of shoots or leaves of certain families, e.g., *Crassulaceae*, *Cactaceae*, etc., where succulence is a common characteristic of the plants. For many years, the term *dark acidification* has been used to describe this increase in acid content. *Malic acid* is often predominant among the acids which accumulate, but not inconsiderable amounts of other acids, e.g., *citric*, *isocitric acids* in *Bryophyllum* may be found at the end of a dark period, i.e., early in the morning. In Crassulacean plants stomata are open at night but closed usully during daytime.

With the return of daylight or when leaves kept in dark are illuminated, the amount of accumulated titrable acid significantly decreases. The synonymous terms *light deacidification* or *photodeacidification* have been used to refer to this decrease in acid content. Titrable acidity often reaches a minimum value in the late afternoon or early evening —the minimum sometimes, but not always, approaches zero. The essential feature of crassulacean acid metabolism is this diurnal fluctuation in titrable acidity. The CO_2 evolved during daytime is refixed by the C_3 pathway.

If the accumulation is chiefly malic acid, the overall reactions, *dark acidification* and *light deacidification* can be represented thus:

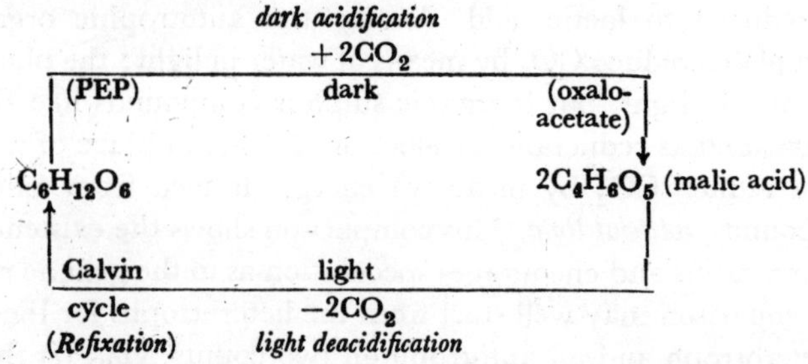

In C_4-plants the C_3 and C_4 pathways are separated in space, in CAM plants they are separated in time.

Three way carbon traffic in vascular plants

It has been evident during the last few years and experimental evidence from several disciplines also supports the conclusion that at least *three distinct photosynthetic mechanisms exist in vascular plants:*

(1) Calvin cycle, primarily in temperate plants; (2) HSK-C_4 dicarboxylic acid pathway or β-carboxylation pathway predominates in tropical plants, i.e., photosynthetic fixation of carbon proceeds either through RUDP (PGA, first measurable product of carboxylation), or β-carboxylation pathways (malate and aspartate are the initial carboxylation products) and (3) Crassulacean acid metabolism (CAM).

Calvin cycle plants, we have seen before, can also be differentiated from HSK plants, at least to some extent, on the basis of leaf-anatomy (Crooksten and Moss, 1970; Downton and Tregunna, 1968) and chloroplast structure—uniform type in Calvin plants (Laetsch, 1969) and dimorphic (more than one type of plastid with more than one function) in HSK plants (Walker, 1970). In tropical grasses (HSK plants, e.g., leaves of maize, sugarcane, etc.) the usual 'grana' stacks of thylakoid membranes are present in some plastids, but entirely absent or at least diminished, in those immediately adjoining the vascular bundles ('bundle sheath' chloroplasts). Normally stacked chloroplasts in the adjacent mesophyll are also much less elongated compared to the bundle sheath ones.

The CAM plants, on the other hand, have a distinct biochemical system for the fixation of CO_2. These plants are differentiated further in that they are probably all non-halophytic succulents. CAM plants, as is well-knwon, are generally adapted to arid environments. In photosynthetic bacteria, like *Chlorobium* and *Chromatium* another pathway operates simultaneously with Calvin cycle, the *reductive carboxylic acid cycle.* This is believed to involve the conversion of acetyl coenzyme A to oxalacetate and then to citrate (analogous to a reversal of Krebs cycle changes). Reduced ferredoxin and NADH provide the reducing power for reversing the reactions and crossing energy for irreversible reactions. Following the cleavage of citrate or isocitrate, the acetate formed can easily serve its purpose by regenerating acetyl coenzyme A and there would be theoretically a net yield of one molecule of C_4-dicarboxylic acid for each turn of the cycle.

PHOSPHORYLATION AND ENERGY TRANSPORT MECHANISM

The most general and important mode of energy transport in living cells appears to occur through the intervention of particular phosphorylated compounds. Evidences from a variety of sources indicate that phosphorus metabolism is linked with photosynthesis. Moreover it is also evident that phosphate must be in some way involved in the production of phosphoglyceric acid, the first detectable product of photosynthetic CO_2-fixation.

The esterification of organic substrate with inorganic phosphates is an integral part of the mechanism, whereby endergonic (energy consuming) reactions are able to utilise the energy made avaible by oxidation reactions, e.g., respiration. It has been pointed out earlier that the primary function of respiration is to provide energy for the energy consuming endergonic reactions of the cell—the synthetic, organisational, mechanical, osmotic and similar activities of the organism. In a biological system, the only known way in which the energy released by an exergonic reaction can be made available to an endergonic one is for the two reactions to be coupled together by means of a substance which participates in both. Thus much of the energy released in respiration or any other oxidative process is not liberated as heat (which is ultimately dissipated and will be of little use to the organism except increasing its temperature) but is retained in the oxidised substrate in association with the incorporated phosphate group. Inorganic phosphate is a necessary component of all metabolic systems; the reaction cannot ordinarily proceed in the absence of phosphate.

The photosynthetic fixation of carbon cannot occur if there is no phosphorylation. No life, including microbial life, is possible without ATP. The key materials in energy transfer are the adenine derivatives; adenylic acid, adenosine diphosphate (ADP) and adenosine triphosphate (ATP). Adenine (6-amino-purine), a nitrogenous base of the group of *purines*, is linked in all these materials with the *pentose sugar*, *ribose* (D-ribofuranose) to form *adenosine*. In adenylic acid, a phosphoric acid molecule is esterified with the ribose portion. ADP, as the name implies, contains two $PO_4^{=}$ residues. The exergonic energy produced by most oxidations, as we know, is not dissipated but are often stored by the simultaneous synthesis of adenosine triphosphate (ATP), *the basic energy currency*, which is the universal fuel for all biochemical work within the cell, plant, or animal, which is also certainly, as we have seen before, one of the major direct products of photosynthesis in plants and the universal energy intermediary of contemporary terrestrial organisms. It has rightly been called the '*energy coin* of the biological realm'. When the third PO_4 group is added to ADP to make ATP, the

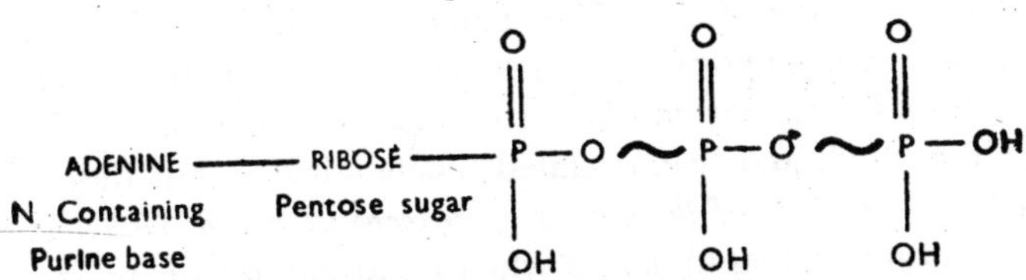

FIG. 711. Adenosine triphosphate (ATP).
Thick dark spiral lines represent high-energy bonds.

phosphate group is loosely held and is rather easily transferred to other metabolic compounds, like sugar and thus the sugar molecules are esterified. When such a transfer happens, the liable energy of the phosphates is not lost but is transferred to the new compound, e.g., the sugar phosphate formed. This is unusual and does not happen when energy is released, for example, in normal oxidation or hydrolysis.

Phosphorylated compounds can be divided into two groups according to the amount of energy released in their hydrolysis; some yield about 3 Cal. per mole., whilst others liberate between 7·5 and 12 Cal. per mole. (7–8 for ATP or ADP). The latter, such as ATP, are known as high-energy or energy-rich phosphate compounds and they contain what is known as high-energy or energy-rich phosphate bonds (the terminal phosphates of ATP or ADP are referred to as energy-rich phosphates).

We have seen that in the oxidative reactions of respiration, phosphorylated compounds such as ATP are formed and part of the energy liberated is stored as the energy of the phosphate bond. The energy-rich phosphate groups may be subsequently used as the source of energy for synthetic reactions. A simple example of such energy coupling is the formation of starch from glucose. The formation of glucosidic bonds in starch requires external source of energy—this energy is derived from respiration or photosynthesis through the mediation of energy-rich phosphate, ATP or UTP.

We have seen before that phosphates are directly linked with photosynthesis and the primary acceptor of CO_2 in photosynthesis is phosphorylated. Thus the continuous and cyclic production of the primary CO_2-acceptor, involves ATP.

Rapid utilisation of ATP has been reported in photosynthesis—but it is certain that though it is directly related to CO_2-fixation, ATP may also function: (*i*) in the production of CO_2-acceptor , and (*ii*) ATP must be present for the conversion of phosphoglyceric acid (PGA) to hexose diphosphate by way of diphosphoglyceric acid and phosphoglyceric aldehyde (G3P) or DHAP (dihydroxy acetone phosphates). The energy requirement to accomplish the reduction of CO_2 to carbohydrate level in photosynthesis cannot be satisfied only by the energy supplied by the reduced hydrogen acceptor molecules (the reduced coenzyme NADPH) produced in the photochemical reaction, *but a source of energy in addition to NADPH or NADH would be required.* As we know, ATP serves as this source.

PRIMARY ENERGY RELATIONS OF THE CELL

We know that the two major pigments of the cell are porphyrins which are related biochemically and arise from a common precursor, protoporphyrin. These are the two prominent pigments of living matter on the earth. One is the green pigment chlorophyll and the other the red pigment haem. The haem pigment is the primary catalyst of biological oxidation; the chlorophyll pigment is the primary catalyst of photosynthesis. Haem is the prosthetic or active group of haemoglobin—the chromoprotein which transports O_2 in the blood stream; it is also the active group of proteins, essential for biological oxidation (respiration). The haem proteins include the cytochromes, cytochrome oxidases, peroxidases, catalases, etc. Chlorophyll, we have seen, is the fundamental pigment for formation of organic matter in the earth. Not only are these two pigments, chlorophyll and haem related biochemically, but they are also related functionally in being *catalysts of the process of decomposition and formation of water.* The functional relationship becomes apparent when we attempt to express and summarise the *primary energy changes* by the following simple chemical equation:

$$O_2 + 4H^+ + 4(e) \underset{\substack{\text{chlorophyll} \\ +nh\nu+\text{ATP}}}{\overset{\substack{-(\Delta H+\text{ATP}) \\ \text{haem enzymes}}}{\rightleftharpoons}} 2H_2O$$

In this equation: (e) represents electrons of high potential energy, forming chemical bonds of organic substrates (the energy contained in the configurations of respiratory substrate such as sugars); the heat produced in oxidation is ΔH; and the number of light quanta of frequency (v), absorbed and stored in protosynthesis is nhv, where h is Planck's constant.

The forward reaction certainly expresses an oxidation with atmospheric oxygen to form water, energy being thereby released. This oxidation is catalysed by a series of haem enzymes in all living cells. The electrons of high potential energy (e) of the sugar molecules combine with protons $4H^+$ and O_2 to form water, where (water) the electrons have a much lower potential energy. This oxidative process, we know, does not take place spontaneously but is catalysed by enzymes which have haem as their prosthetic group (coenzyme). The forward reaction then expresses an overall oxidation in which energy is released when the electrons of high potential energy of organic respiratory substrates combine with protons (positively charged hydrogen ions) and O_2 to form water which has lower potential energy. This released energy is utilised by the protoplasm to make more of itself from the components of its environment. We say again that this release of energy is catalysed by haem enzymes.

What do we know of the reverse reaction? The reverse reaction is essentially the reaction of photosynthesis. It is an energy-storing reaction, the enrgy of sunlight being stored primarily in the electronic configuration of the organic compounds, e.g., sugar. This reaction of photosynthesis, where storage of energy of sunlight occurs, is also catalysed by a porphyrin pigment, more precisely as its another name *magnesium dihydroporphyrin*, called chlorophyll. If four quanta of light are required (i.e., $n=4$) then the energy or reducing capacity of (e) would be in the neighbourhood of hydrogen electrode potential. We have seen before that H_2O is probably the major, if not the exclusive source of O_2, in photosynthesis; water is decomposed in presence of sunlight to yield O_2, protons (H^+) and electrons, where (e) represents the electrons taken up by some electron acceptor in the chloroplasts, which ultimately reduces CO_2 to the carbohydrate level.

From these considerations we see that the release of energy is essentially an oxidation with O_2, water being the end product. The reverse reaction of the removal of hydrogen from water and the consequent storage of energy, due to the reducing ability of the electrons liberated, is the primary reaction of photosynthesis with organic compounds as end products. This is the manner in which haem (enzymes) and chlorophyll are related; they are the catalysts of fundamental energy reactions of the cell.

PRODUCTIVITY

Since life on earth depends on the productivity of plants, plant productivity is of tremendous importance. Life is sustained by solar energy. Surprisingly, only a small fraction of the solar energy—0·1 to 10%—which reaches the leaf surface everyday is utilized by plants. Nevertheless, it has been estimated that total plant dry matter produced every year is about $1{\cdot}64 \times 10^{11}$ tons. The ocean with its rich algal flora utilizes a large fraction of the total energy it receives. N_2-fixing micro-organisms also contribute considerably to productivity.

Crop productivity in India is one of the lowest in the world. Crop productivity depends, in addition to high photosynthetic rates, on the amount of fertilizer, irrigation water, fungicides, pesticides, weedicides and other growth regulating chemicals applied. Most of our farmers cannot afford to purchase them, although situation is gradually improving. Good quality seeds of the high-yielding varieties with proper agricultural practice have increased crop production considerably.

For every kg.cal. of solar energy utilized about 14 mg of carbohydrate is produced. An agricultural land 1m^2 in area receiving 500 kg.cal. of solar energy per day should have a potential productivity of 70 g. World average for rice is only 3·8% of this value; this has been increased to 11·4% with intensive agriculture. Performance of wheat is no better but for sugarcane, sorghum and maize it is as high as 26·3, 31·4 and 38·5%, respectively.

C_4 plants have higher photosynthetic and lower photorespiration rates than the C_3 plants. Unfortunately although sugarcane, maize and sorghum are C_4 plants the major cereals, rice and wheat, are C_3 plants. The high-yielding varieties of wheat and rice require the application of relatively high doses of fertilizers. Most of them are rather short in size and possess erect leaves; maximum solar energy is utilized in this way. However, the popular fine-grained varieties are usually tall in habit, photoperiod sensitive and occupy crop fields for many months, depending on the date of sowing. Their low yield is largely due to poorer translocation rates than the high-yielding varieties. Appropriate breeding experiments may combine useful characters. Crossing C_4 and C_3 plants should theoretically improve many cultivars, but C_4 and C_3 varieties belonging to the same species or genus are rare. Some interesting results have been obtained with *Atriplex*. Several chemicals are available which reduce photorespiration, but this is still expensive. Bright tropical sun is suitable for all the varieties of rice and wheat but in summer and afterwards in India many days are cloudy. Cultivars which can photosynthesise at relatively low light intensities are highly desirable. Cheaper substitutes of nitrogenous fertilizers are essential. In the ocean light intensity is much lower and many marine algae efficiently utilize this low light intensity, in the medium of sea water which includes all essential minerals. As world population increases at a fast rate and there is very little land area left for cultivation, we have to depend, for our subsistence, on the productivity of the oceans, in the not too distant future.

EXPERIMENTS ON PHOTOSYNTHESIS AND PLANT PIGMENTS

I. Test for starch

Take some thin flat leaves from bean, *Cucurbita*, nettle, etc., and boil in water. Note that the green colour does not come out. After throwing away the hot water, place the boiled leaves in organic solvents, e.g., 95% alcohol, acetone or ether, in a water bath and notice that the leaves gradually lose their colour while the solvent turns green. When the leaves are completely bleached, place them in white saucers and pour dilute iodine solution (in potassium iodide) over them. The depth of colour produced indicates roughly how much starch is present. The colour is nearly black when there is an abundance of starch, if little, it is bluish; if none at all, the iodine solution turns the leaf brownish (test for proteins). Test leaves of plants growing in light (abundance of starch) and plants kept in darkness (no starch).

II. Experiments with plant pigments

A. Plastidial Pigments

(*a*) Weigh a number (about 12) of one-gram samples of air-dried suitable leaf powder and place each sample in a test tube. Add 10 ml portions of the following solvents, stopper and allow to stand with occasional shaking to observe solubility of pigments in organic solvents: 80% methyl alcohol, 95% methyl alcohol, absolute methyl alcohol, 80% ethyl alcohol, 95% ethyl alcohol (rectified spirit), absolute ethyl alcohol, benzene, chloroform, acetone, ether, petrol ether and carbon bisulphide. Estimate the relative green (or yellow) colouring in each solvent and arrange them (the solvents) in order of merit as effective pigment extractors.

A concentrated solution of leaf plastidial pigments can be easily obtained from dried powdered leaves by 90% acetone or alcohol, etc. by using a Buchner funnel and sucking the material against the filter paper by a filter pump. A deep blue-green solution is obtained which contains all the *four* pigments. The solution shows strong fluorescence and appears blood red in reflected light; in transmitted light it appears green.

(*b*) Fill a test tube with chlorophyll solution and place it between a direct vision spectroscope and the source of light; observe strong black absorption bands in red and also in blue-violet parts of the spectrum, showing efficient light absorption by chlorophyll solution in those regions of visible light.

(*c*) In an alcoholic extract of chlorophyll, add concentrated HCl, drop by drop, until a brownish-green colour is formed. HCl removes magnesium from chlorophyll molecule and replaces it by hydrogen, forming a *phaeophytin*. This brownish solution if examined with spectroscope will be found to have a new absorption band in the green. Add a crystal of copper, zinc or iron acetate to the phaeophytin solution and heat gently until the colour of the solution changes back to a brilliant green but without fluorescence. The magnesium of the chlorophyll molcule has been replaced by copper, zinc or iron.

(*d*) Chlorophyll, a neutral substance, on treatment with alkalis forms salts of acids, the latter being known as *chlorophyllins*. These salts are soluble in water forming green solutions which are not, however, fluorescent.

$$R<^{COOCH_3}_{COOC_{20}H_{39}}$$

On treatment in the cold, with alkali, the ester is saponified and the alkali salt of the chlorophyllin is formed. During saponification, there is a change in colour of the pigment, the so-called brown phase, followed by a return to green.

Take a little ether solution of chlorophyll into a test tube and in a pipette take a little of 30% KOH in pure methyl alcohol. Place the lower end of the pipette at the bottom of the test tube and allow the KOH solution to run below the chlorophyll solution. The brown-coloured layer formed at the interface between the solutions in 10 minutes changes to pure green. The potassium salt formed is insoluble in ether and if water is added in a separating funnel, the green colour is no longer present in the ethereal layer. This phase test is shown by all compounds containing an intact cyclopentane ring. The phaeophytin and protochlorophyll also give the phase test. During saponification, the pentane ring is opened up by hydrolysis which in absense of oxygen can be converted back again to original compound with the usual closed ring.

$$R<^{COOCH_3}_{COOC_{20}H_{39}} + 2KOH \longrightarrow R<^{COOK}_{COOK} + CH_3OH + C_{20}H_{39}OH$$

If saponification is done with hot alkalis, *isochlorophyllins* are formed which are fluorescent. On heating chlorophyllins with concentrated hot alcoholic alkalis, a series of decomposition products are formed by the removal of carboxyl groups. The final *phyllin* has only one carboxyl group instead of two. When this is removed by further heating, a substance *aetiophyllin* is formed which contains no oxygen.

(*e*) **Separation of the plastidial pigments**

(*i*) **By capillary analysis:** Dip a strip of filter paper hung by suitable means so that the lower edge of the strip just dips into a raw chlorophyll solution. In a few minutes, observe that the solution will have mounted the paper and the yellow zone (chiefly carotenoids) goes in advance of the green along with the solvent, whereas chlorophyll *a* and *b* are fixed at the point of contact. It may be possible to distinguish between the bluish-tinted upper edge of the green zone (chlorophyll *a*) and the lower yellowish-green portion (chlorophyll *b*). Brockmann's alumina can also be used for separation of the pigments by adsorption followed by elution with orgânic solvents

(*ii*) An alcoholic extract of green leaves when treated with an equal volume of benzene and a little water and the mixture inverted several times, separates into two layers, the upper dark-green benzene layer contains the two chlorophylls and carotene while the lower alcoholic layer retains yellow xanthophyll.

(*iii*) **Separation of green and yellow pigments—chlorophylls and carotenoids:** Shake about 5 ml of an ether extract of chlorophyll with 2 ml of 30% KOH in pure methyl alcohol. Note the disappearance of green colour and shortly its reappearance. Now slowly add 10 ml of water and then a little more ether. On shaking the test tube two layers are produced of which the lower aqueous alkaline layer contains the saponified green pigments while the yellow carotenoids are contained in the upper ethereal layer.

(*iv*) **Separation of chlorophyll *a* and chlorophyll *b*.** Shake about 10 ml of freshly prepared acetone extract of chlorophyll with 10 ml of petrol ether in a separating funnel and add water until two layers are formed. Run off the lower acetone layer and wash the petrol ether extract with repeated aliquots of water. Discard the washings and shake the purified petrol ether solution with 10 ml of 90% methyl alcohol. Two layers are formed of which the upper petrol ether contains chlorophyll *a* and the lower methyl alcohol layer, chlorophyll *b*. The colour difference between the two layers is not very distinct owing to the presence of yellow pigments; carotene in petrol ether with chlorophyll *a* and xanthophyll in methyl alcohol with chlorophyll *b*.

(*v*) **Separation of two yellow pigments—carotene and xanthophyll.** The ether layer containing yellow carotenoid pigments obtained in experiment (e) (*iii*) is washed with water in a separating funnel and evaporated down to 1 ml in a water bath. Add 10 ml of petrol ether to it and next mix with 10 ml of 90% methyl alcohol and shake thoroughly. The lower methyl alcohol layer is removed and the upper petrol ether layer is again treated with methyl alcohol and again removed. The process is repeated until the methyl alcohol layer is only faintly coloured. The methyl alcohol layer contains xanthophyll and the upper petrol ether layer, carotene.

(*vi*) Prepare thin sections of fresh leaf samples (*Cucurbita*, bean, nettle, etc.) and mount them in a drop of 90% ethyl alcohol. Leave the slide under a bell jar containing a petri-dish of absolute alcohol. The section will slowly dry in the course of about 12 hours or sometimes a little longer. It is then examined under microscope when there will be observed the characteristic triangular and hexagonal crystals of *ethyl chlorophyllide* (crystalline chlorophyll).

$$R<^{COOCH_3}_{COOC_{20}H_{39}} + C_2H_5OH \longrightarrow R<^{COOCH_2}_{COOC_2H_5} + C_{20}H_{39}OH$$

Methyl or ethyl chlorophyllide crystals may also be prepared by covering a gram of fresh material of *Antirrhinum* (snapdragon) with 4 ml of 70% methyl or ethyl alcohol in a stoppered bottle. Allow to stand. When the leaf material turns yellow, mount in water and look for the crystals.

Carotenoids. (*i*) Make an extract of carotenoids from carrot, yellow flower heads of sunflower, *Calendula*, etc., in 95% alcohol. Treat a portion of the yellow extract separately with concentrated HCl and H_2SO_4. A characteristic deep blue to blue-green colour is obtained.

(*ii*) Cover sections of carrot with a 20% solution of KOH in 40% (by volume) of methyl alcohol and keep them in a stoppered bottle in dark for two days. Wash in distilled water to remove KOH and mount a thin secion in water or glycerine. Carotene appears as slender, sometimes curved, orange-yellow crystals.

B. Pigments of the Cell sap

Anthocyanins. (*a*) When an alcoholic extract of red leaves or flower petals is prepared and treated with an equal volume of benzene and shaken, the lower alcoholic layer appears pink to red owing to the presence of anthocyanin pigments.

(*b*) Boil red or blue flower petals in water and observe that the conspicuous pigments are all soluble in water.

(*c*) Place some blue flowers of *Clitoria* (aparajita) in weak solution of HCl. When the acid has penetrated the tissue, the flowers sometimes (not always) turn red in colour. Similarly, observe that pink or red flowers (e.g., red rose, *Geranium*, etc.) will often turn blue in weak ammonia solution.

(*d*) Grind about 5-10 grams of orange-yellow *Calendula* petals with clean sand or emery and a little acetone. Transfer to a Buchner funnel and extract with 80% acetone, using suction. About 20 ml of the coloured acetone-extract are now transferred to a separating funnel and add an equal volume of ether and water to cause separation of an ether and an aqueous-acetone layer. The upper ether layer contains carotenoids while the water soluble anthocyanins remain in the lower aqueous-acetone layer. Pour portions of each extract into two test tubes and examine their relative colour intensities. Test each test tube with a few drops of dilute alkali (N/10) and also with a drop or two of concentrated HCl.

Betacyanin, the red pigment of sugar beet, however, though water soluble, is not an anthocyanin and thus shows no colour change with acid or alkali.

III. **Exchange of gas in photosynthesis.** Take three wide-mounted conical flasks, three shallow dishes and a large basin of water. Cut a thin flat piece of cork, smaller than the mouth of the flask and stick a small bit of candle on it so that it may float upright in the water in the basin. After lighting the candle on the floating cork, invert the flasks over it one by one, with the mouth of the flask lying below the surface of water. It will be seen that the candle will go out soon. The CO_2 produced by the burning of the candle will extinguish the light.

When the three flasks have about 0·3% of CO_2 of the contained air (indicated by the extinction of the candle flame), push a green fresh leafy shoot up into the flask A and slip a shallow dish under the mouth of the flask in water and lift it out of the basin, keeping the mouth of the flask under water in the dish. Stand the whole thing in bright light for about 24-28 hours. Treat the flasks B and C similarly but place B with leafy shoot inside in the dark. Allow the flask C to act as a control (i.e., do not insert any leafy shoot into it).

After the desired time, return the flasks to the water basin, withdraw the leafy shoots from flasks A and B and slip a plate of glass over the mouth of each flask under water. Take them out of water and now place the flasks, mouth upwards on the table. Test the enclosed gas in the flasks A, B and C by lowering into each a piece of lighted candle or simply a lighted match stick. It will be seen that only in the flask A, which contained the leafy shoot and stood in the light, the candle or the match stick continues to burn proving that CO_2 originally produced by burning of the candle inside the flask has been removed and replaced by oxygen in photosynthesis of the leafy shoot.

In the flasks B and C, the candle or the lighted match stick goes out immediately showing that the original CO_2 could not be consumed as the flask B was in the dark (no photosynthesis) and in C, there was no photosynthesising tissue.

IV. **Demonstration of necessity of chlorophyll in photosynthesis.** Select vigorous and fully grown leaves of *Coleus*, *Croton* or other suitable variegated starch-forming leaves. Outline the leaf and its various coloured areas on a paper. If the leaves show red colouring, heat in boiling water for a minute or two until the water soluble anthocyanin is extracted. Heat the leaf in boiling 95% alcohol in a water bath to remove chlorophyll from the green parts and test the bleached leaf for starch acumulation with iodine in KI solution. Compare the starch pattern with the leaf outline and observe that starch is only formed in the green portions of leaf. Pure white or yellow leaves may also be tested similarly showing no starch accumulation in such leaves.

V. **Demonstration of necessity of light in photosynthesis.** Shade a portion of a starch-free leaf from *Cucurbita*, *Geranium*, *Nasturtium*, etc., with *Ganong's light screen* (Fig. 712). The leaf may be made starch-free by putting the experimental potted plant in a dark room for about a day or two or attaching the light screen on to the leaf early in the morning before the sun rises. The same effect, however, could be obtained as well by covering a portion of the leaf with two uniform pieces of black paper. The advantage of the standard light screen is that, light is cut off from a portion of the leaf, without hindering proper ventilation.

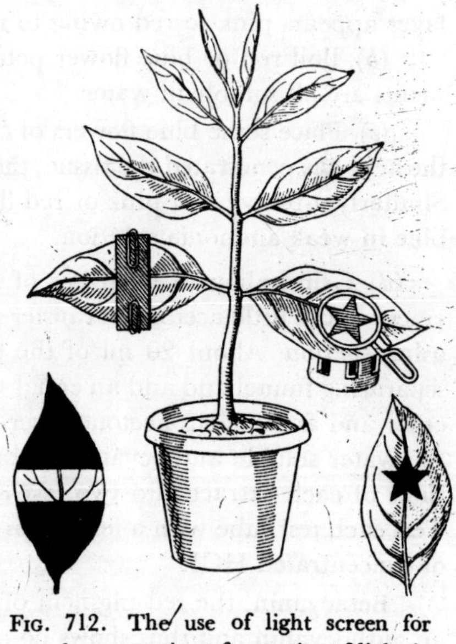

FIG. 712. The use of light screen for demonstration of necessity of light in photosynthesis.

The leaf with the light screen attached, can now be exposed to light in the open or kept near an open window for 24-28 hours. The light screen is then taken off, the leaf plucked and put into boiling 95% alcohol. When the leaf is completely decolourised, wash in water and treat with iodine solution for about a minute. Note that the exposed part turns blue showing formation of starch whereas the covered portion where light could not penetrate is yellowish, showing absence of starch. Instead of a light screen, a suitable photographic negative (negative must be light with clear-cut differentiation into black and colourless areas) can be used. After fixing the negative with meticulous care on a *Nasturtium* leaf (*Geranium* or any other suitable leaf might also do) and exposing the whole in strong light for a day or two and subsequently treating the decolourised leaf with iodine solution, reasonably good positive prints could be obtained.

VI. **Demonstration of necessity of CO_2 in photosynthesis.** Take a suitable leaf from a potted plant which has been kept in darkness for a period of 24-48 hours to free the leaf from starch. Now, take a wide-mouthed bottle and fit a split cork into the mouth. Pour a small quantity of 20% KOH solution in the bottle and clamp it horizontally, resting on a tripod stand. The starch-free leaf is now introduced half inside the bottle and half outside as in Fig. 713. (*Moll's experiment*). Make the corked mouth of the bottle airtight with vaseline and expose the whole apparatus in strong sunlight for about 12 hours. Remove the leaf from the bottle and from the plant, bleach it with alcohol and test with iodine (in KI) solution. The portion of the leaf, which was outside the bottle and which received both light and CO_2, becomes dark blue with iodine showing formation of starch. The other half of the leaf which was inside the bottle, from which all available CO_2 was removed by KOH solution, turned yellowish, showing absence of starch.

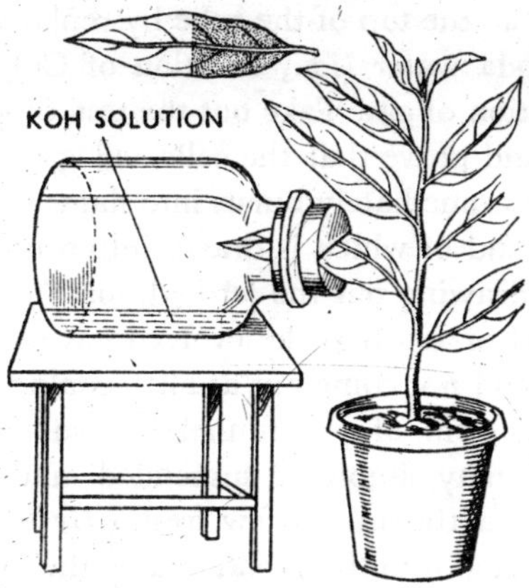

FIG. 713. Moll's experiment.

Take some starch-free leaves of ordinary land plants and after tying them to a stone, sink them under water in a glass jar. Illuminate the jar and after 24 hours, decolourise the leaves with boiling alcohol and test with iodine solution. No starch is found in these leaves. Repeat the experiment with water plants, e.g., *Ceratophyllum*, *Utricularia*, etc.; abundant starch is found in the leaves. It is apparent from this experiment that water plants can effectively utilise the dissolved CO_2 in water for photosynthesis whereas the source of CO_2 for land plants is exclusively atmospheric air.

VII. **Deomonstration of evolution of oxygen in photosynthesis**

(*a*) CHEMICAL PROOF OF PRODUCTION OF OXYGEN IN PHOTOSYNTHESIS. Make a very dilute solution of the dye, indigo carmine, with previously boiled and cooled distilled water. Add very carefully minute quantities of sodium hydrosulphite ($NaHSO_3$) until the blue colour just disappears. Take two test tubes, *a* and *b*, and completely fill them up with the decolourised and reduced dye solution. Put a few *Hydrilla* plants in each and cork carefully to exclude all contact with atomospheric air. Expose one tube *a* to bright light and the tube *b* is kept in the dark. Tube *a* becomes blue soon whereas *b* continues to be colourless (no photosynthesis). Indigo carmine which was reduced to a colourless compound by sodium hydrosulphite is oxidised back to its original state in the tube *a* by oxygen, evolved during photosynthesis.

(*b*) Place some water plants such as *Hydrilla* in a large beaker of tap water and expose the apparatus to bright sunlight. Observe that bubbles of gas are given off. Cover the vessel with a black cloth or replace the tap water with water which has been previously boiled and cooled. Note that bubbling in both cases, ceases. In the former it is due to light being cut off and

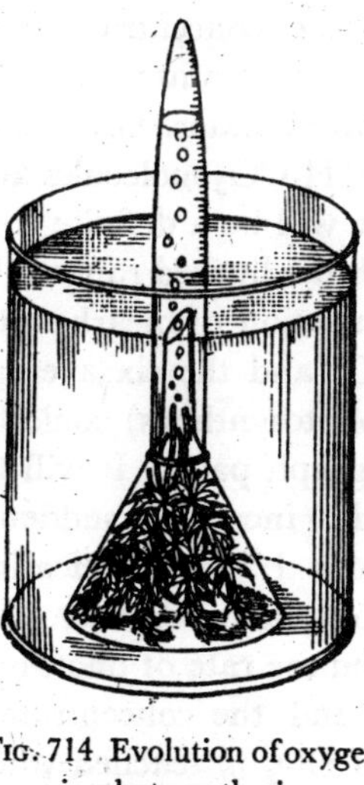
FIG. 714 Evolution of oxygen in photosynthesis.

in the latter, due to dissolved CO_2 being driven off from water by boiling. Press a glass funnel down on the plants, covering them up completely and invert over it a test tube filled with water (Fig. 714). Bubbles of gas collect at the top of the tube by replacement of water. The introduction of a few ml of soda water (i.e., addition of CO_2) improves the experiment by hastening up the evolution of gas. Take out the test tube from the vessel by closing the open end with finger and prove that the collected gas is chiefly oxygen by inserting a glowing splinter, which immediately bursts into flame.

Instead of the test tube, a small glass tube one end of which is drawn into a fine jet and slightly bent (proper care must be taken in drawing out the jet end, for if the bending is not just right or the jet opening too small or too big, the evolved gas may accumulate there and not come out at all) could be fitted to the short stem of the funnel by rubber tubing. Or the jet may simply be suspended under water by means of a thread. A few fresh *Hydrilla* plants are now inserted into the broad end of the jet (Fig. 715, modified *Wilmott's bubbler*) which is dipped in water and the apparatus put in strong sunlight. The water used may be previously boiled for about half an hour to drive off all the dissolved CO_2 and cooled. About 500 ml of this CO_2-free water are taken in a large beaker and six 1·25 gram-weights of $KHCO_3$ added and dissolved in water in the beaker by turn. (Sufficient time must be allowed for the adjustment of plant from one concentration to the next).

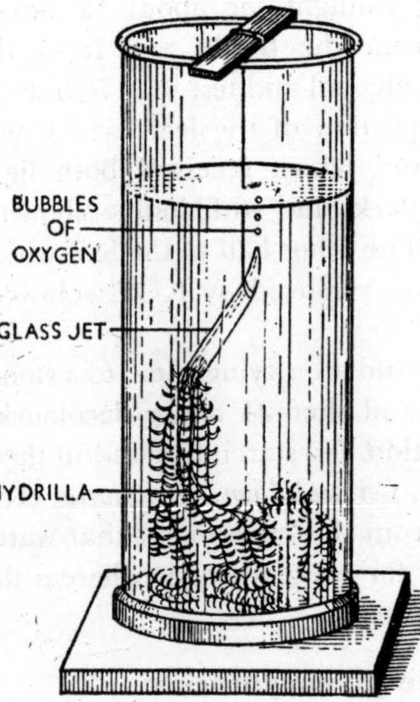

FIG. 715. Quantitative determination of the effect of CO_2^- concentration on the rates of photosynthesis.

Number of oxygen bubbles evolved, coming out through the fine jet end (it will be advisable to discard the very small bubbles coming out in torrents—it will almost be impossible to count them—and select only relatively bigger bubbles of more or less same size) is counted per unit time, preferably a minute in each of the six concentrations of $KHCO_3$ (0·25, 0·50, 0·75, 1·00, 1·25 and 1·50%; $KHCO_3$ solution contains ions of K^+ and HCO_3^- from the bicarbonate as well as H^+ and OH^- ions from water. Since both H^+ and HCO_3^- ions are present in equilibrium with undissociated H_2CO_3 molecules and H_2CO_3 with CO_2, for a definite concentration of $KHCO_3$, there will be a definite concentration of dissolved CO_2). Thus it may be assumed that within a limited range, the CO_2-concentration of the solution is proportional to the concentration of bicarbonate. Three readings should be taken for each concentration of CO_2 and the six average values of number of bubbles (taken as a measure of the rate of photosynthesis) could be plotted against CO_2-concentration (percentage $KHCO_3$) on a graph paper. It will be seen that the rate of evolution of oxygen bubbles which was steadily increasing suddenly shows a rapid decrease at a concentration of $KHCO_3$ of about 1·5-2%, indicating that at this high concentration, CO_2 supply has become toxic for photosynthesis. This clearly demonstrates that there is a direct proportionality between the rate of photosynthesis as measured by the number of oxygen bubbles evolved and the concentration of CO_2 in the external medium until a certain concentration of CO_2 is reached when photosynthesis shows an abrupt fall. Alkalinity at higher concentrations of $KHCO_3$ may also be another cause of the decrease in the rate of photosynthesis.

VIII. Ganong's photosynthometer

This is a standard apparatus for simultaneous quantitative determination of the volumes of CO_2 utilised and O_2 evolved by a green tissue in photosynthesis. With accurate handling an idea about the photosynthetic quotient (O_2/CO_2) can be obtained. The apparatus essentially consists of a glass bulb (Fig. 716) and a graduated measuring tube with stop-cock. The experimental material (about 2 ml of suitable leaf, such as *Geranium*, *Nasturtium*, etc.) is placed in the glass bulb. Inverting the measuring tube and closing the stop-cock it can be filled with water up to a mark corresponding to the desired percentage of CO_2. Placing the hollow stopper on the tube and closing the tap, fill up the hollow with water. Closing it with hand, invert it into a basin of water, clamping it so that the water is level with the bottom of the stop-cock hole. Opening up the lower tap, connect the top of the tube with a Kipps apparatus, generating CO_2. Open the top tap cautiously to allow CO_2 to get into the tube and close it again, when the water has been displaced down in the tube to the level of the water outside. Close both the taps and transfer the whole tube to the top of the glass bulb containing the experimental material. Open the lower stop-cock carefully now and allow the CO_2 gas in the tube to diffuse into the bulb. Place the apparatus in bright sunlight for about 3–4 hours and after noting the time, shut the lower stop-cock and take out the tube from the bulb. Put it into a basin of water and remove the hollow stopper under water. Fix the graduated tube with the zero mark exactly at the surface of water and then open the top tap gently to let water rise inside the tube exactly to the zero mark.

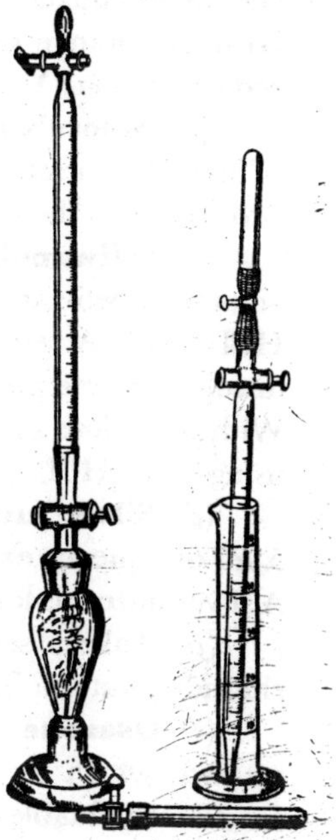

FIG. 716. Ganong's photosynthometer.

Now fill one test tube completely with 30% KOH solution and slip a rubber connection on to the bottom of the graduated tube (as shown in the Fig. 716) thus connecting the two. Lift all out of water, open the clamp and allow the KOH to flow into the graduated tube. Shake thoroughly, drain back the KOH solution into the test tube and shut the clamp. Place the end of the graduated tube under water with zero mark at the surface level and disconnect the test tube. The water will now rush up to a mark corresponding to the volume of CO_2 absorbed by KOH (volume of CO_2 left unutilised by the photosynthetic experimental material after the desired time). Repeat the test with a test tube filled with alkaline pyragallol to absorb the oxygen evolved. The reduction in the volume of CO_2 originally present and the gain in O_2 give us the volumes CO_2 absorbed and O_2 evolved in photosynthesis. These two values resemble closely, thus proving that the photosynthetic quotient is generally near about unity.

QUALITATIVE TESTS FOR CARBOHYDRATES

I. General test for carbohydrates: Molisch test

An aqueous extract of plant tissue is shaken with a few drops of α-napthol solution. (1% α-napthol in 70% alcohol) and conc. H_2SO_4 is carefully poured down the inside of the test tube. A violet colouration, sometimes as a violet ring, is formed at the junction of the two liquids. This is a general test for all carbohydrates.

II. **Reducing sugars**

All monosaccharides, trioses, pentoses and hexoses, etc.; disaccharides, e.g., maltose, etc. Work with extracts of fruits and make aqueous solutions.

(*a*) **Fehling test.** In an aqueous extract of the tissue in a test tube, add an equal volume of Fehling A and B mixture. (Fehling A: 35 gm., $CuSO_4$, $5H_2O$ in 500 ml water; Fehling B: NaOH, 50 gm. and Rochelle salt, 173 gm., dissolved in 500 ml of water. When the two are mixed, the ppt. of $Cu(OH)_2$ is kept in solution by rochelle salt). Boil in water bath. An orange red ppt. of Cu_2O shows the presence of reducing sugars. $Cu(OH_2)$ is reduced to Cu_2O, whereas reducing sugars are oxidised to their corresponding acids.

(*b*) **Moore's test.** Boil a little aqueous extract in a test tube with an equal volume of NaOH solution. If free glucose is present, the solution turns yellow and then reddish-brown due to the formation of a condensation product (caramel) of the sugar.

(*c*) **Seliwanoff's test.** To about 5 ml of aqueous extract, add about an equal volume of Seliwanoff's reagent (0·1 gm of resorcinol in 100 ml of 50% by volume conc. HCl). Boil. A red colouration, often a red ppt. is formed. Addition of a little alcohol dissolves the red ppt. This is a specific test for all reducing *ketoses* ($=C=O$, e.g., fructose). With an aldose sugar, e.g., glucose, galactose, etc., the red colour is produced only by using conc. HCl.

(*d*) **Silver mirror and metallic mercury precipitation tests.** (*i*) Take about 5 ml of aqueous extract with ammoniacal $AgNO_3$ in a test tube. Boil in a steam bath. A silver mirror is formed due to precipatation of silver.

(*ii*) Take a second aliquot of 5 ml of aqueous extract with alkaline mercuric chloride solution. Boil as before. The mercury salts are reduced to metallic mercury.

(*e*) **Osazone test.** All reducing sugars—aldoses or ketoses—form crystalline *osazones*, which by virtue of their melting points and characteristic crystalline forms, constitute valuable identification tests for these sugars. Glucose, mannose and fructose all form identical osazones (bunches of fine, yellow needles, variously aggregated into clusters, sheaves or rossettes) whereas maltose gives osazones of small but broader and flatter needles.

Take about 10 ml of aqueous extract in a test tube, add a few crystals of phenyl-hydrazine hydrochloride and at last twice that amount of solid sodium acetate and also a few drops (about 1 ml) of glacial acetic acid. Place the test tube in a beaker of boiling water for about 15 minutes. Cool slowly under the tap and observe a little of yellow crystalline ppt. under microscope. It is seen that most of the crystals are in the form of clusters of needles.

Make a solution of pure maltose and examine the osazones formed. Compare them with glucosazones.

III. **Specific tests for pentoses, e.g., arabinose**

(*a*) Heat a few ml of a solution of arabinose with half its volume of conc. HCl in a test tube and hold a piece of filter paper soaked in aniline acetate solution near the mouth of the tube. A red colour is developed on the paper.

(*b*) Warm arabinose solution with an equal volume of conc. HCl and add a small quantity of phloroglucinol. A bright red colour is produced.

IV. **Sucrose or cane sugar and other simpler polysaccharides**

Observe that the tests for reducing sugars give *negative* results with sucrose soln.

Hydrolysis of sucrose: Take about 10 ml of pure sucrose soln. Boil for 10 min. with

2·5 ml of N H_2SO_4. Cool, neutralise with N NaOH ending up with N/10 H_2SO_4, using phenolphthalein as indicator. Dilute with water to about 50 ml. Perform all the previous tests for reducing sugar and note that they are all positive now. The positive result with Seliwanoff's reagent proves the presence of fructose as a constituent of the sucrose molecule.

Hydrolysis of simpler polysaccharides in plant extracts: Many polysaccharides are non-reducing carbohydrates. Prepare an extract of fruit and test for free-reducing sugars with Fehling soln. Filter off the ppt. Repeat it once more so as to ensure the removal of all reducing sugars. Now hydrolyse the soln. with N H_2SO_4 as before. Neutralise with N NaOH and perform the tests for reducing sugars, including Seliwanoff's.

QUALITATIVE TESTS FOR TANNINS

Materials: Barks of trees, unripe fruits, e.g., guava, banana, etc., fruits of myrobalan, tea leaves, etc.

10 gm. of material (if dry, powder) are taken in 100 ml water, boiled for half an hour and filtered: make the following experiments with the filtrate:

I. Treat extract with a drop or two of $FeCl_3$ (5%) soln. A deep blue-black colour is produced. After some time the insoluble salts of tannic acid separate out.

II. Dilute a few drops of tannin extract with water. Add a little dil. NH_4OH and then a few drops of potassium ferricyanide soln. A red colouration appears.

III. To about 5 ml of the tannin extract in a test tube, add an equal volume of saturated potassium dichromate soln. Heating might quicken the process of formation of a voluminous red ppt.

IV. In a test tube take a few ml of $AgNO_3$ soln. and a few drops of NH_4OH. If this mixture is boiled for some time with tannin soln., dark colloidal silver separates (silver mirror may be formed) out due to the reducing properties of tannin phenols (*catechol* and *pyragallol*).

V. Add a drop or two of iodine soln. to tannin extract. A transient red colour is seen.

VI. Make an aliquot of tannin soln; carefully make it alkaline with NH_4OH or with lime water. A reddish or blue colouration is seen.

VII. Add $KMnO_4$ to tannin extract. Acidify with HCl or H_2SO_4. $KMnO_4$ is decolourised if pyragallol is present.

VIII. Add HNO_3 to a tannin solution. A reddish to yellow colour is observed.

IX. Tannin extracts, with a few drops of acetic acid, give dark red colour.

X. To 5 ml aliquots of tannin extract in three test tubes, add respectively: (*a*) 5% lead acetate soln. The tannin may be pptd. The ppt. may turn red on the addition of dil. KOH and may dissolve to a red soln. with excess of potash; (*b*) 5% lead nitrate. Tannin is pptd; exception is gallic acid and (*c*) add 10% gelatine, drop by drop, after gentle warming. The *gelatin is pptd;* exception is gallic acid.

CHAPTER XLII

PROTEINS AND THEIR SYNTHESIS

Proteins and other related nitrogenous compounds are the principal constituents of protoplasm and hence are directly or indirectly involved in all the physiological processes taking place in the living cells. Evolution of life itself depended upon the pre-existence of the protein molecule. As soon as the protein molecule was formed from inorganic materials, evolution of life became possible. Proteins often occur in plants in the form of stored foods, particularly in the seeds of many species. Such stored or reserve proteins differ markedly in their physical and chemical properties from the protoplasmic proteins which are highly complex and do not respond to the usual chemical protein tests.

The possible role of nucleic acids in protein synthesis has been analysed recently.

All proteins contain carbon (50–54%), hydrogen (about 7%), oxygen (20–25%) and nitrogen (16–18%). All plant peoteins contain sulphur; sulphurous amino acid, methionine, has an empirical formula, $C_5H_{11}O_2NS$. Phosphorus is also a component of certain fundamental plant proteins.

The dimensions of protein molecules are enormous, the molecular weights of plant proteins usually ranging from a mere 10,000 to 6,000,000. *Insulin*, an animal protein (vertebrate hormone, secreted by pancreatic glands) is, however, the smallest known protein molecule, with a molecular weight of only 5,733 of 2 subunits each, i.e. =11,466. The molecular weights of virus proteins (nucleoproteins) of tobacco mosaic have been found to be in the neighbourhood of 40,000,000. Though the amino acid compositions of some proteins have been worked out, there are very few proteins again (exceptions *insulin* composed of 51 amino acids of 15 different types in two distinct chains and *myoglobin*) whose actual structure—spatial arrangement of amino acids—are known with certainty.

Proteins can be hydrolysed by treating them with acids or with suitable proteolytic enzymes. The end product of complete hydrolysis of any protein is always a mixture of amino acids. During the course of hydrolysis a number of intermediate compounds of varying complexities are formed:

Proteins⟶proteoses⟶peptones⟶polypeptides⟶dipeptides⟶amino acids.

It is evident, therefore, that the amino acids are structural units from which proteins are ultimately synthesised in the living cells.

The classification of proteins

Because the molecular structure of most proteins are so imperfectly understood it is impossible to classify them on a strictly chemical basis. The physical properties such as solubility in acids, alkalis or salt solutions, coagulation, etc., are sometimes taken as basis of classification. The following *three* main groups can be recognised:

I. *Simple peoteins*—yield only amino acids on hydrolysis. The more important simple proteins in plant cells are *albumins*, *globulins*, *histones*, *prolamines*, *glutelins*, etc. Albumins are characterised by their solubility in water; globulins are water-insoluble but readily dissolve in dilute salt solution. Prolamines and glutelins, which are soluble in 70% alcohol, are found in cereals.

II. *Conjugated proteins*—the members of this group are proteins with which are combined some non-protein groups. Examples are afforded by *nucleoproteins* (genes and viruses), ***phosphoproteins*** (casein of milk), ***chromoproteins*** (haemoglobin of red blood cells and phycobilins present in certain algae of the groups, *Cyanophyta*, *Rhodophyta* and *Cryptomonad* algae) and *Lecithoproteins* or *lipoproteins* (combination of protein with phospholipid, *lecithin*)—important constituents of protoplasm.

III. *Derived proteins*—this group includes products that are formed from partial hydrolysis or decomposition of simple proteins. *Proteoses*, *peptones* and *peptides* are representatives of the derived proteins.

Properties of proteins and amino acids

Proteins are colloidal in nature and do not readily diffuse through plasma membranes. Almost all proteins are irreversibly coagulated by heat. Proteins are insoluble in neutral salts such as NaCl, $MgSO_4$, etc., in which they precipitate without change in composition but go into solution on further dilution of the salt. Some are soluble in alkali while basic proteins are soluble in acids.

The amino acids are amphoteric compounds for the amino group is basic in reaction whereas the carboxyl group is acid; hence the amino acids are reactive and unite together to form larger molecules, the so-called *peptide linkage*—CONH—taking place between the amino and carboxyl groups with the elimination of water. Proteins consist of chains of amino acids ranging from a few to many, united by such peptide linkages. Plant and animal protein molecules generally consist of about 200–1000 amino acid units. About 20 amino acids make up all proteins and all of them are characterised by the molecular grouping R—$CHNH_2COOH$, with R standing for benzene or indole nucleus, an aliphatic chain or a sulphur-containing group.

All the amino acids found in plant proteins are of α-amino type.

The amino group, NH_2 replaces hydrogen of one of the alkyl radical and the amino group is always attached to the so-called α-carbon atom which is next to the carboxyl group.

$$\underset{\text{acetic acid}}{CH_3COOH} \xrightarrow{+NH_2-} \underset{\substack{\text{amino acid,} \\ \textit{glycine}}}{CH_2NH_2COOH}$$

$$\underset{\text{propionic acid}}{CH_3CH_2COOH} \xrightarrow{+NH_2-} \underset{\substack{\text{amino acid,} \\ \textit{α-alanine}}}{CH_3CHNH_2COOH}$$

Unlike animals, plants must synthesise all the amino acids necessary for the formation of proteins. In addition, however, plants synthesise about 60 amino acids which, so far as is known, are not incorporated into protein.

Amides are salts of amino acids which correspond to inorganic salts. The two most commonly occurring amides in plants are *asparagine* and *glutamine*. Amides are formed from amino acids by replacement of hydroxyl part of the acid by another NH_2 radical, e.g., asparagine. The amino acid concerned here is *asparatic acid* (derived from succinic acid). The amide glutamine is a derivative of the amino acid, *glutamic acid* which is obtained from the parent, α-oxoglutaric acid.

$$\underset{\text{succinic acid}}{\begin{array}{l} CH_2COOH \\ | \\ CH_2COOH \end{array}} \xrightarrow{+NH_2-} \underset{\substack{\text{amino acid,} \\ \textit{aspartic acid}}}{\begin{array}{l} CH_2COOH \\ | \\ CHNH_2COOH \end{array}} \xrightarrow{+NH_2-} \underset{\substack{\text{amide,} \\ \textit{asparagine}}}{\begin{array}{l} CH_2CONH_2 \\ | \\ CHNH_2COOH \end{array}}$$

It is evident that amides contain more nitrogen compared to amino acids. In the green leaf, breakdown of proteins results directly or indirectly in the production of some amide, asparagine. Proteins are broken down to amino acids and the amino acids thus formed are converted to amides at the expense of respiratory energy (ATP molecules).

Synthesis of proteins

Leaves must synthesise amino acids before they can make any proteins. If plants absorb inorganic nitrogen from soil in the form of NO_3^- ions, the NO_3^- ions must be converted into NH_2 groups before being elaborated into amino acids. This would seem to indicate that NH_4^+ form of nitrogen would be more efficient source of nitrogen for the rapid formation of amino acids, since respiratory energy must be utilised to reduce NO_3^- to NH_4^+ form. Nitrate is reduced by a molybdoflavo-protein, nitrate reductase, the electrons for reduction being provided by NAD(P)H. The reduction of NO_3^- to NH_4^+ form can also be accomplished by NADPH obtained from photolytic decomposition of water by chlorophyll. However, under certain conditions NO_3^- form may be as satisfactory source of nitrogen as NH_4^+ form.

With the formation of NH_4^+ ion, the evidence indicates that it quickly combines with various keto acids—some of which are obtained from the Krebs cycle—to form amino acids. The synthesis of protein as also of fats is intimately linked with carbohydrate metabolism through the PSCR and Krebs cycles.

Specific enzymes are required for these synthesis. Amino acids are synthesised either in light or in darkness. In roots of some species of plants, synthesis of amino acids occurs in dark and they are translocated to the growing shoots where they are built up into proteins. In other plants, the inorganic nitrogen, either as NO_3^- or NH_4^+ form is translocated directly to the leaves. The amino acids are synthesised in the leaves in presence of light and ultimately built up into proteins.

The synthesis of a particular protein requires that its constituent amino acids be arranged in a specific order. Recent theories postulate that the site of protein synthesis in the cell is capable of doing these because it continually receives information about the correct sequences of amino acids from the ultimate controller, the gene, situated in the chromosomes. The site of protein synthesis in prokaryotes is the 70S ribosome, in plants and other eukaryotes it is 80S. Each ribosome has two subunits, 30S and 50S in the case 70S particles, and 40S and 60S in the 80S particles. The formation of peptide linkage involves the expenditure of energy; this is provided by ATP and GTP (guanosine triphosphate). In the first step of protein synthesis amino acids are activated by ATP, to form aminoacyl adenylates—from which each amino acid is transferred to a specific type of RNA-molecules—small in size being about 78 nucleotides long, and soluble in water—each specific for a particular amino acid. These RNA-molecules are transfer or soluble RNA (tRNA or sRNA). Unlike DNA these RNA molecules are single stranded, but are folded in the pattern of a clover leaf. The amino acids are hooked on at one end and the bulge at the opposite end is used for proper association with the messenger RNA at a particular region.

Biological information as is now known is coded in the four letter alphabet of nucleic acid bases—adenine, guanine, cytosine and thymine (See Chapter LXII for structures). The three dimensional chemical structure of adenine is such that it fits snugly only with thymine of DNA (or uracil of RNA), and guanine fits only with cytosine. Adenine is connected to thymine by a pair of hydrogen bonds; three H-bonds are formed between guanine and cytosine. When RNA is synthesized by RNA polymerase, the sequenee of nucleotides in the RNA, becomes exactly complementary to that in DNA, it thus is a negative replica of the nucleotide sequence in DNA. This RNA is considered to carry the message of DNA. This message has now to be translated into another language which is a 20-letter alphabet—in the form of 20 amino acids—permutations of which constitute different protein molecules.

This RNA now associates with a large number of 70S or 80S ribosomes—the polyribosomes or polysomes. For initiation and elongation and termination of the polypeptide chain several factors (which are themselves proteins) are required. The different factors in prokaryotes and eukaryotes are very similar. They are referred to as $1F_1$, $1F_2$, $1F_3$ in so far as initiation of the protein chain is concerned. The mRNA associates with the 30S or 40S subunit first, with the help of 1F-2.

The first amino acid carried by a tRNA which initiates the polypeptide chain formation is methionine—usually formylated (formylated in prokaryotes and chloroplasts, not in plants); its tRNA has a sequence of UAC in the anticodon region and thus associates with the AUG region of mRNA. On each ribosome the tRNA attaches at the sites—one is called the P (peptidyl or donor) site, and the other A (or acceptor) site. The amino acid carried by the tRNA on the P site is transferred to the next amino acid on the tRNA attached to the A site forming a peptide linkage between the free $-NH_2$ and $-COOH$ groups of the two amino acids. The tRNA on the donor site unloading its amino acid becomes free. The mRNA then moves like a conveyor belt bringing in a new amino acyl tRNA to the acceptor site, as the previous one occupies the P position. The peptide chain is then transferred to the newly arrived amino acid, and the process continues.

For elongation, elongation factors Ts and Tu are required in prokaryots, an EF-1 and EF-2 for plant systems. EF-1 stimulates the binding of amino acyl tRNA to ribosomes. The enzyme which forms the peptide bond is known as peptidyl transferase and translocation is achieved by EF-2, presumably a translocase. For translocation of peptidyl tRNA from A site to P site, another factor G is used.

The nucleotide sequence in the mRNA is specific for the nucleotide sequence in the anticodon region of the tRNA carrying a particular amino acid. It is now known that a three nucleotide sequence—a triplet—codes for a particular amino acid. Thus UUU codes for the amino acid phenylalanine, UGC for cysteine, AUG for methionine, and so on. The code is however, degenerate, though universal. Thus, leucine has 6 codons. i.e., there are 6 different tRNA molecules which can recognize leucine and carry it to the site of protein synthesis. There are however, three nucleotide sequences UGA, UAA and UAG, which are not the codons for any amino acid, i.e., no tRNA can recognize it (unless mutated). Consequently when the elongating polypeptide chain comes to a point where one of these codons are there, the chain cannot elongate any further. However, the ribosome with the mRNA keeps moving resulting in the dislodgement of the polypeptide chain from the ribosome–aminoacyl–tRNA complex and it is set free. Some release factors R_1, R_2 and R_3 are believed to be required for this. Then it becomes folded in a particular fashion to contribute to the three dimensional structure of the protein molecule. The different steps involved are given is pp. 616–617.

The genetic code has been worked out largely through the effort or Nirenberg, Mathaei and Khorana. Nucleotide sequence of a tRNA was first achieved by Holley's group. Marcus and associates have contributed considerably to our knowledge of protein synthesis in plants. The wheat germ system has been extensively investigated.

In some bacteria as in *Bacillus*, certain types of polypeptides are synthesized without the participation of ribosomes, mRNA and tRNA. Here the sequence is maintained by association of amino acids to a multienzyme complex in a specific manner followed by the formation of peptide linkage. The peptides may be cyclic and some amino acids are D-amino acids unlike L-amino acids used in ribosomal protein synthesis. Bacterial cyclic polypeptide antibiotics like bacitracin, gramiciden, polymyxin, etc. are synthesised in this way.

Step I **Amino acid activation**

Amino acid +ATP+Ligase⟶Ligase-aminoacyl AMP +Pyrophosphate.
Ligase-aminoacyl−AMP+tRNA (Aminoacyl tRNA synthetase) →Ligase+ AMP+aminoacyl−tRNA

Amino acid +ATP+tRNA⟶Aminoacyl tRNA +AMP +PP.

Step II **Binding of mRNA to ribosome**

mRNA+30S or 40S ribosome $\xrightarrow[\text{GTP}]{\text{1F-2, 1F-3}}$

mRNA bound to ribosome.

Step III. **Binding of initiating aminoacyl tRNA to mRNA ribosome complex and chain initiation**

mRNA bound to (30S or 40S) ribosome +formyl methionyl tRNA (or methionyl tRNA) $\xrightarrow[\text{GTP}]{\text{1F-1, 1F-2}}$ formylmethionyl—tRNA—mRNA—ribosome complex.

Step IV **Binding of 50S or 60S subunit to the initiation complex**

Step V **Chain elongation**

(i) Aminoacyl −tRNA+GTP+Tu−Ts (or EF1-EF_2) (elongation factors).
⟶Aminoacyl tRNA−TU−GTP.

(ii) Aminoacyl tRNA−TU−GTP on P site +peptidyl (n) tRNA on A site ⟶ peptidyl (n+1)—tRNA on A site +TU−GDP+tRNA on P site.

(iii) Tu−GDP+Ts⟶Tu−Ts+GDP.

Peptide bond is formed by peptidyl transferase or peptide synthetase.

Step VI **Translocation**

Deacylated tRNA on P site + peptidyl—tRNA on A site of mRNA → mRNA bound to ribosome $\xrightarrow[\text{GTP}\rightarrow\text{GDP}]{\text{G Factor}}$ Movement of ribosome with respect to mRNA bound to ribosome with peptidyl tRNA on P site+deacylated tRNA (released).

Step VII **Chain propagation**

Repetition of steps V and VI for continued addition of amino acids to the carboxyl end of the growing polypeptide chain.

Step VIII **Chain termination**

When UAA, UAG or UGA codons of mRNA are reached,

$$\text{peptidyl tRNA—mRNA bound to ribosome} \xrightarrow[\text{Factor}]{\text{Release}} \text{peptide chain} + \text{mRNA} +$$

tRNA + 70S or 80S ribosomes.

Step IX. **Dissociation of ribosome**

$$\begin{matrix} \text{70S ribosomes} \\ \text{or} \\ \text{80S ribosomes} \end{matrix} \xrightarrow[\text{Factor D (1F}_3\text{?)}]{\text{Dissociation}} \begin{matrix} \text{30S+50S} \\ \text{or} \\ \text{40S+60S subunits} \end{matrix}$$

Schematic representation of the steps involved in ribosomal polypeptide synthesis (I F, initiation factors, (Peptidyl) *n*, peptide chain containing *n* aminoacids is given above.

An important feature of protein metabolism of the plant as a whole is the fact that the plant proteins appear to be in a state of continuous fluctuation wherein protein breakdown is balanced by resynthesis. Even if the total protein content of a leaf may remain constant, this certainly does not mean that the same individual protein molecules remain in the leaf. On the contrary the decomposition of protein which is taking place continuously is more or less evenly balanced by protein synthesis. If we supply a mature plant, in which the total amount of protein in the leaves is relatively constant, with nitrogen in the form of NH_4^+ ions, it could be shown with the help of tracer technique, using ^{15}N, that entirely new proteins have been synthesised. Since, however, total protein content remains constant, the new synthesis must have been balanced by an equal amount of protein breakdown.

A diagrammatic representation in broad outlines of protein synthesis and protein breakdown is given here:

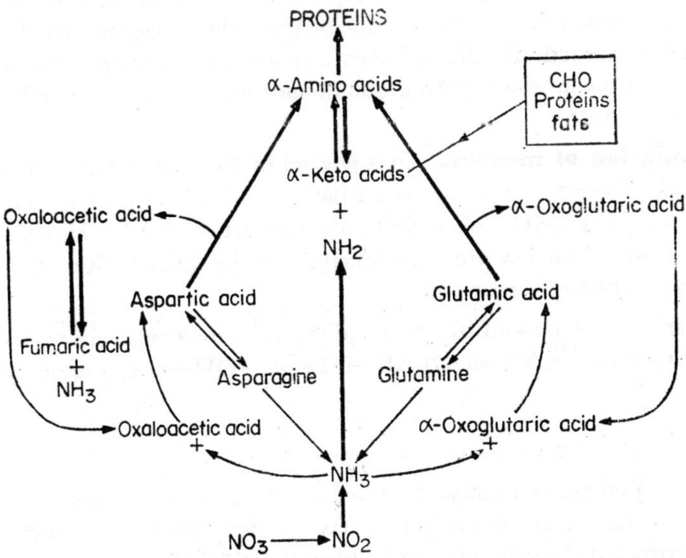

FIG. 717. Simple diagrammatic representation of probable pathways of protein metabolism in green plants.

QUALITATIVE TESTS FOR PLANT PROTEINS AND AMINO ACIDS

A. Proteins

Make extracts from crushed seeds, wheat flour, germinating seeds, etc. Take 10 g. of material, stir in 100 ml water or phosphate buffer and allow to stand for half an hour. Filter and make the following tests with the filtrate:

I. **Biuret test.** To a portion of the aqueous extract add about 1 ml of 4% NaOH soln. and a drop of 1% $CuSO_4$ soln. A violet or pink colour is produced. This is a general test for all proteins containing the—CONH—group.

II. **Xanthoproteic test.** To a portion of the aqueous extract add conc. HNO_3 (about 1/3 the vol.). A white ppt. is formed which on boiling may turn yellow and may also partially dissolve to give a yellow soln. Cool under the tap and add strong NH_4OH till the reaction is alkaline. The yellow colour becomes orange. This test is given only by those proteins which contain the amino acids like tyrosine and tryptophan.

III. **Millon's test.** To about 5 ml of the protein extract, add about 1/2 its vol. of Millon's reagent (5 grams of Hg in 95 ml conc. HNO_3. Stir occasionally until brown fumes are no longer evolved. Add 200 ml water). A white ppt. is formed. On warming, the ppt. turns brick red, or it may dissolve, giving a red soln. Given only by the proteins having a phenolic grouping in the molecule, e.g., tyrosine.

IV. **Sulphur test.** Boil about 5 ml of the plant extract with an equal vol. of 40% NaOH and a drop or two of 10% lead acetate. The soln. turns black or brownish due to the formation of PbS by the action of strong alkali on the sulphur of the protein [$Na_2S + (CH_3COO)_2Pb = PbS + 2CH_3COONa$]. This is a positive test for the sulphur containing amino acids, e.g., cystine, methionine, etc.

V. **Adamkiewicz's reaction.** If conc. H_2SO_4 is added to a soln. of protein to which some acetic, or better, glyoxylic acid has previously been added, a violet colour is produced. The colour will appear as a ring at the juncture of the two liquids if H_2SO_4 is poured carefully down the side of the tube or throughout the mixture if shaken up. It depends upon the interaction of glyoxylic acid, which is generally present as an impurity in acetic acid, with the amino acid tryptophan and thus this test is given by all proteins containing tryptophan.

VI. **Precipitation of proteins.** To a few ml of purified extract, add an excess of absolute alcohol. Protein is precipitated. Take about 5 ml aliquots of the extract in 3 test tubes and add respectively 5% $CuSO_4$, 5% Pb-acetate and 5% $HgCl_2$ soln.; protein is pptd. in each case. Proteins are also precipitated by $(NH_4)_2SO_4$, phosphotungstic acid, protamine sulphate, acetone etc.

Proteins are best separated by gel filtration using molecular sieves like Sephadex columns or chromatography on DEAE—cellulose, CM—cellulose or by gel electrophoresis.

B. Identification of Amino acids

By paper chromatography. Mustard or legume seed, ground and covered with water, is incubated for about three days under toluene (to prevent bacterial infection) or extracted directly with 70% ethanol. 25 ml of this solution is pipetted off and filtered. Acidify the filtrate with acetic acid. Boil and filter off any further ppt. on cooling. Test aliquots of clarified extract for amino acids, *tyrosine*, *tryptophan*, *arginine* and *cystine*. Evaporate a part of the clarified extract nearly to dryness and apply the solution at the corner of a filter paper. Develop the chromatogram with phenol/water and after drying, spray with 0·1% ninhydrin solution (the colour change due to reaction of ninhydrin with amino acids is utilised for the detection of the amino acids). Estimate the RF value (the ratio of the distance traversed by the solute to that of the solvent front) of the spots and compare with spots containing known amino acids.

Specific tests for plant amino acids. (*a*) *Test for tyrosine:* Add a few drops of Millon's reagent to the extract in a test tube. Take care to avoid excess. Heat and observe the gradual development of a dark red colour.

(*b*) *Test for tryptophan:* (*i*) Add a few drops of glyoxylic acid and H_2SO_4 to the extract in a test tube. Development of reddish-violet ring at the juncture of the two liquids is seen (*see* tests for proteins). (*ii*) Add bromine water to another part of the extract. There is pink colouration.

(*c*) *Test for arginine:* To 5 ml of the extract, add a few ml of a saturated solution of flavianic acid in water. A thick yellow ppt. is observed.

(*d*) *Test for cystine:* (*i*) To a part of the extract, add sodium nitroprusside soln. and then a few drops of 40% NaOH. A conspicuous violet colour is developed. (*ii*) Boil about 5 ml of the extract with a little of 40% NaOH and a few drops of 10% lead acetate solution. There is a black ppt. of lead sulphide (*see* test for proteins).

CHAPTER XLIII

FATS AND THEIR SYNTHESIS

Fat and fat-like substances are invariably present in all living cells, plants or animals. They are produced in all actively metabolising plant cells and they serve a number of indispensable roles in plants, particularly as reserve food substance just as does starch.

The fats and certain other compounds, more or less closely related, are often called *lipides* (*lipids*) or *lipoids*.

Lipids have been synthesised in the laboratory by man, but only slowly. In contrast to proteins, lipids are generally molecules of small weight, seldom over 1,000. In general, lipids are insoluble in water, but soluble in the so-called fat-solvents —ether, chloroform, benzene, etc. Chemically they are esters of fatty acids and glycerol. The following simple classification of lipids, found in plants, is universally accepted:

I. Simple lipids—esters of fatty acids with alcohols:

(*a*) *Fats*—esters of fatty acids with *glycerol* (glycerine)—fatty acids esterified with glycerol to form triglycerides. These are true fats and are indispensable as reserve food material and as a source of energy in all living cells.

(*b*) *Waxes*—esters of higher aliphatic fatty acids, with long chain alcohols *other than glycerol*. Waxes are sometimes important components of the cuticle of the epidermal cell walls. Waxy layer sometimes seems to inhibit bacterial, fungal or insect attack. In insectivorous *Nepenthes*, waxy interior of the pitcher effectively helps it to trap insects.

II. Compound lipids—esters of fatty acids containing groups in addition to alcohol and fatty acid radicals:

Phospholipids—fats containing phosphoric acid. *Lecithin*, the best knwon among phospholipids in plant cells, is an essential structural material for all living cells. Sulpholipids contain sulphur.

Galactolipids—are the major lipid constituents of green leaf tissue. The most important types in the chloroplasts, mentioned before, are monogalactosyl diglyceride and digalactosyl diglyceride.

III. Derived lipids and related compounds—certain substances related to lipids or derived from compounds in the above groups by hydrolysis:

(*a*) Fatty acids of various series.

(*b*) *Terpenes and terpenoids*. These are formed from C_5-isoprene units e.g., camphor (monoterpene), gibbereliic acid (diterpene), oleanolic acid (triterpene). Tetra- and pentacyclic terpenoids also occur in plants.

(*c*) *Sterols*—these are mostly complex alcohols of very large molecular weights, soluble in fat-solvents like ether, chloroform, etc. Collectively the sterols found in plants are called *phytosterols*, the most important of which are *cholesterol*, *ergosterol*, *stigmasterol* and *sitosterol*. These are related to the triterpenes and some are polycyclic.

Fats, like proteins and carbohydrates serve as reserve food in seeds. In almost all seeds, fat is found as the principal reserve material. In many seeds, fats make up nearly 35–50% of the total dry weight material while vegetative tissues in general contain less than 5% of fat on a dry weight basis. Fats occur in leaves, stems, fruits, flowers and even in pollen grains but the largest concentration is certainly in the seeds. As fats are insoluble in water, they are found in plant cells in the form of very small droplets or globules, dispersed in the protoplasm. Sometimes they are large enough to be seen under microscope, particularly when strained red with Sudan III.

Fats contain the same elements C, H and O as the carbohydrates but the proportion of carbon to oxygen is much greater in fats than in carbohydrates or in other words carbon in fats is in a more highly reduced form than in carbohydrates.

The principal fatty acids of higher plant fats are those containing 12–26 carbon atoms. The fatty acid *lauric* (12 carbon atoms), *palmitic* (16 carbon atoms), *stearic* (18 carbon atoms) and *oleic* (18 carbon atoms) are the commonly occurring fatty acids in the plant kingdom and almost all fats, identified in plants, contain these fatty acids as their major components. Oleic acid, the most abundantly found fatty acid in plant, constitutes almost 35–40% of the total fatty acids produced in all the world's edible fats. Fatty acids may be saturated or unsaturated having one to few double bonds.

Plant fats may be either liquid or solid at ordinary temperatures. The liquid fats are commonly referred to as *oils* while solid fats are fats proper. The solid fats generally contain a much larger proportion of saturated fatty acid than oil whereas oil contains proportionately a much higher percentage of unsaturated fatty acids (60–75% of linseed oil is unsaturated fatty acids). Conversion of liquid plant oils such as from soybean, peanut, nuts, etc., into solid fat by artificial hydrogenation (direct addition of hydrogen to convert a proportion of the unsaturated fatty acids of oil to solid fat) has become a profitable industrial practice for the manufacturing of solid oils and margarine.

When alcoholic or aqueous solution of fats are boiled with caustic alkali, e.g., NaOH or KOH, they are readily hydrolysed into corresponding fatty acids and glycerol, the alkali immediately combining with fatty acids to form soap. This process is termed *saponification*.

Breakdown and synthesis of fats

Fat is hydrolysed by the enzyme lipase to glycerol and fatty acids. The fatty acid is then oxidised by α- or β-oxidation process. In the former case the carboxyl carbon is oxidised to CO_2, and the α-carbon, i.e. the carbon adjacent to it is oxidised to —CHO by fatty acid peroxidase. An NAD-linked aldehyde dehydrogenase then oxidises the —CHO to —COOH, so that the resulting fatty acid molecule is now one carbon less than the initial molecule. In the next turn similar reactions will follow (See Fig. 717a).

In the β-oxidation process, which is the usual mode of oxidation, the β-carbon is oxidised to =CO, in several steps, and at the end of every turn of the "cyclic" operation, the fatty acid molecule becomes 2-carbon atoms less, because an acetyl coenzyme A (CH_3 COS CoA) molecule is liberated each time. The fatty acid molecule has to be activated first by hooking on a coenzyme A (HS CoA) residue to the carboxyl end, using ATP energy in the process. In subsequent reactions two hydrogens are removed from the β and γ carbons, a water molecule is added to the double bond formed, and two hydrogens removed from the —CHOH group involving the β-carbon, resulting in its oxidation to =CO. One molecule of acetyl CoA is split off by β-ketothiolase.

The reaction sequence is shown in Fig. 717a. These reactions take place on the glyoxysomes and peroxisomes, where the enzymes are present. The $NADH_2$ and the $FADH_2$ formed are oxidised by the electron transport system (See later). The acetyl CoA molecules may enter Krebs cycle or other metabolic pathways. If it enters Krebs cycle, 3 molecules of $NADH_2$ and one molecule of $FADH_2$ will be produced per turn of Krebs cycle resulting in the production of 12 molecules of ATP per molecule of acetyl CoA oxidised through the cycle (one molecule coming from the substrate level phosphorylation during conversion of succinyl CoA to succinate). Thus, during oxidation of stearic acid, which is a saturated fatty acid having 18 carbons, 9 acetyl CoA will be produced, which may contribute 108 molecules of ATP; to this has to be added 40 ($8 \times 3 + 8 \times 2 = 40$) molecules of ATP obtained by the oxidation of one $NADH_2$ and $FADH_2$ produced in each of the 8 times of the β-oxidation process producing 9 molecules of acetyl CoA. Since 1 ATP is used at the commencement of the reaction sequence, the total ATP yield per molecule of stearic acid oxidised this way would be 147. In terms of 6C oxidised the value would be 49 against 38 in the case of glucose. β-oxidation of fatty acids thus, is a more efficient process. α-oxidation however, is less efficient as only one $NADH_2$ ($\equiv$3 ATP) is produced per turn.

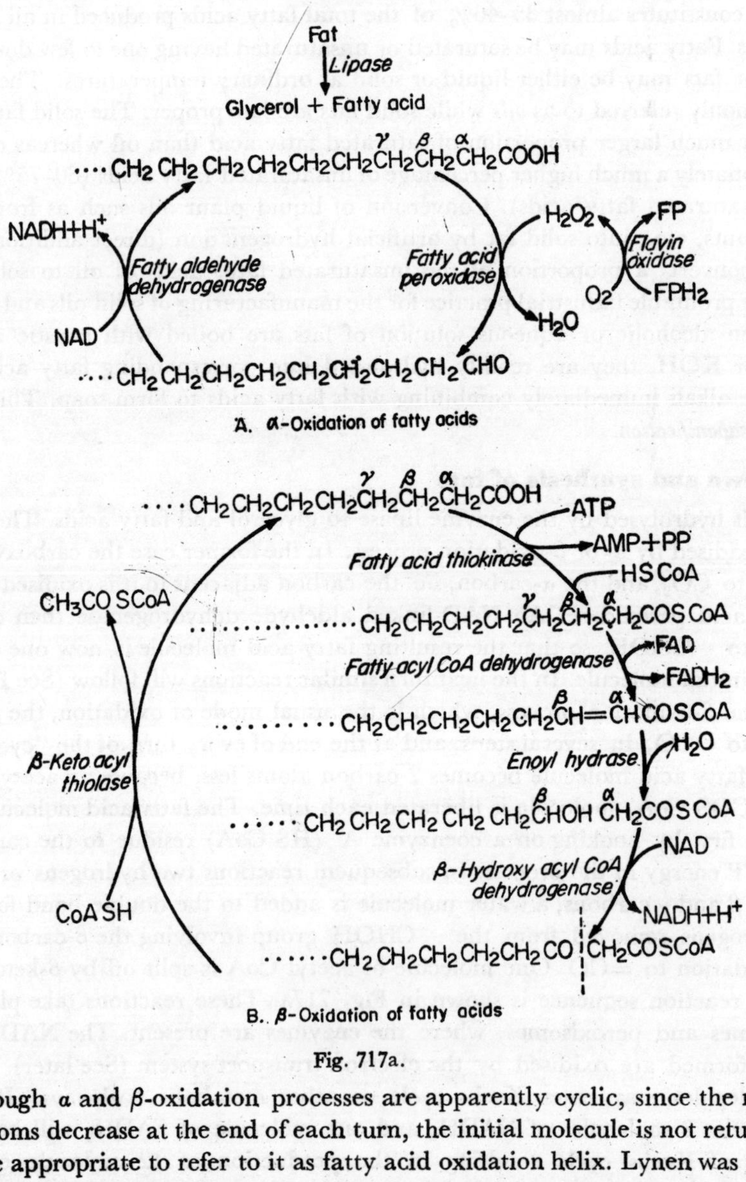

Fig. 717a.

Although α and β-oxidation processes are apparently cyclic, since the number of carbon atoms decrease at the end of each turn, the initial molecule is not returned. It is thus more appropriate to refer to it as fatty acid oxidation helix. Lynen was the first to work out the β-oxidation process.

The glycerol produced after hydrolysis of fat with lipase is broken down through glycolysis and Krebs cycle reaction sequences after conversion to dihydroxyacetone phosphate via oxidation of α-glycerol phosphate which in turn is produced by α-glycerokinase action on glycerol using ATP.

Synthesis of fats. Fats are not synthesized by a reversal of the breakdown process. For synthesis of fatty acids acetyl CoA in the cytoplasm is carboxylated to malonyl CoA in presence of ATP. The malonyl part is then transferred to the SH group of acetyl carrier protein (ACP) also present in the cytoplasm. The malonyl-S-ACP then condenses with acetyl-S-ACP forming acetoacetyl S-ACP and regenerating ACP-SH and CO_2. The acetoacetyl-SACP is then reduced to butyryl-S-ACP through several steps with the help of $NADPH + H^+$. The butyryl-S-ACP then condenses with a malonyl-S-ACP, followed by decarboxylation as before, and the fatty acid chain is elongated by C_2 fragments coming from acetyl coenzyme A and the process is repeated until a long chain fatty acetyl CoA is produced which is then released from the enzyme—a synthetase. In the mitochondria acetyl CoA condenses repeatedly with a short chain fatty acid molecule until the long chain results. The cytoplasmic pathway is the major pathway.

For glycerol synthesis dihydroxy acetone phosphate is reduced by a NADP-linked dehydrogenase to glycerol phosphate, from which the phosphate residue is removed by a phosphatase.

Fats are produced from glycerol and fatty acids by condensation reactions.

Conversion of fat to carbohydrate. In germinating fatty seeds, like castor bean, fat is converted to carbohydrate by glyoxylic acid cycle and reversal of glycolysis.

The reactions take place on glyoxysomes. Fatty acids are first oxidised to acetyl CoA. The acetyl CoA then passes through Krebs cycle until isocitrate is formed, the isocitrate is broken down by isocitratelyase to glyoxylate and succinate. The glyoxylate condenses with acetyl CoA with the help of the enzyme malate synthase to form malate, which is then oxidised to oxalacetate by malic dehydrogenase. The succinate finds its way to the mitochondria where it is also oxidised to oxalacetate. The oxalacetate then is decarboxylated to phosphoenol pyruvate which by reversal of glycolysis produces carbohydrates (Fig. 717b).

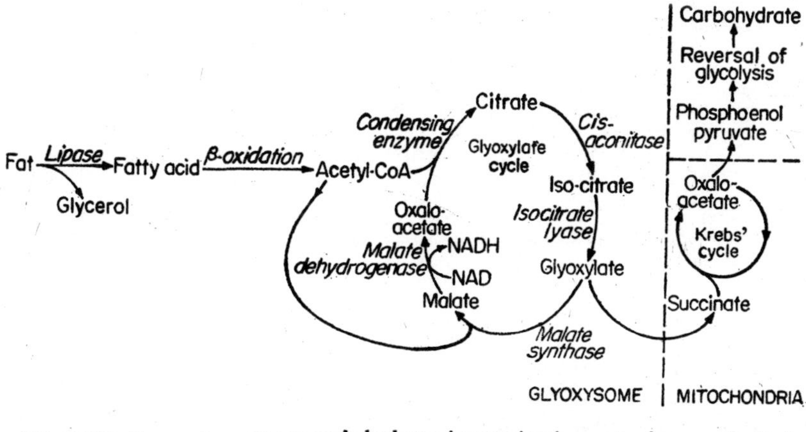

FIG. 717b. Conversion of fat to carbohydrates in germinating castor bean seeds. Only major reactions are shown.

QUALITATIVE TESTS FOR FATS AND OILS

I. Study sections of *Gladiolus* leaves, onion scales and fatty seeds. Place sections in a soln. of Sudan III (0·5% in 70% alcohol) and allow to stand for 20 min. Wash with 50% alcohol and then water and mount in glycerine. Observe the red-stained oil drops which can be seen floating around the section as well as in the cell-inside.

II. (*a*) Test section with 1% osmic acid (OsO_4) and observe the black-stained drops.

(*b*) Add a drop of 1% osmic acid soln. to a little olive soil. A black colour is produced.

III. Try solubilities of olive and other oils in water, ether, alcohol and chloroform. Observe that they are insoluble in water and alcohol but soluble in ether and chloroform.

IV. In a beaker dissolve about 0·5 gm of Sudan III in 50 ml of 95% alcohol. Add 10 ml of olive (or any other oil) oil and stir well. After a short time the oil collects at the bottom and becomes deep orange-red.

V. **Hydrolysis of fats—saponification.** About 100 ml of olive oil are added to about 200 ml of 10% NaOH in a water-bath. Boil for half an hour and after cooling, add Na_2SO_4 (or saturated NaCl soln.). The soap formed separates out and rises to the surface. Filter. Neutralise the filtrate with dilute H_2SO_4. The liquid is evaporated and the residue treated with alcohol. This dissolves the glycerol formed, but not the sulphate. A syrup like liquid is formed which is glycerol. Test for glycerol.

(*a*) To a little of the liquid, add a few crystals of $KHSO_4$ in a test tube and heat strongly. Pungent odour of acrylic aldehyde is easily detected.

(*b*) To a little of the soln. add a few drops of $CuSO_4$ soln. and then some NaOH. A blue, clear soln. is obtained. Glycerol prevents the precipitation of $Cu(OH)_2$ formed.

CHAPTER XLIV

TRANSLOCATION OF SOLUTES IN PLANTS

There are two principal translocatory systems in the vascular bundles of plants, the xylem vessels and the phloem sieve tubes. These two systems are quite different both morphologically and physiologically. Whereas the xylem vessels are lignified, thick walled dead cells, the sieve tubes are thin-walled living cells. Because the sieve tubes are living cells with semipermeable membrane they may develop considerable turgour when water enters into them from outside. Actually osmotic pressures, between 15–30 atm. have been recorded from sieve tubes and consequently they are potentially capable of developing considerable turgour pressure. As a matter of fact, the sieve tubes are normally fully turgid or nearly so due to their nearness to the xylem vessels, which have very low osmotic pressures. Thus in a rapidly transpiring plant, though the xylem vessels may not collapse due to tension, the sieve tubes may still be turgid. Under such circumstances of tension, a cut across the stele may show absorption of water by xylem vessels and exudation of water from the phloem. Of course, in rare cases, when the transpiration rates are very high and absorption of water by roots relatively low, sufficiently severe tension may develop in the xylem vessels to reduce even the turgour of phloem to zero. It is evident no phloem exudation is possible under those exceptional conditions.

According to the classical concept inorganic solute substances are carried in the xylem vessels with the ascending sap of water in the transpiration stream whereas phloem is the pathway of downward translocation of organic solutes (synthesised foods like sugars, amino acids, etc.).

Translocation of inorganic solutes

The classical concept of movement of inorganic solutes upwards through the xylem is unquestionably true but it is difficult to estimate with certainty the proportion of mineral salts thus translocated. Any solute conducted through the xylem is carried along with the ascending streams of water which are pulled up through the plant by transpirational suction. The rates at which inorganic solutes are translocated through the xylem vessels corresponds closely with the rates of translocation of water. There is much indirect evidence for the view—analysis of xylem sap reveals the presence of inorganic salts. By feeding plants with radio-isotopes, it was shown conclusively that inorganic substances move up the plant in the xylem.

Most of the evidence, indicating that organic solutes move downwards from the aerial parts towards the basal portions of stem and that xylem is the main pathway of movement of inorganic solutes upwards, have been obtained by the so-called *ringing experiments*. Ringing really refers to the removal of a narrow continuous band of the tissue external to the xylem. In a tree or herb, such a ringing always produces accumulation of organic solutes, chiefly sugars, in the tissues above the ring, while sugars in the stem and root tissues below the ring, become depleted as a result of utilisation of sugars in respiration. Below the ring the development of buds and new roots takes place very slowly or none at all (scarcity of organic food) whereas above the ring new roots are rapidly and profusely formed (supply of food abundant). Conduction of inorganic solutes from the roots upwards is scarcely affected. (*See* Fig. 686).

The classical concept that mineral substances are carried upwards in the xylem vessels has been shaken, during recent years by some apparently contradictory observations and a re-examination of the classical concept was certainly called for. Although analysis of xylem sap has shown the presence in significant amounts of Ca, Mg and K, 50–90% of the total solutes was sucrose, most of the rest being organic nitrogenous compounds. This even promoted drastic suggestions such as that all substances, both inorganic and organic, move in the xylem. On the other hand, other investigators have championed for phloem as the main path of movement, upward or downward, of all solutes, inorganic as well as organic! It has also been claimed that the path of movement of solutes depends upon the quantity of solutes available. When the quantity of solute to be translocated is high, it may move in the xylem; when the quantity is low, it moves in the phloem!

Later investigators, have, however, confirmed that xylem is the main path of movement of inorganic minerals while others even now stick to phloem.

More recent evidence shows that N, P, S, K, Mg and Cl can be translocated in phloem in the downward direction but certainly not Ca. Thus, we have no other alternative but to admit that inorganic substances may move both in the xylem and phloem but the available evidence as yet favours the view that upward movement of these mineral substances is primarily, if not solely, in the xylem; downward movement probably solely in the phloem. Since by far the greater part of the movement of minerals is in the upward direction from roots to the leaves, this would assign to the xylem, the main path of movement of inorganic solutes in plants whereas the phloem unquestionably is almost the sole path of movement of organic substances synthesised in plants and the path of downward movement of all solutes, inorganic or organic. By feeding single leaves with $^{13}CO_2$, it has been shown that photosynthate, carrying ^{13}C label was transported both up and down and found to accumulate at the points of most active growth. On killing a section of stem with hot wax, ^{13}C-labelled-photosynthate did not move in either direction in such a block. This again, certainly suggests that the phloem is the exclusive path of transport for photosynthate. The classical concept of xylem for inorganic substances and phloem for the organic, is threfore reasonably near the truth.

TRANSPORT THROUGH PHLOEM

We have seen in the previous discussion on the transport mechanism in plants, that phloem is unquestionably the major path of downward movement of all solutes. In some cases, however, particularly in developing fruits which are attached to stem in such a position that some or all of the food translocated into them must move through the stem in the upward direction. Similar examples can be cited in early stages of development of seedlings when upward translocation of foods certainly occurs from the endosperm and cotyledons to the apical shoot meristems in which rapid growth is taking place. Other examples are provided in upward translocation of organic foods during the earlier stages of shoot growth from underground stems such as bulbs, rhizome, tubers, etc Extensive recent investigations favour the idea that these and other similar types of upward translocation of organic foods take place in the phloem and not in the xylem and this certainly contradicts the classical idea that all upward transport occurs in the xylem.

Of the various stem tissues, only the xylem and phloem possess structures which are suitable for a longitudinal movement of the solutes with the important difference that whereas xylem vessels are dead, sieve tubes are very much alive. Both these tissues are characterised by the presence of elongated cells which are joined end to end in such way as to form a continuous system—a sort of pipe line.

Like the xylem, the phloem is continous from the top to the bottom of the plant and ultimate termination of the phloem is in stem tips, lateral organs and in the root tips. Cross-walls (sieve plates), separating the sieve tube cells one from another, are often oblique (present in the side walls of the sieve tubes) and provided with pores. Strands of cytoplasm (*plasmodesmata*) pass from one sieve tube to the next to form a continous living conducting pipe system.

The mechanism of translocation of solutes in phloem

It is evident that the mechanism of the transport of solutes through phloem must be entirely different from the mechanism of transport through the xylem. The fact that the xylem vessels are dead cells points to a fundamental difference of xylem vessels from sieve tubes, with its rich protoplasmic content, highly permeable walls and the consequent development of abnormally high turgour pressure. The xylem vessels, in the conduction of solutes through them, act only passively with its thick incollapsible cell walls. The water is pulled up by tension developed in the aerial organs and by the cohesive forces of water molecules between each and the adhesive force between the water and thick, rigid walls of the vessels.

It has long been recognised that simple diffusion is totally inadquate to account for the known rates or the translocation though the phloem which can, in many cases, be as high as 100 cm per hour from a 10% solution of sugar, for the total cross-sectional area of the sieve tube in potato tuber and grasses. The rate of movement of ^{14}C-labelled photosynthetic products has recently been shown to be unexpectedly fast. In soybean plants (*Glycine soja*) amino acids are known to move through phloem at 370—1,370 cm per hour and sometimes even at 3,000 cm per hour. Later experiments confirmed these results and showed that a small part of the assimilates moved downwards towards the roots, at over 5,000 cm per hour—an incredible feat of food transport through narrow channels like sieve tubes—almost a yard a minute! So diffusion alone is quite insufficient to account for the observed rates of solute movement.

There is general agreement that gradients in concentration are intimately involved in organic food movement in the phloem. Concerning the mechanism of movement, there are many theories; these may be grouped into three categories: (A) The Druckström or mass-flow (literally pressure-flow) or convectional flow hypothesis visualising a flow of solution along the positive gradient—from higher pressure to lower—of hydrostatic pressure developed osmotically, proposed by Ernst Munch, (B) those postulating movement of the solute molecules in, through or upon the surface of the sieve tube protoplasm. By this mechanism the solutes are pictured as moving along concentration gradient (i.e., in differenece in concentration) but independently of each other and of the solvent water and (C) the export of sugar from photosynthesising leaves (e.g., sugar cane, sugar beet, etc.), is nowadays thought to be primarily dependent on light and sugar accumulation. All this diffusion is supposed to be accelerated by protoplasmic streaming, *activated diffusion*, a term implying an accelerated diffusion carried on by the protoplasm and which certainly involves an expenditure of energy.

A. The Druckström or mass-flow theory

The principles involved in this theory are clarified by a reference to a simple physical system:

Two cells with semipermeable walls are connected by a tube to form a closed system and dipped in water. Cell A is assumed to enclose a stronger solution of sucrose than cell B. Water no doubt will enter both the cells but due to greater osmotic pressure in cell A (due to higher conc.) than in B, greater turgour pressure will no doubt be developed in A than in B. The higher hydrostatic pressure developed in A will be transmitted through the closed system resulting in greater diffusion pressure in the water in B than in pure water. The diffusion pressure of water in B will actually be augmented by the imposed hydrostatic pressure transmitted from A. As a result, water will pass out of the cell B and coincidentally there will be a convectional flow of solution along the tube from A to B and then into the external medium. This mass-movement of solution under pressure will continue until the sugar concentration in both the cells are equalised. If, however, one could arrange for continuous addition of sugar in the cell A as rapidly it is moved out of it (the supplying cell) and continuous removal of sugars from the cell B (the receiving cell) by utilisation or by converting them into some insoluble form (i.e., osmotic pressure in the cell B is never allowed to increase more than cell A), a mass-flow of solution under pressure should continue indefinitely.

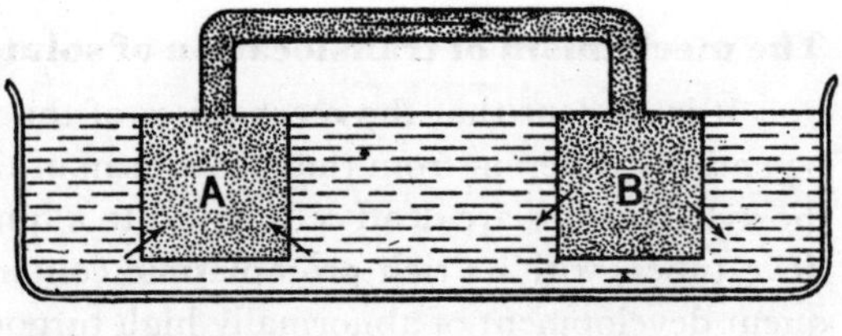

Fig. 718. Diagram of an osmotic system in which mass-flow of solution is supposed to occur.

Some recent investigators have questioned the principle underlying the whole set-up of this experiment, maintaining that this is based on fundamentally questionable concept and the similarity between this experiment and actual transport of organic food in plants, is too far-fetched to be credible.

The Munch hypothesis, introducing the *symplast concept*, visualises the interconnected protoplasm of plant through *plasmodesmata* as constituting a continuous living system—the symplast concept—relatively impermeable on the outer surface but highly permeable throughout its mass.

Fig. 719, shows diagrammatically the downward translocation of solutes. Cells L_1, L_2 and L_3 represent green photosynthetic cells (supplying cells) of the leaf and correspond to the cell A (Fig. 718). Similarly R_1, R_2 and R_3 represent root cells corresponding to (receiving cells) cell B. The continuous system of phloem correspond to the horizonrtal tube connecting A and B in Fig. 718. The osmotic pressure of the leaf cells is maintained relatively high as a result of photosynthesis. In the root cells, the osmotic pressure is supposed to be lower because most of the sugars, translocated to them, are utilised in metabolic activities or are converted into insoluble storage forms. Water is supplied continuously from the roots to the leaf cells through the xylem vessels.

This hypothesis assumes that in the supplying leaf cells higher turgour pressure developed due to higher osmotic concentrations, will cause a convectional flow of solution downward in the phloem. Plasmodesmata connecting adjacent cells are supposed to permit mass-movement of solution from $L_3 \rightarrow L_2 \rightarrow L_1$ and from L_1 into the sieve tubes. Once the convectional flow enters the sieve tubes, the function of sieve tubes is completely passive, allowing the solution under pressure to pass through them. In other words, the sieve tubes are pictured as making up a specialised mechanism of the symplast when water absorbed osmotically brings about high turgour in the regions of synthesis, i.e., the leaves. The turgour promotes a rapid mass-flow of solution through the sieve tubes, through the sieve pores to the region of utilisation, where assimilation, respiration and condensation for storage reduce the osmotic activity of solutes in the roots, thereby reducing its turgour and allowing the water *only* to return to xylem *via* cambium or other receiving root cells. Movement in the sieve tubes would then always have to be from regions of high to one of low turgour pressure. It could, therefore, occur in an upward direction also from storage tissues to the developing aerial shoots for the osmotic pressure, consequently turgour, is evidently much higher in the cells of the storage organs such as underground stems than in the actively growing cells of the developing shoots. The mass-flow mechanism involves a recirculation of water and of the total amount of water used by the plant, only about 5% is involved in this recirculation. Excess of water in the phloem stream is squeezed (?) laterally through the cambium into the xylem as pholem solutes move downward.

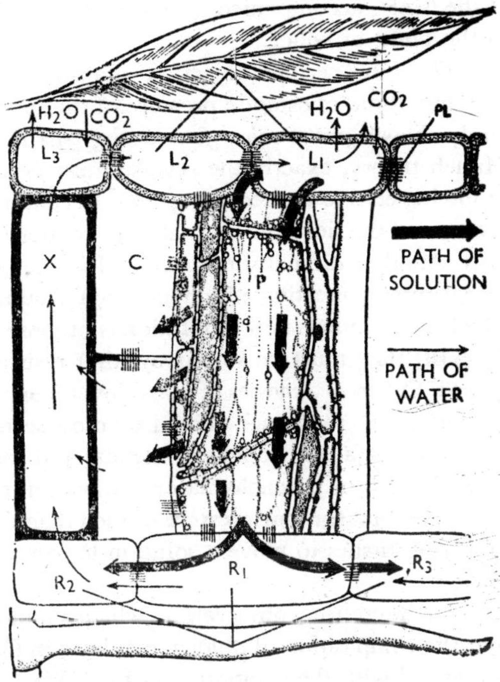

FIG. 719. Diagram to illustrate the mechanism of solute translocation according to Munch hypothesis. L_1, L_2, L_3 represent leaf mesophyll cells (corresponding to membrane A in Fig. 718); R_1, R_2, R_3 represent root cells (membrane B in Fig. 718); PL, plasmodesmata connection; P, continuous system of phloem (sieve tubes) connecting leaf and root cells; X, xylem and C, the cambium. The path of the solution is shown by the thick stout lines and that of water by thin lines.

An elaboration of Munch hypothesis was provided by Crafts from anatomical studies. According to Crafts, movement from leaf cells to the sieve tubes is by diffusion, accelerated by protoplasmic streaming (rotation and circulation of protoplast) *via* the plasmodesmata connection at the expense of released cellular metabolic energy. The sieve tubes allow a ready passage of the solutes in solution from one to another. Removal of the assimilates (carbohydrates, amino acids etc.) from sieve tubes is pictured as active absorption (that too requires respiratory energy) of solutes of the sieve tubes by the adjacent living parenchyma cells. This is followed by symplastic movement, i.e., movement along the interconnected continuous protoplast of the plant. Undoubtedly if this modification of Munch hypothesis by Crafts is correct, it would, as has been pointed out before, require the expenditure of considerable respiratory energy.

The strongest experimental evidence in support of Munch theory is that sap will often exude rapidly from a cut made into the phloem of a stem and this ***phloem exudation*** has been seen to continue for a period of more than 24 hours. This phloem exudate is apparently cell sap with high concentration of normal translocation substances and *not* cytoplasm. This suggests that the phloem exudate is under considerable pressure. Evidently a *turgour push* of considerable magnitude developed originally in the leaf cells is operating in the phloem elements of intact stems. It has been shown that after a sieve cell is punctured by an *aphid*, it continues to exude sucrose solution for a long period, until the exudate exceeds by many thousands of times the volume and sucrose content of single sieve cell. This may be a pathological disturbance of the normal system, but it does certainly suggest that a free flow of sugar solution is theoretically possible through the sequence of interconnected sieve tubes.

On the whole the mass-flow mechanism is supported by evidences, such as permeable condition of the functioning sieve tube, the presence of a concentration gradient of osmotically active materials in the phloem and the last, but not the least, the phenomenon of phloem exudation. The elegant idea of establishment of 'source' and 'sink' by regulating assimilation, with hydrolysis at the 'source' and utilisation and storage at the 'sink', accompanied by recirculation of the solvent component, water, fascinates to the point of belief.

The most serious objection against Munch hypothesis is that the experimental determination, of osmotic pressure of the supplying cells and receiving cells, has consistently shown *greater values for receiving cells than supplying cells.* For normal operation of Munch theory, exactly the reverse must be true, that is a pressure gradient from the supplying cells to the receiving cells. Recent investigations further show that the osmotic pressure of the sieve tubes is actually greater than the osmotic pressure of the mesophyll cells. Another difficulty encountered is that cytological observations do not normally reveal interconnecting plasmodesmata channels from one sieve cell to another along which a continuous flow of solution is supposed to take place. Moreover if such channels do exist, they must be so narrow that resistance to such a flow would be very great indeed. Other weaknesses of mass-flow theory include a complete lack of our knowledge relating to the resistance offered by cross-sectional walls in the phloem and also by the plasmodesmata connections of the leaf and root cells. The resistance offered by passage through phloem parenchyma or companion cell will also be considerable and it seems incredible that turgour pressure of such magnitude could develop in the leaves and other supplying tissues to move a solution for any great distance and at such a speed in the phloem.

An interesting observation which cannot be explained by Munch hypothesis is that the composition of phloem exudates in *Cucurbita* shows fundamental discrepancy as contrasted with the composition of its fruit.

The mass-flow theory will account for translocation in the phloem, not to speak of individual sieve tubes, *always in one direction at any one time* and there is some evidence that at least in some plants (e.g., maize, sugar beet, etc.) the translocation of carbohydrate is *polar;* i.e., occurs only in one direction. There are also some more convincing evidences, however, that translocation in the phloem is simultaneously *bi-directional*, even in individual sieve tubes. By use of isotopic ^{14}C and ^{32}P, strong evidence has been produced of the simultaneous movement of carbohydrates and phosphates in opposite directions, *even in one* individual sieve tube. If this is generally true, this speaks strongly against the pressure-flow hypothesis of Munch.

B. Protoplasmic streaming: Accelerated diffusion theory

The fact that the protoplasm of most living plant cells is frequently in active motion has been widely observed in the cells of many aquatics like *Chara*, *Vallisneria*, *Elodea*, etc., as well as in root hair cells, staminal hairs and also in many kinds of leaf hairs. This is known as *cyclosis*. The mechanism of this movement (shown by the movement of coarse, protoplasmic granular plastids which are actually taken along the stream) is not known but expenditure of energy seems clearly indicated. In the mass-flow hypothesis the function of sieve tubes was taken as completely passive, whereas according to the protoplasmic streaming hypothesis, oxygen is necessary for the sieve tubes to operate in translocation, i.e., living cells are indispensable for translocation of solutes. The functioning sieve tubes are pictured as active elements, rich in protoplasm and capable of continued expenditure of metabolic energy. It has been shown that the conducting tissues of sugar beet and *Heracleum* have high respiratory intensities in both xylem and phloem.

According to this concept, solute molecules move in, through or upon the surface of the sieve tube protoplasm and are carried along the protoplasmic streaming movement from one end of the sieve tube to the other. The solute molecules are usually assumed to pass from one sieve tube to the next by diffusion through the pores of the sieve plates. Clearly the movement is pictured as due to a combination of simple diffusion and protoplasmic streaming. Some investigators have claimed that streaming protoplasm may be continuous from sieve tube to the next through sieve pores. Any way, within a particular sieve tube, the solute molecules which have diffused in from the sieve tube above, would be carried downward by protoplasmic streaming; those that diffused in from the cell below, would be carried upward by the same cyclic streaming. Thus this theory may account for the known movements of substances along concentration gradients and also for the apparrent fact that two substances can be simultaneously translocated in opposite directions—*bi-directional*—in individual sieve tubes. The solutes can thus move along their own concentration gradients but independently of each other and of solvent water.

The most serious objection against the protoplasmic streaming hypothesis is that active streaming, though easily observed in young sieve tube elements, is difficult to observe in mature sieve tubes. Careful technique has, however, succeeded in demonstrating that streaming does occur in mature sieve tubes. Again, the rate of streaming seemingly required to account for the measured rates of transport would be very high and it seems improbable that protoplasmic cyclosis occurs at such rates in any living cell, not to speak of sieve tubes. The possibility, however, remains that streaming of cytoplasmic films or layers which are below the usual range of microscopic visibility might occur at rates, which may be *theoretically* sufficient to account for these observed rates of solute translocation.

Perhaps the most difficult point to reconcile with this theory is the fact that phloem exudate in many plants is apparently cell sap and not cytoplasm and yet the cell sap contains high concentrations of organic foods. This would seem to indicate that organic food translocation is not through or upon the protoplasm but rather *via* the sieve tube vacuole.

To conclude, we may state that the protoplasmic streaming theory is supported by evidence that under certain conditions, solutes appear to move independently of each other in the phloem, that the sieve tubes have high respiration rates and are rich in protoplasmic contents and radioactive ^{32}P absorbed into the phloem move simultaneously in both upward and downward directions in the stem. The greatest weakness of this theory is that no physical mechanism is known or has been pictured by the advocates of this theory that will account for the measured rates and for such rapid transport of large quantities of materials along the protoplasm, independent of the solvent, water.

Since no theory of translocation of solutes is satisfactory, several other theories have been proposed from time to time. We shall discuss only two of them. These are based primarily on electronmicroscopic observations of the internal structures of sieve tubes and some speculations, which seem to be theoretically sound. Through the pores of sieve plates microfibrils are believed to pass; these microfibrils are composed of a protein, named P-protein, which has contractile properties. The microfibrils are 60-280 Å in diameter and the interstices are large enough to accommodate sugar molecules (10 Å in diameter) or hydrated K^+ ions (5 Å in diameter). According to Spanner the surface of the pore walls is negatively charged and positively charged ions (e.g. K^+) are held rather firmly by intermolecular forces; ions in the outer wall are held only weakly and are thus mobile. This results in the movement of a current of K^+ ions from one tube to another. Each K^+ will carry with it the water molecules which hydrate it along with the dissolved uncharged sugar molecules. ATP from bordering companion cells of the sieve elements provide the energy required for driving the K^+ ions through the sieve plate pores overcoming any resistance; the internal surface of the sieve tube cell wall is believed to be rough. So that the adjacent plasma membrane has a large surface area, which further helps in the transport of the K^+ ions. Solute movement through sieve tube thus, according to Spanner is by electro-osmosis. However, Fensom thinks that the swishing of microfibrils results in the development of a fast wave and particles attached to microfibrils were actually found to move with a sort of bouncing motion, as observed in the case of Brownian movement; this results is a slower motion involving mass flow of the fluid in the sieve tube. This explains the bidirectional movement of solute in the same sieve tube. This observation and the theory, as also that of Spanner, await verification by other workers.

CHAPTER XLV

RESPIRATION

The act of breathing in animals was recognised long before the significance of the process was understood. Soon after the discovery of oxygen by Priestly in 1772 it was realised that animals in their breathing absorbed large quantities of oxygen and gave off large quantities of CO_2. Shortly after the discovery that green plants evolve O_2 in light, in about 1779, Ian Ingen-housz demonstrated that two types of gaseous exchange occur in plants, one taking place in the green leaves during day-time in which CO_2 is absorbed and oxygen is given off and another carried on by leaves at night and by all other living tissues at all times in which oxygen is absorbed and CO_2 evolved. Ingen-housz pointed out the similarity of this latter process to the breathing in animals. By about 1865, the confusion was settled by Sachs who pointed out that these gas exchanges were associated with two fundamentally distinct processes, one which is now called photosynthesis associated with manufacture of food and the other associated with a process nearly comparable with breathing of animals.

In the process of photosynthesis it has been found that the gas exchange is only but a small part of the story and that other changes associated with it, particularly the manufacture of carbohydrates, are of much greater significance. The same is true for respiration; it is now recognised that other and more fundamental changes are involved in respiration. The following changes are commonly associated with the process of respiration and can be easily demonstrated: (*i*) an absorption of O_2 gas which (*ii*) disappears during the oxidation of materials within the cell; (*iii*) the oxidation results in a release of energy, part of which is used in some of the energy-consuming life processes and a part of which is lost as heat; (*iv*) during oxidation, many intermediate compounds are formed some of which play an important part in the life of the cell; (*v*) CO_2 and H_2O are usually the end products and (*vi*) as a result of this loss in CO_2 and H_2O, there is usually a reduction in the dry weight of the plant.

We know that when a growing organism is supplied with sugar, a part of that is finally oxidised to CO_2 and H_2O. We also know that if a given amount of sugar is completely burnt to CO_2 and H_2O, a definite amount of energy is released, no matter how many intermediate steps are involved. Thus 1 gram of glucose on complete oxidation produces 3,740 calories of heat. One gram molecule of glucose (180 grams) on complete oxidation will release 673,200 calories as shown by the equation:

$$C_6H_{12}O_6+6O_2 \longrightarrow 6CO_2+6H_2O+673,200 \text{ cal.}$$

In respiration a part of the energy produced appears in the form of heat and the rest of it takes some other form. Part of the energy released takes the form of chemical energy, in the form of ATP molecules, building up more complex compounds out of simpler ones. For example, energy is necessary for the formation of proteins from sugars and nitrates or of fats from sugars.

Thus the light energy which is captured in a few bold chemical steps and stored up in the form of carbohydrates in photosynthesis, is used by the plant for all its further growth and metabolism. This means that thousands of different chemical reactions and syntheses have to be powered by the exergonic energy obtained from the catabolic breakdown of the sugars in respiration, e.g., powering of protein synthesis (where each peptide bond needs some energy for its formation). Energy contained in the sugars is thus changed into smaller and still smaller packets (ATP) for protein synthesis and also for hundreds of other endergonic metabolic reactions taking place every minute in the living cells. This apportioning of the sugar energy takes place in glycolysis and Krebs cycle where the sugar molecule is gradually broken down to CO_2 and water in a large number of steps, yielding small amounts of energy by turn, originally present in the sugar molecule.

Respiration has thus two important functions: (*i*) it is a process that results in the production of highly reactive compounds out of which important cell constituents are formed and (*ii*) it is a process by which energy is released from food, the energy released being harnessed in such a way as to result in the formation of cell structures and in the doing of 'work'.

Aerobic and anaerobic respiration

Respiration in presence of oxygen is knwon as *aerobic respiration* in order to distinguish it from another less frequent, but not unimportant, kind of respiration called *anaerobic respiration* which occurs in higher plants in the absence of a supply of atmospheric oxygen.

$$\text{aerobic} - C_6H_{12}O_6 + 6O_2 \longrightarrow 6CO_2 + 6H_2O + 673{,}200 \text{ cal.}$$

$$\text{anaerobic} - C_6H_{12}O_6 \longrightarrow 2C_2H_5OH + 2CO_2 + 50{,}000 \text{ cal.}$$

Many organisms are knwon to be able to break down carbohydrates to CO_2 and some other substances (e.g., alcohol, lactic acid, etc.) without utilising molecular oxygen. This process has been called by several names; *anaerobic respiration* because it may occur in absence of air, *fermentation* because it is so often controlled by micro-organisms or their enzymes (previously knwon as ferments) and *glycolysis* because it is a breakdown of sugar. It is now known that all higher plants are capable of respiring anaerobically for varying lengths of time but most die within a few days in complete absence of molecular oxygen.

Respiratory quotient (R.Q.)

The molar ratio of CO_2 evolved to oxygen absorbed in respiration is known as the respiratory quotient. This ratio is 1 when glucose or some other carbohydrate is completely oxidised to CO_2 and H_2O as in the equation below:

$$C_6H_{12}O_6 + 6O_2 \longrightarrow 6CO_2 + 6H_2O + \text{energy}$$

$$\text{R.Q.} = \frac{6}{6} = 1$$

There is considerable evidence that hexose sugars are generally or perhaps invariably the substrate which is oxidised in the cells of the higher green plants. When the cells contain both carbohydrates or fats, the former are consumed first as the respiratory substrate before any inroads are made upon the fats. When fatty seeds are allowed to germinate in contact with sugar, it has been found that sugar is oxidised first. When fats serve as respiratory substrate, they are either converted into sugars (which is itself an oxidation process) and in turn oxidised to CO_2 and water or fats may be directly oxidised to CO_2 and H_2O. Utilisation of proteins in respiration apparently occurs only in tissues which have been completely depleted of carbohydrates and fats.

There is considerable evidence that protein may be used in respiration of starved leaves (kept in the dark). The minimum constant respiration rate of starved leaves was termed by Blackman, *protoplasmic respiration* (respiratory substrates, mostly protein) in contradistinction to *floating respiration* or normal respiration in which carbohydrates or other reserve materials are utilised.

Thus if substances other than glucose are respired, a different respiratory quotient from 1 usually results. Thus, a fat would give:

$$2C_{51}H_{98}O_6 + 145O_2 \longrightarrow 102CO_2 + 98H_2O + \text{energy}$$

$$R.Q. = \frac{102}{145} = 0{\cdot}7$$

The respiratory quotient is therefore 0·7. This is to be generally expected from the chemical composition of fats. Fats are compounds which are relatively poor in oxygen compared to carbohydrates and thus would require more oxygen to oxidise them. On the other hand, compounds such as organic acids which are produced from amino acids after hydrolysis of proteins, are relatively richer in oxygen compared to carbohydrates or fats and consequently will require less oxygen to oxidise a molecule of such a substrate, if utilised in respiration. An R.Q. value of 4 could theoretically be expected for complete oxidation of oxalic acid:

$$2\begin{array}{l}COOH\\ |\\ COOH\end{array} + O_2 \longrightarrow 4CO_2 + 2H_2O + \text{energy}$$

$$R.Q. = \frac{4}{1} = 4$$

The values for proteins have actually been found to lie somewhere between 0·7 and 4. The R.Q. can vary from *zero* (no CO_2 evolved) to *infinity* (no oxygen absorbed as in anaerobic respiration). In general, a respiratory quotient of less than 1 indicates that: (*i*) a compound other than carbohydrate is being used as the respiratory substrate, or (*ii*) a carbohydrate is incompletely oxidised, for example in the formation of organic acids that may trap O_2 without the simultaneous release of CO_2 as happens in many succulents or (*iii*) O_2 uptake occurs in processes other than respiration or (*iv*) the CO_2 evolved may be re-assimilated before it can diffuse out of the plant (as observed in acid-metabolising succulents where the respiratory CO_2 is metabolically fixed in the synthesis of malic acid; R.Q. value of 0·03 has been reported from *Opuntia* indicating that only a very small portion of respiratoty CO_2 escaped in these plants). A R.Q. value, if greater than 1 may mean that some high-oxygen compounds, such as organic acids are being utilised in respiration as substrates or that respiration is only partially complete and therefore does not utilise much molecular O_2. In spite of these many exceptions the R.Q. of higher plants is commonly 1 and this certainly indicates that carbohydrates are the primary substrates for respiration.

When leaves are starved (kept in the dark to prevent accumulation of photosynthetic carbohydrate), the stored carbohydrates are used up in a few days. Some carbohydrates are, however, respired more readily than others. Fructose is used up first, then sucrose and starch and finally glucose. But the utilisation of sucrose and starch is at first accompanied by an increase in glucose content. Only after the other carbohydrates have been mainly exhausted does the glucose disappear. This indicates that the decrease in sucrose is initially due to a respiratory breakdown of the fructose component of sucrose.

Comparative rates of respiration

As a rule, the rates of respiration (measured either as absorption of O_2 or evolution of CO_2 or decrease in dry weight) are found to be greatest in the young and actively growing stem or root tips or the embryos of germinating seeds. It is precisely in such tissues that the proportion of protoplasm is greatest in relation to the total dry weight of the tissues. In mature tissues, where the proportion of protoplasm is small compared to the young tissues, the respiration rates are found to be very low. The lowest rates of respiration are found in dormant seeds and spores.

The photosynthetic/respiratory ratio

In leaves and other chlorophyllous tissues the rate of photosynthesis exceeds the rate of respiration during the daylight hours. The CO_2 evolved in respiration is re-utilised by the green cells in photosynthesis but since the latter process is occurring more rapidly than the former during day-time, additional CO_2 is continually diffusing into the green cells from the external atmosphere. Similarly, photosynthesis produces more oxygen than could be utilised in respiration, the surplus O_2 is diffusing out of the plant to the atmosphere. Hence, during daylight hours, as favourable conditions of photosynthesis prevail, there is a net movement of CO_2 into the green cells and a net loss of oxygen. During day-time, therefore, the occurrence of gaseous exchange associated with respiration may be completely masked.

At night or during darkness, the reverse conditions obtain; photosynthesis is totally absent and the only exchange of gas between the plant and the outside environment is entrance of O_2 and exit of CO_2 and this is certainly the characteristic gaseous exchange of all living non-chlorophyllous organs whether in light or in the dark.

Since at low temperature light is the limiting factor in photosynthesis, it is evident that there should be a certain light intensity at which the rate of photosynthesis and the rate of respiration in leaf and other chlorophyllous organs are exactly equal. At this light intensity, often called the *compensation point*, the volume of CO_2 released in respiration is exactly equal to the volume being consumed in photosynthesis while the converse is true for oxygen, i.e., all the oxygen released in photosynthesis is wholly utilised for respiration. At compensation point of a light intensity, it is evident that there would be no exchange of gas between the plant and the environment.

The maintenance of plant life at this theoretical light intensity of compensation point would, however, be impossible for any length of time for: (*i*) light is available only for an average of 12 hours of the 24-hour period—hence there is no photosynthesis to compensate for the respiration at night; (*ii*) compensation point is measured only for green parts of the plant, no allowance is made for respiation of roots and other non-green organs and (*iii*) no plant can survive indefinitely without the consumption of some food in assimilation and at compensation point no food would be available which could be utilised in assimilation.

It seems apparent then that the actual minimum light intensity must be considerably higher than compensation point in order that the plants can grow and develop normally.

At compensation point of light, the rate of photosynthesis is exactly equal to the rate of respiration, i.e., the ratio between them is unity. Under this condition, the sugar synthesised in photosynthesis is completely used up for respiration. Nothing is left over for accumulation. No plant can survive under such conditions which permit a maintenance of respiration equal to or in excess of the rate of photosynthesis. If the P/R ratio is lower than 1 it is evident that respiratory breakdown is in excess of photosynthetic manufacture of food. This results not only a check of accumulation of carbohydrate but may entirely inhibit assimilation. The growth rate of the plant may be greatly slowed down in absence of any food left over for assimilation. Plants subjected to such conditions may survive for a time drawing on its reserves and previously stored foods but eventually they must face starvation. In extreme cases of continued low P/R ratio, death from starvation is the inevitable result. The magnitude of the P/R ratio will be affected by any factor which influences either or both of the processes and temperature is evidently the most important single factor which influences this ratio.

A daily alternation between relatively cool night temperature and moderately high day-time temperature must result in a greater P/R ratio than if both day and night temperatures are relatively high because under former condition the night rates of respiration is comparatively reduced due to low night temperature. A better and more profitable crop of potatoes can thus only be obtained in climates where the night temperature is low compared to warm countries where there is very little difference between the day and night temperatures, for the filling of the potatoes or other tuberous storage organs depends upon the accumulation of starch and other foods. This accumulation can only happen if there is a large excess of food left over after all the requirements of food for consumption in respiration and assimilation have been met with and satisfied. The temperature during daylight hours must, however, be moderately but not too high (for too high temperature will favour high respiration rates) so that the plant can manufacture food sufficiently in excess for utilisation in respiration during night (when there certainly is no photosynthesis) and the net excess, after all consumption requirements have been met, can be utilised for growth and storage.

FACTORS AFFECTING THE RATE OF AEROBIC RESPIRATION IN PLANTS

1. *Protoplasmic conditions.* Young actively growing meristematic tissues have always higher rates of respiration than older and more mature parts. The proportion of protoplasm, both relative and absolute, is always greater in the young cells compared to maturer, vacuolated cells. There seems to be a direct relationship between the amount of protoplasm and the respiration rates—the greater the protoplasm, the higher is the respiration rate. Hydration of protoplasm and the quantity of respiratory enzymes of the mitochondria (all the respiratory Krebs cycle enzymes are known to occur in mitochondria) are important protoplasmic factors which contribute to the effects obtained.

2. *Temperature.* Within certain limits (0-45°C.) increase in temperature leads to an increase in the initial respiration rate. Above 30°C., the respiration rate slows down and the decrease above the optimum may possibly be due to the progressive inactivation of respiratory enzymes. Other causes which might contribute to this lowering of respiration rate may include: (*a*) O_2 may not get access to the cell fast enough, (*b*) CO_2 may accumulate in toxic concentration, (*c*) the supply of oxidisable respiratory substrate may be inadequate to keep pace with the high rates of respiration, etc. At 0°C., respiration rate greatly diminishes and soon becomes imperceptible though there are some records of measurable respiration even at −20°C. The decline of respiration at temperatures below zero can be attributed to the formation of ice and consequent dehydration of protoplasm.

The temperature coefficient (Q_{10}) of respiration within the temperature range of 0-35°C. appears to be about 2·0 to 2·5.

3. *Supply of oxidisable food.* Increase in soluble food content readily available for utilisation as respiratory substrate, generally leads to an increase in the rate of respiration up to a certain point when some other factor becomes limiting.

5. *Oxygen concentration of the atmosphere.* In considering first the effect of oxygen concentration in gas mixtures at N.T.P., the whole range of oxygen concentrations falls in two clear cut divisions, a lower range, in which anaerobic respiration occurs and an upper range, in which the entire volume of CO_2 evolved results from aerobic respiration. The lower range extends from zero to the extinction point (generally

about 4-5% oxygen) which, as we know, is that concentration of oxygen just high enough to completely suppress the anaerobic component. The upper range extends from the extinction point to 100% oxygen.

In general, the rate of respiration is decreased by oxygen concentration smaller than that of air (21%) which drops off very sharply at concentration of oxygen less than 5-10%.

The *oxygen poisoning*, i.e., the significant fall in respiration rate, was observed in many tissues in pure O_2, even at N.T.P. This inhibiting effect was also observed with green peas when they were exposed to pure oxygen exerting a pressure of 5 atm.—the respiration rate fell rapidly. The oxygen poisoning effect was reversible, if the exposure to high oxygen pressure was not too prolonged.

The relation of oxygen concentration to respiration has a particularly important implication in the growth of the roots. Roots must respire vigorously if they are to grow and take up minerals from the soil by active absorption. The vigorous root respiration is only possible if the space surrounding the roots and root hairs has ample supply of oxygen. When the roots are poorly aerated as in water-logged or heavy soils, growth of the plant may be significantly restricted.

5. *Carbon dioxide concentration of the atmosphere.* In general, the higher the concentration of CO_2 of the atmosphere, the lower is the rate of respiration. This fact is made use of in storage of fruit. Air containing 10% CO_2 (in atmosphere it is only 0·03%) retards respiratory breakdown and therefore reduces sugar consumption and thus prolongs the life of the fruit. The oxygen content of the air, however, must be maintained as high as normal to prevent anaerobic respiration. In some plant tissues, however, respiration rate actually increases when exposed to relatively high concentration of CO_2. It is thought now that the effect of CO_2-concentration on the respiration rate is influenced not only by its concentration in the medium but also depends upon the kind of tissue and the period of exposure.

6. *Supply of water.* Increase in the percentage of moisture leads to a general increase in respiration rate. This increase is slow at first but very rapid later. This is very clearly seen in the tissues of many xerophytes. As the water content of such plants is increased, often there is no great immediate effect upon the rate of respiration, until a certain water content (which varies according to the tissue) is attained after which respiration rate shows rapid increase. On the other hand, minor variations in water content of well-hydrated plant cells do not appear to have very great influence upon the rate of respiration.

7. *Light.* As far as we know light has no direct effect on respiration except in bringing about an increase in temperature which certainly, as we know well, influences respiration. The indirect effect of light on respiration is, of course, immense because only in light the primary respiratory substrates are synthesised.

That light has no direct effect on respiratory activity has been shown in some mutant strains of *Chlorella* which, although green, cannot photosynthesise. These algae, however, grow and respire when supplied with a suitable source of carbon. The respiration rate of these forms was entirely unaffected by illumination. With blue-green alga, *Anabaena*, however, respiration rate was found to depend on light, and the effect was also influenced by O_2-concentration.

The term **photorespiration** has attracted a lot of attention during the last few years. It is used to indicate increased respiratory activity in light, regardless of the pathways of respiration, by which CO_2 is released and oxygen consumed.

Some plants, e.g., tobacco, evolve CO_2 when brightly illuminated in CO_2-free air whereas others, e.g.,, maize, do not. This light respiration is stimulated by high oxygen concentration of the medium. The different responses by plants to temperature also suggest that this photorespiration is different from normal mitochondrial respiration. There are strong evidences also for progressive increase in photorespiration with increasing light intensities.

In tropical HSK-plants (e.g., sugar cane leaves), photorespiration is very difficult to detect as these species are extremely efficient in photosynthesis. Compared to these plants, the temperate species, which fix CO_2 primarily by following the classical cycle, have very high respiratory rates.

8. *Inorganic salts.* The chlorides of alkali cations of Na and K, as also the divalent cations of Li, Ca and Mg, generally increase the rate of respiration as measured by the amount of CO_2 evolved, although there is considerable difference in the effects of monovalent and divalent cations. With monovalent chlorides of K and Na, the high respiration rates may be maintained for some 7—10 days whereas with divalent chlorides of Li, Ca and Mg, the increased rate observed generally falls off after about a day. Similar results are obtained with NH_4Cl but in all cases this effect seems to be transitory.

The transitory increased respiration rates on addition of chlorides, particularly of potassium and sodium may be calculated as total respiration in salts *minus* the *ground respiration* (ordinary respiration), unconnected with salt uptake. The ground respiration is essentially cyanide resistant whereas the extra respiration due to salt addition was found to be cyanide sensitive—10^{-4}M cyanide completely abolished the increased respiration. This enhanced respiration has been termed *salt respiration* or *anion respiration* which, according to Lundegardh, was directly related to the total amount of anion absorbed by plant cells rather than to the absorption of cations.

9. *Injury and the effects of mechanical stimulation.* Wounding or injury almost invariably results in an increase in the rate of respiration. Broken and shrivelled seeds and kernels have always higher respiration intensities than clean, intact seeds of the same type and moisture content. It is possible that the increased respiration rate associated with injury may partly be correlated to an increase in the sugar content, actually observed in the cells close to the injury, which may be as high as 70% compared to intact parts. This increased rate of respiration following injury is not, however, maintained for more than 48—72 hours.

A considerably higher respiration rate is observed when leaves are first touched or handled and this effect could also be produced by stroking or bending the leaf by a mechanical arrangement. The possibility that mechanical stimulation actually affects the oxidation process itself is supported by the observation that mechanical stimulation has no effect on respiration rate in an atmosphere of only nitrogen. The effect of two successive stimulation with a short interval between them is interesting. Both the first and second stimulations increase the respiration rate but althought the respiration rate reached as a result of second stimulation may be actually higher than the peak reached by the first stimultion, the net increasein resp ration rate after the second stimulation is always less than that resulting from the first stimulation, if the time between the two stimulations is less than three days. It is evident that the sensitivity of a tissue to respiratory mechanical stimu tion is lowered by a previous stimulation.

10. *Effect of various chemical substances.* Various chemical substances such as chloroform, ether, acetone, morphine, etc., temporarily increase respiration rates if given in small doses but if the dose is large, a rapid decrease in respiration rate is observed.

Ripening fruits produce ethylene and this is accompanied by an increase in respiration rate; other volatile products responsible for the flavour (aroma), e.g., methyl, ethyl, amyl esters of formic, acetic, caproic and caprylic acids also associated with increased respiration rate, reach a maximum during ripening of fruits. Artificial ripening of fruits may be induced by treating unripe fruits with ethylene gas (1 : 1000) and this is always accompanied by an increase rate of respiration.

PASTEUR EFFECT OR PASTEUR REACTION

If in the presence of oxygen the products of respiration are CO_2 and water and in absence of oxygen they are CO_2 and alcohol, the if the rate of hexose consumption is the same under both sets of conditions , the ratio of CO_2 produced under aerobic conditions to that produced under anaerobic will be 3, as an inspection of the equations at once makes clear:

$$C_6H_{12}O_6 + 6O_2 \longrightarrow 6CO_2 + 6H_2O$$
$$C_6H_{12}O_6 \longrightarrow 2CO_2 + 2C_2H_5OH$$

In many tissues, however, a ratio of less than 3 and sometimes very much less, has been observed which means that the presence of O_2 lowers the rate at which sugar breakdown occurs in its absence. This is the *Pasteur effect* or *Pasteur reaction.*

Pasteur effect thus refers to inhibition of sugar breakdown owing to the presence of oxygen and since in aerobic respiration, complete breakdown of a molecule of hexose to CO_2 and H_2O should result in the production of 6 molecules of CO_2, it may be assumed that Pasteur effect is operative when the rate of CO_2 evolution in air is less than 3 times the rate of evolution of this gas in absence of oxygen, i.e., when the ratio (as measured by CO_2 output) is greater than 0·33.

Pasteur effect has also been observed in a number of seeds and also in senescent fruits, e.g., tomatoes, mango and guava during recent years.

Pasteur effect has also been shown to occur in many storage tissues, but not certainly in all plants, and values as high as 1·36 (theoretical value 0·33) have been reported.

The operation of Pasteur effect may be explained as due to: (*i*) the extra CO_2 in anaerobic respiration may come from degradation of some compounds other than carbohydrates; (*ii*) specific inhibition of glycolysis in presence of O_2 and (*iii*) the probable occurrence of *oxidation anabolism* in presence of O_2 *only* in which the intermediate products of glycolysis, e.g., pyruvic acid is built back into the system.

Alcohol/Carbon dioxide ratio

The alcohol/carbon dioxide ratio, theoretically, should be *one* and as the molecular weights of these substances are 46 and 44 respectively, the quantities of the two products produced in anaerobic respiration, should also be almost equal in weight:

$$\frac{C_2H_5OH}{CO_2} = \frac{46}{44} = 1{\cdot}05$$

Actually, however, the quantity of alcohol produced nearly always falls short of this theoretical amount—the divergance may sometimes be considerable. In *Vitis vinifera*, this ratio is almost unity but in potato tuber, only minute quantities, if any at all, of ethyl alcohol was found.

Extinction point

In low concentrations of O_2, respiratory quotients are sometimes observed which are greater than unity indicating the preponderance of anaerobic over aerobic respiration.

As the O_2 concentration is increased the respiratory quotient falls until it reaches a constant value in the neighbourhood of unity indicating the minimisation of anaerobic component of respiration. This oxygen concentration is called *extinction point*.

The extinction point varies widely in different plant material and tissues. Instances are on record (in onion root tips) where some anaerobic respiration occurs even in 100% oxygen of the medium. But although, the operation of anaerobic component of respiration has been observed in high oxygen concentration, for most tissues the extinction point is very often in the neighbourhood of 4–5% oxygen (in wheat seedlings, potato tuber, etc., the extinction point values may be as high as 14% oxygen).

When both aerobic and anaerobic components of respiration are in simultaneous operations in tissues in a low range of oxygen concentration—the range extending from zero to extinction point—the actual values of both the components can be estimated if both the rate of CO_2 evolution and the rate of O_2 absorption are knwon. We know that with a sugar substrate the R.Q. for aerobic respiration is *unity* and for anaerobic respiration, *infinity*. Hence if the rate of O_2 absorption by a tissue is a and the rate of CO_2 evolution is b, the aerobic components R_a is a and the anaerobic component, R_{an} is $b—a$. If the respiratory substrate is other than a carbohydrate, e.g., fats or organic acids which results in a R.Q. having a value other than unity, say x, the CO_2 of R_2 would be xa and that of R_{an}, $b—xa$. Very little, however, is known even now of the anaerobic respiration in which the substrate is not a carbohydrate.

PHOSPHORYLATION AND RESPIRATION

Both anaerobic and aerobic breakdown of sugars involve a release of respiratory energy. Just how this energy is utilised to drive various endergonic processes in the living cells is not completely understood. However, it is known that it is primarily accomplished by a transfer of phosphate groups from the intermediate products of respiration to the energy-consuming processes. The complete oxidation of a molecule of hexose may lead to the synthesis of 38 molecules of ATP. This synthesis of ATP has been shown to occur in the mitochondria—synthesis of specialised anabolic event driven by respiratory energy.

In the oxidative reactions of respiration, phosphorylated compounds such as ATP are formed and part of the energy liberated in the oxidation is stored as labile energy of the phosphate bond. The energy-rich (~) phosphate groups may be subsequently used as the source of energy for the energy consuming synthetic processes. Thus it is evident that inorganic phosphate is a necessary component of the oxidative system; the reaction cannot ordinarily proceed in the absence of phosphate. Furthermore, ADP which acts as a final receptor for the energy released and becomes ATP is an integral component of the oxidative system.

FERMENTATION AND ANAEROBIC RESPIRATION

Of all the many known types of fermentation, by far the most important is the process of alcoholic fermentation. Fermentation is identical with anaerobic respiration. Fermentation is the older name given by Pasteur in 1857 because it is so often controlled by micro-organisms or the enzymes secreted by them (enzymes were previously knwon as *ferments*; hence the name fermentation).

The ancient Egyptians, long before, perhaps a thousand years before Christ, knew how to malt barley and prepare a wine from it and called it fermentation—any process which was accompanied by effervescence. Fermentation processes were knwon even earlier, as early as, 3000 B.C., when the Chinese were fully acquainted with a process, making a kind of wine from rice grain.

For the first time, Pasteur realised that alcoholic fermentation was an anaerobic process and in fact characterised fermentation as 'life without oxygen'. The active agent or the enzyme, *zymase* was extracted from yeast cells in 1897 and it was clearly proved that zymase could catalyse the reaction in total absence of yeast cells. Alcoholic fermentation will occur in almost all sugar solution in which yeast is growing vigorously. Yeast can ferment glucose, fructose, galactose and mannose directly and also several disaccharides after hydrolysing them into monosaccharides.

The following summary equation represents the chemical changes occurring in alcoholic fermentation:

$$C_6H_{12}O_6 \xrightarrow{\textit{zymase}} 2CO_2 + 2C_2H_5OH + 50{,}000 \text{ cal.}$$

Yeast cells cannot ferment starch as they produce no starch breaking enzyme. The quantity of energy released—about 50,000 cal. per mol. of glucose—is much less (about 13 times) than in aerobic respiration in which oxidation of one gram molecule of glucose sets free 686,000 cal. In spite of the relative inefficiency, the process of fermentation is the only method by which yeast cells and a host of other micro-organisms and fungi obtain all their necessary energy.

When sugar solution is fermented by yeast, one of the end products—ethyl alcohol—accumulates in the solution while the other CO_2 escapes as a gas. However, there seems to be generally a definite limit to the accumulation of alcohol in the medium; when the alcohol content reaches a value of about 15%, fermentation process usually, but not invariably, in strains of yeast, comes to a stop.

Any higher animal which is deprived of oxygen will die within a few minutes. Higher plants, however, will continue to live in an atmosphere devoid of oxygen. Some plant or plant organs can survive under such conditions for a long period; others may succumb within a day or two. A number of species of plants produce seeds with coats which are more or less impermeable to oxygen, e.g., pea. During the earlier stages of germination of such seeds, before the seed coats are ruptured, respiration is largely of anaerobic type. Anaerobic respiration occurs naturally in some fleshy fruits such as grapes.

The reasons for believing that the two processes, alcoholic fermentation and anaerobic respiration, are identical, are:

(*i*) hexose sugars are usually the respiratory substrate in both the processes;

(*ii*) the principal end products—CO_2 and C_2H_5OH—are also the same in the two processes;

(*iii*) the widespread and probably of universal occurrence of the enzyme, *zymase complex*, in all the tissues of higher plants;

(*iv*) pyruvic acid, acetaldehyde, etc., are formed as intermediate compounds in both the processes;

(*v*) inorganic phosphates enhance the rates of sugar breakdown in anaerobic respiration as well as in alcoholic fermentation.

Prolonged anaerobic respiration has a detrimental, even lethal effect on tissues of many higher plants. As pointed out earlier, aerobic respiration is more than 25 times as effective as an energy releasing process as anaerobic respiration. It is understandable why many metabolically active plant tissues cannot endure prolonged absence of oxygen for the amount of energy released in anaerobic respiration is not sufficient for the maintenance of complex forms of life for any considerable length of time. Accumulation of alcohol in the medium may also be toxic to the protoplasm.

There are other types of fermentations besides the alcoholic fermentation. Many different organisms bring about *lactic acid* fermentation, some converting nearly 98% of the hexose sugar to lactic acid while others produce smaller amounts of lactic acid and varying amounts of other end products. Lactic acid, as an end product, is also formed by many tissues of flowering plants during anaerobiosis. Production of lactic acid by thin slices of potato tuber, slices of tomato stem, cotyledons of mung bean, etc., in anaerobic respiration can account for as much as 13% of the CO_2 evolved. Perhaps the best knwon organism which produces large quantities of lactic acid is *Streptococcus lactis* which commonly causes the souring of milk.

$$\underset{\text{glucose}}{C_6H_{12}O_6} \xrightarrow{\textit{glycolysis}} \underset{\text{pyruvic acid}}{2CH_3COCOOH} \xrightarrow[\textit{lactic dehydrogenase}]{+2NADH_2} \underset{\text{lactic acid}}{2CH_3CHOH.COOH} + 2NAD$$

There are other types of fermentations such as *butyric acid* fermentation, *citric*, *acetic* (for mation of vinegar) as also *oxalic acid* fermentation. The over-simplified equation of the chemical changes involved in butyric acid fermentation may be written as follows:

$$\underset{\text{glucose}}{C_6H_{12}O_6} \xrightarrow{+2NAD} \underset{\text{butyric acid}}{CH_3CH_2CH_2COOH} + 2CO_2 + 2NADH_2$$

The bacteria concerned are obligate anaerobes such as *Clostridium butyricum.*

THE MECHANISM OF RESPIRATION

Since so many plants accumulate alcohol under anaerobic conditions, early workers suggested that the normal aerobic respiration takes place in two steps; in the first, the anaerobic breakdown of sugars to alcohol and CO_2 and in the second, and oxidation of alcohol into CO_2 and H_2O. This theory soon proved to be untenable. It is now known that alcohol is not produced when the plant is respiring aerobically, though the first

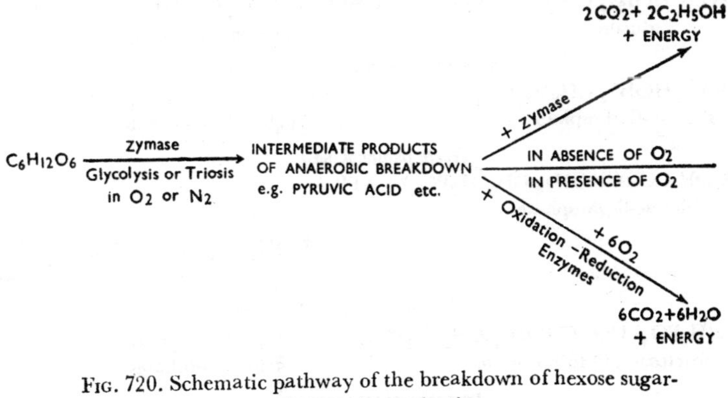

FIG. 720. Schematic pathway of the breakdown of hexose sugar-substrate in respiration.

stage of aerobic respiration is the same as anaerobic up to a point previous to alcohol formation. Alcohol is thus an end product that occurs in a branch line instead of in the main line of reactions. The side branch operates only in absence of oxygen in higher plants and becomes the main line in case of anaerobic micro-organisms. The idea has been illustrated schematically in Fig. 720.

Sugars are broken down by the enzymes of the zymase complex into some intermediate products, this anaerobic phase being the first step in both types of respiration. No oxygen is needed for this anaerobic cleavage of 6-carbon sugar molecules into 3-carbon compound, pyruvic acid, which is the pivot in the respiratory breakdown of sugar. The course of reactions then depends upon whether or not oxygen is available. In absence of oxygen, the reaction proceeds anaerobically and with specific respiratory enzymes, catalysing each and every step, the pyruvic acid is converted into acetaldehyde and CO_2. The acetaldehyde and alcohol are thus on the anaerobic branch line. The acetaldehyde then takes hydrogen from reduced coenzyme $NADH_2$ and is converted into ethyl alcohol. Thus the end products in complete absence of oxygen are alcohol and CO^2. When oxygen is present, pyruvic acid is converted into another organic acid which in its turn is changed to a third one, etc., by various oxidation-reduction enzymes of the zymase complex, ultimately into CO_2 and H_2O (Krebs cycle).

Pfeffer about 75 years ago proposed the above scheme that in plants the first stage of breakdown of carbohydrate is identical in aerobic and in anaerobic respiration. Very many researches since then have led to wide acceptance of this theory and of the view that the series of reactions now known as *Embden-Meyerhof-Parnas* (*EMP*) pathway offers a credible explanation of the way common to both aerobic and anaerobic respiration by which sugar molecules are converted to pyruvic acid.

It must be borne in mind that the series of reactions involved the phosphorylated sugars rather than the sugars themselves. A sucrose or a glucose molecule is perfectly stable inside the cell in spite of its very high energy content. But when these sugar molecules are phosphorylated by means of ATP under the influence of specific enzymes, they become chemically labile, and at each change in the sequence of steps, gives up a small packet of the chemical energy originally present in the sugar molecule.

Each molecule of hexose used for the formation of hexose diphosphate requires two molecules of ATP and the series of reactions and the enzymes involved may be summarised as follows:

(*a*) $$\underset{\text{glucose}}{CHO.(CHOH)_4.CH_2OH} + ATP \xrightarrow{\textit{hexokinase}+Mg.} \underset{\text{glucose-6-phosphate}}{CHO.(CHOH)_4.CH_2(H_2PO_4)} + ADP$$

(*b*) $$\underset{\text{glucose-6-phosphate}}{CHO.(CHOH)_4.CH_2(H_2PO_4)} \overset{\textit{phosphohexo isomerase}}{\rightleftharpoons} \underset{\text{fructose-6-phosphate}}{CH_2OH.CO.(CHOH)_3.CH_2(H_2PO_4)}$$

(*c*) $$\underset{\text{fructose-6-phosphate}}{CH_2OH.CO.(CHOH)_3.CH_2(H_2PO_4)} + ATP \xrightarrow{\textit{fructo phosphokinase}} \underset{\text{fructose-1, 6-diphosphate}}{CH_2(H_2PO_4).CO.(CHOH)_3.CH_2(H_2PO_4)} + ADP$$

(*d*) $$\underset{\text{fructose-1, 6-diphosphate}}{CH_2(H_2PO_4).CO.(CHOH)_3.CH_2(H_2PO_4)} \overset{\textit{aldolase}+Zn(Co\ \&\ Fe)}{\cdots} \underset{\text{triosephosphates}}{2C_3H_5O_2(H_2PO_4)}$$

$$\left(\underset{\text{dihydroxyacetone phosphate}}{CH_2OH.CO.CH_2(H_2PO_4)} \overset{\textit{triosephosphate isomerase}}{\cdots} \underset{\text{phosphoglyceric aldehyde}}{CH_2(H_2PO_4)CHOH\text{-}CHO}\right)$$

(*e*) $$2CH_2(H_2PO_4)CHOH.CHO + 2H_3PO_4 + 2NAD \overset{\textit{triosephosphate dehydrogenase}+Zn}{\cdots} \underset{\text{1, 3-diphosphoglyceric acid}}{2CH_2(H_2PO_4)CHOH.CH:(H_2PO_4)} + 2NADH_2$$

phosphoglyceric kinase

(*f*) $2CH_2(H_2PO_4).CHOH.CO(H_2PO_4)+2ADP \rightleftharpoons$
1, 3-diphosphoglyceric acid

$\rightleftharpoons 2CH_2(H_2PO_4).CHOH.COOH$ + **2ATP**
3-phosphoglyceric acid

phosphoglyceromutase

(*g*) $2CH_2(H_2PO_4).CHOH.COOH \rightleftharpoons 2CH_2OH.CH(H_2PO_4).COOH$
3-phosphoglyceric acid — 2-phosphoglyceric acid

enolase+Mn(Mg & Zn)

(*h*) $2CH_2OH.CH(H_2PO_4).COOH \rightleftharpoons 2CH_2{:}C(H_2PO_4).COOH+2H_2O$
2-phosphoglyceric acid — 2-phosphoenolpyruvic acid

pyruvic kinase

(*i*) $2CH_2{:}C\,(H_2PO_4).COOH+2ADP \rightleftharpoons 2CH_3.CO.COOH$ + **2ATP**
2-phosphoenolpyruvic acid — pyruvic acid

pyruvic decarboxylase+Mn (Mg)

(*j*) $2CH_3CO.COOH \rightleftharpoons 2CH_3CHO+2CO_2$
pyruvic acid — acetaldehyde — carbon dioxide

alcohol dehydrogenase+Zn

(*k*) $2CH_3CHO+2NADH_2 \rightleftharpoons 2CH_3.CH_2OH+2NAD$
acetaldehyde — ethyl alcohol,

Thus, the new overall equation, in which two water molecules and a *net gain* of two high-energy phosphate bonds of ATP in addition to CO_2 and alcohol are end products, can be written thus:

$$C_6H_{12}O_6+2H_3PO_4+2ADP \rightleftharpoons 2CH_3CH_2OH+2CO_2+2ATP+2H_2O$$

The first six steps (*a* to *f*) of partial glycolysis shown here in which hexoses are degraded to 3-phosphoglyceric acid recapitulate the photosynthetic production of hexoses from phosphoglyceric acid but only *in reverse.* Or in other words steps in photosynthetic production of sugar are apparently only a reversal of the reactions taking place in glycolysis and steered by the same enzymes. An undeniable point of difference can, however, be noted here. In photosynthetic production of triosephosphates from phosphoglyceric acid, the reverse of reactions, *e* and *f* can proceed *only in light* as the energy levels of both ATP (photosynthetic phosphorylation) and the reduced coenzyme (obtained from photolysis of water) are essential for supplying the energy requirements of photosynthetic production of sugar. The triosephosphate dehydrogenase of photosynthesis is however NADP-linked.

That phosphate is necessary for fermentation by yeast was shown as long ago as 1904, but the view that the phosphorylation is effected by means of ATP with its energy-rich phosphate bond is comparatively recent. Actually ATP was isolated from higher plant tissues (*Phaseolus aureus*) only in 1950. A number of phosphorylated intermediates of glycolysis, e.g., hexose phosphates and triosephosphates have, however, been recognised in various plant tissues since about 1946.

The existence of most of the enzymes with their coenzymes required for the series of reactions involved in the EMP pathway was demonstrated in yeast more than 25 years ago and since then all of them had also been shown to be present in the cells of higher plants. All these enzymes which have already been isolated as crystalline proteins are now knwon to be among the distinct specific enzymes present in the apozymase complex.

The formation of pyruvic acid completes the initial or anaerobic phase of respiration. Each molecule of hexose which passes through the system yields two molecules of pyruvic acid. Two molecules of ATP are required for the formation of hexose phosphates.

During the course of formation of 2 molecules of pyruvic acid from a hexose, 4 molecules of ATP are produced by substrate level phosphorylation per molecule of hexose utilised. As two molecules of ATP are only utilised for the formation of hexose phosphates, the phosphate energy of the *two surplus ATP molecules* are available for general cellular uses.

When the respiration is aerobic, i.e., when oxygen is present in the medium, the sequence of stages goes only as far as pyruvic acid and then on to Krebs cycle. Acetaldehyde and alchol are only formed in absence of oxygen.

KREBS CYCLE : MITOCHONDRIAL CYCLE

Pyruvic acid is therefore the pivot. When aerobic respiration occurs, pyruvic acid is converted to another organic acid which in turn is changed to a third one, etc. Although these organic acids are of widespread occurrence in plant tissues, it is only relatively recently that we have gained an understanding of their function or the ways in which they are synthesised by the plant. We now realise that the plant organic acids are intermediates in the respiratory oxidation of pyruvic acid. The pathway of the oxidative degradation of pyruvic acid molecule takes place in several steps to CO_2 and H_2O, passing through a series of changes converting it successively into 6, 5 and 4-carbon acids. This cyclic series is known as Krebs cycle [or citric acid or tricarboxylic acid (TCA) cycle], after the English biochemist H.A.Krebs (Nobel Prize winner, 1953) who first formulated and proposed the mechanism. The respiratory significance of this cycle is that it provides a mechanism for the oxidation of pyruvic aid, the glycolytic product of sugar breakdown (Fig. 721).

The specific enzymes of the Krebs cycle are all located on the inner walls of the *cristae of mitochondria* (subcellular organelles). These organelles can be regarded as *efficient low temperature furnaces* where organic molecules are burnt with oxygen. Most of the released energy is converted into high energy bonds of ATP. Within each mitochondrion, there are the particles capable of performing Krebs cycle reactions including the oxidative phosphorylation and containing the complete set of enzymes and co-factors. These particles have rightly been called elementary particles or *oxysomes*.

Recently the cells of higher organisms have been found to contain another type of subcellular organelles, called *peroxisomes*, whose major function is thought to be the *protection* of cells from free molecular oxygen. Tolbert and his colleagues (1968, 1969) isolated peroxisomes from a number of leaves. The peroxisomes contain the enzymes that catalyse the direct reduction of oxygen molecules, forming H_2O_2, through the oxidation of metabolites, such as amino acids and acids of the Krebs cycle. In addition to *catalase*, the leaf peroxisomes contain *glycolate oxidase, glutamate, glyoxylate, amino transferase, malate dehydrogenase*, etc.

The classical dark respiratory pathway may be referred to as *mitochondrial respiration* in contrast to *peroxisomal respiration* to refer to the light-induced or light dependent pathway.

The rate of reduction of oxygen by the peroxisomes increases proportionately with oxygen concentration, so that an excessive amount of oxygen in the cell increases the rate of its reduction to H_2O_2 or H_2O by peroxisomes.

The basic difference between the mitochondrial oxidation reactions and those of peroxisomes, is that in the oxisomes, the steps of oxidation *are not coupled to the synthesis of ATP*. The energy released in peroxisomes is thus lost to the cell; apparently the function of the peroxisome is primarily to protect the cells against the destructive effects of free molecular oxygen which can sometimes be disastrous.

Now leaving aside, peroxisomes and photorespiration, for the present, the mitochondria, in general, perform all of the reactions of Krebs cycle including the formation of ATP molecules, when atmospheric oxygen is finally reduced by electrons obtained from acid substrates.

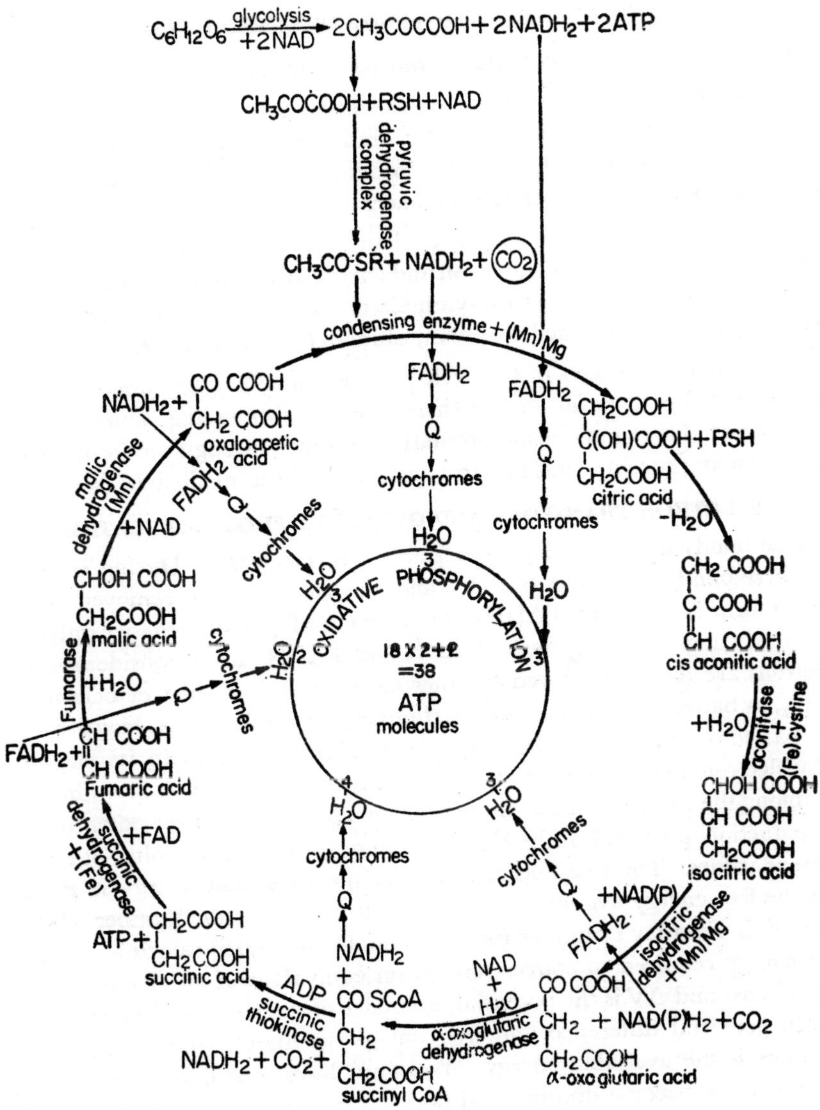

FIG. 721. The operation of Krebs cycle in aerobic respiration. Each time the cycle is traversed, 2 molecules of CO_2 are formed and there is a net gain of 3 molecules of H_2O. $CH_3CO.COOH$, pyruvic acid; RSH, coenzyme A; CH_3COSR, acetyl coenzyme A; *the sequence of acids formed, from the top of the cycle (in clockwise manner) are* citric acid (6C); cis-aconitic acid (6C); isocitric acid (6C); α-oxo glutaric acid (5C); succinyl CoA; succinic acid (4C); fumaric acid (4C); malic acid (4C) and oxalacetic acid (4C) respectively; $FADH_2$, reduced flavin adenine dinucleotide; Q, ubiquinone

The labelling of the mitochondria as the *power-houses of cell* and the classical Krebs cycle as *mitochondrial cycle* seem appropriate enough and totally justified. Mitochondrial cycle or energy-transmutating cycle is certainly the energy transformer of the cell whereby energy inherent in chemical compounds is ultimately modulated to the biological energy of the ATP molecules, which is essential for the organism.

The pyruvic acid formed by glycolysis reacts with CoA, forming acetyl CoA, *before* it enters the cycle, as written below:

$$CH_3CO.COOH+RSH+NAD \text{——} CH_3CORS+NADH_2+CO_2$$

where RSH is coenzyme A; it contains S and pantothenic acid.

A molecule of CO_2 is evolved here and the acetyl coenzyme A, thus formed enters the cycle proper and in presence of water, and condensing enzymes, react with oxalacetic acid obtained from the cycle, to produce 6-carbon citric acid and making the cycle 'go' normally in a clockwise manner.

With appropriate enzyme systems with their activators available for the sequence of cyclic reactions, the TCA cycle moves on and energy, originally contained in the sugar, is apportioned and ATP molecúles are synthesised as the electrons released run through the coupled electron transport redox system, ultimately forming water.

In the sequence of changes in the cycle proper (Fig. 721) ten organic acids take part of which three are *ketoacids—oxalosuccinic* (?) not shown in the figure, *a-oxoglutaric* and *oxalacetic* acids. There may be an alternative pathway through the formation of *glyoxylic acid*, by-passing succinic and fumaric acids, directly to malic acid (see Fig. 721).

ELECTRON TRANSPORT AND OXIDATIVE PHOSPHORYLATION

Chemical bonds are formed by the sharing of electrons. During oxidation of substrates these bonds snap and the H^+ and the electrons are removed by NAD^+ or FAD^+ to form NADH and $FADH_2$, the products of dehydrogenase catalysed reactions. The newly acquired electrons of NADH and $FADH_2$ carry considerable energy. These electrons are then transferred to other electron carriers like cytochromes and finally to O_2, as batons are passed on to participants of a relay race. However, as the electron is passed on from one carrier to another, there is a drop in its energy. The reason why electrons move from one carrier to another is that there is a difference in potential among the carriers. Electrons move from a higher to a lower potential and the oxidation-reduction potential of the $H_2 \rightarrow O_2$ system is about 0·81 volts; other systems have different values. The relationship between potential and energy is $E=nF\Delta V$, where E is the free energy available for work to be done, n is the number of electrons transferred, F is Faraday's number (96494 coulombs or 23,066 calories—which is the amount of energy required to convert one gram equivalent of an element to its one equivalent of ions) and ΔV is the potential difference of the systems. Since the relationship between potential difference and free energy is direct, greater the potential difference more is the available energy. (You must have seen that when line voltage drops electric lamps become dimmer and fans move slowly). As electrons are transported from one carrier to another, there is a drop in potential. At only three points this drop is equivalent to more than 7000 calories, the amount of energy required to form ATP from ADP and inorganic phosphate; consequently during electron movement from NADH to O_2 3 molucules of ATP are formed. In the case of $FADH_2$, the potential drop is enough for formation of only two molecules of ATP. At some points in the electron flow the drop is much less; as the energy equivalent in calories is less than 7000 calories, so no ATP is formed there. The sequence of electron flow is as follows:

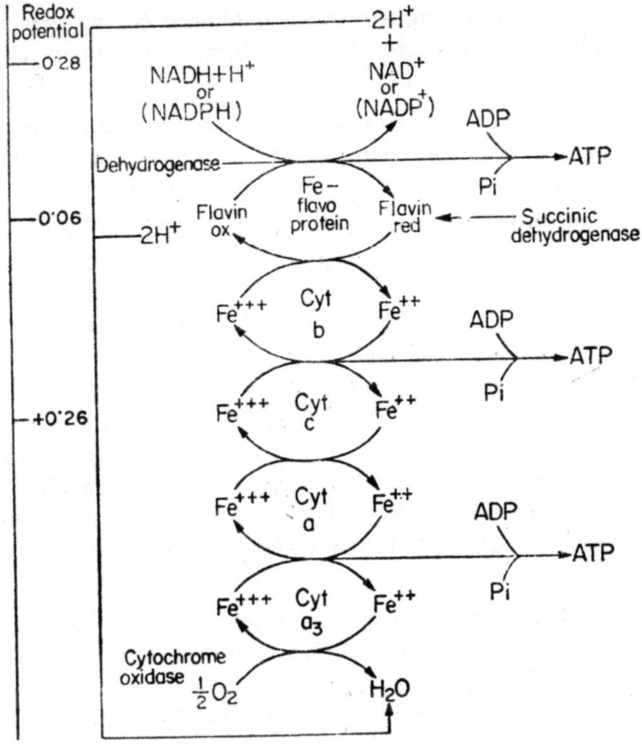

FIG. 722. Terminal electron transport in mitochondria. The redox potential (Eo) values given on the left hand side of the figure are at pH 7.0 and 25°C. Any reduced electron carrier in the upper part of the figure has a higher redox potential than one down below and thus can donate an electron to it thereby reducing it.

Like NADH+H^+, NADPH+H^+ is oxidised similarly producing 3 molecules of ATP per molecule of NADPH+H^+ oxidised to $NADP^+$ and H_2O. In the scheme although one cytochrome *c* is shown, more than one form may be involved. It has been suggested that in some systems ubiquinone (or coenzyme Q) mediates electron transport between FAD and cytochrome b. The sites of ATP synthesis are not completely established.

The mechanism of ATP synthesis is not well understood. It has been suggested that some proteins of the mitochondrial inner membrane are involved, since addition of such purified proteins—the so-called coupling factors, to inactive mitochondrial preparations, results in oxidative phosphorylation. Presumably, high energy phosphorylated proteins are formed as an intermediate. Peter Mitchell, however, does not think that such proteins are actually formed; he has suggested an entirely different theory of ATP synthesis—the *chemiosmotic theory*. On the basis of the observation that H^+ are actually expelled from mitochondria and concentration of OH^- ions increases inside during respiration, he postulated that disappearance of water—a product of the reaction, ADP+Pi⟶ATP+H_2O would result in ATP synthesis by mass action or Le Chatelier's principle. Experiments during the last 19 years have strongly supported this concept and Mitchell was awarded the Nobel prize for this work recently.

Mitchell demonstrated this by treating mitochondria with buffer solutions which resulted in the production of a pH gradient between the two sides of the membrane. Under normal conditions ionization of water is enhanced during movement of electrons. How specifically protons are ejected by membrane components of mitochondria, however, is not understood at present.

The coupling of ATP synthesis to electron transport implies that expenditure of ATP energy would drive the reaction sequence in the backward direction. It would also explain ATP dependence of uptake of solute, including ions, across membranes.

Let us now examine the efficiency of Krebs cycle in relation to ATP synthesis. An examination of Fig. 722 a would show that phosphorylation takes place (*i*) during the oxidation of NADH or NADPH; (*ii*) during the oxidation of a common acceptor cytochrome b and probably, (*iii*) during the oxidation of reduced cytochrome a. Thus for every organic acid substrate oxidised *via* NAD or NADP, *there are three molecules of ATP* formed for every atom of oxygen that gets reduced to form water. In the case of succinate oxidation, where NAD stage is eliminated, *only* 2 *ATP molecules* are formed for every oxygen atom that is reduced. Again when α-oxoglutarate is oxidised by NAD, there is also an additional phosphorylation—*substrate level phosphorylation*—by which one extra ATP molecule (or GTP) is formed. Thus in the case of oxidation of α-oxoglutarate, altogether *four* molecules of ATP are formed for each atom of oxygen reduced (see Fig. 721).

A total of 15 ATP molecules are formed from oxidation of one molecule of pyruvic acid. 3 ATP molecules are also formed from a reduced $NADH_2$ obtained when triosephosphates are oxidised to phosphoglyceric acid in the EMP pathway. A total of 18 ATP molecules are thus obtained from oxidation of one equivalent pyruvic acid. As two pyruvic acids are obtained from one hexose sugar, a total of 36 ATP molecules is clearly visualised. To this must be added the two surplus ATP molecules obtained by substrate phosphorylation in the glycolytic route. So we get a grand total of 38 ATP molecules formed as a hexose sugar substrate runs through and is completely oxidised by the EMP and Krebs cycle routes (see Fig. 721).

We know most of the metabolic pathways along which these Krebs cycle intermediates travel. As a result, we cannot escape the realisation, that the key plan, where the *regulations of the metabolic pathways occur and also intersect each other, may be at the level of Krebs cycle.*

Verification of the cycle

On the basis of many experimental observations evidence seems to be complete and the inescapable conclusion is that the Krebs cycle is operative in the living cell. The primary evidences can be summarised as follows:

(*i*) The widespread and universal occurrence of organic acids, known to take part in the cycle in all types of plant tissues;

(*ii*) The extraction, isolation and purification of enzymes of the Krebs cycle from mitochondrial preparations in species from many families of angiosperms, *Conifers*, as well as from *Bryophyta* and *Pteridophyta;*

(*iii*) If plant mitochondria are suspended in a buffer solution and if any of the acids of the cycle, e.g., citric, α-oxoglutaric or oxalacetic acids, is added to the solution, it is soon converted into the adjacent acid in the cycle;

(*iv*) When pyruvic acid labelled with radioactive tracers such as, ^{14}C is fed to mitochondrial preparation, each of the other ten or eleven acids becomes labelled soon after. If ^{14}C-labelled pyruvic acid is given to whole organisms and if various chemical compounds such as carbohydrates, fats and proteins are isolated and analysed, one can have a fair idea of the fate of the individual carbon atoms in the injected pyruvic acid. On the basis of many such experiments, there seems to be no doubt, whatsoever, about the operation of the Krebs cycle in all living cells.

PENTOSE PHOSPHATE SHUNT OR DIRECT OXIDATION PATHWAY

During the last few years a number of investigators have proposed that hexose may be broken down in plants to carbon dioxide and water by an alternative course which is independent of the EMP pathway and the Krebs cycle (glycolysis is eliminated). That such a more direct pathway might exist in animal tissues and yeast was indicated more than 40 years ago but the possibility of its existence in higher plant tissues has been definitely recognised only during the last 25 years.

It has been known for a long time that when yeast is treated with sodium mono-iodoacetate in proper concentration, activity of the enzyme systems of glycolysis can be completely inhibited. It was noticed, however, that sugar can still be oxidised completely to CO_2 and H_2O by yeast cells. Evidently, all the enzymes of EMP-linked Krebs cycle may not be necessary for the chain of reactions in which sugar is completely oxidised. The chain of reactions of glycolysis culminating in the production of 3-carbon compounds from hexose molecule was supposed, by many workers in this field, to play perhaps no part in preparing the phosphorylated sugar molecules for oxidation.

The *shunt pathway* also known as Hexose Monophosphate Pathway (HMP) or Pentose Phosphate Pathway (PP) involves a cycle of reactions in which phosphorylated hexose is degraded to phosphorylated pentose with release of CO_2 and H_2O and the ultimate reformation of hexose from the pentose through a series of reactions in which other phosphorylated sugars like heptulose, tetrose and triose are the intermediates. In plant systems the occurrence of this pathway was established largely through the work of Gibbs, Axelrod and Beevers. Its presence in microbial and animal systems was earlier demonstrated by Warburg, Christian, Lipmann, Dickens, Horecker and Racker.

Glucose, the respiratory substrate is first phosphorylated as in the first stage of EMP pathway, the enzyme catalysing this reaction is, as we have seen before, *hexokinase*. The cycle of reactions in this direct oxidation pathway can be simplified as shown in Fig. 723.

For every six molecules of hexose entering this cycle of reactions, *five* are reformed. Only *one* hexose molecule is thus consumed, providing 6 molecules of CO_2. There are also 12 molecules of reduced coenzyme ($NADPH_2$) of the dehydrogenases and these are oxidised by 12 atoms or 6 molecules of oxygen to 12 molecules of water by the flavoprotein enzymes and cytochrome oxidase system (36 ATP molecules formed). This leaves a *net* gain of 6 molecules of H_2O (6 molecules of H_2O being utilised in reaction I). The overall equation for the pentose-heptose phosphate shunt is, thus again, the all familiar: $C_6H_{12}O_6 + 6O_2 \rightarrow 6CO_2 + 6H_2O$.

Evidences in support of direct oxidation pathway

The evidence for the existence of the direct oxidation pathway has been further strengthened by the observation that pentose and heptose sugars are widely distributed in plants. By allowing leaves of *Sedum spectabilis* to photosynthesise in an atmosphere

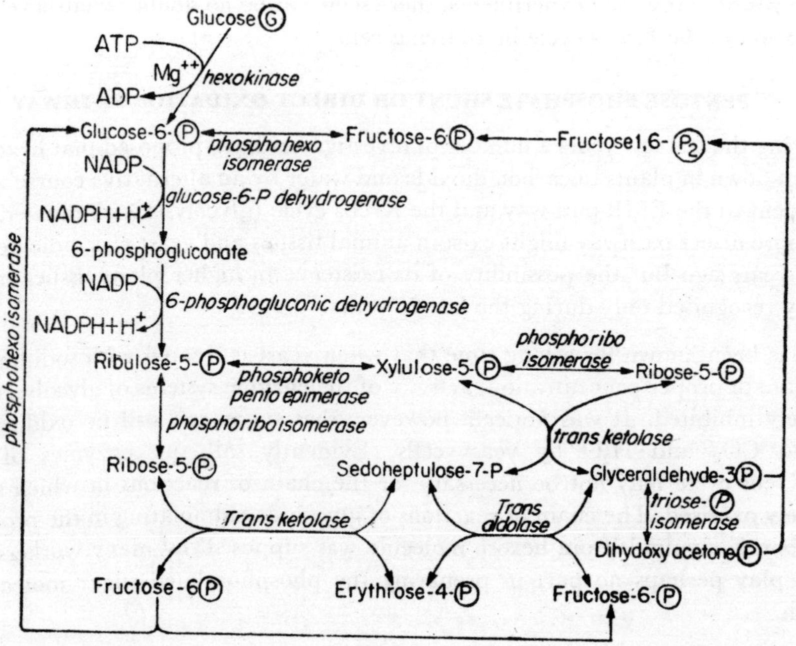

Fig. 723. Schematic representation of the Hexose monophosphate (HMP) or the Pentose Phosphate Pathway. Note that some of the enzymes are common to HMP and EMP pathways, the reactions concerned with the generation of 4-, 5-, 6- and 7- carbon sugars are reversible and that transketolase may transfer a C_2 fragment from the Ketose sugars xylulose-, fructose or sedoheptulose-phosphates to aldose sugars like ribose-, erythrose and glyceraldehyde—phosphate; consequently modification of the reaction suquences as shown, is possible.

containing $^{14}CO_2$, it has been definitely shown that 7-C heptose sugar (*sedoheptulose*), containing ^{14}C can be synthesised by the leaves.

As regards the enzymes supposed to be involved in the direct oxidation pathway almost all of them have been shown to occur in different plant tissues, e.g., leaves of spinach, lucerne, parsley, etc., roots of parsnip and turnip and also in the wheat embryos and fruits of cucumber, etc.

Evidence that plant material is capable of effecting the oxidation of the various sugar intermediates in the direct oxidation pathway has also been recently obtained. The results seem to indicate that the direct oxidation mechanism operated certainly in yeast thus confirming the hypothesis of Lundegardh (1937). With maize roots, evidences seem to be in favour that the direct oxidation was not followed or at most to a very small extent. With other higher plant tissues, e.g., leaves and roots of garden pea, stems and leaves of sunflower, petiole and root slices of carrot, leaves of *Bryophyllum*, maize leaves, etc., a significant, if not the major part, of the respiratory breakdown of hexose was found to be by the direct oxidative pathway.

In tomato fruits, Barbour (1958) and others found that there was a significant degradation of sugar by the direct oxidation pathway—about 84% along EMP-Krebs cycle and about 16% by the pentose phosphate pathway.

The proportion of sugar degradation following the direct oxidation pathway, seems to increase with age—in senescent tissues, more CO_2 is evolved following direct oxidation compared to young and vigorously growing cells, which tend to follow the normal glycolytic pathway.

It has been sometimes seen that tissues of safflower (*Carthamus tinctorius*) infected by rust, *Puccinia*, show significantly higher respiration rate as measured by evolution of CO_2. Comparing healthy with diseased plants, it has been shown by Daly and Sayre (1957) that in the respiration of healthy plants, EMP-Krebs cycle route was preferentially followed whereas most of the increased respiration—observed in the infected tissue—was being maintained by following the direct oxidation pathway.

In tissues which are respiring simultaneously via EMP and HMP pathways, the relative contribution of either pathway can be assessed by feeding glucose labelled with radioactive carbon at selective positions. In the HMP pathway C-1 of glucose is oxidised first to CO_2, which can be trapped in alkali and counted. In the EMP pathway C-3 and C-4 (which equilibrate during triosephosphate isomerase action) is oxidised to CO_2 first; C-1 and C-6 also equilibrate and appear later. From the rate of CO_2 evolution from differently labelled glucose molecules used as substrates for respiration, an idea may be obtained regarding the relative preponderance of the two pathways.

Photorespiration

It is now known through the work of Tolbert that the glycollate produced during photosynthesis is the substrate for photorespiration. Glycollate may be produced by the

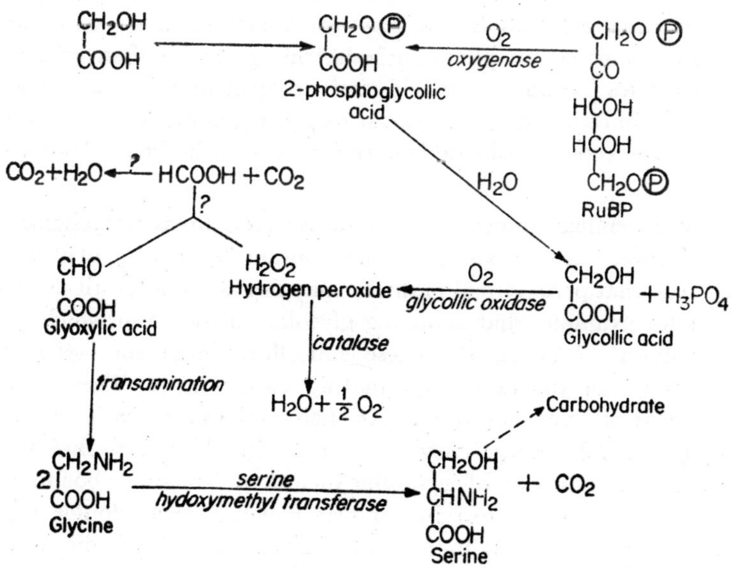

Fig. 724. Schematic representation of the probable reaction sequences in photorespiration.

oxidation of the C_2 active glycolaldehyde fragment carried by transketolase or by the action of oxygenase activity of RuBP-carboxylase preparations on RUBP, producing glycolaldehyde phosphate. The glycollic acid so produced on the chloroplasts (several photosynthetic organisms e.g., *Chlorella* excrete appreciable quantities of glycollate), passes on to the peroxisomes and mitochondria where it is converted to glycine via glyoxylic acid by transamination reactions. From two molecules of glycine one molecule of serine and one molecule of CO_2 are produced in the mitochondria (Fig. 724). Serine is then further metabolised partly to carbohydrate. It has been suggested by Zelitch that CO_2 may also be produced via formic acid.

Photorespiration and microbodies

We have already discussed about peroxisomal-photorespiration. It remains only to state a few more interesting facts about this increased respiration rate in light and also about a number of other newly discovered subcellular organelles—the *microbodies*.

The term photorespiration, we know, is usually used to describe all respiratory activity in light, regardless of the pathways by which CO_2 is released and O_2 consumed, i.e., a light dependent release of CO_2. Some plants, such as temperate Calvin—Benson plants (plants following CO_2-fixation by PSCR cycle of photosynthesis) evolve CO_2 when brightly illuminated in CO_2-free air while others, such as species of tropical Gramineae (plants fixing CO_2 primarily by HSK-pathway) do not. This evolution of CO_2 is stimulated by high concentration of the oxygen of the medium, whereas, we know, that the classical mitochondrial dark respiration is largely unaffected by changes in the O_2-concentration, between 2-20%. This photorespiration is inhibited only by CMU and DCMU (dichlorophenyl dimethyl urea). Different responses to temperature also suggest that photorespiration and dark respiration are separate processes. There are also strong evidences, we repeat, in progressive increase in photorespiration with increasing light intensities.

As we have just mentioned, the group of plants, which follow the HSK-pathway *via* C_4-dicarboxylic acid, have no sizable respiration rate in light; these species are all usually extremely efficient in photosynthesis; this group is often referred to as low compensation species (because of their capacity to deplete the surrounding atmosphere to very low CO_2-concentration) in contrast to plants following Calvin cycle pathway, which have as we know considerable respiratory rate in light—high compensation speceis.

This light-dependent respiration is now referred to as peroxisomal (contain 8 enzymes) respiration—the peroxisomes do not contain the enzymes of glyoxylate cycle—but it is also true that peroxisomal pathway may not account for all this light-induced component as for example, that involving glycolate metabolism which takes place in *glyoxisomes* (contain 17 enzymes). These subcellular organelles—*plant microbodies*—have been detected in tissues of germinating castor bean. They occur in many shapes and appear pliable; most of them in their profiles are nearly circular, with diameter ranging from 0·2 to 1·5μ. They are characterised by a *single* limiting membrane, a granular matrix, a dense core of crystalline inclusion. The single bounding membrane is without cristae, in contrast to mitochondria which have the double membrane with inner membrane showing invagination into cristae. The microbodies differ also from mitochondria in that they have *no cytochromes* and *phosphorylated substrates* have not yet been found there.

Thus, it is evident that while mitochondrial respiration conserves biochemical energy as ATP by oxidative phosphorylation, the peroxisomal and glyoxisomal respiration are characterised by oxidation of H_2O_2—generation and the *loss of energy as heat* during catalase destruction of peroxide.

Leaf glyoxisomes are unique in one respect; they contain β-oxidation enzyme-complex, essential for the conversion of fatty acids to acetyl coenzyme A and ultimately in the synthesis of carbohydrates.

Other terms for particles, similar or identical to microbodies are *phragmosomes, cytosomes* and *crystal containing bodies.* In roots, microbodies have not yet been reported in detail. The root organelles do not contain glyoxylate cycle enzymes—they may represent another biochemically different type of microbody (Tolbert, 1971).

OXIDATIVE ANABOLISM

In anaerobic respiration, only one-third of the carbon acted upon in triosis or glycolytic stage finally appears as CO_2, the other two-thirds as C_2H_5OH.

$$C_6H_{12}O_6 \longrightarrow 2CO_2 + 2C_2H_5OH$$

If then carbon atom is taken as a unit, we can say that anaerobic respiration as measured by the volume of CO_2 evolved is only one-third of glycolysis or using Blackman's nomenclature:

$$GL = 3NR \text{ (respiration in } N_2\text{, i.e., anaerobic).}$$

In Blackman's experiments when air or oxygen was replaced by N_2, i.e., when aerobic was replaced by anaerobic respiration, the respiration rate increased to 1·33 (sometimes 1·5) times that in oxygen. Consequently, using the symbol OR for aerobic respiration (oxygen respiration):

$$NR = 1{\cdot}33 \text{ OR and therefore}$$
$$GL = 4 \text{ OR}$$

This certainly means that under aerobic conditions only *one* atom of carbon out of every *four* subjected to triosis or glycolysis is actually released as CO_2. Consequently for every atom of carbon released as CO_2, *three* atoms of carbon are worked back into the system or again using Blackman's nomenclature:

$$\text{OA (oxidative anabolism)} = 3 \text{ OR (oxygen respiration).}$$

Thus in aerobic respiration, part of carbon is completely oxidised to water and CO_2 but the CO_2 so formed, only accounts for a fraction of the carbon which appears to be

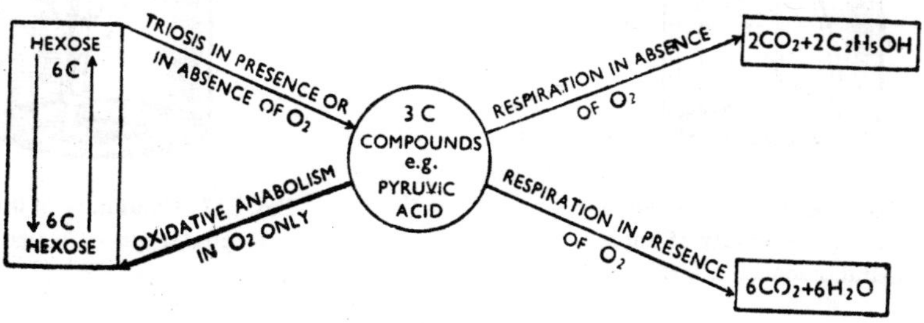

FIG. 725. Pathway of sugar breakdown in respiration and the operation of oxidative anabolism in presence of O_2 only.

involved in glycolysis. The rest of carbon, therefore, must have some other fate. No new substances appear to accumulate in the tissues and the conclusion is therefore drawn by Blackman that only in presence of oxygen, the majority of the products of glycolysis in many tissues of plants, but certainly not all, are worked back into the system to the hexose sugar or to some other substance such as malic acid which can be converted ultimately into hexose. The name *oxidative anabolism* was given to this process. The scheme can be diagrammatically represented as in Fig. 725. Apparently O_2 in some way inhibits respiration.

EXPERIMENTS ON RESPIRATION

I. To prove that the intercellular spaces in plants form a continuous system

Take two wide-mouthed bottles and fill about three-fourths of them with water. Fit in rubber corks, each with two holes and fit them air-tight on the mouths of the bottles. In Fig. 726, insert the petiole of a *Nasturtium* or any other suitable leaf with a long petiole through one of the holes and allow it to dip into water of the bottle. In Fig. 727, insert a freshly cut twig, the cut ends of which are closed. In both the bottles, horizontally bent glass tubes as shown in the figures are inserted whose short vertical ends lie just below the corks above the surface of water.

Draw the air out of the bottles either by sucking through mouth or better by suction through filter pump. Observe the air bubbles coming out in water at the cut end of the petiole in Fig. 726 and throughout the entire surface of the stem twig, in Fig. 727. In

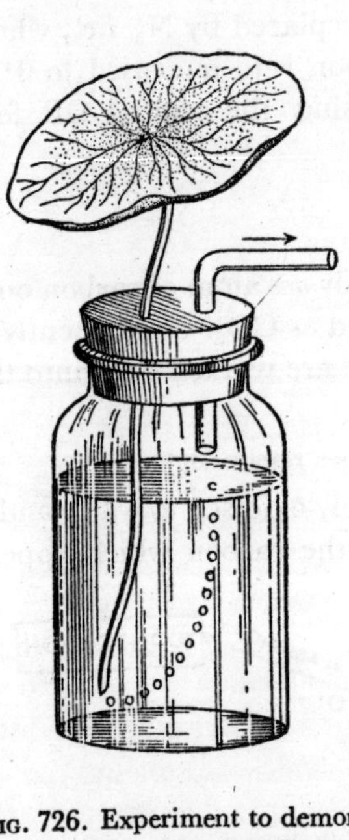

Fig. 726. Experiment to demonstrate the continuity of intercellular spaces in leaf.

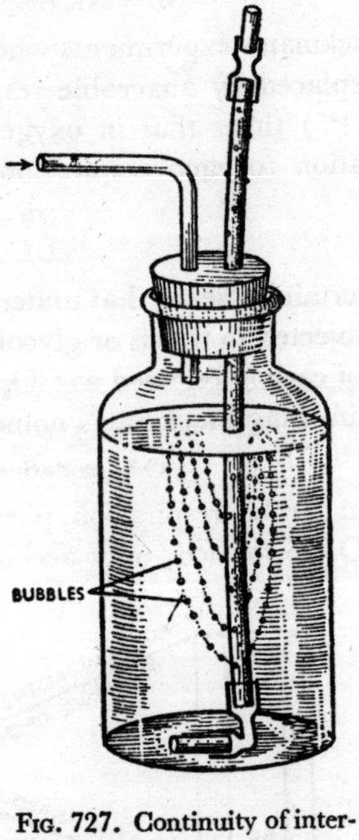

Fig. 727. Continuity of intercellular spaces in stem.

Fig. 723, suction causes air from atmosphere to enter the leaf through the stomata and through the continuous intercellular spaces of the leaf and petiole tissues and ultimately comes out through the cut end dipped in water. In Fig. 727, as both the cut ends are closed and since bubbling may proceed indefinitely it is proved that: (*a*) atmospheric air can enter through the lenticels of the cut stem situated above the cork and (*b*) that the air can travel downwards through the continuous intercellular spaces and bubble out again into water through the lenticels of the submerged portion of the stem.

II. Demonstration of aerobic respiration in plant tissues

Aerobic respiration in plants can be experimentally proved very simply by the use of a respiroscope, (Fig. 728) which consists essentially of a stout vertical tube which is bent into a bulb at the one end. The experiment can be done as well with a long-necked round-bottommed flask or simply a very wide test tube (Fig. 729).

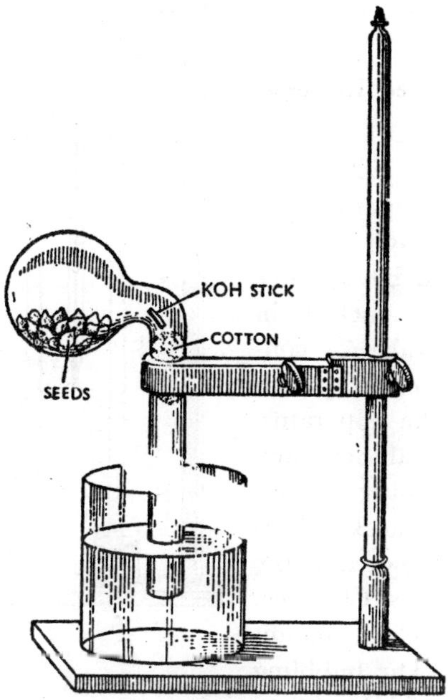

FIG. 728. Respiroscope for demonstration of aerobic respiration in plants.

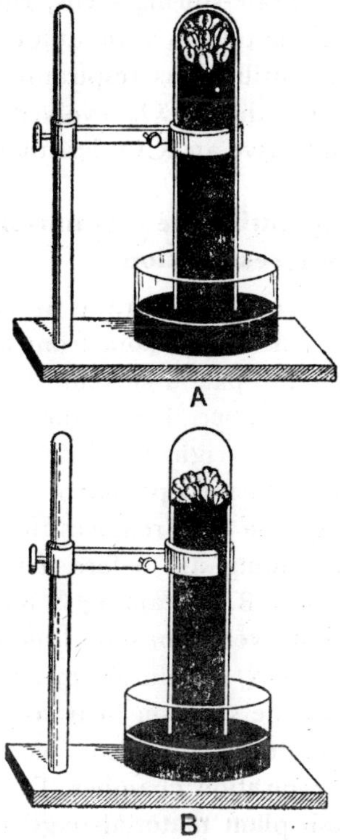

FIG. 729. Simple arrangement for demonstration of aerobic respiration in plant material.

The respiroscope (or the inverted tube) is fixed vertically to a stand and a few germinating seeds or flower buds (from which all green parts have been removed) are placed in the bulb of the respiroscope, the vertical tube of which is dipped just below the surface of water in a beaker (instead of water, mercury may also be used as in Fig. 729). The air enclosed in the apparatus is thus cut off from external air. Introduce a piece or two of caustic potash pellets in the bent portion of the respiroscope, preventing its fall with loosely held cotton wool as shown in the figures. Allow the apparatus to stand for a few hours (2–3 hrs.); it will be seen that water has risen in the vertical tube of the apparatus proving the production of partial vacuum (due to absorption of evolved CO_2 in respiration by KOH) which draws water (or mercury, Fig. 729) upwards into the tube. Instead of water, it is better to use brine for CO_2 is highly soluble in pure water.

In Fig. 730, there is a pair of respiroscopes fixed on a vertical stand. In the tube A, the swollen portion contains germinating seeds but no caustic potash whereas in B, a small tube of pellets of KOH is hung from the top inside the apparatus. Both the tubes are dipped below the surface of water in two vessels. It will be seen after sometimes, that water rises in the respiroscope tube B due to absorption of CO_2 evolved by the seeds by

KOH whereas the level of water both outside and inside the tube in A remains stationary. In the tube A, the volume of oxygen utilised is approximately equal to the volume of CO_2 liberated; hence there is no vacuum inside the tube. If seeds or other materials, storing fats and oils as reserve food in the endosperm, e.g., castor seeds are used in the tube A instead of carbohydrate-storing seeds, rise of level of water could be seen in the tube A without any KOH, indicating that when fats are utilised as respiratory substrate, more oxygen is absorbed than CO_2 evolved in respiration. This will certainly give an R.Q. less than unity.

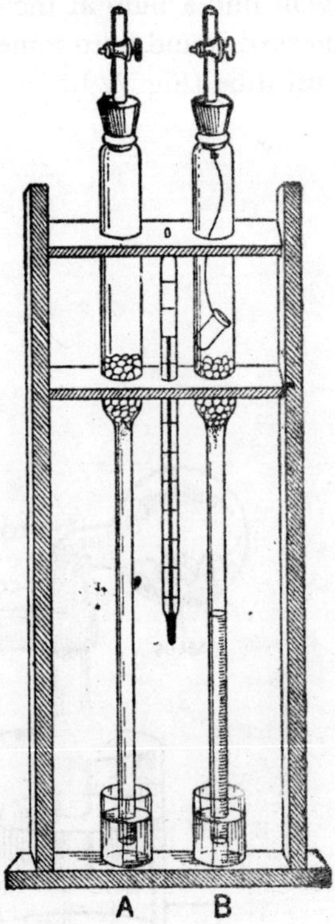

FIG. 730. A pair of respiroscopes.

III. Quantitative estimation of CO_2 released in aerobic respiration

Set up the apparatus as diagrammed in Fig. 728, and test to ensure freedom from all leaks. The apparatus is designed to pass a current of CO_2-free air through the respiring chamber, D. As the suction is applied by a filter pump in the right-hand side, atmospheric air is sucked in by the left-hand opening of the tube. The 'U' tube A, containing soda lime removes the major portion of CO_2 from the incoming air before it passes through the apparatus. The bottle B, containing N $Ba(OH)_2$ solution absorbs any CO_2 left over from soda lime tube, A. The bottle C, containing a N/10 solution of baryta serves as a check and as long as the solution in it remains clear, it shows that the stream of incoming air passing through D is CO_2-free. The respiration chamber, D, contains an weighed amount of fresh plant material (e.g., potatoes, etc.). The bubbling of air is continued for 1–2 hrs. (time should be noted accurately), the respiring material absorbs oxygen from air and releases CO_2 which is absorbed in a known volume (say, 50 ml) of N/10

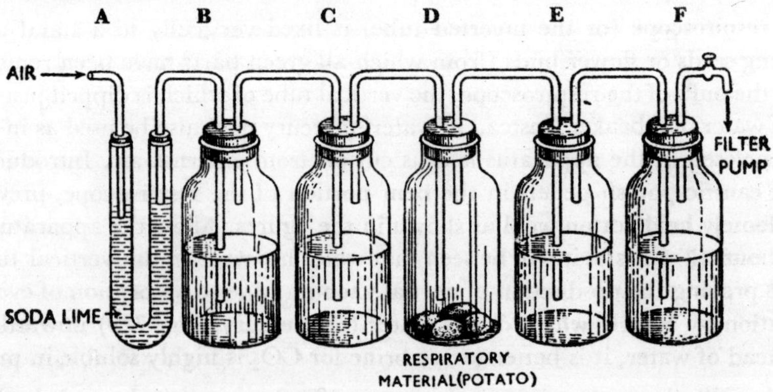

FIG. 731. Arrangement of the apparatus for quantitative determination of CO_2 evolved in aerobic respiration of plant material.

$Ba(OH)_2$ (or of any other known concentration, e.g., N/50, N/100) coloured red with phenolphthalein in the bottle E. The indicator serves as a test for excess alkali. If the solution in E becomes colourless during a run, stopping the gas stream, a further 10—20 ml of N/10 baryta may be added in the bottle E. The bottle F (also containing N/10 baryta) serves as a final safety check and as long as the solution in this bottle remains clear, it shows that all the CO_2 evolved in respiration by the material in the chamber D, is being completely absorbed by the solution in test bottle E. The CO_2 released by the respiring material partially precipitates baryta solution as $BaCO_3$ in the test bottle E. The residual $Ba(OH)_2$ is titrated against N/10 HCl and mgs. of CO_2 evolved are easily calculated for a particular weight of the material in a given time, preferably 1 hour. This titration value should be subtracted from the titration of a fresh 50 ml sample of N/10 baryta with acid. The weight of CO_2 released in respiration could be calculated from the following equation:

mgs. $CO_2 = V \times N \times 22{\cdot}0$ where V is the difference between the blank and experimental titration; N, the normality of the acid used for titration and 22·0 the equivalent weight of CO_2.

IV. Demonstration of anaerobic respiration in plants

(*a*) Take a few soaked gram seeds from which the seed coats have been removed carefully without damaging the embryos (the seed coats are removed to facilitate diffusion of gases). Fill a test tube completely with mercury and invert it just below the surface of mercury contained in a petridish. Pass the seeds under the open end of the tube in the mercury by means of forceps when they will naturally float up to the top of the test tube (Fig. 732A). Observe that in

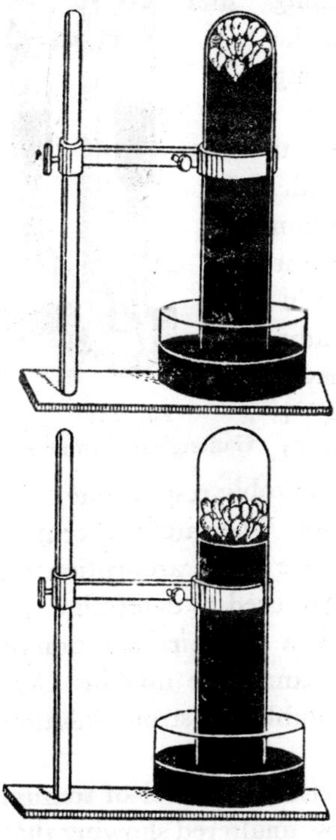
A
B

FIG. 732. Demonstration of anaerobic respiration of seeds.

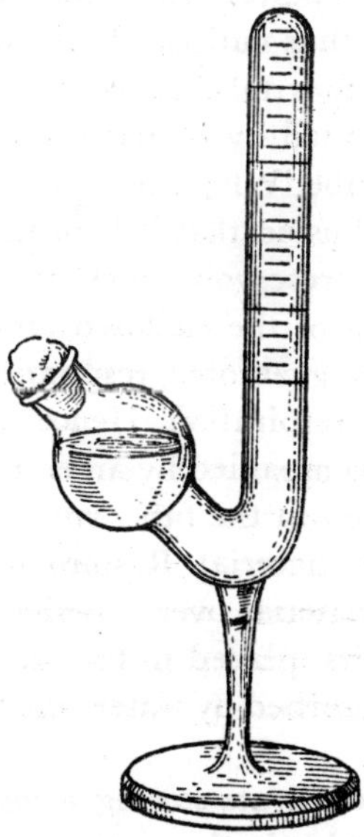

FIG. 733. Fermentation vessel (Kuhne's vessel).

a day or two, the mercury level falls in the test tube in B and the test tube is half or more full of a gas. Introduce a little caustic potash pellet through the open end of the tube. Caustic potash absorbs the gas and the mercury level in the test tube again rises. This proves that the gas is CO_2. That the gas produced by respiration of germinating seeds in complete absence of oxygen supply is CO_2, can also be proved by inserting a lighted match stick in the test tube which is at once extinguished.

(*b*) Fill three fermentation tubes (Kuhne's vessel, Fig. 730) with mixtures of 10% solutions of glucose, sucrose and maltose respectively mixed with small quantities of bakers' yeast (or a suspension of living yeast cells). Plug the open end of the apparatus with cotton wool and observe the occurrence of fermentation (anaerobic respiration) and the collection of CO_2 gas in the back arm of the tube. One with a sensitive nose may even detect faint smell of alcohol produced when the cotton wool is taken off.

V. Ganong's respirometer

The apparatus (Fig. 734) consists of three parts: (*i*) the bulb for the respiring material, with a second smaller bulb at the botton of the first and provided with a stopper with a lateral hole which can be opened or closed by turning the stopper, (*ii*) a graduated manometer tube and (*iii*) a levelling or reservoir tube connected with the manometer tube by a stout rubber tubing. 2 ml of respiring material are put into the bigger bulb of the respirometer and a 10% solution of KOH is placed in the manometer tube. At the outset the air around the material is brought at the atmospheric pressure by turning the stopper of the bulb until the hole coincides with the neck of the tube. The levelling tube on the right is so adjusted that the potash solution in the tube is at the 100 ml mark at the bottom. The 2 ml of respiring material is thus surrounded by 100 ml of air.

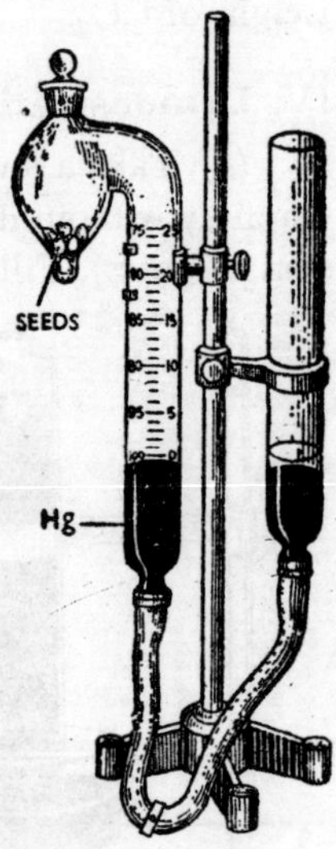

FIG. 734. Ganong's respirometer.

Now the experiment is started by turning the glass stopper at the top and thus cutting off all connections with the external air. Respiration now takes place in a closed space and the absorption of CO_2 liberated by KOH is shown by the rise of solution in the graduated tube. When the solution rises up to 80 mark, the solution is so adjusted that it is in the same level in both the graduated and the reservoir tubes. It remains stationary at this level, i.e., one-fifth of the enclosed air (% O_2 in air is about 21%) is utilised during aerobic respiration. This fraction represents O_2 absorbed in respiration. Hence the only inference is that in aerobic respiration CO_2 output is accompanied by an equal volume of oxygen absorption. The rate of ascent of KOH solution in the tube can be taken as a measure of the rate of O_2 absorption by the respiring material. Respirometers may also be conveniently used to determine the R.Q. of a material over a period of time. In order to do this a saturated solution of NaCl is at first placed in the manometer tube. (Pure water cannot be used as CO_2 evolved is absorbed by water which is, however, only slightly soluble in a strong solution of brine.)

If carbohydrate-storing seeds are used as respiring material, the level of sodium chloride solution in the graduated tube will remain more or less unaltered showing that the volumes of CO_2 evolved and oxygen absorbed are the same and R.Q.$=1$. If solid KOH pellets are now added to the salt solution in the tube, the accumulated CO_2 is absorbed and can be measured from the reading in the tube. This certainly also represents the volume of O_2 utilised.

In another experiment, instead of potash or brine, the reservoir-levelling tube can be filled with mercury (as shown in Fig. 734, depicted here). Observe that after a time the mercury in the two tubes maintains a constant level showing, R.Q.=1. Another experiment may be performed in which *alkaline pyrogallol* solution (which absorbs oxygen readily) is taken in the manometer tube exactly up to 100 mark. As this promptly absorbs all the oxygen in the tube, the volume of oxygen in air in the tube can be found out by the rise of pyrogallate solution which at its maximum may attain 80 mark in the manometer tube, showing that approximately one-fifth of the atmospheric air is oxygen and this is available for respiration of the experimental plant material.

If instead of carbohydrate-storing seeds, fatty seeds, e.g., castor, mustard, linseed, etc., are used as respiring material, and saturated sodium chloride solution is taken in the tube, the salt solution rises in the graduated tube without any addition of KOH, indicating that R.Q. is less than unity, i.e., greater volume of oxygen is absorbed compared to CO_2 liberated. Suppose that this excess oxygen is V_1 ml. KOH pellets are now added to the salt solution in the graduated tube; there is a further rise of the solution in the graduated tube, say about V_2 ml. V_2 ml is really the volume of CO_2 evolved and V_1+V_2 then represent the volume of O_2 absorbed by the fatty seeds in respiration. R.Q. is evidently, $V_2/(V_1+V_2)$.

VI. Demonstration of liberation of heat energy during respiration

Surface sterilized germinating seeds placed in thermos flasks show rise in temperature for a few days in the flask A (Fig. 735). In the flask B, in which seeds have been previously killed by boiling water the temperature did not rise for about 2 days. The rise of temperature was, however, noticed in flask B after 2–3 days as micro-organisms began to grow and multiply in the dead-seed medium in the flask, leading to a rise in temperature due to bacterial respiration. In the flask C, in which the seeds were killed by immersing them in 1% mercuric chloride, the temperature did not show any increase, remaining constant throughout the experiment. The $HgCl_2$ kept the killed seeds sterile. The results of the three experiments have been graphically represented in the lower part of the figure 735.

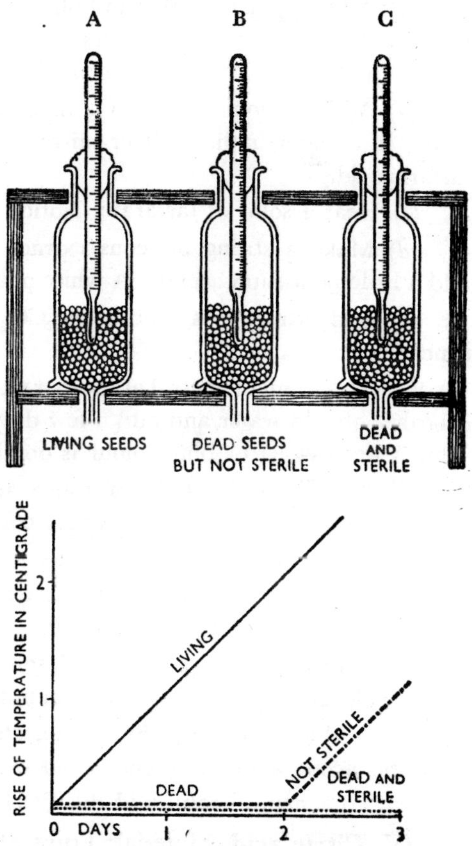

Fig. 735. Arrangements for demonstrating liberation of heat energy during respiration of seeds. The results are graphically represented in the lower figure.

IDENTIFICATION OF COMMON ORGANIC ACIDS IN PLANT TISSUES

1. **Oxalic acid.** Occurs most commonly as calcium oxalate crystals, e.g., raphides, sphaeraphides, etc.

Material: *Oxalis* leaves.

Make an extract by rubbing the leaves in a mortar with sand and dilute HCl. Use litmus paper. Neutralise the filtrate with NH_4OH. Boil off excess NH_3.

(*a*) Treat with 5% $CaCl_2$ soln. Observe that the ppt. of calcium oxalate is insoluble in acetic acid even on heating but soluble in dilute mineral acids.

(*b*) To 5 ml of the neutral extract, add a few drops of 5% lead acetate soln. A white ppt. of lead oxalate is formed. Add an equal quantity of strong acetic acid and warm; the ppt. is insoluble.

(*c*) Add $KMnO_4$ to the neutral extract and then dil. H_2SO_4. The colour of $KMnO_4$ is discharged showing reduction of $KMnO_4$ and the oxidation of oxalic acid.

II. Tartaric acid. Material: Fruits of tamarind, *Ananas*, grapes, tomatoes, etc. Make aqueous extracts. Filter and neutralise the filtrate with NH_4OH and boil off the excess ammonia.

(*a*) Add 5% $CaCl_2$ soln. and observe the characteristic rhombic crystals (prisms) with octahedral faces or as needles of calcium tartarate. Observe that the crystals are soluble in both HCl and acetic acid.

(*b*) Add ammoniacal $AgNO_3$ in a part of the extract in test tube and warm gently in a water bath. A silver mirror is formed showing the reducing properties of tartaric acid.

(*c*) Heat a soln. of tartarate. Notice the odour of burnt sugar.

(*d*) Make a strong aqueous extract of grapes. Acidify with glacial acid and add a little potassium acetate. A white ppt. of potassium hydrogen tartarate is formed.

(*e*) Add extract to a soln. of Na_2CO_3. Observe the liberation of CO_2 by effervescence.

(*f*) To 2–3 ml of neutral grape extract add a few drops of $FeSO_4$. Place the tube in a beaker of cold water and add a few drops of H_2O_2 followed by an excess of NaOH soln. A deep violet or blue colour is obtained.

(*g*) To a neutral extract of grapes add 2 drops of a 2% soln. of resorcinol and 3 ml strong H_2SO_4. Heat gently. A rosy colour is formed which soon turns to violet-red.

II. **Malic acid.** Material: Fruits of tomato, apples, pears, succulents of the family *Crassulaceae*, etc. Make aqueous extracts; filter, and neutralise the filtrate with NH_4OH. Boil off the excess ammonia.

(*a*) Add a few drops of 5% $CaCl_2$ soln. Boil and cool. No ppt. Add an equal volume of 95% alcohol. A white ppt. of calcium malate is seen.

(*b*) Add a few drops of 5% lead acetate soln. A white ppt. of lead malate, soluble in warm acetic acid (difference from oxalic acid) is obtained.

(*c*) To a neutral soln. add dil. $FeCl_3$ soln. Yellow colour is observed.

IV. **Citric acid**. Material: Fruits of the genus, *Citrus*, e.g., orange, lime, lemon, etc., also fruits of tomato. Unripe lemon has about 6% citric acid.

Make an extract of lemon in water. Filter and test the filtrate with litmus paper.

(*a*) In a neutral or slightly alkaline soln. of the extract add 5% of $CaCl_2$ soln. No ppt. Heat to boiling and observe now the thick ppt. of calcium citrate which is soluble in acetic acid.

(*b*) To the aqueous extract add ammoniacal $AgNO_3$. A silver mirror is formed very slowly.

(*c*) Add 5% lead acetate soln. The white ppt. of lead citrate is soluble when an equal volume of acetic acid is added and warmed.

If an unknown solution is given, an idea about the identity of the organic acid in it could be found by simply treating the solution with $CaCl_2$. The soln. must be made neutral (add dilute NH_4OH and test with litmus paper; if ammonia is added in excess, drive off the excess by boiling) before adding $CaCl_2$ in the cold. If there is a white crystalline ppt. with $CaCl_2$ in the cold, the solution must be either *oxalic* or *tartaric acid.* If the ppt. is soluble in acetic acid, it is probably *tartaric acid*; if not, *oxalic.*

If there is no ppt. with $CaCl_2$ in the cold, the neutral solution is boiled. If now there is a thick ppt., the solution contains *citric acid.* If there is no ppt. even on heating, the solution is thoroughly cooled and an equal volume of 95% alcohol added. A thick white ppt. shows the presence of *malic acid.* Further confirmatory tests indicated before should be done to identify the individual acid completely.

Sometimes DENIGE'S REAGENT (dissolve slowly about 5 g. of yellow mercuric oxide in 100 ml of conc. H_2SO_4; make it up to 500 ml with water) has been successfully used in the identification of three of these acids—*oxalic*, *tartaric* and *citric.* In all the cases given below, only neutral solutions, as previously, are to be used.

Oxalic acid treated with Denige's reagent gives a stable, white ppt., which remains insoluble even on heating.

When tartaric acid is treated in a test tube with a few drops of the reagent and a drop or two of 5% $KMnO_4$ solution is added, the colour of the permanganate is completely discharged and the solution remains clear; if citric acid is similarly treated, even to boiling, and the permanganate is added while the solution is still hot, the colour of the permanganate disappears no doubt, but is followed immediately by a heavy white ppt.

CHAPTER XLVI

GROWTH AND DEVELOPMENT

One of the most self-evident of all natural phenomena is the continuous increase in size of plants and production of new organs at least during some stage of their life histories. The term growth is popularly employed to designate this complex of processes, which depends on many other physiological processes. It is impossible to understand plant growth without a sound knowledge of the basic facts of absorption, transport of materials such as minerals to the place of synthesis and metabolism itself. Besides these processes other factors specific to growth and development such as *hormones* must be mentioned. The advances made in this field during the previous 30 or 40 years have really been tremendous.

Growth may be defined as a permanent and irreversible change in any dimension of an organism.

The dry matter which is incorporated into the structure of both protoplasm and cell walls during growth comes almost entirely from the foods synthesised by plants themselves. In the synthesis of new protoplasm, i.e., in the production of new cells, the foods which are assimilated are largely proteins, for protoplasm, as we know, is largely proteinaceous. Foods which are assimilated in the formation of cell walls are mostly carbohydrates, e.g., cellulose and pectins.

As a result of assimilation, a growing region, such as stem tip or root tip increases in dry weight during growth, i.e., new cells are continuously being formed from pre-existing cells.

The mere increase in size of a plant or plant part is not necessarily,the result of growth. We know that dry seeds when soaked in water, imbibe water and certainly increase in volume and in weight. But that cannot be considered as growth for removal of water will also reduce the volume of the seeds to original volume and size and there is also no increase in dry weight.

We know that growth of plant occurs in regions of growing points known as *meristems*. Three main changes occur in meristems: cell division, cell enlargement and cell differentiation. The growth of a plant may be due either to cell division or to cell enlargement. This is possible only because the two daughter cells produced, enlarge to the approximately same size of the mother cell, before they divide again. Thus the growth of a plant or plant organ is essentially an increase in size due to an enlargement of cells and this cell enlargement must be *irreversible* and not simply a reversible increase due to normal increase in hydrostatic pressure, caused by turgour.

In classical concept of cell growth, the causative force is turgour pressure. Yet it is a fact that mature cells have higher turgour pressure than younger meristematic cells. The mature cells, however, have more rigid cell walls but the growing cells stretch more readily. Consequently a small turgour pressure is sufficient to stretch the walls of a young growing cell which may not be sufficient to stretch the wall of a more mature cell. As water enters the cell during cell enlargement, solutes also are absorbed. In this way, the cell maintains a high osmotic pressure of the cell sap, and conditions for the development of sufficiently high turgour pressure to stretch the cell walls, are well maintained.

Besides water and solute absorption other changes also occur during cell growth. New cell wall materials are laid down by *intussusception*, i.e., insertion of new molecules of cellulose on the old wall particles that become softened to permit the stretching the wall undergoes during enlargement. Thickening may also take place by *apposition*, i.e., deposition of new cell wall matter on the inner surface of the wall during the maturation phase of the cell.

The meristematic cell is filled with cytoplasm and has only a few very small vacuoles. As the cell enlarges, the vacuoles increase in number and by coalescing with each other, occupy a much larger proportion of the cell and the cytoplasm as a result becomes a thin layer, lining the large central vacuole.

Thus growth of a cell can be said to involve the following primary processes:

1. absorption of water,
2. absorption of solutes for maintaining higher turgour pressure of the cells,
3. deposition of new cell wall material on the stretched cell walls,
4. formation of new protoplasm by synthesis of proteins, lipids, etc.

Growth is therefore a very complex process involving nearly all the other physiological processes known to occur in the plant. And it requires expenditure of large quantities of energy since absorption of solutes is probably active and not passive (osmotic), completely depending upon energy released in respiration. The synthesis of cell wall materials as well as the synthesis of protoplasmic substances are also endergonic processes.

STAGES OF GROWTH

The growth of an entire plant or only of a plant organ characteristically passes through stages represented by a S-shaped curve. The time during which this occurs has been called *grand period of growth* (Fig. 736). It has been assumed that the rate of production of new materials in growth is proportional to the size of the plant. Thus the growth of an annual plant probably follows a *compound interest law*. The difference between simple interest and compound interest is that when increase takes place at a rate proportional to the initial quantity, it is following a simple interest formula. When the increase occurs at a rate proportional to the new quantity at each stage, it is the compound interest formula. For example, Rs. 100·00 at 5% interest becomes naturally Rs. 110·00 after two years.

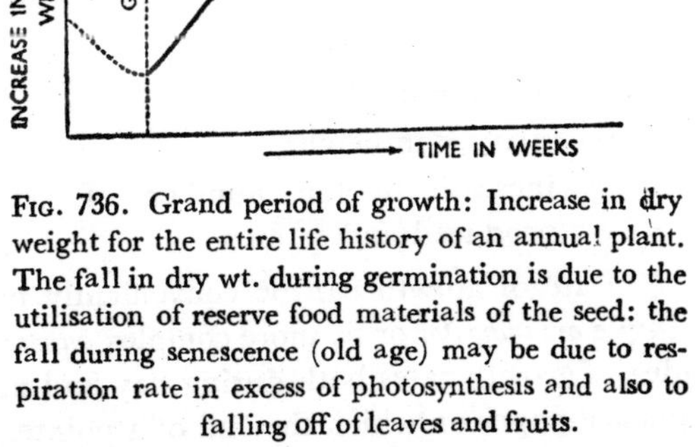

FIG. 736. Grand period of growth: Increase in dry weight for the entire life history of an annual plant. The fall in dry wt. during germination is due to the utilisation of reserve food materials of the seed: the fall during senescence (old age) may be due to respiration rate in excess of photosynthesis and also to falling off of leaves and fruits.

But in compound interest law Rs. 100·00 at the same 5% interest becomes 105 after one year, but at the end of second year where as in simple interest the total becomes 100+5+5, i.e, 110, in compound interest it is 105+5% of 105 and so on.

Blackman applied the compound interest formula to explain growth:

$W_1 = W_0 e^{rt}$ or in logarithmic form, $\log_e (W_1/W_0) = rt$

or $2{\cdot}3026 \log (W_1/W_0) = rt$

where W_1=final dry weight attained by an annual plant.
W_0=initial dry weight,
e=base of natural logarithm,
r=rate of increase in dry weight per unit time, and
t=time interval.

Thus if a plant has doubled itself as regards increase in dry weight in, say, 15 days, i.e.,

$W_1/W_0=2$. Since $\log_e 2=0{\cdot}69315$
the rate of increase in dry weight is:

$r=\log_e (W_1/W_0)\div 15=(\log_e 2)\div 15=0{\cdot}69315\div 15=0{\cdot}04621$ or 4·62 per cent. per day.

If the period of doubling were 10 days, the rate would be 6·93 per cent. per day.

It has been assumed that growth is dependent on chemical transformations of reserve substances into new living protoplasm. The new protoplasm as soon as it is formed begins to participate in the mechanism of transformation by apparently catalysing its own synthesis (*autocatalytic reaction*). Following this principle, the theoretical curve obtained resembles the typical S-shaped curve (Fig. 737).

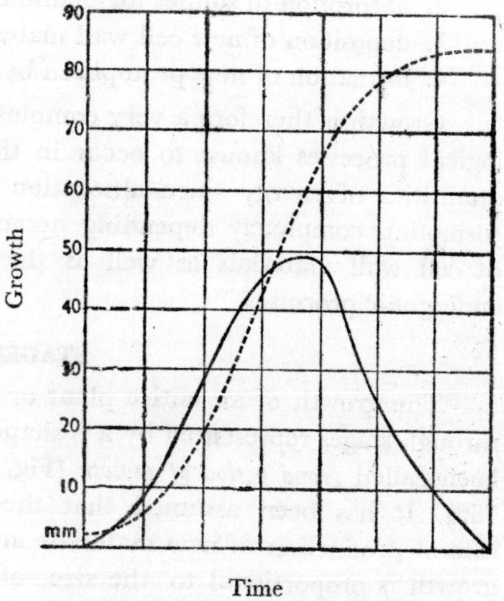

Fig. 737. Curves of course of growth in time. Note the typical S-shaped curve (dotted lines). The other curve represents the *rate of growth* of an organ which is never uniform even under constant external conditions; it gradually increases from very small values to a maximum and then again decreases to zero. The area inside the closed curve below represents the total growth made by the organ concerned.

Measurement of growth

Rate of growth can be measured by various ways. Any one of the following measures can be suitably applied in determining the growth rate and total growth of the individual organs of a plant or the plant taken as a whole:

(*a*) increase in weight,
(*b*) increase in volume,
(*c*) increase in height,
(*d*) increase in the number of cells produced and so on.

Rate of growth can be conveniently measured by standard instruments like the simple *arc indicator* or by more complex *auxanometers*. Auxanometers give a more detailed idea of growth rates both during day and night. The revolving drum, with the pasted smoked paper on which the end of a pointer needle is attached, can be easily made to record the rate of a growth in a particular period, say, 72 hours (Fig. 738).

In the apical meristem of stems and roots, the rate of growth is not uniform throughout its length. If the apical portion of the radicle of a germinating seeds is marked into equal divisions it is found that the growth rate is maximum not at the very apex, but at more basal regions, a few mm from the apex. If the apical root meristem is divided into ten equal divisions from the tip as in Fig. 739, it is seen that the portion of tissues between divisions two and four has increased most. Evidently the growth of the tissue in

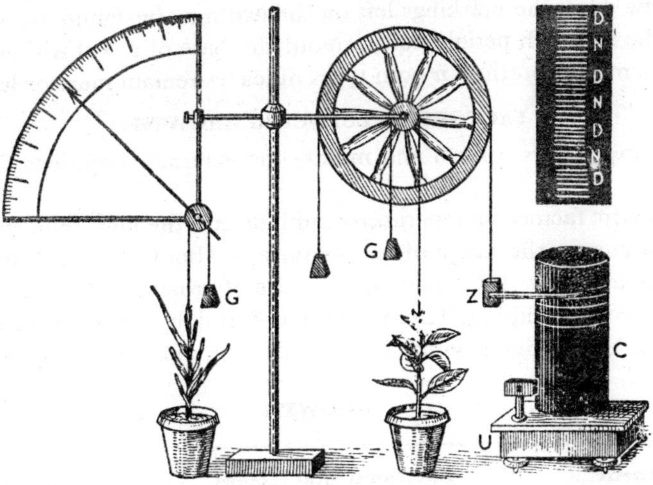

FIG. 738. Simple (arc indicator) and self-registering auxanometers (right). In self-registering auxanometers, the drum, C, with pasted smoked paper is kept in rotation (once every hour) along the vertical axis by a clockwork. D and N represent day and night tracings; the perpendicular distances between tracings on the drum indicate proportional hourly growth. Note that such distances are larger at night compared to day markings, indicating increased growth during the hours of darkness.

the more basal region is controlled and maintained by some sort of stimulus which might be originating at the very tip which is not actually the growing region. The growth of the individual leaf is not also uniforn throughout its entire surface. This can be demonstrated with the help of a *spacemarker disc* (Fig. 740). Two leaves, one sessile and the other with petiole, are marked near the tip, the middle and at the base by the spacemaker

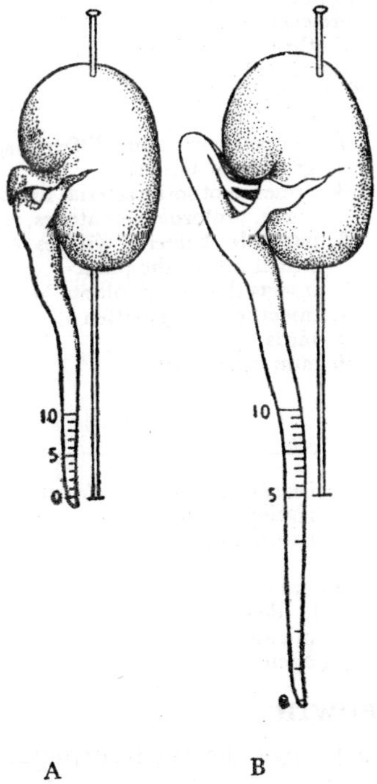

FIG. 739. The distribution of growth in the root tip, A: The root tip is divided by marking with India ink into 10 zones, each 1 mm long. B: the same root after about 24 hrs. Note that by the unequal growth of the different zones, the lines between 0 and 5 have become separated by unequal distances.

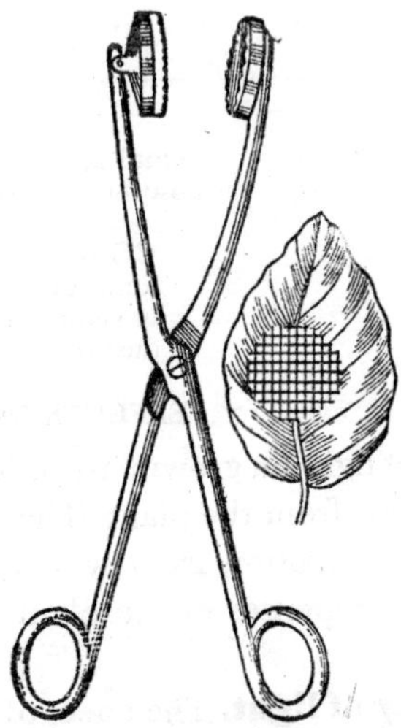

FIG. 740. Space-marker disc.

disc. After a few days, the markings left on the leaf may be found to be wider apart in the tips of the leaf with petiole, and around the base of a leaf without petiole. The markings in the middle portion in both types of leaves remain more or less uniform.

FACTORS INFLUENCING GROWTH

The following factors control and modify the external conditions that affect total plant growth.

Of the climatic factors—temperature and light are the most important. Growth of higher plants occurs in the range of temperature of about 0–35°C. Within this range, raising of temperature by 10°C, increases the rate of growth 2–3 times (Q_{10}=2 or 3) if other factors are not limiting. There are three, certainly not very sharp, temperature zones, known as the cardinal points for growth—the *minimum*, *optimum* and *maximum* (Fig. 741).

GROWTH

- **Hereditary or genetic factors** (the particular set of genes which the plant possesses—genes determine or condition one or more behaviour).
- **Environmental factors**
  - **Climatic factors** (act upon the plant through the atmosphere)
    1. temperature. (air or soil)
    2. light
       - *intensity* quantity or amount of energy received by the plant.
       - *quality* actual spectral composition of light.
       - *duration of light* effect of variation in the length of day and night.
       - *direction*

    CO_2-concentration of the atmosphere humidity precipitation—rainfall, dew or snow wind velocity volatile chemicals, dust particles and other polluting agents.
  - **Chemical factors** (operate through air or soil in which the plant grows).
    1. available water in the soil
    2. goes in air or soil
    3. quantity and nature of soil solutes
    4. pH of the soil.
  - **Biotic factors** (arising from the presence of other plants and animals)
    1. competition among different species
    2. grazing of animals
    3. presence of soil bacteria, fungi
    4. algae, protozoa, nematodes, viruses etc.
    5. parasitic bacteria, viruses and fungi which attack the plant
    6. insects that injure plants
    7. insects carrying pollen
    8. birds
    9. human activities.
- **Internal growth regulating factors**
  1. hormones
  2. vitamins
  3. inhibitors.

EFFECTS OF LIGHT ON GROWTH

Effects of light on growth are at least *four-fold*, each aspect having its own particular type of response from the plant. They are intensity of light, quality of light, duration of light and the direction from which the incident rays fall on a plant. The following characteristic responses of the plant to the various aspects of light on growth can be recognised:

Intensity of light. The effect of high intensity of light on growth is, in general, to retard the growth process (except for photosynthetic function).

Quality of light. Shorter rays beyond the violet and longer rays beyond far red are detrimental to growth. Red and far red light seem to be the most active for

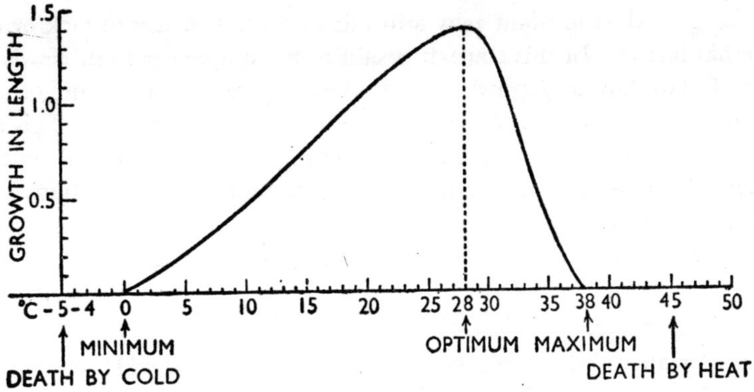

FIG. 741. The growth *rate* in length at various temperatures. The optimum curve. Note that in general the *minimum* temperature for growth is in the neighbourhood of 0°C. The curve then rises to an *optimum* where greatest growth takes place (approx. 25–30°C); from this onwards the curve falls to *maximum* (approx. 38–40°C) beyond which again no growth is found to take place.

regulation of growth. The spectral quality of light is very important, for the plant contains perceptive pigments responsive only to light of particular wave-lengths.

Duration of light. Effect is restricted to certain physiological processes the most remarkable of which is *photoperiodism*.

Direction of light. Shoots, some roots and leaves and also other organs usually show different orientations to the direction of the incident light—the phenomenon of *phototrophic curvature.*

Though the growth of higher plants eventually depends on photosynthesis, nevertheless light is not essential for the growth process itself, so long as sufficient quantities of organic substances are available. But the kind of growth is different when light is absent. In the dark, higher plants show a weak, excessively elongated growth in case of most dicotyledons but the leaves are underdeveloped. Little tissue differentiation, however, occurs in the dark, the tissue remaining mostly parenchymal. Other non-green plants also show a similar stretching in darkness. Monocotyledons as a rule show excessive elongation of the first internode only in darkness and may also then show excessive leaf development. Light actually retards all this excessive growth in the dark. An extreme case is the dwarfing of alpine plants (plants growing in high altitude, above 3,600 metres in tropics) by intense light to which they are subjected to. The light in the alpine region is very rich in violet and ultra-violet radiations and it seems that in general, light of smaller wave-lengths as also light waves longer than far red, have specific inhibitory effect on vegetative growth.

Among the edaphic factors, available water in the soil must take a pre-eminent place since all growth, as we know, depends on the development of a considerable turgour pressure of the cells. Consequently a deficiency of water supply to the regions of meristematic activity will, of course, retard or completely stop growth. Thus, all factors influencing or facilitating absorption of water will naturally influence growth, if other factors are not limiting.

Some of the above factors may completely stop growth. This is most commonly the case in nature when the temperature is below the minimum for growth (e.g., in severe winter) or when its water content is too low (e.g., seeds). In this stage of suspended growth, the plant is said to be dormant. Then, even if environmental conditions are favourable for growth, the plant may still fail to grow because of lack or scarcity of certain internal factors. In this state, it is said to be in a rest period. Newly harvested potato tubers fail to show any growth though the temperature, water and other environmental conditions are optimum for the growth of older potato tubers. This rest period can be broken by artificial means such as treating with hormones or some chemicals or other methods such as removal of the terminal bud (the lateral buds are thereby forced). Other treatments include warm baths and also subjecting the organ at rest period, to alternate freezing and thawing and scarification.

GROWTH REGULATORS

Among substances which markedly influence reactions and metabolism of the animal body are hormones which are internally secreted by glands, (e.g., adrenaline, insulin, thyroxine, etc.) and vitamins. Convincing evidence has accumulated that similar hormone like substances also occur in plants. These compounds are called *growth substances*, *hormones*, *growth hormones* or *phytohormones*. It is characteristic of both plant and animal hormones that they usually exert their physiological effects while present in minute concentrations. Such concentrations are difficult to measure by conventional chemical methods but are bio-assayed using plant materials as test objects, since plant organs exhibit detectable response in presence of low concentrations of the hormones and the response, within limits—has a quantitative relationship with concentration.

Plant hormones, however, as far as is known until now, are not produced in any glands. They may be produced in growing regions like shoot tips or root tips or in leaves and flowers and growing fruit tissues. Like animal hormones they are usually active in target organs some distance away from the site of synthesis. The physiological effects are exhibited throughout growth and developmental phases—from germination to senescence and thus such hormones should not be referred to as "growth hormones" or "growth substances"—because growth is only one of many physiological responses they evoke. The same hormone may be active in different target organs, but the optimum concentrations for the different responses they evoke, vary. Many animal hormones are polypeptide or steroidal in nature. While no plant protein or polypeptide has so far been found to possess hormonal properties, some plant phytosterols and terpenoids —di and tri— have been found to be biologically active. They have not yet been accepted as "hormones" although their regulatory role has been recognised. Plant hormones have considerable similarities with animal hormones—in their basic pattern of mode of action—particularly in relation to the long term effects. The molecules of plant hormones are much smaller, the smallest being ethylene—C_2H_4.

Plant hormones usually exhibit two phase concentration curves, i.e., low concentrations may evoke promotive responses, whereas at higher concentrations the degree of promotion may be reduced or the effect may be inhibitory. They interact with some site (s) in the cell, which apparently become saturated at relatively low concentrations, beyond which there is a competition among the molecules, due to which proper attachment with the site does not occur and the response is decreased or almost non-existent.

Several groups of plant hormones are now known. These are classified according to their physiological activity. The major classes are: (a) auxins, (b) cytokinins,

(c) gibberellins, (d) ethylene and (e) abscisic acid. In addition to these there are the leucoanthocyanins, brassicins, traumatic acid, helminthosporol etc. Phenolics, coumarins and many tetra- and pentacylic tri-terpenoids have inhibitory properties—and they are referred to as "inhibitors"—although at very low concentrations some of them may also promote growth.

Many synthetic compounds have been prepared which stimulate the naturally occurring hormones in their physiological activity; some of them also have some structural similarity with the natural hormones. The synthetic substances, because they can be produced by synthetic means in large quantities—are much cheaper and are thus used in agriculture and horticulture. But they should not be referred to as "hormones", because they do not occur in nature. The natural and synthetic growth regulatory substances together are termed "growth regul..tors"; they include both promoters and inhibitors.

We shall now discuss very briefly the occurrence, properties and physiological activity of the more important classes of growth regulators.

A. Auxins

When *Avena* coleoptile tips are exposed to light coming from one side, the coleoptile bends towards the direction of light. If a coleoptile tip is decapitated and removed by a cut made several mm below the apex, the rate of growth of the stump is immediately retarded. If, however, the tip is replaced on the stump, growth is resumed. Such experiments indicate that the growth of the coleoptile which occurs in the more basal regions is maintained only under the influence of some sort of stimulus originating at the tip and which is transmitted from apex to base through the coleoptile. If the cut-off tips of the oat coleoptiles are placed on a thin layer of agar, after an hour the cut tips are removed and the agar cut into a number of equal sized small blocks, then each one of these blocks, when placed upon the stump of a decapitated coleoptile can accelerate the growth of the stump almost as much as an intact coleoptile.

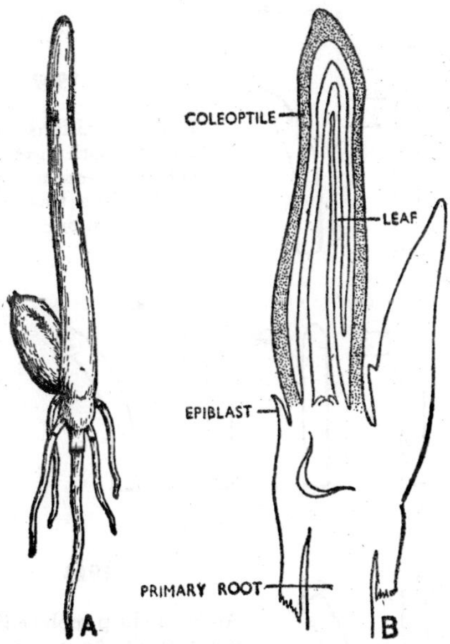

FIG. 742.
A: oat coleoptile and B: a longitudinal section of the same showing coleoptile sheath, enclosing primary leaves.

It seems evident from these results that some substance or substances was transported out of the tip into the agar block and subsequently out of the block into the decapitated stump, when it was translocated downwards to the elongating region (region of response) of the coleoptile. The substances which induce such responses are now classed as *auxins*. Auxins are apparently universally present. They are non-specific in their action, i.e., the same auxin or auxins, chemically speaking, which influences growth phenomenon in one species apparently also influence the same phenomenon in all pieces. The auxins seem to be used up during the physiological process they influence and thus lack this and other characteristics of enzymes. The term auxin was first coined in about 1905.

Auxins are known to be of widespread distribution in plants. They occur in very small quantities and their detection by standard chemical methods is usually difficult and often impossible. With the discovery of a new, easily handled technique early in the last decade—*partition chromatography on paper* as also *thin layer chromatography*—precise identification of all naturally occurring auxins has been possible. While in the past, auxins were estimated by curvatures such as those in the *Avena* test or in pea test, now any critical auxin analysis of plant extracts must involve separation and identification of auxins by the chromatographic techniques before estimation because there are several auxins and auxin precursors as also some non-auxinic substances which may promote or inhibit growth.

The chronological sequence of the development of our early knowledge of auxins in plants is illustrated in the following pages:

1880

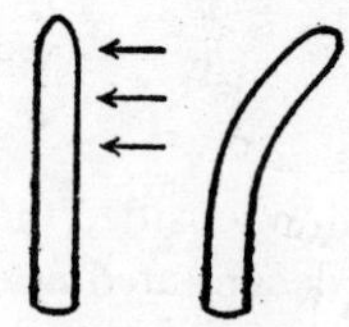

Light falling unilaterally on the tip of a coleoptile causes some influence to be transmitted downwards; the coleoptile curves towards the light.

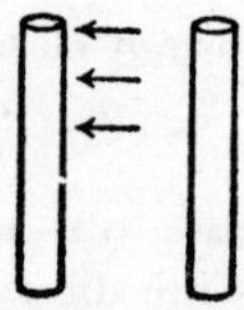

When coleoptile tip is removed, phototropic response does not occur.

1907

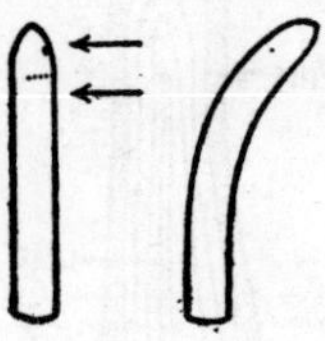

Lateral incisions either on one or both sides of the coleoptile do not prevent its bending towards light from one side.

1910

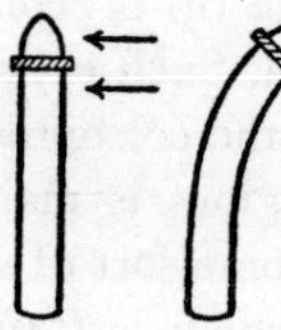

In an excised coleoptile tip replaced with gelatin inserted between it and the stump, phototropic curvature resulted as in normal coleoptiles.

1911

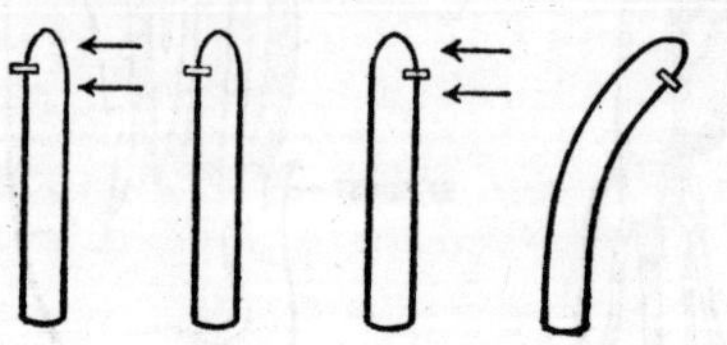

Insertion of mica plates on the shaded side prevents curvature following unilateral illumination of the tip. Mica plates inserted on the illuminated side resulted in curvature in the usual way. Conclusion: a substance migrates down the shaded side, promoting growth curvature towards light.

1918

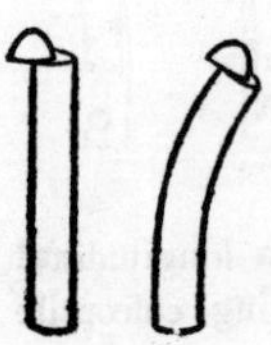

An excised tip replaced on one side of the coleoptile stump causes accelerated growth beneath the tip, resulting in curvature.

1925

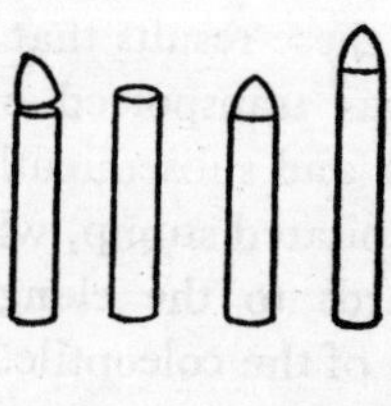

Decapitation results in diminished growth of the coleoptile. Excised tips if replaced on the stump, growth in length is renewed.

1921–1925

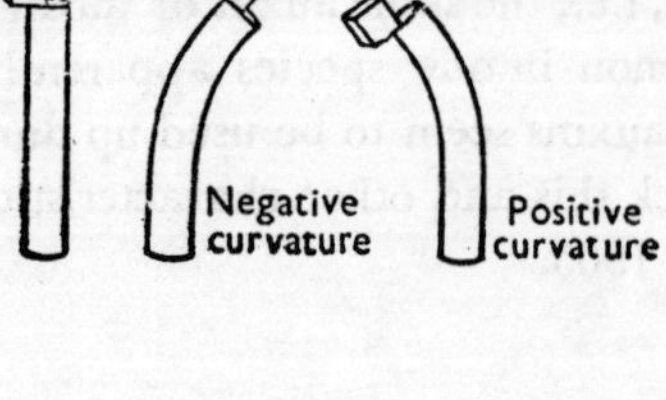

When expressed sap from coleoptile tips was put into agar blocks and applied unilaterally to cut stumps, curvature resulted.

1928

In incised oat coleoptile, growth in length ceases; **a.** The addition of a plain agar block (without any expressed sap from coleoptile tips) has no effect; **b.** Growth is, however, renewed by the addition of a block containing sap from excised coleoptile tips; **c.**

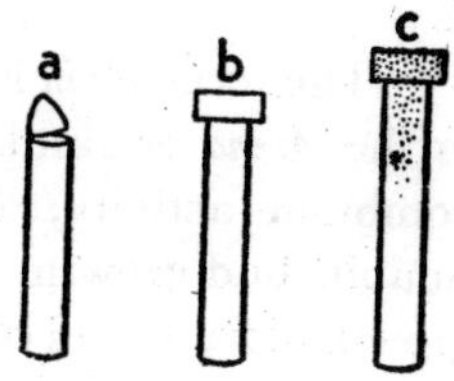

1928

Growth hormone diffuses from the excised tips into agar which is then divided into rectangular blocks (**a, b** and **c**). When such a small block is placed unilaterally on a cut stump, the resulting curvature is approximately proportional to the conc. of the hormone.

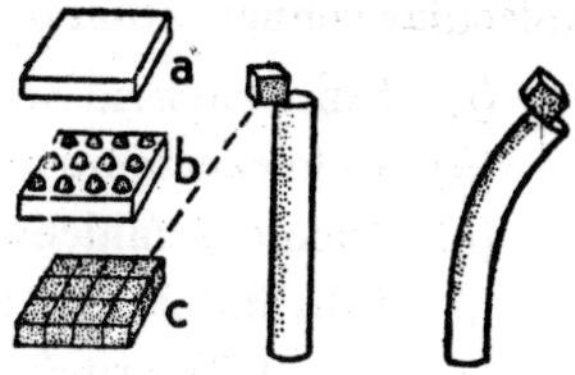

1928

Transport of hormone is polar from morphological apex to base and not *vice versa*. **a,** normal coleoptile; **b,** *inverted;* **f,** plain agar block; **e,** block with hormone. In **a** diffusion from **e→f** (and f′); **b,** no such diffusion.

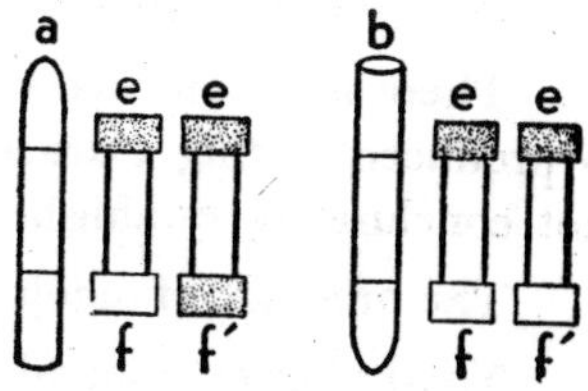

1929

An excised coleoptile tip, **a,** placed in a horizontal position in contact with two plain agar blocks, **b** and **c,** causes displacement of hormone towards the lower side—a greater proportion hormone (67%) accumulates in the lower block, **c.**

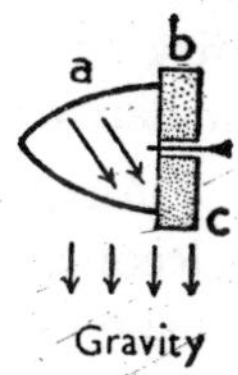

1933

An agar block with hormone when placed upon a coleoptile trunk standing upon two plain agar blocks, **b** and **c** and exposed to one sided illumination, causes displacement of hormone towards the shaded side approximately to a proportion of 70% (shaded) and 30% (illuminated).

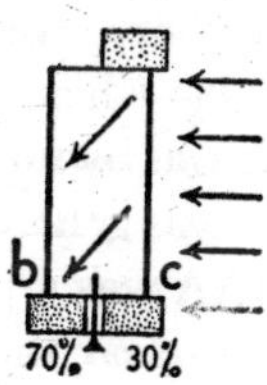

1934

If hormone is supplied in agar block to the cut stump of a coleoptile placed in a horizontal position, greater proportion is transported towards the lower side of the morphological base (65%).

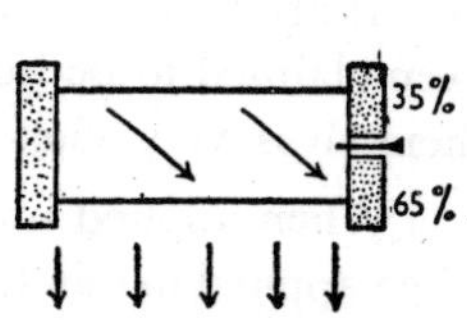

1934

Phototropic bending results from displacement of hormones to the lower side of the plant axis. The shoot curves upwards because its growth is promoted and root end turns downwards because its growth is inhibited by the same conc. of hormone.

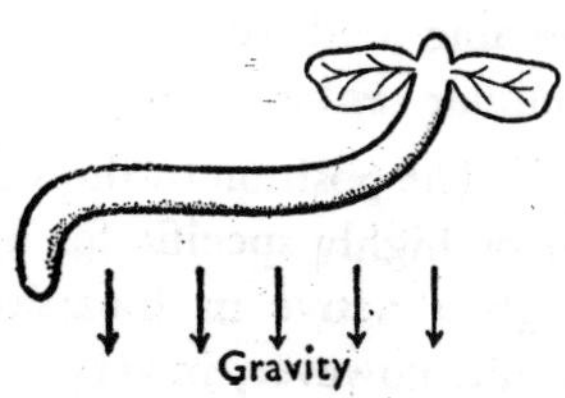

Indole auxins

The term *auxin* is defined here to include only growth promoting substances active in the *Avena* or similar bioassays. However, in addition to this, auxin also promotes combium activity, xylem formation, protoplasmic streaming and root initiation, inhibits bud growth, helps formation of female flowers, and stimulates the setting and growth of fruits and causes abscision depending on concentration. The principal naturally occurring auxins in plants that have been definitely identified, isolated, purified and their molecular configuration determined are all indole derivatives as mentioned below.

(*i*) Indoleacetic acid (IAA—mol. wt. 175)
(*ii*) Indoleacetonitril (IAN—mol. wt. 156)
(*iii*) Indoleacetaldehyde (IAAld—mol. wt. 159).
(*iv*) Ethylindoleacetate (mol. wt. 203)
(*v*) Indole-3-pyruvic acid (IPyA—mol. wt. 203)
(*vi*) Indole-3-ethonol (IEtOH—mol. wt. 161)

There is a controversy regarding whether ethylindoleacetate occurs naturally or is produced during extraction as an artifact. Occurrence of indoleacetamide is also not conclusively established.

The presence of indole ethanol in plants also is questionable.

Non-indole compounds, e.g., some fatty acids may also possess auxin-like properties. But they have not yet been properly characterised. Many synthetic compounds resembling auxins in their biological properties are known; they usually have some structural similarity to indole acetic acid.

Auxins are present in tissues in very low cocentrations and are bioassayed by the coleoptile curvature or straight growth test, internode elongation test, slit pea test, tomato ovary test, leaf epinasty test etc.

Although indoleacetic acid was known to chemistry since 1885, its biological activity was not suspected and it was recognised as a plant growth substance only in 1934 when it was identified and extracted from urine and termed, *heteroauxin*. Even then it was by no means clear that IAA was the native auxin of higher plants; on the contrary it was generally held that '*auxin a and b*' were the primary auxins of higher plants while IAA was a product of fungi and bacteria. By the end of 1950, however, it was becoming clear that indoleacetic acid is the principal naturally occurring auxin in all higher plants. The previously known '*auxin a*' and '*auxin b*' isolated from plant tissues by Kögl (1934) are now considered to be only of theoretical importance for, although they have been claimed to be isolated back in 1934, since then, to say the least, they have proved themselves very elusive indeed.

When applied uniformly to decapitated coleoptiles, IAA causes increased growth; when applied to one side, it causes curvature towards the other side. The greatest puzzle of all seems to be the mechanism whereby traces of auxin such as a relatively simple organic acid, like IAA and its derivatives can so dramatically bring about growth or cellular enlargement.

The position of the side-chain in the ring structure in indole-3-acetic acid appears to be highly specific for activity, since 1-, 2-, and 4-indole acetic acids are only very slightly active in bioassay. Substitution with halogens, like fluorine and chlorine can result, however, in very active derivatives, e.g., the 4-chloro and 6-chloro compounds.

The replacement of an aromatic CH by N can give, 7-*aza* (indole-3) acetic acid (Fig. 743, VII) which is less active in the *Avena* test or indazole-3-acetic acid (Fig. 743, VIII) which has activity equal to IAA.

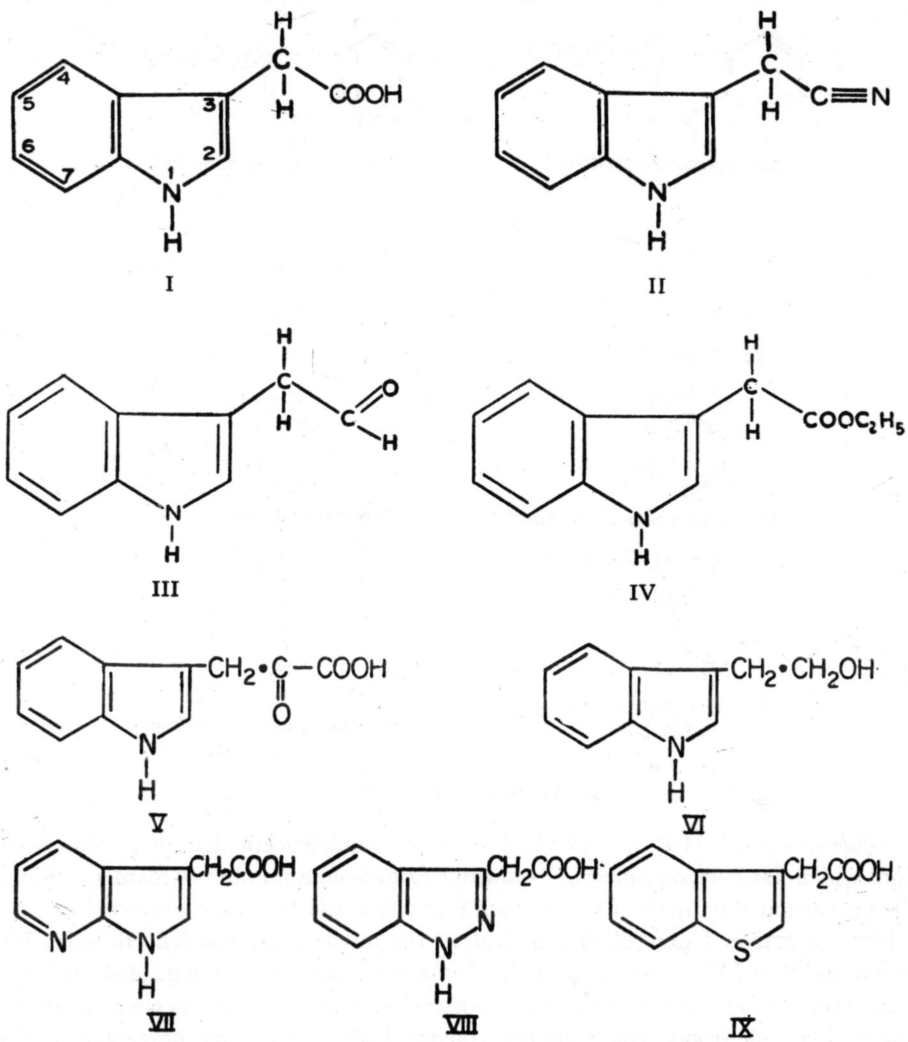

FIG. 743. Molecular structures of naturally occurring IAA and its derivations, as well as a few (the last three) synthetic ones: I: Indole-3-acetic acid; II: Indole acetonitrile; III: Indole acetaldehyde; IV: Ethylindole acetate; V: Indole-3-pyruvic acid; VI: Indole-3-ethanol; VII: 7-*aza* (indole-3) acetic acid; VIII: Indazole-3 acetic acid; IX: Thianaphthen-3-indole acetic acid.

IAN may be converted into either IAC (indole-3-carboxylic acid) or IAA. When the imino group (NH) in IAA is replaced by a sulphur atom (Fig. 739, IX), the resulting compound, thianaphthen-3-indole acetic acid is quite active in various bioassay tests (Allsopp, 1965).

Indole propionic acid (IPA) is less active than IAA, but indole butyric acid (IBA) is more active than IPA; in some tests as in the case of rooting of cuttings, IBA is even as or more active than IAA. Side chains having even number of C-atoms is more active than those having an odd number. It is usually believed that the side chain is oxidised to IAA before the compound shows a biological activity. The chemical structures of some synthetic auxins is given in Fig. 744.

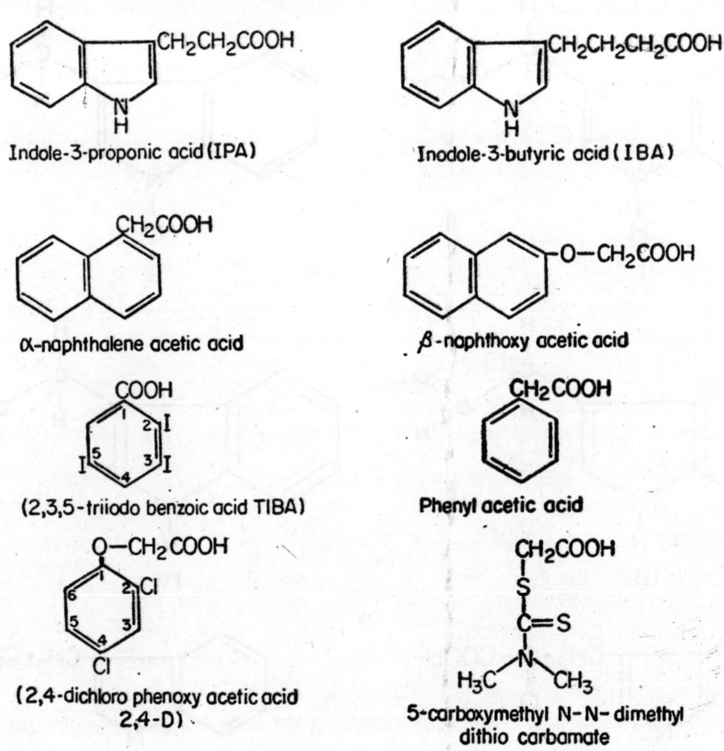

Fig. 744. Some synthetic auxins.

Indoleacetonitrile (IAN), a nitrile derivative of indoleacetic acid, is a neutral substance which has been obtained in crystalline form from a number of plant materials. It is believed that the nitrile, IAN, as such is inactive and has to be converted to IAA in order to be active. Upon alkaline hydrolysis the nitrile yields indoleacetic acid. It has been found that IAN promotes growth of only those plant organs, e.g., *Avena* coleoptile, bean seeds, tomato ovary etc., which possess the enzyme indole aceto nitrilase while tissues, like pea roots which cannot convert IAN to IAA are unreactive to IAN. In higher concentrations IAN is inhibitory like IAA.

An ethyl ester of IAA—*ethylindoleacetate*, $C_8H_6N.CH_2COOC_2H_5$—has also been isolated from plant tissues. Both IAN and this ester seem to be more active than IAA, in those tissues which are capable of converting them into IAA. This may be due to the better penetration of these neutral substances than the acid itself. The aldehyde derivative, *indoleacetaldehyde*, $C_8H_6N.CH_2CHO$, the nitrile, IAN and the ethyl ester all three exist in nature as auxin precursors and all of which can be converted into IAA. In addition to acting as an auxin precursor, IAN is also known to be an activator or booster of IAA-responses. For although IAN is unable to bring about a response in pea test curvature even at 20 p.p.m., its presence in the medium, boosts the response of 0·1 p.p.m. IAA from 2° to nearly 60° curvature.

The work, of Fawcett, however, indicates that the activities of IAN is not always governed by its conversion to IAA.

Indole-3-pyruvic acid (IPyA) has been isolated from corn.

Phototrophic curvature is due to higher concentration of auxin on the side of the coleoptile farther from light. This may be due to either a transverse translocation of auxiin from the illuminted side to the darker side or to a photochemical destruction of auxin (destruction activated by light absorbed by β-carotene or riboflavin) of the lighted side. Geotropism can also be explained in terms of auxin distribution.

Evidence exists that in addition to free auxins, which are present in all growing regions such as germinating seeds, in seeds there are also stored precursors of auxin in which auxin is bound to proteins, carbohydrates, amino acids, ascorbic acid (ascorbigen) etc. which change into IAA proper when seeds absorb water. This change may occur in light or in the dark. The precursors give rise to auxin predominantly in the apical regions. The cyclic amino acid, tryptophan is regarded as a precursor of IAA. When proteins are hydrolysed, tryptophan is liberated and this tryptophan undergoes oxidative deamination and subsequent decarboxylation to yield *indoleacetaldehyde* which is oxidised to the free auxin (IAA). The importance of Zn and cobalt ions in auxin formation is recognised.

There is a difference in sensitivity of roots and stems to the concentraion of auxin. The same concentration (e.g., 10^{-6}M) which causes increased growth of the stem cells may inhibit growth of roots cells. Thus cells may be either stimulated or inhibited, depending on the concentration of auxins. This only applies to increase in elongation. On the other hand for root initiation or new root formation, a relatively higher concentrations of auxin is necessary, which concentration, however, completely inhibits shoot formation or lateral bud growth.

This dual effect of auxin provides the plant with a switch mechanism for turning the growth process on or off. We have seen that when terminal buds grow, the growth of the lateral buds is inhibited. If the terminal bud is removed, the growth of the lateral buds may be stimulated which otherwise would have remained checked. The terminal bud perhaps is able to prevent the growth of the lateral buds (*apical dominance*) by liberating excess auxin to the lateral buds, thereby preventing, (i.e., excess concentration of auxin) the growth of the lateral buds. Thus, the lateral buds may be kept in the dormant state. When the terminal bud is removed, the resumption of growth of the dormant lateral buds may be due to lowering of the concentration of auxin to an effective level, which would now stimulate growth.

In potato tubers and many seeds, auxins are present in growth-stimulating quantity but the tubers and seeds are kept dormant (gibberellins can break this dormancy of potato tubers which action is contrary to that caused by IAA) by the presence of inhibitors which prevents the proper functioning of auxin. The effects of auxins on growth result from an increased or decreased cell enlargement at low or high concentrations respectively. Auxins also affect cell division; callus tissues may be formed when IAA is applied to a cut stump. Renewed activity of cambium and apical meristems seem to be dependent on auxin concentration.

We know that the transport of auxin is usually *polar*, from morphological apex to base. More information is, however, required as to the route and mechanism of transport of these comparatively smaller molecules of IAA throughout the body of the plant.

During recent years, apparantly contradictory effects of auxin, in the phenomena of fruit-set and fruit-drop, have brought out one rather general principle, namely production of auxin is a transient activity of the developing tissue which is soon given up and taken over in time by other tissues. It seems probable that some kind of balance is involved between IAA and gibberellin (see later) and that buds or fruits continue to develop only when the balance is maintained, while excess of either one, inhibits.

Auxins apparently control the growth of only those organisms that show marked cell enlargement, the higher plants. They have no unequivocal effect on the growth of micro-organisms or fungi. No growth response is induced by plant auxins in organisms without cellulose cell wall, e.g., animals. It has been amply proved that the cell enlargement induced by auxin is not due to any increased turgour pressure but perhaps due to the *increase in plasticity of the cell wall* brought about by auxins and which is certainly in some way linked with respiration, for it is known that auxins markedly affect respiration rates in plants. Thus, due to the influence of auxin, plasticity of the cell wall increases, i.e., cell walls soften, thereby allowing osmotic pressure inside the cell to stretch them further as more water enters the cell. During this process, new cell wall material is constantly laid down and relatively less cellulose but more pectin material, is initially laid down. IAA thus causes elongation as well as enlargement of individual cells by reducing the cell wall-pressure, thereby considerably increasing water uptake or suction pressure of the cells.

IAA certainly induces softening of the cell walls. A good deal of efforts have already naturally been devoted to an analysis of cell wall metabolism in connection with auxin actions. It now seems clear, however, that these changes can only be a secondary sequence of events occurring elsewhere in the cell, that is, the changes must be of more fundamental nature.

How does an auxin produce such diverse physiological reponses, as mentioned earlier, in plants? We do not as yet know the initial reaction which causes auxin to set off a chain of events which ultimately expresses itself as a physiological response. Very likely it is a physico-chemical event concerning the whole cell, which changes the enzyme matrix or pattern inside the cells.

Synthetic auxins

Several small molecules have been snthesized (Fig. 744), which exhibit biological properties characteristic of IAA, though not in all respects. They are usually derivatives of benzoic acid, phenoxy acetic acid and naphthalene acetic acid. Some of the potent compounds are: 2, 6-dichloro—and 2,3,5—triodo benzoic acids (TIBA), 2,4—dichloro (2,4—D)—and 2,4,5,-trichloro phenoxy acetic acid (2,4,5—T), α–naphthelene acetic acid (NAA), β–naphthoxy acetic acid etc. Phenyl acetic acid is a weak auxin. An acidic side chain, unsaturation in the ring and an unfilled ortho-position usually characterise many active molecules, but no general rule can be framed. Some thiocarbamates which do not satisfy most of the requirements, are quite active. The D-forms are usually more active than the L-form; *cis*-cinnamic acid is an auxin, whereas *trans*-cinnamic acid behaves as an antiauxin. The three-dimensional structure of the molecule and the spatial relationship between the side chain and the aromatic ring, when present, determine biological activity.

These compounds which give positive tests in *Avena* coleoptile curvature bioassay are now proving to be of immense practical value to the agriculturist and horticulturist for such uses as promotion of flowering, selective weed control, rooting of cuttings, prolonging and breaking of dormancy of cuttings as also of seeds and buds, production of seedless parthenocarpic fruits, thinning of blossoms, prevention of premature, pre-haıvest fruit-drop, etc.

GLUCOBRASSICIN

Although glucobrassicin has been reported to be inactive in the straight-growth test, nevertheless it was found highly active in the *Avena* curvature bioassay. Glucobrassicin was first of all isolated from cabbage leaves. These compounds are members of the so-called mustard oil glucosides of glucosinolates.

B. Cytokinin

The isolation and identification of *kinetin*, (6-furfurylaminoprine) from nucleic acid preparations, was first reported by Miller (1956). In extracts of rapidly growing tissues, such as immature and developing fruits, evidence of kinetin-like activity was obtained by numerous investigators. Assay technique for this type of activity showed these substances' effect of inducing rapid cell division in undifferentiated plant tissues, tissue cultures or callus, with IAA being required as a synergistic cofactor, in most systems. Unlike auxins cytokinins promote bud growth. They also defer senescence and help in the retention of chlorophyll. Cytokinins are usually assayed by their effects on callus growth (soybean, tobacco etc.), expansion of cotyledons (cucumber etc.) and retention of chlorophyll in ageing leaf discs.

The term *cytokinin* has been accepted practically universally, as a generic name for substances which promote cell division and exert other growth regulatory actions in the same manner as *kinetin.*

Two such substances which have been isolated in crystalline form from *Zea mays* immature kernels by Letham (1964) and Koshimizu (1967), have been named *zeatin* and *dihydrozeatin.*

The identification of cytokinins as constituents of t-RNA, suggests that cytokinin moieties in t-RNA may have important role in protein synthesis. Isopentenyl adenine is highly active.

Zeatin riboside (Letham, 1966) and *zeatin ribotide* (Letham 1966, Miller, 1967) also occur naturally in plants; zeatin and zeatin riboside have also been isolated from mycorrhizal fungus (Miller, 1967). A *cis*-ribosyl zeatin and a *ms*-ribosyl zeatin have also been extracted from plant tissue. Diphenyl urea and several of its derivatives have cytokinin activity. Several synthetic cytokinins are available. These include benzyl adenine and benzimidazole. Adenine also possess some cytokinin activity.

The chemical structures of some cytokinins are given in Fig. 745.

C. Gibberellins

The field of gibberellins may be said to be truly a product of both East and West.

This strikingly potent and unusual group of growth substances—the Gibberellins —have been attracting the attention of botanists all over the world since the end of World War II. These compounds stimulate the growth of many plants, promote flowering in some cases and cause a variety of other interesting morphological and physiological responses. They have been known to the Japanese agriculturists during the past 50 years [a farmer in Japan, as far back as 1929, had actually described the symptoms of a

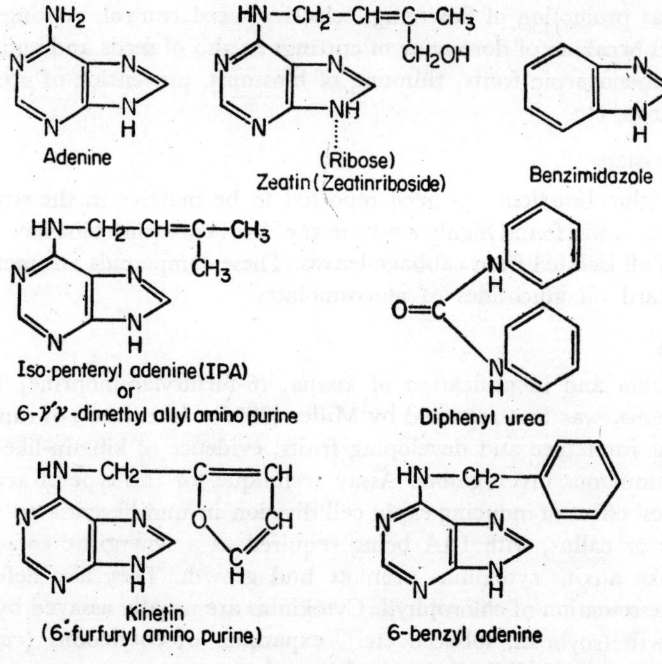

Fig. 745. Chemical structures of several synthetic cytokinins—kinetin, benzyl adenine and benzimidazole.

disease of rice plants (locally known as 'bakanae' meaning foolish) in which there was appearance of very tall and thin plants, sometimes pale, with longer and narrower leaves markedly overgrowing their unaffected neighbours] that a rice-disease causing ascomycetous fungus—*Gibberella fujikuroi* (the perfect form, occurring only occasionally; the imperfect form is *Fusarium moniliforme*) was responsible for the production of these substances.

Gibberellins (the abbreviation GA has already been accepted) have been extracted in the pure state from the culture filtrates of *G. fujikuroi* and higher plants. The basic structure of gibberellins has been well established and found to be derivatives of the aromatic hydrocarbon *fluorene* with bicyclic unsaturated *lactone* systems. More than fifty types of gibberellins possessing biological activity, are now recognised. GA_1, GA_2, GA_3 GA_4 (*gibberellic acid*), GA_5 and GA_7 are highly active. There are some gibberellins with C-20 atoms and some have 19 carbon atoms each. Most gibberellins are monocarboxylic, some are dicarboxylic and very few are tricarboxylic acids. The different gibberellins vary considerably in their physiological activity. Some like GA_3 and GA_7 are highly active, while those like GA_8 are almost inactive. Chemical structures of some gibberellins are given in Fig. 746.

Certain GAs have been found in plant tissues to occur in *bound* or *conjugated forms* with other compounds—free GAs can be released from bound ones by enzymatic (*emulsin*) treatment. In pea seedlings, during the first few days of growth, the necessary GA needed for this phase of growth, may partly be released from 'bound ones'. Among the bound forms of GAs, the following deserve special mention *acetyl* GA_3, D-*glucopyranosyl* -GA_3, -GA_8, -GA_{26}, -GA_{27}, and -GA_{28}.

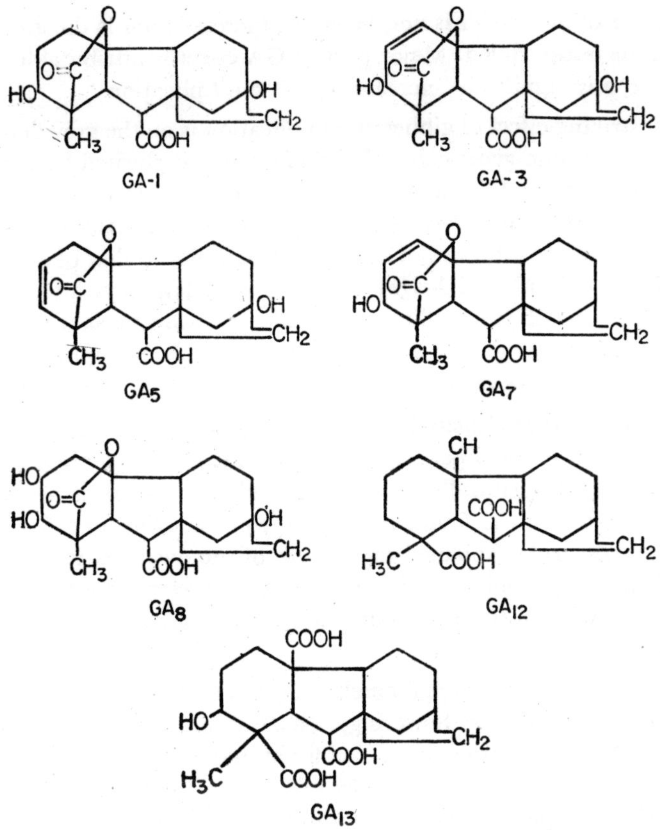

Fig. 746. Chemical structures of some gibberellins. Over 50 gibberellins are known. Note the small differences between any two GAs. Differences in biological actixitp are sometimes associated with presence or absence of double bonds, OH groups in particular positions, presence of the loctone ring and carboxyl groups. Also note that while most GAs, have only one, GA_{12} are GA_{13} have two and three carboxyl groups.

The most typical and striking plant response to gibberellins is stem elongation, actually internodal elongation, limited exclusively to younger tissues which are still growing (mature tissue is not influenced). It is more active on intact plants than on tissue sections. GA is a growth regulator like IAA which brings about cell elongation but does so by a different mechanism than IAA. This response may be greatly influenced by external conditions and may result in the change of habit of the plant (forming a vine from a bushy plant or producing a more bushy habit due to outgrowth of laterals). Even plants like lettuce, which has never been known to show twining habit, have become vines on treating with GA.

There is some evidence that one gibberellin is converted into another within the plant. Thus, there is some evidence to suggest that GA_5 is converted into GA_1 in pea-seedlings but the extent of this conversion is not nearly enough to account for the spectacular effect of GA_5 on growth in this material. Several GAs are probably connected to GA_8 in a bound form.

Gibberellin applied in one part of plant or even to the soil can stimulate growth in all the growing parts of the plant.

The number of internodes is not, however, increased nor is the growth exceptionally abnormal or uncontrolled. Major part of GA-effect is attributable to cell elongation and not to any significant increase in cell multiplication.

The most striking effect of gibberellin application is on the growth of the genetical dwarfs (true also for non-genetic dwarfs) which is not duplicated by auxin, IAA. This apparently specific effect of GA in contrast to all other growth-promoting substances on dwarf varieties of pea, *Vicia*, *Phaseolus*, etc., (also dwarf mutants of maize) can bring about an increase in growth rates of such dwarf varieties to that of a tall veriety. It is possible to obtain differnces in height of 500% in a few days after one application of a few micrograms of gibberellin on a leaflet of dwarf variety of pea. The name gibberellin for this unique growth substance seems singularly appropriate as gibberella in Latin means hunchback or dwarf.

Gibberellins have been isolated from all parts of seedlings of tall and dwarf varieties of pea and also from seeds of pea, french bean, wheat, etc., (there is little doubt now that gibberellins occur naturally in all higher plants). There is no evidence as yet, however, that gibberellin is found in greater concentration in tall varieties than in dwarf, which might have explained the growth of the dwarf varieties on GA application. Even then it is possible, that in dwarfs, GA is present in a form unable to exert any influence on cell elongation or conceivably there may be inhibitors present which prevent their action.

Another exciting property of gibberellins is their ability to induce flowering in some plants (IAA generally inhibits flowering). In *Hyoscyamus*, flower formation could be stimulated by GA in an environmental condition under which flowers otherwise would never have formed. Earlier flowering after gibberellin treatment has been observed in plants in which flowering is normally photo-induced as well as in some species of day-neutral plants. Usually long-day plants flower in response to gibberellin application; the very few short-day plants which exhibit this response include *Impatiens balsamina* and a few grasses. After treatment, chrysanthemums were larger and bloomed earlier. Gibberellin nearly doubles the diameter of some *Geranium* inflorescence. GA usually promotes maleness of flowers.

Gibberellins have been found to substitute for the cold treatment required by biennial long-day plants. *Hyoscyamus* treated with gibberellin, flowers in the first season without any previous cold treatment. Cold requirement for germination of peach seeds can be annulled by gibberellin treatment. It has been suggested that GA does not directly affect the internal reactions which take place during a cold vernalisation and that it is certainly not identical with *vernalin*, the hypothetical product of vernalisation (attempts to vernalise cereal grains with gibberellin have been disappointing). The cold treatment is replaced by GA, most probably through the removal of the blockage which stands in the way of the internal processes which lead to flowering in cold requiring plants, perhaps by switching on a mechanism entirely different from that utilised normally by low temperature.

It is well known that osmotic inhibition of germination of certain types of seeds can be prevented and actually reversed by giving small quantities of red light of specific spectral quality—660 nm. GA has been found to substitute for red light in protecting lettuce and tobacco seeds against this type of inhibition (see later).

One of the most remarkable effects of gibberellins is that it promotes the induction of many hydrolases, particularly α-amylase and proteinases. In seeds of cereals like barley, wheat or rice these enzymes are produced in the aleurone layers and hydrolyse the starch and protein present in the endosperm to soluble monomers (glucose, amino acids etc.), which are utilized in growth. Gibberellins promote the growth of leaf bases or the basal region of the leaves . While gibberellins have no effect on the growth of the coleoptile of cereals it promotes the growth of the first leaf which it encloses; the leaf sheath is more sensitive than lamina. These properties are utilized in the bioassay of gibberellins.

Gibberellins do not show the polar transport (they move freely both in the phloem sieve tubes and also in the xylem) within the plant which is characteristic of auxin. In contrast to auxins, GA does not inhibit root growth nor does it cause the typical IAA-epinasty of leaf petiole. Gross deformities and callus formation in plant tissues brought about by IAA are not duplicated by GA.

Gibberellins, however, strongly promote the growth of tissues during certain growth phases that auxins do not. Evidences seem to indicate that the responses of auxins and gibberellins are conditioned by different genes. In classical auxin bioassay, the *Avena* coleoptile curvature as also in pea curvature tests, gibberellins have so far consistently given only negative results. The most remarkable of all the differences between GA and IAA appear to be the evident indifference of GA-effect to light direction.

Gibberellins are much more potent than auxins in inducing parthenocarpy (fruit formation without flower fertilsation) and enhancement of fruit yield in grapes and tomatoes. In tomato, seedless parthenocarpic fruits were obtained when the styles of emasculated flowers were treated with GA mixed with lanolin. The fruits produced, however, were greatly reduced in size and weight, compared to the normally set, seeded fruits. It is not always necessary that the flower itself be treated; male sterile tomatoes may set fruit when GA is applied to several parts of the plant or even to the soil.

No clear-cut linkage of gibberellin treatment with any metabolic pathway in plants has yet been established but analyses so far have always shown the greatest change among carbohydrate constituents. Sometimes spectacular increase in total sugar concentration has been observed in GA-induced parthenocarpic tomato fruits. Another apparent effect is the reduction in chlorophyll content accompanying chlorosis caused by application of higher dosages of gibberellin. Promotion of respiration of growing parts of treated plants and seeds has also been reported. Though there are variations in the level of certain enzyme activities, gibberellins themselves are not known to activate isolated enzymes.

There is no convincing evidence, perhaps except one, as yet of any GA-effect on any organism except higher plants including mosses and liverworts. GA itself is apparently destroyed by soil micro-orgnisms but some reports indicate that the microbial population is modified on GA-application to soil so that growth of the free-living nitrogen-fixing bacteria, *Azotobacter*, may be favoured in certain types of soil.

Attempts are being made to ascertain the economic importance of gibberellins in agricultural practices. The preliminary results are not very encouraging. Gibberellins have been tried in increasing leaf-expansion in the leafy crops of toabcco and tea. The most promising avenue for further research still may lie in increasing the yield of hay or jute which may be economically profitable. Spectacular increase in the rate of growth of some trees, ornamental or otherwise, may give interesting results.

D. Ethylene

The effects of smoke and illuminating gases on plant growth were recognized since the 19th century. The hormonal role of ethylene was first suggested by Crocker, Hitchcock and Zimmermann about fifty years ago.

The most remarkable physiological effect of ethylene is its effect on fruit ripening. Ripening fruits emit ethylene which helps in the ripening of other fruits. Many fruits including bananas, oranges, lemons and apples are ripened this way; however, there are a few exceptions, e.g., peaches and apricots. Removal of ethylene by a current of air prevents ripenning and helps storage; removal of O_2, introduction of N_2 or CO_2 which interferes with its production, also helps in the retention of greenness of fruits. Other physiological effects include inhibition of stem elongation, increased radial expansion—particularly at the basal regions. Abolition of positive geotropism and promotion of horizontal orientation (presumably due to an interference with polar transport of auxins), stimulation of abscission of leaves and petals, discolouration of flowers, promotion of leaf epinasty (bending of petioles) and hastening of senescence.

Ethylene may be produced from a variety of substances but methionine seems to be the most efficient precursor. Auxin stimulates ethylene production and a number of auxin effects including the early flowering of pine-apple are believed to be mediated through ethylene, ethrel or ethephon, which when sprayed on plants, generate ethylene, are used in agriculture and horticulture.

HELMINTHOSPOROL

Helminthosporol has been isolated and extracted from cultures of the pathogenic fungus, *Helminthosporium sativum* by Tamura and others (1965).

OTHER GROWTH-REGULATING SUBSTANCES FROM PLANTS

A. ABSCISIC ACID (ABA)

The first pure abscission inducing substance, isolated and extracted from petroleum ether extract of cotton fruits (1961), has now got the fully approved name, *abscisic acid* (Fig. 747). It was found to accelerate abscission when applied in very small quantities to the petiole stumps of cotton seedlings. A second crystalline acidic substance was *abscisic acid* II, isolated also from young cotton fruits (1963). It also accelerated abscission when applied to petiole bases of cotton seedlings, more effectively than GA_3, but had no gibberellin-effect on dwarf maize and was also a strong inhibitor of IAA-induced straight-growth in *Avena* curvature test.

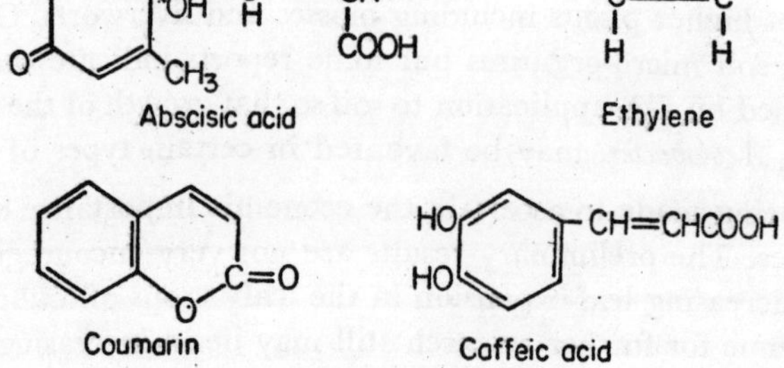

Fig. 747. Some growth regulators other than auxins, cytokinins and gibberellins.

ABA also promotes senescence and bud and seed dormancy and inhibits seed germination, growth of excised embryos mesocotyls, rice seedlings, expansion of leaf discs and *Lemna* fronds, regulates stomatal movement, causes turgour loss, and accelerates abscission of excised abscission zones. It also interacts with other growth hormones. Thus, auxin effects on coleoptile growth and GA_3-induction of α-amylase in aleurone layers of barley grains is antagonized by ABA, and the inhibition is reversed by higher concentrations of the growth substance. ABA-inhibition of the growth of *Lemna* fronds is also countered by a synthetic cytokinin, benzyladenine. The reversion, however, is not observed in all systems. It enhances colour formation in oranges. ABA effects are rapid and may disappear. ABA is present in all plants, usually in microgram quantities. It is produced probably by the isoprenoid pathway via mevalonic acid or by catabolic degradation of xanthophylls like violaxanthin. It may be assayed by leaf abscission and induction of bud dormancy tests, *Lemna* frond growth inhibition and α-amylase induction inhibition tests, but none is specific. Gas liquid chromatographic techniques, which can detect very small quantities, have been used with advantage by many workers.

B. TRAUMATIC ACID

Haberlandt's observation, that in the injured or cut surfaces, some substances were responsible for inducing new cell division to heal up the wound, in 1921 led to the discovery of the so-called, *nekrohormones* or *wound hormones* from the cells bordering the injury. The following formula was assigned to it:

$$HOOC.CH{=}CH.(CH_2)_8.COOH$$

Traumatic acid supplied exogenously is known to induce extensive wound periderm formation in washed potato tuber.

Perhaps Went's (1938) postulated existence of *calines* should be mentioned here. Most of the evidence for their existence in plants is indirect (they have never as yet been identified in plants) and the adenine derivative of cytokinin, *zeatin* has effects on plants analogous to those of the postulated calines. The various postulated calines are: (*i*) *rhizocaline*, synthesised in the leaves or stored in the seeds and necessary for root formation; they are thought to be translocated primarily downwards; (*ii*) *caulocaline;* synthesised in roots or it may be stored in the seeds but necessary for elongation of stems—translocated primarily upwards; and (*iii*) *phyllocaline*, made or stored in the cotyledons and supposed to be necessary for leaf growth.

Synthetic Growth Regulators

A. GROWTH RETARDANTS

Many synthetic organic compounds (Fig. 748) are now known which antagonise or even nullify the longitudinal growth promotion induced by GA in plants. They appear to reduce stem elongation by altering plant metabolism and hence blocking GA biosynthesis (*antimetabolites*), rather than by acting as competitive antagonists to GA as well as to IAA in their growth processes. These compounds inhibit one or more specific steps of GA biosynthesis. Plants become short and bushy but greener. Flowering in some plants, e.g., azaleas and rhododendrons is accelerated. With cycocel biennial mango plants have been found to flower in the off-season. Cycocel also promotes the formation of female flowers in Papaya. They are usually called growth retardants.

Among the specific GA-antagonists, are Phosphon D (tributyl-2,4-dichlorobenzyl phosphonium chloride), AMO 1618, CCC, etc. All of them are complex chlorides.

B. Morphactins

A chemical group of regulator substances, morphactins (morphologically active substances) has been synthesised by 1960 in E. Merck's Laboratories. Since 1964-65, this group, consisting of fluorene-9-carboxylic acids and their derivatives, has been the subject of considerable interest as a representative or new type of growth regulator. Their action is systemic, but slow, and can be transported both basically and acropetally (as in GA). These non-toxic substances inhibit and modify development and new growth. The morphogenic action of morphactins begins only with higher plant groups, starting with highly organised brown algae, the mosses and ferns. Unicellular plants and simple cell filaments are not affected by morphactins.

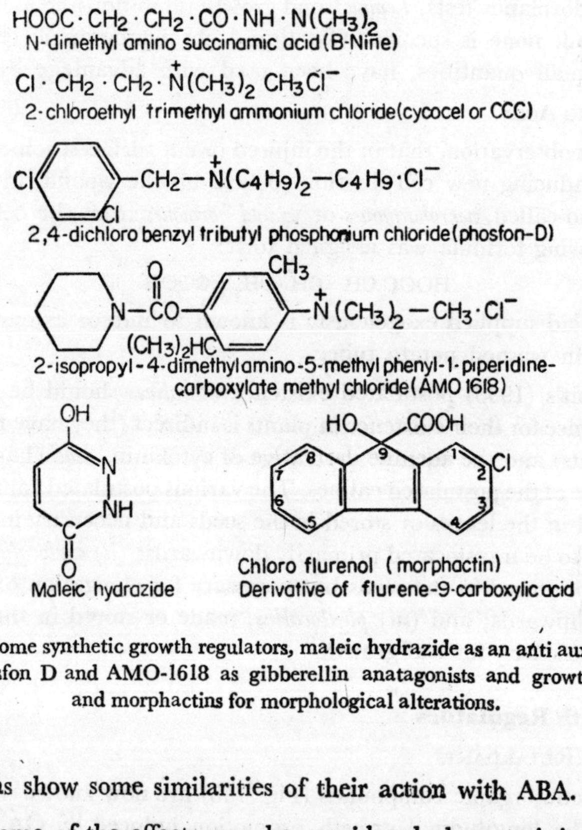

Fig. 748. Some synthetic growth regulators, maleic hydrazide as an anti auxin, B-NINE, CCC, phosfon D and AMO-1618 as gibberellin anatagonists and growth retardants and morphactins for morphological alterations.

Morphactins show some similarities of their action with ABA.

However, some of the effects may be considered characteristic for morphactins, which are not duplicated by any other synthetic or natural growth regulators. They are really *unique* in bringing about these responses, as for example, the complete and striking abolition of polarity phenomena in plants by morphactins; the reduction of apical dominance of the main shoot; promotion of shoot branching; reinforcement of the apical dominance of the tap roots; the abolition of gravity and photoinduced directional growth—geotropism and phototropism; some disturbance, difficult to specify, of the coordination and correlation phenomena and the action on cell division; the inhibition of mitoses in apical meristems; stimulation of cell division in cambium and pericycle of intact plants and cuttings, etc. Morphactin effects may disappear after withdrawal.

C. INHIBITORS

Various phenolics, coumarins (Fig. 748) and cucurbitacins (abundant in the family Cucurbitaceae) are some of the naturally occuring inhibitors which affect several physiological processes including germination, growth, flowering and senescence in addition to metabolic processes.

Mechanism of action of growth regulators

Plant growth regulators have varied physiological effects and each of them produces multiple effects, depending on the concentration used and the organ under study. Some, like auxins, have fast and delayed actions. The morphogenetic changes brought about take quite some time and involve the formation of new enzymes and structural proteins. This implies an interaction with the genomal DNA. Receptor molecules for some hormones have been claimed to have been isolated, as in the case of animal hormones. Hormones are believed to interact with some molecules, presumably proteins, which then interact with the chromatin thereby influencing transcription. Direct interaction between auxin, GA and cytokinin on the one hand and DNA, RNA and protein on the other hand have also been demonstrated, although their biological singnificance has not been adequately explained. In the case of gibberellins, synthesis of the α-amylase enzyme protein in response to GA_3 application has been demonstrated.

The fast action of auxin is evident within a few minutes of application. Since growth is enhanced by dilute acid solution, it has been suggested that auxin in some way influences membrane potential and activates a "proton pump"—the active process concerned with transport of H^+-ions accross membranes. There H^+-ions probably weaken the H-bonds connecting xyloglucan molecules with cellulose fibrils in the primary cell wall. In fact some xyloglucan is liberated in pea stem sections as growth is promoted by IAA. As the wall is loosened, more water is taken up by the cell due to enhanced osmoitc pressure or decreased water potential arising as a consequence of hydrolysis of some cell substance or enhanced ionisation of some electrolyte. With increase in cell volume the membrane presses against the loosened wall. New wall material including polysaccharides and proteins are gradually laid down so that the wall regains its rigidity after some time. The proton pump hypothesis, however, has been questioned recently and a more satisfactory explanation is necessary.

Cytokinins which stimulate cell division and delay senescence are known to promote DNA, RNA and protein synthesis is several tissues, although the mechanism of this stimulation is not understood. Some nucleotides of t-RNA have cytokinin activity; it is not known whether they help in the assembly of amino acids on the m-RNA-ribosome complex. Cytokinins probably influence the activity of certain enzymes—particularly hydrolases and those which synthesise certain aminoacids.

ABA has also been suspected to affect nucleic acid and protein metabolism in some way but no conclusive evidence is available. Its rapid effect or stomatal movement is suggustive of other sites of action.

The effects of various growth substances overlap and they may also interact indirectly. In some cases the effect of one has been found to be duplicated by another. While this may be mediated through the transcription-translation machinery, other mechanisms are also possible, though evidence the present moment is not available.

DEVELOPMENT

If a callus tissue is grown in artificial culture media, it may contunue to grow indefinitely without any tissue differentiation. This is growth without any development. The normal plant, however, shows both growth and development. The fertilised egg cell develops into an embryo, the seed develops into a seedling, the seedling produces vegetative organs for some time and then switches on to its reproductive phase, i.e., flowering, and ultimately the fruit is developed. When the reproductive phase is completed it may be followed in annuals by senescence and death.

The physiological factors, controlling the development of the plant, have only in recent times been critically analysed. The development of a plant from the vegetative to the reproductive stage is one of the commonest of all natural observations. In case of annuals, it is observed that only when the vegetative growth is completed, reproductive phase begins. It is apparent that vegetative growth and reproduction are largely antagonistic to each other. In case of perennials, the plants produce vegetative parts during some period of their life history and reproductive organs in the other. The most important thing is that the two phases—vegetative growth and reproduction—do not usually take place simultaneously. It is only when one stops, the other begins.

Photoperiodism

In all parts of the world except tropics and subtropics marked seasonal variations occur in the length of the daylight period.

The phenomenon of photoperiodism is the development of plants in relation to the length of day and night. It is one of the most notable of all reactions of plants to their environment.

The foundations of our present knowledge on photoperiodism were laid when Garner and Allard showed in 1920 that flowering of many plants could be induced or prevented simply by controlling the length of the daylight period.

In trying to carry out breeding experiments with a giant variety to tobacco, they found that the plants would not flower when growth is in the open in summer, although they would do so and profusely in the greenhouse in autumn or winter. After many unsuccessful attempts Garner and Allard were able to induce these tobacco varieties to flower profusely in the open during summer months by reducing the daylength to correspond with that which the plants get when they flowered in greenhouses in winter. The daylength period was artificially reduced by simply moving the potted plants into a dark room after exposure to the desired number of hours of daylight.

The realisation dawned on them suddenly that growth of the plant, and formation of reproductive organs are strongly influenced by the length of the day, most plants having a particular optimum requirement in this respect—the *photoperiod* (duration of light). Some plants flowered most rapidly when the daylength was about 12 hours or less, others when the daylength was about 12 hours or more. Some plants belonging to the second group were found to flower even in continuous illumination.

As instances multiplied, it was found that plants could be classified into *three* broad groups:

SHORT-DAY PLANTS (SDP). They are really long-night plants (LNP); flowering best when subjected to night periods longer than a critical duration as in many tropical plants, cereals, tobacco, *Crysanthemum* sp., *Xanthium* sp., etc.; also most of the spring and autumn flowering species of temperate regions. It is a varietal character. All *Aman* varieties of rice are short day plants.

LONG-DAY PLANTS (LDP). Flowering best if subjected to light periods longer than a critical duration. Most of the agricultural plants of the temperate regions are long-day plants e.g., wheat, potato; and also many summer-flowering plants of the temperate regions. Species growing in high latitude (60° and further north) are mostly long-day types. Potato is a long-day plant for flower formation but it is a short-day one for tuberisation. Economic cropping of potato can only be obtained if potato is grown in short-days.

INDIFFERENT OR DAY-NEUTRAL PLANTS. The flowering of plants of this group is *independent of daylength.* This group includes tomato, maize and a large number of tropical plants which have no known flowering rhythm, but which produce flowers all the year round. Flowering in tomatoes seems to be controlled by temperature effect—an alternation between night temperature of 15°C and a day temperature of 25°C, appears to be optimal for flowering in tomatoes. This phenomenon has rightly been termed *thermoperiodism* or *thermoperiodicity.* Other plants also frequently show similar responses. *Aush* varieties of rice belong to this group, flowering only after a fixed number ('period') of days.

The photoperiodic effect has been found in all types of plants, herbaceous, woody, annual, biennial, perennial and even in animals. Though it is true that the most important effect of daylength is upon the flowering of plants, the daylength or the nightlength also affects many other phases of the physiology of the plant. Many trees do not enter their rest period in the autumn if long days are maintained instead of short days. As a result, they fail to develop frost resistance and are readily killed by winter cold. Shedding of leaves in many plants (e.g., yellow poplar, *Liriodendron*) can be stopped by artificially illuminating the plants, i.e., giving long days in autumn or in early winter; the plants remain evergreen.

Short-day plants will flower even when exposed to long-day conditions, provided they have already been previously exposed to a sufficient number of short days. Long-day plants will similarly flower in short days if they have been previously exposed to a sufficient period of long days. This carry-over of the effect is called *photoperiodic induction* (or after-effect).

As has been said before long-day plants will flower even in continuous illumination and do not really require any dark intervals whatsoever for flower initiation.

In short-day plants, however, conditions are entirely different—a dark period is obligatory; however it needs a light period for photosynthetic purpose. The flowering of short-day plants in short days thus happens because the dark periods are long and certainly not because the light periods are short.

In conjunction with experiments by Hamner and Bonner and also by Cajlachjan, (1936) who called the hypothetical flower forming hormone, *florigen*, the general assumption was that, in all plants a specific substance is manufactured in the leaves, from whence it is translocated to the growing points, in which it induces the formation of floral initials. The rate of movement of this flowering stimulus from the leaves of certain plants has also bean measured.

In 1961, Lincoln, Mayfield and Cunningham claimed success (?) when they prepared an alcohol-extract at very low temperature from flowering *Xanthium* plants which induced flowering in similar plants under non-induced photoperiods. The degree of flowering, however, was only small, although a significant fraction, of the full potential. On this basis, Lincoln *et al* claimed the first preparation of a natural extract capable of imparting a reproducible flowering stimulus by external application. It was later claimed to have been substantiated by Carr (1964).

The active fraction isolated was acidic in nature and Lincoln and others (1964) felt that the acidic nature of the active fractions warranted assigning the name *florigenic acid* to this compound, which even now remains, by and large, clearly hypothetical.

For the flowering response light stimulus is apparently perceived by a phycocyanin-like pigment (distinct from chlorophyll). Recent findings also indicate that this pigment named *phytochrome* is chemically related to phycobilins of algae. The active spectrum of flower formation is strikingly similar to the absorption spectrum of *allophycocyanin*. It seems certain that two forms of the same pigment are involved and the induction of photoperiodic response depends upon a reversible photoreaction of the photomorphogenic pigment. Phytochrome is a chromoprotein (biliprotein) which exists in two interchangeable forms—one form, R phytochrome, absorbs short-red light in the vicinity of 660 nm and is thereby converted into the other form, FR phytochrome, which absorbs in the far-red region—730 nm—and brings about the reverse reaction. This may enable one response induced by red light or white light containing red be promptly cancelled by an exposure to far-red light or *vice versa*. However, in one direction, this reversion of FR phytochrome to R phytochrome can also proceed in the dark—by a slow, temperature-controlled, metabolic pathway.

660 nm (short-red)

R phytochrome ⇌ FR phytochrome

730 nm (far-red)

dark conversion

At the end of the daylight period it is apparent that all the phytochrome is in the far-red absorbing form and the slow night reversion starts.

For a short-day plant, e.g., chrysanthemum, even a single flash of light (or exposure to red light) will prevent flowering response, for such an exposure to light, however brief, re-converts all the R phytochrome (formed at night prior to exposure) into the FR form thereby altering the effective concentration promoting flowering in short-day plants. The same light-flash treatment, as will readily be understood, induces flowering in long-day plants.

The same perceptive system has been found to be present in certain seeds, e.g., lettuce, tobacco, etc., which require light, particularly short-red light, for germination. Such seeds can be caused to germinate or prevented from germinating by alternate exposure to red and far-red light—the last exposure determining the response—if red, the seeds germinate; if far-red, germination does not take place.

It is imperative in all cases that alternate light exposures follow each other fairly promptly for cancellations to occur. If delayed, i.e., if the 'trigger' or the chain of reactions for a particular response has already been set in motion—it cannot be reversed.

The phytochrome system is also the sensing mechanism that brings about changes in the growth of a developing seedling—in the dark, a great elongation and no opening of leaf-buds, and in light, short stem internodes and expansion and unfurling of leaves. The same response follows on exposure to red light which can again be nullified by a prompt exposure to far-red.

Phytochrome was finally isolated from etiolated plant materials, e.g., pea, corn seedlings, spinach leaves, etc., in the spring of 1959. It is very different from chlorophyll and is largely proteinaceous in nature, most probably, as we already know, a biliprotein like allophycocyanin.

It is a pigment having a molecular weight of about 120,000 and is believed to be associated with membranous structures of all green plants from algae to angiosperms. There is good evidence that it controls permeability of metabolites across membranes. According to Borthwick and Hendricks, the signal of alteration in the level of metabolites is ultimately transmitted to the genome in an unknown manner which then responds by an alteration of its function, i.e., an alteration in the products of transcription and translation. According to Galston phytochrome is associated with the nuclear membrane. Changes in the patterns of RNA synthesised in red and far red light have been observed. Antimetabolites which initiate RNA and protein synthesis also inhibit flowering.

As in the case of auxins, phytochrome exhibits both fast and delayed actions. The fast action is believed to be mediated through the conversion of phytochrome from one form to another due to acceptance or removal of H^+; a membrane potential is developed resulting in the movement of protons. The fast reaction can be demonstrated within a few seconds. Several enzymes are inducible by phytochromes and many are activated. The manifestation of physiological changes like emergence of radicle or plumule, biochemical effects like formation of the red pigment in tomato cuticle takes much longer time. The extreme example of the delayed effect is exhibited by the photoperiodic effect or flowering. The link between the fast and the delayed process is not understood at present.

HOW DOES A PLANT FLOWER?

As in the case of growth itself, there is strong evidence of a hormonal control of flowering. The response certainly occurs in the growing points, yet the growing points themselves (regions of response) do not have to be exposed to any particular daylength. The region of perception of stimulus for flower formation, the leaf, even a single one, if subjected to the necessary photoperiod (daylength) may induce growing points to flower. It has been possible to transfer this floral stimulus from one plant to another by grafting. Certainly then, if the stimulus, is a chemical substance, it is synthesised in the leaf and translocated to the regions where flower formation takes place at the growing points. Some believe that the chemical substance in question is the same one that controls growth—*hormones*. Indoleacetic acid inhibits flowering in concentrations which promotes vegetative growth in general.

Gibberellins promote the flowering of usually long-day plants. In several plants cytokinins have also hastened flowering. Consequently it may appear that the naturally occurring hormones themselves or through their interaction with "flowering hormone" —whatever be its nature—are responsible for flowering. Chailakhyan has modified his original florigen concept now to imply a balance between anthesins (flower forming substance) and gibberellins. More recently he has shown, along with Anton Lang and others that graft transmissible inhibitors, which are produced under non-inductive conditions play an important role in the regulation of flowering. A balance of flower forming and inhibiting substances proposed many years ago by Wellensiek and others has also been visualised. The inhibitor(s) has not yet been isolated. It is interesting to mention that level of abscisic acid is affected remarkably by photoperiods and has been shown to possess marked effects on flowering of plants. Other inhibitors may also be involved. There is a net appearance of florigen in the leaf which can be transported to the region of flower formation through the living tissues such as phloem. And once the flowering hormone or florigen has reached the growing bud, differentiation begins almost immediately, transforming the vegetative bud into a floral one. Although the exact nature of these changes is not clear, we do know that an active bud or its equivalent is necessary. Evidences discussed previously and above thus show the existence of several processes in the photoperiodic response to flowering. The sequence of stages may be summarised as follows:

High-intensity light process ⟶ dark process ⟶ 2nd high intensity light process ⟶ translocation ⟶ differentiation

There is no absolute requirement for the second high-intensity light process since plants will flower even in continuous darkness after the first photoperiodic induction. The suggestion is that florigen appears unstable until it is fixed in some manner by the second high-intensity light process following the dark period.

The existence of a universal, graft-transmissible flower-hormone (florigen) has been questioned in recent years.

It has been realised for some time that there are actually multiple paths to floral induction in many plants and this, perhaps, cannot be readily explained by supposing the universality of florigen concept (Evans, 1971).

On the other hand, the analysis of photochrome action in flowering, suggests that at least two separate photoperiodic processes, one favoured by high and the other by low FR-phytochrome level, are required for flowering in both LDP and SDP plants. The two processes may proceed simultaneously, however, in light sources, giving intermediate proportions of FR-form. Translocation studies (Evans, 1966, King and Evans, 1968) also suggest that there may be at least two primary photoperiodic stimuli for floral induction.

It seems, according to Evans (1971), that we may have to discard our present attractive concept of a universally-active single florigen. And the case for doing so is already pretty strong, according to Evans (1971).

VERNALISATION

Some winter varieties of cereals, e.g., wheat or rye must be sown in temperate regions in the early autumn in order to make them flower in the following summer. If they are sown in the spring they continue to grow vegetatively with abundant tillering but fail to come to ear or flower much later than those sown in autumn of the year preceding. They thus contrast with spring varieties which planted in spring, flower in the same year. If, however, seeds of winter varieties are moistened with enough water to increase their weight to about 20-60%, allowing germination to begin but not such as to encourage rapid growth (when tips are just emerging) and kept at a temperature close to freezing, e.g., 0-5°C. for abour 3 or 4 weeks, they can be sown in the spring to flower at about the same time as the previously autumn-sown seeds. This treatment is known as *vernalisation.* Plants requiring vernalisation are, of course, confined to temperate and polar regions. Thus vernalisation is an agricultural practice of treatment of seeds with low temperature in initial stages of seedling development, accelerating its development. This treatment induces a rapid development towards a physiological older condition and results in shortening of the interval between sowing and flowering. The flowering factor, may be the hypothetical *vernalin*, can actually be translocated from a vernalised plant to a non-vernalised one and cause it to flower. An unvernalised plant of *Hyoscyamus* can also be brought to flower by grafting on to it a scion of the annual variety which does not require vernalisation. By grafting it has been possible to show that 'vernalin' is distinct from photoperiodic florigen. By vernalisation treatment, winter varieties of cereals can be made to behave as spring varieties which planted in the spring, flower and come to ear the same year.

Gregory and Purvis first put out the concept that growing tips perceive the vernalisation stimulus of low temperature. Wellensiek (1961–1970) from his studies with leaves and isolated roots, arrived more or less to the same conclusion; that dividing cells are necessary for the action of low temperature, no matter where they occur in the plant; so it seemed that dividing cells were prerequisites for vernalisation. However, Chakravorti (1970) from his work with *Linum*, concluded that the cells of the fully formed epidermis are capable of perceiving the vernalisation stimulus. This has long been claimed by Efeikin (1958), working with sugar-beet, who showed that the perception of low-temperature stimulus, necessary for flowering, is not restricted to the local changes in meristems.

Lysenko stressed the distinction between growth and development; that they are not identical phenomena and advocated what was once known as the theory of *phasic development.* According to this theory the development processes of an annual plant consists of a series of distinct phases which occur in strict predetermined sequence. The onset of any one phase will only take place when the preceding phase is completed. In winter varieties of wheat, low temperature is necessary for a certain phase of development, the *thermophase,* which must be passed before the next phase, the light-controlled *photophase* can begin (daylength being the controlling factor). Recent researches, however, indicate that the different developmental processes cannot be as clearly separated as suggested by Lysenko. Nevertheless it is true that flowering is undoubtedly influenced both by temperature and daylength to which the plants are exposed and vernalisation may accelerate a part of the course of development.

Vernalisation has been developed in Europe, particularly in Russia where the significance of this treatment has been demonstrated and found to be economically very profitable. In Russia, the severe cold during the winter months sometines kills the autumn-sown winter cereals. Vernalisation method is now used extensively to avoid this severe loss due to killing of seeds. The winter cereal seeds are now vernalised and sown like ordinary spring varieties. Other possibilities arising from the vernalisation treatment, e.g., introduction of a crop plant into a new area with a growing season shorter than that under which it normally grows, are being explored. Very stisfactory results have also been obtained in this line.

The actual temperature requirement varies considerably from species to species. The mechanism of this temperature effect is possibly controlled by a hormone and the name *vernalin* has been suggested for this unknown hypothetical substance. It is thought that vernalin is either a precusor of florigen or a catalyser essential for florigen formation. Alternatively vernalisation reaction may simply consist in the removal of restriction to the production of the largely hypothetical flower hormone, florigen.

Some plants can be vernalised after an appreciable amount of vegetative growth has taken place. Biennial plants such as beet, generally grow vegetatively during the first season and flower only after they have spent the winter in the soil. The low temperature stimulus is perceived by the plant in winter, persists within the plant and subsequently causes flowering in the following summer. The perception of the low-temperature simulus in the biennial is restricted to the apical bud and is independent of the leaves and can be transmitted to another plant across a graft.

The low-temperature requirement for vernalisation of buds or seeds may be determined by a single gene.

Vernalisation effect is, however, reversible. Completely vernalised rye or wheat may entirely revert to the normal condition on being dried for several weeks or simply by storing them under anaerobic conditions (*devernalisation*).

POLARITY, RESTITUTION, CORRELATION AND DIFFERENTIATION

Polarity

In the development of a plant (and animals) growth does not proceed at random producing a formless mass of living cells but is an orderly process that gives rise to specific three dimensional forms of organ or body. Growth in one region or dimension is related to growth in the others and thus the plant becomes an integrated individual. A notable feature of these bodily forms is the presence in them of an *axis* which establishes a longitudinal dimension for an organ or the plant. Along the axis and symmetrically to it, the lateral structures develop. The two ends of *poles* or the axis are usually different both as to structure and physiological activity. Thus a typical vascular plant has a major axis with the root at one end and shoot at the other, with lateral appendages, e.g., leaves, branches and lateral roots arranged symmetrically around it. Growth is usually more rapid parallel to the axis than at right angles, resulting usually, but by no means invariably, in an elongate form. These patterns appear very early in the development of an organism as the result of differences in growth or in planes of cell division. This characteristic orientation of an organism, which is typically *bipolar* and *axial*, is termed *polarity*.

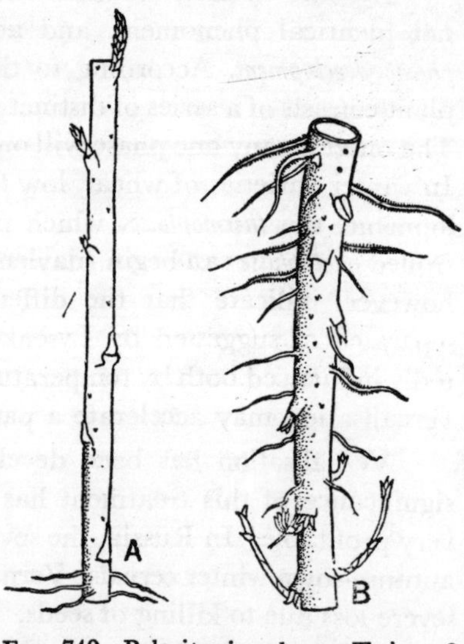

FIG. 749. Polarity in plants. Twigs of willow. A. Suspended in a moist chamber in the normal position. Note rooting from the lower end and buds near the other end, expanding into shoots. B. Twig hung in the inverted position. Note that the corresponding buds at the end which is now lowest, still give rise to shoots while roots are produced near the lower end which is now, of course, uppermost.

Polarity may manifest itself in a variety of ways. The structures at the two ends of the axis are very much unlike as in the cases of root and shoot, stem end and flowering end and petiole and leaf blade. The movement of certain substances may take place in one direction along the axis but not in the other, thus showing polarity in physiological activity. Individual cells show polar behaviour *in the plane. of division* and in the different characters of the two daughter cells formed.

By axiate polarity is then understood as the property of the plant or an organ which determines the contrast between apex and base. In higher plants, this polarity is already determind in the fertilised egg cell and once established it seems to be strictly followed. When a piece of stem is cut off the plant this will usually reform roots and often shoots. The roots are always produced at the basal end of the axis whereas shoots develop near the apical end. This polarity seems to be due to polar transport of auxin in cells and tissues where it always moves from apex to base. This causes auxin accumulation at the base which favours root formation and a deficiency of auxin at the apex which favours shoot initiation. Thus it has never been possible to transform the shoot end into root end and conversely by simply inverting a plant or a piece of stem (except by treatment with morphactins).

Polarity in rare cases may be determined by external factors. in *Fucus* zygotes, rhizoids always form on the side opposite to the direction of the source of light and the first wall is laid down at right angles to this direction. If the zygotes are, however, exposed to polarised light, the rhizoids tend to develop in the plane of polarisation. It has also been found that when *Fucus* zygotes are subjected to centrifugal force, rhizoids seem to grow out from the centrifugal pole. In *Fucus* eggs, even the shape of the egg-cell sometimes modifies egg polarity!

Restitution

By restitution is commonly understood the formation of new organs which as a rule follows the mutilation of a plant and can take place in situations where no active growth would have occurred in an uninjured plant. Two types of restitution are commonly observed: (*i*) new development occurring at the wounded surface and (*ii*) at some distance from the wounded surface:

(*i*) The production of a lost organ from the injured surface though not uncommon in lower plants such as algae and fungi is certainly of very restricted occurrence in higher plants. Only tissues that are still actively dividing—meristematic or embryonic—are capable of this.

In higher plants, formation of new organs at the injured surface, is most frequently seen in the growing points of roots. If the growing tip of a root is removed by a transverse cut, not more than 0·5 mm from the tip, the tip may be formed again. A longitudinally split root tip, tends to complete itself in such a way so as to give rise to two distinct growing points. This type of regeneration is, however, rarely seen in the growing points of shoot or in leaf primordia.

Recent evidences strongly favour the idea that all these regenerations of organs due to injury is controlled by the specific hormone—the wound hormone, *traumatin* (or *traumatic acid*, we have discussed before)—formed at the site of injury.

(*ii*) The formation of organ at some distance from the wounded surface is of widespread occurrence in plant kingdom. In algae and fungi it is commonly seen that an organ which has been lost is soon replaced either by the formation of a new one in the vicinity of the wound or by the outgrowth of one which was in rudimentary and dormant condition.

In higher plants if the growing point is destroyed, a new growing point may be developed from the meristem above the younger leaf primordia. Restitution here is restricted to meristematic cells and is certainly hormone controlled. In other cases, older and fully grown mature cells may return to meristematic conditions and thus tissues may also be regenerated from mature, parenchymatous cells. When water-conducting xylem vessels are sometimes interrupted, new xylem vessels may be formed from the neighbouring thin-walled cells to re-establish connection.

The phenomenon of restitution has great economic importance in horticulture. Plants of horticultural importance may be rapidly multiplied from cuttings without the aid of seeds. In *Begonia* leaves, the origin or restitution of shoots are commonly observed.

Correlation

If roots or shoots are decapitated (the apex is removed) they fail to show any tropic response. If, however, they are exposed to the stimulus, e.g., light, gravity, etc., and then decapitated (before any positive or negative curvature can occur), the stumps show a definite response. Thus the tip is necessary for the perception of stimulus, (of light, gravity, etc.) but not for the response, which takes place further down the tip. This control of the response of one part of a plant by another part (sensitive tip) is known as correlation.

Differentiation

Cells in different organs of a plant though apparently different, are identical in the sense that they are totipotent and each cell theoretically, having the same set of genes as an embryonic cell, is capable of giving rise to a complete plant. When a stem cell of a cutting develops into the root initials or when a leaf cell in *Bryophyllum* develops into a bud capable of developing into a new plant, this becomes evident. The reason why morphology of the different organs of a plant composed of cells containing the same number of chromosomes, having the same nucleotide sequence in its DNA and hence the same genetic information is so different, is that the DNA is differently repressed in different organs. When a young tissue section is transferred to a medium containing all nutrients essential for maintenance of life activity (White's or Murashige and Skoog's mediun), an undifferentiated mass of cells (a callus) grows and will continue to grow indefinitely on such a medium if transferred at regular intervals. However, if the ratio of some nutrients in the medium, particularly hormones like auxin and cytokinin, are altered, differentiation may start and plantlets may be produced, which may later grow into normal plants capable of flowering and fruiting. However, all plants do not respond similarly; thus, while some may respond to kinetin, others prefer zeatin and still others may need benzyladenine. In the cases of auxins also IAA, NAA and 2,4-D are not always equally effective. There are also plants which do not respond to any such treatment. While these studies indicate that hormonal substances play very important roles in differentiation, we have to know a lot more about their interaction and the cell components concerned with differentiation.

In recent years single cell and protoplasts are used extensively in probing the secrets of differetiation and other life processes. Success has been claimed with respect to fusion of protoplasts belonging to same and even different species or genera—although regeneration of intact plants has been difficult. Unicellular algalc ells, plasmids and isolated DNA molecules have been incorporated in protoplasts and plant cells. It has been claimed that the *nif* gene (responsible for nitrogenase activity) has been introduced in eukaryotic cells, but such cells have not survived through several generations. The cause of this failure is not known. Obviously many hurdles have to be crossed before such efforts become successful. We can certainly be optimistic, since progress in this field is very rapid.

CHAPTER XLVII

PLANT MOVEMENTS

Movement is not usually associated with plants and least of all with trees or shrubs. If we think of movement in the narrow sense of an entire organism translocating, then we find very few instances in plant world. Perhaps, the best known cases are when a unicellular alga such as *Chlamydomanas* moves towards light or when spermatozoids move towards the egg cell by means of flagella. Many locomotions are, however, automatic and do not depend on any external stimulus such as shown by plasmodesmata of *Myxomycetes* or amoeboid movements which are really creeping movements due to some internal stimulus. Flagella or cilia are locomotive organs, whip-like in shape which the organism moves to and fro by a process of contraction and expansion which recalls, but is really quite different from, the muscular contraction of animals. Locomotions of this type are shown by ciliated plants, gametes and zoospores while the plasmodia of *Myxomycetes* display creeping, amoeboid movement.

If on the other hand we think movement in the sense of a part of an organism moving due to unequal growth rates of the opposite side, then the examples in the plant world are numerous. Parts of the plants, their roots, their leaves, their stems all are capable of movements and do in fact move continuously.

Plants possess *irritability* and execute movements in response to a variety of external stimuli which act on protoplasm of the cell. Such movements can only be *induced* because the stimuli act on an excitable protoplasm. Some purely mechanical movements, which have nothing to do with protoplasm, are of common occurrence in plant kingdom as when sporangia or fruits dehisce in the dispersal of spores and seeds and in the coiling or uncoiling of elaters as in *Equisetum*. These are purely *hygroscopic movements*, largely determined by the presence or absence of water.

It is often customary to refer to autonomic movements. These are movements which are apparently independent of the surroundings of plants, i.e., of any external stimulus. A rough analogy in a man would be the continuous beating of heart which is quite independent of the environment as opposed to the response of the hand which withdraws when it is scorched. In the latter case, an external stimulus causes the movement. Perhaps the best known example of autonomic movement in plants is the movement described by the apex of the stem of many plants. In plants such as peas or beans, it will be observed that the stem apex regularly describes a circular, spiral type of growth movement. This circular movement which is quite independent of any external conditions is always in the same direction for the same plant species. The more rapid rate of growth travels around the tip which, as it grows upwards, must therefore rotate. The immediate cause is that the growth on the different sides of the stem differs. First one side grows faster, then the other, and this goes on in a continuous regular alternation. This type of movement is referred to as *nutation*, preferably *circumnutation*. Similar type of movements are shown by the tendrils of many plants.

The second type of movement which we usually meet in plants is that which takes place in response to changes in the enyiromental conditions of the plant. The bending of aerial shoots grown near a window towards the light, the downward growth of roots, etc., all these exemplify the responses to changes in the surroundings. These are referred to as *tropisms* because there is a directional relationship between the response and its causative effect—response is either towards the stimulus which is acting *unilaterally* from one direction only or away from it or at a definite angle to the direction of the stimulus. There is another kind of movement (*nastic movement*) in which the stimulus is diffuse, i.e., it does not affect the organ from any definite direction and the movement, thus induced by the stimulus, is determined exclusively by the properties of the irritable cells. The response of the organ here is independent of the direction of the stimulus. Examples are afforded by many flowers which open when temperature rises and close when it falls. Pulvinar movements such as shown by the sensitive leaflets of *Mimosa pudica* are always *nastic* in nature and determined by the properties of the pulvini.

The environmental factors responsible for the growth difference are definitely known and is spoken of as the *stimulus*. The length of time to which the plant or the plant organ must be continuously exposed to a given intensity of stimulus for a visible response to follow is called the *presentation time;* the period from the commencement of stimulation to the commencement of the reaction or response is termed *reaction time* and the time it takes for the plant to recover its original position, after the stimulus has been removed, is the *relaxation time*. An organ may also perceive a stimulus but may be unable to react to it.

Movements can be classified in various ways. The classification which the biologist employs is largely a matter of convenience by which certain groups of plant behaviours can be related to each other. The differences in terminology are again based on convenience rather than on principle.

Classification of plant movements

A. Purely mechanical *movements unrelated to irritability*

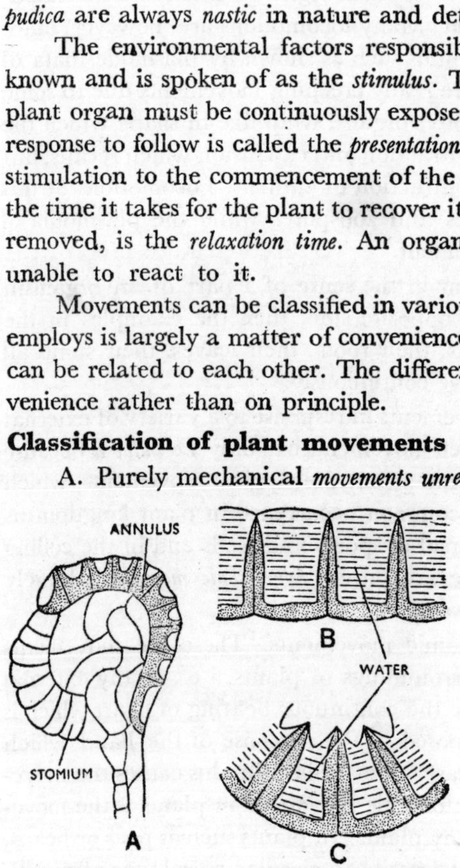

FIG. 750. Dehiscence of a sporangium of fern. A: A sporangium showing annulus and thin-walled stomium. B: Cells of sporangium in original position. C: On exposure to dry air, the cells of the annulus lose their contained water. The watery contents do not, however, separate from the cell walls nor is there a rupture in the liquid since the adhesion to the wall and the cohesive forces between water molecules are very great indeed (about 300-350 atm.). Hence, a deformation of the cell wall, as seen in C, follows with diminishing water content of the cells; the thin outer wall is pulled inwards and the one-sided shortening of the annulus thus produced leads to the opening of the sporangium near the stomium.

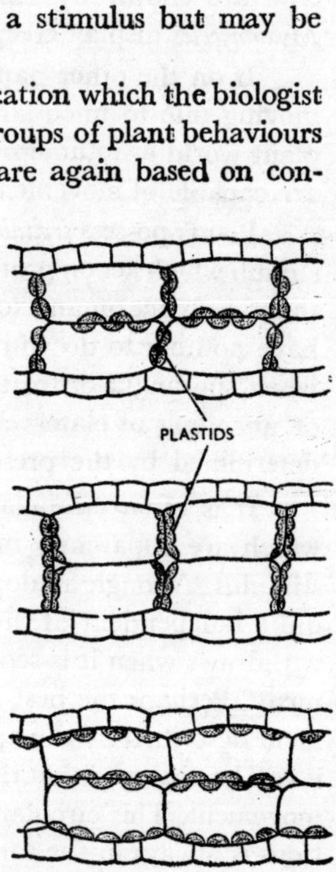

FIG. 751. Phototaxis of chloroplastids in green living cells of *Lemna*. Varying positions of the plastids due to the difference in the intensity of light. The upper figure shows positions of plastids under diffuse daylight (shade conditions). The middle figure, in direct sunlight (note aggregations of plastids along lateral walls only, thus preventing injury). The lower figure shows position at night.

of the protoplasm—movements associated with dehiscence of sporangia (e.g., fern; Fig. 750), with explosive bursting of many fruits, dispersal of spores and seeds, movements of elaters (e.g., *Marchantia* and *Equisetum*) and so forth.

B. Movements as are *induced* when stimuli act on irritable or sensitive protoplasm.

(1) Movement shown by freely motile organisms and free parts of fixed plants. *Movement of locomotion.*

(2) Movement shown by attached organs of fixed plants. *Movement of orientation of curvature.*

Movement of locomotion:

Autonomic or spontaneous locomotion. Movements induced by internal stimulus.

Examples are offered by autonomic movements of some motile algae, fungi, spores and gametes; *movements of protoplasm inside living cells of higher plants; the movements of chromosomes during nuclear division.*

Tactic locomotory movements or Taxes. Movements induced by external stimulus.

Examples are *Phototaxis* of phototropic bacteria and motile algae or of chloroplasts in the cells of green plants (Fig. 751); *Chemotaxis* [movements of spermatozoids towards the ovum induced by chemical substances of definite nature, e.g., malic acid in fern (a conc. of 0·001%), cane sugar in moss, citric acid in lycopods, albumen in *Marchantia*, etc.]. On the other hand, many bacteria move away from or are repelled by substances such as acids and alkalis (negatively chemotactic). If chemotaxis is determined by oxygen, it is termed *Aerotaxis*. *Thermotaxis*; *Rheotaxis* (due to difference in water currents); *Galvanotaxis* (due to difference in electric potential) and so forth.

(2) *Movements of orientation or curvature*

Autonomic or spontaneous movements (movements induced from *within* by internal stimulus).

Paratonic movements (movements induced by *external* stimulus).

Autonomic or spontaneous movements:

Growth movements, e.g., growth itself; *nutations*; *epinasty* and *hyponasty*.

Movements of variations, e.g., pulvinar movement of the lateral leaflets of telegraph plant (*Desmodium gyrans*); terminal leaflets of red clover.

Paratonic movements:

Direction of response bears a definite relation to the direction of the stimulus, generally either towards it or away from it. Stimulus acting on the protoplasm from *one* direction only.

Tropic movements or Tropisms. Phototropism; Geotropism; Hydrotropism; Chemotropism and *Haptotropism* (or *Thigmotropism*) of shoot and root systems.

Response unrelated to the direction of the stimulus, e.g., difference in the intensity of light, temperature, etc. Movements confined to dorsiventral organs. Direction of movement always depends on the reacting organ and not on the environment. Stimulus acting on the protoplasm *from all sides*.

Nastic movements or Nastis. Photonasty; Thermonasty; Haptonasty (*Thigmonasty*) and *Seismonastic* pulvinar movements of leaves of *Mimosa pudica* and other sensitive plants.

Tropic movements

In case of tropic movements, as we have seen before, the direction of movement shows a definite relation towards, away from or at a definite angle, to the direction of the stimulus.

The curvature developed by an organ, when its preferred orientation with respect to an environmental factor is disturbed, may be called a 'tropic response'. The environmental factors only act as triggers which alter the pathway of metabolism in such a manner that the energy-delivering system of metabolism enables the reponse to take place. The environmental or the external factor is called a *stimulus*—the stimulation process ieself, i.e., the very earliest changes inside the organism which are brought about by the effect of the external factor is still little understood. The processes could be subdivided into susception and perception (reception), i.e., purely physical processes and the immediately following physiological ones respectively

A movement towards the stimulus is called a *positive* movement, away from the stimulus, a *negative* movement. The terms *geotropism*, *phototropism*, *hydrotropism*, *chemotropism* and *haptotropism* (or *thigmotropism*) describes movements induced by external stimuli such as gravity, light, water, chemical substances and contact or touch respectively.

GEOTROPISM

The stimulus of gravity must be unilateral since the force is exerted from one direction only. Primary roots are *positively geotropic* (i.e., bend towards the centre of the earth), primary stems are *negatively geotropic;* as a matter of fact all vertically upward-growing organs, whether stems, leaves, flower stalks, parts of flowers and even roots, such as the respiratory pneumatophores of many halophytic plants, are certainly negatively geotropic. When such organs are forced out of their upright position, they may assume it again if they are still capable of growth. Secondary lateral roots and shoots show a weaker response or no response at all and take up a position at an angle to the gravitational force. A spreading fibrous root system as in monocotyledons is indicative of very feeble geotropic response. Secondary lateral roots are said to be *plageotropic* and are directed downwards and laterals of higher orders, both roots and shoots are practically insensitive to the stimulus of gravity (*apogeotropic*). Rhizomes or runners below ground or on the surface, grow horizontally involving only a positional response, that is they take up a position at right angles to the direction of gravity (*diageotropic*). The adventitious,

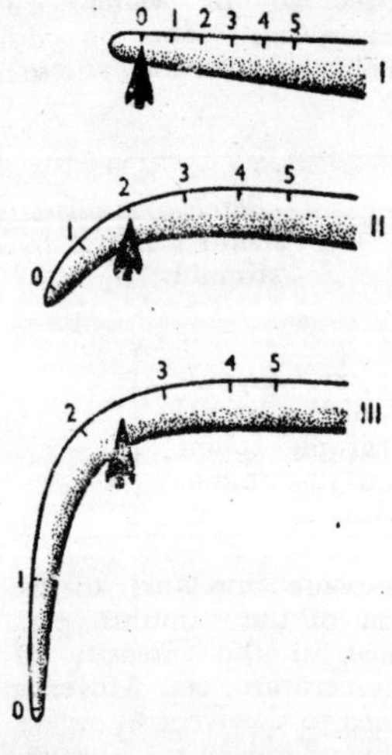

FIG. 752. Course of a geotropic curvature in roots. I: root placed horizontally and divided into 5 equal divisions. II: after more than six hours and III: after twenty-four hours. The arrow signifies a fixed index. Note that bending down from horizontal position occurs at an angle of approx. 90° at a short distance away from the apex, at the region of the greatest growth.

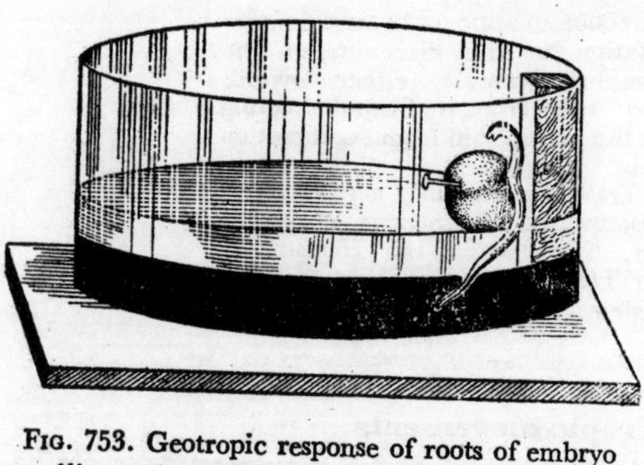

FIG. 753. Geotropic response of roots of embryo seedlings. The radicle goes down and penetrates against considerable pressure exerted by mercury in response to the stimulus of gravity.

pneumatophores of halophytes are negatively geotropic and negatively hydrotropic but seem positively *aerotropic.*

The response of an organ to the gravitational stimulus may change with the stage of development. The peduncle of a poppy flower bud is directed downwards, i.e., positively geotropic, but it gradually becomes negative as the flower opens.

The first sign of geotropic response occurs at a short distance away from the apex (Fig. 752), except in monocotyledons, in which case it occurs at the nodes where the meristems are found.

The geotropic responses are of universal occurrence. When an entire plant is placed horizontally its stem will quickly go upwards and its roots downwards. It was at one time thought that the downward curvature of horizontally placed roots resulted solely from their own weight. But it is now cell established that geotropic curvatures are possible only through growth. Roots when they execute positive geotropic movements penetrate mercury, (Fig. 753) a liquid in which dead roots float. That gravity is the operative stimulus in geotropic curvature can be easily shown if we expose all sides of the plant

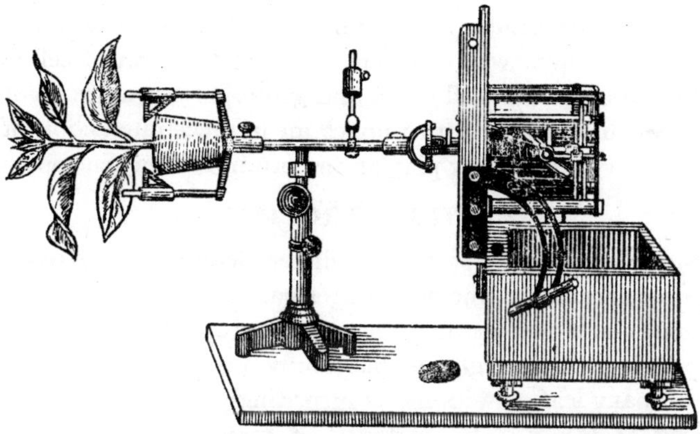

Fig. 754. Clinostat by which the horizontally placed potted plant is rotated in the vertical plane around a horizontal axis. The effect of force of gravity on the plant organs is eliminated by rotation (1-4 per hour) of the pot by the clockwork mechanism.

equally to gravity. This is done by placing a plant horizontally on a wheel which slowly revolves in the vertical plane around a horizontal axis in an apparatus known as *clinostat* (1–4 rotations per hour). In this way, the plant is slowly turned so that each side in turn is exposed to gravity to the same extent (Fig. 754). When this is done, the plant or its parts no longer shows any upward or downward curvature. It just grows as it is placed—straight along (Fig.755).

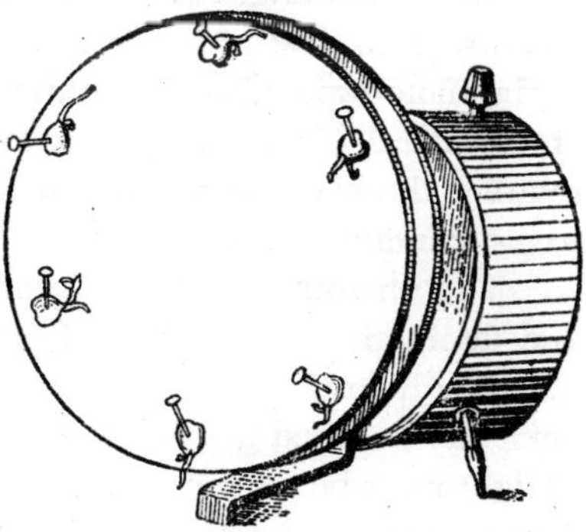

Fig. 755. A clinostat rotated along a horizontal axis with a number of seedlings attached to the wheel. Note that due to the elimination of the forces of gravity, the seedlings grow as they were placed.

If, however, germinating seedlings are fastened in all possible positions at the periphery of a wheel and the wheel rotated very rapidly on a horizontal axis, it is found that the roots grow radially away from and all the shoots radially towards the centre of the wheel. As effects due to gravity were eliminated due to rapid rotation (at the same time considerable centrifugal force was produced) the centrifugal force determined the curvature of the seedlings, as gravity does normally.

The geotropic and phototropic movements are due to unequal growth of the two sides of the organ which is again caused by an unequal distribution of native auxin. The gravitational stimulus must be of certain minimum length of time and intensity in order to cause the response. The exact nature of perception of gravitational stimulus is, however, less clear, but attempts to explain the pheonomenon in terms of *sensory cells* (*statocysts*) and *statoliths* have achieved remarkable results during the last few years.

Many plants contain in certain cells of root tips, large starch grains (*amyloplasts*) which move under the effect of gravity within these sensory cells—statocysts. These movable starch grains were for a long time (since 1858) supposed to be receptors of gravitational stimulus or impulse. Recent observations during the last decade completely confirm the idea inherent in the classical works of Haberlandt and others (1900). Starch grains or amyloplasts normally form an integral part of the graviperception mechanism in plant organs and the basic concept of statolith theory is now generally accepted. However, mitochondria and dictyosomes also respond to gravitational stimulus and the same is also true for other cellular microbodies. Since dictyosomes and vesicles formed by them seem to be in some way connected with metabolism of cell wall the components and the formation of cell wall itself, and geotropic curvature involves differential auxin-induced growth which in turn implies an indirect interaction of auxin on cell wall components, dictyosomes may play an important role in geotropic curvature.

PHOTOTROPISM

As a rule, stems are *positively phototropic* whereas leaves are *plageophototropic* or transversely phototropic (Fig. 756). The advantages gained by these responses are that assimilating parts are brought to a position where light is plentifully received. Study of many leaf-mosaics shows surprisingly little overlapping of leaves. The majority of roots are *insensitive* to light. Only in a few species, one of which is white mustard (*Sinapis alba*) do the roots show the reaction one might usually expect, namely, negative phototropism. The peduncle of some flowers are positively phototropic after fertilisation is over. The effect of unilateral light stimulus may be neutarlised by rotating the plant on a vertical clinostat. No phototropic curvature then occurs.

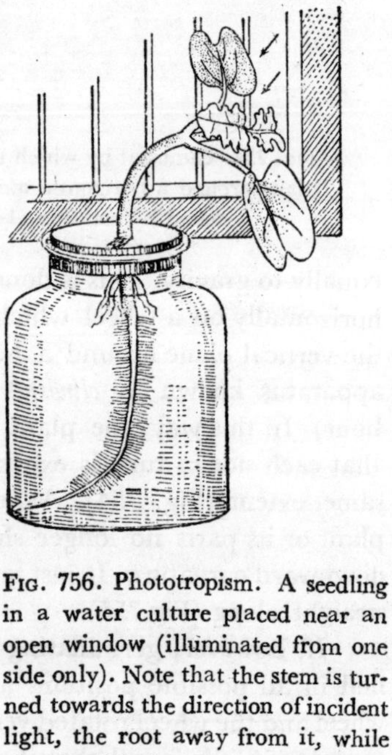

FIG. 756. Phototropism. A seedling in a water culture placed near an open window (illuminated from one side only). Note that the stem is turned towards the direction of incident light, the root away from it, while the leaf-blades are expanded at right angles (transversely phototropic) to the direction oflight).

The phototropic phenomenon is a very intricate and fascinating mechanism. The reaction to light is certainly quantitative. A certain minimum of light is definitely required to cause a response. This minimum of light must be given for a certain period of time. There seems to be a strict relationship between the period of illumination and the light intensity. Illumination causing a definite known response in plant as shown by the bending of a stem or a coleoptile by a measurable number of degrees, will in every case be the same if the illumination is for 100 seconds with

10-foot-candles or for 10 seconds with 100-foot-candles as with 1,000 seconds with one-foot-candle. Thus a given amount of energy produces a given amount of phototropic curvature and the product of the light intensity and the period of illumination needed to bring about a given phototropic response is constant. Time period, however, cannot be reduced indefinitely; there is a minimum *presentation time* during which the light must be given to the plant for it to be perceived. However, both the minimum time and the minimum light needed to cause response are very small. The time may be as little as fraction of a second and the minimum amount of light required not much more than the light from an ordinary torch (one-foot-candle).

These clear-cut and simple light responses do not, however, hold good when high light intensities are reached and then there is no longer a straightforward relationship between response and intensity of illumination as indicated above. Thus if a coleoptile tip is illuminated for 1 min. with 1,000-foot-candles, we certainly do not get 10 times the response given by 100-foot-candles for 1 min. nor again does the latter give 10 times as much as 10-foot-candles for 1 min. In fact the response in the form of the extent of curvature becomes progressively smaller as the light intensity rises till saturation point is reached. There is then no further increase in the response by increasing the intensity of light stimulus. Indeed if the light intensity is very high, the plant or the coleoptile tip till will move *away from* the light and not towards it.

As in the case of geotropism, it is easily shown that the region of perception and the region of response to the light stimulus are not the same. The sensitive zone is the first 1-1·5 mm from the tip.

In the phototropic response of dicotyledonous leaves the question of perception of stimulus is interesting. In the peltate leaf of *Nasturtium*, lateral illumination of the whole leaf produces a strong positive curvature of the petiole. Removal of the lamina, drastically reduces the extent of the response. On the other hand, screening of the lamina with black paper hardly impairs the positive curvature of the petiole while the lateral illumination of the lamina with only the petiole shaded, produces no curvature at all. Thus it is apparent the light stimulus is perceived by the *petiole only*. The leaf blade appears to act as a natural source of auxin for the petiole, enabling the petiole to carry out the growth curvature.

The wave-lengths that induce phototropic response towards a unilateral light are certainly not identical with those involved either in photoperiodism or photosynthesis. The maximum, as we have seen before, is found in the blue, in the neighbourhood of 450 nm with a minor peak at 480 nm. Phototropic curvatures, however, have been observed in response to much shorter wave-lengths, far into violet. The photoreceptive pigment is probably a flavone derivative, e.g., riboflavin (vitamin B_2) or it may conceivably be β-carotene, or some other carotenoid.

EXPLANATION OF PHOTOTROPISM AND GEOTROPISM BY THE HORMONE CONCEPT

If a coleoptile tip is exposed to unilateral light, it results in an unequal distribution of hormones on the two sides of the tip and it can easily be shown, by collecting the hormone diffusing out into agar blocks from the two sides, that the dark side always has more hormone than the side which is illuminated (see Figures in pages 654 and 655).

It seemed pretty obvious that light causes the destruction of hormone on the illuminated side of the plant. If this is so, then in the tip of the oat coleoptile or for that matter in any sensitive region illuminated hormones appear in different amounts on opposite sides of the tissue. Other possibilities of this unequal distribution of hormone in different amounts on the two sides are: (*i*) that there may be a light-induced transverse migration of hormone from the lighted side to the darker side or (*ii*) the sensitivity of the tissue on the lighted side to hormone may be decreased. In any way, in an illuminated coleoptile, the darker side certainly contains more hormone than the lighted side. The hormone then diffuses downwards and reaches the zone of response in unequal amounts, causing different rates of cell elongation which in turn causes curvature to take place.

For light to be effective in any system like this it must first be absorbed to bring about a photochemical reaction, leading to the destruction or redistribution of auxin. Since hormones are colourless compounds, it is clear that hormones cannot be the absorbers of visible light, which is effective in producing phototropic curvature. It has definitely been established, we know, that not all light is equally effective in causing phototropic curvature. Blue light is very effective in causing the movement, while red light is almost ineffective, the response to other colours of light being intermediate of the two. Of the pigments which are present in the oat coleoptile only two are so far known to be active in the absorption of blue light. They are both yellow pigments, vitamin B_2 (riboflavin) and carotene. The candidature of both are strong. The action spectrum (the curve relating action, i.e. phototropic curvature to wavelength of light) of phototropism exhibits peaks at about 360 nm and 445 nm with shoulders or minor peaks at 425 and 472 nm. Riboflavin does have an absorption maximum at about 360 nm, but β-carotene has no absorption in this region; on the other hand the absorption peaks in the region 400 nm and 500 nm are shown by β-carotene whereas riboflavin has only one broad peak at about 460 nm. The absorption spectrum of none of the pigments thus corresponds reasonably with the action spectrum of phototropism. However, the absorption spectra of isolated pigments may not be identical with the native bound form and thus, either of the pigments or both acting co-operatively may be the photoreceptor pigments. However, it should be pointed out that a strain of the fungus *Pilobolus* produces sporangiophores (Fig. 757) which contain no carotene at all but these sporangiophores which are completely carotene-free appear to have retained their full phototropic responsiveness; they seem indeed even more sensitive than the normal ones with carotene. Other flavins may also be involved.

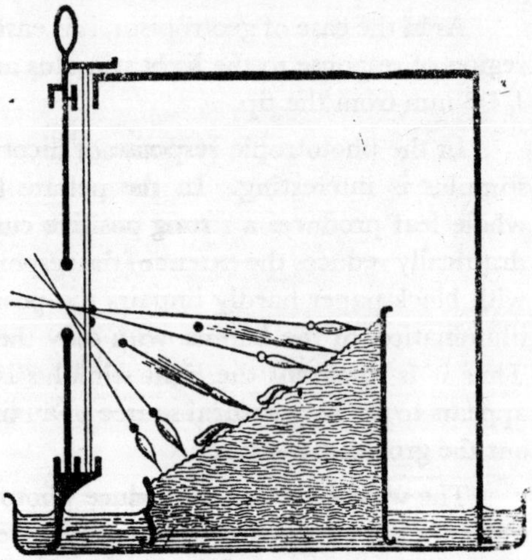

FIG. 757. Sporangiophores of *Pilobolus*. Note that the sporangiophores are positively phototropic and have all turned their black sporangial heads towards the source of light (light is only admitted through the centre of the glass over a small aperture) as shown in the figure. Note also that the sporangia are being shot away from the sporangiophores and the liberated sporangia are collecting only near the aperture through which light is allowed entrance.

It is now generally believed, and there is good evidence for this, that the phototropic curvature is due to lateral transport of auxin from the lighted to the shaded side, and not due to any marked destruction of auxin. *In vitro* studies however, show that riboflavin may catalyse photolysis of IAA.

Photosensitiser substances are of frequent occurrence in the living cells including plant cells. Riboflavin absorbs blue light and the light energy thus absorbed by the photoreceptor pigment for the phototropic reaction, is thought to be utilised for the photochemical destruction of IAA in the lighted side thereby bringing about the unequal distribution of hormone in the two sides of the coleoptile.

What about the other yellow pigment, carotene, whose absorption spectrum considerably coincides with the action spectrum of phototropism? Can it bring about a similar photolysis of IAA as riboflavin? Available evidence indicates that it is not so. Using aqueous mixtures of IAA and colloidal solutions of carotene or using true alcoholic solutions of both, no photolytic destruction of IAA could be discovered in visible light. Another possible effect in a different direction which could be produced by the presence of carotene in the system has been examined and the most impressive discovery was that the addition of carotene to the reaction system, IAA+riboflavin, almost completely inhibits the photocatalytic action of riboflavin on IAA. This protective effect seems to be chiefly due to the similarity of the absorption spectra of both the yellow pigments in the critical region where they largely overlap. Carotene may also act as an internal light-filter screening off the most effective wave-length necessary for photo-inactivation of IAA by riboflavin thereby causing a suitable drop in the light intensity across the coleoptile. It seems certain that there is a strong competition for light between riboflavin and carotene and if sufficient light is diverted for absorption by carotene, riboflavin can no longer mediate the destruction of hormone. Without carotene, light would reach both sides of the coleoptile in equal amounts as the coleoptile sheath is only about 1·5-2 mm thick. Therefore, the presence of protective carotene filter in the coleoptile could reduce the transverse light gradient, thereby progressively increasing the hormone content from the lighted to the dark side.

Many of the observations which have been described for phototropism are also true for geotropism. Thus, the geotropic movement is due to unequal growth which is caused by an unequal distribution of hormones. If an *Avena* coleoptile tip is placed horizontally, the lower part is found to contain as much as 66% of the total hormone content of the tip (see Figures on page 655). But why should gravity cause such an unequal distribution of auxin? It is worth pointing out here again that reaction of roots to hormones differs from that of stems and the important thing to remember is that while in the stem, the side containing most hormone grows fastest, the opposite is the case in the root. Thus it must be postulated that the reaction of tissues to particular concentration of hormone is different in the two types of tissues, the shoot and the root. Whereas a particular concentration of hormone induces acceleration of growth in the shoot end, the same concentration actually produces an inhibitory effect in the growth and enlargement of the root cells. This explains why when a plant is placed horizontally, the stems turn up and the roots down even in absence of light. The hormone in this case concentrates on the lower side of the plant (actually growth inhibitors have been identified by chromatographic separation from the upper part of the horizontally placed mesocotyl—the nodal zone between the grain and the coleoptile of maize) resulting in more *rapid growth of the cells of the lower side of the stem tissue* causing them to bend upwards and *faster growth on the upper side of the root tissue*, causing roots to bend downwards. Hence the overall geotropic response is obtained. It has been suggested that the reason for the unequal distribution of auxin in the horizontally placed plant is partly due to the creation of a potential difference across the plant, the lower side becoming positive. This may cause an accumulation of hormone on the lower side of the plant because of the polar flow towards the positive pole, i.e., lower side.

HYDROTROPISM

In certain localities, soils do not contain water uniformly; some parts may be wetter than others. Primary roots tend to grow towards the wetter regions. The positive

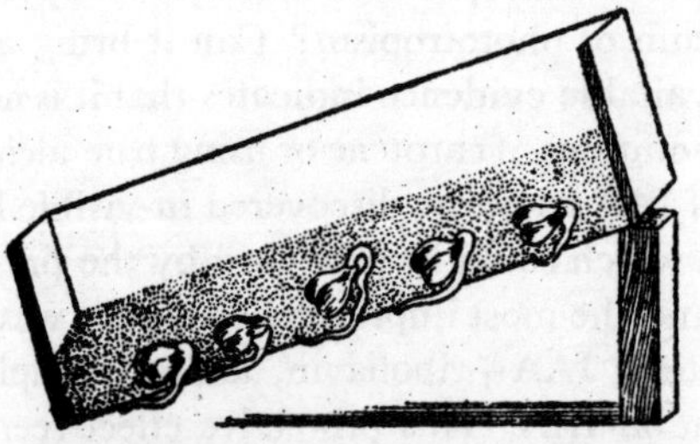

FIG. 758. Experiment on hydrotropism. Germinating seeds growing in moist sawdust placed in an inclined plane in a box. Note the bending back of the root from the vertically down positions and their growth along the surface of the moist sawdust.

hydrotropism of roots may be demonstrated by growing plants in moist sawdust in a shallow sieve, inclined to the vertical (Fig. 758). Hydrotropism of roots can also be demonstrated as in Fig. 759. The roots grow vertically downwards due to the stimulus of gravity until they penetrate the sieve. They then bend back by growth curvature from the dry air outside towards the moist sawdust. Apparently hydrotropic stimulus is more powerful than gravitational impulse.

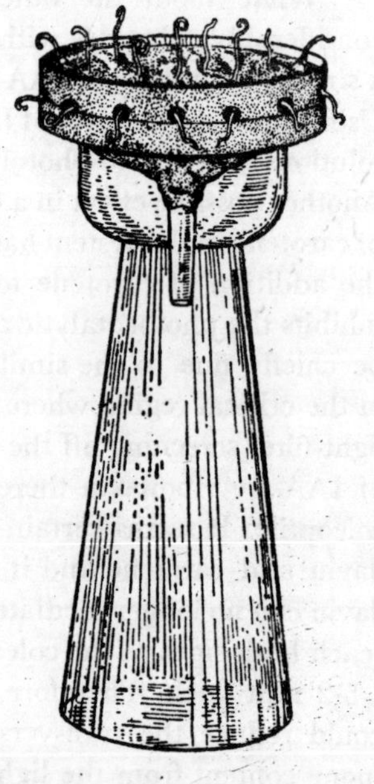

FIG. 759. Experiment to demonstrate hydrotropic movement (induced by water) of roots. The porus funnel filled with sawdust and soaked seeds are arranged in a circle / corresponding to the pores of the funnel. The roots formed at first grow vertically downwards in response to the force of gravity until they penetrate the pores. Note that the roots then bend back by growth curvature from dry air towards the moist sawdust in the funnel proving that hydrotropic stimulus is more powerful than even the stimulus exerted by forces of gravity.

CHEMOTROPISM

Chemical substances in the environment have a directive influence on the growth of certain plant structures. The classical example is afforded by the growth of the pollen tube through stigma and style towards the embryosac cell. It is presumed that chemical substances present in the carpels is the directive force for the growth of the pollen tube. Calcium and boron are the major chemical substances responsible for this movement. When pollen grains germinate in nutrient jelly in which pieces of carpel have been sown, generating pollen tubes grow away from the air and move towards the pieces of carpel; the pollen tubes show *negative aerotropism* and *positive chemotropism.* The first orientation of germinated pollen tubes into the stigma takes place by hydrotropism, followed by a mechanical conduction all along the slimy way. For the final part of the movement of pollen tube through the style, there is no doubt that chemotropic orientation is responsible (Schneider, 1956). The sucking roots of parasites and hyphae of parasitic fungi penetrate into the tissues of the host plant in response to the stimulus of perhaps other chemical substances contained in them. Movements of individual tentacles of *Drosera* in response to various chemical substances, both inorganic and organic (in the form of insects), placed in the leaf blade are also examples of *chemotropism*. A portion of the chemical substance is absorbed by the leaf, protoplasm is stimulated and a semi-motor impulse goes from the leaf to the bases of surrounding tentacles which bend down on the substance or the insect on the leaf blade. Thus glandular heads of the tentacles are brought in contact with the insect and the organic matter can be digested and assimilated by the plant with the help of excreted enzymes.

HAPTOTROPISM OR THIGMOTROPISM

The petioles of many species of plants, the leaf tip of *Gloriosa*, specialised tendrils of garden pea, the stipules of *Smilax*, the branch of vine, etc., are sensitive to contact with the uneven surface of solid bodies and execute haptotropic curvatures. One of the most remarkable facts is that these structures while sensitive to contact with the surface of a glass rod, is *insensitive* to that of a glass rod coated with gelatine gel (i.e., a colloidal emulsion). Raindrops, again a liquid, seem to inhibit haptotropic curvature. It appears that haptotropism results from compression of cells on the stimulated side (i.e., the side in contact with the foreign body) of the sensitive structure while the growth of the opposite side continues. A short curvature is thereby produced and the sensitive structure coils round the support with which it makes contact (Fig. 760).

Though the movement of the individual marginal tentacles of *Drosera* towards the body of the insects has rightly been described previously as chemotropic movement since the inducing stimulus has a directional influence, the first movement of a marginal tentacle towards the small central tentacles, on the other hand, can only be regarded as *chemonastic* or even *haptonastic*. This movement is exclusively determined by the properties of the tentacle for the stimulus exerts no directive influence.

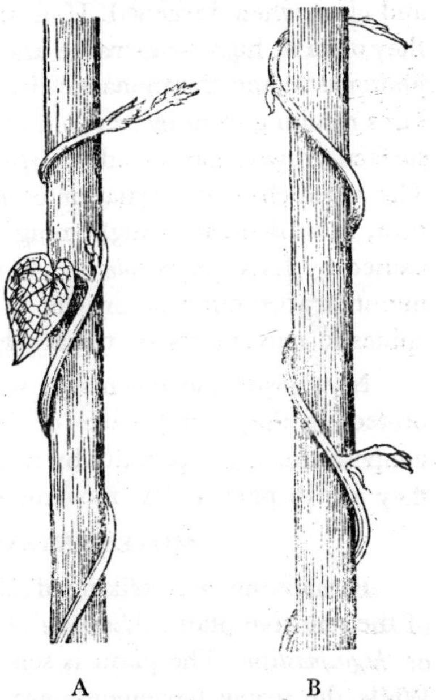

A B

FIG. 760. Haptotropic (twining) movements executed by plants. A. Left-handed climbers —plant circles to the left—anticlockwise. The spiral which is formed around the support mounts from the left to the right. B. Right-handed climbers—plants circle to the right, in the direction of the hands of a clock—clockwise.

NASTIC MOVEMENT

When an external stimulus induces a certain process or influences its intensity without attaching any significance to the direction of the stimulus, we have *nastic* movements. In nastic movements, the direction of curvature is, as we know, *morphologically pre-determined*. As in the case of tropic movement, nastic movements can be brought about either by unequal growth on the opposite halves of the organ or simply by turgour changes.

Diverse forms of biological advantages may result to the organism from *nastic* movements. The *chemonastic* movements shown by insectivorous plants lead to the assimilation of nitrogenous food.

The commonest of nastic movements are *nyctinastic* movements, the day-and-night movements of leaves and flowers. By nyctinastic growth curvatures, certain flowers can perform opening and closing movements. Perianths and many compound leaves may open during the day and close at night. An interesting nastic movement is observed in some species of cactus; the flowers only open in the dark. This is also true for some tobacco flowers. Some of the nastic movements depend on turgour changes rather than on differential rates in growth. Families, typically showing nyctinastic leaf movements are *Leguminoseae*, *Oxalidaceae*, *Euphorbiaceae* and also pteridophytic species of ferns and *Marsilia*. The night position of shoot is always characterised by a vertical position of the leaf blade, the petiole curving either upwards or downwards. In the day position the leaf blades stand horizontally or at right angles to the incident light.

Nyctinastic movements may be controlled by temperature and light. These individual factors acting singly may induce *photonastic* and *thermonastic* movements respectively. The petals of tulip and crocus open at a constant temperature when illuminated and close when darkened. If on the other hand, the intensity of light is kept constant, they open at high temperatue and close at low temperature. They are, therefore, both photonastic and thermonastic. In each case, the opening is the result of growth curvatures resulting from more rapid growth of the upper surface (*epinasty*). When the lower surface grows more rapidly (*hyponasty*), growth curvature resulting in closure follows. The rolled circinate vernation of young fronds of fern is due to epinastic growth curvature; its subsequent straightening up is due to hyponasty. Nastic movements are also caused by IAA (*auxin epinasty*) as well as due to the presence of certain substances in minute traces; ethylene present in concentration of 1 : 500,000 parts of air may cause epinastic movements of tomato leaf petioles.

Nyctinastic movements in some cases appear to have ecological significance in protecting the young leaves or developing flower buds from injury due to light, high temperature and especially by rain. It must be admitted, however, that in most cases, they are in part useless movements.

MOVEMENTS OF LEAFLETS OF MIMOSA PUDICA

Perhaps the most striking of all plant movements is that of the *seismonastic* movements of the sensitive plant, *Mimosa pudica*. This is also sometimes referred to as *haptonastism* or *thigmonastism*. The plant is sensitive to touch, impact, electrical impulses and heat. While the tropic movements are generally, slow taking many minutes, the response of the sensitive plant to any form of shock is almost instantaneous. *Mimosa* has compound leaves made up of leaflets which in turn are made up of pinnules. When the plant is stimulated by touch, the pinnules begin almost immediately, within seconds, to fold up. If the stimulation is vigorous, this is followed up by drooping of the whole leaflets at the junction of the petiole. When a really vigorous stimulation is provided, as for example applying a match to the tip of one of the pinules, the petiole itself droops at its junction with the stem and this is followed by the closing and drooping of all the neighbouring leaves till the leaves of an entire branch have completely folded up. The whole process is extremely fast and within a minute all the leaves of the branch will have closed. The really important thing is the sequence of closing; in the stimulated leaf, first the pinnules, then the leaflets and finally the petiole close up. In the neighbouring leaves, however, the process is completely reversed. Here the petiole drops first followed by the leaflets and only at the last stage do the pinnules fold up in pairs.

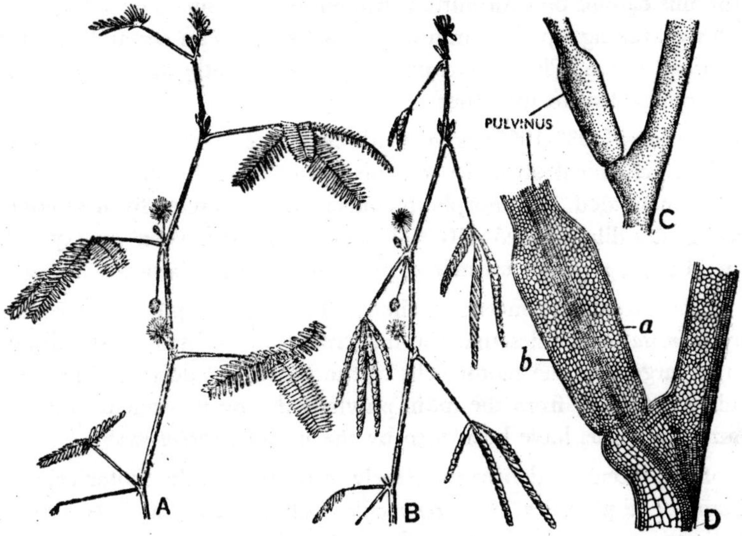

FIG. 761. Seismonasty in *Mimosa pudica*. A: a plant with leaves in normal diurnal position; B: the position attained by the leaflets on stimulation by touch or shock; C: the swollen leaf base, the *pulvinus* which is a sort of semimotor organ causing the folding up or drooping down of the leaflets; D: a section through the pulvinus. *a* & *b* represent upper and lower halves respectively of the pulvinus. On stimulation, cells of the lower half (*b*) of the pulvinus lose their previous rigidity due to excretion of water to the neighbouring cells. The pulvinus, now unable to support the weight of the leaflet or the leaf attached to it, causes the leaflet or the entire leaf to fold up.

The movement of the leaflets of *Mimosa* is affected by external conditions such as temperature. The leaflets pass into a cold rigour at low temperature and into a heat rigour at about 40°C. The leaflets lose their power of movement in continued scarcity of water and also in continued darkness. The power of movement is lost also under anaerobic conditions.

The tropic, nastic and various other movements previously described are, as we have seen, caused by differential growth. The movement of *Mimosa* is, however, very different in mechanism. The pinnules are connected to the leaf-rachis and the rachis is connected to the petiole by swollen leaf bases which are called *pulvini* (Fig. 752). These pulvini contain cells chiefly situated in the lowet half which normally are completely filled with water, i.e., they are fully turgid. These turgid cells of the pulvini, however, may readily lose their water to the neighbouring cells and intercellular spaces, as actually happens, due to shock or touch stimulus. When therefore a sudden loss of water occurs from the cells, the lower half of the pulvinus loses its previous rigidity and can no longer support the weight of the leaflet or leaf attached to it which as a result droops downdards.

In view of the extreme rapidity of the response of *Mimosa* to shock or touch stimulus, which almost suggests the nervous response in animals, it seems that a *chemical substance* is being translocated through the plant very rapidly. There is more than one way of transmitting this shock stimulus. One is very fast and passes in one or two seconds from the pinnule stimulated, to the main petiole pulvinus. This pathway exists only through living cells (primarily through phloem) of the plant and it has been proved that this stimulus cannot be transmitted through dead tissues.The second, however, is somewhat slower, taking up to a minute or two and it can certainly be transmitted through the dead tissue. The second pathway is apparently the mode of transmission from one compound leaf to the other.

Several workers (1938-1957) have made attempts at the purification and determination of the chemical structure of this stimulating substance in *Mimosa*, but so far none has succeeded. An amorphous concentrate, extracted from *Mimosa*, gives the pinna response, at a dilution, $1{\cdot}5 \times 10^{-8}$. This substance behaves as an oxyacid, containing nitrogen (4·5%), with an estimated molecular weight, between 300–450.

Most of the literature concerning the mechanism of rapid movements in plants, particularly *Mimosa*, concludes that the movements are caused by the diminishing or sudden loss of turgour in the motor cells of the pulvini or decrease in their volume or both. Expulsion of water from the main pulvinus during movement and reabsorption of water during recover, have been seen by the use of *micropotometer*.

Much of the recent evidences do clearly indicate that the motor cells in *Mimosa*, either in pinna or in pinnules, have *contractile vacuoles*, whose activity causes liquid—probably cell sap—to be expelled from the motor cells, just like the contractile vacuoles seen in *Protozoa*; this activity might conceivably result in decrease in turgour or the volume of the motor cells. This has nothing to do, however, with the increase in permeability of the plasma membrane.

In 1960, Aimi proved that the sensitive motor cells exist even in the upper-half as well as the lower-half of the main pulvinus. His conclusions were that the magnitude and direction of the petiolar movement in *Mimosa* is expressed by *two forces* antagonising each other, each of which constists of bending force, due to the *contraction of one-half* and *tissue tension in the other;* normally this force, in a downward direction, would be much greater than the upward force. Some differences between the two halves of the pulvinus can actually be observed in the morphology and the activity of the cells—the walls of the cells of the upper-half are actually significantly thicker than that of the lower.

The enzyme ATPase has been identified in the young pulvini but the significance of its presence is still unknown. It may be that the activity of contractile vacuoles or proteins depends on a mechano-chemical reaction caused by an ATP-ATPase system.

How the liquid, which has been expelled into intercellular spaces during response, re-enters the cell interior during recovery, is still unknown. An active accumulation of salts or ions, particularly potassium, has been detected in the intercellular spaces of *Mimosa* petiole and the presumption is that, ATP, working the K-pump takes part in the recovery process of the motor cells, i.e., the re-entrance of the liquid from exterior to the interior of the cells.

It must be admitted that the sensitiveness and the accompanying somewhat complicated motor movement of the leaflets of *Mimosa* do not appear to serve any obvious useful purpose to the plant.

CHAPTER XLVIII

THE PHYSIOLOGY OF REPRODUCTION

Ultimately the end of all living organism is death, either natural or accidental. In order to perpetuate a species, it must reproduce and have descendants of its own kind. It is certainly an essential criterion of life.

The main feature of reproduction consists in a portion of an individual continuing after its death and possessing the power of developing into new individuals.

The ways in which reproduction in vegetable kingdom is carried out may be divided broadly into two categories: *vegetative propagation* and *sexual reproduction.* While in unicellular and simple multicellular plants, every cell can serve for reproduction, in complex highly organised higher plants a division of labour is sometimes apparent between vegetative organs and organs of reproduction. This division of labour culminates in the formation of highly specialised forms—the flowers—which are exclusively set apart for reproduction. The subdivisions of plants into classes, orders and families are sometimes based on the diversity of structure and position of the organs of reproduction.

The formation of reproductive bodies may be considered as a stage or phase in a plant's development. Unlike animals, plants do not bear differentiated reproductive organs at all stages of their life cycle. Rather, they produce their reproductive organs at appropriate seasons in response to primarily environmental stimuli and then consummate the reproductive process fairly promptly. In the century plant, *Agave,* there is vegetative growth for twenty to even hundred years or so before flowers form. The plant immediately dies after the fruits are produced. In many plants with bulbs, vegetative growth is terminated by the formation of both sexual and vegetative structures of reproduction, flowers and bulbs or corms. In other species after a period of vegetative growth, both this and production of flowers may continue together as in some perennials and also in some annuals. In other perennials, however, there appears to be more or less definite vegetative and reproductive phases in each year's growth as in many trees which produce flowers before their vegetative growth is renewed. They usually shed their reproductive organs and produce them again in the subsequent year.

Reproduction commences as a rule when vegetative growth is slackening. It has even been said that vegetative growth and reproduction are antagonistic to each other and in general it is true that all plants must make a certain minimum amount of vegetative growth before they can make preparation for reproducing their kind.

VEGETATIVE PROPAGATION

It is commonly seen in many familiar instances among the horticulturally important plants. Among higher plants, which no longer produce descendants sexually, plants like cultivated bananas, some forms of vines, oranges, strawberries, etc. may be mentioned. With ornamental season flowers, e.g., rose, carnation, chrysanthemum, etc., the commonly prevalent practice is to remove shoots from the parent plant and to place the excised shoots in soil under conditions suitable for the formation of new roots. The modern practice of treating excised shoot ends with artificial hormones to facilitate rooting before putting them into soil has been of great economic importance particularly in plants where such rooting does not take place under normal conditions. The rooted cuttings constitute new individuals which are identical with the parent in genetic constitution. Many kinds of fruit trees, including mangoes, apples, litchies and various citrus species are propagated by *grafting*. In grafting, a bud or short piece of stem containing a bud is removed from the parent plant and its tissue is brought into contact with the exposed meristematic cambial cells of a plant of the same or related species (Fig. 753). When the graft-union is established, a new individual is produced whose upper and lower portions both retain their individual genetic constitution of the parent from which they are derived. *Trichosanthes dioica* (*Palwal*) is propagated by root cuttings. Adventitious buds formed in *Bryophyllum* leaf-margins can grow into new individuals.

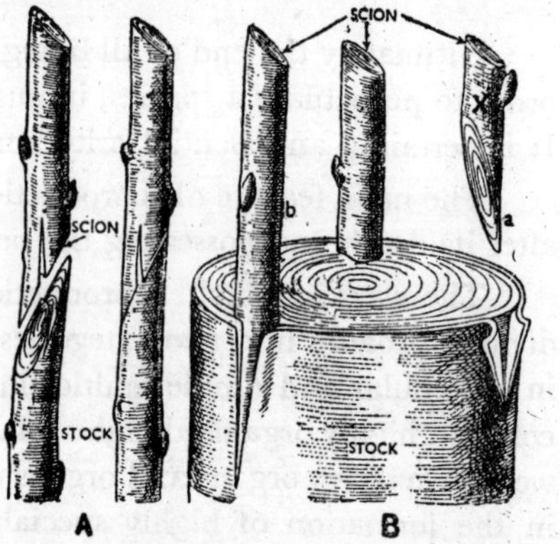

FIG. 762. Propagation by grafting. A: tongue or whip grafting; stock and scion of about same thickness, both trimmed with long sloping cuts. B: bark or crown grafting; diameter of the stock much greater than scion, b; scion already inserted in the vertical slit made in the bark of the stock; a, base of the scion trimmed with a long sloping cut.

In some plants such as *Globba bulbifera*, garlic, etc., flower-buds are modified into *bulbils* which grow into new plants when separated from the parent.

Vegetative propagation by fragmentation is of common occurrence in lower cryptogams. Formation of new individuals by budding in yeast, by simple division of the parent body in many algae and fungi, by formation of advntitious branches and by gemmae in liverworts and mosses, are well known. Vegetative propagation after perennation through unfavourable environmental conditions is a characteristic of underground stems and tuberous roots.

Thus, it is evident that plant species can be perpetuated by simple vegetative propagation and it must be noted here that in many cases no degeneration associated with purely vegetative multiplication has really been observed. Thus, if the monogenic reproduction—where the progeny is identical in genetic constitution with the parent—is sufficient to maintain species, why then the more uncertain and at the same time more complex sexual reproduction has evolved in higher plants? We do not know the precise answer to that yet as we do not know fully the chemical processes associated with the complex control mechanism of sexual reproduction.

SEXUAL REPRODUCTION

If we consider the case of the lower cryptogams such as algae and fungi, it might be supposed that sexual reproduction led to the formation of specially resistant zygotes which could endure a longer period of rest under unfavourable conditions compared to spores produced asexually. But even this is reversed in higher cryptogams such as pteridophytes where the fertilised egg cell must develop immediately after union with the sperm or they perish; here the asexual spores are adapted to endure unfavourable habitat conditions.

It is the rule in higher plants that sexual cells are individually incapable of further development unless they unite in pairs. Why are the individual sexual cells arrested in their growth if no union with its opposite number takes place? Can we say that sexual union removes the arrest of growth and the independent development of sexual cells which was evident before they united, for the united zygote is certainly potentially capable of developing and growing into a new individual? It is true that in the gradual loss of the capacity of independent development of sexual cells, the probability of fusion between pairs was increased? And this union certainly makes the new individual produced from it, different from either of the sex cells for the progeny is formed by the fusion and mingling of characters (genes) of both.

Sexual reproduction certainly brings about a qualitative change in the growth of the plant, a change from the vegetative to the reproductive state.

In higher plants, flowering is controlled chiefly by two environmental factors,e.g., length of the day and the temperature. We have discussed before the striking influence of the length of daylight period on the flowering of plants (*photoperiodism*).. In the short-day and long-day flowering plants, the effects of temperature seem to be closely related with photoperiodism. In plants which flower in short-day lengths, the response to flowering is primarily to the length of dark period. During the course of dark periods the initiation of the qualitative change from the vegetative to reproductive phase is set into motion—ultimately the vegetative buds are converted into flower primordia. As we have seen before, this alteration in habit is theoretically possible through the production and export of one or more hormonal substance, which brings about this conversion. Plants which flower in long daylight periods of summer do not flower in short days; in such plants, long dark periods appear to produce some substance or substances, which being antagonistic to flowering, inactivate the hormone essential for the production of flowers. Flowering response of the plant to the long dark periods is essentially conditioned by the prevailing temperature of long nights. Thus flowering of long-day plants is favoured by short nights. Still other species of plants, e.g., winter cereal varieties, flower only after their seeds or the vegetative buds have received an exposure to low temperature (a range of 0·–5°C) and the appropriate length of day necessary for their flowering (when the temperature treatment is given artificially it is called, *vernalisation*). Tomatoes appear to flower over a wide range of daylengths. They do not seem to have a critical photoperiod.

The initiation of the flower primordia is but the first stage in the sequence of reproductive development. As is well known, this is followed by the growth of the flower, pollination, fertilisation, growth of the fruit and the development of seed. If any one of the sequence of events fails to take place in strict order, the sexual reproduction is totally ineffective. Sometimes, even if all the steps follow one another in the right order, sexual union may not take place. This may be due to incompatibility and incompatibility is determined by some internal factors, e.g., *Rh* factor of blood in higher animals. In flowering plants sometimes, although pollination may be fully effective in securing the deposition of pollen grains on the stigma, the sexual union between male and female cells does not take place due to the arrest of growth of pollen tube which fails to reach the embryo-sac cell. Other cases of incompatibility are known, such as withering away of the pollen grains after their deposition on the stigma, which may be due to the absence of certain substance secreted by the stigma and which is essential for the germination of the pollen grain. Incompatibility may also be due to length of the style. Shortening and successful grafting of style has long been accomplished as a measure against such incompatibility.

The nutritional aspect of sexual reproduction which gained a certain amount of importance a few decades ago may be briefly mentioned here. In lower plants, particularly in fungi and algae, it has been known for long time that they tend to reproduce vegetatively when there is a plentiful supply of assimilable food materials and water in the medium. A familiar example is the vegetative reproduction by budding in yeast which generally takes place when the sugar content is high in the medium. The sexual union in yeast seems only to take place under unfavourable environmental conditions, particularly scarcity in carbohydrate food supply. It is also interesting to note that in hydrophytes which are adapted to an aquatic medium, vegetative propagation is the general rule rather than an exception.

For higher plants, Klebs proposed a sort of nutritional theory which envisaged the initiation of flowering related with nitrogen and carbohydrate contents of the plant—the so-called *carbon/nitrogen ratio.* A high carbon-nitrogen ratio is supposed to promote flowering by bringing about the conditions favouring the qualitative change of the vegetative buds to flower primordia, whereas a low carbon-nitrogen ratio stimulates vegetative growth and propagation. This nutritional theory, however, fails to explain the flowering in short-day plants.

ASEXUAL REPRODUCTION

The formation of asexual reproductive cells, the spores, is an intermediate stage in the life histories of higher plants. In many fungi and in algae, however, spores may be the only reproductive units.

The asexual reproductive spores are really analogous to gametes or sex cells as regards their genetic constitution (genotype). The spores are, however, physiologically very different from the sexual units for spores are capable of further growth (forming the gametophyte) whereas the gametes are totally arrested in their further development unless they unite in pairs. The united diploid cell is only then capable of growing. Thus, physiologically the asexual spores can be compared with this poduct of union of sex cells, the zygote nucleus, which by its growth and development produces the sporophytic organism. Individually either the male sperm or the female egg cell is incapable of any division of its nucleus but the act of union between the two cells evidently supplies a mitogenic factor or factors which stimulate the united cell, i.e., zygote to become meristematic, may be after a short time lag, and initiate such fundamental changes in the nucleus of the cell which lead to division.

Why do the chromosomes continue to exist as resting nucleus in the sex cells for indefinite period and the sex cells even degenerate if they do not chance-unite in pairs? And what make the double chromosomes of the united zygotic nucleus to become shorter and separate out at metaphase? Again, we do not know the precise answers.

PHYSIOLOGICAL ASPECTS OF SEX DETERMINATION IN PLANTS

IAA and other synthetic growth substances like 2, 4-D, NAA, etc., seem to be effective in the determination of the sex of individual plants and also in the development of sex organs.

In dioecious species of hemp (*Cannabis sativa*) genetically male plants could be induced to produce female flowers if during the period of differentiation of flower buds the third and fourth leaves of male plants were treated with napthaleneacetic acid. This probably suggests that sexuality here is determined by the concentration of native auxin during the period of primordium differentiation and that femaleness is associated with a relatively high auxin level. In some *Euphorbiaceae*, e.g., *Mercurialis*, carbon monoxide gas in low concentrations sometimes reduced the number of male flowers in genetically monoecious types. This effect is presumably due to the effect of carbon monoxide on auxin. In some monoecious cucumber, painting of the lower surface of leaves with

NAA caused an increase in the proportion of female flowers, sometimes altogether suppressing the production of male flowers. In general it can be concluded from these observations that female flowers tend to differentiate under higher concentrations of auxins than do male flowers. As has been mentioned earlier, gibberellins promote maleness; cycocel, the gibberellin antagonist, favours femaleness. Animal sex hormones do not elicit comparable responses in plant systems and there is no conclusive evidence that progesterone or steroidal substances which occur naturally in plants determine sex expression in plants.

If sex in dioecious plants is determined by specific substances (in all polypodiaceous ferns there seems to be a specific substance that stimulates the formation of male sex organs of prothalli), these have not yet been isolated nor can they be passed from one plant to another of opposite sex by grafting.

The length of the photoperiod may also affect the differentiation of sex. If seeds of *Mercurialis annua* are grown in short days instead of long days (in long days the plants produced were all staminate only) the percentage of pistillate plants significantly increased. Similar results were also obtained with unisexual heads in *Ambrosia* (*Compositae*). *Cannabis sativa*, growing under 16-hour photoperiod, produced plants which were about half staminate and half pistillate but if the daylength is shortened to 8-hour, the plants were about half bisexual, half females and none males. Again in general then, long days seem to favour production of male sex organs.

SENESCENCE

In annual herbs flowering is a signal for the death of the individual. In perennial plants, death is put off for many years and in *Sequoia* (Californian Red wood) or *Pinus excelsa* life expectancy is prolonged to four or five thousand years. Nevertheless various organs of all plants senesce at various times during ontogeny. There is apparently no reason why a plant (an open system) growing in the field or in the green house where there is no lack of nutrients and environmental conditions are also congenial, should die. Various theories have been put forward for ageing in plants and animals. These include (i) Wear and Tear theory, according to which like the parts of a machine, the various functional components of a cell gradually become less efficient and degenerate; (ii) the Toxin theory which assumes the accumulation with time of a toxic substance capable of interfering with some vital reactions of the cell (there is no tangible evidence for this theory); (iii) Gene repression theory according to which normal functioning of genes is gradually repressed with time resulting in their functioning at suboptimal rates; (iv) Error theory—which takes into consideration the probability of errors being committed in the gene duplication, transcription and translation processes over a period of time—resulting in the formation of abnormal proteins and (v) the Mutation theory according to which after a passage of a certain length of time mutation takes place in the genome, which proves lethal (ionising radiation accelerates such charges).

In the germinating seedling, the cells of the endosperm first degenerate; this is followed by the yellowing of the first and subsequent leaves. Once flowers are formed, the floral parts quickly senesce—the sepal being the last to senesce. In the annual plant after fertilization the entire plant yellows and dries up leaving only the fertilized ovule to continue its life activities. In each cell the endoplasmic reticulum gradually disintegrates accompanied by decreased functioning of chloroplasts, mitochondria and microsomes. As senescence sets in, marked changes take place in metabolic processes like respiration, photosynthesis, and nucleic acid and protein synthesis. While DNA, RNA and protein syntheses decrease, hydrolases and basic proteins like histones are synthesized in larger quantities than previously and new isoenzymes are also produced. Degradative changes predominate due to enhanced activity of nucleases and proteinases. There is a marked loss of chlorophyll due to increase in the activity of chlorophyllase and decreased synthesis of chlorophylls.

In plant systems senescence can be delayed by the application of growth substances like cytokinins, auxins and gibberellins. The effect of cytokinins is most marked; auxin effects are prominent only in some trees and shrubs. Gibberellin effects are less marked. The decrease in DNA, RNA and protein synthesis are checked by all of them and the synthesis of basic proteins is inhibited. As chlorophyll breakdown is also controlled, leaves retain their greenness for a longer period of time.

In the preceding pages we have discussed the effect of environment on plant life. How time affects metabolic processes and ultimately it determines the life span of the individual is unknown at present. It has been suggested that every organism possesses a gene, which is responsible for the termination of the individual. Obviously the expression of such genes is delayed in perennial organisms by years and decades and sometimes centuries. However, in microorganisms which divide by cell fission, apparently there is no death, since the daughter cell divides after a few minutes or hours.

Although individuals die as the fertilized ovule gives rise to a new plant and the same processes of germination, growth, reproduction and senescence are repeated in the same sequence from generation to generation, the information and experience of each generation is passed on to the next until the species is totally exterminated. The first living cells, which originated some three billion years ago in an abstract sense, is still living in all the individuals which are living on this planet to-day.

DORMANCY

The seed, which is a fertilized ovule, usually does not germinate immediataly after its development is completed. It usually needs a period of "rest", in which metabolic activities are largely suspended. This period of "dormant" condition is known as "dormancy". The period varies from a few days to several months or years. Seeds like groundnut have a dormancy period of about two weeks; rice also has a short dormancy period. Some seeds like those of jack-fruit sometimes may germinate within the fruit (viviparous germination).

The causes of dormancy may be varied. This may be due to (i) incomplete development of seed, resulting in immature embryos; (ii) hard seed coat, (iii) accumulation of inhibitors and (iv) sub- or supra-optimal levels of hormones essential for the early phase of seed germination. Some seeds also need exposure to light, temperature etc.

Seed coat or its associated components quite often are impermeable to gases and thus, the respiratory CO_2 cannot diffuse out of the seed, thereby restricting aerobic respiration. Some seeds (e.g., legumes) are surprisingly impermeable to water. Some seed coats are so tough that the radicles and plumules cannot penetrate them. Breaking the seed coat by scarification or even treatment with strong acids (some seeds germinate after treatment with concentrated HNO_3 for 20 min.!) removes these hurdles and the seeds germinate. In nature seed coats are softened by the activity of microorganisms in soil. This is also helped by rain fall, which probably explains why weeds are so abundant in the rainy season.

Many seeds possess considerable quantities of inhibitors, which include abscisic acid, coumarins, phenolics, alkaloids, tannins, unsaturated lactones and cyanide releasing substances. Much of these are leached out during imbibition with water. Supraoptimal levels of growth substances are also responsible sometimes.

When the levels of growth substances are low, sometimes they can be germinated by the application of such substances; gibberellins and cytokinins are particularly useful in this respect. Other chemicals which have been used successfully in breaking dormancy include thiourea and KNO_3. Ethylene chlorohydrin is used for breaking dormancy of potato buds.

Seeds like lettuce, need application of light for germination; red light is most useful since the phytochrome pigment (P_{730}) is involved here. Dormancy of buds of trees is also influenced by photoperiods; short photoperiods in winter months may be responsible for bud dormancy. Some rosaceous seeds need low temperature treatment.

The chemical basis of dormancy is not understood. In dormant potato buds gene function is almost totally repressed. Derepression takes place on treatment of the buds with ethylene chlorhydrin, which, as mentioned earlier, breaks dormancy.

Viability. After a certain length of time, even when the period of dormancy is over, seeds do not germinate. They are then considered to be 'non-viable'. Loss of viability apparently is related to ageing. In rice seeds there is an accumulation of auxins at relatively high concentrations, along with high levels of phenolics and ABA-like substances during storage. Age-induced formation of free radicals may also be a major cause, since chemicals which destroy free radicals increase the viability of seeds. The structural integrity and functional efficiency of organelles may also be irreversibly affected.

Germination

The early reactions in seed germinations are physical. Cellular macromolecules absorb considerable quantities of water. The hydrogen bonding and dielectric properties of water presumably play important roles but how water triggers the initial reaction by which the quiscent seed is activated to resume its normal life activities is not understood. One of the first major events is the release of gibberellins from some bound form which then interacts with the aleuone layer surrounding the endosperm, resulting in the formation and release of α-amylase; other hydrolases like proteinase, nuclease, cellulase, lipase, phosphatase etc. are all similarly produced. This leads to the production of monomers which are utilized for producing the energy required (via respiration) for the synthetic reactions which follow. The monomers like amino acids, nucleotides and monosaccharides are used for synthesis of proteins (structural and enzymic), nucleic acids and cell wall polysaccharides. Release of auxin from some bound form and production of cytokinins (from some t-RNA initially?) help in the elongation and division of cells in the embryonic axis. There is a controversy regarding whether the m-RNA required for the early metabolic reactions is stable and does not have to be synthesised, but there is active synthesis of t-RNA and r-RNA undoubtedly. In photoblastic seeds phytochromes may trigger the early reactions—although how this is achieved, is not understood.

Light is also responsible for the phototropic curvature and chlorophyll synthesis. Gravitational stimuli lead to the differential distribution of auxins resulting in the geotropic curvature. The newly emerged seedling is subjected to practically the same environmental conditions as its parent was and the same events take place in the same sequence. So life continues.

PART FOUR

Plant Ecology

CHAPTER XLIX

INTRODUCTION

As early as 370 B.C., Theophrastus recognised that different plants grew in dissimilar habitats, such as woodland, marsh, lake, river and other distinct plant associations.

The zoologist Reiter was the first man to propose in 1885 the term ecology which is derived like the common word economy, from the Greek *oikos*, meaning home or abode and *logos* meaning study or discourse. Thus the word implies a study of organisms at home, i.e., as they exist in their natural homes or, perhaps even better, the study of their household affairs. It is thus an aspect of botany which deals with the home lives of plants—a study also of the reciprocal relations between plants and their environment.

It is apparent from the wide sense of the term defined above that the field of plant ecology is synthetic and many sided and cannot be properly considerd as a separate branch of botany because ecology is a common meeting ground of all aspects of plant sciences and certain aspects of other sciences as well.

There is no sharp line of division between plant ecology and plant geography. Ecology is concerned, as we have seen, with relationship between plants and their surroundings as also the groupings of plants into kinds of vegetation. On the whole plant ecology is *intensive*, while plant geography is *extensive* in outlook, but both are concerned with plants. The term *geobatany*, often found in literature, should include both plant ecology and plant geography.

Modern plant ecology for the greater part is an outdoor study, for it introduces one from the outset to plants as they actually exist—ecology must be studied primarily in the field. The chief aim of a plant ecologist (the same is true also for plant physiologist) is to know how and why plants grow as they do in the field and how to control their activities.

The critical importance of genetics has largely been overlooked by ecologists as well as plant physiologists for long. Interpretation of much of the recent work in ecology has been rendered extremely difficult by not knowing what part of the observed effects is to be attributed to ecological responses of races, and what part to unknown genetic difference among them.

The comparative physiology of the ecological races is an integrator of diverse disciplines in ecology, primarily genetics and plant physiology. Ecology seems to serve both as a point of departure and focus for tying together the results of genetic and physiological experiments (Hiesey and Milner, 1965). This integration should command the attention of an increasing number of contemporary and active younger investigators in this field, where much remains yet to be done.

Ecological races and species are commonly called *ecotypes* or *ecospecies*. The diverse ecological races have arisen through natural selection by different combinations of environmental and biotic factors. It may be possible to develop in near future an *ecosystem physiology* to account for the *total biotic-community-environment-complex*.

The earth's major *ecosystems* are: the forests, grasslands, oceans, marshes, estuaries, lakes, rivers, tundras and deserts.

The complexity of ecosystems is so great as to preclude any simple, single factor analysis that is both accurate and generally satisfying. Because of the central role of energy in life, however, an examination of the fixation of energy and its flow through ecosystems, help in understanding of the ecosystems themselves. It also reveals somewhat glaringly some of the obscure but fundamental details of the crisis of environment. Forests which cover about a tenth of the earth's surface, fix almost half of the *biosphere's* (*mainly* composed of hydrogen, carbon, nitrogen and oxygen—the essential constituents of living matter) total energy. The biosphere is defined as that part of the earth in which life exists—but this definition certainly needs elaboration. At considerable altitudes above the earth's surface as well as in too hot, too cold or too dry regions the spores of bacteria and fungi can still be traced.

Anyhow this *aeroplankton* does not appear to be engaged in active metabolism. Thus, as a terrestrial envelope, the biosphere obviously has a somewhat irregular shape, in as much as it is surrounded by an indefinite *parabiosphere*, a region in which some dormant life-forms are always present. Ecosystems continue to be developed by migration and continuing evolution. Ecologcal races in taxonomically unrelated species show corresponding distributions over wide climatic transects. Varieties of crop plants are, in effect, ecological races selected by man for economic purposes.

Autecology is the study of ecological relations of individuals or kinds in contradistinction to *synecology*, which is the study of communities, especially their environmental relations and structure. *Genecology* is defined as the ecology of orgnisms or *taxons*—the combination of genetic and ecological methods and concepts, particularly in species and intraspecies.

Whereas plant ecology is usually defined as the study of plants in relation to their habitat, the self-explanatory *phytosociology* is on the other hand, based on a study of interrelationships of plants themselves, within any particular area of vegetation. In this sense, phytosociology could as well mean the study of the interrelationships of the vegetative offshoots of a single plant or to a set of individuals of a single species, growing together in the field. The current definition of the term of phytosociology has been restricted still further so that it now frequently deals only with attempts at classification of plant communities into a generalised scheme.

We shall now consider the implications of the term environment which must be the key word in the definition of ecology. Etymologically the term means surroundings. An analysis of the environment gives us its factors such as soil moisture, wind, temperature, etc. As a matter of fact any condition, force or substance which affects organisms in any way becomes a factor of their environment. The sum total of all such factors constitutes the environment.

The ecological factors may be grouped under *four* major categories : (1) *climatic* or atmospheric such as rainfall, temperature of the air, etc; (2) *edaphic* or factors of the soil such as soil moisture, soil temperature, etc., pertaining to the soil, especially its influence on organisms; edaphic factors may be physical, chemical or biological; (3) *physiographic*, which include those resulting from topography and from soil and (4) *biotic* factors, which result from the presence of other plants or animals, including man, e.g., parasitism, herbivory, etc.

These factors act on existing plant life but they have acted also over entire periods of geological time. Moreover, the plants have themselves evolved from pre-existing kinds; in other words, they are the *historic aspects* of all the external and internal factors to be considered. One is indeed tempted to say that *phytochoria* are the accidents of geological history (Turrill, 1959).

The place where organism or community of organisms lives is its *habitat.* Defined ecologically habitat is the sum total of all the conditions of environments (all the ecological factors) which are effective in determining the continued existence of the individual species of a plant or a community of plants in that particular locality. It may be applied to the particular surroundigs of a certain plant species or a plant community or it may also be used to designate similar combinations of critical factors.

The various ecological factors, as will be readily understood, are to a considerable extent, interdependent upon one another. We know that there are large differences between the vegetation of a mountain summit and that of the slopes and foothills below. The differences in the structure and behaviour of the plant communities in the two places may be largely due to climatic differences but surely, these climatic differences may in turn be due to topographic reasons.

CHAPTER L

ECOLOGICAL FACTORS

The problems relating to the factors of the habitat are extremely complex and a satisfactory solution may be difficult. Nevertheless investigations of the nature and determination of the effects of different ecological factors seem to be of primary importance to the student of plant ecology.

I. Climatic Factors

The climatic factors of the habitat are rainfall, dew (precipitation), atmospheric humidity, velocity of wind, temperature of the air and light.

The term *microclimate* or *microenvironment* refers to strictly local combination of atmospheric factors which owing to local variations in the climate, topography or other factors, differ from the prevalent general climate of the region (*macroclimate*). These microenvironments (microclimates) may be arranged both vertically and horizontally within even a small area. Within each area embraced by one particular macroclimate there may exist several microclimates, some of which may differ so much from the macroclimates as to be of considerable ecological importance; as for examples marked local climates on mountain summits, on steep slopes, in deep ravines, etc., which are relativity cool and damp and also in shallow basins. Very limited cases of local climate are seen in conditions obtaining on the side away from the wind of a rock or even a large stone which may protect the plants in its immediate neighbourhood from strong light and wind, thus creating a microclimate. Similar conditions are also found in the sheltered side of a wood and in the lower strata of the woodland vegetation. In a rooted green plant, the root system is certainly in a different environment compared to the aerial parts. The microclimate or microenvironment of the roots in the soil is really very unlike the general atmospheric climate (macroclimate) in which the shoots occur.

A formation of deciduous trees may establish itself on a wide range of soils if certain more or less definite combination of climatic factors are prevalent in a particular locality characterising definite types of climate. A particular type of climate prevailing over a region thus tends to produce in that region a definite, predominant and stable type of vegetation or, as it is ecologically termed, *climax type of vegetation*. The climate of Western Europe is such that it generally results in deciduous forests as the natural climax vegetation. Where the deciduous forests are lacking in Western Europe, its non-occurrence can be explained by the local operation of some other interfering strong ecological factor, particularly of the soil (edaphic). The deciduous forests are then replaced by other formations adapted to the particular local soil conditions.

RAINFALL

In general rainfall is of little direct significance although rain is of tremendous importance and a factor of prime significance for plants as source of soil moisture. Some absorption of rain water also takes place through cutinised and suberised coverings of shoots. Most of the rain that falls on land results from the condensation of water vapour derived from the surface of the oceans and thus the amount of rain that falls on a particular area depends at least to some extent on its proximity to an ocean or sea. Mountain slopes usually have a relatively more wet side facing the sea from which the moisture is derived than the landward slope and as a result vegetation differs considerably on seaward and landward slopes.

Rainfall operates indirectly through the medium of other habitat factors, its direct importance being connected with its influence on the available water in the soil. Rain is also important in regard to its effect on humidity of the atmosphere. Dry mosses and lichens absorb moisture from a humid atmosphere and the abundance of these plants which grow on rock and bark surfaces is in direct ratio to the humidity of the climate. Quite recently the probable importance of high humidity under exceptional environmental conditions has been further emphasised through the experimental proof that higher plants also can absorb water through their leaves from a saturated atmosphere and this together with contained solutes, can exude through the roots into the soil which (soil) may thereby attain field capacity.

The claims that the presense of a forest causes an increase in rainfall over the amount which would fall on bare ground seem to have been based on insufficient data. Though the presence of forests certainly raises the humidity of the air near the ground level, it is certainly inadequate to cause any heavy rainfall, for only major mass movements of air actually bring about precipitation.

When rainfall increases the atmospheric humidity it exerts the same influence on the plant community as shade conditions (i.e., weak illumination). Structural peculiarities in plants which accompany the weak light climatic conditions are more or less the same as plants grown in full light but in a moisture saturated atmosphere.

The time of the year, when moisture is deficient, is always a critical factor. In areas where rainfall is restricted chiefly to winter months compared to the areas where most of the rain falls during spring and early summer, the floras generally have very few species in common. Nearly every difference in rainfall, either in the annual, total or in the seasonal distribution, brings about a significant difference in the natural plant populations.

The character of a soil depends primarily on climate. A cool wet climate (Eastern Himalaya) means that a large amount of rain water falls on the soil while on the other hand evaporation is slow. The tendency of water will therefore be to move through the soil chiefly in the downward direction. A comparatively dry warm climate, on the other hand, may result in alternate upward and downward movement of soil-water derived from rainfall.

In the tropics, the low rainfall region (12·5 cm) is arid desert with an exceedingly sparse vegetation, highly adapted to scarcity of water. The high rainfall regions (200 cm or above), on the other hand, will bear tropical rain forests, the most complex and highly developed vegetation in the world. The soils at both the regions would be totally different though they may be derived from absolutely identical rock. Heavy rainfall disintegrates the same rock, which remains largely bare in the deserts, to some depths only in high rainfall regions and the deep soil, is covered partly mixed with a layer of humus, which would constantly have a considerable amount of moisture. The nature of the soil, thus distinguished by the difference in rainfall, determines the nature of vegetation.

Heavy rainfall must be one of the determining factors in forming the evergreen forests of tropical regions and the deciduous forests in Western Europe where rainfall, though heavy, is restricted to a very short period only. In rainy climates, above average rainfall is sometimes associated with reduced plant vigour and crop yield.

ATMOSPHERIC HUMIDITY

This invisible water vapour content of the air is usually expressed as relative humidity. It is really the actual moisture content of a volume of air expressed as a percentage of the maximum quantity that the air can hold at the prevailing temperature.

Moisture in the atmosphere intercepts much of the radiant energy before it reaches the earth's surface. Even then, the radiant energy received at the earth's surface, though greatly reduced by such atmospheric absorption, is much more than normal for a great many plants. The detrimental effect of full sunlight upon a vegetation is consequently more pronounced in dry than in humid atmosphere. On the other hand, deficiency of light and heat under shade becomes more critical in humid atmosphere than in dry. Moist air favours the growth of many fungi which readily become serious pests on higher plants. A series of rain showers, within a short period or even cloudy weather (high humidity conditions), favours the growth of rusts and other parasites on plants, commonly the field crops.

VELOCITY OF WIND

The action of wind on plant formation as an ecological factor may be direct or indirect. If the wind velocity is high, the plant communities are frequently in danger of being uprooted and blown away by strong winds. The result is that in those regions, the vegetation is largely composed of species which have prostrate habit of growth and a very tenacious underground root or rhizomatous stem. The occasional nor'westers and gales which sweep through our country cause considerable mechanical damage in breaking off twigs or branches of plants, particularly in tall trees, thus preventing them from growing above a certain height.

A cover of low herbaceous plants strongly reduces the velocity of wind along the ground and a forest cover reduces the velocity still further. By means of this action in slowing down wind currents near the surface of the earth, a plant cover prevents wind-erosion of soil. A *windbreak* is generally a densely planted strip of tall trees usually about a thirty metre wide and placed at right angles to the direction of the prevailing wind, in order to reduce its velocity and this may be a good protection for fields and orchards. A well planned windbreak definitely increases total crop production.

Plants developing under the influence of violent drying winds never attain the degree of hydration essential for the expansion of the maturing cells to normal size. As a result, in general all organs are dwarfed. This is very commonly observed on an approach to sea coast or arctic regions, where the stature of the trees is greatly reduced.

Trees vary greatly in their response to the force of strong wind. Many trees such as pines and oaks often grow flattened against the ground while other trees in the same habitat remain erect. Tree branches often develop only on the side away from the wind. This may simply be a pressure effect. Asymmetrical crowns of trees sometimes result from the death of all buds on the windward surface. Constant high wind velocity may cause woody plants to sway chiefly in one direction, causing an adaptational response which consists in flattening of the trunk branches and roots in a plane parallel to the direction of the wind.

Lodging, a form of wind injury, caused usually by violent winds, peculiar to plants like wheat, maize, sugarcane, etc., frequently flatten the plants against the ground. The prostrated plants, if the stems are not too mature, may become partially erect once more. This adaptation is due to differential growth at still meristematic lower nodes of the plants.

When strong wind carries particles of ice or soil, it is a very effective abrasive force. Bark and buds are sometimes eroded away from the windward sides of woody stems if they are situated in exposed locality. Crops grown in sandy soils in windy climates are very often damaged in this way.

A plant cover is very effective in preventing the blowing away of soil by wind (*soil erosion*). But if the plant cover is thinned or destroyed even at one point, the wind may scoop out the soil to such an extent as to expose the roots of the adjacent plants, bringing about death and devastation over a wide area. The eroded and blown sand may be deposited over wide areas on other plants near their roots on the ground, bringing about a sharp reduction in aeration about their roots following deposition of new soil upon the old. Very few plants can tolerate inadequate aeration and as a result many perish. Those that survive this change adapt themselves by developing adventitious roots at successively higher levels on the stem as sand deposition, caused by wind, occurs.

Along the sea coasts, the salt sprays of the waves and breakers are carried ashore by wind and certain sensitive species of plants are sometimes injured considerably by these seawind-borne salt sprays. Fortunately, air borne salt content diminishes with increasing distance from the shore and it has been found that only those species, which are tolerant of salt, grow nearest the sea. Salt damage from violent storms has been seen as far as 80 kilometres or so inland.

In areas of uneven topography snow is swept from the windward slopes of the peaks and deposited on the away slopes and sometimes in hollows and cracks, winter after winter. Vegetation types on slopes where accumulation of snow is little and those where it is excessive are fundamentally different and their boundaries very sharp.

The indirect general effects of wind in increasing the rate or cuticular transpiration in plants by constantly bringing unsaturated air into contact with the leaves is well known. Wind velocity increases with height above the surface of the soil, so the nearer the leaves of a plant are to the ground, the less certainly will be the danger of excessive transpiration caused by wind.

The wind velocity seems to determine the limit of the height to which a particular plant may grow, for theoretically the higher the plant is, the greater will be the rates of transpiration and consequently a plant can only attain such a height where the loss of water from the leaves is lower than the absorption of water from the soil. The adaptive response is seen in many tall trees where the height is limited by the decay of those upper shoots which are above the critical level where transpiration is in excess of absorption.

TEMPERATURE

The effect of temperature on plants, as in all living organisms, becomes obvious when we compare arctic with tropical vegetation which shows large morphological and anatomical differences. The active growth and reproduction in summer in plants is in marked contrast with the dormant state in winter, particularly in plants of the temperate regions. Such marked differences in the kinds of vegetation and their seasonal relations are brought about only at narrow range of temperatures, between 0–50°C, for there is little living-activity below 0° or above 50°C.

All physiological activities of living cells including reproduction are greatly influenced by temperature and as a result temperature as an uniformly operated ecological factor significantly affects the vegetation of a locality.

The landward migration of plants from their ancestral environment of the open seas certainly necessitated marked adaptations for enduring wide variations in temperature that characterised the newer environment. It is true that as there is no region on earth which is so cold or so hot that some plants cannot live there, adaptations acquired by plants are not so perfectly adjusted and as a result extremes of temperature frequently cause permanent injury or even death.

Considering the vegetation of the earth as a whole it is possible to correlate to some extent certain vegetation characters with annual or seasonal temperatures. It must be borne in mind that a consideration of temperature as a separate habitat factor is difficult for action of wind and atmospheric humidity (more definite and particular in their operation) cannot be separated from the temperature effect.

Some investigators believe that temperature, i.e., geographical latitude is probably of the greatest importance in determining distribution of plants on the earth. According to them the entire vegetation of the earth can be divided into *four* following classes:

(i) *megatherms*—plants which require more or less constant high temperature throughout the year for their optimal growth and development. These plants occur mainly in the tropical zones.

(ii) *mesotherms*—these plants are capable of enduring considerably lower temperature during some period of the year, such as winter months. Many plants of the tropical and subtropical regions of the world can be included in this class.

(iii) *microtherms*—plants of the temperate regions of the earth demand much lower temperature for their growth and development. These plants are incapable of enduring high temperatures, even for some part of the year, such as summer months. All high altitude plants (up to about 3,600 metres) of the tropical and subtropical regions can also be included in this group.

(iv) *hekistotherms*—these plants are restricted only to arctic and to alpine regions, above 4,800 metres in tropics and about 3,600 metres in the temperate zone regions of the world. The hekistotherms have the lowest temperature demand of all plants and they are also adapted to short summer which prevails in the extreme temperate regions of the world. These plants seem to endure long and extremely cold winter months without any permanent injury.

Temporary protoplasmic adaptation affording a certain measure of immunity to low temperature injury is called *low temperature-hardening* and the perennial vegetation of cold climates regularly develops this particular hardiness in autumn. In spring this hardiness seems to be lost and it is not uncommon in summer conditions that the same plants are killed by temperatures far above those which they endure easily in winter.

Considering high temperature injury of plants, aside from its role in desiccation and in bringing about disparity in the balance of respiration and photosynthesis, (respiration is greatly in excess of photosynthesis) high temperature injures and kills protoplasm. The principal adaptational features which protect plants against high temperature injury are: (*a*) high rate of transpiration which may cool the leaves, at least to some extent, (*b*) a vertical orientation of the leaf blades compared to the leaves at right angles to the sun's rays, (*c*) white incrustation on the surface of the leaves which reflects rays that would otherwise be absorbed increasing the temperature of the leaf surface, (*d*) a covering of dead hairs shading living cells, (*e*) a bark insulating phloem and cambium and (*f*) protoplasm with a low moisture content.

Stems are more frequently injured by high temperature than leaves or roots. *Stem girdle* (presence of discoloured and shrunk bands of tissue in stems) is due to scorching of tender stems at the soil surface when the soil temperature becomes very high. Seedlings and herbs may be killed by contact with such hot soil.

As we have seen before, interaction between various factors, climatic, edaphic or biotic, frequently prevents a plant from attaining its theoretical maximum growth with respect to temperature.

It is well known that when we consider equal areas, the number of plant species increased in an equatorial direction from both north and south poles, although other factors such as stable climate and the land continents play important parts in bringing about such distribution. The longitudinal variations in temperature may have greatly influenced the rate of evolution, for the rate of mutation has generally been found to increase directly with increase in temperature.

LIGHT

The energy necessary to sustain life on earth is derived from sunlight directly by green plants or indirectly by other heterotrophic organisms which must eventually depend upon various organic compounds synthesise by the green plants only. The essential connecting link between living organisms is provided by the ability of chlorophyll to absorb radiant energy and convert it into chemical energy of carbohydrates. Light also exerts much stimulating influence upon plants, especially on the differentiation of tissues and organs. It can be truly said that with the exception of water, light is unrivalled in its influence upon the morphological and anatomical characteristics of vegetative life.

To an ecologist, however, the importance of light as a factor determining the formation of plant communities is not so great as its physiological importance, on account of the fact that the action of light is comparatively uniform over wide areas. The general features of plant communities are not usually affected by light except in the case of relatively small plant communities, growing in deeply shaded situations, such as caves, etc. However, individual species within a plant community, shows particular internal structures which are modified in relation to the operation of the light factor. The majority of plants have definite requirements with regard to light. Some thrive only in full intensity of sunlight, others favour a moderate amount of shade while still others can tolerate deep and continued shade. Most plants can adapt themselves to moderate variations in the intensities of light. There are some plants, however, which exhibit a maximum range of adaptabillty: they are apparently equally at home in both strong light and in the deep shade. Plants may be classified ecologically according to their relative requirements of sunlight and shade; those growing best in full sunlight are called *heliophytes* and those preferring shade (low light intensities) are *sciophytes*. Among heliophytes, some species can grow fairly well under shade (*facultative sciophytes*) compared to those sun plants which cannot do so (*obligate heliophytes*). Sciophytes can be similarly divided in two groups depending upon their ability to tolerate high light intensities. One of the very common and disagreeable weeds in India, *Elephantopus* (*Compositae*) seems to be a strict sciophyte growing luxuriantly only where there is deep and continued shade.

In comparison to plants grown in shade (sciophytes), species thriving best under full sunlight conditions (heliophytes) usually exhibit some characteristic features, e.g.: (*a*) thicker stems with well-developed xylem, (*b*) shorter internodes with more prolific branching, (*c*) usually smaller, but thicker leaf blades with stomata and veins smaller and closer together. (*d*) leaves generally hairy with more hairs per unit area, (*e*) thicker cuticle and cell walls, (*f*) better developed palisade tissue with weakly developed spongy cells, (*g*) intercellular spaces reduced in size and number, (*h*) leaf blades not flat and oriented at other than right angles, (*i*) roots longer and more numerous, resulting in higher root/shoot ratio, (*j*) greater fresh and dry weights of both roots and shoots, (*k*) lower photosynthetic (proportionately greater chlorophyll *a*/*b* ratio) rate but higher respiration rate, (*l*) more rapid transpiration, (*m*) higher salt content and consequently higher osmotic pressure of the cells, (*n*) cell sap more acidic, (*o*) higher carbohydrate/nitrogen ratio, (*p*) pronounced vigour in flowering and fruiting, (*q*) earlier blossoming and (*r*) a significantly greater resistance to injuries due to temperature, drought or parasitic infection (Fig. 763).

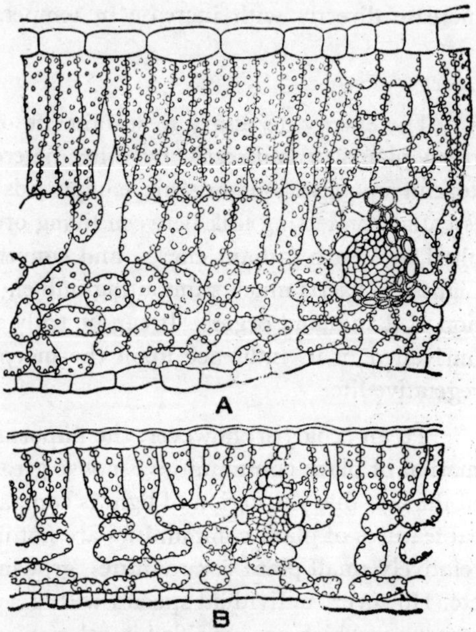

Fig. 763. Changes in leaf structure produced by environmental conditions. Transverse sections of two leaves of a single sugar maple (*Acer* sp.) tree. A: fully exposed to sun and B: from the centre of the much-shaded leafy crown. Note that the differences in both light and rate of transpiration are responsible for the marked differences in structure.

In moist climates, vegetation tends to be a complex series of tall trees, low trees, shrubs, herbs, mosses, etc., and a high percentage of the flora, except perhaps the trees, must needs be facultative or obligate sciophytes. In dry climates, however, the reverse is true but even here there are many plants that thrive only in the shade of larger ones. Shade conditions are mostly brought about by plants themselves—forest trees produce the shade conditions under which the undergrowth community of vegetation exists. Different types of woodlands cast different amount of shades, as for example, undergrowth in the very deep shades of pine woods is almost absent (deep shade) compared to the luxuriant ground flora of an oak wood (shade not so deep).

Underwater differences in light values are very important factors which determine the depth to which submerged plants can live. Red algae as a group have lower light requirements than any other algae. This may be a case of chromatic adaptation, for it is well known that phycobilin pigments in *Rhodophyta* can make exceptionally efficient use of weak light of relatively shorter wave-lengths, e.g., green. Has the presence of non-green pigments then any bearing upon the ability of aquatic flora to grow in deep water?

II. Edaphic Factors

THE SOIL

In a seed plant though the root systems usually comprise not more than a quarter of the whole plant on a dry weight basis, they are so finely divided that they may occupy a mass of soil greater than the volume of the atmosphere surrounding the shoot. There is no doubt then of the tremendous amount of surface contact between the soil and the plant. The plants, more or less completely dependent upon the soil for anchorage, water and nutrients, cannot be over-emphasised. Thus the factors of the soil, the edaphic factors, are of supreme importance to the ecologist not only due to this intimacy of contact between the plant and the soil, which are strongly influenced by each other, but also due to the tremendous complexity of the structure of the soil. While climatic factors determine the general layout of vegetation, the edaphic factors are primarily responsible for local differences found within the same climatic regions.

With the very beginning of agriculture, the important part which soil plays in plant growth, was recognised and appreciated. Man soon discovered that in his attempt to control environment for the benefit of his good crops, little can be done in altering climatic factor but much can be accomplished with the soil.

In the first text-book on plant ecology written by Warming in 1895, plants were classified according to the characteristics of the soil, into *oxylophytes* (on acid soils), *halophytes* (on saline soils), *psammophytes*(on sandy soil), *lithophytes* (on rocks) and *chasmophytes* (on rock fissures and crevices, etc.).

Soil study now ranks as a highly technical independent branch of science and is known as *Pedology*. Soil is the surface material of the earth in which plants grow containing the underground parts of higher plants as well as algae, fungi and bacteria (soil flora) and also a soil fauna, ranging from protozoa, worms, insects to small mammals, such as moles, mice, etc.

Soil may be defined as the weathered superficial skin covering of the earth's crust with which are intimately mixed living oragnisms and the products of their decay (humus). The earth's crust is created and continuously affected by surface agencies—primarily heat, frost, wind and above all water. These agencies act upon the rock from which the soil is derived and continually also on the soil itself. The soil is also essentially affected, sometimes even newly created soil, e.g., peat, by the vegetation growing upon it. The spaces in the soil remaining between the solid particles are filled with water and gases. The continuing action of heat, wind, water, etc., naturally tends to the development of *stratification*—differentiation of soil into distinct layers from the surface down to the parent rock. This is the natural soil.

The basic framework of most soils conists of small fragments of mineral matters which have been derived from the parent rock by disintegration by weathering agencies, both physical and chemical. Besides these inorganic particles, an organic constituent (humus) is nearly always present. According to the particular rock from which the soil is derived, the nature of the inorganic soil particle differs both chemically and physically; but all soils contain basically three main inorganic chemical substances—*alumino-silicates*, *silica* and *calcium carbonate*. With alumino-silicates are associated various elements which form alkaline salts, e.g., calcium, magnesium, potassium and sodium —all except sodium, being elements on plant nutrition. Iron salts, mostly as ferric ions are also always present in all soils and which give many soils the typical brown or red colour. Besides these, phosphates and sulphates are always present. Nitrogen is partly derived from the organic humus and also from fixation of atmospheric nitrogen by certain soil micro-organisms. Traces of other elements in extremely minute quantities are found in most soils.

Almost all rocks liberate considerable quantities of soluble salts as they weather and the marine sediments also yield considerable quantities of sodium chloride by leaching. Salt accumulation on the surface of the soil is sometimes called '*alkali*' though the deposits are not always basic. There are two major types of 'alkali' in saline soils. '*White alkali*' consists chiefly of chlorides and sulphates of sodium, potassium, calcium and magnesium. During dry seasons, the concentration of these salts produces a thin, white incrustation on the surface of such soils. The white alkali soils are fairly porous. The growth of most crop plants is adversely affected, except halophytes (can thrive even in soil with white alkali content of more than 5–8%), when the 'white alkali' content rises above 0·5%. '*Black alkali*' soils are dominated by carbonate or bicarbonate of sodium and sometimes potassium—about 10–15%. These salts yield strong bases like caustic soda or caustic potash by hydrolysis, which, as is well known, has a strong corrosive action on plant tissue and humus. As a result, large amounts of organic matters are dissolved into such soil which give the soil surface a blackish colour. Black alkali soils are structureless and more or less impervious to water and gases. Owing to these characteristics of the black alkali soil most plants cannot tolerate as much of it as they can of the white alkali.

An interesting type of soil is found in the regions of high rainfall, especially on sandy ground. Rainfall here exceeds evaporation and iron and other minerals tend to be washed down from the surface layers. The minerals are deposited lower down as a *hard pan* which is naturally dark in colour, in contrast to the pale colour of the leached surface layer. The leached surface is acidic, and as the hard pan is impervious to water, bog vegetation often develops there.

Calcium is the dominant basic ion in the soil and when present in quantity, imparts the physical and chemical stability to the formation of the soil. These ions sometimes aggregate into fine colloidal particles and sometimes also into compound articles, thus giving a structure to the soil.

The basic reaction of the soil is usually due to the presence of free calcium carbonate. The extremely acid soils with pH values between 3 and 4 can support only specialised natural vegetation (*oxylophytes*). These soils are sometimes referred to as '*sour*' and they need the addition of lime to render them fertile.

The terms *calcicoles* or *calciphiles* for plants growing exclusively on calcareous soils and *calcifuges* or *calciphobes* (lime-hating soils) for those found only on acid soils, have long been recognised by ecologists. Calcifuge plants are typical of sandy soils. Some plants may occur indiscriminately on either types of soil but the reality of the above ecological classification may not be overlooked. Species of one genus may differ and even physiological types adapted to either type of soil, may occur within one species, e.g., '*edaphic ecotypes*' of *Trifolium repens* (Snaydon, 1962).

The speeies of moss genera, *Tortella* and *Neckera* can thrive only in the soil or rock surface which are rich in calcium or lime. These soils are apparently alkaline in nature and the species of such plants are *calcicoles* (*acidophobic*); other moss species such as *Polytrichum* and all *Sphagnum* grow exclusively in acid soils and these *calcifuges* consequently strictly avoid lime-rich soil or rock (*acidophilous*). The growth of different moss species may thus reveal something of the potentialities and nature of the land on which they grow. Consequently in recent years the importance of moss as '*indicator plants*' (indicating whether the soil is basic or acidic in nature) is coming to be appreciated in increasing extent.

Sometimes growth of natural flora of a particular habitat may act as an index for judging the quality of land from the point of view of fertility and deterioration. The following species of plants serve as such indicator plants in some dry or semi-dry regions of the Punjab: *Salvadora oleoides*, *Zizyphus nummularia*, etc., are indicators of good productive land; *Calotropis procera*, *Saccharum munja*, etc., indicate sandy soils; *Tamarix articulata*, *Butea monosperma* and *Suaeda fruticosa* indicate saline soil in the process of deterioration.

The timber yielding *Shorea robusta* communities in India, the conifer, *Pinus roxburghii* seem invariably to avoid calcium-rich soils (alkaline soils) and they can thus only be classed as *calcifuge* or *calciphobe*, i.e., acidophilous (best growth between pH 6·0–6·5).

Use of indicator plants in geobotanical prospecting is being extensively employed for indicating the presence of heavy metal ore deposits. Indicator plants, designed '*universal*' are those species which always indicate the presence of a definite element and are confined to soils over certain ore deposits—mapping the distribution of a particular indicator species, most affected by the minerals sought, has been economically very profitable. Examples are afforded by *Astragalus* sp. for uranium prospecting in Colorado (U. S. A.), *Viscaria alpina* for copper in Norway, *Silene cobalticola* for cobalt in Katanga (Africa), *Equisetum arvense* for gold in Czechoslovakia, etc.

The four essential constituents of the soil are: (*a*) mineral particles constituting the inorganic framework of the soil, (*b*) organic matter or humus arising from the death and decay of plants, and animals such as protoza, insects, etc., living on the surface of the soil, (*c*) soil water containing dissolved substances and (*d*) air in the soil.

(*a*) MINERAL PARTICLES IN THE SOIL

The nature and size of the inorganic particles resulting from the disintegration of rocks not only influence the chemical processes but also directly determine the physical nature of the soil and its effects upon plants. The texture of a soil depends primarily on the sizes of its mineral particles and this feature is of considerable importanc to vegetation because the texture of the soil controls aeration and water-holding capacity of a particular soil. The mechanical composition of the soil, i.e., the relative proportions of the different sized particles determines to a considerable extent the availability of water for absorption by plants. By mechanical analysis in the laboratory the physical composition of a soil may be determined. The mineral part of the soil is separated into a number of generally accepted by arbitrary fractions, each consisting of particles whose size lies between definite limits, which are as follows:

coarse gravel (stones) particles—above			5·0	mm	dia.
gravel	,,	from	5·0—2·0	,,	,,
coarse sand	,,	,,	2·0—0·2	,,	,,
fine sand	,,	,,	0·2—0·02	,,	,,
silt	,,	,,	0·02—0·002	,,	,,
clay	,,	below	0·002	,,	,,

All soils contain, in fact, particles belonging to more than one of these categories and a soil is conveniently named after the fraction which is preponderant. A *loam* is a soil which consists of a thorough mixture of particles of widely different sizes with an adequate humus fraction (Fig. 764).

Gravels are soils with a preponderance of large mineral particles above 2 mm in diameter and always mixed with coarse sand as well as some fine sand and silt. Aeration and percolation of water in such soils are extremely free. Thus rain falling on this coarse-textured soil penetrates almost immediately so that very little water is lost. Gravel soils are not very suitable for plant life in general because of their dryness (they lose their water too quickly) and dearth of nutrients.

Sandy soils have a preponderance of particles between 2 mm (coarse sand) and 0·02 mm (fine sand) in diameter. The mineral particles are mainly silicates. As in gravels, percolation of water and aeration are unrestricted and as a result sandy soils are generally dry, warm soils, warming up quickly in spring. As in gravels, they are poor in nutrients because of the deficiency in the finer particles of silk and clay with which the bases are associated. Because they are always liable to thorough leaching, the nutrients are usually washed away. The sandy soils are referred to as *light soils*.

Sandy soils with sufficient proportions of the finer particles of silt and clay and with a little humus readily support deciduous summer forests. Sandy soils make good agricultural soils for certain food crops.

Silt soils are intermediate between sand and clay in the size of their particles. They are the most favourable soils for all types of vegetation as they have considerable water-holding capacity accompanying easy percolation and aeration. The name is derived from the texture of alluvial soil (silt) laid down in the plains by flooding of rivers. Fine silts when wet form *heavy stiff soils* but unlike clay, they are easily crumbled, when dry.

Clay soils differ very strikingly both chemically and physically from the three types of soil constituents mentioned before. The mineral particles, below 0·002 mm in diameter of hydrated aluminium silicates are numerous in such soils and when any soil has more than 30–40% of these particles, it may be called a clay soil. Clay soils may in some measure be secondary in origin as they are not entirely composed of particles of rock-forming minerals but in addition also contain precipitations of dissolved CO_2 and organic acids, derived from the weathering of rocks. Clay, owing to smallness of its particles, is essentially colloidal in nature and is thus able to take up and retain considerable quantities of water, swelling up and increasing markedly in volume during the process. Clay soils have the qualities exactly opposite to those of sand; extremely slow percolation of water, imperfect aeration and high water-holding capctity. Clay soils are *wet*, *heavy* (compared to sandy soils which are *light and warm*) *and cold* as they warm up only very slowly.

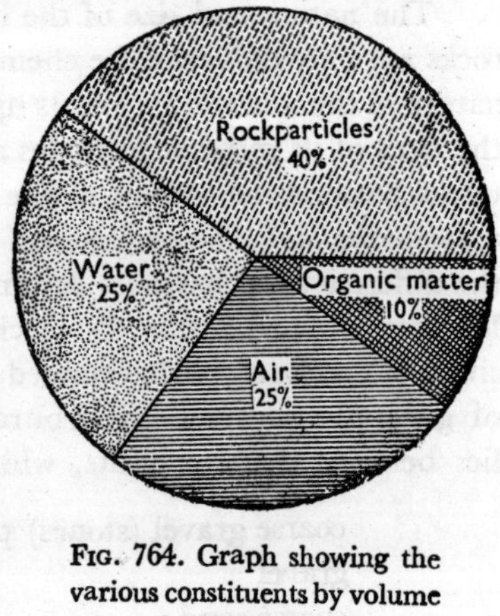

FIG. 764. Graph showing the various constituents by volume of a rich loam soil.

The very fine silt and clay offer considerable resistance to penetration by roots and hence clay soils are unsuitable for many plants. In plants adapted to these heavy soils (fine silt and clay), the root systems tend to be shallow because of the difficulty of aeration at greater depths. The extent and the degree of branching of the roots are decreased. This becomes very important in those regions where relatively long dry periods intervene between rains. This effect can be seen on clay grass land where the grasses are mostly shallow-rooted species. Clay soils are sometimes much improved as regards their fertility by an abundance of humus, which make them lighter and secures better aeration and freer water movement.

Loams are soils which have a good mixture of particles of all the different sizes and are the most suitable soils for the great majority of plants as they combine the good qualities of all the types of soils—clay and humus fractions give consistency and water-holding capacity and also supply nutrients essential for plant growth; silts permit capillary rise of water and sand particles facilitate good aeration. Loams can be classified as heavy (dominant particle, clay), medium and light (dominant particle, sand) (Fig. 764).

Lime soils are those which are derived directly from limestones containing 90–100% of calcium carbonate. The lime soils are shallow and contain surface-humus saturated with calcium (calcareous soil). These soils support a herbaceous vegetation, e.g., *Clematis* (*calcicole*) which tolerates drought exceedingly well.

(*b*) ORGANIC MATTER OR HUMUS IN THE SOIL

All soils contain organic material derived from the dead remains of plants or parts of plants such as dead leaves, roots, rhizomes, etc., with a small addition of animal excreta or dead bodies. When the dead remains are returned to the soil, the complex organic compounds are broken down by the activity of micro-organisms into humus and ultimately mineralised (the process, an oxidative decomposition, takes a long period for completion) into molecular or ionic forms (the chief of which are CO_2 and water) which then become available to the future generation of plants. The term *humus* is usually applied to the whole complex of continually disintegrating organic material in the soil. Sometimes the term humus is restricted to the intermediate conversion of humus into brown substance, soluble in acidulated water (*humification*). Humification takes place only when oxygen supply is deficient.

The soils of deserts in which vegetation is scanty and new soils freshly formed from rocks, apparently contain very little humus, while mature soils with vegetation, the most.

Peat is the term given to a soil which is really a deposit of partially decomposed humified organic matter which accumulates as a layer on a soil. Peat formed on wet soil or under water is sometimes referred to the type of vegetation from which it is derived. If the peat is strongly acid in reaction, due to oxygen deficiency and shortage of calcium, it may be called *acid peat* or *acid humus*. In shallow lakes, where temperature and calcium supply are suitable and oxygen supply plentiful, organic matter is decomposed into what is known as *neutral peat* or *neutral humus*, which is usually neutral but may also be slightly basic in reaction.

In general humus is an important source of mineral nutrients, and the fertility of a soil is usually correlated with its humus content. Peat soil is, however, notoriously low in phosphorus and potassium. Humus, like clay and perhaps, even more efficiently than clay, helps to make aggregate structure of the soil possible.

Humus is the source of food for most soil organisms. Green plants certainly add much more to the soil than what they extract from it. While they absorb only small amounts of soluble minerals they return these (the minerals) *plus* even greater quantities of synthesised organic materials such as starch, protein, cellulose, etc., when they die. The vast and complex groups of saprophytic soil organisms and organisms which grow as parasites on these organic substances thrive only in humus.

(*c*) WATER CONTENT OF THE SOIL

The amount of water contained in a soil at any particular moment apparently depends upon various factors. Water-holding capacities of soils of different textures are very different; the amount of water held in the soil depends, as mentiond before, on the size of soil particles and the amount of humus present. The coarser the particles are, the larger are the air-spaces between them. A clay, because of its constituent fine particles, takes longer to dry and holds water so very strongly that plant roots can scarcely extract from it nearly all the water present (plants can use nearly all water in sand which is relatively loosely held)—a small percentage of water remains unavailable to the roots all the time (*hygroscopic water*).

The water content of the soil is more important than any other single ecological factor in causing differnces between plant communities. Soil water is not only important in connection with direct water requirements of plants, it is, as far as it is known, also the medium by which mineral salts essential for plant nutrition enter the plants. Water is also responsible, as we have seen before, for most of the actual soil-forming processes.

A heavy carpet of vegetation sometimes dries out a soil although a close vegetation decreases direct evaporation from the soil surface. The reason for this anomaly is clearly understood if we consider that a soil may lose water more rapidly by absorption through the roots and transpiration from the extensive leaf surface during hot weather than by direct evaporation from the soil surface.

(*d*) AIR IN THE SOIL

The presence of a continuous air system which ramifies all but waterlogged soils is essential to the respiratory activity of underground plant organs and of the soil microorganisms. Decay of roots, leaves the soil full of minute channels and burrowings also by worms and insects create minute passages through which oxygen can enter from atmosphere and the respiratory CO_2 passes out readily. Composition of soil air differs only very slightly from that of the atmosphere proving an active gaseous exchange between the two—the soil and the atmosphere. Good aeration supplies oxygen for root respiration and the energy derived seems to be essential for active absorption of minerals from soil by the root.

In marshy waterlogged soils, the regular aerating process of other soils seems to be inoperative, for water fills up all the places between the soil particles thereby excluding the air. Under such circumstances oxygen deficiency and an excess of CO_2 (from anaerobic root respiration) may be accompanied by production of other gases in the soil, such as marsh gas or methane. The composition of soil air may easily be an important factor in determining the characteristics of a community, such as a marshland vegetation.

(*e*) SOIL TEMPERATURE

The porosity of coarse-textured, light sandy soils and even of heavy soils such as clay, which are well aggregated tends to favour a condition of equilibrium between the soil and the atmospheric temperature. But due to lower moisture-holding capacity of light sandy soils they become warm much sooner in spring than the clay. Root injury due to low temperature in winter is more pronounced in sandy soil than in clay.

The rate at which a soil surface is warmed up, is considerably more rapid than the rate of cooling—on an average the temperature of the soil is higher than that of the air immediately above at all seasons.

Temperature of the soil is one of the most important factors which influence the rate at which absorption of water and nutrients from the soil can take place and as such, as we have seen before, plays a conspicuous role in determining the geographical distribution of plants on earth.

III. Physiographic Factors

Strong topographical relief (steep hills and deep valleys) has a profound effect on vegetation largely because it may produce characteristic local climates or even microclimates and may also modify the edaphic factors of the habitat. These physiographic factors originate from behaviour and the structure of earth's surface. Mountains, hills and valleys result from irregularities in the form of earth's surface. This behaviour of the surface of the earth definitely also includes such processes as erosion, silting up of lakes and river mouths and the deposition of sand along the sea coast by wind.

Steepness of slope especially in comparatively high latitudes increases the exposure to the sun. A steep southern slope receives rays of the midday sun almost perpendicularly while the northern slope receives only oblique rays during morning and evening and sometimes none at all, except perhaps for a short period at summer. This difference in light and heat sometimes affects the vegetation profoundly. Some times striking and almost incredible effects on vegetation are seen due to this difference in as much as the steep northern slope protected from the sun (where fog may often form) may bear virgin forest with hygrophilous ground vegetation, whereas the southern face constantly heated by the sun, can support only a xerophytic vegetation of shrubs and herbs, strongly protected against excessive transpiration. In fact, there may not be a single species in common in the two slopes separated only, say by 90–120 metres of rock face. This complete contrast in vegetation types may entirely be due to difference in the climates, produced locally in the two slopes.

Slopes may play a very important part in determining the character of the soil. Due to gravity, ground water always moves downwards and in a slope of a mountain, this downward movement of water removes soil from a slope and carries it down and may deposit in the valley. Sometimes, if the slope is steep, this results in a soil surface of unstable material and the vegetation may be wholly unable to establish there or is just able to maintain a precarious foothold. Water content of a sloping soil surface is less compared to a horizontal one and consequently the soil is drier, for rain water runs off before it can be retained by the sloping soil. This may result in almost complete absence of higher green plants from a sloping hill side.

In general, northern slopes usually bear a vegetation adjusted to wetter conditions than southern slopes in the Northern Hemisphere (opposite is true in case of Southern Hemisphere) and often the same communities of vegetation establish in the southern slopes at higher altitudes, as in general, greater the altitude, wetter is the climate.

The vegetation of a deep sheltered slope sometimes differs considerably from that of an exposed slope, even in the immediate neighbourhood. This effect of shelter on windy sea coasts and mountains is sometimes very pronounced.

Local climates may vary considerably within a few metres or even centimetres on an uneven ground (microclimate). The shelter from wind and heat provided by a hillock or even by a small rock may enable a plant community to establish itself which could not exist if it were not so sheltered.

Eroded rock faces, hills, steep slopes and river banks produce a special class of habitat in which only a certain species of plants can precariously maintain themselves. If the erosion is rapid enough, the habitat may be absolutely devoid of any plant community. Gravel, sand or clay, eroded by water and wind from rock surfaces, may be brought down by rivers and streams and deposited as silt at their mouths, providing new habitats (like salt marshes) for colonisation by plants of different types.

Wind-blown sand along the sea coast accumulates near the coast on which sometimes a grass vegetation with a creeping underground stem and root system develops and establishes itself on the loose sand. These grasses eventually bind the loose sand together and ultimately render it partially suitable for the other species of plants which ordinarily require a heavier stable soil.

IV Biotic Factors

The biotic factors are those involving animals and plants other than the organism being considered. The biotic factors are often overlooked, perhaps because they are the most complex and we know least about them.

The interaction of organisms, the reciprocal effects of plants and animals are sometimes termed *Coaction*.

The average green plant is commonly referred to and thought of as an autotrophic or independent organism in contrast to fungi, bacteria and animals which cannot synthesise their own food. In reality, however, not even green plants are independent in the strict sense of the word for they are also considerably influenced by other living organisms. For example, a flowering plant may depend upon insects for pollination, upon birds for seed dispersal and upon fungi and bacteria associated with its roots for absorption of nutrients from soil. Even the CO_2 used in green plant photosynthesis has been released in respiration by other organisms and respiratory oxygen has accumulated in the air by generations of green plants. The amounts of heat, light, moisture and nutrients available to one plant are all conditioned and determined by the proximity and competition of other living organisms. Environment of any organism is in reality mostly *biologic* and *only partly physical*.

Thus the biotic factors of the habitat (so-called because they have origin in the activities of plants and animals) depend directly on the action of living organisms on the vegetation. Plants in any community have a profound effect upon one another just as mutual effects of the individuals of a human society. In the theoretical sense, a particular plant community, adapted to a habitat must also include the soil bacteria, either free living or in symbiotic association with other organisms, earthworm and other soil fauna, parasitic fungi, the snails or the insects which live on the plants or in the soil and the birds that live in the trees which may be destroying as well as distributing fruits and seeds. The name *biome*, a major climax community, has been suggested for such a complex of organisms, both plants and animals, naturally living together and forming a sociological unit. It is said to be equivalent to climax or formation. Biome should also include the sheep or cattle which graze on the semi-natural pasture communities and help to keep the plant community in equilibrium.

Out of man's diverse activities have arisen to a large group of biotic factors which operate on vegetation—farming largely consists of modifying vegetation in certain definite directions. Actually cultivation of the soil and growing of food crops are biotic factors due to intense human interference. Grazing by domestic animals such as sheep may result in a particular vegetation being dominated by certain species of plants, whereas in its absence, perhaps some other species would have dominated the field. Modification of vegetation by destruction by fire (an indirect biotic effect due to man's activity) may cause a temporary replacement of one dominant species by another—in extreme case it might even lead to the permanent alteration of the whole character of the vegetation.

Animals including man himself react on and determine the nature of vegetation in many other ways. In areas where the soil is loose and sandy, grazing factors may cause serious results, leading to total destruction of the soil surface. Trampling by man and animals may often prevent any vegetation except a few mosses to grow in such soils.

Relics of an ancient era, survivors of a former prevailing flora, are referred to as *Epibiotic*. When regions harbour such relic species and have not been drastically altered, these areas are termed *Refuginum.*

Interactions occurring between various plants which grow together in a community do not seem to affect the community as a whole but perhaps only individual species. In any plant community, the dominant and also the largest species seem to make the greatest demands on the resources of the habitat. As such, the life form of the dominant species is chiefly the outcome of climatic conditions, physiographic and biotic conditions seem to be only of minor importance.

It is evident that in many cases biotic factors cannot be looked upon as definite factors in the same sense as climatic or edaphic factors—biotic factors only modify or limit the expression of other primary important factors of the habitat.

CHAPTER LI

LIFE FORMS OR GROWTH FORMS OF PLANTS

Various types of plant communities resemble or differ from each other in different ways. In various communities of trees such as pine, oak, beech, etc., the communities resemble each other in that in each, the dominant and the characteristic species are trees, but they differ from each other in that these species are pine, oak, and beech respectively. Thus plant communities may differ from each other not only in respect to their dominant species, but may also at the same time resemble each other in that the dominant species in all cases may be trees, shrubs, or herbs or in other words, are of the same life or growth form. Thus, if we recognise behaviour of units of the vegetation, the *life forms* (the vegetative form of the plant body, thought usually to be a hereditary adjustment to environment; growth form) of the plants comprising these units must also be taken into account. The life forms of the various plants comprising a community play an important part, as would be readily seen, in determining its general structure.

Any classification of life forms must take into account the characteristics of a plant adjusting it to its habitat. Different classification of life forms based on different vegetative features have been proposed from time to time but the earliest one was proposed by Humboldt (1806). He attempted to group vegetation types on a *physiognomic* basis, with at least some effort to express relations between environment and certain life-form groups. Humboldt named about 15 groups of plants whose physiognomy was important in the study of landscapes:

1. Banana form, *Musa*, *Strelitzia*, etc.
2. Palm form, *Cocos*, etc.
3. Tree fern form , *Alsophila*, *Cyathea*, etc.
4. Aroid form, *Arum*, *Pothos*, etc.
5. Conifer form, *Taxus*, *Pinus*, *Picea*, etc.
6. All sharp-leaf form, *Araucaria*, *Juniperus*.
7. Tamarisk form, *Mimosa*, etc.
8. Mallow form, *Stêrculia*, *Hibiscus*.
9. Liana form, *Bauhinia*, etc.
10. Orchid form, *Epidendron*, *Cattleya*, etc.
11. Cactus form, *Opuntia*, *Cereus*, etc.
12. Casuarina form, *Casuarina*, *Equisetum*.
13. Grass form, *Andropogon*, *Panicun*, *Scirpus*, etc.
14. Moss form, *Bryum*, *Sphagnum*, *Polytrichum*, etc.
15. Lichen form, *Cladonia*, *Gyrophora*, etc.

The following classification (Raunkiaer), which came much later, based on a single principle—the position of the perennating vegetative bud in relation to the ground level—has proved to be of great value and has received considerable world-wide use:

1. Parasites and Epiphytes.
2. Plants with succulent habit.
3. Climbers and Twiners.

4. Trees and shrubs—*phanerophytes* (Ph) (from *phaneros*—visible):
   (*a*) tall trees (more than 30 metres in height)—*megaphanerophytes* (Mg) (from *mega*—large);
   (*b*) trees (between 30 metres and 8 metres)—*mesophanerophytes* (Ms);
   (*c*) trees and shrubs (between 8 metres and 2 metres)—*microphanerophytes* (Mc) (from *micro*—small);
   (*d*) shrubs (between 2 metres and 25 cm)—*nanophanerophytes* (N) (from *nanus*—dwarf);
   (*e*) undershrubs or herbs with vegetative buds not more than 25 cm above the surface of the soil—*chamaephytes* (Ch) (on the ground);
   (*f*) plants with buds on the surface of the soil—*hemicryptophytes* (H) (from *hemicrypto*—half-hidden);
   (*g*) plants with buds buried deep in the soil on underground stems—*geophytes* (G) (from *geo*—in the earth).
5. Annual herbs—*therophytes* (Th) (from *theros*—summer).
6. Water and marsh plants—plants with buds under water—*hydrophytes* (from *hydro*—wet) and *helophytes* (from *helo*—marsh).

All forest communities of the tropical regions are dominated by *phanerophytes*, the deserts by *nanophanerophytes* and *chamaephytes* (ephemerals in some deserts are *therophytes*), grassland by *hemicryptophytes* and much of the cold arctic and alpine vegetation by *chamaephytes*. Communities inhabiting loose and soft soil are marked by the abundance of *geophytes*. *Therophytes* are abundant on all open soil habitats, on seashores and in all watery places are early phases of plant succession.

An interesting aspect of the distribution of life forms—the list of percentages of different classes of life forms in the total flora of a climatic region—*biological spectrum*—is clearly related to the prevailing climate.

A very interesting example of variations in life form adopted by some species of plants due to changes in the environmental conditions, particularly climate, is found in the Eastern Himalayas. The flaming red-flowered *Rhododendrons* which are tall trees (*megaphanerophytes*?) at altitudes between 1800–2400 metres become strikingly dwarf, *microphanerophytes* at alpine altitudes of 3600–4200 metres. The effect is mostly thermal, no doubt, for near freezing temperatures generally prevail at high altitudes for most of the year but preponderance of ultra-violet radiations in sunlight (compared to light at lower altitudes) and high wind velocity might also be significant contributory factors.

CHAPTER LII

THE UNITS OF VEGETATION

PLANT COMMUNITIES

Individual plants (the same is true for animals, including man) growing in the open under more or less natural conditions tend always to associate together in communities—*plant communities*. Each community has its special features which differentiates it from another. The edge of a river may be lined by tall grasses forming a community while in the water are other plants with floating leaves or others again, completely submerged. All these are communities and they are essentially natural, though sometimes modified to a considerable extent by human agency. We often find plants of various degrees of heights associated together forming a stratified community, one above another. The difference in the growth form actually enables them to grow together as members of the same community.

It is also very easy to see that the various units of vegetation are not all of equal rank. In an oakwood, as an example, in addition to the dominant species—oak, there are certain other characteristic plants. Lichens or moss covering the bark of a tree in a wood cannot really be considered as a community of the same status to the woodland itself. Yet all these are units of vegetation and the term plant community is applied to any such unit of whatever size or rank. A plant community is any collection of plants living together which has, as a whole, a certain distinctiveness or better, individuality. Plant communities have thus a characteristic composition as regards the species which they contain, each having their own definite structure—they are the units, which together make up the vegetation of the earth.

Transect studies are sometimes indispensable for knowing the variation of vegetation which occurs with changing environment, e.g., caused by irregularities in the topography of a soil. These studies also help us in determining correctly the succession of plant communities—how one community is replaced by another—in a particular vegetation.

A *transect* is a long rectangular sample plot—a continuous strip (the strip may also be divided into squares equal to their width) that gives a rough cross-section of a particular vegetation.

The *belt transect* is a strip of vegetation of uniform width (1-10 metres) in woodlands (largely determined by the character of vegetation) and usually of considerable length —determined by its purpose. Transects can be profitably utilised in determining primarily the biotic influence of trees upon other plants, as affected through water content and available light.

The transect method is also a commonly used one in studying forest trees.

Isolation transects are frequently used in grazing experiments. They usually consist of two strips with a protected (from cattle and rodents) strip in between, acting as a control. One of the two strips is grazed, the other left ungrazed. A history of the development of vegetation as affected by time of grazing may thus be obtained by keeping alternately grazed and ungrazed areas, open or closed.

While *belt transect* studies give us the structure of a vegetation in two dimensions—relative height and the lateral spread of plants (and the interrelation of one plant to the other)—the *bisect* studies supplement this by supplying the third—the relative position, depth and the extent of underground parts such as rhizomes or corms as well as the root system of each plant. Bisect of *profile* is essentially a line transect along which a trench is dug to a depth greater than that of the deepest root system. It is truly a representation of a vertical section of a community showing the roots and shoots in their natural positions.

Other methods of studying vegetation can be mentioned. A *quadrat* study is essential for a knowledge of the structure of vegetation in a particular habitat. The quadrat (loosely, any sample plot, whether or not it is square), simply square sampling plots, is a square area of varying size (maximum about 100 sq. m.), marked off for the purpose of detailed study, which in its simplest form is the counting the individuals of each species to determine their relative abundance. Among the kinds of quadrats are, *list quadrat*, (all the species in a sample unit are listed and their number counted), *basal area* or *diameter quadrat* (the measurement of basal area occupied by each species is estimated), *clip quadrat* (clippings are removed and dry weights, volumes and fresh weights of the tops of each individual determined), *chart quadrats* are those in which the position of each plant is indicated accurately in its proper place on a chart by a planimeter, a *cover quadrat* is one on which the actual or relative coverage or shading of these species is recorded and a *count quadrat* is a sample plot of which the density of plant species is determined. Some quadrats are marked in such a way as to permit study of plants from year to year (*permanent quadrat*). Other quadrats are devised from which vegetation is removed artificially in order to follow critically the manner in which the plants re-enter the particular quadrat and the course and outcome of competition (*denuded quadrat*) among plants.

Within the last thirty years, wide and various *application of mathematics* have progressively invaded the study of vegetation. In order to put this study on an increasingly objective basis, hence scientifically more respectable, efforts are being made in closely re-examining certain basic assumptions with regard to the nature of the data and the modern probabilistic, statistical methods to be profitably employed. Use of such methods eliminates personal bias and produces easily reproducible results for a given set of data, analysed by a given standard method. It will be imprudent, however, to eliminate the subjective element. In recent times, attempt are made to evaluate a given set of ecological data obtained from bisect, transect or quadrat studies, on an objective basis certainly, without losing sight of the subjective at any time.

The size of the quardrat is generally adapted to the characteristics of the community being analysed. In general, greater the floristic richness and the more heterogeneous the life forms of the community, the larger is the quadrat size.

Strictly speaking, the shape of the quadrat is square, but, for all practical purposes, sample plots can be of any regular shape—square, circular, short rectangular or even strips.

Plant communities because they possess a certain degree of individuality can be compared as a whole with a living organism in that they also have developmental history and only finally attaining a stable maturity, which is called a *climax*. We have seen before that plant communities are of a number of types, the nature of each being determined by the local habitat conditions. A similar combination of the same habitat factors in different areas always produces more or less similar type of plant communities.

A *climatic climax* is an expression of climate and is dominated by highest type of growth forms. As we have seen before, besides climate, which is probably the most stable factor, there are edaphic, physiographic, biotic and other factors which may influence the climatic factor. These factors, often termed *pro-factors*, help in the development of *pro-climaxes*. Pro-climaxes include: (1) *sub-climaxes* where the development of typical vegetation for a particular climate type has been arrested before attainment of the final stage due to an influence other than climate, (2) *pre-climaxes* and *post-climaxes* where the climax formation deviates from the normal type on account of some local physiographic factor and (3) *dis-climaxes* which may result from some disturbance due to biotic modifications or change of typical climax.

Another common phenomenon seen in a tropical rain forest—*stratification* or *storeys of vegetation*—refers to the growth of shrubs, herbs, climbers, epiphytes and sometimes parasites and saprophytes as well in such forests in addition to the tall trees. These diverse life forms according to their stature and nature of existence have been called *synusiae*. A *synusia* is thus a group of plants of similar life forms and floristic composition, making similar demands on similar habitats. The synusiae appear to have the common property of representing tightly-knit groups of species of similar ecological requirements. Thus, the species constituting a synusia may be of different taxonomical origin but can be regarded as ecologically equivalent. The total assemblage of plants living at a particular station; the community as a whole, including all its synusiae, is sometimes termed *Phytocoenosis*.

The linean system of taxonomy appears to require existence in a group of variational unity, conforming satisfactorily in their properties with those required of the taxonomic species. These units constitute the *biological species*—groups of actually or potentially interbreeding populations—which are reproductively isolated from other such groups.

PLANT FORMATIONS

The largest and the most comprehensive kind of plant community, equivalent to climax and biome, is the *plant formation*. Plant formations may be determined primarily either by the climatic factors of the habitat—climatic plant formations—or by factors of the soil—edaphic plant formation. There are, however, some formations which are determined partly by soil and partly by climate. In suitable soils they can only exist within a narrow range of variations in the climate. Plant formations may be determined not only by habitat factors of soil and climate but also in some cases of specific and continuous operations of man himself—*anthropogenic formations*.

The great climatic plan formation types of the world are determined by well-marked types of climate prevailing over wide areas of the earth's surface. Thus a number of major vegetation types associated with particular climatic conditions can be clearly recognised such as the deciduous forests in areas, where a favourable climate alternates with an unfavourable one, due to low temperature or inadequate humidity and evergreen coniferous forests with areas where the length of the flowering season is restricted. The tropical rain forest are conspicuous as these formation types occur all over the equatorial regions of the world. The deciduous summer forests in Europe, North America and parts of Eastern Asia and subalpine forests of coniferous trees are also among the most important climatic plant formations. Each is characterised by a specific kind of vegetation formed by the prevailing climate.

The edaphic plant formation types are determined not by climate but by the general nature of the soil. The formation types which grow in shallow water on the edges of lakes and rivers all over the world sometimes show similar growth forms, although differing in the particular species of which they consist. Other formation types such as the mangrove formations are determined by saline soil and they are to a large extent independent of the climate.

A world formation of high latitudes (northern region) and high altitudes (high mountain side or top), cold desert, (*frigorideserta*) is widely known as a *tundra*. They are really the vegetation represented in a cold desert. The (*deserta*) regions are frozen for most of the year but which support a certain amount of plant life. The tundra is beyond the tree line and the plants are all small. A few stunted creeping willows and birches are found and perhaps also a few flowering plants, which make the tundra a brilliant mass of colour for a short period each year. Most of the species are, however, mosses and lichens. Their growing season is very short and plants growing there must produce reproductive structures very quickly while the ground surface is still unfrozen.

A formation of completely treeless grasslands, especially short grasses of warm temperate or subtropical regions, is usually called *steppe* or *terriherbosa* (steppe and prairie vegetation).

Plants with which ants or termites are associated, often with special adaptive features, are referred to as *myrmecophilous*.

A general feature of all formation types is that the dominant species—those plants which give the formation its characteristic appearance—belonging to each are characterised by specific life forms. The following recognised climax forest formation types cover the whole of India, Bangladesh, Pakistan and Burma (Fig. 765). All these forests are determined by well-marked climate types (mainly rainfall and the accompanying humidity) prevailing over wide areas. The tidal forest formations are, however, determined by the general nature of the soil (saline); they are edaphic formations:

1. *Tropical wet evergreen*—Burma, Bangladesh, Assam and a long belt in Western India.
2. *Tropical wet semi-evergreen*—Orissa and Andhra coasts, West Bengal, Assam, Bihar, Bombay, Mysore, Kerala and Bangladesh.
3. *Tropical moist deciduous*—Central and Northern Burma.
4. *Tropical moist deciduous* (Sal)—Northern India belt; Orissa extending up to Madhya Pradesh.
5. *Tropical dry deciduous*—almost all Central India extending up to extreme South.
6. *Tropical thorn forest*—Rajasthan deserts, Hyderabad, Sind and Pakistan.
7. *Tropical dry evergreen*—a belt on the coasts of Andhra and Madras and up to extreme South.
8. *Subtropical wet forest*—Mysore, Madras, North Assam, North Burma and Bangladesh.
9. *Subtropical pine forest*—a thin belt from Kashmir in the north to Assam and Burma.
10. *Subtropical dry*—Pakistan and Rajasthan (small area).
11. *Wet temperate*—Subtropical Himalayas from North Bihar to North Assam and North Burma.
12. *Moist temperate*—a subtropical Himalayan belt extending from Kashmir to North Assam.

13. *Dry temperate*—a Himalayan belt from Pakistan (N.W.F.P.) right up to North Assam, immediately below the Alpine forest.

14. *Alpine*—Himalayan region, above 4800 metres.

15. *Tidal*—West Bengal (Sunderbans, Sagar Islands); Bangladesh; Godavari delta in Andhra; a very thin belt on the Bombay coast; Burma (Irawaddy delta); and Andaman and Nicobar Islands.

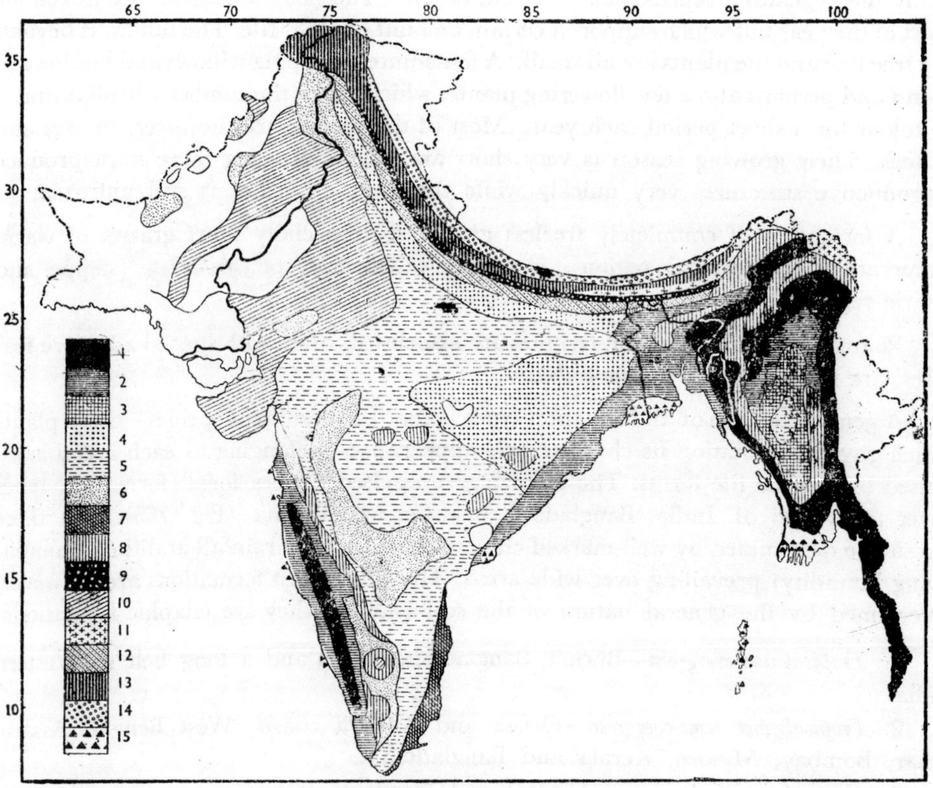

FIG. 765. Map showing the main climax forest formation types of India, Pakistan, Bangladesh and Burma (after *Champion*). 1. Tropical wet evergeen; 2. Tropical wet semi-evergreen; 3. Tropical moist deciduous; 4. Tropical moist deciduous, Sal forest; 5. Tropical dry deciduous; 6. Tropical thorn forest; 7. Tropical dry evergreen; 8. Subtropical wet forest; 9. Subtropical pine forest; 10. Subtropical dry; 11. Wet temperate; 12. Moist temperate; 13. Dry temperate; 14. Alpine and 15. Tidal.

The plant association and consociation

The actual units of natural vegetation belonging to one of the plant formations are of various nature and status. The largest of these vegetation classes, apart from the climatic types dominated by one or more species, are called the *plant associations*. Thus the term association refers to the major subdivisions of a formation.

A concrete community of a size smaller than minimal area and not having typical composition and structure of the associated, is called *fragmentary association*.

A plant association has commonly more than one dominant species because it is usually composed of a number of smaller unit communities each with its own dominant. An association dominated by a single species of plant is called *consociation*. A large plant association may, therefore, consist of several consociations. Thus an oakwood dominated by one of the species of oak or a beechwood dominated by the common beech, is a consociation whereas deciduous summer forest though characterised by deciduous tree, is a formation but again each formation with its particular species of tree is an association. Certain groups or pairs of moss species are habitually associated together, as for example the associations of *Philonotis fontana* and *Bryum pseudotriquetrum* of bog springs on mountains. *Typha* and *Scirpus*, although showing individually typical life forms, sometimes grow intermingled (association), but more often, separately in consociations.

If an association is characterised not by a single species but by the grouping of dominant or controlling species, it is called a *faciation*—the entire area of an association is thus composed of its various faciations. Each faciation corresponds to a particular regional climate—small differences in rainfall, evaporation and temperature. Temperature appears to play the leading part in the arrangement of grassland faciations.

A community dominated by shrubs and bushes is called a *Scrub*.

In nearly bare or recently populated areas, embankments, abandoned fields, etc., the vegetation usually consists of individuals belonging to a single species; such a community of a single species is called a *family* in the ecological sense. The family is usually a small unit and is typical of early stages of development, rarely being found in a stabilised vegetation.

If the individuals of a family become more numerous, adjacent families may then merge or one family may invade another. It may also be that two or more species may invade an area simultaneously. In either case, such a small initial community is generally termed a *colony*—a characteristic of early developmental stage of a vegetation.

There are seldom sharp lines of demarcation between the mergings of two communities of higher rank. Such transitional overlapping areas of mixed community are termed *ecotones;* ecotones contain characteristic species of each and presumably is intermediate in environment. Ecotones may be broad and indefinite, but on mountains, or steep hill slopes or in the associations of mangroves or in associations around lakes and ponds communities may be rather strikingly delimited.

All vegetation shows more or less striking differences, very few centimetres or metres. Differences in habitat, particularly microclimate, are reflected both in number and in kinds of plants. Where these differences are continuous, usually around swamps or ponds, the structure of the vegetation is clearly shown in *zonation*.

A vegetation shows a definite structure with changes in temperature and available light—seasonal changes. The spring vegetation, with abundance of herbaceous growth and blossoms, changes to locally taller species with the advent of summer which again gives way to another structure in autumn. The adjustment of various species of these seasonal changes results in what are called, *aspects*. Thus aspect is a seasonal impress on a community, such as the spring aspect in deciduous forests produced by perennating (geophytic) herbs or the winter leafless aspect.

Huxley (1938) introduced the word *cline* as a taxonomic concept indicating the existence of a variational gradient of biotypes. Cline is actually a series of form changes i.e., a gradient of biotypes along an environmental transition. Gregor (1939) used this word with prefix *eco* or *topo* to define certain terms concerned with intraspecific variation within a community of plants; a *topocline* is a cline which only follows a geographical gradient and an *ecocline*, a cline, apparently correlated with an observable ecological gradient or environmental transition. Ecoclines can be regarded as ecological varieties of plant species. A *topotype* has been defined as a population of plant species in a geographical region, differing from those of another region. A topotype may be *extraclinal* if it does not fall within a single geographical gradient in character expression or *intraclinal*, if it has a reference to a particular, variable range on a geogaphical gradient.

An *ecotype* is a particular range of characters on an ecocline, related to their respective environments.

All species found in associations or consociations other than the *dominants* may be called *subordinate species* whose growths are largely determined by the presence of the dominant species (by the humus produced by the decay of the dominant species). A community of such subdominant species within an association is called a *society*, the lowest category in a plant community. Subordinate species which are invariably present with an association or a consociation (may also be found outside or not) are known as *constants*. The exclusives and constants are the characteristic species of a plant community. In eastern tropical Himalaya, the dominant species of plants such as *Shorea*, *Terminalia*, and *Garuga*, each forming its own consociation, produce what is known as *Garuga-Shorea-Terminalia* association. In western Himalaya, the chief associates of *Shorea robusta* in the upper Doon valley are *Eugenia jambolana* and *Terminalia tomentosa* or *bellerica*, forming a *Shorea-Eugenia-Terminalia* association.

CHAPTER LIII

PLANT SUCCESSION

We have seen before that a plant community can be compared to an individual living organism with a developmental history this developmental history is called *plant succession*. Among the plant groups some are more stable than the others. Some plant groups remain essentially the same for a long time, perhaps for centuries or even more, others are changeable, giving place, in course of a short time, to other communities of different composition. But generally there is a definite trend in one particular direction, towards a position of equilibrium; all such progressive change is referred to as plant succession which is really the developmental process of plant communities from birth to maturity.

If an area of ground is bare of vegetation or if new bare ground is formed by emergence of land from the sea or river, by drying up of a lake or by a deposit of sand blown by wind or by landslide or by biotic agency, plants will begin to appear in it sooner or later. The first vegetation to occupy the area as pioneer colonisers are few and soon give way to others and these again to others, until a stable and intimate relationship between the vegetation and the habitat is established. This type of development of vegetation is called—*autogenic succession.*

Change in habitat may also be brought about by continually acting outside factors such as a gradual change in climate, filling up of a lake by deposition of silt, increasing concentration of solutes in soil water, leaching or gradual washing out of nutrients from surface layers of soils, etc. This may result in the habitat becoming less suitable for the first colonisers and more suitable for others. Plant has nothing to do with the changes in the habitat which thus takes place and in this type of succession the first colonisers pack up, being unable to cope with changes in the habitat and others more adapted to the new circumstances take their place. A change in vegetation is thus brought about—*allogenic succession.*

During the first stage of the formation of plant communities on bare rock areas denuded of all vegetation, individuals which appear from seeds and spores blown in by winds, with fewer demands on the soil, have more or less wide unoccupied spaces between them—*open community*. Later the spaces become occupied either by new arrivals from outside the area or by the next generation seedlings of the original colonizers. Eventually all available soil surface becomes occupied by plants forming a *closed community*. When a community of plants becomes closed, the habitat conditions change due to the reaction of the plants forming the community upon the habitat. As the individuals die, the products of their decay accumulate in the surfac layers of the soil as humus, thereby altering physical structure as also the chemical nature of the soil leading to an increase of its water-holding capacity. Thus the inhospitable soil now becomes more favourable and allows plant life, much more complex and more suited to the changed habitat conditions than the original pioneers, as the regenerated soil can now satisfactorily meet the higher demands made on the soil by these more evolved newcomers. Thus bigger plants progressively replace smaller ones, the original dominants are replaced by successful newcomers and ultimately woody plants appear on the scene and occupy the area.

With the arrival of subordinate species, a highly complex climax forest vegetation is finally established. A stage of equilibrium is reached in time between the plant community and the climatic as well as edaphic factors of the locality. Thus the plant community becomes permanent.

Any particular and concrete example of plant succession is termed a *sere* and the various changes, a particular sere undergoes are termed *seral stages* or *seral communities.*

The early course of initiation of vegetation on bare grounds, wet or dry, differs entirely according to the nature of the initial habitat, for the plants which can colonise such different habitats differ very widely indeed. If the habitat be wet, on submerged ground on the edge of a lake or a pond, the pioneer stages in succession will consist of species of hydrophytic nature, from submerged aquatic plants (i.e., *Hydrilla*, *Ceratophyllum*, etc.) to aquatics with floating leaves (e.g., water-lily). As succession proceeds the soil level is built up by accumulation of decaying plant remains till it reaches the surface of water. Shrubs and trees which can tolerate waterlogged soil around their roots—the *helophytes* (marsh plants, like *Typha*)—often follow and as the level of soil is built higher above the water-level by accumulation of decomposing plant remains (humus), the habitat becomes drier by the addition of organic matter and is now certainly better aerated. The trees can now come in and the series of communities may be ultimately completed by climatic forest formation; such a sere beginning on a wet area is called *hydrosere.*

On dry bare habitats with extreme deficiency of water such as exposed rock surface, rocky slopes, wind-blown sand and sometimes pumice and volcanic ash, the early stages of plant succession are totally different. Lichens, (the algal partner is very often a member of the group, *Cyanophyta*) both crustose and foliaceous and terrestrial algae, predominantly the blue-green *Cyanophyta* (*Nostocaceae*) together with certain xerophytic rock mosses are the pioneer colonisers on such areas. Blue-green algae seem to be pre-eminently suited as the most impressive plant-pioneering community for all types of dry, inhospitable soils. Their success as pioneering colonisers is, no doubt, due to their marked ability to withstand adverse conditions of the habitat such as desiccation, extreme temperatures, ability to hold rainwater, etc., but a capacity for considerable nitrogen fixation, as shown by many species of their group of algae in addition to photosynthesis, cannot be discounted as a possible factor determining their high status among all pioneer colonisers. The lichens, mosses and the algae disintegrate the surface of the substratum which with decayed parts of their bodies (humus) form a thin soil. Further colonisation by other mosses follows which form thicker cushions and the amount of soil gradually increases enough to support xerophytic herbs followed by xerophilous shrubs and trees and eventually with the formation of reasonably thick layer of soil and humus, the climatic forest formations. Successions initiated on such areas where there is an extreme deficiency of water are termed *xerach* and such a sere, i.e., different stages of the developmental process, is a *xerosere.* If the sere takes its origin only on rock surface, it is termed *lithosere.*

A sere taking its origin on saline soil can be termed *halosere.* The pioneer communities in early stages of plant succession in a halosere in the Ganges and Godavari delta are the true salt-resisting mangroves such as *Avicennia* followed by other true mangroves, e.g., *Bruguiera*, *Rhizophora*, *Ceriops*, etc. These are the plants which are best adapted to those inhospitable surroundings of high salt conditions of the soil which is largely unsuitable for other plants. As the soil becomes less saline, other species which are not really true mangroves, such as *Excoecaria*, *Agriceras*, *Acanthus ilicifolius* and succulents like *Suaeda* arrive and adapt themselves ultimately forming mixed mangrove forests. With the arrival of salt tolerant perennial grasses, halophytic herbs and shrubs come in, and the series is completed and a new climax forest formed.

A sere starting on sand is a *psammosere*. The pioneer communities to begin succession on such a sere are generally sand-binding grasses such as *Spinifex*, *Ipomea*, etc. Varieties of shrubs and trees follow in due succession and thereby the sand is fixed and stabilised. The organic content of the sand increases with the simultaneous increase in the moisture-holding capacity. The sandy soil may now be fit to support a climax type of forest formation.

Any seral community below the climax association is termed *associes*.

A miniature of vegetative successsion of microscopic communities such as bacteria and fungi and followed occasionally by terrestrial algae on fallen logs, twigs, etc., has been often termed a *serule*.

The development of vegetation in an era is called an *eosere*, the major developmental process of climatic climax forest formations in a particular geological era. The term *geosere* has also been sometimes used to designate the total plant succession in time of the geological past.

A series of climaxes, following one another in any given area as a result of climatic change, is sometimes referred to as *clisere*. Contiguous climaxes move together in a common direction because of widespread climatic change that induces regional parallelism.

If the development of a vegetation is initiated not on a new ground, but on an area whose pre-existing vegetation has been destroyed by fire, earthquake or by other means, it is known as *secondary succession* (*subseres*). The classic case of recolonisation of Krakatoa group of islands in the Pacific which was completely destroyed by a volcanic eruption denuding the islands of all traces of visible plant life is an example of secondary succession. The pioneering was started by a few members of the family *Nostocaceae* (*Cyanophyta*) within a few years which appeared first as a dark-green gelatinous layer on the volcanic ash. Members of this group of algae are equally important pioneers in secondary plant successions under less inhospitable circumstances, e.g., eroded soil. The course of a subsere is necessarily different from either a hydrosere or xerosere on new ground because the point of commencement is different and the time taken to complete it is also less.

An enumeration of the various stages of succession which are characterised by similar series of stages, whether in primary or in secondary, hydrosere, xerosere or halosere, may be advantageous:

(*a*) *nudation*—the formation of the bare areas by natural causes or artificially.

(*b*) *colonisation* and *aggregation*—after the establishment of the first scattered invaders, the individuals come to be grouped as a result of propagation.

(*c*) *ecesis*—adjustment, establishment and final attainment of maturity of the colonising species of plants that have migrated from neighbouring areas to new situations. It essentialy consists of three well-known physiological processes, e.g., germination, growth and reproduction. Ecesis is the most decisive factor in *invasion*, and *migration* without it, is totally ineffective.

(*d*) *reaction*—between the colonising plant species and the habitat causing change in habitat conditions of atmosphere and the soil, e.g., area once fully lighted becomes more or less shaded (temperature becomes lower and the air is more humid); a wet area becomes drier for the migrants absorb large amounts of water from the soil which is lost through transpiration; water-retaining capacity of a dry soil increases and it becomes richer by the accumulation of humus caused by the decay of dead roots, stems and leaves; the dry area gradually becomes more moist and certainly more favourable to plant growth.

(*e*) *competition*—as a result of ecesis, competition among different colonising species ultimately leading to the survival of the species, fittest or luckiest for that particular habitat.

(*f*) *stabilisation*—achievement of complete emancipation and stability of the dominant species.

(*g*) *climax*—final formation of climax-type forests.

CHAPTER LIV

ECOLOGICAL CLASSIFICATION OF PLANTS

ECOLOGICAL TYPES

In the first comprehensive text book on plant ecology written by Warming in 1895, plants were classified into ecological groups characterised by one type of substratum with special emphasis on the edaphic factors of soil. No attempt was, however, made then on the taxonomic relationship of the plants forming the ecological groups.

Taking Warming's classification, based on edaphic factor, as a starting point, the following ecological groups can be broadly recognised:

(*i*) plants usually found on acid soils —*oxylophytes*
(*i*) ,, ,, ,, ,, saline soils —*halophytes*
(*iii*) ,, ,, ,, ,, sand —*psammophytes*
(*iv*) ,, ,, ,, ,, rock surfaces —*lithophytes*
(*v*) ,, ,, ,, ,, rocks, tree trunks, etc. —*saxicolous*
(*vi*) ,, ,, ,, ,, rock crevices —*chasmophytes*
(*vii*) ,, ,, ,, ,, ice and snow —*cryophytes*
(*viii*) ,, ,, ,, ,, barks of trees, as in the case of bryophytic communities—*corticolous*
(*ix*) ,, ,, ,, ,, in wasteland or land, subject to disturbance by man, road sides, rubbish dumps—*ruderal*

That the soil factor is at best equally as important to aquatics (perhaps only those rooted in the mud of ponds and lakes) as to land plants, was not fully recognised until recently. Differences in soil conditions under water certainly strongly influence the distribution of submerged or floating, rooted hydrophytes. *Epiphytes*, as also saxicolous and corticolous, some of the very characteristic of all the ecological types, perhaps should not be included in this classification for they may grow upon other plants without any connections with the soil whatsoever—they have sometimes no need for the soil in the accepted sense of the term.

Another fact was recognised early by Warming that plants growing in habitats with closely similar water relations show many similarities in their structure, both morphological and physiological. Since many of these plants may be only remotely related taxonomically, the structural similarities must necessarily indicate parallel evolutionary adaptations of different species to the same environment. The three major groups of this classification based on water relations are *hydrophytes*, *xerophytes* and *mesophytes* which refer to plants of wet, dry and intermediate (neither too dry nor too wet) habitats respectively.

The limits between the above three ecological groups, as in all other biological classifications are, to say the least, very ill-defined.

Giving major importance to water balance in climatic and edaphic ecology, Warming later on (1909) erected four major categories of vegetation: *Hydrophytenverine*, *Xerophytenverine*, *Halophytenverine* and *Mesophytenverine*.

It has been found in recent years that mutations, either spontaneous or induced, sometimes alter the reaction of plant to its normal environmental conditions and thus it seems possible to produce new *ecotypes*. Ecotypes of barley are fairly well-known which react differently to different levels of nitrate fertilisers in the soil. Application of this principle of mutation in the case of crop plants such as rice, wheat, etc., may have considerable potentialities, as desired strains, suitable for a particular habitat condition, can be produced artificially which otherwise may have proved economically useless under that particular condition. Thus effective land reclamation can be achieved for profitable cropping.

HYDROPHYTES

Vascular plants inheriting from remote marine ancestors a multicellular body with unlimited apical growth and a dimorphic life cycle, slowly conquered the land and spread themselves over much of the land surface. Yet, despite this flourishing conquest of the land, some angiosperms, most of them herbs and a few pteridophytes, ventured back into fresh-waters and even further, to their ancestral habitat, the oceans. They carried with them the relics of their terrestrial heritage and the advanced reproductive methods of their terrestrial relatives which, to say the least, seem more of an anachronistic obstacle to them now—for from whatever depth at which they grow, most angiospermic hydrophytes still aim at raising their flowers above the surface of water for wind or insect pollination.

Hydrophytes are plants normally growing in water and also include plants inhabiting swampy or marshy habitats containing a quantity of water which would prove much more than optimal for the average land plant. It will be evident that hydrophytes are subject to less extremes of temperature than land plants for the watery habitat in which the plants grow certainly takes longer to be heated and also longer to cool. Conditions thus are more uniform compared to plants growing in soil and as a result, many of the hydrophytes are very widely distributed geographically. Submerged aquatics, because they are screened from too high intensities of light show many of the characteristics of shade plants, *sciophytes*. Habitat water *per se* is not harmful but because of extremely low solubility of oxygen in water or in water-saturated soils, (oxygen readily dissolves in water when the surface is in contact with air but subsequent diffusion downwards is so slow that a permanent oxygen deficiency exists all the time for submerged hydrophytes; on the other hand the availability of CO_2 in water for photosynthesis in submerged plants is certainly as much or even somewhat more, compared to land plants; at 20°C, one litre of water can hold 0·3 ml of CO_2) a complex of critical environmental conditions is produced and only specialised forms of plants can suitably cope with it. The upper sunlit zone of the sea where the light intensity is great enough to permit photosynthesis, is referred to as *euphotic zone*, usually regarded as being to a depth of a hundred metres, though there might be differences depending on such factor as the amount of solid carried in suspension. River estuary obviously is less transparent than the mid-oceans.

Roots of vascular hydrophytes are evidently able to live and grow normally in very low oxygen tension.

The ability to respire anaerobically is especially well developed in certain hydrophytes growing in still water or in wet soil. The dispersal of seeds and fruits by water is a common characteristic of hydrophytes, for seeds of most xerophytes or mesophytes are killed when wetted (due to lack of aeration). The streams and ocean-currents carry the spores of hydrophytic cryptogams, the floating fruits of *Cocos nucifera* and *Xanthium*, the floating seedlings of mangroves and entire floating plants such as *Lemma*, *Salvinia*, etc.

Land plants which live in situations in which atmosphere is permanently moist are spoken of as *hygrophilous*. Many hygrophilous plants, living in almost saturated atmospheric conditions of tropical forests, show considerable resemblance to hydrophytes in form and structure.

Without water, there is no active metabolism nor any possible development. This is sometimes termed *anabiosis*, when no respiration nor any other life-activities occur. In barley, seeds hardly consume any O_2 in respiration, below 8% humidity of the environment.

Many hygrophytes show adaptations to excessive water absorption in these plants by developing *hydathodes* or similar structures as means of getting rid of surplus water (exudation) in the liquid form.

Among the principal morphological and anatomical characteristics of hydrophytes are the following:

(*i*) Because of the suboptimal aeration of tissues due to the watery habitat, one of the outstanding structural peculiarities shared by one and all hydrophytes is the *sponginess* of their tissues. In hydrophytes, owing to disintegration of groups of cells or their separation, large intercellular spaces or cavities develop in all parts of the plant which remain filled with air. Vascular hydrophytes with a terrestrial ancestry became so adjusted to high oxygen content of the air during their sojourn in land that when they again take to water they must make special provision for aeration; hence the development of *aerating tissue*. On the other hand most thallophytes do not possess any special aerating tissue.

The remarkable absence of air cavities or *lacunae* from the thalli of angiospermic *Podostemaceae*, must be mentioned here. It is possible that lacunae may have been absent in *Podostemaceae* from ancestral stocks and the family has been necessarily restricted to well-aerated watery medium throughout its evolution. Alternately, it may have been that the ancestral stocks had lacunae but the ability to develop them has subsequently been lost in the course of evolution, for the lacunae become superfluous in more favourable environment, where gaseous exchange is facilitated usually by a swift current and

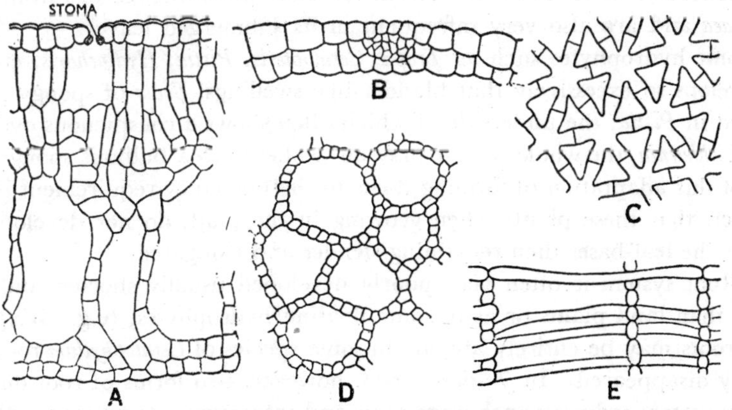

FIG. 766. Some structural features of hydrophytes. A: Section through a floating leaf-blade showing thin cuticle and stomata confined to the upper surface. B: Section through a delicate submerged leaf showing again lack of cuticle and stomata. The lamina is thin and has direct contact with water. C: Aerenchyma in the stems. D: Aerenchyma in the stem of *Hydrocotyle*. E: Aerenchyma with two rows of normal cork cells in the stem of *Jussiaea*

oxygen-saturated running water (Sculthorpe, 1967). It is thus probable that because the thallophytes originated in water medium they have been fully adjusted to oxygen deficiency condition in water from the very beginning. An interesting case of the development of so-called *aerenchyma* (Figs. 766, 767, 768, 770, 772 and 773) is seen in the spongy and enlarged petioles of water hyacinth. The air storage in such cells gives buoyancy to the plants which helps to keep them afloat. In lotus, even the fruits contain large air cavities rendering them buoyant and also facilitating their dispersal by water. On the other hand, *Azolla* owes its buoyancy primarily to the minute epidermal hairs on the upper surface of the dorsal lobes of the leaves; the layer of air trapped by the hairs keeps the surface of the leaves dry, at the same time preventing the plant from being immersed. In *Jussiaea repens* and *J. suffruticosa* the bulk of the tissue consists of aerenchyma if growing in water but if *J. suffruticosa* be grown on land no development of aerenchyma occurs. In some marine brown algae, e.g, *Fucus*, special swellings, like bladder serve to keep the shoots erect in water. The air cavities, in the floating leaves of aquatics such as water-lily, generally form a continuous air-communicating system by means of which the submerged organs can easily exchange gases with the air outside through the ever open stomata.

The ability to develop air cavities (*lacunae*) is probably genetically endowed and in submerged hydrophytes, the seemingly apparent reaction to the aquatic environment, may very well be no more than the full realisation of this ever present potential. The number of air cavities is sometimes specific and may vary in a characteristic way between different species of a genus; it may be of some value as a criterion for identification.

The fragility of lacunate tissue is counterbalanced by radial plates and in many species of vascular hydrophytes, e.g., *Pontederia*, *Potamogeton*, *Sagittaria* sp., etc., by the development of water-tight transverse *diaphragms* which interrupt the air cavities at intervals. The diaphragms are groups of very thin-walled cells, 1–3-cell thick, cell walls of which swell into the cavities, branching freely and rapidly, usually forming a spongy mass of loosely arranged cells.

Diaphragms are commonly found at the nodes of submerged stems and are usually scattered in the petioles. They are characteristic of *Nymphaeceae* and perhaps of all aquatic nonocotyledons. They are absent for some unknown reasons from the petiole of *Nymphaea* and are also very infrequent in its submerged leaves.

In some hydrophytes such as *Trapa*, *Ceratopteris*, *Pistia*, *Hydrocharis*, etc., lacunate tissue develops so excessively that bladder-like swellings, *floats* of spongy parenchyma are formed in *Pistia*, the underside of which often shows a conspicuous ovoid swelling, while in *Eichhornia*, the whole petiole may be swollen into a bulbous spongy float. The relation of this adaptation of forming floats to the buoyancy requirement is confirmed by the fact, that these plants when growing in the mud, do not develop floats, the petioles or the leaf-bases then remaining slender and elongated.

(*ii*) Root system is often very poorly developed usually shorter and much less branched than land plants or as in some floating hydrophytes, (e.g., *Wolffia*, *Salvinia*, etc.) the roots may be entirely absent (in some species of *Ceratophyllum* root hairs have completely disappeared). In *Jussiaea repens*, however, two forms of root develop when growing in water; ordinary anchorage roots and erect very spongy negatively geotropic roots which grow upwards often till they reach the surface of water.

The above generalisation is, however, quite unjustified. Many hydrophytes, it must be apparent to even a casual observer, have exceedingly well-developed root systems or rhizomes which show few anatomical modifications compared with those of land plants, e.g, *Nymphaea*, *Pontederia*, etc., have stout spongy rhizomes. The tough and sometimes woody-rhizomes of *Cyperus*, *Scirpus*, *Typha*, etc., branch freely, covering vast areas in growing season. The root stock of *Vallisneria* which is bulbous in *Sagittaria* and hard and corm-like in *Aponogeton* and *Isoetes*, are well-known.

All *free-floating rosette plants*, e.g., *Eichhornia*, *Pistia*, etc., have supremely well-developed adventitious roots. Relieved of the function of anchorage (the floating plants do not need any anchorage), the roots of these plants are at least partly responsible in preserving the stability of their rosette leaves. The roots of these genera can develop chloroplastids in their epidermal and cortical cells when they get sufficient light and may contribute something, however small that may be, to the net assimilation of the plant.

Ceratopteris has a shallow system of quite freely branched roots. It has been found that a significant proportion of *biomass* is represented in vascular hydrophytes, by the underground parts, e.g., rhizome, roots, bases of the aerial shoots—in *Typha* sp. more than 50% of the biomass can be accounted for by the root system on fresh weight basis. Biomass, is actually the quantity of organic substance produced in a given area, as in the weight of the vegetable matter, removed by a *dipping quadrat* (a sample plot from which the vegetation is removed) or the plankton of a given volume of water. Contrary to popular idea, the roots of most submerged hydrophytes do develop abundant root hairs.

The notion of wholly or mainly foliar absorption of water and nutrients by floating or submerged hydrophytes thus rests on quite unwarranted evidences. These include, as mentioned before, (*a*) rudimentary nature of the root system. We know that it cannot be true, for far from being rudimentary, root system of many species is remarkably well developed; (*b*) lack of root hairs; again this is certainly not applicable to the vascular hydrophytes for root hairs are not only present but may be conspicuously abundant; moreover, the absence of dearth or root hairs are not real indications of the roots' ability to absorb water and nutrients; (*c*) vascular reduction; it is true that, there is reduction in the extent of xylem vessels, but is the uptake of water and inorganic ions by roots directly related to it?

An interesting anatomical anomaly must be mentioned here. Except in *Nymphaea*, and a few other hydrophytic species, the stele of roots is surrounded invariably, in all manner of watery habitats, by a very conspicuous endodermis, bearing typical *casparian bands* (*Myriophyllum*) or heavily thickened transverse, radial and inner tangential walls (*Potamogeton* sp.). This cannot conceivably be a trend towards structural reduction of roots—may not also be dismissed so easily as mere ancestral relics. We know it for certain that endodermis is functional in regulating lateral movement of water and ions in terrestrial roots; how could we now reconcile logically the prominence of endodermis in aquatic roots with the classical notion that in vascular hydrophytes there is scarcely any significant absorption and transport? (Sculthorpe, 1967).

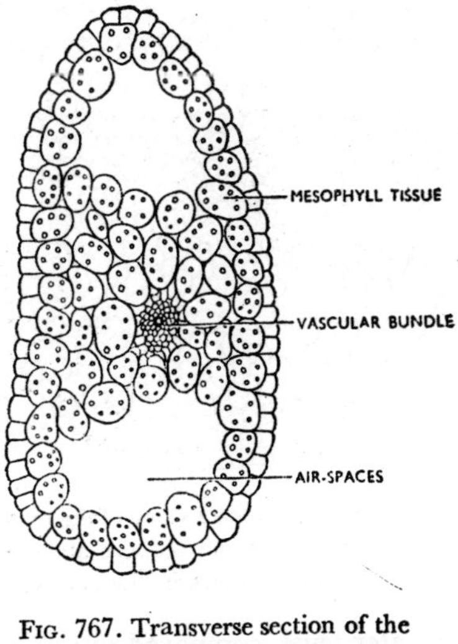

FIG. 767. Transverse section of the leaf of *Zannichellia* sp. showing large air-spaces and feeble development of vascular tissue.

The root caps of vascular hydrophytes are typically elongated and sheath-like (*root pockets*) or may also be totally absent (*e.g.*, *Azolla*). Root cell walls usually remain abnormally thin.

(*iii*) Epidermis usually lacks cuticle or periderm (suberised cork cells) and stomata are generally absent in the submerged organs. If present they are usually functionless (Figs. 766 and 767). This certainly adds to the evidence of the terrestrial ancestry of the vascular aquatics. In the floating leaves, stomata, if present, are generally restricted to the upper epidermis. The lack of cutin and suberin from the cell walls of submerged organs, theoretically makes these organs capable of absorbing water and nutrients throughout their surface directly, rather than by means of root absorption.

(*iv*) Mechanical tissue, although very feebly developed, is amply compensated by the buoyancy due to the presence of aerenchyma and the support afforded by surrounding water, renders the formation of mechanical tissue seemingly unnecessary.

(*v*) Conducting tissue is poorly developed but there is evidence that a transpiration current operates in the xylem and there is also conclusive evidence that at least in vascular aquatics, root absorption is indispensable and which, as we have seen just now, may be quite considerable for optimal growth and development of vascular hydrophytes (Figs. 767, 768, 769 and 770).

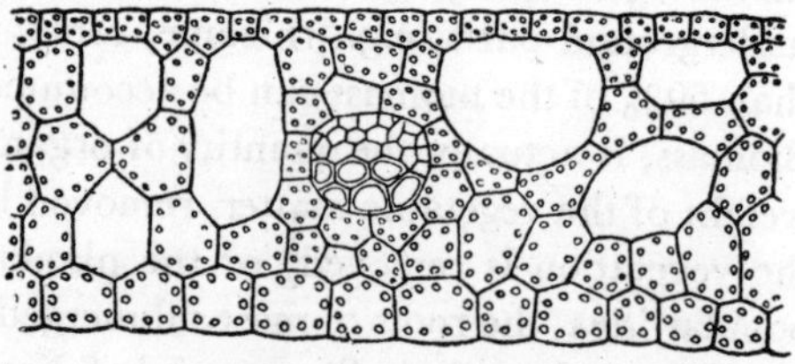

FIG. 768. Transverse section of a submerged leaf of *Potamogeton*. Note epidermal chloroplasts, absence of cuticle and stomata and reduced vascular elements.

Vascular tissues are usually found towards the centre of the stele with very few bundles towards the periphery for the pulling forces, exerted in quickly flowing water, are more economically met by aggregation of vascular bundles towards the centre. The almost complete absence of secondary growth in thickness of stems and roots is also an important feature of hydrophytes (Figs. 769, 770 and 772).

(*iv*) In floating hydrophytes, rooted in the mud, e.g., water-lily, etc., with the usually rounded leaves lying flat on the surface of water, certain mechanisms, such as waxy surface causing water-drops to roll off quickly or as in *Salvinia*, leaf hairs holding water-drops above the surface of the leaf, protect the leaf surface from getting wet. Closely packed hairs, particularly those with hydrophobic tips can also protect the leaf surface from getting too wet. A homogeneous wax covering on the leaf surface might also be expected to provide an unwettable surface. Much of the strength of floating leaves results, however, from their leathery texture. The presence of copious mucilage accumulation on the aerial organs seems also an adaptation for protecting them from getting too wet. Mucilage may also prevent too rapid diffusion of water through the cell wall. This certainly is an obvious benefit to the plants, whose stomata usually are confined to the upper surface of leaf lamina.

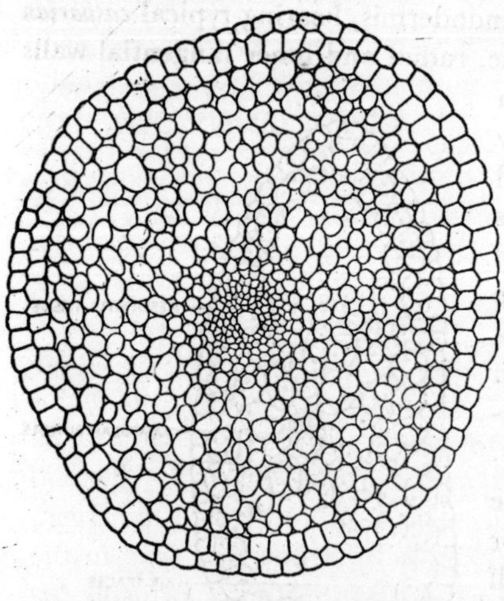

FIG. 769. Transverse section of the root of *Elodea* showing much reduced stele and large air-passages in the cortex.

(*vii*) In the 'amphibious' group of aquatics, growing in shallow water-logged soils with their shoots well above the surface of watery medium, striking *dimorphism* is sometimes exhibited by the leaves that diverge from the stem below the water level, compared to those that diverge above water.

(*viii*) *Heterophylly*—Heterophylly is the phenomenon which is commonly observed in some hydrophytes, as for example, *Sagittaria sagittifolia*, *Limnophila heterophylla*, etc. It indicates in contrast to *homophylly* (only one form of leaf) production of completely different forms of leaves in the same plant (transition forms are also sometimes seen). In submerged aquatics with floating aerial leaves, the submerged leaves are generally linear, ribbon-shaped or finely dissected while the aerial leaves are usually entire and rounded or slightly lobed. This difference in the two or more forms of leaf structure may be due to the characteristic physiological features of water plants, such as: (1) quantitative reduction in transpiration, (2) reduction of light intensity (in the completely submerged parts), (3) plants subjected to much hydrostatic pressure, (4) unaffected by drought and lastly (5) variation in life form and habitat.

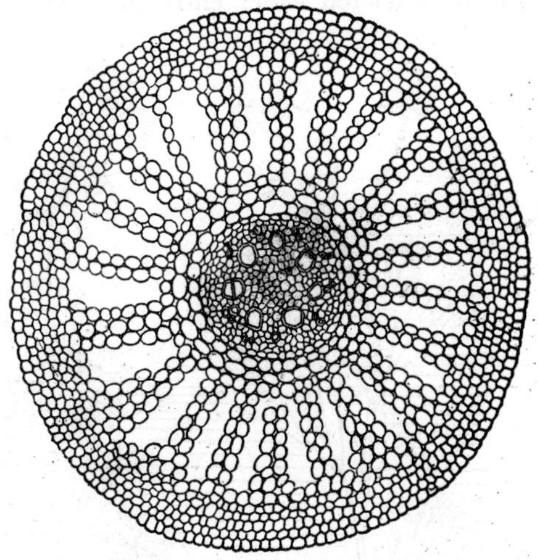

FIG. 770. Transverse section of the root of marshland *Typha latifolia* showing abundance of aerenchyma.

In *Sagittaria* (Fig. 771) the heterophylly seems to be due to the difference in the intensity of light in the submerged and aerial parts (more light intensity in the aerial parts apparently favouring formation of entire leaves). In others, e.g., *Proserpinaca palustris*, the aquatic leaves are finely serrate on horizontal stem whether in air or in water while the aerial leaves are linear but entire on erect vertical stems. In the last case, the heterophylly is most probably determined by genotype.

In *Limnophila heterophylla*, the case is even more interesting. On aerial stems, transition forms of leaves are found, from finely dissected leaves to the fully aerial entire types; in the submerged leaves, however, only the veins of the aerial lamina are developed, the remaining development of lamina (the intervenous mesophyll tissue) is totally arrested, (are these leaves vestigial?) perhaps as an evolutionary adaptation (Fig. 771).

Heterophylly, it must be pointed out here, is not unique to hydrophytes. It also occurs in many terrestrial plants, both woody and herbaceous. Thus heterophylly has been attributed both in the internal causes, e.g., age, genotype and nutrient status of the plant as well as to the influence of environmental factors, notably the photoperiod, temperature and moisture conditions.

Heterophylly, as mentioned before, means the presence of two or more distinct types of leaf in a single individual. The leaf types may differ markedly in shape and in anatomical organisation or they may differ in habit and anatomy, yet be of comparable shape (e.g., the floating leaves of some *Nymphaea*). In extreme case of heterophylly, the leaf types differ in all the above three aspects and a full-grown plant may bear submerged, floating and aerial leaves at the same time. The change from one leaf type to the other may be quite abrupt or it may be more gradual with transitional forms.

In some hydrophytes, a *heteroblastic* sequence (an ontogenic sequence in which early formed leaves are markedly different from later-formed ones) is apparent regardless of the environment. In *Sagittaria sagittifolia*, the first formed 'juvenile' leaves are invariably of the dissected or ribbon-shaped submerged type, irrespective of whether the plant is in water or on land, and the later formed ones on the other hand, are the floating or aerial type (usually associated with the reproductive phase).

Heterophyllous polymorphism in many species has confused their taxonomic significance and nomenclature within several hydrophyte groups. The close morphological similarities of ecological forms, which really belong to different species, has brought out clearly the unreliability of any identification based wholly or primarily on leaf characters (attention should be diverted to fruit characters instead). Allsopp (1965) believes that the principle involved in the morphogenesis of hydrophytes, is the highly unrestricted water entry into the developing tissue.

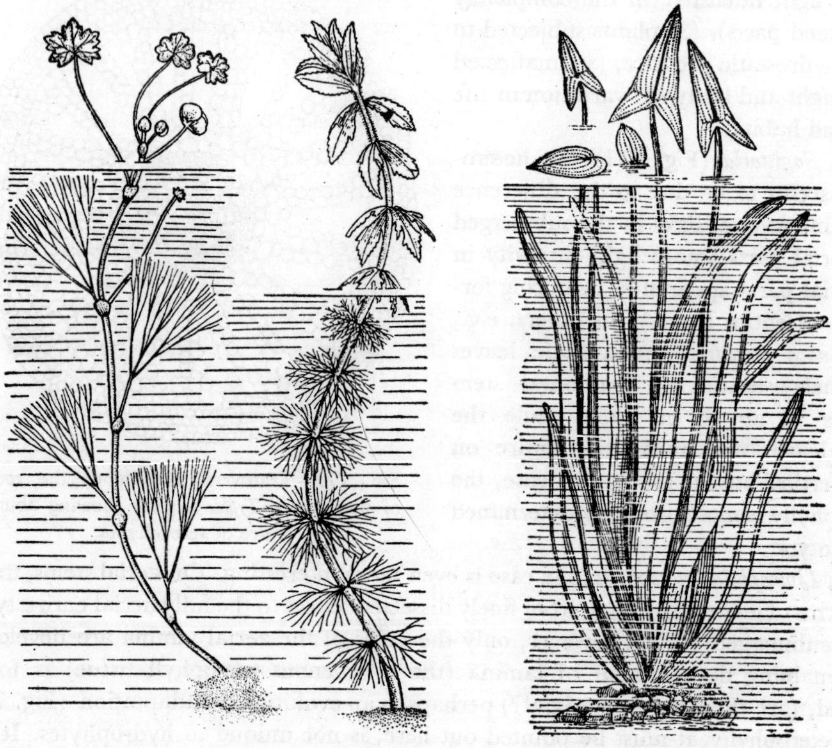

FIG. 771. Heterophylly: A, *Rananculus aquatilis*; B, *Limnophila heterophylla* and C, *Sagittaria sagittifolia.*

Allsopp (1956) found that in sporelings of amphibious *Marsilea drummondi*, the heteroblastic sequence, from the juvenile subulate first leaves through spathulate and bifid intermediate types to the quadrifid or quadrifoliate adult form, could be affected by the composition of the media. Carbohydrate or nitrate starvation induces formation of only juvenile submerged type of leaf and also to a reduction from solenostele to protostelic condition. Wardlaw (1965) also found that in ferns, higher sugar concentration favours the aerial adult type of leaf-formation while in low carbohydrate media, only the juvenile types are formed. Allsopp suggested that the structural differentiation, distinctive of land and water forms, is perhaps mediated through the carbohydrate balance of the plants.

The heteroblastic manifestations in many heterophyllous plants is certainly correlated with the progressive enlargement of the shoot apex, the degree of which is certainly determined by protein synthesis (Allsopp, 1954–1965). Further evidence of the importance of protein synthesis in the heterophyllous plants is afforded by White (1966). He found that if *thiouracyl*, a long-known inhibitor of protein synthesis, is added to the media, the rate of leaf formation in some *Marsilea* sp., is retarded and leaves are only of one type, the typical quadrifoliate natural land-forms. Sometimes fully formed water-forms are converted into land-forms in presence of thiouracyl

The linear and the finely divided forms of leaves in the submerged parts may be of considerable ecological importance to plants for the dissected leaves certainly offer less resistance to under-currents and the stress and strain of water which, if the leaves were entire and broad, might have been torn to shreds and permanently damaged. Dissection of the leaf is not a direct reaction of plant to flowing water, since it occurs in many hydrophytes of stagnant habitats. Dissected leaves certainly have a higher surface area/volume ratio than most entire aerial leaves but the advantage gained by this is debatable. Sometimes mechanical tissues are found along the margins of such finely divided or ribbon-shaped leaves. In the aerial leaf of *Eichhornia*, angular collenchyma along the margins of blade certainly offers resistance to tearing stress.

(*ix*) In lotus, the long petioles seem to adapt themselves to the depth of water where they grow, always keeping the leaf lamina on the surface of water. This is perhaps due to the fact that petioles of floating-leaved hydrophytes have always a great capacity for renewed growth, (as much as 17 mm/hr in *Sagittaria*) which is perhaps regulated by auxins. (1 p.p.m. IAA, IBA, IPA, NAA, etc., actually promoted cell expansion in *Nymphaea* and *Sagittaria*). An interesting observation is the geotropic growth curvature exhibited by some *Nymphaea* petioles.

(*x*) The propagation of most aquatics is essentially vegetative, e.g., by rapid fragmentation in *Elodea*, by runners in *Eichhornia*, by rapid production of thalli in *Lemna*, etc. They usually flower much less freely than land plants. When flowers are present fertilisation takes place at or above the surface of water. Setting of fruits and seeds and their dispersal are always uncertain in the watery habitat.

(*xi*) The majority of hydrophytes are perennial. The relative uniformity of the aquatic environment encourages good vegetative growth. The insulation against intense illumination, afforded by water, abundance of CO_2 in the watery medium and freedom from violent fluctuations of temperature, all seem to maintain active photosynthesis of submerged hydrophytes at a high optimal level, leading to a sustained and sometimes luxuriant development of foliage. This "excessive vegetative activity of water plants, acts in all probability as a deterrent to sexual reproduction" (Mrs. Arber). It may be that the increasing number of vegetative meristems make a greater demand on the plants' nutrients, leaving very little for the production of floral meristems.

There is thus very often a tendency against hydrophytes towards the replacement of sexual by vegetative reproduction. This is quite in line with the fact that the flowers of many angiospermic hydrophytes are singularly ill-adapted to aquatic life and have still to be raised into the air above the surface of water for pollination. A capacity for sustained vegetative propagation, evading the hazards of pulling up the flowers above the water level, may be advantageous especially to the plants living in deeper waters.

Although the frequency of vegetative reproduction is higher among the hydrophytes, the organs actually employed as *propagules* are essentially similar to those of terrestrial herbs. Fragmentation of the plant body, followed by regeneration from any small part, bearing a vegetative bud, is a common occurrence in some free-floating plants and also in submerged hydrophytes with long and delicate stems. Regeneration from any fragment of the body containing meristematic tissue is also common among members of the family *Podostemaceae*. *Gemmipary* the development of new plants from vegetative buds on the parent body, occurs in many wild hydrophytes.

Colonisation by means of rhizomes, stolons, runners, etc., is widespread amongst hydrophytes and in numerous taxa, these organs also store food reserves and serve as *hibernacula*, enabling plants to survive through unfavourable environmental conditions.

Towards the ends of the growing season, some species of *Sagittaria* and *Potamogeton* form small tubers on the ends of stolons or lateral branches, whilst some species of *Hydrocharis*, *Myriophyllum* and *Utricularia* develop specialised dwarf-shoots known as *turions*. Among other specialised structures, *Ceratophyllum* has dense shoot apices, protected by mucilage or a thick cuticle which are full of food reserves. These remain dormant through the winter, germinating again with the advent of favourable weather in the spring. Throughout large sectors of their geographical distribution. *Elodea canadensis* and *Acorus calamus*, propagate only by vegetative reproduction—in those regions they do not produce flowers at all.

There are numerous examples amongst hydrophytes of a phenomenon known as *pseudovivipary* (it also occurs in many terrestrial monocotyledons and dicotyledons). Here, the vegetative propagules replace some or all of the normal sexual flowers in the inflorescence. It has been known for a long time that if an inflorescence of *Myriophyllum verticillatum* became submerged, a turion was formed at the apex. In an *Aponogeton* sp., the digitately branched axis may bear at the same time the much reduced inflorescence as well as young plants (de Wit, 1958).

The hydrophytes of the family *Alismaceae* are rich in pseudoviviparous species. *Caldesia grandis*, an Indian species has an inflorescence-axis, bearing whorls of turions instead of flowers in submerged plants growing in water of 50 cm or greater depth (Den Hartog, 1957). Several American species of submerged hydrophytes (e.g., *Echinodorus*) bear plantlets in place of flowers all along the axis of their inflorescence (de Wit, 1958).

Gemmiparous buds may also be produced from foliar tissue in some hydrophytes. These buds arise at specific sites of meristematic *cell-aggregates* in the leaf-lamina. Since a single leaf may bear several such buds, each capable of developing into a plantlet (or bulbil) which may either drop off or become independent when the leaf decays, gemmipary is certainly a prolific method of vegetative multiplication. We do not know much as yet of the physiology and cytology of these sites of cell-aggregates and how they differ from the remaining cells of the leaf tissue. In the aquatic fern, *Ceratopteris*, which produces plantlets in the vicinity of vein-endings at the base of the marginal furrows of the mature leaf, the excised meristems of those specific sites will produce healthy plants on a nutrient medium, *only* in presence of either adenine or IAA (Gottlieb, 1963).

(*xii*) *Physiology of sexual reproduction in vascular hydrophytes*. There is some evidence that in some hydrophytes, the initiation of the flowering phase may depend upon nutrition. The carnivorous *Utricularia* flowers when grown aseptically in inorganic nutrient medium but only when supplemented with organic nitrogenous compounds, e.g., a mixture of peptone and meat extract (Harder, 1963). The carnivorous habit does seem likely to be indispensable in natural habitats for the fulfilment of the life cycle of *Utricularia*. A photoperiodic response may also be involved in *Utricularia*. Pringsheim (1962) and Harder (1963) found that the plant flowered in artificial light in supplemented nutrient medium, under 11–12 hr. photoperiod.

Eichhornia crassipes, on the other hand, seems to be a day-neutral plant—flowering seems to be influenced by temperature rather than daylength. Species of the genus *Lemna* show opposite photoperiodic behaviour in that, some species flower in short days while others in long days or even in continuous illumination. Some do not flower under any photoperiodic condition. The floral effect in *Lemna* is indeed complex with complications arising from interaction of different external factors, e.g., temperature, CO_2-concentration, chelators, micronutrients (e.g., Cu and Fe), etc., with intensity and duration of light.

Wolffia microscopica in artificial culture behaves as a short-day plant for flowering—a minimal dark period of 12–14 hr. of at least one cycle, seems to be the essential requirement. The recently isolated cytokinin, *zeatin*, when added to the medium, however, promotes flowering in *Wolffia* even under a long photoperiod (Maheswari and Venkataraman, 1966).

Photoperiodic responses are probably involved in the promotion of reproductive phase in some aquatic pteridophytes such as *Salvinia*. *Salvinia natans* appears to be a short-day plant—in nature, sporocarps appear in daylength of 13 hr. or less and even a 7 hr. photoperiod seems to be sufficient to induce 100% response, if the treatment is given for a week (Nakayama, 1952).

(*xiii*) The seeds of an overwhelming majority of aquatic vascular angiosperms exhibit prolonged dormancy. In most species, this is due to the mechanical imprisonment of the embryo within a hard, heavily cuticularised testa or pericarp. Germination is extremely erratic and uncertain—some way remain dormant but viable for four or five years. Seeds of lotus show good viability, ranging from two to four years, irrespective of the temperature and humidity of the medium.

(*xiv*) In some hydrophytes, we have noted before, food is stored up in the rhizome, (e.g., water-lily) in others such as *Sagittaria*, tubers are formed.

CLASSIFICATION OF HYDROPHYTES

Mrs. Arber (1920) recognised two primary groups of aquatic angiosperms: *rooted* and *non-rooted*, which she again subdivided according to the type of foliage and inflorescence produced and the position of these organs with respect to the water level. She realised that this distinction cannot be a strict one as the lines of demarcation between them were blurred by transitional types. Penfound (1952), following the outline of Hess and Hall (1945), split hydrophytes into *wet-land types*—in soils saturated with water and *aquatic types*—in soils covered with water for most of the growing season. He recognised three forms among the aquatic forms—*emergent*, *floating* and *submerged*.

Danserau (1945) developed a new classification of the biological forms of hydrophytes. This is a rather sensitive system with attributes of correlating the zonation of communities along streams and lakes.

Danserau classified all aquatic forms into two broad types—*helophyta* (paludous) and *hydrophyta*. The hydrophyta is further subdivided into *natantia* (not fixed to the substratum; e.g. *Lemna*, *Ceratophyllum*, etc.), *radicantia* (fixed to the substratum) and *adnata* (fixed on rocks, or on other plants; e.g., *Podostemon*, *Fontinalis*, etc.). Radicantia is divided into *emersa* (at least partially emerged) and *subemersa* (none or few leaves floating). Emersa is further divided into *foliacea* (with a grand leaf development; e.g, *Sagittaria*, *Pontederia*, etc.), *Junciformia* (reduced leaf development; e.g., *Scirpus*, etc.) and *nymphoidea* (leaves floating ; e.g., *Nymphoea*, etc.). Subemersa is put into 3 classes: *vittata* (with long stem, leaves soft; e.g., *Potamogeton*, *Vallisneria*, etc.), *rosulata* (leaves reduced, basal; e.g., *Isoetes*, *Lobelia*, etc.) and *annua* (therophytic annuals; e.g., *Najas*, etc.).

Hejny (1957, 1960) simply recognised three groups: *euhydatophytes*—with vegetaive organs completely submerged and the reproductive organs, submerged or aerial: *hydatoaerophytes*—with partly submerged and partly floating vegetative body and aerial inflorescence, and *tenagophytes*—amphibious plants occurring in habitats with marked fluctuations of the water level. On the basis of their attachment to the soil, Luther (1949) classified hydrophytes into: *haptophytes*—plants which are attached to, but do not penetrate much to the substrate, e.g., many algae, lichens, bryophytes and among angiosperms, only the *Podostemaceae; rhizophytes*—whose basal parts actually penetrate the soil or the substrate in which they grow and *planophytes*—freely floating plants with submerged or surface-floating photosynthetic organs. The third group includes microscopic *planktophytes* and macroscopic *pleustophytes* (e.g., larger floating algae, liverworts, ferns and some angiosperms). Luther further divided pleustophytes into three groups according to the level at which they float—a differentiation which can hardly be sustained as species of *Lemna*, *Ceratophyllum*, *Utricularia*, etc., seem to rise and fall according to the season or to their stage of development. Den Hartog and Segal (1964) further elaborated and then supplemented Luther's division of rhizophytes and pleustophytes. Eleven basic types of growth habit were recognised among rhizophytes, e.g., rhizophytes possessing a short stem and a rosette of stiff leaves, with or without stolons, distinguished as *isoetids* (*Isoetes*, etc.); rhizophytes with long stems, entire submerged leaves, no floating leaves, and aerial or submerged reproductive organs as *elodeids*, (e.g., *Elodea*, *Najas*, etc.) and so on.

The British ecologists, Tansley (1949), Spence (1964) and Sculthorpe (1967) recognised principal life and growth forms of vascular hydrophytes and adopted a scheme of classification based on this recognition:

A. *Hydrophytes attached to the substratum*

(1) Emergent hydrophytes:

Occur on exposed or submerged soils; mainly rhizomatous or cormous perennials; in heterophyllous species submerged and/or floating leaves *precede* the mature aerial leaves; many submerged forms (usually sterile); all produce aerial reproductive organs, e.g., *Typha*, *Phragmites*, etc.

(2) Floating-leaved hydrophytes:

Occur in submerged soils in water depths up to approximately 4 m.; some species may exist as reduced land forms; in heterophyllous species, submerged leaves precede or accompany the floating leaves; many species produce aerial leaves in crowded habitats; reproductive organs, floating or aerial;

(*i*) rhizomatous or cormous type, with floating leaves on long flexible petioles, e.g., *Aponogeton* sp., *Nymphaea*, etc.

(*ii*) stoloniferous type, with trailing stems and producing floating leaves on relatively short petioles, e.g., *Potamogeton natans*, *Nymphoides*, etc.

(3) Submerged hydrophytes:

Occur on submerged soils at all water depths to a maximum of about 12 m.; foliage entirely submerged; leaves often filiform or ribbon-shaped or even finely divided; a few species may produce land forms; reproductive organs aerial floating or submerged.

(*i*) *caulescent type* with or without a rhizome, the long leafy stems rooting from nodes, e.g., *Elodea*, *Hydrilla*, *Najas*, *Potamogeton* sp., etc.

(*ii*) *rosette type*, with radical leaves arising often from a condensed, tuberous stock or a rhizome; often stoloniferous, e.g., *Aponogeton* sp., *Isoetes*, *Sagittaria* sp., *Vallisneria*, etc.

(*iii*) *thalloid type*, with plant body reduced to a cylindrical or flattened, creeping or floating , polymorphic thallus, with erect or trailing secondary branches, e.g., the *Podostemaceae* such as species of *Hydrobryum*, *Podostemon*, etc.

B. *Free-floating hydrophytes*

All are unattached in slow-flowing water, but some species with extensive root system may become attached to the substrate in shallow water; numerous species produce land forms; very diverse in form and habit, ranging from stoloniferous plant, with rosettes of aerial and/or floating leaves and well-developed submerged roots (e.g., *Ceratopteris* sp., *Eichhornia crassipes*, *Pistia*, *Trapa*, etc.), to minute surface-floating or submerged plants, with reduced assimilatory thallus, having few or no roots (e.g., *Lemna* and *Wolffia*); reproductive organs floating or aerial, very rarely submerged (e.g., *Ceratophyllum* and *Salvinia*); numerous submerged species rise to the surface to flower, and may sink to the bottom of the substratum to perennate (e.g., *Lemna trisulca*, *Utricularia*, etc.).

For convenience, the hydrophytic vegetation has been grouped here under the following five morpho-ecological groups:

(a) FLOATING HYDROPHYTES—Include all plants that float or are in contact with water and air but *not* rooted in the soil. Examples—*Pistia*, *Lemna* (in the family *Lemnaceae*, where the leaf and stem, as mentioned before, are represented by a small thallus, the roots are in the process of disappearing; *Lemna* has only one strand of root), *Wolffia* (entirely rootless, one of the minutest of the flowering plants), *Aldrovanda*, *Eichhornia* (water hyacinth), *Limnanthemum*, the water fern, *Salvinia*, etc.

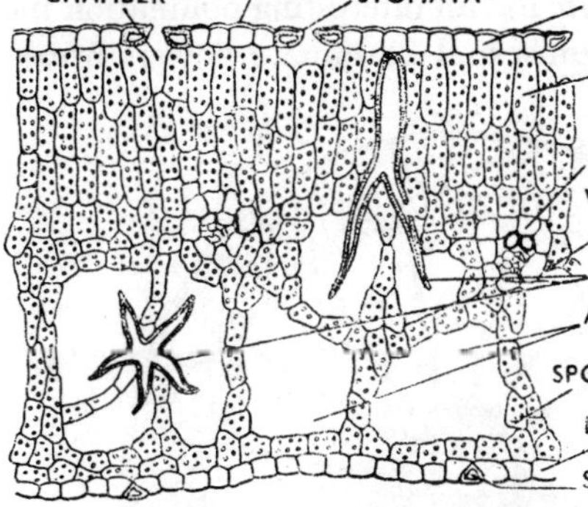

FIG. 772. Transverse section of a floating leaf of water-lily. Note larger air-chambers, mucilaginous glands and very much reduced vascular tissue.

(b) SUSPENDED HYDROPHYTES—Plants such as the brown alga, *Sargassum* and also *Lemna trisulca*, etc., are in contact with water alone. They are not on the surface but just below the surface of water. The plants under this group are absolutely free from any necessity of transpiration, yet for some quaint reason or other, they are found in the best-lighted and best-aerated part of the water.

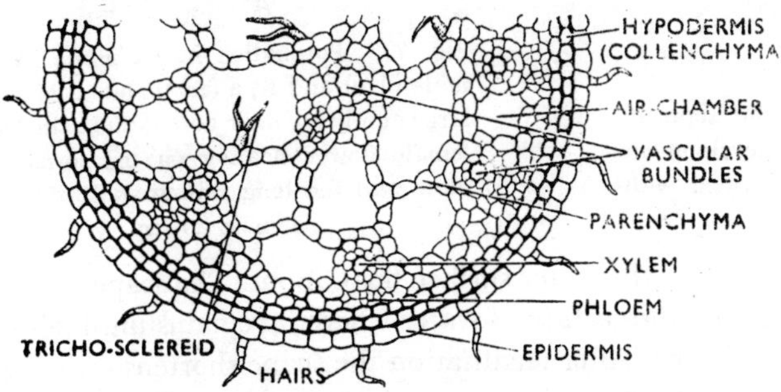

FIG. 773. Transverse section of the petiole of water-lily. Note ground tissue with large air-chambers and very poorly developed vascular bundles.

(c) FLOATING BUT ROOTED HYDROPHYTES—The plants in this category are anchored in the muddy soil at the bottom of the ponds and lakes. Examples—water-lily, lotus, *Potamogeton indicus*, *P. crispus*, *P. mucronatum*, *Aponogeton crispum*, *Neptunia oleracea*, etc. (Figs. 772 and 773). The hydrophytic fern, *Ceratopteris thalictroides* (*Parkeriaceae*) which is of world-wide distribution throughout tropical and subtropical regions may grow on damp soil, but usually it is a true aquatic, either rooted in the mud or free-floating.

(d) SUBMERGED AND ROOTED HYDROPHYTES—Plants in this groups are completely submerged and are also anchored to the substratum. Extensive development of blooms of floating blue-green algae such as *Anabaena*, *Oscillatoria*, etc., (sometimes the growth is so luxuriant that more than 9,000,000 filaments may be found per litre of water!) on the water surface particularly under calm conditions may cut down light penetration to such an extent that the growth of submerged hydrophytes may be severely reduced. Because of this shade effect compared to light available to land plants (light both in quantity and quality is already altered for deeply submerged plants due to passage of light through several feet of water) many submerged hydrophytes show features which are associated with shade-loving land plants, *sciophytes*, e.g., long internodes, no palisade parenchyma, presence of chloroplastids in the epidermal cells, etc. Three genera or blue-green algae, *Anabaena*, *Aphanizomenon* and *Microcystis*, which produce the most spectacular water blooms bear the specific name of '*flos aquae*'. Examples—Mostly monocotyledons such as *Vallisneria*, *Hydrilla*, some *Potamogeton* species, *Najas*, etc., and dicotyledons such as *Ceratophyllum*, *Myriophyllum* (Fig. 774), *Utricularia* (insectivorous bladderwort), etc. *Chara* and almost all macroscopic algae. *Vallisneria* has an interesting pollination mechanism. The unisexual staminate flowers break off from the inflorescence and after rising to the water surface float about. The scape of the pistillate flower, elongates and reaches the surface of water where it remains until pollinated by pollens of the staminate flower. After fertilisation the scape shortens drawing and developing young plant under water where it completes its development (Fig. 775). In *Myriophyllum*, however, inflorescences are always maintained above water and flowers are always(?) wind pollinated.

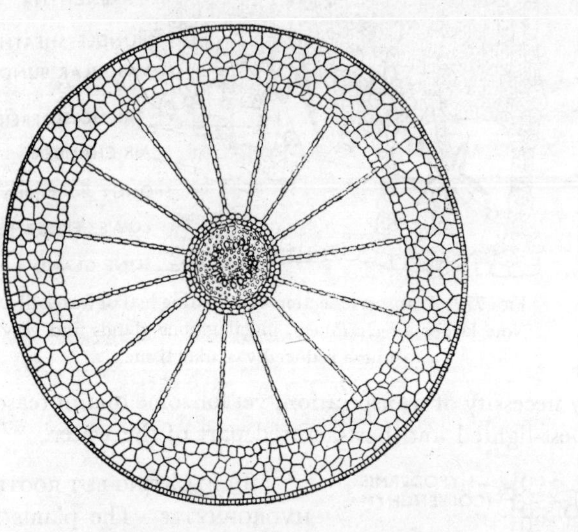

FIG. 774. Transverse section of the stem of *Myriophyllum*. Note the poor development of vascular bundles and cortex provided with large air-chambers.

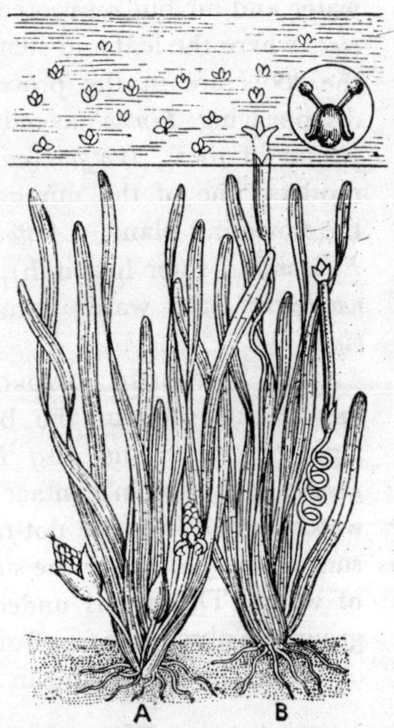

FIG. 775. Pollination in *Vallisneria*. A: male plant and B: a female plant. Note large number of male flowers and a single female flower floating on the surface. Note also the long coiling stalk of the female flower.

(e) ROOTED AMPHIBIOUS HYDROPHYTES—These are so named because the plants in this group grow in shallow water with their underground parts in water or in water-saturated muddy soil, while extending their shoots well above the surface of water. They are also known as marsh plants or *helophytes.* The characteristic feature of marsh plants is that their lower parts, buried in muddy water are adapted to aquatic life, while the air-exposed upper parts resemble ordinary mesophytes or sometimes may even be adapted to withstand atmospheric drought. However, the reduced leaf surface and the presence of the cuticle in aerial parts, observed in some marsh plants, may only be xeromorphic in nature, i.e., the persistence of ancestral features—they must have been land plants at the beginning before they started their watery career—in spite of a striking change of habitat. These so-called "marsh xerophytes are really typical hydrophytes, wearing a false xerophytic mask."

EXAMPLES—Some varieties of rice, *Oryza sativa* (rice plants which live with their roots partly submerged in water, i.e., under oxygen-poor conditions, usually have large air-channels in these roots); the typical marshland plants, e.g., *Scirpus*, the grass *Typha*; the mangroves *Rhizophora*, *Avicennia*, etc. (ecological adaptations not in the least hydrophytic; sometimes striking xeromorphic characteristics); the heterophyllous *Sagittaria, Proserpinaca palustris; Enhydra fluctuans* (Fig. 776), *Polygonum barbatum*, *Jussiaea repens* (may also grow completely submerged), *Jussiaea suffruticosa*, *Commelina* sp., etc.

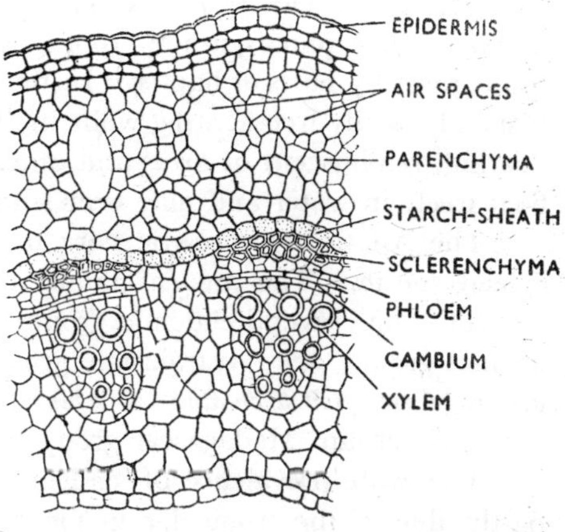

FIG. 776. Transverse section of a portion of the stem of marshland hydrophyte, *Enhydra*. Note again (*i*) feeble cuticle and (*ii*) conspicuous air-spaces in the cortex.

Among other common amphibian hydrophytes from water-logged rice fields of southern part of West Bengal are: *Cardenthera triflora*—usually heterophyllous with leaves above the water level entire, submerged leaves in water are much dissected. The plant remains heterophyllous even when water is drained off from the paddy fields after the harvesting of rice. When the land becomes completely dry, sometimes homophyllous plants are seen; *Hygrophila spinosa*, common in rice fields with only one type of leaves; *Limnophila gratioloides*, a small herbaceous plant, with pink leaves and stems appear in the wet rice fields after rice harvest, etc.

Many of these amphibious hydrophytes grow comfortably in this type of muddy soil because the oxygen concentration is low in such habitats and these plants have developed especially low oxygen requirement for germination, as for example oxygen requirement for the most satisfactory germination of some rice varieties is only about one-sixth of that of wheat.

In many of this type of marshy hydrophytes such as *Rhizophora*, *Ceriops*, etc., special root branches grow erect until they project above the poorly aerated muddy soil. These structures called *pneumatophores* usually have well-developed intercellular system of air-spaces continuous with the stomata of the leaf and must be of unquestioned value in supplying oxygen to the plants.

Somewhat loose associations of two hydrophytes are sometimes observed in many watery-habitats of India as for example, *Eichhornia-Potamogeton association*, *Hydrilla-*

Vallisneria-association and *Ceratophyllum-Vallisneria association*. Similar associations are sometimes observed between moisture-loving, insectivorous *Drosera burmani* (also *Drosera indica*) and the marshland hydrophyte, *Eriocaulon*. In certain localities of North Bengal the presence of *Eriocaulon* indicates the probable presence of *Drosera* in the neighbourhood.

The proportion of *endemic species* is pretty high among the hydrophytes—about 25–30 per cent. The high frequency of endemism is probably related to the edaphic characteristics of the soil and is a possible relic status. Most endemic hydrophytes are naturally to be found in the tropics; numerous ecologically restricted members of the *Podostemaceae* (perhaps the strangest and most provocative family of angiosperms) being prominent amongst them.

In some members of the family *Podostemaceae*, where the vegetative body is reduced to a creeping or floating thallus, closely resembling certain fresh-water and marine algae or may even be strongly reminiscent of hepatics and lichens,—this resemblance of the thalloid types to far distant cryptogams is not easily interpreted—astonishing polymorphism is clearly shown. The rock-creeping anomalous thallus is most probably a rhizome-like photosynthetic root (the presence of root cap supports this), adhering to the rocks by hairs or haptera. In the species *Dicraea dichotoma*, the thallus is ribbon-like and distally free-floating while in *D. stylosa*, it is broad, freely branched, and often floating, reminiscent of *Fucus* and other sea weeds. The endogenous secondary shoots float freely in water and may bear simple, subulate or linear leaves.

The Asiatic hydrophytic flora includes many endemics and relatively few wide-spread species. The ancient family of hydrophytic *Aponogetonaceae* contains species which fall into several endemic groups, which display morphological affinities. These distinct groups are Australasian group, South-East Asiatic group, Indo-Sri Lanka group (*Aponogeton undulatus*, *A. echinatus*, *A. microphyllus*, *A. crispus* and *A. rigidifolius*), Sri Lanka group, Madagascar group and continental African group.

The wide prevalence of endemic hydrophytes in India and Sri Lanka may be partly due to the tropical rain forest and suitable coastal low lands being isolated from those of Africa and of continental and oceanic south-east Asia.

The total number of families of vascular hydrophytes which have gone back to water and is represented at present in the aquatic medium, perhaps does not exceed 40. This is certainly a small total. Of these families, thirty have less than 10 genera and seventeen are monogeneric. The number of aquatic species probably exceeds 100 in only two families, *Podostemaceae* and the continental, *Haloragaceae*.

XEROPHYTES

There have been many interpretations of the term *xerophyte*. If we use the term in a loose qualitative way, xerophytes are plants of relatively dry habitats—dry in soil and most often also climatically. A truly ecological definition approaching as near as possible a quantitative basis is that xerophytes are plants which grow on substrata which usually become greatly depleted of gravitational ground water to a depth of at least 20–25 cm during the course of a normal season. In dry regions all plants not confined to the sides of streams or lakes are considered xerophytes whereas in the regions of high rainfall, xerophytes would likely be represented only by some shallow-rooted plants in light sandy soils, by the vegetation on dry hill-tops and the cryptogamic flora of lichens, algae and mosses.

The environmental adaptations achieved by xerophytes to escape drought or to endure recurrent drought (drought is not easily defined; it refers to periods during which the soil contains little or no available water) can be either morphological (structural) or physiological or both. In order to endure dry conditions xerophytic plants must adopt all the means at their disposal, aiming constantly at two definite objectives: (*i*) to procure as much water as they can get from the inhospitable soil and (*ii*) to economise on this supply of water by conservative use and keeping the loss of water from the aerial parts as also the water requirement of the plant to a minimum.

This is, however, drought resistance in the narrow sense of the term. During drought, the plant suffers, in addition to dehydration of its cells and tissues, from a considerable increase in the temperature of the body, i.e., from overheating. Thus, the lack of water caused by drought in usually accompanied by high body temperatures. Hence drought resistance of the plant includes both the abilities—first, ability to withstand dehydration and secondly, the ability to withstand overheating. High heat-resistance is, however, not always linked with high drought rsistance. It is necessary also to note that drought not only affects plants in different ways, but even individual species may have different types of response.

According to many authors, such as Henckel (1954, 1960) drought resistance "is a property which is formed and developed in the process of ontogenesis and is based on the whole preceding phylogeny of the plant" (Henckel, 1964). This property thus is not unchangeable and may be markedly affected by individual development of the plant.

The following definition of drought-resistance seems to satisfy reason and the results of experimental observations: "Drought-resistant plants are those which in the process of ontogenesis are able to adapt to the effect of drought and which can normally grow, develop and reproduce under drought conditions because of a number of properties acquired, in the process of evolution, under the influence of environmental conditions and natural selection."

We must try to understand how drought injures and kills plants. Soil drought first of all dehydrates the cells and tissues. A slight water deficiency is generally a normal phenomenon which does not greatly impair the functioning of the plant. Only when the deficiency exceeds a certain limit, its effect on the plant becomes fatal. This fatal level is not the same for different plants—drought-resistant plants can lose great amounts of water without any harmful effects; drought-susceptible plants, however, can only endure a slight water deficiency, which sometimes may be enough to disturb the balance between their various physiological processes.

Interesting facts with regard to the development of *xeromorphism* have been obtained during recent years. It was found that phosphorus nutrition, light and water deficiency, all stimulate the formation and development of xeromorphic morphological characteristics. All these factors affect the same aspect of plant metabolism, namely morphogenesis, through nucleic acid metabolism—phosphorus-stimulating morphogenesis through increased nucleic acid synthesis, while water or nitrogen deficiency, inhibiting nucleic acid metabolism, may hasten tissue differentiation.

The nature of injury and death of plants caused by drought may be considered now. Ilyin (1930-1957) thought that during dehydration, cells are injured and die as a result of sudden and violent mechanical disruption of the protoplasm. This concept was, however, modified later and it was strongly suspected that actually the cause of death was the destruction of protein molecules. The injury and the subsequent death was attributed at first to excessive protein decomposition but later, it was found that *decreased protein is not a cause but a result of the injury.* Stocker (1950-1960) attributes the main cause of death during drought to changes in the submicroscopic structure of protoplasm—loosening or modifying specific molecular connections in the submicroscopic protoplasmic network—accompanied by changes in viscosity, permeability, electric charge as well as due to inactivation of enzymes.

Heat-resistant plants respond to high temperatures with intensive protein synthesis and intensive restorative processes. These are the results of high levels of nucleic acids, RNA in particular (complexed with protein), in these plants which stimulate protein synthesis and high respiration rates. The hardened plants as a rule maintain all their synthetic processes at a high level during drought. Leaves of hardened plants are usually richer in starch and all tissues are higher in organic high-energy bond phosphates (ADP and ATP). The protoplasmic colloidal-chemical properties play an essential role in the resistance of plants to high temperatures—high hydrophilic viscosity, the degree of hydration of colloids and increased content of bound water are also important aspects of heat-resistance.

Hardened plants have a greater total absorbing surface in the root system. Thus the absorption of mineral elements is more intense in hardened plants—in addition to this, the plants also bear more primary roots. As a rule, roots are more resistant than leaves to high temperatures and accumulate more starch (Petinov, 1961).

It has been observed that kinetin placed on the leaf blades increases the heat resistance of tobacco plants. This effect is attributed to the ability of kinetin to promote the mobilization of amino acids (Englebrecht and Mothes, 1960).

Hardened plants have a different and specific type of xeromorphism in that the foliage area is larger with denser network of veins and ribs. Usually smaller leaves are correlated with xeromorphism but because of extensive increases in nucleoproteins, enhanced protein synthesis, cell division and differentiation of cells—the hardened plants develop this specific xeromorphic morphology concomitant with an increase in leaf area.

It is quite possible that the nature of high temperature-killing of plants is different from drought injury. A slow increase in temperature may result in ammonia poisoning; a fast increase, however, disrupts protoplasmic submicroscopic structure, culminating in the coagulation of protoplasmic proteins. Protective adaptive reactions against high temperatures generally consist of more rapid respiration, increased amounts of organic acids, which may bind released ammonia, and increased hydrophilic viscosity of protoplasm.

Hardening of plants against drought and high temperature falls sharply during the development of reproductive organs (crucial period). Reproductive organs are particularly sensitive to drought; among them, androecium is much more sensitive to drought than the gynaecium. In mustard, cucumbers and others, drought, however, injures the gynaecium which also leads to a reduced fruit set. Under drought conditions, pollination and fertilisation are in general poor and in cereals, result in partially or entirely empty heads. The most severe injuries to a plant resulting from drought occurs at the sexual-cell-formation stage—a stage at which a plant undergoes extensive protoplasmic reorganisation because growth processes are now, as mentioned before, crucially directed towards the formation of reproductive organs. The change is reflected in the modification of colloidal-chemical properties of the cell—the decrease in elasticity and protoplasmic viscosity and respiration of the plant at this stage, certainly requires much more of water and is naturally very sensitive to drought.

Plants possessing structural features and adaptations similar to evergreen xerophytes are as we know, referred to as *xeromorphic*. Xeromorphic characters are inherited (genetically fixed) under any environmental conditions and as such xeromorphy is found outside the desert as many evergreen trees, e.g., hydrophytic mangroves. *Xeroplastic* features are, however, only induced by drought conditions (characters are acquired) and are in most cases associated with dry habitats (characters not heritable).

Three general types of xerophytes having practically nothing in common morphologically, physiologically and perhaps even more truly taxonomically, may be recognised:

EPHEMERAL ANNUALS

The ephemerals are a prominent feature of vegetation of all semi-arid regions which are characterised by definite and regular, however brief, rainy seasons. These regions support a considerable flora of small annuals which complete their entire life cycle within a few weeks after the first rains have fallen. With the onset of rains seeds of such annuals germinate, quickly grow to maturity, flower and set seed, i.e., the entire life cycle is completed before the soil dries out again. The new crop of seeds set, survives through intervening dry season until the next advent of rains.

Among seeds of these desert annuals mechanisms are found which delay germination of seeds until a rain of sufficient intensity has wetted the soil to such depths that seedlings can complete their life cycle with the water. This is accomplished by *water soluble inhibitors* which are leached out of the seeds only by a heavy rainfall. Such inhibitors may either be simple inorganic salts like NaCl or they may be organic substances.

The principal morphological adaptations of ephemerals are their small size (they cannot possibly grow to a big size for the duration of life cycle is so short) and large shoots in relation to root system.

Such plants have been termed '*drought escaping*' and not true xerophytes as they do not really endure drought (they cannot endure a severe reduction of water content for extended periods without permanent injury to the cells) rather escape it. Yet it is a significant fact that by no means all mesophytic annuals can grow in the desert. It is also true that the percentage of such annuals in the flora varies directly with increasing dryness of the soil.

A great number of these desert annuals have been investigated and in most cases no xerophytic characteristics could be discovered in these plants. Most of these plants are small roundish, dense shrubs represented among others by species of *Papilionaceae* (*Astragalus* sp., etc.), some inconspicuous *Compositae* (e.g., *Artemesia*), a few *Zygophyllaceae*, *Boraginaceae* and some grasses.

SUCCULENTS

Succulents also contribute a large percentage of vegetation to the flora of the most semi-arid regions. They are also frequently found in locally dry habitats such as sandy soils, sea beaches, etc., of more humid climate.

Succulence is due to the proliferation of parenchymatous cells accompanied by an enlargement of vacuoles of mature cells and a considerable reduction in the size of intercellular spaces. The succulents are a distinct group of plants, not only in structure but also in metabolism and water-economy as well (the accumulation of organic acids, pentosans and mucilage by succulents is well-known).

This structural modification enables the plant to accumulate in the proliferated tissue, depending upon the extensibility of individual cells, large reserves of water during brief rainy seasons. The cells of the water-storage tissues are rich in glue-like mucilage. This holds water very efficiently, and as *cacti* have got no normal leaves, there are no large stomata-covered leaf surface, through which much water might have been lost in transpiration. A few stomata are seen in the epidermis of the fleshy stems and it is probable that they are *open only at night*. The succulents shrivel during the periods of drought as they become depleted of water, again swelling up with the advent of rain.

Succulence, to be effective, must necessarily be accompanied by reduced transpiration rates. They must also economically use the water stored in their tissue. A relatively thick cuticle and the fact that in many succulents, stomata are open only during night and closed during day, are important factors permitting conservation of water. In some cases, the form of the stem succulents such as cacti may approach that of a sphere; this for a given volume certainly exposes the minimum transpiring surface and is thus largely helpful in reducing transpiration rates (with the same supply of soil water, it has been estimated that a spherical cactus may lose more than 500 times less water than from an equally heavy plant of the same volume but with greater transpiring surface).

The succulents, like cacti can survive periods of many rainless months and replenish their water supply again during the following rain. To decrease their water loss, as mentioned earlier, they have evolved an interesting type of metabolism by which they open their stomata during night when they absorb CO_2 (dark fixation) which is stored in the cells in the form of dicarboxylic acids (e.g., malic acid). During the day, they keep their stomata closed thereby losing very little water and also transforming the dicarboxylic acids formed at night into sugars (*Crassulacean acid metabolism*).

The most conspicuous succulents of all semi-desert regions belong to the cactus family (*Cactaceae*). The other important families of plants which include a number of succulent species are: *Euphorbiaceae*, *Crassulaceae*, *Liliaceae*, *Amaryllidaceae*, *Portulacaceae*, etc. Succulence may occur in roots (in species of desert *Pelargonium*, *Oxalis*, etc.) but is rare. Stem succulents are the cacti (desert plants but shallow rooted) and the leaf succulents, *Agave*, *Aloe*, *Sedum*, etc. Cacti are unique among all types of desert plants in having shallow, but vast spreading roots, enabling them to absorb superficial water quickly, with rootlets which are drought-deciduous. The rootlets wither away when supply of water in the surface soil is exhausted, reappearing within a very short time when there is again a supply of water in the soil.

Some ecologists have excluded succulents from the group of true xerophytes as they avoid drought by means of their water reserves. It seems that the tissues of succulents do not have any intrinsic resistance to the harmful effects of droughts: survival of plant depends wholly on its outer line of defences—xeromorphic characters (e.g., leaves in many cases reduced to spines, ridges or protuberances, so that when the storage tissue shrinks due to depletion of water, contraction can take place without damage; thick and waxy epidermis, etc.) and its water-storage capacity. But succulents are by far the best represented plants in the desert vegetation all over the world. It seems perfectly justifiable to regard succulence, found in certain true xerophytes simply as a unique mode of adaptation to cope with extreme water-scarcity conditions of the environment.

NON-SUCCULENT PERENNIALS

Members of this category are the true xerophytes (*euxerophytes*)—the *drought-enduring plants*—which can successfully endure long and continuous dearth of water in the soil. Soil-drought conditions are usually accompanied by dry atmospheric conditions such as high temperature, low humidity and often high wind velocity, all of which favour high transpiration rates. In *euxerophytes*, water deficiency usually reaches 60–70% of their gross fresh weight. As a rule, water deficiency is harmful; first of all it decreases growth process as a whole especially during cell elongation. The plants which survive drought are as a rule small and weak. The morphological and physiological characteristics that enable these plants to withstand drought are many but the principal ones are primarily as follows:

(1) Plants respond to progressive dehydration by changing the colloidal chemical state of the protoplasm, namely by increased hydration and hydrophily of the colloids of the protoplasm (1959, 1960). It has been noted that dehydration causes the same changes in the colloidal system as does cell-aging, i.e., it lowers the water-holding capacity and ability to swell.

(2) RAPID ELONGATION OF TAP ROOT. In arid and semi-arid regions the surface soil dries out no deeper than few cm. So all plants whose tap root system is long enough to penetrate into deeper soil and whose rate of penetration can keep ahead of the progressive drying of the soil from the surface downwards, are potentially capable of growing in such habitats.

(3) One of the most important adaptations of true xerophytes seems to be the production of extensive root system in proportion to shoot system. It not only increases the total absorptive capacity of the plant but also exposes relatively only a small proportion of the plant to the atmosphere. Many desert plants have roots, sometimes adventitious, deep enough to absorb whatever little amount of available water (during and immediately after rains) there may be from the moist subsoil. Well-aerated and dry soils are a boon to most xerophytes for there they allow roots to penetrate to great depths where there may be permanently available water.

(4) HIGH OSMOTIC PRESSURE OF THE CELLS

It is well known that osmotic pressure of plant cells vary inversely with their water supply and consequently high osmotic pressure of the cells (which produce such high suction pressure, i.e., diffusion pressure deficit that an equilibrium between loss of water and water uptake can probably be maintained by the cells) is a characteristic feature of the true xerophytes but how effective, this factor really is in helping plants to extract more water from the arid soil, is debatable. It may only be an expression of too high solute content of the soil.

In xerophytes, the leavel usually develop a very large water-potential by increasing the osmotic pressure of the cells—osmotic values, as high as 100–150 atm. are sometimes exhibited by such cells.

Desert perennials can withstand considerable dehydration; some of these plants being quite able to tolerate desiccation to about 50% of their dry weight (in mesophyte this value averages about 200%)—that is they can tolerate protoplasmic desiccation to a much larger extent compared to ordinary mesophytic plants. The gelatinous coating on the cell walls of the blue-green algae perhaps plays an important role here.

The cell walls of some true xerophytes exhibit little or no elasticity and which characteristic may greatly influence the water-holding capacity of the cell. In cells with inelastic walls there is essentially no change in the volume of the cell from the turgid state to incipient plasmolysis. This is believed to enable xerophytes with inelastic cell walls to adapt themselves to drought conditions as such a condition would greatly enhance the plant's resistance to water loss through transpiration.

Thus the xerophytic habit of a plant, which may not, necessarily, involve greatly reduced transpiration, seems to be closely linked with rigid and inelastic cell-wall structure, preventing cellular collapse under high water-stress. Consequently it also allows the development of highly negative water-potential for water uptake, even from very dry soils. Thus the reduction of *negative* water-potential gradient during the day-light hours and the production of a *positive* potential gradient during the night, allow water vapour uptake as long as the relative humidity of the surrounding air is high.

(5) APILITY TO REDUCE TRANSPIRATION RATES TO EXTREMELY LOW LEVELS DURING PERMANENT WILTING

The organ that is most strikingly modified in xerophytes is the leaf. Although some desert plants may have low transpiration rates, conclusive evidence points to the fact that true xerophytes transpire more freely and vigorously (actually there is a great number of stoma per unit area in xerophytic leaves) than ordinary mesophytes when the availability of water is equal. In many desert plants, stomata very often close during the hottest hours of the day when the cooling effect of transpiration would be most advantageous. The control of water loss by transpiration is, however, strikingly different in a xerophyte compared to a mesophyte when the condition of permanent wilting sets in.

The leaf-shedding habit of many xerophytic deciduous perennials is a very efficient means of enduring transpiration for the branches, devoid of leaves, lose very little water.

Many xerophytes which retain their leaves througout the season, decrease transpiration rates by special structural adaptations, of course, chiefly effective only during peiods of permanent wilting. Plants with evergreen leaves whose epidermis (sometimes muitiple or more than one layer of epidermis is observed) is heavily cutinised or with

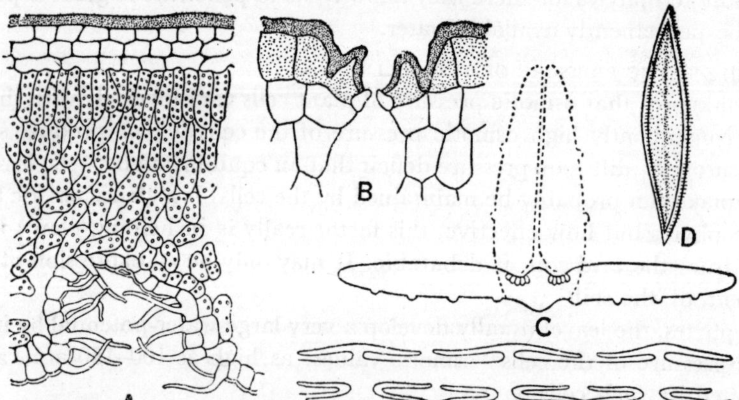

FIG. 777. Some structural features of xerophytes. A: Cross-section of a xerophytic leaf showing thick cuticle and sunken stomata in pits with hairs; B: Section of a succulent leaf of *Agave* with thick cuticle and sunken stomata; C: Outline of a section through a leaf showing method of folding; D: Underside of a leaf showing stomatal surface protected by hairs and permanent revolute margin of the lamina and E: section through a leaf showing peltate hairs.

waxy cell walls, show greater resistance of desiccation under conditions of extreme soil dryness. This is most likely due to closure of stomata coupled with highly efficient cutinisation of the epidermal cells. Plants of this type are sometimes referred to as *sclerophylls* (hard leaved). This thick cuticle may also be advantageous in preventing the breakage and consequent damage to the leaves caused by bending of the blades by high wind generally prevalent in a desert. A shiny, heavily cutinised leaf surface reflects much of the sun's rays, thereby reducing transpiration through reduced heat absorption. Transpiration rates may also be reduced as a result of protection variously given to the stomata. A greater density of epidermal hairs is a prominent feature of one or both surfaces of the leaves of many xerophytes (Fig. 777). But living hairs probably increase the rate of cuticular transpiration by increasing the transpiring surface of the leaf but if the hairs are white in colour it might actually reduce the transpiration by reflection of large proportion of incident light. Dead hairs, on the other hand, perhaps reduce the rate of stomatal transpiration particularly under conditions of intense sunlight or strong wind for it is supposed that such hairs impose a mechanical barrier to the effect of wind in transpiration. Dead hairs projecting outwards may keep air currents well above the stomatal surface. Sometimes the stomatal surface is protected by a dense coating of such hairs and also by permanent revolute margins of the blade. Peltate hairs sometimes thinly shade the upper leaf surface of some xerophytes to such an extent as to create a dead air space just above the stomata-bearing under surface of the leaf, thereby reducing transpiration rates. But it must also be pointed out that conclusive experimental evidence about the efficiency of a dense coating of hairs on the leaves in the reduction of transpiration rates is lacking.

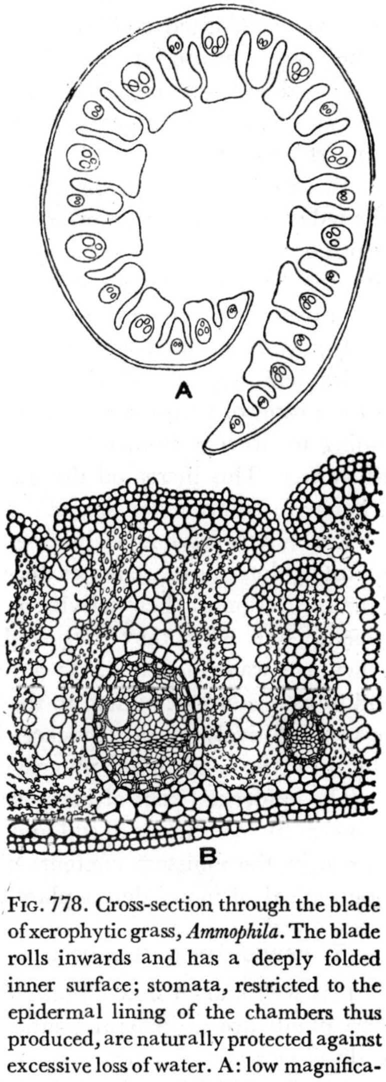

FIG. 778. Cross-section through the blade of xerophytic grass, *Ammophila*. The blade rolls inwards and has a deeply folded inner surface; stomata, restricted to the epidermal lining of the chambers thus produced, are naturally protected against excessive loss of water. A: low magnification and B: higher.

SUNKEN STOMATA—Stomata, sunken in cavities below the level of epidermal surface and thus also well below the level of atmospheric stress and tensions, have generally been found to be well-capable of reducing the transpirational water loss (Fig. 777). Sometimes stomata are over-arched by adjoining cells so they come to be situated in cavities protected from the wind. In some other plants stomata are sometimes plugged, however temporarily, with a deposition of wax, or resin. The mesophyll of most xerophytes is characterised by the small size of the intercellular spaces especially those which are continuous with the pores of the stomata. There is often no spongy tissue but only very compactly packed palisade cells beneath both upper and lower epidermis. All these certainly contribute largely towards an actual reduction in transpiration rates.

When the drought conditions are first felt by xerophytic plants, many of them change form or position so that the amount of light received per unit area generally becomes less. Often part of aerial transpiring surface is thereby protected from direct contact with outside air. The leaflets of many desert legumes fold upwards in such a manner that only approximately half the leaf-surface is exposed to air. In case of plants that grow in mats, a very common form in xerophytic habitats, aggregation of stems brings about protection of the leaves. In addition, plants may often have erect or oblique leaves. Some desert grass leaves roll or fold lengthwise along the longitudinal furrows in their upper surface; these turgour movements are due to loss in turgidity of the rows of enlarged colourless cells, often termed *bulliform cells* (*motor cells*), situated at the bottom of the furrows (Figs. 778 and 779). The free surface of leaf is reduced in many desert grasses (*Stipa*) and in certain species of the family *Ericaceae*, by the inrolling of the leaf surface. The leaves of moss, *Polytrichum*, also fold longitudinally in a similar manner.

In certain erect and shrubby xerophytic plants, the leaf blades are permanently oriented in a vertical position so they are never subjected to full incident sunrays. The greyish or sometimes light green colour of most desert plants is thought to be of value in reflecting light rays which, otherwise, would have been absorbed and converted into heat, raising the temperature of the leaf thus promoting rapid transpiration.

It must be pointed out here clearly that anatomical features, e.g., dead hairs, sunken stomata, leaf-roll, etc., which had formerly been considered as helping to reduce transpiration rates of xerophytes, actually benefit the plant in a completely different way by tending to prevent stomatal closure by maintaining high humidity conditions at the leaf surface. This increased the duration of active photosynthesis by allowing gaseous exchange to continue for longer time.

Xerophytes have more intense assimilation rate than other plants for their palisade cells and chloroplasts are better developed and also can only function part-time because of the restriction of their activity during periods of water-deficit. Green tissues function efficiently only when they are turgid and their photosynthetic efficiency falls off rapidly if they are water-deficient. As a result, the photosynthesis of plants under drought conditions is usually confined to the early morning and the late evening when the cells are comparatively turgid (Stocker, 1960). In extreme water-deficient conditions of the habitat on the other hand photosynthetic efficiency is greatly decreased.

The epidermal cells of many xerophytic leaves are in a collapsed state in extreme water scarcity conditions of soil and air but the cells straighten up whenever there is an increase in the moisture content—this bellow-like action of the epidermal cells seems to be typical of many desert plants.

(6) REDUCTION IN SIZE OF LEAF BLADES

It is well known that all leaf blades, pinnae and pinnules of compound leaves of desert plants are in general smaller in size and more compact, i.e., the desert flora is mostly *microphyllous*. Though the size of leaf blades is reduced, the blades contain a denser network of veins. The area of the leaf usually does not exceed 1 sq. cm. Theoretically, presence of microphyllous leaves should reduce the amount of transpiration due to this decrease of the aggregate area of the total transpiring surface. But very often in microphyllous plants, the small size of the individual leaf units is compensated by greater numbers with the result that nothing is really gained by the plants in the way of reducing the total transpiring surface. Thus it is evident that reduction in the size of individual leaf-lamina has no positive significance with respect to reducing transpiration.

Still microphylly is encountered everywhere in the arid desert regions and this omnipresence must be indicative of some fundamental role played by the minute leaf blades of shrubby desert plants. That is, why microphylly? When average-sized leaf blades are injured by severe atmospheric drought it is commonly seen that the marginal parts, especially those between the larger veins die first, while those in the immediate neighbourhood of the veins, degenerate last. This pattern of injury can be explained by the facts: (*i*) that because the water diffusion through the mesophyll is very slow (diffusion is cell by cell), only those cells which are situated near the veins can obtain sufficient water supply during the crisis brought about by drought and (*ii*) more stomata are found towards the midrib than the margins of the leaves and thus in a linear leaf blade the tip, which is the remotest part of the blade from the water supply (transpiration stream), is the first to die. If we closely examine a leaf blade which has almost been killed, but not quite, by desiccation, we find that the pattern of tissue still remaining alive in the injured leaf blade is strikingly similar to a much dissected pinnatifid or pinnatipartite blade. It is interesting to speculate whether this evolutionary breaking up of leaf blades or reducing the size of individual leaf units by discarding much of the intervening mesophyll has been to reduce the likelihood of severe drought degeneration of leaf tissues of the desert plants.

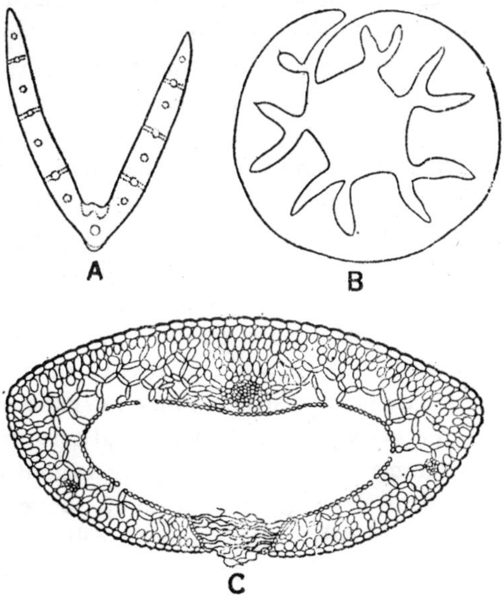

FIG. 779. Leaf-folding devices in xerophytes. A: diagram of cross-section of a half-folded leaf of *Poa pratensis* showing folding device and position of vascular bundles and accompanying mechanical tissue; B: outline of rolled-leaf of wheat grass (*Agropyron* sp.); C: cross-section of the blade of the xerophytic, *Empetrum*. The blade is rolled back so that the margins almost meet thus preventing excessive transpiration from the stomata which lie on the inner surface. A felt of wooly hairs, developed at the leaf margins also contributes to the prevention of excessive loss of water.

Microphylly is sometimes a definite advantage in the desert environment in that, as is evident, the smaller the leaf blade, there is less likelihood of its being overheated when exposed to strong intensity of light commonly prevalent in deserts.

Desert plants often lack green leaf altogether. In certain plants such as *Ephedra*, the reduction in the size of the leaf blades has progressed so far that the leaves in *Ephedra* are probably vestigial (only a trace, the remaining sign that the leaf blades were there). In *Australian acacia*, leaf blades have been lost entirely; the photosynthetic function is taken over by the expanded petioles (*phyllodes*) or stems or both.

(7) CHANGE IN THE SIZE AND SHAPE OF THE CELLS

The characteristics of the cells of xerophytic plants are that they are relatively small in size with small vacuoles.

Mosses, algae, lichens and other lower organisms adapted to dry habitats have always cells with small volume, some being only 50 or 300 or 700 cu. microns in size. The cell volume of higher plants not adapted to drought is between 2–3 million cu. microns. Consequently when desiccated small cells decrease their volumes to nearly half, compared to larger cells, where the reduction may be as high as 5–10 times. As a result, the former suffer much less disturbance compared to the latter (large cells) when a cell is desiccated and water is lost from the vacuole. There is thus a lesser readjustment of walls to each other in smaller cells than in larger ones.

Buds of higher plants are quite resistant to drought. Their cells are completely devoid of vacuoles. As these buds develop, vacuoles are gradually formed and simultaneously the natural resistance to drought of buds decreases. Germinated seeds behave in a similar way.

Cells having a larger proportion of protoplasm and consequently smaller vacuole are least disturbed by loss of water and are also protected against injury. Cells of hibernating or storage organs and reproductive structures, e.g., spores, zygotes and seeds generally lack vacuoles which help them to survive through drought conditions.

The *shape* of the cell is also a factor in resisting drought. The spherical and cubical cells of plants not adapted to desiccation have their walls further apart compared to elongated cells of many xerophytes. Certain xerophytic mosses have cells which are much elongated with narrow cell cavity; their opposite walls are almost in contact. Thus there may be almost no change in shape of such cells when water is lost from them.

Vegetative organs of ferns and certain angiosperms can remain in a dry state and can again revive in water supply wthout any apparent injury to their cells. Though the cells of these organs have comparatively larger vacuoles, the protoplast of these cells is never violently disturbed as the cells are actually protected against such injury by the extraordinarily firm nature of the cell vacuoles which resist desiccation.

(8) METABOLISM

Xerophytism is expressed in many ways due to changed metabolism of the cells owing to desiccating conditions. These result, as we know, in thickened cell walls and in the formation of protective coverings and these may again be due to accelerated conversion of polysaccharides into their anhydrous forms, e.g., cellulose, etc. Desiccation may also promote development of suberised cork cells. Desiccation actually promotes the development of cork—thickness of bark is greatest in desert plants compared to plants grown in moist conditions. In xerophytes conducting vessels are, as a rule, well developed and the heavily thickened vessels are more numerous, larger in diameter and longer.

As with cutin, lignification of the vessels in xerophytes begins much earlier than in mesophytic vessels. Annual rings of growth are more pronounced in xerophytes than in mesophytes. Lignified bast fibres and sclerenchyma sometimes reach their highest achievement in the xerophytes. In xerophytic leaves often even the palisade cells are partly lignified, and accumulation of gums and tannins may be of frequent occurrence in the lignified fibres or cutinised cells, constituting the hypodermis of many xerophytes.

Desert plants, as mentioned before, usually have a dull greyish colour quite in contrast to the bright green of mesophytes and hydrophytes which may probably be due to thicker epidermal coverings and deep-seated chloroplasts or it may also be due to chloroplasts being paler and fewer in number.

Enzymatic activity is generally greatly affected by the water-deficit conditions of the tissues of xerophytic plants. The activities of enzymes such as *catalase* and *peroxidase* have been found to be relatively higher in the tissues of xerophytes (is much of the respiration of these plants peroxisomal?) compared to plants growing in normal supply of water. The same is perhaps also true of the hydrolytic activity of the starch-splitting enzyme, *amylase*.

In general, drought-resistant plants have higher respiratory rates than plants growing in abundant supply of water. It is probable that they maintain the high water-content of the protoplasmic colloids with water molecules obtained from intensified respiration(?).

There is perhaps an inhibition of normal carbohydrate-phosphorus metabolism—corroborating the conclusion of Zholkevich (1960) that drought first of all disturbs phosphorylation process.

There seems to be a relative acceleration of hydrolytic processes (in some cases leading to accumulation of reducing sugars) with a simultaneous decrease in synthetic processes in the tissues of many xerophytes investigated. Xerophytes seem to use a lesser quantity of respiratory substrates than do mesophytes. The enzymatic equilibrium is perhaps not quite normal in xerophytic plants.

Dry air and soil conditions seem to favour production of male flowers, and a ratio of 10 : 1 of male and female, is sometimes exhibited.

(9) GENETICAL

In the cells of many xerophytic plants there seems to be considerable reduction in the number of terminal junctions and chiasmata during the conjugating phase of reduction division. The conjugation phase itself may also be inhibited to a certain extent in water-deficient xerophytic plants.

Many plants enduring water scarcity conditions of the soil have been found to be polyploids. It is probable that polyploids which show a wider range of variability in osmotic pressure values of their tissues are more suitably adapted to water-scarcity conditions than diploids.

In concluding this discussion of plants adapted to conditions of great scarcity of water, it becomes apparent that there are no definite anatomical or physiological characteristics common to all members of this ecological class. Each species seems to have solved its own water requirement (really water balance) problem by its own peculiar and even almost specific combination of adaptive characters.

Ecological literature concerning adaptation of plants for resistance to drought contains many statements which have proved to be quite untenable in the light of modern critical experimental evidence. Early views were essentially abstract, philosophical interpretations of various morphological and anatomical characters associated with xerophytes. It was sublimely reasoned that because the xerophytes had to conserve on their limited supply of water they must have developed structural adaptations to enable them to transpire very slowly. Nothing is really further from the truth for most xerophytes have higher rates of transpiration (if there is any water to lose) with a greater number of stomata per unit area than mesophytes.

Another misconception, long current in ecological literature, was that xerophytes are more efficient in reducing the soil water-content—could deplete it to a lower percentage before showing permanent wilting—than mesophytes. Critical experimentations, however, have conclusively proved that there is no valid evidence for such a belief.

MESOPHYTES

Mesophytes are ordinary land plants growing in situations which are neither too wet nor too dry. This third group of ecological class of plants essentially based on water relations includes all normal terrestrial plants, which on the one hand cannot grow in water or in wet soil nor on the other can endure habitats where the supply of available water is significantly scanty. The structural adaptation common to xerophytes and hydrophytes are generally absent from mesophytes except in certain speeies derived in the course of evolution from either xerophytes or hydrophytes.

Most species of mesophyte, however, can survive short dry periods—in more humid climates of the world, such periods are relatively infrequent and seldom last very long —without serious injury as there is not at the same time prolonged periods of soil water deficiency as well. Among the mesophytes also there are plants which thrive best in dry regions compared to wet. Cropping of some varieties of wheat in the dry regions seems to be more productive and profitable as well.

Tropophytes (changing plants) are an interesting group of plants which can be included under mesophytes. In most tropical regions weather condition remains more or less uniform throughout the year. But in others there is an alternation between a period favourable for growth and a more or less unfavourable season brought about either by dryness in summer or by dryness and cold in winter. Plants living in such climates necessarily show differences in structural modifications compared with those of uniformly moist tropical regions. Only the tropophytes succeed in these regions because they can endure in one way or another the regular cycle of favourable and unfavourable habitat and climatic conditions.

Tropophytic flora behaves as a xerophyte during summer and winter months and as a mesophyte or even a hygrophyte during the rest of the year. Shedding of leaves during winter—*deciduous habit*—is an important characteristic of many tropophyllous plants; with the advent of spring there is a luxuriant reappearance of foliage on the bare shoots. The shedding of leaves in summer may be due to drought conditions and in winter due to winter cold. If there is no shedding of leaves, the persisting xeromorphous foliage (e.g., conifers) is amply protected against drought and cold. In herbaceous plants, sometimes only the flowering shoot dies in winter but in rhizomatous herbs, all the aerial parts may die, reappearing again with the increase of warmth and water supply in spring. The formation of underground stems is also an adaptation for the perennating buds to survive through unfavourable summer and winter conditions.

Other structural modifications tropophytes adopt in dry weather conditions include: (a) better protection of winter buds, (b) thick cork covering of stems and scars left by shed leaves, (c) efficient closing up of lenticels by the formation of cork cells, etc.

The cold, alpine or subalpine temperate flora is mostly tropophytic in nature. Other examples: silk cotton tree, *Bombax*, *Erythrina*, banyan, *Ficus benghalensis; Shorea robusta; Tectona grandis*, etc.; in contrast to general winter shedding of leaves, the last two shed their leaves in summer.

HALOPHYTES

Halophytes are plants which grow in saline soils, soils in which the concentration of simple inorganic salts is so great that only specially adaped plants such as halophytes grow naurally upon them. True halophytes are those plants that normally occur only on saline soils or in water too salty for average mesophyte. Sugar beet and almond can be made to grow on saline soils and therefore can be called *facultative halophytes*. When soil contains approximately 0·5% to 1% of sodium salt, either Cl^- or SO_4^{--} ions, or more by weight of soluble salts, chemical properties of soil seem to determine its ecological character, physical differences being only by subsidiary importance.

The action of dissolved inorganic salts on plants is partly osmotic (physical) and partly, specifically chemical, depending upon the types of kinds of ions.

If a plant cell is to maintain turgidity the concentration of osmotically active substance within the cell sap must exceed that its water supply, i.e., the concentration of the soil solution. Most non-halophytic plants (*glykophytes*) begin to show reduced growth when the osmotic pressure of the soil solution rises above 2 atm. Beyond this, ordinary plants have very limited ability to increase the osmotic concentration of their sap. But all halophytes possess this ability to a remarkable degree. The osmotic pressures of halophytic cells, due to the presence of chloride ion, frequently equal 40 atm. or more (values as high as 150 atm. have been frequently reported) and as a result sometimes their tissue fluids are so impregnated with salt that they taste salty. Osmotic pressure, even as high as 200 atm., have been recorded in mangrove trees, such as *Avicennia*. Sometimes in some of them (e.g., *Spartina*, *Glaux maritina*, *Cressa cretica*, *Frankenia* sp., *Limonium* sp., etc.) superfluous salts are excreted by guttation through special glands. All these salt-excreting of halophytes, are sometimes referred to *Crinohalophytes*. Arisz *et al* (1955) have shown that the osmotic pressure of the exuded fluid is higher than that of cell sap, so that mediation by ATP is necessary for the removal process; oxygen is indeed necessary as the exudation process can be inhibited by respiratory inhibitors, e.g., KCN, iodoacetate, fluoride, etc.

It is said that anthocyanin develops more and more strongly in the plants with increasing salt content, but so far as *Salicornia* sp. is concerned, anthocyanin development appears rather to be determined genetically. It seems that one perhaps should recognise *physiological races*.

Halophytes certainly take in large quantities of sodium chloride from the media to which they are adapted. In order that sodium ions in the cells may not reach toxic concentration, halophytic plants are known to adopt mechanisms for active extrusion of sodium and chloride ions. In submerged marine and brackish-water algae, sodium *out-pumps* have actually been located in the *plasmalemna* (lipid membrane bounding the cytoplasm; lipid molecules are stabilised by protein on each side) and in the chloroplasts. The true mangrove species of *Aegialitis* and *Avicennia* are known to eliminate large quantities of sodium and potassium salts through special glands on the leaves, a property which, however, other species such as *Rhizophora* and *Sonneratia* do not seem to posses (Scholander, 1962; Eppley, 1962).

When crop plants are grown in saline soil, their root systems are generally dwarfed, transpiration is considerably reduced and there is a general tendency in the part of the plant towards water economy and conservative use of it. Because the presence of soluble inorganic salts in soil so evidently interferes with the absorption of water by non-halophytic plants, saline soils have long been considered '*physiologically dry*' (Schimper, 1898), though certainly physically wet, sometimes very wet for non-halophytic plants. Even in thirties of this century, the concept of physiological drought embraced halophytes but subsequent experimental research has shown that this old concept does not apply to them. That halophytes experience absolutely no difficulty whatsoever in absorbing water from highly concentrated salt solutions of such soils is demonstrated by their high transpiration rates (per unit leaf-area, shown in three halophytic species of *Plantago*), exudation of guttation water containing dissolved salts, production of shallow absorbing roots, etc. This shows that in halophytes water absorption is considerably in excess of water loss and considering the high rates of transpiration observed in most halophytes, one of our most strongly accepted and favourite ideas is shattered. It must be admitted then that halophytes form a very interesting ecological group of plant characterised not only to endure high concentration of salts (true, certain ions only) but also to absorb water with perfect ease and composure under these adverse conditions.

Halophytic herbs grow chiefly during the rainy season owing to dilution of the soil solution and also due to the movement of salt down below the root zone. The characteristic shallow roots of any halophytes is a decided advantage in this connection and may also be an important factor from the standpoint of aeration, for saline soils frequently become water-logged during at least a part of the year. To counteract this lack of aeration of the root zone, many halophytes produce special negatively geotropic root branches which grow erect until they project above the poorly aerated muddy soil surface. These structures are called *pneumatophores* (Figs. 780, 781 and also Fig. 787).

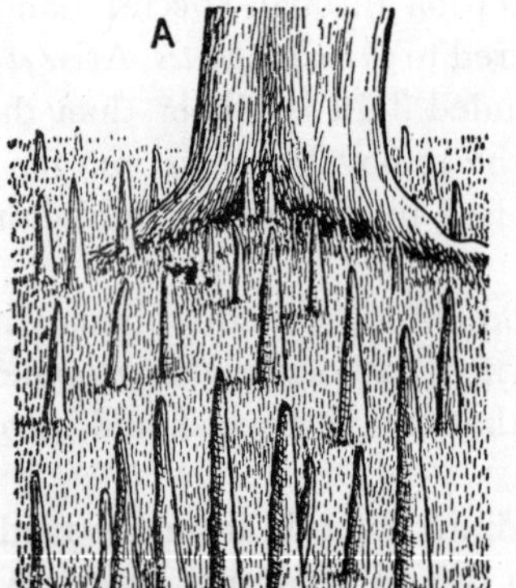

Fig. 780. Pneumatophores. A: pneumatophores at the base of a mangrove tree; B: section through the soil showing the origin of pneumatophores from normal horizontal roots.

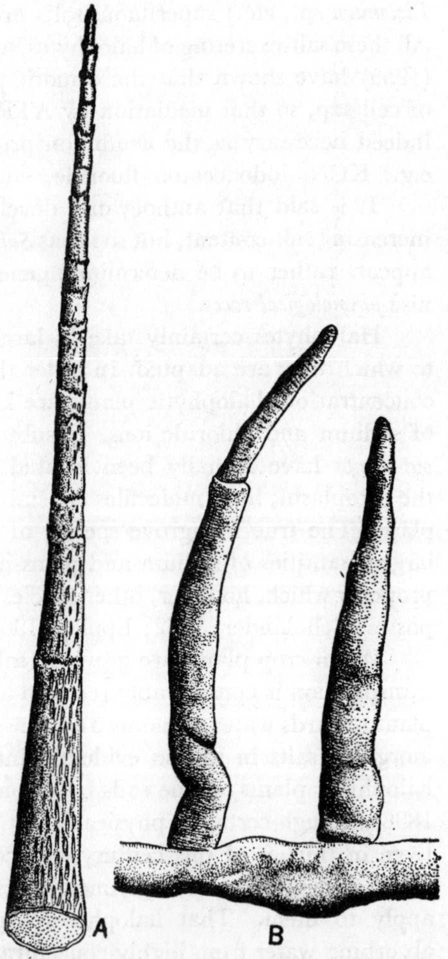

Fig. 781. A and B, pneumatophores of *Avicennia officinalis*. Note lenticel-like scars in A.

In *Carapa obovata*, the pneumatophores are displayed in their simplest form—the serpentine creeping roots projecting above the mud with their upper edge, shaped like the blade of a thick knife, studded with lenticel-like pores. In the species of *Bruguiera*, the horizontal parent roots, here and there bend out of the mud into *knee-like* structures, which in *Bruguiera gymnorhiza*, bear large lenticels while in *B. caryophylloides*, lenticels are wanting due to shedding of cortex. *Avicennia officinalis, Sonneratia alba, Ceriops* sp., etc., all bear negatively geotropic lateral roots, protruding from the ground like asparagus (Figs. 780 and 781). These are sometimes as long as 10 cm or as in *Sonneratia*, 45–50 cm. The species of *Rhizophora* do not possess any special pneumatophores, yet the upper parts of their descending stilt roots that are above mud, probably perform the same function.

As has been mentioned before the pneumatophore supply the underground parts of these halophytic trees with oxygen. All pneumatophores are provided with devices for absorbing oxygen from the atmosphere such as lenticel-like pores and thin cork, and also for transporting the absorbed oxygen; there are highly developed intercellular passages continuous with the stomata of the leaves in the primary cortex and also in the phloem.

It is interesting to point out here that in their apparent anatomical structures such as collateral and endarch arrangement of the vascular bundles, mature pneumatophores resemble normal stem and not root (Fig. 782). In very young pneumatophores, however, sometimes the radial and exarch arrangements of bundles, typical of roots, can be discerned.

A large percentage of halophytic flora, e.g., *Suaeda maritima, Mesembryanthemum, Salsola kali*, etc., exhibit succulence, the significance of which is obscure (Fig. 783).

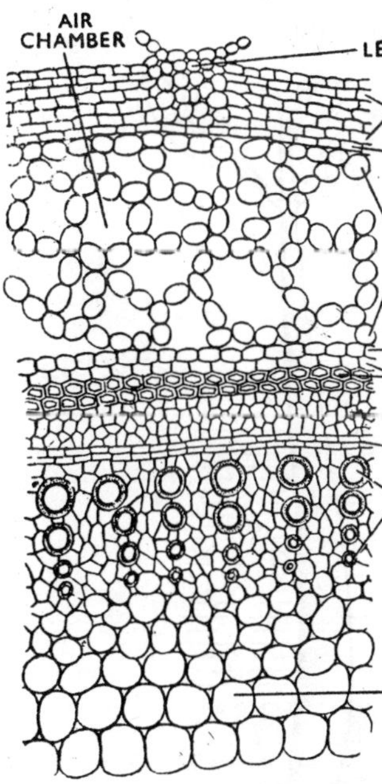

FIG. 782. A portion of the cross-section of a pneumatophore. Note lenticel-like openings, air-spaces in the cortex and *endarch* xylem.

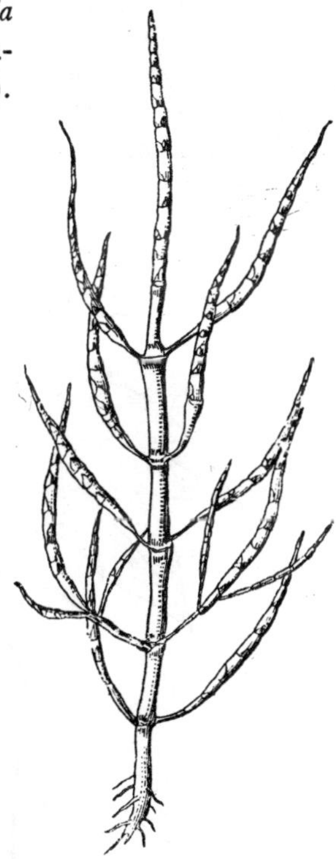

FIG. 783. A characteristic succulent halophyte, *Salicornia herbacea* (*Chenopodiaceae*).

Suaeda maritima may only be an *aggregate species*—the species will tolerate high salinities, though optimum growth takes place at about 0·95–1·00% (not more) NaCl in the soil. *Suaeda maritima* perhaps represents a distinct variety or *ecotype*. The amount of sodium cloride in the tissue can be considerable and may even exceed that of *Salicornia* (2·5%

NaCl). It is *par excellence*, the pioneer colonists of secondary bared areas such as pans, ditch spoil, etc.

There is abundant seed production in this genus; a single plant of *Suaeda macrocarpa* producing up to 1000 seeds, but the viability of the seeds is low and it also decreases with increasing salt concentration of the medium. Dissolved salts in soil solutions especially chlorides of sodium, as pointed out before, are definitely known to stimulate succulence in many non-halophytes.

Even in halophytes, in which sodium is present in greatest quantity, it does not seem to be indispensable for them. Halophytes live in soils rich in sodium, not because they love it, but because they can tolerate it better than other plants. The greater ability to accumulate large quantities of sodium ions in their cells without any apparent ill effect, perhaps leads to their succulence which concentration would have been injurious to non-halophytes. The characteristic succulence of some halophytes may be lost, more or less completely, if grown in soil without common salt.

The salt tolerant *Asparagus* is more succulent when grown in saline soil (rich in sodium chloride) than in non-saline soil.

Cryptogams are very poorly represented in the halophytic flora. Only a few moss, e.g., *Pottia salina*, *Pottia crinita*, *Bryum salinum*, *Grimmia maritima*, etc. (the single moss that grows in the sundarbans is *Calymperes sundarbanense*, and ferns e.g., the polypodiaceous *Acrostichum aureum* and *Ceratopteris thalictroides*, perhaps the only ferns which occur in number in the Gangetic estuarine Bengal and *Drymoglossum carnosum*, with succulent leaves) seem to have been able to adjust to the true saline habitat.

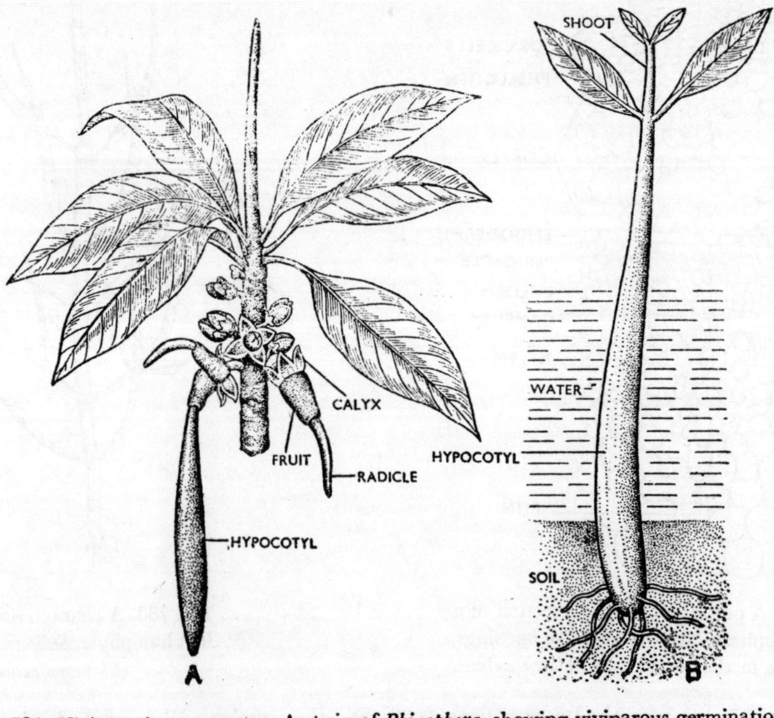

Fig. 784. Vivipary in mangroves. A: twig of *Rhizophora*, showing viviparous germination and B: a seedling established on tidal water-logged soil.

In non-halophytes, germination of seeds is greatly retarded and the developing seedling-survival is almost impossible in the saline medium. Certain halophytes have more than overcome this difficulty by some physiological adaptation for in many such plants, the presence of certain minimum concentration of salt is much more favourable for germination and seedling-survival than its absence. Consequently successful germination of halophytic seeds is usually restricted to rainy seasons when the salt solution is more dilute.

A number of halophytes of sea coasts or other saline soils bordering the seas, in the deltas and estuaries, has developed a peculiar type of germination known as ***vivipary***, which is defined as germination of seeds before the fruits break off from the plant (Fig. 784). The large embryo, without endosperm, grows and develops in the seed within the fruit when it is still attached to the plant. In *Rhizophora mucronata* the fruit which is leathery and indehiscent, soon after the completion of its growth, it pierced at its summit by the green hypocotyl, as the embryo does not require any period of rest, continuing its development without interruption. The hypocotyl is club-chaped and attains a length of 50–75 cm (sometimes even a metre) before it falls down, leaving behind the fused cotyledons which served as absorbing organs of the developing seedling. As the lower end of the hypocotyl is thicker, the embryo seedling falls vertically with its root-tip downwards into the soft saline mud and within a few hours develops lateral roots that fix the seedling firmly. Sometimes, if the water level be high, the embryo floats with the hypocotyl (radicle) downwards and waits for a good chance of catching a suitable crack in the soft mud as the water level falls. Vivipary is also exhibited by *Aegiceras majus*, *Aegialitis rotundifolia*, *Kandellia reedi*, *Ceriops* sp. and *Brugiera* sp. An incipient type of vivipary is also seen in *Avicennia* sp.

Some interesting cases in the relationship of vivipary and non-vivipary is exhibited by certain halophytic plants; for example some typical non-viviparous plants in their native habitats, e.g., Sweden and Finland, when transplanted to California become viviparous. It was presumed that photoperiod was a controlling factor in the development of vivipary. Similarly, it was found that viviparous plants, e.g., *Poa alpina*, from eastern Greenland, within a short time turned non-viviparous when grown in Zurich (Switzerland). All these seem to confirm the fact that both *light duration* and *intensity* are important factors in controlling vivipary.

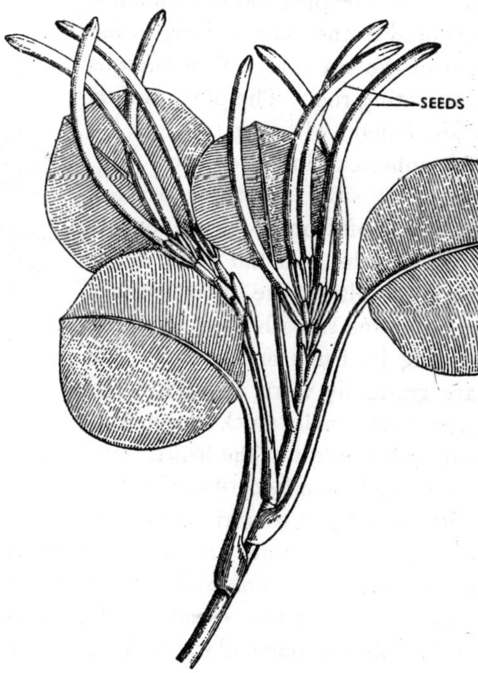

FIG. 785. A cluster of viviparous seeds with elongated hypocotyl before they are shed, in *Aegialitis rotundifolia.*

A distinctive viviparous germination of seeds is seen in *Aegialitis rotundifolia*, a striking and widely occurring mangrove in the Sunderban region of West Bengal. Beautiful, green clusters (about 15–20) of viviparous seeds slightly curved at the top (7–8 cm long) are seen in a cymose umbellate type of arrangement (Fig. 785) all pointing upwards. When dry, the green outer-skin of the thickened hypocotyl is seen to rupture in five longitudinal strips, leaving a solid central core behind.

The embryo-seedlings of *Aegiceros* are curved like horns and remain enclosed in the thin pericarp of the fruit and are in general shorter than those of the *Rhizophoraceae*. Those of *Avicennia*, which sometimes fall still surrounded by the leathery pericarps that dehisce subsequently, and at other times without the pericarp, are provided with bent and densely hairy hypocotyl attached to two large cotyledons. The hairs are stiff and curve upwards and fix the seedling to the mud. In the remaining mangrove plants, vivipary does not occur but the embryo-seedlings of some species, particularly those of *Acanthus ilicifolius*, a thistle-like herb, are always further developed in the fruit than in non-halophytes.

The phenomenon of vivipary can be looked upon as an adaptation by which the retarding influence of high salt concentration on seed germination is conveniently avoided by some halophytic plants.

A littoral bluish-green species of *Graminae*, *Spinifex squarrosus* (the other 4 species of *Spinifex* are all Australian desert xerophytes) growing in sandy saline seashores of peninsular India, particularly in Andhra sea coasts and further down south has an interesting type of fruit dispersal. This dioecious plant with female spikelets massed together in globular inflorescence and infructescence (which latter appears to be composed of long radially arranged stiff, spiny bracts) has creeping shoots which are firmly anchored in the loose shifting saline substratum of the sea coasts by numerous deeply penetrating roots. The plant is thus in much less danger of being torn out by wind. The spherical infructescence (Fig. 786) formed by stiff, bracteate bristles, when ripe breaks off from its dry stalk and becomes the plaything of the wind. Dancing and somersaulting, the infructescence is hurried along the smooth surface of the sand dropping its fruits on its trail. The bristles are gradually worn away and the infructescence gets finally stuck in the sand and buried with the rest of the fruits.

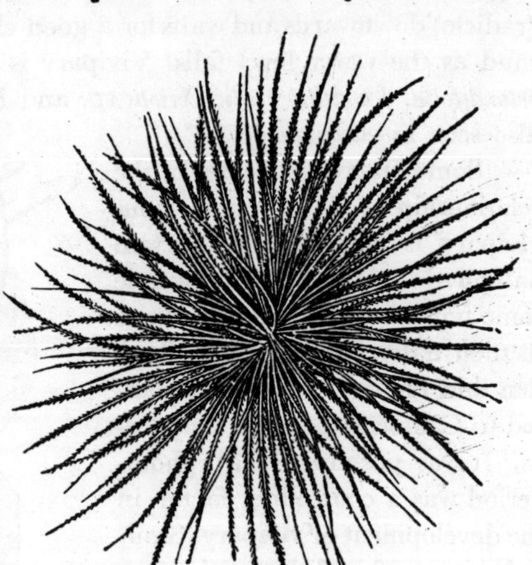

FIG. 786. Infructescence of *Spinifex squarrosus*. Note long radially arranged stiff, spiny bracts.

It is true that the effects of concentration of salts are largely osmotic in nature but the carbonates of sodium and potassium seem to have special physiological effects which are not found with other salts or ions. Each of the salt responsible for the salinity of soil seems to have certain specific effects upon plants.

Some halophytes, particularly halophytic annuals, may be completely lacking in any special regulatory mechanism for their survival, except of course, the increase in the osmotic pressure of the cells in leaves which gradually rises to a maximum, at which point (end of the season) the leaves die. In these plants, chloride content of their root cells is regarded as rising to lethal values at the end of vegetative period or during long dry periods.

It has been suggested by Kursanov (1956) that the increase in salinity of the medium, may alter or determine the occurrence of sex in plants. In dioecious species, salinity may sometimes be responsible for an increase in staminate flowers.

In halophytes growing in northern latitudes, incidence of polyploidy was found to be significantly higher. It may be that polyploidy in these plants is associated with tolerance of extreme climatic and edaphic conditions.

Mycorrhizas, mostly endophytic types, are sometimes associated with plants adapted to saline soils—only the smallest roots are infected and the fungus seems to be restricted to a specific zone of the cortex.

MANGROVES

A special halophytic vegetation represented by the well-known *mangrove formations*, some of them typically Indian (Eastern Asia) are found on the saline soils of the Gangetic estuarine deltas near the Bay of Bengal—the Sunderban areas, in Western India near the sea coasts of Bombay and Kerala (barely a mile in width) and in the banks and creeks of the Gautami Godavari in South India where the river meets the sea. The mangrove forests are highly developed in both the Andaman and Nicobar Islands (21 sp. in about 1200 sq. km). In the Andamans they form a distinctive feature on the borders and river estuaries and lagoons. These luxuriant mangroves are really the guardians and builders of the Andamans for the seaward growing forests of mangroves, spreading roots into the shallow water, hold the mud and the sand together, which otherwise would have been washed down the rivers. All these regions are continually washed by tidal rivers and as a result the saline soil everywhere is sandy, loose and swampy. Such waterlogging by saline water, relativly high rainfall, high humidity, more or less equable temperature and complete absence of ground frost are some of the essential conditions necessary for the development of mangroves. Dense and frequently repeated cloudiness apparently is also an important climatic factor for the occurrence of mangroves in the tropics. During the recent years the mangrove forests have undergone considerable modifications in various regions due to biotic influences, i.e., man-made activities.

The typical Indian true mangrove flora comprises about 20 species of flowering plants belonging to about 10 genera. The characteristic adaptive features of these mangroves can be summarised as follows:

(1) production of respiratory roots, *pneumatophores* for efficient aeration—the soil is water-logged and the aeration in the root medium is consequently much below optimal (Figs. 780–782 and also Fig. 787);

(2) formation of *stilt-like prop roots* arising from the main trunk of the trees for efficient anchorage of plants in the loose sandy soil where fixation by the shallow ordinary roots is often difficult.

The stilt roots which give *Rhizophora mucronata* such a peculiar appearance are relatively strongly developed in the herbaceous *Acanthus ilicifolius* and to a lesser extent in *Rhizophora conjugata*. In the other mangrove trees they are either absent or only very slightly developed, particularly those stilt roots descending from the branches. Sometimes a large number of aerial adventitious roots, *flying buttress*, are developed and these combined with pneumatophores, stilt roots, *knee roots* and sometimes, *plank buttresses* (these structures are not true roots but are outgrowths of roots and stems which may or may not have a continuous development. They are all more or less triangular and may extend up the trunk to 75 cm or more; others may be comparatively low but extend laterally from the base for about 50 cm or even over. Buttresses, when present, add strngth to the shallow-rooted trees) may make a mangrove forest an impenetrable tangle of roots;

(3) viviparous method of germination of seeds;

(4) plants are generally shrubs; herbs are somewhat rare, except perhaps, *Acanthus ilicifolius*, *Heliotropium ovalifolium*, etc. Some shrubby species become dome-shaped be repeated cymose branching;

(5) *succulence* of stem is a common characteistic. Leaves are xerophilous, generally small in size, thick and are very often glossy. There are also thick cuticle, large mucilage cells, protected stomata but *never truly sunken*, and sometimes large-celled thin-walled aqueous tissue. Sometimes old leaves serve as water reserviors for younger leaves (e.g., *R. mucronata*). These adaptations certainly seem xeromorphic in nature. But why xeromorphy in halophytes? We know that as regards availability and absorption of water these plants live in conditions of luxury. Thus xeromorphy in halophytes is apparently an example of purposeless adaptation and, as has been pointed out before, is only a remainder of their xerophytic ancestry. The halophytic sp. of *Spartina* which grows abundantly on the eastern sea coasts of England is definitely known to be derived from two parental forms which are themselves not halophytic at all.

The mangrove flora of Gangetic W. Bengal shows different varieties of life forms correlated with substrata. They belong to several different families but present many points of similarity in their adaptations. Most of them are very much isolated in their systematic position and consequently appear to be very old associations. Recent determination of chromosome numbers of a few mangroves (*Ceriops*, *Sonneratia*, *Avicennia*, etc.) may throw interesting light on the taxonomic positions of the mangroves. Names of a few species, classified according to the soil substrata in which they grow are given here. Some of them are really non-mangroves but are seen in association with mangrove plants at various levels. The mangrove communities in these regions, as in all others, are primarily edaphic in nature and can be classed as edaphic climaxes. These mangroves generally develop in sheltered tidal areas where the alluvial substratum is subjected to periodic flooding:

Open sea—planktonic algae.

Deep sea bottom—hold-fast fixed benthic algae.

Terrestrial but not submerged.

SANDY SOILS OF RIVER ESTUARIES

Prostrate succulent herbs or shrubs—*Suaeda maritima* (Chenopodiaceae); *Mesembryanthemum* (Ficoideae); *Abronia martima* (Nyctaginaceae); the last two are not common in the Gangetic estuary.

Shrubs—*Ceriops roxburghiana* (Rhizophoraceae).

HEAVIER SOILS: SANDY SOILS MIXED WITH RIVER SILT AND HUMUS

Tall, evergreen shrubs or low trees—*Avicennia alba* (Verbenaceae); *Rhizophora nucronata*, *Rhizophora conjugata*, *Bruguiera gymnorhiza* (all Rhizophoraceae). *Rhizophoraceae* may have originated from *Theaceae* or even *Tiliaceae*. *Aegiceras majus* (Myrsinaceae); *Sonneratia apetala* (Sonneratiaceae), *Heritiera* sp., *macrophylla*, *minor*, etc., (Sterculiaceae). The Rhodophytan alga, *Coelarthrum* sp. (Sphaeroccocaceae), epiphytic on the mangrove tree trunks and also on pneumatophores are plentiful in these regions. These algae are completely submerged at high tides, getting exposed, however, when the tide is low. The moss *Calymperes sundarbanense* is an epiphyte growing just at the tidal level.

An interesting point to note here is that almost all large and small fishes around

FIG. A photograph of mangrove forest in the Sunderban region. Note stilt roots and pneumatophores. (Photo by Zainab Begum)

these regions do have enormous masses of this *Coelarthrum* in their alimentary canal; these red algae look like making nutritive food for fishes. Is it possible that these insignificant algae around here, may be rich in protein content? Certainly, an examination of their total nitrogen content—inorganic and organic—is called for.

Succulent herbs, shrubs or evergreen small trees—The succulent fern, *Ceratopteris thalictroides* (Parkeriaceae); *Excoecaria agallocha* (Euphorbiaceae). All *Excoecaria* sp. of Sunderban region seem to harbour a hemiparasitic species of *Loranthus articulata*. The host parasite relationship is very strong for all *Excoecaria* sp., so far examined are found to have the green *Loranthus* on them. The parasitic selection of host seems specific indeed and certainly points to an intimate physiological relationship between them. *Sesuvium portulacastrum* (Ficoideae), the succulent twiner, *Basella rubra* (Chenopodiaceae), *Salicornia brachiata*(Chenopodiaceae), etc. In some species of *Salicornia*, they are capable of absorbing water directly through cuticle when they are flooded by the tide; indeed as much as 60% or more of water lost in the preceding exposed period can be reabsorbed in a comparable period of submergence.

Low, non-succulent perennial herbs or shrubs—*Acanthus ilicifolius* (Acanthaceae); *Aegialitis rotundifolia* (Plumbaginaceae); *Carapa obovata* (Meliaceae); the saline palms, *Nipa fruticans* and *Phoenix paludosa* (both Palmae); the polypodiaceous fern, *Acrostichum aureum*.

Grasses—Truly salt-tolerant *Oryza sativa*, var *Patnai*, *Achra*, etc., and the wild rice species of *Oryza coarctata; Myriostachya* (Graminaceae); *Scirpodendron* (Cryperaceae).

A number of succulent species of the order *Chenopodiales*, growing abundantly in saline soil, has been grouped under a special family—*Halophytaceae* (Hutchinson, 1946).

In addition to the soils in which sodium chloride is predominant there are other soils with much Na_2SO_4 and Na_2CO_3. They carry a special flora, and the species and zonations are different from those of true mangroves in NaCl- soil. In the waters of these "*soda lakes*", species of *Enteromorpha*, *Spirogyra*, *Zannichellia*, etc., flourish.

Prain divided mangroves of Sunderban region localised into three plant geographical zones: (1) southern coastal strip and the south-western part, (2) the central zone of *Heritiera* and (3) the north-eastern part. Other regional divisions may be: (*i*) the fresh-water forests, (*ii*) the moderately salt-water forests and (*iii*) the salt-water forests.

Ceriops forms pure forest on higher grounds near the sea, one of the dominant species of the Sunderban region.

The stages in ecological plant successions in the formation of mangrove forests in the Sunderban region may be of two different progressional types—one starting from the sand dune and the other conceivably from a saline swamp, both progressions ending up with the same climax forest of *Heritiera fomes*.

The zonations of mangroves from sea-water to inland in South-Eastern Asia may be: (1) *Avicennia*, *Sonneratia* type; (2) *Bruguiera caryophylloides* type; (3) *Rhizophora* type; (4) *Bruguiera parviflora* type and (5) *Bruguiera gymnorhiza* type. Similar zonal distribution is also likely to be met in the Sunderban region. This frequency and distribution of mangroves species are undoubtedly correlated with the degree of salinity of the soil—the salinity evidently gradually decreasing from the sea-water line to landward side.

The mangrove species of the western sea coasts of India are not as rich compared to the mangroves of Ganges estuary, both in distribution and in density. The six communities which are common in the western sea coasts are: (*i*) *Avicennia alba*, *Avicennia officinalis*, (*ii*) *Avicennia alba*, *Acanthus ilicifolius*,, (*iii*) *Sesuvium portulacastrum*, (*iv*) *Suaeda fruticosa*, (*v*) *Aleuropus repens* and (*vi*) *Salvadora* sp. The mangroves of Salsette islands near Bombay contain a few characteristic species such as *Ceriops tagal* (Rhizophoraceae).

The ecological succession of plants in mangrove forests in western coasts in India may be, from the pioneer species of mangroves of swampy regions, e.g., *Rhizophora*, *Acanthus*, *Ceriops*, *Bruguiera*, *Sonneratia*, etc., followed in succession by tidal mangroves, e.g., *Avicennia*, *Suaeda*, etc., then by non-littoral mangrove species, e.g., *Thespesia*, and *Tamarix* and finally colonisation is completed by xerophilous non-mangroves.

The mangrove vegetation of the Godavari delta in Andhra (South India) is rich and the pioneer species of the muddy flat regions, constituting about 75% of the mangrove forest flora are *Avicennia alba* and *Avicennia marina*, the last one seems to be one of the characteristic mangrove species of these regions. Further inland, *Rhizophora*, *Bruguiera*, *Ceriops* and *Sonneratia* and still further to it, *Acanthus ilicifolius*, *Myriostachya wightiana*, *Clerodendron inerme*, etc., form dense thickets. On the landward higher land, true mangroves seem to disappear, their place being taken by almost pure crops of *Aegiceras* and *Excoecaria*, growing side by side with really non-mangrove climbers such as *Dalbergia spinosa*, *Derris uliginosa* and *Sarcobolus carinatus*.

Communities of plants, common on the saline soil of Pondicherry sea coasts are distinctive in their features. A few species, cited here, are specific for these regions and are not usually found in the Sunderban delta, e.g., two *Suaeda* species, *Suaeda nudiflora* and *Suaeda monoica*. Among the interesting trees are *Salvadora persiea*. Other common species in the saline marshes and of coastal sands are *Cressa cretica*, *Aeluropus lagopoides*, *Heliotropium curassavicum*, etc. A *Spinifex* species, *Spinifex littoreus* is abundantly found on the seashore and on the newer sand dunes.

The composition of any mangrove vegetation undergoes changes with the gradual elevation of the land from the sea-level. The vegetation is thus truly transitory in nature. The succesion of vegetation on a virgin land arising from the sea-level consists of (*i*) pure crops of pioneering true mangroves such as *Avicennia*, followed by (*ii*) other mangrove species such as *Rhizophora*, *Bruguiera*, *Ceriops*, etc., forming the so-called mixed mangrove forests. As the ground rises above the spring high-tide level, the true mangroves disappear, followed by (*iii*) invasion and colonisation of such lands by halophytic plants such as species of *Excoecaria*, *Aegiceras*, etc. which are specially adapted to such saline soils forming climax forest. The *Excoecaria*, etc., forests are interspersed with (*iv*) extensive barren lands with no other important vegetation except perhaps the succulent and prostrate, *Suaeda maritima*. Finally (*v*) the salt-tolerant grasses arrive, colonise and fully stabilise themselves in the area and ultimately make the land fit for cultivation of cereals like varieties of paddy such as the salt-enduring *Oryza sativa*, var *Patnai*, *Achra*, etc. (A hybrid selection from the two has been raised and found to be highly salt-tolerant). The evolution and decline of a mangrove forest are thus completed.

It may be of interest to point out here that high saline soils are not necessarily restricted to sea or ocean coasts only. In many regions such as in certain parts of the arid Rajasthan and other deserts of India, far away from any sea coasts, the soil is distinctly saline due to the presence of salts like NaCl, $CaSO_4$, Na_2CO_3, etc. particularly NaCl. The range of pH in these soils varies from near neutral (7·5) to appreciably alkaline (9·5) . The origin of high salinity of these soils is not quite clear, but may be due to: (*i*) weathering of rocks, (*ii*) salts brought by wind from neighbouring salt concentrations and (*iii*) regional leaching and concentrations. A characteristic halophytic vegetation, beautifully adapted to both high salinity and water scarcity conditions, represented among others by following species which seem to thrive well in these regions:

Tamarix dioica, *Tamarix articulata*, *Chenopodium album*, *Artiplex cressifolia*, *Suaeda fruticosa*, *Haloxylon salicorneum*, *Haloxylon multiflorum*, *Salsola foetida*, etc.

These communities of desert halophytes can very appropriately be described as *microedaphic formations*.

EPIPHYTES

Epiphytes (or *aerophytes*) are plants that grow upon other plants, differing from parasites in not deriving nutrients and water from the supporting host plant and from *lianas* (rooted vascular plants which maintain their stems in more or less erect position by making use of convenient objects for support) in not having any connections with the soil. As an ecological class, epiphytes are the most directly dependent upon rainfall, snow and dew for their water requirements. They are sometimes capable for enduring drought conditions in absence of rain or dew. They derive their nutrient supply mainly from rainfall, accumulated wind-borne dust, smoke and also from the decaying bark surfaces of the supporting host plant.

Epiphytes are common on trunks and limbs of trees, shrubs and even on submerged plants. Rarely, they also occur on the upper leaf surface of woody plants—*epiphylls*. They are generally abundant in the forks of trees, on horizontal or nearly horizontal trunks and branches but are seldom found on vertical and smooth surfaces.

The association between the supporting host plant and the epiphyte must be very loose, for neither partner gets much benefit out of this association. The epiphyte certainly is benefitted to a small degree by the food it can derive from the surfac of host plant as mentioned before but what can the host plant possibly get from the epiphyte? Perhaps nothing. Thus these associations cannot, by any stretch of imagination, be compared to symbioses, even if the water absorbed by aerial epiphytic roots could contribute (which is very doubtful) to the water requirement of the supporting plant. The epiphytic associations between terrestrial *Cyanophyta* and a few species of flowering plants (e.g., *Gunnera*) and *Cycads* can, however, be some sort of *symbiosis* for [there is certain amount of evidence that the alga (*Nostoc punctiforme*) separated from *Cycad* is scarcely capable of autotrophic nutrution] the atmospheric nitrogen fixed by the alga (*Nostoc*) does possibly contribute to the nitrogen requirements of the host plant. Even if we consider these associations as symbioses these are neither so highly developed nor so successful as those between legume and *Rhizobium*.

The greatest single factor operating against the epiphytic habit is drought and consequently, terrestrial flora is very rich in epiphytes in situations where drought is never of long duration. In cold and dry climates, epiphytes are few and chiefly represented by cryptogams, e.g., terrestrial algae, lichens, liverworts and mosses. In warm wet climates, epiphytic flora is particularly rich and a variety of vascular plants such as ferns and angiospermic families of *Bromeliaceae* and *Orchidaceae* predominate. The luxuriant epiphytic flora of the tropical rain forests include not only species which are shade and moisture loving (confined only to lower trunks of trees) but also xerophytic *Cactaceae* (e.g., *Cereus*) which are sometimes found on tree tops. Apparently requirement of light ranks higher to these tree-top *Cereus* than occasional dryness due to wind which they are adapted to endure; they sacrifice moist soil conditions in preference to bright light.

Some epiphytes seem to show marked preference of particular supporting plant, i.e., they are very selective in their choice of host partner. An interesting case is that of the moss, *Tortula pogorum*, which is found only on the barks of trees *within the city limits*.

This certainly intrigues one's imagination. What can a plant utilise from city life? Is it higher temperature, compared to that of the wide open spaces, they crave for or do they require some constituents from air-borne dust and smoke available inside the towns, which may be essential for the growth of this epiphytic moss?

Some epiphytes grow on rocks and some even on horizontal telegraph wires as well as upon plants.

Some epiphytic ferns start their lives as rooted plants in the soil but soon, by extending their rhizomes up a readily available tree trunk, lose all connections with the ground and eventually become true epiphytes. Others such as hemlock (*Tsuga*) and certain figs germinate on trees and live as epiphytes during the early part of their life cycle until their roots grow down the hosts' trunks and establish connections with the soil, when they cease to be true epiphytes.

Study of symbiotic relations between higher plants has been largely neglected. A number of observations, however, indicate that there may be something in it. It is well-known that various epiphytic orchids and some species of *Ericaceae* (a family which provides many true epiphytes in the tropics) in tropical rain-forests occur exclusively on certain host trees in such constant ratios, that the host trees can be identified by their specific epiphytic communities.

As some epiphytes have no soil connections during their entire life history, xeromorphic adaptations such as thickened cuticles, sunken stomata, succulence (*pseudobulbs* or shoot tubers may often serve for water storage) and reduction in the number of leaves (in many orchids, there is only a single leaf) are common. In *nest epiphytes* the roots accumulate large masses of dust and debris that have water holding capacities of a soil. The aerial roots of many tropical epiphytes belonging to *Araceae* or *Orchidaceae* extend outward into the air (Fig. 788). They are thick, unbranched, whitish structures covered on the surface by a peculiar tissue called *velamen*. This velamen is a special absorptive tissue that can take up water rapidly from a brief shower of rain and even from a near-saturated atmosphere, like a piece of blotting-paper soaking up water. It is essentially a modification of many-layered epidermis, a parchment-like sheath of considerable thickness, sorrounding the aerial roots. The cells of the velamen are empty (dead) and they are generally spirally or reticulately thickened. A saturated velamen root looks pale-greenish, whereas when air is dry, it looks white (Fig. 789).

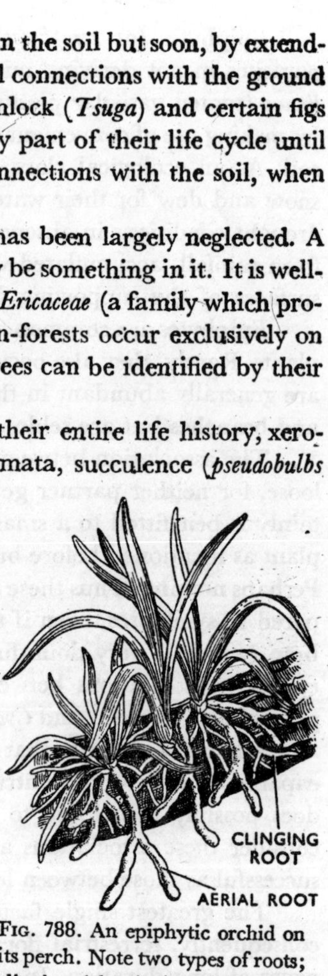

Fig. 788. An epiphytic orchid on its perch. Note two types of roots; clinging (clasping) and the hanging aerial roots.

Sometimes greenish white aerial roots, 0·75–1·5 metre in length, with typical velamen-like properties are developed on the branches of some non-epiphytic species such as the showy, *Vernonia monosis* (*Compositae*). The absorbed water is stored in the velamen until absorbed by the living cells of the aerial roots.

Among epiphytic species of the family *Bromeliaceae* (e.g., *Tillandsia*), thigmotropic and often positively phototropic roots principally serve as organs of anchor for supporting the plant (*clasping roots*) or they may also be completely wanting, when the function of absorption of water is taken over by leaves and stems. The clasping roots consist almost exclusively of thick-walled fibres and take absolutely no part in nutrition. In some other *Tillandsia*, the peltate-scale-covered finely divided shoot systems collect capillary water with the help of the absorptive scales. The water is finally absorbed by uncutinised spots on epidermis of the leaves. The leaf bases of a few *Tillandsia* sp. are shaped like furrows or cavities and can collect water in their axils—*tank* or *cistern epiphytes.*

In *Dischidia nummularia* (*Asclepiadeceae*), one of the most specialised of all epiphytes inhabiting the alluvial Sunderban region, one of the leaves forms pitchers that collect

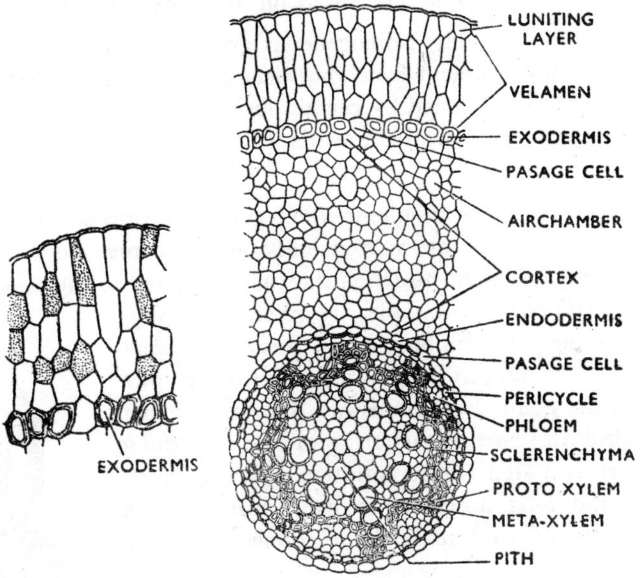

FIG. 789. Transverse sections of aerial root of epiphytic orchid, A; B: *velamen* with exodermis and reticulately thickened cells.

and accumulate rain water and into which modified, special and freely branching absorptive roots extend. The pitchers sometimes harbour ants and other insects, the excreta and the dead remains of which may serve as a source of additional organic nitrogenous food for the plant. In *Dischidia rafflesiana* (Fig. 790) and in epiphytic fern.

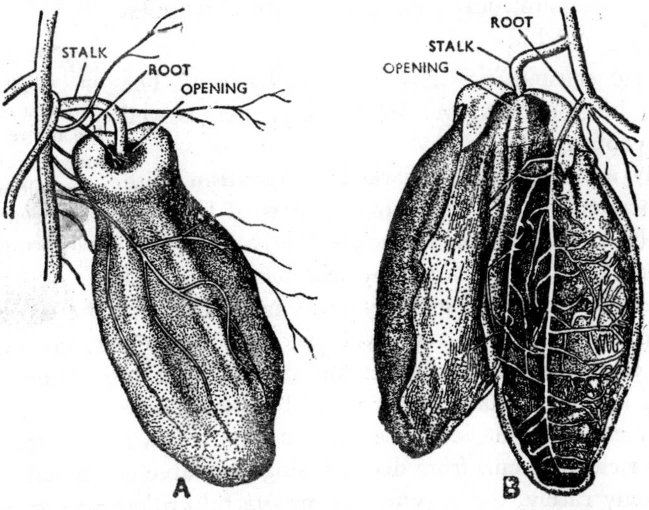

FIG. 790. Pitchers formed by the leaves of *Dischidia* sp. A: entire pitcher. B: pitcher cut open to show the modified freely branching absorptive roots extending into the pitcher.

Asplenium nidus (bird-nest epiphyte), leaves specialised into leaf pockets, (leaves forming a rosette, enclosing a funnel-shaped cavity) help the plants to absorb humus and mineral nutrients which accumulate in the bark-surface crevices of the host support.

Only plants whose seeds are usually dispersed by wind, insects or birds to the tops and branches of trees can adopt an epiphytic mode of life. According to their mode of life, the following *four* groups (mainly based on Schimper) can be distinguished:

I. PROTOEPIPHYTES—A really homogeneous group and includes species that are compelled to acquire nourishment from the surface of the supporting host and from atmosphere. Examples are afforded by certain ferns, *Peperomia* and a number of *Asclepiadaceae*. They frequently lack any definite adaptive features except perhaps aerial roots with velamen.

II. HEMIEPIPHYTES—These are plants which germinate and pass through their early development on host structures but subsequently become connected with the ground by their own roots. As regards their nutrition they are subjected to the same conditions as are ordinary terrestrial plants. Examples are given by epiphytic fig trees and some root-climbing large *Aroid*, *Scindapsus officinalis* (Fig. 791). The largest hemiepiphytes in the tropics are the species of the genus *Ficus* (banyan). Its roots at first are of one kind and all serve as anchoring or/and absorbing organs; later there is a sharp differentiation between short, anchoring roots and long absorbing roots. Ultimately the absorbing roots descend and make final connections with the soil.

There are some *lianes*—stem climbers that germinate in the soil but their stems gradually die from below upwards, so that at maturity, they exist just like hemiepiphytes. They have been termed *pseudoepiphytes*.

III. NEST EPIPHYTES—This group is composed of some characteristic species, which by appropriate and efficient devices collect, large quantities of humus and water. Examples are given by many orchids.

FIG. 791. *Scindapsus officinalis*, an epiphytic aroid with climbing roots. It begins its life as a terrestrial plant but after climbing up the host plant ultimately gets completely detached from the soil and becomes an epiphyte.

IV. EPIPHYTIC BROMELIADS—This is a class by itself. The *Bromeliaceae* is a large family (over 1600 species) that is well known for its epiphytism.

The family as a whole has marked xerophytism with which are associated various absorbing systems of leaves—in particular, (1) absorbing peltate scales that act as one-way valves and (2) tank formation by leaf bases.

Again, *four* principal ecological types of bromeliads can be recognised:

type I—*Soil root*: The soil root type has well developed root sytems and soil constitutes the exclusive source of water and mineral nutrients. They have no specialised absorbing organs. Tanks are not formed by leaf bases.

type II—*Tank root*: The leaf bases in this type retain rain water and dew. The water becomes rich in humus from decomposing vegetable or animal remains. Roots enter the soil only rarely, but a system of *mycorrhizal axillary roots* grow upward between expanded and overlapping leaf bases, utilising the collected water and nutrients. Perhaps most of them are not exclusively dependent upon the interfoliar roots.

type III—*Tank-absorbing trichomes:* They constitute the great majority of the bromeliads. The tank is much better equipped and is also better developed than in the preceding type. With leaves rosulate and upright, the rain and falling leaves are effectively collected in the tank. Roots merely serve to anchor the plants, absorption occuring mostly through epidermal scales. Some of the epiphytic plants in tropical rain forests become gigantic in size—in some plants of this type, the inflorescence may be as high as 3 metres and its tank may have as much as 5 litres of water! The tank contains a *microcosm*, rich in microorganisms and even hydrophytic insectivorous plants, i.e., *Utricularia* which flourish there, catching numerous small creatures, e.g., ants, worms, numerous protozoa, etc., in such traps.

Several sub-types are recognised:

(*a*) species wholly dependent on the *single* tank produced:

(*b*) other species produce *nests* as a result of vigorous vegetative propagation and the additional humic material entrapped in the nests, is utilised as well as that in the tanks;

(*c*) others, such as some *Tillandsia* sp. have ephemeral tanks, and

(*d*) there are species associated with ants that may be regarded as myrmecophilous; some of them are never found except in ant gardens.

type IV—*Atmospheric vapour, dew and dust-absorbing trichomes:* In these plants the tanks and the adventitious absorbing roots are both lacking; in *Tillandsia usneoides* all roots are absent. Rain and dew are absorbed by scales that usually cover the entire plant and the *only* source of mineral nutrients for them must be dust in the air. There is a suggestion that some species of this type have symbiotic fungi which may conceivably be nitrogen-fixers although direct evidence is lacking.

The type IV may again be broadly divided into three categories:

(1) certain forms intermediate with type III,

(2) those that absorb everything wholly from the atmosphere and

(3) those that are myromecophilous.

Some authors studied the occurrence of bromeliads in the forests and divided them again into 3 main groups:

(*i*) *exposure group:* consists of species that occur at the top, i.e., in the highest levels of the forest with complete exposure to the sky. All members of this group, almost exclusively are tank-forming species but with no nest formation. Under drier conditions, the exposure group becomes predominantly atmospheric type (type IV);

(*ii*) *sun group:* contains the largest number of species. These plants include both the true tank-type and the tank-nest type; they are amongst the most massive growth-type of bromeliads, and

(*iii*) *shade-tolerant group:* occurs predominantly in the lower-levels of forests, even near the ground but they increase in density with increased illumination. Most of these are of the ephemeral tank-type, containing only very small amounts of water. The nest-type is absent, as it is in the exposure group.

LITHOPHYTES

According to Warming's classification based on edaphic factors of the soil, *lithophytes* are a group of plants which are adapted to the rock surface habitats and naturally grow upon them. Of all the ecological factors which affect the vegetation of a particular place, physiographic factors, e.g, steep hills are the most important in determining the vegetation on rock surfaces.

Two types of lithophytes are generally recognised: *true lithophytes* growing on bare, inclined rock surfaces and *chasmophytes* which grow on cracks, fissures and crevices of rocks.

True lithophytic flora is mainly composed of some terrestrial algae, lichens and mosses. The habitat is dry and as a result the life forms of individuals comprising the lithophytic flora are adapted to derive water from rain, dew and melted snow (in higher altitudes). Many of the cryptogams included among lithophytes have devices for collecting water (such as rhizoids, which also help to attach the plants on the precipitous incline). Even then, lithophytes must endure considerable desiccation due to the topography or due to dry cold wind or both, high insolation, etc., and like epiphytes, must obtain their water requirements from dust and precipitations such as rain, snow and dew, or from trickles of water sometimes running down the hills.

Chasmophytes are lithophytes that grow on rock crevices and fissures where *micro-climate conditions are often attained* (rock crevices and fissures are sheltered from wind and are thus comparatively free from extreme variations in temperature, thereby becoming shady and moist compared to open rock surface). Chasmophytes, like true lithophytes, consist mainly of algae and pteridophytes (e.g., *Equisetum*) but considerable spermatophytic population is also observed. The chasmophytic spermatophytes are generally tufted or rosette in appearance and may be both mesophytic and xerophytic in structural adaptations. The prostrate forms of vegetation of high altitude, and rosette of other lithophytes may be of considerable advantage in allowing these plants to utilise the comparatively warm layers of air near the surface of the soil. This may permit the basal leaves and shoots carry on photosynthesis at times when it may be too cold for photosynthesis by taller green shoots, which are up in the air. Succulence is often met with, chiefly represented by the family *Crassulaceae.*

Families, *Saxifragaceae* and *Orchidaceae* are among the more widely distributed chasmophytes. In eastern Himalayas, chasmophytic grasses (e.g., *Erianthus* sp. and *Miscanthus nepalensis*), with flattened and tufted roots penetrating deep into the crevices of the rock side, are often met with. Other higher spermatophytic chasmophytes are *Anaphalis* sp. *Aster*, wild daisy, *Digitalis* and many orchids.

The developmental process (the succession of vegetation) of a community of lithophytic plants on virgin rock surface is called a *lithosere.* The pioneer communities are the lichens and mosses and the higher angiosperms seem to come in reluctantly and the series is somehow completed.

CHAPTER LV

ENVIRONMENTAL POLLUTION AND PLANTS

The environment we live in is polluted greatly by biological activity. If the equilibrium among the activities of different organisms—plant, animal and micro-organisms is seriously disturbed, new equilibria have to be attained sooner or later but that may not be favourable to man or the plant or animal world. Although pollution has been existent since time immemorial, it has assumed disturbing proportions since the industrial revolution and now it is almost alarming.

Since industries are usually situated in or near the cities, the environment of the cities and suburbs are most polluted. The effect gradually decreases with increase in distance. It has been estimated that Calcutta and Howrah air contains about 5·57 tons of dust, 122 tons of SO_2, 440 tons of CO, 70 tons of oxides of nitrogen and 102 tons of hydrocarbons everyday. The smoke produced by coal and oil combustion and automobile fumes combine with fog to produce the smog characteristic of most cities. Aerosols, which have both solid and liquid components are suspended in air or gaseous environments and quite often contain toxic ingredients which include heavy metals like Zn, Pb, Cu and Fe.

Sewage, agricultural and industrial waste discharges are the main causes of water pollution. Sewage discharges in the Dal lake of Kashmir and Loklak lake of Manipur has resulted in the development of extensive vegetation. The Sagar lake near Hyderabad has been polluted by inorganic effluents. The river Ganges receiving the sewage and industrial discharge from the cities on her banks is now one of the most polluted rivers of the world; the water is even unfit for bathing, not to speak of drinking purposes. The water of the river Yamuna near Delhi is so polluted that it has to be diluted about 8000-fold to reduce its toxicity below permissible level. Important pollutants for river and oceans include heavy metals, halogenated hydrocarbons, petroleum, artificial radioactivity and litter.

Many of the pollutants mentioned above have mutagenic and carcinogenic properties. Respiratory difficulties, nervous disorders, disabilities of various sorts and several other diseases have been reported. Concentration of population in cities, urbanisation, lack of consciousness of pollution hazards have further helped in the spread of pollution. The pathogens responsible for various diseases of man, viruses, bacteria, fungi and protozoa are multiplying and spreading biological pollution. Much less work has been done with plants.

The industries are producing huge quantities of gases which are escaping into the atmosphere of our earth every day. The gases include CO_2, CO, SO_2, H_2S, NH_3 C_2H_4, oxides of nitrogen and various hydrocarbons. The CO_2 concentration according to some estimates has increased considerably; although this should increase crop productivity the heat loss due to radiation will decrease, as CO_2 does not allow heat to pass through, resulting in an increase in temperature. If this increase continues, a day will come when the snow caps on North and South Poles will melt, increasing the sea level to an extent that many coastal cities and lowlying areas will be inundated. Increase in CO_2 levels will also affect respiratory processes in both animals and plants. Carbon monoxide, which is considered to be a poisonous gas, may give rise to several diseases; it is poison for plants also, where it affects the cytochrome system. The oxides of nitrogen, in addition to their harmful effects react with the ozone of our atmosphere, which provides a protective shield from the lethal and mutagenic ultraviolet radiation coming from the sun. This will affect both plant, animal and microbial life. The application of heavy doses of synthetic nitrogenous fertilizers is encouraging the activity of nitrifying and denitrifying bacteria; the latter group of microorganisms produce N_2O (in addition to N_2) which is dangerous in this respect. The jet aircrafts are also discharging considerable quantities of nitrogen oxides into our atmosphere. The hydrocarbons in the exhaust fumes of automobiles and the discharges of petrochemical industries are causing serious health hazards to men, plants and animals also. The industries are also throwing into our atmosphere, in addition to soot and dust particles, various compounds of metals like lead, mercury, arsenic, cadmium, etc. Plant life, particularly, trees are visibly affected in some parts of the Durgapur region.

Equally dangerous are the discharges from various industries, paper, leather, jute, cotton, iron and chemicals (see Table below). A variety of chemicals used in these industries is thrown into river water as industrial waste. This is seriously affecting aquatic life—both flora and fauna. During the past decade the aquatic flora of many places have changed. One of the reasons why Hilsa fish is becoming more scarce in the Ganga is pollution of Ganga water, particularly in certain regions and they are migrating to new places. The aquatic flora on which they live are gradually disappearing. Chemical wastes from industries are consumed by fish etc., which ultimately produce toxicity symptoms in man. The havoc caused by industrial wastes containing mercury in the coastal region of Japan is well known. The oil refineries in our country are also polluting our waters. The various petroleum products float on water and plant life,

The Pollutants of some industrial effluents

Industries	Pollutants discharged
Paper, sugar, distilleries, vanaspati, textiles, tanneries	Various organic material in large quantities
Fertilizer industries	Alkalis, chromates, ammonia, cyanides, phenols, naphthalenes, etc.
Coal, iron and steel	Phenol, harmful gases, smoke
Petroleum processing	Oil and phenols
Sulphuric acid	Sulphurous gases
Caustic Soda	Mercury
Rayon, pesticides, petroleum products	Zinc and many other toxic chemicals.
Thermal Power plants	Various gases including SO_2, unburnt particles, soot, etc.
Nuclear Power plants	Radioactive wastes

both in water and on the river banks is seriously affected. At Haldia tree planting on the strand is posing a problem. Polluted river water is dangerous for drinking and irrigation purposes.

Modern agricultual practices are also adding to pollution hazards. Thus, application of excessively high doses of fertilizers are encouraging the growth of unwanted plants and algal blooms, which produce organic matter via photosynthesis, and settle at the bottom of tanks and ponds. Drainage water from agricultural lands contain inorganic salts, which are difficult to remove by conventional procedures and toxic compounds including pesticides used for plant protection purposes. Faulty irrigation of crops is also quite often responsible for salt and alkali accumulation; inadequate drainage leads to increased salinity of ground water.

The extensive use of fungicides, insecticides and agricultural chemicals has created serious problems. Many of the chemicals, though in small quantities, have found their way to human bodies through the crops and vegetables we use. It has been estimated that everyone of us is carrying in his adipose tissues about 20 mg. of the insecticide DDT, the use of which has now been banned in many Western countries.

Consumption of crabs collected from rice fields receiving very high doses of pesticides caused unusual paralytic attacks in man. The agricultural use of seed dressing by mercuty and industrial effluents resulted in the accumulation of mercury in fish and birds resulting in the thinning of eggs. Several cases of poisoning have also been reported in green house cultivations in Japan.

Many of the chemicals are not easily decomposed and are deposited in our soils to be carried by rain water to various regions. Burying of idustrial water in the soil reduces water pollution but it is no solution since sooner or later it will be carried into river or sea water. Some of them may be reabsorbed by plants. The soil microbial life is also affected.

The extensive use of antibiotics in food processing, fodder and crop protection has contributed to the selection of microorganisms resistant to these antbiotics. Intensive cultivation of the so-called high-yielding cultivars, developed in foreign countries, where common pathogens of our country are non-existent, is replacing our less-yielding but relatively resistant cultivars and crop epidemics are becoming common. Development of the Nagarjunasagar dam in Andhra Pradesh has increased molybdenum concentration of soil to such an extent that cereals like sorghum are experiencing toxic conditions. Air and other pollutants quite often lead to chlorosis, decreased photosynthesis and necrosis. Particulate matter containing magnesium and lime decrease crop yield, smoke soot suppresses growth, sulphuric acid droplets with moisture, blacken leaf surfaces and automobile fumes cause lead injury in crops. Lethal effects of pollutants on aquatic flora lead to a decrease in oxygen concentration of water, resulting in asphyxia of fish an other aquatic fauna.

The radioactive wastes from nuclear power plants cause radiation hazard to both plants and animals, whether buried in the soil or dumped in the sea. The radioactive fall-out from atomic or hydrogen bomb explosions is polluting air, land and water. ^{90}Sr which is usually present in such fall-outs is absorbed by plants which are used as fodder for cattle and the milk of such cattle contains the ^{90}Sr which ultimately replaces Ca in our bones becoming a potential carcinogenic source.

Sharma has suggcted that induction of plants resistant to pollution may be very rapid, particularly when they are exposed to acute and chronic pollution hazards. The range of tolerance of the plant species may enlarge enabling them to survive in a polluted situation. This is of special ecological significance and plants showing such capability may be used effectively as indicators and pollutant scanvengers.

Plants are also helping considerably in reducing the pollution hazards. Thus, the relatively higher CO_2 content of our atmosphere is gradually replaced by O_2 via photosynthesis. A 50 ton tree may produce 1 ton O_2 per year. Many harmful gases are also absorbed by plants. Soot dust particles (in India 40% of air pollutants are dust particles) and industrial wastes in the particle form—including compounds of heavy metals are collected by leaf surfaces—the total area of which is considerable. Many trees collect 1–5 gm dust per m^2 leaf surface. As they absorb various polluting substances from soil and water, they reduce their concentrations in the environment. Some plants are particularly remarkable in this respect e.g., water hyacinth which has been found to grow luxuriantly in polluted water in several places, as it can remove many toxic compounds including metals, and the alga *Porphyra*, which accumulates ^{106}Ru. But plants are probably suffering as much from pollution of our air, water and soil as animals do. Several microorganisms are now used for detoxification of water. The products of degradation may even be useful in some cases—microorganisms capable of producing hydrocarbons which may be used as fuel or for utilising hydrocarbon—producing energy and food etc. will be particularly useful in this respect. Some plants introduced in this country have helped in the biological control of certain pests Eradication of alternate hosts of pathogens may also eliminate certain diseases.

Some plants may cause pollution themselves. The fungal spores and pollen grains of many plants are responsible for allergic diseases, particularly respiratory ones. The cause of hay fever and several types of asthma has been traced to pollen grains. Grass pollen grains are particularly remarkable in this respect, but many other plants are equally dangerous. Plants like *Parthenium* which are multiplying rapidly are highly toxic and so are several other poisonous plants. Seeds of *Argemone mexicana* which are used as adulterants of mustard seeds contain toxic principles. Many plants produce the so-called allelochemicals which inhibit the growth of other plants. Since this is done on a large scale for establishment of a community, their implications are quite serious for plant life. The chemicals include abscissic acids, phenolics, alkaloids, terpenes and a variety of substances. Many bacteria and fungi produce toxins. Aflatoxin produced by *Aspergillus flavus* is posing a serious problem.

Pollution thus, is causing a serious problem, in as much as it is upsetting the ecological balance. While industrialization cannot be avoided, the polluting agents, whatever be their source, have to be identified and inactivated as far as possible. Fortunately people have become conscious of pollution hazards now and scientists in many countries of the world, including India, are studying the extent of various types of pollution and carrying on research for formulation of remedy measures to check or reduce pollution in our environment.

Meanwhile if pollution is to be controlled, it is essential that antipollution acts are strictly enforced, environmental education is encouraged in schools and colleges, agricultural and industrial wastes are adequately recycled and afforestation and tree planting are pursued intensively. Use of biofertilizers and biological control methods, establishment of bioshpere reserves, conservation of resources in the Himalayas, Arunachal Pradesh, Himachal Pradesh etc., utilization of biomass as a source of energy, conservation of flora and fauna and social forestry are some of the other measures which have been suggested for the control of pollution in our country.

CHAPTER LVI

FOOD CHAIN, ENERGY FLOW AND ECOLOGICAL BALANCNE

Any interference with ecological principles, as has been exemplified in the discourse on pollution, in the previous chapter, upsets the ecological balance among the organisms which populate the earth. Ecological balance, and in fact, the continuity of life, depends on the coupling of energy-producing and energy-consuming processes. Energy production is restricted only to a group of autotrophic organisms, many of which are endowed with the capacity of converting light energy into chemical energy. Some non-photosynthetic organisms also can oxidise reduced inorganic compounds and utilize the chemical energy liberated in the process. The ultimate source of energy, in all cases, is the sun.

Biomass production

Every minute, each cm^2 of the earth's surface receives about 1.34 K Cal of solar energy. The process of photosynthesis, on which life depends, however, can utilize no more than 2.5% of this energy. As light intensity decreases with increase in the depth of water or as sunlight filters through the canopy of plants, the net effect decreases still further. Nevertheless, about 10^{11} tons of carbon are fixed from the atmosphere per year. The photosynthetic bacteria which are usually present in anaerobic environments in aquatic sediments, where light intensity is low, fix small amounts of CO_2 and N_2 from the surrounding environment. In many lakes the contribution of the photosynthetic sulphur bacteria is no more than 5% of the total organic matter production per unit time; however, in places where there is a good supply of H_2S, naturally or through discharges of industries, about 25% of photosynthesis can be attributed to them. The non-sulphur photosynthetic bacteria also make significant contribution. The cyanobacteria or the blue green algae are much more efficient. *Oscillatoria* is a major component of the Caribbean Sea plankton, comprising about 60% of the total chlorophyll in the upper 50 metres and is responsible for about 20% of the primary production. The *Oscillatoria* formation has been estimated to fix about 1.3 mg of N_2 per m^2 per day. Primary productivity of our oceans is about 1000 K Cal/m^2/yr, the total gross production being 10^{16} K Cal/yr. In coastal zones and in the estuaries and reefs the primary productivity is 2,000–20,000 K Cal/m^2/yr. About 18–36 million tons of N_2 are also fixed in the ocean/yr. The blue-green algae also fix large amounts of nitrogen in association with plants like *Azolla.* In Eastern India *Azolla* has been found to fix as much as 37 mg N_2/cm^2/day.

Terrestrial primary production has been estimated to be 57.4×10^{16} K Cal/yr, about half of which come from the evergreen forests in the tropical and subtropical regions and about one-fifth is contributed by grasslands and pastures. Total primary production/yr. is about 10^{18} K Cal. About 175 million tons of N_2 are fixed by microorganisms independently or in association with other organisms, particularly plants; the permanent grasslands contribute about one-fourth of this amount and forests and woodlands about one-fifth. A part of the carbon fixed is lost in respiration, a small amount may also be lost as ammonia and excreta. The difference of the two is the net biomass.

The carbohydrates, lipids and proteins are the major organic constituents of living organisms. CO_2 and N_2 of our atmosphere, through a series of metabolic reactions, pass through these compounds, which in addition to being responsible for the structure and form of almost all organisms, also provide energy for activities of the living forms, depending on the level of reduction of carbon and nitrogen.

Non-photosynthetic microorganisms like the sulphur-reducing or methane-producing bacteria produce these compounds as products of anaerobic metabolism. Several microorganisms can also utilize H_2. Their overall contribution to energy economy of the biosphere, however, is not very large. The anaerobic microorganisms derive energy by oxidation of reduced compounds using compounds other than O_2 as terminal electron acceptors e.g., nitrate, sulphide, sulphate etc. Organic compounds may also be reduced as in fermentation reactions.

Water was used as a source of electrons first about 2.6 billion years ago and since then the atmosphere is becoming richer in O_2. When oxygen accumulated in sufficient concentration, mechanisms were developed for utilization of O_2 as the terminal acceptor of electrons. This resulted in aerobic respiration accompanied by oxidative phosphorylation. ATP production in aerobic respiration is much higher than that in anaerobic respiration. Since green plants are capable of both photosynthetic and oxidative phosphorylation, the autotrophic forms were more efficient for biomass production than the heterotrophic forms, even though the primitive living organisms were undoubtedly heterotrophic. The aerobic environment encouraged the growth and multiplication of non-photosynthetic aerobic autotrophs like the nitrifying bacteria, the emergence of which made the operation of nitrogen cycle in nature possible and which together with carbon cycle are essential for continuity of life on earth.

Producers and consumers

The photosynthetic organisms are the primary producers and the heterotrophs are the consumers. In the lichens, which is an association between an alga and a fungus, the alga is the primary producer and the fungus the consumer; the algal cells however are not consumed by the fungi. The fungi grow parasitically or saprophytically and thus, are always consumers, although sometimes they may produce growth factors or hormones which may be beneficial for the autotrophic hosts. When *Anabaena* lives symbiotically in the leaf cavities of *Azolla*, it does not photosynthesize and utilizes the organic compounds of the host plant for N_2 fixation. It is thus, a consumer; however, it returns much of the organic compounds in the form of nitrogeneous substances to the host plant and so far as nitrogen nutrition is concerned it may be considered as a primary-producer. In water-logged rice fields where *Azolla* multiplies rapidly, the products of decomposition of *Azolla* are utilized by the rice plant, which although, itself a primary producer, should be considered as a consumer when it receives the nitrogenous substances from *Azolla*, though indirectly.

In ponds and other aquatic environments, the blue-green and the green algae and the aquatic plants are the primary producers. The phytoplanktons are consumed by the zooplanktons, the zooplanktons by other small animals, which are utilized by fish and the fish may be consumed by man, birds or some other animals. The goats feed on the grasses which are the primary-producers. Man becomes a consumer when he takes the mutton. When a man is killed by a tiger, the tiger is the consumer. The larger is the number of organisms which intervene between the primary-producer and the ultimate consumer, the longer is the food chain. When a man eats rice or potato, the chain is shortest, when he eats fish or some animal product, the chain is longer. In the case of heterotrophic consumers also the food chain may be long. Thus, protozoa may be preyed upon by insects, insects by bacteria and bacteria by *Bdellovibrios* and viruses. The viruses in this food chain are the ultimate consumers. In the food chain there may be interlocking i.e., cross utilization at several points, resulting in what is known as a food web!

In every step of the food chain, however, there is a loss of energy, which may be as much as 90%. With every passing step, there is a decrease in size and volume and increase in protein content.

Decomposers

When organisms—autotrophs or heterotrophs—die, microorganisms start decomposing the corpse or detritus, as also the excreta in various ways. Plants are also known to leach or exude organic and inorganic matter.

Insects and other small animals help in splitting the larger particles into smaller ones. The degradation continues depending on the enzyme-makeup of the decomposers. Many such decomposers produce hydrolases, which degrade cellulose, pectins, proteins, and lipids to simple monomers like sugars, amino acids, organic acids etc., which are then metabolized rapidly by other microorganisms. In the case of plant remains, in many cases the fungi are the first attackers because many of them are active producers of enzymes which degrade cellulose, pectin and lignin. However, the degradation is scarcely complete because aromatic substances resulting from degradation of lignin and other large molecular weight aromatic substances cannot be degraded and metabolized by most microorganisms. These undefined aromatic substances along with polysaccharides etc. constitute "humus", which is so important for soil formation and its physical structure. A good amount of nitrogen is also locked up in humus. Animal remains are degraded much rapidly. In the case of invertebrates it is quite rapid; in the case of vertebrates the bones are much difficult to degrade and they may be recognised in soil even after hundreds of years. The decomposers utilize the products of decomposition and are thus consumers. The CO_2 evolved and the heat generated provide a measure of decomposition.

Biogeochemical cycles

The ultimate products of degradation are usually small molecules, gases, oxides and salts. The CO_2 liberated in respiration or decomposition of organic matter returns to the food chain *via* photosynthesis, the primary producers and the consumers. The ammonia produced is usually oxidised to nitrate by the nitrifiers and denitrified by denitrifying bacteria to N_2 or N_2O. N_2 comes back to the biosphere through N_2 fixation by bacteria, freely or symbiotically with other plants. H_2S is oxidized to S° or SO_4'' which being solids are retained in the biosphere. The phosphates present in plants or animal bones are slowly solubilized by phosphate-solubilizing bacteria and recycled through plants, animals and microorganisms. Similar cycles also exist for Mn, Fe, and other minerals.

The turnover time depends on the nature of the compound being recycled. In the case of gases it may be of the order of a million years or more.

The consequences of interference with the ecological balance

Human activity is interfering seriously with the operation of these—cyclic processes endangering the life of man in the distant future. Because the rate of photosynthesis is higher than the rate of respiration, there has been an accumulation of photosynthetically evolved O_2 in our atmosphere. Its utilization in respiration is preventing the attainment of high oxidation levels. The atmosphere has changed from a reducing to a moderately oxidizing one to which the primitive and more evolved forms of life have adopted themselves.. Industrial processes are consuming considerable O_2 and producing large amounts of CO_2. By 1959 CO_2 concentration in our atmosphere increased by 13% and by the year 2000 the increase may be as much as 25%, which in turn will increase the temperature of many places affecting the life of various organisms. Alteration of CO_2 concentrations had profoundly influenced the climate in the geological past. Application of heavy doses of nitrogenous fertilizers and the use of jet aircrafts are enhancing the concentration of nitrogen oxides in our atmosphere, which in turn is depleting the ozone shield of our atmosphere.. The excessive application of fungicides and insecticides etc. has now come to a point that several species have totally disappeared from certain places and the resistant forms, which have gradually become selected, are now becoming predominant. Every man now carries about 21 mg of DDT in his adipose tissues; even in the breast milk of mothers, the concentration is not far from the value lethal for the babies. Malarial parasite which was controlled by DDT, is now reappearing with vengeance, being resistant to many insecticides. Excessive use of antibiotics has selected the drug resistant forms to an extent that unless new and more powerful drugs are discovered continually, disastrous epidemics may occur in the near future. Extensive deforestation has been suggested to be the cause of decrease in rainfall in several places; attempts to convert arid areas into green belts by irrigation etc. have also been suspected to change the location of low pressure areas resulting in diversion of clouds to other regions. The prudence of the introduction of crops and other plants to distant places has also been questioned. Some of the high-yielding varieties of crops developed in Phillipines or Mexico have ousted the local low-yielding disease resistant cultivars and the consequences may well be disastrous, since the introduced varieties are not always resistant to the indigenous pathogens. Alongwith the food grains imported from foreign countries, we have also "imported" weeds like *Parthenium* and several pathogens which have already started causing serious problems in many places. *Parthenium* is spreading at a fast rate changing the composition of the weed flora and causing considerable health hazards. Man has to pay heavily for upsetting the ecological balance. He still has a long way to go if he has to understand the ways of nature and what she has up her sleeves, if she is disturbed.

PART FIVE

Evolution, Genetics and Plant Breeding

CHAPTER LVII

WHAT IS EVOLUTION?

The word *evolution*, arising from the Latin word *evolvere* (to unroll) means the development of something by natural processes. Thus, we may talk of the evolution of the Earth, of the evolution of Man or of the evolution of a political dogma. When we talk of the **Evolution Theory** or about **Organic Evolution** we mean the evolution of life on this Earth, the greatest exponent of which idea was Charles Darwin.

If there can be a story based on facts there can be none more interesting than that of the evolution of life on our Earth. The ideas of Evolution once began as poetic dreams, then as whispered theories, finally condensing into the bold assertions of Charles Darwin (discussed in next Chapter). Since then, evidences have piled up to such an extent that Evolution is no longer a mere theory but an established fact. Let us see how it took place.

In talking about the evolution of life we begin from the beginning. Astronomers and geologists have concluded that this Earth first evolved out of celestial bodies as a fiery spinning ball of extremely hot gases. Gradually, through hundreds of millions of years, the gases cooled down and a solid core was formed. An atmosphere of air (possibly somewhat different in composition from what it is today) formed a covering round it and the hot steam poured down as water forming the first seas and lagoons that were rather shallow. Life first began on this shallow water when it cooled down a bit. How did it begin? We have not yet got the answer but, perhaps, someday we shall. In all probabilities its origin was merely accidental and was possible only in the peculiar conditions of that time. Still there are some scientists who believe that life originated outside our Earth and then it got transferred here. Whatever may be the cause of its origin, life began on this water by the formation of a speck of that mysterious substance—protoplasm. But, possibly, before the appearance of protoplasm there was a long history of natural synthesis of the most interesting organic substance *protein*, which is the basis of all living matter.

Proteins are formed of a number of common elements—C, H, O and N. But, protein molecules are very complicated and very large. As compared to the molecular weight 18 of water, that of *insulin*, a very simple protein and the only protein whose constitution is known (the structure was first established by Sanger in 1958 for which

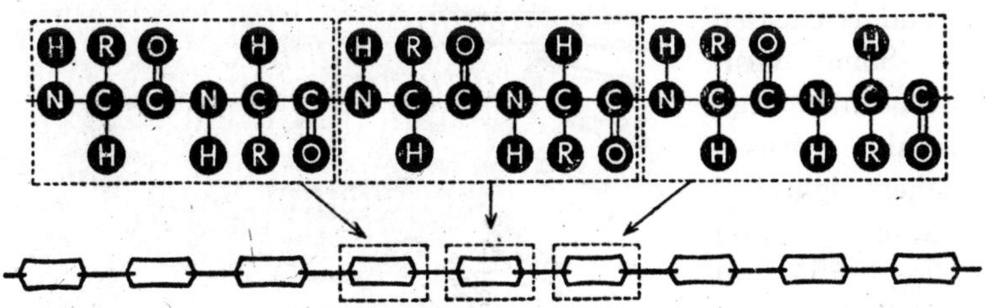

FIG. 792. How amino acid molecules combine to form peptide chains. Three amino acid molecules with their radical side chains symbolised by R are shown above. The peptide chain formed by a number of such molecules is shown below.

he was awarded the Nobel Prize), is about 5,000 while most protein molecules weight over 100,000 and some are known weighing about 10,000,000. These complex molecules are formed of simpler *amino acid* molecules which have the peculiar characteristic of having both acidic (CO · OH) and basic (NH · H) parts sc that the acid part of one molecule can unite with the basic part of another and, thus, a number of amino acid molecules may form what is known as a *peptide chain* (Fig. 792). These chains turn and twist to form the complex protein molecules (Fig. 793). The atmosphere on our very primitive Earth had not been the same through the ages. About 3500 million years ago, when 'life' was possibly being evolved, it was much warmer, there was no free oxygen in the air which, on the other hand, was possibly composed of free hydrogen, ammonia, methane and water vapour. It has been shown by Miller and Urey in 1955 that such a mixture of gases in a closed chamber when subjected to electric discharges gives rise to many of the known amino acids by artificial synthesis. Such synthesis of amino acids might have taken place naturally in those days as result of atmospheric electric discharges or natural radiation. As this type of synthesis was possibly only in the atmosphere prevalent at that time, one may suppose that the origin of 'life' was possible only in those days. At least 21 types of amino acids are well-known today and they may combine in endless ways forming hundreds of protein types. Some of the proteins, like the egg albumin, have been obtained in crystalline forms.

We are now aware of a very interesting group of substances—the **Viruses** which are somewhat intermediate between the living and the non-living. The virus substances had been known since their discovery by Ivanovsky in 1892 as peculiar things which cause dreadful epidemic diseases in plants (e.g., tobacco mosaic virus) as well as animals (chicken-pox, smallpox, measles, influenza, poliomyelitis, rabies, etc.). The infective principles were known but they were not visible under the microscope and they could not be assigned to any type of living beings. In 1953 Stanley obtained tobacco mosaic virus in the form of crystals which could infect other tobacco plants. Further researches have shown that these virus particles are *nucleoproteins*, i.e., compounds of nucleic acid and some protein. Each virus particle has a central core of ribonucleic acid (*RNA*) or, sometimes, deoxyribonucleic acid (*DNA*) and is encased by proteins. Some virus particles are rod-like (e.g., tobacco mosaic virus—Fig. 794), others spherical (e.g., influenza, smallpox and polio virus—Figs. 795 and 796). Figure 797 shows the comparative sizes of some big protein molecules, some known genes, some virus particles, the X-chromosome of *Drosophila*, a bacterium and a red blood corpuscle. It should be remembered

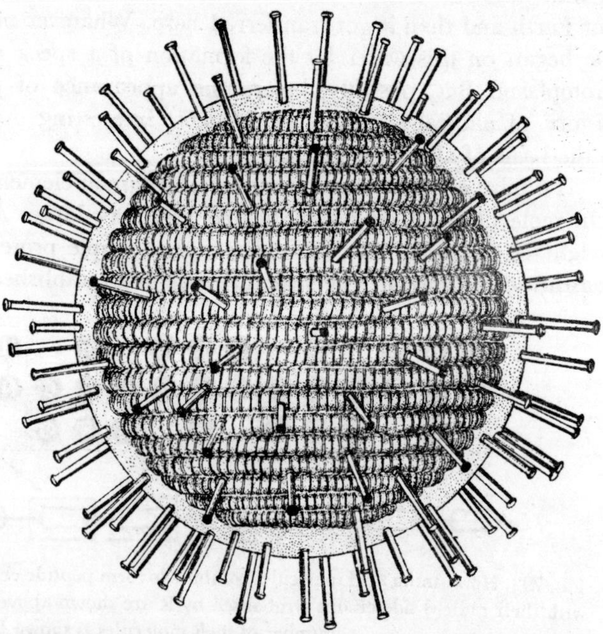

Fig. 793. Influenza virus magnified one million times. The thicker core strands are protein. The thinner threads twisted round them are *RNA*. The outer cover is a shell of protein and fatty material with spinous emergences which enable the 'flu' virus to get a foothold on the slimy surface of the body's respiratory tract. (Redrawn from *Life*).

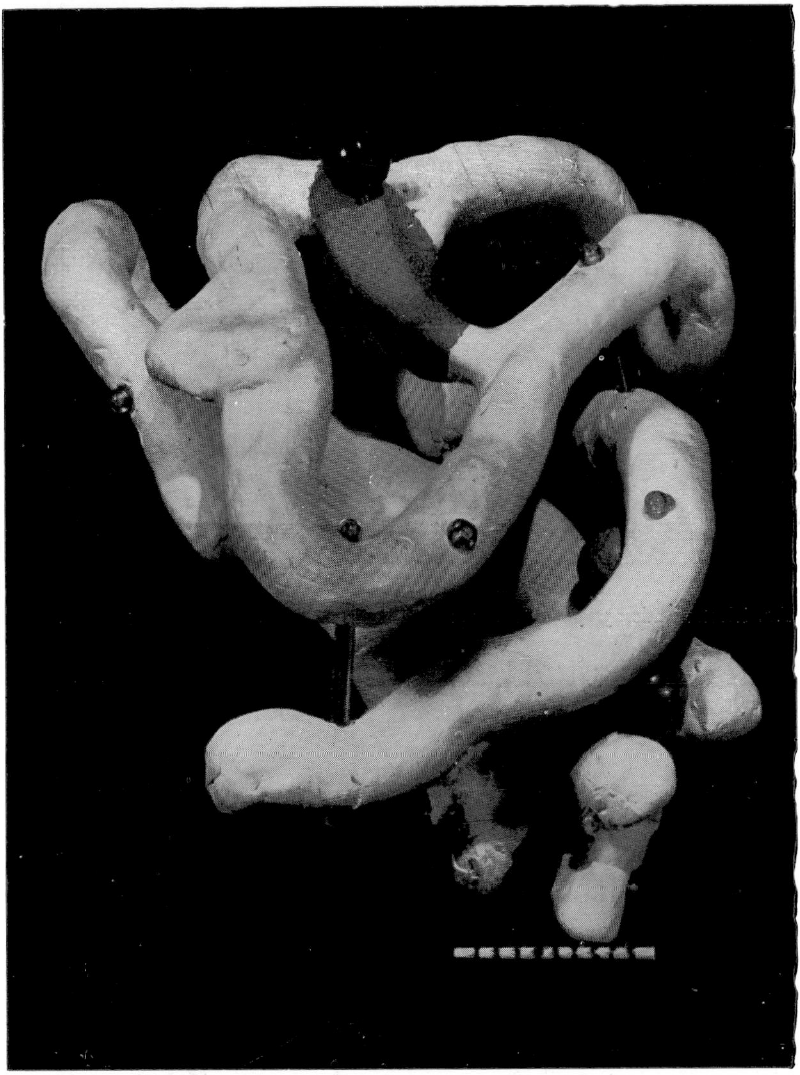

FIG. Model of a protein molecule—*myoglobin* (present in muscles and other tissues). Note the peculiar twisted form and the great size of such molecules. The *polypeptide chains* are shown white. Each mark in the scale equals 1 Angstrom unit (Å), i.e., one-tenth millimicron ($\frac{1}{10}$ $\mu\mu$). (Photo through the courtesy of Dr. J. C. Kendrew of the Cavendish Laboratory, University of Cambridge, England.)

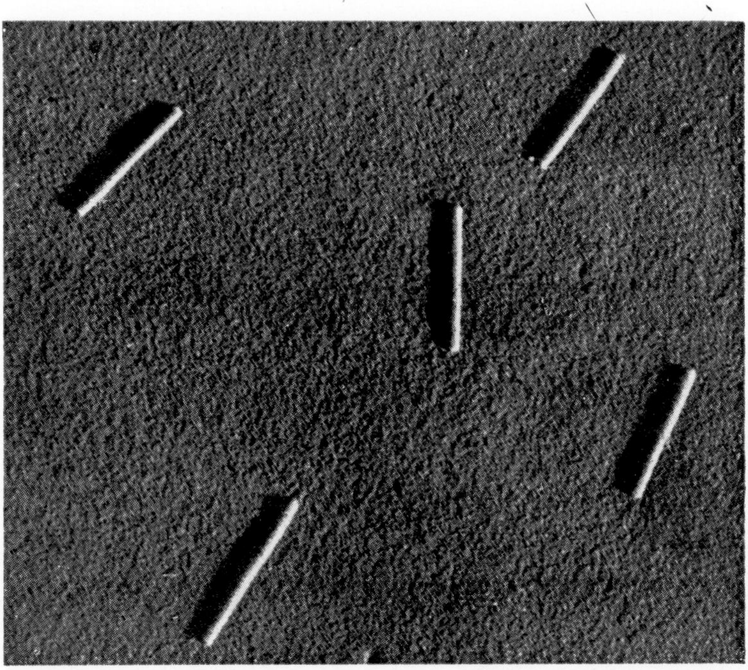

Fig. Electron micrograph of tobacco mosaic virus particles. ×72000.

Fig. Electron micrograph of polio virus particles. ×90000.
(Both the above electron micrographs were obtained through the courtesy of Dr. R. C. Williams, Associate Director, Virus Laboratory, University of California.)

that it is not possible to see anything smaller than 450 $\mu\mu$ with the optical microscope while the largest virus particle, the rod-like tobacco mosaic virus (Fig. 794), is only about 300 $\mu\mu$ long. If these virus particles are injected into the cells of plants or animals of the specific type, they multiply enormously. This multiplication is a characteristic of living substances. But virus particles as those of tobacco mosaic possess neither water nor enzymes without which no vital function can go on. The virus particles do not multiply in the normal way. When these particles enter the living cell the protein casing is first destroyed. Then the protoplasm of the living cell forgets its normal way of metabolism and begins to produce innumerable new nucleic acid particles instead—the first virus particle somehow induces it to do it. The new nucleic acid particles get invested with new protein casings and then the virus particles increase rapidly.

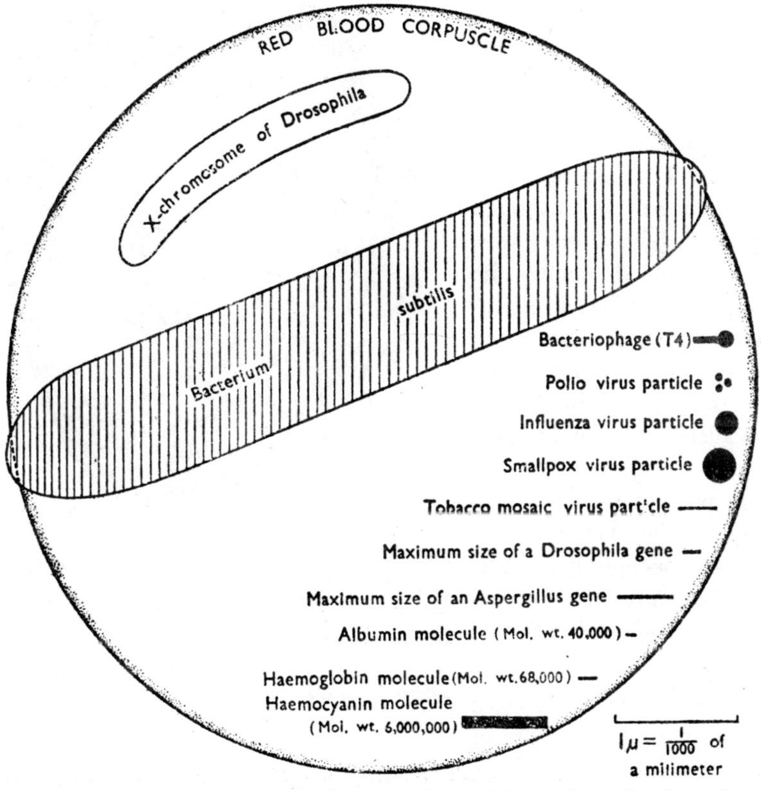

FIG. 797. Relative sizes of some known virus particles, genes, protein molecules, a bacterium and the X-chromosome of *Drosophila* as compared with a red blood corpuscle (the outer circle).

The minute bacteria also have their virus enemies, the *bacteriophages.* These have been extensively used by doctors, because they are known to destroy bacteria causing diseases. The bacteriophages have tadpole-like structures (Fig. 797), their head part containing a nucleic acid (deoxyribonucleic acid or *DNA*) and the tail part containing some protein. It has been found that when a bacteriophage attacks a bacterium, it is only the *DNA* which enters the bacterial cells. The whole structure at the time of infection may be compared to that of a doctor's hypodermic syringe from which the injection fluid (here, *DNA*) is pushed into the bacterium and the syringe proper is represented by the protein shell of the bacteriophage. Very soon after, the protoplasm inside begins to synthesise new bacteriophages until the cell is full of these and is destroyed. In the tobacco mosaic virus also, it has been found that the *RNA* alone (without the protein) can carry on the infection. So the nucleic acid is the more important and we shall again note its importance as we study chromosomes (Chapter LXIII).

Further studies of virus types have shown that there are various degrees of complexity within them and, even though they cannot be considered as 'living', they surely possess many characteristics of the'living'. Recently, another group of pathogens have been detected and named **Mycopalsma.** Its structure (Fig. 798) is somewhat more complex than that of the virus. It contains *DNA* strands as well as *RNA* granules encased in a membrane of proteins, fatty mattef and cholesterol. It develops 'baby' particles inside it which bud off as new particles. **Rickettsiae** form another group of pathogenes (causing a type of typhus) which, like the mycoplasma come nearer to the bacteria. The peculiarity of these particles suggest that although what is known as 'virus' today was, perhaps, not evolved before the living protoplasm (it has been suggested that virus might have been evolved by the degeneration of intracellular parasites which lost their cell walls), there might have been substances intermediate between living protoplasm and non-living inorganic life. Whatever be it, our definite conception of 'life' begins with the evolution of protoplasm with all its complexities of metabolism.

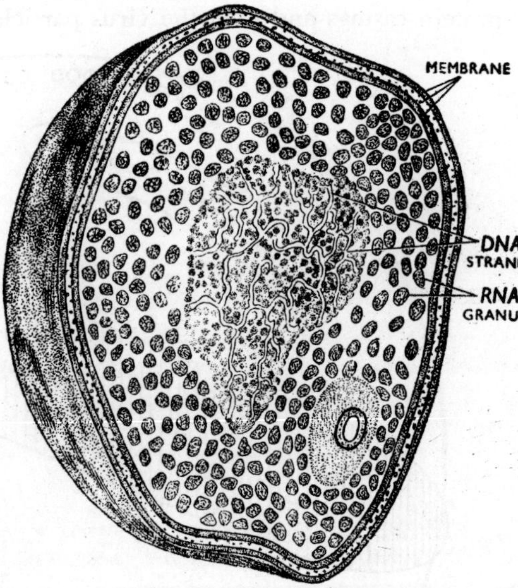

FIG. 798. Mycoplasma. The outer membrane shows outer and inner layers of protein and a middle layer of fatty matter within which last are particles of cholesterol (shown as dots). In the central mass are reticulated *DNA* strands (shown white). All round the central mass are big *RNA* granules. The oval on side is a bud or 'baby' mycoplasma. (Redrawn from *Life*).

Only recently, Dr. Theodore O. Diener, a plant pathologist of the U.S.D.A., has discovered yet another virus-like particle about one-eightieth the size of the smallest virus. This also causes plant diseases but in its constitution there is merely a speck of nucleic acid without the protein coating which is the rule in viruses. This has been named a **viroid.**

The impact of the discovery of these sub-living objects may be seen in a new biological classification of the living world proposed recently by Hu Hsen-Hsu (1965) which divides the living beings into (i) *Protobiota* (virus, etc.) and (ii) *Cytobiota* (cellular plants and animals).

Is there the existence of life anywhere outside our Earth? We have not yet got any definite evidence as to that. But, this Universe is big—much bigger than we can imagine. In it there are thousands of planetary bodies with exactly the same conditions as prevailing on our Earth. If that be so, and if the first origin of life was accidental, we can well imagine that there is a possibility of the co-existence of life on some other celestial bodies. It is also likely that such living bodies are different from what we are familiar with.

The first speck of protoplasm was different from all other inorganic matter hitherto known. It possessed 'life', i.e., it could grow, carry on metabolism, respire and reproduce. Reproduction was rapid and out of a single speck there were millions. The first formed living bodies were all alike and they depended solely on inorganic substances for their nutrition as the organic world was just born. At that time there was no differentiation

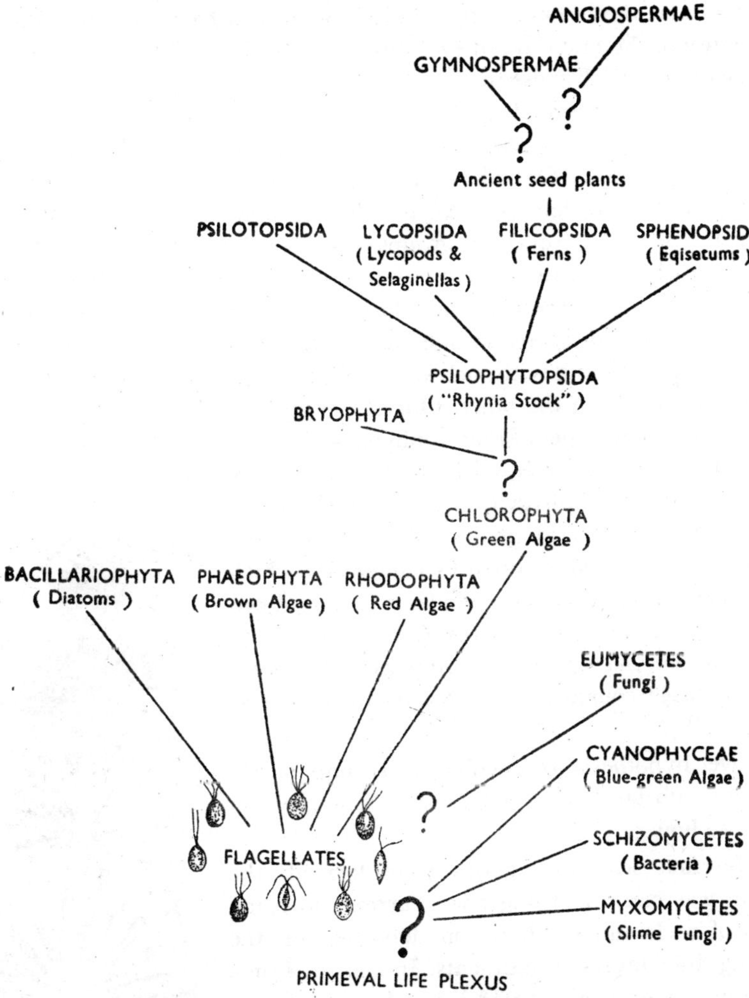

FIG. 799. The possible course of evolution of the plant groups.

into plants and animals. But, through thousands of years, while the first formed speck of protoplasm was passing through millions of generations, very insignificant *variations* in the descendants were accumulating into large differences so that one could now distinguish different forms. Meanwhile, the world was changing and certain new variations were found to be better *adapted* to the newer conditions. These two facts of *variation* and *adaptation* were gradually combining into a great moulding force causing the evolution of new forms of life. Very soon it was found that life was evolving in two directions; firstly, towards, the development or animals which is beyond the field of

Botany and secondly, to develop plants where the most striking achievement is the development of chlorophyll and, later, of the erect terrestrial plant habit with vascular bundles. There is no reason to suppose that evolution went on only along one or two particular lines. Plant and animal forms were changing in all direction. Most of the lines were being abandoned later on as the forms failed to survive the rigour of their environments while a few lines survived and again evolved. Evolution went on in this way.

The course of evolution may be traced with reference to the different plant groups that exist today and those which are definitely known to have existed in the ancient days. The picture of the first days of evolution is rather hazy. There was some form of primordial life out of which arose the bacteria (*Schizomycetes*), the peculiar slime moulds (*Myxomycetes*), the blue-green algae (*Cyanophyceae*) and above all, the *Flagellatae*—an interesting group of unicellular naked organisms which move about in water and out of which the true algae arose. Out of the different types of flagellates arose the diatoms (*Bacillariophyta*), the brown algae (*Phaeophyta*), the red algae (*Rhodophyta*) and the green algae (*Chlorophyta*). Out of these, all the groups ended blindly evolving no farther, but the Chlorophyta line which, possibly, gave rise to the land plants. The fungi (*Eumycetes*) arose by degeneration at some early and hazy stage of evolution. The Chlorophyta possibly gave rise to the *Bryophyta* on one hand and the *Pteridophyta* (sometimes called *Vascular Cryptogams*, beginning with such simple forms as the extinct *Psilophytopsida*—the so-called 'Rhynia stock') on the other. Out of the ancestral Pteridophyte as simple as the Psilophytes arose four types of Pteridophytes—(i) the *Psilotopsida;* (ii) the Lycopods, Selaginella and Isoetes (*Lycopsida*); (iii) the Equisetums (*Sphenopsida*) and (iv) the ferns (*Filicopsida*). It is the Pterophyta line represented by the Filicopsida among the Pteridophytes, or probably, the *Progymnosperms*, a specialised group of Pteridophytes, which developed further giving rise to the seed plants (*Spermaphyta*). The Spermaphytes developed into the *Gymnosperms* at an early stage and the *Angiosperms* at comparatively recent times. This probable course may be traced in a plan as shown in Fig. 799.

The above hypothesis on evolution is not purely imaginary but is substantiated by the invisible writing on rocks. Let us see how.

When the first seas an lagoons were formed, the torrential rain was causing strong river currents to bring down debris of rock and earth which deposited on the seabed forming thick layers of sediments. In course of time these layers were being consolidated into what are known as *sedimentary rocks*. Such sedimentary rocks are being continually formed and are being deposited one layer above another. These sedimentary rocks contain not only mineral matter but also remains of plants and animals that were present in that particular period of time. Such remains are known as *fossils*. Geologists can ascertain the ages of the different strata of sedimentary rocks. Thus, if a fossil is discovered, it is possible to know at what time that type of plant or animal flourished. This fact has been of the greatest help in tracing the evolution of the plant and animal king-

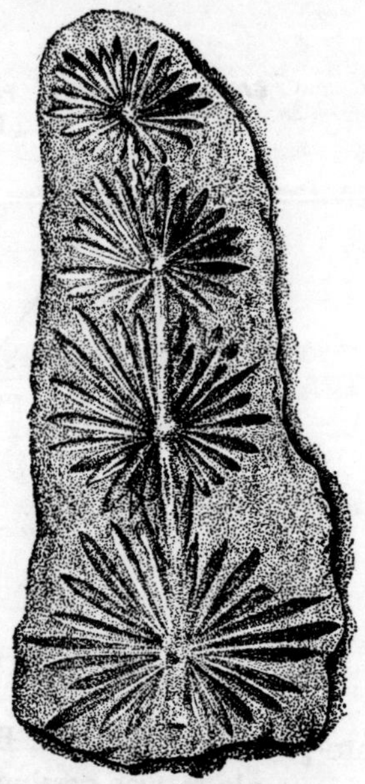

FIG. 800. A fossil The age of such fossils has to be counted in millions of years.

FIG. 801. Dense forest of the Permo-Carboniferous that has become coal today. *It is difficult to think of the stillness that reigned in these luxuriant forests.*

doms through ages as, it is found, rocks of different ages contain remains of different types of organisms.

It is almost impossible to make a correct estimate of the ages of the rocks in terms of years. Yet, different estimates have been made, and, naturally, they vary widely. One such estimate puts the total age of all the rocks under survey as about 4500 million

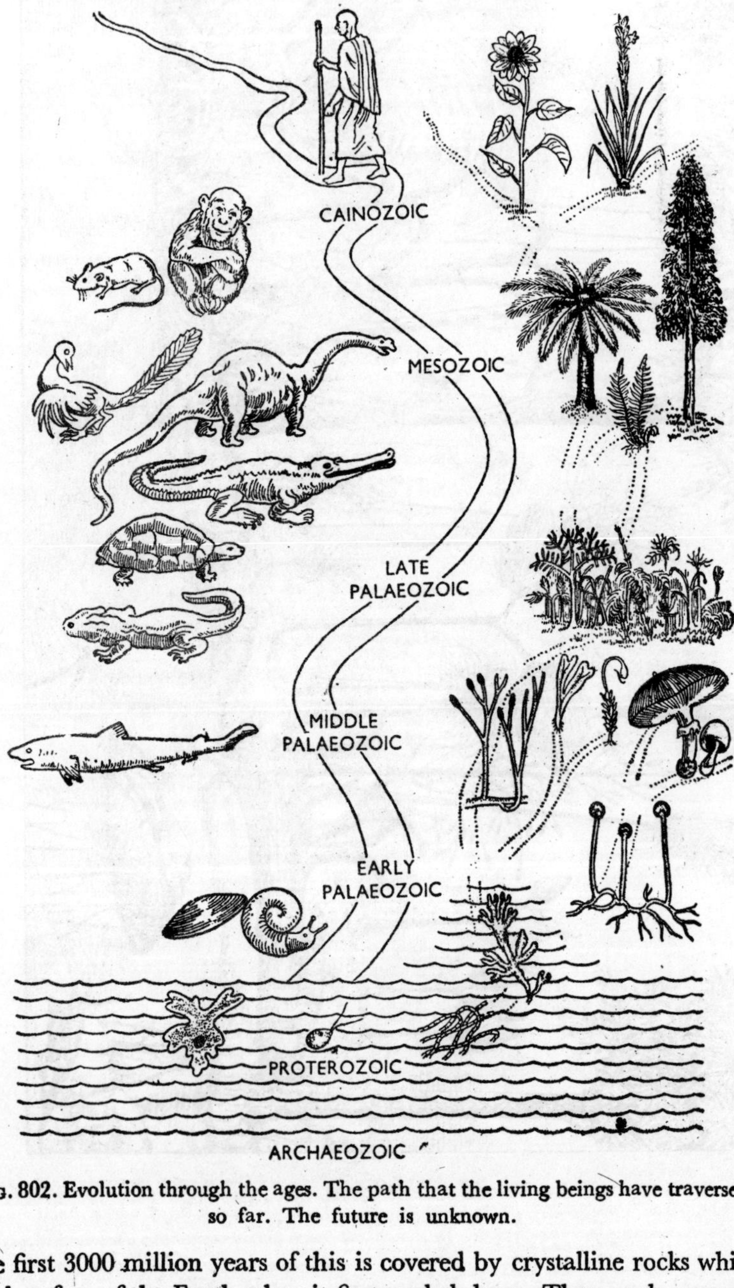

Fig. 802. Evolution through the ages. The path that the living beings have traversed so far. The future is unknown.

years. The first 3000 million years of this is covered by crystalline rocks which formed the original surface of the Earth when it first cooled down. These rocks scarcely contain any living remains and are called **Azoic** (lifeless) or **Archaeozoic** (primitive life) rocks. The latter name is because evidences are accumulating that life did exist even in

TABLE I

HISTORY OF ORGANIC EVOLUTION

ERA	PERIOD	CLIMATE	ORGANIC LIFE
Cainozoic 60 m. years		Global climatic zones with cooling of poles as found today. Previously there were alternating periods of extreme cold (glaciation) and warmer spells. Formation of mountains.	Zonation of plants as seen today. Predominance of Angiosperms and Mammals and rise of the Primates culminating in the human race.
Mesozoic 120 m. years	*Cretaceous* 65 m.	Varying	Dwindling of Gymnosperms and increase of Angiosperms. Reptiles decreasing and Mammals increasing.
	Jurassic 25 m.	from temperate	Age of Gymnosperms and Reptiles. First Angiosperms and first birds.
	Triassic 30 m.	to warm.	Mostly Cycadophyte Gymnosperms. Plant types fewer. Reptiles like dinosaurs. Mammals appear.
Palaeozoic 330 m. years.	*Permian* 25 m.	Rigorous climate and glaciation.	Plant types fewer. Early Gymnosperms. Dwindling of Pteridophytes. Reptiles.
	Carboniferous 50 m.	Warm and	Age of giant lycopods, equisetums and cycadofilicales forming great coal forests and swamps. Early reptiles. Amphibians.
	Devonian 60 m.	humid	First recognised land flora of vascular plants (Pteridophytes — psilophytes and lycopods). Early Broyphytes. Fishes, Early amphibians.
	Silurian 35 m.	Warm with	Definitely known migration of life from water to land. Early Pteridophytes (Psilophytales). Marine animals.
	Ordovician 80 m.	local drought.	Marine algae and animals. Evidences are accumulating that land plants might have originated even before Cambrian.
	Cambrian 80 m.	Probably warm.	
Proterozoic 1000 m. years			Early sedimentary rocks with traces of algae and elementary animal forms.
Archaeozoic 3000 m. years			Life began towards close of this Era. There is evidence of bacterial & algal life. Igneous and metamorphosed rocks.

these very ancient days. Evidences are accumulating that algal life existed 2500 million years ago and bacterial life started even before them (3400 million years old fossils are claimed from South Africa). During the next 1000 million years we get the **Proterozoic** (beginning of life) rocks which also show very little evidence of life. But, on the whole, there is evidence that the thallophytes and the protozoans had already developed and so plant and animal kingdoms had differentiated.

Evolution of life gained a new momemtum after the Proterozoic days. The next 330 million years are covered by the **Palaeozoic** (ancient life) rocks. As the lowermost Palaezoic rocks are called *Cambrian* all the rocks before that (Archaeozoic and Proterozoic) are sometimes called **Pre-Cambrian** during which life developed rather slowly and fossils are practically unknown. During the first half of the Palaezoic (*Early Palaezoic*—**Cambrian** and **Ordovician**) probably all life was confined to the seas and lagoons as only the Thallophytes are known. A dreadful bleakness and stillness reigned on land. Then, gradually, a change occurred and life moved from water to land. This acquisition of the terrestrial habit was a big event in both the plant and the animal kingdoms. It is certainly difficult to tell today when exactly the migration of plants from water to land took place. Recently, some woody elements (tracheides) and spores of land plants have been discovered from Indian Cambrian rocks as also in some other countries. It is then justifiable to say that the first land plant (Pteridophytes) existed during Cambrian, i.e., in the Early Palaeozoic and the migration might have taken place even immediately before the Cambrian. During the *Middle Palaeozoic* (**Silurian** and **Devonian**) we get evidence of the existence of Bryophytes (*Hepaticites devonicus*) and of well-developed Pteridophytes (*Psilophytales*, *Protolepidodendron*—allied to Lycopodium, *Hyenia*—allied to Equisetum and some *Primofilices*—primitive ferns). In the *Late Palaeozoic* (**Carboniferous** and **Permian**) we get the maximum development of the Pteridophytes. The Permo-Carboniferous is known as the **Coal Age** as at this period the first big forests were formed on the edges of lagoons at river estuaries and these forests have carbonised and have given rise to the present-day coal. These forests were formed by giant trees which were allied to the Lycopods and the Equisetums and are now extinct. There were also some ancient, ferns and the first Gymnosperms—the *Cycadofilicales* or *Pteridosperms* and the *Cordaitales*. Perhaps the atmosphere differed in its constitution and was warmer at that time.

At the end of the Palaeozoic (that is, after the Permian) there was some climatic change on this Earth. There was a period of desolation. All the older forms of life died and when climatic condition again became favourable, newer plants were evolved. This period is covered by the **Mesozoic** (middle life—**Triassic, Jurassic** and **Cretaceous**) rocks covering about 120 million years. This was the age of gigantic reptiles like the Dinosaurs and of the Gymnosperms. The ferns also attained the maximum height of their development in this age.

Towards the end of the Mesozoic (Cretaceous) another great change came over this Earth. Perhaps there was some catastrophic change and the older forms of life became unfit for survival. Most of the older forms vanished, the Angiosperms among the plants and the Mammals among the animals raised their heads.

The rocks since the Mesozoic to the present day (about 60 million years) form the **Cainozoic** or **Cenozoic** (recent life). The world during this period has not varied much although there have been periods of cooling down of the Earth involving the approach of glaciers. It is during this period that the present-day flora (predominated by the Angiosperms) and the present-day fauna (predominated by the Mammals) evolved.

Evolution during these ages was by no means uniform. It was always being moulded by the frequent climatic and geographic variations of this Earth. During all this time there had been alternating periods of extreme cold when glaciers advanced to areas which are very warm today, and of warm spells when a luxuriant vegetation covered the face of the Earth including the poles which are frozen today. There were also alternating phases of humid and arid climates. Changes were sometimes catastrophic causing sudden extinction of whole races of plants and animals. The geography also was always changing. Geologists think that during the late Palaeozoic Coal Age (and persisting in the Mesozoic) there was a vast continent in the south covering what are India, Africa, Australia, South America and Antarctica today. This continent is named **Gondwana land** (Figs. 803 and 804) after the *Gond* tribe of Madhya Pradesh. To the

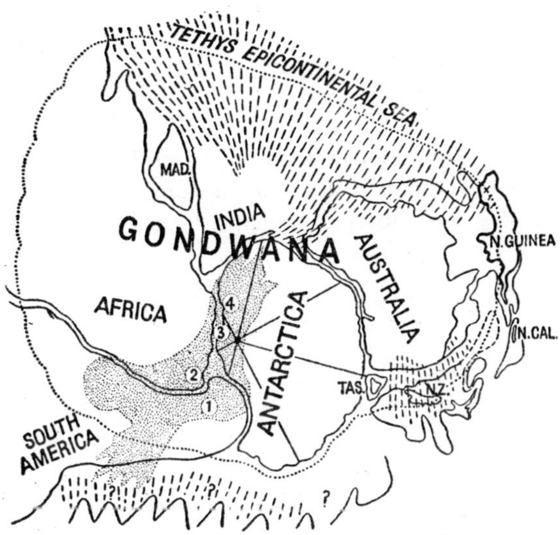

FIG. 803. The Gondwana land as it was during the Permian long before the modern continents took shape after drifting apart.

north of it were two continents—Angaraland (modern Siberia and North China) and the European-North American land mass. In between was the sea of Tethys. This geography explains the fact that at this age the same types of plants and animals flourished in places like India, Africa, Australia and South America, which are so distant today. The particular Gondwana flora of the Permo-carboniferous has been named the **Glossopteris Flora** from the preponderance of a leaf-type called *Glossopteris* (Fig. 805) which was probably a pteridosperm (seed-bearing fern). The Gondwana fossils are represented in our country in two groups of rocks called *Lower Gondwana* (Permo-Carboniferous, i.e., Late Palaeozoic—predominated by *Glossopteris*), and *Upper Gondwana* (Mesozoic—predominated by *Ptilophyllum* leaves of the *Cycadeoidales* group of Gymnosperms). The former can be seen in the Talchir-Barakar Raniganj-Panchet coal mine areas and the latter is Panchmarhi, Rajmahal and Jabalpur. The Gondwana land was possibly formed during the Carboniferous and began to split up during the Cretaceous. The Gondwana Flora is treated more elaborately in College Botany Vol. II.

This is the story how on a hot fiery mass life first came into being and then how that speck of protoplasm developed into human beings and lovely flowers of today. But,

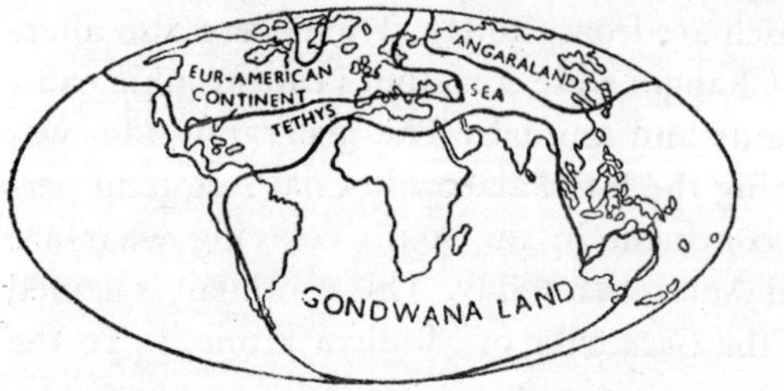

FIG. 804. How the geography of the world has changed. The thick line shows the map of the world during the late Palaeozioc while the thinner line shows the world today.

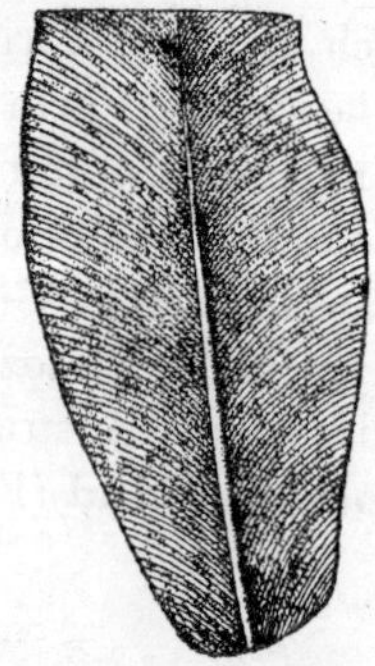

FIG. 805. A fossilised leaf of *Glossopteris.*

scientists cannot take any theory for granted unless solid evidences are advanced proving the theory. The above is a statement of the phenomenon of Evolution while the next chapter will explain how this theory obtained a firm foothold in science.

CHAPTER LVIII

DEVELOPMENT OF THE IDEA OF EVOLUTION: DARWINISM

Ever since the human race became intelligent enough to ask whys and hows, the most baffling and persistent question was about their own origin. Accordingly all civilised races had their respective stories of Creation and, naturally, all such speculations were based on pure imagination. Nevertheless, some ancient philosophers of the more civilised races, the Hindus, the Chinese, the Hebrews and the Greeks, imagined some sort of Evolution. But, not being based on facts, such theories had no strong foothold. The idea of Evolution, as we understand it now, grew along with modern science. It is surprising that the first notable contribution in this direction came from a great man who is widely known as a painter and not as a scientist. This versatile genius was Leonardo da Vinci (1452-1519) who asserted in the Florentine courts that fossils were nothing but the remains of organisms that had existed in ancient times but were no more. The next great contribution came more than three centuries later—it was the publication of *Philosophie Zoologique* in 1809 by the French zoologist Lamarck. He gave his ideas as to how species of organisms change their appearances giving rise to new species and thereby causing Evolution (see Lamarckism—next chapter). About the same time an English clergyman, T. R. Malthus, caused a sensation by pointing out the great overproduction of human beings and the resulting struggle for supplies.

FIG. 806. Leonardo da Vinci (1452–1519)—*the versatile genius.*

It was in this world, when people were already thinking about evolution, variation and overproduction that Charles Darwin (1809-1882) was born. He was not supposed to be a very successful student when, after graduation, he got the opportunity to accompany an exploration ship, the Beagle, as a naturalist. He spent three years on this ship and visited the south of Southern America. The visit to the Galapagos Islands impressed him greatly and he made a very thorough study of the overproduction of organisms, the variations occurring spontaneously and the influence of nature on the future generations. A study of these made him formulate his theory of Evolution.

Friends of Darwin knew that Darwin was working on his theory and in 1858 Darwin was surprised to receive from A. R. Wallace, then in

FIG. 807. Charles Darwin (1809–1882). *He fought against odds and established the theory of Organic Evolution.*

Malaya, an essay for his opinion. This essay contained all the main points of the theory that Darwin was putting in shape and this made Darwin hesitate before publishing his own theory. Later on, however, both the papers were read together at the Linnaean Society, London. Because of the two authors, the theory is often called *Darwin and Wallace's theory.* Next year, in 1859, Darwin published his **Origin of Species.**

The above will show that Darwin was neither the 'discoverer' nor the 'inventor' of the Theory of Evolution. His genius was employed in giving a concrete form to certain ideas which were already agitating some thinking men. But the publication of the Theory caused a great furore. European culture of that period was dominated by religious missionaries who, again, were fanatic followers of every word in the Bible. The Bible said *God created man after His owm image* and everything around us was created in six days. So, talking about a slow process of evolution culminating in human beings was declared irreligious. For a long time Darwinism had to fight such fanatics. It should be mentioned, however, that every missionary was not a fanatic; there were quite a few progressive ones among them who, at least privately, supported Darwinism in some form.

Darwin's Theory, as explained in the *Origin of Species*, may be summarised as follows:

1. *Overproduction:* Animals and plants in nature give rise to more offsprings than could possibly survive. This is also the doctrine of Malthus.

2. *Struggle for existence:* Overproduction causes a struggle as to which of the offsprings would survive. The struggle may be active, as in the case of animals or passive, as in the case of plants. The victors survive and the vanquished become extinct.

3. *Variation:* Offsprings do not entirely resemble one another but differ among themselves in details. Such variations sometimes help an individual in the struggle for existence while another variation may be a drawback. Thus, some are better fitted for the struggle while others are ill fitted.

4. *Natural selection, or Survival of the fittest:* It is logical to suppose that the one which is the best fitted for the struggle, ultimately survives eliminating others which are not so well fitted. Thus, in the long run, all the variants that are not well fitted die out leaving only the fittest. Ultimately, it seems as if Nature, acting like a Super-gardener, has exercised some selection during which she has eliminated all the unfit.

5. *Heredity* and *Origin of Species:* When the fit ones survive, the variations that cause their conquest, i.e., the *advantageous variations*, are passed on to their offsprings. Thus, these variations become a part of their character and they become better species than their parents.

This is how Nature is always evolving newer and newer species which are better than their immediate ancestors and thus causing Progress in the course of Evolution. But, environment may often so change that what was once thought 'progress' becomes a drawback. That is why many gigantic plants and animals of the Palaeozoic and the Mesozoic became extinct.

The **evidences** on which Darwin relied in giving shape to his theory are enumerated as follows by T. H. Huxley, his ablest champion:

1. *The gradation of organisms:* It is well known how the living plants may be arranged from the simple Thallophytes to the more and more complex types—the Bryophytes, the Pteridophytes, the Gymnosperms and the Angiosperms. A comparative study of the different plant groups easily shows a gradation which can only be explained by Evolution. Thus, we find:

(a) *Evolution in structure and migration to land.* From the primitive unicellular plants the multicellular (filamentous, thalloid, and then solid forms of land plants) ones have arisen. These have developed mechanical tissues and a strong conducting system ultimately becoming the higher land plants.

(b) *Change in the alternation of generations, evolution of sex and ultimately the seed habit.* The type of sexuality with typical flower structure and the alternation of generations that we see in the Angiosperms (page 190) shows gradation down to lower and lower plant groups. Sexuality and alternation of generations are first established in the algae and then the gametophyte becomes progressively more and more reduced while the sporophyte becomes ascendant. The evolution of the seed habit is the climax attained so far.

Similar gradation due to evolution is also seen among the animals.

3. *Embryology:* A comparative study of the embryos of different groups of plants shows the great resemblances among these groups which is not clear when the adult plants are considered. The difference between the embryo of a bean and that of a tamarind is not as great as that between a bean plant and a tamarind tree. The embryos of animals are even more characteristic as the embryos of such widely different groups

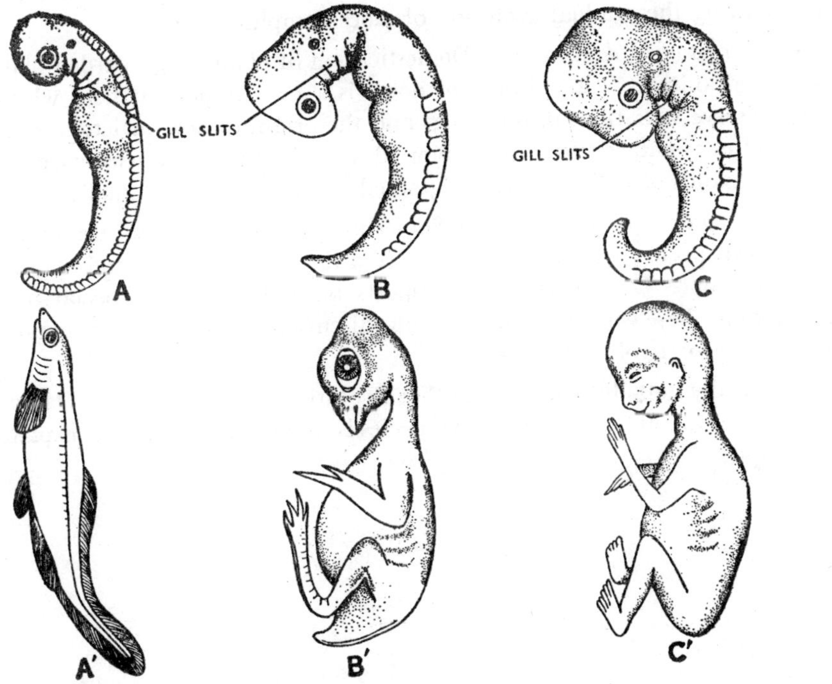

Fig. 808. Embryos of widely different animals from different groups showing a similarity that suggests organic evolution. Early and late stages of fish (A and A′), chick (B and B′) and human (C and C′) embryos are shown.

as fishes, frogs, reptiles, birds and mammals resemble one another very strongly. This led some scientists to state that the embryonic development of animals rapidly recapitulates the stages of evolution of that group (*recapitulation theory*). Thus, the human as well as the chick embryo show vestigeal gills (Fig. 808) which represent fishes, i.e., a lower stage of evolution. *Acacia moniliformis*, in which leaf development has been suppressed by the development of phyllodes, shows normal leaves in the seedling stage (Fig. 182).

3. *Comparative Morphology:* On studying the morphology of the same organ in different plant and animal groups one cannot but be impressed by their similarity. One may thus easily trace the evolution of the leaf, the flower, the brain, the heart or the hands and legs.

4. *Geographical distribution:* Allied groups of plants are found to live together. An impassable barrier like an ocean or a high mountain range shows different flora and fauna on either side. It is explained later that *isolation* has been a very important factor in the origin of new species. These facts point out that related species must have a common origin.

Mention may be made in this connection of the much more recent **Age and Area Law** of Willis. The more widely distributed area a species occupies the older it is in age. The widely distributed species occupy a more primitive place in the scale of evolution while the more localised species are of comparatively recent origin.

5. *Geological succession of organisms:* This is one of the most glaring evidences and has already been pointed out in the previous Chapter. Older rocks show fossils of simpler and more primitive organisms. As one comes to newer and newer rocks one finds remnants showing the gradual evolution of more complex forms.

6. *Changes under domestication:* Domestication of plants (e.g., garden plants or crop plants) and animals (e.g., pigeons or dogs) is a very important field where one can closely observe variations in offsprings and the origin of new varieties. Here one sees the processes of nature taking place more quickly by the added selective efforts of men. Thus, thousands of varieties have originated in a comparatively smaller period of time.

7. *The observed facts of mutation:* Sometimes individual plants are found, specially among wild species, which greatly vary from the parent species and may form a new species all at once. Darwin called them 'sports' but did not lay any special stress on them. But, later on, these variants, called *mutants* (discussed later), became as important as the whole theory of Evolution.

Practically nobody today denies that Evolution did take place as Darwin said but Darwin's idea about the origin of new species is subject to criticism as explained in the next Chapter.

CHAPTER LIX

DEVELOPMENT OF THE IDEA OF EVOLUTION: THE ORIGIN OF NEW SPECIES

Once evolution is taken for granted, the big question comes in—How did it take place? The first real tackling of the question came from the French zoologist Jean Baptiste Lamarck (1744-1829).

LAMARCKISM

The genius of Lamarck would be appreciated all the more today if one remembers the tragedy of his life. He published his views on evolution at a late age, in *Philosophie zoologique* in 1809 but it brought little luck to him in his old age. He was snubbed and ignored by Napoleon himself, became blind seven years before his death, lived a life of loneliness and poverty until he was buried in some unknown grave. Yet, he was an all round scientist of his age—a botanist, a zoologist, a chemist and a physicist. He postulated theories on evolution which, although they may be greatly disputed, were something unknown before his time. His theories tried to explain how new species could have evolved. These are as follows:

1. *The direct effect of environment:* It is common knowledge that environment may influence the form of organisms. Thus, a plant grown in fertile soil will have a luxuriant form while the same plant in a barren tract will be sickly and stunted. In the case of animals the effect of environment may also be indirect as the habits of animals, may change in a new environment and a change of habits, according to Lamarck, causes a change of form. Lamarck assumed that such changes of form caused by the direct or indirect effect of environment are inherited and are of cumulative nature. Thus, in course of generations, new species may be formed.

FIG. 809. Jean Baptiste Lamarck (1744–1829). *He postulated the first theory on the mechanism of evolution.*

2. *Conscious effort:* Lamarck thought that when an animal feels the necessity of an organ he can develop it by conscious effort. The 'need' of an organ causes an organ to be produced. Thus, the need of horns to fight with and teeth to chew with would cause the growth of horns and teeth. This idea of Lamarck is not at all taken seriously today as it baffles scientific reasoning.

3. *Use and disuse:* Excessive use of an organ will cause increase in size or strength of that organ while its disuse will cause its disappearance in course of generations. Thus, horse-like animals transferred to an area where there were only trees and no grass had to stretch their necks for food. This excessive use of the neck caused its excessive development resulting in the evolution of a race of giraffe-like animals. Similarly, disuse of the tails of monkeys has caused their loss in human beings.

4. *Inheritance of acquired characters:* All such characters acquired during the lifetime of an organism will automatically be inherited by its offsprings. The character will gradually increase in successive generations and such cumulative effect will ultimately give rise to a new species. This is the main as well as the most disputed assumption of Lamarck.

DARWINIAN VIEW—THE THEORY OF PANGENESIS

Darwin assumed in his Theory of Evolution that *variations* occur spontaneously and nature selects out the suitable variants. Thus, new species are evolved. At first he did not bother much as to how the variations arise. But, later on, this problem engaged his attention more and more. His views were a modification of Lamarckism. He also thought that variations arise in both plants and animals as a direct effect of the environment and as a result of use and disuse in the case of animals. He did not think much of the indirect effect of environment on animals by a change in their habits. Similarly, he never thought that the 'need' of an organ might cause it to grow. Darwin also thought that some variations might arise due to unknown causes which variations he called *chance variations*. Like Lamarck, Darwin thought that all new variations, i.e., acquired characters, are inherited.

Darwin realised that the last assumption is not quite easy to accept. A plant is propagated by its seed and any variation, in order to be inherited, must influence the seed or the reproducitve parts. Supposing that a leaf or a root is slightly modified by a variation—how can that affect the seed? This problem racked the brain of a genius like Darwin and he gave a purely hypothetical explanation. This is embodied in one of his later books, *The Variation of Animals under Domestication*, as the **Theory of Pangenesis.** This supposes that every cell of the body of an organism is producing heredity-bearing particles called *pangenes*. These pangenes are passing on to the germ cells (cells that will give rise to the reproductive parts) where these would accumulate and ultimately pass on to the next generation. Thus, any change in an organ during the lifetime of an organism is duly registered in the germ cells and is thus inherited. In the case of animals the blood stream is constantly receiving these pangenes or *gemmules* which, later, accumulate in the germ cells.

The Theory of the pangenes, however, was based purely on imagination and there was no experimental evidence to prove it. Darwin proposed this theory only 'tentatively'. Followers of Darwin like Galton noted that blood transfusion did not change heredity and, finding nothing in favour of this theory, ultimately dropped it.

WEISMANN

August Weismann (1834–1914) was a devoted disciple of Darwin who was strongly against the Lamarckian doctrine of inheritance of acquired characters. He could not see eye to eye with the Darwinian support to this doctrine and thought that these views of Darwin needed modification. This revised view of Darwinism formed the basis of the *neo-Darwinian* or modern-Darwinian school as opposed to the *neo-Lamarckian* or modern-Lamarckian school.

Fig. 810. August Weismann (1834–1914). *He denied the inheritance of acquired characters.*

Within two years of the death of Darwin, Weismann began by analysing variations and gradually developed this new idea of Heredity. He said that variations are of two types. Some are congenital, i.e., organisms are born with them. Others are acquired during the lifetime of a particular plant or animal. It was with this latter type of variations, the *acquired characters*, that Weismann was very much concerned. He gradually developed his **Theory of the Germplasm** to explain that acquired characters could not be inherited.

The *Germplasm Theory* is based on the concept that living bodies are formed of two fundamentally different kinds of material—the *germplasm* and the *somatoplasm*. The germplasm is responsible for the development of reproductive parts and remains in the organism, sealed and secluded from the somatoplasm or the general body of the organism which develops out of the germplasm. In course of time, the germplasm gives rise to the offspring in which again the germplasm remains separate from the new somatoplasm that it develops. Thus, the germplasm is a continuous stream from generation to generation while the somatoplasm is formed anew at every generation and is discontinuous. In bisexual plants and animals, at every generation the germplasm stream becomes a confluence of the maternal and the paternal streams (Fig. 811).

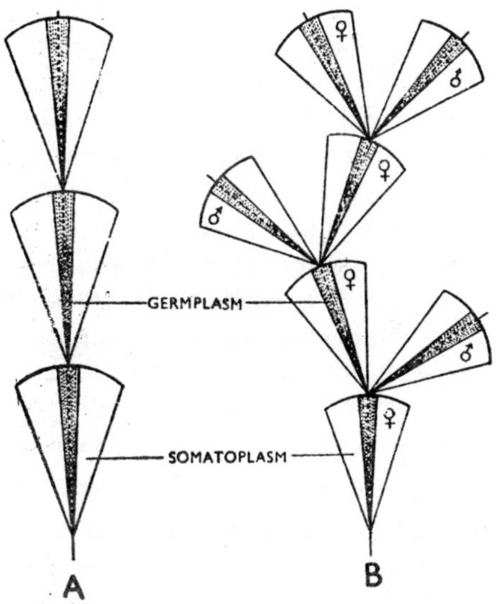

The germplasm forms a continuous stream from generation to generation while the somatoplasm is discontinuous being formed anew in each generation.

In sexual organisms the germplasm in each generation is a confluence of maternal (♀) and paternal (♂) streams.

FIG. 811. Diagrams to illustrate the continuity of germplams. Each cone represents a generation. The shaded part represents the germplasm while the rest forms the somatoplasm.

Weismann's conclusion from the above was that the somatoplasm is a discontinuous entity and so, any variation in it cannot be inherited. The somatoplasm does not influence the germplasm. Weismann did not believe in Darwin's Theory of pangenesis which tried to explain the influencing of the germplasm by the somatoplasm. The independence of the germplasm from the somatoplasm is further shown by the modern technique of grafting the ovary from a guinea-pig of one variety to that of another variety. It is found that such a grafted ovary passes its original characters to the progeny and is not influenced by the body of the new guinea-pig to which it had been grafted and on which it is growing.

WEISMANN'S THEORIES OF HEREDITY

After a clear statement of his theory of the germplasm, Weismann tried to find out the mechanism of inheritance. Fortunately, at the time of Weismann, more was known about the structure and division of cells than during the lifetime of Mendel. It was known that chromosomes are stable bodies within the nucleus, their number for a species or variety is fixed that they are equally divided during mitosis. Weismann declared that heredity is determined by certain *determiners* which are located within complicated bodies called *ids*. The *ids* again are components of the chromosomes. These determiners cannot be changed so that the heredity of an organism does not change.

A complete conviction in the theory of the germplasm led Weismann to believe that no acquired character, as such, could be inherited. But, he was confronted with certain difficulties in the form of variations which he knew to be heritable. These he explained by supplementary hypotheses. Firstly, he admitted that there are certain variations in organisms that could be inherited. He explained this by his **hypothesis of parallel modification of germplasm and soma.** He said that under certain circumstances there is a variation of the organic body (soma) which also involves a parellel variation of the germplasm substance. This variation is inherited. The second difficulty was even greater. Certain organisms have different forms under different circumstances. The amphibious plant *Ranunculus aquatilis* (Fig. 161) develops dissected leaves when grown in water and broad entire leaves when grown on land. This is called *polymorphism*. Some butterflies and moths have different forms according to the environment and the nutrition they receive. Among bees, it is well-known that the same egg may develop into a queen bee or a worker bee according to the nutrition supplied to it. These phenomena cannot be explained by Weismann's theory of the determiners. To avoid this difficulty Weismann developed another **hypothesis**—that **of germinal selection.** He now said that the germplasm has got not a single set of determiners but a number of different sets of determiners. It is the environment that decides which set of determiners will express themselves. Thus, there are any number of determiners struggling for expression within the germplasm and finally one set is selected out by the environment.

While the earlier theories of Weismann (germplasm and chromosomes) were well received and still form the basis of Genetics, his last theory about germinal selection met a different fate. This last theory died out in course of time like Darwin's pangenesis as, likewise, this also was based on mere speculation. While Darwin thought of a centripetal theory (pangenes from soma to germplasm), this was a centrifugal (determiners from germplasm to soma) one. Weismann gave the first concrete form to the modern conception of inheritance through chromosomes but went too far in his speculation with determiners.

DARWIN TO DE VRIES

The period from the death of Darwin to de Vries and the rediscovery of Mendelism was occupied by extreme speculations of the Weismannian type. But even in this period, some people were not satisfied with speculative theories alone but were bent upon observing more closely what was actually happening in nature. Galton and Bàteson closely studied variations and concluded that there are two kinds of variations—continuous and discontinuous. *Continuous variations* are the common graded fluctuating variations such as one finds when measuring the lengths and other measurable (quantitative) attributes of organs like leaves, fruits, etc. Here the extremes are connected by a graded series of intermediates. The other type of variations is *discontinuous*. These arise all on a sudden and are separated from normal ones by big gaps—there being no intermediate form. Darwin called these *sports* but he did not consider them to be of much significance. On the other hand, even before de Vries, Galton, Bateson and Korjinsky believed that it is the discontinuous type of variations that plays the real part in species formation and evolution. The continuous types are unstable and not strictly inherited while the discontinuous ones are very stable as well as hereditary. Galton sought to explain this by his famous analogy of a polyhedron. If a polyhedron is slightly tilted it comes back to its original position as soon as the hand is removed. This is the case of continuous variations. But, if the polyhedron is moved too far it settles on another of its faces and cannot come back to its previous face. This is what happens in discontinuous variations. The importance of discontinuous variations forms the core of the mutation theory of de Vries.

Along with the realisation of the importance of discontinuous variations another science was being developed by Galton and Pearson. This was the science of **Biometry** which involves actual measurement and statistical study of variations and heredity (Chapter LXII).

DE VRIES—MUTATION THEORY

Hugo de Vries (1848–1935), a Dutch botanist, pointed out the line of evolution which has now become the most important. Although Darwin was aware of the '*sports*' he did not stress their importance. Both Galton and Bateson realised how important they were. But, it was de Vries who was their real champion.

Fig. 812. Hugo de Vries (1848–1935). *He pointed out that mutation is the real cause of evolution.*

Fig. 813. *Oenothera lamarckiana.* The plant that was made famous by de Vries.

De Vries made an exhaustive study of variations and in course of this he used to collect variations growing wild and to plant them in his own nursery. His study revealed two important facts:

1. *The occurrence of elementary species:* Among many wild and domesticated species there are varieties with minute differences that breed true. These differences are not very marked and de Vries called these varieties *elementary species*. Even the 'fluctuating' variations really included some 'elementary species' of this type.

2. *Mutation in Oenothera: Oenothera lamarckiana* or Evening Primrose is a garden plant (Fig. 804) with yellow flowers which open towards the evening. In Holland, this plant often grows in a wild state. In an abandoned potato field, de Vries found certain variants of this plant which looked very different from the normal. De Vries brought these to his home and on cultivation found seven types that bred true. These he called *mutants.* After breeding for eight generations de Vries classified his mutants as follows:

(a) *Progressive species:* due to addition of certain characters: to be considered as elementary species.

(1) *O. gigas* (stouter with larger leaves and flowers)
(2) *O. rubrinervis* (red veined leaves and stems).

(b) *Retrogressive species:* due to loss of some characters.

(3) *O. nanella* (dwarf)
(4) *O. laevifolia* (smooth, narrower leaves)
(5) *O. brevistylis* (shorter style).

(c) *Degressive species:* very weak forms; not likely to survive in nature.

(6) *O. albida* (pale whitish)
(7) *O. oblonga* (oblong leaves on weak, dwarfish plants).

Besides these seven, de Vries also got two other forms which did not breed true and were giving rise to mutants like the parent *O. lamarckiana.* These he grouped together as:

(d) *Inconstant species:*

(1) *O. lamarckiana* (the parent form)
(2) *O. lata* (with broad leaves)
(3) *O. scintillans.*

Starting with nine plants from the field and growing them for eight years de Vries found that the same seven mutants arise from the stock parental forms year after year. De Vries, therefore, concluded as follows:

1. *New elementary species appear suddenly and attain full constancy at once.*

2. *The same new species are produced in a large number of individuals.* We get the same mutants again and again so that there is a greater chance of selection by nature.

3. *Mutability is something fundamentally different from fluctuating variability.*

4. *Mutations take place in nearly all directions.* This is because a mutation may mean either a gain or a loss and may involve any character.

Although de Vries formulated an extremely important theory which stands well established today, the material on which his thesis was based was later on found to be rather abnormal. Even de Vries had noticed that *Oenothera lamarckiana* was an 'inconstant species' unlike the more usual true species. The reason of this inconstancy has now been found out cytologically. The chromosome set of *O. lamarckiana* ($2n=14$) is derived from two sets—the *gaudens* complex and the *velans* complex. These plants give rise to two types of gametes—*gaudens* and *velans.* Among the zygotes formed, the *gaudens-gaudens* and *velans-velans* homozygous true breeding forms fail to survive while the *gaudens-velans* 'balanced heterozygotes' only live to form the common *O. lamarckiana* plants. Thus, *O. lamarckiana* is a hydrid, now known as a 'complex heterozygote' (also called 'structural hybrid') which is unstable and is prone to throw out chromosomal mutations. Many of the mutants obtained by de Vries were not mutants in the de

Vriesian sense (gene mutations) but mere unstable chromosomal 'aberrations'. Of his mutations, *O. gigas* is now known to be a tetraploid ($4n=28$); *O. rubrinervis* and *O. nanella* (a balanced lethal) are recombination plants ($2n=14$); *O. albida*, *O. oblonga*, *O. lata* and *O. scintillans* are unstable trisomics ($2n+1=15$), while *O. brevistylis* and *O. laevifolia* are true gene mutants. So, de Vries proved his theory with a very unnatural species but, nevertheless, these facts do not in any way diminish the importance of the theory.

PRESENT POSITION OF DARWINISM AND THE ROLE OF MUTATION IN THE MECHANISM OF EVOLUTION

Darwin named his book *Origin of Species* but really sensible criticism are levelled today only against that part of his theory which tries to explain the origin of new species. Darwin's Theory established the fact of Evolution but still there was a serious gap. Darwin said that Evolution was the result of natural selection of variations but he could not explain how these variations arose. Darwin's variations were, mainly, acquired characters. But, Weismann convinced most geneticists that acquired characters as such could not be inherited. This assertion caused a great confusion. Then where were those heritable variations which were necessary for the evolution of new species? In the words of de Vries, "*Natural selection may explain the survival of the fittest but it cannot explain the arrival of the fittest.*" The explanation came after the mutation theory became established. It is now understood that some of the variations, which form the raw material on which natural selection works, may be selections from natural hybrids but the more important variations are the mutants. Thus mutation has assumed a great importance today as the principal mechanism moulding evolution. The present-day idea of mutation is somewhat different from the idea of de Vries. Different types of mutation are recognised today and these are discussed in Chapter LXVI. Many such mutations, like the trisomic aberrants, are unstable. Even de Vriesian mutations (gene mutations) need not be very prominent but may involve slight changes of characters looking like ordinary fluctuating variations. These mutations are just the variations mentioned by Darwin. Such mutated variations are inherited and an accumulation of a number of such insignificant mutations may give rise to a new species. Environment may be a factor is causing and selecting such mutations. Thus, in the mutation theory we have found a way of explaining the Darwinian view of Evolution. The modern concept of origin of species are discussed again in Chapter LXVIII.

The second point in Darwinism which has been subject to criticism in his doctrine of the survival of the fittest. This part of his theory did not take into sufficient account catastrophic destructions of whole races of organisms (when both the 'fit' and the 'unfit' are destroyed) and the role played by *chance* when some so-called 'unfit' organisms may also survive.

CHAPTER LX

GENETICS AND VARIATION

With the growth of this science, Bateson coined the word **Genetics** in 1906 for that branch of Biology which is concerned wih the "elucidation of the phenomena of heredity and variation". The word itself is of Latin origin and means the genesis or origination of organisms. Genetics seeks to account for the resemblances and differences which are found among organisms related by descent. **Heredity** is the transmission of parental qualities, expressed or latent, to the progeny. Thus Genetics is the critical study of heredity enabling one to understand the mechanism of Evolution. *Eugenics* is the science of Genetics applied to human beings with special reference to race betterment.

The science of Genetics began with a study of variations to find out which of them were inherited.

VARIATION

The actual build up of any organism is determined by two factors—heredity and environment, i.e., **nature** and nurture. We have seen that while Lamarck stressed the importance of environment (in addition to the 'conscious efforts' of animals) alone, Weismann went to the opposite extreme and, at first, completely ignored the effect of environment. But it is clear that if heredity be so constant and immutable then new species cannot be formed and no evolution can take place. To explain evolution there must be *variation*. Variation is of the greatest importance in evolution. Nature selects out those hereditary variations which render the organism more suitable to its environment. Thus, the selected variants become better adapted to their surroundings and, in course of time, progressive selection forms a new species. So, *variation*, *selection* and *adaptation* principally determine the course of evolution. Adaptations are special modifications which suit the particular environment of the organism.

The most bitter arguments in Genetics and Evolution center around the problem of the origin of variations. This is a matter of contention between the neo-Lamarckians and the Weismannians or neo-Darwinians. Some **vitalist** philosophers (Bergson, Smuts, Bernard Shaw, etc.) talked of some supernatural force or some Life Force causing evolution in certain directions. But these are merely speculative and not based on actual facts. At present, all arguments have condensed down to two main channels—either, heritable variations are caused by accidental changes in the germplasm or, environment moulds such variations. Most geneticists today believe in **accidental evolution** caused by the accidental variations or mutations. It is possible that some of the mutations are due to disturbances caused by enviromental changes. On the other hand, there are others who believe that new true breeding variations arise as a result of the direct effect of the environment. The Russian Lysenko school of Genetics, who claimed that they alone were the true followers of Darwin, were strong advocates of the latter view.

Certain phenomena in evolution have been differently explained by the vitalists, the mutationists and the environmentalists. **Orthogenesis** means evolution in a definite direction. That is how the seed habit in plants and the brain in amimals developed. The vitalists argue that a directive force caused this evolution and 'predetermined' the direction of orthogenesis. The mutationists, on the other hand, say that nature selected out such mutations while the enviromentalists would say that environment moulded out these forms. **Epharmosis** means that evolution of similar forms in different groups by the same environment. Thus, cushion and rosette forms have been developed in alpine plants of different types. Xerophytic cacti and xerophytic Euphorbias look much alike though they come from widely different groups. This phenomenon is also known as *homoplasy*. This *convergence in evolution caused by epharmonic adaptation* has given rise to much speculation. *Parallel evolution* of heterospory in different Pteriodophytic groups may also be ascribed to similar reasons. Vavilov formulated an interesting *Law of Homologous Series in Variation* by showing that the different cultivated crops follow the same course in the evolution of new varieties. This also provokes arguments as to whether such parallel evolution can be accidental.

Organisms with the same heredity have been found to show the following types of variation:

(1) The *fluctuating* or *continuous variations* are very common. It is very difficult to find two leaves on the same plant which are exactly alike. This variation is caused more by 'nurture' than by hereditary 'nature'. Internal and external environment causes such variations and these are not inherited.

The original concept of *fluctuating variations*, as understood by Darwin and others, however, included something more than what is stated above. This includes the segregates of *polygenes*, i.e., the *multiple genes* and such minor *polygenes* and *modifying genes* (explained in Chapter LXIV) which are constantly arising in nature by mutation.

(2) *Mutations* or *discontinuous variations* appear suddenly. They are caused accidentally, or as an effect of environmental changes. Different types of mutation are discussed in Chapter LXVI as a better knowledge of chromosomes and genes is needed before they can be understood.

(3) *Variation by hybridisation and recombination of the germinal material:* Mendel is called the Father of Genetics as he was the first scientist who demonstrated how hybridisation could be utilised in understanding the heredity of organisms. This is taken up in the next chapter.

(4) *Chimaeras:* In grafts between two plants sometimes some peculiar buds occur at the point of union of the *stock* and the *scion* (Fig. 814A). The tissue of such a bud is

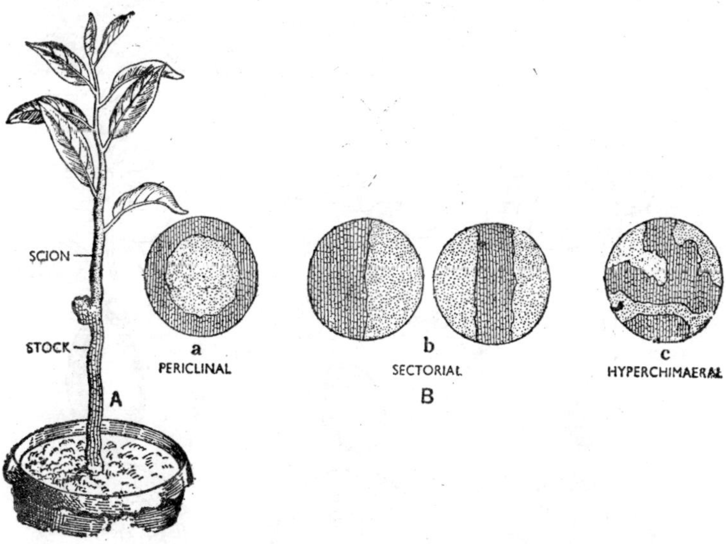

FIG. 814. A. A graft showing a bud divided between the stock and the scion which turns out to be a chimaera. B. Different types of chimaeras.

actually a mixture of the tissue of the two plants. Branches arising from such buds and plants vegetatively propagated from them show some mixtures of the two parent plants simulating true hybrids and are known as **chimaeras.** Chimaeras are also known to arise on normal plants without grafting, being caused by mutation of somatic cells, i.e., *bud mutations*. Chromosomal mutations (explained in Chapter LXVI) are known to take place in the somatic tissues of plants so that two tissues, with cells which are genetically different (i.e., with difference in chromosomes), come to lie side by side. This is called a *chromosomal chimaera*. A bud developing out of such a tissue will be a chimaeral bud mutation. A very interesting chimaera is the *gynandromorph* form of insects, half of which is male and half female as one-half of the insect is developed of male cells, and the other

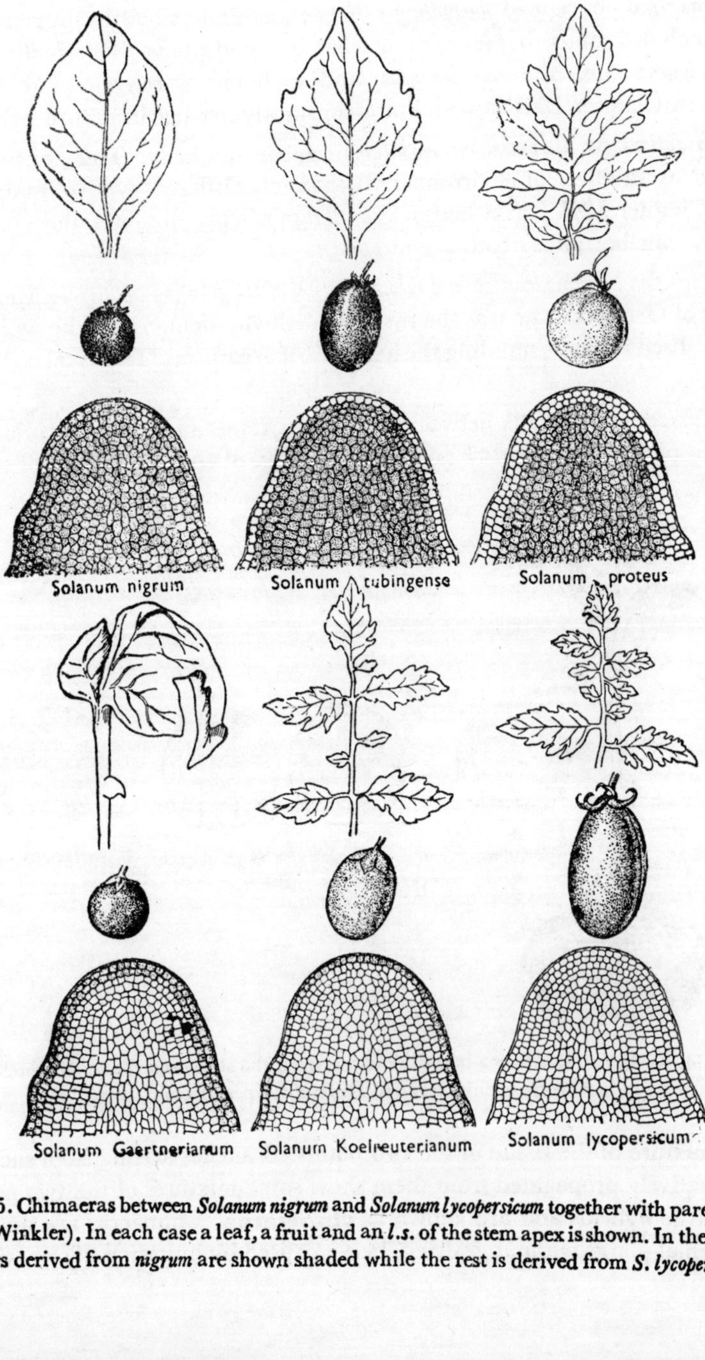

Fig. 815. Chimaeras between *Solanum nigrum* and *Solanum lycopersicum* together with parent forms (after Winkler). In each case a leaf, a fruit and an *l.s.* of the stem apex is shown. In the last, the layers derived from *nigrum* are shown shaded while the rest is derived from *S. lycopersicum*.

half out of female. There may also be a chimaera involving more than two types of genetically different tissues. This is termed a *policlinal chimaera.*

The two types of parental tissues in a chimaera may be arranged differently and different types of chimaeras are recognised according to this. Fig. 814B shows the arrangements in (a) *periclinal* (one tissue on periphery only), (b) *sectorial* (two sections occupied by two tissues) and (c) *hyperchimaeral* (two tissues mixed up indiscriminately) types.

The term **graft hybrid** has been used for similar variations by some classical authors, but, in a different sense. The 'graft hybird' hypothesis supposes that there is actual hybridisation or, rather, nuclear fusion between the cells of the stock and the scion. Tissues may separate out later by the segregation of these fused cells. The same term 'graft hybrid' had, again, been used in a different sense by the Russian Lysenko school (discussed in a later chapter).

Winkler (1911) obtained remarkable chimaeras from *Solanum nigrum* (woody nightshade) × *Solanum lycopersicum* (tomato) grafts (Fig. 815) all of which could pass as true hybrids. But, on closer examination, these 'graft hybrids' were found to be periclinal chimaeras. *Cytisus adami*, a periclinal chimaera between *Laburnum vulgare* and *Cytisus purpureus*, is also famous.

CHAPTER LXI

MENDELISM

PIONEERS OF HYBRIDISATION

As early as 1760, the German Botanist Joseph Kölreuter (1733–1806) hybridised two species of tobacco and found the hybrid to be intermediate between the two. He was followed by Knight (1799) and Goss (1822) in England, Gärtner (1849) in Germany, Naudin (1862) in France and others. The idea of all these pioneers was that hybridisation means something like mixing of two fluids—one maternal and one paternal. The particulate nature came to be realised only later, after Mendel. In the animal world the common concept was that fertilisation means the mixing of two bloods. In 1879, Galton even declared a *Law of Heredity* on this basis. He based his law on his study of pedigreed dogs. The law stated that in the heritage of an individual the parents, on the average contribute one-half the grand-parents one-fourth, the great-grandparents one-eighth and so on, the total heritage being taken as unity.

All such speculations were set at rest by Mendel's experiments which placed the conception of inheritance on a sound footing. The particulate nature of this inheritance became clear later, when the nature of chromosomes and gene became known.

MENDELISM

Gregor Johann Mendel (1822–1884), now famous as the Father of Genetics, was all through his lifetime an obscure monk at the Austrian township of Brünn (now Brno in Czechoslovakia). He possessed a very clear logical brain and carried on his experiments for eight years with a thoroughness unprecedented in his time. He published his work in a local paper in 1866 but this did not attract the slightest attention. It is surprising that even Nägeli, another great Botanist to whom Mendel used to communicate all his papers, failed to appreciate the importance of his work. Unrecognised and bitterly disappointed, Mendel died in 1884. Sixteen years after his death, in 1900, de Vries arrived at certain conclusions from his own observations which he called the '*law of the splitting of hybrids*'. In digging up old literature to find out if anybody had worked on this line, he rediscovered the paper by Mendel. The same discovery was simultaneously made by von Tschermark of Austria and Correns of Germany in the same year. Thus the genius of Mendel came to be discovered long after his death.

Fig. 816. Gregor Johann Mendel (1822–1884). *The Father of Genetics.*

Mendel was familiar with the work of his predecessor plant breeders and surmised that their failure to attain a greater success was due to a want to precise and continued analysis. He realised the necessity of starting with pure-breeding material, of considering each character separately (i.e., not considering all the characters of the plant together) and of always keeping the generations separate. He also realised that the progeny from each individual must be separately recorded. When such precautions had been observed, he anticipated that regular and reliable results would be obtained if the experiments were carried on a sufficient scale. Mendel left clear data of his experiments.

SEGREGATION AND DOMINANCE : MONOHYBRID EXPERIMENTS

After several preliminary trials, Mendel selected the edible pea (*Pisum sativum*) for his subject. Following his idea that the heredity of each character must be separately

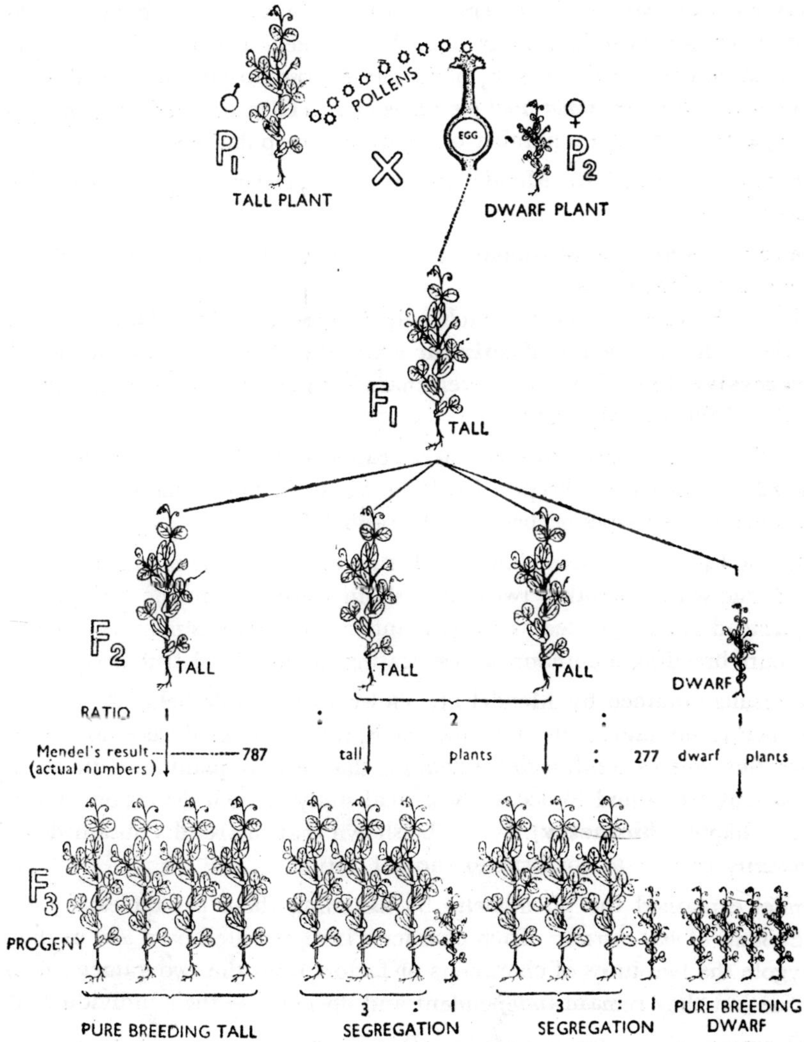

Fig. 817. Diagram illustrating the result obtained by Mendel in his Tall×Dwarf cross.

investigated in order to arrive at a clear and uncomplicated conclusion, he chose a number of *pairs* of characters and made separate crosses for each pair between varieties differing markedly in respect of the particular characters. As the hybrids involve single pairs of characters these are known as **Monohybrid Experiments.** He chose seven such character pairs, namely, (i) form of ripe seed—round or wrinkled; (ii) colour of cotyledons—yellow or green; (iii) colour of seedcoats and flowers—coloured or white; (iv) form of pod—ripe pod simply inflated or constricted between pods; (v) colour of pod—unripe pod green or yellow; (vi) position of flowers—arranged along axis or bunched at top; and (vii) length of stem (i.e., height of plant)—tall (6 to 7ft.) or dwarf ($\frac{3}{4}$ to $1\frac{1}{2}$ ft.). It will be found out later that Mendel was specially lucky in getting perfect dominance in all these cases and in having the seven characters located in seven different chromosomes. Otherwise, his results could not have been so clear.

The pea plants are naturally self-fertilised. Mendel first carried on his experiment separately for each pair of characters. For each pair he crossed between two suitable plants of the proper varieties and grew the hybrid seed. The hybrid plant represents the *first filial generation* or the F_1. The F_1 plant self-fertilised itself, all its seed was collected and planted to form the *second filial generation* or F_2. Then the seed of each F_2 plant was planted separately to form a number of separate F_2 populations.

The result obtained by Mendel in the Tall×Dwarf cross is diagrammatically represented in Fig. 817.

The results were exactly similar in all the seven different crosses. Mendel found that in the case of each cross—

1. In the F_1, one character of each pair disappeared. The character that showed itself in the F_1 he signified as **dominant** while the character that did not show was termed **recessive.** In the list of the seven character *pairs* given above, the first character is *dominant* and the second *recessive.*

2. In the F_1, the parental characters reappeared. Thus, the dominant F_1 plant **segregated** into the parental types. In all these cases he found that in the F_2, the dominant and the recessive types occurred in the ratio 3 : 1.

3. In the F_3, the recessive plants bred true. But, of the dominant plants only one-third bred true while the other two-thirds again segregated in the ratio 3 : 1. So, the F_2 ratio, instead of being stated as 3 dominants : 1 recessive, is more correctly stated as 1 true or pure breeding dominant : 2 segregating dominant : 1 true breeding recessive.

The results obtained by Mendel are shown in the table below:

A glance at the results will show that the F_1 ratios obtained were not exactly 3:1 in most cases. But, when a result is dependent on chance (any plant may die accidentally) such deviations are bound to occur. A critical study of such deviations will be found in the next chapter (**Biometry**). It will be shown that the results obtained by Mendel are sufficiently correct to conform to the 3:1 ratio.

Mendel explained the phenomena by assuming these principles:

1. *Existence of unit character*. Each character is represented by a *unit* in the gamete. In the zygote the two units of characters or factors from the two gametes come to lie side by side but they remain independent and do not lose their individual identities.

2. *Occurrence of characters in pairs* and *Dominance of characters*. There are pairs of all types of characters, e.g., tallness and dwarfness for height, round and wrinkled for seed, etc. In a hybrid zygote one such unit comes from each parent and the two units lie together. When both the character units are present, only one of them shows itself and is called *dominant* while the other lies latent and is called *recessive*. Such pairs of characters are called *allelomorphs* or alleles.

TABLE II

SUMMARY OF F_2 RESULTS OF MENDEL'S MONOHYBRID EXPERIMENTS WITH PEAS

CHARACTER	DOMINANT CHARACTER	NO. OB-TAINED	RECESSIVE CHARACTER	NO. OB-TAINED	TOTAL PLANTS	PHENOTYPIC PERCENTAGE AND RATIO
1. Form of seed	Round	5474	Wrinkled	1850	7324	74·75:25·25 2·96:1
2. Colour of cotyledon	Yellow	6022	Green	2001	8023	75.00:25·00 3·00:1
3. Colour of seedcoat	Coloured	705	White	224	929	76·00:24·00 3·15:1
4. Form of pod	Simply inflated	882	Constricted	299	1181	74·75:25·25 2·95:1
5. Colour of pod	Green	428	Yellow	152	580	73·75:26·25 2·82:1
6. Position of flowers	Axial	651	Terminal	207	858	75·75:24·25 3·14:1
7. Height of plant	Tall	787	Dwarf	277	1064	74·00:26·00 2·84:1
TOTAL		14949		5010	19959	75·00:25·00 3·00:1

3. *Eventual segregation of units contributed by respective parents.* The hybrid plants will again form gametes of two types and these gametes will pair to form zygotes of the next generation. The zygotes will be diffeent as gametes may associate in differnt ways. So, the F_2 plants will be of parental as well as of hybrid types. This breaking down of the hybrid type into parental and hybrid types is called *segregation* which is sometimes called the **First Law of Mendel.** Mendel declared" *it is now clear that the hybrids (F_1) form seeds having one or other of the two differentiating characters and of these one half develop again the hybrid form, while the other half yields plants which remain constant and receive the dominant or the recessive characters (respectively) in equal numbers.*"

Segregation can be explained easily if we consider the association of the gametes. A gamete can contain only one unit character of each allelomorphic pair. Thus, if **T** and **d** represent the unit characters for tallness and dwarfness respectively, there can be male gametes (formed out of pollens) and female gametes (eggs) of either **T** or **d** types. Thus, when the first cross is made, one of the parents gives rise to only **T** gametes (within pollens) and the other to only **d** gametes (eggs). The F_1 zygote is **Td**. This **Td** plant now gives rise to two types of pollens resulting in two types of gametes **T** and **d** in aproximately equal numbers. The same parent will also give rise to equal numbers of **T** and **d** eggs. These two types of gametes are now capable of fertilising one another freely . The above facts are diagrammatically represented in Fig. 818.

The diagram explains that the F_2 segregation ratio is

1 **TT** : 2 **Td** : 1 **dd**

or, 1 true breedling tall : 2 segregaing tall : 1 true breeding dwarf.

The **TT** plants, when self-fertilised, can form only **T** gametes—so they will be true breeding talls. Similarly **dd** plants will be true breeding dwarfs. But, the **Td** plants will again give rise to **T** and **d** male and female gametes so that the descendants of these plants will again segregate. Hence, the **Td** plants are segregating talls.

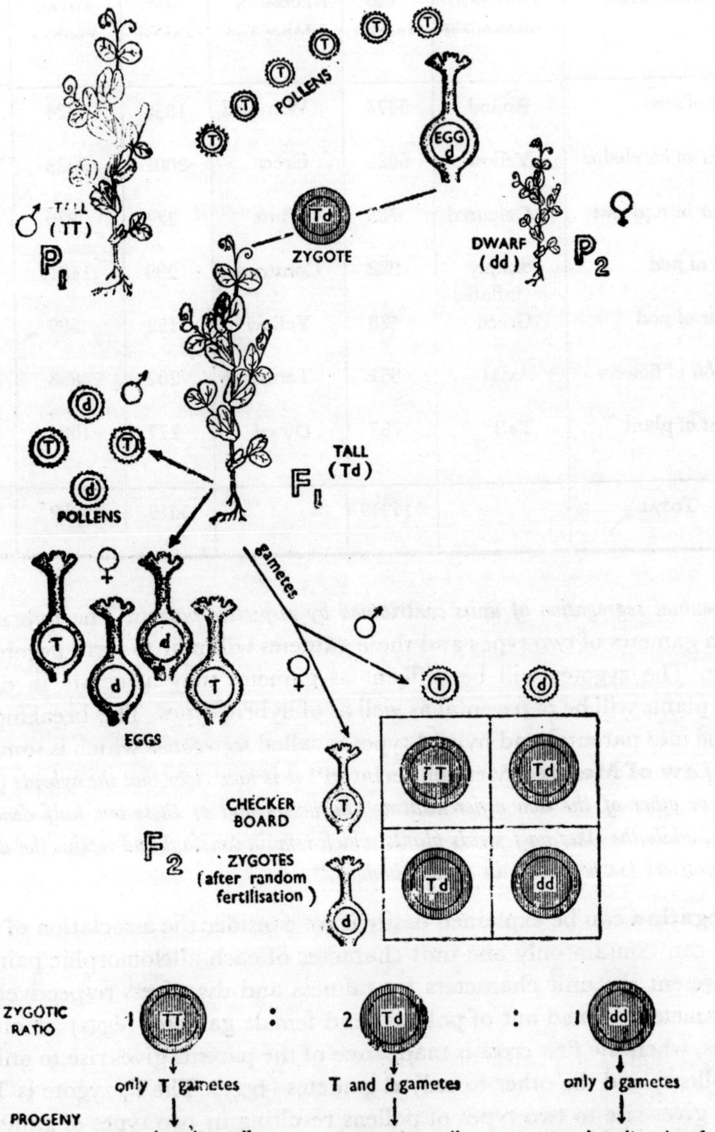

FIG. 818. Diagram explaining monohybrid segregation.

We find that among the tall F_2 plants there are two types, **TT** and **Td**. These are constitutionally different as their future segregation shows but, outwardly, their appearances are the same—i.e., tall. When a group of organisms are similar in appearance but not necessarily similar in their genetic constitution, they are said to form a **phenotype.**

On the other hand, organisms of the same genetic constitution form a **genotype.** Thus, the F_1 segregation shows two phenotypes—tall and dwarf, but three genotypes—**TT, Td** and **dd.**

The zygote formed by the union of two gametes always contains two factors of an allelomorphic pair. If the two factors are the same, e.g., **TT** or **dd**, then the zygote is called a **homozygote** and the plant from it is *homozygous*. On the other hand, if the two factors are different, e.g., **Td,** then it is a **heterozygote** and the plant from it is called *heterozygous*. *Homozygous* plants are always true breeding while *heterozygous* ones always segregate.

In order to find out the segregation ratio, the gametes and the zygotes have been represented in the diagram in the form of a checker-board or chess-board. This method of finding out the segregation ratio is known as the **Checker-board** or **Chess-board Method.**

INDEPENDENT ASSORTMENT: DIHYBRID & TRIHYBRID EXPERIMENTS

After considering the character pairs singly, Mendel now began his experiments with two pairs of characters simultaneously and thus obtained the **Dihybrid Ratio.** Here again Mendel had another bit of luck in experiencing no linkage (discussed in Chapter LXV).

For a dihybrid experiment, Mendel crossed a variety having yellow cotyledons and round seeds with one having green cotyledons and wrinkled seeds. The result obtained is shown in Fig. 819.

The F_2 ratio can be explained in the same way as was done in the case of the monohybrid ratio. Let **Y** represent yellow cotyledon, **g** green cotyledon, **R** round seed and **w** wrinkled seed. Thus, a gamete will contain either **Y** or **g** and, at the sametime either **R** or **w.** The eventual segregation is explained by Fig. 820. The checker-board shows that only four phenotypes in the ratio 9 **Y-R-** : 3 **Y-ww** : 3 **ggR-** : 1 **ggww** are possible because **Y** is dominant over **g** and **R** over **w.** The actual ratio obtained by Mendel is very close to its expected ratio. It should also be noted that the four phenotypes are composed of the following genotypes:

The 9 of **Y-R-** phenotype include	1 of **YYRR** genotype
	2 of **YYRw** ”
	2 of **YgRR** ”
	4 of **YgRw** ”
3 of **Y-ww** phenotype include	1 of **YYww** genotype:
	2 of **Ygww** ”
3 of **ggR-** phenotype include	1 of **ggRR** genotype:
	2 of **ggRw** ”

The **ggww** phenotype is a genotype by itself.

Next, Mendel worked with three pairs of characters and thus obtained a **Trihybrid Ratio.** He considered the same **Yg** and **Rw** alleles along with the characters coloured or white seedcoat. Putting **C** and **c** for coloured and white seedcoat respectively, his results may be shown as in Fig. 821.

The F_2 segregation can be easily explained as has been done in the cases of monohybrid and dihybrid ratios. Naturally, the checker-board will be a little more com-

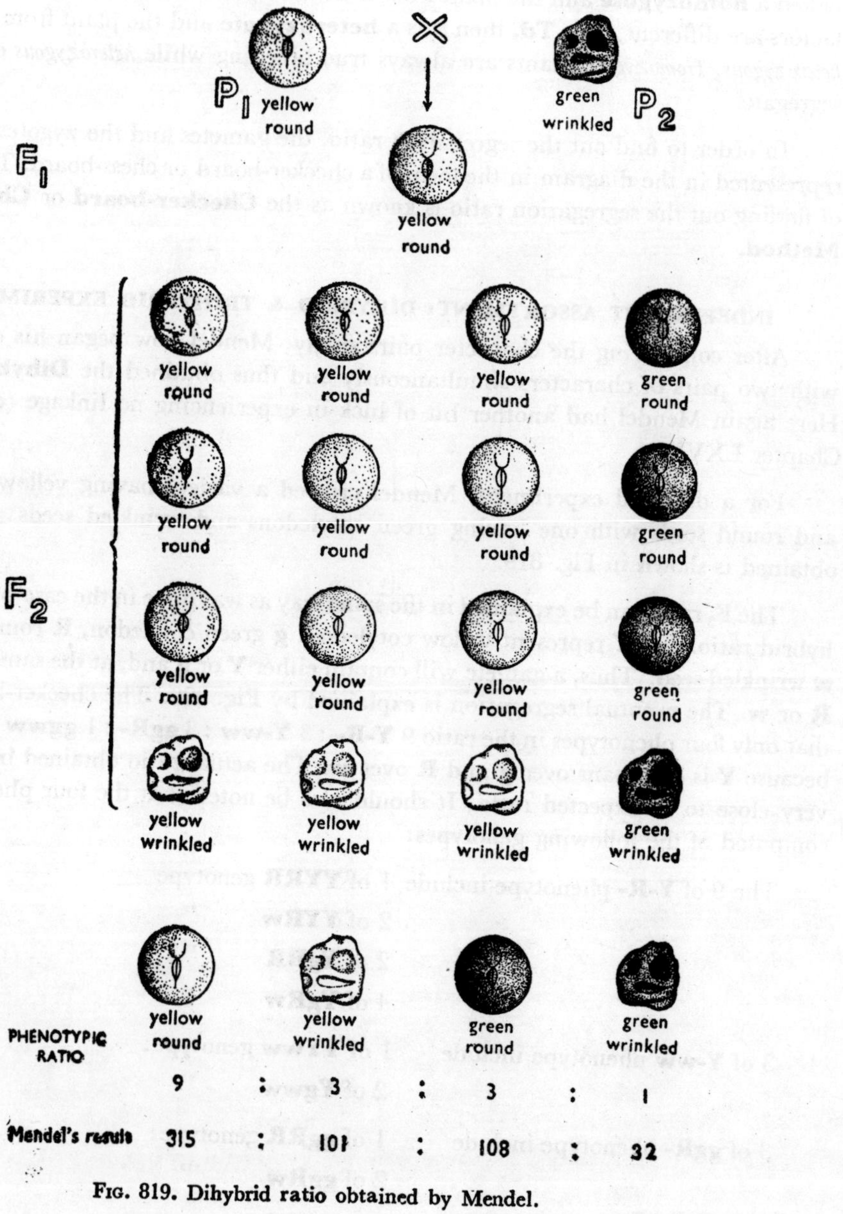

FIG. 819. Dihybrid ratio obtained by Mendel.

plicated, there being 8 possible types of gametes so that the checker-board will have 8×8=64 squares.

Fom the dihybrid and trihybrid ratios obtained, Mendel concluded his principle of **independent assortment of characters** which is sometimes known as **Mendel's Second Law.** This states that the different factors or allelomorphic pairs in the gametes

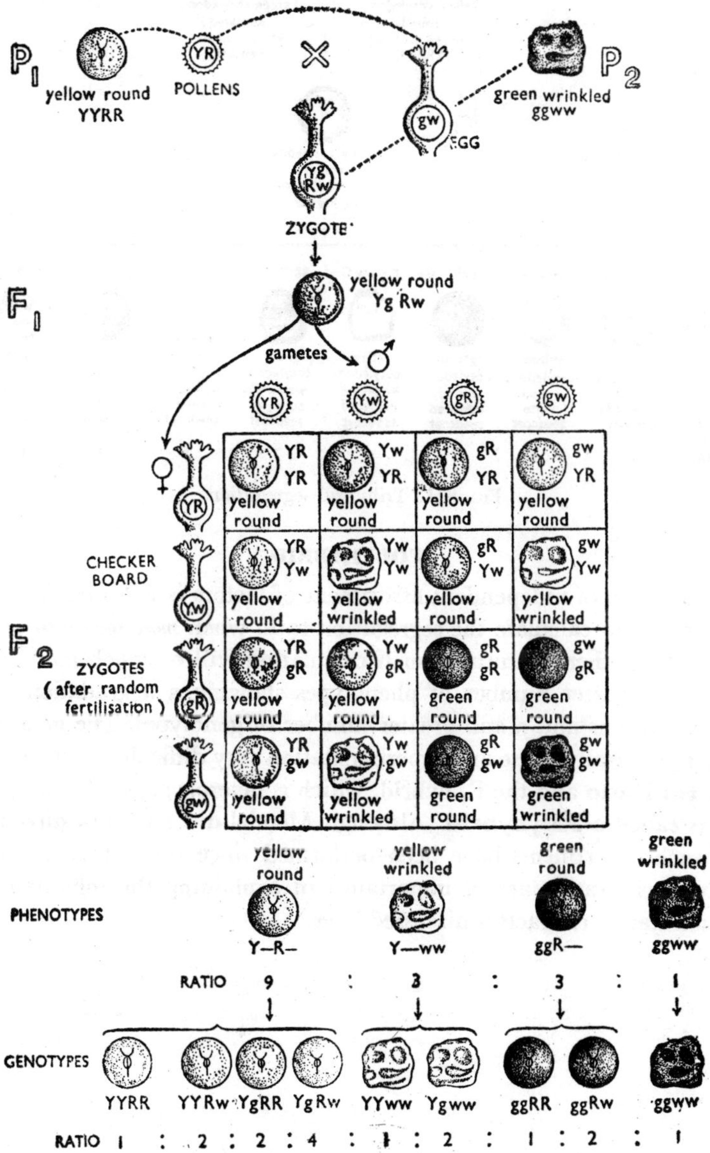

Fig. 820. Explanation of dihybrid ratio by checker-board.

and zygotes assort themselves and segregate independetly of one another. Mendel stated "*the offsprings of the hybrids in which several essentially different characters are combined exhibit the terms of a series of combinations, in which the developmental series for each pair of differentiating characters are united. It is demonstrated at the same time that the relation of each pair of different characters in hybrid union is independent of the other differences in the two original parental stocks.*"

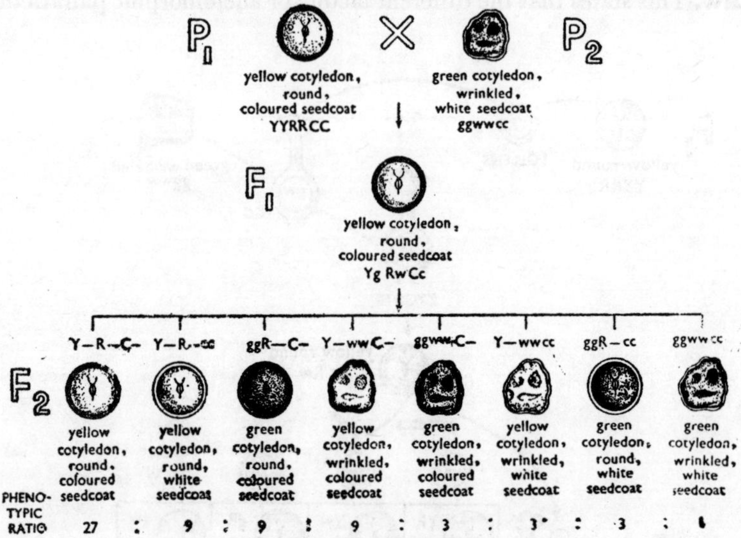

FIG. 821. Trihybrid segregation.

POLYHYBRID

If the principle of independent assortment of factors is followed, it will be found that it is possible to calculate the segregation ratio when *more than three pairs of factors* are considered together. Naturally, such calculation will be much more complicated. There will be a greater number of phenotypes showing a a gradation almost as in fluctuating variations and a still greater number of genotypes. The great variation of segregates will naturally cause any experimentation very difficult. Such a ratio is known as **Polyhybrid Ratio** and the F_1 hybrid, which is heterozygous for a large number of characters is called a **Polyhybrid.** Although Mendel did not work directly on this, a number of such experiments have been performed since the rediscovery of Mendel's paper. Polyhybrid ratios are of importance in explaining the inheritance of multi-factorial quantitative characters discussed later.

CHAPTER LXII

BIOMETRY

It has already been stated that during the post-Weismannian period, the tendency of studying variations closely gave birth to a new science—**Biometry**. *Biometry* or **Quantitative Biology** is the application of statistical methods to biological facts. The science was begun by Sir Francis Galton (1822-1911) and was developed by Karl Pearson (1857-1936) and his disciples. Through the researches of R. A. Fisher, S. Wright, J. B. S. Haldane, K. Mather and others, Biometry has today become a very exact tool in the hands of geneticists. While some such work is too mathematical to be well understood by the non-mathematical scientists, the formulae usually employed are so useful that nowadays very few experimental evidences are much relied upon unless the results are statistically tested. This is true not only for genetics but for most branches of science where living organisms are concerned or, where *chance* plays a role.

The scope of the present work is too limited to give any real idea of the different aspects of Biometry. Only a few topics of immediate importance are being taken up.

MEASUREMENT OF VARIATION

Biometry developed with measurement of variations. It deals with masses and not with individuals. Thus, in stating that the percentage of sugar in a variety of sugar beet if 15.5, one has got to examine a whole population of beets or a suitable number (called

FIG. 822. Graph showing variation in the sugar content of 40,000 sugar beets. (After de Vries). The broken line is the theoretical curve obtained from the binomial formula $(a+b)^{20}$.

Percent								*Data*							
sugar	12	12.5	13	13.5	14	14.5	15	15.5	16	16.5	17	17.5	18	18.5	19
Frequency	340	635	1192	2205	3597	5561	7178	7829	6925	4458	2233	692	133	14	5

a *sample*) of them, and to get the *mean* or *average* value. Fig. 822 shows a *variation curve* plotting the percentage of sugar in forty thousand beets as obtained by de Vries. The curve shows how variations are evenly distributed about a mean in case of fluctuating variations. It should be noted that while considering the data of quantitative variations, the individuals are not considered separately but are grouped under different **classes**. Thus, in this case the classes were 12% (i.e., 11·75 to 12·25%), 12·5% (12·25–12·75%), 13% (12·75–13·25%), etc. The number of individuals showing the measurement of a class is known as the **frequency** of the class. Thus, 635 beets showed the sugar content of the 12·5% class (i.e., 12·25–12·75%).

In case of ordinary fluctuating variations, it is found that deviations from the average lie equally distributed on both sides of the average. A curve for this (as the curve in Fig. 822) is a **normal curve** which is symmetrical about a vertical axis called the **mode** which represents the highest frequency. In a perfectly normal curve the mode should also show the *mean* value. The **mean** or **average** may be obtained by adding up all the values and dividing by the total number of by the formula.

$$\text{Mean} = \frac{\text{sum of (each class value} \times \text{its frequency)}}{\text{total number}}$$

$$\text{Or, symbolically, } M = \frac{\Sigma f.v}{n}$$

A measure of the amount of deviation of individuals from the mean is expressed by **standard deviation** (σ or s) which is equal to

$$\sqrt{\frac{\text{sum of } \{(\text{deviation of each class from mean})^2 \times \text{class frequency}\}}{\text{total number} - 1}}$$

$$\text{Or, symbolically, } \sqrt{\frac{\Sigma\,(v\text{-means})^2 \times f}{n-1}} \text{ or } \sqrt{\frac{\Sigma\,d^2.f}{n-1}}$$

When a computing machine is at hand the following formula (which is mathematically derived from the above formula) would be found more useful:

$$\text{Standard deviation} = \sqrt{\frac{\Sigma X^2 - \frac{(\Sigma X)^2}{n}}{n-1}}$$

where X is the value for each sample and n is the total number. Standard deviation is the square root of *variance* (s^2), which is a measure of the variation in a population. Deviation is also expressed by the *Coefficient of Variation* given by the formula

$$\text{Coefficient of Variation (C.V.)} = \frac{\sigma \times 100}{\text{mean}} \text{ per cent.}$$

If, instead of taking one sample, several samples are taken it will be found that the *standard deviations* of the different samples also vary. This variation is measured by the **standard error** (or, S.E. of a mean) which is given by the formula $\frac{\sigma}{\sqrt{n}}$ which gives an idea as to how any mean obtained from a sample may differ from the true hypothetical mean of the population. Sometimes, instead of the Standard Error, **Probable Error** is used which is equal to 0.6745 times the standard error. In making a statement on some measurement one should state the average $\pm$ the standard error. Thus, one may say that the leaves of a particular plant measure 2·5 $\pm$ 0·15 cm.

Correlation Coefficient (r) is another measure which shows how one type of variation may be related with another type of variation. If X and Y be the two variables, their correlation r is given by the formula

$$r_{XY} = \frac{\Sigma xy}{\sqrt{\Sigma x^2} \times \sqrt{\Sigma y^2}}$$

where x is the difference if each X from its mean and y is the same for each Y.

For computation, however, the following formula is easier to use

$$r_{XY} = \frac{\Sigma XY - \frac{\Sigma X . \Sigma Y}{n}}{\sqrt{\Sigma X^2 - \frac{(\Sigma X)^2}{n}} \times \sqrt{\Sigma Y^2 - \frac{(\Sigma Y)^2}{n}}}$$

n being the total number.

When $r = +1$, there is perfect positive correlation, perfect negative correlation when $r = -1$, there being no correlation when $r = 0$. Intermediate values mean correlations of intermediate degrees. The following illustration shows how the formula may be used:

TABLE III

HYPOTHETICAL DATA ON MUTATION INDUCTION BY X-RAYS

Experiment No.	Relative dosage of X-rays	Relative amount of mutation obtained
1	1	1
2	2	2
3	3	3
4	4	4
5	5	5

Using the above formula

$$r_{\text{X-ray. mutation}} = \frac{55-45}{\sqrt{55-45} \times \sqrt{55-45}} = \frac{10}{10} = +1$$

which means perfect positive correlation, i.e., mutation rate varies directly with intensity of X-rays.

The 't test' for significance

If the mesurements of two different samples (e.g., the yields of two different crop varieties) vary there is a method of testing whether the difference is **significant** enough to enable one to assert that the two samples are really different or if the difference is merely due to some ordinary cause of fluctuation or some error in the experiment. This is known as the **'t test'.**

$$\text{The statistic } \mathbf{t} = \frac{\text{mean difference}}{\text{standard error of differences}}$$

Let us take up a concrete example to understand this method. The table below shows some hypothetical data on the yield trials of rice grown in a year at 10 different farms in different localities:

TABLE IV
RICE YIELD TRIAL DATA

Locality	Yield: kilogram per acre		Difference	Square of Difference
	Variety A	Variety B		
1	810	800	10	100
2	855	825	30	900
3	810	800	10	10
4	840	810	30	900
5	730	700	30	900
6	820	800	20	400
7	770	760	10	100
8	760	730	30	900
9	700	700	0	0
10	700	670	30	900
Sum			200	5200

We are going to test whether the yield of the variety A is *significantly* superior to the yield of the variety B.

$$\text{Mean difference} = \frac{\text{sum of differences}}{\text{number of trials}(n)} = \frac{200}{10} = 20$$

Standard error of difference

$$= \sqrt{\frac{(\text{sum of square of differences}) - \dfrac{(\text{sum of differences})^2}{n}}{(n-1).n}}$$

$$= \sqrt{\frac{5200 - \dfrac{40000}{10}}{9 \times 10}} = \sqrt{13{\cdot}3} = \pm 3{\cdot}65$$

$$\therefore \mathbf{t} = \frac{20}{3{\cdot}65} = 5{\cdot}48^{**} \text{ (sign ignored).}$$

TABLE V
SHORT-CUT **t** TABLE UP TO 20 *d.f.*

Degree of freedom (*d.f.*)	Probability of a larger value of **t** (*P*)		*d.f.*	*P*	
	0·05	0·01		0·05	0·01
1	12·71	63·66	11	2·20	3·11
2	4.30	9·92	12	2·18	3·06
3	3·18	5·84	13	2·16	3·01
4	2·78	4·60	14	2·14	2·98
5	2·57	4·03	15	2·13	2·95
6	2·45	3·71	16	2·12	2·92
7	2·36	3·50	17	2·11	2·90
8	2·31	3·36	18	2·10	2·88
9	2·26	3·25	19	2·09	2·86
10	2·23	3·17	20	2·09	2·84

Now let us look at the table of **t.** For 9 degrees of freedom (or *d.f.* which is one less than the number of trials or comparisons) **t**=2·26 vor *P* of 0·05 and 3·25 for *P* of 0·01. So in this case *P* is much$<$0·01. Stated in words, this means that chances are much less than 0·01 (or 1% or 1:99) that this difference is due to mere chance and not due to some real difference between the two varieties. In other words, the yield difference observed is *highly significant* which fact may be noted by two asterisks**. In **t** tests $P<0·05$ only is considered *significant.*

Analysis of Variance

Finally, there is an elaborate method of **Analysis of Variance** which enables one to measure critically the effect of every source of variation. This is rather complicated and not discussed in this book.

PROBABILITY

Laws of Probability enable one to forecast the occurrence of any specified event. Thus if a coin is tossed 100 times it is likely that it will fall 'head' 50 times and 'tail' 50 times because the chances of a coin falling head or tail are equal i.e., 1:1. The result will be more accurately 1:1 the greater number of times one tosses the coin. Accuracy in Biometry always depends on the large number of individuals one considers. In order to get reliable statistics the sample size must be fairly large.

If instead of taking a single coin one takes two coins at a time, it will be found that the ratio of Two Heads : One Head the Other Tail : Two Tails will be as 1:2:1 (Fig. 823). Probability in such cases is given by the coefficient of the expansion of the binomial formula $(a+b)^n$ where *n* is the number of objects. If such coefficients are plotted on a graph paper, a *normal curve* will be obtained. Thus, if four coins are tossed at a time

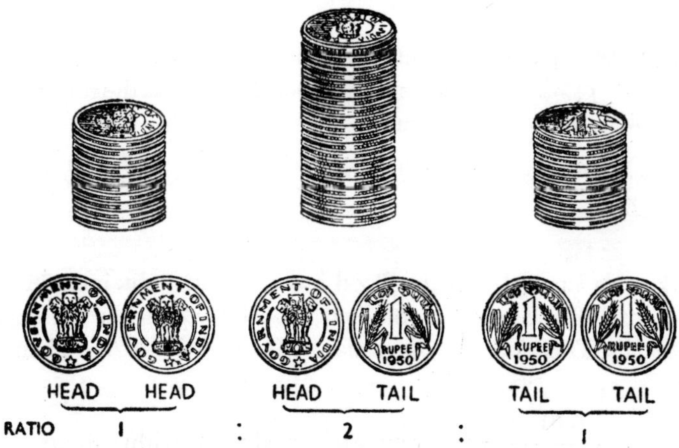

FIG. 823. If two coins are tossed together *large number of times,* probabilities are that the ratio of two heads: head and tail: two tails falling together would be 1 : 2 : 1.

the probabilities will be given by the coefficients in $(a+b)^4$, i.e., $a^4+4a^3b+6a^2b^2+4ab^3+b^4$. This means that the proportion of 4 Heads : 3 Heads and 1 Tail : 2 Heads and 2 Tails : 1 Head and 3 Tails : 4 Tails will be as 1 : 4 : 6 : 4 : 1. In the beet sugar content normal curve (Fig. 813) the broken line is obtained by plotting the coefficients of the expansion of $(a+b)^{20}$ and the two curves match very well indeed!

There can be an interesting application of this law in forecasting the composition of human families. The probability of any particular child of a family being a boy or a girl is 1 : 1 just as in the case of head or tail of a coin. So, if we collect the statistics of families having four children each, we are likely to find the proportion of boys and girls in the same way. It is probable that out of every 16 families having 4 children each, 1 will have all boys, 4 will have 3 boys and 1 girl, 6 will have 2 boys and 2 girls, 4 will have 1 boy and 3 girls and the last family will have all girls. The numbers are obtained from the coefficient of the expansion of $(a+b)^4$ just as in the case of the coins.

TESTING 'GOODNESS OF FIT'

A very important use of Biometry in the field of Genetics is for testing whether a particular ratio of observations may be taken to fit to an expected ratio. For example, we expect that a monohybrid ratio should be 3:1 but perhaps the actual result obtained from an experiment is 2·85 : 1. Should we accept this different ratio to be within ordinary limits of error? Such deviations from the expected ratio are bound to occur in all such experiments as accidents will always happen with living organisms. It is, therefore, always necessary to test if the deviation is within reasonable limits. This is known as **Testing Goodness of Fit** which may be done by two methods:

1. The Standard Error Method

This is quite a good process when the ratio between two classes only (e.g., the monohybrid 3 : 1 ratio) is to be tested. The standard error of the ratio is obtained by the formula

$$\text{S.E.}_r=\sqrt{\frac{p.q}{n}}$$

when p is one of the expected percentages (conveniently expressed as a decimal) and q the other (i.e., $1-p$) while n is the total number. *Deviation* is the difference between an expected value and the value obtained.

Results obtained are usually taken to be a good fit to a ratio so long as $\frac{\text{deviation}}{\text{S.E.}_r}$ is less than 2.

Let us taken an example. In the **T×d** cross, Mendel obtained 787 talls and 277 dwarfs—a total of 1064. The ratio obtained is 0·74 : 0·26 while the expected ratio is 0·75 : 0·25 (3 : 1). Therefore, the deviation is 0·75—0·74=0·01. Then by the above formula

$$\frac{\text{deviation}}{\text{S.E.}_r}=\frac{0{\cdot}01}{\sqrt{\frac{0{\cdot}74\times0{\cdot}26}{1064}}}=\text{approximately 1 which is within limits}$$

(i.e., less than 2) so that the result may be taken to conform to the 3:1 ratio.

2. The Chi-square method

Chi-square (named after the Greek letter χ) is a method which can be used for various types of biological problems and is, therefore, much more useful than the standard error method. Without going into the reasonings, it is quite convenient to use the following formula mechanically:

$\chi^2 = \Sigma \frac{d^2}{e}$ which is a summation of the square of the deviation in each class (d^2) divided by the expected number (e) for each class.

An example will clarify its use. Let us take Mendel's **T** × **d** cross again. The observed numbers were 787 snd 277 while the expected numbers, inorder to conform exactly to the 3:1 ratio, ought to be 798 and 266. Then, considering the two classes (3 and 1, i.e., **T** and **d**)—

$$\chi^2 = \frac{(787-798)^2}{798} + \frac{(277-266)^2}{266}$$

$$= \frac{121}{798} + \frac{121}{266} = 0{\cdot}61^*$$

Now it has to be tested if the value of χ^2 obtained fits to a 3 : 1 ratio. For that we are to consult the short table below:

TABLE VI

DEGREE OF FREEDOM (*d.f.*)	PROBABILITY OF A LARGER VALUE OF CHI-SQUARE (*P*)			DEGREE OF FREEDOM (*d.f.*)	PROBABILITY OF A LARGER VALUE OF CHI-SQUARE (*P*)		
	0·99	0·50	0·05		0·99	0·50	0·05
1	0·0002	0·46	3·84	6	0·87	5·35	12·59
2	0·02	1·39	5·99	7	1·24	6·35	14·07
3	0·12	2·37	7·82	8	1·65	7·34	15·51
4	0·30	3·36	9·49	9	2·09	8·34	16·92
5	0·55	4·35	11·07	10	2·56	9·34	18·31

Degree of freedom means the number of alternative classes which will be one less than the number of actual classes. So, for a 3 : 1 ratio the *d.f.* is 1; for 9 : 3 : 3 : 1 *d.f.* is 3. The right-hand column shows the probability of obtaining a higher value of χ^2 as a result of chance alone if the sample is from a population conforming to the particular ratio. Looking at the table we find that the χ^2 of 0·61 obtained has P between 0.50 and 0·05 that is, the probability of obtaining a value higher than this is much greater than 0·05 or 5% or 5 times in 100 trials. So, the result obtained is a *very good fit* (marked by an asterisk) to the 3 : 1 ratio. Values showing $P > 0{\cdot}05$. (i.e., above 5% level)are considered *good fit* and higher the value of P (i.e., smaller the chi-square) better is the fit. It should be remembered that in statistical experiments, the conclusions are reliable only when the number of individuals (n) considered is reasonably large. Segregation ratios involving fewer individuals may show good fit to 2 or 3 different ratios rendering the conclusion doubtful.

For the dihybrid ratio, Mendel obtained 315, 101, 108 and 32 plants, while, for a 9 : 3 : 3 : 1 ratio the expected number are 312·75, 104·25, 104·25, and 34·75.

$$\text{So, } \chi^2 = \frac{(2{\cdot}25)^2}{312{\cdot}75} + \frac{(3{\cdot}25)^2}{104{\cdot}25} + \frac{(3{\cdot}75)^2}{104{\cdot}25} + \frac{(2{\cdot}75)^2}{34{\cdot}75}$$

$$= 0{\cdot}44^*$$

which is much less than the maximum limit for 3 *d.f.* So this is also a very good fit.

CHAPTER LXIII

THE PHYSICAL BASIS OF INHERITANCE: CHROMOSOMES AND GENES

The earlier geneticists had a conception of *heredity* quite different from the modern idea based on chromosomes. Early in the eighteenth century the idea of *performation* prevailed. It was supposed that the future human body was already present in a miniature form as the '*homonculus*' (also spelt *homunculus*) either in the spermatozoa or in the ovum and then in the embryo. The future development of the human body was only an enlargement of this. The idea of preformation gave place to the idea of *epigenesis* in the nineteenth century. This does not believe in any preformation but believes that the future organs are somehow evolved out of the predecessor. The modern version of epigenesis is the study of *epigenetics* and *morphogenesis* by Waddington and others which try to explain the development of different organs not disregarding the chromosomes and the gene theory.

While the gemmules or pangenes of Darwin were purely speculative, Weismann made a very near approach to the genes by his 'ids' and even located them in the chromosomes although he ascribed peculiar imaginary characters to his ids. When Mendel published his paper, he was completely ignorant of the existence of chromosomes. It is surprising how he could get so near the facts without having any knowledge of these bearers of heredity. Since then, the researches of a number of workers like Janssens, Morgan (Fig. 824), Muller, Bridges followed by a host of others have so clarified the behaviour of chromosomes that it is now difficult to dispute that chromosomes are the bearers of hereditary units. It is now widely accepted after Morgan that certain particles called **genes** are present along the whole length of the chromosomes and these *genes* are the controllers of heredity.

FIG. 824. Thomas Hunt Morgan (1866–1945). *He was awarded a Nobel Prize in 1933 for his contribution towards the development of the Theory of the Gene.*

The eukaryotic chromosomes were discovered in the 1870s and 1880s (Strasburger 1875, Bütschli 1876, Balbiani 1876, Pfitzner 1881, van Beneden 1883, Boveri 1887, etc.,) the term being coined by Flemming (literally, "coloured bodies", the rod-like structures becoming coloured when treated with certain chemicals), and it was found that the number and structures of chromosomes in each species of plants and animals were constant and, gradually, geneticists of the late nineteenth century realised that the chromosomes were the bearers of hereditary characters. As early as 1902, Sutton concluded that the chromosome number of the zygote is *diploid*, one *haploid* set of chromosomes being contributed by each of the maternal egg and the paternal gamete.

To understand the role of chromosomes and before taking up their detailed physical and chemical nature, it is first necessary to make oneself very clear about the mitotic and meiotic methods of cell division. The phases of mitosis (pages 259–261) and meiosis (pages 263–266) have already been explained but, it is worthwhile to recapitulate as the understanding of chromosomes is essential for an understanding of genetics.

Mitosis

A zygote is formed by the union of syngamy of an egg and a male gamete, each of which contains one set of chromosomes of *genomes* and the chromosome number in a gamete or egg is represented by the symbol n or termed *haploid.* Thus, the zygote has two almost identical sets of chromosomes or *genomes*, one coming from the maternal egg and the other from the paternal gamete, so that the zygotic chromosome number is *diploid* and is symbolically expressed as $2n$. This is the first *somatic* cell of the plant body. So in the somatic nucleus there are two chromosomes of each type. Such pairs are known as **homologues**. The *homologues*, although morphologically alike, may contain different allelomorphic pairs of characters. The essence of *mitosis* is that the exact

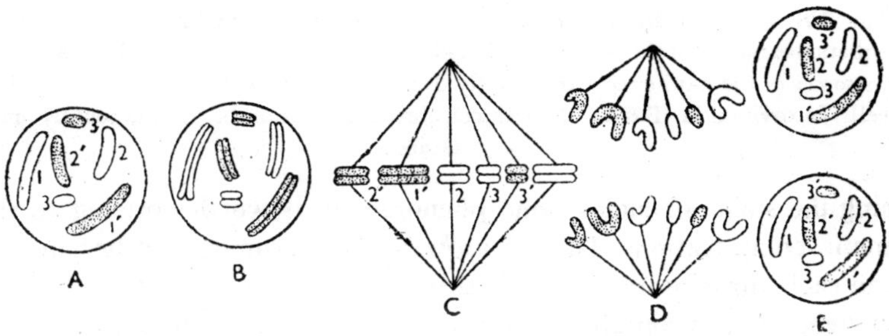

FIG. 825. Diagram showing equational nature of mitosis. Stages A to E represent successive stages. The $2n$ chromosomes (1, 2, 3 and their homologues 1′, 2′, 3′) of the mother cell A are equationally divided between the daughter cells E.

chromosome complement is passed on from the dividing mother nucleus to the daughter nuclei so that all the cells bear the same hereditary characters. This happens by an exact longitudinal doubling of each mother chromosome into two daughter *chromatids* which then pass on to the two daughter nuclei as chromosomes. The mitotic division is, therefore, equational (Fig. 825) so that both the daughter nuclei receive exactly the same double genome ($2n$). The number of chromonema threads in each chromosome or chromatid is not so important from the genetic point of view.

Meiosis

All somatic cells arise by mitosis so that they have the same $2n$ chromoomes. But in the life-history of evey plant, a stage comes when what is known as *meiosis* or *reduction division* takes place. This takes place only in one stage of its life—during the formation of the spores. In the formation of male (i.e., microspore or pollen) and female (i.e., megaspore) spores, the spore mother cells divide by the method of meiosis. The essential feature of the first (heterotypic) division of meiosis is that during prophase and metaphase the homologous pairs of chromosomes come to be associated side by side. At this stage each chromosome doubles into two chromatids. Then, during anaphase, *one of each pair of chromsomes* (not of each pair of chromatids) pass to the opposite poles. Thus, each daughter nucleus receives only *one* genome or exactly half the number (i.e., n) of chromosomes that are found in somatic cells. The chromosome number in the spores, is therefore, *reduced* to half. This reduced number now passes on to the gametes. This reduction is shown diagrammatically in Fig. 826.

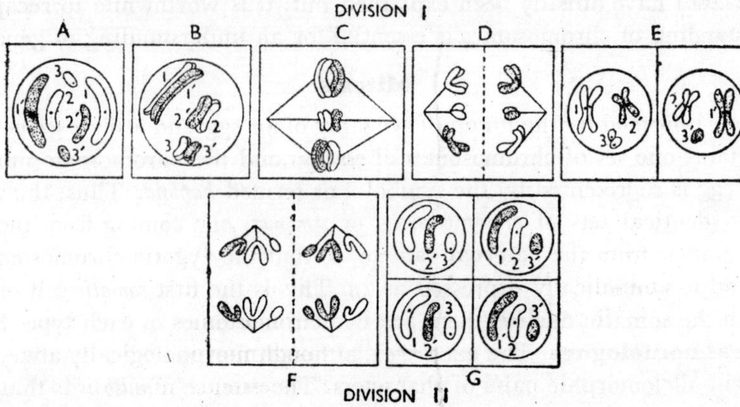

Fig. 826. Diagram illustrating the reductional nature of the first division and the equational nature of the second division of meiosis. In division I (A to E) the 6 chromosomes of the mother cell are equally divided (3 each) between the two daughter cells. In division II (F & G) this reduced chromosomes number is equationally maintained.

During the stages of meiosis one phenomenon is worth noticing as this explains many genetic complications. During *pachytene* the chromosomes become shorter and thicker and each chromosome is clearly seen to be of a double nature. Every one of the thick pachytene threads contains two chromosomes, each containing two chromonemata so that a thread is constituted of *four chromatids* (*not* half-chromatids as in the case if mitosis). *Early pachytene* chromosome is two stranded whereas *late pachytene* chromosome is four stranded. This proves that chromosome doubling takes place in pachytene. The two chromosomes coil round each other (*relational coiling*). The coiling tightens from the prophase to metaphase, so that the length of the chromosome decreases. This coiling is further complicated by the coiling of the two chromosome threads or chromatids round each other. The amount of twisting certainly cause stress on the chromatids. During the next stage, *diplotene*, the two chromosomes of each pachytene pair, or rather, the

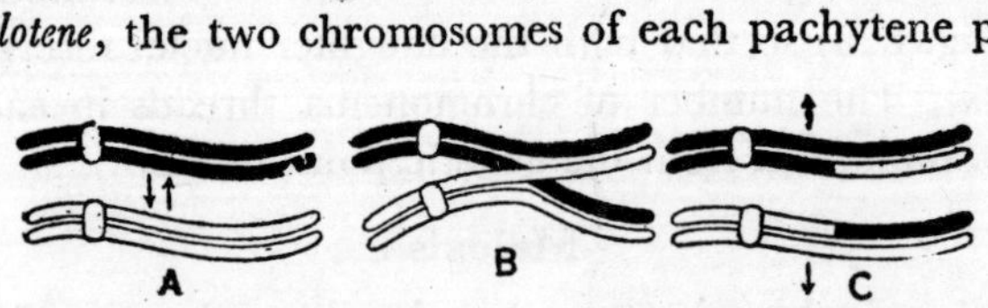

Fig. 827. Diagrammatic representation of chiasma formation and crossing-over. A—pachytene pair of chromosomes; B—chiasma during diplotene and diakinesis; C—dissociation during anaphase showing two crossover and two non-crossover (normal) chromatids.

pairs of sister-chromatids, begin to dissociate from one another because of a force of repulsion noticed at this stage. At this time,a disturbance is very often seen in the dissociation of the two chromosomes. Sometimes, two chromosomes exchange bits of chromatids and this causes the formation of what are known as **chiasmata** (singular *chiasma*). When the two chromosomes try to move away from each other they appear to be sticking to one another at these *chiasmata* points giving the appearance of an **X** when there is a single *interstitial chiasma* or showing loops when there are several. Two *terminal chiasmata* will make the bivalent look like a ring. Often it is found that the chiasmata shift from an interstitial position to a terminal position during pachytene to diakinesis. This is known as *terminalisation* of chiasmata. During diplotene, the chromosomes shorten greatly by the tightening of the coils within the matrix which later

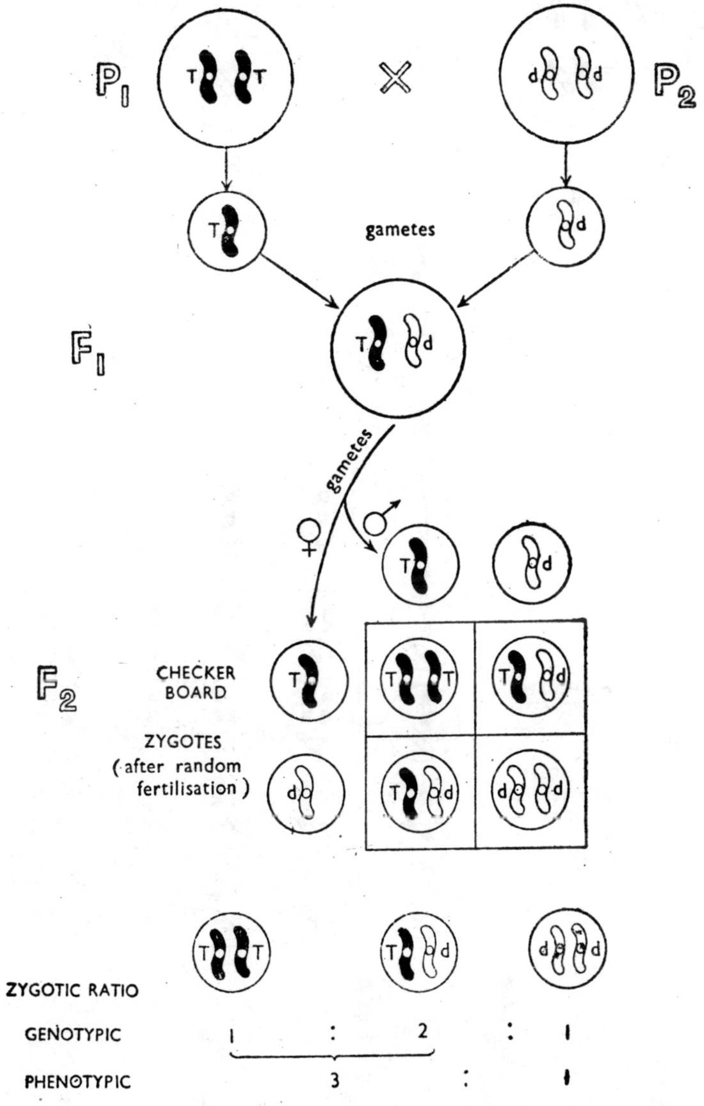

FIG. 828. Diagram illustrating segregation of **T** and **d** genes in Mendel's **T** × **d** monohybrid experiment.

also gradually becomes more and more stainable and renders the chromonema threads inside invisible. At the next stage, *diakinesis*, the chromatids are compact and well spaced out in the nucleus, usually in the peripheral region. This is a stage suitable for counting the chromosomes and the chiasmata. During *anaphase* there is a strong force of repulsion between the centromeres of the chromosomes so that the chromosomes dissociate and this dissociation causes breakage of the chromatids at the points of the interstitial chiasmata so that *crossing-overs* become permanent. Each chiasma (which always involves two of the four chromatids) and resultant crossing-over gives rise to two *crossover chromatids* and two *non-crossover* (i.e., of the original parental types—one

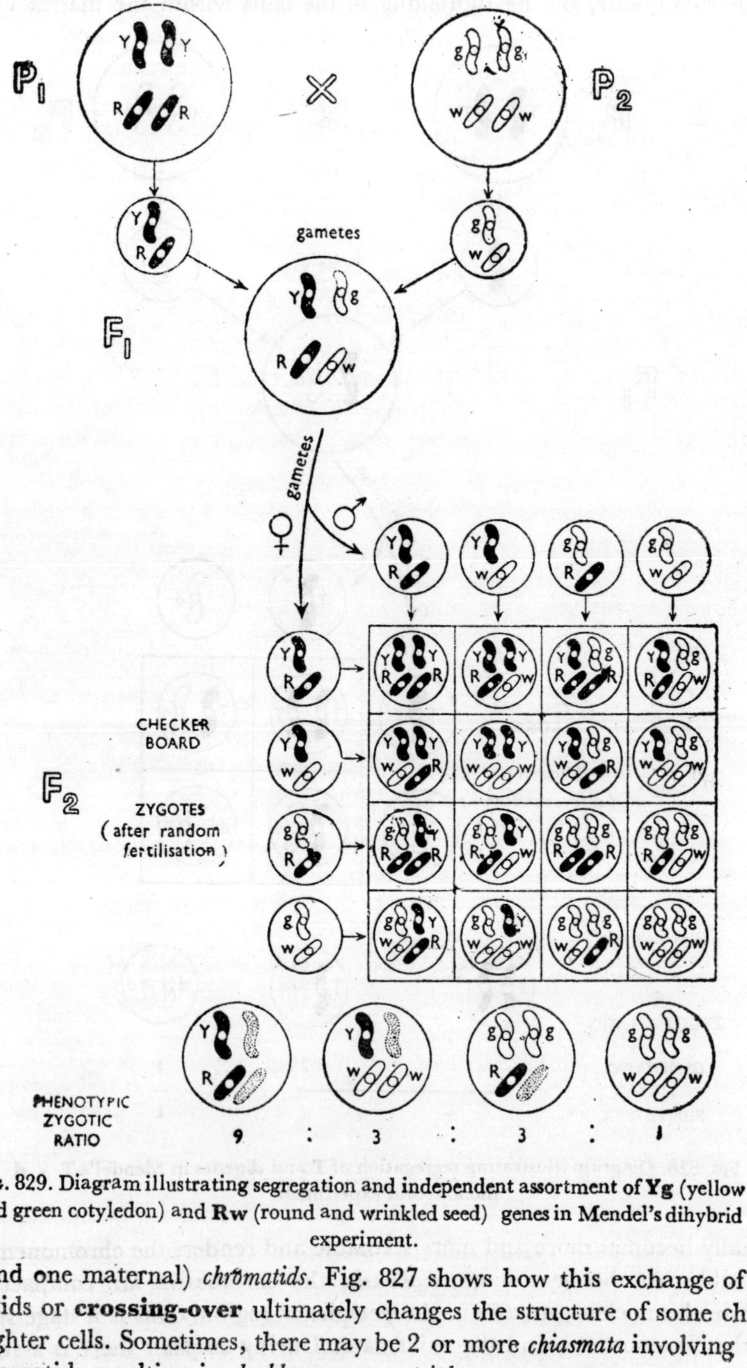

FIG. 829. Diagram illustrating segregation and independent assortment of **Yg** (yellow and green cotyledon) and **Rw** (round and wrinkled seed) genes in Mendel's dihybrid experiment.

paternal and one maternal) *chromatids*. Fig. 827 shows how this exchange of portions of chromatids or **crossing-over** ultimately changes the structure of some chromatids in the daughter cells. Sometimes, there may be 2 or more *chiasmata* involving the same pair of chromatids resulting in *double crossovers*, *triple crossovers*, etc.

Thus, as a result of meiosis, the spore mother cell containing two genomes of chromosomes (2*n*) gives rise to a tetrad of cells each one of which has got only a single genome (*n*) of chromosome. The four cells will ordinarily be of two types, but, if there be a crossing-over, then they will be all different (Fig. 851).

The chromosome as bearer of hereditary characters

If one supposes that alleles are located in homologous chromosomes then the meiotic behaviour of chromosomes exactly explains segregation and independent assortment of chromosomes as found by Mendel. Fig. 828 explains monohybrid segregation.

Fig. 829 similarly explains the segregation and independent assortment in a dihybrid ratio.

The chromosome

By the word chromosome we usually mean the chromosomes of the **eukaryotes.** But, besides the eukaryotes, there is a vast group of **prokaryotes** (virus, bacteria, blue-green algae) which do not have any properly organised nucleus or chromosomes. Nevertheless DNA molecules are present there in some form or other to control heredity.

As already explained, the eukaryotic chromosome shows spirally twisted *chormonema* threads enveloped by a *matrix* (Fig. 830). The matrix is in a highly diffused state in the metabolic and early prophasic stages and is scarcely stainable at this state. But during late prophase and metaphase it is a condensed structure and becomes deeply stainable because of changes in the chemical constitution so that the chromonemata get masked. The chromonema threads are the more important components of the chromosome as the genes are located on these while the matrix is non-genic. As early as 1875 it was noticed by Balbiani that the chromonema presents a beaded appearance—the beads or knots being called *chromomeres.* The chromomeres have been supposed to be particular regions on an otherwise uniform chromonema thread where more nucleic acid or nucleoprotein is synthesised or accumulated (Kaufmann 1948). There is also evidence that chromomeres are caused by the denser coiling of the chromonema thread (Ris 1945) and if the chromonema be stretched to its full length the chromomeres disappear. Chromomeres have also been supposed to be the loci of the genes (Belling 1923) but this is no longer supposed to be always true. Some genes may be located on some of the chromomeres while some genes may be located on the interchromomeric parts of the chromonema thread. *Polytene chromosomes* (p. 866) show characteristic bands formed by the identical chromomeres on parallel chromatids (Fig. 841.).

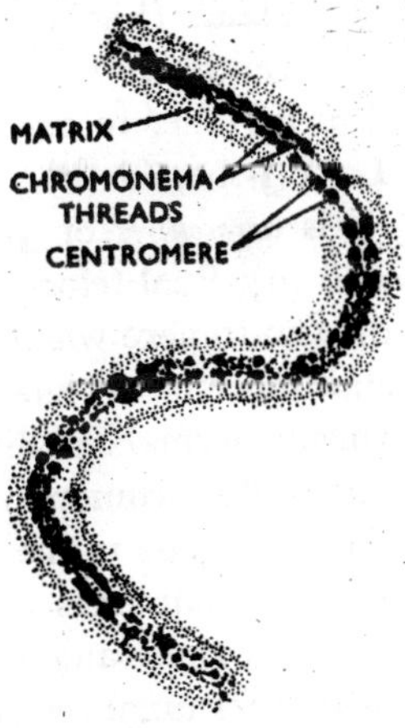

FIG. 830. The structure of a chromosome.

Every chromosome shows a specialised region called the *centromere* or the *kinetochore* which is associated with its movement to the two poles in the spindle during anaphase being the point of spindle attachment. It is usually localised at what is known as the *primary constriction* region. Sometimes, however, it is not localised at all but diffused along the chromosome when the entire chromosome shows the property of spindle attachment. This diffused state is noted in *Luzula* (Juncaceae) and some members of Cyperaceae. *Luzula* chromosomes may be broken by X-rays when each broken bit behaves as a normal chromosome as it is capable of getting attached to the spindle fibres by its diffused centromere. With diffuse centromere, the chromosomes are rod-shaped at anaphase and they show parallel movement to the poles as they are not pulled by a localised centromere.

In a localised centromere, it is a region which absorbs very little stain and the genes in this region govern the formation of the spindle and guide the movement of the chromosomes during cell division.

Tjio and Levan (1950) noted in each centromere four large chromomere granules (two in each chromomere thread) joined with each other by a thin interchromatic thread and again similarly joined on two sides to the two chromosome arms presenting the appearance of  . Recent electron microscopic studies have shown that a bunch of microtubules originate from the centromere connecting it with the spindle. The main function of the centromeres seem to be its attachment to the spindle during cell division. During somatic mitosis, the centromere usually appears as a non-staining constriction. In maize, during the pachytene of meiosis it can be seen as a distinct unstained oval body, somewhat larger in diameter than the remainder of the chromosome (McClintock 1933). In *Tradescantia* and other plants with large chromosomes it may appear at somatic metaphase or early anaphase as a complex with stainable granules connected to the remainder of the chromosomes by thinly stretched threads. It is a specialised segment of the chromosome strand.

The position of the centromere varies in different plants (Fig. 831). If the centromere is at the centre, it is V-shaped and *metacentric* (a). If it is rather to one side, it is *submetacentric* and **L-** or **J**-shaped (b). It is very near one end, it is *acrocentric* or **J**-shaped (c). If it is at the extreme end it is rod-shaped or *telocentric* (d). Real telocentric chromosomes are rare. Accidentally a chromosome may lose the centromere when it is *acentric*. Such a chromosome is useless and gets lost as it cannot move or orientate properly during cell division.

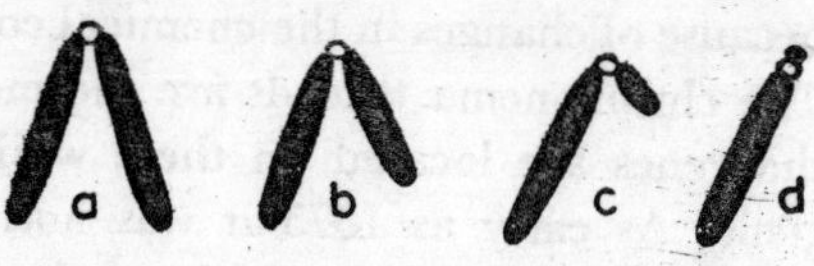

Fig. 831. Position of centromere.

During normal cell division, the centromere splits along the line of the chromonomea threads giving rise to identical chromatids to pass to the two poles. But, Darlington (1940) showed that sometimes there is a *misdivision* and the centromere splits at right angles completely separating the arms, each with a terminal centromere and two chromatids. These new telocentric chromosomes may continue as members of the chromosome complement or sometimes the two fragments again join up forming the original chromosome. Sometimes, however, the new telocentric chromosomes split into normal chromatids and one of these split arms rotate 180° giving rise to a new metacentric chromosome of which both the arms are identical and completely homologous. The two arms tend to pair during meiosis and form a ring. These new types of chromosomes are called **isochromosomes.** (Fig. 832).

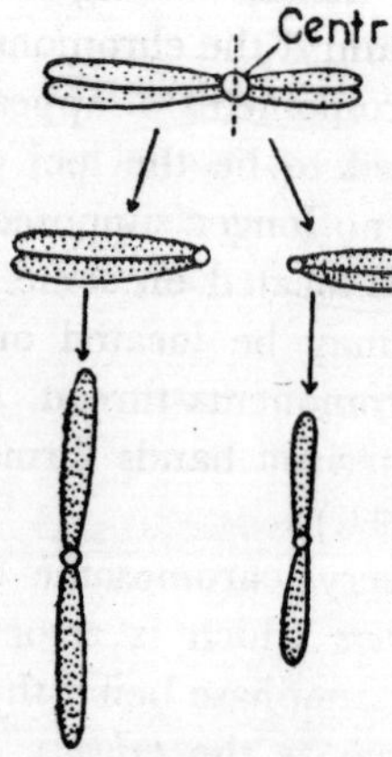

Fig. 832. Origin of isochromosomes.

Most chromosomes have single primary centromeres but a secondary one is also sometimes found. Such chromosomes are *dicentric*. *Polycentric* chromosomes with several centromeres have also been noted in *Ascaris*.

Certain chromosomes show *secondary constrictions* which are usually associated with the organisation of nucleoli. It now appears that the so-called 'constriction' is an optical illusion—the chromosomes being as stout at this region as elsewhere; only, this part remains unstained. Usually each genome set has one chromosome with a *nucleolus organiser* (Fig. 833). The *nucleolus* (the spherical body containing nuclcie acid and proteins in the metabolic nucleus) remains attached to the organiser (a knob-like structure in maize) during early prophase. It gradually diminishes and disappears during prophase when the point where it was attached is marked by an unstainable part in the chromosome and looks like a 'secondary constriction'. The part of the chromosome beyond this point looks like a satellite—hence the name *sat-chromosome* (Fig. 522). The terms 'secondary constriction' and 'satellite' are based on their topography. If the secondary constriction is located at nearly the end of a chromosome so that the distal end looks like a dot the term *satellite* is used. If, on the other hand, it is located at any other portion, the term *secondary constriction* is applied. The functions of both are identical.

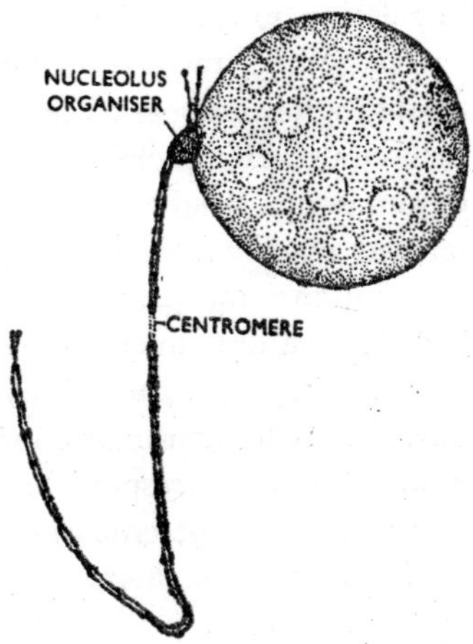

FIG. 833. Chromosome VI of maize (*Zea*) in meiotic prophase, showing its *nucleolus organiser* in contact with the large nucleolus (after McClintock).

Each chromosome ends in a structure called *telomere* which, though not always morphologically discernible, has a distinct and characteristic behaviour. When a chromosome breaks (Chapter LXVI), the broken end of the chromosome attaches itself to some other broken end. But, intact chromosomes with telomeres do not show this characteristic; plant chromosomes are not known to join end to end or a telomere does not get attached to another chromosome.

The chemistry of chromosomes has been and is being studied very closely in recent days. The most important constituent of the chromosome is *nucleic acid* and this is actually the *chromatic* material which causes the staining of the chromosome. Chromosomes often show differential staining in different regions and this means difference in the chemical nature. Certain regions are known as *euchromatic* and others *heterochromatic*. Heterochromatin may be *constitutive* or *facultative*. When *constitutive* it is obligatory and is inherited in the same position in the two homologues. It may occur (1) as unstained areas on the two sides of the centromere (appearing as the *primary constriction*) but becomes brightly stained as **prochromosomes** during telophase onwards, (2) at the telomeric regions in some plant chromosomes, (3) in the nucleolus organising regions, (4) as intercalary heterochromatin forming bands, (5) in the entire sex chromosome, (6) in the entire *supernumerary* or *accessory* chromosomes. Good examples of *facultitative* heterochromatin are found in the animal world. (1) In human sex chromosomes (XX-XY), the X in the male and one of the X's in the female are always euchromatic. But, the other X in the female becomes heterochromatic at a later stage and is called the *hot X*. This phenomenon may be an effort to balance the inert Y-chromosomes of the male. (2) In male mealy bugs the entire chromosome set (the genome) obtained from the male parent becomes heterochromatic while the natural genome remain euchromatic. Thus, in the next generation, the male bug can contribute only the maternal, euchromatic compliment as the heterochromatic gets lost before gametogenesis. The euchromatic zone is considered as the active site having the genes while the heterochromatic regions contain some minor *polygenes* (Chapter LXIII) and as such may appear as *inert*. But Muller has shown that the heterochromatic Y-sex chromosome of *Drosophila* contains some genes while Gates has proved the gene for hairy ear in the human heterochromatic Y-chromosome. Heterochromatin is usually concentrated near the centromere. It has been found that the stainability of these regions changes in relation to the accumulation of nucleic acid which is different in different phases. In metaphase stages the heterochromatic regions mostly do not stain while the euchromatic regions appear as bright. The sex chromosomes too occasionally show the same behaviour. Heterochromatic regions may be lighter or darker than the general chromosome. Brighter staining is termed *positive heteropycnosis* whereas fainter staining is termed *negative heteropycnosis*. Heteropycnosis is also called *allophylly*. Heterochromatin is hereditary and passes on from chromosomes to daughter chromosomes.

The resting nuclei show a number of these deeply staining *prochromosomes* or *chromocentric regions*, which contain much nucleic acid. Since each chromosome shows only one heterochromatic region (about the centromere), the number of prochromosomes usually correspond to the number of chromosomes but it may be more when there are more heterochromatic regions in the chromosomes or it may be less by the fusion of some of them.

Banding Technique

In the 1970's a spectacular technique has been developed for identifying any chromosome. It was discovered that chromosomes stained with fluorochrome quinacrine or its derivatives fluoresce in alternate bright and dark brands when examined under ultraviolet light. The banding pattern in each chromosome is fixed and usually consistent in a taxon. Thus, it is as easy to identify a chromosome with its banding pattern as human beings are identified by their finger prints. The differences in the banding patterns is obviously due to the differences in the DNA in the different parts of the chromosomes. Different types of bands are obtained by using different types of techniques and these are made visible through the low and high intensity regions under the fluorescence microscope or even as differentially stained bands under the optical microscope. The main types of bands and techniques may be named as follows according to the International Convention :

Q-band: as obtained by the fluorochrome quinacrine.

G-band: Obtained by using the Giemsa technique

R-band (banding, reverse to Q band): when stained with Giemsa after heating to 87°C.

C-band: demonstrated by the denaturation-reassociation technique showing the constitutive heterochromatin.

E-band: obtained by enzymic digestion.

CT-band: centromeric and telomeric bands after barium hydroxide and "stains all" staining in mammals.

N-band: at nucleolus-organising regions in mammals.

O-band: with orcein staining. Intercalary and heterochromatin bands (mainly in plant chromosomes).

Construction of the chromosomes

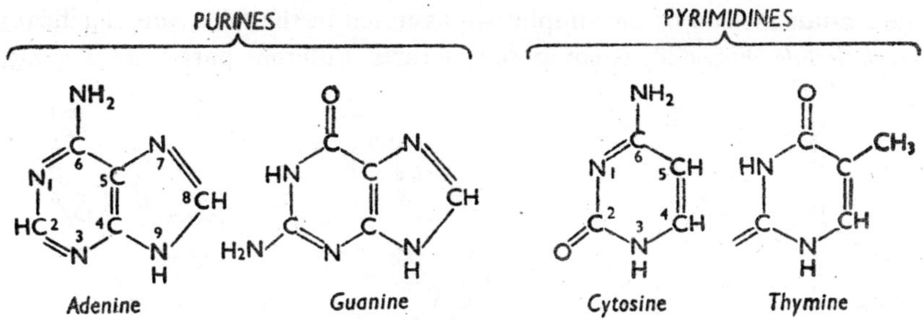

FIG. 834. Molecular structures of Purine and Pyrimidine bases.

The chromosome is constructed of nucleoprotein fibrils. The constituents are *deoxyribose-nucleic acid* (*DNA*), protein (mainly *histone* and *protamine*) and calcium. It is

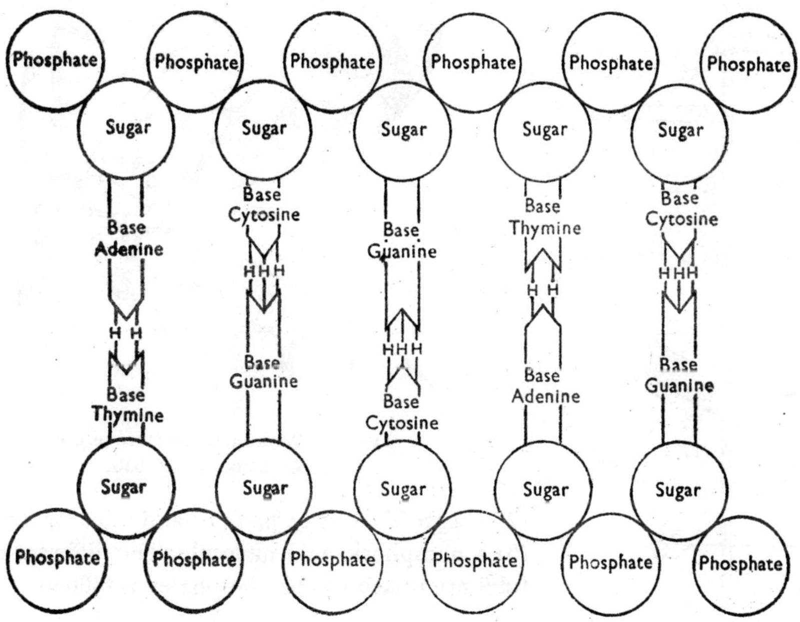

FIG. 835. Paired polynucleotide chains in *DNA* molecule according to Watson and Crick.

the nucleic acid and not the protein which plays the main role in heredity. Two types of nucleic acids are important: The deoxyribose-nucleic acid or *DNA* in the formation of which *deoxyribose* (*d-2-deoxypentose*) sugar takes part and ribose-nucleic acid or *RNA* in the formation of which *ribose* sugar takes part. The nucleic acid molecules are long, spirally twisted and thread-like. These are formed by the end-to-end linking to a large number (may range from a few thousand to millions) of *nucleotides* forming what are known as *polynucleotide chains.* The *DNA* plays the most important role in heredity. The probable structure of the *DNA* molecule was shown in a model by Watson and Crick in 1953. The British scientists T.H.C.Crick and M.H.F.Wilkins and the American scientist J.D.Watson were awarded the Nobel Prize in Physiology and Medicine in 1962 for their researches on the constitution of nucleic acids. Although partly hypothetical, this concept is considered very reasonable. Even then many modifications have been suggested. For example, the *DNA* molecule probably very often contains several polynucleotide strands instead of the simple two assumed in the following clarification.

Each *nucleotide* (Fig. 838) is composed of three different parts: (1) a *nitrogen base,*

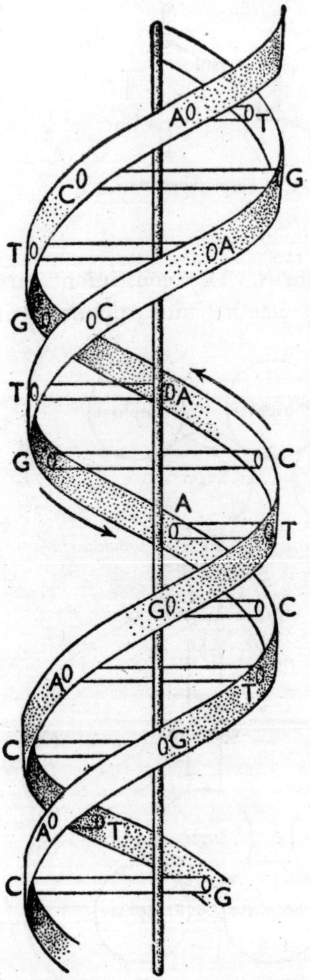

Fig. 836. The double helical structure of the *DNA* molecule (Diagrammatic figure after Watson and Crick).

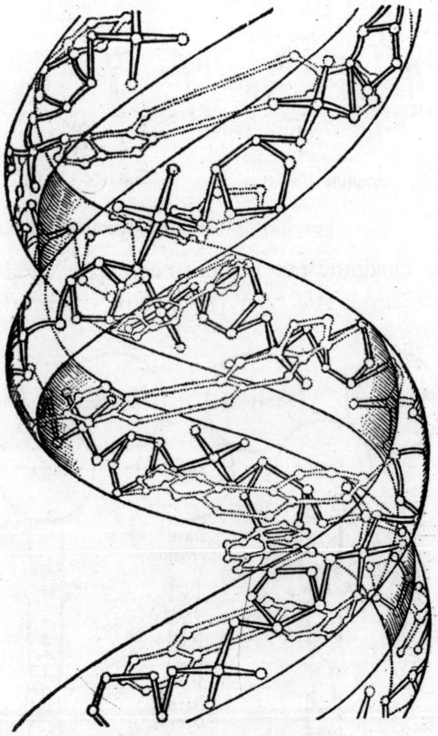

Fig. 837. Molecular structure of the *DNA* helix in Fig. 836.

(2) a *sugar* (*deoxyribose* for *DNA* and *ribose* for *RNA*) and (3) a phosphoric acid molecule. The different nucleotides are linked by the phosphates as follows:

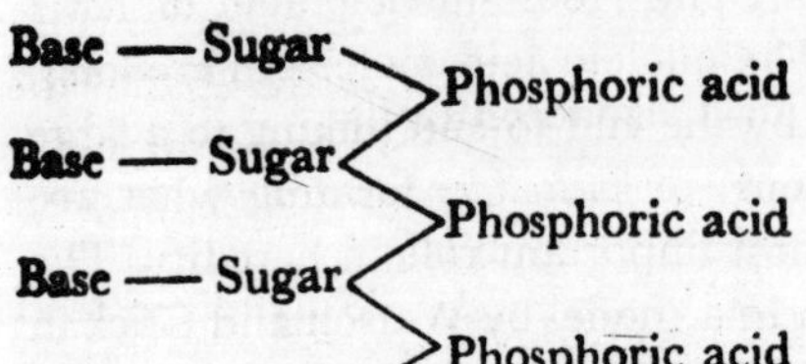

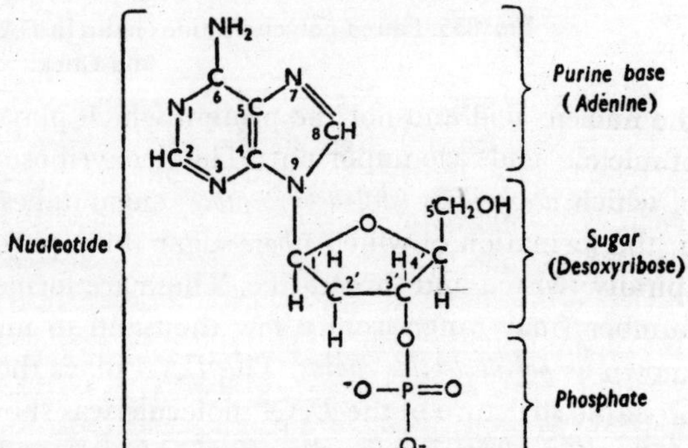

Fig. 838. Molecular structure of a Nucleotide.

In the *DNA* molecule only four nitrogenous bases (Fig. 835) are concerned—two *Purines* (*Adenine* and *Guanine*) and two *Pyrimidines* (*Cytosine* and *Thymine*). The four bases are symbolised as A, G, C and T respectively.

Although there are only four bases, the bases may be arranged in different orders in a polynucleotide chains (e.g., CCAGTTAC.... or CAGTCA....) giving rise to hundreds of types of nucleic acid chains and thus causing different types of inherited characters. In *RNA* the bases are the same except that in place of *Thymine* there is *Uracil*, the bases being A, G, C & U. The *RNA* nucleotide is sometimes called *ribotide* and the chain *polyribotide* chain. The big difference between *RNA* and *DNA* molecules is that while the *RNA* molecule is formed simply of single chains in *DNA* there are always two such parallel polynucleotide chains. Here, two such chains get linked—each adenine being linked with a thymine of the other chain through two H-atom bonds and similarly

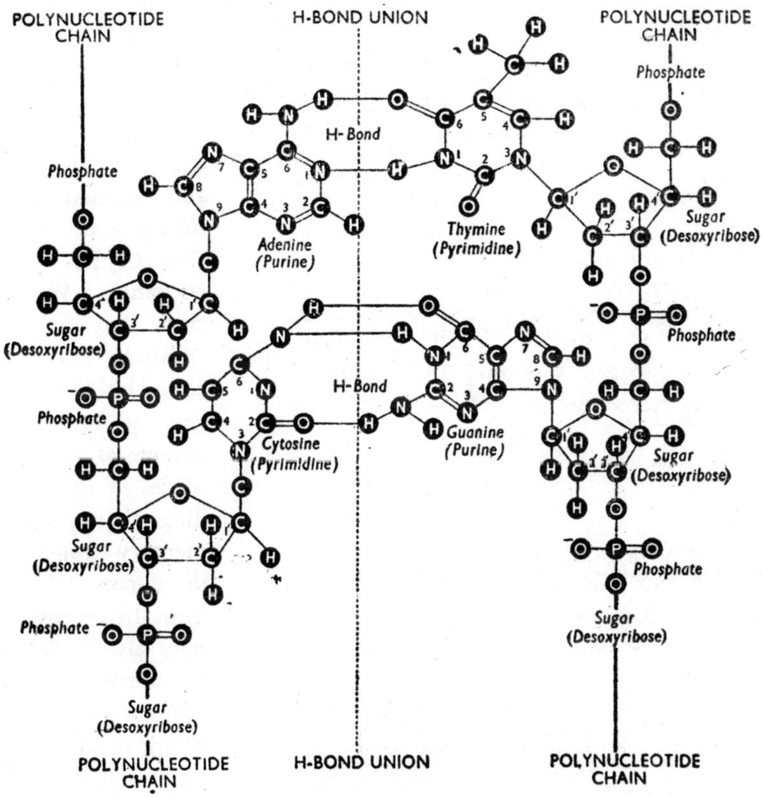

FIG. 839. Molecular structure of a bit of *DNA* helix shown more clearly.

each cytosine getting linked to a guanine of the opposite chain by three H-atoms (Fig. 835). Figs. 836, 837 and 839 also show this linking of two polynucleotide chains, the first diagrammatically and the latter two showing the actual molecular structure of a bit of the united chains. The molecular structure shows that the two chains cannot be straight so that they twist round each other giving the *DNA* molecule a double helical structure shown diagrammatically in Fig. 837 and with the molecular structure in Figs. 838 and 839. Such a molecule is very large and has a molecular weight of 5 to 10 million. A number of such *DNA* molecules get joined lengthwise. The actual method of joining is not clear but, it is possibly through the proteins (histone and protamine) which are of basic nature and calcium also may play a role here. Thus, the long nucleoprotein fibrils are formed. The electron microscope has actually shown the presence of 'microfibrils', 100 to 200Å (i.e., 10 to 20$\mu\mu$) wide and coiled like a cork-screw inside the chromosome (Ambrose 1956).

The splitting and doubling of these polynucleotide paired threads of the chromosomes may also be explained on the basis of the *DNA* molecular structure. Splitting of the composite thread means somehow the hydrogen bonds between the nitrogenous bases break so that the two threads go apart (Fig. 840B) and immediately each nucleotide in the split threads attracts its opposite nucleotide from the free nucleotides present in the vicinity and the linked double chain is reformed (Fig. 840C) and ultimately, the whole chromosome gets doubled (Fig. 840D). This is called *replication.* We may, conclude that each fibril attracts identical material to itself, chain by chain and gene by gene, so that the original two strands of the helix take the places of two *templates* moulding the two new complementary strands. The fixation of a complementary strand from the original double helix may also be compared with the developing of a photographic print from a negative (Fig. 840). When sufficient raw material accumulates, a new mitosis starts. During the mitotic cycle, the chromosomes lose water and thereby acquire a fibrous coat of nucleic acid (stainable matrix) as it coils and assumes the metaphasic condensed appearance. Then it gains water, uncoils and ceases to be separately visible (loses stainability), i.e, reverts to the resting phase.

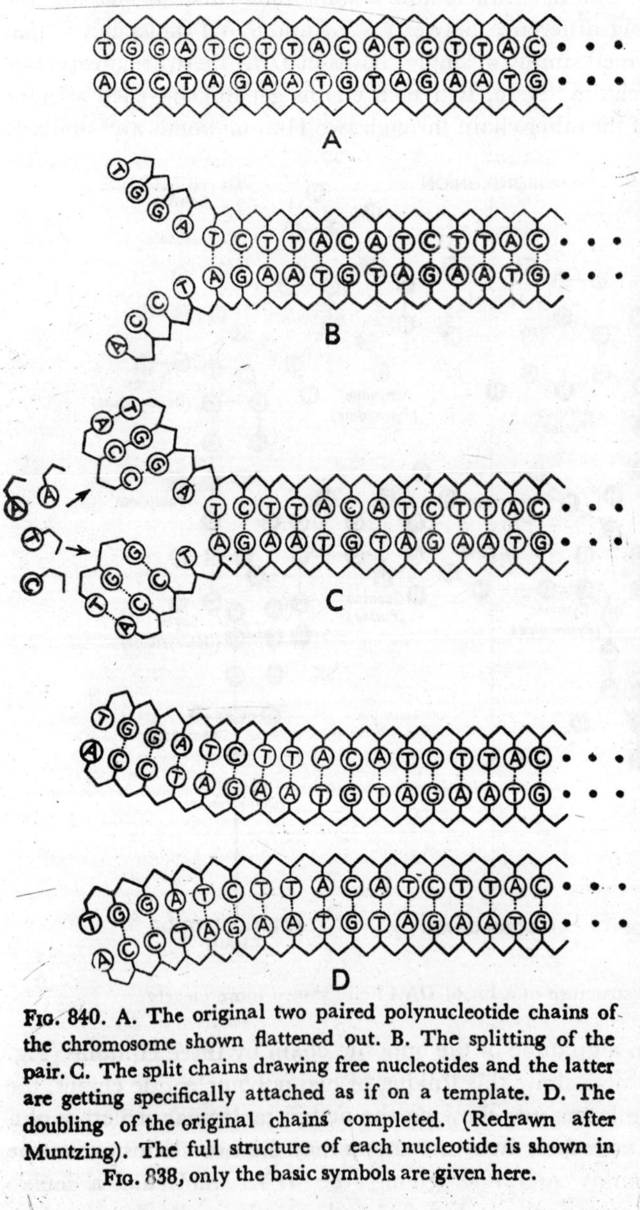

FIG. 840. A. The original two paired polynucleotide chains of the chromosome shown flattened out. B. The splitting of the pair. C. The split chains drawing free nucleotides and the latter are getting specifically attached as if on a template. D. The doubling of the original chains completed. (Redrawn after Muntzing). The full structure of each nucleotide is shown in FIG. 838, only the basic symbols are given here.

Endomitosis

The chromosome number in a species of plants or animals is expected to be fixed. But, since the beginning of the current century spectacular discoveries have been made where it has been found that in the same plant there are patches of cells which show a different chromosome number. If a bud develops from such a plant, the new branch (which may become a separate plant) would show a different chromosome number. This is more common in plants which propagate vegetatively. This difference may be due to non-desjunction (some chromosome pairs do not actually divide), somatic reduction etc. but most interesting is the case of *endomitosis*. In endomitosis, the chromosomes of a somatic cell divide but the daughter chromosomes do not separate into daughter cells and the nuclear membrane does not break so that the ultimate cell contains double the chromsomes (in case of complete endomitosis) or a different number (in case of partial endomitosis). The somatic patch may be said to be formed of *endopolyploid* cells and when perfect may contain $2n$, $3n$, $4n$ etc., cells.

Operon Concept

There are thousands of genes in a cell every one of which secretes enzymes which give rise to different proteins. But, all these genes never act together and so no chaotic situation arises. The reason of this coordination had been considered a problem. Apparently, there must be some *switching* mechanism in the chromosomes which induces or stops the action of the genes. A hypothesis to explain this phenomenon has been presented by F. Jacob and J. Monod of the Paris Pasteur Institute in 1961 for which they were awarded a Nobel Prize (Biochemistry) in 1965. They have suggested the **operon** model which pictures the *operon* which includes the gene *codon* along with similar genes which regulate, operate, repress or induce the action of the *codon*. The best worked out of the *operons* is the **lac operon** (lactose forming) of *Escherechia coli* which is a system of the *codon* TAC along with all its complicated controlling genes.

THE GENES

The ultimate unit which controls heredity is now called the **gene.** The word **gene** had been used by earlier Botanists but a clear picture came out only after the work by Morgan for which he was awarded a Nobel Prize. Morgan stated his *Theory of the Gene* as follows:

"*The theory states that the characters of the individual are referable to paired elements (genes) in the germinal material that are held together in a definite number of linkage groups; it states that the members of each pair of genes separate when the germ cells mature in accordance with Mendel's first law, and in consequence each germ cell comes to contain one set only; it states that the members belonging to different linkage groups assort independently in accordance with Mendel's second law; it states that an orderly interchange—crossing-over—also takes place, at times, between the elements in corresponding linkage groups; and it states that the frequency of crossing-over furnishes evidence of the linear order of the elements in each linkage group and of the relative position of the elements with respect to each other.*" (the aspects of linkage and crossing-over are explained in Chapter LXV).

What is the gene? It can be located on the chromosome and assigned to some particular locus (Chapter LXV), it can mutate (Chapter LXVI) and it can be hit by radiation rays like any other particle. The conclusion is that it is a particle and these particles are arranged on the chromonema thread in a linear order. Naturally, the chromomeres on the chromonema threads were first supposed to be the genes or the seats of the genes. When homologous chromosomes pair, exactly identical chromomeres come to lie side by side. But further studies have shown that while some genes may be located in the chromomeres there are others in the interchromomeral regions. Moreover, recent workers doubt the suitability of considering the gene as a single particle. The structures of the chromosome as now understood clearly defines the gene as a *small segment of the polynucleotide chain* consisting of hundreds or thousands of nucleotides, thus the gene can no longer be considered as a single unit—as a matter of fact, every nucleotide bit of it may act as a *subunit*. Besides these subunits with the gene, there may be more elements within the genetic material affecting heredity which cannot be expressed by the assumption of a simple gene particle as shown by a linkage map. Nevertheless, although it has not yet been possible to identify these genes clearly under the optical microscope, their behaviour has made it possible to locate them and to speculate as to their dimensions and weights. Muller (1947) located four genes on one of the *Drosophila* salivary gland chromosomes (Fig. 841) and calculated their mean length to be 1250Å, i.e., 125$\mu\mu$. Pontecarvo (1952) considered 4500Å (450 $\mu\mu$) to be the gene size in the ascomycete *Aspergillus nidulens*. These are sizes of large virus particles (Fig. 797). The importance of nucleic acid in heredity has already been discussed (Chapter LVII). It has been found that in some bacteria the bare nucleic acid may cause hereditary changes. The role of the protein in the chromosome is at yet uncertain. The nucleic acid of chromosomes is DNA but it has already been stated that in certain viruses (Chapter LVII) *RNA* (*ribose-nucleic acid*—another type having the same characteristic as DNA) plays the same role and acts like genes. RNA, which is present in plant cytoplasm and nucleoli, is known to be connected with protein synthesis as explained later.

It has now been established that the nucleic acid DNA constitution the gene. In support of this the following facts may be pointed out

(i) The DNA content of a cell varies directly with the number of sets of chromosomes (haploid, diploid, triploid etc.), i.e., with the chromosome volume.

(ii) Different genetic problems, as will be seen later in this chapter, can be best explained on the basis of the DNA constitution of the gene.

(iii) In virus (e.g. bacteriophage) it is the DNA core which cause hereditary transmission and not the protein casing.

(iv) **Bacterial transduction:** In certain cases when the DNA material of a bacteriophage, on entering a bacterium does not destroy the latter but remains as a host inside the bacterium as a host side by side with the bacterial chromosome. The host DNA is then called a *prophage*. This prophage, being smaller than the bacterial chromosome, behave as a gene and divides simultaneously with the bacterial chromosome but does not lose its identity. In this way the prophage is transferred from bacterium to bacterium and the phenomenon is called *transduction*. This behaviour of the prophage proves that DNA is the gene substance. Transduction has been put to further use in genetic engineering.

(v) **Bacterial transformation:** *Pneumococci* producing disease have capsulated cells. Certain strains of *Pneumococci* do not possess capsules as they cannot synthesis the polysaccharide needed for it. These cannot cause any disease. If a noncapsulated strain be hybridised with a capsulated one, the offspring is capsulated—capsulation being dominant. It has been found that if an extract of dead capsulated pneumococci be added to the noncapsulated strain the latter become capsulated. This is called **transformation.** Other instances of transformation have also been discovered. It has been found that the dead extract is at least 98% DNA. Hence, DNA is responsible for this transformation.

Considerable work has been done lately on the structure of the gene. Mendel's concept was that the factors (now known as genes) were the ultimate units of heredity. But, lately it has been shown that the genes are divisible into subunits, i.e., units of mutation, units of recombination, etc. The *DNA* molecule of the chromosome contains a large number of genes which, again in turn, are quite large and may be divided into finer subunits as the nucleotides or neucleotide *triplets* as explained later. Subunits are referred in terms of the different functions of the genes. Detailed studies of them have been done in the bacterium *Escherichia coli*.

Mention must be made of a very interesting and abnormal type of chromosomes which throws light on the structure of chromosomes and on genes. These are the giant **salivary gland chromosomes** of *Diptera* flies. The larvae of these flies (*Drosophila, Chironoma, Bibio, Sciara, Rhyncosciara, Camptomya,* etc) show abnormally large cells with large nuclei (about 25μ in diamater) in their salivary glands. Similar cells are also found in their rectal epithelia and Malphighian tubes but the salivary glands are easier to dissect and study. These nuclei contain large chromosome, 50 to 200 times the normal size—some of them as much as half a millimetre long and visible under a simple dissecting microscope. Such chromosomes were noted by Balbiani in *Chironoma* as early as 1881 but their study gained importance only in the '30s of the present century. It is found that each such 'chromosome' is actually formed by a complete synapsis of the two homologous chromosomes along their entire lengths in the somatic cells. After this abnormal *somatic synapsis* cell division does not progress any further, the cells remain permanently in the prophasis stage but the chromosomes increase in length enormously. Moreover, it is seen that every chromosome shows distinct *bands* which are heavily stainable but less so in intercalary zones throughout the entire length (Fig. 841). The bands always occur at distinct

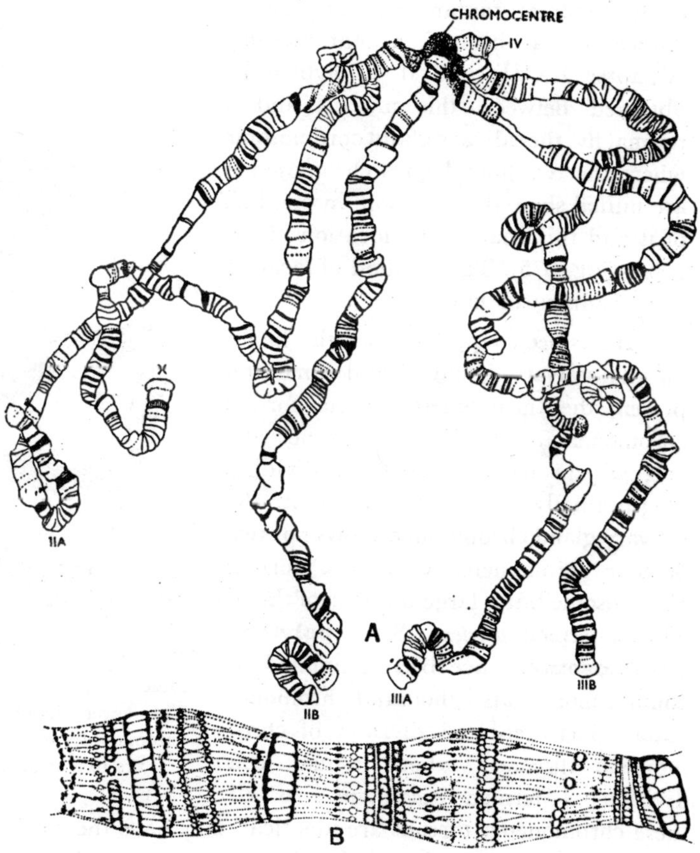

FIG. 841. A. Salivary gland chromosomes of a female *Drosophila melanogaster* (after Painter)—the small chromosome IV, the sex chromosome X and the two long chromosomes II and III (each with two arms A and B) are joined at the *chromocentre*. Each of these 'chromosomes' is actually a pair of homologues in synapsis. B. A bit of salivary gland chromosome much enlarged to show bands and chromomeres (after Painter and Griffin).

places in a particular chromosome and are of definite widths. If a bit of a chromosome somehow gets lost, the lost zone can be easily detected by examining the pattern of the zones. Thus each chromosome has a definite and constant pattern which may be called a *biological spectrum.* On the large chromosome of *Drosophila* there are more than 2000 distinct bands along its length. In *Drosophila*, the four chromosomes remain adhering to each other at the centromeres. All the heterochromatic material from near the centromeres of all the chromosomes (as also the whole Y-chromosome of the male fly, which is formed almost entirely of heterochromatin) fuse together to form a *chromocentre.* From this chromocentre the four chromsomes spread out like the arms of an octopus (Fig. 842)—the two long chromosomes II and III, having centromeres in the middle, show two arms each. In other *Diptera* flies the chromosomes remain separate from one another in the nucleus.

Naturally, these bands have been studied closely. Each band seems to be formed of one or more Lumps of chromatic granules extending across the cylindrical double-chromosomes. These are the regions where nucleic acid accumulates greatly and they may be the seats of the genes. If a fly becomes mutant by the loss or *deletion* of a bit of a chromosome (Chapter LXVI) and if a hybrid be obtained between this mutant and a normal fly, the salivary gland chromosome where the two homologous chromosomes are united show the 'deletion' in one half of it and this enables the location of the genes (Fig. 875). This method of locating genes is discussed later.

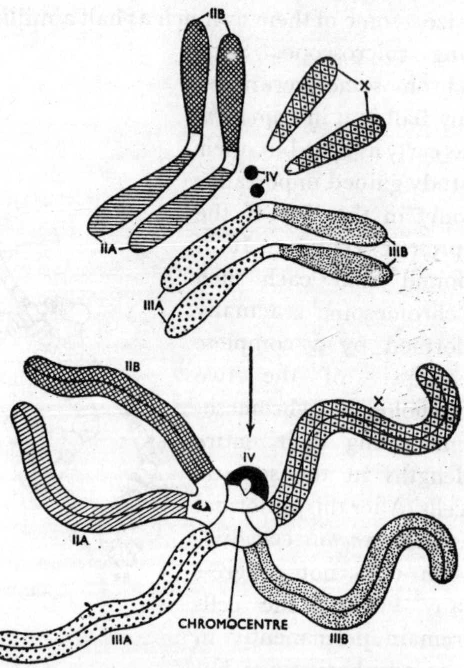

FIG. 842. The upper figure shows the chromosome complement in a female *Drosophila.* The lower figure shows the salivary gland chromosomes in which the above chromosomes are in a synaptic condition. (See text).

It has been supposed that the bands on the salivary gland chromosomes represent the chromomeres of the normal chromonema. Somehow or other the original chromonema goes on dividing longitudinally repeatedly so that the salivary gland chromosome shows not two or four chromonemata as in a normal chromosome but a large number of them. This is termed *polyteny*. Some **polytene chromosomes** have been estimated to contain more than a thousand chromonemata. Such repeated division of chromonema threads without actual cell division is termed *endomitosis*. As all these chromonema threads are identical, they have the chromomeres at identical spots which press on one another as the chromonemata are closely pressed together. So many chromomeres at one place appear as a band or disc on the chromosome which is completely uncoiled. Some authorities, however, think that such a structure is also possible by the swelling of the four normal chromonemata instead of undergoing actual polyteny.

In Genetics, genes are symbolised in the same way as unit characters of Mendel by letters, e.g., **T** and **d**. It is customary to use capital letters for dominant genes and small letters for recessive genes. Nowadays, the same letter or letters are used for pairs of allelomorphic genes, e.g., **A** and **a** form one pair and **Pr** with **pr** form another. Dominant genes found in the wild type are frequently donoted by the common symbol +.

PHYSIOLOGICAL GENETICS: THE GENETIC CODE

When the existence of the genes is agreed, the still greater problem is confronted—How are the different characters actually caused by the genes? Experiments and speculations as to this are of recent origin and come under the field of *Physiological Genetics*. It is possible that the genes, though insignificant particles, start certain physico-chemical reactions in the living body which cause the development of the characters. That some chemicals come out of the nucleus (i.e., the genes, according to the gene theory) is beautifully illustrated by the siphonaceous alga, *Acetabularia*. *Acetabularia* is a large but unicellular alga. The single cell is differentiated into a holdfast, a stipe and a mushroom-like cap (Fig.843). If the unicellular plant be cut into bits, the bits survive for a good period of time and, also, one bit from one plant may be grafted to another bit from another. It is found that if a bit contains the nucleus, which is located in the holdfast, it not only survives but also regenerates the other parts. A bit without a nucleus cannot do so. Two species of this alga, *A. mediterranea* and *A. wettsteinii* have two different types of cap. If a stipe of *A. mediterranea* be grafted on a holdfast of *A. wettsteinii* with a nucleus, it is found that a cap develops on the *mediterranea* stipe but this is not of the *mediterranea* type but of the *wettsteinii* type evidently because the chemicals controlling growth are coming from the *wettsteinii* nucleus.

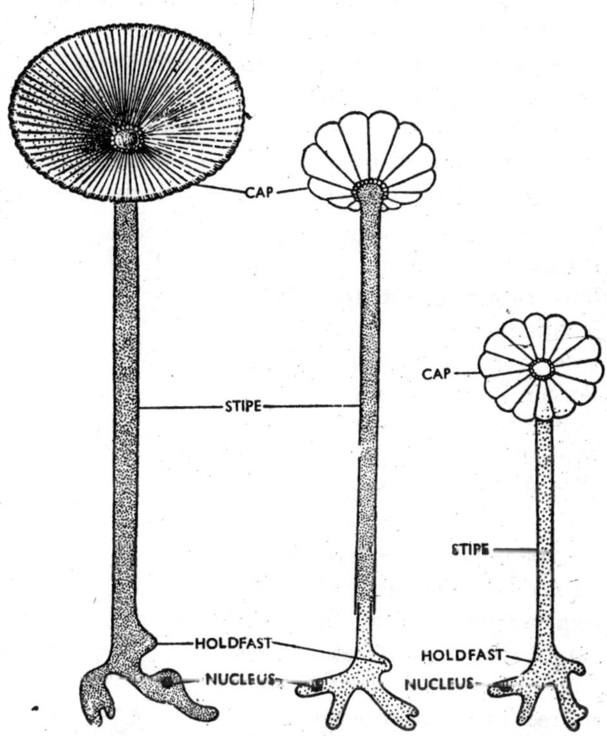

FIG. 843. A—*A.mediterranea*. B—A stipe of *A. mediterranea* grafted on a holdfast of *A. wettsteinii*. The regenerated cap is of the *wettsteinii* type. C—*A. wettsteinii*. (After Hämmerling).

Another very recent spectacular experiment also demonstrates the role of the nucleus in heredity. In recent days great successes have been achieved in the asexual reproduction achieved from the culture of single cells of higher plants and animals in artificial media resulting in the generation of complete plants out of single cells. In such an experiment Dr. J. B. Gurdon of Oxford University destroyed by radiation the nucleus of an unfertilised egg cell of an African clawed frog and transplanted within this cell a nucleus from a body cell of another frog. The new cell combination began dividing and regenerated a complete tadpole which matured into a frog which was nothing but an identical twin of the frog which donated the nucleus and not the frog which originally donated the egg!

Interesting types of gene action, that of controlling flower and seed colour, are described in the next chapter.

Light is thrown on gene action by the study of **phenocopies**. It has been found that there are various races among butterflies showing different types of designs on wings. Some of these natural races may be produced in the laboratory by treating a different type by different ranges of temperature during growth. *Drosophila melanogaster* normally shows a light brown body. In 1910 Morgan obtained a yellow-bodied mutant. In 1939 Rapopport found that when larvae of normal flies are fed on food containing silver salts, yellow-bodied *Drosophilas* may be raised. Such an artificially raised butterfly or *Drosophila* fly is termed a *phenocopy* and is not the same as the original race or mutant as these copied characters are not hereditary. Actually the artificial treatments had the same effects as the genes which cause these characters. While the genes pass on from generation to generation these artificial treatments do not.

Gene action is sometimes affected by some unknown cause so that, although present, it may show no action on the individual. When a gene shows its full action it is said to have *complete* **penetrance**. When it fails to act completely in some cases it shows *incomplete penetrance*. Persons inheriting genes of some serious disease may not be actually affected by the disease because the gene does not show penetrance. The expression of a gene in the phenotype, i.e., the actual appearance of the organism is its **expressivity**, Reduced penetrance will cause reduced expressivity. That genes show different degrees of penetrance and expressivity does not go against the gene theory. The gene action is caused by complicated, chemical reactions resulting in the control of enzyme synthesis. There may be some other factor or even some other gene which hinders some phase of this reaction causing an incomplete penetrance.

FIG. 844. HarGobind Khorana—*of whom we are proud.*

In recent years there has been phenomenal progress in the field of biochemical genetics involving the synthesis of *DNA* and *RNA*, which has thrown much light on the action of the genes. Three awards of Nobel Prizes have already made on this subject. First, in 1959, S. Ochoa (New York) and A. Kornberg (Washington) were awarded a Nobel Prize for the synthesis of nucleic acid. Then, in 1962, Crick, Williams and Watson got the award as already mentioned. Finally, in 1968, the award has gone to M. W. Nirenberg, R. Holley and HarGobind Khorana (Fig. 844) for the elucidation of the *genetic code*. Khorana is now well known in India for he is an Indian who has settled in the U.S.A. The conclusions that arise from their extensive work is as follows:

The genes control the synthesis of enzymes and other proteins which again control the development of characters. It has been seen that in a polynucleotide chain the bases are arranged in different orders. Three such consecutive bases (say AAC—i.e., Adenine—Adenine—Cytosine) forms a *nucleotide triplet* which is the unit for the synthesis of a particular amino acid. This unit is named **codon** as the unit has got a code on it which causes the particular amino acid to be synthesised. Thus, a number of codons arranged in a particular order will cause the synthesis of the particular amino acids in a particular order. These amino acids will get linked in that particular order forming a polypeptide chain which forms a molecule of the particular protein. But, protein is

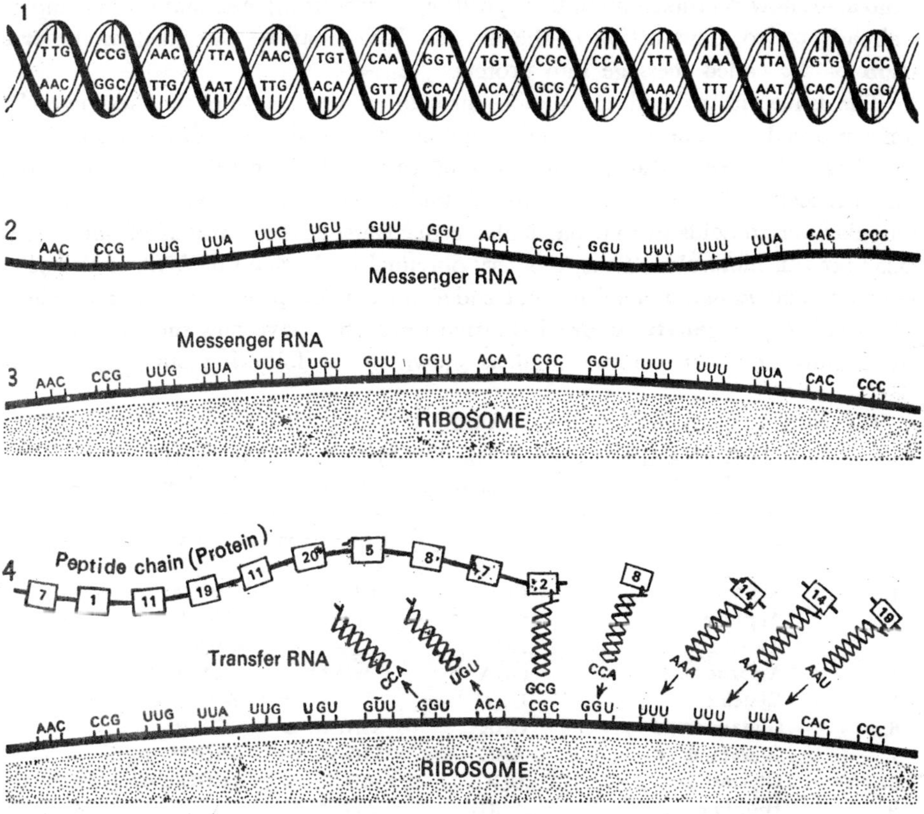

FIG. 845. Synthesis of proteins according to genetic code. 1. The DNA double helix. 2. Messenger RNA formed by one of the polynucleotide chains. (Thymine of DNA changes to Uracil in RNA). 3. The messenger RNA fixed on a ribosome. 4. The messenger RNA giving rise to transfer RNAs which are fixing the amino acids (numbered as shown in Table VII) in position. A polypeptide chain of a protein molecule (exactly what was *coded*) is being developed. (After Nirenberg and Müntzing).

synthesised in particular places called *ribosomes* which are ultramicroscopic bodies in the cytoplasm. So, what actually happens is that of the two polynucleotide *DNA* chains in a double helix gene (Fig. 845-1) *one* procreates an *RNA* replica (Fig. 845-2) which is called the **Messenger RNA**. This may be termed as the **transcription** of ıe **code** on the RNA. This RNA contains the nucleotides in the same order only the sugar is changed from deoxyribose to ribose by oxidation, the bases are the same excepting that Thymine (T) is changed to Uracil (U). The messenger RNA now moves out of the nucleus into the cytoplasm and gets attached to a ribosome (Fig. 845-3) which is to manufacture the protein as *coded* on this messenger RNA. Even this is not done directly. Each codon on the Messenger RNA produces soluble **Transfer RNA** (Fig. 845-4). Usually, in RNA, there is a single helix. But, in the transfer RNA the single thread is folded over forming a double helix of which one end is closed and the other end is open. The transfer RNA holds the codon which is the opposite (C for G, U in place of T in DNA for A or *vice versa* as explained earlier) while in the open end it draws out the specific amino acid fixed for that codon. This amino acid is fixed in that place. The codons on the sides also draw out the respective amino acids. All these amino acids now get linked up in their positions synthesising the exact protein molecule as already coded. This action of the transfer RNA may be described as the **translation** of the coded message into protein.

The transcription and translation throughout the length of the messenger RNA do not proceed all at once but different codons are translated at different speeds. This control is called **modulation** and may be responsible for differentiation of organs. It is thus seen that the *codons* or *nucleotide triplets* may cause the synthesis only of the particular amino acids in that particular order causing the synthesis of only the particular protein molecules. Thus, the very polynucleotide chain of DNA or the gene or the consequent *messenger RNA* has got coded on it what protein it can manufacture. This is called the **genetic code**. Extensive researches have now shown which codon may manufacture which amino acid as shown in the following table for 20 amino acids.

TABLE VII

THE RNA CODES FOR THE DIFFERENT AMINO ACIDS
(After Müntzing summarising the work of Nirenberg and others)

No.	Amino acid	RNA code words or codons			
1.	Alanine	CCG	UCG	ACG*	
2.	Arginine	CGC	AGA	UGC*	CGA*
3.	Asparagine	ACA	AUA	ACU*	
4.	Aspartic acid	GUA	GCA*	GAA*	
5.	Cysteine	UUG			
6.	Glutamic acid	GAA	GAU*	GAC*	
7.	Glutamine	AAC	AGA	AGU*	
8.	Glycine	UGG	AGG	CGG	
9.	Histidine	ACC	ACU*		
10.	Isoleucine	UAU	UAA		
11.	Leucine	UUG	UUC	UCC	UUA
12.	Lysine	AAA	AAU		
13.	Methionine	UGA			
14.	Phenylalanine	UUU	CUU		
15.	Proline	CCC	CCU	CCA	CCG0
16.	Serine	UCU	UCC	UCG*	ACG
17.	Threonine	CAC	CAA		
18.	Tryptophan	CGU			
19.	Tyrosine	AUU			
20.	Valine	UGU	UGA*		

* Not definite.

Khorana and his students now have even succeeded in creating artificially a yeast gene of 77 nucleotides. This is, of course, the first step, as a complete chromosome contains millions of nucleotides—each single human cell has been calculated to contain chains involving some 6000 million nucleotides. Later (1976), Khorana with his associates has been able to synthesise the tyrosol suppressor **t**RNA gene of **Escherichia coli** complete with it's control elements. This gene was able to function inside the bacterial cell when inserted. This is a spectacular achievement and leads us on to **Genetic Engineering.**

GENETIC ENGINEERING

The experiments of Khorana and others gave rise to wild hopes and 'dreams'. If it be possible to synthesise genes and to introduce them into bacteria, is it not possible to **create** new genes and introduce them into the cells of other animals giving rise to unknown characters? It may be possible to create new antibiotics and medicines, to cure incurable diseases or even to create a 'god' ! But, with unknown factors, it may also work the other way. Instead of beneficial, dangerous and uncontrollable organisms (say, those causing epidemics or even a Frankenstern) may arise. The idea may be the source of hair-raising science fictions but, in actuality, may not be palatable at all. Since the '70s of this century much experimentation has been done and some very hopeful results have been obtained. A new science called **Genetic Engineering** has developed. The aim of this science is to introduce foreign DNA into living cells by special techniques and thereby changing the gene complex and the behaviour of these organisms. The scientists are aware of the dangers, they have been warned and all precautions are taken so that no evil may result. In the U.S.A., where the most work of this type is being done, the National Institute of Health has formulated certain guide lines for safety and these are followed in all laboratories.

It is easy to understand that synthesising all genes is not an easy job. Secondly, it is much safer to work with known genes already present in organisms without trying to 'create' new genes. There is enough scope to achieve remarkable success this way. The method of genetic engineering that has been most widely used is what is known as the **Recombinant DNA Method.**

RECOMBINANT DNA METHOD

In understanding the Recombinant DNA Method one says of the *tools* required for this type of engineering. But actually no mechanical tool is meant as this is not a case of mechanical repair of a machine or that of a surgical operation with the doctor's kit. Actually, these are cytochemical techniques which are performed *in vitro*. Certainly, the methods are not easy but the general principles may be explained as follows.

(1) The gene that has to be transferred is to be isolated. This gene is called the **passenger DNA** as this is a gene foreign to the receiving organism and has to be transferred passively to a strand of DNA of the recipient.

(2) The passenger DNA is now *cloned* to the **vector** (also called *vehicle*) **DNA** which has been very suitably found in the *plasmids* of bacteria or even in bacteriophages (Fig. 846). In bacteria, extrachromosomal DNA occurs in the cytoplasm as small circular rings. They can exist independently of the main genome set of the bacteria and multiply independently. These are called the

plasmids. Being very small, they can be isolated and transferred out of the bacterium.

After, reintroducing the 'recombined' or the 'hybrid' plasmid into the bacterium it becomes a part of it and goes on dividing and multiplying the new characters normally.

The recombination needs very delicate manipulation which cannot be done by any apparatus but is done by biochemical techniques. Special *enzymes*, which have also been isolated from bacteria—*exonucleases*, *endonucleases* and, above all, the *restriction endonucleases*. All these enzymes are capable of cleaving by digestion a DNA (a single or paired double chain) into bits.

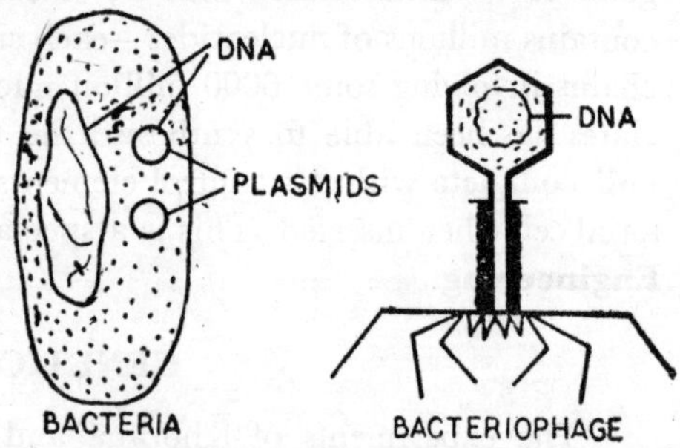

Fig. 846. Vectors suitable for cloning.

Exonucleases can digest or break the base of any single chain of DNA. *Endonucleases* cleave the DNA at any point of a DNA chain except the ends. But, the *restriction endonuclease* cause breaks, in particular points of nucleotide sequence of both the strands of a pair. Since this 'particular point' is not in the same place in the two chains, the cut becomes uneven so that the broken ends form something like a mortise cut and become 'sticky' (Fig. 847). These sticky ends make it easy for the subsequent splicing into the *vector* plastid chain.

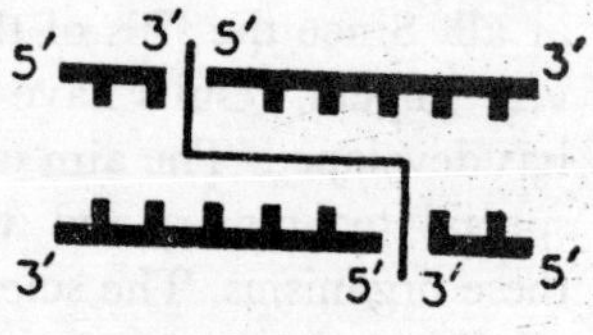

Fig. 847. Uneven break of DNA chain by restriction endonuclease.

The *passenger DNA* or the gene to be implanted (or *cloned*) may be (1) a synthetic gene as first synthesised by Khorana or (ii) a complementary DNA (cDNA) which is synthesised on an RNA template with the help of *reverse transcriptase* on the DNA-RNA complex and is then subjected to alkaline digestion to isolate single-stranded DNA; the complementary DNA is actually synthesised on this single-stranded DNA by *DNA polymerases*. The cDNA so produced is specific for a gene and can be linked to a vector DNA for transfer. If none of these techniques work then a gene may be isolated by (iii) a 'shot gun' experiment. In this method, the whole DNA of the organism from which a desirable gene is to be isolated, is broken up into fragments by restriction endonucleases. Each fragment is now recombined with suitable vector DNA and each recombination tested for the required gene after transfer to a host.

Experiments following these lines have not only given deeper access to the organisation of chromosomes and the working of the genes but have also given us some very useful practical products and there are possibilities of more benefits. Some of the fields where they have been applied are:

(i) Synthesis of polypeptides and proteins of medical importance, specially production of insulin and other medically important hormones, proteins and peptides.

(ii) Production of vaccines.

(iii) Production of *interferons* and specific antibodies.

(iv) Production of new strains of organisms with beneficial properties.

(v) Determination of defects and replacement therapy for genetic diseases.

(iv) Digestion of wastes, specially removal of hydrocarbon waste and its conversion into protein.

Some of the outstanding achievements are being mentioned below:

(1) *Insulin production* (Fig. 848)

The protein, *insulin* is secreted from the pancreas for normal sugar metabolism. The absence of this secretion causes the very common disease *diabetes mellitus*. It may be treated by injection of insulin which is normally obtained only from animal sources. But this source is extremely limited and insufficient Hence, this attempt at genetic engineering. If the gene which causes insulin secretion could be introduced within the patient, it would have been the ideal treatment. The second best way has been found by transferring an insulin gene into a bacterium which would grow very fast and would be a cheap source of insulin.

The bacterium *Escherichia coli* has proved to be a gold mine for gene transfer. The strain *E. coli* X 1776 was isolated by Curtiss in 1976 (U.S. Bicentennial year—named 1776 for this). This strain was found to be perfectly harmless and was certified by the N.I.H. for use in this type of experiments. The actual transfer was made by Ullrich and associates in 1977 in the University of California, San Fransisco campus (Biochemistry & Biophysics Division). The steps followed are as follows (See Fig. 848):

(i) Messenger RNA (mRNA) was isolated from rat pancreas and complementary DNA (cDNA) was synthesised with the help of the enzyme *reverse transcriptase*. A *restriction endonuclease* was now used which cut the DNA strands between the 3′ and 5′ nucleotides (also see Fig. 847) leaving sticky ends. *E. coli* X 1776 bacteria was treated with a detergent in a test tube which dissolved the outer membrane of the cells exposing the DNA strands. The plasmid rings are now separated by a centrifuge and then treated with the same *restriction endonuclease* so that the ring is broken at the same points and a part cut off. (The ring may now become linear.)

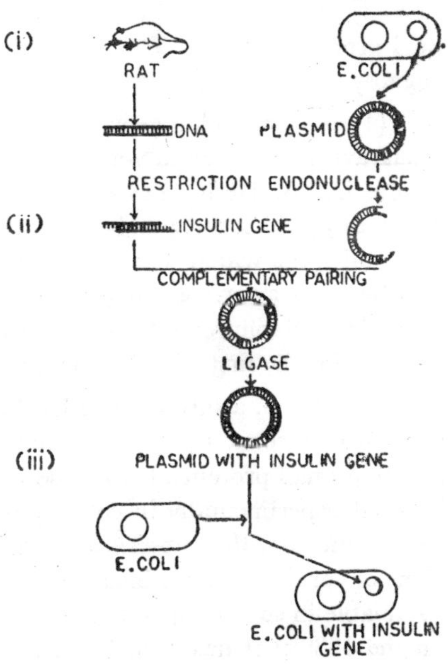

Fig. 848. Showing diagrammatically the utilisation of Recombinant DNA technique in closing Insulin gene (compare with text for explanations).

(ii) The open rings of the plastid DNA (which may be linear) are new placed together with the broken mice insulin DNA in the presence of the enzyme *T4 DNA ligase*. This results in the recombination of the broken plasmid DNA ring when the rat insulin DNA gets attached like a chimaera.

(iii) The recombinant DNA is placed in a solution of cold calcium chloride containing normal *E. coli* X 1776 cells. This solution is suddenly heated so that the bacterial outer membrane becomes permeable allowing entrance to the hybrid recombined plastids.

The recombined plasmid now divides along with the bacteria duplicating the genes as in normal cases.

The experiments have been simply described but they are by no means that simple. There are possibilities of footfall at every stage. Actually, Ullrich *et al* did not show that the new cloned bacteria do secrete insulin. This was shown next year (1978) by Gilbert (who performed the cloning by a different technique) who found that minute quantities of proinsulin (precursor of biologically active insulin) was secreted by the recombined plasmids. But, still, this technique needs lot of improvement to be of practical use.

(2) *Transfer of nitrogen fixation*

Certain bacteria fix atmospheric nitrogen by their *nif* genes. It has been found possible to transfer this *nif* from nitrogen-fixer to non-fixer bacteria. The aim now is to transfer the *nif* gene to the crop plants themselves—first to the symbiotic legume plants and then to crop plants like rice or wheat when all such crop plants would be able to utilise atmospheric nitrogen without the help of bacteria. Such a goal does not seem beyond reach today.

(3) It has been found possible to transfer the human hormone hGH (an amino acid protein) to *E.coli*. This gives rise to great hopes in the field medicine and pharmacognosy.

(4) A spectacular genetic engineering success has been achieved by Dr. A.M. Chakravarty of Calcutta who now works in the General Electric Research Laboratory in the U.S.A. He has developed by genetic engineering a new type of the bacterium *Pseudomonas* which can gulp hydrocarbons. This is to be used for clearing accidental petroleum oil spills which sometimes spread over hundreds of kilometres of ocean surface and becomes a serious danger to the littoral countries. So important is this discovery that the U. S. Supreme Court has ruled that a patent should be granted for this bacterium—for the first time for the 'manufacture' of any living organism.

(5) Of late, a substance called **interferon** (IFN) has been discovered as a soluble substance secreted by cells infected by virus. This substance is now known to be a group of small proteins. It can also be produced by sensitised lymphocytes and may be induced experimentally in cells by treating them with double stranded RNA. It has been found that this substance has immense possibility in the immunisation and treatment of various hitherto incurable virus diseases and even cancer. But the substance is obtainable in so little quantity that it is almost impossible to use it commonly. Now (in the 1980's) it has been found possible to produce it more easily by genetic engineering. The interferon mRNA has been isolated from leucocytes, converted into cDNA by *reverse transcriptase*, cloned to *E. coli* and this has been shown to be active *in vitro* as well as *in vivo*. So, it may be possible to produce *interferon* in greater amount in future. By 1982 interferon is being used on patients in cancer hospitals. But, some fatal results have already stopped its use in French Government programmes.

(6) Recently (1982), closed DNA molecules contaning genes causing secretion of enzymes have been injected into the eggs of mice and such genes have after been found to be transmitted to the offsprings. But, is most cases it is doubtful if the transferred genes have at all been able to generate the same enzymes. These are problems to be solved in future.

CHAPTER LXIV

POST-MENDELIAN DEVELOPMENTS.

I. GENE ACTION AND INTERACTION

The impetus of the rediscovery of Mendel's work caused hundreds and thousands of genetical experiments to be performed and the word 'Mendelism' or the term 'Mendelian genetics' today does not mean merely what Mendel discovered but includes many post-Mendelian modifications based on the gene concept. Mendel was extremely lucky to have found rather simple inheritance in all the cases that he studied. But, today, sufficient post-Mendelian evidences have accumulated which show that inheritance is not always as simple as Mendel found it to be. Very often, as in the many cases cited below, inheritance of characters cannot be explained on the basis of single independent units showing dominance in allelomorphic pairs, So, at the first sight, these inheritances may appear non-Mendelian.

FACTOR HYPOTHESIS AND THE GENE CONCEPT

To explain this apparent anomaly, Bateson presented his **factor hypothesis** as a supplement to original Mendelism. This stated that characters are not always caused by 'units' but by *factors* of a fractional nature which may combine in different ways to cause the different characters. At that time the concept of particulate genes and the true structure of chromosomes were not very clear. Clarification of these facts has greatly changed the original concept of factors. It is now becoming more and more clear that *genes* or *factors* behave as unit particles which remain quite separate and yet they may have combined effect on a character. Thus, although these are not cases of 'unit characters' yet they are being controlled by units of genes. It has already been stated that genes are complex molecules which are able to cause a number of physico-chemical activities which develop different characters. The behaviour of the genes is being studied very closely and this new science of *physiological genetics* has advanced rapidly. With the modern concept of gene action, it is now quite easy to visualise that different gene actions may interact with one another, a number of genes may combine to form a character, a single gene may affect different characters, and so on. Thus, Mendel's idea of 'unit characters' has to be modified. Moreover, the particulate concept of the 'gene' has induced many geneticists to use the word 'factor' as a synonym of it which was not its original meaning.

Presence or Absence Hypothesis—Explanation of Dominance

In this connection, mention may be made of the Mendelian concept of dominance. After Mendel had clarified his idea about dominance some attempts were made to explain why this takes place. Bateson and Punnett tried to explain this by their **Presence or absence hypothesis** which was, later on, further elaborated by others. This hypothesis supposes that *alleles* do not mean that two determiners are present. There is only one determiner whose *presence* is dominance and whose *absence* causes the recessive characters. Thus, the recessive determiner does not exist. That means that in the monohybrid ratio the true-breeding dominant tall has two doses of the determiner (say, **TT**) and is called *duplex*, the segregating dominant tall has one dose of the determiner (**Tx** instead of **Td**, x signifying absence), and is called *simplex* while the true breeding recessive dwarf has no determiner at all (xx or *nil* instead of **dd**) and should be called *nulliplex*.

The above explanation is no longer popular among genticists as it does not explain many other experimental results (e.g., that of *multiple alleles*, discussed later in connection with *mutations*) and is, therefore, untenable. Modern explanation of dominance is different and rather more complicated but it is always based on the assumption that there are actually two different genes representing an allelomorphic pair. Some cases of dominance may be explained rather simply by supposing that a character is caused by certain reactions and that while the dominant enbales these reactions to be carried on to the fullest extent, the recessive gene *limits* or checks the reactions so that the character is not fully developed or is developed differently, or, in some cases, the recessive gene may have no effect at all.

TYPES OF INHERITANCE

The important types of inheritance that have now become clear may be discussed as follows:

1. Simple inheritance

These are the cases where inheritance is found to be as simple as observed by Mendel. Hundreds of examples of this type have been found after Mendel's own classical experiments. *Drosophila* or the fruitfly is an insect which has proved to be a gold mine showing numerous instances of this and other types of inheritance. Among human beings, brown eye and blue eye form an allelomorphic pair of characters where brown eye is dominant.

2. Incomplete dominance or blending inheritance (1 : 2 : 1 ratio)

The classic example of this is that of the *Mirabilis jalapa* flower (four-o' clock plant) found by Correns (1903). When the red-flowered variety (*rosea*) is crossed with the white-flowered variety (*alba*), the

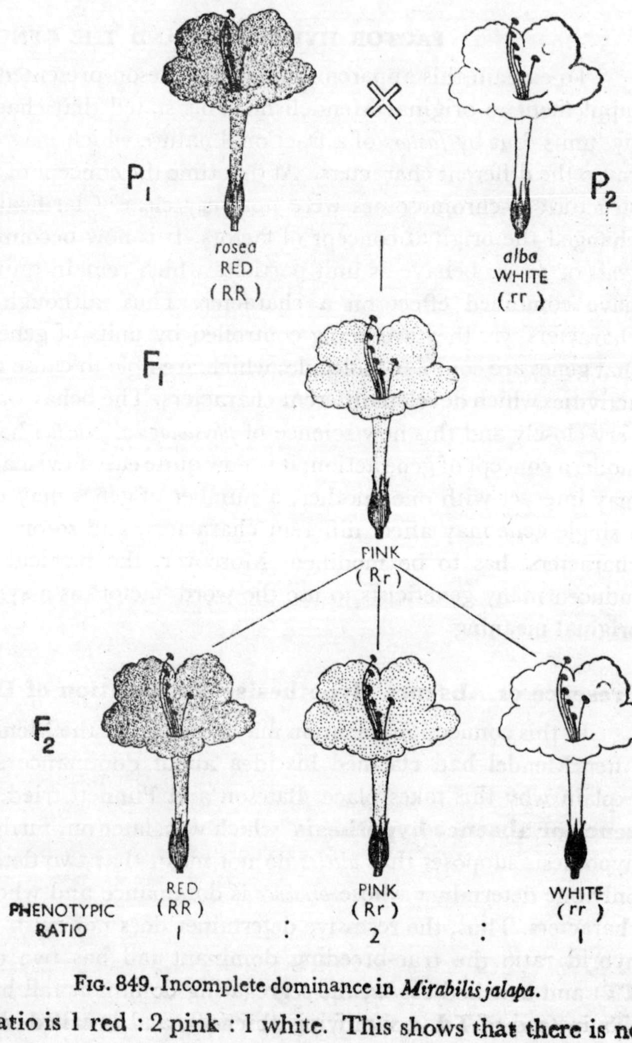

FIG. 849. Incomplete dominance in *Mirabilis jalapa*.

F_2 hybrid is pink and the F_2 ratio is 1 red : 2 pink : 1 white. This shows that there is no complete dominance in the heterozygous plant which bears pink flowers. If the pair of alleles controlling flower colour be named **Rr**, then red flowers develop on **RR** plants, white on **rr** plants, while the heterozygous **Rr** plants develop pink flowers and not red ones which would have been the case if there was complete dominance of **R** over **r** (Fig. 849). This is also called **blending inheritance** as if the two colours have blended in the heterozygote. But this does not necessarily mean actual *blending* of two gene particles and may well be caused by the simultaneous action of two genes, one of which may 'dilute' the other.

There are many other examples of this type. In some rice, when a short-glumed variety is crossed with a long-glumed one, the F_1 shows glumes of intermediate length. This is a case of quantitative inheritance where we may suppose that the dominant gene increases the length while the recessive retards it. Then the heterozygous F_1 should be intermediate between the double dominant and the double recessive.

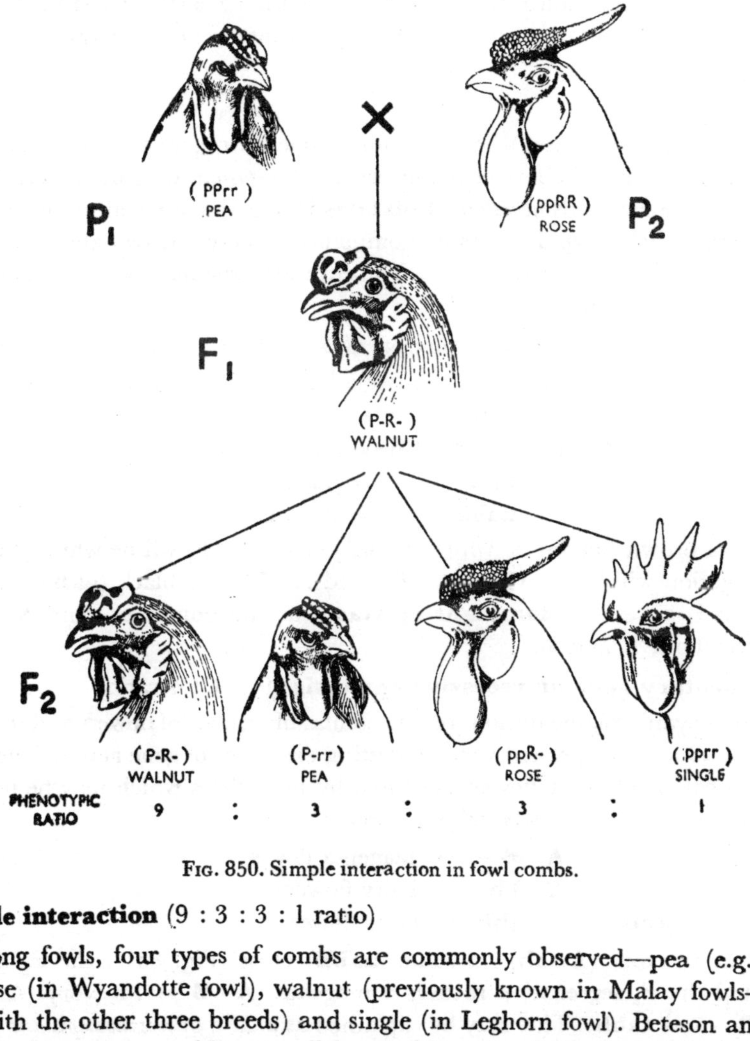

FIG. 850. Simple interaction in fowl combs.

3. **Simple interaction** (9 : 3 : 3 : 1 ratio)

Among fowls, four types of combs are commonly observed—pea (e.g., Brahma fowl), rose (in Wyandotte fowl), walnut (previously known in Malay fowls—not connected with the other three breeds) and single (in Leghorn fowl). Beteson and Punnett found that these are caused by two allelomorphic pairs of genes which may be symbolized as **Pp** and **Rr** (capital letters signify dominant genes while small letters signify their recessive alleles). When the dominant gene **P** is present alone (i.e., without the simultaneous presence of the other dominant gene **R**) the comb is of the pea type. When **R** is present alone the comb is rose. But when both **P** and **R** are present, there is an interaction and the comb becomes walnut. When no dominant gene is present, the comb is of the double recessive single type. Thus if one crosses a pure-breeding pea-combed fowl with a pure-breeding rose-combed one (Fig. 850) the F_1 will show a comb of walnut type and, when selfed (sib-crossing), the F_2 will segregate in the ratio 9 walnut : 3 pea : 3 rose : 1 single.

The above is a simple case of interaction where the normal dihybrid ratio (though with unexpected phenotypes) 9 : 3 : 3 : 1 is obtained.

The modes of interaction discussed below are more complex and involve complex ratios. The F_2 segregation in all such cases of interaction, however, will be understood if one remembers that in a dihybrid ratio where two allelomorphic pairs, **Aa** and **Bb,** are involved, the segregation ratio is 9 **A-B-** : 3 **A-bb** : 3 **aaB** : 1 **aabb,** The phenotypic expression will vary according to the type of interaction and the ratio will vary accordingly.

4. **Epistasis**

There are numerous cases where a factor of one pair masks the expression of the factors of another pair. The former is then said to be *epistatic* over the latter. The latter may, then, be designated as *hypostatic*. **Epistasis** is thus the same effect between factors of two different allelomorphic pairs as dominance is between two alleles of the same pair. Epistasis may be of different types. The epistatic gene may be a dominant gene or a recessive one or both.

Epistasis of Dominant Gene (12 : 3 : 1 ratio)

In the coat colour of dogs, there are two allelomorphic pairs which may be named **Aa** and **Bb**. They interact as follows:

A—B— and **A—bb** = white
aaB— = black
aabb = brown

Thus in a brown (**aabb**) × white (**AABB**) cross, the F_1 wil be white (**AaBb**) and the F_2 segregation will be 12 white (**A—B—**and **A—bb**) : 3 black (**aaB—**) : 1 brown (**aabb**). This is a case of **dominant epistasis** as the dominant gene **A** (white) is epistatic over both **B** and **b**.

5. **Supplementary gene or recessive epistasis** (9 : 3 : 4 ratio)

The anthocyanin pigmentation (red, magenta, blue, etc.) of flowers and the melanin pigmentation of animals present very interesting examples of gene action. The red-type anthocyanin colour of many flowers is caused by two alleles which may be termed **Aa** and **Bb**. In the snapdragon (*Anrirrhinum*) flower

A—B— = magenta flower
A—bb = ivory flower
aaB— and **aabb** =white flower

Thus, in an ivory (**AAbb**) × white (**aaBB**) cross (Fig. 851), the F_1 (**AaBb**) is magenta and the F_2 segregation ratio is 9 magenta (**A—B—**) : 3 ivory (**A—bb**) : 4 white (**aaB—** and **aabb**). This may be explained as a case of *recessive epistasis* by saying that the recessive factor **aa** suppresses the expression of all colour. But a much more plausible explanation is the *supplementary factor hypothesis*. It has been experimentally shown that the ivory flowers contain anthoxanthin and magenta flowers anthocyanin pigments. It is known that anthoxanthin and anthocyanin are synthesised out of the same raw materials, the former in the first stage and the latter when some additional

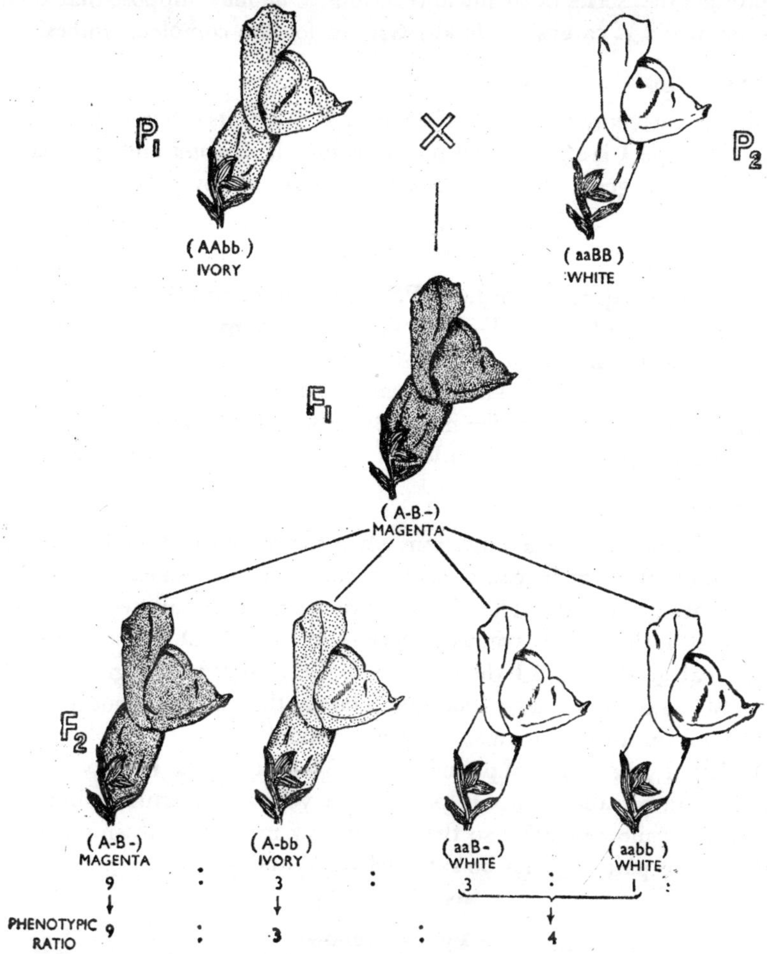

FIG. 851. Inheritance of flower colour in *Antirrhinum majus*. (Supplementary genes).

chemical reactions take place. It is, therefore, reasonable to suppose that the pigment-producing chemicals in the flower are brought to the anthoxanthin stage by the gene **A**. When the gene **A** is supplemented by the gene **B**, this reaction proceeds to a higher stage producing the anthocyanin pigment. Thus, the magenta anthocyanin pigment may only be produced in two stages. The first stage is completed by **A** and the end product is ivory (anthoxanthin). The second stage (anthocyanin) is brought about by **B**. Thus **B** is a *supplementary gene* having no effect by itself (i.e., without **A**). Flowers with neither **A** nor **B** (**aabb**) are white.

An exactly similar situation is met with in rat colour. Here

R—C— = black
rrC— = cream
R—cc and **rrcc** = albino white

This also may be explained in the same way by considering **R** as a supplementary gene. The development of pigments in animal skin is now known to follow the same pattern as the development of anthocyanin pigments. The black pigment *melanin* is the final oxidation product of a series of chemical reactions. One may suppose that **C** causes the basic reactions while **R** causes the final reactions for the complete synthesis of *melanin*.

6. **Inhibitory genes** (13 : 3 ratio)

In some rice variety the presence of the gene **P** causes its leaves to be coloured deep purple. But if a gene **I** is present the purple colour is *inhibited* and the leaf becomes normal green. The **I** gene may be considered as epistatic over **P**.

I—P—, I—pp, iipp = green
iiP— = purple

Thus in a green (**IIpp**) × purple (**iiPP**) cross the F_1 will be green (**IiPp**) and the F_2 segregation will be 13 green (**I—P—, I—pp** and **iipp**) : 3 purple (**iiP—**) as the presence of **P** in the absence of **I** only may cause the purple colour.

Similar inhibitory genes are known in many other plants and animals. For example, the purple grain colour of maize may be inhibited by an **I** gene and the white leghorn fowl owes its colour to the inhibition of a colour gene by an **I** gene.

7. **Lethal genes**

These usually occur as gene mutations when the presence of the mutated gene **l** or **L** (more commonly the former) causes death of the organism. An example is the '*xantha*' (yellowness) character of rice. The character develops as a mutated recessive gene **l**. The mutant plant is then heterozygous with the constitution **Ll**. The seedlings segregate as 1 **LL** (normal green) : 2 **Ll** (heterozygous, i.e., segregating green): 1 **ll** (lethal yellow or *xantha*), i.e., 3 normal : 1 xantha. The xantha plants cannot develop chlorophyll and die within a week or two after germination.

Another lethal yellow type of pigmentation is found in rats. Here, yellow and black pigmentation form an allelomorphic pair where yellow is dominant but the double dominant dose of yellow is lethal so that the yellow rats exist only in the heterozygous hybrid state. If the pair of genes be signified by **Y** and **y**.

YY = lethal
Yy = yellow
yy = black

Thus if two yellow rats be crossed the offspring will show the F_2 ratio of 1 **YY** (dies): 2 **Yy** (yellow) : 1 **yy** (black), i.e., only yellow and black rats in the ratio 2 : 1.

8. **Complementary genes** (9 : 7 and 9 : 6 : 1 ratios)

In some rice varieties, the following interaction of two pairs of genes has been noted:

A—B— = red grain
aaB—, A—bb and **aabb** = grey grain

Thus in a grey (**AAbb**) × grey (**aaBB**) cross, the F_1 is red (**A—B—**) and the F_2 segregation is 9 red (**A—B—**) : 7 grey (**aaB—, A—bb** and **aabb**).

The phenomenon may be explained by the *hypothesis of complementary factors*. This supposes that **A** and **B** are two *complementary* genes which have no effects by themselves but when they combine together, the red colour is produced. As in the case of supplementary factors, one may suppose in this case that the production of the red pigment requires two stages brought about independently by the factors **A** and **B**, But, the effect of any one factor is not outwardly manifested.

Similar instances are known in many other cases. Bateson obtained hybrid pea plants with red flowers on crossing two white-flowered plants because of similar gene action:

C—P— = purple flower
ccP—, C—pp and **ccpp** =white flower

Thus, Bateson obtained red F_2 plants (**CcPp**) from a white (**CCpp**) × white (**ccPP**) cross.

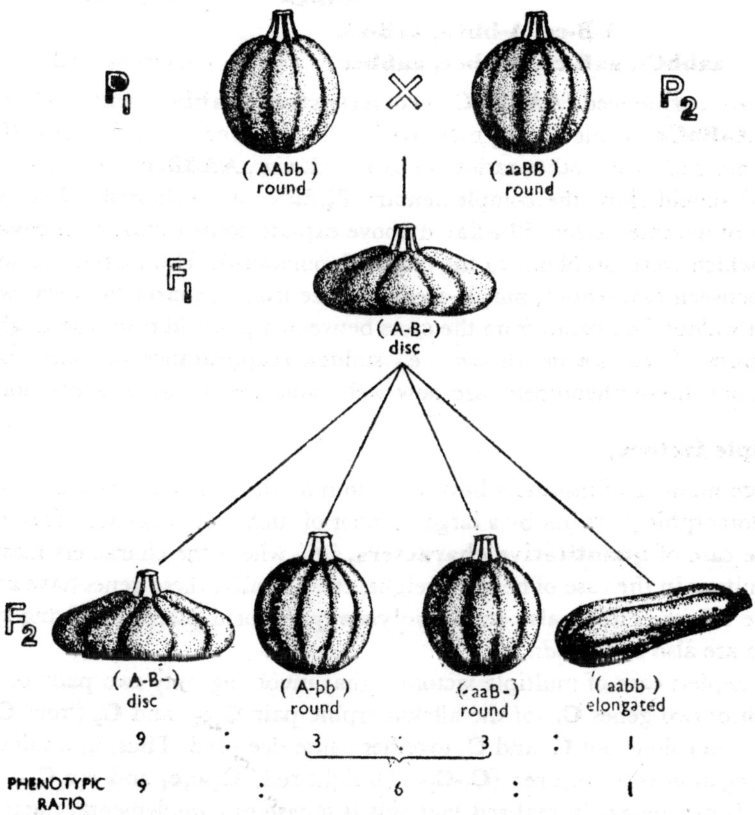

FIG. 852. Inheritance of fruit shape in *Cucurbita pepo* showing complementary genes.

There are other types of complementary gene which have some expression even singly. For example, the shape of the common gourd (*Cucurbita pepo*—also called squash in some countries) is controlled by two pairs of genes:

A—bb and **aaB—** = rounded or spherical
A—B— = discoid
aabb = elongated

Fig. 852 shows the segregation resulting after crossing two round type plants of composition **AAbb** and **aaBB**. The F_2 segregation is 9 discoid : 6 rounded : 1 elongated and not 9 : 7 as we are concerned here with three phenotypes and not two.

9. **Supplementary and complementary genes** (27 : 37 ratio)

We may now consider an interesting case of gene interaction involving three pairs of genes instead of two. Emerson found that seed-coat colour in maize is controlled by three pairs of genes. **C** and **R** are two complementary genes which have no action independently but together cause a brownish green colouration of the stem while the grain remains colourless. If the gene **A** is added to **C** and **R** some plant parts as well as the grain-coat become purple by the development of anthocyanin. We may suppose that the anthocyanin pigment is developed in two stages : **C** and **R** are complementary in forming the brown pigment and when **A** is supplemented the full anthocyanin pigment is developed. **A** is, therefore, a supplementary gene.

A-B-C- = purple seed

A-B-cc, A-bbC-, aaB-C-,
aabbC-, aaB-cc, A-bbcc, aabbcc = colourless seed

Thus, a cross, purple seed (**AABBCC**) × colourless seed (**aabbcc**), will produce a purple F_1 plant (**AaBbCc**) while the F_2 ratio will be 27 coloured : 37 colourless. If the cross involves other genotypes, other ratios are expected, viz., **AABBcc** (colourless) × **aabbcc** (colourless) should show the complementary F_2 ratio of 9 coloured : 7 colourless.

Many of the interactions illustrated above explain some instances of **reversion** or **atavism** which were problems to the classical geneticists. Thus, a red pea flower from the cross between two whites, purple-seeded maize from the cross between two colourless ones, a walnut fowl comb from the cross between a pea and rose type might well be taken as cases of *reversion* or *atavism*, i.e., sudden reappearance of some lost ancient character, but, these phenomena are now well explained by gene interaction.

10. **Multiple factors**

A large number of instances have been found where characters are caused not by single allelomorphic pairs but by a large number of such pairs of genes. This is specially true in the case of **quantitative characters,** i.e., where the characters may be measured in units as in the case of height, weight, etc. Usually, these genes have additive or cumulative effects. This is also called **polymery.** Hybrid plants involving such multiple genes are also *polyhybrids*.

The simplest case of multiple factors is that involving only two pairs of genes. In wheat each of two genes $\mathbf{C_1}$ (of the allelomorphic pair $\mathbf{C_1c_1}$) and $\mathbf{C_2}$ (from $\mathbf{C_2c_2}$) may cause light red colour but $\mathbf{C_1}$ and $\mathbf{C_2}$ together cause deep red. Thus, in a suitable cross, the $\mathbf{F_2}$ segregation is 9 deep red ($\mathbf{C_1\text{-}C_2\text{-}}$) : 6 light red ($\mathbf{C_1\text{-}c_2c_2}$ and $\mathbf{c_1c_1C_2\text{-}}$) : 1 white ($\mathbf{c_1c_1c_2c_2}$). It can be easily realised that this is a case of complementary action of two genes as in the case of *Cucurbita pepo* shape but this is considered under multiple factors as this is a case of quantitative inheritance involving measurement, *viz.*, the intensity of colouring.

The above is a case of only two factors having an additive effect. There may be more factors having such cumulative effects. For example, there are cases of plant height being controlled by a number of genes each of which add, say, 2 cms to the height. When a large number of such factor pairs are involved, there will be a large number of segregated phenotypes and there will be such gradation among them that, sometimes, it may be impossible to distinguish between them. Multiple factors, which are caused by a large number of factors, each having a small effect, may also be affected by minor *polygenes* (explained below) and even by the environment, e.g., plant size, controlled by a large number of genes, may be affected by the availability of nutrition; human skin colour, also controlled by a large number of genes, may also be affected by exposure to sunlight.

11. Multiplication of factors

In some cases there may be two or more allelomorphic pairs causing the same character. Such genes have no additive effect.

In the case of a rice variety, the awn is caused either by the gene $\mathbf{A_1}$ (from $\mathbf{A_1a_1}$) or by its **duplicate gene** $\mathbf{A_2}$ (from $\mathbf{A_2a_2}$). Thus, in a suitable awned × awnless cross the $\mathbf{F_1}$ will be awned while the $\mathbf{F_2}$ segregation will be 15 awned ($\mathbf{A_1\text{-}A_2\text{-}}$, $\mathbf{A_1\text{-}a_2a_2}$ and $\mathbf{a_1a_1A_2\text{-}}$) : 1 awnless ($\mathbf{a_1a_1a_2a_2}$).

Similar duplicate genes are known for rice apiculus (lemma and palea tips) colour where the presence of two pairs of genes, $\mathbf{Ap_1ap_1}$ and $\mathbf{Ap_2ap_2}$, caused a 15 coloured : 1 green $\mathbf{F_2}$ ratio. Duplicate genes point out the presence of duplicate chromosomes (present in *polyploids*—explained in connection with mutation) and, therefore, suggest a polyploid origin of rice.

Instances of **triplicate genes** (also suggesting polyploidy) are also known. Here the expected $\mathbf{F_2}$ ratio is 63 : 1.

12. Pleiotropy

It has now come to light, specially in studies with *Drosophila*, that one gene may control two or more different characters simultaneously. This is called *pleiotropy*. For example, anthocyanin genes is plants cause colouration in different parts; the 'vestigial wing' gene in *Drosophila* also causes development of bristles.

13. Modifying genes and Polygenes

Modifying genes are minor genes which modify other characters. *Polygenes** are similar minor genes which act together in modifying quantitative characters. Ordinarily, the individual effect of such genes is very insignificant and may even go undetected. But, in the long run, they may play an important role in the evolution of new species.

14. Cytoplasmic inheritance

In this book, it has hitherto been taken for granted that hereditary characters are controlled by genes located in the chromosomes. But in the formation of a zygote a big amount of egg cytoplasm takes part. This cytoplasm (along with the plastids contained in it) may cause the transmission of certain maternal characters just as it may transmit any virus disease particle contained in it. Some chlorophyll deficiencies or variegations are known to be inherited in this way. Correns showed long ago that the *albomaculatus* type leaf variegation in *Mirabilis jalapa* is inherited in this way.

Recently, some instances have been found where cytoplasm seems to play a more complicated part in inheritance. For example, in the unicellular animal *Paramecium*, Sonneborn has shown that the inheritance of a certain character ('killer' character) is partly controlled by some particles contained in the cytoplasm. The 'killer' *Paramecium* contains a gene **K** in its nucleus and also particles called *kappa particles* in its cytoplasm which are capable of regeneration. The kappa particles are made of *DNA*

* The term *polygene* has also been used by geneticists in other senses. The multiple factors or genes controlling quantitative characters are sometimes called *polygenes* in the literal sense. Again, there are some very small genes located in the heterochromatic region of the chromosome showing less complicated and less specific action which are also called *polygenes*.

and they secrete a substance called *paramecin* which kills all 'sensitive' *Paramecia.* i.e., those possessing **kk** genes. The 'killer' character is ineffective unless both the **K** gene and the kappa particles are present. These kappa particles must come through the physical inheritance of some cytoplasm containing these. Such heredity controlling particles in the cytoplasm are termed **plasmagenes** (sometimes also called *cytogenes*) as opposed to the common nuclear genes.

Other instances of inheritance through maternal cytoplasm are well-known. The dextral and sinistral coiling of the snail *Limnaea* is determined not by its own genes but by the genes of its mother. Wettstein noted that when the moss *Funaria hygrometrica* is crossed with *F. mediterranea* the appearance of the $\mathbf{F_1}$ plant is like its mother whichever way the cross was made.

XENIA

When, in course of hybridisation, the pollen changes the appearance of the hybrid seed still on the mother plant, the phenomenon is known as **xenia**. This is apparent in the case of endosperm characters. Formerly, it was thought to be a direct effect of the male on the female exerted through the pollen and the term 'xenia' was used by Focke as early as 1881 in this sense.

While the zygote is formed by the union of a male and a female gamete so that it is *diploid* with two sets of genomes, the endosperm mother cell in Angiosperms is formed by the fusion of three nuclei—first two embryo-sac nuclei and then a male gamete. So the endosperm mother cell is *triploid* having *three* sets of genomes. The endosperm will then have three genes of an allelomorphic pair which may be symbolically represented as **Aaa, Ttt, Ppp,** etc. It is found that, in most cases where strict dominance is observed, if only one of these three genes is a dominant gene then the endosperm shows the dominant character inspite of the simultaneous presence of two recessive genes.

The second important point in xenia is that endosperm characters are evident one generation earlier. In ordinary hybridisation the hybrid seeds and ordinary self-fertilised seeds on the mother plant all look alike. The hybrid character becomes apparent only when the hybrid seed grows into the F_1 plant. On the other hand, in the case of xenia, the hybrid seed become at once distinct from the self-fertilised seeds because of their endosperm characters. This is because the endosperm develops fully while it is still located on its mother plant, i.e., on the plant representing the previous generation.

The phenomenon can be clerly seen when a maize plant with white grains is crossed with one with purple grains (Fig. 853). If some of the flowers in a female inflorescence of a pure breeding white grained maize plant are intentionally or accidentally pollinated with pollens from a pure breeding purple grained maize plant, the mature maize cob will show the hybrid grains as purple (since purple is dominant over white) while the self-fertilised grains will remain white. Thus, this purple F_1 character will be seen on the mother parent itself as an expression of xenia. Similarly the F_1 plant, when fully grown and self-fertilised, will show the F_2 segregation of endosperm colour on the cobs that will still be on the F_1 plants. The proportion of purple to white grains on such a cob will be 3 : 1 (of course, allowing statistical limits of variations). i.e., in the F_2 ratio.

Similar endosperm characters are common specially in the different cereal crops which are mainly endosperm crops. Starchy (dominant) and glutinous (recessive) grains of rice, as also the starchy (dominant) and sugary (recessive) grains of maize, are similar pairs of endosperm characters which show xenia.

Metaxenia is another phenomenon sometimes included within *xenia*. This means the effect of the pollen (or the post-pollination effect induced by the growing endosperm and embryo) on the maternal tissue enclosing the seed. This effect is specially visible in

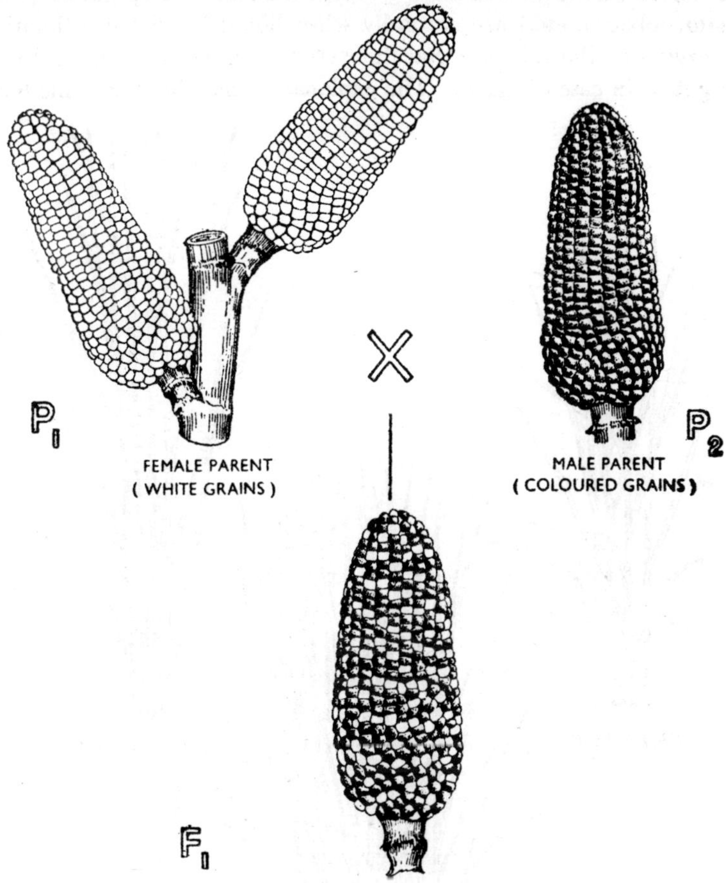

FIG. 853. Xenia in maize grains. The purple endosperm colour and its segregation is seen one generation earlier than in the case of ordinary characters. The female parent shows the purple hybrid grain on the otherwise white cob.

the development of apples. This type of development, however, is better explained differently than in the case of xenia.

HETEROSIS OR HYBRID VIGOUR

Ever since the art of hybridisation became known it was also clear that the F_1 hybrid plant is much more vigorous than either parent. F_1 shows the limit of such vigour and usually it decreases in the future generations. Shull (1914) proposed the name **heterosis** for this hybrid vigour. Jones (1917) suggested that the reason of this vigour is that the maximum number of dominant genes come together in the F_1 when they counteract any ill effect of the recessive genes. East (1936) pointed out that because of linkage (discussed in next chapter) it is not possible to get all the dominants in any pure line form and the maximum association of the dominants is possible only in the heterozygous form of the F_1. Hence the increase in vigour in hybrids as seen in the F_1 hybrid rice plant in Fig. 854.

If the cause of increased vitality of hybrids is heterozygosis, it follows that the homozygous condition means weakness. A large number of crop-plants (wheat, rice, pulses, tomato, tobacco, etc.) are naturally self-pollinated and self-pollination always leads to homozygosity. But it is seen that these crops do not show so much loss of vigour as inbreeding does in case of maize. The explanation may be that while harmful and

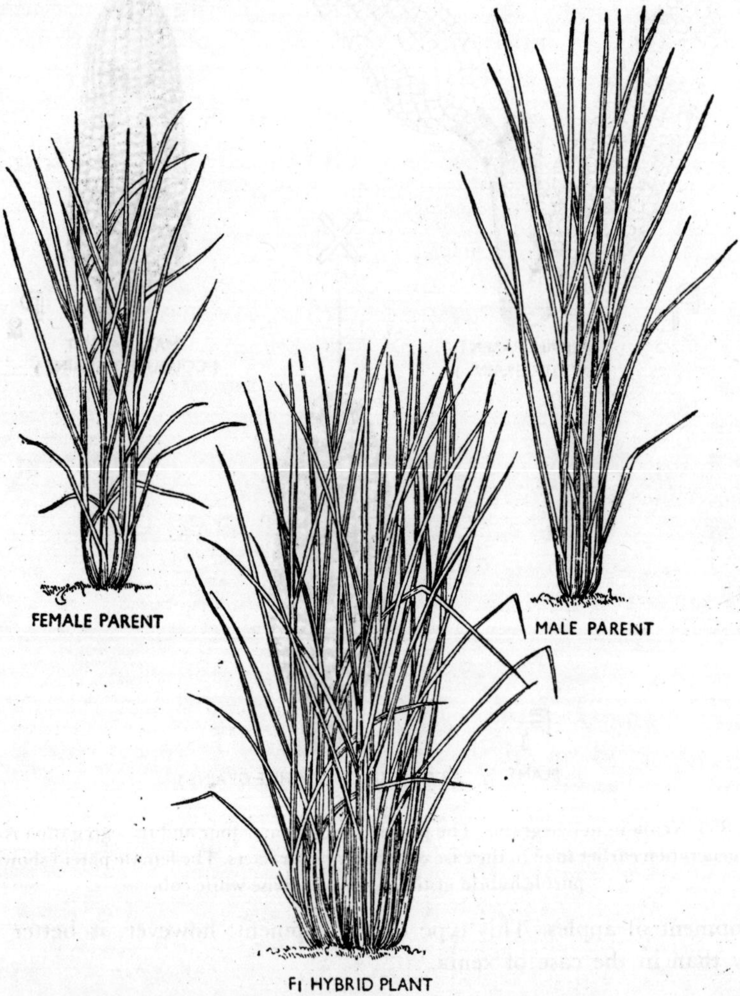

FIG. 854. Hybrid vigour in rice. The hybrid F_1 rice plant, grown side by side with the parents, is seen to be much more vigorous than either of them.

lethal mutant genes are constantly arising in plants, natural inbreeding results in homozygosity bringing out the harmful characters and their eventual elimination in course of natural selection. But, in normally cross-bred plants, these harmful genes remain in a heterozygous recessive condition and are not eliminated causing weakness of the strain.

In recent years, the phenomenon of heterosis has been utilised commercially in the use of hybrid maize grains ('hybrid corn') directly for seed (discussed in Chapter LXX).

CHAPTER LXV

POST-MENDELIAN DEVELOPMENTS. II. LINKAGE; SEX CHROMOSOMES

LINKAGE AND CROSSING-OVER

About five years after the rediscovery of Mendel's paper, in 1906, Bateson and Punnett reported a dihybrid experiment in sweet pea (*Lathyrus odoratus*) where the segregation ratio could not be explained by Mendel's Law of Independent Assortment. They crossed purple flowered-long pollen grained sweet peas with red flowered-round pollen grained ones. It was known before, that purple-red flower colour and long-round

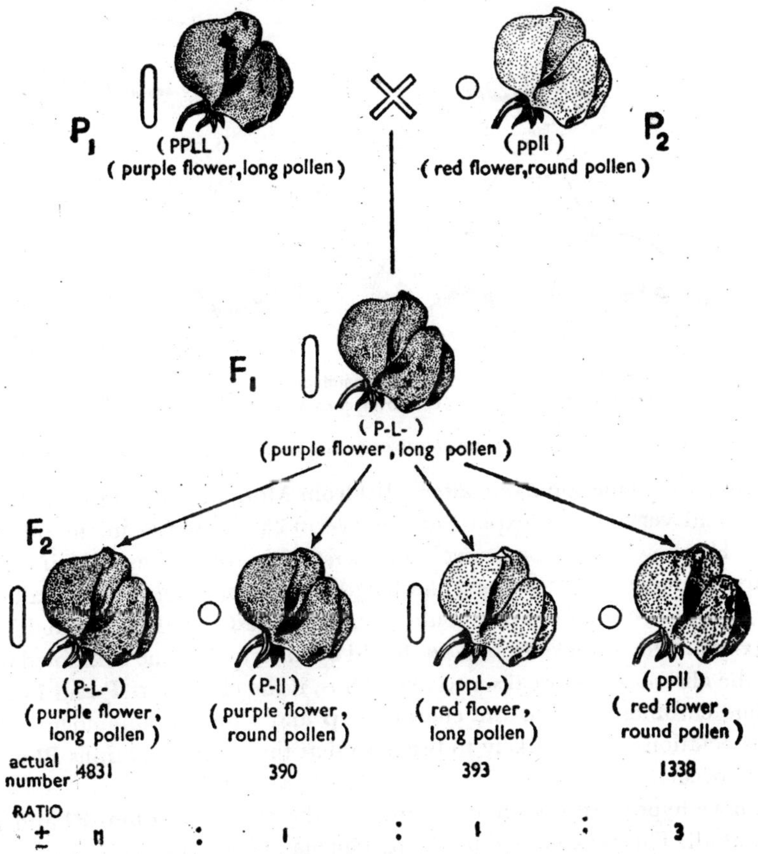

Fig. 855. Linkage in sweet pea as found by Bateson and Punnett.

pollen characters behave as allelomorphs. The F_1 was purple flowered-long pollen grained. But the F_2 segregation was not as would be expected from a 9 : 3 : 3 : 1 ratio. The actual numbers obtained are shown in Figure 855. Clearly, there were too many of the parental types—purple flowered-long pollened (say, **P-L-**) and red flowered-round pollened (**ppll**) and too few of the new associations **P-ll** and **ppL-**. Bateson and Punnett explained this by saying that there is a **coupling** between the factors **P** and **L** on the one hand and between **p** and **l** on the other so that they tend to remain together.

Secondly, on examining the **Pl** abd **pL** types they found that in these again, in future generations, the **Pl** and **pL** characters tend to come together more often than they should. This was called the **repulsion phase** as, it seemed, the parental associations refuse to be formed again and the new associations tend to persist.

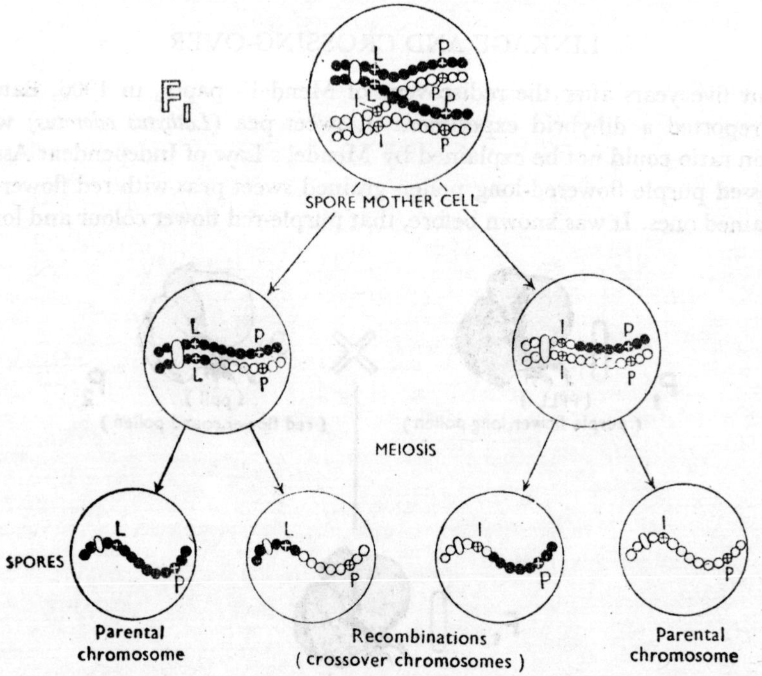

FIG. 856. Diagram illustrating crossing-over giving rise to the **Pl** and **pL** recombinations.

The proper explanation came after 1910 from Morgan who established the theory of the gene and very clearly explained the role of chromosomes in the inheritance of characters. He showed that both *coupling* and *repulsion* are well explain if one assumes a linear arrangement of genes along the lengths of chromosomes. Then, all the genes located in a single chromosomes should remain **linked** together or should form one **linkage group.** The whole *linkage group* should be inherited as one combined unit unless and until the chromosome breaks into two. Thus, if the characters **P** and **L** are located on the same chromosome and the characters **p** and **l** on the homologue, then these parental associations are not likely to break so that the new associations **Pl** and **PL** are not likely to occur.

The above hypothesis does not explain how the new associations **Pl** and **pL** could then occur at all. This is explained by the penomena of chiasma formation and crossing-over. We have already seen how bits of chromonema threads may crossover from one chromosome to its homologue. This easily explains the translocation of some genes to the homologous chromosomes. The new combinations **Pl** nd **pL** occur in this way (Fig. 856) and represent the **crossovers** or **recombinations.**

It is easy to follow that if the above hypothesis is true, the number of *linkage groups* is an organism should coincide with the number of homologous pairs (i.e., the *n* number) when the pairs of alleles and not single genes are considered. Exhaustive experiments have shown this to be true and four linkage groups have been clearly demonstrated in *Drosophila* (n=4) and ten linkage groups in maize (n=10).

Detection of Crossovers : The Testcross

Crossing-over takes place during meiosis which takes place only in the spore mother cells. Fig. 856 shows that for each crossover in a spore mother cell there would be 2 *crossover spores* (or *recombinations*) and 2 *non-crossover spores* (original parental combinations). Thus, if there be 1 crossover in 100 spore mother cells (i.e., 1%) there would be

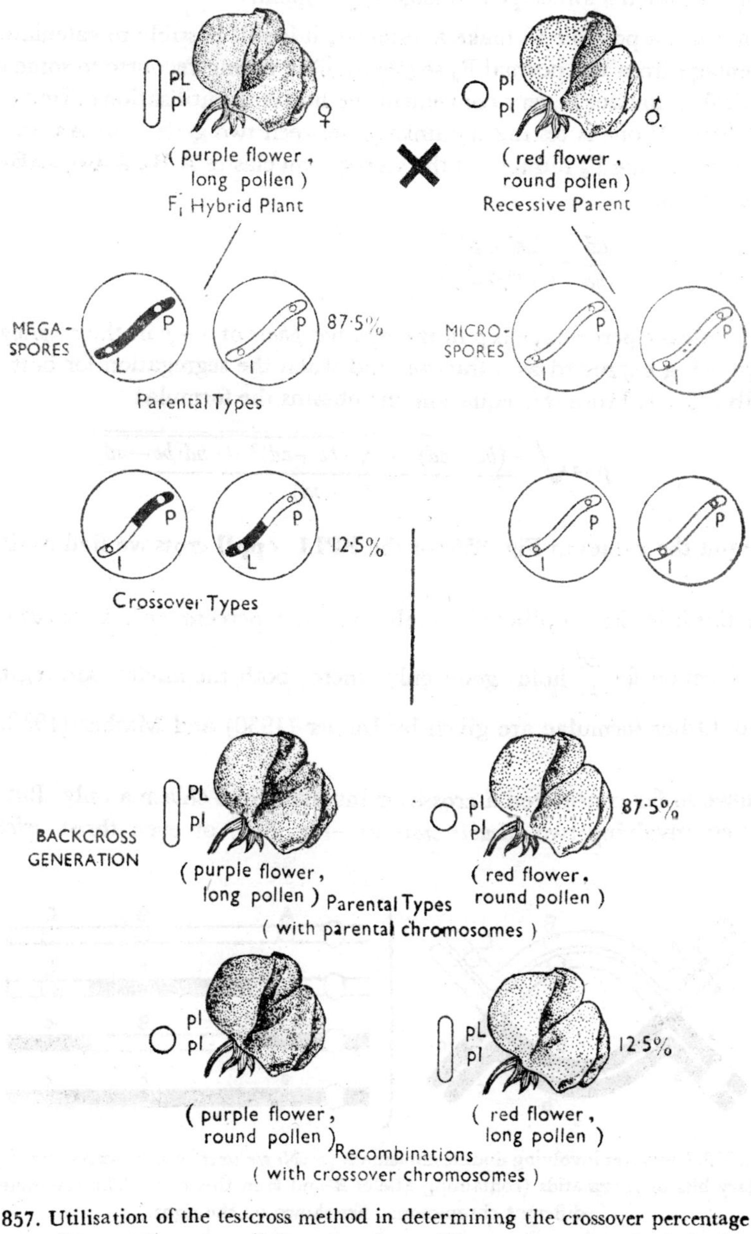

Fig. 857. Utilisation of the testcross method in determining the crossover percentage in Bateson and Punnett's sweet pea linkage experiment. The crossover plants in the backcross (or testcross) generation gives the crossover percentage.

2 crossover spores among 400 spores, i.e., only 0.5%. If there be one crossover in every spore mother cell there would be 50% crossovers.

To detect the percentage of crossovers, the simplest way is to take recourse to the **Testcross**. *Testcross* is crossing the F_2 hybrid plant back to the recessive parent. If a large number of *backcross* plants be obtained in this way, the percentage of recombination plants in the backcross population gives the percentage of crossovers. Fig. 857 shows how the testcross method was utilised in finding the crossover percentage in Bateson and Punnett's sweet pea linkage experiment.

When it is not possible to make a testcross, it is still possible to calculate the crossover percentage from the normal F_2 segregation by taking recourse to some improvised mathamatical formulae. A very convenient method is the utilisation of Immer's Product Method (1930). If one is testing the linkage between two gene pairs **Aa** and **Bb** and if *a, b, c, d* represent the F_2 number of the segregation classes **A-B-, A-bb, aaB-** and **aabb** respectively, then,

$$\frac{ad}{bc}=\frac{2p^2+p^4}{1-2p^2+p^4}$$

when the crossover percentage is p in the *repulsion phase* or $1-p$ in the *coupling phase*, the percentage being expressed as a fraction and when the segregation for both the alleles **Aa** and **Bb** is 3 : 1. From this equation one obtains the formula:

$$p=\sqrt{\frac{-(bc+ad)+\sqrt{(bc+ad)^2+ad(bc-ad)}}{(bc-ad)}}$$

If we put the values in Fig. 855 for the **PPLL** × **ppll** cross we find $p=0{\cdot}87$ approximately.

Since this is in the coupling phase, the crossover percentage is approximately 13%.

This formula for $\frac{ad}{bc}$ holds good only where both the alleles are segregating in 3 : 1 ratio. Other formulae are given by Immer (1930) and Mather (1938) for other ratios.

We have so far considered a crossover involving one chiasma only. But there may be crossovers involving two (*double crossover*—Fig. 858) or even three (*triple crossover*)

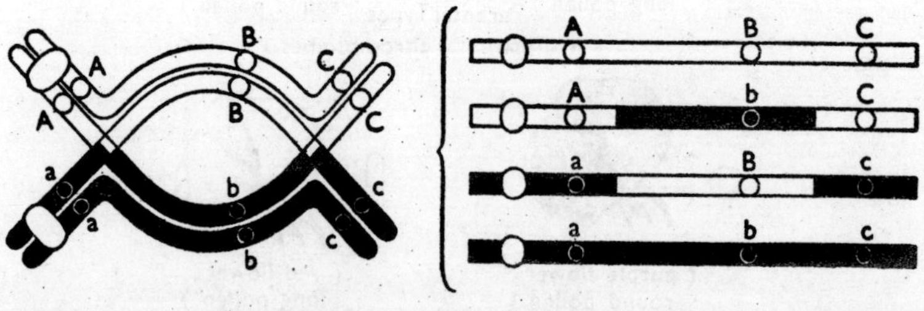

FIG. 858. Crossover involving double chiasmata (*double crossover*) causing crossover of intercalary bits of chromatids (containing alleles **B** and **b** in this case). The resultant four different chromosomes are shown on the right.

chiasmata. Moreover, when there are more than one chiasmata, all the chiasmata may not be between the same two chromatids so that complicated *compound crossovers* are met with.

Chromosome maps

As crossing-over is a matter of chance it may be assumed that there is a greater chance of crossover between two genes of the same linkage group the farther apart they are. On this basis it is possible to locate the different character genes on a chromosome or to 'map the chromosome'.

The method is to effect a crossing involving two pairs of linked genes and to calculate the percentage of crossover obtained. The distance between the genes is taken to be as many units (this unit has been named *morgan* in honour of the exponent of the gene theory) as the percentage of crossovers. Suppose **A, B, C** are three genes in one linkage group and, therefore, located in one chromosome. There is a crossover percentage of 10 between **A** and **B,** 10 between **B** and **C** and 20 between **A** and **C**. It is justifiable to assume that the distance between **A** and **B** is 10 units and that between **B** and **C** another 10 units. In this manner, if sufficient data be available, it is possible to

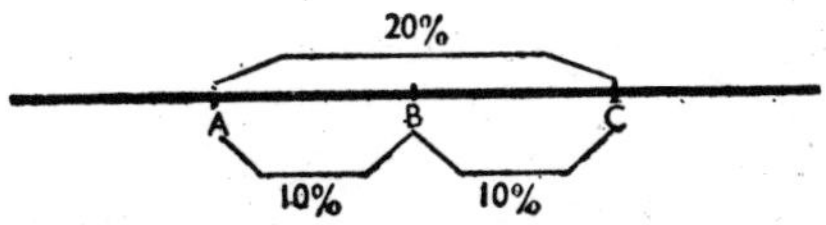

FIG. 859. Diagram illustrating location of genes **A, B,** and **C** on a chromosome.

locate all the genes. Complete chromosome maps have now been worked out for the four chromosomes of *Drosophila* (Fig. 860) and the mapping of the ten chromosomes of maize is almost complete. It should be noted that the relative lengths of the linkage groups correspond to the relative lengths of the actual chromosomes.

It should be pointed out as a necessary precaution that every crossover percentage should not be considered as an absolute measure of the distance between two genes. Cytologists are aware of environmental and other factors which may affect the crossover percentage between any two genes. A correct linkage map as that of *Drosophila* can only be completed after repeated experiments and consideration of all factors which may interfere with the result.

The salivary gland chromosomes have made it possible to obtain partial linkage maps of *Drosophila* by an altogether different cytological technique—that of studying these chromosomes in deleted mutants (explained in connection with *deletion* in the next chapter). The positions of the genes in the two types of maps have been found to correspond after making allowances for certain explainable discrepancies.

Another point should be noted in the chromosome maps. It has already been pointed out that the maximum percentage of crossover possible is 50. But the map shows more than 100 units. This is because such maps, in order to be more accurate, are made up by determining the crossovers of genes placed very closely (say, not showing more than 10% crossover between them in case of *Drosophila*). When all the genes are located and the distances added up, it may go to more than 100. So, in case of genes placed wide apart, the unit distance is to be considered as unit distance only and not as percentage of crossover.

Certainly, mapping of chromosomes is a major evidence in support of the chromosome and gene theory.

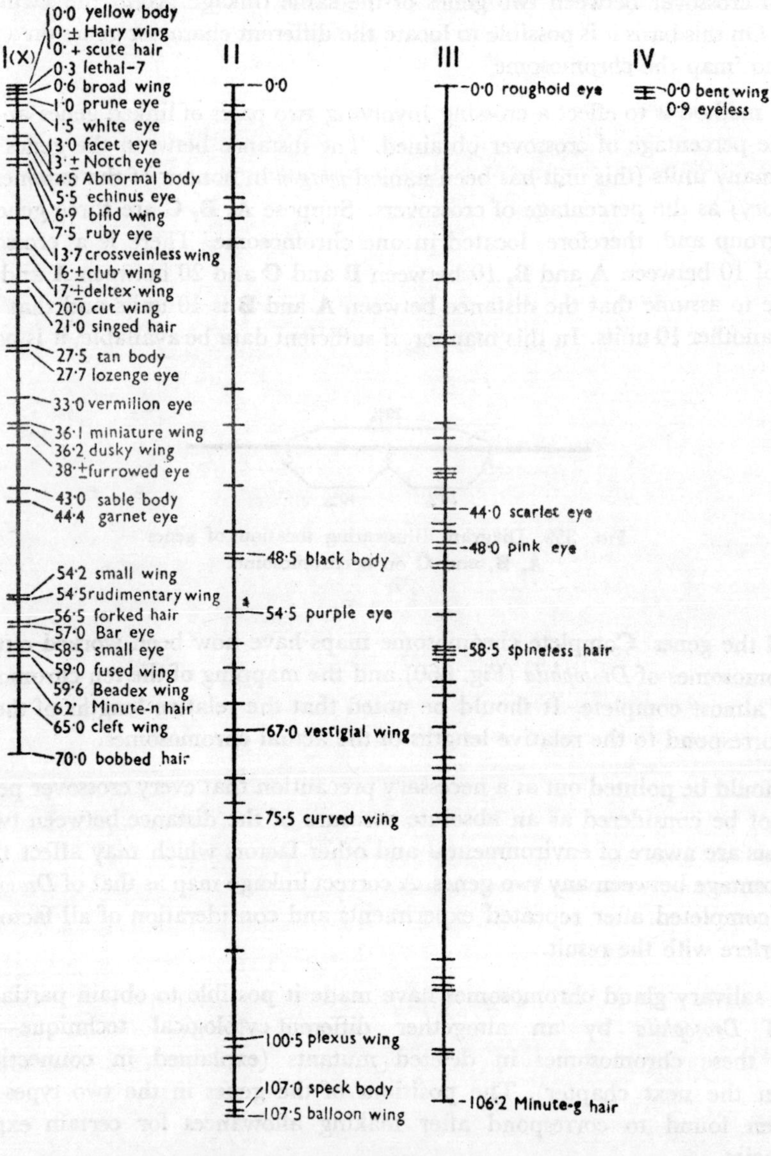

FIG. 860. The four linkage groups of *Drosophila* showing the locations of the genes on the four chromosomes. Mapping of Chromosome I (X-chromosome) has been shown in detail. In the others, only a few genes have been shown. The relative lengths of the linkage groups are proportional to the lengths of the chromosomes as seen in Fig. 861A.

SEX DETERMINATION, SEX CHROMOSOMES AND SEX LINKAGE

While in most plants male and female organs are located on the same individual, most animals and a good number of plants show a differentiation into separate male and female organisms. The factors determining this difference in sex has always attracted the attention of biologists. Numerous experiments have been performed and volumes have been written on this problem but still there are many points that are not yet clear. The immediate causes of sex difference are physiological (c.f. p. 714) but this physiology is often controlled by genes. These sex determining genes may be located in ordinary chromosomes but, sometimes, they are located in special **sex chromosomes** which are often quite different from the normal chromosomes. The normal chromosomes are termed **autosomes** (symbolised by **A**) while the special sex chromosomes form the **allosomes**. Sex chromosomes have been found to be of different types. The most important types are as follows:

XX (**female**)—**XY** (**male**) type

This is the '*Insect type*' found in many insects, the most familiar example being *Drosophila*. Here, in addition to the ordinary chromosomes (autosome), the female shows two similar sex chromosomes (**XX**) while the male shows two dissimilar ones (one **X** and one **Y**). Thus, the **XY** pair of the male is not homologous. The female forms

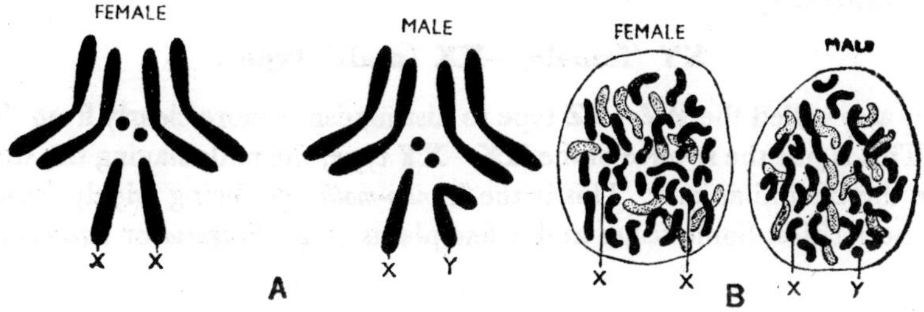

FIG. 861. **X** and **Y** sex chromosomes in *Drosophila melanogaster* (A) and human beings (b).

gametes of only one type (**X**) while the male gametes are of two types (**X** and **Y**) in equal proportions. Offsprings, therefore, should be equally male and female. Human beings show the same type of sex chromosomes. A good number of plants like *Coccinia*, *Elodea*, *Hydrilla*, *Smilax*, *Cannabis*, *Melandrium*, etc., are also of the same type.

As to the actual sex genes in the sex chromosomes of this important group, two very important cases are known. In *Drosophila*, it seems that the female genes are present in the **X** while maleness is contributed by the autosome **A**. The **Y** is *inert* or useless in this sense. Thus, the sex of a *Drosophila* is determined by the Autosome/**X** chromosome balance. 1·5 **X** chromosomes seem to balance the maleness of the autosome. That is

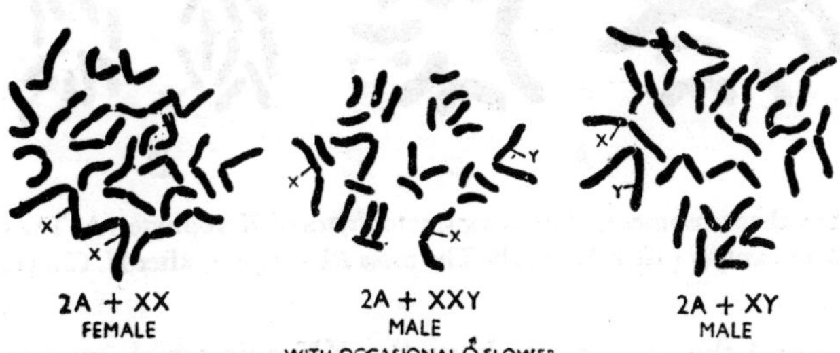

FIG. 862. Sex chromosome balance in *Silene dioica*. Root tip cells show that **2A+XX** plants are female, **2A+XXY** ones are male and hermaphrodite while **2A+XY** plants are male.

why **A**+1**X** is male, **A**+2**X** is female and 2**A**+3**X** individuals are *intersexes*. The other well-known example is that of the plant *Silene dioica* (formerly put under *Melandrium* or *Lychnis*) where the autosome does not seem to be of any importance, the sex being determined by the **X/Y** balance. Plants with **X**'s only are female while it is male when **X** is balanced by **Y** ($\frac{2}{3}$**Y** or more for each **X**). If **X** be incompletely balanced ($\frac{1}{3}$ to $\frac{2}{3}$**Y** for each **X**) then the plant is bisexual. Thus a plant having **XX** is female, one having **XY** is male but one having **XXY** is partially hermaphrodite (Fig. 862). In human beings the **Y** chromosome has been found to be very strong. Abnormal individuals with **XXXXXY** chromosomes have been found to be male, i.e., one **Y** is sufficient to outbalance 5**X** chromosomes. An abnormal **XO** (i.e., without **Y**) individual becomes a weak female or a eunuch.

XX (female)—XO (male) type

This may be taken as a modification of the former in which the **Y** chromosome is missing. The female has two **X** chromosomes while the male has only one which is not paired. This is also found in some lower animals like the grasshoppers and among plants like some *Dioscorea* sp.

XY (female)—XX (male) type

This is also called the **WZ—ZZ** type to distinguish it more clearly from the **XX—XY** type. This is just the reverse of the **XX—XY** type, the male having the similar and the female the dissimilar pair. This is the '*Avian-moth type*' being widely found among birds and butterflies. Some fishes and a few plants (e.g., *Fragaria* or strawberry) show this type.

XO (female)—XX (male)

This is just a development from the previous type, the **Y** being dropped from the female leaving the **X** without a pair. The moth *Fumea* is an example.

Bryophytes

Among Bryophytes the conspicuous plants are the haploid gametophytes and sex differentiation is often shown in these plants. In many instances it has been found that the haploid female contains an **X** chromosome while the male contains an **Y** chromosome. The zygote and the sporophyte have the **XY** pair which on reduction division forms **X** and **Y** spores in equal numbers resulting in an equal number of male and female gametophytic plants.

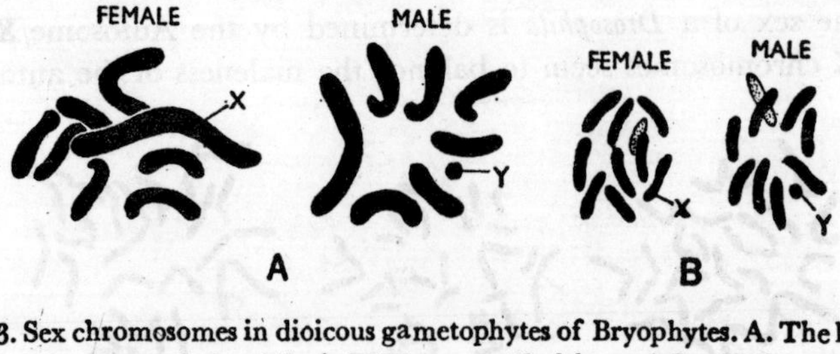

Fig. 863. Sex chromosomes in dioicous gametophytes of Bryophytes. A. The liverwort *Sphaerocarpos* (after Allen). B. The moss *Rhodobryum* (after N. Chopra).

Special types

Certain other plants show variations in the number and composition of the sex chromosomes. There are various types of them, e.g., different species of *Rumex* show the **XX** (female)—$\mathbf{XY_1Y_2}$ (male) type (i.e., the female contains two **X** chromosomes while the male contains one **X** chromosome and two different **Y** chromosomes).

When plants are dioecious although sex chromosomes are not found, one may suppose that there are sex genes within the *autosome* which control sex. In the papaw (*Carica papaya*), Hofmeyr explained the existence of male, female and hemaphrodite plants by assuming the presence of an allelic series of three sex genes, $\mathbf{M_1}$, $\mathbf{M_2}$ and **m**, **mm** plants are female, $\mathbf{M_1m}$ male and $\mathbf{M_2m}$ hermaphrodite while $\mathbf{M_1M_1}$, $\mathbf{M_2M_2}$ and $\mathbf{M_1M_2}$ zygotes are non-viable. There are, however, modifying genes which may affect the sex expression.

While sex is found to be controlled by sex chromosomes or by genes located in chromosomes which cannot be differentiated from ordinary chromosomes, the importance of environmental and physiological factors must not be ignored. The latter is often stronger than the fomer which may even cause a sex reversal (i.e., male to female or femle to male) in organisms controlled by genes (c.f. p. 714).

As the sex genes are located in some chromosomes, the other genes on such a chromosome will certainly be inherited together with the sex gene. Thus, the characters controlled by these genes will be **sex-linked**, i.e., they will occur either with the male or with the female. This is why certain characters are always found in the male and others in the female. It should be noted in this connection that the **Y** chromosome of the males is often *inert*, i.e., it contains very few genes. Only two genes are known to be present on the *Drosophila* **Y** chromosome and four on the human **Y** chromosome. Neverthelesss, the human **Y** chromosome, as already stated, is very strong in sex determination while the *Drosophila* **Y** chromosome is ineffective from this standpoint.

CHAPTER LXVI

POST-MENDELIAN DEVELOPMENTS. III. MUTATION

Mutation of a plant or an animal means *a sudden change in its hereditary make-up*. Suddenly a mutated organism arises and the changed or mutated appearance is usually found to be hereditary. i.e., it breeds true. Since heredity is controlled by genes, it follows that the genes somehow change their behaviour. While mutations are always taking place is nature, very few of them are of a major nature to be readily apparent. In many cases, the changes are so minor that they are not readily apparent while in most cases they are recessive and remain latent. The de Vriesian concept of *mutation* has changed now as many changes which were considered as *fluctuating variations* by Darwin and de Vries have been found to be true mutations of a minor nature.

The actual change in the genes may be of two types: (I) *intragenic* or (II) *intergenic*.

I. *Intragenic change* means something has happened *within* the gene itself or the *gene has mutated*, i.e., it has undergone some chemical or structural change. Since the gene itself has mutated this may be called **gene mutation** which is sometimes considered as the true type of mutation. Gene mutation is also called **point mutation** since the mutation has taken place at a point or locus on the chromosome.

II. *Intergenic changes* involve merely the rearrangement of the genes already existing or the loss of some of them. So, this type of mutation does not involve the formation of mutated genes. They are **chromosomal mutations** or **chromosomal aberrations** which are caused either (A) by a change in the number of chromosomes which may involve either individual chromosomes (**aneuploidy**) or whole sets of genomes (**euploidy**), or (B) by a change in the structure of chromosomes.

I. GENE MUTATION (INTRAGENIC CHANGE)

As we come across a really mutated gene only in this type of mutation, sometimes the word *mutation* is applied to *gene mutation* or *point mutation* alone. In nature, genes are constantly duplicating themselves. Every cell division means the formation of new genes—daughter genes from mother genes. In this duplication the original gene is generating an exact replica of itself. But it is possible that while this process is going on for millions of times, once a while something may go wrong and the new gene may not be an exact replica of the original. The gene, formed of nucleoproteins, must have a complex molecular structure and there may be a slight change somewhere. This change, however small it may be, gives rise to a mutated gene whose behaviour is different from its predecessors in some degree. Such spontaneous gene mutations are always occuring in nature. Among some 10 million specimens of *Drosophila melanogaster* examined, a few hundreds of gene mutants were observed. These mutants give rise to true-breeding stable genotypes varying in some characters (eye colour, wing type, etc.) from the original wild type. Mutations like these are always observed in all types of organisms from human beings to bacteria. They arise all on a sudden. Most mutations bring about such a minute change that they evade common observation. Moreover, most mutations are recessive as well as lethal. They produce no effect before appearing in a homozygous form in some future generation. They are, therefore, undetectable. However, there are some special cytogenetical methods (e.g., the *ClB method*) for detecting lethal mutations in **X** chromosomes of *Drosophila*.

ClB Method

In this method of detecting lethal mutations in the **X** chromosomes of *Drosophila* the irradiated male fly is first crossed with a female of a special stock called *ClB*. In this, of the two **X**'s of the female, one is normal (shown by the thick bar in the figure) while the other contains C (an inverted segment of chromosome), l (a lethal gene which cannot act being dominated by the normal **X** homologue which has no lethal) and B (bar-eye character) one fourth of the offspring would be female with one treated **X** (shown by a thick bar in the figure) and one *ClB* **X**, one fourth would be male with one *ClB* **X** which would die and half would be normal. Only the female from the first quarter is crossed with a normal male in a second cross. Among the offsprings all males would die if a lethal gene has been induced. If no lethal gene has been induced there would be 2 : 1 female to male (Fig. 864).

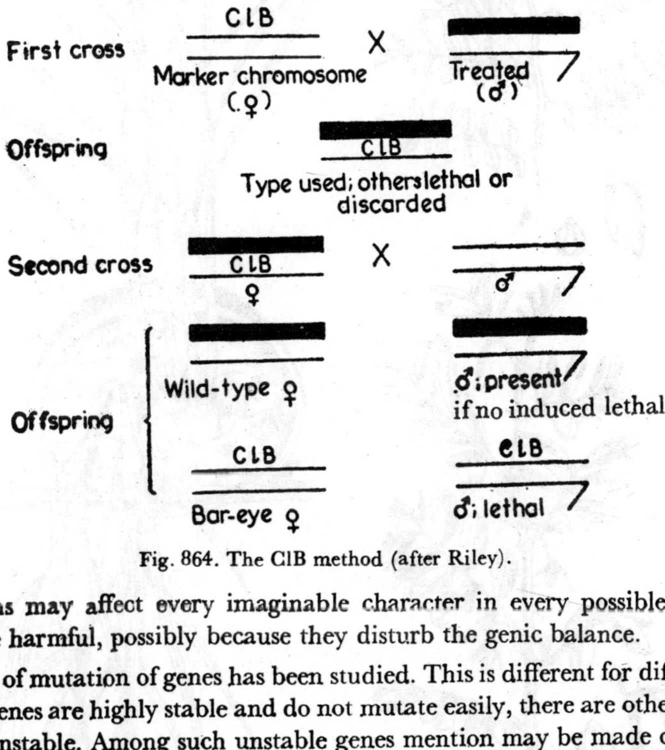

Fig. 864. The ClB method (after Riley).

Mutations may affect every imaginable character in every possible way. Most mutations are harmful, possibly because they disturb the genic balance.

The rate of mutation of genes has been studied. This is different for different genes. While some genes are highly stable and do not mutate easily, there are others which are remarkably unstable. Among such unstable genes mention may be made of the gene **c** in chromosome IX of maize causing a colourless aleurone in the grain which was found by Jones to easily mutate into the dominant **C** developing colour so that such mutations in the somatic aleurone cells give rise to spots on the grain. Similarly the *rose-a* and *lavender-a* genes in *Delphinium* (Larkspur) were found by Demerec to mutate into purple giving rise to purple patches on the otherwise rose or lavendar petals. This mutation may also take place on the germ cell giving rise to a fully purple-flowered mutation. The stability of genes seem to be different in different species. In *Drosophila* the possibility of mutation of a gene has been calculated as 1 in 10^5, in higher organisms 1 in 10^6. In maize the rate of mutation of the **R** gene is of the order 492 in $1 \cdot 5 \times 10^6$ while the same for the **I** gene is 106, for the **S** gene 1 and for the **Wx** gene none. Haldane had calculated the rate of mutation in human beings as 1 to 5 in 10^5. If this be the rate of mutation of a single gene, the possibility of a mutation in an individual is great considering that there are thousands of genes in a cell. It has been suggested by Muller that there is some mutation in one out of every 20 gametes in *Drosophila* and in human beings there is a possibility of having one mutated gamete out of every 10. The rate of mutation has been found to be affected by environmental and other factors viz., it increases with temperature and with age. The effect of X-rays and other ionising radiations is discussed later. The rate of mutation may even be affected by other genes in the same cell.

FIG. 865.

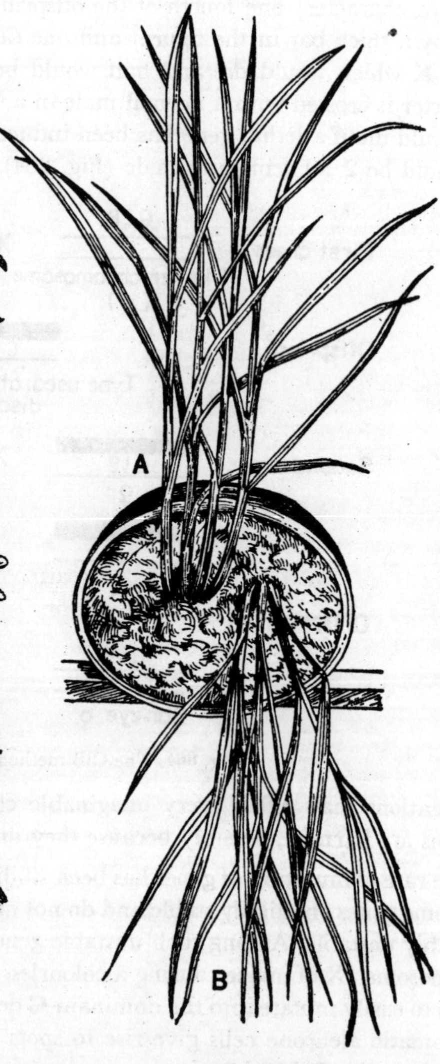

FIG. 866.

MUTANTS. FIG. 865. *Cicer arietinum* (A) and its mutated form *Cicer gigas* (B) (*after Dixit*). FIG. 866. In a tilted pot, a normal rice plant (A) is growing erect while the 'lazy' mutant (B) is growing in a pendant manner (*after Jones and Adair*).

Rhoades has shown that the **Dt** gene in chromosome IX of maize causes the $\mathbf{a_1}$ gene in chromosome III to mutate into $\mathbf{A_1}$ so that colourless plants show purple streaks on leaf and purple dots in aleurone.

Mutations may take place in somatic cells as well as in germinal cells. While the latter gives rise to mutated seeds and mutated progenies, the former, as seen above, gives rise to patches of tissues with mutated cells. Some such tissue may develop buds and shoots and *bud sports* or *bud mutations* arise in this way. When vegetatively propagated, these buds may give rise to clones of mutated plants. Many agricultural and horticultural varieties have arisen in this way. Washington Navel orange plants are known to give rise to single mutated fruits as a result of somatic mutation just before the development of the flower bud. The origin of *chimaeras* as a result of somatic mutation has already been discussed.

Although most of the mutations are useless and even harmful, sometimes a mutant may be of some biological value and they certainly play a great part in the evolution of new species. Many new breeds of domesticated animals and strains of cultivated crops must have arisen by gene mutation. Different breeds of sheep (e.g., the short-footed Ancon sheep), swine, rats, horses, dogs, pigeons, etc., are known to have arisen in this way. The Shirley poppy, the dwarf 'Cupid' sweet pea and many varieties of showy flowers also arose by gene mutation. The cultivated *Cicer gigas* arose as a mutation of the common gram, *Cicer arietinum* (Fig. 865). Among cereals, an ageotropic mutant called 'lazy' (Fig. 866) is very common. Different types of lethals, part lethals, chlorophyll deficiencies and albinos (human as well as in many animals like elephants, tigers, snakes etc.) are common and well-known instances of gene mutation.

INDUCTION OF GENE MUTATION

Since environmental factors are known to affect the rate of gene mutation. Naturally, they have been employed for the artificial induction of mutations. Although attempts at induction of such mutation had been made before and Morgan had obtained some *Drosophila* wing mutants by radium treatment, the first concentrated work was by Muller who published in 1927 his work on X-ray induced mutations in *Drosophila*, a work for which he was awarded a Nobel Prize in 1946. This was followed by Stadler's work on barley and then spread to other fields.

The most important mutagenic agents have been found to be ionising radiations, viz., α- (projected nuclei of helium atoms), β- and γ- radiations of radioactive substances; X-rays ; neutrons (projected electrically neutral particles, mainly from hydrogen nuclei) and protons. All these act by ionising the atoms of the matter through which they pass so that they eject electrons. The ejected electrons, in their passage through matter, cause further ionisation losing energy with each collision until they finally halt. The actual cause of mutation is still a matter of speculation. One theory is that mutation is caused by the direct 'hit' of the 'target' by electrons. The other view holds the cause of mutation as indirect being resultant of the chemical and physical changes brought about in the molecules surrounding the genes. The α-, β-, γ-and X-rays are measured in *roentgen* (r) units which are products of the strength of the rays and the duration through which they work. It has been found that the rate of mutation is directly proportional to the amount of total radiation received (in r-units) by the cells and not to the strength or the duration (Fig. 867) alone and not even to the type of radiation employed.

Induction of mutation by radiation has been tried on all types of organisms. The resultant mutations are sometimes visible, sometimes latent, being apparent only in future generations as they mainly occur as recessive genes. In *Drosophila* 80% of the mutated genes have been found to be of the lethal type. The same may be true for other organisms. Mutagenic treatments may ensure one of getting mutants but the type of resultant mutations cannot be predicted.

Induction of mutation has been tried on plant viruses and on bacteria with variable degrees of success. Among fungi, mutations have been induced in *Neurospora*. There are numerous instances in higher plants—barley (Stadler and Gustafsson found hundreds of chlorophyll deficiencies and a few other types like the *erectoid* form), maize, tobacco,

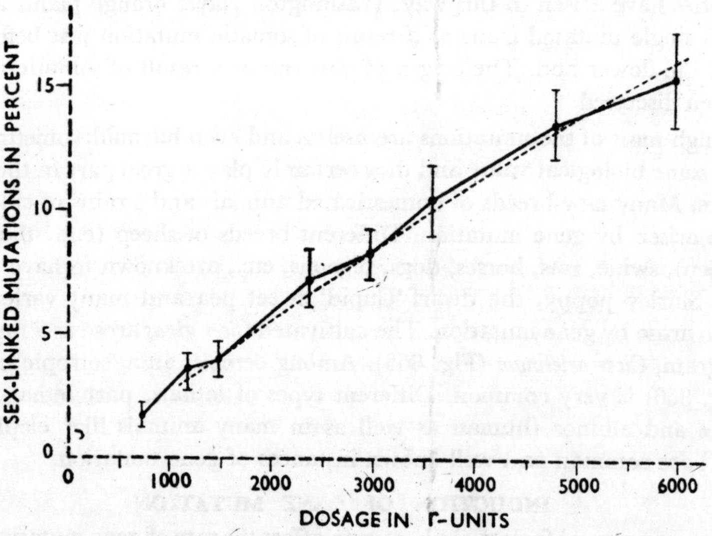

FIG. 867. Graph after Timoféeff-Resovsky showing that the rate of sex-linked mutations in *Drosophila melanogaster* is directly proportional to the amount of radiation applied.

Antirrhinum, *Oenothera*, *Datura*, etc., and they have all been found to be equally responsive. However, most spectacular results have been obtained in *Drosophila*. Besides the numerous recessive lethal genes (the X chromosome shows 1·5 lethal mutations in every thousand chromosomes per generation and this rate increases 70 times when the flies are subjected to a dosage of 4800 r-units) which are detectable only by adapting special cytogenetic techniques (c.f. *ClB* method), a large number of visible mutants have been obtained. The most interesting of these are the eye colour mutants. White eye ($\mathbf{w}^{+}$) mutates to eosin ($\mathbf{w}^{e}$) and apricot ($\mathbf{w}_{a}$) while coral ($\mathbf{w}^{co}$), buff ($\mathbf{w}^{bf}$), cherry ($\mathbf{w}^{ch}$), apricot ($\mathbf{w}^{a}$), eosin ($\mathbf{w}^{e}$) and tinge ($\mathbf{w}^{t}$) have been found to mutate to white. So the recessive white may also mutate to its dominants like eosin. These mutants are similar to the spontaneous mutations at the same locus.

Besides ionising radiations, non-ionising ultraviolet rays, chemical mutagens like those of the mustard gas group, and temperature have been found to be effective in inducing mutations but not to the same extent as the first. Mustard gas (tried on *Drosophila*, *Neurospora* and bacteria) has sometimes been found to be as effective as X-rays. Mutation rate generally increases with temperature but the rate of increase is different in different cases and there are some genes (specially the *unstable* ones) in which the mutation rate is lower in higher temperatures.

MULTIPLE ALLELOMORPHS

Since the same gene may mutate again and again, and all such genes will remain at the same locus of the same homologous chromosomes, it is apparent that there may be a series of genes occupying the same position. Then they will form a series of *multiple allelomorphs*, any two of which may act as an ordinary allelomorphic pair. The presence of multiple allelomorphs helps to understand the nature of alleles and disproves the *presence and absence hypothesis*. Common examples of multiple allelomorphs are found in the anthocyanin genes of cotton, rice, maize, jute and in the eye colour genes of *Drosophila*. The eye colour of *Drosophila* has already been mentioned. It is controlled by an allelomorphic series of at least twelve genes—Red wild type (**W** or +), white ($\mathbf{w}^+$), ivory ($\mathbf{w}^i$), pearl $\mathbf{w}^p$), tinged ($\mathbf{w}^t$), buff ($\mathbf{w}^{bf}$), honey ($\mathbf{w}^h$), apricot ($\mathbf{w}^a$), cherry ($\mathbf{w}^{ch}$), eosin ($\mathbf{w}^e$), blood ($\mathbf{w}^{bl}$) and coral ($\mathbf{w}^{co}$). In cotton, the cultivated Old World cottons *Gossypium arboreum* and *G. herbaceum* show an allelic series of some 20 genes at the $\mathbf{R_2}$ locus causing pigmentation of different plant organs and petals. The New World cottons *G. hirsutum* and *G. barbadense* contain another series of alleles also controlling anthocyanin colouration at the $\mathbf{R_1}$ locus in addition to $\mathbf{R_2}$. This proves the hypothesis that the New World cotton originated after hybridisation with the Old World cotton. A third similar allelic series of genes are known at the $\mathbf{R_3}$ locus in *G. anomalum*. In maize, the allelic series **P** in chromosome I, $\mathbf{A_1}$ in chromosome III and **R** in chromosome X control anthocyanin colouration. In rice, at least two different allelic series **C** and **Sp**, located in two different chromosomes, seem to control anthocyanin colouration. In Jute, 5 alleles at the **A** locus, also controlling anthocyanin colouration, have become known.

II. CHROMOSOMAL ABERRATIONS (INTERGENIC CHANGES)

As already explained, these result from *chromosomal aberration* disturbing the normal number or structure of the chromosomes. Thus, they may also be called *chromosomal mutations*. In these, the gene disturbance is *intergenic* not involving their internal structures.

According as the aberration involves the number or the structure of the chromosomes, these may be of two types:

A. Changes involving change in the number of chromosomes

The number of chromosomes for every species of organism is fixed. In higher plants and animals the ordinary cells contain $2n$ number of chromosomes (i.e., *two* sets of the same *genome*) and the organisms are, therefore, called *diploid*. It is found that this fixed number may change causing a mutation into a new type. If such a change be sufficiently great and hereditary, then the mutant forms a new species.

Change in the number of chromosomes sometimes involves multiplication of the whole chromosome sets (*genomes*), i.e., the number, instead of being $2n$, becomes $3n$, $4n$, $6n$, etc. These are generally called **Polyploids** or *true plolyploids* (**Euploids**). *Haploids* (n) caused by a decrease in genomes, is usually considered along with the Euploids. In other cases, the change involves increase or decrease in the number of single chromosomes and not of whole sets. Thus numbers like $2n-1$ or $2n+1$ may be obtained. These are *irregular polyploids* or **Heteroploids** or **Aneuploids**.

Besides the above, some changes in the chromosome number may be caused by fusion of chromosomes, by breaking down of chromosomes or by formation of chromosome-like bodies by inert material which may not contain any gene (**supernumerary chromosomes, e.g.,** *B-chromosomes* of maize and *m-chromosomes* of mosses).

(a) EUPLOIDS OR TRUE POLYPLOIDS

These again are of two types. Some involve the multiplication of exactly the same genome (**Autopolyploid**) while in others different types of genomes are involved (**Allopolyploid**). These polyploids have played an important role in the evolution of new species. Many cultivated crops have developed from wild plants by means of allopolyploidy. Sometimes a polyploid may be sterile but it may still be a very important agricultural or horticultural crop by vegetative propagation. Among animals, polyploids are rare.

Autopolyploids

Cytological investigations have shown the presence of polyploid series among some cultivated plants. Thus, Fig. 868 shows the chromosomes of diploid ($2n$), triploid ($3n$),

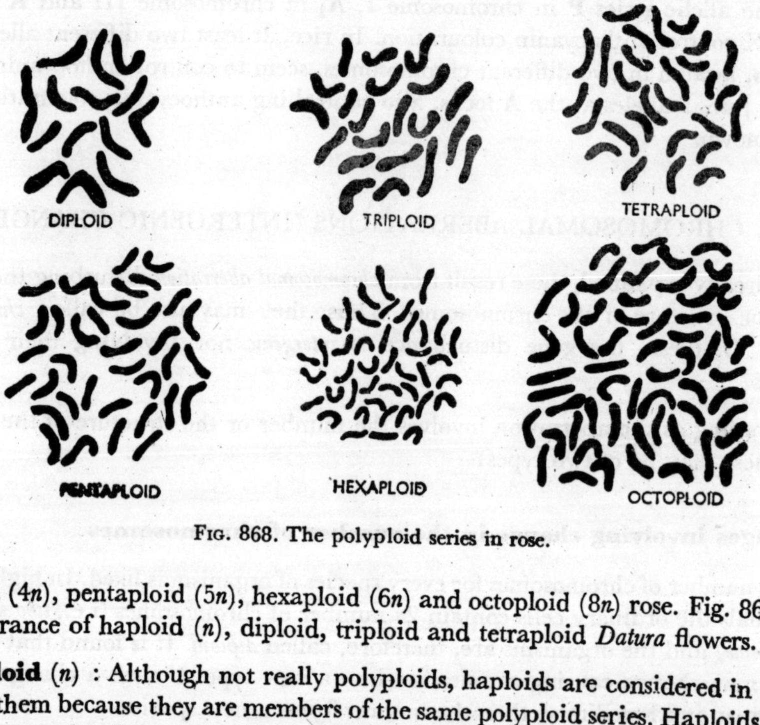

FIG. 868. The polyploid series in rose.

tetraploid ($4n$), pentaploid ($5n$), hexaploid ($6n$) and octoploid ($8n$) rose. Fig. 869 shows the appearance of haploid (n), diploid, triploid and tetraploid *Datura* flowers.

Haploid (n) : Although not really polyploids, haploids are considered in connection with them because they are members of the same polyploid series. Haploids usually occur as a result of parthenogenesis—whether spontaneous or induced. They have been found in rice containing double embryos of which one is normal and the other is parthenogenetic. The stimulus of parthenogenesis may come from ineffective pollination when foreign pollen or X-rayed pollen is used in pollination. Haploid plants have been found in a number of species like *Datura stramonium*, *Oryza sativa*, *Oenothera* sp., *Triticum* sp., *Zea mays*, etc. Haploid plants are all-round weak and small but there is a report of a haploid plant in pepper as healthy as a normal diploid. As there is only one set of chromosomes, there should be no pairing during meiosis so that normal pollens or eggs are not formed. A normal gamete is possible only if by chance all the chromosomes remain at the equator and form a single nucleus during irregular meiosis. Haploids are, therefore, usually sterile. In some cases, however, bivalents are noticed during meiosis. This indicates that in the original genome duplication of chromosomes was already present and the basic number of the species is actually less than the haploid number. This throws a light on the origin of the species. Haploids may be utilised in cytogenetics in another way. By applying colchicine (discussed below) the chromosome set in the haploid may be doubled and normal diploid may be obtained. The diploid will be perfectly homozygous—a condition rarely obtained in nature. Such a diploid plant is of great use in cytogenetics and plant breeding.

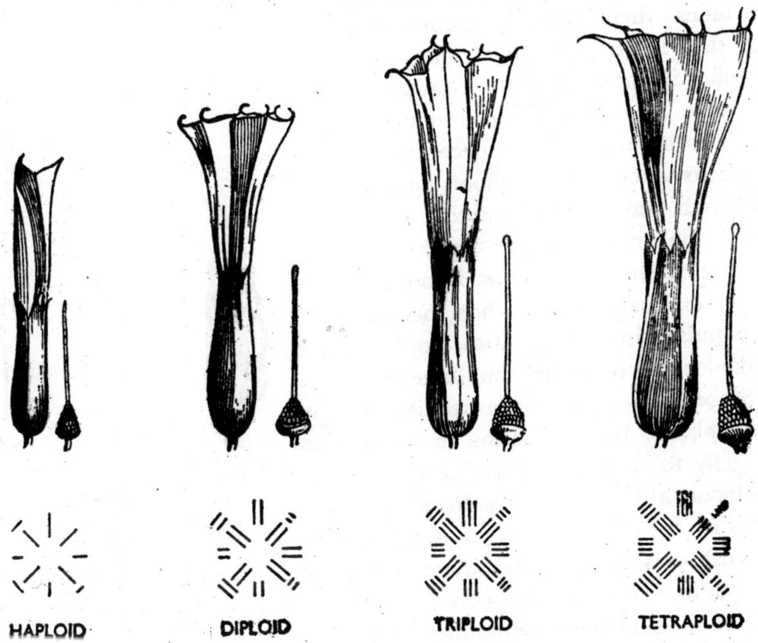

FIG. 869. *Datura* polyploid series (*after Blakeslee*).

In the animal world, the males of bees, wasps and other *Hymenoptera* are normally haploid. They have adjusted their lives to this nature. There is no reduction during meiosis. There is no synapsis and either during the first division all the chromosomes normally pass to one pole or the first division is completely eliminated. The meiosis, therefore, is equational and normal sperms are formed.

Triploid(3*n*) : Autotriploids have been observed or obtained like the haploids in a number of plants like *Oenothera*, *Datura*, rose, rice and many others. *Drosophila* triploids are well known. Triploids are usually formed as a result of the fusion of a diploid gamete (formed abnormally) with a normal haploid one. Diploid pollens or eggs may come from tetraploid plants or sometimes occur abnormally under the influence of X-rays, etc. As in the case of haploids, the three sets of genomes are unable to pair during meiosis so that functional gametes are rarely formed. As a result, triploid plants are usually sterile like the haploids. But, triploids may be useful in horticulture as, when propagated vegetatively, they have been found to develop seedless fruits in many cases. Triploid plants may or may not be bigger than normal diploids.

Tetraploids (4*n*) : Tetraploids are the most important of the polyploids. They are found widely in nature and may also be induced artificially Numerous cultivated varieties have autotetraploid varieties along with normal diploids. Since the four sets of chromosomes may easily pair between themselves normal diploid gametes are formed which form normal tetraploid seeds. During diakinesis one often sees tetravalents of chromosomes instead of the bivalents seen in normal diploids. Nevertheless, varying degrees of sterility are observed among tetraploids. Tetraploid plants are usually larger, more succulent and have bigger pollens, stomata, fruits, etc. The succulence may render the tetraploid more susceptible to disease. Their relative sterility often precludes their use as commercial crops but the large size of fruits, etc., may be useful in horticulture. Tetraploids sometimes arise as somatic bud mutations. They have been obtained by the decapitation of tomato plants when the new shoots become tetraploid. High temperature induces tetraploidy. But, the most popular method is treatment with chemicals, the most important of which is *colchicine* which has become extremely popular since its practical application by Blakeslee and by Nebel simultaneously in 1937. Since then it has been applied to thousands of plants. Its application in liquid form or as jelly to seeds and growing tips is not very difficult although the results are not always certain. In the simplest cases, seeds soaked in dilute solutions germinate into tetraploid plants. Colchicine prevents the separation of daughter chromosomes after they have divided by arresting the formation of the spindle and forms nuclei with double the number of chromosomes ($4n$) during mitosis. As a result, a $4n$ tissue arises forming a diploid shoot.

Higher-ploids: *Pentaploids* ($5n$), *Hexaploids* ($6n$), *Octoploids* ($8n$), etc., have also been found among wild and cultivated plants (e.g., rose in Fig. 868). All these show how the formation of new species is facilitated by polyploidy.

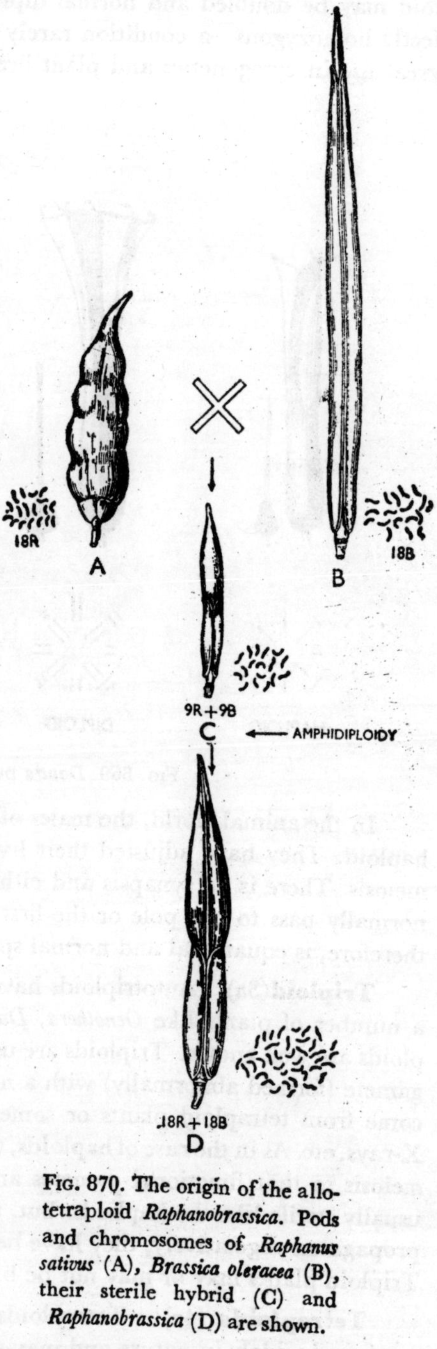

FIG. 870. The origin of the allotetraploid *Raphanobrassica*. Pods and chromosomes of *Raphanus sativus* (A), *Brassica oleracea* (B), their sterile hybrid (C) and *Raphanobrassica* (D) are shown.

Allopolyploids

While the autopolyploids are formed by the multiplication of the same set of chromosomes (genome) all the allopolyploids are formed by the union of different genomes from different plants.

A very interesting example of an allopolyploid obtained artificially is the *Raphanobrassica* (Fig. 870) obtained by Karpechenko (1927). He crossed *Raphanus sativus* (radish, $2n=18$) with *Brassica oleracea* (cabbage, $2n=18$). The hybrid was intermediate between radish and cabbage and had 9+9 (9R of *Raphanus* and 9B of *Brassica*) chromosomes which did not pair during meiosis as they were from different plants. The hybrid was sterile. Accidentally, polyploidy occurred. The 9R+9B chromosomed hybrid doubled its chromosomes (18R+18B). Now, the chromosomes could pair between themselves and perfect seeds were formed. This pairing of chromosomes is *autosyndesis* as, actually, chromosomes of similar genomes (R or B) are pairing with their respective homologues. The 36 chromosomed final plant is a tetraploid but it is an *allotetraploid* as the 36 chromosomes are coming from two different genomes. This sudden doubling of chromosomes which formed the *allotetraploid* is called **amphidiploidy.**

Subsequently, it has been found that such amphidiploidy may be induced artificially by colchicine. As a result, some hybrids which were sterile have been transformed into fertile allotetraploids by the induction of amphidiploidy.

An examination of the cultivated plants shows that allopolyploidy has played a very important role in the evolution of these species. Cytogenetical analysis of cultivated wheat plants have yielded very valuable results. Investigations by Kihara and others show that the wheat species fall under three groups—diploid, allotetraploid and allohexaploid. These were formed from three different genomes.

TABLE III

GENOMES IN WHEAT (*c.f.* FIG. 871)

		Diploid chromosome No.	Genome Constitution
EINKORN GROUP	*Triticum monococcum*	14	2A (diploid)
EMMER GROUP	*T. drurum* (hard wheat), etc.	28	2A+2B (allotetraploid)
VULGARE GROUP	*T. aestivum* (=*vulgare*) (bread wheat), etc.	42	2A+2B+2C (allohexaploid)

The genome C is present in the grass *Aegilops squarrosa* ($2n=14$) which grows wild in the region from Armenia to Afghanistan. Thus, in the evolution of the wheat species, one may suppose that *Triticum monococcum* (genome A) hybridised with another plant having genome B and also having $2n=14$. There was amphidiploidy. *T. durum* (2A+2B, $2n=28$) resulted. Again, a member of the *durum* group hybridised with *Aegilops squarrosa* (genome C, $2n=14$) and an amphidiploidy gave *T. aestivum* (=*vulgare*) (2A+2B+2C, $2n=42$). Actually two such hybridisations have been successfully done. Mcfaden and Sears obtained a wheat-like *Triticum spelta* (a member of the 42 chr. *vulgare* group) by inducing amphidiploidy on a *T. dicoccoides* (a wild member of *emmer* group from the region Armenia to Palestine) × *Aegilops speltoides* hybrid. Kihara has actually obtained a bread wheat similar to the cultivated *T. aestivum* (=*vulgare*) var. *erythospermum* by amphidiploidising a *T. persicum* (a cultivated *emmer* wheat) × *Aegilops squarrosa* hybrid. It has now been established that the B genome has been derived from *Aegilops speltoides*. It is reasonable to suppose that the bread wheat originated in some place between Armenia, Persia and Afghanistan.

If the genome constitution of a natural species shows that it is an amphidiploid or allopolyploid, it may be possible to *synthesise* the species artificially. A very interesting example is that of *Brassica juncea*. Among cultivated mustards there are three species —*Brassica campestris* (rape, $2n=20$), *Brassica nigra* (black mustard, $2n=16$) and *Brassica juncea* (rai, $2n=36$). Genome analysis showed that the third species was an amphidiploid of the first two. Subsequently, a plant has been obtained synthetically by inducing amphidiploidy in the sterile *Brassica campestris* × *Brassica nigra* hybrid (Fig. 872). This synthetic plant differs only slightly from natural *Brassica juncea*.

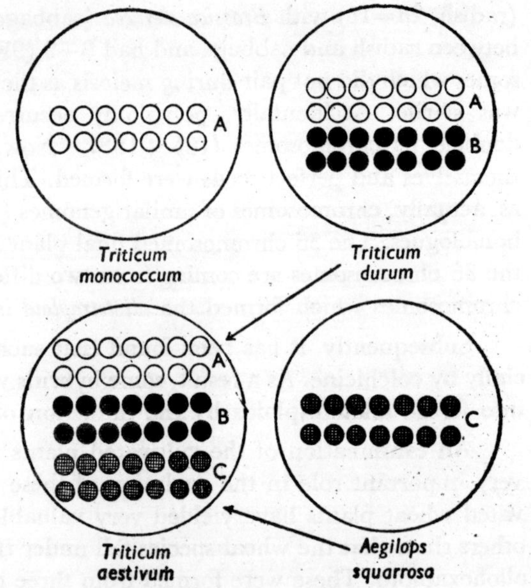

FIG. 871. Genomes in wheat.

Segmental allopolyploids

In the allopolyploids discussed above it has been presumed that the chromosomes of the two parents have no affinities between themselves and so, there is absolutely no pairing between themselves. But, Stebbins is of opinion that in nature, in most allopolyploids there is some affinity between the chromosomes of the two different genomes resulting in *allosyndesis* for some chromosomes. During meiosis of these allopolyploids some tetravalents as well as some bivalents are seen and this results in some sterility. These have been named *segmental allopolyploids* and examples are found in *Primula kewensis* (derived from *P. floribunda* × *P. verticillata*), *Tradescantia canaliculata-himulis*, *Delphinum gyposophilum* (derived from *D. hesperium* × *D. recurvatum*), etc.

(b) ANEUPLOIDS OR HETEROPLOIDS OR IRREGULAR POLYPLOIDS

In the **aneuploids**, one is concerned not with the multiplication of whole sets of chromosomes but with the increase or decrease of the number of homologues. Ordinarily, two homologues of each chromosome are present in the diploid so that the chromosome number is $2n$. But, in exceptional cases there may be disturbances in the division of chromosomes so that the number of homologues instead of being two is changed to three (*trisomic*), four (*tetrasomic*), one (*monosomic*) or none (*nullisomic*). Such disturbances may be spontaneous due to crossing between polyploids and heteroploids, different types of non-disjunction (*aberration*), incompatibility due to hybridisation between distant plants, etc., or may be induced by X-rays, etc. Aneuploids are usually sterile. When aneuploids show an *increase* in chromosome number (e.g., $2n+1$) it may be called a **hyperploid**. When there is a decrease (e.g., $2n-1$) it is a **hypoploid**.

Trisomics ($2n+1$): These are formed by the addition of an extra homologue of one chromosome. Many of the *Oenothera* 'mutations' found by de Vries are trisomics. Trisomics are widespread in nature and have been extensively studied in *Datura* and also in maize, tomato, wheat, *Nicotiana* and *Drosophila*. They are readily obtained by selfing a triploid or by crossing diploids with triploids.

Blakeslee and his students studied the *Datura* trisomics very closely and obtained 12 trisomics (Fig. 873) for the triplication of every one of the 12 chromosomes in *Datura*

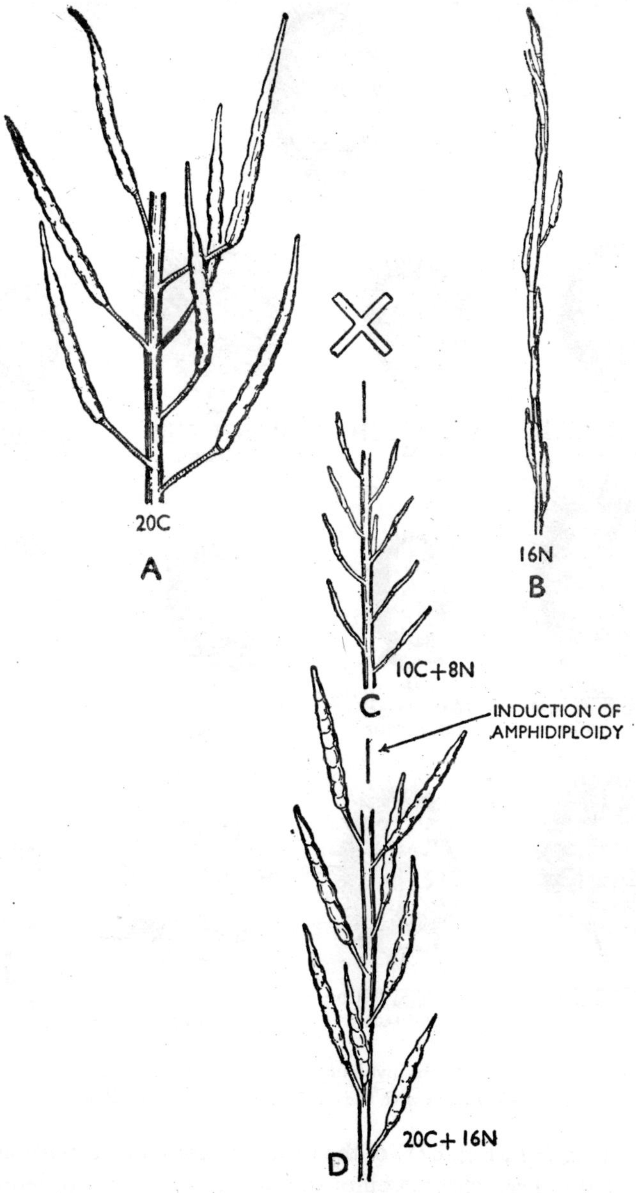

Fig. 872. Synthesis of *Brassica juncea*. A. *Brassica campestris* with 20C chromosomes. B. *Brassica nigra* with 16N chromosomes. C. Hybrid A×B with 10C+8N chromosomes. D. Synthetic *Brassica juncea* obtained by doubling the chromosomes in C.

stramonium. As can be seen from the figure, every one of the trisomics is distinct from the other. Besides these *normal* ones they also obtained other trisomics in which the three chromosomes were exactly alike, there being some segmental interchanges or translocation. In a **secondary trisomic** the translocations are from the same chromosome.

Thus, while three 1·2 chromosomes give the *rolled* normal trisomic (Fig. 873), two 1·2 and one 1·1 gives *sugarloaf* and two 1·2 with one 2·2 gives *polycarpic*. In a secondary

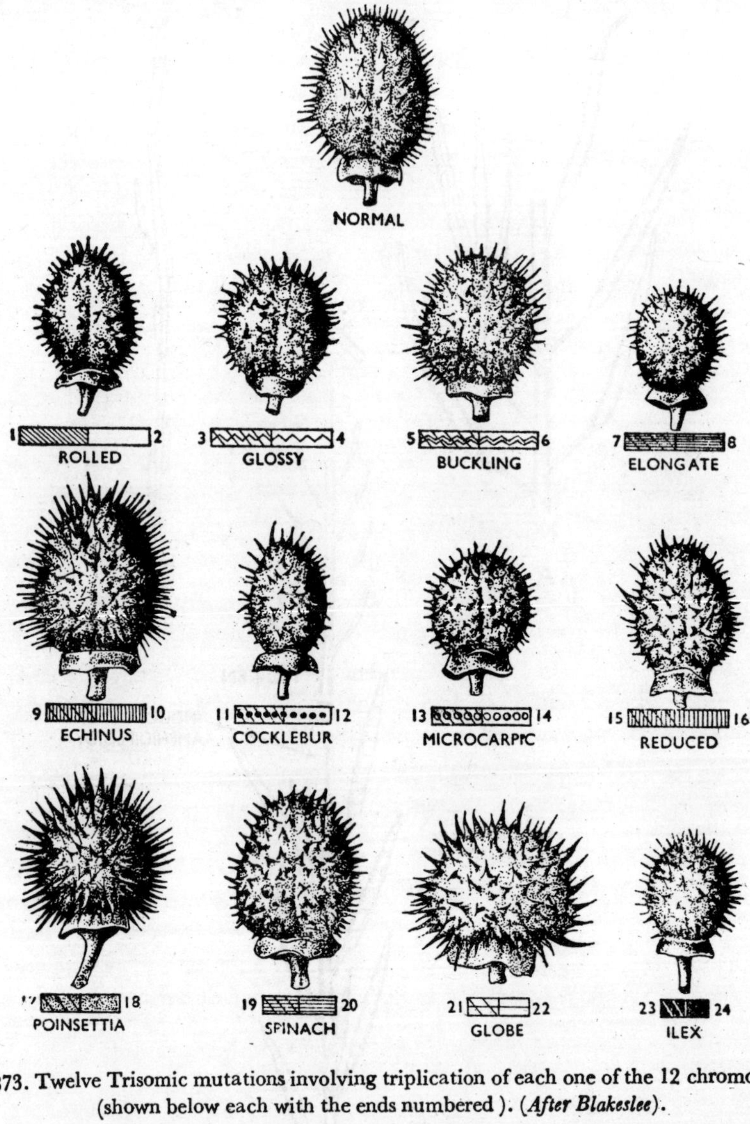

FIG. 873. Twelve Trisomic mutations involving triplication of each one of the 12 chromosomes (shown below each with the ends numbered). (*After Blakeslee*).

trisomic a closed ring (*c.f.*, Fig. 877) of the three chromosomes is possible. In a **tertiary trisomic**, one of the three chromosomes contains a bit from a different chromosome. Thus, the trisomic *hedge* has one 1·9 chromosome (produced by translocation between 1·2 and 9·10) in addition to two 1·2 chromosomes.

Sears obtained trisomics in bread wheat (*Triticum aestivum*), but these do not differ much from the normal plants, possibly because the species is an allohexaploid.

A trisomic (47 instead of the $2n=46$ chromosomes) mutation in human beings is known to cause the disease 'Down's syndrome' (mongoloid features combined with mental slowness).

Double trisomic ($2n+1+1$) shows three homologues of each of two *different* chromosomes.

Tetrasomic ($2n+2$) has four homologues of the same chromosome. Sears obtained these as well in *Triticum aestivum* but these also do not differ from normal plants.

Monosomic ($2n-1$) : There are two homologues of each chromosome excepting one of which there is one only. Apparently, the homologue of this chromosome is somehow lost. Monosomics arise in the same way as the trisomics but they are not usually viable. *Datura* monosomics are not viable. But, while the monosomic are not viable in true diploids, they have been successfully obtained in some allopolyploids. In the allopolyploid *Nicotiana tabacum* ($2n=48$) Clausen and his colleagues obtained all the 24 possible monosomics and these are different from one another. Similarly Sears obtained all the 21 possible monosomics in *Triticum aestivum* (=*vulgare*) but these do not differ much from the normal. Monosomic analysis greatly facilitates assignment of genes to the linkage groups.

Nullisomic ($2n-2$): In these plants, both the homologues of a particular chromosome somehow get lost so that the chromosome is completely missing from the plant. Such a plant should clearly show what genes were contained in the missing chromosome. Nullisomic analysis, combined with monosomic, has greatly helped in determining linkage groups. Nullisomics are obtained by selfing monosomics. They are usually inviable like the monosomics. They are, however, known in maize and in *Triticum aestivum* (=*vulgare*) Sears obtained 17 of the 21 possible nullisomics.

B. Changes involving change in the structure of chromosomes

Some accidents sometimes occur which end in the breaking-down of chromosomes. The broken bits may get healed up or get re-attached in a wrong way or may even get lost. These accidents are not to be confused with the normal crossing-overs. Such incidents cause structural modifications of chromosomes involving a re-arrangement or loss of genes which may influence heredity by causing aberrations and certainly influence linkage and crossing-over. In some cases, it has been found that a gene located at one position on a chromosome behaves differently when placed on a different position. This is known as the **position effect,** A very striking example of this is the '*bar-eyed*' character of *Drosophila*. While structural modifications of chromosome occur in nature it has been possible to get a great number of them by subjecting dividing cells to harsh treatment, chiefly by X-rays and other ionising radiations in the same way as explained for gene mutation. Structural modifications may be of several types (Fig. 874).

Deficiency: A *deficiency* has a *bit* of a chromosome lost altogether. Some genes are, therefore lost. A deficiency may be *terminal* when it involves the end of a chromosome, or *intercalary* when it is an intermediate part that is deficient. Intercalary deficiency is also called **deletion.** Both terminal and intercalary deficiencies are known in maize. They may arise spontaneously or as a result of artificial radiation.

Deletions have been extensively studied in *Drosophila* where the salivary glands have enabled another method of location of genes. It has been explained that in a salivary gland chromosome the two homologues remain in a stage of synapsis. If a mutation arises by deletion and if this mutated individual be crossed with a normal, the hybrid will be a heterozygote and in its salivary gland cells there will be the pair of chromosomes of which one homologue is normal and the other deleted. Close examination of the band of this chromosome shows exactly which bands are missing and, since the two homologues pair band by band, there will be a short curvature at the point of deletion

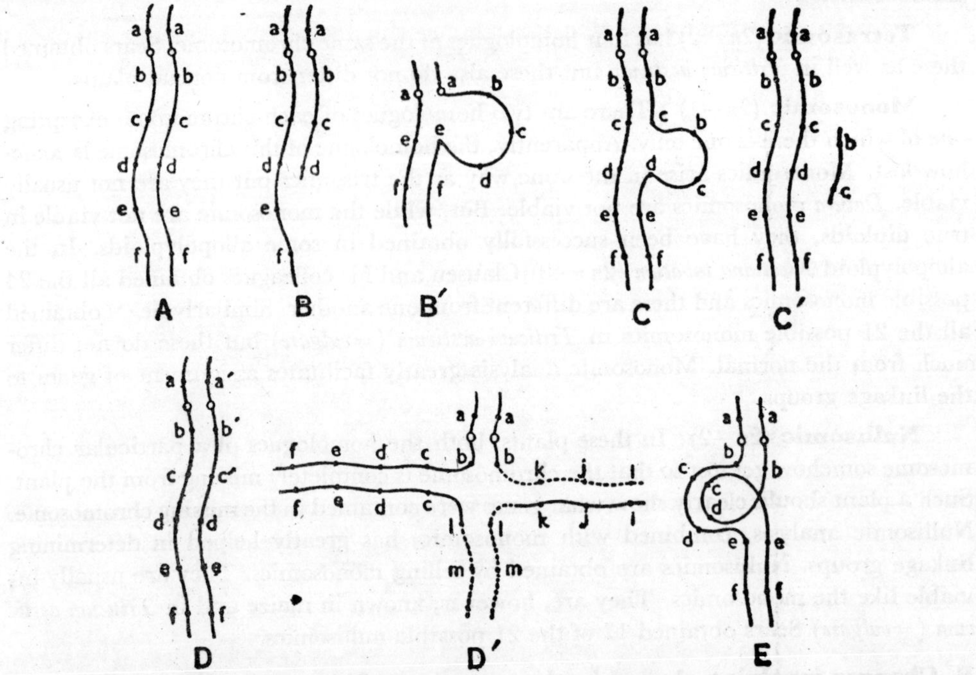

FIG. 874. Structural modifications of chromosomes. A. Pachytene pairing of normal homologues with genes *a, b, c, d,* etc. The other diagrams show pairing involving one homologue with terminal deficiency (B), intercalary deficiency (B′) duplication (C), duplication in a fragment (C′), translocation involving homologues (D), translocation involving non-homologues (D′) and inversion (E)

since one homologue is shorter by a few bands there (Fig. 875). By this observation, it is possible to locate the gene on the particular chromosome and by studying a number of such deleted mutants in the same way it is possible to get a sort of chromosome map as obtained by studying crossover percentages (Chapter LXV). It is found that the two types of chromosome maps tally with each other keeping in mind that the crossover values at certain regions of the chromosomes are known not to be exactly proportional to the distance.

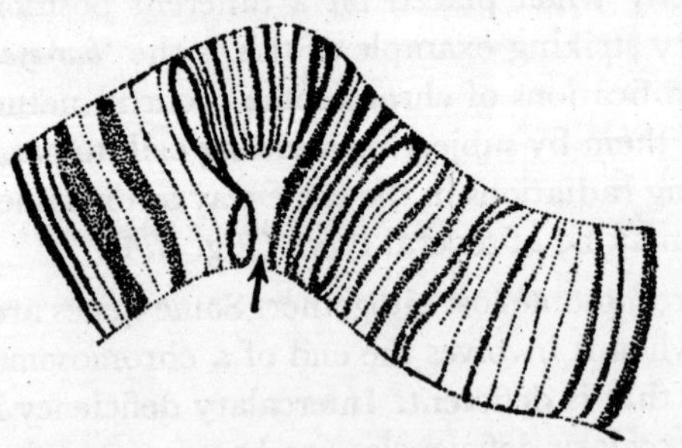

FIG. 875. A salivary gland chromosome of a hybrid *Drosophila melanogaster* showing a deletion in the lower chromosome of the synapsed pair (see text). The arrow shows the region of deletion.

Deletion causes a disturbance in the genic balance and gametes with deleted chromosomes are often inviable. In *Drosophila,* deletion of a bit containing more than 50 bands has lethal effect.

Duplication: A broken bit of a chromosome may remain free in the nucleus as a *fragment* in addition to two complete homologues. However, no such fragment can survive if it does not contain a centromere. Thus, some alleles will be represented thrice. The broken bit, instead of remaining free may also remain attached to some other broken chromosome (which may or may not be its homologue) at an intercalary position. It should be remembered that there can be no attachment to the unbroken telomere end.

Translocation: A broken bit of a chromosome may get attached to some other chromosome. Translocations are usually *reciprocal*—somewhat resembling crossing-over—but very different from the latter as whole chromosomes are involved here. Such reciprocal translocation may involve homologous or non-homologous chromosomes. *Simple translocation* of only one bit of a chromosome to another is extremely rare. If that rare event happens, the broken bit may even get re-attached in a different position on the mother chromosome.

Inversion: A segment of a chromosome gets inverted during reattachment. Thus, a chromosome having the genes *abcdef* in linear order may get the segment *cd* inverted. Then the new arrangement will be *abdcef*.

Fig. 874 shows that structural modifications cause typical appearances during pachytene pairing as the allelic genes tend to come side by side.

Inversions also are widely found. An inversion not containing a centromere (called a *paracentric* inversion) and followed by a crossover may cause an anaphasic *inversion chromatid bridge* and an *acentric fragment* as explained in Fig. 876.

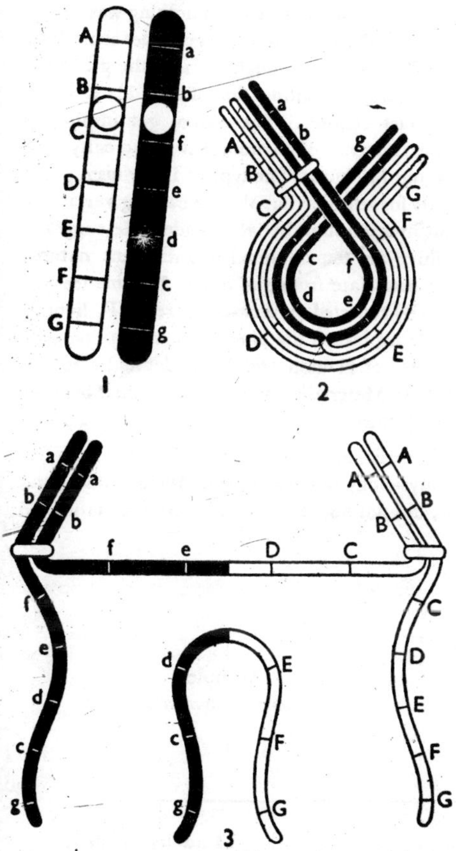

FIG. 876. An inversion not including a centromere giving rise to a *chromatid bridge* and an *acentric fragment*. (1) The two homologues, one showing a *paracentric inversion* of the *cdef* region not involving a centromere. (2) Diplotene pairing showing a chiasma in the inverted region which ends in a crossover. (3) Anaphase showing the *dicentric chromosome* forming a *chromatid bridge* and an *acentric chromosome* which is usually cast off later.

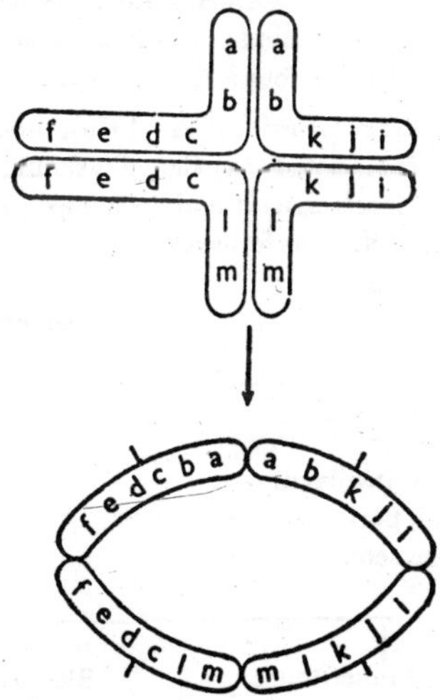

FIG. 877. Four chromosomes involving translocation of segments forming a ring of four chromosomes at metaphase.

Translocation is widely present in plants and animals—in *Datura*, maize, pea, wheat, *Tradescantia*, *Rhoeo*, etc. Segmental interchanges between chromosomes (translocation) lead to *ring formation* (Fig. 877).

CHAPTER LXVII

LYSENKO GENETICS

For twenty years some new ideas dominated Genetics in Soviet Russia. These ideas caused a great sensation and were variously named 'New Genetics', 'Soviet Genetics' or 'Lysenko Genetics' as opposed to 'Mendelian Genetics'. Although the most sensational doctrines of this new Genetics have lost their force even in Russia, it is worth while to know what it was all about as a new angle of vision is good for any growing science.

Soon after 1930, new ideas about Genetics and Heredity began to be expressed in Russia. By 1939 this new school of Genetics was in the ascendency and gradually the Mendelian school almost disappeared from the U.S.S.R. The leader of this new school was T.D. Lysenko, an eminent plant physiologist whose reputation was established by his theory of phasic development and his experiments on vernalisation.

The thoughts guiding this new school was a link from the Darwinian days when nothing was known about chromosomes and the physical basis of Mendelian Genetics. The particulate idea of genes controlling hereditary characters was unknown in those days and hybridisation was thought as something like the mixing together of maternal and paternal fluids. Lysenko Genetics is a continuation of this idea ignoring entirely the development of Mendelian Genetics during the last century. The school got inspiration from the views expressed by Lamarck and Darwin about the influence of environment on heredity and from the works of Naudin, Millardet, Timiriazev, Burbank and Michurin. The type of Darwinism popular in Russia holds environment or Nature responsible not only for selection but also for causing variations. Naudin (France, 1863) gave the first clear hypothesis about hybridisation before Mendel. He suggested that the hybrid should be regarded as an individual in which are found united two different natures, each having its own mode of life and trying to assert on and dissociate from the other. Dominance and segregation is the result of this struggle. Millardet (France, 1894) reported some cases where the F_1 hybrids bred true and did not show any segregation in the F_2. He explained the phenomenon as a complete annihilation of the properties of the recessive parent. The ideas of Naudin and Millardet had a special appeal to the Soviet mind as it is in accordance with Dialectical Materialism in which a conflict between *thesis* and *antithesis* before a final establishemnt of *synthesis* is pictured.

Timiriazev (Russia, 1843—1920), though a plant physiologist, made full study of Genetics from this new angle. He insisted that Genetics should be studied as physiology of development. Weismannism and Mendelian Genetics, as it had developed in the early days, did not appeal to him. Timiriazev tabulated types of inheritance as follows:

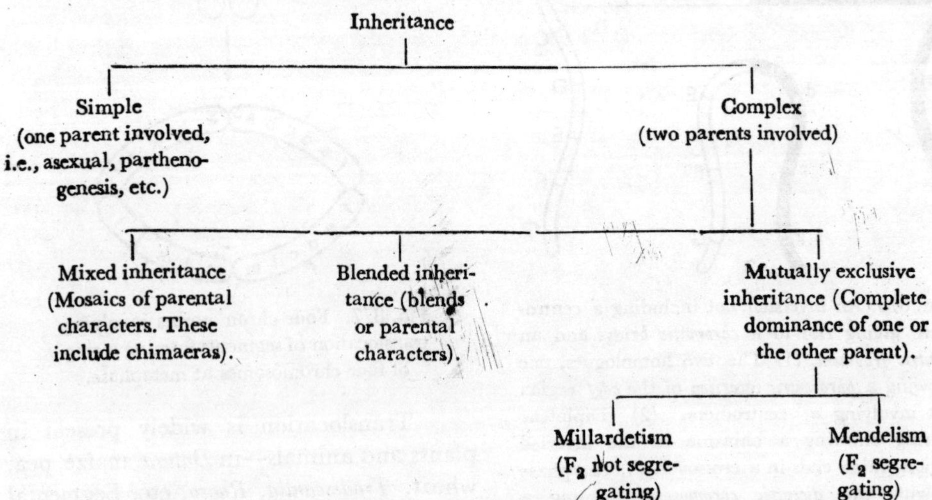

Luther Burbank (U.S.A., 1849—1926) and I.V.Michurin (Russia, 1855—1935) were practical plant breeders rather than geneticists and they were convinced about the influence of environment and nutrition on heredity. Burbank declared "*Heredity is the sum of all environment of a complex ancestry back to the beginning.*" Burbank developed and introduced thousands of new plant varieties. Michurin spent a lifetime in the same pursuit as Burbank. He was similarly convinced about the all important influence of environment in changing plants. He developed some spectacular techniques in horticulture and some novel theories. One is his **'mentor'** theory which states that if an F_1 hybrid is used as a stock on which another plant is used as a scion, the latter plant may act as a *mentor* in changing the genetical constitution of the hybrid so as to bring it nearer to the *mentor*. Mendelism, according to him, is valid only under particular environmental conditions. Under other conditions results may be entirely different, e.g., the dominant may become the recessive.

Trofym Denisovitch Lysenko (Ukranian, born 1898) and his followers gave shape and idea to this new line of thought. This school does not believe in the Mendelian concept of Genetics controlled by chromosomes or genes or in any such control. According to them Genetics and Heredity are purely physiological phenomena. Lysenko defines heredity as "*the property of a living body to require definite conditions for its life, its development and to react definitely to various conditions.*" Heredity is thus, not a result of interaction of factors or genes but response of the plant as a whole to a complex set of internal and external environmental factors most of which are not well understood.

Of special importance in this new Genetics is the role of environment in giving expression to the heredity of the plant. There is nothing like *pure line* as heredity is being constantly changed. If the environment can be changed the expression of heredity will change. Environment is the all important thing and plants gradually change because of changes in the environment whose effect on a plant is direct. Most astounding were the claims that *Triticum durum* gets converted into *Triticum aestivum* (=*vulgare*), rye changes into wheat while oats and barleys develop from wheat in one or two generations. Very interesting in this connection is the claim put forward about the formation of **true graft hybrids** as a result of grafting. The reasoning is that as the scion is getting part of its nutrition from the stock and *vice versa*, it is natural that each of them will be influenced by the other. Thus, it is possible to get true hybrid seeds (not chimaeras or mixtures of tissues) from either of them. A hybrid combines the adaptations to the environments of two plants. All environmental modifications are inherited. Frequent crossings *within* a variety are necessary to keep up the vigour of a race. Biometry was thought to be a very faulty science by this school of geneticists.

FIG. 878. Trofym Denisovitch Lysenko (Born 1898)—*the man behind this new Genetics.*

Lysenko lost his supremacy since the setback of orthodox Stalinism in Russia (about 1957) and the so-called 'Mendelian geneticists' regained their status.

During the twenty years of the dominance of this school of thought in Russia there were much harsh words, specially between the Lysenkoists in Russia and the Mendelians in the U.S.A. but, not considering this, a geneticist should consider whether there is anything in this Lysenko Genetics worth consideration. There may be something in the Russian claim of the importance of environment in causing some plasticity of the gene. Already the idea of the evolution of modifying genes and polygenes has rendered our concept of the gene a little more fluid. The idea of *pure lines* also has undergone some modifications as it is impossible to get homozygosity with so many insignificant genes. Mendelian genetics, based on genes and chromosomes, is now very well established. but, this has been modified in the past in the light of fresh discoveries and may again infuse something from the Lysenkoist idea of the plasticity of the gene. Some modern Russian geneticists (who are the products of the Lysenko School) seem to be doing wonderful work. In 1954, Academecian Tsitsin gave an account in India of how they had succeeded in getting hybrids between very distant plants following Michurin's technique and how they were progressing towards developing perennial wheat plant and peas growing on perennial trees.

CHAPTER LXVIII

ORIGIN OF SPECIES : MODERN CONCEPT

With a clear idea of the mechanism of heredity based on genes and chromosomes we may now re-examine the question of the origin of new species or that of speciation.

Lamarck and Darwin based their ideas of origin of species on the *inheritance of acquired character*. Lamarck believed that variations arise as a direct effect of environment or conscious effort or use and disuse. Darwin was vague about the actual origin of the variations. He pointed to the small fluctuating variations in nature and supposed that because of natural selection some of these variations survive. There was some confusion about this very conception of **acquired characters.** With the definition that *an acquired character is any variation acquired by an organism during its own lifetime and not inherited from its progenitors*, we include many variations which, *prima facie*, cannot be inherited. It was, therefore, very easy for Weismann to completely deny the inheritance of acquired characters because a soldier's war scars cannot be inherited, the tails of mice or dogs cannot be shortened by generations of docking, the mutilated feet of Chinese ladies did not become smaller although they were kept bandaged for generations, and so on. Even the facts that the queen bees become abnormally large because of the special food 'royal jelly', that *Ranunculus aquatilis* produces two types of leaves under aquatic and terrestrial conditions and that *phenocopies* (Chapter LXIII) can be obtained artificially, are not questions of inheritance but of *morphogenesis*. But, while the neo-Lamarckians were wrong in making a dogmatic assertion that all acquired characters were inherited the Weismannians also were equally wrong in categorically stating that no acquired character could be inherited. As a matter of fact, all variations should not be placed under the same category. This was realised at the beginning of the present century and so de Vries and others classified the variations in nature. De Vries made a distinction between 'fluctuating variations' and 'mutations' and contended that only the latter are hereditary variations and they alone give rise to new species.

The gene theory today tells us that a variation, in order to be inherited must be a function of the genes and the chromosomes. A new species cannot originate without a change in the chromosomes. The different ways of the mutation of the gene and the chromosome have been explained in the last chapter. Today, the only possible method of the origin of new species seems to be either through these muations or through fresh natural selections from the segregates of hybridisation. Mutation and hybridisation supply the raw material out of which new species arise as a result of natural selection. Mutant genotypes undergo combinations in a population and favourable recombinations survive. But, these mutations are slightly different from the mutations as understood by de Vries. Although de Vries gave us the 'mutation theory' his idea of mutation was not as clear as it is to the geneticists today. Even among what he considered as mere 'fluctuating variations' were some mutants—mutations caused by very minor *modifying genes* or *polygenes* whose effects are almost insignificant though hereditary. Mutations of the three types viz., (a) gene mutations, (b) variations in chromosome numbers—mainly polyploidy and (c) structural variations in chromosomes, have all been found very important in the origin of new species.

The gene mutation, if sufficiently pronounced, or when a number of gene mutations accumulate, can easily give rise to new species as they always have done. The importance of polyploidy in giving rise to new species has already been stated. All the advanced plant groups are dominated by species arising through polyploidy. Müntzing and others are of opinion that more than half the species today are of polyploid origin. If polyploidy is important, structural changes in chromosomes are not less so. An idea may be made about the origin of a number of related species by a close study of their *karyotypes* through hybridisation, etc. *Karyotype* means the basic chromosomes of a plant keeping in view their structure, size and number. These chromosomes may be put in the

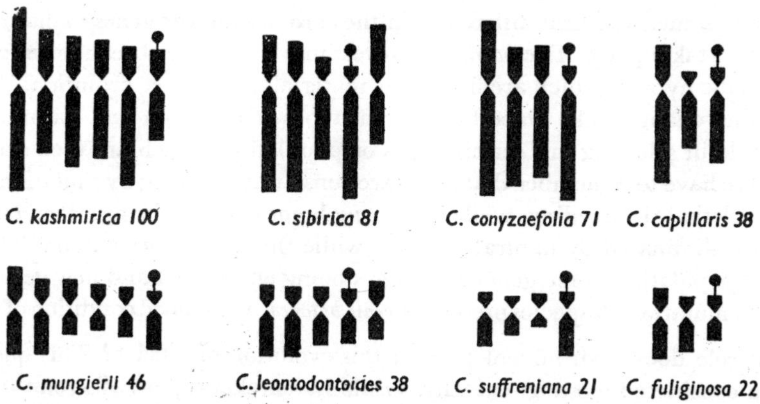

FIG. 879. Idiograms showing karyotypes of eight species of *Crepis* with their relative total chromosome lengths shown at the end of each name (After Babcock, Stebbins and Jenkins).

form of a diagram called the *idiogram*. Karyotype analysis was originally done thoroughly in the genus *Drosophila* as also in the genera *Crepis* and *Oenothera* among plants. Later, however, a large number of plants have been extensively analysed from the point of view of karyotypes. It has been shown that not only the species but even strains of the same species may differ in chromosome morphology. The role of structural alterations of chromosomes in the evolution of species was not properly assessed in the past. In *Crepis*, Babcock and co-workers have studied 113 of the 196 recognised species of the genus. Chromosomes of 96 of them have been examined. The haploid numbers of 3 of them is 7, 14 have 6, 19 have 5, 57 have 4 and 3 have 3. Figure 879 shows the karyotypes of eight of them with the total chromosome length of each. Babcock has shown that the lower chromosome numbers have evolved from the higher, mostly by translocation and loss of some chromosome parts. Figure 880 shows how Tobgy explained the origin of *C. fuliginosa* ($n=3$, total chromosome length 22) out of *C. neglecta* ($n=4$, total chromosome length 31).

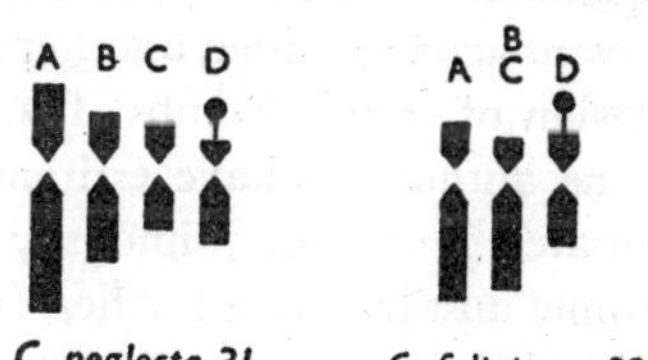

FIG. 880. Origin of *Crepis fuliginosa* out of *C. neglecta*. Translocations involving B and C chromosomes of the latter, possibly, caused the origin of the BC chromosome of the former. (After Babcock, Stebbins and Jenkins).

Taxonomists say that the idea of 'species' is rather fluid. They often change their opinion whether a group of plants is to be considered a 'species' or a 'subspecies' or a 'variety'. A cytotaxonomist will be satisfied that a 'species' has evolved when the population shows a definite chromosomal difference in number or structure and, above all, if it fails to hybridise naturally with other species. Thus, Mayr defines a species as a *group of actually or potentially inbreeding natural populations which are reproductively isolated from other such groups*. Emerson defines it as *a genetically distinctive, reproductively isolated, natural population.* Unfortunately, in the taxonomic description of a species, all one has is some herbarium specimen of unknown genetic constitution. The delimitation of a species cannot be ascertained unless it is studied cytogenetically as well. *Cytotaxonomy* is more accurate in describing a species than taxonomy alone.

So, in speciation the main factor is *isolation.* This isolation may be due to a geographical, ecological or seasonal (two populations flowering at different seasons) *barrier* or genic, due to some structural difference, in the chromosomes or genes so that no normal hybridisation takes place. The *barrier*, however, may sometimes break down giving rise to interspecific hybrids which are fresh 'raw material' for further evolution. Within the *barrier* further changes take place by different types of mutation and natural selection plays its role in selecting out certain types or populations which are then stabilised as 'species'. We have to remember that genes are constantly evolving by mutation and most of them are lethal. Inbreeding, while increasing homozygosity, brings them out and the harmful are eliminated by natural selection while the useful ones remain. Thus, in an inbreeding population, some genes are always being eliminated and new genes are arising by fresh mutation. A species presents a *genic balance*, a balanced condition of the genes.

What role does environment play in this evolution of species? This question has often been considered too dogmatically. While Lamarckians paid all credit to environment, the Weismannians completely ruled it out. But, environment has certainly a role —if not actually in bringing out the mutation at least in natural selection and stabilisation of the 'species'. Some philosophic ideas on the direction of evolution have been mentioned in Chapter LX. Mutations occur spontaneously in nature. There is no proof that these mutations are being caused by any particular environmental agent. It was suggested that some natural radiation, like cosmic rays, might cause it. Vavilov's concept of Centres of Origin of crop plants also hinted at this possibility. But, Muller and others have calculated that the natural radiation in atmosphare is too little for mutations. Yet, there is the question that there may be certain regions where radioactive gases may accumulate, viz., near radioactive mines, where more mutations may take place. The present age has added another great factor in the atomic dust or 'fall out' caused by explosion of atomic bombs. Direct exposure to atom bombs may be even more serious. Some authorities have estimated that every survivor at Hiroshima carries at least one mutated gene. One cannot say whether the direction of evolution may not change if this atomic dust increases further. The role of temperature in causing mutations is, however, more definite. With more mutations in higher temperature, there is more 'raw material' for selection. This is possibly the reason why there are more species of plants in the tropics than elsewhere. Among animals, 80% of reptile species and 58% of mammal species are confined to the tropical region.

Whether the environment is directly responsible for mutations or not, there is little doubt that environment has a role in natural selection. It has been observed that the same species growing in different climatic areas may give rise to different *subspecies* or *ecospecies* or *races* because of ecological variations. It one of two populations of the same species be located on the Himalaya and another on the warmer plains, the environments will be different in the two different places. The inbred genotype-segregates and the mutants will be selected differently in the two places, genotypes suitable for the particular environment will survive, and ultimately, in course of time two different *ecotypes* or *subspecies* (if the differences become great) will appear in the two places. We may easily imagine that if the two ecotypes continue to thrive in the two different ecological surroundings, they may change further by chromosomal mutations and later form two *species* which will perhaps, by then, fail to cross when brought together. Mention may be made in this connection of the two subspecies *indica* and *japonica* within the cultivated rice species *Oryza sativa*. The former developed in tropical India and South-East Asia while the latter flourishes in temperate North China and Japan. It is found that although there is no apparent chromosomal difference between the two subspecies, hybrids between them show a high degree of sterility.

A very beautiful example of the action of environment on evolution is the change in the peppered moth types in England. The normal form of this moth has light mottled wings and this was the prevalent type one century ago. The normal type often threw out black 'melanic' type (Fig. 881) mutants which, being more conspicuous, would soon be

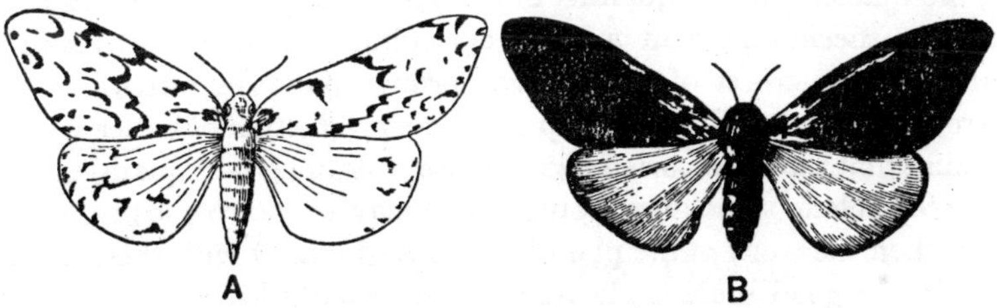

FIG. 881. Normal (A) and melanic (B) types of peppered moth.

exterminated by ravaging birds. But, with the Industrial Revolution, the sooty atmosphere of the industrial towns provided a very adequate camouflage for the survival of the malenic type. The normal type in agricultural England of one century ago has been completely replaced by the black melanic type in modern industrial England.

At present it is beyond doubt that for the origin of species, both the environment and the genotype must act in a co-ordinated way. By environment, both extracellular and intracellular environments are implied. Genes, although containing the raw materials of evolution as well as the essence of all characters, cannot function unless they get the proper background for action. Every gene can only function if a proper cytoplasmic background is provided. Naturally, therefore, one cannot eliminate the important role of environment in the functions of genes. Evolution represents the ultimate manifestation of a co-ordinated action of genes and environment, the former supplying the necessary characteristics and the latter helping in their expression.

CHAPTER LXIX

PLANT BREEDING

The art of Plant Breeding is perhaps older than the science of Genetics. Ever since men took to agriculture they came to find that better seed means better harvest. There was, therefore, a natural tendency to *select* the best ear of grains or the best fruit for seed. With the improvement of communication facilities, the importance of *introduction* of good and novel varieties from foreign countries was realised and finally, some pioneer plant breeders like Burbank and Michurin showed what miracles *hybridisation* might perform in the direction of getting new and better varieties.

It is an experience of farmers that if only a part of the harvest is used as seed year after year, the quality and the quantity of the harvest deteriorates. *Production of seed* is a technique more specialised than ordinary agriculture and that is why seed farms with capitals running into crores of rupees have sprung up in progressive countries like the U.S.A. Here, trained plant breeders supervise the production of better seed. Moreover, it is often found that the local varieties grown are poor in certain qualities. For example, it may be susceptible to some disease, the grains may be too coarse or the yield may be too low. It is then the work of the plant breeder to do away with these defects. Even if the local variety is good there is always scone to make it better.

The pioneer plant breeders were often vigorous workers with no scientific training. But an automobile is always better in the hand of a driver when the driver is also a mechanic and knows the ins and outs of the machine in his charge. Similarly, a modern plant breeder must have *a thorough knowledge of genetics* in all its aspects. Moreover, he must be sufficiently conversant in the statistical methods of designing and laying out his experiments as well as those of interpreting his data (*field plot technique* and *biometry*). He should be well aware of *special techniques* adapted to meet special ends. With this training he must tackle the particular problem. For this, he must know all about the *characteristics of the crop* (*Crop Botany*) he wants to improve and then must know what exactly are the *improvements needed* by the growers and the country. The improved seed is now to be distributed. Here comes the need of seed farms like that at the Panth Agricultural Institute.

The geneticist must have a clear idea as to what is a **Pure Line.** A *pure line* is a strain of plants which is homozygous and will not segregate any further. Ordinarily, cultivated crops become contaminated by natural crossing, etc. So, a plant breeder has to segregate pure lines. But, it should be borne in mind that the term *pure line* is only relative. With innumerable insignificant genes, it is impossible to get perfect homozygosity. But a pure line is one which is homozygous for the characters usually studied in economic plants.

In studying the characters of crops the plant breeder has to study the **modes of propagation** of plants. Cultivated plants fall under the following groups according to the mode or propagation:

A. The asexual group: Many horticultural crops and some agricultural crops are propagated vegetatively. Most fruit trees and many flower plants are propagated by graftings, cuttings, layerings, etc. Sugarcane is cultivated by nodes of stems (*sets*) and potato by seed tubers. A group of plants propagated from a single original stock is called **a clone**. Ordinarily , there can be no change in the genotypic constitution of plants propagated through clones and there is no segregation. Heterozygotes, may be propagated in this way. It may, however, be necessary to get the seed of these plants for experimental purposes like obtaining new hybrids, etc.

B, The sexual group: Most agricultural crops and some horticultural crops are propagated sexually.

(*a*) *Naturally self-pollinated group:* Some plants are found to be naturally self-pollinated. Cross-pollination is rare (less than 4%) even if no precaution is taken to stop it. Under this group come rice, wheat, barley, oats, potato, peas, beans, tobacco, flax, tomato, etc. Most jute varieties, specially *Corchorus capsularis*, also come under this group.

(*b*) *often cross-pollinated group:* Though self-pollination is the rule, cross-pollination takes place very often unless one is very careful about it. Cotton and sorghum come under this group.

(*c*) *Naturally cross-pollinated group:* Cross-pollination is the rule among these plants so that strict measures are to be taken if selfing is desired. Some plants in this group have unisexual flowers which simplify the technique to be followed in hybridisation. In this group are maize, rye, many grasses, cucurbits, *Brassica* species, sugar beets, clover, many fruits and root vegetables.

(*d*) *Dioecious plants:* The male and female plants being separate, there can be nothing but cross-pollination. Here come hemp, dates and palms, spinach, etc. Naturally, under this group breeding must be undertaken both for male and female lines.

The most important aim of field experimentation technique is the avoidance of *soil heterogeneity*. However careful one may be, a plot of land is never uniform in the fertility of the soil. As a result, plants at one end may vary from plants at another end because of the difference in soil and not because of any genetic cause. Statistical methods are employed to minimise this serious error and also other errors due to competition, etc.

CHAPTER LXX

METHODS EMPLOYED IN PLANT BREEDING

A plant breeder usually plans his work along one or more of the following lines:

I. PLANT INTRODUCTION

Foreign varieties are secured and cultivated in research stations. In advanced countries foreign introductions are usually screened through quarantine stations so that foreign disease pests may not be introduced along with the plants. The introduced varieties are then selected and tested. Some varieties may be found to be good from the beginning or they may be improved after acclimatisation or by hybridisation. Usually, introduced plants react differently to their new country of adaptation. They may die outright or may somehow survive as a poor straggler. In some other cases it is found that although they look weak during the first two or three years, after that they revive and become as healthy as in their native land. The converse is also found—that is, they thrive for a year or two and then suddenly deteriorate. For these reasons an introduced variety has to be *acclimatised* first. It is usually grown in different regions of the country of adaptation and their performances are recorded and tested. Potato, tobacco and tomato are some of the well-known introduced crops in this country. Introduced varieties are always kept in a *variety nursery* in modern plant breeding stations so that they may be utilised as future breeding stock. The spectacular *IR* rice varieties are introductions from the International Rice Research Institute in Manilla, Philippines and the *Sonora* wheat varieties are from the International Maize & Wheat Improvement Centre at Sonora, Mexico.

II. SELECTION

The oldest method of plant improvement is by the selection of the best plants and by growing only the seed from them. This is useful only so long as the population of plants is a mixture of pure lines. Selection within a pure line is useless according to Mendelian Genetics.

The Pure Line Concept: Pure Line Selection

Louis de Vilmorin (France, 1856) developed the method of *progeny test*. Individual plants are isolated and their progeny tested to find out if all the offsprings show uniform characters (e.g., sugar content of sugar beets), i.e., if characters are segregating. By strict progeny tests W. Johannsen (Denmark, 1857–1927) established the *Pure Line Theory*. He took a commercial variety of beans, selected for the weight of beans and found that a single variety could be broken up into 19 pure lines. Each one of the 19 pure lines had a constant average weight of beans (different for the different lines) and this average within a line could not be altered by further selection. This is because the pure line is homozygous for the character or characters studied, no further segregation takes place and, therefore, any further selection within it is futile.

The Lysenko school denied the existence of pure lines since, according to them, heredity is not anything fixed but varies with a variable and unpredictable environment. Even in the Mendelian sense it is not possible to get a perfectly homozygous plant for all characters although one may get a plant homozygous for all predominant characters.

The case becomes even more complicated if polygenes or modifying genes be present. The term *pure line* should, therefore, be used in a relative sense.

Pure line selection is important whenever a new variety of uncertain origin is obtained, or when investigations are begun on a new crop. It loses its importance, as soon as all the pure lines are isolated. But, it has already been pointed out that the word *pure line* is relative. Whenever investigations are taken in hand for a new character, fresh pure line selection is necessary. Thus, one may even now carry on pure line selection for protein content or vitamin content of rice.

In the methods of selection two courses are usually followed:

1. **Mass selection:**

Not individual plants but whole groups of plants are selected out. This is the simpler method.

2. **Individual Plant selection or Pedigree selection:**

Individual plants are selected out, isolated, and its seed grown separately (*Progeny test*). The same process may be repeated for a few generations. The process is naturally more rigorous but yields better result. A very well-known method is the *ear-to-row* or *panicle-row method*. Ears or panicles of cereal crops are selected out and each ear or panicle is grown in a separate row for future selection.

CLONAL SELECTION

Clones are plants propagated vegetatively from single original stocks and it has already been pointed out that the genotypic constitution of plants propagated in this way is not likely to change. They are as stable as pure lines and no segregation or variation is expected among them. So, selection within a clone is not likely to be fruitful. But, in nature, *bud mutations* have been found to occur occasionally and the selection of such bud mutations has played an important role in the breeding of clonal crops like potato, sugarcane, pineapple, apple, citrus, etc. In clonal crops, even better results are expected if clonal propagation be combined with actual hybridisation—sometimes in places far away from the actual fields. That is why there are special sugarcane and potato breeding stations on the hills where very important work is done on the hybridisation of such clonal crops.

VARIETY TEST

It can now be realised that all plant breeding stations must maintain different sections for different purposes. First, there must be the *variety plots* where hundreds and thousands of varieties (introductions and selections of varieties usually grown) are grown year after year as a living herbarium from which seed of any variety may be obtained for further selection or hybridisation. Secondly, there are the *pedigree culture* or *progeny test plots* where the pure lines are found out. Thirdly, there must be the *variety test plots*. In whatever way may a variety be obtained (by introduction or selection or hybridisation), its performances must be tested before it can be recommended to the farmers. For this, the varieties under test are grown side by side with standard varieties and their qualities compared. There may be different plots for testing different qualities, viz., *yield trial plots*, *disease nursery plots*, etc.

Selection played a very important part in the early days of plant breeding. It was largely employed in the selection of Indian wheat varieties by Mr. and Mrs. Howard. Valuable strains have similarly been obtained in India by pure line selection in rice, millets, cotton, etc.

III. HYBRIDISATION

Hybridisation is a very important method in plant breeding. It has played a big role in the development of improved sugarcane strains at Coimbatore where sugarcane has been experimentally crossed with sorghum, maize and even with *arundinacia* bamboo. Hybridisation has also been usefully employed in getting good strains of wheat, rice, cotton, etc. There was phenomenal improvement of wheat varieties of England by hybridisation. There may be a local variety which is good in all respects but inferior in one particular character, e.g., the grain may have an unwelcome red colour. Let it have the constitution **AABBCCRR** (**A, B & C** are good qualities and **R** for red grain). The plant breeder will then find out another pure line which may not be a good variety but it will have white grains. Its constitution may be **aabbccrr** (**a, b & c** are bad recessives and **r** white recessive). If a cross be made, the F_1 hybrid plant will show the dominant characters and will have the constitution **AaBbCcRr**. In the F_2 and subsequent generations the characters will segregate and recombine in all different ways. Some will be inferior types with white grains, some superior with red grains but there will be a very few superior types with white grains. The plant breeder will now go on selecting for a few generations only the desirable combinations, i.e., good qualities with white grains (phenotype **A-B-C-rr**). The progeny will be all white as it is homozygous for **rr** but there will be segregation for the **A, B** and **C** characters as most of the selected plants will be heterozygous for these. If he goes on selecting for a few more generations he will ultimately find out an **AABBCCrr** plant which will be a pure line, being homozygous, and this will be the desired combination. The plant breeder may carry on this selection of the F_1 progeny by two well established methods:

A. Pedigree method

The plant breeder grows every one of the F_2 selected plants separately. In the F_3 he again selects the suitable plants (**A-B-C-rr**) and grows every one of the selected plants separately keeping clear pedigree records of each plant. The advantage of this method lies in the quickness with which he will get the true breeding **AABBCCrr** plant. Among the F_2 plants that he selects (**A-B-C-rr**), there may be some **AABBCCrr** plants which he will be able to identify in 3 or 4 generations if he is lucky enough from the simple fact that these pedigreed lines will not show any segregation.

While the pedigree method is advantageous in being very quick it is disadvantageous in demanding much more attention and labour. In plant breeding stations, usually a large number of crosses are handled simultaneously and it is impossible to carry on with all those crosses by the pedigree method. The second method (*Bulk method*) is, therefore, used to save labour.

B. Bulk method

This is a mass selection method. The F_2 selected plants are not grown separately but are bulked together to form a single F_3 population. In the F_3 again, the suitable (**A-B-C-rr**) plants are collected and bulked together. This goes on for a few years after which the **A-B-C-rr** plants are tested separately to find out the true breeding **AABBCCrr** plant. This method involves more time but minimises labour as the plants do not need individual attention. Very often the plant breeder has no other option but to follow this method.

The table below shows two programmes, one according to pedigree method and the other according to bulk method, which are meant for rice but may also be followed for other small grains like wheat or barley.

Pedigree method	*Year*	*Bulk Method*
Cross is made.	1	Cross is made.
F_1 is grown.	2	F_1 is grown.
F_2 is grown and panicles from desirable segregates selected	3	Desirable segregates selected from F_1 plants and grown (F_2).
F_3 panicle rows grown and desirable panicles selected.	4	F_3 grown in bulk plots. 100 desirable panicles selected.
F_4 panicle rows grown and desirable panicles selected.	5	Selected F_4 grown in bulk plot. 100 panicles selected.
Panicle rows or blocks of three rows. Checks every fifth row or every third block.	6	F_5 in bulk plot. 100 panicles selected.
5 three-row blocks or each selection. Checks as above.	7	Panicle rows and selection of desired plants.
10 three-row blocks of each selection. Checks as above.	8	Twenty-row blocks from 2 or 3 selections.
As in the 8th year.	9	Selections in small plots with checks.
Replicated yield trial plots with checks in different localities.	10	Selections in small plots with checks.
Increase plots.	11	Replicated yield trial plots with checks.
Distribution to growers.	12	Yield trials on larger plots and in different localities.
	13	Increase plots.
	14	Distribution to growers.

HYBRIDISATION TECHNIQUE

When a plant breeder wants to hybridise between two varieties he must first gather all informations about the flowers, viz., the time of flowering, the exact time when the anther and the stigma become mature for pollination, which flowers give healthy seeds, how long do the pollens remain viable, etc. He must take all precautions so that

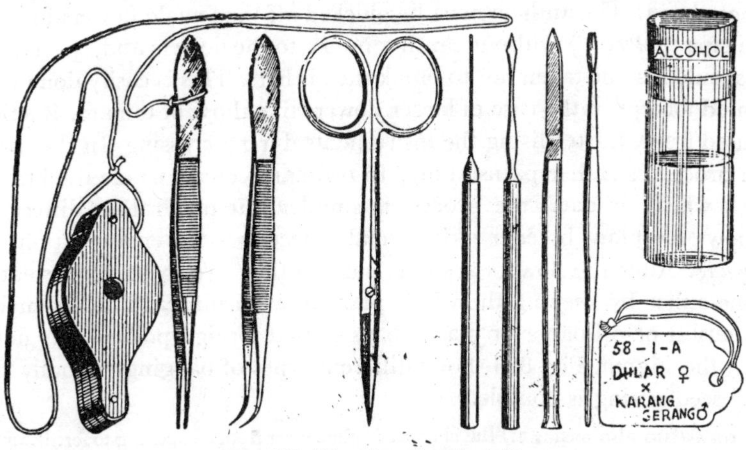

FIG. 882. The plant breeder's kit.

hybridisation takes place only in the way he desires, precluding all chances of self-pollination and must ensure that no foreign pollen can contaminate the result. He should follow the following stages:

1. *Selection and preparation of parents: Isolation:* The plant breeder first selects the plant that he will use as the mother parent and keeps the male parent ready so that the anther will be ripe just at the desired time. If there are too many flowers on the branch of the mother parent he clips off a number of them. This is specially true in the cereals (wheat or rice) where there are a big number of flowers on the spike or panicle. In rice, about ten or twelve flowers of the same age are kept. It is necessary to *isolate* the female parent and, sometimes, even the male parent, by growing on isolated plots or by *bagging* or *caging*. Necessity of isolation increases with the percentage of natural cross-pollination.

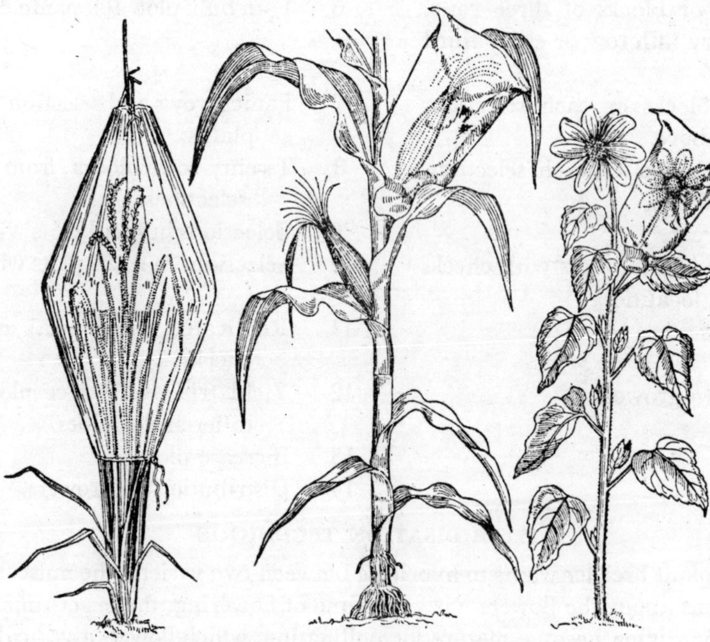

FIG. 883. Different method of bagging.

2. *Emasculation:* The anthers must be plucked off the female flowers just before the anthers are ripe (*anthesis**) without causing injury to the flowers and, specially, the carpels. Care should also be taken not to break the anthers. This is easily done with a pair of fine pointed forceps in the case of larger flowers like those of tomato. Rectified spirit should be used freely in sterilising the instruments during crossing. In the case of small flowers the process is rather painstaking. In ordinary cereals where the bracts are not brittle (e.g., wheat or oats) the process is simpler but it is rather different in rice. Fig. 882 shows the plant breeder's kit specially necessary for emasculation.

3. *Bagging:* After the flowers are emasculated they are to be kept isolated which may be done either by keeping the whole plant in a muslin cage or by enclosing the flowers in muslin or oil paper or plastic bags so that foreign pollens may not come in contact with the stigma. Fig. 883 shows different types of bagging. Usually these bags are kept till seed-setting is complete.

* The term *anthesis* also means all the changes in the flower from its opening to fertilisation, i.e., the period during which it attains the climax of its duration.

4. *Pollination:* When the stigma of the emasculated flower is mature the bag is temporarily removed and the stigma pollinated by dusting with complete broken anthers or pollens from the male parent. Special study should be made as to the viability of the pollens. Flowers are bagged again after pollination.

Care must always be taken to keep the crossed flowers properly labelled or tagged. The label should be as brief as possible but complete. It should bear the names of the parents (female parent first) and, at least, a number referring to the field record book as shown in Fig. 882. All orher necessary particulars should be entered in a handy field record book with a number and the number referred to on the tag.

The hybridisation technique must be adapted to the particular crop on which work is being done. Techniques usually followed for three important crops—rice, wheat and cotton are give below:

RICE

In rice, the lemma and palea are rather hard and the flowers remain open for only about half an hour, sometimes after the sun warms up. Emasculation must be done before the time of anthesis which can be easily ascertained by placing the closed flowers against the sun (Fig. 884) and looking through the semitransparent lemma-palea.

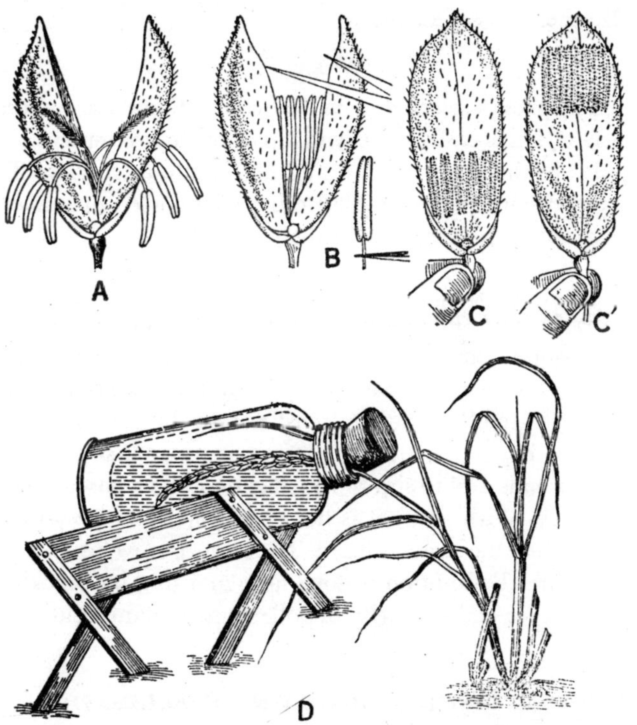

FIG. 884. Hybridisation of rice. A. A spikelet. B. Forcing the flower open and removing anthers by forceps. C. A spikelet against sun showing immature anthers. C′, Same spikelet just before anthesis. D. Hot water emasculation (*after Gangulee*).

Just before anthesis the anthers rise from the base to the top of the closed spikelet. Emasculation may be done in four ways: (1) By slightly forcing open the lemma and palea just before the opening of the flower. This is the method for wheat and other cereals and this is the method usually adapted for rice in India. (2) By clipping off the tip of the open flower with a pair of scissors and taking out the anthers through the opening by a pair of fine pointed forceps or a needle. Care must taken not to break the anthers and to take out all the six anthers of rice. (3) The panicle is covered in the early morning, before blooming, by a dark or brown paper bag (Ramiah). The heat inside forces the flowers open and emasculation is carried on as usual. (4) The last method of hot water emasculation, developed by N. E. Jodon, is very interesting and is now widely used in the U.S.A. It has also been found very fruitful in India (Gangulee 1959). Warm water (about 43°C.) is taken in a thermoflask, a mature tiller of rice is tilted and a whole panicle kept immersed wthin the warm water for about 10 minutes (Fig. 884). Within a few minutes of taking the panicle out of the water, just the *mature flowers* open out. Not only that, the temperature renders all the anthers sterile while the carpels are not injured, thus causing automatic emasculation. Pollination is to be completed within the next half an hour after which the flowers automatically close. For pollination whole broken anthers, from flowers in which anthesis is about to take place, are used. When emasculation is carried on by forcing the flowers open, the latter are kept closed by small rubber rings so that bagging is unnecessary. Bagging is necessary for the second and the third methods of emasculation in rice and necessary only for the half an hour while the flowers are open after hot water treatment. Only a few flowers in a panicle should be pollinated and all others scissored off.

Wheat

After the female spike is selected, 5 to 10 healthy flowers are chosen and the rest scissored off. Emasculation is easily done with a pair of narrow forceps by gently opening the lemma and palea as the latter are soft and not brittle as in rice. The emasculated spikes are kept bagged with paper or plastic bags supported on stakes. Pollination is done after two days with whole broken anthers.

Cotton

Emasculation is done on the afternoon of the day previous to the normal opening of the flower. An incision is carefully made round the base of the corrolla and the latter is removed taking care not to injure the pistil. The anthers are removed by plucking or scraping them off very carefully. Anthers must not be broken in the process. Emasculated flowers are now bagged. A still simpler process is to take a small bit of drinking straw, to seal one of its ends and to slip the open end on the emasculated pistil so that it fits tightly on the ovary. With straw it is not necessary to scrape off the lower anthers which get rubbed off when the straw is fitted. The straw is now fastened to the stem with soft and fine copper wire. Pollination is done next morning by plucking whole flowers from previously bagged (to ensure purity of pollen) male plants and dusting the pollens on the stigma.

SPECIAL METHODS INVOLVING HYBRIDISATION

1. Breeding for disease resistance

Disease often cause serious havoc among plants. Effective control of such diseases by germicides, etc., is often too expensive and inconvenient. On the other hand, strains of crop plants (which may even be wild, distantly related varieties or species) are sometimes found to be naturally immune (resistant) while other strains are normally susceptible for specific diseases. This resistance and suceptibility to diseases (e.g., rust in wheat or *Helminthosporium*=*Ophiobolus* in rice) are usually genetic factors. Sometimes this resistance is found in wild varieties which are otherwise useless. When this is so, it is possible to get the resistance factor from the resistant variety combined with the good qualities of some suitable cultivated variety. The cultivation of the new resistant variety will then be an efficient as well as cheap measure in controlling the disease. In breeding disease resistant strains, every hybrid generation is to be subjected to artificially created disease producing conditions. Special *inoculation tents* are sometimes found useful for this purpose. All plants are artificially inoculated with fungal spores, etc., and the proper environment (specially, humidity) created. The resistant segregates are then easily spotted in every generation. Selection is carried on for a number of years till the homozygous strain is obtained. Breeding for disease resistance is not always simple as it is difficult to get strains resistant under all conditions and the inheritance of disease resistance is often multifactorial. Moreover, most important diseases have a large number of strains of the pathogens and it is difficult to get any variety of the crop resistant to all the strains. Some success has been attained in breeding disease resistant strains of various Indian crop plants. One of the serious limitations of the present method of raising disease resistant varieties is that it does not take into account the potentiality of the disease producing organisms to undergo mutation. That is why it has often been found that a disease resistant variety once obtained does no remain so for a long time. Later, it becomes susceptible due to mutation of the organism causing the disease, which can then infect the so-called resistant plant.

2. Backcross and Testcross methods

It has already been shown how the crossover percentage of two linked genes may be determined by the *testcross*, i.e., by *backcrossing* the hybrid F_1 plant back to the recessive parent. This testcross method is very usefully utilised by the plant breeder and the geneticist in determining the genotypic constitution of any plant. As backcrossing takes place with the recessive parent, the latter does not show itself in the progeny and the backcross segregation ratio represents the gametic ratio of the plant in question and, hence, its genotypic constitution. Thus, Fig. 885 shows the result of a testcross with a tall ×dwarf heterozygous pea plant. The backcross ratio can only be explained by assuming that there are **T** and **d** gametes on the plant in equal proportions, i.e., it is a **Td** plant. If the backcross plant is a pure line (with only one type of gametes), the backcross generation must show plants of one type only.

Another use of the *backcross* method in plant breeding is introduction of a character from either (i.e., recessive or dominant) of the parents more quickly. Thus in breeding for disease resistance, the hybrid disease resistant plants of successive generations are repeatedly backcrossed with the disease resistant parent. In this way homozygosity for disease resistance may be attained a few years earlier. Backcrossing is rather easy when a monogenic character is to be transferred while it may also be adapted for multifactorial characters.

3. The use of hybrid seed

F_1 plants are always more vigorous because of heterosis. Yield of a crop can be greatly increased if F_1 seeds can be directly used as seed by the farmers. But, production of such seed is usually so costly that it can only be used for experimental purposes and is hopelessly uneconomic if used as the farmers' seed. However, an exception has been found in the use of *hybrid corn* (maize) as seed. Methods have been developed in the U.S.A. for obtaining hybrid corn seed in large scales at low cost. Production of maize has been greatly increased in recent years by this process. Production of hybrid corn has now taken hold in India. The author's own experiments show that it is possible to cultivate hybrid rice on a small plot and obtain a much higher yield.

Maize is a naturally cross-fertilised crop. So, the first step towards producing hybrid corn is isolating the pure lines by inbreeding. The inbred pure lines are rather weak

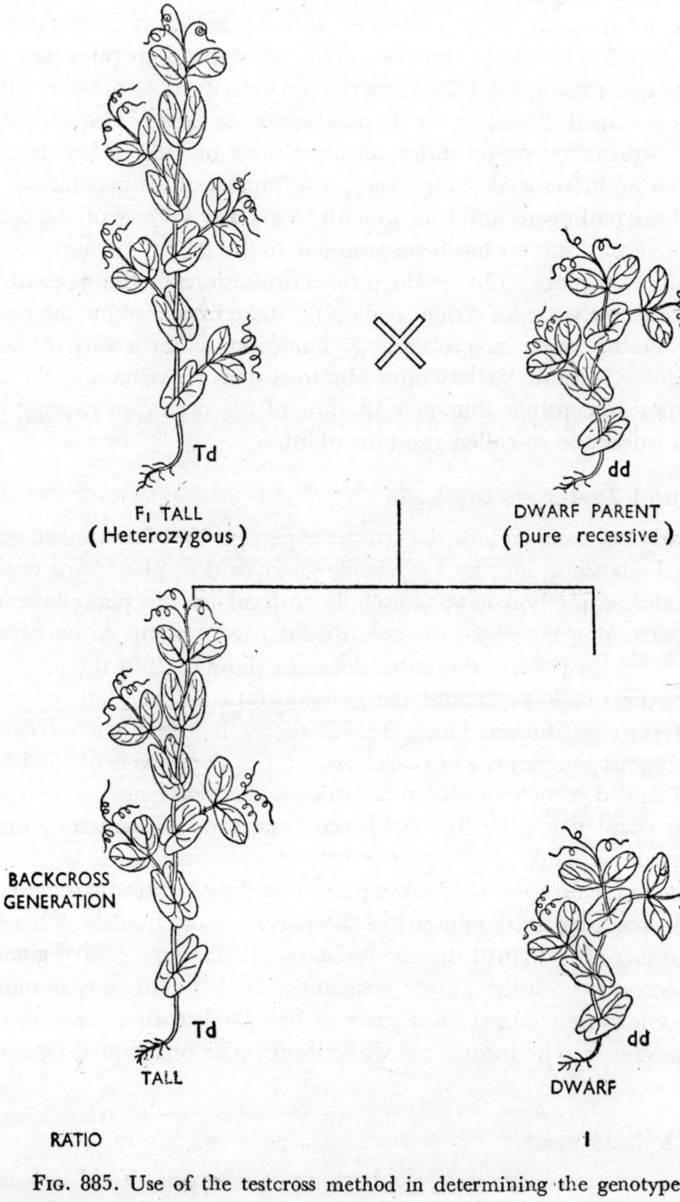

FIG. 885. Use of the testcross method in determining the genotype.

Two such pure lines are planted in the field alternately, one as the male stock or pollinator and the other as the feamle stock or the seed producer. Such fields must be isolated from all other varieties of maize to get good uncontaminated hybrid corn. The usual method is to plant one pollinator row for evey two seed rows. Male flowers of maize are borne on apical panicles (Fig. 274) or 'tassels' while female cobs are borne on the axils. So, emasculation is rather easy by simply lopping off these panicles and it is now possible to carry on large-scale emasculation or 'detasseling' mechanically. All kernels developing on the detasseled variety are hybrids, pollination being possible only from the other variety. But, as the inbred pure line plants are weak, the production of *single-cross grains* on them is rather low. So, double-crossing is resorted to by making a hybrid

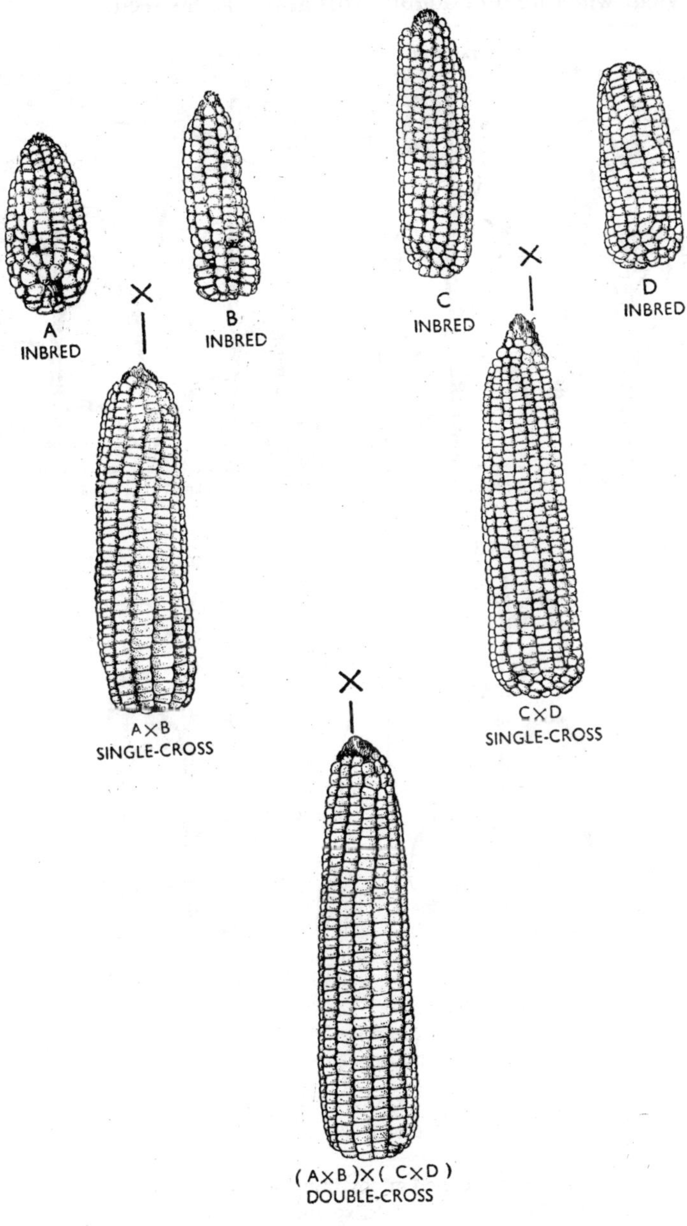

FIG. 886. Maize cobs become increasingly larger from the inbred varieties to the single-cross and from the single-cross to the double-cross.

of the second order out of two F_1 single-cross hybrids in order to get a large quantity of *double-cross grains*. The cobs on the double hybrids (Fig. 886) are even larger than on the single hybrids. For double crossing, two rows of pollinator single-cross plants are planted for six to eight rows of single-cross seed rows. As the single-cross plants are much more vigorous than the pure line plants, the production of double-cross grains is much more bountiful. Four pure lines are involved in producing a double-cross hybrid and the method of such production is diagrammatically represented in Fig. 887. The farmer gets a heavy yield when he uses double-cross grains as his seed.

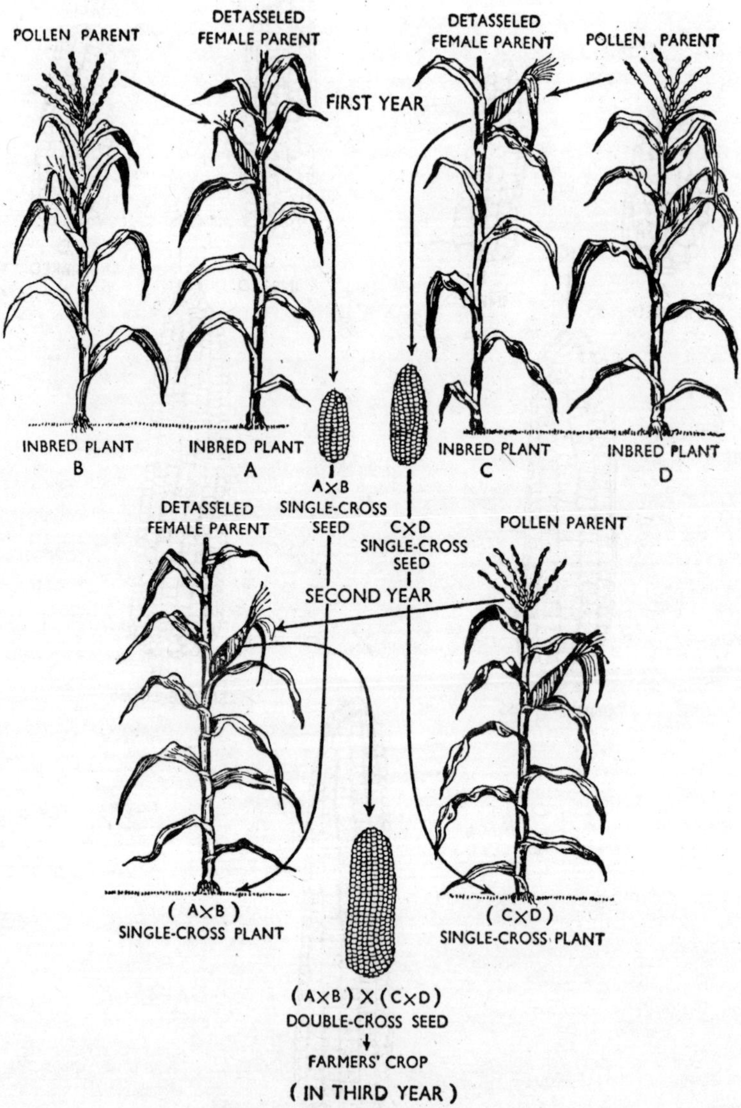

Fig. 887. Method of obtaining double-cross seed.
(After Texas Agricultural Experiment Station Bulletin).

A new development in the production of hybrid corn is the utilisation of the '*male-sterile*' character. Plants having this gene or character have sterile pollens so that they are naturally emasculated. If a strain having this character is used as the seed parent then it is no longer necessary to emasculate or detassel it. If it be possible to get such suitable 'male-sterile' genes in other crops it may be possible to get cheap hybrid F_1 seeds as the labour involved in emasculating will be reduced to the minimum.

IV. MUTATION BREEDING

A new line of plant breeding has opened up in recent years—that of mutation breeding. Mutations and their role in evolution of new species have been dealt with in the chapter on mutations. Important crop varieties are known to be mutants. Müntzing thinks that more than half the species of flowering plants are polyploids. A plant breeder has to keep his eyes open to select out any naturally occurring mutant (point mutation or intergenic mutation) that may look promising. Selection within clones and pure lines is possible only when such mutants occur.

Artificial induction of mutations has been extensively employed in recent years. The most important agents for such mutation induction are (*i*) X-rays and other types of ionising radiation explained in Chapter LXVI (giving rise to the science of Radiation Biology) and (*ii*) Chemical mutagens like mustard gas or colchicine. All growing organs, seeds, pollens, eggs, etc., may be subjected by such irradiation or chemical treatment. There is no fundamental difference between natural and artificial mutations. It is true that most mutations are of no practical importance or even harmful and also that there is no way of predicting what type of mutation one is going to get. But, this should not give rise to any scepticism and a plant breeder should be satisfied if he gets a new beneficial character in a million. Some very useful radiation-induced mutations have already come to the use of agriculturists and horticulturists. Sweden has greatly advanced in mutation breeding since 1929 with workers like Gustafsson and Müntzing at their Svaloff station. They have obtained improved 'erectoid' stiff-strawed barley varieties by X-ray treatment and similar treatment has yielded better varieties of white mustard, *Phaseolus*, ground-nut, oats, peas, etc. Similarly, there are reports of improved barley from Germany; rust resistant wheat from Austria; improved barley, peanut and short-strawed rice (by Beachell) from the U.S.A. and improved wheat varieties in India. Radiation induction has been even more fruitful in horticulture. New flower varieties have been raised and then propagated vegetatively. In India attempts have also been made to get better varieties of rice, sesame and jute in this way. A short-strawed, high-tillered rice mutant has been reported by Ramanujam and Parthasarathy. Other results are still under investigation.

Induction of mutation with chemical mutagens has become even more popular as the method is accessible to all types of workers. Colchicine has been extensively used in the production of tetraploids and amphidiploids involving hundreds of species. Although most of these polyploids are of no practical value, they have been found to be very useful material which can possibly be improved by further breeding. Thus, some useful *Triticale* (wheat × rye, i.e., *Triticum* × *Secale* amphidiploid) varieties have been obtained by hybridising different types of *Triticales*. Müntzing has obtained tetraploid winter rye. Kihara and his associates in Japan have reported triploid sugar beet with higher sugar content, triploid water-melon, tetraploid radish, etc. Tetraploid grapes, also, have been produced.

Cytogenetics has placed more synthetic breeding material in the hand of the plant breeders in the form of plants with altered genomes or substituted chromosomes and the study of trisomics, monosomics, nullisomics, etc., has enabled them to work on the definite positions of the desirable genes in the chromosomes.

Finally, the latest attempt of inducing mutations in a novel way is well worth mentioning. It has been mentioned that the most important chemical constituents of genes are nucleic acids and there is evidence that free nucleic acid may control heredity. There has been a claim from France that deoxyribose-nucleic acid (known so cytogeneticists as *DNA*) extracted from the eggs of the Khaki Campbell variety of ducks when injected into the Peking variety, has induced some Khaki Campbell characters to the Peking ducks and this change has been found to be hereditary. As *DNA* has been proved to be the gene substance, it is quite likely that in near future it may be possible to use just the *DNA* extract from one of the desirable parents in the hybridisation of higher plants as has already been done in the case of *Pneumococcus* strains of bacteria. If this method of changing heredity be established and improved, possibly it will be the most important tool in the hands of the plant and animal breeders. (c.f. Genetic Engineering).

CONCLUSION

The importance of plant breeding in modern agriculture

With the present population explosion the cry in the densely populated underdeveloped countries, specially of Asia, is for food and more food. Many such countries like India does not produce enough cereals to feed their own populations. The need is to develop more amount of food on the same area of land. For this, besides improved methods of agriculture, better and high yielding varieties and strains must be developed. Fortunately, great success has been achieved in this direction in recent years. A *green revolution* has taken place in India and some other countries so that these countries would soon produce more food that what they presently need.

In this connection one must mention Norman E. Borlaug (Fig. 888), the plant pathologist plant breeder devoting his life at the International Maize and Wheat Improvement Centre at Sonora in Mexico. His splendid work in developing new high yielding, rust resistant, non-lodging dwarf wheat varieties and strains, which are now being cultivated in many countries, is the basis of the present 'green revolution'. His *Sonora 64*, with *Sonalika*, *Kalyansona* and other varieties developed in India, is working a miracle. Most deservingly, Borlaug was awarded a Nobel Prize for Peace in 1970. Nothing contributes more to peace than self-sufficiency in food.

FIG. 888. Norman E. Borlaug—*the man behind the green revolution.*

Similarly, in the field of rice also high-yielding varieties have been developed by plant breeders. *IR*–8, developed in the International Rice Research Institute located in the Philippines and the strains *Padma* and *Jaya* developed in India are contributing towards 'green revolution' here.

INDEX

F

H

I

J

K

M

N

O

P

Q

S

T

U

V

Z